Manual of Pediatric Hematology and Oncology

Fifth Edition

Philip Lanzkowsky, M.B., Ch.B., M.D.,
Sc.D. (honoris causa), F.R.C.P., D.C.H., F.A.A.P.

Chief Emeritus, Pediatric Hematology-Oncology
Chairman Emeritus, Department of Pediatrics
Executive Director and Chief-of-Staff (Retired)
Steven and Alexandra Cohen Children's Medical Center
of New York, New Hyde Park, New York
Vice President, Children's Health Network (Retired)
North Shore-Long Island Jewish Health System
Consultant, Steven and Alexandra Cohen Children's
Medical Center of New York
Professor of Pediatrics
Hofstra North Shore-LIJ School of Medicine, Hempstead, New York

AMSTERDAM • BOSTON • HEIDELBERG • LONDON • NEW YORK • OXFORD
PARIS • SAN DIEGO • SAN FRANCISCO • SINGAPORE • SYDNEY • TOKYO

ELSEVIER   Academic Press is an imprint of Elsevier

Academic Press is an imprint of Elsevier
32 Jamestown Road, London NW1 7BY, UK
30 Corporate Drive, Suite 400, Burlington, MA 01803, USA
525 B Street, Suite 1800, San Diego, CA 92101-4495, USA

First edition 1989
Second Edition 1994
Third Edition 1999
Fourth Edition 2005
Fifth Edition 2011

Notice

British Library Cataloguing-in-Publication Data
A catalogue record for this book is available from the British Library

Library of Congress Cataloging-in-Publication Data
A catalog record for this book is available from the Library of Congress

ISBN : 978-0-12-375154-6

For information on all Academic Press publications
visit our website at www.elsevierdirect.com

Typeset by MPS Limited, a Macmillan Company, Chennai, India
www.macmillansolutions.com

Printed and bound in the United States of America

11 12 13 14   10 9 8 7 6 5 4 3 2 1

Contents

Contributors

Robert J. Arceci, M.D., Ph.D.   King Fahd Professor of Pediatric Oncology, Professor of Pediatrics, Oncology and the Cellular and Molecular Medicine Graduate Program, Kimmel Comprehensive Cancer Center, Johns Hopkins University, Baltimore, Maryland

Histiocytosis Syndromes

Suchitra S. Acharya, M.D.   Associate Professor of Pediatrics, Hofstra North Shore-LIJ School of Medicine, Hempstead, New York; Attending, Division of Pediatric Hematology-Oncology and Stem Cell Transplantation, Section Head, Bleeding Disorders and Thrombosis, Department of Pediatrics, Steven and Alexandra Cohen Children's Medical Center of New York, New Hyde Park, New York

Hemostatic Disorders; Thrombotic Disorders

Melissa A. Alderfer, Ph.D.   Assistant Professor of Pediatrics, University of Pennsylvania School of Medicine, Philadelphia, Pennsylvania, Psychologist, The Cancer Center at the Children's Hospital of Philadelphia, Philadelphia, Pennsylvania

Psychosocial Aspects of Cancer for Children and Their Families

Mark Atlas, M.D.   Assistant Professor of Pediatrics, Hofstra North Shore-LIJ School of Medicine, Hempstead, New York; Attending, Pediatric Hematology-Oncology and Stem Cell Transplantation, Section Head, Childhood Brain and Spinal Cord Tumor Program, Department of Pediatrics, Steven and Alexandra Cohen Children's Medical Center of New York, New Hyde Park, New York

Central Nervous System Malignancies

Rochelle Bagatell, M.D.   Assistant Professor of Pediatrics, University of Pennsylvania; Attending Physician, Division of Oncology, The Children's Hospital of Philadelphia, Philadelphia, Pennsylvania

Neuroblastoma

Mary Ann Bonilla, M.D.   Assistant Professor of Pediatrics, Columbia University College of Physicians and Surgeons, Attending Pediatric Hematologist-Oncologist, St. Joseph's Children's Hospital, Peterson, New Jersey

Disorders of White Blood Cells

James Bussel, M.D.   Professor of Pediatrics and Professor of Pediatrics in Obstetrics and Gynecology and in Medicine, Weill Cornell Medical College; Director, Platelet Research & Treatment Program, Division of Pediatric Hematology-Oncology. Department of Pediatrics Weill Cornell Medical Center and New York-Presbyterian Hospital, New York, New York

Disorders of Platelets

Mitchell S. Cairo, M.D.   Professor of Pediatrics, Medicine, Pathology and Cell Biology, Columbia University, Chief, Division of Blood and Marrow Transplantation, New York-Presbyterian, Morgan Stanley Children's Hospital, New York, New York

Non-Hodgkin Lymphoma

Andrew Chen, D.O.   Fellow, Hematology and Medical Oncology, University of Utah, Salt Lake City, Utah

Polycythemia

Jeffrey Dome, M.D., Ph.D.   Associate Professor of Pediatrics, George Washington University School of Medicine and Health Sciences, Chief, Division of Oncology, Center for Cancer and Blood Disorders, Children's National Medical Center, Washington, D.C.

Renal Tumors

Steven DuBois, M.D.   Assistant Professor of Pediatrics, University of California, San Francisco School of Medicine, Attending Physician, Hematology-Oncology, University of California at San Francisco Children's Hospital, San Francisco, California

Malignant Bone Tumors

Carolyn Fein Levy, M.D.   Assistant Professor of Pediatrics, Hofstra North Shore-LIJ School of Medicine, Hempstead, New York; Attending, Division of Pediatric Hematology-Oncology and Stem Cell Transplantation, Department of Pediatrics, Steven and Alexandra Cohen Children's Medical Center of New York, New Hyde Park, New York

Rhabdomyosarcoma and Other Soft-Tissue Sarcomas

Jonathan Fish, M.D.   Assistant Professor of Pediatrics, Hofstra North Shore-LIJ School of Medicine, Hempstead, New York; Attending, Division of Pediatric Hematology-Oncology and Stem Cell Transplantation, Section Head, Center for Survivors of Childhood Cancer, Department of Pediatrics, Steven and Alexandra Cohen Children's Medical Center of New York, New Hyde Park, New York

Evaluation, Investigations and Management of Late Effects of Childhood Cancer

Debra L. Friedman, M.D., M.S.   Associate Professor of Pediatrics, Vanderbilt University School of Medicine, Nashville, Tennessee, E. Bronson Ingram Chair in Pediatric Oncology, Director, Division of Pediatric Hematology/Oncology, Co-Leader, Cancer Epidemiology, Control and Prevention Program, Vanderbilt Ingram Cancer Center, Nashville, Tennessee

Retinoblastoma, Hodgkin Lymphoma

Richard Gorlick, M.D.   Associate Professor of Pediatrics and Molecular Pharmacology, The Albert Einstein College of Medicine of Yeshiva University, Bronx, New York, Vice Chairman, Division Chief of Hematology-Oncology, Department of Pediatrics, The Children's Hospital at Montefiore, Bronx, New York

Malignant Bone Tumors

Eric Gratias, M.D.   Assistant Professor of Pediatrics, University of Tennessee College of Medicine, Chattanooga, Tennessee, Division of Pediatric Hematology/Oncology, T.C. Thompson Children's Hospital, Chattanooga, Tennessee

Renal Tumors

Jessica Hochberg, M.D.   Assistant Professor of Pediatrics, New York Medical College, Attending, Maria Fareri Children's Hospital, Valhalla, New York

Non-Hodgkin Lymphoma

Katherine A. Janeway, M.D.   Instructor of Pediatrics, Harvard Medical School, Boston, Massachusetts, Attending Physician, Pediatric Hematology-Oncology, Dana Farber Cancer Institute and Children's Hospital, Boston, Massachusetts

Malignant Bone Tumors

Janet L. Kwiatkowski, M.D.   Associate Professor of Pediatrics, University of Pennsylvania School of Medicine, Director, Thalassemia Program, Children's Hospital of Philadelphia, Philadelphia, Pennsylvania

Hemoglobinopathies

Philip Lanzkowsky, M.B., Ch.B., M.D., Sc.D. (honoris causa), F.R.C.P., D.C.H., F.A.A.P. Professor of Pediatrics, Hofstra North Shore-LIJ School of Medicine, Hempstead, New York; Chief Emeritus, Division of Pediatric Hematology-Oncology, Department of Pediatrics, Steven and Alexandra Cohen Children's Medical Center of New York, New Hyde Park, New York. Chairman Emeritus, Department of Pediatrics, Chief-of-Staff and Executive Director (Retired), Steven and Alexandra Cohen Children's Medical Center of New York, New Hyde Park, New York

Classification and Diagnosis of Anemia In Children, Anemia During the Neonatal Period, Iron Deficiency Anemia, Megaloblastic Anemia, Lymphadenopathy and Splenomegaly

Jeffrey M. Lipton, M.D., Ph.D. Professor of Pediatrics, Hofstra North Shore-LIJ School of Medicine, Hempstead, New York; Chief, Division of Pediatric Hematology-Oncology and Stem Cell Transplantation, Department of Pediatrics, Steven and Alexandra Cohen Children's Medical Center of New York, New Hyde Park, New York

Bone Marrow Failure

Neyssa Marina, M.D. Professor of Pediatrics, Stanford University School of Medicine, Associate Chief of Clinical Affairs, Division of Hematology-Oncology, Stanford University and Lucile Packard Children's Hospital, Palo Alto, California

Malignant Bone Tumors

Jill S. Menell, M.D. Assistant Professor of Pediatrics, Columbia University College of Physicians and Surgeons, Chief, Pediatric Hematology-Oncology, St. Joseph's Children's Hospital, Paterson, New Jersey

Disorders of White Blood Cells

Thomas A. Olson, M.D. Associate Professor of Pediatrics, Emory University School of Medicine, Atlanta, Georgia, Attending Physician, Aflac Cancer Center and Blood Disorders Service, Children's Healthcare of Atlanta, Atlanta, Georgia

Germ Cell Tumors, Hepatic Tumors

Pinki Prasad, M.D. Research Fellow, Division of Pediatric Oncology, Vanderbilt University Nashville, Tennessee

Hodgkin Lymphoma

Julie R. Park, M.D.   Associate Professor of Pediatrics, University of Washington School of Medicine, Program Director, Hematology-Oncology Education, Pediatric Hematology-Oncology Specialist, Seattle Children's Hospital, Seattle, Washington

Neuroblastoma

Josef T. Prchal, M.D.   Professor of Medicine, Pathology and Genetics, University of Utah, Director of Huntsman Cancer Hospital Myeloproliferative Disorders Clinic and the George A. Wahlen Veterans Administration Medical Center Myeloproliferative Disorders Clinic, Salt Lake City, Utah

Polycythemia

Arlene Redner, M.D.   Associate Professor of Pediatrics, Hofstra North Shore-LIJ School of Medicine, Hempstead, New York; Attending, Division of Pediatric Hematology-Oncology and Stem Cell Transplantation, Department of Pediatrics, Steven and Alexandra Cohen Children's Medical Center of New York, New Hyde Park, New York

Leukemias

Thomas Renaud, M.D.   Fellow, Pediatric Hematology-Oncology, Division of Pediatric Hematology-Oncology, Department of Pediatrics, Weill Cornell Medical College and New York-Presbyterian Hospital and Memorial Sloan-Kettering Cancer Center, New York, New York

Disorders of Platelets

Susan Rheingold, M.D.   Assistant Professor of Pediatrics, University of Pennsylvania School of Medicine, Philadelphia, Pennsylvania, Medical Director, Outpatient Oncology Program, The Cancer Center at the Children's Hospital of Philadelphia, Philadelphia, Pennsylvania

Management of Oncologic Emergencies

Lorry Glen Rubin, M.D.   Professor of Pediatrics, Hofstra North Shore-LIJ School of Medicine, Hempstead, New York, Chief, Division of Infectious Disease, Department of Pediatrics, Steven and Alexandra Cohen Children's Medical Center of New York, New Hyde Park, New York

Supportive Care of Patients with Cancer

Indira Sahdev, M.D.   Associate Professor of Pediatrics, Hofstra North Shore-LIJ School of Medicine, Hempstead, New York; Attending, Division of Pediatric Hematology-Oncology and Stem Cell Transplantation, Section Head, Stem Cell Transplantation, Department of Pediatrics, Steven and Alexandra Cohen Children's Medical Center of New York, New Hyde Park, New York

Hematopoietic Stem Cell Transplantation

Jessica Scerbo, M.D   Fellow, Division of Pediatric Hematology-Oncology, Department of Pediatrics, Steven and Alexandra Cohen Children's Medical Center of New York, New Hyde Park, New York

Management of Oncologic Emergencies, Supportive Care of Patients with Cancer

David T. Teachey, M.D.   Assistant Professor of Pediatrics, University of Pennsylvania School of Medicine, Attending Physician, Pediatric Hematology-Oncology, Children's Hospital of Philadelphia, Philadelphia, Pennsylvania

Lymphoproliferative Disorders, Myeldysplastic Syndrome and Myeloproliferative Disorders

M. Issai Vanan. M.D., MPH   Research Fellow, Division of Pediatric Hematology-Oncology and Stem Cell Transplantation and Oncology and Cell Biology, Department of Pediatrics, Steven and Alexandra Cohen Children's Medical Center of New York, New Hyde Park, New York

Hematologic Manifestations of Systemic Illness

Leonard H. Wexler, M.D.   Associate Professor of Clinical Pediatrics, Weill Cornell Medical College; Associate Attending Physician, Department of Pediatrics, Memorial Sloan-Kettering Cancer Center, New York, New York

Rhabdomyosarcoma and Other Soft-Tissue Sarcomas

Lori S. Wiener, Ph.D.   Coordinator, Psychosocial Support and Research Program, Pediatric Oncology Branch, Co-Director, Behavioral Science Core, National Cancer Institute, National Institutes of Health, Bethesda, Maryland

Psychosocial Aspects of Cancer for Children and their Families

Lawrence Wolfe, M.D.   Associate Professor of Pediatrics, Hofstra North Shore-LIJ School of Medicine, Hempstead, New York; Section Head, Hematology, Attending, Division of Pediatric Hematology-Oncology and Stem Cell Transplantation, Department of Pediatrics, Steven and Alexandra Cohen Children's Medical Center of New York, New Hyde Park, New York

Hematologic Manifestations of Systemic Illness, Red Cell Membrane and Enzyme Defects, Extracorpuscular Hemolytic Disease, Management of Oncologic Emergencies, Supportive Care of Patients with Cancer

Introduction

Reflection on 50 Years of Progress in Pediatric Hematology-Oncology

As the fifth edition of the *Manual of Pediatric Hematology-Oncology* is published, I have reflected on the advances that have occurred since I began practicing hematology-oncology over 50 years ago and since my first book on the subject was published by McGraw Hill in 1980. The present edition is more than double the size of the original book.

Our understanding of hematologic conditions has advanced considerably with the explosion of molecular biology and the management of most hematologic conditions has kept pace with these scientific advances. Our understanding of the basic science of oncology, molecular biology, genetics and the management of oncologic conditions has undergone a seismic change. The previous age of dismal and almost consistent fatal outcomes for most childhood cancers has been replaced by an era in which most childhood cancers are cured. This has been made possible not only because of advances in oncology but because of the parallel development of radiology, radiologic oncology and surgery as well as supportive care such as the pre-emptive use of antibiotics and blood component therapy. It has been a privilege to be a witness and participant in this great evolution over the past 50 years. Yet we still have a long way to go as current advances are superseded by therapy based upon the application of knowledge garnered from an accurate understanding of the fundamental biology of cancer.

In the early days of hematology-oncology practice, hematology dominated and occupied most of the practitioner's time because most patients with cancer had a short life span and limited therapeutic modalities were available.

Automated electronic blood-counting equipment has enabled valuable red cell parameters such as mean corpuscular volume (MCV) and red cell distribution width (RDW) to be applied in routine clinical practice. This advance permitted the reclassification of anemias based on MCV and RDW. Previously these parameters were determined by microscopy with considerable observer variability. The attempt at a more accurate determination of any one of these parameters was a laborious, time-consuming enterprise relegated only as a demonstration in physiology laboratories.

Rh hemolytic disease of the newborn and its management by exchange transfusion, which occupied a major place in the hematologists' domain, has now become almost extinct in developed countries due to the use of Rh immunoglobulin.

The description of the various genetic differences in patients with vitamin B_{12} deficiency has opened up new vistas of our understanding of cobalamin transport and metabolism. Similar advances have occurred with reference to folate transport and metabolism.

Gaucher disease has been converted from a crippling and often disabling disease to one where patients can live a normal and productive life thanks to the advent of enzyme replacement therapy. Replacement therapy has also been developed for other inborn errors of metabolism.

Aplastic anemia has been transformed from a near death sentence to a disease with hope and cure in 90% of patients thanks to immunosuppressive therapies, hematopoietic stem cell transplantation and advanced supportive care. The emergence of clonal disease years later in patients treated medically with immunosuppressive therapy, however, does present a challenge. The discovery of the various genes responsible for Fanconi anemia and other inherited bone marrow failure syndromes has revealed heretofore unimaginable advances in our understanding of DNA repair, telomere maintenance, ribosome biology and other new fields of biology. The relationship of these syndromes to the development of various cancers may hold the key to our better understanding of the etiology of cancer as well as birth defects.

The hemolytic anemias, previously lumped together as a group of congenital hemolytic anemias, can now be identified as separate and distinct enzyme defects of the Embden–Meyerhof and hexose monophosphate pathways in intracellular red cell metabolism as well as various well-defined defects of red cell skeletal proteins due to advances in molecular biology and genetics. With improvement in electrophoretic and other biochemical techniques, hemoglobinopathies are being identified which were not previously possible.

Diseases requiring a chronic transfusion program to maintain a hemoglobin level for hemodynamic stability such as in thalassemia major frequently had marked facial characteristics with broad cheekbones and developed what was called "bronze diabetes" a bronzing of the skin along with organ damage and failure, particularly of the heart, liver, beta cells of the pancreas and other tissues due to secondary hemachromatosis because of excessive iron deposition. The clinical findings attributed to extramedullary hematopoiesis are essentially of historic interest because of the development and widespread use of proper transfusion and chelation regimens. However, the full potential of the role of intravenous and oral chelating agents is yet to be realized due to the problems of compliance with difficult treatment regimens and also due to failure of some patients to respond adequately. Advances in our understanding of the biology of iron absorption and transport at the molecular level hold out promise for further improvement in the management of these conditions. Curative therapy in thalassemia major and other conditions by hematopoietic stem cell transplantation in suitable cases is widely available today.

In the treatment of idiopathic thrombocytopenic purpura, intravenous gammaglobulin and anti-D immunoglobulin have been added to the armamentarium of management and are useful in specific indications in patients with this disorder.

Major advances in the management of hemophilia have included the introduction of commercially available products for replacement therapy which has saved these patients from a life threatened by hemorrhage into joints, muscles and vital organs. Surgery has become possible in hemophilia without the fear of being unable to control massive hemorrhage during or after surgery. The devastating clinical history of tragic hemophilia outcomes has been relegated to the pages of medical history. Patients with inhibitors, however, still remain a clinical challenge. The whole subject of factors associated with inherited thrombophilia such as mutations of factor V, prothrombin G20210A and 5,10-methylenetetrahydrofolate reductase as well as the roles of antithrombin, protein C and S deficiency and antiphospholipid antibodies in the development of thrombosis has opened new vistas of understanding of thrombotic disorders. Notwithstanding these advances, the management of these patients still presents a clinical challenge.

There are few diseases in which advances in therapy have been as dramatic as in the treatment of childhood leukemia. In my early days as a medical student, the only available treatment for leukemia was blood transfusion. Patients never benefitted from a remission and died within a few months. Steroids and single-agent chemotherapy, first with aminopterin, demonstrated the first remissions in leukemia and raised hope of a potential cure; however, relapse ensued in almost all cases and most patients died within the first year of diagnosis. In most large pediatric oncology centers there were few patients with leukemia as the disease was like a revolving door – diagnosis and death. The development of multiple-agent chemotherapy for induction, consolidation and maintenance, CNS prophylaxis and supportive care ushered in a new era of cure for patients with leukemia. These principles were refined over time by more accurate classification of acute leukemia using morphological, cytochemical, immunological, cytogenetic and molecular criteria which replaced the crude microscopic and highly subjective characteristics previously utilized for the classification of leukemia cells. These advances paved the way for the development of specific protocols of treatment for different types of leukemia. The management of leukemia was further refined by risk stratification, response-based therapy and identification of minimal residual disease, all of which have led to additional chemotherapy or different chemotherapy protocols, resulting in an enormous improvement in the cure rate of acute leukemia. The results have been enhanced by modern supportive care including antibiotic, antifungal, antiviral therapy and blood component therapy. Those patients whose leukemia is resistant to treatment or who have recurrences can be successfully treated by advances that have occurred with the development of hematopoietic stem cell transplantation. The challenge of finding appropriate, unrelated transplantation donors has been ameliorated by molecular HLA-typing techniques and the development of large, international donor registries. Emerging targeted and pharmacogenetic therapies hold great promise for the future.

Hodgkin disease, originally defined as a "fatal illness of the lymphatics," is a disease that is cured in most cases today. Initially, Hodgkin disease was treated with high-dose radiation to the sites of identifiable disease resulting in some cures but with major life-long radiation damage to normal tissues because of the use of cobalt machines and higher doses of radiation than is currently used. The introduction of nitrogen mustard early on, as a single-agent chemotherapy, improved the prognosis somewhat. A major breakthrough occurred with the staging of Hodgkin disease and the use of radiation therapy coupled with multiple-agent chemotherapy (MOPP). With time this therapeutic approach was considerably refined to include reduction in radiation dosage and field and a modification of the chemotherapy regimens designed to reduce toxicity of high-dose radiation and of some of the chemotherapeutic agents. These major advances in treatment ushered in a new era in the management and cure of most patients with this disease. The management of Hodgkin disease, however, did go through a phase of staging laparotomy and splenectomy with a great deal of unnecessary surgery and splenectomies being performed. There were considerable surgical morbidity and post-splenectomy sepsis, occasionally fatal, that occurred in some cases. With the advent of MRI and PET scans, surgical staging, splenectomy and lymphangiography have become unnecessary.

Non-Hodgkin lymphoma, previously considered a dismal disease, is another success story. Improvement in histologic, immunologic and cytogenetic techniques has made the diagnosis and classification more accurate. The development of a staging system and multiagent chemotherapy was a major step forward in the management of this disease. This, together with enhanced supportive care including the successful management of tumor lysis syndrome, have all contributed to the excellent results that occur today.

Brain tumors were treated by surgery and radiation therapy with devastating results due to primitive neurosurgical techniques and radiation damage. The advent of MRI scans has made the diagnosis and the determination of the extent of disease more accurate. Major technical advances in neurosurgery such as image guidance, which allows 3D mapping of tumors, functional mapping and electrocorticography, which allow pre- and intraoperative differentiation of normal and tumor tissue, the use of ultrasonic aspirators and neuroendoscopy, have all improved the results of neurosurgical intervention and has resulted in less surgical damage to normal brain tissue. These neurosurgical advances, coupled with the use of various chemotherapy regimens, have resulted in considerable improvements in outcome for some. This field, however, still remains an area begging for a better understanding of the optimum management of these devastating and often fatal tumors.

In the early days of pediatric oncology Wilms tumor in its early stages was cured with surgery followed by radiation therapy. The diagnosis was made with an intravenous pyelogram and inferior venocavogram and chest radiography was employed to detect pulmonary metastases. The diagnosis and extent of disease were better defined when CT of the abdomen and chest became available. The development of the clinicopathological staging system and the more accurate definition of the histology into favorable and unfavorable histologic types, allowed for more focused treatment with radiation and multiple chemotherapy agents, for different stages and histology of Wilms tumor, resulting in the excellent outcomes observed today. The success of the National Wilms Tumor Study Group (NWTSG), more than any other effort, provided the model for cooperative group therapeutic cancer trials, which in large measure have been responsible for advances in treatment of Wilms tumor.

The diagnosis of neuroblastoma and its differentiation histologically from other round blue cell tumors such as rhabdomyosarcoma, Ewing sarcoma and non-Hodgkin lymphoma was difficult before neurone-specific enolase cytochemical staining, Shimada histopathology classification, N-myc gene status, VMA and HVA determinations and MIBG scintigraphy were introduced. In the future, new molecular approaches will offer diagnostic tools to provide even greater precision for diagnosis. The existing markers coupled with a staging system have enabled neuroblastoma to be assigned to various risk group categories with specific multimodality treatment protocols for each risk group which has improved the prognosis in this disease. Improvements in diagnostic radiology determining extent of disease and modern surgical techniques have enhanced the advances in chemotherapy in this condition. However, despite all the advances that have occurred, disseminated neuroblastoma still has a poor prognosis.

Major advances have occurred in rhabdomyosarcoma treatment over the years. Early on treatment of this disease was characterized by mutilating surgery including amputation and a generally poor outcome. More accurate histologic diagnosis, careful staging, judicious surgery, combination chemotherapy and radiotherapy have all contributed a great deal to the improved cure rates with significantly less disability.

Malignant bone tumors had a terrible prognosis. They were generally treated by amputation of the limb with the primary tumor; however, this was usually followed by pulmonary metastases and death. The major advance in the treatment of this disease came with the use of high-dose methotrexate and leukovorin rescue which, coupled with limb salvage treatment, has resulted in improved survival and quality-of-life outcomes.

The advances in the treatment of hepatoblastoma were made possible by safer anesthesia, more radical surgery, intensive postoperative management together with multiagent chemotherapy and more recently the increased use of liver transplantation. These advances have allowed many patients to be cured compared to past years.

Histiocytosis is a disease that has undergone many name changes from Letter-Siwe disease, Hand-Schüller-Christian disease and Eosinophilic Granuloma to the realization that these entities are one disease, re-named histiocytosis X (to include all three entities) to its present name of Langerhans Cell Histiocytosis (LCH) due to the realization that these entities have one pathognomonic pathologic feature that is the immunohistochemical presence of Langerhans cells defined in part by expression of CD1a or langerin (CD207), which induces the formation of Birbeck granules. Advances have occurred in the management of this disease by an appreciation of risk stratification depending on number and type of organs involved in this disease process as well as by early response to therapy. Once this was established, systemic therapy was developed for the various risk groups which led to appropriate and improved therapy with better overall results.

Until a final prevention or cure for cancer in children is at hand, hematopoietic stem cell transplantation must be viewed as a major advance. Improved methods for tissue typing, the use of umbilical and peripheral blood stem cells, improved preparative regimens, including intensity-reduced approaches and better management of graft-versus-host disease has made this an almost routine treatment modality for many metabolic disorders, hemoglobinopathies and malignant diseases following ablative chemotherapy in chemotherapy-sensitive tumors. Post-transplantation support with antibiotic, antifungal, antiviral, hematopoietic growth factors and judicious use of blood component therapy has made this procedure safer than it was in years gone by.

The recognition of severe and often permanent damage to organs and life-threatening complications from chemotherapy and radiation therapy has, over the years, led to regimens consisting of combination chemotherapy at reduced doses and reduction in dose and field of radiation with improved outcome. An entire new scientific discipline, Survivorship, has arisen because of the near 80% overall cure rate for childhood cancer. Focusing on the improvement of the quality of life of survivors coupled with research in this new discipline gives hope that many of the remaining long-term effects of cancer chemotherapy in children will be mitigated and possibly eliminated.

Major advances have occurred in the management of chemotherapy-induced vomiting and pain management because of the greater recognition and attention to these issues and the discovery of many new, effective drugs to deal with these symptoms. The availability of symptom control and palliative care has provided a degree of comfort for children undergoing chemotherapy, radiation and surgery that did not exist only a few years ago.

Hematologist-oncologists today are privileged to practice their specialty in an era in which most oncologic diseases in children are curable and at a time when national and international cooperative groups are making major advances in the management of these diseases and when basic research is at the threshold of making major breakthroughs. The present practice is grounded in evidence-based research that has been and is still being performed by hematologist-oncologists and researchers that form the foundation for ongoing advances. Today we stand on the shoulders of others, which permits us to see future advances unfold to benefit generations of children. While we bask in the glory of past achievements, we should always be cognizant that much work remains to be done until the permanent cure of all childhood malignancies and blood diseases is at hand.

This book encompasses the advances in the management of childhood cancer which have been accomplished to date and which have become the standard of care.

Philip Lanzkowsky, M.B., Ch.B., M.D.,
Sc.D. (honoris causa), F.R.C.P., D.C.H., F.A.A.P.

Preface to the Fifth Edition

The fifth edition of the *Manual of Pediatric Hematology and Oncology* differs considerably from previous editions but has retained the original intent of the author to offer a concise manual of predominantly clinical material culled from personal experience and to be an immediate reference for the diagnosis and management of hematologic and oncologic diseases. I have resisted succumbing to the common tendency of writing a comprehensive tome which is not helpful to the practicing hematologist-oncologist at the bedside. The book has remained true to its original intent.

The information included at all times keeps "the eye on the ball" to ensure that pertinent, up-to-date, practical clinical advice is presented without extraneous information, however interesting or pertinent this information may be in a different context.

The book differs from previous editions in many respects. The number of contributors has been considerably expanded drawing on the expertise of leaders in different subjects from various institutions in the United States. Increased specialization within the field of hematology and oncology has necessitated including this large a number of contributors in order to bring to the reader balanced and up-to-date information for the care of patients. In addition, the number of chapters has increased from 27, in the previous edition, to 33. The reason for this is that many of the chapters, such as hemolytic anemia and coagulation, had become so large and the subject so extensive that they were better handled by subdividing the chapter into a number of smaller chapters. An additional chapter on the psychosocial aspects of cancer for children and their families, not present in previous editions, has been added.

Some chapters have been extensively revised and re-written where advancement in knowledge has dictated this approach, e.g., Hodgkin lymphoma, neuroblastoma and rhabdomyosarcoma and other soft-tissue sarcomas, whereas other chapters have been only slightly modified. In nearly all the chapters there has been significant change in the management and treatment section reflecting advances that have occurred in these areas.

This edition has retained the essential format written and developed decades ago by the author and, with usage over the years, has proven to be highly effective as a concise, practical, up-to-date guide replete with detailed tables, algorithms and flow diagrams for investigation and management of hematologic and oncologic conditions. The tables and flow diagrams included in the book have been updated using the latest information and the most recent protocols of treatment, which have received general acceptance and have become the standard of care, have been included. In a book with so many details, errors inevitably occur. I do not know where they are because if I did they would have been corrected. I apologize in advance for any inaccuracies that may have crept in inadvertently.

The four previous editions of this book were published when the name of the hospital was the Schneider Children's Hospital. Effective April 1, 2010 the name of the hospital was changed to the Steven and Alexandra Cohen Children's Medical Center of New York.

I would like to acknowledge Morris Edelman, MB, BCh, B.Sc (Laboratory Medicine) for his contribution in reviewing the pathology on Hodgkin disease.

I thank Rose Grosso for her untiring efforts in the typing and coordination of the various phases of the development of this edition.

Philip Lanzkowsky, M.B., Ch.B., M.D.,
Sc.D. (honoris causa), F.R.C.P., D.C.H., F.A.A.P.

Preface to the Fourth Edition

This edition of the *Manual of Pediatric Hematology and Oncology* is the fourth edition and the sixth book written by the author on pediatric hematology and oncology. The first book written by the author 25 years ago was exclusively on pediatric hematology and its companion book, exclusively on pediatric oncology, was written 3 years later. The book reviewers at the time suggested that these two books be combined into a single book on pediatric hematology and oncology and the first edition of the *Manual of Pediatric Hematology and Oncology* was published by the author in 1989.

It is from these origins that this 4th edition arises – the original book written in its entirety by the author was 456 pages – has more than doubled in size. The basic format and content of the clinical manifestations, diagnosis and differential diagnosis has persisted with little change as originally written by the author. The management and treatment of various diseases have undergone profound changes over time and these aspects of the book have been brought up-to-date by the subspecialists in the various disease entities. The increase in the size of the book is reflective of the advances that have occurred in both hematology and oncology over the past 25 years. Despite the size of the book, the philosophy has remained unchanged over the past quarter century. The author and his contributors have retained this book as a concise manual of personal experiences on the subject over these decades rather than developing a comprehensive tome culled from the literature. Its central theme remains clinical as an immediate reference for the practicing pediatric hematologist-oncologist concerned with the diagnosis and management of hematologic and oncologic diseases. It is extremely useful for students, residents, fellows and pediatric hematologists and oncologists as a basic reference assembling in one place, essential knowledge required for clinical practice.

This edition has retained the essential format written and developed decades ago by the author and, with usage over the years, has proven to be highly effective as a concise, practical, up-to-date guide replete with detailed tables, algorithms and flow diagrams for investigation and management of hematologic and oncologic conditions. The tables and flow diagrams have been updated with the latest information and the most recent protocols of treatment, that have received general acceptance and have produced the best results, have been included in the book.

Since the previous edition, some five years ago, there have been considerable advances particularly in the management of oncologic disease in children and these sections of the book have been completely rewritten. In addition, advances in certain areas have required that other sections of the book be updated. There has been extensive revision of certain chapters such as on Diseases of the White Cells, Lymphoproliferative Disorders, Myeloproliferative Disorders and Myelodysplastic Syndromes and Bone Marrow Failure. Because of the extensive advances in thrombosis we have rewritten that entire section contained in the chapter on Disorders of Coagulation to encompass recent advances in that area. The book, like its previous editions, reflects the practical experience of the author and his colleagues based on half a century of clinical experience. The number of contributors has been expanded but consists essentially of the faculty of the Division of Hematology Oncology at the Schneider Children's Hospital, all working together to provide the readers of the manual with a practical guide to the management of the wide spectrum of diseases within the discipline of pediatric hematology-oncology.

I would like to thank Laurie Locastro for her editorial assistance, cover design and for her untiring efforts in the coordination of the various phases of the production of this edition. I also appreciate the efforts of Lawrence Tavnier for his expert typing of parts of the manuscript and would like to thank Elizabeth Dowling and Patrician Mastrolembo for proof reading of the book to ensure its accuracy.

Philip Lanzkowsky, M.B., Ch.B., M.D.,
Sc.D. (honoris causa), F.R.C.P., D.C.H., F.A.A.P.

Preface to the Third Edition

This edition of the *Manual of Pediatric Hematology and Oncology,* published five years after the second edition, has been written with the original philosophy in mind. It presents the synthesis of experience of four decades of clinical practice in pediatric hematology and oncology and is designed to be of paramount use to the practicing hematologist and oncologist. The book, like its previous editions, contains the most recent information from the literature coupled with the practical experience of the author and his colleagues to provide a guide to the practicing clinician in the investigation and up-to-date treatment of hematologic and oncologic diseases in childhood.

The past five years have seen considerable advances in the management of oncologic diseases in children. Most of the advances have been designed to reduce the immediate and long-term toxicity of therapy without influencing the excellent results that have been achieved in the past. This has been accomplished by reducing dosages, varying the schedules of chemotherapy, and reducing the field and volume of radiation.

The book is designed to be a concise, practical, up-to-date guide and is replete with detailed tables, algorithms, and flow diagrams for investigation and management of hematologic and oncologic conditions. The tables and flow diagrams have been updated with the latest information, and the most recent protocols that have received general acceptance and have produced the best results have been included in the book.

Certain parts of the book have been totally rewritten because our understanding of the pathogenesis of various diseases has been altered in the light of modern biological investigations. Once again, we have included only those basic science advances that have been universally accepted and impinge on clinical practice.

I thank Ms. Christine Grabowski, Ms. Lisa Phelps, Ms. Ellen Healy and Ms. Patricia Mastrolembo for their untiring efforts in the coordination of the writing and various phases of the development of this edition. Additionally, I acknowledge our fellows, Drs. Banu Aygun, Samuel Bangug, Mahmut Celiker, Naghma Husain, Youssef Khabbase, Stacey Rifkin-Zenenberg, and Rosa Ana Gonzalez, for their assistance in culling the literature.

I also thank Dr. Bhoomi Mehrotra for reviewing the chapter on bone marrow transplantation, Dr. Lorry Rubin for reviewing the sections of the book dealing with infection, and Dr. Leonard Kahn for reviewing the pathology.

Philip Lanzkowsky, M.B., Ch.B., M.D.,
Sc.D. (honoris causa), F.R.C.P., D.C.H., F.A.A.P.

Preface to the Second Edition

This edition of the *Manual of Pediatric Hematology and Oncology*, published five years after the first edition, has been written with a similar philosophy in mind. The basic objective of the book is to present useful clinical information from the recent literature in pediatric hematology and oncology and to temper it with experience derived from an active clinical practice.

The manual is designed to be a concise, practical, up-to-date book for practitioners responsible for the care of children with hematologic and oncologic diseases by presenting them with detailed tables and flow diagrams for investigation and clinical management.

Since the publication of the first edition, major advances have occurred, particularly in the management of oncologic diseases in children, including major advances in recombinant human growth factors and bone marrow transplantation. We have included only those basic science advances that have been universally accepted and impinge on clinical practice.

I would like to thank Dr. Raj Pahwa for his contributions on bone marrow transplantation, Drs. Alan Diamond and Leora Lanzkowsky-Diamond for their assistance with the neuro-radiology section, and Christine Grabowski and Lisa Phelps for their expert typing of the manuscript and for their untiring assistance in the various phases of the development of this book.

Philip Lanzkowsky, M.B., Ch.B., M.D.,
Sc.D. (honoris causa), F.R.C.P., D.C.H., F.A.A.P.

Preface to the First Edition

The *Manual of Pediatric Hematology and Oncology* represents the synthesis of personal experience of three decades of active clinical and research endeavors in pediatric hematology and oncology. The basic orientation and intent of the book is clinical, and the book reflects a uniform systematic approach to the diagnosis and management of hematologic and oncologic diseases in children. The book is designed to cover the entire spectrum of these diseases, and although emphasis is placed on relatively common disorders, rare disorders are included for the sake of completion. Recent developments in hematology-oncology based on pertinent advances in molecular genetics, cytogenetics, immunology, transplantation, and biochemistry are included if the issues have proven value and applicability to clinical practice.

Our aim in writing this manual was to cull pertinent and useful clinical information from the recent literature in pediatric hematology and oncology and to temper it with experience derived from active clinical practice. The result, we hope, is a concise, practical, readable, up-to-date book for practitioners responsible for the care of children with hematologic and oncologic diseases. It is specifically designed for the medical student and practitioner seeking more detailed information on the subject, the pediatric house officer responsible for the care of patients with these disorders, the fellow in pediatric hematology-oncology seeking a systemic approach to these diseases and a guide in preparation for the board examinations, and the practicing pediatric hematologist-oncologist seeking another opinion and approach to these disorders. As with all brief texts, some dogmatism and "matters of opinion" have been unavoidable in the interests of clarity. The opinions expressed on management are prudent clinical opinions; and although they may not be accepted by all, pediatric hematologists-oncologists will certainly find a consensus. The reader is presented with a consistency of approach and philosophy describing the management of various diseases rather than with different managements derived from various approaches described in the literature. Where there are divergent or currently unresolved views on the investigation or management of a particular disease, we have attempted to state our own opinion and practice so as to provide some guidance rather than to leave the reader perplexed.

The manual is not designed as a tome containing the minutiae of basic physiology, biochemistry, genetics, molecular biology, cellular kinetics, and other esoteric and abstruse detail. These subjects are covered extensively in larger works. Only those basic science advances that impinge on clinical practice have been included here. Each chapter stresses the pathogenesis, pathology, diagnosis, differential diagnosis, investigations, and detailed therapy of hematologic and oncologic diseases seen in children.

I would like to thank Ms. Joan Dowdell and Ms. Helen Witkowski for their expert typing and for their untiring assistance in the various phases of the development of this book.

Philip Lanzkowsky, M.D.,
F.R.C.P., D.C.H., F.A.A.P.

CHAPTER 1

Classification and Diagnosis of Anemia in Children

Anemia can be defined as a reduction in hemoglobin concentration, hematocrit, or number of red blood cells per cubic millimeter. The lower limit of the normal range is set at two standard deviations below the mean for age and sex for the normal population.[*]

The first step in diagnosis of anemia is to establish whether the abnormality is isolated to a single cell line (red blood cells only) or whether it is part of a multiple cell line abnormality (red cells, white cells and platelets). Abnormalities of two or three cell lines usually indicate one of the following:

- bone marrow involvement, (e.g., aplastic anemia, leukemia), or
- an immunologic disorder (e.g., connective tissue disease or immunoneutropenia, idiopathic thrombocytopenic purpura [ITP] or immune hemolytic anemia singly or in combination) or
- sequestration of cells (e.g., hypersplenism).

Table 1-1 presents an etiologic classification of anemia and the diagnostic features in each case.

The *blood smear* is very helpful in the diagnosis of anemia. It establishes whether the anemia is hypochromic, microcytic, normocytic, macrocytic or shows spezcific morphologic abnormalities suggestive of red cell membrane disorders (e.g., spherocytes, stomatocytosis or elliptocytosis) or hemoglobinopathies (e.g. sickle cell disease, thalassemia).

The mean corpuscular volume (MCV) confirms the findings on the smear with reference to the red cell size, e.g., microcytic (<70 fl), macrocytic (>85 fl) or normocytic (72–79 fl). Figure 1-1 delineates diagnosis of anemia by examination of the smear and Table 1-2 lists the differential diagnostic considerations based on specific red cell morphologic abnormalities. The mean corpuscular hemoglobin (MCH) and mean corpuscular hemoglobin concentration (MCHC) are calculated values and generally of less diagnostic

---

[*]Children with cyanotic congenital heart disease, respiratory insufficiency, arteriovenous pulmonary shunts or hemoglobinopathies that alter oxygen affinity can be functionally anemic with hemoglobin levels in the normal range.

Manual of Pediatric Hematology and Oncology. DOI: 10.1016/B978-0-12-375154-6.00001-X

Table 1-1   Etiologic Classification and Major Diagnostic Features of Anemia in Children

Etiologic Classification	Diagnostic Features
I. Impaired red cell formation	
A. Deficiency	
Decreased dietary intake (e.g., excessive cows' milk [iron-deficiency anemia], vegan [vitamin B_{12} deficiency])	
Increased demand, e.g., Growth (iron) hemolysis (folic acid)	
Decreased absorption	
Specific: intrinsic factor lack (Vitamin B_{12})	
Generalized: malabsorption syndrome (e.g., folic acid, iron)	
Increased loss	
Acute: hemorrhage (iron)	
Chronic: gut bleeding (iron)	
Impairment in red cell formation can result from one of the following deficiencies:	
1. Iron deficiency	Hypochromic, microcytic red cells; low MCV, low MCH, low MCHC, high RDW,[a] low serum ferritin, high FEP, guaiac positivity
2. Folate deficiency	Macrocytic red cells, high MCV, high RDW, megaloblastic marrow, low serum and red cell folate
3. Vitamin B_{12} deficiency	Macrocytic red cells, high MCV, high RDW, megaloblastic marrow, low serum B_{12}, decreased gastric acidity; Schilling test positive
4. Vitamin C deficiency	Clinical scurvy
5. Protein deficiency	Kwashiorkor
6. Vitamin B_6 deficiency	Hypochromic red cells, sideroblastic bone marrow, high serum ferritin
7. Thyroxine deficiency	Clinical hypothyroidism, low T_4, high TSH
B. Bone marrow failure	
1. Failure of a single cell line	
a. Megakaryocytes[b]	
(1) Amegakaryocytic thrombocytopenic purpura with absent radii (TAR)	Limb abnormalities, thrombocytopenic purpura absent megakaryocytes
b. Red cell precursors	
(1) Congenital red cell aplasia (Diamond–Blackfan anemia)	Absent red cell precursors
(2) Acquired red cell aplasia (transient erythroblastopenia of childhood – TEC)	Absent red cell precursors
c. White cell precursors[b]	
(1) Congenital neutropenias	Neutropenia, recurrent infection

(Continued)

Table 1-1   (Continued)

Etiologic Classification	Diagnostic Features
2. Failure of all cell lines (characterized by pancytopenia and acellular or hypocellular marrow)	
a. Congenital	
(1) Fanconi anemia	Multiple congenital anomalies, chromosomal breakage
(2) Familial without anomalies	Familial history, no congenital anomalies
(3) Dyskeratosis congenita	Marked mucosal and cutaneous abnormalities
b. Acquired	
(1) Idiopathic	No identifiable cause
(2) Secondary	History of exposure to drugs, radiation, household toxins, infections; (parvovirus B19, HIV) associated immunologic disease
3. Infiltration	
a. Benign (e.g., osteopetrosis, storage diseases)	
b. Malignant	
Primary (e.g., leukemia, myelofibrosis)	Bone marrow: morphology, cytochemistry, immunologic markers, cytogenetics, molecular features
c. Secondary (e.g., neuroblastoma, lymphoma)	VMA, skeletal survey, bone marrow
4. Dyshematopoietic anemias (decreased erythropoiesis, decreased iron utilization)	
(1) Anemia of chronic disease	Evidence of systemic illness
(2) Renal failure and hepatic disease	BUN and liver-function tests
(3) Disseminated malignancy	Clinical evidence
(4) Connective tissue diseases	Rheumatoid arthritis
(5) Malnutrition	Clinical evidence
(6) Sideroblastic anemias	Hypochromic anemia, Ring sideroblasts
II. Blood loss	Overt or occult guaiac positive
III. Hemolytic anemia	
A. Corpuscular	Splenomegaly, jaundice
1. Membrane defects (spherocytosis, elliptocytosis)	Morphology, osmotic fragility
2. Enzymatic defects (pyruvate kinase, G6PD)	Autohemolysis, enzyme assays
3. Hemoglobin defects	
a. Heme	
b. Globin	
(1) Qualitative (e.g., sickle cell)	Hb electrophoresis
(2) Quantitative (e.g., thalassemia)	Quantitative HbF, A_2 content

(Continued)

Table 1-1   (Continued)

Etiologic Classification	Diagnostic Features
B. Extracorpuscular	
1. Immune	Direct antiglobulin test (Coombs' test)
a. Isoimmune	
b. Autoimmune	
(1) Idiopathic	Direct antiglobulin test, antibody
(2) Secondary	identification
Immunologic disorder (e.g., lupus)	Decreased C_3, C_4, CH_{50}-positive ANA
One cell line (e.g., red cells)	Anemia – direct antiglobulin test positive
Multiple cell line (e.g., white blood cells, platelets)	Neutropenia – immunotropenia,
2. Nonimmune (idiopathic, secondary)	thrombocytopenia – ITP

[a]RDW = coefficient of variation of the RBC distribution width (normal between 11.5% and 14.5%).
[b]Not associated with anemia.
Abbreviations: FEP, free erythrocyte protoporphyrin; G6PD, glucose-6-phosphate dehydrogenase; Hb, hemoglobin; ITP, idiopathic thrombocytopenic purpura, MCH, mean corpuscular hemoglobin; MCHC, mean corpuscular hemoglobin concentration; MCV, mean corpuscular volume; RBC, red blood cell; RDW, red cell distribution width (see definition); VMA, vanillylmandelic acid.

value. The MCH usually parallels the MCV. The MCHC is a measure of cellular hydration status. A high value (>35 g/dL) is characteristic of spherocytosis and a low value is commonly associated with iron deficiency.

MCV and red cell distribution width (RDW) indices, available from automated electronic blood-counting equipment, are extremely helpful in defining the morphology and the nature of the anemia and have led to a classification based on these indices (Table 1-3).

The MCV and reticulocyte count are helpful in the differential diagnosis of anemia (Figure 1-2). An elevated reticulocyte count suggests chronic blood loss or hemolysis; a normal or depressed count suggests impaired red cell formation. The reticulocyte count must be adjusted for the level of anemia to obtain the reticulocyte index,[*] a more accurate reflection of erythropoiesis. In patients with bleeding or hemolysis, the reticulocyte index should be at least 3%, whereas in patients with anemia due to decreased production of red cells, the reticulocyte index is less than 3% and frequently less than 1.5%.

In more refractory cases of anemia, bone marrow examination may be indicated. A bone marrow smear should be stained for iron, where indicated, to estimate iron stores and to diagnose the presence of a sideroblastic anemia. Bone marrow examination may indicate a

---

[*]Reticulocyte index = reticulocyte count × patient's hematocrit/normal hematocrit. Example: reticulocyte count 6%, hematocrit 15%, reticulocyte index = 6 × 15/45 = 2%.

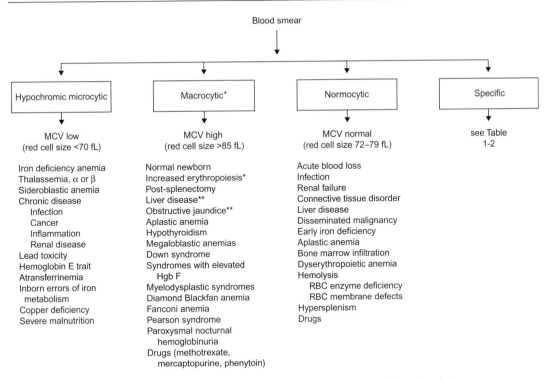

Figure 1-1 An Approach to the Diagnosis of Anemia by Examination of the Blood Smear.
⁺Spurious macrocytosis (high MCV) may be caused by macroagglutinated red cells (e.g., Mycoplasma pneumonia and autoimmune hemolytic anemia).
*Increased number of reticulocytes.
**On the basis of increased membrane resulting in an increased membrane/volume ratio. Increased membrane results from exchanges between red cell lipids and altered lipid balance in these conditions.

Table 1-2 Specific Red Cell Morphologic Abnormalities

I. Target cells
Increased surface/volume ratio (generally does not effect red cell survival)
    Thalassemic syndromes
    Hemoglobinopathies
        Hb AC or CC
        Hb SS, SC, S-Thal
        HbE (heterozygote and homozygote)
        HbD
    Obstructive liver disease
    Postsplenectomy or hyposplenic states
    Severe iron deficiency
    LCAT deficiency: congenital disorder of lecithin/cholesterol acyltransferase deficiency (corneal opacifications, proteinuria, target cells, moderately severe anemia)
    Abetalipoproteinemia

(Continued)

Table 1-2   (Continued)

II. Spherocytes

Decreased surface/volume ratio, hyperdense (>MCHC)

Hereditary spherocytosis

ABO incompatibility: antibody-coated fragment of RBC membrane removed

Autoimmune hemolytic anemia: antibody-coated fragment of RBC membrane removed

G-6-PD Deficiency

Microangiopathic hemolytic anemia (MAHA): fragment of RBC lost after impact with abnormal surface

SS disease: fragment of RBC removed in reticuloendothelial system

Hypersplenism

Burns: fragment of damaged RBC removed by spleen

Posttransfusion

Pyruvate kinase deficiency

Water-dilution hemolysis: fragment of damaged RBC removed by spleen

III. Acanthocytes (spur cells)[a]

Cells with 5–10 spicules of varying length; spicules irregular in space and thickness, with wide bases; appear smaller than normal cells because they assume a spheroid shape

Liver disease

Disseminated intravascular coagulation (and other MAHA)

Postsplenectomy or hyposplenic state

Vitamin E deficiency

Hypothyroidism

Abetalipoproteinemia: rare congenital disorder; 50–100% of cells acanthocytes; associated abnormalities (fat malabsorption, retinitis pigmentosa, neurologic abnormalities)

Malabsorptive states

IV. Echinocytes (burr cells)[a]

10–30 spicules equal in size and evenly distributed over RBC surface; caused by alteration in extracellular or intracellular environment

Artifact

Uremia

Dehydration

Liver disease

Pyruvate kinase deficiency

Peptic ulcer disease or gastric carcinoma

Immediately after red cell transfusion

Rare congenital anemias due to decreased intracellular potassium

V. Pyknocytes[a]

Distorted, hyperchromic, contracted RBC; can be similar to echinocytes and acanthocytes

VI. Schistocytes

Helmet, triangular shapes, or small fragments. Caused by fragmentation upon impact with abnormal vascular surface (e.g., fibrin strand, vasculitis, artificial surface in circulation)

Disseminated intravascular coagulation (DIC)

Severe hemolytic anemia (e.g., G6PD deficiency)

Microangiopathic hemolytic anemia

Hemolytic uremic syndrome

Prosthetic cardiac valve, abnormal cardiac valve, cardiac patch, coarctation of the aorta

Connective tissue disorder (e.g., SLE)

Kasabach–Merritt syndrome

[a]May be morphologically indistinguishable.   *(Continued)*

Table 1-2 (Continued)

Purpura fulminans
Renal vein thrombosis
Burns (spheroschistocytes as a result of heat)
Thrombotic thrombocytopenia purpura
Homograft rejection
Uremia, acute tubular necrosis, glomerulonephritis
Malignant hypertension
Systemic amyloidosis
Liver cirrhosis
Disseminated carcinomatosis
Chronic relapsing schistocytic hemolytic anemia

VII. Elliptocytes

Elliptical cells, normochromic; seen normally in less than 1% of RBCs; larger numbers occasionally seen in a normal patient

Hereditary elliptocytosis
Iron deficiency (increased with severity, hypochromic)
SS disease
Thalassemia major
Severe bacterial infection
SA trait
Leukoerythroblastic reaction
Megaloblastic anemias
Any anemia may occasionally present with up to 10% elliptocytes
Malaria

VIII. Teardrop cells

Shape of drop, usually microcytic, often also hypochromic

Newborn
Thalassemia major
Leukoerythroblastic reaction
Myeloproliferative syndromes

IX. Stomatocytes

Has a slit-like area of central pallor

Normal (in small numbers)
Hereditary stomatocytosis
Artifact
Thalassemia
Acute alcoholism
Rh null disease (absence of Rh complex)
Liver disease
Malignancies

X. Nucleated red blood cells

Not normal in the peripheral blood beyond the first week of life

Newborn (first 3–4 days)
Intense bone marrow stimulation
　Hypoxia (especially postcardiac arrest)
　Acute bleeding
　Severe hemolytic anemia (e.g., thalassemia, SS hemoglobinopathy)
Congenital infections (e.g., sepsis, congenital syphilis, CMV, rubella)

(Continued)

Table 1-2   (Continued)

Postsplenectomy or hyposplenic states: spleen normally removes nucleated RBC

Leukoerythroblastic reaction: seen with extramedullary hematopoiesis and bone marrow replacement; most commonly leukemia or solid tumor – fungal and mycobacterial infection may also do this; leukoerythroblastic reaction is also associated with teardrop red cells, 10,000–20,000 WBC with small to moderate numbers of metamyelocytes, myelocytes and promyelocytes; thrombocytosis with large bizarre platelets

Megaloblastic anemia

Dyserythropoietic anemias

XI. Blister cells

Red cell area under membrane, free of hemoglobin, appearing like a blister

G6PD deficiency (during hemolytic episode)

SS disease

Pulmonary emboli

XII. Basophilic stippling

Coarse or fine punctate basophilic inclusions that represent aggregates of ribosomal RNA

Hemolytic anemias (e.g., thalassemia trait)

Iron-deficiency anemia

Lead poisoning

XIII. Howell–Jolly bodies

Small, well-defined, round, densely stained nuclear-remnant inclusions; 1 μm in diameter; centric in location

Postsplenectomy or hyposplenia

Newborn

Megaloblastic anemias

Dyserythropoietic anemias

A variety of types of anemias (rarely iron-deficiency anemia, hereditary spherocytosis)

XIV. Cabot's Ring bodies

Nuclear remnant ring configuration inclusions

Pernicious anemia

Lead toxicity

XV. Heinz bodies

Denatured aggregated hemoglobin

Normal in newborn

Thalassemia

Asplenia

Chronic liver disease

Heinz body hemolytic anemia

normoblastic, megaloblastic, or sideroblastic morphology. Figure 1-3 presents the causes of each of these findings.

Table 1-5 lists various laboratory studies helpful in the investigation of a patient with anemia.

Table 1-3 Classification of Nature of the Anemia Based on MCV and RDW

	MCV Low	MCV Normal	MCV High
RDW Normal	*Microcytic Homogeneous*	*Normocytic Homogeneous*	*Macrocytic Homogeneous*

	Heterozygous thalassemia
	Chronic disease
		Normal
		Chronic disease
		Chronic liver disease
		Nonanemic hemoglobinopathy
		  (e.g., AS, AC)
		Chemotherapy
		Chronic myelocytic leukemia
		Hemorrhage
		Hereditary spherocytosis
			Inherited bone marrow
			  failure syndromes
			Preleukemia

RDW High	*Microcytic Heterogeneous*	*Normocytic Heterogeneous*	*Macrocytic Heterogeneous*

	Iron deficiency
	S β-thalassemia
	Hemoglobin H
	Red cell
	  Fragmentation
	  disorders
		Early iron or folate
		  deficiency
		Mixed deficiencies
		Hemoglobinopathy
		  (e.g., SS)
		Myelofibrosis
		Sideroblastic anemia
			Folate deficiency
			Vitamin B_{12}
			  deficiency
			Immune hemolytic
			  anemia
			Cold agglutinins

Abbreviations: MCV, mean corpuscular volume; RDW, red cell distribution width, which is coefficient of variation of RBC distribution width (normal, 11.5–14.5%).

The investigation of anemia entails the following steps:

- Detailed **history** and **physical examination** (see Table 1-1)
- **Complete blood count**, to establish whether the anemia is only due to a one-cell line (red cell line) or part of a three-cell line abnormality (abnormality of red cell count, white blood cell count and platelet count)
- Determination of the **morphologic characteristics** of the anemia based on blood smear (Table 1-2) and consideration of the **MCV** (Figures 1-1, 1-2 and Table 1-3) and **RDW** (Table 1-3) and morphologic consideration of **white blood cell and platelet morphology**
- Reticulocyte count as a reflection of erythropoiesis (Figure 1-2)
- Determination if there is evidence of a hemolytic process by:
  - Consideration of the clinical features suggesting hemolytic disease (Table 1-4)
  - Laboratory demonstration of the presence of a hemolytic process (Table 1-5)
  - Determination of the precise cause of the hemolytic anemia by special hematologic investigations (Table 1-5).

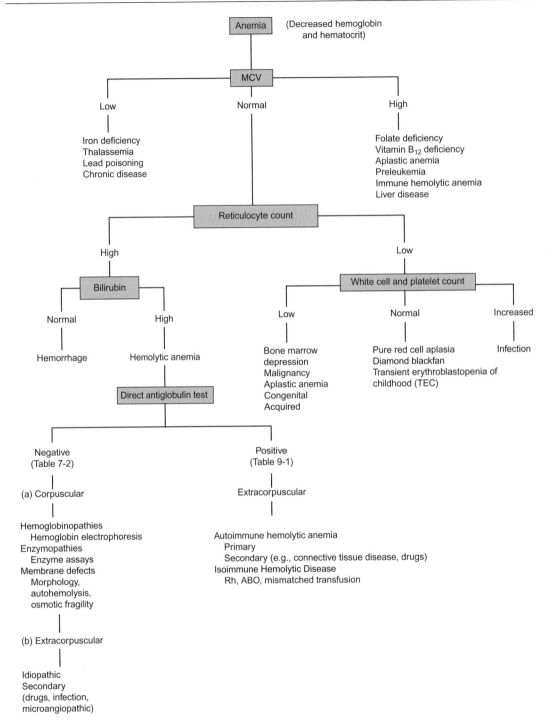

Figure 1-2 Approach to the Diagnosis of Anemia by MCV and Reticulocyte Count.

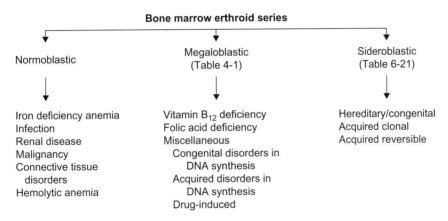

Figure 1-3 Causes of Normoblastic, Megaloblastic and Sideroblastic Bone Marrow Morphology.

Table 1-4   The Clinical Features Suggestive of a Hemolytic Process

- Ethnic factors – incidence of sickle gene carrier in the black population (8%), high incidence of thalassemia trait in people of Mediterranean ancestry and high incidence of glucose-6-phosphate dehydrogenase (G6PD) deficiency among Sephardic Jews
- Age factors – anemia and jaundice in an Rh-positive infant born to a mother who is Rh negative or a group A or group B infant born to a group O mother (setting for a hemolytic anemia)
- History of anemia, jaundice or gallstones in family
- Persistent or recurrent anemia associated with reticulocytosis
- Anemia unresponsive to hematinics
- Intermittent bouts or persistent indirect hyperbilirubinemia
- Splenomegaly
- Hemoglobinuria
- Presence of multiple gallstones
- Chronic leg ulcers
- Development of anemia or hemoglobinuria after exposure to certain drugs
- Cyanosis without cardiorespiratory distress
- Polycythemia (2,3-diphosphoglycerate mutase deficiency)
- Dark urine due to dipyrroluria (unstable hemoglobins, thalassemia and ineffective erythropoiesis)

- Bone marrow aspiration, if required, to examine erythroid, myeloid and megakaryocytic morphology to determine whether there is normoblastic, megaloblastic or sideroblastic erythropoiesis and to exclude marrow pathology (e.g. aplastic anemia, leukemia and benign or malignant infiltration of the bone marrow) (Figure 1-3)
- Determination of underlying cause of anemia by additional tests (Table 1-5).

Table 1-5   Laboratory Studies in the Investigation of a Patient with Anemia

Usual initial studies
  Hemoglobin and hematocrit determination
  Erythrocyte count and red cell indices, including MCV and RDW
  Reticulocyte count
  Study of stained blood smear
  Leukocyte count and differential count
  Platelet count
Suspected iron deficiency
  Free erythrocyte protoporphyrin
  Serum ferritin levels
  Stool for occult blood
  ^{99m}Tc pertechnetate scan for Meckel's diverticulum – if indicated
  Endoscopy (upper and lower bowel) – if indicated
Suspected vitamin B$_{12}$ or folic acid deficiency
  Bone marrow
  Serum vitamin B$_{12}$ level
  Serum folate level
  Gastric analysis after histamine injection
  Vitamin B$_{12}$ absorption test (radioactive cobalt) (Schilling test)
Suspected hemolytic anemia
Evidence of red cell breakdown
    Blood smear – red cell fragments (schistocytes), spherocytes, target cells
    Serum bilirubin level
    Serum haptoglobin
    Plasma hemoglobin level
    Urinary urobilinogen
    Hemoglobinuria
Evidence of increased erythropoeisis (in response to hemoglobin reduction)
    Reticulocytosis – frequently up to 10–20%; rarely, as high as 80%
    Increased mean corpuscular volume (MCV) due to the presence of reticulocytosis and increased RDW
      as the hemoglobin level falls
    Increased normoblasts in blood smear
    Specific morphologic abnormalities – sickle cells, target cells, basophilic stippling, irregularly
      contracted cells (schistocytes) and spherocytes
    Erythroid hyperplasia of the bone marrow – erthroid/myeloid ratio in the marrow increasing from 1:5
      to 1:1
    Expansion of marrow space in chronic hemolysis resulting in:
      a. prominence of frontal bones
      b. broad cheekbones
      c. widened intra-trabecular spaces, hair-on-end appearance of skull radiographs
      d. biconcave vertebrae with fish-mouth intervertebral spaces
Evidence of type of corpuscular hemolytic anemia
    Membrane defects
      Blood smear: spherocytes, ovalocytes, pyknocytes, stomatocytes
      Osmotic fragility test (fresh and incubated)
      Autohemolysis test

(Continued)

Table 1-5 (Continued)

Hemoglobin defects
    Blood smear: sickle cells, target cells
    Sickling test
    Hemoglobin electrophoresis
    Quantitative hemoglobin F determination
    Kleihauer–Betke smear
    Heat-stability test for unstable hemoglobin
Enzymes defects
    Heinz-body preparation
    Autohemolysis test
    Specific enzyme assay
Evidence of type of extracorpuscular hemolytic anemia
Immune
    Direct antiglobulin test: IgG (gamma), C′3 (complement), broad-spectrum both gamma and
        complement
    Flow cytometric analysis of red cells with monoclonal antibodies to GP1-linked surface antigens for
        PNH
    Donath-Landsteiner antibody
    ANA
Suspected aplastic anemia or leukemia
Bone marrow (aspiration and biopsy) – cytochemistry, immunologic markers, chromosome analysis
Skeletal radiographs
Other tests often used especially to diagnose the primary disease
Viral serology, e.g., HIV
ANA, complement, CH_{50}
Blood urea, creatinine, T_4, TSH
Tissue biopsy (skin, lymph node, liver)

Suggested Reading

Bessman JD, Gilmer PR, Gardner FH. Improved classification of anemias by MCV and RDW. Am J Clin
    Pathol. 1983;80:322.
Blanchette V, Zipursky A. Assessment of anemia in newborn infants. Clin Perinatol. 1984;11:489.
Lanzkowsky P. Diagnosis of anemia in the neonatal period and during childhood. p. 3. In: Pediatric
    Hematology-Oncology: A Treatise for the Clinician, 1980; McGraw-Hill, New York.

CHAPTER 2

Anemia During the Neonatal Period

Anemia during the neonatal period is caused by:

*   *Hemorrhage*: acute or chronic
*   *Hemolysis*: congenital hemolytic anemias or due to isoimmunization
*   *Failure of red cell production*: inherited bone marrow failure syndromes, e.g., Diamond–Blackfan anemia (pure red cell aplasia).

Table 2-1 lists the causes of anemia in the newborn.

HEMORRHAGE

Blood loss may occur during the prenatal, intranatal, or postnatal period. Prenatal blood loss may be transplacental, intraplacental, or retroplacental or may be due to a twin-to-twin transfusion.

Prenatal Blood Loss

Transplacental Fetomaternal

In 50% of pregnancies fetal cells can be demonstrated in the maternal circulation, in 8% the transfer of blood is estimated to be between 0.5 ml and 40 ml and in 1% of cases exceeds 40 ml and is of sufficient magnitude to produce anemia in the infant. Transplacental blood loss may be acute or chronic. Table 2-2 lists the characteristics of acute and chronic blood loss in the newborn. It may be secondary to procedures such as diagnostic amniocentesis or external cephalic version. Fetomaternal hemorrhage is diagnosed by demonstrating fetal red cells by the acid-elution method of staining for fetal hemoglobin (Kleihauer–Betke technique) in the maternal circulation. Diagnosis of fetomaternal hemorrhage may be missed in situations in which red cells of the mother and infant have incompatible ABO blood groups. In such instances the infant's A and B cells are rapidly cleared from maternal circulation by maternal anti-A or anti-B antibodies. The optimal timing for demonstrating fetal cells in maternal blood is within 2 hours of delivery and no later than the first 24 hours following delivery.

Manual of Pediatric Hematology and Oncology. DOI: 10.1016/B978-0-12-375154-6.00002-1

Table 2-1 Causes of Anemia in the Newborn

I. Hemorrhage
  A. Prenatal
    1. Transplacental fetomaternal (spontaneous, traumatic-amniocentesis, external cephalic version)
    2. Intraplacental
    3. Retroplacental
    4. Twin-to-twin transfusion
  B. Intranatal
    1. Umbilical cord abnormalities
      a. Rupture of normal cord (unattended precipitous labor)
      b. Rupture of varix or aneurysm of cord
      c. Hematomas of cord or placenta
      d. Rupture of anomalous aberrant vessels of cord (not protected by Wharton's jelly)
      e. Vasa previa (umbilical cord is presenting part)
      f. Inadequate cord tying
    2. Placental abnormalities
      a. Multilobular placenta (fragile communicating veins to main placenta)
      b. Placenta previa – fetal blood loss predominantly
      c. Abruptio placentae – maternal blood loss predominantly
      d. Accidental incision of placenta during cesarean section
      e. Traumatic amniocentesis
      f. Placental chorioangioma
    3. Hemorrhagic diathesis
      a. Plasma factor deficiency
      b. Thrombocytopenia
  C. Postnatal
    1. External
      a. Bleeding from umbilicus
      b. Bleeding from gut
      c. Iatrogenic (diagnostic venipuncture, post-exchange transfusion)
    2. Internal
      a. Cephalhematomata
      b. Subgaleal (Subaponeurotic) hemorrhage
      c. Subdural or subarachnoid hemorrhage
      d. Intracerebral hemorrhage
      e. Intraventricular hemorrhage
      f. Intra-abdominal hemorrhage
      g. Retroperitoneal hemorrhage (may involve adrenals)
      h. Subcapsular hematoma or rupture of liver
      i. Ruptured spleen
      j. Pulmonary hemorrhage
II. Hemolytic anemia (see Chapters 7, 8 and 9)
  A. Congenital erythrocyte defects
    1. Membrane defects (with characteristic morphology)
      a. Hereditary spherocytosis (p. 175)
      b. Hereditary elliptocytosis (p. 179)
      c. Hereditary stomatocytosis (p 182)
      d. Hereditary xerocytosis
      e. Infantile pyknocytosis[a]
      f. Pyropoikilocytosis

(Continued)

Table 2-1   (Continued)

 2. Hemoglobin defects[b]
   a. α-Thalassemia Syndromes[c]
      - single α-globin gene deletion (asymptomatic carrier state)
      - Two α-globin gene deletion (α thalassemia trait)
      - Three α-globin gene deletion (Hemoglobin Hβ_4 and Hemoglobin Barts γ_4)
      - Four α-globin gene deletion (death in utero or shortly after birth)
   b. γβ-Thalassemia
   c. εγ δβ Thalassemia
   d. β-Thalassemia[d]
   e. Unstable hemoglobins (Hb Köln[c], Hg Zürich[c], Hb F Poole[e], Hb Hasharon[e]) (p. 229–230)
 3. Enzyme defects
   a. Embden–Meyerhof glycolytic pathway
      (1) Pyruvate kinase
      (2) Other enzymes, e.g. 5′-nucleotidase deficiency, glucose phosphate isomerase deficiency
   b. Hexose-monophosphate shunt
      (1) G6PD (Caucasian and Oriental) with or without drug exposure[c]
      (2) Enzymes concerned with glutathione reduction or synthesis[c]
 B. Acquired erythrocyte defects
   1. Immune
      a. Maternal autoimmune hemolytic anemia
      b. Iso-immune hemolytic anemia: Rh disease, ABO, minor blood groups (M, S, Kell, Duffy, Luther)
   2. Non-immune
      a. Infections (cytomegalovirus, toxoplasmosis, herpes simplex, rubella, adenovirus, malaria, syphilis, bacterial sepsis, e.g., *Escherichia coli*)
      b. Microangiopathic hemolytic anemia with or without disseminated intravascular coagulation: disseminated herpes simplex, coxsackie B infections, gram-negative septicemia, renal vein thrombosis
      c. Toxic exposure (drugs, chemicals) ± G6PD ± prematurity[c]: synthetic vitamin K analogues, maternal thiazide diuretics, antimalarial agents, sulfonamides, naphthalene, aniline-dye marking ink, penicillin
      d. Vitamin E deficiency
      e. Metabolic disease (galactosemia, osteopetrosis)
III. **Failure of red cell production**
   1. Congenital (Chapter 6)
      a. Diamond–Blackfan anemia (Pure red cell aplasia)
      b. Dyskeratosis congenita
      c. Fanconi anemia
      d. Aase syndrome
      e. Pearson Syndrome
      f. Sideroblastic anemia
      g. Congenital dyserythropoietic anemia
   2. Acquired
      a. Viral infection (hepatitis, HIV, CMV, rubella, syphilis, parvovirus, B$_{19}$)
      b. Malaria
      c. Anemia of prematurity

[a]Not permanent membrane defect but has characteristic morphology.
[b]β chain mutations (e.g., sickle cell) uncommonly produce clinical symptomatology in the newborn. In homozygous sickle cell disease, the HbS concentration at birth is usually about 20%.
[c]All these conditions can be associated with Heinz-body formation and in the past were grouped together as congenital Heinz-body anemia.
[d]β-thalassemia syndromes only become clinically apparent after 2 or 3 months of age. The first sign is the presence of nucleated red cells on smear or continued high HgbF concentration.
[e]Hemolysis subsides after the first few months of life as fetal hemoglobin ($\alpha^2 \gamma^2$) is replaced by adult hemoglobin ($\alpha_2 \beta_2$).

Table 2-2   The Characteristics of Acute and Chronic Blood Loss in the Newborn

Characteristic	Acute Blood Loss	Chronic Blood Loss
Clinical	Acute distress; pallor; shallow, rapid and often irregular respiration; tachycardia; weak or absent peripheral pulses; low or absent blood pressure; no hepatosplenomegaly	Marked pallor disproportionate to evidence of distress. On occasion signs of congestive heart failure may be present, including hepatomegaly
Venous pressure	Low	Normal or elevated
Laboratory Hemoglobin concentration	May be normal initially; then drops quickly during the first 24 hours of life	Low at birth
Red cell morphology	Normochromic and macrocytic	Hypochromic and microcytic anisocytosis and poikilocytosis
Serum iron	Normal at birth	Low at birth
Course	Prompt treatment of anemia and shock necessary to prevent death	Generally uneventful
Treatment	Normal saline bolus or packed red blood cells. If indicated, iron therapy	Iron therapy. Packed red blood cells on occasion

From: Oski FA, Naiman JL. Hematologic problems in the newborn, 3rd ed. Philadelphia: Saunders, 1982, with permission.

Intraplacental and Retroplacental

Occasionally, fetal blood accumulates in the substance of the placenta (intraplacental) or retroplacentally and the infant is born anemic. Intraplacental blood loss from the fetus may occur when there is a tight umbilical cord around the neck or body or there is delayed cord clamping. Retroplacental bleeding from abruptio placenta is diagnosed by ultrasound or at surgery.

Twin-to-Twin Transfusion

Significant twin-to-twin transfusion occurs in at least 15% of monochorionic twins. Velamentous cord insertions are associated with increased risk of twin-to-twin transfusion. The hemoglobin level differs by 5 g/dl and the hematocrit by 15% or more between individual twins (maximal discrepancy in cord blood hemoglobin in dyzogotic twins is 3.3 gm/dl). The donor twin is smaller, pale, may have evidence of oligohydramnios and show evidence of shock. The recipient is larger and polycythemic with evidence of polyhydramnios and may show signs of hyperviscosity syndrome, disseminated intravascular coagulation, hyperbilirubinemia and congestive heart failure (Chapter 10).

Intranatal Blood Loss

Hemorrhage may occur during the process of birth as a result of various obstetric accidents, malformations of the umbilical cord or the placenta or a hemorrhagic diathesis (due to a plasma factor deficiency or thrombocytopenia) (Table 2-1).

Postnatal Blood Loss

Postnatal hemorrhage may occur from a number of sites and may be internal (enclosed) or external.

Hemorrhage may be due to:

- Traumatic deliveries (resulting in intracranial or intra-abdominal hemorrhage)
- Plasma factor deficiencies (see Chapter 13)
    - Congenital – hemophilia or other plasma factor deficiencies
    - Acquired – vitamin K deficiency, disseminated intravascular coagulation
- Thrombocytopenia (see Chapter 12)
    - Congenital – Wiskott–Aldrich syndrome, Fanconi anemia, thrombocytopenia absent radius syndrome
    - Acquired – isoimmune thrombocytopenia, sepsis.

Clinical and Laboratory Findings of Anemia Due to Hemorrhage

1. Anemia – pallor, tachycardia and hypotension (if severe e.g. $\geq$20 ml/kg blood loss).
2. Liver and spleen not enlarged (except in chronic transplacental bleed).
3. Jaundice absent.
4. Laboratory findings:
    - Direct antiglobulin test (DAT) negative
    - Increased reticulocyte count
    - Polychromatophilia
    - Nucleated RBCs raised
    - Fetal cells in maternal blood (in fetomaternal bleed).

Treatment

1. Severely affected
    a. Administer 10–20 ml/kg packed red blood cells (hematocrit usually 50–60%) via an umbilical catheter
    b. Cross-match blood with the mother. If unavailable, use group O Rh-negative blood or saline boluses (temporarily for shock)
    c. Use partial exchange transfusion with packed red cells for infants in incipient heart failure.
2. Mild anemia due to chronic blood loss
    a. Ferrous sulfate (2 mg elemental iron/kg body weight three times a day) for 3 months.

HEMOLYTIC ANEMIA

Hemolytic anemia in the newborn is usually associated with an abnormally low hemoglobin level, an increase in the reticulocyte count and with unconjugated hyperbilirubinemia. The hemolytic process is often first detected as a result of investigation for jaundice during the first week of life. The causes of hemolytic anemia in the newborn are listed in Table 2-1.

Congenital Erythrocyte Defects

Congenital erythrocyte defects involving the red cell membrane, hemoglobin and enzymes are listed in Table 2-1 and discussed in Chapters 7 and 8. Any of these conditions may occur in the newborn and manifest clinically as follows:

- Hemolytic anemia (low hemoglobin, reticulocytosis, increased nucleated red cells, morphologic changes)
- Unconjugated hyperbilirubinemia
- Direct antiglobulin test negative.

Infantile Pyknocytosis

Infantile pyknocytosis is characterized by:

- Hemolytic anemia – Direct antiglobulin test negative (non-immune)
- Distortion of as many as 50% of red cells with several to many spiny projections (up to 6% of cells may be distorted in normal infants). Abnormal morphology is extracorpuscular in origin
- Disappearance of pyknocytes and hemolysis by the age of 6 months. This is a self-limiting condition
- Hepatosplenomegaly
- Pyknocytosis may occur in glucose-6-phosphate dehydrogenase (G6PD) deficiency, pyruvate kinase deficiency, vitamin E deficiency, neonatal infections and hemolysis caused by drugs and toxic agents.

Anemia in the Newborn Associated with Heinz-Body Formation

Red cells of the newborn are highly susceptible to oxidative insult and Heinz-body formation. This may be congenital or acquired and transient.

Congenital

Hemolytic anemia associated with Heinz-body formation occurs in the following conditions:

- Unstable hemoglobinopathies (e.g., Hb Köln or Hb Zürich)

- α-Thalassemia, for example, hemoglobin H (α-chain Tetramers)[*]
- Deficiency of G6PD, 6-phosphogluconic dehydrogenase, glutathione reductase, glutathione peroxidase.

Acquired

Hemolytic anemia associated with Heinz-body formation occurs transiently in normal full-term infants without red cell enzyme deficiencies if the dose of certain drugs or chemicals is large enough. The following have been associated with toxic Heinz-body formation: synthetic water-soluble vitamin K preparations (Synkayvite), sulfonamides, chloramphenicol, aniline dyes used for marking diapers and naphthalene used as mothballs.

Diagnosis

1. Demonstrate Heinz bodies on a supravital preparation.
2. Perform specific tests to exclude the various congenital causes of Heinz-body formation mentioned above.

Acquired Erythrocyte Defects

Acquired erythrocyte defects may be due to immune (direct antiglobulin test-positive) or nonimmune (direct antiglobulin test-negative) causes. The immune causes are due to blood group incompatibility between the fetus and the mother, for example, Rh (D), ABO, or minor blood group incompatibilities (such as anti-c, Kell, Duffy, Luther, anti-C and anti-E) causing isoimmunization. Kell antigen is second to Rh(D) in its immunizing potential and occurs in about 9% of whites and 2% of blacks.

Immune Hemolytic Anemia

Rh Isoimmunization

Clinical Features

1. Anemia, mild to severe (if severe, may be associated with hydrops fetalis[**]).
2. Jaundice (unconjugated hyperbilirubinemia)
   a. Presents during first 24 hours

---

[*] α-Chain hemoglobinopathies are evident during fetal life and at birth whereas β-chain hemoglobinopathies such as sickle cell disease or β-thalassemia are generally not apparent until 3–6 months of age when synthesis of the β globin chain increases.

[**] Infants have ascites, pleural and pericardial effusions and marked edema. Pathogenesis of hydrops may be due to heart failure, hypoalbuminemia, distortion and dysfunction of hepatic architecture and circulation due to islets of extramedullary erythropoiesis.

    b. May cause kernicterus
- (1) Exchange transfusion should be carried out whenever the bilirubin level in full-term infants rises to, or exceeds, 20 mg/dl.
- (2) Factors that predispose to the development of kernicterus at lower levels of bilirubin, such as prematurity, hypoproteinemia, metabolic acidosis, drugs (sulfonamides, caffeine, sodium benzoate) and hypoglycemia, require exchange transfusions below 20 mg/dl.

    c. See Table 2-3 for a list of various causes of unconjugated hyperbilirubinemia. Figure 2-1 outlines an approach to the diagnosis of both unconjugated and conjugated hyperbilirubinemia.

3. Hepatosplenomegaly; varies with severity.

4. Petechiae (only in severely affected infants). Hyporegenerative thrombocytopenia and neutropenia may occur during the first week.

5. Severe illness with birth of infant with hydrops fetalis, stillbirth, or death in utero and delivery of a macerated fetus.

6. Late hyporegenerative anemia with absent reticulocytes. This occurs occasionally during the second to the fifth week and is due to a diminished population of erythroid progenitors (serum concentration of erythropoietin is low and the marrow concentrations of BFU-E and CFU-E are not elevated).

Laboratory Findings

1. Serologic abnormalities (incompatibility between blood group of infant and mother); direct antiglobulin test positive in infant. Mother's serum has the presence of immune antibodies detected by the indirect antiglobulin test.

2. Decreased hemoglobin level, elevated reticulocyte count, smear-increased nucleated red cells, marked polychromasia and anisocytosis.

3. Raised indirect bilirubin level.

Severity of disease is predicted by:

- History indicating the severity of hemolytic disease of the newborn in previous infants
- Maternal antibody titers
- Amniotic fluid spectrophotometry
- Fetal ultrasound
- Percutaneous fetal blood sampling.

Management

Antenatal

Patients should be screened at their first antenatal visit for Rh and non-Rh antibodies. Figure 2-2 shows a schema of the antenatal management of Rh disease.

Table 2-3   Causes of Unconjugated Hyperbilirubinemia

I. **"Physiologic" jaundice:** jaundice of hepatic immaturity
II. **Hemolytic anemia (Chapters 7 and 9 for more complete list of causes)**
 A. Congenital erythrocyte defect
  1. Membrane defects: hereditary spherocytosis, ovalocytosis, stomatocytosis, infantile pyknocytosis
  2. Enzyme defects (nonspherocytic)
   a. Embden–Meyerhof glycolytic pathway (energy potential): pyruvate kinase, triose phosphate isomerase, etc. (see p. 191)
   b. Hexose monophosphate shunt (reduction potential): G6PD (see p. 194)
  3. Hemoglobin defects
   Sickle cell hemoglobinopathy[a]
 B. Acquired erythrocyte defect
  1. Immune: allo-immunization (Rh, ABO, Kell, Duffy, Lutheran)
  2. Nonimmune
   a. Infection
    (1) Bacterial: *Escherichia coli,* streptococcal septicemia
    (2) Viral: cytomegalovirus, rubella, herpes simplex
    (3) Protozoal: toxoplasmosis
    (4) Spirochetal: syphilis
   b. Drugs: penicillin
   c. Metabolic: asphyxia, hypoxia, shock, acidosis, vitamin E deficiency in premature infants, hypoglycemia
III. **Polycythemia (Table 10-1 for more complete list of causes)**
 A. Placental hypertransfusion
  1. Twin-to-twin transfusion
  2. Maternal–fetal transfusion
  3. Delayed cord clamping
 B. Placental insufficiency
  1. Small for gestational age
  2. Postmaturity
  3. Toxemia of pregnancy
  4. Placenta previa
 C. Endocrinal
  1. Congenital adrenal hyperplasia
  2. Neonatal thyrotoxicosis
  3. Maternal diabetes mellitus
 D. Miscellaneous
  1. Down syndrome
  2. Hyperplastic visceromegaly (Beckwith–Wiedemann syndrome), associated with hypoglycemia
IV. **Hematoma**
 Cephalhematoma, subgaleal, subdural, intraventricular, intracerebral, subcapsular hematoma of liver; bleeding into gut
V. **Conjugation defects**
 A. Reduction in bilirubin glucuronyl transferase
  1. Severe (type I): Crigler–Najjar (autosomal-recessive)
  2. Mild (type II): Crigler–Najjar (autosomal-dominant)
  3. Gilbert disease

(Continued)

Table 2-3   **(Continued)**

B. Inhibitors of bilirubin glucuronyl transferase
1. Drugs: novobiocin
2. Breast milk: pregnane-3α, 20β-diol
3. Familial: transient familial hyperbilirubinemia
VI. **Metabolic**
Hypothyroidism, maternal diabetes mellitus, galactosemia
VII. **Gut obstruction** (due to enterohepatic recirculation of bilirubin)
(e.g., pyloric stenosis, annular pancreas, duodenal atresia)
VIII. **Maternal indirect hyperbilirubinemia**
(e.g., homozygous sickle cell hemoglobinopathy)
IX. **Idiopathic**

[a]Not usually a cause of jaundice in the newborn because of the predominance of Hgb F (unless associated with concomitant G-6PD deficiency).

If an immune antibody is detected in the mother's serum, proper management includes the following:

• Past obstetric history and outcome of previous pregnancies. History of prior blood transfusions

• Blood group and indirect antiglobulin test (to determine the presence and titer of irregular antibodies). Most irregular antibodies can cause erythroblastosis fetalis; therefore, screening of maternal serum is important. Titers should be determined at various weeks of gestation (Figure 2-2). The frequency depends on the initial or subsequent rise in titers. Theoretically, any blood group antigen (with the exception of Lewis and I, which are not present on fetal erythrocytes) may cause erythroblastosis fetalis. Anti-Le[a], Le[b], M, H, P, S and I are IgM antibodies and rarely, if ever, cause erythroblastosis fetalis and need not cause concern

• Zygosity of the father: If the mother is Rh negative and the father is Rh positive, the father's zygosity becomes critical. If he is homozygous, all his future children will be Rh positive. If the father is heterozygous, there is a 50% chance that the fetus will be Rh negative and unaffected. The Rh genotype can be accurately determined by the use of polymerase chain reaction (PCR) of chorionic villus tissue, amniotic cells and fetal blood when the father is heterozygous or his zygosity is unknown. Mothers with fetuses found to be Rh D negative (dd) can be reassured and further serologic testing and invasive procedures can be avoided. Fetal zygosity can thus be determined by molecular genetic techniques. Fetal Rh D genotyping can be performed rapidly on maternal plasma in the second trimester of pregnancy without invading the fetomaternal circulation. This is performed by extracting DNA from maternal plasma and analyzing it for the Rh D gene with a fluorescent-based PCR test sensitive enough to detect the Rh D gene in a single cell. The advantage of this test is that neither the mother nor the fetus is exposed to the risks of amniocentesis or chorionic villus sampling

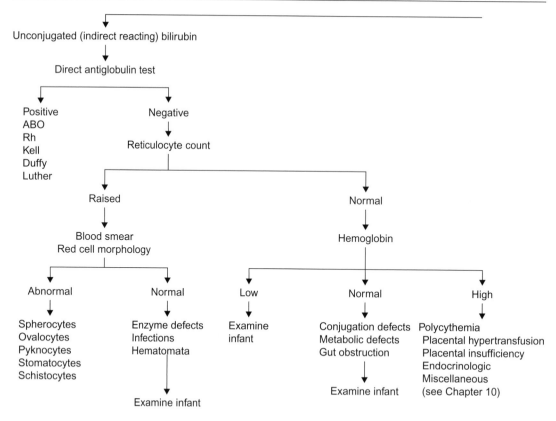

Figure 2-1 Approach to Investigation of Jaundice in the Newborn.

- Examination of the amniotic fluid for spectrophotometric analysis of bilirubin. Past obstetric history and antibody titer are indications for serial amniocentesis and spectrophotometric analyses of amniotic fluid to determine the condition of the fetus. Amniotic fluid analysis correlates well with the hemoglobin and hematocrit at birth ($r = 0.9$) but does not predict whether the fetus will require an exchange transfusion after birth. The following are indications for amniocentesis:
  - History of previous Rh disease severe enough to require an exchange transfusion or to cause stillbirth
  - Maternal titer of anti-D, anti-c, or anti-Kell (or other irregular antibodies) of 1:8 to 1:64 or greater by indirect antiglobulin test or albumin titration and depending on previous history. An assessment of the optical density difference at 450 μm (ΔOD_{450}) at a given gestational age permits reasonable prediction of the fetal outcome (Figure 2-3). Determination of the appropriate treatment depends on the ΔOD_{450} of the amniotic fluid, the results of the fetal biophysical profile scoring and the assessment of the presence or absence of

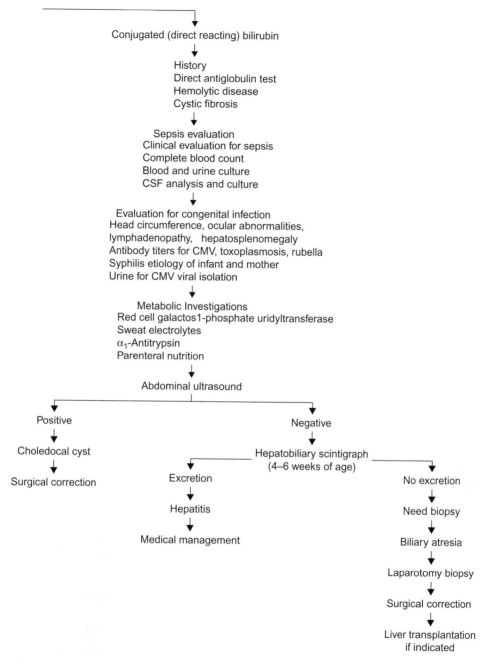

Figure 2-1 (*Continued*)

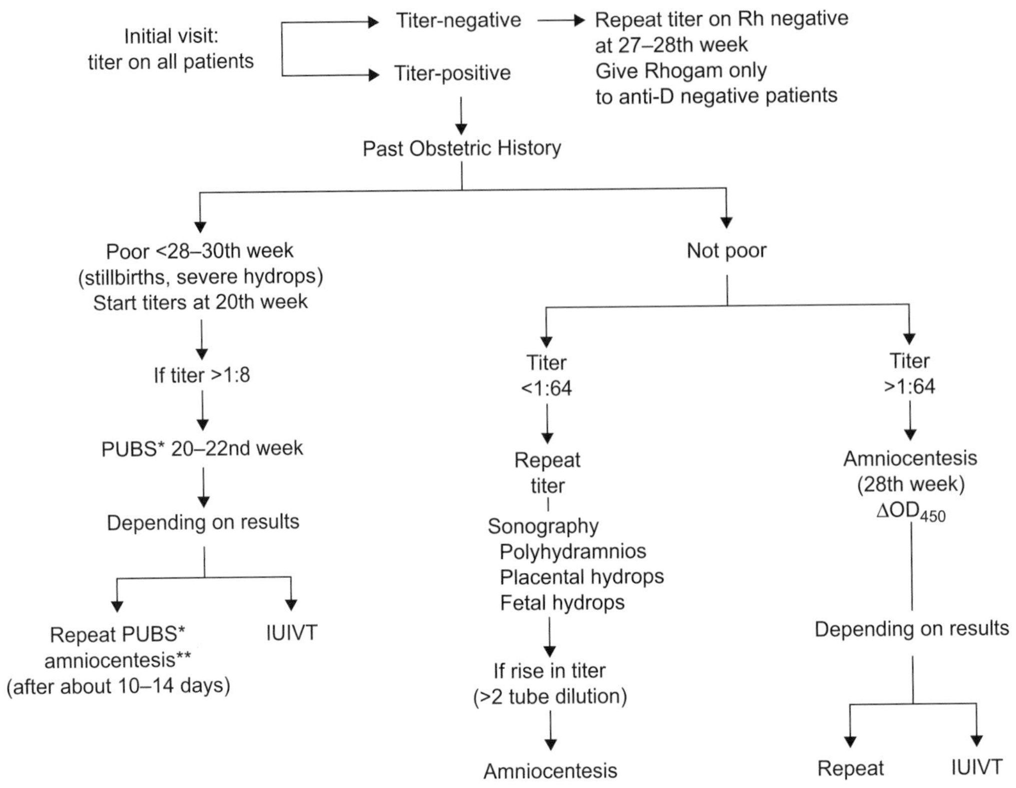

Figure 2-2 Schema of Antenatal Management of Rh Disease.
*Percutaneous umbilical vein blood sampling.
**Amniotic fluid analysis is less reliable prior to the 26th week of gestation and PUBS is recommended. IUIVT, Intrauterine intravenous transfusion.

fetal hydrops (seens on ultrasound) and amniotic phospholipid determinations (lung profile).*

Features of Lung Profile	Immature Fetus	Mature Fetus
Lecithin/sphingomyelin ratio	<2.0	>2.0
Acetone-precipitable fraction	<45%	>50%
Phosphatidylinositol	Absent	Present (small amounts)
Phosphatidylglycerol	Absent	Present (prominent)

*Ultrasound for the assessment of gestational age must be done early in pregnancy. The fetal biophysical profile scoring uses multiple variables: fetal breathing movements, gross body movements, fetal tone, reactive fetal heart rate and quantitative amniotic fluid volume. This scoring system provides a good short-term assessment of fetal risk for death or damage *in utero*.

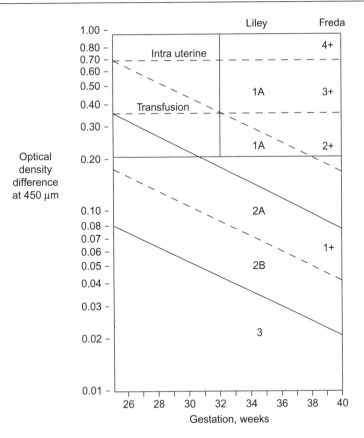

Figure 2-3 Assessment of Fetal Prognosis by the Methods of Liley and of Freda. Liley's Method of Prediction.
Zone 1A: Condition desperate, immediate delivery or intrauterine transfusion required, depending on gestational age. *Zone 1B*: Hemoglobin less than 8 g/dl, delivery or intrauterine transfusion urgent, depending on gestational age. *Zone 2A*: Hemoglobin 8–10 g/dl, delivery at 36–37 weeks. *Zone 2B*: Hemoglobin 11.0–13.9 g/dl, delivery at 37–39 weeks. *Zone 3*: Not anemic, deliver at term. Freda's method of prediction: *Zone 4+*: Fetal death imminent, immediate delivery or intra-uterine transfusion, depending on gestational age. *Zone 3+*: Fetus in jeopardy, death within 3 weeks, delivery or intrauterine transfusion as soon as possible, depending on gestational age. *Zone 2+*: Fetal survival for at least 7–10 days, repeat amniocentesis indicated, possible indication for intrauterine transfusion, depending on gestational age. *Zone 1+*: Fetus in no immediate danger.
From: Robertson JG. Evaluation of the reported methods of interpreting spectrophotometric tracings of amniotic fluid analysis in Rhesus isoimmunization. Am J Obstet Gynecol 1966;95:120, with permission.

If the amniotic fluid optical density difference at 450 μm (ΔOD_{450}) indicates a severely affected fetus and phospholipid estimations indicate lung maturity, the infant should be delivered. If the ΔOD_{450} indicates a severely affected fetus and the phospholipid estimations indicate marked immaturity, maternal plasmapheresis and/or intrauterine intra-vascular transfusion (IUIVT) should be carried out. IUIVT has many advantages over intra-peritoneal fetal transfusions and is the procedure of choice. This decision is made in conjunction with the biophysical profile score.

Intensive maternal plasmapheresis antenatally using a continuous-flow cell separator can significantly reduce Rh antibody levels, reduce fetal hemolysis and improve fetal survival in those mothers carrying highly sensitized Rh-positive fetuses. This procedure together with IUIVT should be carried out when a high antibody titer exists early before a time that the infant could be safely delivered.

If the risk of perinatal death resulting from complications of prematurity is high, then an IUIVT should be carried out. Percutaneously, the umbilical vein is used for blood sampling (PUBS) and venous access and permits a fetal transfusion via the intravascular route (IUIVT). With the availability of high-resolution ultrasound guidance, a fine (20 gauge) needle is inserted directly into the umbilical cord, either at the insertion site into the pla-centa or into a free loop of cord. This allows the same blood sampling as is available post-natally in the neonate. Temporary paralysis of the fetus with the use of pancuronium bromide (Pavulon) facilitates the procedure, which may be applied to fetuses from 18 weeks' gestation until the gestational age when fetal lung maturity is confirmed. The inter-val between procedures ranges from 1 to 3 weeks.

Blood used for IUIVT should be cytomegalovirus-negative packed RBCs with a packed cell volume of 85–88%. Cells should be fresh, leukocyte-depleted and irradiated to prevent the low risk of graft-versus-host disease. The use of kell antigen-negative blood is optimal, if available.

The risks of IUIVT include:

- Fetal loss (2%)
- Premature labor and rupture of membranes
- Chorioamnionitis
- Fetal bradycardia
- Cord hematoma or laceration
- Fetomaternal hemorrhage.

The overall survival rate is 88%. Intraperitoneal transfusion can be performed in addition to IUIVT to increase the amount of blood transfused and to extend the interval between transfusions.

Modern neonatal care, including attention to metabolic, nutritional and ventilatory needs and the use of artificial surfactant insufflation, makes successful earlier delivery possible. The need for IUIVT and intraperitoneal transfusion is rarely indicated.

Postnatal

Hydropic infant at birth. In addition to phototherapy[*] the following measures are employed:

- Adequate ventilation must be established
- Drainage of pleural effusions and ascites to improve ventilation
- Use of resuscitation fluids and drugs, surfactant and glucose infusions to counteract hyperinsulinemic hypoglycemia should be employed
- Partial exchange transfusion may be necessary to correct severe anemia
- Double-volume exchange transfusion may be required later.

Hyperbilirubinemia is the most frequent problem and can be managed by exchange transfusion. Phototherapy is an adjunct rather than the first line of therapy in hyperbilirubinemia due to erythroblastosis fetalis. Postnatal management and criteria for exchange transfusion have changed over the years and still remain somewhat controversial. We currently use the following indications for exchange transfusion:

- A rapid increase in the bilirubin level of greater than 1.0 mg/h and/or a bilirubin level approaching 20 mg/dl at any time during the first few days of life in the full-term infant is an indication for exchange transfusion. In preterm or high-risk infants, exchange transfusion should be carried out at lower levels of bilirubin (e.g., 15 mg/dl)
- Clinical signs suggesting kernicterus at any time at any bilirubin level are an indication for exchange transfusion.

The blood for exchange transfusion should be ABO compatible and for anti-D hemolytic disease of the newborn, Rh negative. If the mother is alloimmunized to an antigen other than D, the blood should not have that antigen. It should be crossmatched compatible with the mother's serum. Ideally, the blood should be leukocyte-depleted and be negative for Kell antigen (to avoid sensitizing the infant) and be hemoglobin S negative.[**] If the initial exchange transfusion is carried out using the group O blood, any further exchange transfusions should use O blood. Otherwise, brisk hemolysis and jaundice due to ABO incompatibility may become a further complication. Graft-versus-host disease occurs rarely after exchange transfusion, but blood should be irradiated if possible, especially for premature infants.

Prevention of Rh Hemolytic Disease

Rh hemolytic disease can be prevented by the use of Rh immunoglobulin at a dose of 300 μg, which is indicated in the following circumstances:

---

[*]Intensive phototherapy implies the use of high levels of irradiance [430–490 nm, i.e., usually 30 μW/cm^2 per nm or higher].

[**]In hypoxic infants sickle cell trait blood could cause an iatrogenic sickle cell crisis or death.

- For all Rh-negative, Rh0 (D")-negative mothers who are unimmunized to the Rh factor. In these patients Rh immunoglobulin is given at 28 weeks' and 34 weeks' gestation and within 72 hours of delivery. Antenatal administration of Rh immunoglobulin is safe for the mother and the fetus
- For all unimmunized Rh-negative mothers who have undergone spontaneous or induced abortion, particularly beyond the seventh or eighth week of gestation
- After ruptured tubal pregnancies in unimmunized Rh-negative mothers
- Following any event during pregnancy that may lead to transplacental hemorrhage, such as external version, amniocentesis, or antepartum hemorrhage in unimmunized Rh-negative women
- Following tubal ligation or hysterotomy after the birth of a Rh-positive child in unimmunized Rh-negative women
- Following chorionic villus sampling at 10–12 weeks' gestation. In these patients 50 μg of Rh immunoglobulin should be given.

ABO Isoimmunization

ABO incompatibility is milder than hemolytic disease of the newborn caused by other antibodies.

Clinical Features

1. Jaundice (indirect hyperbilirubinemia) usually within first 24 hours; may be of sufficient severity to cause kernicterus.
2. Anemia at birth is usually absent or moderate and late anemia is rare.
3. Hepatosplenomegaly.

Table 2-4 lists the clinical and laboratory features of isoimmune hemolysis due to Rh and ABO incompatibility.

Diagnosis

1. Hemoglobin decreased.
2. Smear: spherocytosis in 80% of infants, reticulocytosis, marked polychromasia.
3. Elevated indirect bilirubin level.*
4. Demonstration of incompatible blood group
   a. Group O mother and an infant who is group A or B
   b. Rarely, mother may be A and baby B or AB or mother may be B and baby A or AB.

---

*In the era of early discharge of newborns the use of the critical bilirubin level of 4 mg/dL at the 6th hour of life predicts significant hyperbilirubinemia and 6 mg/dL at the 6th hour will predict severe hemolytic disease of the newborn. The reticulocyte count, a positive antiglobulin test and a sibling with neonatal jaundice are additional predictors of significant hyperbilirubinemia and reason for careful surveillance of the newborn.

Table 2-4 Clinical and Laboratory Features of Isoimmune Hemolysis Caused by Rh and ABO Incompatibility

Feature	Rh Disease	ABO Incompatibility
Clinical evaluation		
Frequency	Unusual	Common
Occurrence in first born	5%	40–50%
Predictably severe in subsequent pregnancies	Usually	No
Stillbirth and/or hydrops	Occasional	Rare
Pallor	Marked	Minimal
Jaundice	Marked	Minimal (occasionally marked)
Hepatosplenomegaly	Marked	Minimal
Incidence of late anemia	Common	Uncommon
Laboratory findings		
Blood type, mother	Rh-negative	O
Blood type, infant	Rh-positive	A or B or AB
Antibody type	Incomplete (7S)	Immune (7S)
Direct antiglobulin test	Positive	Usually positive
Indirect antiglobulin test	Positive	Usually positive
Hemoglobin level	Very low	Moderately low
Serum bilirubin	Markedly elevated	Variably elevated
Red cell morphology	Nucleated RBCs	Spherocytes
Treatment		
Need for antenatal management	Yes	No
Exchange transfusion		
Frequency	~2:3	~1:10
Donor blood type	Rh-negative group specific, when possible	Rh same as infant group O only

5. Direct antiglobulin test on infant's red cells usually positive.
6. Demonstration of antibody in infant's serum
   a. When free anti-A is present in a group A infant or anti-B is present in a group B infant, ABO hemolytic disease may be presumed. These antibodies can be demonstrated by the indirect antiglobulin test in the infant's serum using adult erythrocytes possessing the corresponding A or B antigen. This is proof that the antibody has crossed from the mother's to the baby's circulation
   b. Antibody can be eluted from the infant's red cells and identified.
7. Demonstration of antibodies in maternal serum. When an infant has signs of hemolytic disease, the mother's serum may show the presence of immune agglutinins persisting after neutralization with A and B substance and hemolysins.

Treatment

In ABO hemolytic disease, unlike Rh disease, antenatal management or premature delivery is not required. After delivery, management of an infant with ABO hemolytic disease is directed toward controlling the hyperbilirubinemia by frequent determination of unconjugated bilirubin levels, with a view to the need for phototherapy or exchange transfusion. The principles and methods are the same as those described for Rh hemolytic disease. Group O blood of the same Rh type as that of the infant should be used. Whole blood is used to permit maximum bilirubin removal by albumin.

Late-Onset Anemia in Immune Hemolytic Anemia

Infants not requiring an exchange transfusion for hyperbilirubinemia following immune hemolytic anemia may develop significant anemia during the first 6 weeks of life because of persistent maternal IgG antibodies hemolyzing the infant's red blood cells associated with a reticulocytopenia (antibodies destroy the reticulocytes as well as the red blood cells). For this reason, follow-up hemoglobin levels weekly for 4–6 weeks should be done in those infants.

Nonimmune Hemolytic Anemia

The causes of nonimmune hemolytic anemia are listed in Table 2-1.

Vitamin E Deficiency

1. Vitamin E is one of several free-radical scavengers that serve as antioxidants to protect cellular components against peroxidative damage. Serum levels of 1.5 mg/dl are adequate and levels greater than 3.0 mg/dl should be avoided because they may be associated with serious morbidity and mortality.
2. Vitamin E protects double bonds of lipids in the membranes of all tissues, including blood cells.
3. Vitamin E requirements increase with exposure to oxidant stress and increase as dietary polyunsaturated fatty acid (PUFA) content increases. Vitamin E is now supplemented in infant formulas in proportion to their PUFA content in a ratio of E:PUFA >0.6.
4. The lipoproteins that transport and bind vitamin E are low in neonates.

Clinical Findings

1. Hemolytic anemia and reticulocytosis.
2. Thrombocytosis.
3. Acanthocytosis.
4. Peripheral edema.
5. Neurologic signs:
   - Cerebellar degeneration
   - Ataxia
   - Peripheral neuropathy.
6. Hemolytic anemia develops under the following conditions:
   - Diets high in PUFA supplemented with iron, which is a powerful oxidant

- Prematurity
- Oxygen administration, a powerful oxidant.

Diagnosis

Peroxide hemolysis test: Red cells are incubated with small amounts of hydrogen peroxide and the amount of hemolysis is measured.

FAILURE OF RED CELL PRODUCTION

Congenital

The inherited bone marrow failure syndromes (Chapter 6) are listed in Table 2-1 and discussed in Chapter 6.

Acquired

Viral Diseases

Viral interference (e.g., CMV, HIV) with fetal hematopoiesis may cause anemia, leukopenia and thrombocytopenia in the newborn. HIV disease may be associated with a number of hematologic abnormalities (Chapter 5).

ANEMIA OF PREMATURITY

This anemia is normocytic and normochromic and characterized by reduced bone marrow erythropoietic activity (hypoproliferative anemia) with low reticulocyte count and low serum erythropoietin (EPO) levels. It may be compounded by folic acid, vitamin E and iron availability and frequent blood sampling.

The low hemoglobin concentration is due to:

- Preterm infants deprived of third trimester hematopoiesis and iron transport
- Decreased red cell production (premature infants have low EPO levels which reach a nadir between days 7 and 50, independently of weight at birth and are less responsive to EPO)
- Shorter red cell life span
- Increased blood volume with growth
- Marked blood sampling in relation to their weight.

The nadir of the hemoglobin level is 4–8 weeks and is 8 g/dl in infants weighing less than 1,500 g. However, small-for-gestational-age infants who have had intrauterine hypoxia exhibit increased erythropoiesis. The anemia of prematurity rarely occurs in association with cyanotic congenital heart disease or with respiratory insufficiency; indicating that higher oxygen-carrying capacities can be maintained in infants in the first few weeks of life if the need arises.

Clinical Features

Tachycardia, increased apnea and bradycardia, increased oxygen requirement, poor weight gain.

Treatment

Delaying cord clamping for 30–60 seconds in infants who do not require immediate resuscitation may reduce the severity of anemia of prematurity. In addition, limiting blood loss by phlebotomy is important.

Recombinant human erythropoietin (rHuEPO) in a dose of 75–300 units/kg/week subcutaneously for 4 weeks starting at 3–4 weeks of age has been employed to increase reticulocyte counts and raise hemoglobin. It takes about 2 weeks to raise the hemoglobin to a biologically significant degree, which limits its usefulness when a prompt response is needed. Despite extensive studies, many of which have shown a reduction in need for transfusions in premature infants, particularly less than 1,000 g who have significant phlebotomy losses, there is still no definite consensus as to whether rHuEPO minimizes the need for blood transfusion. A potential advantage of rHuEPO is the associated right shift in the oxyhemoglobin dissociation curve, most likely due to the increased erythrocyte 2,3-DPG content. The incidence of necrotizing enterocolitis has been shown to be lower in very-low-birthweight infants treated with rHuEPO. However, the risk of severe retinopathy of prematurity (stage 3 or higher) is increased with early rHuEPO treatment compared with placebo.

Supplemental oral iron in a dose of at least 2 mg/kg/day or intravenous iron supplementation may also be required to prevent the development of iron deficiency. Adequate intake of folate, vitamin E and protein are important to support erythropoiesis.

The criteria for transfusion of preterm infants vary considerably among different institutions. All transfusions should be provided from a single donor and be less than 7–10 days old and be leukodepleted. Packed red cells should be adjusted to a hematocrit of 60–79 percent with normal saline or 5% albumin solution. Low-risk cytomegalovirus blood products (cytomegalovirus-negative or leukodepleted red cells) should be used only for neonates with birth weight less than 1,200 g who are cytomegalovirus-negative or have unknown cytomegalovirus status. As a general rule, hemoglobin values should be maintained above 12 g/dl during the first 2 weeks of life. After that period indication for transfusion should not be based on hemoglobin concentration alone but on available tissue oxygen which is determined by:

- Hemoglobin concentration
- Position of the oxyhemoglobin dissociation curve
- Arterial oxygen saturation
- Infants clinical condition which includes:
  - Weight gain
  - Fatigue during feeding

Table 2-5   Indications for Small-Volume RBC Transfusions in Preterm Infants

Transfuse well infant at hematocrit ≤20% or ≤25% with low reticulocyte count and tachycardia, tachypnea, poor weight gain, poor suck or apnea

Transfuse infants at hematocrit ≤30%
   a. If receiving <35% supplemental hood oxygen
   b. If on CPAP or mechanical ventilation with mean airway pressure <6 cm H_2O
   c. If significant apnea (>6/day) and bradycardia are noted while receiving therapeutic doses of methylxanthines
   d. If heart rate >180 beats/min or respiratory rate >80 breaths/min persists for 24 hours
   e. If weight gain <10 g/day is observed over 4 days while receiving ≥100 kcal/kg/day
   f. If undergoing surgery

Transfuse for hematocrit ≤35%
   a. If receiving >35% supplemental hood oxygen
   b. If intubated on CPAP or mechanical ventilation with mean airway pressure >6 cmH_2O

Do not transfuse
   a. To replace blood removed for laboratory tests alone
   b. For low hematocrit value alone

Abbreviations: CPAP, continuous positive airway pressure by nasal or endotracheal route.
Modified from: Hume, H. Red blood cell transfusions for preterm infants: the role of evidence-based medicine. Semin Perinatol 1997;21:8–19, with permission.

- Tachycardia
- Tachypnea
- Evidence of hypoxemia by an increase in blood lactic acid concentration.

Table 2-5 gives indications for small-volume red cell transfusions in preterm infants.

PHYSIOLOGIC ANEMIA

In utero the oxygen saturation of the fetus is 70% (hypoxic levels) and this stimulates erythropoietin, produces a reticulocytosis (3–7%) and increases red cell production causing a high hemoglobin at birth.

After birth the oxygen saturation is 95%, erythropoietin is undetectable and red cell production by day 7 is 10% of the level in utero. As a result of this, the hemoglobin level falls to 11.4 + 0.9 g/dl at the nadir at 8–12 weeks (physiologic anemia). At this point oxygen delivery is impaired, erythropoietin stimulated and red cell production increases. Infants born prematurely experience a more marked decrease in hemoglobin concentration. Premature infants weighing less than 1,500 g have a hemoglobin level of 8.0 g/dl at age 4–8 weeks.

DIAGNOSTIC APPROACH TO ANEMIA IN THE NEWBORN

Figure 2-4 is a flow diagram of the investigation of anemia in the newborn and stresses the importance of the direct antiglobulin test, the reticulocyte count, the mean corpuscular

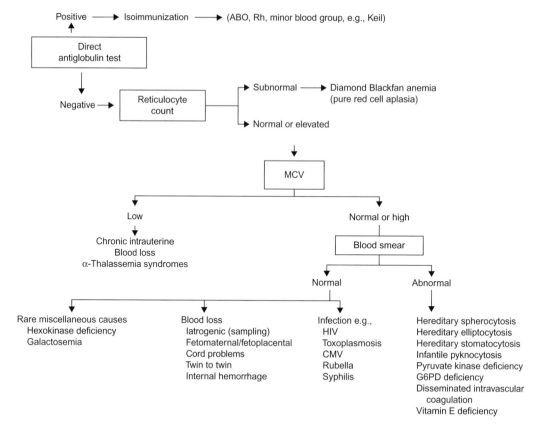

Figure 2-4 Approach to the Diagnosis of Anemia in the Newborn.

Table 2-6   Clinical and Laboratory Evaluation in Anemia in the Newborn

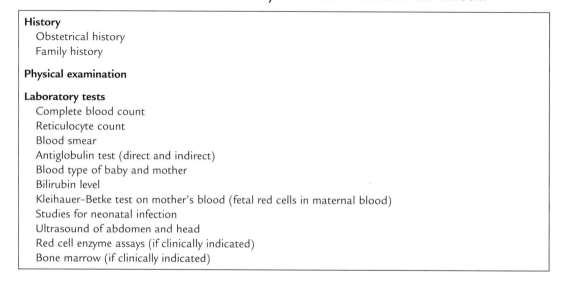

History
  Obstetrical history
  Family history

Physical examination

Laboratory tests
  Complete blood count
  Reticulocyte count
  Blood smear
  Antiglobulin test (direct and indirect)
  Blood type of baby and mother
  Bilirubin level
  Kleihauer–Betke test on mother's blood (fetal red cells in maternal blood)
  Studies for neonatal infection
  Ultrasound of abdomen and head
  Red cell enzyme assays (if clinically indicated)
  Bone marrow (if clinically indicated)

volume (MCV) and the blood smear as key investigative tools in elucidating the cause of the anemia and Table 2-6 lists the clinical and laboratory evaluations required in anemia in the newborn.

Suggested Reading

Academy of Pediatrics. Provisional Committee for Quality Improvement and Subcommittee on Hyperbilirubinemia. Pediatrics. 1994;94:558–565.

Aher S, Malwatkar K, Kadam S. Neonatal anemia. Seminars in Fetal & Neonatal Medicine. 2008;13(4):239–247.

Bishara N, Ohls RK. Current controversies in the management of the anemia of prematurity. Seminars in Perinatology. 2009;33(1):29–34.

Bowman J. The management of hemolytic disease in the fetus and newborn. Semin Perinatol. 1997;21:39–44.

Halperin DS, Wacker P, LaCourt G, et al. Effects of recombinant human erythropoietin in infants with the anemia of prematurity: a pilot study. J Pediatr. 1990;116:779–796.

Hann IM, Gibson BES, Letsky EA. Fetal and neonatal haematology. London: Bailliere Tindall; 1991.

Maier RH, Obladen M, Müller-Hansen I, et al. Early treatment with erythropoietin [beta] ameliorates anemia and reduces transfusion requirements in infants with birth weights below 1000g. Journal of Pediatrics. 2002;141:8–15.

Messer J, Haddad J, Donato L, et al. Early treatment of premature infants with recombinant human erythro poietin. Pediatrics. 1993;92:519–523.

Murray NA, Roberts IAG. Haemolytic disease of the newborn. Archives of Disease in Childhood Fetal & Neonatal Edition. 2007;92(2):F83–F88.

Pilgrim H, Lloyd-Jones M, Rees A. Routine antenatal anti-D prophylaxis for Rh-D-negative women: a systematic review and economic evaluation. Health Technology Assessment (Winchester, England). 2009;13(No. 10):1–103.

Smits-Wintjens VEHG, Walther FJ, Lopriore E. Rhesus haemolytic disease of the newborn: Postnatal management, associated morbidity and long-term outcome. Seminars in Fetal & Neonatal Medicine. 2008;13(4):265–271.

Steiner LA, Gallagher PG. Erthrocyte disorders in the perinatal period. Seminars in Perinatology. 2007;31(4):254–261.

Urbaniak SJ, Greiss MA. RhD haemolytic disease of the fetus and the newborn. Blood Reviews. 2000;14(1):44–61.

CHAPTER 3

Iron-Deficiency Anemia

Iron deficiency is the most common nutritional deficiency in children and is worldwide in distribution. The incidence of iron-deficiency anemia is high in infancy. It is estimated that 40–50% of children under 5 years of age in developing countries are iron deficient. The incidence is 5.5% in inner-city school children ranging in age from 5 to 8 years, 2.6% in pre-adolescent children and 25% in pregnant teenage girls.

PREVALENCE

There is a higher prevalence of iron-deficiency anemia in African-American children than in Caucasian children. Although no socioeconomic group is spared, the incidence of iron-deficiency anemia is inversely proportional to economic status.

Peak prevalence occurs during late infancy and early childhood when the following may occur:

- Rapid growth with exhaustion of gestational iron
- Low levels of dietary iron
- Complicating effect of cow's milk-induced exudative enteropathy due to whole cow's milk ingestion (page 42).

A second peak is seen during adolescence due to rapid growth and suboptimal iron intake. This is amplified in females due to menstrual blood loss.

Table 3-1 lists causes of iron deficiency and Table 3-2 lists infants at high risk for iron deficiency.

ETIOLOGIC FACTORS

Diet

1. One mg/kg/day to a maximum of 15 mg/day (assuming 10% absorption) is required in normal infants.

Manual of Pediatric Hematology and Oncology. DOI: 10.1016/B978-0-12-375154-6.00003-3

Table 3-1   Causes of Iron-Deficiency Anemia

I. Deficient intake
   Dietary (milk, 0.75 mg iron/l)
II. Inadequate absorption
   Poor bioavailability: absorption of heme $Fe>Fe^{2+}>Fe^{3+}$; breast milk iron>cow's milk
   Antacid therapy or high gastric pH (gastric acid assists in increasing solubility of inorganic iron)
   Bran, phytates, starch ingestion (contain organic polyphosphates which bind iron avidly)
   loss or dysfunction of absorptive enterocytes (inflammatory bowel disease, celiac disease)
   Cobalt, lead ingestion (share the iron absorption pathways)
III. Increased demand
   Growth (low birth weight, prematurity, low-birth-weight, twins or multiple births, adolescence, pregnancy), cyanotic congenital heart disease
IV. Blood loss (Chapter 2)
   A. Perinatal
      1. Placental
         a. Transplacental bleeding into maternal circulation
         b. Retroplacental (e.g., premature placental separation)
         c. Intraplacental
         d. Fetal blood loss at or before birth (e.g., placenta previa)
         e. Feto-fetal bleeding in monochorionic twins
         f. Placental abnormalitites (Table 2-1)
      2. Umbilicus
         a. Ruptured umbilical cord (e.g., vasa previa) and other umbilical cord abnormalities (Table 2-1)
         b. Inadequate cord tying
         c. Post exchange transfusion
   B. Postnatal
      1. Gastrointestinal tract
         a. Primary iron-deficiency anemia resulting in gut alteration with blood loss aggravating existing iron deficiency: 50% of iron-deficient children have positive guaiac stools
         b. Hypersensitivity to whole cow's milk ? due to heat-labile protein, resulting in blood loss and exudative enteropathy (leaky gut syndrome) (Table 3-4)
         c. Anatomic gut lesions (e.g., esophageal varices, hiatus hernia, peptic ulcer disease, leiomyomata, Meckel's diverticulum, duplication of gut, hereditary hemorrhagic telangiectasia, arteriovenous malformation, polyps, hemorrhoids); exudative enteropathy caused by underlying bowel disease (e.g., allergic gastroenteropathy, intestinal lymphangiectasia); inflammatory bowel disease; substantial intestinal resection.
         d. Gastritis from aspirin, adrenocortical steroids, indomethacin, phenylbutazone
         e. Intestinal parasites (e.g., hookworm [*Necator americanus* or *Ancylostoma duodenale*] and whipworm [*Trichuris Trichiura*])
         f. Henoch-Schönlein purpura
      2. Hepato-biliary system: hematobilia
      3. Lung: idiopathic pulmonary hemosiderosis, Goodpasture syndrome, defective iron mobilization with IgA deficiency, tuberculosis, bronchiectasis

(Continued)

Table 3-1   (Continued)

     4. Nose: recurrent epistaxis

     5. Uterus: menstrual loss

     6. Heart: intracardiac myxomata, valvular prostheses or patches

     7. Kidney[a]: infectious cystitis, microangiopathic hemolytic anemia, nephritic syndrome (urinary loss of transferrin), Berger disease, Goodpasture syndrome, hemosiderinurias-chronic intravascular hemolysis (e.g., paroxysmal nocturnal hemoglobinuria, paroxysmal cold hemoglobinuria, march hemoglobinuria)

     8. Extracorporeal: hemodialysis, trauma

V. Impaired absorption

Malabsorption syndrome, celiac disease, severe prolonged diarrhea, postgastrectomy, inflammatory bowel disease, *helicobacter pylori* infection-associated chronic gastritis.

VI. Inadequate presentation to erythroid precursors

Atransferrinemia

Anti-transferrin receptor antibodies

VII. Abnormal intracellular transport/utilization

Erythroid iron trafficking defects

Defects of heme biosynthesis

[a]Hematuria to the point of iron deficiency is extremely uncommon.

Table 3-2   Infants at High Risk for Iron Deficiency

Increased iron needs:
   Low birth weight
   Prematurity
   Multiple gestation
   High growth rate
   Chronic hypoxia-high altitude, cyanotic heart disease
   Low hemoglobin level at birth
Blood loss:
   Perinatal bleeding
Dietary factors:
   Early cow milk intake
   Early solid food intake
   Rate of weight gain greater than average
   Low-iron formula
   Frequent tea intake[a]
   Low vitamin C intake[b]
   Low meat intake
   Breast-feeding >6 months without iron supplements
   Low socio-economic status (frequent infections)

[a] Tea inhibits iron absorption.
[b] Vitamin C enhances iron absorption.

2. Two mg/kg/day to a maximum of 15 mg/kg/day is required in low-birth-weight infants, infants with low initial hemoglobin values and those who have experienced significant blood loss.

Food Iron Content

A newborn infant is fed predominantly on milk. Breast milk and cow's milk contain less than 1.5 mg iron per 1,000 calories (0.5–1.5 mg/L). Although cow's milk and breast milk are equally poor in iron, breast-fed infants absorb 49% of the iron, in contrast to about 10% absorbed from cow's milk. The bioavailability of iron in breast milk is much greater than in cow's milk.

Table 3-3 lists iron content of infant foods.

Growth

Growth is particularly rapid during infancy and during puberty. Blood volume and body iron are directly related to body weight throughout life. Iron-deficiency anemia can occur at any time when rapid growth outstrips the ability of diet and body stores to supply iron requirements. Each kilogram gain in weight requires an increase of 35–45 mg body iron.

The amount of iron in the newborn is 75 mg/kg. If no iron is present in the diet or blood loss occurs the iron stores present at birth will be depleted by 6 months in a full-term infant and by 3–4 months in a premature infant.

The commonest cause of iron-deficiency anemia is inadequate intake during the rapidly growing years of infancy and childhood.

Table 3-3   Iron Content of Infant Foods

Food	Iron, mg	Unit
Milk	0.5–1.5	liter
Eggs	1.2	each
Cereal, fortified	3.0–5.0	ounce
Vegetables (starched)		
Yellow	0.1–0.3	ounce
Green	0.3–0.4	ounce
Meats (strained)		
Beef, lamb, liver	0.4–2.0	ounce
Pork, liver, bacon	6.6	ounce
Fruits (strained)	0.2–0.4	ounce

Blood Loss

Blood loss, an important cause of iron-deficiency anemia, may be due to prenatal, intranatal, or postnatal causes (see Chapter 2, Table 2-1). Hemorrhage occurring later in infancy and childhood may be either occult or apparent (Table 3-1).

Iron deficiency by itself, irrespective of its cause, may result in occult blood loss from the gut. More than 50% of iron-deficient infants have guaiac-positive stools. This blood loss is due to the effects of iron deficiency on the mucosal lining (e.g., deficiency of iron-containing enzymes in the gut), leading to mucosal blood loss. This sets up a vicious cycle in which iron deficiency results in mucosal change, which leads to blood loss and further aggravates the anemia. The bleeding due to iron deficiency is corrected with iron treatment. In addition to iron deficiency per se causing blood loss it may also induce an enteropathy, or leaky gut syndrome. In this condition, a number of blood constituents, in addition to red cells, are lost in the gut (Table 3-4).

Cow's milk can result in an exudative enteropathy associated with chronic gastrointestinal (GI) blood loss resulting in iron deficiency. Whole cow's milk should be considered the cause of iron-deficiency anemia under the following clinical circumstances:

*   One quart or more of whole cow's milk consumed per day
*   Iron deficiency accompanied by hypoproteinemia (with or without edema) and hypocupremia (dietary iron-deficiency anemia not associated with exudative enteropathy is usually associated with an elevated serum copper level). It is also associated with hypocalcemia, hypotransferrinemia and low serum immunoglobulins due to the leakage of these substances from the gut
*   Iron-deficiency anemia unexplained by low birth weight, poor iron intake, or excessively rapid growth
*   Iron-deficiency anemia recurring after a satisfactory hematologic response following iron therapy
*   Rapidly developing or severe iron-deficiency anemia
*   Suboptimal response to oral iron in iron-deficiency anemia
*   Consistently positive stool guaiac tests in the absence of gross bleeding and other evidence of organic lesions in the gut
*   Return of GI function and prompt correction of anemia on cessation of cow's milk and substitution by formula.

Blood loss can thus occur as a result of gut involvement due to primary iron-deficiency anemia (Table 3-4) or secondary iron-deficiency anemia as a result of gut abnormalities induced by hypersensitivity to cow's milk, or as a result of demonstrable anatomic lesions of the bowel, e.g. Meckel's diverticulum.

Table 3-4 Classification of Iron-Deficiency Anemia in Relationship to Gut Involvement

Primary Iron Deficiency (Dietary, Rapid Growth)

	Mild or Severe			Severe[a]	
	None	Leaky Gut Syndrome		Malabsorption Syndrome	
Gut Changes	None	Leaky Gut Syndrome		Malabsorption Syndrome	
Effect	No blood loss	Loss of: Red cells only	Loss of: Red cells, Plasma protein, Albumin, Immune globulin, Copper, Calcium	Impaired absorption of iron only	Impaired absorption of xylose, fat and vitamin A; Duodenitis
Result	Iron-deficiency anemia (IDA)	IDA, guiac-positive	IDA, exudative enteropathy	IDA, refractory to oral iron	IDA, transient enteropathy
Treatment	Oral iron	Oral iron	Oral iron	Oral iron	IM iron-dextran complex

Secondary Iron Deficiency

	Mild or Severe	Severe
Pathogenesis	Cow's milk-induced? Heat-labile protein	Anatomic lesion (e.g., Meckel's diverticulum, polyp, intestinal duplication, peptic ulcer); Blood loss
Effect	Leaky gut syndrome; Loss of: Red cells, Plasma protein, Albumin, Immune globulin, Copper, Calcium	
Retuls	Recurrent IDA, exudative enteropathy	Recurrent IDA
Treatment	Discontinue whole cow's milk; soya milk formula; oral iron	Surgery, specific medical management, iron PO or IM iron dextran

[a]Can occur in severe chronic iron-deficiency anemia from any cause.

Table 3-5   Important Iron-Containing Compounds and their Function

Compound	Function
α-Glycerophosphate dehydrogenase	Work capacity
Catalase	RBC peroxide breakdown
Cytochromes	ATP production, protein synthesis, drug metabolism, electron transport
Ferritin	Iron storage
Hemoglobin	Oxygen delivery
Hemosiderin	Iron storage
Mitochondrial dehydrogenase	Electron transport
Monoamine oxidase	Catecholamine metabolism
Myoglobin	Oxygen storage for muscle contraction
Peroxidase	Bacterial killing
Ribonucleotide reductase	Lymphocyte DNA synthesis, tissue growth
Transferrin	Iron transport
Xanthine oxidase	Uric acid metabolism

Impaired Absorption

Impaired iron absorption due to a generalized malabsorption syndrome is an uncommon cause of iron-deficiency anemia. Severe iron deficiency because of its effect on the bowel mucosa, may induce a secondary malabsorption of iron as well as malabsorption of xylose, fat and vitamin A (Table 3-4).

NON-HEMATOLOGICAL MANIFESTATIONS

Iron deficiency is a systemic disorder involving multiple systems rather than exclusively a hematologic condition associated with anemia. Table 3-5 lists important iron-containing compounds in the body and their function and Table 3-6 lists the tissue effects of iron deficiency.

DIAGNOSIS

1. *Hemoglobin*: Hemoglobin is below the acceptable level for age (Appendix 1).
2. *Red cell indices*: Lower than normal MCV, MCH and MCHC for age. Widened red cell distribution width (RDW) in association with a low MCV is one of the best screening tests for iron deficiency.
3. *Blood smear*: Red cells are hypochromic and microcytic with anisocytosis and poikilocytosis, generally occurring only when hemoglobin level falls below 10 g/dl. Basophilic stippling can also be present but not as frequently as is present in

Table 3-6 Tissue Effects of Iron Deficiency

I. **Gastrointestinal tract**
   A. Anorexia-common and an early symptom
      1. Increased incidence of low-weight percentiles
      2. Depression of growth
   B. Pica-pagophagia (ice) geophagia (sand)
   C. Atrophic glossitis with flattened, atrophic, lingual papillae which makes the tongue smooth and shiny
   D. Dysphagia
   E. Esophageal webs (Kelly-Paterson syndrome)
   F. Reduced gastric acidity
   G. Leaky gut syndrome
      1. Guaiac-positive stools-isolated
      2. Exudative enteropathy: gastrointestinal loss of protein, albumin, immuno-globulins, copper, calcium, red cells
   H. Malabsorption syndrome
      1. Iron only
      2. Generalized malabsorption: xylose, fat, vitamin A, duodenojejunal mucosal atrophy
   I. Beeturia
   J. Decreased cytochrome oxidase activity and succinic dehydrogenase
   K. Decreased disaccharidases especially lactase with abnormal lactose tolerance tests
   L. Increased absorption of cadmium and lead (iron deficient children have increased lead absorption)
   M. Increased intestinal permeability index

II. **Central nervous system**
   A. Irritability
   B. Fatigue and decreased activity
   C. Conduct disorders
   D. Lower mental and motor developmental test scores on the Bayley Scale which may be long-lasting
   E. Decreased attentiveness, shorter attention span
   F. Significantly lower scholastic performance
   G. Reduced cognitive performance
   H. Breath-holding spells
   I. Papilledema

III. **Cardiovascular system**
   A. Increase in exercise and recovery heart rate and cardiac output
   B. Cardiac hypertrophy
   C. Increase in plasma volume
   D. Increased minute ventilation values
   E. Increased tolerance to digitalis

IV. **Musculoskeletal system**
   A. Deficiency of myoglobin and cytochrome C
   B. Impaired performance of a brief intense exercise task
   C. Decreased physical performance in prolonged endurance work
   D. Rapid development of tissue lactic acidosis on exercise and a decrease in mitochondrial alpha-glycerophosphate oxidase activity
   E. Radiographic changes in bone-widening of diploeic spaces
   F. Adverse effect on fracture healing

V. **Immunologic system**
   There is conflicting information as to the effect on the immunologic system of iron deficiency anemia.

(Continued)

Table 3-6 (Continued)

A. Evidence of increased propensity for infection
  1. Clinical
    a. Reduction of acute illness and improved rate of recovery in iron-replete compared to iron-deficient children
    b. Increased frequency of respiratory infection in iron deficiency
  2. Laboratory
    a. Impaired leukocyte transformation
    b. Impaired granulocyte killing and nitroblue tetrazolium (NBT) reduction by granulocytes
    c. Decreased myeloperoxidase in leukocytes and small intestine
    d. Decreased cutaneous hypersensitivity
    e. Increased susceptibility to infection in iron-deficient animals
B. Evidence of decreased propensity for infection
  1. Clinical
    a. Lower frequency of bacterial infection
    b. Increased frequency of infection in iron overload conditions
  2. Laboratory
    a. Transferrin inhibition of bacterial growth by binding iron so that no free iron is available for growth of microorganisms
    b. Enhancement of growth of nonpathogenic bacteria by iron
VI. **Cellular changes**
  A. Red cells
    1. Ineffective erythropoiesis
    2. Decreased red cell survival (normal when injected into asplenic subjects)
    3. Increased autohemolysis
    4. Increased red cell rigidity
    5. Increased susceptibility to sulfhydryl inhibitors
    6. Decreased heme production
    7. Decreased globin and α-chain synthesis
    8. ? Precipitation of α-globin monomers to cell membrane
    9. Decreased glutathione peroxidase and catalase activity
      a. Inefficient H_2O_2 detoxification
      b. Greater susceptibility to H_2O_2 hemolysis
      c. Oxidative damage to cell membrane
      d. Increased cellular rigidity
    10. Increased rate of glycolysis-glucose 6-phosphate dehydrogenase, 6-phosphogluconate dehydrogenase, 2,3-diphosphoglycerate (2,3-DPG) and glutathione
    11. Increase in NADH-methemoglobin reductase
    12. Increase in erythrocyte glutamic oxaloacetic transaminase (EGOT)
    13. Increase in free erythrocyte protoporphyrin
    14. Impairment of DNA and RNA synthesis in bone marrow cells
  B. Other tissues
    1. Reduction in heme-containing enzymes (cytochrome C, cytochrome oxidase)
    2. Reduction in iron-dependent enzymes (succinic dehydrogenase, aconitase)
    3. Reduction in monoamine oxidase (MAO)
    4. Increased excretion of urinary norepinephrine
    5. ? Reduction in tyrosine hydroxylase (enzyme converting tyrosine to di-hyroxyphenylalanine)
    6. Alterations in cellular growth, DNA, RNA and protein synthesis in animals
    7. Persistent deficiency of brain iron following short-term deprivation
    8. Reduction in plasma zinc

Table 3-7   **Causes of Elevated Levels of Free Erythrocyte Protoporphyrin (FEP) and Advantages of FEP Compared to Transferrin Saturation as a Diagnostic Tool**

Causes of raised levels of FEP:
1. Iron-deficiency anemia
2. Conditions with high reticulocyte count[a]
3. Lead poisoning (very high levels)
4. Chronic infection
5. Erythropoietic protoporphyria
6. Acute myelogenous leukemia
7. Rare cases of dyserythropoietic and sideroblastic anemias

Advantages of FEP compared with transferrin saturation:
1. FEP is not subject to daily fluctuations
2. FEP remains elevated during iron treatment (returns to normal after cells with excess FEP are replaced)[b]
3. FEP is not elevated in α- and β-thalassemia

[a]Reticulocytes have a slightly higher concentration of FEP. It occurs in hemolytic anemias (e.g., hemoglobin SS disease).
[b]Useful to know whether a patient who is in the process of receiving iron treatment was iron deficient before commencement of iron therapy.

thalassemia trait. The RDW is high ($>14.5\%$) in iron deficiency and normal in thalassemia ($<13\%$).

4. *Reticulocyte count*: The reticulocyte count is usually normal but, in severe iron-deficiency anemia associated with bleeding, a reticulocyte count of 3–4% may occur.

5. *Platelet count*: The platelet count varies from thrombocytopenia to thrombocytosis. Thrombocytopenia is more common in severe iron-deficiency anemia; thrombocytosis is present when there is associated bleeding from the gut.

6. *Free erythrocyte protoporphyrin*: The incorporation of iron into protoporphyrin represents the ultimate stage in the biosynthetic pathway of heme. Failure of iron supply will result in an accumulation of free protoporphyrin not incorporated into heme synthesis in the normoblast and the release of erythrocytes into the circulation with high free erythrocyte protoporphyrin (FEP) levels.

   a. The normal FEP level is 15.5 ± 8.3 mg/dl. The upper limit of normal is 40 mg/dl. Table 3-7 gives the causes of elevated levels of FEP and its advantages over transferrin saturation levels as a diagnostic tool.

   b. In both iron deficiency and lead poisoning, the FEP level is elevated. It is much higher in lead poisoning than in iron deficiency. The FEP is normal in α- and β-thalassemia minor. FEP elevation occurs as soon as the body stores of iron are depleted, before microcytic anemia develops. An elevated FEP level is therefore an indication for iron therapy even when anemia and microcytosis have not yet developed.

7. *Serum ferritin*: The level of serum ferritin reflects the level of body iron stores; it is quantitative, reproducible, specific and sensitive; and requires only a small blood sample. A concentration of less than 12 ng/ml is considered diagnostic of iron

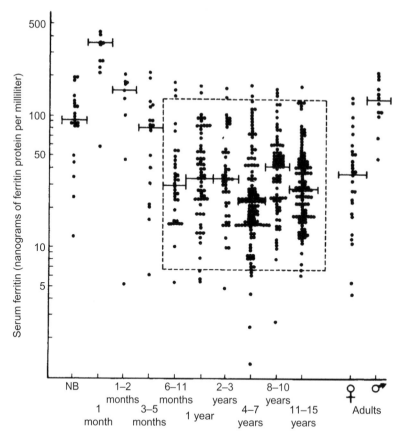

Figure 3-1 Serum Ferritin Concentrations During Development in the Healthy Nonanemic Newborn, in Infants and in Children of Various Age Groups, together with Adult Male and Female Values.
The median value in each age group is indicated by a horizontal line. The dashed line encloses a square, which includes the 95% confidence levels of the values between the ages of 6 months and 15 years.
Note: Normal ferritin levels can occur in iron deficiency in the presence of bacterial or parasitic infection, malignancy or chronic inflammatory conditions because ferritin is an acute-phase reactant.
From: Siimes MA, Addrego JE, Dallman PR. Ferritin in serum: diagnosis of iron deficiency and iron overload in infants and children. Blood 1974;43:581, with permission.

deficiency. Normal ferritin levels, however, can exist in iron deficiency when bacterial or parasitic infection, malignancy or chronic inflammatory conditions co-exist because ferritin is an acute-phase reactant and its synthesis increases in acute or chronic infection or inflammation. Figure 3-1 depicts the normal range of serum ferritin concentrations at different ages.

8. *Serum iron and iron saturation percentage*: Serum iron estimation as a measure of iron deficiency has serious limitations. It reflects the balance between several factors, including iron absorbed, iron used for hemoglobin synthesis, iron released by red cell

destruction and the size of iron stores. The serum iron concentration represents an equilibrium between the iron entering and leaving the circulation. Serum iron has a wide range of normal, varies significantly with age (see Appendix 1) and is subject to marked circadian changes (as much as 100 µg/dl during the day). The author has abandoned the use of serum iron for the routine diagnosis of iron deficiency (in favor of MCV, RDW, FEP and serum ferritin) because of the following limitations:

- Wide normal variations (age, sex, laboratory methodology)
- Time consuming
- Subject to error from iron ingestion
- Diurnal variation
- Falls in mild or transient infection.

9. *Therapeutic trial*: The most reliable criterion of iron-deficiency anemia is the hemoglobin response to an adequate therapeutic trial of oral iron. Ferrous sulfate, in a dose of 3 mg/kg per day is given for one month. A reticulocytosis with a peak occurring between the fifth and tenth days followed by a significant rise in hemoglobin level occurs (a hemoglobin rise of more than 1 g/dl in one month). The absence of these changes implies that iron deficiency is not the cause of the anemia. Iron therapy should then be discontinued and further diagnostic studies implemented. Table 3-8 summarizes the diagnostic tests in the investigation of iron-deficiency anemia.

Other tests for iron deficiency not in common usage include:

- *Serum transferrin receptor levels (STfR)*: This is a sensitive measure of iron deficiency and correlates with hemoglobin and other laboratory parameters of iron status. The STfR is increased in instances of hyperplasia of erythroid precursors such as iron-deficiency anemia and thalassemia. It is unaffected by infection and inflammation. With erythroid hypoplasia or aplasia, e.g., aplastic anemia, red cell aplasia or chronic renal failure, the STfR concentration is decreased. It is therefore of great value in distinguishing iron deficiency from the anemia of chronic disease and in identifying iron deficiency in the presence of chronic inflammation or infection. It can be measured by a sensitive enzyme-linked immunosorbent assay (ELISA) technique
- *STfR/log ferritin ratio*: Calculating the ratio of serum transferring receptor concentration to the logarithm of the serum ferritin concentration provides the highest sensitivity and specificity in the presence of chronic inflammation or infection
- *Red blood cell zinc protoporphyrin/heme ratio*: When available bone marrow iron is insufficient to support heme synthesis, zinc substitutes for iron in protoporphyrin IX and the concentration of zinc protoporphyrin relative to heme increases. This is more sensitive than plasma ferritin levels, is inexpensive and simple and is not altered in chronic inflammatory diseases or acute infections.

Table 3-8   Diagnostic Tests for Iron-Deficiency Anemia

1. Blood smear
   a. Hypochromic microcytic red cells, confirmed by RBC indices:
      (1) MCV less than acceptable normal for age (see Appendix 1)
      (2) MCH less than 27.0 pg
      (3) MCHC less than 30%
   b. Wide red cell distribution width (RDW) greater than 14.5%
2. Free erythrocyte protoporphyrin: elevated
3. Serum ferritin: decreased
4. Serum iron and iron binding capacity
   a. Decreased serum iron
   b. Increased iron binding capacity
   c. Decreased iron saturation (16% or less)
5. Therapeutic responses to oral iron
   a. Reticulocytosis with peak 5–10 days after institution of therapy
   b. Following peak reticulocytosis hemoglobin level rises on average by 0.25–0.4 g/dl/day
      or hematocrit rises 1%/day
6. Serum transferrin receptor level[a]
7. Red blood cell Zinc protoporphyrin/heme ratio[a]
8. Bone marrow[b]
   a. Delayed cytoplasmic maturation
   b. Decreased or absent stainable iron

[a]Rarely required or readily available.
[b]Used only if difficulty is experienced in elucidating cause of anemia.

Stages of Iron Depletion

1. Iron depletion: This occurs when tissue stores are decreased without a change in hematocrit or serum iron levels. This stage may be detected by low serum ferritin measurements.
2. Iron-deficient erythropoiesis: This occurs when reticuloendothelial macrophage iron stores are completely depleted. The serum iron level drops and the total iron-binding capacity increases without a change in the hematocrit. Erythropoiesis begins to be limited by a lack of available iron and serum transferrin receptor levels increase.
3. Iron-deficiency anemia: This is associated with erythrocyte microcytosis, hypochromia, increased RDW and elevated levels of FEP. It is detected when iron deficiency has persisted long enough that a large proportion of circulating erythrocytes were produced after iron became limiting.

Differential Diagnosis

Although hypochromic anemia in children is usually due to iron deficiency, it is not necessarily attributable to this condition. A list of the causes of hypochromia is given

Table 3-9   Disorders Associated with Hypochromia

1. Iron deficiency
2. Hemoglobinopathies
    a. Thalassemia (α and β)
    b. Hemoglobin Köln
    c. Hemoglobin Lepore
    d. Hemoglobin H
    e. Hemoglobin E
3. Disorders of heme synthesis caused by a chemical
    a. Lead
    b. Pyrazinamide
    c. Isoniazid
4. Sideroblastic anemias (Table 6-20)
5. Chronic infections or other inflammatory states
6. Malignancy
7. Hereditary orotic aciduria
8. Hypo- or atransferrinemia
    a. Congenital
    b. Acquired (e.g., hepatic disorders); malignant disease, protein malnutrition
       (decreased transferrin synthesis), nephrotic syndrome (urinary transferrin loss)
9. Copper deficiency
10. Inborn error of iron metabolism
    Congenital defect of iron transport to red cells

in Table 3-9. In some of these cases, there is an inability to synthesize hemoglobin normally in spite of adequate iron (e.g., thalassemia, lead poisoning). The red cell distribution width is normal in patients with thalassemia but high in those with iron deficiency. The plasma in iron deficiency is watery and in thalassemia it is straw-colored. In unusual or obscure cases of hypochromic anemia, it is necessary to do additional investigations, such as determination of serum ferritin, serum transferrin receptor levels, hemoglobin electrophoresis and examination of the bone marrow for stained iron, in order to establish the cause of the hypochromia.

Table 3-10 lists the investigations employed in the differential diagnosis of microcytic anemias and Figure 3-2 depicts a flow chart for the diagnosis of microcytic anemia using MCV and RDW.

In addition to making a diagnosis of iron-deficiency anemia, its pathogenesis must be established. The history should include conditions resulting in low iron stores at birth, dietary history and consideration of all factors leading to blood loss. The most common site of bleeding is into the bowel and the most important investigation is examination of the stool for occult blood. If occult blood is found, its cause should be established by examination of stools for ova, rectal examination, barium enema, upper GI series, ^{99m}Tc-pertechnetate scan for detection of a Meckel's diverticulum, upper endoscopy and colonoscopy.

Table 3-10 Summary of Laboratory Studies in Microcytic Anemias

	Ethnic origin	Hb	MCV	MCV in Parents	RDW	FEP	Ferritin	Serum Iron	TIBC	Bone Marrow Iron Status	Hb Electro Phoresis	Other Features
Iron deficiency	Any	↓	↓	N	↑	↑	↓	↓	↑	↓	Normal	Dietary deficiency Normal examination
β-Thalassemia β⁺ trait (heterozygous)	Mediterranean	Slight↓	↓	One parent ↓	N	N	N or ↑	N	N	N	A₂ raised F normal or ↑	
β⁰ (homozygous)	Mediterranean	↓	↓	Both parents ↓	↑	↑	↑	↑	↑	↑	F raised (60–90%)	Hepatosplenomegaly Transfusion dependent
α-Thalassemia Silent carrier (α-thal-2)	Asians, blacks, Mediterranean	N	N	N	N	N	N	N	N	N	Normal	No hematologic abnormalities
Trait (α-thal-1)	Asians, blacks, Mediterranean	N or slightly ↓	↓	One parent ↓	N	N	N or ↑	N	N	N	Normal	
Hemoglobin H disease		↓	↓		↑	N	N or ↑	N or ↑	N	↑	Hemoglobin H (2–40%)	Hemolytic anemia of variable severity Inclusion bodies in RBCs
Anemia of chronic infection	Any	↓	N	N	N	↑	N or ↑	↓	N or ↑	N or ↑	Normal	
Sideroblastic	Any	↓	N	N	↑	N or ↑	N or ↑	N or ↑	N or ↓	↑	Normal	

Abbreviations: FEP, free erythrocyte protoporphyrin; Hb, hemoglobin; MCV, mean corpuscular volume; RDW, red cell distribution width; TIBC, total iron-binding capacity; ↑, abnormally high; ↓, abnormally low; N, normal.

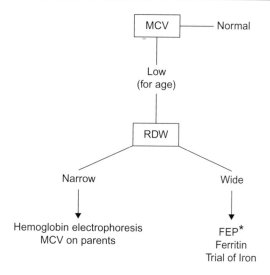

Figure 3-2 Flow chart depicting the diagnosis of microcytic anemia using MCV and RDW.

Negative guaiac tests for occult bleeding may occur if bleeding is intermittent; for this reason, occult bleeding should be tested for on at least five occasions when gastrointestinal bleeding is suspected. The guaiac test is only sensitive enough to pick up more than 5 ml occult blood. Excessive uterine bleeding, epistaxis, renal blood loss (hematuria) and, on rare occasions, bleeding into the lung (idiopathic pulmonary hemosiderosis and Goodpasture's syndrome) may all be causes of iron-deficiency anemia. Bleeding into these areas requires specific investigations designed to detect the cause of bleeding.

TREATMENT

Nutritional Counseling

1. Maintain breastfeeding for at least 6 months, if possible.
2. Use an iron-fortified (6–12 mg/L) infant formula until 1 year of age (formula is preferred to whole cow's milk). Restrict milk to 1 pint/day. Avoid cow's milk until after the first year of age because of the poor bio-availability of iron in cow's milk and because the protein in cow's milk can cause occult gastrointestinal bleeding.
3. Use iron-fortified cereal from 6 months–1 year.

4. Evaporated milk or soy-based formula should be used when iron-deficiency is due to hypersensitivity to cow's milk.
5. Provide supplemental iron for low birth weight infants:
    *   Infants 1.5–2.0 kg: 2 mg/kg/day supplemental iron
    *   Infants 1.0–1.5 kg: 3 mg/kg/day supplemental iron
    *   Infants <1 kg: 4 mg/kg/day supplemental iron.
6. Facilitators of iron absorption such as vitamin C-rich foods (citrus, tomatoes and potatoes), meat, fish and poultry should be included in the diet and inhibitors of iron absorption such as tea, phosphate and phytates common in vegetarian diets should be eliminated.

Oral Iron Medication

The goal of therapy for iron deficiency is both correction of the hemoglobin level and replenishment of body iron stores.

1. *Product*: Ferrous iron (e.g., ferrous sulfate, ferrous gluconate, ferrous ascorbate, ferrous lactate, ferrous succinate, ferrous fumarate, or ferrous glycine sulfate) is effective. Ferric irons and heavily chelated iron should not be used as they are poorly and inefficiently absorbed.
2. *Dose*: 1.5–2.0 mg/kg elemental iron three times daily. Older children: ferrous sulfate (0.2 g) or ferrous gluconate (0.3 g) given three times daily, to provide 100–200 mg elemental iron. In children with gastrointestinal side effects, iron once every other day may be better tolerated with good effect.
3. *Duration*: 6–8 weeks after hemoglobin level and the red cell indices return to normal.
4. *Response*: The lower the hemoglobin to start the higher the reticulocyte response and rise in hemoglobin.
    a. Peak reticulocyte count on days 5–10 following initiation of iron therapy.
    b. Following peak reticulocyte level, hemoglobin rises on average by 0.25–0.4 g/dl/day or hematocrit rises 1%/day during first 7–10 days.
    c. Thereafter, hemoglobin rises slower: 0.1–0.15 g/dl/day.
5. *Failure to respond to oral iron*: The following reasons should be considered:
    *   Poor compliance – failure or irregular administration of oral iron; administration can be verified by change in stool color to gray-black or by testing stool for iron
    *   Inadequate iron dose
    *   Ineffective iron preparation
    *   Insufficient duration
    *   Persistent or unrecognized blood loss
    *   Incorrect diagnosis – thalassemia, sideroblastic anemia
    *   Coexistent disease that interferes with absorption or utilization of iron (e.g., chronic inflammation, inflammatory bowel disease, malignant disease, hepatic or renal

disease, concomitant deficiencies [vitamin B_{12}, folic acid, thyroid, associated lead poisoning])
- Impaired gastrointestinal absorption due to high gastric pH (e.g., antacids, histamine-2 blockers, gastric acid pump inhibitors).

Parenteral Therapy

Intramuscular

Iron-dextran, a parenteral form of elemental iron, is available for intramuscular use. It is safe, effective and well tolerated even in infants with a variety of acute illnesses, including acute diarrheal disorders.

An increased risk for clinical attacks of malaria and other infections have been demonstrated in malarious regions, particularly with parenteral or high-dose oral iron supplementation.

Indications

1. Noncompliance or poor tolerance of oral iron.
2. Severe bowel disease (e.g., inflammatory bowel disease) – use of oral iron might aggravate the underlying disease of the bowel or iron absorption is compromised.
3. Chronic hemorrhage (e.g., hereditary telangiectasia, menorrhagia, chronic hemoglobinuria from prosthetic heart valves).
4. Acute diarrheal disorder in underprivileged populations with iron-deficiency anemia.
5. Rapid replacement of iron stores is needed.
6. Erythropoietin therapy is necessary, e.g. renal dialysis.

Dose

For intramuscular iron-dextran the following formula is used to raise the hemoglobin level to normal and to replenish iron stores:

$$\frac{\text{Normal hemoglobin} - \text{initial hemoglobin}}{100} \times \text{Blood volume (mL)} \times 3.4 \times 1.5$$

1. Normal hemoglobin (see Appendix 1).
2. Blood volume – 80 ml/kg or 40 ml/lb body weight.
3. Multiplication by 3.4 – converts grams of hemoglobin into milligrams of iron.
4. Factor 1.5 – provides extra iron to replace depleted tissue stores.

Iron-dextran complex provides 50 mg elemental iron/ml.

Side Effects

Staining at the site of intramuscular injection may occur especially in cases in which the solution is accidentally administered into the superficial tissues. Staining is of a transient type, disappearing after a few weeks or months. A "Z-track" injection into the muscle minimizes the chance of a subcutaneous leak. The local inflammatory reaction is slight. Nausea and dizziness have been reported in occasional cases. Because of the painful nature and the skin discoloration that occurs with intramuscular injection the preferred route for parenteral iron administration is intravenous.

Intravenous

Sodium ferric gluconate (Ferrlecit) or iron (III) hydroxide sucrose complex (Venofer) for intravenous use is effective and has a superior safety profile when compared with intravenous iron dextran. They are especially useful in anemia associated with renal failure and hemodialysis. Dosage ranges from 1–4 mg/kg per week.

A small test dose should be given and the patient observed for 30 minutes to rule out an anaphylactoid reaction.

Contraindications to Parenteral Iron Therapy

1. Anemias not due to iron deficiency.
2. Iron overload.
3. History of hypersensitivity to parenteral iron preparations.
4. History of severe allergy or anaphylactic reactions.
5. Clinical or biochemical evidence of liver damage.
6. Acute or chronic infection.
7. Neonates.

Blood Transfusion

A packed red cell transfusion should be given in severe anemia requiring correction more rapidly than is possible with oral iron or parenteral iron or because of the presence of certain complicating factors. This should be reserved for debilitated children with infection, especially when signs of cardiac dysfunction are present and the hemoglobin level is 4 g/dl or less.

Partial Exchange Transfusion

A partial exchange transfusion has been recommended in the management of a severely anemic child under two circumstances:

- In a surgical emergency, when a final hemoglobin of 9–10 g/dl should be attained to permit safe anesthesia
- When anemia is associated with congestive heart failure, in which case it is sufficient to raise the hemoglobin to 4–5 g/dl to correct the immediate anoxia.

Suggested Reading

Ballin A, Berar M, Rubinstein U, et al. Iron state in female adolescents. Am. J. Dis. Chil. 1992;146:803–805.

Chaparro CM. Setting the stage for child health and development: prevention of iron deficiency in early infancy. Journal of Nutrition. 2008;138(12):2529–2533.

Clark SF. Iron Deficiency Anemia. Nutrition in Clinical Practice. 2008;23(2):128–141.

Committee on Nutrition. Iron supplementation for infants. Pediatrics, 1976, 58:765–768.

Dallman PR. Iron deficiency and related nutritional anemias. In: Nathan DG, Oski FA, eds. Hematology of Infancy and Childhood. Philadelphia: WB Saunders; 1987.

Dallman PR, Reeves JD. Laboratory diagnosis of iron deficiency. In: Steckel A, ed. Iron Nutrition in Infancy and Childhood. New York: Raven Press; 1984:11.

Dallman PR. Progress in the prevention of iron deficiency in infants. Acta Paediatr. Scan. 1990;365 (Suppl.):28–37.

Grant CC, Wall CR, Brewster D. et al., Policy statement on iron deficiency in pre-school-aged children. Journal of Paediatrics & Child Health. 2007;43(7–8):513–521.

Lanzkowsky P: Iron deficiency anemias: A systemic disease. Transactions of the College of Medicine of South Africa, July–December, 1982, pp. 67–113

Lanzkowsky P. Iron-deficiency anemia. Pediatric Hematology-Oncology: A Treatise for the Clinician. New York: McGraw-Hill; 1980.

Lanzkowsky P. Iron metabolism and iron deficiency anemia. In: Miller DR, Pearson MA, Baehner RL, McMillan CW, eds. Blood Diseases of Infancy and Childhood. 4th Ed. Saint Louis: CV Mosby; 1978.

Lozoff B. Iron deficiency and child development. Food & Nutrition Bulletin. 2007;28(4 suppl):S560–S571.

Lukens JN. Iron metabolism and iron deficiency anemia. In: Miller DR, Baehner RL, McMillan CW, eds. Blood Diseases of Infancy and Childhood. St. Louis: CV Mosby; 1984.

Lynch S, Stoltzfus R, Rawat R. Critical review of strategies to prevent and control iron deficiency in children. Food & Nutrition Bulletin. 2007;28(4 Suppl):S610–S620.

Oski FA. Iron deficiency in infancy and childhood. N. Engl. J. Med. 1993;199s(329):190–193.

CHAPTER 4

Megaloblastic Anemia

Megaloblastic anemias are characterized by the presence of megaloblasts in the bone marrow and macrocytes in the blood. In more than 95% of cases, megaloblastic anemia is as a result of folate and vitamin B_{12} deficiency. Megaloblastic anemia may also result from rare inborn errors of metabolism of folate or vitamin B_{12}. In addition, deficiencies of ascorbic acid, tocopherol and thiamine may be related to megaloblastic anemia. The causes of megaloblastosis are listed in Table 4-1.

VITAMIN B_{12} (COBALAMIN) DEFICIENCY

Dietary vitamin B_{12} (Cb1),[*] acquired mostly from animal sources, including meat and milk, is absorbed in a series of steps that include:

- Proteolytic release of Cb1 from its associated proteins and Cb1 binds to haptocorrin, a cobalamin-binding protein, produced by salivary and esophageal glands
- In the duodenum after exposure to pancreatic proteases, Cb1 is released from haptocorrin
- In the proximal ileum Cb1 binds to intrinsic factor (IF), a gastric secretory protein, to form IF–Cb1 complex
- Recognition of the IF–Cb1 complex by specific receptors on ileal mucosal cells, which is taken into lysosomes where the IF–Cb1 complex is released and intrinsic factor is degraded
- Transport across ileal cells in the presence of calcium ions
- Release into the portal circulation bound to transcobalamin II (TC II) – the serum protein that carries newly absorbed Cbl throughout the body.

Figure 4-1 shows the pathway of cobalamin absorption, transport and cellular uptake.

---

[*] For the purposes of this chapter, vitamin B_{12}, cobalamin and cbl are used interchangeably. Vitamin B_{12} contains a metal ion in the form of cobalt and therefore is also known as cobalamin.

Manual of Pediatric Hematology and Oncology. DOI: 10.1016/B978-0-12-375154-6.00004-5

Table 4-1   Causes of Megaloblastosis

I. **Vitamin B$_{12}$(cobalamin) deficiency** (see Table 4-2)
II. **Folate deficiency** (see Table 4-6)
III. **Miscellaneous**
    A. Congenital disorders in DNA synthesis
        1. Orotic aciduria (uridine responsive)-pyrimidine biosynthesis is interrupted
        2. Thiamine-responsive megaloblastic anemia[a]
        3. Congenital familial megaloblastic anemia requiring massive doses of vitamin B$_{12}$ and folate
        4. Associated with congenital dyserythropoietic anemia (Tables 6-16 and 6-17)
        5. ? Lesch–Nyhan syndrome (adenine-responsive)-purine nucleotide regeneration is blocked
    B. Acquired defects in DNA synthesis
        1. Liver disease
        2. Sideroblastic anemias (Table 6-20)
        3. Leukemia, especially acute myeloid leukemia (M6) (Chapter 17)
        4. Aplastic anemia (congenital or acquired)
        5. Refractory megaloblastic anemia
    C. Drug-induced megaloblastosis
        1. Purine analogs (e.g., 6-mercaptopurine, aza- thioprine and thioguanine)
        2. Pyrimidine analogs (5-fluorouracil, 6-azauridine)
        3. Inhibitors of ribonucleotide reductase (cytosine arabinoside, hydroxyurea)

[a]Associated in some cases with diabetes and sensorineural hearing impairment and in others with the DIDMOAD syndrome (p. 80). There is wide clinical heterogeneity of this rare disorder. Only the anemia is responsive to high doses of thiamine.

Cobalamin is converted into the two required coenzyme forms, adenosylcobalamin (AdoCbl) and methylcobalamin (MeCbl). The cellular metabolism by which the coenzymes are formed involves the following:

- Receptor-mediated binding of the TC II–Cbl complex to the cell surface
- Adsorptive endocytosis of the complex
- Intralysosomal degradation of the TC II
- Release of Cbl into cytoplasm
- Enzyme-mediated reduction of the central cobalt atom and
- Cytosolic methylation to form MeCbl or mitochondrial adenosylation to form AdoCbl.

The causes of vitamin B$_{12}$ deficiency are listed in Table 4-2.

Clinical Manifestations

Vitamin B$_{12}$ deficiency is characterized by the following:

- Failure to thrive, anorexia, weakness, glossitis
- Pallor, scleral icterus

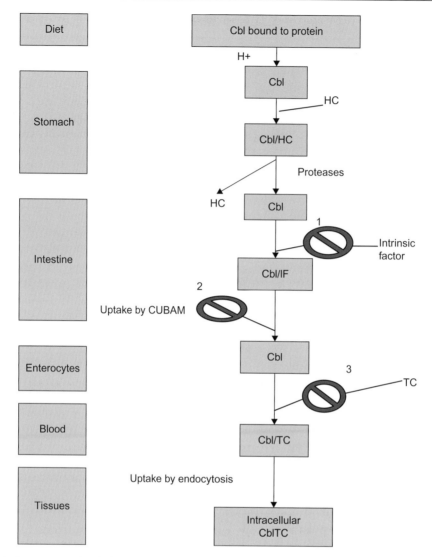

Figure 4-1 Summary of Cobalamin Absorption, Transport and Cellular Uptake.
Abbreviations: Cbl, cobalamin; Cbl/HC, cobalamin haptocorrin complex; Cbl/IF, cobalamin intrinsic factor complex; Cbl/TC, cobalamin transcobalamin complex; CUBAM, ileal receptors made up of cubilin and amnionless proteins; HC, haptocorrin; TC, transcobalamin; 1, intrinsic factor deficiency; 2, Imerslund–Gräsbeck syndrome; 3, transcobalamin deficiency.
Adapted from: Morel and Rosenblatt, British Journal of Haematology. 2006;134:125–136, with permission.

Table 4-2   Causes of Vitamin B_{12} Deficiency

I. Inadequate vitamin B_{12} intake

    A. Dietary ($<2\ \mu g$/day): food fads, lacto-ovo vegetarianism, low animal-source food intake, veganism, malnutrition, poorly controlled phenylketonuria diet

    B. Maternal deficiency leading to B_{12} deficiency in breast milk

II. Defective vitamin B_{12} absorption (Table 4-3)

    A. Failure to secrete intrinsic factor

        1. Congenital intrinsic factor deficiency (gastric mucosa normal) (OMIM 261000)

            a. Quantitative

            b. Qualitative (biologically inert)[a]

        2. Juvenile pernicious anemia (autoimmune) (gastric atrophy)[b]

        3. Juvenile pernicious anemia (gastric autoantibodies) with autoimmune polyendocrinopathies (OMIM 240300)

        4. Juvenile pernicious anemia with IgA deficiency

        5. Gastric mucosal disease

            a. Chronic gastritis, gastric atrophy (elevated serum gastrin and/or low serum pepsinogen 1 concentrations) often caused by *Helicobacter pylori*

            b. Corrosives

            c. Gastrectomy (partial/total)

    B. Failure of absorption in small intestine

        1. Specific vitamin B_{12} malabsorption

            a. Abnormal intrinsic factor[a]

            b. Defective cobalamin transport by enterocytes-abnormal ileal uptake (Imerslund–Gräsbeck syndrome) (OMIM 261100)

            c. Ingestion of chelating agents (phytates, EDTA) (binds calcium and interferes with vitamin B_{12} absorption)

        2. Intestinal disease causing generalized malabsorption, including vitamin-B_{12} malabsorption:

            a. Intestinal resection (e.g., congenital stenosis, volvulus, trauma)

            b. Crohn's disease

            c. Tuberculosis of terminal ileum

            d. Lymphosarcoma of terminal ileum

            e. Pancreatic insufficiency[c]

            f. Zollinger–Ellison syndrome (caused by gastrinoma in duodenum or pancreas)

            g. Celiac disease (gluten enteropathy), tropical sprue

            h. Other less specific malabsorption syndromes

            i. HIV infection

            j. Long-standing medication that decreases gastric acidity (H_2-receptor antagonists and proton pump inhibitors)

            k. Parasites (*Giardia, Lamblia, Diphyllobothrium latum*)

            l. Neonatal necrotizing enterocolitis

        3. Competition for vitamin B_{12}

            a. Small-bowel bacterial overgrowth (e.g., small-bowel diverticulosis, anastomoses and fistulas, blind loops and pouches, multiple strictures, scleroderma, achlorhydria, gastric trichobezoar)

            b. *Diphyllobothrium latum,* the fish tapeworm, (takes up free B_{12} and B_{12}-intrinsic factor complex), giardia lamblia, plasmodium falciparum, strongyloides stercoralis

III. Defective vitamin B_{12} transport

    A. Congenital TC II deficiency (OMIM 275350)

    B. Transient deficiency of TC II

    C. Partial deficiency of TC I (haptocorrin deficiency) (OMIM 193090)

IV. Disorders of vitamin B_{12} metabolism

    A. Congenital

(Continued)

Table 4-2   (Continued)

> 1. Adenosylcobalamin deficiency *Cb1A* (OMIM 251100) and *Cb1B* diseases (OMIM 251100)
> 2. Deficiency of methylmalonyl-CoA mutase (mut°, mut⁻)
> 3. Methylcobalamin deficiency *Cb1E* (OMIM 236270) and *Cb1G* diseases (OMIM 250940)
> 4. Combined adenosylcobalamin and methylcobalamin deficiencies: *Cb1C* (OMIM 277400), *Cb1D* (OMIM 277410) and *Cb1F* diseases (OMIM 277380)
>
> B. Acquired
>   1. Liver disease
>   2. Protein malnutrition (kwashiorkor, marasmus)
>   3. Drugs associated with impaired absorption and/or utilization of vitamin B_{12} (e.g., *p*-aminosalicylic acid, colchicine, neomycin, ethanol, oral contraceptive agents? Metformin)

[a]Same condition.
[b]Pernicious anemia is the final stage of an autoimmune disorder in which autoantibodies against H^+K^+-adenosine triphosphatase destroy parietal cells in the stomach.
[c]Because of lack of the enzymes needed to liberate B_{12} from haptocorrin, the protein that initially binds ingested vitamin B_{12}.
OMIM, Online Mendelian Inheritance in Man (see p. 65).

- Anemia with high MCV, hypersegmented neutrophils, leukopenia, thrombocytopenia
- Megaloblastic bone marrow
- Elevated urinary and plasma methylmalonic acid and homocysteine
- Muscle hypotonia, tremor, myoclonus.

Nutritional Deficiency

Recommended dietary allowance of vitamin B_{12} for children is 0.9–2.4 µg/day. The most common cause of Cb1 deficiency in infants is dietary deficiency in the mother. Mothers following vegetarian, vegan, macrobiotic and other special diets are at particular risk. Cb1 in breast milk parallels that in serum and is deficient when the mother is a vegan or has unrecognized pernicious anemia, has had previous gastric bypass surgery or short gut syndrome.

Defective Absorption

Table 4-3 lists the features of congenital and acquired defects of vitamin B_{12} absorption, Table 4-4 lists the main features of genetic defects in processing of vitamin B_{12}.

Food Cobalamin Malabsorption

Some patients suffer from an inability to release cobalamin from the protein-bound state in which it is normally encountered in food. This process requires both an acid pH and peptic activity. Impaired absorption occurs when there is impaired gastric function, e.g. atrophic gastritis, partial gastrectomy. In this condition, there is a low serum cobalamin, mild

Table 4-3 Features of Congenital and Acquired Defects of Vitamin B_{12} Absorption

Condition	Stomach			Schilling Test		Serum Antibodies		
	Histology	Intrinsic Factor[a]	Hydrochloric Acid (HCL)	Without IF	With IF	Intrinsic Factor	Parietal Cell	Associated Features
Congenital pernicious anemia	Normal	Absent	Normal	Decreased	Normal	Absent	Absent	None; relative of patient may exhibit defective vitamin B_{12} malabsorption
Juvenile pernicious anemia (autoimmune)	Atrophy	Absent	Achlorhydria	Decreased	Normal	Present (90%)	Present (10%)	Occasional lupus erythematosus, IgA deficiency, moniliasis, endocrinopathy in siblings
Juvenile pernicious anemia with polyendocrino-pathies or selective IgA deficiency	Atrophy	Absent	Achlorhydria	Decreased	Normal	Present	Present	Hypothyroidism (chronic auto-immune thyroiditis–Hashimoto's thyroiditis) insulin-dependent diabetes mellitus, primary ovarian failure, myasthenia gravis, hypoparathyroidism, Addison's disease, moniliasis, or selective IgA deficiency
Enterocyte vitamin B_{12} malabsorption (Imerslund-Gräsbeck)	Normal	Present	Normal	Decreased	Decreased	Absent	Absent	Benign proteinuria, amino-aciduria, no generalized malabsorption[b]
Generalized malabsorption	Normal	Present	Normal	Decreased	Decreased	Absent	Absent	Malabsorption; syndrome; history of ileal resection, Crohns disease, lymphoma

[a]Either absent secretion of immunologically recognizable IF or secretes immunologically reactive protein that is inactive physiologically. The latter group includes patients whose IF has reduced affinity for the ileal IF receptor, reduced affinity for cobalamin or increased susceptibility for proteolysis.

[b]Rare cases have been described of this syndrome associated with generalized malabsorption reversed by vitamin B_{12} administration and rare cases have been described without proteinuria or aminoaciduria.

IF, Intrinsic factor.

Table 4-4   **Main Features of Genetic Defects in Processing of Vitamin B$_{12}$**

Defect	Serum B$_{12}$	Clinical/Biochemical
Food cobalamin malabsorption	Low	N.A. ± Anaemia, mild ↑ MMA/tHcy
Intrinsic factor deficiency	Low	Anaemia, delayed development, mild ↑ MMA/tHcy
Enterocyte cobalamin malabsorption (Imerslund–Gräsbeck)	Low	Anaemia, proteinuria, delayed development, mild ↑ MMA/tHcy
Transcobalamin I (R-Binder) deficiency	Low	No abnormality, No ↑ MMA/tHcy
Transcobalamin II deficiency	Normal	N.A. ± Anaemia, failure to thrive, mild ↑ MMA/tHCy
Intracellular defects of cobalamin	Normal	Severe disease, ↑ MMA/tHcy

Abbreviations: N.A., neurologic abnormalities; MMA, methylmalonic acid; tHcy, total homocysteine; ↑, increased.

increase in methylmalonic acid and homocysteine and a normal Schilling test (see page 83).

Intrinsic Factor Deficiency

Patients with absent or defective intrinsic factor (also known as S-binder) have low serum B$_{12,}$ megaloblastic anemia, developmental delay and myelopathy. Patients have a mild increase in methylmalonic acid and homocysteine. This autosomal recessive disorder usually appears early in the second year of life, but may be delayed until adolescence or adulthood. The abnormal absorption of cobalamin is corrected by mixing the vitamin with a source of normal intrinsic factor. Some patients have no detectable intrinsic factor, whereas others have intrinsic factor that can be detected immunologically but lacks function. The gene for human intrinsic factor (GIF gene) has been cloned and localized to chromosome 11. Mutations have been identified in the GIF gene, together with a polymorphism (68A→G) which may be a marker for this inheritance. Homozygous GIF mutations result in complete loss of intrinsic factor function.

Defective Cobalamin Transport by Ileal Enterocyte Receptors for the Intrinsic-Factor–Cobalamin Complex (Imerslund-Gräsbeck Syndrome)

The Imerslund–Gräsbeck syndrome is an autosomal recessive disorder which has a low serum B$_{12}$ due to selective defect in cobalamin absorption that is not corrected by treatment with intrinsic factor. It usually presents with pallor, weakness, anorexia, failure to thrive, delayed development, recurrent infections and gastrointestinal symptoms within the first two years of life but has been reported up to 15 years of age. In many patients, proteinuria of the tubular type is found that is not corrected by systemic cobalamin. Most of the known patients reside in Norway, Finland, Saudi Arabia and among Sephardic Jews in Israel. In these patients, intrinsic factor level is normal, they do not have antibodies to intrinsic factor and the intestinal morphology is normal. They have a mild increase in methylmalonic acid and homocysteine. In some cases the ileal receptor for intrinsic factor–cobalamin complex is absent, whereas in other patients it is present.

There has been a decrease in the number of new cases suggesting that dietary or other factors may influence the expression of this disease. The locus for Imerslund–Gräsbeck syndrome has been assigned to chromosome 10. Imerslund–Gräsbeck-causing mutations are found in either of two genes encoding the epithelium proteins: cubilin (CUBN) and amnionless (AMN). The gene receptor, cubilin P1297L (OMIM 602997)[*] is a 640-kDa protein which recognizes intrinsic factor–cobalamin and various other proteins to be endocytosed in the intestine and kidney. The exact function of AMN is unknown but mutations affecting either of the two proteins may cause Imerslund–Gräsbeck syndrome.

Defective Transport

Patients have a low serum B_{12} due to selective defect in cobalamin absorption that is not corrected by treatment with intrinsic factor.

Table 4-5 lists clinical manifestations, laboratory finding and treatment of inborn errors of cobalamin transport and metabolism and Figure 4-2 shows the pathways of vitamin B_{12} metabolism and sites of inborn errors of vitamin B_{12} metabolism.

Transcobalamin II Deficiency (OMIM 275350)

Transcobalamin II (TC II) is the principal transport carrier protein system of cobalamin. The TC II gene is located on chromosome 22. In the absence of TC II, a serious and potentially fatal condition occurs. It presents clinically as follows:

- Age 3–5 weeks
- Autosomal-recessive inheritance
- Failure to thrive, weakness
- Vomiting and diarrhea
- Hematologically: severe megaloblastic anemia, some patients present with progressive pancytopenia or isolated erythroid hypoplasia. Defective granulocyte function has been described
- Immunologic deficiency both cellular and humoral
- Neurologic disease (appears 6 to 30 months after onset of symptoms)
- Hyperhomocysteinemia, Homocystinuria and methylmalonic aciduria
- Normal serum cobalamin levels (most of the cobalamin in serum is bound to transcobalamin I).

---

[*]The 6-digit number is the entry number for the disorder in Online Mendelian Inheritance in Man [OMIM] a continuously updated electronic catalog of human genes and genetic disorders. The online version is accessible through the world wide web [http://www.ncbi.nlm.nih.gov/omim/].

Table 4-5 Clinical Manifestations, Laboratory Findings and Treatment of the Autosomal Recessive Inborn Errors of Cobalamin Transport and Metabolism

Condition (OMIM no.)	Defect	Typical Clinical Manifestations	Typical Onset	Laboratory Findings	Treatment and Response
TC II deficiency (OMIM 275350)	Defective/absent TCII	Failure to thrive, megaloblastic anemia, later neurologic features and immunodeficiency	Early infancy 3–5 weeks	Usually normal serum Cb1; elevated serum MMA, homocysteine; absent/defective TCII	High doses of Cb1 by injection; good response to treatment if begun early
TC I (R-binder) deficiency (OMIM 193090)	Deficiency/absence of TCI in plasma, saliva, leukocytes	Neurologic symptoms (myelopathy) reported, but unclear if these are related to condition	Unclear if observed symptoms are related to condition	Low serum Cbl, normal TCII-Cb1 levels. No increase in MMA or homocysteine	Cbl therapy does not appear to be of benefit
Defective synthesis of AdoCb1: cblA (OMIM 251100) cblB (OMIM 251110)	Defective synthesis of AdoCb1	Lethargy, failure to thrive, recurrent vomiting, dehydration, hypotonia, keto-acidosis hypoglycemia	First weeks or months of life	Normal serum Cbl, homocysteine and methionine; elevated MMA, ketones, glycine, ammonia; leukopenia, thrombocytopenia, anemia	Pharmacologic doses of Cb1, dietary protein restriction, oral antibiotics. Treatment response for cblA better than for cblB

Defective synthesis of MeCb1: cblE (OMIM 236270) cblG (OMIM 250940)	Defective synthesis of MeCb1	Most in first 2 years of life	Normal serum Cb1 and folate; homocystinuria, hypomethioninemia	Pharmacologic doses of Cb1, betaine; good treatment response in some patients treated early
Defective synthesis of AdoCb1 and MeCb1: cblC (OMIM 277400) cblD (OMIM 277410) cblF (OMIM 277380)	Impaired synthesis of both AdoCb1 and MeCb1	Variable from neonatal period to adolescence majority with neonatal onset	Normal serum Cb1, TCII; methylmalonic aciduria, homcystinuria, hypomethioninemia	Pharmacologic doses of hydroxocobalamin, moderate protein restriction, betaine treatment. Response often not optimum

Abbreviations: TCII, Transcobalamin II; OMIM, Online Mendelian Inheritance in Man; Cb1, cobalamin; MMA, methylmalonic acid; TCI, transcobalamin I; AdoCb1, Adenosylcobalamin; MeCb1, methylcobalamin.

Modified from: Rasmussen SA, Fernhoff PM and Scanlon KS. Vitamin B12 deficiency in children and adolescents. Journal of Pediatrics, 2001;138:110, with permission.

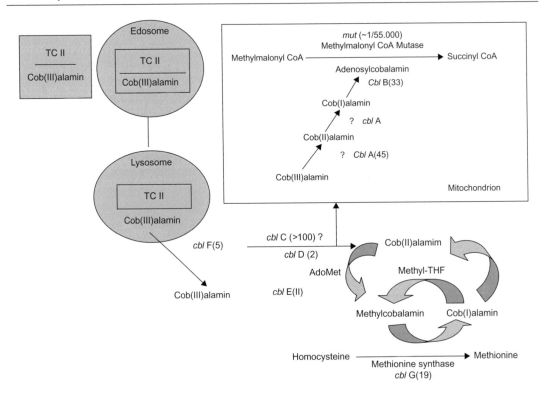

Figure 4-2 Cobalamin Metabolism in Cultured Mammalian Cells and the Sites of the Known Inborn Errors of Cobalamin Metabolism.

AdoMet, S-adenosylmethionine, cob(III)alamin, cob(II)alamin, cob(II)represent cobalamin with its cobalt in the 3+, 2+ or 1+ oxidation state, methyl-THF is 5-methyltetrahydrofolate. The incidence or minimum numbers of patients with a given diseases are shown in parentheses.

Adapted from: Rosenblatt DS and Whitehead VM (1999). Cobalamin and Folate Deficiency: Acquired and Hereditary Disorders in Children. Seminars in Hematology 36:19, with permission.

Diagnosis

Absence of protein capable of binding radiolabeled cobalamin and migrating with TC II on chromatography or gel electrophoresis, or by immunologic techniques. An abnormal Schilling test result is usually found. TC II is synthesized by amniocytes, permitting prenatal diagnosis.

Treatment

One thousand μg vitamin B_{12} intramuscularly 1–2 times weekly. Serum cobalamin levels must be kept very high (1,000–10,000 pg/ml) in order to treat TC II patients successfully.

Partial Deficiency of Transcobalamin I (Haptocorrin Deficiency) (OMIM 193090)

Partial deficiency of transcobalamin I (also known as haptocorrin or R-binder) has been reported. Serum vitamin B_{12} concentrations are persistently low and patients show no signs of vitamin B_{12} deficiency (normal values for hemocysteine and methylmolonic acid and show no megaloblastic hematologic features) because their TC II-cobalamin levels are normal and patients are not clinically deficient in vitamin B_{12}. TC I concentrations range from 25 to 54% of the mean normal concentration.

Clinically this syndrome is characterized by a myelopathy, not attributable to other causes and the etiology of these symptoms remains unclear.

Disorders of Metabolism

Congenital

The conversion of a vitamin to its active co-enzyme and subsequent binding to an apo-enzyme producing active holo-enzyme are fundamental biochemical processes. Therefore deficient activity of an enzyme can result not only from a defect of the enzyme protein itself, which may involve interaction of a co-enzyme with an apo-enzyme, but also from a defect in the conversion of the vitamin to a co-enzyme.

Once vitamin B_{12} has been taken up into cells, it must be converted to an active co-enzyme in order to act as a co-catalyst with vitamin B_{12}-dependent apoenzymes. Two enzymes known to depend for activity on vitamin B_{12} derivatives are:

- Methylmalonyl Coenzyme A (CoA) mutase, which requires adenosylcobalamin. Methylmalonyl CoA mutase catalyzes the conversion of methylmalonyl CoA to succinyl CoA. A decreased activity of methylmalonyl CoA mutase is reflected by the excretion of elevated amounts of methylmalonic acid
- N^5-methyltetrahydrofolate homocysteine methyltransferase which requires methylcobalamin. Lack of methylcobalamin leads to deficient activity of N^5-methyltetrahydrofolate homocysteine methyltransferase, with reduced ability to methylate homocysteine, resulting in hyperhomocysteinemia and homocysteinuria.

Patients with inborn errors of cobalamin utilization present with methylmalonic acidemia and hyperhomocysteinemia, either alone or in combination. Methylmalonic acidemia occurs as a result of a functional defect in the mitochondrial methylmalonyl CoA mutase or its cofactor adenosylcobalamin which catalyzes the conversion of L-methylmalonyl CoA to succinyl CoA. Hyperhomocysteinemia occurs as a result of a functional defect in the cytoplasmic methionine synthase or its cofactor methylcobalamin. Those disorders causing methylmalonic aciduria are characterized by severe metabolic acidosis, with the accumulation of large amounts of methylmalonic acid in blood, urine and cerebrospinal fluid.

The incidence is estimated at 1:61,000. All the disorders of Cb1 metabolism are inherited as autosomal recessive traits and prenatal diagnosis is possible. Classification has relied on somatic cell complementation studies in cultured fibroblasts. Prenatal detection of fetuses with defects in the complementation groups *cblA, cblB, cblC, cblE* and *cblF* has been accomplished using cultured amniotic cells and chemical determinations on amniotic fluid or maternal urine. In several cases, *in utero* cbl therapy has been attempted with apparent success.

Adenosylcobalamin Deficiency CblA (OMIM 251100) and CblB (OMIM 251110) Diseases

Deficiency of adenosylcobalamin synthesis leads to impaired methylmalonyl CoA mutase activity and results in methylmalonic acidemia. Cobalamin-responsive methylmalonic aciduria characterizes both *CblA* and *CblB* diseases. Intact cells from both *CblA* and *CblB* patients fail to synthesize adenosylcobalamin. However, cell extracts from *CblA* patients can synthesize adenosylcobalamin when provided with an appropriate reducing system, whereas extracts from *CblB* patients cannot. The defect in *CblA* may be related to a deficiency of a mitochondrial nicotinamide adenine dinucleotide phosphate (NADPH)-linked aquacobalamin reductase. The defect in *CblB* affects adenosyltransferase, which is involved in the final step in adenosylcobalamin synthesis.

This group of patients presents with:

- Life-threatening or fatal ketoacidosis in the first few weeks or months of life
- Hypoglycemia and hyperglycinemia
- Failure to thrive or developmental retardation (may be a consequence of the acidosis and reversed by relief of the ketoacidosis)
- Serum cobalamin concentrations are normal
- Both *CblA* and *CblB* are autosomal recessive diseases.

Studies of these patients have shown that intact cells fail to oxidize propionate normally. Methylmalonyl CoA arises chiefly through the carboxylation of propionate, which in turn derives largely from degradation of valine, isoleucine, methionine and threonine.

Treatment

Ninety percent of *CblA* patients respond to therapy with systemic hydroxocobalamin or cyanocobalamin whereas only 40% of *CblB* patients respond to this therapy. Only 30% have long-term survival.

Deficiency of Methylmalonyl-CoA Mutase (mut°, mut⁻)

Defects in methylmalonyl CoA mutase apoenzyme formation can occur and results in methylmalonic aciduria, which is accompanied by life-threatening or fatal ketoacidosis, unresponsive to vitamin B_{12}.

Clinical Findings

Infants are well at birth but become rapidly symptomatic on protein feeding. Symptoms include lethargy, failure to thrive, muscular hypotonia, respiratory distress and recurrent vomiting and dehydration. Children normally excrete <15 to 20 μg of methylmalonic acid per gram of creatinine, whereas patients with methylmalonyl CoA mutase deficiency excrete >100 mg up to several grams daily. Patients may have elevated levels of ketones, glycine and ammonia in the blood and urine. Many also have hypoglycemia, leukopenia and thrombocytopenia.

It is an autosomal recessive disease and prenatal diagnosis is possible.

Treatment

Treatment is protein restriction using a formula deficient in valine, isoleucine, methionine and threonine, with the goal of limiting amino acids that use the propionate pathway. Therapy with carnitine has been advocated for those patients who are carnitine deficient. Lincomycin and metronidazole have been used to reduce enteric propionate production by anaerobic bacteria. These patients do not respond to vitamin B_{12} therapy. Despite therapy, a number of patients have experienced basal ganglia infarcts, tubulointerstitial nephritis, acute pancreatitis and cardiomyopathy as complications. Liver transplantation has been attempted.

Culture of patients' fibroblasts show two classes of mutase deficiency: those having no detectable enzyme activity are designated mut°, whereas those with residual activity, which can be stimulated by high levels of cobalamin, are called mut⁻. Some mut° cell lines synthesize no detectable protein.

Methylcobalamin Synthesis Deficiency: CblE (OMIM 236270) and CblG (OMIM 250940) Diseases

Abnormalities in methylcobalamin synthesis result in reduced N^5-methyltetrahydrofolate: homocysteine methyltransferase and consequently lead to homocysteinuria with hypomethioninemia. Thus homocysteinuria and hypomethioninemia, usually without methylmalonic aciduria, characterize functional methionine synthase deficiency (*CblE, CblG*), although one *CblE* patient had transient methylmalonic aciduria. Fibroblasts from *CblE* and *CblG* patients show a decreased accumulation of methylcobalamin with a normal accumulation of adenosylcobalamin after incubation with cyanocobalamin. Their fibroblasts show decreased incorporation of labeled methyltetrahydrofolate as well. Cyanocobalamin uptake and binding to both cobalamin-dependent enzymes is normal in *CblE* fibroblasts and in most *CblG* fibroblasts.

Clinical Findings
1. Most patients become ill within the first 2 years of life, but a number have been diagnosed in adulthood.
2. Megaloblastic anemia.
3. Various neurological deficits including developmental delay, cerebral atrophy, EEG abnormalities, nystagmus, hypotonia, hypertonia, seizures, blindness and ataxia.
4. Failure to thrive.

Treatment
Hydroxocobalamin administered systemically, daily at first, then once or twice weekly. Usually this corrects the anemia and the metabolic abnormalities. Betaine supplementation may be helpful to reduce the homocysteine further. The neurological findings are more difficult to reverse once established, particularly in *CblG* disease. There has been successful prenatal diagnosis of *CblE* disease in amniocytes and the mother with an affected fetus can be treated with twice weekly hydroxocobalamin after the second trimester.

Combined Adenosylcobalamin and Methylcobalamin Deficiencies CblC (OMIM 277400), CblD (OMIM 277410) and CblF (OMIM 277380) Diseases

These disorders result in failure of cells to synthesize both methylcobalamin (resulting in homocysteinuria and hypomethioninemia) and adenosylcobalamin (resulting in methylmalonic aciduria) and accordingly, deficient activity of methylmalonyl CoA mutase (leading to homocystinuria and hypomethioninemia with methylmalonic aciduria) and N^5-methyltetrahydrofolate: homocysteine methyltransferase. Fibroblasts from *CblC* and *CblD* patients accumulate virtually no adenosylcobalamin or methylcobalamin when incubated with labeled cyanocobalamin. In contrast, fibroblasts from *CblF* patients accumulate excess cobalamin, but it is all unmetabolized cyanocobalamin, nonprotein bound and localized to lysosomes. In *CblC* and *CblD*, the defect is believed to involve cob(III)alamin* reductase or reductases, whereas in *CblF*, the defect involves the exit of cobalamin from the lysosome. Partial deficiencies of cyanocobalamin beta-ligand transferase and microsomal cob(III)alamin reductase have been described in *CblC* and *CblD* fibroblasts as well.

These patients present in the first month or before the end of the first year of life with:

- Poor feeding, failure to thrive and lethargy
- Macrocytosis, hypersegmented neutrophils, thrombocytopenia and megaloblastic anemia

---

*In this form of cobalamin the cobalt atom is trivalent (cob [III]) and must be reduced before it can bind to the respective enzyme.

- Developmental retardation
- Spasticity, delirium and psychosis (in older children and adolescence)
- Hydrocephalus, cor pulmonale and hepatic failure have been described, as well as a pigmentary retinopathy with perimacular degeneration
- Methymalonic acid levels are less than in methylmalonyl CoA mutase deficiency but greater than in defects of cobalamin transport
- Many patients with the onset of symptoms in the first month of life die whereas those with a later onset have a better prognosis.

CblC, CblD and *CblF* diseases can be differentiated using cultured fibroblasts. Failure of uptake of labeled cyanocobalamin distinguishes *CblC* and *CblD* from all other cbl mutations. There is reduced incorporation of propionate and methyltetrahydrofolate into macromolecules in all three disorders and reduced synthesis of adenosylcobalamin and methylcobalamin. Complementation analysis between an unknown cell line and previously defined groups establishes the specific diagnosis. Prenatal diagnosis has been successfully accomplished in *CblC* disease using chorionic villus biopsy material and cells.

Treatment

The treatment of *CblC* disease can be difficult. Daily therapy with oral betaine and twice-weekly injections of hydroxocobalamin improve lethargy, irritability and failure to thrive, reduce methylmalonic aciduria and return serum methionine and homocysteine concentrations to normal. There has been incomplete reversal of the neurological and retinal findings. Surviving patients usually have moderate to severe developmental delay, even with good metabolic control.

Acquired

In protein malnutrition (kwashiorkor, marasmus) and liver disease impaired utilization of vitamin B_{12} has been reported. Certain drugs are associated with impaired absorption or utilization of vitamin B_{12} (see Table 4-2).

FOLIC ACID DEFICIENCY

Food folate occurs in the polyglutamate form which must be hydrolyzed by conjugase in the brush border of the intestine to folate monoglutamates. These are absorbed in the duodenum and upper small intestine and transported to the liver becoming 5-methyl-tetrahydrofolate, the principal circulating folate form.

Folate binds to and acts as a coenzyme for enzymes that mediate single-carbon transfer reactions. They accept and donate single-carbon atoms at different states of oxidation. 5,10-methylene tetrahydrofolate is used unchanged for the synthesis of thymidylate, reduced to 5-methyl tetrahydrofolate for the synthesis of methionine, or oxidized to 10-formyl tetra-hydrofolate for the synthesis of purines.

The recommended dietary allowance of folate increases from 150 to 400 µg/day from age 1 year to 18 years.

The causes of folic acid deficiency are listed in Table 4-6.

Acquired Folate Deficiency

Folate deficiency, next to iron deficiency, is one of the commonest micronutrient deficiencies worldwide. It is a component of malnutrition and starvation. Women are more frequently affected than men. Folate deficiency is common in mothers, particularly where poverty or malnutrition is prevalent and dietary supplements are not provided. Folate stores are depleted after 3 months or sooner when the growing fetus and lactation impose increased demands for folate. The major benefit of folate sufficiency for the fetus is the prevention of neural tube defects. This is currently best achieved by administering folate (and cobalamin) to mothers during the periconceptional period.

In addition, low daily folate intake is associated with a twofold increased risk for preterm delivery and low infant birth weight. These findings suggest that maternal folate status may affect birth outcome in ways other than neural tube defects.

Clinical folate deficiency is seldom present at birth. However, rapid growth in the first few weeks of life demands increased folate. There is a need for folate supplements at this time, particularly for premature infants in doses of 0.05 to 0.2 mg daily.

There is a greater than normal folate requirement under the following conditions:

- Diseases of the small intestine causing malabsorption
- Medication ingestion, e.g., antiepileptic medication, oral contraceptive pills
- Hemolytic anemia with rapid red cell turnover, e.g., sickle cell anemia, thalassemia
- Pregnant women and lactating women
- Goat's milk diet (goat's milk contains almost no folate)
- Chronic infections, e.g., diarrhea, hepatitis (may disturb folate stores), HIV infection.

Inborn Errors of Folate Transport and Metabolism

Inborn errors include hereditary folate malabsorption, methylene-tetrahydrofolate reductase (MTHFR) deficiency and glutamate formiminotransferase deficiency. In addition to these rare severe deficiencies, polymorphisms in the MTHFR gene have been implicated with neural defects and vascular thrombosis. Table 4-7 lists the clinical and biochemical features of inherited defects of folate metabolism.

Table 4-6   Causes of Folic Acid Deficiency

I. Inadequate intake
  A. Poverty, ignorance, faddism
  B. Method of cooking (sustained boiling loses 40% folate)
  C. Goat's-milk feeding (6 μg folate/L)
  D. Malnutrition (marasmus, kwashiorkor)
  E. Special diets for phenylketonuria or maple syrup urine disease
  F. Prematurity
  G. Post bone marrow transplantation (heat-sterilized food)
II. Defective absorption
  A. Congenital, isolated defect of folate malabsorption[a]
  B. Acquired
       1. Idiopathic steatorrhea
       2. Tropical sprue
       3. Partial or total gastrectomy
       4. Multiple diverticula of small intestine
       5. Jejunal resection
       6. Regional ileitis
       7. Whipple's disease
       8. Intestinal lymphoma
       9. Broad-spectrum antibiotics
      10. Drugs associated with impaired absorption and/or utilization of folic acid, e.g., methotrexate, diphenylhydantoin (Dilantin), primidone, barbiturates, oral contraceptive agents, cycloserine, metformin, ethanol, dietary amino acids (glycine, methionine), sulfasalazine and pyrimethamine
      11. Post bone marrow transplantation (total body irradiation, drugs, intestinal GVH disease)
III. Increased requirements
  A. Rapid growth (e.g., prematurity, pregnancy)
  B. Chronic hemolytic anemia, especially with ineffective erythropoiesis (e.g., thalassemia major)
  C. Dyserythropoietic anemias
  D. Malignant disease (e.g., lymphoma, leukemia)
  E. Hypermetabolic states (e.g., infection, hyperthyroidism)
  F. Extensive skin disease (e.g., dermatitis herpetiformis, psoriasis, exfoliative dermatitis)
  G. Cirrhosis
  H. Post bone marrow transplantation (bone marrow and epithelial cell regeneration)
IV. Disorders of folic acid metabolism
  A. Congenital[b]
       1. Methylenetetrahydrofolate reductase deficiency (OMIM 236250)
       2. Glutamate formiminotransferase deficiency (OMIM 229100)
       3. Functional N^5-methyltetrahydrofolate: homocysteine methyltransferase deficiency caused by *cblE* (OMIM 236270) or *cblG* (OMIM 250940) disease
       4. Dihydrofolate reductase deficiency (less well established)
       5. Methenyl-tetrahydrofolate cyclohydrolase (less well established)
       6. Primary methyl-tetrahydrofolate: homocysteine methyltransferase deficiency (less well established)
  B. Acquired
       1. Impaired utilization of folate
            a. Folate antagonists (drugs that are dihydrofolate reductase inhibitors, e.g., methotrexate, pyrimethamine, trimethoprim, pentamidine)
            b. Vitamin B_{12} deficiency

(Continued)

Table 4-6   (Continued)

   c. Alcoholism
   d. Liver disease (acute and chronic)
   e. Other drugs (IIB10 above)
V. Increased excretion (e.g., chronic dialysis, vitamin B_{12} deficiency, liver disease, heart disease)

[a]Rare disorder. Isolated disorder of folate transport resulting in low CSF folate and mental retardation. The ability to absorb all other nutrients is normal. Defect is overcome by pharmacologic oral doses of folic acid or intramuscular folic acid (Lanzkowsky P, Erlandson, ME, Bezan AI. Isolated defect of folic acid absorption associated with mental retardation and cerebral calcifications, Blood 34:452–465, 1969; Amer J Med 48:580–583, 1970).
[b]These disorders are associated with megaloblastic anemia, mental retardation, disorders in gait and both peripheral and central nervous system disease.
Abbreviations: OMIM, Online Mendelian Inheritance in Man (p. 65).

Hereditary Folate Malabsorption (OMIM 229050)

Hereditary folate malabsorption (congenital malabsorption of folate) is due to a rare autosomal recessive trait and is characterized by megaloblastic anemia, chronic or recurrent diarrhea, mouth ulcers, failure to thrive and usually loss of developmental milestones, seizures and progressive neurological deterioration. The most important diagnostic feature is megaloblastic anemia in the first few months of life, associated with low serum, red cell and cerebrospinal fluid folate levels.

All patients have an abnormality in the absorption of oral folic acid or of reduced folates (5-methyltetrahydrofolate or 5-formyltetrahydrofolate [folinic acid]). They may have an elevated excretion of formiminoglutamate (FIGLU) and of orotic acid. This disease indicates that there is a specific transport system for folates across both the intestine and the choroid plexus and that this carrier system is coded by a single gene. Even when blood folate levels are increased sufficiently to correct the anemia, folate levels in the cerebrospinal fluid may remain low. These patients are unable to achieve the normal 3:1 CSF:serum folate ratio indicative of failure to transport folates across the choroid plexus. The uptake of folate into other cells is probably not defective and the uptake of folate into cultured cells is not abnormal.

Treatment

It is essential to maintain levels of folate in the blood and in the cerebrospinal fluid in the range associated with folate sufficiency. Oral folic acid in doses of 5–40 mg daily and lower parenteral doses correct the hematologic abnormality, but cerebrospinal fluid folate levels may remain low. Oral doses of folates may be increased to 100 mg or more daily if necessary. Oral methyltetrahydrofolate and folinic acid can increase cerebrospinal fluid folate levels, but only slightly. If oral therapy is not effective, systemic therapy with reduced

Table 4-7   Clinical and Biochemical Features of Inherited Defects of Folate Metabolism

	Hereditary Folate Malabsorption	Methylene-H$_4$ Folate Reductase Deficiency	Glutamate Formimino-Transferase Deficiency	Functional Methionine Synthase Deficiency	
				Cb1E	*Cb1G*
Clinical Signs					
Prevalence	13 cases	>30 cases	13 cases	8 cases	12 cases
Megaloblastic anemia	A	N	N[a]	A	A
Developmental delay	A	A	N[a]	A	A
Seizures	A	A	N[a]	A	A
Speech abnormalities	N	N	A[a]	N	N
Gait abnormalities	N	A	N[a]	N	A[a]
Peripheral neuropathy	N[a]	A	N[a]	N	A[a]
Apnea	N	A	N[a]	N[a]	N
Biochemical Findings					
Homocystinuria/ Homocysteinemia	N	A	N	A	A
Hypomethioninemia	N	A	N	A	A
Formininoglutamic aciduria	A[a]	N	A	N	N[a]
Folate absorption	A	N	N	N	N
Serum Cbl	N	N	N[a]	N	N
Serum folate	A	A	N[a]	N	N
Red blood cell folate	A	A[a]	N[a]	N	N
Defects detectable in cultured fibroblasts					
Whole cells					
CH$_3$H$_4$ folate uptake	N	N	N	A	A
CH$_3$H$_4$ folate content	N	A	N	N	N
CH$_3$B$_{12}$ content	N	N[a]	N	A	A
Extracts					
Activity of holoenzyme of methionine synthase	N	N[a]	N	N[b]	A
Glutamate Formininotransferase				Activity undetectable in cultured cells ?Abnormal in liver and erythrocytes	
Methylene-H$_4$ folate reductase	N	A	N	N	N
Treatment	Folic acid or reduced folates in pharmacologic doses	Folates, betaine methionine	? Folates	OH-Cb1, folinic acid, betaine	

[a]Exceptions described in some cases.
[b]Abnormal activity with low concentrations of reducing agent in assay.
Abbreviations: N, normal; A, abnormal; (i.e. clinical or laboratory findings). From Rosenblatt, DS. Inherited disorders of folate transport and metabolism. In: Scriver CR, Beaudet A, Sly WS, Valle D, editors. The metabolic and molecular bases of inherited disease 7th ed. New York, McGraw-Hill, 1995, with permission.

folates should be tried. It may be necessary to give intrathecal reduced folates if cerebrospinal fluid levels cannot be normalized.

Methylene-Tetrahydrofolate Reductase Deficiency (OMIM 236250)
Clinical Findings
Clinically asymptomatic but biochemically affected individuals have been reported. The condition is a rare autosomal recessive disorder and can present severely in early infancy (first month of life) or much more mildly as late as 16 years of age. Clinical symptoms vary and consist of developmental delay which is the most common clinical manifestation, hypotonia, motor and gait abnormalities, recurrent strokes, seizures, mental retardation, psychiatric manifestations and microcephaly. Autopsy findings including internal hydrocephalus, microgyria, perivascular changes, demyelination, macrophage infiltration, gliosis, astrocytosis and subacute combined degeneration of the spinal cord have been reported. By interfering with methylation, methionine deficiency may cause demyelination. The vascular changes include thrombosis of both cerebral arteries and veins. Megaloblastic anemia is uncommon in patients with this disease because reduced folates are still available for purine and pyrimidine synthesis. MTHFR deficiency results in elevated plasma homocysteine and homocystinuria and decreased plasma methionine levels because levels of methyltetrahydrofolate serves as one of three methyl donors for the conversion of homocysteine to methionine. The gene for MTHFR is located on chromosome 1p36.3 and there are more than 30 mutations.

Diagnosis
MTHFR deficiency can be diagnosed by measuring enzyme activity in liver, white blood cells and cultured fibroblasts. In fibroblasts, the specific activity of MTHFR is dependent on the stage of the culture cycle. There is a rough correlation between the degree of enzyme deficiency and clinical severity. The proportion of total folate in fibroblasts that is methyltetrahydrofolate and the extent of labeled formate incorporated into methionine are better indicators of clinical severity.

Prognosis
Prognosis is poor in early-onset severe MTHFR deficiency.

Treatment
MTHFR deficiency is resistant to treatment. Regimens have included folic acid, methyltetrahydrofolate, methionine, pyridoxine, various cobalamins, carnitine and betaine. Betaine therapy after prenatal diagnosis has resulted in the best outcome to date since it has the theoretical advantage of both lowering homocysteine levels and supplementing methionine levels.

Prenatal diagnosis is possible by enzyme assay in amniocytes, chorionic villus biopsy samples, or cultured chorionic villus cells. The phenotypic heterogenicity in MTHFR deficiency is reflected by genotypic heterogeneity.

Glutamate Formiminotransferase Deficiency (OMIM 229100)

Glutamate formiminotransferase and formiminotetrahydrofolate cyclodeaminase are involved in the transfer of a formimino group to tetrahydrofolate followed by the release of ammonia and the formation of 5,10-methyltetrahydrofolate. These activities are found only in the liver and kidneys and are performed by a single octameric enzyme. It is not clear that glutamate formiminotransferase deficiency is associated with disease, even though formiminoglutamic acid (FIGLU) excretion is the one constant finding. There have been 20 patients described, with ages ranging from 3 months to 42 years at diagnosis. Some have been asymptomatic and several patients have macrocytosis and hypersegmentation of neutrophils.

Mild and severe phenotypes have been described. Patients with the severe form show mental and physical retardation, abnormal EEG activity and dilatation of the cerebral ventricles with cortical atrophy. In the mild form, there is no mental retardation but massive excretion of FIGLU.

Liver-specific activity ranges from 14 to 54% of control values. It is not possible to confirm the diagnosis using cultured cells because the enzyme is not expressed. There is dispute as to whether the enzyme is expressed in red cells.

Patients may have elevated to normal serum folate levels and elevated FIGLU levels in the blood and urine after a histidine load. Plasma amino acid levels are usually normal, but hyperhistidinemia, hyperhistidinuria and hypomethioninemia have been found. The excretion of hydantoin-5-propionate, the stable oxidation product of the FIGLU precursor, 4-imidazolone-5-propionate and 4-amino-5-imidazolecarboxamide, an intermediate in purine synthesis, has been seen in some patients.

Autosomal recessive inheritance is the probable means of transmission because there have been affected individuals of both sexes with unaffected parents.

Functional Methionine Synthase Deficiency (OMIM 250940)

Functional methionine synthase deficiency due to the *cblE* and *cblG* mutations is characterized by homocystinuria and defective biosynthesis of methionine. Most patients present in the first few months of life with megaloblastic anemia and developmental delay. The distribution of cobalamin derivates is altered in cultured cells, with decreased levels of methylcobalamin as compared with normal fibroblasts. The *cblE* mutation is associated with low methionine synthase activity when the assay is performed with low levels of thiol, whereas the *cblG* mutation is associated with low activity under all assay conditions. *cblE* and *cblG* represent distinct complementation classes. Both diseases respond to treatment with hydroxycobalamin (OH-cbl).

Other Megaloblastic Anemias

1. *Thiamine-responsive anemia in DIDMOAD (Wolfram) syndrome*: It is a rare autosomal recessive disorder of thiamine transport, possibly deficient thiamine pyrophosphokinase activity, due to mutations in a gene on chromosome 1q23. Megaloblastic anemia and sideroblastic anemia with ringed sideroblasts may be present. Neutropenia and thrombocytopenia are present. It is accompanied by diabetes insipidus (DI), diabetes mellitus (DM), optic atrophy (OA) and deafness (D). *Treatment*: Anemia responds to 100 mg thiamine daily but megaloblastic changes persist. Insulin requirements decrease.

2. *Hereditary orotic aciduria*: Rare autosomal recessive defect of pyrimidine synthesis with failure to convert orotic acid to uridine and excretion of large amounts of orotic acid in the urine, sometimes with crystals. It is associated with severe megaloblastic anemia, neutropenia, failure to thrive and physical and mental retardation are frequently present. *Treatment*: Oral uridine in a dose of 100–200 mg/kg/day. The anemia is refractory to vitamin B_{12} and folic acid.

3. *Lesch-Nyhan syndrome*: Mental retardation, self-mutilation and choreoathetosis result from impaired synthesis of purines due to lack of hypoxanthine phosphoribosyltransferase. Some patients have megaloblastic anemia. *Treatment*: Megaloblastic anemia responds to adenine therapy (1.5 g daily).

GENERAL CLINICAL FEATURES OF COBALAMIN AND FOLATE DEFICIENCY

1. *Insidious onset*: Pallor, lethargy, fatigability and anorexia, sore red tongue and glossitis, diarrhea – episodic or continuous.

2. *History*: Similarly affected sibling or of a sibling who died, maternal vitamin B_{12} deficiency or poor maternal diet.

3. *Vitamin B_{12} deficiency*: All infants show signs of developmental delay, apathy, weakness, irritability or evidence of neurodevelopmental delay, loss of developmental milestones, particularly motor achievements (head control, sitting and turning). Athetoid movements, hypotonia and loss of reflexes occur. In older children signs of subacute dorsolateral degeneration of the spinal cord may occur. The usual symptoms are paresthesias in the hands or feet and difficulty in walking and use of the hands. Symptoms arise because of a peripheral neuropathy (especially parasthesias and numbness) associated with degeneration of posterior and lateral tracts of the spinal cord. Loss of vibration and position sense with an ataxic gait and positive Romberg's

sign are features of posterior column and peripheral nerve loss. Spastic paresis may occur, with knee and ankle reflexes increased because of lateral tract loss, but flaccid weakness may also occur when these reflexes are lost but the Babinski sign remains extensor. MRI findings include increased signals on T2-weighted images of the spinal cord, brain atrophy and retarded myelination. Long-term cognitive and developmental retardation are irreversible following B_{12} treatment.

4. Deleterious effects of cobalamin or folate deficiency (apart from neurologic complications) include increased risk of vascular thrombosis due to hyperhomocysteinemia.

5. Maternal folate deficiency results in neural tube defects, prematurity, fetal growth retardation and fetal loss.

6. Inborn errors of metabolism of cobalamin and folate result in failure to thrive, neurologic disorders, unexplained anemias or cytopenias. Plasma levels of methylmalonic acid and homocysteine should be done in these cases to elucidate the precise diagnosis. Elevation of these levels reflects a functional lack of cobalamin and/ or folate by tissues even when plasma vitamin levels are at the lower level of normal.

DIAGNOSIS

The age of presentation may help to focus on the most likely diagnosis. Table 4-8 lists disorders giving rise to megaloblastic anemia in early life and their likely age at presentation.

1. Red cell changes
    a. Hemoglobin: usually reduced, may be marked
    b. Red cell indices: MCV increased for age and may be raised to levels of 110–140 fl; MCHC normal
    c. Red cell distribution width (RDW): increased
    d. Blood smear: many macrocytes[*] and macro-ovalocytes; marked anisocytosis and poikilocytosis; presence of Cabot rings, Howell-Jolly bodies and punctuate basophilia.

2. White blood cells: count reduced to 1,500–4,000/mm³; neutrophils show hypersegmentation, i.e., nuclei of more than five lobes.

3. Platelet count: moderately reduced to 50,000–180,000/mm³.

4. Bone marrow: megaloblastic appearance
    a. The cells are large and the nucleus has an open, stippled, or lacy appearance. The cytoplasm is comparatively more mature than the nucleus and this dissociation (nuclear-cytoplasmic dissociation) is best seen in the later cells. Orthochromatic cells may be present with nuclei that are still not fully condensed

---

[*]Macrocytosis can be masked by associated iron deficiency and thalassemia.

Table 4-8   Disorders Giving Rise to Megaloblastic Anemia in Early Life
and their Likely Age at Presentation

Disease	Likely Age at Presentation (Months)		
	2–6	7–24	>24
Folate deficiency			
Inadequate supply			
Prematurity	+		
Dietary (e.g., goat's milk)	+		
Chronic hemolysis			+
Defective absorption			
Celiac disease/sprue			+
Anticonvulsant drugs			+
Congenital	+		
Cobalamin deficiency			
Inadequate supply			
Maternal cobalamin deficiency		+	
Nutritional			+
Defective absorption			
Juvenile pernicious anemia			+
Congenital malabsorption		+	±
Congenital absence of intrinsic factor		+	±
Defective metabolism			
Transcobalamin II deficiency	+		
Inborn errors of cobalamin utilization	+		
Other			
Thiamine responsive			+
Orotic aciduria	+		
Lesch–Nyhan syndrome			+

b. Mitoses are frequent and sometimes abnormal[*]; nuclear remnants, Howell–Jolly bodies, bi- and trinucleated cells and dying cells are evidence of gross dyserythropoiesis

c. The metamyelocytes are abnormally large (giant) and have a horseshoe-shaped nucleus

d. Hypersegmented polymorphs may be seen and the megakaryocytes show an increase in nuclear lobes.

5. Ineffective erythropoiesis manifest by: increased levels of lactate dehydrogenase, bilirubin, serum iron and transferrin saturation.

6. Serum vitamin B_{12} level: normal values 200–800 pg/ml (levels <80 pg/ml are almost always indicative of vitamin B_{12} deficiency).

---

[*]Megaloblastic cells exhibit increased frequency of chromosomal abnormalities, especially random beaks, gaps and centromere spreading. A rare case of nonrandom, transient 7q has been described in acquired megaloblastic anemia.

7. Serum and red cell folate levels: wide variation in normal range; serum levels less than 3 ng/ml low, 3–5 ng/ml borderline and greater than 5–6 ng/ml normal. Red cell folate levels 74–640 ng/ml.
8. Urinary excretion of orotic acid to exclude orotic aciduria.
9. Deoxyuridine suppression test: This test can discriminate between folate and cobalamin deficiencies.

If vitamin B_{12} is suspected proceed as follows:

- Detailed dietary history, history of previous surgery
- Schilling urinary excretion test[*]: It measures both intrinsic factor availability and intestinal absorption of vitamin B_{12}. However, there are increasing difficulties in obtaining labeled vitamin B_{12} and intrinsic factor so that the Schilling test is no longer readily available
- If the Schilling test is abnormal, repeat with commercial intrinsic factor. If absorption occurs, abnormality is due to lack of intrinsic factor. If no absorption occurs then there is specific ileal vitamin B_{12} malabsorption (Imerslund–Gräsbeck) or transcobalamin II deficiency. When bacterial competition (blind-loop syndrome) is suspected, the test may be repeated after treatment with tetracycline and will often revert to normal
- Gastric acidity after histamine stimulation, intrinsic factor content in gastric juice and serum antibodies to intrinsic factor and parietal cells and gastric biopsy help to establish a precise diagnosis
- Measurement of serum holo-transcobalamin II (cobalamin bound to transcobalamin II). In patients with vitamin B_{12} deficiency holo-transcobalamin II falls below the normal range before total serum cobalamin does
- Ileal disease should be investigated by barium studies and small bowel biopsy
- Disorders of vitamin B_{12} metabolism should be excluded by serum and urinary levels of excessive methylmalonic acid and of total homocysteine as well as by other sophisticated enzymatic assays. In folate deficiency, serum methylmalonic acid is normal whereas homocysteine is increased. Therefore evaluation of both methylmalonic acid and total homocysteine is helpful in distinguishing between folate and vitamin B_{12} deficiency
- Persistent proteinuria is a feature of specific ileal vitamin B_{12} malabsorption.

---

[*]The test is performed by administering 0.5–2.0 μg radioactive ^{57}cobalt-labeled vitamin B_{12} PO. This is followed in 2 hours by an intramuscular injection of 1,000 μg nonradioactive vitamin B_{12} to saturate the B_{12}-binding proteins and allow the subsequently absorbed oral radioactive vitamin B_{12} to be excreted in the urine. All urine is collected for 24 hours and may be collected for a second 24 hours, especially if there is renal disease. Normal subjects excrete 10–35% of the administered dose; those with severe malabsorption of vitamin B_{12}, because of lack of intrinsic factor or intestinal malabsorption, excrete less than 3%.

If folic acid is suspected, proceed as follows:

- Detailed dietary and drug history (e.g., antibiotics, anticonvulsants) gastroenterologic symptoms (e.g., malabsorption, diarrhea, dietary history)
- Tests for malabsorption:
  - Oral doses of 5 mg pteroylglutamic acid should yield a plasma level in excess of 100 ng/ml in 1 hour. If there is no rise in plasma level, congenital folate malabsorption should be considered
  - A 24 hour stool fat should be done to exclude generalized malabsorption.
- Upper gastrointestinal barium study and follow through
- Upper gut endoscopy and jejunal biopsy
- Sophisticated enzyme assays to diagnose congenital disorders of folate metabolism.

TREATMENT
Vitamin B_{12} Deficiency

Prevention

In conditions in which there is a risk of developing vitamin B_{12} deficiency (e.g., total gastrectomy, ileal resection), prophylactic vitamin B_{12} should be prescribed.

Active Treatment

Once the diagnosis has been accurately determined, several daily doses of 25 to 100 µg cyanocobalamin or hydroxycobalamin may be used to initiate therapy as well as potassium supplements.[*] Alternatively, in view of the ability of the body to store vitamin B_{12} for long periods, maintenance therapy can be started with monthly intramuscular injections in doses between 200 and 1,000 µg cyanocobalamin or hydroxycobalamin. Most cases of vitamin B_{12} deficiency require treatment throughout life.

Patients with defects affecting the intestinal absorption of vitamin B_{12} (abnormalities of IF or of ileal uptake) will respond to 100 µg of B_{12} injected subcutaneously monthly. This bypasses the defective step completely.

Patients with complete TC II deficiency respond only to large amounts of vitamin B_{12} and the serum cobalamin level must be kept very high. Doses of 1,000 µg IM two or three times weekly are required to maintain adequate control.

Patients with methylmalonic aciduria with defects in the synthesis of cobalamin coenzymes are likely to benefit from massive doses of vitamin B_{12}. These children may require 1–2 mg vitamin B_{12} parenterally daily. However, not all patients in this group are benefited by

---

[*]Hypokalemia has been observed during B_{12} initiation treatment in adults who are severely anemic.

vitamin B_{12}. It may be possible to treat vitamin B_{12}-responsive patients *in utero*. Congenital methylmalonic aciduria has been diagnosed *in utero* by measurements of methylmalonate in amniotic fluid or maternal urine.

In vitamin B_{12}-responsive megaloblastic anemia, the reticulocytes begin to increase on the third or fourth day, rise to a maximum on the sixth to eighth day and fall gradually to normal about the twentieth day. The height of the reticulocyte count is inversely proportional to the degree of anemia. Beginning bone marrow reversal from megaloblastic to normoblastic cells is obvious within 6 hours and is complete in 72 hours. Neurologically, the level of alertness and responsiveness improves within 48 hours and developmental delays may catch up in several months in young infants. Permanent neurologic sequelae often occur. Prompt hematologic responses are also obtained with the use of oral folic acid, but it is contraindicated since it has no effect on neurologic manifestations and may precipitate or accelerate their development.

Folic Acid Deficiency

Successful treatment of patients with folate deficiency involves:

- Correction of the folate deficiency
- Treating the underlying causative disorder
- Improvement of the diet to increase folate intake
- Follow-up evaluations at intervals to monitor the patient's clinical status.

Optimal response occurs in most patients with 100–200 µg folic acid per day. Since the usual commercially available preparations include a tablet (0.3–1.0 mg) and an elixir (1.0 mg/ml) these available preparations are utilized. Before folic acid is given, it is necessary to exclude vitamin B_{12} deficiency.

The clinical and hematologic response to folic acid is prompt. Within 1–2 days, the appetite improves and a sense of well-being returns. There is a fall in serum iron (often to low levels) in 24–48 hours and a rise in reticulocytes in 2–4 days, reaching a peak at 4–7 days, followed by a return of hemoglobin levels to normal in 2–6 weeks. The leukocytes and platelets increase with reticulocytes and the megaloblastic changes in the marrow diminish within 24–48 hours, but large myelocytes, metamyelocytes and band forms may be present for several days.

Folic acid is usually administered for several months until a new population of red cells has been formed. Folinic acid is reserved for treating the toxic effects of dihydrofolate reductase inhibitors (e.g., methotrexate, pyrimethamine).

It is often possible to correct the cause of the deficiency and thus prevent its recurrence, e.g. improved diet, a gluten-free diet in celiac disease, or treatment of an inflammatory disease such as tuberculosis or Crohn's disease. In these cases, there is no need to continue folic acid for life. In other situations, it is advisable to continue the folic acid to prevent recurrence, e.g. chronic hemolytic anemia such as thalassemia or in patients with malabsorption who do not respond to a gluten-free diet.

Cases of hereditary dihydrofolate reductase deficiency respond to N-5-formyl tetrahydrofolic acid and not to folic acid.

Suggested Reading

Allen LH. Causes of vitamin B12 and folate deficiency. Food & Nutrition Bulletin. 2008;29(2 Suppl):S20–34.

Dror DK, Allen LH. Effects of vitamin B_{12} deficiency on neurodevelopment in infants: Current knowledge and possible mechanisms. Nutritional Reviews. 2008;66(5):250–255.

Gordon N. Cerebral folate deficiency. Developmental Medicine & Child Neurology. 2009;51(3):180–182.

Hvas AM, Nexo E. Diagnosis and treatment of vitaemin B_{12} deficiency – An update. Haematologica. 2006; 91(11):1506–1512.

Lanzkowsky P. The megaloblastic anemias: Vitamin B_{12} Cobalamin deficiency and other congenital and acquired disorders. Clinical, pathogenetic and diagnostic considerations of vitamin B_{12} (Cobalamin) deficiency and other congenital and acquired disorders. In: Nathan DG, Oski FA, eds. Hematology of Infancy and Childhood. Philadelphia: WB Saunders; 1987.

Lanzkowsky P. The megaloblastic anemias: Folate deficiency II. Clinical, pathogenetic and diagnostic considerations in folate deficiency. In: Nathan DG, Oski FA, eds. Hematology of Infancy and Childhood. Philadelphia: WB Saunders; 1987.

Whitehead VM. Acquired and inherited disorders of cobalamin and folate in children. British Journal of Haematology. 2006;134(2):125–136.

Hematologic Manifestations of Systemic Illness

A variety of systemic illnesses including acute and chronic infections, neoplastic diseases, connective tissue disorders and storage diseases are associated with hematologic manifestations. The hematologic manifestations are the result of the following mechanisms:

- Bone marrow dysfunction
  - Anemia or polycythemia
  - Thrombocytopenia or thrombocytosis
  - Leukopenia or leukocytosis.
- Hemolysis
- Immune cytopenias
- Alterations in hemostasis
  - Acquired inhibitors to coagulation factors
  - Acquired von Willebrand disease
  - Acquired platelet dysfunction.
- Alterations in leukocyte function.

HEMATOLOGIC MANIFESTATIONS OF DISEASES OF VARIOUS ORGANS

Heart

Microangiopathic hemolysis occurs with prosthetic valves or synthetic patches utilized for correction of cardiac defects (particularly when there is failure of endothelialization, "Waring blender" syndrome) or rarely after endoluminal closure of patent ductus arteriosus. It has the following characteristics:

- Hemolysis is secondary to fragmentation of the red cells as they are damaged against a distorted vascular surface

Manual of Pediatric Hematology and Oncology. DOI: 10.1016/B978-0-12-375154-6.00005-7

- Hemolysis is intravascular and may be associated with hemoglobinemia and hemoglobinuria
- Iron deficiency occurs secondary to the shedding of hemosiderin within renal tubular cells into the urine (hemosiderinuria)
- Thrombocytopenia secondary to platelet adhesion to abnormal surfaces
- Autoimmune hemolytic anemia may occasionally occur after cardiac surgery with the placement of foreign material within the vascular system.

Cardiac Anomalies and Hyposplenism

Cardiac anomalies, particularly situs inversus, may be associated with hyposplenism and the blood film may show Howell–Jolly bodies, Pappenheimer bodies and elevated platelet counts.

Infective Endocarditis

Hematologic manifestations include anemia (due to immune hemolysis or chronic infection), leucopenia or leukocytosis and rarely thrombocytopenia and pancytopenia.

Coagulation Abnormalities

1. A coagulopathy exists in some patients with cyanotic heart disease. The coagulation abnormalities correlate with the extent of the polycythemia. Hyperviscosity may lead to tissue hypoxemia, which could trigger disseminated intravascular coagulation.
2. Marked derangements in coagulation (such as disseminated intravascular coagulation (DIC), thrombocytopenia, thrombosis and fibrinolysis) can accompany surgery involving cardiopulmonary bypass. Heparinization must be strictly monitored.

Platelet Abnormalities

Quantitative and qualitative platelet abnormalities are associated with cardiac disease:

- Thrombocytopenia occurs secondary to microangiopathic hemolysis associated with prosthetic valves
- Cyanotic heart disease can produce thrombocytopenia, prolonged bleeding time and abnormal platelet aggregation (see Chapter 12)
- Patients with chromosome 22q11.2 deletion (DiGeorge syndrome) can have platelet abnormalities including the Bernard–Soulier-like syndrome due to haploinsufficiency of the gene for GpIbβ and thrombocytopenia due to autoimmunity.

Polycythemia

1. The hypoxemia of cyanotic heart disease produces a compensatory elevation in erythropoietin and secondary polycythemia.
2. Patients are at increased risk for cerebrovascular accidents secondary to hyperviscosity.

Gastrointestinal Tract

Esophagus

1. Iron-deficiency anemia may occur as a manifestation of gastroesophageal reflux.
2. Endoscopy may be required in unexplained iron deficiency.

Stomach

1. The gastric mucosa is important in both vitamin B_{12} and iron absorption.
2. Chronic atrophic gastritis produces iron deficiency. There may be an associated vitamin B_{12} malabsorption.
3. Gastric resection may result in iron deficiency or in vitamin B_{12} deficiency due to lack of intrinsic factor.
4. Zollinger–Ellison syndrome (increased parietal cell production of hydrochloric acid) may cause iron deficiency through mucosal ulceration.
5. *Helicobacter pylori* infection in addition to causing chronic gastritis has been implicated in the initiation of iron-deficiency anemia, pernicious anemia, auto-immune thrombocytopenia and platelet aggregation defects (ADP-like defect).

Small Bowel

1. Celiac disease or tropical sprue may cause malabsorption of iron and folate. Table 5-1 lists the various hematologic manifestations of celiac disease.
2. Inflammatory bowel disease may cause iron deficiency from blood loss.
3. Eosinophilic gastroenteritis can produce peripheral eosinophilia.
4. Diarrheal illnesses of infancy can produce life-threatening methemoglobinemia.

Lower Gastrointestinal Tract

1. Ulcerative colitis is often associated with iron-deficiency anemia.
2. Peutz–Jeghers syndrome (intestinal polyposis and mucocutaneous pigmentation) predisposes to adenocarcinoma of the colon.

Table 5-1   Hematologic Manifestations of Celiac Disease

Problem	Frequency	Comments
Anemia: iron deficiency Folate deficiency, Vitamin B_{12} deficiency and other nutritional deficiencies	Common	The anemia is most commonly secondary to iron deficiency but may be multifactorial in etiology. Low serum levels of folate and vitamin B_{12} without anemia are frequently seen. Anemia due to other deficiencies appears to be rare
Thrombocytopenia	Rare	May be associated with other autoimmune phenomena
Thombocytosis	Common	May be secondary to iron deficiency or hyposplenism
Thromboembolism	Uncommon	Etiology is unknown but may be related to elevated levels of homocysteine or other procoagulants
Leukopenia/neutropenia	Uncommon	Can be autoimmune or secondary to deficiencies of folate, vitamin B_{12} or copper
Coagulopathy	Uncommon	Malabsorption of vitamin K
Hyposplenism	Common	Rarely associated with infections
IgA deficiency	Common	May be related to anaphylactic transfusion reactions
Lymphoma	Uncommon	The risk is highest for intestinal T-cell lymphomas

From: Halfdanarson TR, Litzow MR, Murray JA. Hematologic manifestations of celiac disease. Blood 2007;109:412–421, with permission.

3. Hereditary hemorrhagic telangiectasia (Osler–Weber–Rendu disease) may produce iron deficiency, platelet dysfunction and hemostatic defects.

Pancreas

1. Hemorrhagic pancreatitis produces acute normocytic, normochromic anemia. It may also be associated with DIC.
2. Shwachman–Diamond syndrome is characterized by congenital exocrine pancreatic insufficiency, metaphyseal bone abnormalities and neutropenia. There may also be some degree of anemia and thrombocytopenia.
3. Cystic fibrosis produces malabsorption of fat-soluble vitamins (e.g., vitamin K) with impaired prothrombin production.
4. Pearson syndrome is characterized by exocrine pancreatic insufficiency and severe sideroblastic anemia (Chapter 6).

Liver Disease

Anemia

Anemias of diverse etiologies occur in acute and chronic liver disease. Red cells are frequently macrocytic (mean corpuscular volume [MCV] of 100–110 fl). Target cells and acanthocytes (spur cells) are frequently seen. Some of the pathogenetic mechanisms of anemia include:

- Shortened red cell survival and red cell fragmentation (spur cell anemia) in cirrhosis
- Hypersplenism with splenic sequestration in the presence of secondary portal hypertension
- Iron-deficiency anemia secondary to blood loss from esophageal varices in portal hypertenion
- Chronic hemolytic anemia in Wilson disease secondary to copper accumulation in red cells
- Aplastic anemia resulting from acute viral hepatitis (particularly hepatitis B) in certain immunologically predisposed hosts
- Megaloblastic anemia secondary to folate deficiency in malnourished individuals.

Coagulation Abnormalities

The liver is involved in the synthesis of most of the coagulation factors. Liver dysfunction can be associated with either hyper- or hypocoagulable states because both procoagulant and anticoagulant synthesis are impaired. Table 5-2 lists the various coagulation abnormalities seen in liver disease and Table 5-3 lists the tests to differentiate between the coagulopathy of liver disease and other etiologies.

Table 5-2 Coagulation Abnormalities in Liver Disease

Hemorrhage	Thrombosis
(1) Thombocytopenia/Platelet dysfunction due to hypersplenism, altered TPO production (2) Decreased liver synthesis of procoagulant factors (3) Impaired carboxylation of vitamin K factors (4) Dysfibrinogenemia (5) Hyperfibrinolysis due to increased tPA and decreased PAI, alpha2 anti-plasmin	(1) Decreased anticoagulant – AT-III Protein C and S (2) Portal hypertension-portal vein thrombosis

Abbreviations: TPO, thrombopoietin; tPA, tissue plasminogen activator; PAI, plasminogen activator inhibitor.

Table 5-3 Tests to Differentiate Coagulopathies of Different Etiologies

Pro-coagulant factors	Liver	Vitamin K	DIC
F V	Decreased (late)	Normal	Decreased
F VII	Decreased (early)	Decreased	Decreased
F VIII	Normal/increased	Normal	Decreased

Factor I (Fibrinogen)

Fibrinogen levels are generally normal in liver disease. Low levels may be seen in fulminant acute liver failure.

Factors II, VII, IX and X (Vitamin K-Dependent Factors)

These factors are reduced in liver disease secondary to impaired synthesis. Factor VII is the most sensitive.

Factor V

Levels generally parallel factors II and X. If there is associated DIC, factor V level is markedly depressed.

In cholestatic liver disease, factor V may be markedly elevated as an acute-phase reactant. However, factor V synthesis is dramatically impaired in severe liver disease and is used as a more sensitive indicator for the need for liver transplantation.

Factor VIII

Procoagulant activity is generally normal in liver disease. If there is associated DIC, factor VIII will be markedly depressed.

Plasminogen and Antithrombin

Levels are depressed in acute and chronic liver disease. Liver disease results in a hypercoagulable state and may be associated with an increased incidence of DIC.

α_2-Macroglobulin and Plasmin

Inhibitor of thrombin α_2-macroglobulin and plasmin are elevated in cirrhosis.

Tests for Coagulation Disturbances

Prothrombin time (PT) is the most convenient for monitoring liver function.

Kidneys

Renal disease may affect red cells, white cells, platelets and coagulation.

Severe renal disease with renal insufficiency is frequently associated with chronic anemia (and sometimes pancytopenia). This type of anemia is characterized by:

- Hemoglobin as low as 4–5 g/dl
- Normochromic and normocytic red cell morphology unless there is associated microangiopathic hemolytic anemia (as in the hemolytic uremic syndrome), in which case schistocytes and thrombocytopenia are seen

- Reticulocyte count low
- Decreased erythroid precursors in bone marrow aspirate.

The following mechanisms are involved in the pathogenesis of this type of anemia:

- Erythropoietin deficiency is the most important factor (90% of erythropoietin synthesis occurs in the kidney)
- Shortened red cell survival is secondary to uremic toxins or in hemolytic uremic syndrome (HUS) secondary to microangiopathic hemolysis
- Uremia itself inhibits erythropoiesis and in conjunction with decreased erythropoietin levels produces a hypoplastic marrow
- Increased blood loss from a hemorrhagic uremic state and into a hemodialysis circuit causes iron deficiency
- Dialysis can lead to folic acid deficiency.

Treatment

1. Recombinant human erythropoietin (rHuEPO):[*]
   - Determine the baseline serum erythropoietin and ferritin levels prior to starting rHuEPO therapy. If ferritin is less than 100 ng/ml, give ferrous sulfate 6 mg/kg/day aimed at maintaining a serum ferritin level above 100 ng/ml and a threshold transferrin saturation of 20%
   - Start with rHuEPO treatment in a dose of 50–100 units/kg/day SC three times a week
   - Monitor blood pressure closely (increased viscosity produces hypertension in 30% of cases) and perform complete blood count (CBC) weekly
   - Titrate the dose:
     - If no response, increase rHuEPO up to 300 units/kg/day subcutaneously (SC) 3 times a week
     - If hematocrit (Hct) reaches 40%, stop rHuEPO until Hct is 36% and then restart at 25% dose
     - If Hct increases very rapidly (>4% in 2 weeks), reduce dose by 25%.

   Figure 5-1 shows a flow diagram, in greater detail, for the use of erythropoietin-stimulating agents in patients with chronic kidney disease.
2. Folic acid 1 mg/day is recommended because folate is dialyzable.
3. Packed red cell transfusion is rarely required.

Endocrine Glands

Thyroid

Anemia is frequently present in hypothyroidism. It is usually normochromic and normocytic. The anemia is sometimes hypochromic because of associated iron deficiency and

---

[*]Thrombosis of vascular access occurs in 10% of cases treated with rHuEPO.

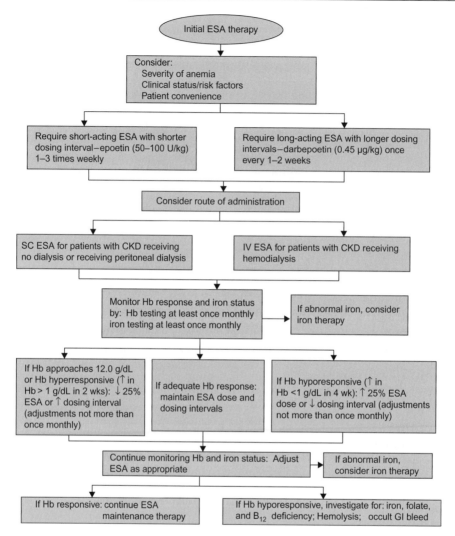

Figure 5-1 Recommended Erythropoietin-stimulating Agent (ESA) Treatment in Patients with Chronic Kidney Disease.
Abbreviations: SC=subcutaneous; IV=intravenous; CKD=chronic kidney disease; ↑, increase; ↓, decrease. From: Wish JB, Coyne DW. Use of erythropoiesis stimulating agents in patients with anemia of chronic kidney disease: Overcoming the pharmacologic and pharmacoeconomic limitations of existing therapies. Mayo Clin Pro 2007;82(11):1372–1380, with permission.

occasionally macrocytic because of vitamin B_{12} deficiency. The bone marrow is usually fatty and hypocellular and erythropoiesis is usually normoblastic. The finding of a macrocytic anemia and megaloblastic marrow in children with hypothyroidism should raise the possibility of an autoimmune disease with antibodies against parietal cells as well as against the thyroid, leading to vitamin B_{12} deficiency (juvenile pernicious anemia with polyendocrinopathies).

Adrenal Glands

1. Androgens stimulate erythropoiesis.
2. Conditions of androgen excess such as Cushing syndrome and congenital adrenal hyperplasia can produce secondary polycythemia.
3. In Addison disease, some degree of anemia is also present but may be masked by coexisting hemoconcentration. The association between Addison disease and megaloblastic anemia raises the possibility of an inherited autoimmune disease directed against multiple tissues, including parietal cells (juvenile pernicious anemia with polyendocrinopathies) (see Chapter 4).

Lungs

1. Hypoxia secondary to pulmonary disease results in secondary polycythemia.
2. Idiopathic pulmonary hemosiderosis is a chronic disease characterized by recurrent intra-alveolar microhemorrhages with pulmonary dysfunction, hemoptysis and hemosiderin-laden macrophages, resulting in iron-deficiency anemia. A precise diagnosis can be established by the presence of siderophages in the gastric aspirate. A lung biopsy may be necessary. Apart from a primary idiopathic type, there is also a variant associated with hypersensitivity to cows' milk and one that occurs with a progressive glomerulonephritis (Goodpasture's syndrome).

Treatment is controversial and may involve:

- Corticosteroids
- Withdrawal of cow's milk
- Packed red cell transfusions when indicated.

Skin

Mast Cell Disease

Mast cell disease or mastocytosis is associated with abnormal accumulation of mastocytes (closely related to monocytes or macrophages rather than to basophils) that occur in the dermis (cutaneous mastocytosis) or in an internal organ (systemic mastocytosis). Systemic form is rare in children. In children, this condition is more common under 2 years of age. It usually presents either as a solitary cutaneous mastocytoma or more commonly, as urticaria pigmentosa. Involvement beyond the skin is unusual in children, bone lesions are the most common, but bone marrow involvement is rare.

Eczema and Psoriasis

Patients with extensive eczema and psoriasis commonly have a mild anemia. The anemia is usually normochromic and normocytic (anemia of chronic disease).

Dermatitis Herpetiformis

1. Macrocytic anemia secondary to malabsorption.
2. Hyposplenism: Howell–Jolly bodies maybe present on blood smear.

Dyskeratosis Congenita (Chapter 6)

This disease is characterized by ectodermal dysplasia and aplastic anemia. The aplastic anemia is associated with high MCV, thrombocytopenia and elevated fetal hemoglobin. This may occur before the onset of skin manifestations.

Hereditary Hemorrhagic Telangiectasia

This autosomal dominant disorder is associated with bleeding disorder. Easy bruisability, epistaxis and respiratory and gastrointestinal bleeding may be caused by telangiectatic lesions.

Ehlers–Danlos Syndrome

This condition may be associated with platelet dysfunction: reduced aggregation with ADP, epinephrine and collagen. An unusual sensitivity to aspirin is described in type IV syndrome (see Chapter 12).

CHRONIC ILLNESS

Chronic illnesses such as cancer, connective tissue disease and chronic infection are associated with anemia. The anemia has the following characteristics:

- Normochromic, normocytic, occasionally microcytic
- Usually mild, characterized by decreased plasma iron and normal or increased reticuloendothelial iron
- Impaired flow of iron from reticuloendothelial cells to the bone marrow
- Decreased sideroblasts in the bone marrow.

The tests to differentiate the anemia of chronic illness from iron deficiency anemia are listed in Table 5-4 and therapeutic options for the treatment of anemia in chronic disease are outlined in Table 5-5.

In inflammatory diseases, cytokines released by activated leucocytes and other cells exert multiple effects that contribute to the reduction in hemoglobin levels. The pathophysiology of anemia of chronic disease is shown in Figure 5-2.

1. Interleukins (IL), especially IL-6 along with endotoxin, induce hepcidin synthesis in the liver. Hepcidin in turn binds to Ferroportin, which leads to internalization and

Table 5-4   Laboratory Tests to Differentiate Anemia of Chronic Disease from Iron-Deficiency anemia[a]

Variable (serum levels)	Anemia of Chronic Disease	Iron-Deficiency Anemia	Both Conditions[b]
Iron	Reduced	Reduced	Reduced
Transferrin	Reduced to normal	Increased	Reduced
Transferrin saturation	Normal to mildly reduced	Reduced	Reduced
Ferritin	Normal to increased	Reduced	Reduced to normal
Soluble transferrin receptor	Normal	Increased	Normal to increased
Cytokine levels	Increased	Normal	Increased

[a]Relative changes are given in relation to the respective normal values.
[b]Patients with both conditions include those with anemia of chronic disease and true iron deficiency.
Modified from: Weiss G, Goodnough LT. Anemia of chronic disease. N Engl J Med 2005;352:1011–1023, with permission.

Table 5-5   Therapeutic Options for the Treatment of Anemia of Chronic Disease

Treatment	Anemia of Chronic Disease	Anemia or Chronic Disease with True Iron Deficiency
Treatment of underlying disease	Yes	Yes
Transfusions[a]	Yes	Yes
Iron supplementation	No[b]	Yes[c]
Erythropoietin agents	Yes	Yes, in patients who do not have a response to iron therapy

[a]This treatment is for the short-term correction of severe or life-threatening anemia. Potentially adverse immunomodulatory effects of blood transfusions are controversial.
[b]Although iron therapy is indicated for the correction of anemia of chronic disease in association with absolute iron deficiency, no data from prospective studies are available on the effects of iron therapy on the course of underlying chronic disease.
[c]Overcorrection of anemia (hemoglobin >12 g per deciliter) may be potentially harmful to patients; the clinical significance of erythropoietin-receptor expression on certain tumor cells needs to be investigated.
From: Weiss G, Goodnough LT. Anemia of Chronic Disease. N Engl J Med 2005;352:1011–23, with permission.

degradation of Ferroportin; the corresponding sequestration of iron within the macrophages limits iron availability to erythroid precursors.

2. Inhibition of erythropoietin release from the kidney (especially by IL-1β and tumor necrosis factor-alpha [TNF-α]) leads to reduced erythropoietin-stimulated hematopoietic proliferation.

3. Director inhibition of the proliferation of erythroid progenitors (especially by TNF-α, interferon-gamma [IFN-γ] and IL-1β).

4. Augmentation of erythrophagocytosis by reticulo-endothelial macrophages.

Treatment involves treating the underlying illness.

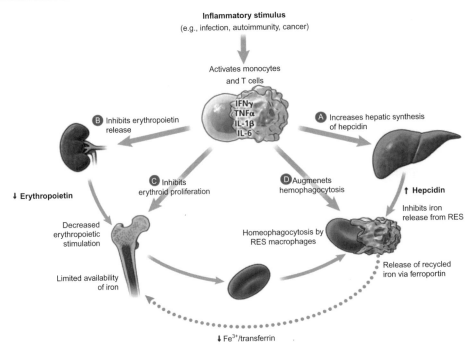

Figure 5-2 Pathophysiology of Anemia of Chronic Disease. RES, Reticuloendothelial system. From: Zarychanski R, Houston DS. Anemia of chronic disease. A harmful or beneficial response. CMAJ 2008;179(4):333–337, with permission.

Connective Tissue Diseases

Rheumatoid Arthritis

1. Anemia of chronic illness (normocytic, normochromic).
2. High incidence of iron deficiency.
3. Leukocytosis and neutropenia common in exacerbations of juvenile rheumatoid arthritis (JRA).
4. Thrombocytosis associated with a high level of IL-6 occurs in many patients, although there may be transient episodes of thrombocytopenia.

Felty's Syndrome

1. Triad of rheumatoid arthritis, splenomegaly and neutropenia.
2. Granulocyte colony-stimulating factor (G-CSF) is effective treatment in some cases.

Systemic Lupus Erythematous

1. Two types of anemia are common: anemia of chronic illness (normocytic, normochromic) and acquired autoimmune hemolytic anemia. Direct antiglobulin test (DAT) positive.

2. Neutropenia is common as a result of decreased marrow production and immune-mediated destruction.
3. Lymphopenia with abnormalities of T-cell function.
4. Immune thrombocytopenia occurs.
5. A circulating anticoagulant (antiphospholipid antibody) may be present and is associated with thrombosis.

Polyarteritis Nodosa

1. Microangiopathic hemolytic anemia may be associated with renal disease or hypertensive crises.
2. Prominent eosinophilia.

Wegener Granulomatosis

This autoimmune disorder is rare in children. Hematological features include:

- Anemia: normocytic; RBC fragmentation with microangiopathic hemolytic anemia
- Leukocytosis with neutrophilia
- Eosinophilia
- Thrombocytosis.

Kawasaki Syndrome

1. Mild normochromic, normocytic anemia with reticulocytopenia.
2. Leukocytosis with neutrophilia and toxic granulation of neutrophils and vacuoles.
3. Decreased T-suppressor cells.
4. High C_3 levels.
5. Increased cytokines IL-1, IL-6, IL-8, interferon-α and TNF.
6. Marked thrombocytosis (mean platelet count 700,000/mm^3).
7. DIC.

Henoch–Schönlein Purpura (HSP)

Henoch–Schönlein purpura (HSP), also called anaphylactoid purpura, is associated with systemic vasculitis characterized by unique purpuric lesions, transient arthralgias or arthritis (especially affecting knees and ankles), colicky abdominal pain and nephritis (see page 375, Chapter 12).

- Anemia occasionally occurs as a result of GI bleeding or decreased RBC production caused by renal failure
- Transient decreased Factor XIII activity may occur
- Vitamin K deficiency from severe vasculitis-induced intestinal malabsorption has been reported.

Infections

Anemia

1. Chronic infection is associated with the anemia of chronic illness.
2. Acute infection, particularly viral infection, can produce transient bone marrow aplasia or selective transient erythrocytopenia.
3. Parvovirus B19 infection in people with an underlying hemolytic disorder (such as sickle cell disease, hereditary spherocytosis) can produce a rapid fall in hemoglobin and an erythroblastopenic crisis marked by anemia and reticulocytopenia. There may be an associated neutropenia.
4. Many viral and bacterial illnesses may be associated with hemolysis.

White Cell Alterations

1. Viral infections can produce leukopenia and neutropenia. Neutrophilia with an increased band count and left shift frequently results from bacterial infection.
2. Neonates, particularly premature infants, may not develop an increase in white cell count in response to infection.
3. Eosinophilia may develop in response to parasitic infections.

Clotting Abnormalities

Severe infections, for example Gram-negative sepsis, can produce disseminated intravascular coagulation (DIC). Polymicrobial sepsis (including both aerobic and anaerobic organisms) in the head and neck region may cause thrombosis of major vessels. When this occurs in the jugular veins it leads to a constellation of findings called Lemierre's syndrome (suppurative thrombophlebitis with inflammation starting in the pharynx and spreading to the lateral parapharyngeal tissues in association with jugular vein thrombosis).

Thrombocytopenia

Infection can produce thrombocytopenia through decreased marrow production, immune destruction or DIC.

Viral and Bacterial Illnesses Associated with Marked Hematologic Sequelae

Parvovirus

Parvovirus B19 has a peculiar predilection for rapidly growing cells, particularly red cell precursors in the bone marrow. It has preference for the red cell precursors because it uses P antigen as a receptor. This viral infection is associated with a transient erythroblastopenic crisis, particularly in individuals with an underlying hemolytic disorder. In addition, it can produce thrombocytopenia, neutropenia and a hemophagocytic syndrome. In immunocompromised individuals, parvovirus B19 infection can produce prolonged aplasia. Bone marrow aspirate shows decreased or arrested maturation of erythroid precursors and the pathognomonic "giant pronormoblasts."

Epstein–Barr Virus (EBV)

EBV infection is associated with the following hematologic manifestations:

- Atypical lymphocytosis
- Acquired immune hemolytic anemia
- Agranulocytosis
- Aplastic anemia
- Lymphadenopathy and splenomegaly
- Immune thrombocytopenia.

EBV infection also has immunologic and oncologic associations (see Chapter 16). Some of the EBV-associated lymphoproliferative disorders are given in Table 5-6.

Human Immunodeficiency Virus (HIV)

The main pathophysiology of human immunodeficiency virus (HIV) infection is a constant decline in CD4+ lymphocytes, leading to immune collapse and death. The other bone marrow cell lines also decline in concert with CD4+ cell numbers as HIV disease (acquired immunodeficiency syndrome [AIDS]) progresses.

HIV infection has the following hematologic manifestations.

Table 5-6   EBV-associated Lymphoproliferative Disorders

EBV-associated B-cell lymphoproliferative disorders
1. Classic Hodgkin lymphoma
2. Burkitt lymphoma
3. Post-transplantation lymphoproliferative disorders
4. HIV-associated lymphoproliferative disorders
– Primary CNS lymphoma
– Diffuse large B-cell lymphoma, immunoblastic
– HHV-8-positive primary effusion lymphoma
– Plasmablastic lymphoma
EBV-associated T/NK-cell lymphoproliferative disorders
1. Peripheral T-cell lymphoma, unspecified
2. Angioimmunoblastic T-cell lymphoma
3. Extranodal nasal T/NK-cell lymphoma

Abbreviations: EBV, Epstein–Barr virus; HHV-8, human herpes virus-8; NK, natural killer.
Modified from: Carbone A, Gloghini A, Dotti G. EBV associated Lymphoproliferative disorders: classification and treatment. Oncologist 2008;13(5):577–585, with permission.

Thrombocytopenia

Thrombocytopenia occurs in about 40% of patients with AIDS. Initially, the clinical findings resemble those of immune thrombocytopenic purpura (ITP). Some degree of splenomegaly is common and the platelet-associated antibodies are often in the form of immune complexes that may contain antibodies with anti-HIV specificity. Megakaryocytes are normal or increased and production of platelets is reduced in the bone marrow (see Chapter 12).

Thrombotic thrombocytopenic purpura (TTP) is also associated with HIV disease. This occurs in advanced AIDS.

Anemia and Neutropenia

HIV-infected individuals develop progressive cytopenia as immunosuppression advances. Anemia occurs in approximately 70–80% of patients and neutropenia in 50%. Cytopenias in advanced HIV disease are often of complex etiology and include the following:

- A production defect appears to be most common
- Antibody and immune complexes associated with red and white cell surfaces may contribute. Up to 40% have erythrocyte-associated antibodies. Specific antibodies against i and U antigens have occasionally been noted. About 70% of patients with AIDS have neutrophil-associated antibodies.

The pathogenesis of the hematologic disorders includes:

- *Infections:* Myelosuppression is frequently caused by involvement of the bone marrow by infecting organisms (e.g., mycobacteria, cytomegalovirus (CMV), parvovirus, fungi and, rarely, *Pneumocystis jeroveci*)
- *Neoplasms:* Non-Hodgkin lymphoma (NHL) in AIDS patients is associated with infiltration of the bone marrow in up to 30% of cases. It is particularly prominent in the small noncleaved histologic subtype of NHL
- *Medications:* Widely used antiviral agents in AIDS patients are myelotoxic, for example, zidovudine (AZT) causes anemia in approximately 29% of patients. Ganciclovir and trimethoprim/sulfamethoxazole or pyrimethamine/sulfadiazine cause neutropenia. In general, bone marrow suppression is related to the dosage and to the stage of HIV disease. Importantly, the other nucleoside analogs of antiHIV compounds [dideoxycytidine (ddC), dideoxyinosine (ddI), stavudine (d4T), or lamivudine (3TC)], are usually not associated with significant myelotoxicity
- *Nutrition:* Poor intake is common in advanced HIV disease and is occasionally accompanied by poor absorption. Vitamin B_{12} levels may be significantly decreased in HIV infection although vitamin B_{12} is not effective in treatment. The reduction in serum vitamin B_{12} levels is due to vitamin B_{12} malabsorption and abnormalities in vitamin B_{12}-binding proteins.

Coagulation Abnormalities

The following abnormalities occur:

- Dysregulation of immunoglobulin production may affect the coagulation cascade. The dysregulation of immunoglobulin production may also occasionally result in beneficial effects, as in the resolution of anti-factor VIII antibodies in HIV-infected hemophiliacs
- Lupus-like anticoagulant (antiphospholipid antibodies) or anticardiolipin antibodies occur in 82% of patients. This is not associated with thrombosis in AIDS patients
- Thrombosis may occur secondary to protein S deficiency. Low levels of protein S occur in 73% of patients.

Role of Hematopoietic Growth Factors in Acquired Immunodeficiency Syndrome (AIDS)

1. rHuEPO results in a significant improvement in hematocrit and reduces transfusion requirements while the patient is receiving zidovudine. rHuEPO therapy should be initiated if the erythropoietin threshold is less than 500 IU/L.
2. G-CSF in a dose of 5 μg/kg/day SC is the most widely used growth factor in neutropenia.
3. Granulocytic-macrophage colony-stimulating factor (GM-CSF) improves neutrophil counts in drug-induced neutropenia. The effects of GM-CSF are seen within 24–48 hours with relatively low doses of GM-CSF (0.1 mg/kg/day).
4. Interleukin-3 (IL-3) given in doses of 0.5–5 mg/kg/day increases neutrophil counts.

Cancers in Children with Human Immunodeficiency Virus Infection

Malignancies in children with HIV infection are not as common as those in adults. Table 5-7 lists the AIDS-related neoplasms in children with HIV infection and Table 5-8 lists the spectrum of lymphoproliferative lesions in children with AIDS.

Table 5-7   AIDS-Related Neoplasms in Children

1. Classic Hodgkin lymphoma (lymphocyte depleted)
2. Non-Hodgkin lymphoma
   - Burkitt lymphoma
   - Central nervous system lymphoma
   - Diffuse large B-cell lymphoma
   - Mucosa-associated lymphoid tissue (MALT)-type lymphoma
3. Leiomyoma and leiomyosarcoma
4. Kaposi's sarcoma
5. Acute leukemias
6. Miscellaneous tumors – isolated cases of hepatoblastoma, fibrosarcoma of liver, embryonal rhabdomyosarcoma of biliary tree, Ewing's tumor of the bone

Modified from: Balarezo FS, Joshi VV. Proliferative and Neoplastic disorders in children with AIDS. Adv Anat Pathol 2002; 9(6):360–370, with permission.

Table 5-8 Spectrum of Systemic Lymphoproliferative Lesions in Children With AIDS

A. Hyperplasia involving
  1. Lymph nodes
  2. Peyer's patches of ileum
  3. Lymphoid nodules in esophagus and colon
  4. Thymus
  5. Pulmonary lymphoid hyperplasia (PLH)
B. Lymphoplasmacytic infiltrates in
  1. Lungs [lymphoid interstitial pneumonitis (LIP)]
  2. Salivary glands
  3. Liver
  4. Thymitis and multilocular thymic cyst
C. Polyclonal polymorphic B-cell lymphoproliferative disorder (PBLD) involving
  1. Lungs
  2. Liver, spleen, lymph nodes
  3. Kidneys
  4. Salivary glands
  5. Muscle, periadrenal fat
D. Myoepithelial sialadenitis
E. Myoepithelial sialadenitis with focal lymphoma
F. MALT lymphoma (involving nodal and extra nodal sites)
G. Non-MALT lymphoma (involving nodal and extra nodal sites)

Modified from: Balarezo FS, Joshi VV. Proliferative and Neoplastic disorders in children with AIDS. Adv Anat Pathol 2002;9(6):360–370, with permission.

Non-Hodgkin Lymphoma (NHL)

NHL is the most common malignancy secondary to HIV infection in children. It is usually of B-cell origin as in Burkitt's (small noncleaved cell) or immunoblastic (large cell) NHL. The mean age of presentation of malignancy in congenitally transmitted disease is 35 months, with a range of 6 to 62 months. In transfusion-transmitted disease, the latency from the time of HIV seroconversion to the onset of lymphoma is 22–88 months. The CD4 lymphocyte count is less than $50/mm^3$ at the time of diagnosis of the malignancy.

The presenting manifestations include:

• Fever
• Weight loss
• Extranodal manifestations (e.g., hepatomegaly, jaundice, abdominal distention, bone marrow involvement, or central nervous system [CNS] symptoms).

Some patients will already have had lymphoproliferative diseases such as lymphocytic interstitial pneumonitis or pulmonary lymphoid hyperplasia. These children usually have advanced (stage III or IV) disease at the time of presentation.

Central Nervous System Lymphomas

Children with central nervous system lymphomas present with developmental delay or loss of developmental milestones encephalopathy (dementia, cranial nerve palsies, seizures, or hemiparesis).

Differential diagnosis includes infections such as toxoplasmosis, cryptococcosis, or tuberculosis. Contrast-enhanced computed tomography (CT) studies of the brain show hyperdense mass lesions that are usually multicentric or periventricular. CNS lymphomas in AIDS are fast-growing and often have central necrosis and a "rim of enhancement" as in an infectious lesion. A stereotactic biopsy will give a definitive diagnosis.

Treatment of HIV Infection-Related Lymphomas

Treatment consists of standard protocols as described in Chapter 20 on non-Hodgkin lymphoma. Treatment of CNS lymphomas is more difficult. Intrathecal therapy is indicated even for those without evidence of meningeal or mass lesions at diagnosis of NHL. Radiation therapy may be a helpful adjunct for CNS involvement.

The following are more favorable prognostic features in NHL secondary to AIDS:

- CD4 lymphocyte count above $100/mm^3$
- Normal serum LDH level
- No prior AIDS-related symptoms
- Good Karnofsky score (80–100).

Proliferative Lesions of Mucosa-Associated Lymphoid Tissue

Mucosa-associated lymphoid tissue (MALT) shows reactive lymphoid follicles with prominent marginal zones containing centrocyte-like cells, lymphocytic infiltration of the epithelium (lymphoepithelial lesion) and the presence of plasma cells under the surface epithelium. These lesions may be associated with the mucosa of the gastrointestinal tract, Waldeyer's ring, salivary glands, respiratory tract, thyroid and thymus. Proliferative lesions of MALT can be benign or malignant (such as lymphomas).

The proliferative lesions arising from MALT form a spectrum or a continuum extending from reactive to neoplastic lesions. The neoplastic lesions are usually low grade but may progress into high-grade MALT lymphomas. MALT lymphomas characteristically remain localized, but if dissemination occurs, they are usually confined to the regional lymph nodes and other MALT sites. MALT lesions represent a category of pediatric HIV-associated disease that may arise from a combination of viral etiologies, including HIV, EBV and CMV.

Treatment of Low-Grade MALT Lymphoma

1. α-Interferon: 1 million units/m^2 SC three times a week (continued until regression of disease or severe toxicity occurs).
2. Rituxan (monoclonal antibody-anti-CD20): 375 mg/m^2 IV weekly for 4 weeks (courses may be repeated as clinically indicated).

Some patients may not require any treatment because of the indolent nature of the disease.

Leiomyosarcomas and Leiomyomas

Malignant or benign smooth muscle tumors, leiomyosarcoma (LS) and leiomyoma (LM), are the second most common type of tumor in children with HIV infection. The incidence in HIV patients is 4.8% (in non-HIV children, it is 2 per million). The most common sites of presentation are the lungs, spleen and gastrointestinal tract. Patients with endobronchial LM or LS often have multiple nodules in the pulmonary parenchyma. Bloody diarrhea, abdominal pain, or signs of obstruction may signal intraluminal bowel lesions. These tumors are clearly associated with EBV infection. In situ hybridization and quantitative polymerase chain reaction studies of LM and LS demonstrated that high copy numbers of EBV are present in every tumor cell. The EBV receptor (CD21/C3d) is present on tumor tissue at very high concentrations but it is present at lower concentrations in normal smooth muscle or control leiomyomas/leiomyosarcomas that had no EBV DNA in them. In AIDS patients, the EBV receptor may be unregulated, allowing EBV to enter the muscle cells and cause their transformation.

Treatment involves:

*   Chemotherapy, including doxorubicin or α-interferon
*   Radiotherapy
*   Complete surgical resection prior to chemotherapy, where feasible.

Despite surgery and chemotherapy, the disease tends to recur.

Kaposi Sarcoma (KS)

KS is rare in children and constitutes the third most common malignancy in pediatric AIDS patients; it occurs in 25% of adults with AIDS. KS occurs only in those HIV-infected children who were born to mothers with HIV. The lymphadenopathic form of KS is seen mostly in Haitian and African children and may represent the epidemic form of KS unrelated to AIDS. The cutaneous form is a true indicator of the disease related to AIDS. Visceral involvement has not been pathologically documented in children with AIDS.

Leukemias

Almost all leukemias are of B-cell origin. They represent the fourth most common malignancy in children with AIDS. The clinical presentation and biologic features are similar to those found in non-HIV children. Treatment involves chemotherapy designed for B-cell leukemias and lymphomas.

Miscellaneous Tumors

There is no increase in Hodgkin disease in children with AIDS as compared to adult patients. Children with AIDS rarely develop hepatoblastoma, embryonal rhabdomyosarcoma, fibrosarcoma and papillary carcinoma of the thyroid. The occurrence of these tumors is probably unrelated to the HIV infection.

TORCH Infections

This is a group of congenital infections including toxoplasma, rubella, cytomegalovirus (CMV), herpes simplex virus (HSV) and syphilis. They can all cause neonatal anemia, jaundice, thrombocytopenia and hepatosplenomegaly. They have significant sequelae so prevention, early identification and treatment are required.

Salmonella Typhii

Typhoid fever usually produces profound leukopenia and neutropenia in the initial stages of the illness and is often accompanied by thrombocytopenia.

Acute infectious Lymphocytosis

Acute infectious lymphocytosis is caused by a *Coxsackie* virus and is a rare benign, self-limiting childhood condition. It is associated with a low-grade fever, diarrhea and marked lymphocytosis ($50,000/mm^3$). Lymphocytes are mainly CD4 T cells. The condition resolves in 2–3 weeks without treatment (p. 302).

Bartonellosis

Bartonellosis is caused by a Gram-negative bacillus *Bartonella bacilliformis* confined to the mountain valleys of the Andes. The vector is a local sand fly. Infection from this organism causes a fatal syndrome of severe hemolytic anemia with fever (Oroya fever). Another species of Bartonella, *B. hensele* causes "cat scratch fever." It is associated with a regional (following a scratch by a cat) lymphadenitis. Thrombocytopenia may occur in this condition.

Tuberculosis

Tuberculosis is caused by *Mycobacterium tuberculosis*. Hematologic manifestations include leukemoid reaction mimicking CML, monocytosis and rarely pancytopenia.

Leptospirosis (Weil Disease)

This disease is caused by a leptospira, *L. icterohemorrhagiae*. A coagulopathy occurs which is complex and can be corrected with vitamin K administration. Thrombocytopenia commonly occurs but DIC is rare.

Parasitic Illnesses Associated with Marked Hematologic Sequelae

Malaria

Acute infections cause anemia which is multifactorial:

- Intracellular parasite metabolism alters negative charges on the RBC membrane which causes altered permeability with increased osmotic fragility. Spleen removes the damaged RBC or the parasites are "pitted" during the passage from spleen which results in microspherocytes of RBC
- Autoimmune hemolytic anemia may also occur. An IgG antibody is formed against the parasite and resulting immune complex attaches nonspecifically to RBC, complement is activated and cell destruction occurs. Positive Coombs' test due to IgG is found in 50% of patients with *P. falcifarum* malariae
- Thrombocytopenia without DIC is common. IgG antimalarial antibody bonds to the platelet bound malaria antigen and the IgG platelet parasite complex is removed by the R-E system.

Babesiosis

Babesiosis is caused by several species from the genus *Babesia* that colonize erythrocytes. It is a zoonotic disease transmitted by the *Ixodid* tick. There are similar clinical features to malaria. The clinical features include fever, myalgia and arthralgia with hepatosplenomegaly and hemolysis. Peripheral blood film may reveal intraerythrocytic trophozoites arranged in the form of a "maltese cross."

Leishmaniasis

The protozoal species *Leishmania* causes progressive splenomegaly, pancytopenia (anemia, neutropenia and thrombocytopenia). The bone marrow is usually hypercellular with hemophagocytosis. Some children may show coagulopathy.

Hookworm

Worldwide hookworm is a major cause of anemia. Two species infest humans:

- *Ancylostoma duodenale* is found in the Mediterranean region, North Africa and the west coast of South America

- *Necator Americanus* is found in most of Africa, Southeast Asia, Pacific islands and Australia.

Hookworms penetrate exposed skin, usually soles of bare feet and migrate through the circulation to the right side of the heart, then lungs (causing hypereosinophilic syndrome), through the airway down to the esophagus. They mature in the small intestine and attach their mouthparts to the mucosa. They suck blood, with each adult *A. duodenale* consuming about 0.2 ml/day. Heavily infested children may present with profound iron-deficiency anemia, hypoproteinemia and marked eosinophilia.

Tape Worm

Diphyllobothrium latum is a fish tape worm. It is acquired by eating uncooked freshwater fish. This worm infestation in the intestine results in vitamin B_{12} deficiency.

Trypanosomiasis

The diagnosis can be made by finding trypanosomes in blood and bone marrow smear.

LEAD INTOXICATION

One of the most striking hematologic features of lead intoxication is basophilic stippling (coarse basophilia) of red cells. It is caused by precipitation of denatured mitochondria secondary to inhibition of prymidine-5′-nucleotidase. Lead also produces ring sideroblasts in the marrow and it is associated with hypochromic microcytic anemia and markedly elevated free erythrocyte protoporphyrin levels.

NUTRITIONAL DISORDERS

Protein-Calorie Malnutrition

Protein deficiency in the presence of adequate carbohydrate caloric intake (kwashiorkor) is associated with mild normochromic, normocytic anemia secondary to reduced RBC production despite normal or increased erythropoietin levels as well as reduced red cell survival. Protein calorie malnutrition is also associated with impaired leukocyte function.

Scurvy

Mild anemia is common. There is a bleeding tendency due to loss of vascular integrity which may result in petechiae, subperiosteal, orbital or subdural hemorrhages. Hematuria and melena may occur.

Anorexia Nervosa

It produces the following hematologic changes in more advanced stages:

- Gelatinous changes of bone marrow which may become severely hypoplastic
- Mild anemia (macrocytic), neutropenia and thrombocytopenia
- Predisposition to infection associated with neutropenia
- Irregularly contracted red cells are seen (as in hypothyroidism) secondary to a disturbance in the composition of membrane lipids.

BONE MARROW INFILTRATION

The bone marrow may be infiltrated by non-neoplastic disease (storage disease) or neoplastic disease. In storage disease, a diagnosis is established on the basis of the clinical picture, enzyme assays of white cells or cultured fibroblasts and bone marrow aspiration revealing the characteristic cells of the disorder. Neoplastic disease may arise *de novo* in the marrow (leukemias) or invade the marrow as metastases from solid tumors (neuroblastoma or rhabdomyosarcoma). Table 5-9 lists the diseases that may infiltrate the marrow.

Gaucher Disease

Gaucher disease is the most common lysosomal storage disease, resulting from deficient activity of β-glucocerebrosidase. It is inherited in an autosomal-recessive manner. There are more than 200 mutations identified in the β-glucocerebrosidase gene located on 1q21

Table 5-9   Diseases Invading Bone Marrow

I. **Non-neoplastic**
  A. Storage diseases
    1. Gaucher disease
    2. Niemann–Pick disease
    3. Cystine storage disease
  B. Marble bone disease (osteopetrosis)
  C. Langerhans cell histiocytosis (Chapter 18)
II. **Neoplastic**
  A. Primary
    1. Leukemia (Chapter 17)
  B. Secondary
    1. Neuroblastoma (Chapter 22)
    2. Non-Hodgkin lymphoma (Chapter 20)
    3. Hodgkin lymphoma (Chapter 19)
    4. Wilms tumor (rarely) (Chapter 23)
    5. Retinoblastoma (Chapter 26)
    6. Rhabdomyosarcoma (Chapter 24)

including point mutations, crossovers and recombinations, yet prediction of clinical course can only be broadly ascribed on the basis of genotyping. Generally, the presence of the 1226G (N370S) mutation on one allele is synonymous with type-I disease (i.e., is apparently protective against neurologic involvement), whereas homozygosity for the allele 1488C (L444P) is invariably correlated with neurological disease. The degree of clinical involvement differs greatly in individual patients, even those with the same genotype and those affected within the same family.

Pathogenesis

Glucocerebrosidase is necessary for the catabolism of glucocerebroside. Deficiency of glucocerebrosidase leads to accumulation of glucocerebroside in the lysosomes of macrophages in tissues of the reticuloendothelial system. Figure 5-3 shows a diagram of the cellular pathophysiology of Gaucher disease. Accumulation in splenic macrophages and in the Kupffer cells of the liver produces hepatosplenomegaly.

Hypersplenism produces anemia and thrombocytopenia. Glucocerebroside accumulation in the bone marrow results in osteopenia, lytic lesions, pathologic fractures, chronic bone pain, bone infarcts, osteonecrosis and acute excruciating bone crises.

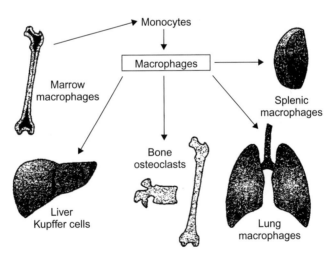

Figure 5-3 Diagram of the Cellular Pathophysiology of Gaucher Disease.
Monocytes are produced in the bone marrow and mature to macrophages in the marrow or in specific sites of distribution as liver Kupffer cells, bone osteoclasts and lung and tissue macrophages. Once resident, they accumulate glucosylceramide by phagocytosis and become end-stage Gaucher cells.
From: Grabowski GA, Leslie N. Lysosomal storage diseases: Perspectives and Principles. In: Hoffmann R, Benz EJ, Shattil SJ, Furie B, Cohen HJ, Silberstein LE, McGlave P, editors. Hematology Basic Principles. 3rd ed. Philadelphia: Lippincott-Raven, 2000, with permission.

Gaucher disease is classified into three types based on the presence and degree of neuronal involvement. Table 5-10 outlines the clinical manifestations of the three types of Gaucher disease.

Patients with type 1 Gaucher disease (non-neuropathic) which accounts for 90% of all cases of Gaucher disease, present with:

- Asymptomatic splenomegaly (rarely, portal hypertension develops). Splenic infarction is common and presents with pain, rigid abdomen and fever. Splenic nuclide scanning is helpful in the presence of an acute abdomen
- Pancytopenia secondary to hypersplenism (rarely from infiltration of the bone marrow with Gaucher cells)
- Skeletal manifestations include bone marrow infiltration with Erlenmeyer flask deformity from bone marrow expansion, generalized bone mineral loss and infarction on radiographs. The resultant osteopenia and infarction can lead to pathologic fractures
- Bone crises characterized by fever and excruciating local pain most frequently along femurs
- Growth delay: 50% of the symptomatic children are at or below the third percentile for height and another 25% are shorter than expected based on their mid-parental height.
- Typical Gaucher cells in the bone marrow[*]
- Decreased glucocerebrosidase activity of white cells
- Characteristic mutations of the glucocerebrosidase gene on chromosome 1 on DNA analysis.

Diagnosis

Glucocerebrosidase assay on leukocytes or cultured skin fibroblasts is the most efficient method of diagnosis. The typical child with type 1 Gaucher disease will have enzyme activity that is 10–30% of normal.

Further Evaluation

1. DNA evaluation for glucocerebrosidase gene abnormalities in patient, parents and siblings.
2. Complete blood count.
3. Serum chemistry with liver function tests.
4. Acid phosphatase level.
5. Angiotensin-converting enzyme.
6. Chitotriosidase.

---

[*]Large tissue cells of macrophage origin. Cytoplasm is pale gray-blue and they have eccentrically placed nuclei. Morphologically "Gaucher-like" cells are observed in chronic granulomatous disease, thalassemia, multiple myeloma, Hodgkin disease, AIDS and acute lymphoblastic leukemia, but can be readily distinguished from true Gaucher cells.

Table 5-10 Clinical Manifestations of Subtypes of Gaucher Disease

Characteristic	Type I		Type II		Type III		
	Asymptomatic	Symptomatic	Infantile	Neonatal	IIIa	IIIb	IIIc
Most common genotype	*1226G (IN370S)* homozygous	*1226G* compound heterozygous	None	Two null mutations	None	*1448C (L444P)* homozygous	*1342C (D409H)* homozygous
Ethnic predilection	Ashkenazi Jews	Ashkenazi Jews	None	None	None	Norbottnians (Northern Sweden)	Palestinian Arab Japanese
Common presenting features	None	Hepatosplenomegaly Hypersplenism Bleeding Bone Pains	OMA Strabismus Opisthotonus Trismus	Hydrops fetalis Congenital icthyosis	OMA Myoclonic seizures	OMA Hepatospleno-megaly Growth retardation	OMA Cardiac valve calcification
Central inervous system involvement	None	None	Severe	Lethal	Slow progressive neurological deterioration	OMA Slow cognitive deterioration	OMA
Bone involvement	None	Mild to severe	None	None	Mild	Moderate to severe	Small
Lung involvement	None	None to severe	Severe	Severe	Mild to moderate	Moderate to severe	Small
Enzyme replacement therapy	Not indicated	Indicated and efficient	Ethically problematic	Not relevant	Recommended for visceral features only		
Life expectancy	Normal	Normal	Death before age 2 years	Neonatal death	Death during childhood	Possible survival to adulthood	Survival to teenage

Abbreviations: OMA, oculomotor apraxia.

From: Elstein D, Abrahamov A, Hadas-Halpern I, Zimarn A. Gaucher's disease. Lancet 2001;358:324–327, with permission.

6. Liver/spleen volume with magnetic resonance imaging (MRI) or CT radiographs of femora and lateral spine.
7. MRI of femora.
8. Bone density of the spine and hips (DEXA).
9. Chest radiograph.

Treatment

Enzyme replacement therapy is recommended for the treatment of symptomatic type 1 patients. Recombinant human macrophage-targeted human glucocerebrosidase [imiglucerase, Cerezyme]* is used for enzyme replacement therapy. The initial dose is 30–60 units/kg IV every 2 weeks. The initial dose must be individualized for each patient based on disease severity and rate of progression. Maintenance dose is 15–60 units/kg IV every 2 weeks. Children who require treatment need to continue therapy indefinitely to maintain their clinical improvement. Prolonged periods without therapy are not appropriate.

Substrate reduction therapy (SRT) is available using N-butyldeoxynojirimycin (NB-DNJ) (Zavesca; Actelion Pharmaceuticals, Allschwill, Switzerland). NB-DNJ is an inhibitor of glucosylceramide (GlcCer) synthase, the enzyme responsible for GlcCer synthesis and hence synthesis of all GlcCer-based glycolipids. Unlike Cerezyme, Zavesca is given orally and does cross the blood–brain barrier. It is important to note that Zavesca causes a number of side effects and long-term reduction in glycolipid levels could affect a variety of cell functions because of the essential roles that these lipids play in normal physiology. Currently, Zavesca is only approved for adult patients with mild disease for whom enzyme replacement therapy (ERT) is unsuitable or not a therapeutic option, or as a supplement to ERT in severe cases. See Table 5-10 for ERT in the other types of Gaucher disease.

Recommendations for monitoring of children with type 1 Gaucher disease receiving and not receiving enzyme replacement therapy are outlined in Table 5-11.

Iron therapy in Gaucher patients with anemia is not recommended, because Gaucher cells avidly take up iron, which leads to hemochromatosis and decreased iron availability for erythropoiesis.

Response to Therapy

The earliest response is an improvement in hematologic parameters. A progressive decrease in liver/spleen size is regarded as a positive response. Skeletal response occurs more slowly (after 2–4 years), along with a decrease in pain and bone crises.

Approximately 5% of patients develop hypersensitivity to enzyme replacement therapy. These reactions respond to interruption of infusion and administration of antihistamine and glucocorticoids. Subsequent reactions can usually be prevented by reducing the initial rate

---

*Manufactured by Genzyme, Cambridge, MA.

Table 5-11 Recommendations for Monitoring Children with Type 1 Gaucher Disease (Minimal Evaluations Only)

	All patients, baseline	Patients not Receiving Enzyme Therapy		Patients Receiving Enzyme Therapy		At time of dose change
		Every 12 months	Every 12–24 months	Every 3 months[a]	Every 12 months[a]	
Hematologic						
Hemoglobin	X	X		X		X
Platelet count	X	X		X		X
Acid phosphatase (total, non-prostatic), Angiotensin converting enzyme, chitotriosidase[b]	X	X		X		X
Visceral[c]						
Spleen volume (volumetric MRI or CT)	X		X		X	X
Liver volume (volumetric MRI or CT)	X		X		X	X
Skeletal[d]						
MRI (coronal; T1- and T2-weighted) of entire femora[e]	X		X		X	X
Radiograph: AP view of entire femora[e] and view of lateral spine	X		X		X	X
Bone density (DEXA): spine and hips	X		X		Every 12–24 mo	
Quality of life[f]						
Patient reported functional health and well-being	X	X			X	

[a]For patients who have reached clinical goals and for whom there has been no change in dose, the frequency of monitoring can be decreased to every 12–24 months.

[b]One or more of these markers should be consistently monitored (at least once every 12 months) in conjunction with other clinical assessments of disease activity and response to treatment.
Of the three currently recommended biochemical markers, chitotriosidase activity, when available as a validated procedure from an experienced laboratory, may be the most sensitive indicator of changing disease activity and is therefore preferred.

[c]Obtain contiguous transaxial 10-mm-thick sections for sum of region of interest.

[d]Additional skeletal assessments that are optional include bone age for patients ≤14 years old. Follow-up is recommended if baseline is abnormal.

[e]Optimally, obtain hips to below knees. As an alternative, obtain hips to distal femur.

[f]Ideally, quality of life should be assessed every 6 months using a standard and valid instrument.

Abbreviations: DEXA, dual energy X-ray absorptiometry.

From: Charrow J, Anderson HC, Kaplan P, et al. Enzyme replacement therapy and monitoring for children with Type 1 Gaucher disease: Consensus recommendations. J Pediatr 2004;144:112–120, with permission.

of infusion, so that no more than 10 units/min are administered. These reactions commonly occur during the first 12 months of treatment. For this reason, the first year of treatment should be administered under the direct supervision of a physician. Following one year therapy can be administered at home by home nursing services. The non-neutralizing IgG antibodies that develop in up to 13% of patients are not clinically relevant.

Niemann–Pick Disease

Niemann–Pick disease types A and B result from deficient activity of acid sphingomyelinase, encoded by a gene on chromosome 11. The defect results in accumulation of sphingomyelin in the monocyte–macrophage system. The progressive deposition of sphingomyelin in the central nervous system leads to type A and in non-neuronal tissues leads to type B. Type C is a neuronopathic form that results from the defective cholesterol transport.

Clinical Manifestations

Depending on the type, Niemann–Pick disease has classic signs, including:

- Hepatosplenomegaly
- Cherry red spot in macula
- Psychomotor deterioration
- Reticular pulmonary infiltrates
- Foamy cells in the bone marrow.

Table 5-12 lists the clinical features of the different types of Niemann–Pick disease.

Table 5-12   Classification of Niemann–Pick disease

	Type		
	A (acute infantile with CNS involvement)	**B** (chronic visceral)	**C/D** (chronic neuropathic)
Age at presentation	3–6 months	Infancy/childhood	Infancy to early adulthood
Inheritance	Autosomal recessive	Autosomal recessive	Autosomal recessive
Ethnicity	Mainly Ashkenazi Jews	Pan-ethnic	Nova Scotia (D)
Neurologic symptoms	Developmental delay Neurologic regression	None	Psychomotor retardation Down-gaze paralysis Ataxia
Hepatosplenomegaly	Present	Present	Present/Absent
Cherry red macula	50% of cases	Absent	Absent
Lymphocyte vacuoles	Present	None	Present
Niemann–Pick cells in marrow	Present	Present	Present
Sphingomyelinase activity in tissue	Marked reduction (<10% of controls)	Marked reduction (<10% of controls)	Normal range[a]
Storage Product	Sphingomyelin	Sphingomyelin	Sphingomyelin and cholesterol

[a]Deficiency in cultured fibroblasts to esterify exogenous cholesterol.

Diagnosis

Diagnosis involves examining leukocytes or cultured fibroblasts to determine sphingomyelinase activity.

Treatment

There is no specific treatment for Niemann–Pick disease. Bone marrow transplantation in type B patients has been successful in reducing spleen and liver volumes, the sphingomyelin content in the liver, the Niemann–Pick cells in the bone marrow and the radiologic infiltration of the lungs.

Splenectomy in type B patients frequently causes progression of pulmonary disease and should be avoided if possible.

"Foam Cells" in Bone Marrow

Foam cells, with numerous uniform vacuoles often described as having a "honeycomb" appearance, are seen in the bone marrow in the following conditions:

- Neimann–Pick disease (types A, B, C, D)
- G_{m1} gangliosidosis (type 1)
- G_{m2} gangliosidosis (Sandhoff variant)
- Lactosyl ceramidosis
- Sialidosis I
- Sialidosis II, late infantile type
- Mucolipidosis II
- Mucolipidosis III
- Mucolipidosis IV
- Fucosidosis
- Mannosidosis
- Neuronal ceroid-lipofuscinosis
- Farber's disease
- Wolman's disease
- Cholesteryl ester storage disease
- Cerebrotendinous xanthomatosis
- Chronic hyperlipidemia
- Chronic corticosteroid therapy
- Hematologic malignancies (e.g., Hodgkin disease, leukemia, myeloma)
- Hematologic disease (e.g., aplastic anemia, ITP).

Careful history (including ethnic and family history), physical examination, examination of bone marrow using a phase electron microscopy and special stains and enzyme assays on white blood cells or cultured skin fibroblasts and liver biopsy for biochemical analysis can assist in making a specific diagnosis of these storage diseases.

Cystinosis

An autosomal-recessive defect, cystinosis is associated with generalized deposits of cystine in the tissues. Cystinosis occurs in the first year of life with the following manifestations:

- Thermal instability, polydipsia, polyuria
- Failure to thrive
- Recurrent episodes of vomiting and dehydration
- Dwarfism and rickets often prominent
- Early renal involvement with tubular dysfunction manifesting as a secondary Fanconi syndrome, leading to chronic renal failure.

Diagnosis
- Cystine crystals in the bone marrow
- Elevated cystine levels in leukocytes or fibroblasts.

Infantile Malignant Osteopetrosis (Marble Bone Disease)

Osteopetrosis is a hereditary disorder that may be present in either a severe or a mild form.

Severe Form (Autosomal Recessive)

The marrow space is progressively obliterated by excessive osseous growth. The difficulty in obtaining marrow by aspiration is a diagnostic clue. Radiologic changes are characteristic and diagnostic, consisting of generalized osteosclerosis. The cranial foramina progressively narrow resulting in blindness due to optic atrophy, deafness and other cranial nerve lesions.

The hematologic characteristics include the following:

- Progressive pancytopenia due to encroachment on the hematopoietic marrow by the overgrowth of bone
- Compensatory extramedullary hematopoiesis with resultant leukoerythroblastic anemia (circulating normoblasts, tear-drop-shaped poikilocytosis and early myelocytes), hepatosplenomegaly and lymphadenopathy
- Bone marrow hypoplasia
- Hemolysis due to splenic sequestration of red cells and perhaps general overactivity of the reticuloendothelial system.

Treatment

Allogeneic stem cell transplantation provides multipotent hematopoietic stem cells, which serve as a source of normal osteoclasts.

Mild Form (Autosomal Dominant)

Pathologic fractures occur in sclerotic bone. Nerve entrapment syndromes may also be present.

Neoplastic Disease

Neoplastic disease can be associated with the following hematologic alterations:

- Hemorrhage
- Nutritional deficiency states
- Dyserythropoietic anemias (including erythroid hypoplasia, sideroblastic anemia and anemia similar to that seen in chronic inflammation)
- Defect in erythropoietin production
- Hemodilution
- Hemolysis
- Pancytopenia secondary to marrow invasion or to cytotoxic therapy
- Acquired von Willebrand disease as in Wilms tumor
- Hypercoagulable states as in non-Hodgkin lymphoma
- Coagulopathy as in acute promyelocytic leukemia
- Leukoerythroblastic anemia and marrow
- Infiltration
- Cytotoxic drug therapy.

Marrow infiltration is suspected when leukoerythroblastic anemia develops. This term signifies the presence of myelocytes and normoblasts with anemia, thrombocytopenia and neutropenia. The explanation of this blood picture is that extramedullary erythropoiesis occurs when the marrow is infiltrated, permitting the escape of early myeloid and erythroid cells into the circulation. Normal blood findings, however, do not exclude marrow infiltration.

Bone marrow examination frequently demonstrates infiltration with tumor cells in the presence of pancytopenia. Because metastatic bone marrow involvement from solid tumors may be patchy, a single aspiration is not diagnostic. At least two aspirates and two biopsies should be performed.

The hematologic alterations associated with malignancy should be managed supportively and resolve if the underlying neoplasms can be successfully treated.

Table 5-13 summarizes some of the peripheral blood manifestations of systemic illness.

Table 5-13 Peripheral Blood Manifestations of Systemic Illness

Condition	Red Blood Cell (RBC)	White Blood Cells (WBC)	Platelets	Comments
Hypersplenism	Spherocytes, schistocytes	Leucopenia	Thrombocytopenia	Splenectomy usually corrects the peripheral blood changes
Hyposplenism	Target cells, Howell-Jolly bodies			
Leuco-erythroblastosis (marrow infiltration)	Tear drop cells, tailed RBC's, nucleated RBC	Leucocytosis (increased immature granulocytes)	Thrombocytosis Thrombocytopenia	Triad of NRBC, tear drop cells and immature granulocytes
Megaloblastosis	Macrocytosis, fragmentation	Leucopenia, Hypersegmented granulocytes	Thrombocytopenia	Deficiency of B_{12}, folic acid
Malignancy				
Acute lymphoblastic leukemia (ALL)	Anemia	Lymphoblasts, leucopenia, Hyperleucocytosis	Thrombocytopenia	Pancytopenia due to marrow infiltration
Acute Myeloid leukemia (AML)		Myeloblasts, Hyperleucocytosis, increased promyelocytes (M3), monocytes (M5) and eosinophils (M5eo)	Thrombocytopenia	Pancytopenia due to marrow infiltration
Hodgkin disease	Immune hemolytic anemia	Eosinophilia, neutrophilia	Thrombocytopenia	These paraneoplastic manifestations can precede the illness
Infections				
Sepsis (especially bacterial infections):	Agglutination, rouleaux formation, hemolytic anemia	Neutrophilic changes include toxic granulation, Döhle bodies, cytoplasmic vacuolation	Thrombocytosis, thrombocytopenia	Presence of immature myeloid precursors indicate leukemoid reaction

	Red cell changes	White cell changes	Platelet changes	Comments
• AIH, warm antibody type	Nucleated RBCs, spherocytes, schistocytes	Leucocytosis, rarely leucopenia	Thrombocytopenia	Common causes include pneumococcal infection, typhoid fever, hepatitis C
• AIH, cold agglutinin type	Agglutination, rarely erythrophagocytosis	Reactive lymphocytes	Thrombocytopenia	Mycoplasma, parvo virus, legionella, Chlamydia, EBV, CMV, VZV, HIV
• AIH, cold IgG antibody type	Spherocytes. Intravascular hemolysis		Thrombocytopenia	Measles, mumps, influenza-A, adenovirus
Bacterial infections		Granulocytes may contain *Staph. aureus*, *Streptococcus*, *Pneumococcus*, meningococci, *Clostridia* and *Bartonella*). Ehrlichiosis (morula within neutrophils/monocytes)		Relapsing fever (Borellia recurrentis), legionella and *Klebsiella* can have extracellular organisms in the smear
Parasitic infections	Intracellular parasitic forms seen in Malaria, Babesiosis	Eosinophilia, Trophozoite forms of Toxoplasma in neutrophils and monocytes		Extra-cellular forms include Filariasis (microfilariae), Trypanosomiasis (trypomastigote forms)
Fungal infections		*Candida*, *Histoplasma* and *Cryptococcus* in neutrophils and monocytes		*Candida* and *Cryptococcus* can also be found extracellularly
Viral infections	Rouleaux formation and acquired Pelger Heut anomaly in HIV. Agglutination in EBV infection	Reactive lymphocytosis (Downey type-II) in EBV infection	Thrombocytopenia	Rarely plasma cells seen in Hepatitis-B, C infections

Abbreviations: NRBC, nucleated red blood cells; AIH, auto-immune hemolytic anemia.

Suggested Reading

Aird WC. The hematologic system as a marker of organ dysfunction in sepsis. Mayo Clin Proc. 2003;78:869–881.

Balarezo FS, Joshi VV. Proliferative and Neoplastic disorders in children with AIDS. Adv Anat Pathol. 2002;9 (6):360–370.

Charrow J, Anderson HC, Kaplan P, et al. Enzyme replacement therapy and monitoring for children wth Type 1 Gaucher Disease: Consensus Recommendations. J Pediatr. 2004;144:112–120.

D'Azzo A, Kolodney EH, Bonten E, Annunziata I. Storage diseases of the reticuloendothelial system. In: Orkin S, Nathan D, et al., eds. Nathan and Oski's Hematology of Infancy and Childhood. 7th ed. Philadelphia: Saunders Elsevier; 2009:1301–1379.

Elstein D, Abrahamov A, Hadas-Halpern I, Zimran A. Gaucher's Disease. Lancet. 2001;358:324–327.

Ezekowitz RA. Hematologic manifestations of system disease. In: Orkin S, Nathan D, et al., eds. Nathan and Oski's Hematology of Infancy and Childhood. 7th ed. Philadelphia: Saunders Elsevier; 2009:1679–1739.

Grabowski GA, Leslie ND. Lysosomal storage diseases: perspectives and principles. In: Hoffman R, Benz EJ, Shattil SJ, Furi B, Silberstein LE, McGlave P, Heslop HE, eds. Hematology Basic Principles and Practice. 5th ed. Philadelphia: Churchill Livingstone Elsevier; 2008.

Granovsky MO, Mueller BU, Nicholson HS, Rosenberg PS, Rabkin CS. Cancer in human immunodeficiency virus-infected children: a case series from the Children's Cancer Group and the National Cancer Institute. J Clin Oncol. 1998;16:1729.

McGovern MM, Desnick RJ. Lipidoses. In: Kliegman RM, Behrman RE, Jenson HB, Stanton BF, eds. Nelsons Textbook of Pediatrics. 18th ed. Philadelphia: WB Saunders; 2007.

Mueller BU, Pizza PA. Cancer in children with primary or secondary immunodeficiencies. J. Pediatr. 1995;126:1.

Volberding PA, Baker KR, Levine AM. Human Immunodeficiency Virus Hematology. Hematology. 2003;294–313.

Weiss G, Goodnough LT. Anemia of chronic disease. N Engl J Med. 2005;352:1011–1023.

Wish JB, Coyne DW. Use of erythropoiesis stimulating agents in patients with anemia of chronic kidney disease: Overcoming the pharmacologic and pharmacoeconomic limitations of existing therapies. Mayo Clinic Proc. 2007;82(11):1372–1380.

CHAPTER 6

Bone Marrow Failure

Bone marrow failure may manifest as an isolated quantitative failure of one cell line, a single cytopenia (e.g., erythroid, myeloid, or megakaryocytic), or as pancytopenia, a failure of all three cell lines with a hypoplastic or aplastic marrow. These disorders may be congenital or acquired (Table 6-1).

Bone marrow failure may also be due to invasion of the bone marrow by non-neoplastic (e.g. storage cells) or neoplastic conditions, primary or metastatic. Table 6-7, later in this chapter, lists the inherited bone marrow failure syndromes (IBMFS) with their known and presumed genes. IBMFS consist of diseases resulting in pancytopenia (Fanconi anemia and dyskeratosis congenita) and those apparently restricted to a single hematopoietic lineage (Diamond Blackfan anemia [DBA]), congenital neutropenias (Shwachman Diamond syndrome, severe congenital neutropenia [SCN], Kostmann syndrome, cyclic neutropenia and other even less common disorders) (see Chapter 11), congenital amegakaryocytic thrombocytopenia and thrombocytopenia absent radii (TAR) syndrome (see Chapter 12). However it has become evident that most of these "single cell cytopenias" may manifest abnormalities in other hematopoietic cell lines, e.g. in Shwachman Diamond syndrome and congenital amegakaryocytic thrombocytopenia, pancytopenia is fairly common, while in DBA it is observed occasionally (Table 6-1).

The qualitative marrow failure disorders known as congenital dyserythropoietic anemia (CDA) are a unique set of disorders resulting in a moderate erythroid failure due to ineffective erythropoiesis with characteristic morphological abnormalities of erythroblasts.

In addition there are mitochrondrial diseases with bone marrow failure syndromes (Pearson syndrome, Wolfram syndrome and various types of sideroblastic anemias).

APLASTIC ANEMIA

Aplastic anemia is a physiologic and anatomic failure of the bone marrow characterized by a marked decrease or absence of blood-forming elements in the marrow and

Manual of Pediatric Hematology and Oncology. DOI: 10.1016/B978-0-12-375154-6.00006-9

Table 6-1   Causes of Single- and Three-Cell Line Bone Marrow Failure

Failure of single cell line (single cytopenia)
Red cells
  Congenital
    Diamond Blackfan anemia (inherited pure red cell aplasia)
    Congenital dyserythropoietic anemia (CDA)
    Pearson syndrome
  Acquired
    Idiopathic
      Transient erythroblastopenia of childhood (TEC)
    Secondary
      Drugs or toxins
      Infection – parvovirus B19 infection in immunodeficiency (chronic bone marrow failure)
      Malnutrition
      Thymoma
      Chronic hemolytic anemia with associated parvovirus B19 infection (transient bone marrow failure)
      Connective tissue disease and autoimmune associated
      Malignancy associated
White blood cells (Chapter 11)
  Shwachman Diamond syndrome
  Severe congenital neutropenia
    ELA2 – autosomal dominant
    HAX1 (Kostmann syndrome) – autosomal recessive
    G6PC3 – autosomal recessive
    Other rare neutropenias
  Reticular dysgenesis (congenital aleukosis)
  Other rare genetic disorders
Platelets (Chapter 12)
  Congenital amegakaryocytic thrombocytopenia
  Thrombocytopenia absent radii (TAR) syndrome
Failure of all three cell lines (generalized pancytopenia)
Inherited
  Fanconi anemia (associated with chromosomal breakages induced by clastogens, e.g. diepoxybutane [DEB])
  Dyskeratosis congenita (associated with short telomeres)
  Shwachman Diamond syndrome (predominantly neutropenia)[a]
  Congenital amegakaryocytic thrombocytopenia (predominantly thrombocytopenia)[a]
  Diamond Blackfan anemia (predominantly anemia)[a]
  Aplastic anemia with constitutional chromosomal abnormalities
  Dubowitz syndrome (congenital abnormalities, mental retardation, aplastic anemia)
Acquired
  Idiopathic (more than 70% of cases)
  Secondary
  A. Drugs[b]
    1. Predictable, dose dependent, rapidly reversible (affects rapidly dividing maturing hematopoietic cells rather than pluripotent stem cells)

(Continued)

Table 6-1   (Continued)

    a. 6-Mercaptopurine
    b. Methotrexate
    c. Cyclophosphamide
    d. Busulfan
    e. Chloramphenicol
  2. Unpredictable, normal doses (defect or damage to pluripotent stem cells)
    a. Antibiotics: chloramphenicol, sulfonamides
    b. Anticonvulsants: mephenytoin (Mesantoin), hydantoin
    c. Antirheumatics: phenylbutazone, gold
    d. Antidiabetics: tolbutamide, chlorpropamide
    e. Antimalarial: quinacrine
B. Chemicals: insecticides (e.g., DDT, Parathion, Chlordane)
C. Toxins (e.g., benzene, carbon tetrachloride, glue, toluene)
D. Radiation
E. Infections
  1. Viral hepatitis (hepatitis A, B and C and non-A, non-B, non-C and non-G hepatitis)
  2. HIV infection (AIDS)
  3. Infectious mononucleosis (Epstein–Barr virus)
  4. Rubella[c]
  5. Influenza[c]
  6. Parainfluenza[c]
  7. Measles[c]
  8. Mumps[c]
  9. Venezuelan equine encephalitis
  10. Rocky Mountain spotted fever[c]
  11. Cytomegalovirus (in newborn)
  12. Herpes virus (in newborn)
  13. Chronic parvovirus
F. Immunologic disorders
  1. Graft versus host reaction in transfused immunologically incompetent subjects
  2. X-linked lymphoproliferative syndrome (Chapter 16)
  3. Eosinophilic faciitis
  4. Hypogammagloblinemia
G. Aplastic anemia preceding acute leukemia (hypoplastic preleukemia)
H. Myelodysplastic syndromes (Chapter 16)
I. Thymoma
J. Paroxysmal nocturnal hemoglobinuria (Chapter 7)
K. Malnutrition
  1. Kwashiorkor
  2. Marasmus[c]
  3. Anorexia nervosa[c]
L. Pregnancy

[a]Can have reduction in other cell lines.
[b]Partial listing.
[c]Pancytopenia with temporary marrow hypoplasia.

pancytopenia (decreased red cells, white blood cells and platelets). Splenomegaly, hepatomegaly and lymphadenopathy do not occur in this condition. Aplastic anemia may be congenital or acquired.

Figure 6-1 delineates in schematic form an approach to the differential diagnosis of pancytopenia and Table 6-2 lists the investigations to be carried out in a patient with pancytopenia.

ACQUIRED APLASTIC ANEMIA
Definition

Severe aplastic anemia is defined as having a bone marrow cellularity of less than 25% and at least two of the following cytopenias: granulocyte count $<500/mm^3$ (<200 mm^3 defines very

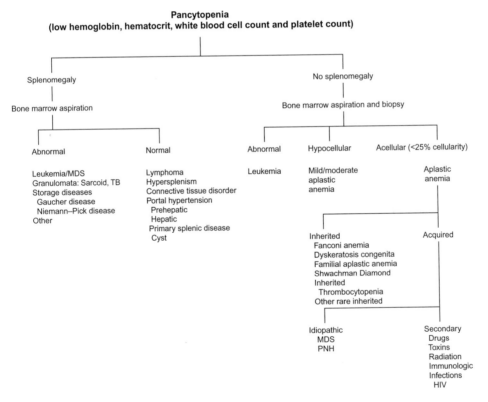

Figure 6-1 Approach to the Differential Diagnosis of Pancytopenia.
Abbreviations: MDS, myelodysplastic syndrome; PNH, paroxysmal nocturnal hemoglobinuria.

Table 6-2   Investigations in Patients with Pancytopenia

1. Detailed drug history, toxin and radiation exposure, family history of aplastic anemia, MDS or leukemia, physical examination for congenital anomalies
2. Blood count: absolute reticulocyte count, granulocyte count, Hb, Hct, MCV, platelet count
3. ANA and DNA titer, direct antiglobulin (DAT) test, rheumatoid factor, liver function tests, tuberculin test
4. Viral serology: HIV, EBV, parvovirus, hepatitis A, B, C. PCR for virus when indicated
5. Serum vitamin B_{12}, red cell and serum folate levels
6. Bone marrow aspirate and trephine biopsy – because of patchiness of bone marrow involvement, biopsy at multiple sites may be required
7. Chromosome breakage assay on blood lymphocytes or skin fibroblasts using clastogen stimulation (e.g., diepoxybutane or mitomycin C) to diagnose Fanconi anemia. Skeletal radiographs, renal, cardiac, abdominal ultrasound, chest radiograph to determine congenital anomalies in Fanconi anemia
8. Telomere length determination for dyskeratosis congenita
9. Cytogenetic studies on bone marrow to exclude myelodysplastic syndromes
10. Flow cytometric immunophenotypic analysis of erythrocytes for deficiency of GPI-linked surface protein (e.g. CD59) to exclude paroxysmal nocturnal hemoglobinuria
11. Diagnostic tests to rule out Shwachman Diamond syndrome (see Chapter 11) such as skeletal radiograph, chest radiograph, pancreatic CAT scan. Pancreatic function tests (72 hour fecal fat, serum trypsinogen and isoamylase)
12. Mutation analysis for inherited bone marrow failure syndromes when suspected.

Abbreviations: Hb, hemoglobin; Hct, hematocrit; MCV, mean corpuscular volume; HIV, human immunodeficiency virus; EBV, Epstein–Barr virus; PCR, polymerase chain reaction.

severe aplastic anemia); platelet count $<20,000/\text{mm}^3$; and reticulocyte count $<20,000/\text{mm}^3$. The definition of mild and moderate aplastic anemia varies among institutions.

Pathophysiology

The effectiveness of immunosuppressive therapy implies that in many patients with acquired aplastic anemia bone marrow failure results from an immunologically mediated, tissue-specific, organ-destructive mechanism. The fact that 50% of identical twins with severe aplastic anemia will not engraft with no conditioning after the infusion of syngeneic stem cells supports this notion at least in half the cases. A reasonable theory suggests that exposure to an inciting antigen, cells and cytokines of the immune system destroy stem cells in the marrow resulting in pancytopenia. Treatment with immunosuppressive modalities leads to marrow recovery. In some cases heretofore unrecognized genetic causes for aplastic anemia are being identified, resulting is a more precise classification of patients.

Clinical and laboratory studies have suggested that γ-interferon (γ-IFN) plays a central role in the pathophysiology of aplastic anemia.

In vitro studies show that the T cells from aplastic anemia patients secrete γ-IFN and tumor necrosis factor (TNF). Long-term bone marrow cultures (LTBMCs) have shown that γ-IFN and TNF are potent inhibitors of both early and late hematopoietic progenitor cells. Both of these cytokines suppress hematopoiesis by their effects on the mitotic cycle and, more importantly, by the mechanism of cell killing. The mechanism of cell killing involves the pathway of apoptosis (i.e., γ-IFN and TNF upregulate each other's cellular receptors, as well as the Fas receptors in hematopoietic stem cells). Cytotoxic T cells also secrete interleukin-2 (IL-2), which causes polyclonal expansion of the T cells. Activation of the Fas receptor on the hematopoietic stem cell by the Fas ligand present on the lymphocytes leads to apoptosis of the targeted hematopoietic progenitor cells. Additionally, γ-IFN mediates its hematopoietic suppressive activity through interferon regulatory factor 1 (IRF-1), which inhibits the transcription of cellular genes and their entry into the cell cycle. γ-IFN also induces the production of the toxic gas nitric oxide, diffusion of which causes additional toxic effects on the hematopoietic progenitor cells. Direct cell–cell interactions between effective lymphocytes and targeted hematopoietic cells probably also occur. The oligoclonal expansion of CD4+ and CD8+ T cells that fluctuate with disease activity further supports an immune etiology. If not overtly genetic it is likely that many affected patients have a genetic predisposition to marrow failure.

The importance of immunosuppressive therapy was recognized when: (a) an unexpected improvement in pancytopenia was observed in aplastic anemia patients following failure of engraftment in allogeneic bone marrow transplantation; and (b) the need for immunosuppressive preparative therapy was realized for successful engraftment in about half of hematopoietic stem cells in identical twin bone marrow transplantation performed for aplastic anemia.

Table 6-1 lists the various causes of acquired aplastic anemia.

Clinical Findings

Acquired aplastic anemia may be idiopathic or secondary. At least 70% of cases are considered idiopathic, without an identifiable cause. The incidence is 2 cases per million per year and the male:female ratio is 1:1. The onset of acquired aplastic anemia is usually in retrospect gradual and the symptoms are related to the pancytopenia:

* Anemia results in pallor, easy fatigability, weakness and loss of appetite
* Thrombocytopenia leads to petechiae, easy bruising, severe nosebleeds and bleeding into the gastrointestinal and renal tracts
* Leukopenia leads to increased susceptibility to infections and oral ulcerations and gingivitis that respond poorly to antibiotic therapy
* Hepatosplenomegaly and lymphadenopathy do not occur; their presence suggests an underlying leukemia.

Laboratory Investigations

1. *Anemia*: normocytic, normochromic or macrocytic.
2. *Reticulocytopenia*: absolute count more reliable.
3. *Leukopenia*: granulocytopenia often less than 1,500/mm^3.
4. *Thrombocytopenia*: platelets often less than 30,000/mm^3.
5. *Fetal hemoglobin*: may be slightly to moderately elevated.
6. *Bone marrow*:
   - Marked depression or absence of hematopoietic cells and replacement by fatty-tissue-containing reticulum cells, lymphocytes, plasma cells and usually tissue mast cells
   - Megaloblastic changes and other features indicative of dyserythropoiesis frequently seen in the erythroid precursors present
   - Bone marrow biopsy essential (only way to assess cellularity) for diagnosis to exclude the possibility of poor aspiration technique or poor bone marrow sampling; additionally, will help to rule out granulomas, myelofibrosis, or leukemia
   - Chromosomal analysis including breakage assay normal; rules out Fanconi anemia and myelodysplastic syndromes
   - Bone marrow cultures for infectious agent and/or DNA; antigen-based evaluation for infectious agent when indicated.
7. *Chromosome breakage assay*: performed on peripheral blood to rule out Fanconi anemia.
8. *Flow cytometry (CD59)*: to exclude paroxysmal nocturnal hemoglobinuria.
9. Telomere length to screen for dyskeratosis congenita.
10. Physical examination, appropriate laboratory and imaging studies and if warranted mutation analysis to rule out other inherited bone marrow failure syndrome (DC, DBA, SDS, CAT).
11. *Liver function chemistries*: to exclude hepatitis.
12. *Renal function chemistries*: to exclude renal disease.
13. *Viral serology testing*: hepatitis A, B and C antibody panel; Epstein–Barr virus antibody panel; parvovirus B19, IgG and IgM antibodies; varicella antibody titer; cytomegalovirus antibody titer.
14. *Quantitative immunoglobulins, C3, C4 and complement*.
15. *Autoimmune disease evaluation*: Antinuclear antibody (ANA), total hemolytic complement (CH50), direct antiglobulin test.
16. *HLA typing*: patient and family done at the diagnosis of severe aplastic anemia to ensure a timely transplantation.

Table 6-3 lists the recommendations for the treatment of moderate and severe aplastic anemia.

Table 6-3 **Recommendations for Treatment of Children with Aplastic Anemia**

1. Moderate aplastic anemia:
   Observe with close follow-up and supportive care
   If the patient develops:
   a. Severe aplastic anemia
      and/or
   b. Severe thrombocytopenia with significant bleeding
      and/or
   c. Chronic anemia requiring transfusion treatments
      and/or
   d. Serious infections.
   Then treat as severe aplastic anemia.
2. Severe aplastic anemia:
   Allogeneic bone marrow transplantation when HLA-matched sibling donor available
   In the absence of an HLA-matched sibling marrow donor:
   Treat the patient with ATG, cyclosporine A (CSA), methylprednisolone and growth factors such as
      G-CSF or GM-CCF (see Table 6-4)[a]
   If no response or waning of response and recurrence of severe aplastic anemia a second course of
      immunosuppressive therapy is controversial. The following is recommended:
   a. HLA-matched matched unrelated bone marrow, peripheral blood or umbilical cord blood transplantation
      if a suitable donor is available. If not available a second course of immunosuppression is warranted
   b. High-dose cyclophosphamide and cyclosporine therapy without stem cell transplantation is carried
      out in some institutions but its use remains controversial.

[a]Partial response: absence of infections and transfusion dependency and sustained increase in all cell counts as follows: reticulocyte count, $\geq 20,000/mm^3$; platelet count, $\geq 20,000/mm^3$; absolute neutrophil count, $\geq 500/mm^3$. Complete response: Normal counts. Partial response and complete response are considered as responses for the evaluation of the success of immunosuppressive therapy.

Treatment

Supportive Care

1. Avoid exposure to hazardous drugs and toxins.
2. The risk of serious bleeding and symptomatic anemia must be balanced against the risk of transfusion sensitization and iron overload. Transfusion of red cells and platelets should be performed judiciously but should not be withheld if clearly indicated. Prior to any transfusion, perform complete blood group typing to minimize the risk of sensitization to minor blood group antigens and to permit identification of antibodies should they subsequently develop. To avoid sensitization to transplantation antigens there should be no transfusions from blood relatives and transfusions should be restricted, if possible, to single unrelated donors to decrease the likelihood of sensitization to donor antigens. In all patients blood products should be leukocyte-depleted to reduce the risk of sensitization and CMV infection. The use of CMV-negative blood products versus CMV-safe blood products is somewhat controversial. Patients receiving chronic red cell transfusion should be followed for evidence of iron overload and chelated appropriately.

The use of single donor platelets, when available, is recommended. In females, menses should be suppressed by the use of oral contraceptives.

3. Drugs that impair platelet function, such as aspirin, should be avoided.

4. Intramuscular injections should be given carefully, followed by ice pack application to injection sites.

5. The antifibrinolytic agent, epsilon aminocaproic acid (Amicar) can be used to reduce mucosal bleeding in thrombocytopenic patients with good hepatic and renal function. Hematuria is a contraindication to its use. A dose of 100 mg/kg/dose every 6 hours is used. The maximum daily dose is 24 grams. Teeth should be brushed with a cloth or soft toothbrush.

6. Avoid infection. Keep patients out of the hospital as much as possible. Good dental care is important. Rectal temperatures should not be taken and the rectal areas should be kept clean and free of fissures. If a patient is febrile:
   - Culture possible sources, including blood, sputum, urine, stool, skin and sometimes spinal fluid and bone marrow, for aerobes, anaerobes, fungi and tubercle bacilli
   - Patients with fever and neutropenia should be treated with broad-spectrum antibiotic coverage (Chapter 31). The specific therapy depends upon the clinical status of the patient, the presence of indwelling vascular access devices and knowledge of the local flora pending specific culture results and antibiotic sensitivities. Patients who remain febrile from 4–7 days, in the face of broad antibacterial coverage, should be started on antifungal therapy empirically. Therapy should be continued until the patient is afebrile and cultures are negative or a specific organism is identified. An appropriate course of therapy is administered if an organism is identified. Specific infections require appropriate coverage, such as anaerobic coverage with clindamycin or Flagyl for a perirectal infection.

7. Patients previously treated with immunosuppressive therapy should receive irradiated cellular blood products to prevent complications of graft versus host disease (GvHD). Patients receiving immunosuppressive therapy should also receive *pneumocystis jeroveci* prophylaxis with trimethoprim and sulfamethoxazole (Bactrim/Septra). No antibacterial prophylaxis should be administered to afebrile, neutropenic patients.

Patients with mild to moderate aplastic anemia should be observed for spontaneous improvement or complete resolution. The treatment of choice for severe aplastic anemia (SAA), for patients who have an HLA-matched related donor, is hematopoietic stem cell transplantation. An increasing number of centers are treating moderate aplastic anemia in a fashion similar to SAA.

Specific Therapy

Hematopoietic stem cell transplantation (HSCT)

As soon as the diagnosis of SAA is suspected in children, HLA typing should be performed where potential donors exist. Patients with related histocompatible donors should have an HSCT (complete investigations to exclude Fanconi anemia [FA], dyskeratosis congenita [DC], paroxysmal nocturnal hemoglobinuria [PNH] or other inherited bone marrow failure syndromes should be carried out). Prolonged neutropenia and multiple transfusions increase the risk of transplantation-related morbidity and mortality. See Chapter 29 for detailed of preparatory regimens employed pre-transplantation.

Immunosuppressive Therapy

Patients unable to undergo HSCT (because no suitable donor is present) should have immunosuppressive therapy consisting of antithymocyte globulin (ATG) and cyclosporine (CSA) which have become the treatment of choice for these patients. In addition to ATG and CSA corticosteroids, methylprednisolone (Solu-medrol) or prednisone (Prednisone), are added to prevent serum sickness. Granulocyte colony-stimulating factor (G-CSF; Neupogen) is used to achieve a more rapid increment in the granulocyte count. Short term, the survival using this approach is in the range of 85%.

The regimen of ATG, methylprednisolone, GM-CSF and CSA treatment is listed on Table 6-4.

Table 6-4   Immunosuppressive Therapy for Severe Aplastic Anemia

1. Antithymocyte globulin: ATGAM anti-thymocyte globulin (equine) (Pharmacia) 20 mg/kg/d IV once daily, or Thymoglobulin (antithymocyte globulin [rabbit], Sang stat) 2.0 mg/kg/d IV once daily days 1 to 8
2. Methylprednisolone, 2 mg/kg/d IV days 1 to 8. Divide into 0.5 mg/kg/dose IV every 6 hours
3. Prednisone taper following an 8-day course of IV methylprednisolone. On days 9 and 10, prednisone, 1.5 mg/kg/d PO to be divided into two equal daily doses. On days 11 and 12, prednisone, 1 mg/kg/d PO to be divided into two equal daily doses. On days 13 and 14, prednisone, 0.5 mg/kg/d PO to be divided into two equal daily doses. On day 15, prednisone, 0.25 mg/kg/d PO to be given in one dose
4. G-CSF, 5 µg/kg SC once daily before bedtime starting on day 5. G-CSF is to be continued until patient has been transfusion independent for 2 months, absolute neutrophil count $>1,000/mm^3$, hematocrit $\geq25\%$ and platelet count $\geq40,000/mm^3$. At that point, taper G-CSF guided by the absolute neutrophil count
5. CSA, 10 mg/kg/d PO initially starting on day 1. Divide into two equal daily doses. Serum drug levels should be monitored as needed with the first level at 72 hours post initiation of therapy. CSA dose to be adjusted to keep serum trough levels between 100 and 300 ng/mL. CSA should be continued for one year to reduce the likelihood of relapse and then decrease the dose by 2.0 mg/kg every 2 weeks

Abbreviations: CSA, cyclosporine (formerly cyclosporin A); G-CSF, granulocyte colony-stimulating factor.
Modified from: Vlachos A, Lipton JM. In Conn's Current Therapy, W.B. Saunders Company, 2002, with permission.

Contraindications to the use of immunosuppressive drugs include:

- Serum creatinine, >2 mg%
- Concurrent pregnancy
- Sexually active females who refuse contraceptives
- Concurrent hepatic, renal, cardiac, or metabolic problems of such severity that death is likely to occur within 7–10 days or moribund patients.

Antithymocyte Globulin (ATG)

Test dose: An intradermal ATG test dose consisting of an injection of 0.1 ml of a 1:1,000 dilution of ATG in 0.9% sodium chloride solution for injection (5 µg equine IgG) should be carried out prior to ATG treatment. A control using 0.9% sodium chloride injection is administered on the contralateral side. Allergy is indicated by erythema greater than 5 mm compared to the saline control, developing after the observation at 15–20 minutes during the first hour of the skin test. The patient should also be observed for signs and symptoms of systemic allergic reaction.

Doses of ATG are listed on Table 6-4.

Usual adverse reactions to ATG include:

- Thrombocytopenia: All patients should receive a daily platelet transfusion on a prophylactic basis to maintain a platelet count of more than 20,000/mm^3 (during administration of ATG). Only irradiated and leukocyte-filtered cellular blood products should be used
- Headache, myalgia
- Arthralgia, chills and fever: Treatment with an antihistamine and corticosteroid is indicated
- Chemical phlebitis: A central line (high flow vein) for infusion of ATG should be used and peripheral veins should be avoided
- Itching and erythema: Treatment with an antihistamine with or without corticosteroids is indicated
- Leukopenia
- Serum sickness – approximately 7–10 days following ATG administration: many patients develop serum sickness. This should be treated by increasing the daily dose of solumedrol until the symptoms abate.

Uncommon adverse reactions to ATG include: Dyspnea, chest, back and flank pain, diarrhea, nausea, vomiting, hypertension, herpes simplex infection, stomatitis, laryngospasm, anaphylaxis, tachycardia, edema, localized infection, malaise, seizures, gastrointestinal bleeding/perforation, thrombophlebitis, lymphadenopathy, hepatosplenomegaly, renal function impairment, liver function abnormalities, myocarditis and congestive heart failure.

Cyclosporine (CSA) Preparations

1.   Neoral oral solution, 100 mg/ml.
2.   Neoral capsule or Sandimmune capsule, 25 mg and 100 mg/capsule.

Oral CSA solution may be mixed with milk, chocolate milk, or orange juice preferably at room temperature. It should be stirred well and drunk at once.

Cyclosporine levels should be performed once a week for the first 2 weeks and then once every 2 weeks for the remainder of the treatment or as necessary to maintain a whole-blood CSA level between 200 and 400 ng/ml. Changes in serum creatinine levels are the principal criteria for dose change. An increase in creatinine level of more than 30% above baseline warrants a reduction in the dose of CSA by 2 mg/kg/day each week until the creatinine level has returned to normal. A serum CSA level of less than 100 ng/ml is evidence of inadequate absorption and a CSA level above 500 ng/ml is considered an excessive dose. If the CSA level is greater than 500 ng/ml, CSA should be discontinued. Levels should be repeated daily or every other day. When the level returns to 200 ng/ml or less, CSA should be resumed at a 20% reduced dose. In responders CSA should be tapered very slowly, with some hematologists beginning to taper the CSA dose only after a year.

Principal side effects of CSA: Renal dysfunction, tremor, hirsutism, hypertension and gingival hyperplasia.

Uncommon side effects of CSA: Significant hyperkalemia, hyperuricemia, hypomagnesemia, hepatotoxicity, lipemia, central nervous system toxicity and gynecomastia. An increase of more than 100% in the bilirubin level or of liver enzymes is treated in the same way as an increase of more than 30% in creatinine and warrants a reduction in the dose of CSA by 2 mg/kg/day each week until the bilirubin and/or liver enzymes return to the normal range.

Contraindications to the use of CSA: Hypersensitivity to CSA.

Pharmacokinetic interactions with CSA:

*   Carbamazepine, phenobarbital, phenytoin, rifampin – decreases half-life and blood levels of CSA
*   Sulfamethoxazole/trimethoprim IV – decreases serum levels of CSA
*   Erythromycin, fluconazole, ketoconazole, nifedipine – increases blood levels of CSA
*   Imipenem-cilastatin – increases blood levels of CSA and central nervous system toxicity
*   Methylprednisolone (high dose), prednisolone – increases plasma levels of CSA
*   Metoclopramide (Reglan) – increases absorption and increases plasma levels of CSA.

Pharmacologic interactions with CSA:

- Aminoglycosides, amphotericin B, nonsteroidal anti-inflammatory drugs, trimethoprim/ sulfamethoxazole – nephrotoxicity
- Melphalan, quinolones – nephrotoxicity
- Methylprednisolone – convulsions
- Azathioprine, corticosteroids, cyclophosphamide – increases immunosuppression, infections, malignancy
- Verapamil – increases immunosuppression
- Digoxin – elevates digoxin level with toxicity
- Nondepolarizing muscle relaxants – prolongs neuromuscular blockade.

Hematopoietic Growth Factors

The addition of human recombinant granulocyte colony stimulating factor (G-CSF) to a regimen of ATG, cyclosporine and corticosteroids theoretically provides improved protection from infectious complications by stimulating granulopoiesis.

Treatment Choices and Long-term Follow-up

Although the short-term outcome with immunosuppressive therapy is comparable to that obtained with HLA-matched related HSCT, the decision to choose HSCT for younger patients, who have a histocompatible donor, is based on the result of long-term follow-up. Although there is some late mortality, due to chronic GvHD and therapy-related cancer, in patients undergoing HSCT for SAA the survival curves are relatively flat. Improved GvHD prophylaxis and safer preparative regimens should further improve these results. In contrast, the risk of clonal hematopoietic disorders MDS, AML and PNH is unacceptably high relative to both the short- and long-term risks of HSCT. Those undergoing immunosuppressive therapy must be closely followed for the development of clonal disorders. In terms of unrelated or poorly matched related donor HSCT, current risk favors the use of immunosuppressive therapy in those patients with SAA who cannot receive a matched related HSCT.

Salvage Therapy

For the patient who fails HSCT, has a partial response (ANC $\geq$500/mm^3, but is red cell and platelet transfusion-dependent) or relapses following immunosuppressive therapy management choices include alternative donor HSCT or further immunosuppressive therapy. The use of HSCT, if an appropriate alternative donor is available, is preferred. Children and teenagers for whom a fully HLA-matched unrelated donor, determined by high-resolution typing, exists are good candidates for an alternative donor HSCT. A delay in transplantation, along with the associated risk of infection and additional transfusions attendant to a second course of immune therapy, seems unwarranted in this setting. For older patients (>40 years) and those without a good alternative donor, a second course of ATG/ CSA/G-CSF is warranted. Androgens and alternative cytokines are being evaluated and should be considered experimental.

High-dose Cyclophosphamide Therapy

Complete remission in severe aplastic anemia after high-dose cyclophosphamide therapy without bone marrow transplantation has been reported. The rationale for the use of high-dose cyclophosphamide is as follows:

- The majority of patients with severe aplastic anemia lack an HLA-identical sibling for treatment with bone marrow transplantation
- Although the majority (80%) of children with severe aplastic anemia benefit from the use of treatment with ATG and cyclosporine, many do not attain completely normal counts and some patients treated successfully with immunosuppressive therapy either relapse or develop late clonal diseases such as paroxysmal nocturnal hemoglobinuria, myelodysplastic syndrome, or acute leukemia
- After preparation with cyclophosphamide, most allografts persist indefinitely; however, in several cases, a complete autologous reconstitution of hematopoiesis has occurred
- Patients with very severe aplastic anemia (i.e., severe aplastic anemia patients with an absolute neutrophil count of less than 200/mm^3 at diagnosis) respond to immunosuppressive therapy, but have greater morbidity and mortality due to the profound neutropenia.

On this basis, patients with severe aplastic anemia who lack an HLA-identical sibling donor have been treated rarely on high-dose cyclophosphamide as a single course. Most experts believe that this regimen is too toxic and that standard immunosuppressive therapy or unrelated stem cell transplantation should be used as salvage therapy rather than high-dose cyclophosphamide treatment.

Long-term Sequelae and Outcomes for Severe Aplastic Anemia

Table 6-5 lists the long-term sequelae following treatment of aplastic anemia.

Outcomes for both immunosuppressive therapy and HSCT have improved considerably.

1. Survival rates are greater than 90% with either immunosuppressive therapy or stem cell transplantation. However, stem cell transplantation is curative for most patients.
2. Immunosuppressive therapy improves hematopoiesis and achieves transfusion independence in the majority of patients, but the time to response is long, hematopoietic response may be partial and relapses are relatively common.

Table 6-5   Long-Term Sequelae Following Treatment of Aplastic Anemia[a]

Sequelae	Type of Therapy and Incidence of Complications	
	Immunosuppressive Therapy (%)	Bone Marrow Transplantation (%)
10-year cumulative cancer incidence	18.8	3.1
10-year cumulative myelodysplastic syndrome (MDS) incidence	9.6	0.0
10-year cumulative acute leukemia (AL) incidence	6.6	0.25
10-year cumulative solid tumor (ST) incidence	2.2	2.9
Conclusion: Survivors of aplastic anemia are at high risk of developing late malignancies. Incidence of MDS and AL is higher in patients treated with immunosuppressive therapies; however, the incidence of solid tumors is the same in both transplantation and immunosuppressive treated patients.		

[a]Report of European Bone Marrow Transplantation (EBMT) working party on severe aplastic anemia.

3. Clonal hematopoietic disorders including PNH, myelodysplasia and leukemia may develop in up to 10% of patients treated with immunosuppressive therapy (IST). An analysis of 1765 patients with acquired aplastic anemia treated with either sibling transplantation ($n = 583$) or immunosuppressive therapy ($n = 1182$) produced the following results:
   - Matched sibling donor HSCT is always superior in young patients (<20 years of age) at any neutrophil count
   - Immunosuppression is superior in older patients (41–50 years) with a neutrophil count greater than 500/mm^3
   - For the 21–40-years age group, the differences are less clear
   - In all age groups there is a higher percentage of late failures in the immuno-suppression-treated patients
   - The difference in survival between patients treated with HSCT and immunosuppression is not linear, but increases with time. For the younger group of patients, a 10% advantage in favor of HSCT at one year became a 19% advantage at 5 years
   - There is a higher risk of late death in patients treated with immunosuppressive therapy due to complications including relapse and evolution to clonal disorders.

The European Bone Marrow Transplantation Working Party compared the rate of secondary malignancies following HSCT and immunosuppressive therapy (IST). Forty-two malignancies developed in 860 patients receiving IST, compared to 9 in 748 patients who underwent HSCT. In this study, acute leukemia and myelodysplasia were seen exclusively in IST-treated patients while the incidence of solid tumors was similar in the two groups of patients.

Treatment of Moderate Aplastic Anemia

The natural history of moderate aplastic anemia is uncertain and clinical experience varies widely. For this reason, it is generally thought that these patients should be treated initially with supportive therapy with very close follow-up. Patients who progress, as the majority appear to do, to develop severe aplastic anemia and/or significant and severe thrombocytopenia and bleeding, serious infections, or a chronic red blood transfusion requirement should be treated with the same treatment options as described for severe aplastic anemia.

CONGENITAL APLASTIC ANEMIAS

Fanconi Anemia

The key shared clinical manifestations of inherited bone marrow failure syndromes are:

* Bone marrow failure
* Congenital anomalies
* Cancer predisposition
* May present in adulthood.

The common pathophysiology is low apoptotic threshold of mutant cells.

Fanconi anemia (FA), is a rare (heterozygote frequency in the general population of 1/300; 1/90 in Ashkenazi Jews (FANCC) and 1/80 in South African Afrikaners (FANCA) due to "founder effect"). It is an autosomal recessive and rarely X-linked recessive inherited bone marrow failure syndrome generally associated with multiple congenital anomalies and a predisposition to cancer.

The details of guidelines for the diagnosis and management of FA are beyond the scope of the book but are available from the Fanconi Anemia Research Fund (FARF) (Eiler M, Frohnmayer, D (Eds). Fanconi Anemia: Guidelines for Diagnosis and Management, 3rd edition, Fanconi Anemia Research Fund, Inc., 2009). It offers the most up-to-date and comprehensive information available.

Pathophysiology of Fanconi Anemia

Somatic cell hybridization studies have thus far defined 13 FA complementation groups. All 13 FA genes have been cloned (Table 6-6). Complementation groups FANCA, C and G represent ≈90% of the cases. The gene products of these 13 genes have been shown to cooperate in a common pathway. Eight of the FA proteins (FANCA, B, C, E, F, G, L and M) assemble in a nuclear complex that is required to monoubiquinate and activate FANCD2 and FANCI. Ubiquinated FANCD2 and FANCI co-localize to a nuclear DNA repair focus containing FANCJ and FANCN as well as BRCA1 and other repair proteins. Of note FANCD1 has been identified as BRCA2. The exact mechanism of FANCD2 and FANCI monoubiquitination and the role of FANCD2, BRCA2 (FANCD1), BCRA1 and the other proteins in the repair DNA complex is being unraveled. Despite the identification of this pathway the manner in which disruption in this cascade of events results in a faulty

Table 6-6 Inherited Bone Marrow Failure Syndrome Genes, Known and Presumed

Disorder	Gene	% of Cases	Locus	Genetics	Gene Product
Fanconi anemia	*FANCA*	66%	16q24.3	Autosomal recessive	FANCA
	FANCB	<1%	Xp22.31	X-linked recessive	FANCB
	FANCC	9.5%	9q22.3	Autosomal recessive	FANCC
	FANCD1	3.3%	13q12.3	Autosomal recessive	FANCD1/BRCA2
	FANCD2	3.3%	3p25.3	Autosomal recessive	FANCD2
	FANCE	2.5%	6p21.3	Autosomal recessive	FANCE
	FANCF	2.1%	11p15	Autosomal recessive	FANCF
	FANCG	8.7%	9p13	Autosomal recessive	FANCG/XRCC9
	FANCI	1.6%	15q25-26	Autosomal recessive	FANCI/KIAA1794
	FANCJ	1.6%	17q22-24	Autosomal recessive	FANCJ/BRIP1/BACH1
	FANCL	<1%	2p16.1	Autosomal recessive	FANCL/PHF9
	FANCM[a]		14q21.3	Autosomal recessive	FANCM
	FANCN	<1%	16p12	Autosomal recessive	FANCN/PALB2
Dyskeratosis congenital	*DKC1*	35%	Xq28	X-linked recessive	Dyskerin
	TERC	5%	3q21-28	Autosomal dominant	Telomerase RNA component
	TERT	1%	5p15.33	Autosomal dominant	Telomerase
	TINF2	10%	14q12	Autosomal dominant	T1N 2
	NOLA3	<1%	15q14-15	Autosomal recessive	NOP10
	NOLA2	<1%	5q35.3	Autosomal recessive	NHP2
Diamond–Blackfan anemia	*RPS19*	25%	19q13.2	Autosomal dominant	RPS19
	RPS17	2%	10q22-23	Autosomal dominant	RPS17
	RPS24	2%	15q25.2	Autosomal dominant	RPS24
	RPL35A	2%	3q29-qter	Autosomal dominant	RPL35a
	RPL5	7%	1p22.1	Autosomal dominant	RPL5
	RPL11	10%	1p36-35	Autosomal dominant	RPL11
Shwachman–Diamond syndrome	*SBDS*		7q11	Autosomal recessive	SBDS
Severe congenital neutropenia (SCN)	*ELA2*	60%	19p13.3	Autosomal dominant	Neutrophil elastase
	HAX1 (Kostmann syndrome)	Rare	1q21.3	Autosomal recessive	HAX1 protein
	G6PC3[b]	Rare	17q21	Autosomal recessive	G6PC3
Amegakaryocytic thrombocytopenia	c-mpl		1p34	Autosomal recessive	Thrombopoietin receptor
Thrombocytopenia absent radii (TAR) syndrome	?		?	Autosomal recessive	?

[a]FANCM is a member of the "core complex" but homozygosity not yet identified in patients with FA.
[b]Other very rare mutated genes resulting in SCN have been described (*GFI-1, WASP, p14*).

DNA-damaged response and genomic instability leading to hematopoietic failure, birth defects and cancer predisposition remains to be determined.

Fanconi anemia cells are characterized by hypersensitivity to chromosomal breakage as well as hypersensitivity to G2/M cell cycle arrest induced by DNA cross-linking agents. In addition there is sensitivity to oxygen free radicals and to ionizing radiation.

Clinical Features

1. FA is inherited as an autosomal recessive disorder (>99%) and rarely as an X-linked recessive (FANCB, <1%) and is the most frequently inherited aplastic anemia. FANCA is the most common complementation group, representing about 70% of cases. FANC and G are the next most common representing 10% of cases each. The remaining ten complementation groups are quite rare, representing the remainder of cases (Table 6-6).
2. Genotype–phenotype correlations are complex and probably relate as much to the nature of the gene product and other factors as to the specific complementation group. However, certain associations relating genotype to specific congenital anomalies, early onset aplastic anemia, leukemia, as well as Wilms tumor and medulloblastoma are emerging.
3. All racial and ethnic groups are affected.
4. Pancytopenia is the usual finding.
   a. The median age at hematologic presentation of patients with aplastic anemia is approximately 8–10 years. Leukemia tends to appear later in the teenage years and solid tumors appear in young adulthood and continue to occur as patients age.
   b. Hematologic dysfunction usually presents with macrocytosis, followed by thrombocytopenia, often leading to progressive pancytopenia and severe aplastic anemia (SAA). FA frequently terminates in myelodysplastic syndrome (MDS) and/or acute myeloid leukemia (AML).
   c. FA cells are hypersensitive to chromosomal breaks induced by DNA cross-linking agents. This observation is the basis for the commonly used chromosome breakage test for FA. The clastogens diepoxybutane (DEB) and mitomycin C (MMC) are the agents most frequently used *in vitro* to induce chromosome breaks, gaps, rearrangements, quadriradii and other structural abnormalities. Clastogens also induce cell cycle arrest in G2/M. The hypersensitivity of FA lymphocytes to G2/M arrest, detected using cell cycle analysis by flow cytometry either *de novo* or clastogen induced, has more recently been used as a screening tool for FA.
5. Bone marrow examination reveals hypocellularity and fatty replacement consistent with the degree of peripheral pancytopenia. Residual hematopoiesis may reveal dysplastic erythroid (megaloblastoid changes, multinuclearity) and myeloid (abnormal granulation) precursors and abnormal megakaryocytes.

Table 6-7   Congenital Anomalies and Frequency in Fanconi Anemia

Anomaly	Approximate Frequency
Skin	55%
Skeletal	51%
Reproductive organs	35%
Small head or eyes	26%
Renal	21%
Low birth weight	11%
Cardiopulmonary	6%
Gastrointestinal	5%

Modified from: Alter B, Lipton J. Anemia, Fanconi. EMedicine Journal [serial online]. 2009. (Available at http://www .emedicine.com/ped/topic3022.htm).

6. Congenital anomalies include increased pigmentation of the skin along with café au lait and hypopigmented areas, short stature (impaired growth hormone secretion), skeletal anomalies (especially involving the thumb, radius and long bones), male hypogenitalism, microcephaly, abnormalities of the eyes (microphthalmia, strabismus, ptosis, nystagmus) and ears including deafness, hyperreflexia, developmental delay and renal and cardiac anomalies. Forty percent of patients lack obvious physical abnormalities. There is great clinical heterogeneity even within a genotype (affected sibling may be phenotypically different). Table 6-7 lists the anomalies and frequency in FA.

7. There is a nearly 800-fold increased relative risk of developing AML in Fanconi anemia and perhaps an even greater relative risk of nonhematologic tumors (e.g. squamous cell carcinoma of head and neck, cancer of the breast, kidney, lung, colon, bone, retinoblastoma and female gynecologic) in patients with FA at much younger ages than that seen in the general population. A relatively large number of patients only become aware that they have FA when they are diagnosed with cancer. Androgen-related, usually benign liver neoplasia may also occur. The risk of solid tumors may become even higher as death from aplastic anemia is reduced and as post-hematopoietic stem cell transplantation (HSCT) patients survive longer. These data must be considered in the context of HSCT, in particular when the risk of non-hematologic malignancy is likely to increase as a result of HSCT conditioning regimens and chronic GvHD. Treatment for cancer is generally ineffective.

8. Prenatal diagnosis is possible in amniotic fluid cell cultures and chorionic villus biopsy.

Diagnosis

Table 6-8 lists the indications for Fanconi anemia screening studies. Table 6-9 lists the laboratory studies required to make the diagnosis of Fanconi anemia. Table 6-10 lists the initial and follow-up investigations to be performed in a patient with an established diagnosis of Fanconi anemia.

Table 6-8 Indications for Fanconi Anemia Screening Studies

All children with aplastic anemia or unexplained cytopenias
All children with MDS or AML
Patients with classic birth defects suggestive of FA
  VATER/VACTRL (vertebral anomalies, anal atresia, cardiac anomalies, tracheoesophageal fistula, renal anomalies and limb anomalies)
  Structural anomalies of the upper extremity and/or genitourinary system
Patients with:
  Excessive café au lait spots, hypo- or hyperpigmentation of skin
  Microcephaly
  Micro-ophthalmia
  Growth failure
Development of FA associated cancers at a young age (e.g. squamous cell carcinoma in esophagus, head and neck <50 years of age, vulvar cancer <40 years of age and uterine cervical cancer <30 years of age or liver tumors)
Patient with leukemia or solid tumor with unusual sensitivity to chemotherapy
Karyotype with spontaneous chromosome breaks
Patients with unexplained macrocytosis and an elevated HbF
Patients with non-immune thrombocytopenia
Males with unexplained infertility
Siblings of known FA patients

Table 6-9 Laboratory Studies for Diagnosis of Fanconi Anemia

1. Screening tests:
   a. Demonstration of the presence of increased chromosomal breakage in T-lymphocytes cultured in the presence of DNA cross-linking agents such as mitomycin C (MMC) or diepoxybutane (DEB). DEB test is used more widely. Chromosome fragility includes breaks, gaps, re-arrangements, radials, exchanges and endoreduplication
      Fibroblast should be studied in patients for whom mosaicism is suspected[a]
   b. Flow cytometry study:
      A flow cytometric technique for the analysis of alkylating agent – treated cells can determine the percentage of cells arrested in G2/M because a characteristic distribution clearly distinguishes FA cells from normal cells[b]
   c. Western blot for D2-L (long protein formed by ubiquitination of FANC D2[b])
2. Definitive test:
   Complementation group analysis[b]
   Mutation analysis (see Table 6-6 for cloned FANC genes)[b]
3. Prenatal diagnosis of FA: DEB test can be used in either chorionic villus or amniocentesis derived samples
4. Detection of carrier state:
   In a FA family, if proband has been identified to have a defect in one of the eight cloned genes, molecular testing is available for the extended family members
   Population-based screening is only done in the at risk populations

[a]Some patients with FA may have two populations of cells exhibiting either a normal or an FA phenotype. Such mosaicism may result in a false-negative chromosome breakage study if the percentage of normal cells is high. The study of fibroblasts is useful in this circumstance.
[b]Done only in specialized laboratories.

Table 6-10 Initial and Follow-up Investigations to be Performed in a Patient with Established Diagnosis of Fanconi Anemia

Endocrine studies for:
  Short stature (growth hormone deficiency)
  Glucose intolerance (diabetes mellitus)
  Hypothyroidism
  Pubertal delay
  Evaluation of undescended testes
  Reduced fertility
Imaging studies[a] and evaluation of:
  Orthopedic anomalies
  Genitourinary abnormalities
Hepatic ultrasound every 6 months while taking androgens
Serum chemistries for:
  Liver function
  Kidney function
Hearing test
Monitoring for iron overload for patients on red cell transfusion therapy:
  1. Ferritin
  2. Liver enzymes
  3. Liver biopsy
  4. Superconductivity quantum interference device biosuseptometry (SQUID)
Survey of family members:
  1. Exclude diagnosis of FA in any other family members
  2. Type family members for the potential availability of an HLA-matched sibling for future consideration of bone marrow transplantation
  3. Provide genetic counseling to parents and patient
Prospective counseling and screening:
  1. Avoid exposure to potential mutagens or carcinogens (e.g., insecticides, organic solvents, hair dye, papilloma virus)
  2. Cancer surveillance:
    a. Examine bone marrow yearly with histologic and cytogenetic studies for evidence of myelodysplasia or leukemia
    b. Yearly head and neck examination over age 7 years
    c. Yearly gynecologic examination beginning at age 16 years
    d. Breast self-examination beginning at age 16 years
    e. Periodic oral cancer screening
Mutation analysis:
  These studies are performed in specialized laboratories only. Mutation analysis may help predict the phenotype as more data become available

[a]Limit exposure to radiation by using appropriate restraint and non-radiologic imaging studies.

Complications

Table 6-11 describes the complications of malignancy and liver disease associated with Fanconi anemia.

Table 6-11   Malignancy and Liver Disease in Fanconi Anemia

	Leukemia	Myelodysplastic Syndrome (MDS)	Cancer[a]	Liver Disease
Number of patients (%)	84 (9)	32 (3)	47 (5)	37 (4)
Male:female	1.3:1	1.1:1	0.3:1	1.6:1
Age (in years) at diagnosis				
Mean	10	13	13	9
Median	9	12	10	6
Range	0.1–28	1–31	0.1–34	1–48
Percentage ≥16 years old	20	32	31	11
Age (in years) at complication				
Mean	14	17	23	16
Median	14	17	26	13
Range	0.1–29	5–31	0.31–38	3–48
Number without pancytopenia (%)	21 (25)	14 (44)	8 (17)	1 (3)
Number died (%)	40 (48)	20 (63)	18 (38)	1 (3)
Number reported deceased (%)	66 (79)	24 (75)	28 (60)	32 (86)

[a]More recent data describes an actuarial risk of hematologic and non-hematologic cancer of 33 and 28%, respectively, by 40 years of age.

Note: 150 patients had one or more malignancies; the number of malignancies was 157. MDS cases included seven who developed leukemia.

From: Alter BP. Arms and the man or hands and the child: congenital anomalies and hematologic syndromes. J Pediatr Hematol/Oncol 1997;19:287–291, with permission.

Differential Diagnosis

1. The differential diagnosis of FA generally includes acquired aplastic anemia, congenital amegakaryocytic thrombocytopenia (CAT), TAR syndrome, as well as VATER/VACTRL (vertebral anomalies, anal atresia, cardiac anomalies, tracheoesophageal fistula, renal anomalies, limb anomalies) syndromes. FA is easily distinguished from TAR syndrome. There is an intercalary defect in TAR consisting of absent radii with normal thumbs, whereas in FA the defect is terminal, an abnormal radius always being associated with anomalies of the thumb. Table 6-12 lists the features differentiating FA from TAR syndrome.

2. FA testing is warranted in any child who presents with hematologic cytopenias, unexplained macrocytosis, aplastic anemia or AML, as well as representative congenital abnormalities or solid tumors typical of FA such as head and neck, esophageal or gynecologic tumors presenting at an early age (Table 6-8).

3. The critical investigations are aspiration and biopsy of the bone marrow and demonstration in peripheral blood of increased chromosomal fragility or G2/M arrest induced by clastogens (e.g. DEB, MMC). Complementation group analysis and/or mutation analysis are helpful after the demonstration of a positive screening test and should, if possible, be obtained.

Table 6-12 Features Differentiating Fanconi Anemia from Amegakaryocytic Thrombocytopenic Purpura (TAR Syndrome)

Feature	Fanconi Anemia	TAR
Age of onset of aplastic anemia symptoms	Median of 8–10 years	Birth to infancy
Low birth weight	~10%	~10%
Stature	Short	Short
Skeletal deformities	66%	100%
Absent radii with fingers and thumbs present	0%	100%
Other hand deformities	~40%	~40%
Lower extremity deformities	~40%	<10%
Cardiovascular anomalies	5–10%	5–10%
Anomalous pigmentation of skin	77%	0%
Hemangiomas	0%	~10%
Mental retardation	17%	7%
Peripheral blood	Pancytopenia Macrocytosis (high MCV)	Thrombocytopenia, eosinophilia, leukemoid reactions, anemia
Bone marrow	Aplastic	Absent or abnormal megakaryocytes, normal myeloid and erythroid precursors
Marrow CFU-GM, CFU-E	Decreased	Normal (decreased CFU-megakaryocytes)
HbF	Increased	Normal
Hexokinase in blood cells	Decreased in some	?
Chromosomal breaks in Leukocytes	Present	None
Malignancy	Common	Rare (leukemia only)
Sex ratio (male/female)	~1:1	~1:1
Inheritance pattern	Autosomal-recessive	Autosomal-recessive
Associated leukemia	Yes	Rare
Prognosis	Poor	Good if patient survives first year when platelet count improves

Abbreviation: ~, Approximately.
Modified from: Hall JG, Levin J, Kuhn JP, et al. Thrombocytopenia with absent radius (TAR). Medicine (Baltimore) 1969;48:411, with permission.

4. FA somatic mosaics with DEB-positive and DEB-negative (double population) cells belong to distinct groups based upon the degree of mosaicism and may present diagnostic problems. Mosaicism leading to a "normal" T-cell that is resistant to the less dose-intense HSCT conditioning, used for FA, may result in graft rejection.

Management

Serial assessment of the bone marrow should be performed to provide evidence of progression and the development or evolution of cytogenetic abnormalities.

Bone marrow aspiration should be performed for cytology, cytogenetics with FISH analysis for cytogenetic abnormalities that may be predictive of leukemia (e.g. 3q26q29 amplification and 7q deletion) approximately yearly and more often if indicated by the emergence of specific clonal or morphological abnormalities.

Bone marrow biopsy should be done for cellularity.

The patient's complete blood counts should be monitored. The degree of cytopenia guides management as follows:

	Mild	Moderate	Severe
Hemoglobin level	$\geq$8.0 gm/dl	<8.0 gm/dl	<8.0 gm/dl
ANC	<1,500/mm^3	<1,000/mm^3	<500/mm^3
Platelet count	150,000–50,000/mm^3	<50,000/mm^3	<30,000/mm^3

When cytopenias are in the mild-to-moderate range and in the absence of cytogenetic abnormalities, counts should be monitored every 3–4 months and bone marrow aspiration should be performed yearly. Monitoring of blood counts and bone marrow should be increased to every 1–2 months and every 1–6 months, respectively, for cytopenia in the presence of cytogenetic abnormalities or more significant dysplasia without frank MDS. With falling (or in some cases rising) counts surveillance must be increased.

Treatment

Androgen therapy: Androgen therapy (oxymetholone 2–5 mg/kg/day and tapered to the lowest effective dose). Approximately 50% of patients will respond to androgens.

Cytokines: G-CSF in a dose of 5 µg/kg every other day or GM-CSF in a dose of 250 µg/kg/m^2 every other day should be administered when moderate to severe cytopenias are present.

Transfusions: Treatment with packed red blood cells and platelets should be minimized and reserved for patients who fail androgen therapy. Blood products should be irradiated, leukocyte-depleted and of single donor origin, when possible. Blood relatives should not be used as blood donors until a matched allogeneic-related donor transplantation is ruled out. Iron status should be monitored at regular intervals to determine the degree of iron overload and the institution of chelation treatment in chronically transfused patients.

Allogeneic hematopoietic stem cell transplantation: HLA typing should be done at diagnosis to facilitate therapeutic planning. If an HLA-matched related donor is available, stem cell transplantation should be carried out. Evidence of true MDS (as opposed to benign clonal abnormalities) or evolution to leukemia are clear indications for transplantation.
The sensitivity of FA patients to traditional transplantation conditioning regimens requires the use of lower dosages of chemotherapy and radiation therapy (Chapter 29).

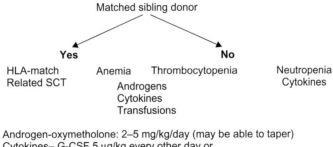

Androgen-oxymetholone: 2–5 mg/kg/day (may be able to taper)
Cytokines– G-CSF 5 µg/kg every other day or
GM-CSF 250 µg/M^2 every other day

Figure 6-2 Treatment of Fanconi Anemia.

Before a family member is used as a donor, the donor should be evaluated to exclude a diagnosis of Fanconi anemia.

HPV vaccination: Vaccination is recommended in patients with FA.

Growth hormone therapy: The majority of patients with FA have short stature. Up to 50% of them have deficient growth hormone. Because of a theoretical association of growth hormone and leukemia, growth hormone should be used with that understanding in patients with FA.

Gene therapy: This approach is experimental and will only be performed in approved clinical trials.

Figure 6-2 summarizes the treatment of FA.

Prognosis

Current results of matched sibling transplantation prior to development of overt leukemia show a long-term disease-free survival of 80–90%. However, the long-term risks of late sequelae from hematopoietic stem cell transplantation, although not sufficiently understood, probably include an increase in cancer risk. Unrelated donor transplantations have generally been reserved for androgen refractory patients and those with MDS or leukemia. However, improvements in HLA-typing, conditioning regimens and overall care as well as the experience in certain transplantation units has improved outcome considerably. Thus every patient should be evaluated for hematopoietic stem cell transplantation.

DYSKERATOSIS CONGENITA

1. Dyskeratosis congenita (DC) is characterized by the classic triad of ectodermal dysplasia consisting of:
   - abnormal skin pigmentation of the upper chest and neck

- • dysplastic nails
- • leukoplakia of oral mucous membranes. leukemia, myelodysplasia and epithelial cancers.

2. Predisposition to bone marrow failure.
3. Predisposition to cancer – hematologic (leukemia, myelodysplasia) and epithelial cancers.
4. Somatic findings in DC include: epiphora (tearing due to obstructed tear ducts), blepharitis, developmental delay, pulmonary disease (fibrosis), short stature, esophageal webs, liver fibrosis, dental carries, tooth loss, premature gray hair and hair loss, ocular, dental, skeletal, cutaneous, genitourinary, gastrointestinal, neurologic abnormalities and immunodeficiency have been reported.
5. The diagnosis of DC requires two of the three elements of the classical diagnostic triad and any other associated abnormality in patients with a known mutation or very short telomeres. The presence of short telomeres in a member of a pedigree with definitive DC is sufficient for the diagnosis.
6. The median age at diagnosis is 15 years. The median age for the onset of mucocutaneous abnormalities is 6–8 years. Nail changes occur first but hematologic abnormalities may precede mucocutaneous changes. The median age for the onset of pancytopenia is 10 years. Approximately 50% of patients develop severe aplastic anemia and greater than 90% develop at least a single cytopenia by 40 years of age. The anemia is associated with a high MCV and elevated fetal hemoglobin. As with FA it is the non-hematologic manifestations of DC that are of particular concern, especially when hematopoietic stem cell transplantation for bone marrow failure is considered.

Pathophysiology

Research establishes DC to be the result of deficient telomerase activity. Telomerase adds DNA sequence back to the ends of chromosomes that are eroded with each DNA replication. Telomerase activity is found in tissues with rapid turnover such as the basal layer of the epidermis, squamous epithelium of the oral cavity, hematopoietic stem cells and progenitors and in other tissues affected in DC. The lack of telomerase activity may also give rise to chromosome instability resulting in the high rate of premature cancer observed in these tissues. Table 6-13 shows the cells in various organs expressing telomerase and the defects that occur in telomerase failure. Epithelial malignancies develop at or beyond the third decade of life. About one in five patients develop progressive pulmonary disease characterized by fibrosis, resulting in diminished diffusion capacity and/or restrictive lung disease. Of note, type 2 alveolar epithelial cells express telomerase. It is likely that more pulmonary disease would be evident if patients did not succumb earlier to the complications of severe aplastic anemia and cancer.

Table 6-13 Cells Expressing Telomerase and Defects Occurring in Telomerase Failure

Organ Systems	Cells Expressing Telomerase	Defect
Hair	Hair follicle	Alopecia
Oral cavity	Squamous epithelium	Leukoplakia
Skin	Epidermis, basal layer	Abnormal pigmentation and dyskeratotic nails
Lungs	Type II alveolar cells	Pulmonary fibrosis
Liver	?	Cirrhosis
Intestines	Crypt cells	Enteropathy
Testes	Spermatogonia	Hypogonadism
Bone marrow	Progenitors	Bone marrow failure

Genetics

Mutations in six genes in the telomerase maintenance pathway have been associated with DC. Dyskeratosis congenita is most commonly inherited as an X-linked recessive but may also be autosomal dominant or recessive. The gene responsible for the X-linked form was mapped to Xq28 and subsequently identified as DKC1. DKC1 codes for dyskerin, a nucleolar protein associated with nucleolar RNAs. Dyskerin is associated with the telomerase complex. This latter function appears to be the one involved in the pathophysiology of DC, as all the genes found to date to be mutated in DC (Table 6-13) are involved in telomere biology. There are many features in common to all three genetic subtypes; however, the clinical phenotypes may vary widely in severity even within different mutations of the same allele. Affected members within the same family may exhibit wide variability in clinical presentation suggesting the influence of modifying genes and environmental factors.

Clinical Course

The clinical course is quite variable.

Bone marrow failure: The incidence of bone marrow failure is 86%. The majority of deaths (67%) are a result of bone marrow failure and a significant number who do not die of bone marrow failure die as a result of lung disease (pulmonary fibrosis) with or without HSCT.

Malignancy: Almost 9% of DC develop cancer (MDS, AML and Hodgkin disease). The most common cancers are squamous cell carcinoma of the head and neck followed by anorectal, stomach and lung. All of these cancers occur at younger ages than these cancers occur in the population at large.

Neurological: Patients with a severe form of DC known as Hoyeraal-Hreidarsson (HH) syndrome have symptomatic cerebellar hypoplasia, microcephaly and developmental delay. Revesz syndrome (RS) is associated with CNS calcification, occasionally cerebellar hypoplasia and exudative retinopathy. Multiple DC genes have been implicated in HH and TINF2 gene has been implicated in RS.

Immunodeficiency: Significant progressive immunodeficiency occurs in DC. Although DC is predominantly a cellular immune defect, humoral immunodeficiency as well as neutropenia probably play a significant role in the infectious morbidity and mortality in DC.

Outcome: The median survival is approximately 40–45 years for patients with DC. In HH it is approximately 5 years of age and in RS the median has not yet been defined. The prognosis for patients with DC is generally poor.

Therapy

Supportive care: Blood products, antibiotics and antifibrinolytic agents are similar to those used for idiopathic aplastic anemia.

Hematopoietic stem cell transplantation (HSCT) should be considered for those patients with an HLA-matched related donor or an acceptable alternative donor and no DC-related contraindications. The results of HSCT have been poor predominantly due to pulmonary complications. All DC patients are at a high risk of interstitial pulmonary disease when undergoing HSCT. There have been too few transplantation survivors to determine whether an increase in the prevalence of cancer will follow as a consequence of HSCT. An immunoablative rather than a myeloablative approach may reduce the incremental risk of pulmonary toxicity as well as the potential for nonhematologic cancer risk.

Responses to androgens, G-CSF or GM-CSF, as well as erythropoietin and rarely splenectomy have been documented. However, these responses have been transient. Immunosuppressive therapy is ineffective.

CONGENITAL APLASTIC ANEMIAS OF UNKNOWN INHERITANCE

Rare cases of aplastic anemia have been associated with Down syndrome; congenital trisomy-8 mosaicism; familial Robertsonian translocation (13;14); nonfamilial translocation in a male with t(1;20); (p22;q13.3) and cerebellar ataxia; bone marrow monosomy-7 manifesting prior to pancytopenia (familial ataxia–pancytopenia syndrome); and increased spontaneous chromosomal breakage without further increase in breakage with mitomycin C as well as other very rare cases with familial associations. Many of these cases were reported before the discovery of the many genes associated with the inherited bone marrow failure syndromes and can now be categorized. However, a large number of cases are yet to be genetically diagnosed and await the identification of new mutated genes.

DIAMOND–BLACKFAN ANEMIA (CONGENITAL PURE RED CELL APLASIA)

Pathophysiology

Diamond–Blackfan anemia (DBA), is a rare pure red cell aplasia predominantly, but not exclusively, of infancy and childhood resulting from defective ribose biosynthesis as a consequence of a ribosome biosynthesis resulting in erythroid progenitors and precursors that are highly sensitive to death by apoptosis. All of the genetically known cases are the consequence of either small or large subunit-associated ribosomal protein haploinsufficiency. Readers are referred to a recent article, Diagnosing and Treating Diamond Blackfan Anemia: Results of an International Clinical Consensus Conference. Br J Haematol 2008, 142: 859–876, for the most comprehensive discussion of the diagnosis and treatment of DBA. The detailed information provided in this paper is beyond the scope of the book.

Genetics

1. Dominant inheritance:
   - The first "DBA gene" was cloned in 1997 and identified as *RPS 19*, a gene that codes for a ribosomal protein, located at chromosome 19q13.2. Studies showed that *RPS 19* mutations accounted for only 20–25% of both sporadic and familial cases. Since that time an additional five genes have been identified (Table 6-6), comprising approximately 50% of DBA cases analyzed. Mutations leading to ribosomal protein (rp) haploinsufficiency, both of the small and large subunit-associated proteins, account for all of the mutations (published and unpublished) to date. The functions of the rp are not fully understood
   - Laboratory studies used for identification of dominant inheritance in family members of a proband with DBA include: hemoglobin level, mean corpuscular volume (MCV), erythrocyte adenosine deaminase activity (the absence of these markers clearly does not exclude dominant inheritance) and mutation analysis when available. By carefully evaluating families it appears that at least 40–50% of cases of DBA may be dominantly inherited. There is no firm evidence of autosomal recessive inheritance of DBA.

To provide genetic counseling, it is important to perform the previously mentioned laboratory studies to reduce the possibility of missing dominant inheritance in presumed recessive or sporadic cases. It is also important to perform these laboratory studies in potential family stem cell donors to increase the likelihood of detection of a silent phenotype. When there is a known mutation the parents should be evaluated and extended family members evaluated as indicated.

Clinical Features[*]

1. Uncommon familial disorder; autosomal-dominant mode of inheritance in almost 50% of cases.
2. The median age at presentation of anemia is 2 months and the median age at diagnosis of DBA is 3–4 months. Over 90% of the patients present during the first year of life. A small percentage of affected infants may be anemic at birth.
3. Platelet and white cell count are usually normal; thrombocytosis occurs rarely, neutropenia and/or thrombocytopenia may occur. Instances of significant cytopenias including aplastic anemia are emerging.
4. Physical anomalies, excluding short stature, are found in 47% of the patients. Of these, 50% are of the face and head (microcephaly, eye anomalies), 38% upper limb and hand (thumb deformity, triphalangeal thumb duplication of thumb and bifid thumb), 39% genitourinary and 30% cardiac. Twenty-one percent of the patients have more than one anomaly.
5. Low birth weight occurs in approximately 10% of all affected patients, with about half of this group being small for gestational age. Over 60% are below the 25th percentile for height. There appears to be a slight increase in the incidence of miscarriages, stillbirths and complications of pregnancy among the mothers who have given birth to infants with this syndrome.
6. Karyotype generally normal.
7. No hepatosplenomegaly.
8. Malignant potential; DBA has been recognized as a cancer predisposition syndrome. The precise incidence of cancer is unknown. Of the approximately 30 reported case of malignancy the most common have been hematopoietic [AML, myelodysplastic syndrome (MDS), lymphoma]. Osteogenic sarcoma is next most common and cases of breast, colon and other solid tumors have been reported, all occurring at a younger age than expected for these malignancies.

Diagnosis

A number of rp gene mutations have been identified in DBA and a number of genetically defined individuals have been identified who lack some or all of the classical clinical criteria.

The following laboratory findings occur in DBA:

- Macrocytosis associated with reticulocytopenia. The white cell count and platelet count are usually normal at presentation but trilineage marrow failure may become evident with increasing age

---

[*] Includes findings of the DBA Registry (DBAR) of North America.

Table 6-14   Diagnostic Criteria for Diamond–Blackfan Anemia[a]

Diagnostic criteria:
Classical
  Normochromic, usually macrocytic anemia, relative to patient's age and occasionally normocytic anemia
    developing in early childhood with no other significant cytopenias
  Reticulocytopenia
  Normocellular marrow with selective paucity of erythroid precursors
  Age less than 1 year
Supporting criteria:
Definitive but not essential
  Presence of mutation described in classical DBA
Major
  Positive family history
Minor
  Congenital abnormalities described in classical DBA
  Macrocytosis
  Elevated fetal hemoglobin
  Elevated erythrocyte adenosine deaminase (eADA) activity

[a]These criteria are under constant analysis and may be modified as new DBA genes are identified. The diagnosis becomes less certain when there are fewer diagnostic criteria and the patient does not have a positive family history or a known mutation.

- An elevated erythrocyte adenosine deaminase (eADA) activity is found in approximately 85% of patients. It may also be raised in leukemia and myelodysplastic syndromes
- Elevated fetal hemoglobin. These parameters may be useful in avoiding potential matched related HSCT donors with genotypic DBA and have been helpful in distinguishing DBA from transient erythroblastopenia of childhood (TEC)
- Bone marrow with virtual absence of normoblasts, in some cases with a relative increase in proerythroblasts or normal numbers of proerythroblasts with a maturation arrest; normal myeloid and megakaryocytic series.

Table 6-14 lists the diagnostic criteria for Diamond–Blackfan anemia.

Differential Diagnosis

This condition must be differentiated from:

- Transient erythroblastopenia of childhood (TEC). Table 6-15 lists the differentiating features of TEC from DBA
- Congenital hypoplastic anemia due to transplacental infection with parvovirus B19 can be differentiated from DBA by performing reverse transcriptase polymerase chain reaction (RT-PCR) for parvovirus B19 on a bone marrow sample. Parvovirus may result in transient red cell failure in a patient with underlying hemolytic anemia or chronic red cell failure in a patient with underlying immune deficiency

Table 6-15   Differentiating Transient Erythroblastopenia from Diamond–Blackfan Anemia

Feature	Transient Erythroblastopenia	Diamond Blackfan Anemia
Frequency	Common (? Increasing)	Rare (5–10 per 10^6 live births)
Etiology	Acquired (viral, idiopathic)	Genetic
Age at diagnosis	6 months–4 years, occasionally older	90%, by 1 year 25%, at birth or within first 2 months
Familial	No	Yes (in at least 10–20% of cases)
Antecedent history	Viral illness	None
Congenital abnormalities	Absent	Present $\sim$50% cases (heart, kidneys, musculoskeletal system)
Course	Spontaneous recovery in weeks to months	Prolonged, 20% actuarial probability of remission
Transfusion dependence	Not dependent	Transfusion or steroid dependent
MCV increased		
At Diagnosis	20%	80%
During Recovery	90%	100%
In Remission	0%	100%
Hemoglobin F increased		
At Diagnosis	25%	100%
During Recovery	100%	100%
In Remission	0%	85%
i Antigen	Usually normal	Elevated
Erythrocyte adenosine deaminase activity	Not elevated	Elevated ($\sim$85% of cases)
Treatment	Packed cell transfusion, if required	Packed red cell transfusion until 1 year of age Prednisone 2 mg/kg/day and taper to lowest effective dose Stem cell transplantation

- Late hyporegenerative anemia due to severe Rh or ABO hemolytic disease of the newborn. This may rarely last for a few months and should be considered in the differential diagnosis of DBA
- Pearson syndrome, which is characterized by refractory aregenerative macrocytic sideroblastic anemia, neutropenia, vacuolization of bone marrow precursors with sideroblasts (usually ring sideroblasts), exocrine pancreatic dysfunction and metabolic acidosis. The anemia presents at 1 month of age in 25% and at 6 months of age in 70% of affected individuals. A deletion in mitochondrial DNA has been found in Pearson syndrome. In many instances the cytopenia may resolve with age. However, many patients will develop neurodegenerative disease (Kearns–Sayre Syndrome) later in childhood. The natural history of Pearson syndrome is not well characterized
- Thymoma – not described in infancy but has been reported in a 5-year-old
- Viral infections
- Medications.

Treatment

1. *Prednisone*: In a dose of 2 mg/kg/day in a single or divided dose is used to initiate therapy. Reticulocytosis usually occurs in 1–2 weeks but may take slightly longer. When the hemoglobin level reaches $\approx$10.0 g/dl, the prednisone dose should be reduced to the minimum dose necessary to maintain a reasonable hemoglobin level on an alternate-day schedule. A dose equivalent of 1 mg/kg/every other day (0.5 mg/kg/day) is generally safe but the corticosteroid dose must be individualized. Any patient who experiences significant steroid-related side effects including growth failure should have steroid medication temporarily discontinued and should be placed on a red cell transfusion regimen. Patients with DBA on low-dose alternate-day therapy of long duration, starting in early infancy, may manifest significant steroid toxicity. Steroid-related side effects have been observed in most patients, 40% manifest cushingoid features, 12% pathologic fractures and 6.8% have cataracts. Corticosteroids should be withheld for the first year of life (during this period consideration should be given to the safety of local blood products and vascular access) to reduce these and other side-effects and to allow for safe and effective immunization.

2. *Packed red cell transfusion*: Leukocyte-depleted packed red cell transfusion should be used to reduce the incidence of nonhemolytic, febrile transfusion reactions, as well as the risk of transmission of cytomegalovirus (CMV) and the risk of human leukocyte antigen (HLA) alloimmunization. Patients who are receiving or who have recently been treated with immunosuppressive drugs should receive irradiated blood products. Patients in whom stem cell transplantation is contemplated should receive CMV-safe blood products. Effective iron chelation must accompany a transfusion protocol.

3. *Hematopoietic stem cell transplantation*: HLA-matched sibling donor transplantation should be considered for any patient with DBA, particularly those who are transfusion-dependent. Consideration should be given to the fact that 20% of all patients attain spontaneous remission, balanced by the risk of hematologic malignancy, myelodysplasia or severe aplastic anemia. A family marrow donor must be tested for the presence of a "silent phenotype." Matched unrelated or incompletely matched related donor transplantations have proven to be more risky and should be reserved for patients with leukemia, MDS, severe aplastic anemia or patients with clinically significant neutropenia or thrombocytopenia. The results for alternative donor transplantations have improved considerably and these recommendations may change for selected patients.

4. *Alternative therapy*: A number of treatments, including erythropoietin, immunoglobulin, megadose corticosteroids and androgens have been utilized in DBA patients with little success. Cyclosporine, interleuken-3 (IL-3), metoclopramide and leucine have resulted in occasional responses in DBA. The toxicity of cyclosporine and the lack of

availability of IL-3 preclude their use for most patients. A more extensive trial with leucine is required to determine whether it has a place in the treatment of DBA. These agents should be explored on a case-by-case basis as an alternative to corticosteroids, transfusion or stem cell transplantation when the risk associated with these proven modalities warrants consideration of alternate therapy.

Prognosis

1. Approximately 80% of DBA patients respond initially to corticosteroid therapy. The remaining 20% require transfusion therapy.
2. The actuarial remission rate in DBA is approximately 20% by age 25, irrespective of their pattern of response to treatment, with the majority remitting during the first decade.
3. The major complication of transfusion is iron overload, the consequences of which include diabetes mellitus, cardiac and hepatic dysfunction, growth failure as well as endocrine dysfunction. Iron chelation with either desferoxamine or deferasirox is therefore an essential component of a transfusion program. New oral iron chelators are in development; however, the oral chelator deferiprone (L1) has caused significant neutropenia in DBA and should not be used. Many patients, however, find nearly daily subcutaneous and even oral chelation therapy onerous and compliance is often poor. Sustained hematologic remissions defined as stable hemoglobin levels without transfusion or steroid requirement for 6 months may occur. Only about half of steroid-responsive patients remain on prednisone for long periods of time. In summary, both chronic corticosteroid therapy and chronic transfusion therapy may lead to a number of significant immediate and long-term complications, supporting a role for HSCT. Survival of patients into adulthood in remission or sustainable on steroids is in the range of 85–100%. Only about 60% of transfusion-dependent patients currently survive to middle age. The overall actuarial survival for DBA at 40 years of age is 75.1±4.8%.
4. HLA-matched-sibling stem cell transplantation patients have long-term survival of over 90% if performed at age 9 years or younger. Well-matched unrelated donor transplantations carried out have resulted in a survival rate in the range of 80%. Favorable transplantation outcomes are most likely if the patient is in good health at the time of HSCT without complications of iron overload and allosensitization. Improvements in supportive care, GvHD prophylaxis and infection control have resulted in a marked decrease in HLA-matched related HSCT transplantation-related morbidity and mortality. Sibling HSCT is recommended for young DBA patients, prior to development of significant allosensitization or iron overload, when there is an available HLA-matched related donor.
5. Death in DBA is due to treatment-related causes (iron overload, infection, complications of stem cell transplantation) in 67% of cases, related to the disease

(severe aplastic anemia, malignancies) in 22% of cases and unknown or unrelated causes in 11%.

6. Patients with DBA who become pregnant may develop either an increased requirement for steroid therapy or red cell transfusions due to worsening anemia and should be considered high risk and require appropriate follow-up. This appears to be a hormonally induced problem because oral contraceptives may cause the same problem in patients with DBA.

7. Fetal hydrops secondary to fetal DBA has been reported.

TRANSIENT ERYTHROBLASTOPENIA

Transient erythroblastopenia of childhood (TEC) is much more common than Diamond–Blackfan anemia (DBA) and must be differentiated from DBA (Table 6-15) in order to avoid unnecessary corticosteroid use. TEC has the following features:

- *Pathophysiology*: The following clinical and laboratory observations have shed light on the basic mechanisms of the pathogenesis of TEC:
  - Viral: There is usually a history of a preceding non-specific viral illness 1–2 months prior to TEC
  - Erythropoietin levels: Serum erythropoietin levels are high in keeping with the degree of anemia
  - CFU-E and BFU-E: Both are decreased in 30–50% of patients, suggesting that the defect might be at the CFU-E and BFU-E levels
  - Serum inhibitors of erythropoiesis: Immunoglobulin G (IgG) inhibitors of normal progenitor cells have been found in 60–80% of patients with TEC
  - Cellular inhibitors of erythropoiesis: Inhibitory mononuclear cells have been observed in approximately 25% of patients with TEC.

  On the basis of these observations, it has been speculated that a nonspecific virus is cleared as the host develops IgG antibody. This IgG antibody probably recognizes shared viral and erythroid progenitor epitopes.
- *Age*: Usually between 6 months and 4 years of age. With more children attending daycare programs younger patients with TEC are being identified.
- *Sex*: Equal frequency in boys and girls.
- *Hematologic values*:
  - Hemoglobin falls to levels ranging from 3 to 8 g/dl
  - Reticulocyte count is 0%
  - White blood cell and platelet count are usually normal.

  However, approximately 10% of patients may have significant neutropenia [absolute neutrophil count (ANC), $<1,000/mm^3$] and 5% have thrombocytopenia (platelet count $<100,000/mm^3$) (Table 6-15 lists the hematologic characteristics). An analysis of

50 patients presenting with TEC at our institution revealed a high incidence of neutropenia (64% with an ANC of less than 1,500/mm^3)

- *Bone marrow*: Absence of red cell precursors, except when the diagnostic bone marrow is performed during early recovery (prior to a reticulocytosis) when variable degrees of erythroid maturation may be observed
- *Prognosis*: Spontaneous recovery occurs within weeks to months with the vast majority of patients recovering within 1 month. Recurrent TEC occurs only rarely
- *Treatment*: Transfusion of packed red blood cells if there is impending cardiovascular compromise. As recovery is usually prompt restraint should be exercised with regard to red cell transfusions.

Other instances of transient red cell failure may occur secondary to:

- *Drugs* – chloramphenicol, penicillin, phenobarbital and diphenylhydantoin
- *Infections* – viral infections (e.g., mumps, Epstein–Barr virus (EBV), parvovirus B19, atypical pneumonia) and bacterial sepsis
- *Malnutrition* – kwashiorkor and other disorders
- *Chronic hemolytic anemia* – hereditary spherocytosis, sickle cell anemia, ß-thalassemia and other congenital or acquired hemolytic anemias. The etiologic agent is human parvovirus B19.

CONGENITAL DYSERYTHROPOIETIC ANEMIA (CDA)

CDA are a group of conditions characterized by ineffective erythropoiesis (intramedullary red cell death, i.e. anemia with reticulocytopenia and marrow erythroid hyperplasia) and by specific morphologic abnormalities in the bone marrow consisting of increased numbers of morphologically abnormal red cell precursors. There are three major types of CDA (I–III) although other variants (IV–VIII plus other variants) have been described.

Clinical Manifestations

CDA has the following clinical manifestations:

- Chronic mild congenital anemia (red cells have nonspecific abnormalities; basophilic stippling, occasional normoblasts) usually presenting in childhood
- Reticulocyte response insufficient for the degree of anemia in the context of erythroid hyperplasia in marrow
- Normal granulopoiesis and thrombopoiesis
- Chronic or intermittent mild jaundice
- Splenomegaly
- High plasma iron turnover rate and low iron utilization by erythrocyte (ineffective erythropoiesis) resulting in hemosiderosis

- Red cell survival time shortened
- Progressive iron overload leading to hemosiderosis
- Marrow with abnormal erythroid morphology that can usually distinguish the three types of CDA.

Other clinical manifestations of CDA include the following:

- CDA associated with atypical hereditary ovalocytosis
- CDA of neonatal onset (with severe anemia at birth, hepatosplenomegaly, jaundice, syndactyly and small for gestational age)
- CDA associated with hydrops fetalis and hypoproteinemia.

Table 6-16 lists the clinical and laboratory features of congenital dyserythropoietic anemia, types I–III and Table 6-17 lists types IV–VI. Cases with clinical manifestations that do not fit the classical categories of CDA have been described. These types share the common features of a congenital, perhaps hereditary, anemia with an inappropriately low reticulocyte count for the degree of anemia and ineffective marrow dyserythropoiesis. Thalassemia and other metabolic abnormalities must be excluded. Table 6-18 lists the myeloid/erythroid (M/E) ratios and percentages of erythroblasts showing various dysplastic changes in ten healthy adults and 12 patients with CDA type I. Table 6-19 lists the diagnostic tests necessary when CDA is suspected.

Differential Diagnosis

The diagnosis of CDA can only be made after the exclusion of other causes of congenital hemolytic anemias associated with ineffective erythropoiesis such as thalassemia syndromes and hereditary sideroblastic anemias. Familial dyserythropoietic anemia with thrombocytopenia has been shown to be associated with mutations in GATA-1.

Treatment

1. Splenectomy performed in severely affected patients results in moderate to marked improvement with CDA type I having the poorest response.
2. Transfusion program with the use of desferoxamine, deferasirox or deferiprone (L1) to ameliorate the effects of iron overload may be required to maintain an acceptable hemoglobin level.
3. Folic acid 1 mg per week should be administered. Iron therapy is contraindicated.
4. Vitamin E has been used in the treatment of CDA type II, with an apparent improvement in red cell survival and a reduction in serum bilirubin and reticulocyte count.
5. Recombinant α-interferon 2a has been used in CDA type I, resulting in an increase in hemoglobin level, a decrease in MCV and red cell distribution width (RDW), a

Table 6-16 **Clinical and Laboratory Features of Congenital Dyserythropoietic Anemia,**
Types I–III

Feature	Type I	Type II (HEMPAS)[a]	Type III
Inheritance	Autosomal recessive	Autosomal recessive	Autosomal dominant
Clinical	Hepatosplenomegaly	Hepatosplenomegaly	Hepatosplenomegaly
	Jaundice	Variable jaundice	Hair-on-end
		Gallstones	appearance on
		Hemochromatosis	skull radiograph
			Increased prevalence of
			lymphoproliferative
			disorders
Gene	*CDAN1*	Unknown	Unknown
Gene locus (in some cases)	15q15.1–15.3	20q11.2	15q22
Red cell size	Macrocytic	Normo- or macrocytic	Macrocytic
Anemia	Mild to moderate (usually presenting in neonatal period) Hemoglobin 8–12 g/dl	Moderate Hemoglobin 6–7 g/dl	Mild to moderate Hemoglobin 7–8.5 g/dl
Reticulocytes	1.5%	±2%	2–4%
Smear	Macrocytic: Marked anisocytosis and poikilocytosis; basophilic stippling	Normocytic: Anisocytosis and poikilocytosis; basophilic stippling; "tear drop" cells; irregular contracted cells; occasionally, normoblasts	Macrocytic: Anisocytosis and poikilocytosis; basophilic stippling
Marrow normoblasts	Megaloblastoid: Binucleated, 2–5%; internuclear chromatin bridges, 1–2%	Normoblastic: Bi- and multinucleated 10–50% Binuclearity predominates	Megaloblastic: Multinuclearity (up to 12 nuclei gigantoblasts), 10–50%
Marrow iron	Scant increase	Increased	Increased
Serum bilirubin and urine urobilinogen	Elevated	Elevated	Elevated
Treatment	Some patients respond to α interferon 2a treatment or undergo HSCT	Splenectomy HSCT	HSCT possibly

[a]Pathognomonic finding in CDA type II is that the patient's red cells are lysed by approximately 30% of acidified sera from normal individuals, but not from patient's own acidified serum. The red cells contain a specific HEMPAS (hereditary erythroblastic multinuclearity associated with a positive acidified-serum test) antigen; many normal sera contain an IgM that is anti-HEMPAS.

Abbreviations: HSCT, Hematopoietic stem cell transplantation.

Modified from: Alter BP. Inherited bone marrow failure syndromes. In Nathan and Oski's, Hematology of infancy and childhood, 6th Ed, Eds. Nathan DG, Orkin SH, Ginsburg D, Look AT, Saunders; 2003.

Table 6-17   Clinical and Laboratory Features of Congenital Dyserythropoetic Anemia, Type IV–VI

	Type IV	Type V	Type VI
Clinical	Mild to moderate splenomegaly	Spleen palpable in few cases. Unconjugated hyperbilirubinemia due to intramedullary destruction of morphologically normal, but functionally abnormal erythroblasts/marrow reticulocytes	Spleen not palpable
Hemoglobin	Very low, transfusion dependent	Normal or near normal	Normal or near normal
MCV	Normal or mildly elevated	Normal or mildly elevated	Very high (119–125) without vitamin B_{12}, folic acid, or other causes of megaloblastic anemia
Erythropoiesis	Normoblastic or mildly to moderately megaloblastic	Normoblastic	Grossly megaloblastic
Nonspecific erythroblast dysplasia	Present	Absent or little	Present

From: Wickramasinghe SN. Dyserythropoiesis and congenital dyserythropoietic anemias. Br J Haematol 1997;98:785-797.

Table 6-18   Myeloid/erythroid (M/E) Ratios and Percentages of Erythroblasts Showing Various Changes in 10 Healthy Adults and 12 Patients with CDA Type I

	Healthy Volunteers		CDA Type I	
	Mean	Range	Mean	Range
M/E ratio	3.1	2–8.3	0.54	0.20–1.30
Cytoplasmic stippling (%)	0.24	0–0.91	7.10	1.02–15.04
Cytoplasmic vacuolation (%)	0.39	0–0.70		
Intererythroblastic cytoplasmic bridges (%)	2.38	0.72–4.77		
Markedly irregular or karyorrhectic nuclei (%)	0.22	0–0.55	3.00	1.32–5.03
Howell-Jolly bodies (%)	0.18	0–0.39	0.97	0.41–1.58
Binuclearity (%)	0.31	0–0.57	4.87	3.50–7.02
Internuclear chromatin bridges (%)	0	0	1.59	0.60–2.83
Number of erythroblasts assessed per subject	713	548–1022	817	500–1185

From: Wickramasinghe SN. Dyserythropoiesis and congenital dyserythropoietic anemias. Br J Haematol 1997;98:785-797.

Table 6-19   Diagnostic Tests for Congenital Dyserythropoietic Anemia[a]

1. Complete blood count, including MCV, red cell distribution width (RDW), blood smear examination
2. Absolute reticulocyte count
3. Quantitative light and, if needed, electron microscope analysis of the bone marrow
4. Serum vitamin B_{12} and red cell folate measurements
5. Parvovirus B_{19}
6. Serum bilirubin levels
7. Hemoglobin (Hb) electrophoresis: Hb A2, Hb F assays
8. Red cell enzyme assays [pyruvate kinase, glucose-6-phosphate dehydrogenase (G6PD)]
9. Sodium dodecyl sulfate polyacrylamide gel electrophoresis of red cell membranes
10. Test for urinary hemosiderin
11. Cytogenetic studies of bone marrow cells
12. Mutation analysis for known CDA genes
13. Studies of globin chain synthesis
14. Studies of globin gene analysis

[a]This list is not exhaustive nor is it required in all patients. These tests may be required when there is a need to rule out ß-thalassemia, thiamine responsive sideroblastic anemia, megaloblastic (B_{12}, folate) anemia, iron deficiency and other causes of ineffective erythropoiesis.

reduction in serum bilirubin and lactic dehydrogenase (LDH) levels, an improvement in morphology of erythroblasts and a reduction in ineffective erythropoiesis.

6. Successful stem cell transplantation has been performed in types I and II CDA.

SIDEROBLASTIC ANEMIAS (MITOCHONDRIAL DISEASES WITH BONE MARROW FAILURE SYNDROMES)

The sideroblastic anemias are a heterogeneous group of disorders characterized by iron deposition in erythroblast mitochondrial.

Laboratory Findings

1. Anemia that may be normocytic, normochromic or microcytic and hypochromic except in Pearson syndrome, which is characterized by macrocytic anemia probably due to fetal-like erythropoiesis.
2. Reticulocytopenia.
3. Ineffective erythropoiesis (i.e., erythroid hyperplasia in bone marrow despite anemia).
4. Presence of iron-loaded normoblasts demonstrated as ring sideroblasts (greater than 10% of erythroid precursor) by Pearls' Prussian blue stain (this stain serves as a surrogate technique for electron microscopy or energy-dispersive X-ray analysis used for the demonstration of iron-loaded mitochondria in normoblasts).
5. Mild to moderate hemolysis due to peripheral red blood cell destruction of unknown etiology.

Table 6-20   Classification of the Sideroblastic Anemias (SAs)

Hereditary/congenital SA
  Isolated heritable
    X-linked (XLSA)
    Glutaredoxin 5 deficiency
    Associated with erythropoietic protoporphyria
    Presumed autosomal
    Suggested maternal
    Sporadic congenital
  Associated with genetic syndromes
    X-linked with ataxia (XLSA/A)
    Thiamine-responsive megaloblastic anemia (TRMA)
    Myopathy, lactic acidosis and sideroblastic anemia (MLASA)
    Mitochondrial cytopathy (Pearson syndrome)
Acquired clonal SA (see Chapter 16, page 497)
  Refractory anemia with ring sideroblasts (RARS)/Pure SA (PSA)
  Refractory anemia with ring sideroblasts and thrombocytosis (RARS-T)
  Refractory cytopenia with multilineage dysplasia and ring sideroblasts (RCMD-RS)
Acquired Reversible SA:
These are associated with:
  Alcoholism
  Certain drugs (isoniazid, chloramphenicol)
  Copper deficiency (idiopathic, zinc ingestion, copper chelation, nutritional, ? malabsorption)
  Hypothermia

From: Bottomly SS. Sideroblastic anemia In Wintrobe's Clinical Hematology, 12th Edition, Lippincott, Williams & Wilkins 2009, with permission.

The sideroblastic anemias can arise from the primary or secondary defects of mitochondria.

In congenital sideroblastic anemias, iron rings are predominantly seen in late normoblasts (i.e., orthochromatic and polychromatophilic normoblasts), whereas they are seen in earlier erythroid cells (i.e., basophilic normoblasts) in the acquired form.

Table 6-20 shows a classification of the sideroblastic anemias.

Pathophysiology

Heme biosynthesis involves eight enzymes, four of which are cytoplasmic and four that are localized in the mitochondria.

δ-aminolevulinic acid synthase (ALA-S): There are two distinct types of ALA-S. ALA-S1 (housekeeping form) occurs in nonerythroid cells and its gene maps on the autosome and ALA-S2 (erythroid-specific form) occurs in erythroid cells and its gene maps on the X chromosome. Distinct aspects of heme synthesis regulation in nonerythroid and erythroid cells are related to the differences between these two ALA-S enzymes. In nonerythroid cells, the synthesis and activity of ALA-S1 is subject to

Table 6-21 **Distinct Features of Iron and Heme Metabolism in Erythroid and Nonerythroid Cells**

	Erythroid	Non-erythroid[a]
Iron		
Iron source	Exclusively Tf	Tf + non-Tf Fe
Tf receptors	Differentiation f ↑	
	Proliferation ↑	
	Differentiation ↓	
+Fe	Little change	↓↓
Regulation	Transcriptional	Primarily mRNA stability
Effect of heme on	Inhibits	No effect
Fe uptake from Tf		
Fe overload	Mitochondria	Cytosol ferritin (never mitochondria)
Heme	Noncovalent assoc. with globin	Covalent binding to cytochromes
Content	Very high	Trace
Major function	O_2 transport	Electron transport
Control of synthesis	Fe from Tf	ALA-S
Effect of ALA-S	Translational induction	Feedback repression
	by Fe (IRE in 5′UTA)	by heme (no IRE)
Heme oxygenase	mRNA ↓ during	Induced by heme
	erythroid differentiation	

[a]"Nonerythroid" cells, in this context, are represented by transformed cells grown in tissue cultures and hepatocytes; it is possible that some specialized cells (e.g., macrophages) have other specific iron/heme metabolism characteristics.
Abbreviations: Tf, transferrin.
From: Ponka, P. Tissue-specific regulation of iron metabolism and heme synthesis: distinct control mechanism in erythroid cells. Blood 1997;89:1–25, with permission.

feedback inhibition by heme, thus making ALA-S1 the rate-limiting enzyme for the heme pathway. In erythroid cells, heme does not inhibit either the activity or the synthesis of ALA-S2 but it does inhibit cellular iron uptake from transferrin without affecting its utilization for heme synthesis.

Table 6-21 shows the distinct features of iron and heme metabolism in erythroid and nonerythroid cells.

These differences explain the large amount of heme production by erythroid cells compared to the low amount produced by nonerythroid cells. They also explain the mitochondrial deposition of iron in iron-loaded erythroid precursors.

Sideroblastic anemias result from injury to the mitochondria. Defects attributed to the mitochondrial pathways of heme synthesis result in sideroblastic anemias.

Mitochondrial injury results from:

- Defective heme synthesis and the accumulation of iron, especially in erythroid precursors. This iron accumulation causes oxidative damage to the mitochondrial

```
┌─────────────────────────────┐  ┌──────────────────────────────┐
│ Erythroid cell normoblast   │  │ Nonerythroid cell            │
│                             │  │                              │
│ Damaged mitochondria, e.g., │  │ Damaged mitochondria due to  │
│ as a Result of mitochondrial│  │ mitochondrial DNA mutation   │
│ DNA damage, ALAS deficiency,│  │            ↓                 │
│ Antibiotic toxins           │  │ Decreased ATP production     │
│            ↓                │  │            ↓                 │
│ Increased accumulation of   │  │ Damage to cellular organelles│
│ Non-utilized iron           │  │ and cell membranes           │
│            ↓                │  │            B                 │
│ Formation of hydroxyl radical│ └──────────────────────────────┘
│ through Fenton reaction     │
│            ↓                │
│ Cross-linking of OH radical │
│ to DNA, proteins and lipids │
│            ↓                │
│ Damage to cellular organelles│
│ and cell membranes          │
│            A                │
└─────────────────────────────┘
```

Involvement of A alone: Sideroblastic anemia without mitochondrial cytopathy
Involvement of B alone: Mitochondrial cytopathies without sideroblastic anemia
Involvement of A and B: Sideroblastic anemias with mitochondrial cytopathies, e.g., Pearson syndrome, Pearson syndrome with Kearns–Sayre syndrome, Wolfram syndrome

Figure 6-3 Simplified View of Pathophysiologic Consequences of Mitochondrial Diseases.

machinery through a Fenton reaction (i.e., the formation of a hydroxyl radical catalyzed by iron and reactive oxygen species damaging mitochondrial DNA by cross-linking DNA strands or by promoting the formation of DNA protein cross links)

• Congenital deletions of mitochondrial DNA.

As a result of mitochondrial damage, there is increased deposition of iron in heme-containing cells (e.g., erythroid cells). Additionally, there is decreased oxidative phosphorylation and decreased adenosine triphosphate (ATP) synthesis in many organs as observed in Pearson syndrome. Figure 6-3 shows a simplified view of the pathophysiologic relationship of various mitochondrial diseases in the context of sideroblastic anemias, bone marrow failure and/or mitochondrial cytopathies.

Sideroblastic anemia in children is often secondary to defects in the enzymes of the heme biosynthetic pathway, namely, ALA-S deficiency. Impaired production of heme resulting from defects in these enzymes results in mitochondrial iron accumulation, damage to the mitochondrial machinery and formation of ring sideroblasts. Porphyrias, however, do not display sideroblastic anemia because they are characterized by defects in the cytoplasmic steps of heme synthesis.

Treatment

1. Oral pyridoxine used in some patients with either congenital or acquired sideroblastic anemia with partial response.
2. Removal of toxin/drug responsible for causing sideroblastic anemia.
3. Stem cell transplantation is employed for the treatment of sideroblastic anemia secondary to myelodysplastic syndrome.

Treatment of *Pearson syndrome* is largely palliative and consists of the following:

- Correction of the metabolic acidosis (e.g., avoidance of fasting, administration of thiamine, riboflavin, carnitine and coenzyme Q to bypass deleted respiratory enzymes)
- Removal of reactive oxygen radical by the use of ascorbate, vitamin E, or lipoic acid. The efficacy of these therapies is not clear at this time
- Anemia is treated with red cell transfusions. G-CSF may be used to support clinically significant neutropenia. If patients do not succumb to metabolic acidosis and organ failure the majority will improve within the first decade of life
- HSCT has been performed and although engraftment occurred the patient succumbed to nonhematopoietic manifestations of the disease.

Suggested Reading

Alter B, Lipton J. Anemia, Fanconi. EMedicine Journal [serial online]. 2009 (Available at: http://www.emedicine.com/ped/topic3022.htm).

Bacigalupo A, Brand R, Oneto R, et al. Treatment of acquired severe aplastic anemia: bone marrow transplantation compared with immunosuppressive therapy – the European Group for Blood and Marrow Transplantation Experience. Semin Hematol. 2000;37:69–80.

Bagby GC, Meyers G. Bone Marrow Failure Syndromes in Hematology/Oncology Clinics of North America. Philadelphia: Saunders; 2009.

Ball SE. The modern management of severe aplastic anemia. Br J Haematol. 2001;110:41–53.

Bridges KR. Sideroblastic anemia: a mitochondrial disorder. J Pediatr Hematol/Oncol. 1997;19:274–278.

Brodsky RA, Sensenbrenner LL, Jones RJ. Complete remission in severe aplastic anemia after high dose cyclophosphamide without bone marrow transplantation. Blood. 1996;87:491–494.

Brown KE, Tisdale J, Barnett J, et al. Hepatitis associated aplastic anemia. N Engl J Med. 1997;336:1059–1064.

Camitta B, Thomas ED, Nathan DG, et al. A prospective study of androgens and bone marrow transplantation for treatment of severe aplastic anemia. Blood. 1979;53:504–515.

Cherrick I, Karayalcin G, Lanzkowsky P. Transient erythroblastopenia of childhood. Amer J Pediatr Hematol/Oncol. 1994;16:320–324.

Connor JM, Gatherer D, Gray FC, et al. Assignment of the gene for dyskeratosis congenita to Xq28. Human Genet. 1986;72:348–351.

De Plauque M, Bacigalupo A, Wursch A, et al. Long term follow-up of severe aplastic anemia patients treated with antithymocytic globulin. Br J Haematol. 1989;73:121–126.

Draptchinskaia N, Gustavsson P, Andersson B, et al. The gene encoding ribosomal protein S19 is mutated in Diamond-Blackfan anemia. Nat Genet. 1999;21:169–175.

Eiler M, Frohnmayer D, eds. Fanconi Anemia: Guidelines for Diagnosis and Management. 3rd edition. Fanconi Anemia Research Fund, Inc.; 2009.

Frickhofen N, Kaltwasser JP, Schrezenmeir H, et al. Treatment of aplastic anemia with antilymphocyte globulin and methylprednisolone with or without cyclosporine. N Engl J Med. 1991;324:1297–1304.

Grompe M. FANCD2: a branch-point in DNA damage response? Nat Med. 2002;8:555–556.

Hedberg VA, Lipton M. Thrombocytopenia with absent radii. A review of 100 cases. Am J Pediatr Hematol/Oncol. 1988;10:51–64.

Joenje H, Patel KJ. The emerging genetic and molecular basis of Fanconi anemia. Nat Rev Genet. 2001;2:446–457.

Kerr DS. Protein manifestations of mitochondrial disease: a mini-review. J Pediatr Hematol/Oncol. 1997;19:279–286.

Koenig JM, Ashton D, DeVore GR, Christensen RD. Late hyporegenerative anemia in Rh-hemolytic disease. J Pediatr. 1989;115:315–318.

Kojima S, Horibe K, Inaba J, et al. Long term outcome of acquired aplastic anemia in children: comparison between immunosuppressive therapy and bone marrow transplantation. Br J Hematol. 2000;11:321–328.

Lipton JM, Federman N, Khabbaze Y, et al. Osteogenic sarcoma associated with Diamond Blackfan anemia: a report from the Diamond Blackfan Anemia Registry. J Pediatr Hematol/Oncol. 2001;23:39–44.

Nathan DG, Orkin SH. Hematology of Infancy and Childhood. 5th ed. Philadelphia: Saunders; 1998.

Perdahl EB, Naprstek BL, Wallace WC, Lipton JM. Erythroid failure in Diamond-Blackfan anemia is characterized by apoptosis. Blood. 1994;83:645–650.

Ponka P. Tissue-specific regulation of iron metabolism and heme synthesis: distinct control mechanism in erythroid cells. Blood. 1997;89:1–25.

Rosenberg PS, Greene MH, Alter BP. Cancer incidence in persons with Fanconi's anemia. Blood. 2003;101:822–826.

Savage SA, Alter, BP in Bagby, GC, Meyers, G. Bone Marrow Failure Syndromes in Hematology/Oncology Clinics of North America. Philadelphia: Saunders, 2009.

Shalev H, Tamary H, Shaft D, et al. Neonatal manifestations of congenital dyserythropoietic anemia type I. J Pediatr. 1997;131:95–137.

Socie G, Henry-Amar M, Bacigalupo A, et al. Malignant tumors occurring after treatment of aplastic anemia. European Bone Marrow Transplantation – Severe Aplastic Anemia Working Party. N Eng J Med. 1993;329:1152–1157.

Teo JT, Klassen R, Fernandez CV, et al. Clinical and genetic analysis of unclassifiable inherited bone marrow failure syndromes. Pediatr. 2008;122:e139–e148.

Vlachos A, Klein GW, Lipton JM. The Diamond Blackfan Anemia Registry: tool for investigating the epidemiology and biology of Diamond Blackfan anemia. J Pediatr Hematol/Oncol. 2001;23:377–382.

Vlachos A, Ball S, Dahl N, et al. Diagnosing and Treating Diamond Blackfan Anemia: Results of an International Clinical Consensus Conference. Br J Haematol. 2008;142:859–876.

Vulliamy T, Dokal I. Dyskeratosis congenita. Semin Hematol. 2006;43:157–166.

Vulliamy T, Marron A, Goldman F, et al. The RNA component of telomerase is mutated in autosomal dominant dyskeratosis congenita. Nature. 2001;413:432–435.

Young NS, Alter BP. Aplastic anemia acquired and inherited. Philadelphia: Saunders; 1994.

Wickramasinghe SN. Dyserythropoiesis and congenital dyserythropoietic anemias. Br J Haematol. 1997;98:785–797.

CHAPTER 7

Red Cell Membrane and Enzyme Defects

GENERAL APPROACH TO DIAGNOSIS OF HEMOLYTIC ANEMIA

An essential feature of hemolytic anemia is a reduction in the normal red cell survival of 120 days. Premature destruction of red cells may result from corpuscular abnormalities (within the red cell corpuscle), that is, abnormalities of membrane, enzymes, or hemoglobin; or from extracorpuscular abnormalities, that is, immune or nonimmune mechanisms.

The approach to the diagnosis of hemolytic anemia should include:

- Consideration of the clinical features suggesting hemolytic disease
- Laboratory demonstration of the presence of a hemolytic process
- Determination of the precise cause of the hemolytic anemia by special hematologic investigations.

Clinical Features

The following clinical features suggest a hemolytic process:

- *Ethnic factors*: Incidence of sickle gene carrier in the African-American population (8%), high incidence of thalassemia trait in people of Mediterranean ancestry and high incidence of glucose-6-phosphate dehydrogenase (G6PD) deficiency among Sephardic Jews
- *Age factors*: Anemia and jaundice in an Rh-positive infant born to a mother who is Rh negative or a group A or group B infant born to a group O mother (setting for a hemolytic anemia)
- History of anemia, jaundice, or gallstones in family
- Persistent or recurrent anemia associated with reticulocytosis
- Anemia unresponsive to hematinics
- Intermittent bouts or persistent indirect hyperbilirubinemia/jaundice
- Splenomegaly

Manual of Pediatric Hematology and Oncology. DOI: 10.1016/B978-0-12-375154-6.00007-0

- Hemoglobinuria
- Presence of multiple gallstones
- Chronic leg ulcers
- Development of anemia or hemoglobinuria after exposure to certain drugs
- Cyanosis without cardiorespiratory distress
- Polycythemia (2,3 Diphosphoglycerate mutase deficiency)
- Dark urine due to dipyrroluria (unstable hemoglobins, thalassemia and ineffective erythropoiesis).

Laboratory Findings

Laboratory findings of hemolytic anemia consist of:

- Evidence of accelerated hemoglobin catabolism due to reduced red cell survival
- Evidence of increased erythropoiesis.

Accelerated Hemoglobin Catabolism

Accelerated hemoglobin catabolism varies with the type of hemolysis as follows:

- Extravascular hemoglobin catabolism (see Fig. 7-1)
- Intravascular hemoglobin catabolism (see Fig. 7-2).

The two may not be easily distinguished if the cause for hemolysis is not obvious, hence the long lists of markers of testing indicated below. The presence of hemoglobinuria and hemosidenuria and the absence of haptoglobin are the major markers of intravascular hemolysis in practice.

Markers of Extravascular Hemolysis

1. Increased unconjugated bilirubin.
2. Increased lactic acid dehydrogenase in serum.
3. Decreased plasma haptoglobin (normal level, 128 ± 25 mg/dl).
4. Increased fecal and urinary urobilinogen.
5. Increased rate of carbon monoxide production.

Markers of Intravascular Hemolysis

1. Increased unconjugated bilirubin.
2. Increased lactic acid dehydrogenase in serum.
3. Hemoglobinuria (Table 7-1 lists the causes of hemoglobinuria).
4. Low or absent plasma haptoglobin.
5. Hemosiderinuria (due to sloughing of iron-laden tubular cells into urine).

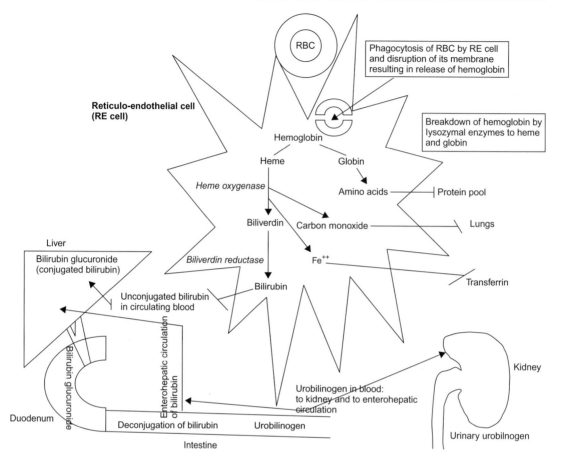

Figure 7-1 Extravascular Hemoglobin Catabolism Following Extravascular Hemolysis.

6. Raised plasma hemoglobin level (normal value <1 mg hemoglobin/dl plasma, visibly red plasma contains >50 mg hemoglobin/dl plasma).
7. Raised plasma methemalbumin (albumin bound to heme; unlike haptoglobin, albumin does not bind intact hemoglobin).
8. Raised plasma methemoglobin (oxidized free plasma hemoglobin) and raised levels of hemopexin–heme complex in plasma.

Increased Erythropoiesis

Erythropoiesis increases in response to a reduction in hemoglobin and is manifested by:

- *Reticulocytosis*: Frequently up to 10–20%; rarely, as high as 80%
- Increased mean corpuscular volume (MCV) due to the presence of reticulocytosis and increased red cell distribution width (RDW) as the hemoglobin level falls

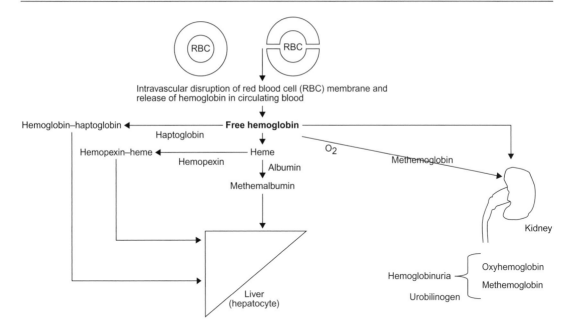

Figure 7-2 Intravascular Hemoglobin Catabolism Following Intravascular Hemolysis. Hemoglobin-haptoglobin, Hemopexin-heme and Methemalbumin are cleared by hepatocytes. Heme is converted to iron and bilirubin. The common pathway for both extravascular and intra- vascular hemolysis is the conjugation of bilirubin (bilirubin glucuronide) by the hepatocytes, its excretion in bile and ultimately formation of urobilinogen by the bacteria in the gut. Part of urobil- nogen enters in entero-hepatic circulation and part is excreted by the kidney in urine and the remainder of urobilinogen is excreted in stool.

- Increased normoblasts in peripheral blood
- *Specific morphologic abnormalities*: Sickle cells, target cells, basophilic stippling, irregularly contracted cells or fragments (schistocytes), eliptocytes, acanthocytes and spherocytes
- *Erythroid hyperplasia of the bone marrow*: Erythroid:myeloid ratio in the marrow increasing from 1:5 to 1:1
- Expansion of marrow space in chronic hemolysis resulting in:
  - Prominence of frontal bones
  - Broad cheekbones
  - Widened intratrabecular spaces, hair-on-end appearance of skull radiographs
  - Biconcave vertebrae with fish-mouth intervertebral spaces.
- Decreased red cell survival demonstrated by 51Cr red cell labeling
- Red cell creatine levels increased.

Table 7-1   Causes of Hemoglobinuria

I. Acute
  A. Mismatched blood transfusions
  B. Warm antibody-induced autoimmune hemolytic anemia
  C. Drugs and chemicals
    1. Regularly causing hemolytic anemia
      a. Drugs: phenylhydrazine, sulfones (dapsone), phenacetin, acetanilid (large doses)
      b. Chemicals: nitrobenzene, lead, inadvertent infusion of water
      c. Toxins: snake and spider bites
    2. Occasionally causing hemolytic anemia
      a. Associated with G6PD deficiency: antimalarials (primaquine, chloroquine), antipyretics (aspirin, phenacetin), sulfonamides (Gantrisin, lederkyn), nitrofurans (Furadantin, Furacin), miscellaneous (naphthalene, vitamin K, British antilewisite [BAL], favism)
      b. Associated with Hb Zürich: sulfonamides
      c. Hypersensitivity: quinine, quinidine, *para*-aminosalicylic acid (PAS), phenacetin
  D. Infections
    1. Bacterial: *Clostridium perfringens, Bartonella bacilliformis* (Oroya fever)
    2. Parasitic: malaria
  E. Burns
  F. Mechanical (e.g., prosthetic valves)
II. Chronic
  A. Paroxysmal cold hemoglobinuria; syphilis; idiopathic
  B. Paroxysmal nocturnal hemoglobinuria
  C. March hemoglobinuria
  D. Cold agglutinin hemolysis

Precise Cause of Hemolytic Anemia

Table 7-2 lists the tests used to establish the cause of the hemolytic anemia.

MEMBRANE DEFECTS

Table 7-3 lists causes of hemolytic anemia due to corpuscular defects.

Hereditary spherocytosis, elliptocytosis, stomatocytosis, acanthocytosis, xerocytosis and pyropoikilocytosis can be diagnosed on the basis of their characteristic morphologic abnormalities.

Spectrin, the major red cell membrane protein, is largely responsible for maintaining the normal red cell shape and overall morphology. It is composed of two subunits, α- and β-spectrin, which are structurally distinct and are encoded by separate genes. Spectrin is integrated vertically into the lipid bilayer of the red cell membrane through the intercession of smaller proteins (ankyrin and protein 4.1) to integral membrane-spanning proteins (Band 3, Rh Antigen and Glycophorin C). These vertical interactions seem to maintain red cell

Table 7-2   Tests Used to Establish a Specific Cause of Hemolytic Anemia

Corpuscular defects
Membrane defects:
> Blood smear: spherocytes, ovalocytes, pyknocytes, stomatocytes[a]
> Osmotic fragility (fresh and incubated)[a]
> Eosin-5-maleimide dye staining with flow cytometry[b]
> Ektacytometry
> Autohemolysis[a]
> Cation permeability studies
> Membrane phospholipid composition
> Scanning electron microscopy

Hemoglobin defects:
> Blood smear: sickle cells, target cells (Hb C)[a]
> Sickling test[a]
> Hemoglobin electrophoresis[a]
> Quantitative fetal hemoglobin determination[a]
> Kleihauer–Betke smear[a]
> Heat stability test for unstable hemoglobin
> Oxygen dissociation curves
> Rates of synthesis of polypeptide chain production
> Fingerprinting of hemoglobin

Enzyme defects:
> Heinz-body preparation[a]
> Osmotic fragility[a]
> Autohemolysis test[a]
> Screening test for enzyme deficiencies[a]
> Specific enzyme assays[a]

Extracorpuscular defects
> Direct antiglobulin test: IgG (gamma), C′3 (complement), broad-spectrum (both gamma and complement)[a]

Serological testing for unusual immune defects:
> IgA induced hemolysis, DAT negative hemolytic anemia
> Donath–Landsteiner test[a]
> Flow cytometric analysis of red cells with monoclonal antibodies to GP1-linked surface antigens (for PNH)[a]

[a]Tests commonly employed and most useful in establishing a diagnosis.
[b]Test available in reference laboratories. Test of choice for hereditary spherocytosis.
Abbreviation: DAT, direct antiglobulin test.

membrane cohesion and abnormalities of these vertical interactions predominantly lead to the varying syndromes of hereditary spherocytosis. Spectrin associates with itself head to head, while the tail of this flexible protein associates with actin and supporting proteins. These horizontal interactions maintain membrane stability as defects in these lateral relationships lead primarily to the varied hereditary elliptocytosis syndromes. The red cell membrane is semipermeable and must maintain its volume in order for the erythrocyte to negotiate the narrower spaces in the circulatory system. Red cell volume is maintained by a number of passive, gradient-driven cation and anion channels as well as active transporters. Errors in these functions lead to the syndromes of over-hydrated and dehydrated stomatocytosis.

Table 7-3   Causes of Hemolytic Anemia Due to Corpuscular Defects

I. Membrane defects
  A. Primary membrane defects with specific morphologic abnormalities
      1. Hereditary spherocytosis
      2. Hereditary elliptocytosis/pyropoikilocytosis
      3. Hereditary stomatocytosis with:
          a. Increased red cell volume (over hydrated stomatocytosis)
          b. Decreased red cell volume (dehydrated stomatocytosis/xerocytosis/desicytosis)
          c. Normal osmotic fragility/normal volume
          d. Rh-null
  B. Secondary membrane defects: abetalipoproteinemia/neuroacanthocytosis
II. Enzyme defects
  A. Energy potential defects (Embden–Meyerhof: anaerobic; ATP-producing pathway deficiencies)
      1. Hexokinase
      2. Glucose phosphate isomerase
      3. Phosphofructokinase
      4. Triosephosphate isomerase
      5. Phosphoglycerate kinase
      6. 2,3-Diphosphoglyceromutase (polycythemia and no hemolysis)
      7. Pyruvate kinase
  B. Reduction potential defects (hexose monophosphate: aerobic; NADPH-producing pathway deficiencies)
      1. G6PD[a]
      2. 6-Phosphogluconate dehydrogenase (6PGD)
      3. Glutathione reductase
      4. Glutathione synthetase
      5. 2,3-Glutamyl-cysteine synthetase
  C. Abnormalities of erythrocyte nucleotide metabolism
      1. Adenosine triphosphatase deficiency
      2. Adenylate kinase deficiency
      3. Pyrimidine 5′-nucleotidase (P5N) deficiency
      4. Adenosine deaminase excess
III. Hemoglobin defects (see Chapter 8)
  A. Heme: congenital erythropoietic porphyria
  B. Globin
      1. Qualitative: hemoglobinopathies (e.g., Hb S, C, M)
      2. Quantitative: α- and β-thalassemias
IV. Congenital dyserythropoietic anemias (see Chapter 6)
  A. Type I
  B. Type II
  C. Type III
  D. Type IV

[a]World Health Organization (WHO) classification of G6PD variant: Class I variant: Chronic hemolysis due to severe G6PD deficiency, e.g., G6PD deficiency Harilaou. Class II variant: Intermittent hemolysis in spite of severe G6PD deficiency, e.g., G6PD Mediterranean. Class III variant: Intermittent hemolysis associated usually with drugs/infections and moderate G6PD deficiency, e.g., G6PDA variant. Class IV variant: No hemolysis, no G6PD deficiency, e.g., normal G6PD (B variant).
Abbreviations: ATPase, adenosine triphosphatase; G6PD, glucose-6-phosphate dehydrogenase.

Enzyme defects and many hemoglobinopathies have nonspecific morphologic abnormalities related to secondary effects on red cell membrane proteins and pumps (e.g. ATP depletion and the Gardos effect).

Hereditary Spherocytosis

Genetics

1. Autosomal dominant inheritance (75% of cases). The severity of anemia and the degree of spherocytosis may not be uniform within an affected family.
2. No family history in 25% of cases. Some show minor laboratory abnormalities, suggesting a carrier (recessive) state. Others are due to a *de novo* mutation.
3. Most common in people of northern European heritage, with an incidence of 1 in 5,000.

Pathogenesis

In hereditary spherocytosis (HS), the primary defect is membrane instability due to dysfunction or deficiency of a red cell skeletal protein. A variety of membrane skeletal protein defects have been found in different families. These include:

- *Ankyrin mutations*: Account for 50–67% of HS. In many patients, both spectrin and ankyrin proteins are deficient. Mutations of ankyrin occur in both dominant and recessive forms of HS. Clinically, the course varies from mild to severe. Red cells are typically spherocytes
- α-Spectrin mutations occur in recessive HS and account for less than 5% of HS. Clinical course is severe. Contracted cells, poikilocytes and spherocytes are seen
- β-Spectrin mutations occur in dominant HS and account for 15–20% of HS. Clinical course is mild to moderate. Acanthocytes, spherocytic elliptocytes and spherocytes are seen
- Protein 4.2 mutations occur in the recessive form of HS and account for less than 5% of HS. Clinical course is mild to moderate. Spherocytes, acanthocytes and ovalocytes are seen
- Band 3 mutations occur in the dominant form of HS and account for 15–20% of HS. Clinical course can be mild to moderate. Spherocytes are occasionally mushroom-shaped or pincered cells.

Deficiency of these membrane skeletal proteins in HS results in a vertical defect, which causes progressive loss of membrane lipid and surface area. The loss of surface area results in characteristic microspherocytic morphology of HS red cells. The sequelae are as follows:

- Sequestration of red cells in the spleen (due to reduced erythrocyte deformability)
- Depletion of membrane lipid
- Decrease in membrane surface area relative to volume, resulting in a decrease in surface area-to-volume ratio
- Tendency to spherocytosis

- Influx and efflux of sodium increased; cell dehydration
- Rapid adenosine triphosphate (ATP) utilization and increased glycolysis leading to increased loss of surface area under ATP-depleted conditions. This leads to the observation of splenic conditioning where the changes in glucose utilization as well as cell volume control are dramatically exacerbated with each circulatory passage through the spleen
- Premature red cell destruction.

Hematologic Features

1. *Anemia*: Mild to moderate in compensated cases. In erythroblastopenic (aplastic or hypoplastic) crisis, hemoglobin may drop to 2–3 g/dl.
2. MCV usually decreased; mean corpuscular hemoglobin concentration (MCHC) raised and RDW elevated.[*]
3. Reticulocytosis (3–15%).
4. *Blood film*: *Spherocytes*, microspherocytes[**] (vary in number); hyperdense cells[***] with or without polychromasia.
5. Direct antiglobulin test negative.
6. Increased red cell osmotic fragility (spherocytes lyse in higher concentrations of saline than normal red cells) occasionally only demonstrated after incubation of blood sample at 37°C for 24 hours (always do this test incubated). In spite of normal osmotic fragility, increased MCHC or an increase of hyperdense red cells is highly suggestive of HS.
7. Autohemolysis at 24 and 48 hours increased, corrected by the addition of glucose.
8. Survival of 51Cr-labeled cells reduced with increased splenic sequestration.
9. *Marrow*: Normoblastic hyperplasia; increased iron.
10. Eosin-5-maleimide dye staining with flow cytometry is the test of choice to diagnose HS but is only available in special reference laboratories.

Biochemical Features

1. Raised bilirubin, mainly indirect reacting.
2. Obstructive jaundice with increased direct-reacting bilirubin; may develop due to gallstones, a consequence of increased pigment excretion.

---

[*] The MCHC is only raised in hereditary spherocytosis, hereditary xerocytosis, hereditary pyropoikilocytosis and cold agglutinin disease. The presence of elevated RDW and MCHC [performed by aperture impedance instruments, e.g., Coulter] makes the likelihood of hereditary spherocytosis very high, because these two tests used together are very specific for hereditary spherocytosis.

[**] The percentage of microspherocytes is the best indicator of the severity of the disease but not a good discriminator of the HS genotype.

[***] Hyperdense cells are seen in HbSC disease, HbCC disease and xerocytosis. In HS, hyperdense cells are a poor indicator of disease severity but an effective discriminating feature of the HS phenotype.

Clinical Features

1. Anemia and jaundice: Severity depends on rate of hemolysis, degree of compensation of anemia by reticulocytosis and ability of liver to conjugate and excrete indirect hyperbilirubinemia.
2. Splenomegaly.
3. Presents in newborn (50% of cases) with hyperbilirubinemia, reticulocytosis, normoblastosis, spherocytosis, negative direct antiglobulin test and splenomegaly. Patients may present with a more aggressive hemolysis in the first 8 weeks of life which may not be reflective of their ultimate clinical severity.
4. Presents before puberty in most patients.
5. Diagnosis sometimes made much later in life, often after the birth of an infant with neonatal jaundice caused by HS.
6. Co-inheritance of HS with hemoglobin S-C disease may increase the risk of splenic sequestration crisis.
7. Co-inheritance of β-thalassemia trait and HS may worsen, improve, or have no effect on the clinical course of HS.
8. Iron deficiency may correct the laboratory values but not the red cell life span in HS patients.
9. HS with other system involvement:
   - Interstitial deletion of chromosome 8p11.1–8p21.1 causes ankyrin deficiency, psychomotor retardation and hypogonadism
   - HS may be associated with neurologic abnormalities such as cerebellar disturbances, muscle atrophy and a tabes-like syndrome.

Classification

Table 7-4 lists a classification of hereditary spherocytosis in accordance with clinical severity and indications for splenectomy.

Diagnosis

1. Clinical features and family history.
2. Hematologic features.

Complications

1. *Hemolytic crisis*: With more pronounced jaundice due to accelerated hemolysis (may be precipitated by viral infection).
2. *Erythroblastopenic crisis (Hypoplastic crisis)*: Dramatic fall in hemoglobin level (and reticulocyte count); usually due to maturation arrest and often associated with giant

Table 7-4 Classification of Spherocytosis and Indications for Splenectomy

Classification	Trait	Mild Spherocytosis	Moderate Spherocytosis	Severe Spherocytosis[a]
Hemoglobin (g/dl)	Normal	11–15	8–12	6–8
Reticulocyte count (%)	≤3	3.1–6	≥6	≥10
Bilirubin (mg/dl)	≤1.0	1.0–2.0	≥2.0	≥3.0
Reticulocyte production index	<1.8	1.8–3	>3	
Spectrin per erythrocyte[b] (percentage of normal)	100	80–100	50–80	40–60
Osmotic fragility				
Fresh blood	Normal	Normal to slightly increased	Distinctly increased	Distinctly increased
Incubated blood	Slightly increased	Distinctly increased	Distinctly increased	Distinctly increased
Autohemolysis				
Without glucose (%)	>60	>60	0–80	50
With glucose (%)	<10	≥10	≥10	≥10
Splenectomy	Not necessary	Usually not necessary during childhood and adolescence	Necessary during school age before puberty	Necessary, not before 5 years of age
Symptoms	None	None	Pallor, erythroblastopenic crises, splenomegaly, gallstones	Pallor, erythroblastopenic crises, splenomegaly, gallstones

[a]Value before transfusion.
[b]Normal (mean±SD): $226\pm54\times10^3$ molecules per cell.
From: Eber SW, Armburst R, Schröter W. J Pediat 1990;117:409.

pronormoblasts in the recovery phase; often associated with parvovirus B19 infection.[*]

3. *Folate deficiency*: Caused by increased red cell turnover; may lead to superimposed megaloblastic anemia. Megaloblastic anemia may mask HS morphology as well as its diagnosis by osmotic fragility.

4. *Gallstones*: In approximately one-half of untreated patients; increased incidence with age, can occur as early as 4–5 years of age. Occasionally, HS may be masked or improved in obstructive jaundice due to increase in surface area of red cells and

---

[*]Parvovirus B19 infects developing normoblasts, causing a transient cessation of production. The virus specifically infects CFU-E and prevents their maturation. Giant pronormoblasts are seen in bone marrow. Diagnosis is made by increased IgM antibody titer against parvovirus and PCR for parvovirus on bone marrow.

formation of targets cells in obstructive jaundice. The co-inheritance of Gilbert syndrome markedly increases the incidence of gallstones.

5. *Complications of chronic anemia.* Patients with more severe HS (see Table 7-4) may suffer growth retardation, anemic heart failure and failure to thrive necessitating intermittent or chronic transfusion.

6. *Hemochromatosis:* Rarely.

Treatment

1. Folic acid supplement (1 mg/day).
2. Leukocyte-depleted packed red cell transfusion for severe erythroblastopenic crisis.
3. Splenectomy for moderate to severe cases. Most patients with less than 80% of normal spectrin content require splenectomy.* Splenectomy should be carried out early in severe cases but not before 5 years of age, if possible. The management of the splenectomized patient is detailed in Chapter 31. Although spherocytosis persists post splenectomy, the red cell life span becomes essentially normal and complications are prevented, especially transient erythroblastopenia and persistent hyperbilirubinemia, which leads to gallstones. Attitudes towards the use and timing of splenectomy continue to evolve. Case reports and retrospective studies suggest the possibility of an increased risk of arterial and venous thrombosis in later life as well as the possibility of enhancing the risk of idiopathic pulmonary hypertension. Most evidence-based reviews still suggest splenectomy is appropriate for severely affected patients (see Table 7-4). These data have also increased interest in the technique of partial splenectomy. In these surgical techniques, up to 90% of the splenic mass is removed leaving enough splenic tissue to protect against infection. Whether this mitigates the risk of subsequent thrombosis is not clear. Also unclear is the rate of reconstitution of the splenic mass and the return of severe hemolysis. The technique is not widely available and current recommendations are to consider its use primarily in transfusion-dependent patients who are under 5 years of age.
4. Ultrasound should be carried out before splenectomy to exclude the presence of gallstones. If present, cholecystectomy is also indicated.

Hereditary Elliptocytosis

Hereditary elliptocytosis (HE) is clinically and genetically a heterogeneous disorder.

Pathogenesis

HE is due to various defects in the skeletal proteins, spectrin and protein 4.1. The basic membrane defects consist of:

---

*Laparoscopic splenectomy is safe in children. Although it requires more operative time than open splenectomy, it is superior with regard to postoperative analgesia, smaller abdominal wall scars, duration of hospital stay and more rapid return to a regular diet and daily activities. It is not known if accessory spleens are readily identified with the laparoscope although the magnification afforded by the laparoscope might be advantageous in some cases.

- Defects of spectrin self-association involving the α-chains (65%)
- Defects of spectrin self-association involving the β-chains (30%)
- Deficiency of protein 4.1
- Deficiency of glycophorin
- "Silent carrier" effect: alpha-spectrin mutant genes which produce less α-spectrin when paired with an α-spectrin structural mutant. They lead to more severe disease (see below).

The mechanically unstable membrane of hereditary eliptocytosis leads to shape change from discocyte to eliptocyte as the membrane is buffeted by sheer stress in the circulation.

Patients who are heterozygotes for these defects have milder disease while double heterozygotes and homozygotes for these mutants have progressively more severe syndromes.

Genetics

HE is characterized by an autosomal dominant or codominant mode of inheritance (with variable penetrance), affecting about 1 in 25,000 of the population.

Occasionally, patients who are severely affected appear to be the offspring of a family with only a single affected parent. In this case a "silent carrier"-like mutation in an α-spectrin gene of the unaffected parent may be the cause. This α-spectrin gene produces less of a normal alpha spectrin, which emphasizes the effect of the output of the mutant α-spectrin gene.

Clinical Features

1. Varies from patients who are symptom-free to severe anemia requiring blood transfusions. The percentage of microcytes best reflects the severity of the disease.
2. About 12% have symptoms indistinguishable from hereditary spherocytosis.
3. The percentage of elliptocytes varies from 50 to 90%. No correlation has been established between the degree of elliptocytosis and the severity of the anemia.
4. HE has been classified into the following clinical subtypes:
   - Common HE, which is divided into several groups: silent carrier state, mild HE, HE with infantile pyknocytosis
   - Common HE with chronic hemolysis, which is divided into two groups: HE with dyserythropoiesis and homozygous common HE, which is clinically indistinguishable from hereditary pyropoikilocytosis (see later discussion)
   - Spherocytic HE, which clinically resembles HS; however, a family member usually has evidence of HE
   - Southeast Asian ovalocytosis, in which the majority of cells are oval; however, some red cells contain either a longitudinal or transverse ridge
   - Infantile hemolytic elliptocytosis of infancy: These patients present with hemolytic elliptocytosis (even occasionally mimicking hereditary pyropoikilocytosis) which changes over the first two years of life to a clinical picture of mild hereditary

elliptocytosis as fetal hemoglobin changes to adult hemoglobin. Usually there is a single affected parent with HE.

Laboratory Findings

1. Blood smear: 25–90% of cells elongated oval elliptocytes.
2. Osmotic fragility normal or increased.
3. Autohemolysis usually normal but may be increased and usually corrected by the addition of glucose or ATP.

Treatment

The indications and considerations for transfusion, splenectomy and prophylactic folic acid are the same as for hereditary spherocytosis.

Hereditary Pyropoikilocytosis

Definition

Hereditary pyropoikilocytosis (HPP) is a congenital hemolytic anemia associated with *in vivo* red cell fragmentation and marked *in vitro* fragmentation of red cells at 45°C. Because of the similarities in the membrane defect in this condition and HE, it is viewed as a subtype of HE.

Genetics

1. Homozygous or doubly heterozygous for spectrin chain mutants (e.g., Sp-$\alpha^{1/74}$ and Sp-$\alpha^{1/76}$. The spectrin chain defects found in HPP are similar to those found in HE.
2. Increased ratio of cholesterol to membrane protein.
3. Decreased cell deformability.

Clinical Features

1. Anemia characterized by extreme anisocytosis and poikilocytosis:
   - Red cell fragments, spherocytes and budding red cells (the red cells are exquisitely sensitive to temperature and fragment after 10 minutes of incubation time at 45–46°C *in vitro*; heating for 6 hours at 37°C explains *in vivo* formation of fragmented red cells and chronic hemolysis)
   - Hemoglobin level, 7–9 g/dl
   - Marked reduction in MCV and elevated MCHC.
2. Jaundice.
3. Splenomegaly.
4. Osmotic fragility and autohemolysis increased.
5. Mild HE present in one of the parents or siblings.

Differential Diagnosis

Similar cells are seen in microangiopathic hemolytic anemias, after severe burns or oxidant stress and in pyruvate kinase deficiency.

Treatment

In infancy these patients require intermittent transfusion for hypoplastic crises. Patients respond well to splenectomy with a rise in hemoglobin to 12 g/dl. Following splenectomy, hemolysis is decreased but not totally eliminated.

Hereditary Stomatocytosis

Definition and Genetics

The stomatocyte has a linear slit-like area of central pallor rather than a circular area. When suspended in plasma, the cells assume a bowl-shaped form. This hereditary hemolytic anemia of variable severity is characterized by an autosomal dominant mode of inheritance. There are two forms of this inherited disorder related to failure to maintain normal red cell volume: over-hydrated stomatocytosis (previously referred to as "hereditary stomatocytosis") and dehydrated stomatocytosis (previously referred to as "hereditary desicytosis or xerocytosis"). The latter is characterized by a relative paucity of stomatocytes with the appearance of cells that appear very hyperchromic.

Etiology

The cells contain high Na^+ and low K^+ concentrations. The disorder is probably due to a membrane and protein defect. Although both forms share the relative increase in red cell sodium, over-hydrated stomatocytosis is associated with an increase in red cell volume as the total cation content increases from unbridled sodium entry while dehydrated stomatocytosis has a reduced red cell volume as the potassium cation loss is not matched by sodium accumulation. The cells are abnormally rigid and poorly deformable, contributing to their rapid rate of destruction. There are many biochemical variants. The properties of the stomatocytosis syndromes are listed in Table 7-5.

Clinical Features

Over-Hydrated Stomatocytosis
1. Very variable.
2. Jaundice at birth.
3. Pallor: marked variability depending on severity of anemia.
4. Splenomegaly.
5. Hematology
   a. Anemia
   b. Smear, 10–50% stomatocytes

Table 7-5   Properties of the Stomatocytosis Syndromes

	Severe Stomatocytosis	Mild Stomatocytosis	Cryothdrocytosis	Xerocytosis
Hemolysis	Severe	Mild–moderate	Mild–moderate	Moderate
Smear	Stomatocytes	Stomatocytes	Stomatocytes	Target cells
MCV fl.	110–150	95–130	90–105	85–125
MCHC %	24–30	26–29	34–38	34–38
Osmotic fragility	Very increased	Increased	Normal/slightly increased	Very decreased
RBC Na^+	60–100	30–60	6–25 at 20°C	10–30
RBC K^+	20–55	40–85	55–90 at 20°C	60–90
Cation leak[a]	10–50	~3–10	2–10 at 20°C	2–4
Cold lysis	No	No	Yes	No
Pseudohyper K^+	? Yes	? Yes	Yes	Occasionally
Perinatal ascites	No	No	No	Occasionally
Genetics	AD	AD	AD	AD

[a]Times normal value.
Table provided by Dr. Samuel Lux, personal communication, 2009.
Abbreviation: AD, autosomal dominant.

   c. Reticulocytosis
   d. Increased MCV
   e. Decreased MCHC
   f. Increased osmotic fragility and autohemolysis.

Dehydrated Stomatocytosis
1. Mild anemia.
2. Variable neonatal presentation.
3. Splenomegaly and gallstones.
4. Mild increase of MCV.
5. Increased MCHC.
6. Decreased osmotic fragility (i.e. osmotic resistance).
7. Increased heat stability (46 and 49°C for 60 minutes).

Differential Diagnosis

Stomatocytosis morphology may occur with thalassemia, some red cell enzyme defects (glutathione peroxidase deficiency, glucose phosphate isomerase deficiency), Rh_{null} red cells, viral infections, lead poisoning, some drugs (e.g., quinidine and chlorpromazine), some malignancies, liver disease and alcoholism. Dehydrated stomatocytosis syndrome resembles pyruvate kinase deficiency and infantile pyknocytosis on blood smear.

Treatment

Most patients have mild to moderate hemolysis that occasionally requires transfusion. Currently splenectomy should be avoided in these syndromes as there seems to be a consistent finding of significant venous thromboembolic complications post splenectomy in these disorders.

Hereditary Acanthocytosis

Definition

Acanthocytes have thorn-like projections that vary in length and width and are irregularly distributed over the surface of red cells. There are apparently a number of genetic syndromes associated with acanthocytosis and their molecular basis is not yet well defined.

Genetics

The mode of inheritance is autosomal recessive.

Clinical Features

1. *Steatorrhea*: In cases when acanthocytosis is associated with severe fat malabsorption.
2. *Neurologic symptoms*: Weakness, ataxia and nystagmus, atypical retinitis pigmentosa with macular atrophy, blindness.
3. *Anemia*: Mild hemolytic anemia; 10–80% acanthocytes; slight reticulocytosis.

Diagnosis

1. Inherited acanthocytosis is associated with the following clinical syndromes:
   • Abetalipoproteinemia (absent beta-lipoprotein in blood)
   • Chorea-acanthocytosis
   • Huntington disease-like 2
   • Pantothenate kinase-associated neurodegeneration
   • HARP syndrome (hypo-betaliproteinemia, acanthocytosis, retinitis pigmentosa and pallidal degeneration)
   • McLeod syndrome (X-linked anomaly of Kell blood group syndrome).
2. Acquired acanthocytosis is associated with the following clinical conditions:
   • Anorexia nervosa
   • Renal failure
   • Microangiopathic hemolytic anemia
   • Subgroup of hereditary spherocytosis
   • Thyroid disease
   • Liver disease: When associated with liver disease, the acanthocytosis is due to an imbalanced loading of cholesterol and phospholipid on to the red cell membrane. Hemolysis may be more brisk in this situation.

Differential Diagnosis

During the neonatal period, hereditary acanthocytosis may have to be distinguished from the benign nonhereditary disorder of infantile pyknocytosis. Later, acquired causes of acanthocytosis must be considered.

PAROXYSMAL NOCTURNAL HEMOGLOGINURIA

Paroxysmal nocturnal hemoglobinuria (PNH) is characterized by a nonmalignant clonal expansion of hematopoietic stem cells that are mutated at *PIGA*. *PIGA* encodes the glycosyl phosphatidlinositol (GPI) anchor, the mutation of which results in a deficiency of GPI-anchor proteins. Many of these are complement regulatory surface proteins, a deficiency of which results in hemolytic anemia by increasing sensitivity to complement-induced hemolysis.

Pathogenesis

Patients with PNH have a somatic mutation in the PIG-A gene (phosphatidylinositol glycan complementation group A).

This mutation occurs in primitive hematopoietic stem cells.

A protein product (probably $\propto$-1,6N-acetylglucosamine transferase) of the PIG-A gene is normally responsible for the transfer of N-acetylglucosamine to phosphatidylinositol. In patients with PNH, there is a mutation in the PIG-A gene, which results in a decrease in its protein product and leads to a metabolic block in the biosynthesis of the glycolipid (i.e., GPI) anchor. This anchoring molecule is required for several surface proteins of the hematopoietic cells.

Table 7-6 lists the surface proteins missing on PNH blood cells as a result of a deficiency in the GPI anchor. Thus, the primary defect in PNH resides in the deficient assembly of the GPI anchor and, as a result, all GPI-linked antigens are absent on the surface of PNH cells.

Mechanism of Hemolysis and Hemoglobinuria in Paroxysmal Nocturnal Hemoglobinuria

The absence of surface complement-regulatory proteins, namely CD55 and CD59, allows deposition of complement factors and C3 convertase complexes, which leads to chronic complement-mediated intravascular hemolysis, resulting in hemoglobinuria.

Mechanism of Hypercoagulable State

The mechanism of a hypercoagulable state in PNH is not well understood. A theory is that complement deposition on platelets results in vesiculations of their plasma membranes,

Table 7-6   Surface Proteins Missing on Paroxysmal Nocturnal Hemoglobinuria Blood Cells

Protein	Expression Pattern
Enzymes	
Acetylcholinesterase (AChE)	Red blood cells
5′-ectonucleotidase (CD73)	Some B and T lymphocytes
Leukocyte alkaline phosphatase	Neutrophils
Adhesion molecules	
Blast-1/CD48	Lymphocytes
Lymphocyte function-associated antigen-3 (LFA-3 or CD58)	All blood cells[a]
Complement regulating surface proteins	
Decay accelerating factor (DAF or CD55)	All blood cells[b]
Homologous restriction factor (HRF or C8bp)	All blood cells[c]
Membrane inhibitor of reactive lysis (MIRL or CD59)	All blood cells
Receptors	
Fcγ receptor III (Fcγ III or CD16)	Neutrophils, NK cells[d], Macrophages[d], some T lymphocytes[d]
Endotoxin binding protein (CD14)	Monocytes, macrophages, granulocytes
Urokinase-type plasminogen activator receptor (CD87)	Monocytes, granulocytes
Blood group antigens	
Comer antigens (DAF)	Red blood cells
Yt antigens (AChE)	Red blood cells
Holley Gregory antigen	Red blood cells
John Milton Hagen (JMH) bearing protein (CD 108)	Red blood cells, lymphocytes
Dombrock residue	Red blood cells
Neutrophil antigens	
NA1/NA2 (CD16)	Neutrophils
NB1/NB2	Neutrophils
Other surface proteins	Various
CD52 (CAMPATH)　　CD109	
CD24　　　　　　　CD157	
CD48　　　　　　　GP500	
CD66c　　　　　　　GP175	
CD67　　　　　　　Folate receptor	
CD90	

[a]On lymphocytes expressed in GPI-linked and transmembrane form.
[b]Level of expression on T lymphocytes varies.
[c]Expression of C8bp on human blood cells is controversial (personal communication, Taroh Kinoshita).
[d]Expressed in a transmembrane form.
From: Young NS, Bressler M, Casper JT, Liu J. Biology and therapy of aplastic anemia. In: Schacter GP, McArthur TR, editors. Hematology 1996. American Society of Hematology, 1996; with permission; and Ware RE. Autoimmune hemolytic anemia. In: Nathan and Oski's Hematology of Infancy and Childhood 6th Edition, Eds Nathan DG, Orkin SH, Ginsburg D, Look TA, Saunders 2003, with permission.

which leads to increased procoagulant activity of the platelets. The monocytes and granulocytes of PNH cells lack the receptor for the GPI-linked urokinase plasminogen activator and this deficiency may lead to impaired fibrinolysis.

The anti-thrombin (AT), protein C and protein S levels are normal in PNH patients.

Mechanism of Defective Hematopoiesis

The mechanism of defective hematopoiesis (macrocytosis with bone marrow erythroid dysplasia) evolving to severe aplastic anemia in some patients is not well understood. However, the following explanations have been considered:

- The initial step is the development of the PIG-A mutation. This is followed by a bone marrow insult
- Resistance of PNH clones to injury by the insulting agents compared with susceptibility of normal hematopoietic stem cells
- Intrinsic proliferation advantage of PNH stem cells compared with normal hematopoietic stem cells results in selection of abnormal stem cells followed by clonal expansion
- Suppression of normal hematopoietic stem cells by PNH cells and evolution to MDS or AML.

In the preceding explanation, it is assumed that two populations of stem cells normally reside in bone marrow: (1) a large population of normal stem cells; and (2) a minor population of PNH stem cells.

Clinical Manifestations

The three main clinical features of PNH are:

- Paroxysmal intravascular hemolysis more frequent at night associated with hemoglobinuria and abdominal and back pain. In most cases hemolytic episodes occur every few weeks although some patients have chronic unrelenting hemolysis with severe anemia
- Bone marrow failure (macrocytosis, pancytopenia to severe aplastic anemia)
- Tendency to venous thrombosis.

PNH can present as a primary "classic" intravascular hemolysis or it may arise during the course of aplastic anemia (AA) as AA-PNH syndrome. The nature of the pathogenetic link between the two conditions remains unknown. They may be differentiated from each other by the clinical findings shown in Table 7-7.

Many patients have an overlap of the aforementioned findings and do not fit precisely into one of these two groups.

Course of the Disease

The onset of PNH is insidious. There is no familial tendency. Venous thrombosis is more often responsible for death than bone marrow failure in patients with PNH. Spontaneous long-term remission or leukemia transformation or aplastic anemia may occur in some

Table 7-7   Clinical Findings in Classic PNH Syndrome and in Aplastic Anemia–PNH Syndrome

Findings	Classic PNH Syndrome	Aplastic Anemia–PNH Syndrome

Hemolysis
Thrombotic
   complications
	Chronic with acute exacerbation
	More often present
	   Acute hemolysis may be preceded by
	   abdominal pain, thought to be due to
	   temporary occlusion of the
	   gastrointestinal veins. Thrombosis of
	   larger abdominal veins may be present
		Hemolysis clinically subtle
		Occurs less frequently
		   Bone marrow failure
		   predominant clinical finding

Abnormal erythrocyte
   or granulocyte
   CD55/CD59
	Positive from the time of diagnosis	Positive in 20–50% of patients
		with SAA. May evolve post
		immunosuppressive therapy

Abbreviation: SAA, Severe aplastic anemia.

patients. Anemia is the most common finding and aplastic anemia is found in approximately 10% of patients.

Patients with classic PNH may have cytopenia of one or all blood cell lineages and the degree of bone marrow failure may vary from mild to severe. About 15% of patients with aplastic anemia develop overt PNH; however, 35–50% of aplastic anemia patients may have flow cytometric evidence of deficiency of GPI-linked molecules at some stage of their disease as evidence of subclinical PNH.

Complications

Intravascular Hemolysis (DAT Negative)

- Hemoglobinuria (dark urine)
- Iron deficiency
- Acute renal failure.

Venous Thrombosis

- Peripheral veins
- Superior and inferior vena cava
- Hepatic veins (Budd–Chiari syndrome)
- Mesenteric veins
- Sagittal sinus
- Splenic vein
- Abdominal wall veins
- Intrathoracic veins.

Defective Hematopoiesis

- Aplastic anemia
- Macrocytosis
- Evolution to myelodysplasic or AML.

Infectious

- Sinopulmonary
- Blood borne.

Other

- Dysphagia.

Table 7-8 lists the laboratory findings in PNH.

Diagnosis

Flow cytometric analysis of GPI-linked molecules: Flow cytometric analysis of blood cells with the use of monoclonal antibodies to GPI-linked surface antigens is a very sensitive method for the diagnosis of PNH and has replaced the Ham test.

All blood cell lineages (i.e., red blood cells, lymphocytes, monocytes, granulocytes) can be analyzed by the flow cytometric technique. Heterogeneous patterns of the phenotypic expressions of various blood cells can be identified with the flow cytometric technique. For example, red blood cell phenotypes can be identified by their CD59 expression:

Table 7-8 Laboratory Findings in Paroxysmal Nocturnal Hemoglobinuria

Nonspecific findings:	Cytopenia involving one or more cell lineages
	Macrocytosis, anisocytosis, polychromasia
	Reticulocytosis
	Decreased neutrophil alkaline phosphatase
	Increased level of lactate dehydrogenase
	Decreased haptoglobin
	Hemoglobinuria, hemosiderinuria
	Iron deficiency, folate deficiency
Bone marrow findings:	Varies from hyperplastic with predominant erythropoiesis to hypoplastic with little or patchy hematopoiesis
	Hypoplasia or aplasia of one or more hematopoietic lineages
	Increased number of mast cells
Cytogenesis:	Usually normal
Specific test for PNH:	Flow cytometric analysis for glycosyl phosphatidylinositol (GPI)-linked cell surface proteins (e.g., CD59) on peripheral blood or bone marrow cells

Adapted from: Young NS, Bressler M, Casper JT, Liu J. Biology and therapy of aplastic anemia. In: Schacter GP, McArthur TR, editors. Hematology 1996. American Society of Hematology, 1996; with permission.

PNH type I=Normal expression of CD59
PNH type II=Partially deficient or residual expression of CD59
PNH type III=Complete absence of expression of CD59.

The proportion of the three different phenotypes may vary from patient to patient.

Because other blood cell lineages can be analyzed, the transfusion of red blood cells to a patient does not interfere with the diagnosis of PNH.

The percentage of granulocytes with a PNH phenotype is usually higher than the percentage of red cells lacking CD59. Thus, flow cytometric analysis of the granulocytes increases sensitivity in the diagnosis of PNH.

Management

The most common manifestation of PNH is hemolytic anemia but thromboembolism is the leading cause of death. The etiology of bone marrow failure in PNH appears to either result in or be the consequence of a selective advantage of the PNH clone. So, aside from stem cell transplantation (for bone marrow failure or severe hemolytic anemia or life-threatening thromboembolic disease) or immunosuppressive therapy for bone marrow failure the therapy is now directed to the resolution of hemolysis.

Ecluzumad

Recently, ecluzumab, a humanized monoclonal antibody that blocks complement activation at C5 preventing the formation of C5a, has been shown to dramatically reduce hemolysis and thromboembolism and dramatically improve the quality of life for patients with PNH and has become the standard of care for PNH. Due to the importance of complement in immunity against *Neisseria meningitidis* patients receiving ecluzumab must be vaccinated.

Corticosteroids

Prednisone 1–2 mg/kg daily can ameliorate hemolysis and is often recommended for 24–72 hours around the time of a hemolytic episode.

Hematopoietic Stem Cell Transplantation

Hematopoietic stem cell transplantation (HSCT) is the only curative treatment for PNH. If a fully matched family donor is available, then HSCT is the treatment of choice, especially for patients who develop bone marrow failure. In the absence of a matched unrelated donor alternative donor transplantations can be considered based on the quality of the available alternative donor and the severity of the PNH.

Immunosuppressive Therapy

Therapy with cyclosporine and ATG is indicated in the setting of PNH-associated aplastic anemia. This treatment may lead to improvement in aplastic anemia but not in the hemolysis of PNH.

Use of Hematopoietic Growth Factor

The use of G-CSF may be attempted in the setting of a pertinent cytopenia.

Supportive Therapy

- Long-term anticoagulant therapy (e.g., with warfarin) is indicated for patients with venous thrombosis. Also, women with PNH should be discouraged from using birth control pills
- Iron and folate supplements are indicated due to chronic hemoglobinuria accompanied by iron loss and chronic hemolysis with increased erythroid marrow activity requiring supplementation of additional folate
- Sidenafil may be effective in treating dysphagia and intestinal spasm and impotence, which are the consequence of decreased nitric oxide consumed by passive quantities of plasma free hemoglobin
- Red blood cell transfusion as needed for symptomatic anemic patients.

ENZYME DEFECTS

There are two major biochemical pathways in the red cell: the Embden–Myerhof anaerobic pathway (energy potential of the cell) and the hexose monophosphate shunt (reduction potential of the cell). Figure 7-3 illustrates the enzyme reactions in the red cell.

Pyruvate Kinase Deficiency

Pyruvate kinase (PK) is an enzyme active in the penultimate conversion in the Embden–Meyerhof pathway. Although deficiency is rare, it is the most common enzyme abnormality in the Embden–Meyerhof pathway.

Genetics

1. Autosomal recessive inheritance.
2. Significant hemolysis seen in homozygotes.
3. Found predominantly in people of northern European origin.
4. Deficiency not simply quantitative; probably often reflects the production of PK variants with abnormal characteristics.

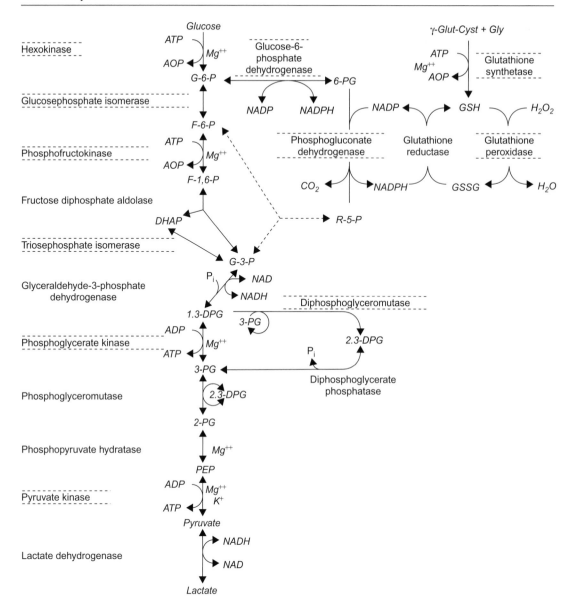

Figure 7-3 Enzyme Reactions of Embden–Meyerhof and Hexose Monophosphate Pathways of Metabolism.
Documented Hereditary Deficiency Diseases are Indicated by Enclosing Dotted Lines.

Pathogenesis

1. Defective red cell glycolysis with reduced ATP formation.
2. Red cells rigid, deformed and metabolically and physically vulnerable (reticulocytes less vulnerable because of ability to generate ATP by oxidative phosphorylation).

Hematology

1. Features of nonspherocytic hemolytic anemia: macrocytes, oval forms, polychroma-tophilia, anisocytosis, occasional spherocytes, contracted red cells with multiple projecting spicules, rather like echinocytes or pyknocytes.
2. Erythrocyte PK activity decreased to 5–20% of normal; 2,3-diphosphoglycerate (2,3-DPG) and other glycolytic intermediary metabolites increased (because of two- to threefold increase in 2,3-DPG, there is a shift to the right in P_{50}).[*]
3. Autohemolysis markedly increased, showing marked correction with ATP but not with glucose.

Clinical Features

1. Variable severity; can cause moderately severe anemia (not drug induced). Patients may tolerate their anemia better because of the increase in 2,3-DPG shifting the hemoglobin oxygen dissociation curve to the right. This leads to superior off-loading of oxygen to the tissues and may mitigate the anemia.
2. Usually presents with neonatal jaundice.
3. Splenomegaly common but not invariable.
4. Late: gallstones, bone changes of chronic hemolytic anemia, cardiomegaly secondary to severe anemia.
5. Erythroblastopenic crisis due to parvovirus B19 infection.
6. Hemochromatosis. These patients seem to have a risk of hemochromatosis beyond the number of transfusions they received. Careful attention should be paid to their iron loading.

Treatment

1. Folic acid supplementation.
2. Transfusions as required.
3. Splenectomy (if transfusion requirements increase); splenectomy does not arrest hemolysis, but decreases transfusion requirements. Note that there is a paradoxical increase in reticulocytosis after splenectomy even as transfusion requirement and hemolytic rate abate.
4. Surveillance for iron overload.

Other Enzyme Deficiencies

1. Hexokinase deficiency, with many variants.
2. Glucose phosphate isomerase deficiency.

---

[*] Because of the right shift of P50, patients do not exhibit fatigue and exercise intolerance proportionate to the degree of anemia.

3. Phosphofructokinase deficiency, with variants.
4. Aldolase.
5. Triosephosphate isomerase deficiency.
6. Phosphoglycerate kinase deficiency.
7. 2,3-DPG deficiency due to deficiency of diphosphoglycerate mutase.
8. Adenosine triphosphatase deficiency.
9. Enolase deficiency.
10. Pyrimidine 5′-nucleotidase deficiency.
11. Adenosine deaminase overexpression.
12. Adenylate kinase deficiency.

These enzyme deficiencies have the following features:

1. General hematologic features:
   - Autosomal recessive disorders except phosphoglycerate kinase deficiency, which is sex linked and Enolase deficiency which presents as an autosomal dominant
   - Chronic nonspherocytic hemolytic anemias (CNSHAs) of variable severity
   - Osmotic fragility and autohemolysis normal or increased
   - Improvement in anemia after splenectomy
   - Diagnosed by specific red cell assays.
2. Specific nonhematologic features:
   - Phosphofructokinase deficiency associated with type VII glycogen storage disease. Hematologic symptoms are mild compared to the significant myopathy
   - Triosephosphate isomerase deficiency associated with progressive debilitating neuromuscular disease with generalized spasticity and recurrent infections (some patients have died of sudden cardiac arrest)
   - Phosphoglycerate kinase deficiency associated with mental retardation, myopathy and a behavioral disorder.

Note the three exceptions to the general hematologic features listed above:

- Adenosine deaminase excess (i.e., not an enzyme deficiency) is an autosomal dominant disorder
- Pyrimidine 5′-nucleotidase deficiency is characterized by marked basophilic stippling, although the other chronic nonspherocytic hemolytic anemias lack any specific morphologic abnormalities
- Deficiency of diphosphoglycerate mutase results in polycythemia.

Glucose-6-Phosphate Dehydrogenase Deficiency

Glucose-6-phosphate dehydrogenase (G6PD) is the first enzyme in the pentose phosphate pathway of glucose metabolism. Deficiency diminishes the reductive energy of the red cell and may result in hemolysis, the severity of which depends on the quantity and type of G6PD and the nature of the hemolytic agent (usually an oxidation mediator that can oxidize NADPH, generated in the pentose phosphate pathway in red cells).

Genetics

1. Sex-linked recessive mode of inheritance by a gene located on the X chromosome (similar to hemophilia).
2. Disease is fully expressed in hemizygous males and homozygous females.
3. Variable intermediate expression is shown by heterozygous females (due to random deletion of X chromosome, according to Lyon hypothesis).
4. As many as 3% of the world's population is affected; most frequent among African-American, Asian and Mediterranean peoples.

The molecular basis of G6PD deficiency and its clinical implications follow:

- Deletions of G6PD genes are incompatible with life because it is a housekeeping gene and complete absence of G6PD activity, called hydeletions, will result in death of the embryo
- Point mutations are responsible for G6PD deficiencies. They result in:
  - *Sporadic mutations*: They are not specific to any geographic areas. The same mutation may be encountered in different parts of the world that have no causal (e.g., encountering G6PD Guadalajara in Belfast) relationship with malarial selection. These patients manifest with chronic nonspherocytic hemolytic anemia (CNSHA WHO Class I)
  - *Polymorphic mutations*: These mutations have resulted from malaria selection; hence, they correlate with specific geographic areas. They are usually WHO Class II or III and not Class I.

The World Health Organization (WHO) classifies G6PD variants on the basis of magnitude of the enzyme deficiency and the severity of hemolysis (Table 7-9).

Table 7-9 WHO Classification of G6PD Variants

WHO Class	Variant	Magnitude of Enzyme Deficiency	Severity of Hemolysis
I	Harilaou, Tokyo, Guadalajara, Stonybrook, Minnesota	2% of normal activity	Chronic non-spherocytic hemolytic anemia
II	Mediterranean	3% of normal activity	Intermittent hemolysis
III	A⁻	10–60% of normal activity	Intermittent hemolysis usually associated with infections or drugs
IV	B (Normal)	100% normal activity	No hemolysis

Pathogenesis

1. Red cell G6PD activity falls rapidly and prematurely as red cells age.
2. Decreased glucose metabolism.
3. Diminished NADPH/NADP and GSH/GSSG ratios.
4. Impaired elimination of oxidants (e.g., H_2O_2).
5. Oxidation of hemoglobin and of sulfhydryl groups in the membrane.
6. Red cell integrity impaired, especially on exposure to oxidant drugs, oxidant response to infection and chemicals.
7. Oxidized hemoglobin precipitates to form Heinz bodies which are plucked out of the red cell leading to hemolysis and "bite cell" and "blister cell" morphology.

Clinical Features

Episodes of hemolysis may be produced by:

- Drugs. Table 7-10 lists the agents capable of inducing hemolysis in G6PD-deficient subjects
- Fava bean (broad bean, *Vicia fava*): ingestion or exposure to pollen from the bean's flower (hence favism)
- Infection.

1. Drug-induced hemolysis
   a. Typically in African-Americans but also in Mediterranean and Canton types
   b. List of drugs (see Table 7-6); occasionally need additional stress of infection or the neonatal state
   c. Acute self-limiting hemolytic anemia with hemoglobinuria
   d. Heinz bodies in circulating red cells
   e. Blister cells, fragmented cells and spherocytes
   f. Reticulocytosis
   g. Hemoglobin normal between episodes.
2. Favism
   a. Acute life-threatening hemolysis, often leading to acute renal failure caused by ingestion of fava beans
   b. Associated with Mediterranean and Canton varieties
   c. Blood transfusion required.
3. Neonatal jaundice
   a. Usually associated with Mediterranean and Canton varieties but can occur with all variants
   b. Infants may present with pallor, jaundice (can be severe and produce kernicterus[*]) and dark urine.

---

[*] The excessive jaundice is not only due to hemolysis but may be due to reduced glucuronidation of bilirubin caused by defective G6PD activity in the hepatocytes.

Table 7-10 Agents Capable of Inducing Hemolysis in G6PD-deficient Subjects[a]

Clinically Significant Hemolysis	Usually not Clinically Significant Hemolysis
Analgesics and antipyretics	
Acetanilid	Acetophenetidin (phenacetin)
	Acetylsalicylic acid (large doses)
	Antipyrine[a,b]
	Aminopyrine[b]
	p-Aminosalicylic acid
Antimalarial agents	
Pentaquine	Quinacrine (Atabrine)
Pamaquine	Quinine[b]
Primaquine	Chloroquine[c]
Quinocide	Pyrimethamine (Daraprim)
	Plasmoquine
Flouroquinones	
Ciprofloxacin	
Sulfonamides	
Sulfanilamide	Sulfadiazine
N-Acetylsulfanilamide	Sulfamerazine
Sulfapyridine	Sulfisoxazole (Gantrisin)[c]
Sulfamethoxypyridazine (Kynex)	Sulfathiazole
Salicylazosulfapyridine (Azulfidine)	Sulfacetamide
Nitrofurans	
Nitrofurazone (Furacin)	
Nitrofurantoin (Furadantin)	
Furaltadone (Altafur)	
Furazolidone (Furoxone)	
Sulfones	
Thiazolsulfone (Promizole)	
Diaminodiphenylsulfone (DDS, dapsone)	Sulfoxone sodium (Diasone)
Miscellaneous	
Naphthalene	
Phenylhydrazine	Menadione
Acetylphenylhydrazine	Dimercaprol (BAL)
Toluidine blue	Methylene blue
Nalidixic acid (NegGram)	Chloramphenicol[b]
Neoarsphenamine (Neosalvarsan)	Probenecid (Benemid)
Infections	Quinidine[b]
Diabetic ketoacidosis	Fava beans[b]
Doxorubicin	
Urate Oxidase (Rasburicase)	
Foods	

[a]Many other compounds have been tested but are free of hemolytic activity. Penicillin, the tetracyclines and erythromycin, for example, will not cause hemolysis and the incidence of allergic reactions in G6PD deficient persons is not any greater than that observed in others.

[b]Hemolysis in whites only.

[c]Mild hemolysis in African-Americans, if given in large doses.

Note: Drugs associated with hemolysis in any WHO class are listed as clinically significant.

Often no exposure to drugs; occasionally exposure to naphthalene (mothballs), aniline dye, marking ink, or a drug. In a majority of neonates, the jaundice is not hemolytic but hepatic in origin.

4. Chronic nonspherocytic hemolytic anemia
    a. Occurs mainly with sporadic inheritance
    b. Clinical picture
        (1) Chronic nonspherocytic anemia variable but can be severe with transfusion dependence
        (2) Reticulocytosis
        (3) Intense neonatal presentation
        (4) Shortened red cell survival
        (5) Increased autohemolysis with only partial correction by glucose
        (6) Slight jaundice
        (7) Mild splenomegaly.

Treatment

1. Avoidance of agents that are deleterious in G6PD deficiency. (For a consistent, up-to-date list of drug susceptibilities visit: www.favism.org).
2. Education of families and patients in recognition of food prohibition (fava beans), drug avoidance, heightened vigilance during infection and the symptoms and signs of hemolytic crisis (orange/dark urine, lethargy, fatigue, jaundice).
3. Indication for transfusion of packed red blood cell in children presenting with acute hemolytic anemia:
    a. Hemoglobin (Hb) level below 7 g/dl
    b. Persistent hemoglobinuria and Hb below 9 g/dl.
4. Chronic nonspherocytic hemolytic anemia (NSHA):
    *   In patients with severe chronic anemia: transfuse red blood cells to maintain Hb level 8–10 g/dl and iron chelation, when needed
    *   Splenectomy has only occasionally ameliorated severe anemia in this disease. Indications for splenectomy are as follows:
        -   Hypersplenism
        -   Severe chronic anemia
        -   Splenomegaly causing physical impediment
    *   Genetic counseling and prenatal diagnosis for severe CNSHA if the mother is a heterozygote.

Other Defects of Glutathione Metabolism

Glutathione Reductase

In this autosomal dominant disorder, hemolytic anemia is precipitated by drugs having an oxidant action. Thrombocytopenia has occasionally been reported. Neurologic symptoms occur in some patients. The disease is mimicked by riboflavin deficiency.

Glutamyl Cysteine Synthetase

In this autosomal recessive disorder, there is a well-compensated hemolytic anemia. This very rare disease has been associated with spinocerebellar degeneration in one patient.

Glutathione Synthetase

In this autosomal recessive disorder, there is a well-compensated hemolytic anemia, exacerbated by drugs having an oxidant action. This is the most common disorder of the group and can also present as a systemic metabolic disorder with acidosis, hemolysis and susceptibility to infection.

Glutathione Peroxidase

In this autosomal recessive disorder, acute hemolytic episodes occur after exposure to drugs having an oxidant action.

Suggested Reading

An X, Mohandas N. Disorders of red cell membrane. Br J Haematol. 2008;141:367–375.

Becker P, Lux S. Disorders of the red cell membrane. In: Nathan D, Oski F, eds. Hematology of Infancy and Childhood. Philadelphia: WB Saunders; 1993.

Bolton-Maggs PH, Stevens RF, et al. on behalf of the General Haematology Task Force of the British Committee for the Standards in Haematology. Guidelines for the diagnosis and management of Hereditary Spherocytosis. Br J Haematol. 2004;126:455–474.

Dacie J. The Haemolytic Anaemias 3. The Auto-Immune Haemolytic Anaemias. 3rd ed Edinburgh: Churchill Livingstone; 1992.

Hillmen P, Lewis SM, Bressler M, et al. Natural history of paroxysmal nocturnal hemoglobinuria. N Engl J Med. 1995;333:1253–1258.

King M-J, Behrens, et al. Rapid flow cytometric test for the diagnosis of membrane cytoskeleton-associated haemolytic anemia. Br J Haematol. 2000;111:924–933.

Lanzkowsky P. Hemolytic anemia. Pediatric Hematology Oncology: A Treatise for the Clinician. New York: McGraw-Hill; 1980.

Tracy E, Rice H. Partial Splenectomy for Hereditary Spherocytosis. Ped Clin North Am. 2008;55:503–519.

Wolfe L, Manley P. Disorders of erythrocyte metabolism including porphyria. In: Arceci R, Hann I, Smith O, eds. Pediatric Hematology. Boston: Blackwell Publishing Ltd.; 2006.

CHAPTER 8

Hemoglobinopathies

SICKLE CELL DISEASE

Incidence

Sickle hemoglobin is the most common abnormal hemoglobin found in the United States (approximately 8% of the African-American population has sickle cell trait). The incidence of sickle cell disease (SCD) at birth is approximately 1 in 600 African-Americans.

Genetics

1. Sickle cell disease is transmitted as an autosomal co-dominant trait.
2. Homozygotes (two abnormal genes, SS) do not synthesize hemoglobin A (Hb A); beyond infancy, red cells contain >75% hemoglobin S (Hb S).
3. Heterozygotes (one abnormal gene), sickle cell trait, have red cells containing 20–45% Hb S.
4. Sickle cell trait provides selective advantage against *Plasmodium falciparum* malaria (balanced polymorphism).
5. $\propto$-thalassemia (frequency of one or two $\propto$ gene deletions is 35% in African-Americans) may be co-inherited with sickle cell trait or disease. Individuals who have both $\propto$-thalassemia and sickle cell disease-SS tend to be less anemic than those who have sickle cell disease-SS alone. The co-inheritance of sickle cell disease and alpha thalassemia trait is associated with a reduction in the risk of some complications, such as stroke, but has no effect on the frequency or severity of vaso-occlusive pain.

Results of DNA polymorphisms linked to the β^s gene suggest that it arose from five independent mutations, four in tropical Africa and one in the Arabian-Indian sub-continent:

- Benin-Central West African haplotype (the most common haplotype)
- Senegal-African West Coast haplotype

Manual of Pediatric Hematology and Oncology. DOI: 10.1016/B978-0-12-375154-6.00008-2

- Bantu–Central African Republic (CAR) haplotype
- Cameroon haplotype
- Arab-Indian haplotype.

Pathophysiology

Hemoglobin S arises as a result of a point mutation (A–T) in the sixth codon of the β-globin gene on chromosome 11, which causes a single amino acid substitution (glutamic acid to valine at position 6 of the β-globin chain). Hemoglobin S is more positively charged than Hb A and hence has a different electrophoretic mobility. Deoxygenated hemoglobin S polymerizes, leading to cellular alterations that distort the red cell into a rigid, sickled form. Vaso-occlusion with ischemia–reperfusion injury is the central event, but the underlying pathophysiology is complex, involving a number of factors including hemolysis-associated reduction in nitric oxide bioavailability, chronic inflammation, oxidative stress, altered red cell adhesive properties, activated white blood cells and platelets and increased viscosity. The following mechanisms are thought to be involved:

- Sickle cells are prematurely destroyed, causing hemolytic anemia
- Intravascular hemolysis reduces nitric oxide (NO) bioavailability by the following mechanisms (Fig. 8-1):
    - Release of arginase from the red cells consumes plasma L-arginine, a substrate for NO production
    - Free plasma hemoglobin reacts with NO, producing methemoglobin and nitrate, thereby depleting NO
    - Increased xanthine oxidase and NADPH oxidase activity in sickle cell disease leads to production of free oxygen radicals that consume NO.
- NO normally regulates vasodilation, causing increased blood flow and inhibits platelet aggregation. Thus, reduced NO bioavailability is thought to contribute to vaso-constriction and platelet activation
- Adhesion molecules are overexpressed on sickle reticulocytes and mature red cells. Increased red cell adhesion reduces flow rate in the microvasculature, trapping red cells contributing to vaso-occlusion
- Sickle cells increase blood viscosity, which also contributes to vaso-occlusion
- Sickle red cells may damage the endothelium leading to production of inflammatory mediators. Ischemia–reperfusion also causes inflammation
- White blood cell counts are often elevated in sickle cell disease and these white cells have increased adhesive properties. White blood cells adhere to endothelial cells and may further trap sickled red cells, contributing to stasis

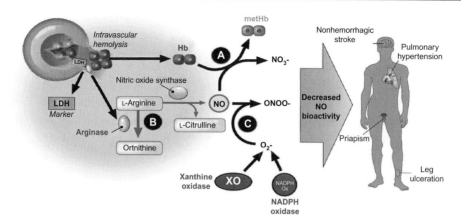

Figure 8-1 Intravascular Hemolysis Reduces Nitric Oxide Bioactivity.
Nitric oxide is produced by isoforms of nitric oxide (NO) synthase, using the substrate L-arginine. Intravascular hemolysis simultaneously releases hemoglobin, arginase and lactate dehydrogenase (LDH) from red cells into blood plasma. Cell-free plasma hemoglobin stochiometrically inactivates NO, generating methemoglobin and inert nitrate (A). Plasma arginase consumes plasma L-arginine to ornithine, depleting its availability for NO production (B). LDH also released from the red cell into blood serum serves as a surrogate marker for the magnitude of hemoglobin and arginase release. NO is also consumed by reactions with reactive oxygen species (O_2) produced by the high levels of xanthine oxidase activity and NADPH oxidase activity seen in sickle cell disease, producing oxygen radicals like peroxynitrite (ONOO−) (C). The resulting decreased NO bioactivity in sickle cell disease is associated with pulmonary hypertension, priapism, leg ulceration and possibly with nonhemorrhagic stroke. A similar pathobiology is seen in other chronic intravascular hemolytic anemias.
From: Kato GJ, Gladwin MT, Steinberg MH. Deconstructing sickle cell disease: Reappraisal of the role of hemolysis in the development of clinical subphenotypes. Blood Reviews. 2007;21, with permission.

- Activated platelets may interact with abnormal red cells, causing aggregation and vaso-occlusion
- Hemoglobin F affects HbS by decreasing polymer content in cells. The effect of HbF on HbS may have direct and indirect effects on other RBC characteristics (i.e. percentage of HbF affects the RBC adhesive properties in patients with SCD). Elevated HbF concentration is associated with a reduction in certain complications of sickle cell disease.

The relative role of hemolysis or viscosity/vaso-occlusion is postulated to differ among different subphenotypes of sickle cell disease (Figure 8-2). In particular, hemolysis and NO depletion are thought to play an important role in priapism, leg ulcers and pulmonary hypertension, while viscosity/vaso-occlusion is thought to be more central in the pathophysiology of vaso-occlusive pain and acute chest syndrome; however, considerable overlap exists.

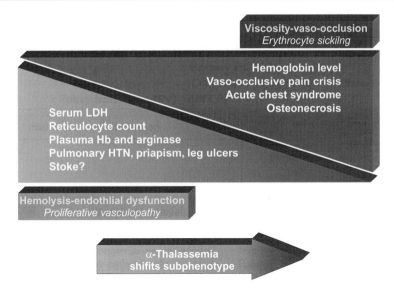

Figure 8-2 Model of Overlapping Subphenotypes of Sickle Cell Disease.
Published data suggest that patients with sickle cell disease with higher hemoglobin levels have a higher frequency of viscosity-vaso-occlusive complications closely related to polymerization of sickle hemoglobin, resulting in erythrocyte sickling and adhesion. Such complications include vaso-occlusive pain crisis, acute chest syndrome and osteonecrosis. In contrast, a distinct set of hemolysis-endothelial dysfunction complications involving a proliferative vasculopathy and dysregulated vasomotor function, including leg ulcers, priapism, pulmonary hypertension and possibly nonhemorrhagic stroke, is associated with low hemoglobin levels and high levels of hemolytic markers such as reticulocyte counts, serum lactate dehydrogenase, plasma hemoglobin and arginase, producing a state of impaired nitric oxide bioavailability. The spectrum of prevalence and severity of each of these subphenotypes overlap with each other. Patients with alpha-thalassemia trait tend to have less hemolysis and higher hemoglobin levels, tending to decrease the prevalence of hemolysis-endothelial dysfunction and tending to increase the prevalence of viscosity-vaso-occlusion. The effect of fetal hemoglobin expression or chronic red cell transfusion is more complex, simultaneously increasing hemoglobin level, but reducing sickling and hemolysis.
From: Kato GJ, Gladwin MT, Steinberg MH. Deconstructing sickle cell disease: Reappraisal of the role of hemolysis in the development of clinical subphenotypes. Blood Reviews. 2007;21, with permission.

Clinical Features

Hematology

1. Anemia – moderate to severe in SS and S-β^0thalassemia, milder with SC or Sβ^+ thalassemia.
2. MCV is normal with SS; MCV is reduced (microcytic cells) with concomitant α-thalassemia or with S-β thalassemia.
3. Reticulocytosis.

4. Neutrophilia common.
5. Platelet count often increased.
6. Blood smear – sickle cells (not in infants or others with high Hb F) increased polychromasia, nucleated red cells and target cells (Howell–Jolly bodies may indicate hyposplenism).
7. Erythrocyte sedimentation rate (ESR) – low (sickle cells fail to form rouleaux).
8. Hemoglobin electrophoresis – hemoglobin S migrates slower than hemoglobin A.

Newborn screening shows FS, FSC, or FSA pattern depending on genotype.

Acute Complications

1. **Vaso-occlusive pain event (VOE)**
   a. Episodic microvascular occlusion at one or more sites resulting in pain and inflammation. Common locations and manifestations of VOE are shown in Table 8-1. Symptoms of fever, erythema, swelling and focal bone pain may accompany VOE, making it difficult to distinguish from osteomyelitis. Unfortunately, no test clearly distinguishes these two entities. Table 8-2 describes clinical, laboratory and radiological features that may aid in differentiating bone infarction from osteomyelitis
   b. The average rate of VOE prompting medical evaluation in sickle cell disease-SS is 0.8 events/year. Approximately 40% of patients never seek medical attention for pain, while about 5% of patients account for a third of all VOE seeking medical attention. These numbers underestimate the true incidence of VOE because many episodes are managed at home

Table 8-1   Common Location of Vaso-Occlusive Pain

Site	Manifestations
Hands/feet (dactylitis)	Most common in children younger than 3 years old. Painful swelling of the dorsum of the hands and/or feet. Fever may be present. Often can be managed with acetaminophen or nonsteroidal anti-inflammatory medication. Unusual in older children because as the child ages, the sites of hematopoiesis move from peripheral location such as the fingers and toes to more central locations such as arms, legs, ribs and sternum
Bone	More common after age 3 years. Often involves long bones, sternum, ribs, spine and pelvis. May involve more than one site during a single episode. Swelling and erythema may be present. May be difficult to differentiate from osteomyelitis because clinical symptoms, laboratory studies and radiological imaging may be similar. Features that may aid in distinguishing these two diagnoses are shown in Table 8-2
Abdomen	Caused by microvascular occlusion of mesenteric blood supply and infarction in the liver, spleen, or lymph nodes that results in capsular stretching. Symptoms of abdominal pain and distension mimic acute abdomen

Table 8-2   Clinical, Laboratory and Radiological Features Differentiating
Bone Infarction from Osteomyelitis

Features	Favoring Osteomyelitis	Favoring Vaso-Occlusion
History	No previous history	Preceding painful episode
Pain, tenderness, erythema, swelling	Single site	Multiple sites
Fever	Present	Present
Leukocytosis	Elevated band count ($>1,000/mm^3$)	Present
ESR	Elevated	Normal to low
Magnetic resonance imaging	Abnormal	Abnormal
Bone scan[a]	Abnormal ^{99m}Tc-diphosphonate Normal ^{99m}Tc-colloid Marrow uptake	Abnormal ^{99m}Tc-diphosphonate Decreased ^{99m}Tc-colloid marrow uptake
Blood culture	Positive (*Salmonella*, *Staphylococcus*)	Negative
Recovery	Only with appropriate antibiotic therapy	Spontaneous

[a]Obtained within three days of symptom onset.

   c. Risk factors for pain include high baseline hemoglobin level, low hemoglobin F levels, nocturnal hypoxemia and asthma

   d. The approach to pain management involves a stepwise progression, beginning with a nonsteroidal anti-inflammatory pain medication and adding an opioid pain medication for moderate to severe pain. The management of vaso-occlusive pain is shown in Table 8-3 and a guideline for dosing of commonly utilized pain medications is provided in Table 8-4. Higher opioid dosing may be required for patients who have developed tolerance.

2. **Acute chest syndrome**

   a. Acute chest syndrome (ACS) is the most common cause of death and the second most common cause of hospitalization. It is defined as the development of a new pulmonary infiltrate accompanied by symptoms including fever, chest pain, tachypnea, cough, hypoxemia and wheezing

   b. Acute chest syndrome is caused by infection, infarction and/or fat embolization. About 50% of ACS events are associated with infections, including viruses, atypical bacteria including *Mycoplasma* and *Chlamydia* and less frequently with *Streptococcus pneumoniae*. Parvovirus B19 infection can also result in ACS. In about half of the cases, ACS develops during hospitalization (often for vaso-occlusive pain) where fat embolization and hypoventilation contribute to the pathophysiology

   c. The incidence of ACS in sickle cell disease-SS is about 24 events per 100 patients in children. The incidence in other sickle cell genotypes is lower (SS>Sβ^0-thalassemia>SC>Sβ$^+$-thalassemia) and concomitant α-thalassemia does not appear to affect ACS rates

Table 8-3 **Management of Vaso-Occlusive Pain Episodes**

At home:
  Ibuprofen and/or acetaminophen
  If continued pain, add oral opiod
    Mild pain – codeine
    Moderate pain – oxycodone, hydrocodone, morphine
  Supportive measures
    Heating pad
    Fluids
    Stool softeners and/or laxative if taking opiods for more than 1–2 days
    If pain persists or worsens, patient should be evaluated and treated in an acute care setting
In Emergency Department/Acute Care Unit:
  Rapid triage and administration of pain medication
  If no pain medications were taken prior to arrival and pain not severe, may use ibuprofen and oral opioid
  If prior pain medications were taken or pain is severe
    Ketorolac tromethamine (non-steroidal anti-inflammatory drug)
    IV opioid
  Fluids to maintain euvolemia. IV normal saline bolus should only be used if evidence of decreased oral
    intake/dehydration
Inpatient:
  Continue nonsteroidal anti-inflammatory agent
  Continue IV opioids. Should be given as scheduled medication rather than "as needed"
  Consider patient controlled analgesia (PCA) pump if pain not adequately controlled
  Ongoing evaluation of adequacy of pain control is essential – utilize pain scales
  Supportive care
    Fluids (oral + IV) to maintain euvolemia
    Incentive spirometry
    Heating pad – must be used carefully to avoid burns
    Bowel regimen to prevent/treat constipation secondary to opiod use
      Stool softeners (e.g. docusate)
      Laxative (e.g. senna)
    Antihistamines (e.g. diphenhydramine, hydroxyzine) for pruritis
  Transition to oral non-steroidal and oral opioid as pain level improves. Addition of long-acting opioid
    (e.g. sustained release morphine)

d. The risk of ACS is directly proportional to the hemoglobin level and white blood cell count; increased levels of cytokines and/or white cell adhesion to the endothelium may play a role. Rates of ACS are also higher in children with asthma. Higher hemoglobin F levels appear to be protective

e. Laboratory findings:
   • White blood cell count is often elevated
   • Hemoglobin level often falls to below baseline values
   • Thrombocytosis may be present and often follows an episode of ACS
   • Secretory phospholipase 2 (an inflammatory mediator) levels are elevated in ACS. The combination of fever and elevated phospholipase 2 levels predicts a high risk of developing ACS in patients hospitalized with vaso-occlusive pain

Table 8-4   Dosages of Commonly Utilized Analgesics for Management of Sickle Cell Vaso-Occlusive Pain

Medication	Usual Dose	Maximum Dose	Route	Interval
Non-steroidal anti-inflammatory medications (NSAIDs)				
Ibuprofen	10 mg/kg	800 mg	PO	Q 6-8 hours
Naproxen	5-7 mg/kg	500 mg	PO	Q 12 hours
Ketorolac	0.5 mg/kg	30 mg	IV, IM	Q 6-8 hours[a]
Opiod pain medications				
Codeine	0.5-1 mg/kg	60 mg	PO	Q 4-6 hours
Oxycodone				
<6 years	0.05-0.15 mg/kg	2.5 mg	PO	Q 4-6 hours
6-12 years	0.05-0.2 mg/kg	5 mg	PO	Q 4-6 hours
>12 years	0.05-0.2 mg/kg	10 mg	PO	Q 4-6 hours
Hydromorphone				
<12.5 kg	0.03-0.08 mg/kg		PO	Q 3-4 hours
≥12.5 kg	1-4 mg/dose	8 mg	PO	Q 3-4 hours
<33 kg	0.015-0.03 mg/kg		IV, IM, SQ	Q 3-4 hours
≥33 kg	1-4 mg/dose	4 mg	IV, IM, SQ	Q 3-4 hours
Morphine (immediate release)				
<6 months	0.1-0.3 mg/kg		PO	Q 3-4 hours
6 months-18 years	0.2-0.5 mg/kg		PO	Q 3-4 hours
Adults	10-30 mg/dose		PO	Q 3-4 hours
Morphine (controlled release)				
>30 kg	0.3-0.6 mg/kg	60 mg	PO	Q 8-12 hour
Morphine				
<6 months	0.05-0.1 mg/kg		IV, IM, SQ	Q 3-4 hours
≥6 months	0.1-0.2 mg/kg	15 mg	IV, IM, SQ	Q 3-4 hours

[a]Duration of therapy should not exceed 5 days.
Abbreviations: PO, orally; IV, intravenously; IM, intramuscularly; SQ, subcutaneously.

- The management of ACS is described in Table 8-5
- Prevention of ACS: Patients with a history of recurrent ACS are candidates for preventative/curative therapies including:
  - Hydroxyurea
  - Prophylactic red cell transfusions. Optimal target HbS level is not known, but usually a goal of 30–50% is used
  - Stem cell transplantation.
3. **Stroke**
   a. Acute symptomatic stroke is usually infarctive in children, although hemorrhagic stroke may occur, particularly in older children
   b. The most common underlying lesion is intracranial arterial stenosis or obstruction, usually involving the large arteries of the circle of Willis, particularly the distal internal carotid artery (ICA) and the middle (MCA) and anterior cerebral arteries (ACA)

Table 8-5   Management of Acute Chest Syndrome

Evaluations:
  Chest radiograph
  Complete blood count and reticulocyte count
  Blood type and screen
  Blood culture
  Viral studies
  Pulse oximetry
  Consider arterial blood gas in room air
Treatment:
  Antibiotics: Broad-spectrum intravenous antibiotic such as cefuroxime plus an oral macrolide
    (erythromycin or azithromycin) to cover atypical bacteria
  Supplemental oxygen if hypoxemic
  Fluids: Intravenous and oral fluids should be kept at maintenance. Avoid overhydration
  Pain control: Must be carefully monitored. Goal is to relieve pain to reduce splinting/poor aeration but
    avoid oversedation with hypoventilation
  Transfusion:
    Simple transfusion (10–15 cc/kg) – do not exceed post transfusion hemoglobin level of ~ 10 g/dL
    Exchange transfusion – if no improvement with simple transfusion or with severe hypoxemia/
    respiratory distress
  Bronchodilators – particular if history of reactive airways disease or if wheezing present
  Steroids may be beneficial for severe acute chest syndrome or if reactive airways disease component.
    There is a risk of rebound pain with discontinuation of the steroids
  Incentive spirometry to reduce atelectasis
  Mechanical ventilation as needed
  Consider thoracentesis if significant pleural effusion

c. Chronic injury to the endothelium of vessels by sickled red blood cells results in changes in the intima with proliferation of fibroblasts and smooth muscle; the lumen is narrowed or completely obliterated. Small friable collateral blood vessels known as *moyamoya* may develop. Infarction of brain tissue occurs acutely as a result of in situ occlusion of the damaged vessel or distal embolization of a thrombus. Perfusional and/or oxygen delivery deficits related to changes in blood pressure or other factors also may contribute to infarction, particularly in watershed zones

d. Stroke is most common in sickle cell disease-SS. Prior to transcranial Doppler ultrasound screening with transfusions for high-risk children, stroke prevalence in children with sickle cell disease-SS is estimated at 11% with the highest incidence rates occurring in the first decade of life (1.02 per 100 patient years in 2–5-year-olds and 0.79 per 100 patient-years in 6–9-year-olds)

e. A number of clinical, laboratory and radiological factors associated with increased risk of overt stroke have been identified (Table 8-6)

f. Symptoms of stroke include:
  • Focal motor deficits (e.g., hemiparesis, gait dysfunction)
  • Speech defects
  • Altered mental status

Table 8-6   Clinical, Laboratory and Radiological Factors Associated with Increased Risk of Overt Stroke in Sickle Cell Disease

Clinical
History of transient ischemic attacks
History of bacterial meningitis
Sibling with SCD-SS and stroke
Recent episode of acute chest syndrome (within 2 weeks)
Frequent acute chest syndrome
Systolic hypertension
Nocturnal hypoxemia
Laboratory
Low steady-state hemoglobin level
No alpha gene deletion
Certain HLA haplotypes
Radiological
Abnormal transcranial Doppler ultrasound
Silent infarct

- Seizures
- Headache.

g. Gross neurological recovery occurs in approximately two-thirds of children, but neurocognitive deficits are common.

h. In untreated patients, about 70% of patients experience a recurrence within 3 years. Outcome after recurrent stroke is worse.

i. Any child with SCD who develops acute neurological symptoms requires immediate medical evaluation. A guideline for management is presented in Figure 8-3. The acute management involves prompt diagnosis and treatment.

  (1) Diagnosis: Physical examination with detailed neurological examination Head computerized tomography (CT) scan – useful for detecting intracranial hemorrhage and often more readily available than magnetic resonance imaging (MRI). May not be positive for acute infarction within the first 6 hours. Brain MRI with diffusion-weighted imaging is more sensitive to early ischemic changes and may be abnormal within one hour. Should be performed as soon as possible in a child with sickle cell disease presenting with acute neurological symptoms, but should not delay empiric treatment. Magnetic resonance angiography (MRA) demonstrates large vessel disease

  (2) Treatment:

- Transfusion. Exchange transfusion, either automated or manual, should be performed as soon as possible. The goal is to reduce the amount of Hb S to less than 30% and to raise the hemoglobin level to approximately 10 g/dl. If exchange transfusion is not readily available, a simple transfusion to raise the hemoglobin level to no greater than 10 g/dl may be used.

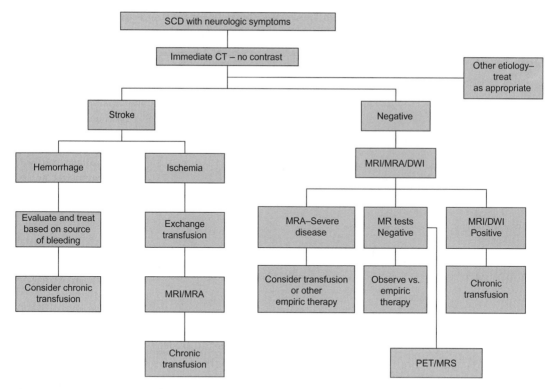

Figure 8-3 Management of the Child with Sickle Cell Disease and Neurological Symptoms.
From The Management Of Sickle Cell Disease, 4th Ed, 2002, National Heart, Lung and Blood Institute of the National Institutes of Healthy and The U.S. Department of Health and Human Services, NIH Publication No. 02-2117, with permission.
Abbreviations: PET, Positron Emission Tomography; MRA, Magnetic Resonance Arterial Angiography; DWI, Diffusion-Weighted Imaging; MRS, Magnetic Resonance Spectroscopy.

- Supportive therapy including avoiding hypotension and maintaining adequate oxygenation and euthermia should be initiated as adjunctive therapy
  j. Long-term management:
    (1) Prevention of recurrent stroke:
      - A chronic red cell transfusion program should be instituted, with the goal of maintaining the pre-transfusion Hb S level at less than 30%. Transfusions must be continued indefinitely, due to the high risk of stroke recurrence after discontinuation of therapy. After a period of 3–4 years following the initial stroke, it may be possible to allow the pre-transfusion HbS level to rise to less than 50% in low-risk patients, without increased risk of stroke recurrence. This approach is associated with decreased iron loading
      - Fetal hemoglobin stimulating agents (e.g., hydroxyurea) may prevent further stroke and are currently under study
      - Stem cell transplantation

- Revascularization procedures such as encephalodurosynangiosis may be beneficial in children with significant vasculopathy, particularly if symptomatic (transient ischemic attacks, recurrent stroke) although published data on use in sickle cell disease are limited
- Prophylactic aspirin may also be useful in children with progressive vasculopathy, but the risks of hemorrhage must be weighed against the potential benefit.

(2) Rehabilitation
- Physical and occupational therapy as needed
- Neuropsychological testing should be performed with educational interventions if indicated.

k. Primary stroke prevention
(1) Screening
- Transcranial Doppler (TCD) ultrasonography is a noninvasive study used to measure the blood flow velocity in the large intracranial vessels of the circle of Willis
- The highest time-averaged mean velocity (TAMMvel) in the distal internal carotid artery (ICA), its bifurcation and the middle cerebral artery (MCA) are used to categorize studies into the following risk groups.
  - Normal (velocity <170 cm/s), low risk
  - Conditional (170–199 cm/s), moderate risk
  - Abnormal (≥200 cm/s), high risk
  - Inadequate – unable to obtain velocity in the ICA or MCA on either side, in the absence of a clearly abnormal value in another vessel. Inadequate TCD may be due to technique, skull thickness, or severely stenosed vessel
- Very low velocity (ICA/MCA velocity <70 cm/s) may indicate vessel stenosis and increased risk of stroke
- Elevated velocity in the anterior cerebral artery (ACA) is associated with increased stroke risk. Treatment of children with isolated high ACA velocities has not been established. Brain MRI/A should be obtained. Transfusion should be considered for children with ACA velocity ≥200 cm/s, especially if silent infarcts and/or cerebral blood vessel stenosis are present on MRI/A
- TCD screening is recommended for children with SCD-SS or SCD-Sβ^0-thalassemia ages 2 to 16 years. Screening is performed annually, unless the prior study was not normal. An approach to screening is shown in Table 8-7. In addition, more frequent screening should also be considered if other known stroke risk factors are present (such as a sibling with SCD-SS and stroke or abnormal TCD)
- Brain MRI/MRA should be obtained in children with abnormal TCD and should be considered for children with conditional TCD

Table 8-7 Transcranial Doppler Ultrasonography Screening Protocol

Last TCD Result (TAMMvel in ICA/MCA)	Screening Interval
Normal (<170 cm/s)	Annual
Low conditional (170–184 cm/s)	3–6 months[a]
High conditional (185–199 cm/s)	6 weeks–3 months[a]
Abnormal (200–219 cm/s)	Within 2 weeks
High abnormal (220 cm/s or higher)	No confirmation needed – recommend treatment

[a]Use the shorter time interval for children <10 years of age.
Abbreviations: TCD, transcranial doppler; TAMMvel, time-averaged mean velocity; ICA, interval carotid artery; MCA, middle cerebral artery.

- Brain MRA is helpful to evaluate cerebral vasculature in children with repeatedly inadequate TCD or with very low velocity.

(2) Treatment
- Chronic transfusion to maintain the hemoglobin S level <30% reduces the risk of stroke by >90% in children with abnormal TCD
- Discontinuation of transfusion therapy after at least 30 months of transfusion with normalization of TCD results is associated with a high risk of reversion to abnormal TCD and stroke. Thus, transfusions are continued indefinitely
- Stem cell transplantation with an HLA-identical sibling donor may be considered
- Hydroxyurea therapy is associated with a lowering of TCD velocities and is currently under study for primary stroke prevention.

4. **Priapism**
   a. Priapism is a sustained, painful erection of the penis. Priapism may be prolonged (lasts more than 3 hours), or stuttering (lasts less than 3 hours). Stuttering episodes often recur or may develop into a prolonged episode
   b. Occurs in 30–45% of patients with sickle cell disease, most commonly in the SS type. The prevalence is likely underestimated due to underreporting by patients
   c. Mean age at the first episode of priapism in patients with SCD is about 12 years
   d. Priapism often occurs during the early morning, when normal erections occur and is probably related to nocturnal acidosis and dehydration. The normal slow blood flow pattern in the penis is similar to the blood flow in the spleen and renal medulla. Failure of detumescence is due to venous outflow obstruction or to prolonged smooth muscle relaxation, either singly or in combination
   e. A history of priapism in childhood is associated with later sexual dysfunction, with 10–50% of adults with SCD and a history of priapism reporting impotence
   f. Treatment
      (1) At home, patients may try warm baths, oral analgesics, increased oral hydration and pseudophedrine

    (2) Patients should be evaluated in an emergency room for episodes lasting over 2 hours

    (3) Initial treatment includes intravenous hydration and parenteral analgesia

    (4) Episodes lasting >4 hours are associated with an increased risk of irreversible ischemic injury and thus warrant more aggressive management. Urological consultation should be obtained. Treatment involves aspiration of the corpus cavernosum followed by irrigation with a dilute (1:1,000,000) epinephrine solution. A dilute solution of phenylephrine, an alpha adrenergic agent, rather than epinephrine, has also been utilized in some centers

    (5) The role of transfusion for the management of priapism is controversial and the clinical response is variable. Exchange transfusion for priapism is associated with the development of acute neurological events

    (6) Surgical shunting procedures (cavernosaspongiousum or cavernosaphenous vein) may be considered if the above treatments fail

  g. Prevention of priapism

    (1) Pseudophedrine, 30–60 mg orally at bedtime

    (2) Hydroxyurea therapy has been employed, although this treatment has not been studied for this indication

    (3) Leuprolide injections, a gonadotrophin-releasing hormone analog that suppresses the hypothalamic–pituitary access, reducing testosterone production

    (4) Transfusion protocol for 6–12 months following an episode of priapism requiring irrigation and injection

5. **Splenic sequestration**

  a. Highest prevalence between 5 and 24 months of age in sickle cell disease-SS (may occur at older ages in other sickling syndromes)

  b. May be seen in association with fever or infection

  c. Splenomegaly due to pooling of large amounts of blood in the spleen

  d. Rapid onset of pallor and fatigue. Abdominal pain is often present

  e. Hemoglobin level may drop precipitously, followed by hypovolemic shock and death

  f. Reticulocytosis and nucleated red blood cells often present

  g. Platelet and white blood cell count also usually fall from baseline

  h. Treatment of splenic sequestration is shown in Table 8-8

6. **Transient pure red cell aplasia**

  a. Cessation of red cell production that may persist for 7–14 days with profound drop in hemoglobin (as low as 1 g/dl)

  b. Reticulocyte count and the number of nucleated red cells in the marrow sharply decrease; platelet and white blood cell counts are generally unaffected

  c. May occur in several members of a family and can occur at any age

  d. Almost invariably associated with parvovirus B19 infection

Table 8-8 Management of Splenic Sequestration

Treatment of acute splenic sequestration episode
Monitor cardiovascular status, spleen size and hemoglobin level closely
Normal saline bolus of 10–20 cc/kg
Red cell transfusion. Administer in small aliquots because transfusion often results in reduction in spleen size with "autotransfusion" of previously trapped red cells. Rapid infusion used for cardiovascular instability
Pain management
Prevention of recurrent splenic sequestration
Splenectomy if history of one major or two minor acute splenic sequestration episodes
For children <2 years of age, chronic transfusion therapy may be employed to postpone splenectomy

    e. Terminates spontaneously usually after about 10 days (recovery occurs with reticulocytosis and nucleated red cells in the blood)

    f. Treatment

      (1) Close monitoring of CBC and reticulocyte count

      (2) Red cell transfusion to raise hemoglobin level to no greater than 9–10 g/dl

      (3) Monitor siblings with sickle cell disease closely (CBC, reticulocyte count, parvovirus PCR and/or titers).

Chronic Complications and End-Organ Damage

1. **Central nervous system – Silent stroke**

    a. Defined as one or more focal T2-weighted signal hyperintensities demonstrated on brain MRI, in the absence of a focal neurological deficit corresponding to the anatomical distribution of the brain lesion

    b. Present in 20–35% of children with sickle cell disease-SS and occur less commonly in other sickle cell genotypes

    c. Associated with neuropsychologic deficits and impaired school performance

    d. Silent infarcts may progress in size and number over time and are associated with an increased risk of overt stroke

    e. Treatment

      (1) Management of children with silent infarcts includes neuropsychological testing and monitoring of academic performance

      (2) Chronic transfusion therapy for this complication is under study.

2. **Cardiovascular system**

    a. Abnormal cardiac findings are present in most patients as a result of chronic anemia and the compensatory increased cardiac output

    b. Cardiomegaly is found in most patients and left ventricular hypertrophy occurs in about 50%

    c. A moderate intensity systolic flow murmur is often present

    d. Echocardiogram may show left and right ventricular dilatation; increased stroke volume and abnormal septal motion

    e. Prolonged QTc

    f. Pulmonary hypertension

        (1) Defined by a pulmonary artery systolic pressure greater than 35 mmHg (tricuspid regurgitant jet velocity (TRV) higher than 2.5 m/s)

        (2) Prevalence of pulmonary hypertension in adults is estimated at 20%, with 10% of these adults having moderate to severe pulmonary hypertension (pressure above 45 mmHg). The prevalence of pulmonary hypertension in children appears to be about 10% and is most common with the SS genotype. Determination of the diagnosis of pulmonary hypertension by TRV alone has been questioned. Children with elevated TRV should be managed along with a cardiologist

        (3) A central role for hemolysis and altered NO bioavailability has been postulated

        (4) The optimal treatment is unknown, but red cell transfusions or hydroxyurea have been used. Treatment with sildenafil, an agent used to treat pulmonary hypertension in other patient groups, was associated with an increased risk of vaso-occlusive pain episodes in adults with sickle cell disease.

3. **Pulmonary**

    a. Reduced PaO_2

    b. Reduced PaO_2 saturation. Pulse oximetry may not correlate with PaO_2 in steady state. Changes in pulse oximetry are useful for monitoring children with ACS. Daytime and/or nocturnal hypoxemia may be present

    c. Pulmonary fibrosis – chronic lung disease: Early identification of progressive lung disease using pulmonary function testing is imperative. Aggressive treatment has little benefit in end-stage lung disease and this should be avoided by prophylactic transfusions

    d. Asthma – Prevalence appears to be higher in children with SCD than in the general population. Asthma is associated with complications of SCD including pain, acute chest syndrome, stroke and pulmonary hypertension. Aggressive management is warranted.

4. **Kidney**

    a. Increased renal flow

    b. Increased glomerular filtration rate

    c. Enlargement of kidneys; distortion of collecting system on intravenous pyelogram

    d. Hyposthenuria (urine concentration defect): Hyposthenuria is the first manifestation of sickle cell-induced obliteration of the vasa recta of the renal medulla. Edema in the medullary vasculature is followed by focal scarring, interstitial fibrosis and destruction of the countercurrent mechanism. Hyposthenuria results in a concentration capacity of more than 400–450 mOsmol/kg and an obligatory urinary output as high as 2,000 ml/m^2/day, causing the patient to be particularly susceptible to dehydration. The increased urine output is associated with nocturia, often manifesting as enuresis. Treatment of nocturnal enuresis includes behavioral modifications such as bedwetting alarm and intranasal 1-deamino-8-D-arginine vasopressin (DDAVP) (0.01%): 10–40 µg at bed time

e. Hematuria: papillary necrosis is usually the underlying anatomic defect. Treatment of papillary necrosis is IV hydration and rest. Frank hematuria usually resolves, although bleeding can be prolonged. Antifibrinolytic agents such as episilon-amino caproic acid have been used for recalcitrant bleeding with variable success. However, caution must be taken when using this drug because of the risk of thrombosis and urinary obstruction. Evaluation for other causes of hematuria (i.e., renal medullary carcinoma) is indicated for the first episode of hematuria

f. Renal tubular acidification defect

g. Increased urinary sodium loss (may result in hyponatremia). Hyporeninemic hypoaldosteronism and impaired potassium excretion are a result of renal vasodilating prostaglandin increase in patients with SCD

h. Proteinuria: Persistent increasing proteinuria is an indication of glomerular insufficiency, perihilar focal segmental sclerosis and renal failure. Intraglomerular hypertension with sustained elevations of pressure and flow is the prime etiology of the hemodynamic changes and subsequent proteinuria. If proteinuria persists for more than 4–8 weeks, angiotensin-converting enzyme (ACE) inhibitors (i.e., enalapril) are recommended

i. Nephrotic syndrome: A 24-hour urine protein of more than 2 g/day, edema, hypoalbuminemia and hyperlipidemia may indicate progressive renal insufficiency. The efficacy of steroid therapy in the management of nephrotic syndrome in SCD is not clear. Carefully monitored use of diuretics is indicated to control edema

j. Chronic renal failure: Renal failure can be managed with peritoneal dialysis, hemodialysis and transplantation.

5. **Liver and biliary system**
   a. Chronic hepatomegaly
   b. Liver function tests: Increased serum glutamic-oxaloacetic transaminase (SGOT) and serum glutamic pyruvic transaminase (SGPT)
   c. Cholelithiasis
      (1) Chronic hemolysis with increased bilirubin turnover causes pigmented stones
      (2) Occurs as early as 2 years old and affects at least 30% by age 18 years
      (3) Sonographic examinations of the gall bladder should be performed in children with symptoms. The treatment for symptomatic cholelithiasis is laparoscopic cholecystectomy. The role of screening and treatment of asymptomatic patients is unclear

d. Transfusion-related hepatitis. Hepatitis C is more common in older patients who received red cell transfusions prior to the availability of screening of blood products

e. Intrahepatic crisis: Intrahepatic sickling can result in massive hyperbilirubinemia, elevated liver enzyme values and a painful syndrome mimicking acute cholecystitis or viral hepatitis. Progression to multiorgan system failure may occur. Early exchange transfusion is indicated

f. Hepatic necrosis, portal fibrosis, regenerative nodules and cirrhosis are common post mortem findings that may be a consequence of recurrent vascular obstruction and repair

g. Transfusional iron overload, secondary to repeated intermittent or chronic transfusions may cause hepatic fibrosis.

6. **Bones**

Skeletal changes in SCD are common because of expansion of the marrow cavity, bone infarcts, or both.

a. Avascular necrosis (AVN): The most common cause of AVN of the femoral head is sickle cell disease. The incidence is much higher with coexistent α-thalassemia, in patients who have frequent painful events and in those with the highest hematocrits. The pathophysiology is sludging in marrow sinusoids, marrow necrosis, healing with increased intramedullary pressure, bone resorption and eventually collapse. About 50% of patients are asymptomatic. Symptomatic patients have significant chronic pain and limited joint mobility. The diagnosis is made radiographically and shows:

   • Subepiphyseal lucency and widened joint space
   • Flattening or fragmentation and scarring of the epiphysis
   • On MRI, avascular necrosis of femoral head can be detected before deformities are apparent on radiograph.

   Treatment: Therapy for AVN is largely supportive, with bed rest, NSAIDs and limitation of movement during the acute painful episode. Transfusion therapy does not seem to delay progression of AVN. Physical therapy is helpful and may reduce the risk of progression. Core decompression of the affected hip has been reported to reduce pain and stop progression of the disease. In this procedure, avascularized bone is removed to decompress the area with the potential for subsequent new bone formation. This procedure seems to be beneficial only in the early stages of AVN and before loss of the integrity of the femoral head. AVN of the hip may have its onset in childhood, so thorough musculoskeletal examination with concentration on the hips should be performed at least yearly in children with SCD. This ensures that AVN is detected early when it is in its most treatable form. Total hip replacement may be the only option for severely compromised patients; 30% of replaced hips require surgical revision within 4.5 years and more than 60% of patients continue to have pain and limited mobility postoperatively. Avascular necrosis of the humeral head is less common. Patients are less symptomatic and arthroplasty is exceedingly rare.

b. Widening of medullary cavity and cortical thinning: Hair-on-end appearance of skull on radiograph.

c. Fish-mouth vertebra sign on radiograph.

7. **Eyes**

a. Retinopathy: sickle retinopathy is common in all forms of SCD, but particularly in those patients with SCD-SC

Nonproliferative: Occlusion of small blood vessels of the eye detected on dilated ophthalmological examination and usually not associated with defects in visual acuity. Treatment not usually needed.

Proliferative: Occlusion of small blood vessels in the peripheral retina may be followed by enlargement of existing capillaries or development of new vessels. Clusters of neovascular tissue "sea fans" grow into vitreous and along the surface of the retina. Sea fans may cause vitreous hemorrhage, which results in transient or prolonged loss of vision. Small hemorrhages resorb, but repeated leaks cause formation of fibrous strands. Shrinkage of these strands can cause retinal detachment.

Neovascularization may not progress or may even regress spontaneously. Indications for treatment include bilateral proliferative disease, rapid growth of neovascularization and large elevated neovascular fronds. Laser photocoagulation and other methods are used to induce regression of neovascularization. With proper screening and new methods such as laser surgery most of the complications of retinopathy can be avoided. Annual ophthalmologic examinations including inspection of the retina are indicated for children from the age of 5 years for children with SCD-SC and 8 years for children with SCD-SS.

b. Angioid streaks: These are pigmented striae in the fundus caused by abnormalities in Baruch's membrane due to iron or calcium deposits or both. They usually produce no problems for the patient, but occasionally they can lead to neovascularization that can bleed into the macula and decrease vision

c. Hyphema: Blood in the anterior chamber (hyphema) rarely occurs secondary to sickling in the aqueous humor, because of its low pH and pO_2. Anterior chamber paracentesis may be performed if pressure is increased

d. Conjunctivae: Comma-shaped blood vessels, seemingly disconnected from other vasculature, can be seen in the bulbar conjunctiva of patients with SCD and variants (SS > SC > Sβ-thalassemia). These produce no clinical disability. Their frequency may be related to the number of irreversibly sickled cells in the blood. This abnormality can be identified by using the +40 lens of an ophthalmoscope.

8. **Ears**

Up to 12% of patients have high-frequency sensorineural hearing loss. The pathophysiology may involve sickling in the cochlear vasculature with destruction of hair cells.

9. **Adenotonsillar hypertrophy**

Adenotonsillar hypertrophy giving rise to upper airway obstruction can become a problem from the age of 18 months. The marked hypertrophy is compensation for the loss of lymphoid tissue in the spleen. It occurs in at least 18% of patients. In severe cases, this can cause hypoxemia at night with consequent sickling. Early tonsillectomy and adenoidectomy may be indicated in these patients.

10. **Skin**

Cutaneous ulcers of the legs occur over the external or internal malleoli. Leg ulcers occur less commonly in children and rarely before age 10 years. Ulcers are most common in homozygous SCD. Ulceration may result from increased venous pressure in the legs caused by the expanded blood volume in the hypertrophied bone marrow.

Treatment:

* Rest; elevation of the leg
* Protection of the ulcer by the application of a soft sponge–rubber doughnut
* Debridement and scrupulous hygiene
* Low-pressure elastic bandage and above-the-knee elastic stockings to improve venous circulation
* Transfusion therapy for a 3–6-month course if ulcers persist despite optimal care
* Antibiotic therapy if acutely infected (typical organisms are *Staphylococcus*, *Streptococcus* and pseudomonal species)
* Oral administration of zinc sulfate (220 mg three times a day) may promote healing of leg ulcers
* Split-thickness skin grafts.

11. **Growth and development**

a. Birth weight is normal. However, by 2–6 years of age, the height and weight are significantly delayed. The weight is more affected than the height and patients with sickle cell disease-SS and $S\beta^0$-thalassemia experience more delay in growth than patients with sickle cell disease-SC and $S\beta^+$-thalassemia. In general, by the end of adolescence, patients with sickle cell disease have caught up with controls in height but not weight. The poor weight gain is likely to represent increased caloric requirements in anemic patients with increased bone marrow activity and cardiovascular compensation. Zinc deficiency may be a cause of poor growth. In these patients, zinc supplementation (dose of 220 mg three times a day) at about 10 years of age should be administered. Growth hormone levels and growth hormone stimulation studies appear to be normal in most children who have impaired growth

b. Delayed sexual maturation: Tanner 5 is not achieved until the median ages of 17.3 and 17.6 years for girls and boys, respectively. In males, decreased fertility with abnormal sperm motility, morphology and numbers is prominent. Zinc sulfate

220 mg three times a day may be effective for sexual maturity in these patients; females are more responsive than males.

12. **Functional hyposplenism**
    a. By 6 months of age, splenomegaly is apparent and persists during early childhood, after which the spleen undergoes progressive fibrosis (autosplenectomy)
    b. Functional reduction of splenic activity occurs in early life. This is the consequence of altered intrasplenic circulation caused by intrasplenic sickling. It can be temporarily reversed by transfusion of normal red cells. Children with functional hyposplenia are 300–600 times more likely to develop overwhelming pneumococcal and *Haemophilus influenzae* sepsis and meningitis than are normal children; other organisms involved are Gram-negative enteric organisms and *Salmonella*. The period of greatest risk of death from severe infection occurs during the first 5 years of life
    c. Functional hyposplenism may be demonstrated by the following:
       - Presence of Howell–Jolly bodies on blood smear
       - ^{99m}Tc-gelatin sulfur colloid spleen scan – no uptake of the radioactive colloid by enlarged spleen
       - Pitted red blood cell count >3.5%.

Diagnosis

1. In utero: Sickle cell disease can be diagnosed accurately *in utero* by mutation analysis of DNA prepared from chorionic villus biopsy or fetal fibroblasts (obtained by amniocentesis). With the advent of polymerase chain reaction (PCR) amplification of specific DNA sequences, sufficient DNA can be obtained from a very small number of fetal cells, thereby eliminating the necessity of culturing fetal fibroblasts from amniotic fluid. These techniques should be employed before 10 weeks' gestation.
2. During newborn period: The diagnosis of sickle cell disease can be established by electrophoresis using:
   - Isoelectric focusing (most commonly used in screening programs)
   - High-performance liquid chromatography
   - Citrate agar with a pH of 6.2, a system that provides distinct separation of hemoglobins S, A and F
   - DNA-based mutation analysis.
   These tests are commonly performed on a dried blood specimen blotted on filter paper (Guthrie cards) used in newborn screening programs.
3. In older children: Table 8-9 lists the diagnosis and differential diagnosis of various sickle cell syndromes.

Prognosis

The survival time is unpredictable and is related in part to the severity of the disease and its complications (with active management, 85% survive to 20 years of age).

Table 8-9 Differential Diagnosis in Sickle Cell Syndromes

Syndrome[a]	Clinical Severity	Splenomegaly	Mean Hemoglobin (g/dl)	Mean Hematocrit (%)	Mean Corpuscular Volume (fl)	Reticulocytes (%)	Red Cell Morphology	Electrophoresis
AS	Asymptomatic	(−)	Normal	Normal	Normal	Normal	Few target cells	35–45% S; 55–60% A; F[b]
SS	Severe	YC(+) OC(−)	7.5	22	85	5–30	Many target cells, ISCs (4+) and NRBCs	80–96% S; 2–20% F[b]
SC	Mild/ moderate	(+)	11	33	80	2–6	Many target cells, few ISCs (1+)	50–55% S; 45–50% C; F[b]
S/β-thalassemia	Moderate/ severe	(+)	8.5	28	65	3–20	Marked hypochromia and microcytosis; many target cells, ISCs (3+) and NRBCs	50–85% S; 2–30% F[b], >3.5% A2
S/β+-thalassemia	Mild/ moderate	(+)	10	32	72	2–6	Mild microcytosis and hypochromia; many target cells few ISCs (1+)	50–80; S; 10–30% A; 0–20% F[b], <3.5% A2
SS/ α-thalassemia-1	Mild/ moderate	(+)	10	27	70	5–10	Mild hypochromia and microcytosis; few ISCs (2+)	80–100% S; 0–20% F[b]
S/HPFH	Asymptomatic	(−)	14	40	85	1–3	Occasional target cells, no ICSs	60–80% S; 15–35% F[c]

[a] All syndromes have positive sickle preparations.
[b] Hemoglobin F distribution; heterogeneous.
[c] Hemoglobin F distribution; homogeneous.

Abbreviations: HPFH, high persistent fetal hemoglobin; ISC, irreversible sickle cell; NRBC, nucleated red blood cell; OC, older child; YC, young child.
(−) absent; (+) present.

Causes of death include:

• infection (sepsis, meningitis) with a peak incidence between 1 and 3 years of age
• acute chest syndrome/respiratory failure
• stroke (especially hemorrhagic) and
• organ failure including heart, liver and renal failure.

Management

1. *Comprehensive care*: Prevention of complications is as important as treatment. Optimal care is best provided in a comprehensive setting. Recommended screening studies are shown in Table 8-10.

2. *Infection*: Because of a marked incidence of bacterial sepsis and meningitis and fatal outcome under 5 years of age, the following management is recommended.
   All children with sickle cell disease should receive oral penicillin prophylaxis starting by 3–4 months of age:
   • 125 mg bid (<3 years)
   • 250 mg bid (3 years and older).

Table 8-10   Routine Health Maintenance Related Laboratory and Special Studies in Patients with Sickle Cell Disease

Laboratory Studies	Starting Age	Frequency
Complete blood count/ reticulocyte count	At diagnosis	Quarterly to yearly with differential white cell count monthly if receiving hydroxyurea
Hemoglobin quantitation	At diagnosis	Yearly (SCD-SS) pre-transfusion for children receiving chronic transfusion therapy
Red cell antigen typing	At diagnosis	–
Liver and renal functions	At diagnosis	Yearly
Urinalysis	1 year	Yearly
HIV, Hepatitis B, C		Yearly if receiving transfusions
Special Studies		
Pulse oximetry	At diagnosis	Quarterly to yearly
Pulmonary function	5 years	Every 3 years
Sleep study		If symptoms present
Eye examinations	5 years for SCD-SC	Yearly
	8 years for SCD-SS	Yearly
Transcranial Doppler	2 years	Based on prior results
Brain MRI/A		If school difficulties, abnormal or repeatedly conditional TCD, neurological symptoms
Abdominal ultrasound		If symptoms of cholelithiasis
Hip radiograph/MRI		If symptoms of AVN
Echocardiogram	10 years	Every 3 years or more frequent if abnormal

Abbreviations: TCD, transcranial doppler; AVN, avascular necrosis.

In patients allergic to penicillin erythromycin ethyl succinate 10 mg/kg orally twice a day should be prescribed.

Penicillin prophylaxis should be continued at least through age five. Because the incidence of invasive bacterial infection declines with age, it may be reasonable to discontinue penicillin in older children. However, given that the rate of infection remains higher than the rate in individuals with spleens, some centers advocate continuing penicillin indefinitely.

All children with SCD should receive routine childhood immunizations including conjugate *H. influenza*. The 24-valent pneumococcal vaccine should be administered at 2 years of age with revaccination at 5 years old; conjugate 7-valent pneumococcal vaccine and hepatitis B according to the routine childhood schedule. Meningococcal vaccination should also be administered. Influenza virus vaccine should be given yearly, each fall.

Early diagnosis of infections requires: Education of the family to identify a child with fever. Families should be instructed to call their physician immediately if their child develops a single temperature greater than 38.5°C (by mouth) or three elevations between 38°C and 38.5°C. The child should be seen immediately by a physician.

The patient should be investigated to determine the etiology of the fever, which should include a CBC with differential and reticulocyte count and blood culture in all children. Chest radiograph is obtained in children younger than 5 years old and in older children with respiratory symptoms. Urinalysis and culture are indicated in children <3 years or older children with symptoms. Lumbar puncture is performed in young infants (<2–3 months) and in older infants and children with symptoms of meningitis. Other studies such as viral studies, stool cultures and sputum cultures are performed based on symptoms.

Prompt antibiotic treatment with a broad-spectrum intravenous antibiotic that covers encapsulated organisms, such as a third-generation cephalosporin should be given.

Many centers recommend inpatient hospitalization for all children younger than 5 years because this group is at highest risk of infection. In addition, all children, regardless of age, with the following high-risk features should be admitted:

- Ill appearance
- High fever (>39.5°C)
- Acute chest syndrome
- Meningeal signs
- Enlarging spleen
- Elevated leukocyte count (>30,000/mm^3)
- Falling blood counts or low reticulocyte count.

A subset of lower-risk children, over age 12 months and without the above high-risk features, may be considered for discharge after a shorter period of observation

(4–18 hours) after having received a long-acting antibiotic such as ceftriaxone. This option should only be considered if the family can be contacted readily, follow-up is ensured and continuous blood culture monitoring is available.

3. *Treatment of specific complications*: This is provided in the acute and chronic complication sections above.

4. *Transfusion therapy*: Transfusion therapy is used to manage acute and chronic complications of sickle cell disease. Indications for transfusions in sickle cell disease are shown in Table 8-11. Risks of transfusion include infection (hepatitis B virus, hepatitis C virus, HIV, bacterial), alloimmunization and iron overload.

The incidence of alloimmunization is 17.6%: mostly Kell (26%) and Rh (E, 24% and C, 16%, respectively) antibodies. Other antibodies also occur in the following order of frequency: Jk^b (10%), Fy^a (6%), M (4%), Le^a (4%), S (3%), Fy^b(3%), e (2%) and Jk^a (2%). All children with SCD should have a red cell phenotype identified at diagnosis. This allows determination of the child's red cell antigen phenotype before any transfusion. The patients should receive blood that is leukocyte-depleted and phenotypically matched to the patient for the Rh and Kell antigens. These measures decrease the incidence of transfusion reactions and alloimmunization. Sickle negative blood should be administered to children receiving chronic transfusion therapy to allow accurate monitoring of Hb S levels.

Chronic red cell transfusion therapy or repeated intermittent transfusions leads to iron overload. Complications of iron overload include hepatic fibrosis, endocrinopathies and cardiac disease and are best defined for thalassemia. The prevalence of certain complications such as heart disease may be lower in SCD than in thalassemia. The

Table 8-11   Indications for Transfusions in Sickle Cell Disease

Episodic Transfusion
  Overt stroke
  Transient pure red cell aplastic episode
  Splenic sequestration
  Acute chest syndrome
  Pre-operatively for surgical procedure with general anesthesia[a]
  Acute multiorgan failure
  Retinal artery occlusion
Chronic Transfusion
  Stroke
  Abnormal transcranial Doppler ultrasound
  Recurrent acute chest syndrome
  Pulmonary hypertension
  Recurrent severe pain

[a]Moderate to high-risk surgical procedures. Controversial for low-risk procedures.

treatment is similar to the approach used for thalassemia described later in this chapter. In addition, in SCD, exchange transfusion limits or prevents iron loading and should be utilized when possible for chronic transfusion therapy.

5. *Induction of fetal hemoglobin*: Sustained elevations in Hb F ($\geq$20%) are associated with reduced clinical severity in sickle cell disease. Hydroxyurea (HU) is the most commonly used drug for Hb F modulatory therapy. HU results in the upregulation of fetal hemoglobin (HbF). HbF, within the red cell, interferes with polymerization of HbS and therefore decreases the propensity of the red cell to sickle. Other effects of HU include increased red cell hydration and decreased expression of red cell adhesion molecules, increased NO production and lowering of white blood cell count, reticulocytes and platelets. Numerous studies in adults and children have shown the following beneficial effect of HU in SCD:

   • Reduces number of vaso-occlusive pain events (VOE)
   • Reduces incidence of acute chest syndrome (ACS)
   • Reduces mortality.

   HU is not yet approved in the US by the FDA for use in children with SCD. Efficacy and safety in children as young as 2 years of age have been as good as adult experience. However, the use of HU in children should be considered investigational, particularly because of concerns regarding potential leukemogenesis, teratogenesis and adverse effect on growth and development.

Dose

The starting dose of HU is 15–20 mg/kg/day. It is increased every 8 weeks by 5 mg/kg/day until a total dose of 35 mg/kg/day is reached or until a favorable response is obtained or until signs of toxicity appear. Evidence of toxicity includes:

• Neutrophil count <1,000/mm^3
• Platelet count <80,000/mm^3
• Hemoglobin drop of 2 g/dl
• Absolute reticulocyte count <80,000/mm^3.

Response is indicated by clinical improvement (reduction in VOE, ACS, etc.) and by laboratory response including rise in Hb F (10–20% is typical), a rise in total hemoglobin of 1–2 g/dl and increased MCV and reduced numbers of sickled red cells on blood smear.

Follow-up

The patient should be monitored with a complete blood count every 2 to 4 weeks and Hb F level every other month. Once a stable and maximum tolerated dose is obtained, the patient can be monitored with CBCs monthly.

Indications

- Three or more VOE in one year
- Recurrent ACS
- Chronic leg ulcers that fail conventional therapy
- Persistent occurrences of priapism despite standard therapy
- Cerebrovascular accident (CVA) with significant alloimmunization. HU therapy is currently under study for secondary stroke prevention.

Side Effects

- Myelosuppression
- Hair loss; skin pigment changes
- Gastrointestinal (GI) disturbance
- Potential birth defects (female patients on hydroxyurea should not become pregnant because of the potential for birth defects)
- Reduced sperm count and motility.

Contraindications

- Creatinine level more than twice the upper limit of normal for age, or greater than 1.5 mg/dl
- Active liver disease.

6. *Other fetal hemoglobin modulating agents*:
   a. Butyrates are short chain fatty acids that raise Hb F levels through modulation of histone acetylation. They have a short half-life and also must be given intravenously. Their effect wanes if they are given continuously, but appear to be more effective if given in pulse courses. New short-chain fatty acid derivatives with better oral bioavailability are under study
   b. Decitabine increases Hb F production by inducing hypomethylation of the gamma globin gene, promoting expression. This drug has been used subcutaneously with or without erythropoietin in a small number of patients with SCD with improvement in Hb F and hemoglobin levels. Further study is underway. Long-term effects such as infertility and malignancy risk are unknown. An oral form of the drug is also being developed.

7. *Hematopoietic stem cell transplantation (HSCT)*: Currently HSCT (including umbilical cord blood) is the only curative therapy. The results of transplantation are best when performed in children with a sibling donor who is HLA-identical. Eligibility criteria for HSCT for SCD are:
   - Availability of a fully HLA-matched sibling donor
   - Sickle cell disease (SCD-SS or SCD-Sβ^0-thalassemia)
   - One or more of the following complications:
     - Stroke or CNS event lasting longer than 24 hours
     - Recurrent ACS (at least 2 episodes in the last 2 years)

  * Recurrent severe, debilitating VOE (three or more severe pain events per year for the past 2 years)
  * Sickle nephropathy
  * Avascular necrosis of multiple joints.

Over 150 patients have undergone HSCT from HLA-identical siblings worldwide. Transplantation morbidity is about 5% and more than 90% of patients survive. Approximately 85% survive free from SCD after HSCT. About 10% of patients experience recurrence. Neurologic complications such as seizures may occur after transplantation. Patients who have stable engraftment of donor cells experience no subsequent sickle-cell-related events and stabilization of pre-existing organ damage. There is also splenic function recovery. Linear growth is normal or accelerated after transplantation in the majority of patients. About 5% of the patients develop clinical grade III acute or extensive GVHD (see Chapter 29). The risk of secondary cancers is estimated to be less than 5%.

Only about 15% of patients with SCD are likely to have an HLA-identical sibling donor. Unrelated donor stem cell transplantation and reduced-intensity conditioning protocols are under development.

Recommendations

* Children with SCD who experience significant sickle cell complications should be considered for HSCT
* HLA typing should be performed on all siblings
* Families should be counseled about the collection of umbilical cord blood from prospective siblings
* For severely affected children who have HLA-identical sibling donors, families should be informed about the benefits, risks and treatment alternatives regarding HSCT.

8. *Psychological support* See Chapter 33. As for any chronic disease, patients require psychological support. Major problems that occur are:
   * Coping with chronic pain
   * Inability to keep up with peers
   * Fears of premature death
   * Delayed sexual maturity
   * Increased doubts about self-worth.

Sickle Cell Trait (Heterozygous form, AS)

The concentration of Hb S in red cells is low and sickling does not occur under normal conditions.

Hematology

1. Indices – usually normal.
2. Blood smears – normal with few target cells.
3. Sickle cell preparation – reducing agents (e.g., sodium metabisulfite) to induce sickling *in vitro*.
4. Hemoglobin electrophoresis – AS pattern (Hb A 55–60%; Hb S, 35–45%).

Clinical Features

1. Usually asymptomatic.
2. Hematuria rarely.
3. Increased propensity for renal medullary cancer.
4. Exertional rhabdomyolysis/exercise-related sudden death. Ensure adequate hydration with sports activities.
5. Complicated hyphema – with secondary hemorrhage, increased intraocular pressure, central retinal artery occlusion. Requires evaluation/treatment by ophthalmologist.
6. Infarction rare, occurring during flights in unpressurized aircraft.

Significance

The genetic implications mandate counseling. Table 8-9 lists the differential diagnosis of sickle cell syndromes.

Hemoglobin C

Basic Features and Pathology

1. Carrier state – 2% in African-Americans.
2. Amino acid substitution (the same codon in the β-chain as in hemoglobin S) – lysine for glutamic acid.
3. Hemoglobin C tendency to form rhomboidal crystals with increases in osmolality – red cell deformability impaired and splenic sequestration increased.

Hemoglobin C Disease (Homozygous CC)

Hematology

1. Anemia – usually mild, hemolytic.
2. Blood smear – numerous target cells, as well as some spherocytes (the result of membrane loss in the spleen); a bar of crystalline hemoglobin across cell due to alteration in intracellular hemoglobin is a frequent finding.
3. Hemoglobin electrophoresis – CC pattern.

Clinical Features

1. Usually clinically asymptomatic.
2. Splenomegaly.
3. Mild hemolysis, cholelithiasis, retinopathy may occur.

Hemoglobin C Trait (Heterozygous Form, AC)

Asymptomatic with only genetic significance.

Hemoglobin SC Disease (Sickle Cell Disease-SC)

Combination of hemoglobin S and hemoglobin C.

Hematology

1. Anemia – if present, usually mild, hemolytic.
2. Blood smear – many target cells; sickle cells occasionally seen.
3. Sickle cell preparations – positive.
4. Hemoglobin electrophoresis – SC pattern (Hb S $\sim 50\%$; Hb C $\sim 50\%$).

Clinical Features

1. Similar to, but usually less severe than, sickle cell disease-SS.
2. Severe infarctions on occasion (e.g., during pregnancy or the puerperium).

Hemoglobin S/β-Thalassemia

1. Combination of hemoglobin S and β-thalassemia trait.
2. Hematology and clinical features vary; severity depends on the amount of normal adult hemoglobin synthesized (0–30%).
3. With no hemoglobin A (S-β-thalassemia), disease comparable to sickle cell disease-SS.

Hemoglobin E

1. Mutation in β-globin gene that creates an alternate splice site which leads to decreased production.
2. Heterozygotes (hemoglobin E trait) and homozygotes (hemglobin E disease) are asymptomatic. The MCV is reduced and target cells are seen on peripheral blood smear. Mild anemia is seen with hemoglobin E disease and less commonly with hemoglobin E trait. Important to distinguish hemoglobin E disease from hemoglobin E/β-thalassemia as the latter is clinically significant.
3. Hemoglobin E/β-thalassemia – causes a thalassemia intermedia or thalassemia major phenotype (see later in chapter).

UNSTABLE HEMOGLOBINS

Unlike the amino acid substitutions in hemoglobin S and hemoglobin C, which affect the polarity of the external surface of the hemoglobin molecule, resulting in polymerization (Hb S) or crystallization (Hb C), the substitutions in unstable hemoglobins occur within the heme cavity or pocket of the α- or β-polypeptide chain. Substitution in the region of heme attachment causes gross molecular instability.

Changes in the oxygen affinity have also been found in some of the unstable hemoglobins and some of the M hemoglobins. An increase in oxygen affinity results in greater tissue anoxia and greater erythropoietin stimulation for a given level of anemia. In at least one hemoglobinopathy, hemoglobin Chesapeake, the only clinical manifestation is mild polycythemia.

Table 8-12 lists the various clinical manifestations that suggest unstable hemoglobinopathies. Table 8-13 presents laboratory data that suggest unstable hemoglobinopathies.

The hereditary methemoglobinopathies are closely related to the unstable hemoglobins. The substitution in these cases is also in the region of heme attachment, but it results in increased susceptibility to oxidation of heme Fe^{2+} to Fe^{3+} with consequent methemoglobin accumulation and cyanosis rather than hemolysis. There is some overlap between these two disorders, insofar as there is an increase in methemoglobin formation in most types of unstable hemoglobinopathies.

Table 8-12   Clinical Manifestations of Unstable Hemoglobins

Chronic nonspherocytic hemolytic anemia, varying from mild to severe
Intraerythrocyte inclusions (Heinz bodies) demonstrable by incubation of the cells with brilliant cresyl blue
    or methyl violet
Urinary dipyrrolic pigment excretion
Drug-induced hemolytic anemia
Methemoglobinemia
Cyanosis
Polycythemia
Chronic hemolytic anemia with normal hemoglobin electrophoresis
Variable response of hemolytic anemia to splenectomy

Table 8-13   Laboratory Data in Unstable Hemoglobinopathies

Chronic hemolytic anemia with normal red cell morphology, red cell enzymes and hemoglobin
    electrophoresis
Abnormal heat stability test; tendency to precipitate on heating at 50°C
Presence of Heinz bodies
Raised methemoglobin levels
Dipyrroluria

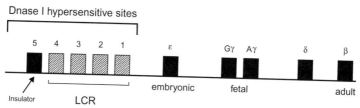

β-globin gene transcription is regulated by activation of the genes of the locus control region (LCR) and repression of the early genes

Figure 8-4 The Structure of the Human β-globin Locus in Chromosome 11. From: Nathan D, Orkin S. Nathan and Oski's hematology of infancy and childhood. 5th ed. Philadelphia: Saunders, 1998:817, with permission.

THALASSEMIAS
Basic Features

Thalassemia syndromes are characterized by varying degrees of ineffective hematopoiesis and increased hemolysis. Clinical syndromes are divided into α- and β-thalassemias, each with varying numbers of their respective globin genes mutated. There is a wide array of genetic defects and a corresponding diversity of clinical syndromes. Most β-thalassemias are due to point mutations in one or both of the two β-globin genes (chromosome 11), which can affect every step in the pathway of β-globin expression from initiation of transcription to messenger RNA synthesis to translation and post-translation modification. Figure 8-4 shows the organization of the genes (i.e., ε and γ, which are active in embryonic and fetal life, respectively) and activation of the genes in the locus control region (LCR), which promote transcription of the β-globin gene.

There are four genes for α-globin synthesis (two on each chromosome 16). Most α-thalassemia syndromes are due to deletion of one or more of the α-globin genes rather than to point mutations.

Mutations of β-globin genes occur predominantly in children of Mediterranean, Southern and Southeast Asian ancestry. Those of α-globin are most common in those of Southeast Asian and African ancestry.

The main genetic variants include:

β-Thalassemia

1. β^0-Thalassemia: no detectable β-chain synthesis due to absent β-chain messenger RNA (mRNA).
2. β^+-Thalassemia: reduced β-chain synthesis due to reduced or nonfunctional β-chain mRNA.

3. δβ-Thalassemia: δ- and β-chain genes deleted.
4. Eβ-Thalassemia: Hemoglobin E (lysine → glutamic acid at 26) combined with β-thalassemia mutation. May be β^0 or β^+.
5. Hb Lepore: a fusion globin due to unequal crossover of the β- and δ-globin genes (the globin is produced at a low level because it is under δ-globin regulation).

α-Thalassemia

1. Silent carrier α-thalassemia: deletion of one α-globin gene.
2. α-Thalassemia trait: deletion of two α-globin genes.
3. Hb Constant Spring: abnormal α-chain variant produced in very small amounts, thereby mimicking deficiency of the gene.
4. Hb H disease: deletion of 3 α-globin genes resulting in significant reduction of α-chain synthesis.
5. Hydrops fetalis: deletion of all 4 α-globin genes; no normal adult or fetal hemoglobin production.

In many populations, α- and β-thalassemia and structural hemoglobin variants (hemoglobin-opathies) exist together, resulting in a wide spectrum of clinical disorders.

Tables 8-14 and 8-15 list some features of the heterozygous and homozygous states of β-thalassemia and its variants. Table 8-16 lists the α-thalassemia syndromes.

Table 8-14   Heterozygous States of β-Thalassemia and Variants

Type	Hb A$_2$	Hb F
β^+-Thalassemia	Increased	Normal to slightly increased
β^0-Thalassemia	Increased	Normal to slightly increased
δβ-Thalassemia	Normal	Increased (5–15%)
HPFH	Normal	Increased (15–30%)

Table 8-15   Homozygous or Doubly Heterozygous States of β-Thalassemia and Variants

Type	Anemia	δ-Globin Chain	β-Globin Chain	β-Globin MRNA	β-Globin Gene Mutation
β^+-Thalassemia	Severe	Present	Decreased	Decreased	Point mutations or deletions
β^0-Thalassemia	Severe	Present	Absent	Absent or abnormal	Point mutations or deletions
δβ-Thalassemia	Mild	Absent	Absent	Absent	Deletion mutation
HPFH	None	Absent	Absent	Absent	Point mutations or deletions

Table 8-16 α-Thalassemia Syndromes

Syndrome	Genetics	Number of α-Genes Deleted	Newborn Hb Barts (δ4) (%)	α/β Synthesis Ratio	Comments
Silent carrier of α-thalassemia	Heterozygous silent carrier	1	1–2	0.8–0.9	No anemia; no microcytosis; detectable by genetic interaction (i.e., two silent carriers can produce a child with α-thalassemia trait; a silent carrier and a person with α-thalassemia trait can produce a child with Hb H disease); also detectable by molecular studies
α-Thalassemia trait	Heterozygous α-thalassemia trait OR Homozygous silent carrier OR Homozygous Hb Constant Spring	2	3–10	0.7–0.8	Microcytosis; hypochromia; mild anemia
Hemoglobin H disease	Heterozygous α-thalassemia trait/silent carrier OR α-thalassemia trait/ Constant Spring	3	25	0.3–0.6	Hemolytic anemia of variable severity; relatively little ineffective erythropoiesis; no transfusion requirement; Hb H (β4) present
Hydrops fetalis	Homozygous α-thalassemia trait Hb Barts (δ4)	4	80–100	0	Death in utero or shortly after birth

β-THALASSEMIA: HOMOZYGOUS OR DOUBLY HETEROZYGOUS FORMS (MAJOR AND INTERMEDIA)

Pathogenesis

1. Variable reduction of β-chain synthesis (β^0, $\beta+$ and variants).
2. Relative α-globin chain excess resulting in intracellular precipitation of insoluble α-chains.
3. Increased but ineffective erythropoiesis with many red cell precursors prematurely destroyed; related to α-chain excess.
4. Shortened red cell life span; variable splenic trapping.

Sequelae

1. Hyperplastic marrow (bone marrow expansion with cortical thinning and bony abnormalities).
2. Increased iron absorption and iron overload (especially with repeated blood transfusion), resulting in:
    * Fibrosis/cirrhosis of the liver
    * Endocrine disturbances (e.g., diabetes mellitus, hypothyroidism, hypogonadism, hypoparathyroidism, hypopituitrism)
    * Skin hyperpigmentation
    * Cardiac hemochromatosis causing arrhythmias and cardiac failure.
3. Hypersplenism:
    * Plasma volume expansion
    * Shortened red cell life (of autologous and donor cells)
    * Leukopenia
    * Thrombocytopenia.

Hematology

1. Anemia – hypochromic, microcytic.
2. Reticulocytosis.
3. Leukopenia and thrombocytopenia (may develop with hypersplenism).
4. Blood smear – target cells and nucleated red cells, extreme anisocytosis, contracted red cells, polychromasia, punctate basophilia, circulating normoblasts.
5. ^{51}Cr-labeled red cell life span reduced (but the ineffective erythropoiesis is more important in the production of anemia).
6. Hemoglobin F raised; hemoglobin A2 increased.
7. Bone marrow – may be megaloblastic (due to folate depletion); erythroid hyperplasia.
8. Osmotic fragility – decreased.
9. Serum ferritin – raised.

Biochemistry

1. Raised bilirubin (chiefly indirect).
2. Evidence of liver dysfunction (late, as cirrhosis develops).
3. Evidence of endocrine abnormalities (e.g., diabetes [typically late], hypogonadism [low estrogen and testosterone], hypothyroidism [elevated thyroid stimulating hormone]).

Clinical Features

Because of the variability in the severity of the fundamental defect, there is a spectrum of clinical severity (major to intermedia), which considerably influences management. β-Thalassemia intermedia is defined as homozygous or doubly heterogeneous thalassemia, which is not transfusion-dependent. β-Thalassemia major is defined as homozygous or doubly heterogeneous thalassemia (β^0 or β^+), which requires regular transfusions to manage clinical complications. Clinical manifestations of beta thalassemia major include:

- Failure to thrive in early childhood
- Anemia
- Jaundice, usually slight; gallstones
- Hepatosplenomegaly, which may be massive; hypersplenism
- Bone abnormalities:
  - Abnormal facies, prominence of malar eminences, frontal bossing, depression of bridge of the nose and exposure of upper central teeth
  - Skull radiographs showing hair-on-end appearance due to widening of diploic spaces
  - Fractures due to marrow expansion and abnormal bone structure
  - Generalized skeletal osteoporosis.
- Growth retardation, delayed puberty, primary amenorrhea in females and other endocrine disturbances secondary to chronic anemia and iron overload
- Leg ulcers
- Skin bronzing.

If untreated, 80% of patients with beta thalassemia major die in the first decade of life. With current management, the life expectancy has dramatically increased. Patients now reach the fifth decade of life and are expected to live even longer.

Complications

Complications develop as a result of:

- Chronic anemia (in patients who are undertransfused or in untransfused thalassemia intermedia patients)
- Iron overload – Due to repeated red cell transfusions in β-thalassemia major. In patients not treated with chelation therapy, cardiac disease from iron loading typically develops

in late teens and early 20s. Iron overload also develops in β-thalassemia intermedia due to increased absorption of dietary iron.

Even in carefully managed patients, the following complications may develop:

- Endocrine disturbances (e.g., growth retardation, pituitary failure with impaired gonadotropins, hypogonadism, insulin-dependent diabetes mellitus, adrenal insufficiency, hypothyroidism, hypoparathyroidism)
- Cirrhosis of the liver and liver failure (exacerbated if concomitant hepatitis B or C infection is present)
- Cardiac failure due to myocardial iron overload (often associated with arrhythmias and pericarditis may occur)
- Osteopenia and osteoporosis are common and the risk is directly proportional to age (the prevalence of osteoporosis is about 60% in patients 20 years and older). The causes of this include medullary expansion, deficiency of estrogen and testosterone, nutritional deficiency (including calcium, vitamin D and zinc) and chelator toxicity. Genetic factors likely also contribute
- Pulmonary hypertension (tricuspid regurgitant jet velocity greater than 2.5 m/s) occurs in both β-thalassemia major and β-thalassemia intermedia. Splenectomy may exacerbate this risk, particularly in patients who are not regularly transfused.

Causes of Death

1. Congestive heart failure.
2. Arrhythmia.
3. Sepsis secondary to increased susceptibility to infection post-splenectomy.
4. Multiple organ failure due to hemochromatosis.

Management

Transfusion Therapy

Indications for initiation of regular red cell transfusions include:

- Hemoglobin level <7 g/dl (on at least 2 measurements)
- Poor growth
- Facial bone changes
- Fractures
- Development of other complications (pulmonary hypertension, extramedullary hematopoiesis, etc.).

The goal of transfusions is to maintain a pretransfusion hemoglobin greater than 9–9.5 g/dl. Typical programs involve transfusion of 10–15 cc/kg of packed leukodepleted red cells. Blood should be matched for ABO, C, E and kell antigens to reduce the risk of alloimmunization (some centers perform extended red cell antigen matching). Post-transfusion hemoglobin falls roughly 1 g per week, necessitating transfusions every 3–4 weeks.

Transfusions result in:

- Maximizing growth and development
- Minimizing extramedullary hematopoiesis and decreasing facial and skeletal abnormalities
- Reducing excessive iron absorption from gut
- Retarding the development of splenomegaly and hypersplenism by reducing the number of red cells containing α-chain precipitates that reach the spleen
- Reducing and/or delaying the onset of complications (e.g., cardiac).

Iron overload results from:

- Ongoing transfusion therapy
- Increased gut absorption of iron (more important in β-thalassemia intermedia).

Monitoring Iron Overload
A number of tests are available to monitor iron loading, including:

- Serum ferritin – particularly useful to follow trends. Value may be altered by infection, inflammation and vitamin C deficiency
- Liver iron concentration (LIC) may be measured by different techniques. Liver iron concentration ≥15 mg/g dry weight of liver is associated with an increased risk of cardiac disease and death. Methods to measure LIC include:
  - MRI: R2 methodology is most common but other techniques including R2*, T2 and T2* are available
  - Superconducting quantum interference device (SQUID): highly specialized equipment available in few centers worldwide
  - Liver biopsy: the gold standard, but invasive. This is the method of choice if histopathological examination is needed.
- Cardiac iron measurment by T2* MRI. Cardiac iron may be high even if the liver iron concentration is low, particularly in patients with a history of high iron levels in the past with recent intensification of chelation
  - T2* ≥ 20 ms indicates minimal cardiac iron loading
  - T2* of 10–19 ms indicates some cardiac iron loading. This result should prompt a discussion with patient/family about adherence with chelation. Intensification of chelation may be warranted
  - T2* <10 ms is associated with a high risk of cardiac disease (arrhythmias, congestive heart failure). Improved adherence and/or intensification of chelation therapy is indicated.

Chelation Therapy

The objectives of chelation therapy are:

- To bind and detoxify free (non-transferrin bound) extracellular iron
- To remove excess intracellular iron
- To maintain a safe level of body iron burden:
  - Reduce previous iron loading
  - Reverse organ dysfunction
  - Prevent new iron loading.

Chelation therapy typically is not used in children younger than 2 years old and is often deferred until age 3 to 4 years. Indications for chelation therapy in patients receiving chronic transfusions include:

- Cumulative transfusion load of 120 ml/kg or greater
- Serum ferritin level persistently >1,000 ng/ml
- Liver iron concentration >5–7 mg/g dry weight.

Transfusion requirements and iron burden should be monitored closely and doses of chelation adjusted to maintain liver iron concentration at 3–7 mg/g dry weight and serum ferritin level between 500–1,500 ng/ml.

Currently available options for chelation therapy in the United States include deferoxamine and deferasirox. A third chelator, deferiprone, is licensed for use in Europe for patients unable to tolerate deferoxamine. The properties of the common chelators are summarized in Table 8-17.

Deferoxamine was the first available chelator, in clinical use for about 40 years. Due to its poor oral bioavailability, this drug must be administered parenterally, usually as a subcutaneous infusion over 8–24 hours. Potential complications of deferoxamine are listed in Table 8-17. Audiological and ophthalmological toxicities are more common when the iron burden is low relative to the chelator dose. Similarly, bone changes including metaphyseal dysplasia, are more common in young children with lower iron burden. Thus, it is important to avoid "over-chelation" in all patients and lower doses of chelation therapy should be used in young children to avoid toxicity.

Nightly subcutaneous administration of deferoxamine is time-consuming and painful and interferes in many ways with the lifestyle of the patient. For this reason, treatment adherence is often suboptimal and patients develop iron overload. The availability of oral chelation may help improve adherence to therapy. Two oral chelators have undergone extensive study:

- Deferasirox (Exjade, Novartis) is supplied as orally dispersible tablets, which are dissolved in a glass of water or apple juice and administered ½ hour before meals

Table 8-17   Properties of Common Chelators[a]

Property	Deferoxamine	Deferiprone	Deferasirox
Chelator:iron binding	1:1	3:1	2:1
Route of administration	Subcutaneous or intravenous	Oral	Oral
Usual dosage	25–50 mg/kg/day	75 mg/kg/day	20–40 mg/kg/day
Schedule	Administered over 8–24 hours, 5–7 days/wk	Three times a day	Daily
Route of excretion	Urine/feces	Urine	Feces
Adverse effects	Local reactions – swelling, rash Ophthalmologic – cataracts, reduction of visual fields and visual acuity and night vision Hearing impairment Bone abnormalities Pulmonary Neurologic Allergic reactions	Gastrointestinal disturbances Transaminase elevations Agranuloctyosis/ neutropenia Arthralgia	Gastrointestinal disturbances Transaminase elevations Hepatic failure Gastrointestinal bleeding Rise in serum creatinine Proteinuria Rash
Advantages	Long-term data available	May be superior in removal of cardiac iron	The only oral chelator licensed for use in United States
Disadvantages	Compliance problems may be greater	Not licensed for use in United States Variable efficacy in removal of hepatic iron	Long-term data lacking Efficacy at cardiac iron removal not well studied
Special monitoring considerations		Weekly complete blood count with differential	Monthly blood urea nitrogen, creatinine, hepatic transaminases (also obtain 2 weeks after starting the medication) and urinalysis

[a]Adapted from: Kwiatkowski, JL, "Oral Iron Chelators" in Pediatr Clin N Am 55 (2008) with permission.

Studies have shown efficacy similar to that of deferoxamine. Gastrointestinal disturbances including abdominal pain, nausea, vomiting and diarrhea are common and may improve with continued administration of the drug. The gastrointestinal effects may be related to lactose intolerance as lactose is present in the drug preparation. Elevations in hepatic transaminases to more than five times above normal can occur and fulminant hepatic failure has been reported rarely. Liver function tests should be measured every 2 weeks for the first month after starting the medication and tested monthly thereafter. Elevations in serum creatinine are also common, although renal insufficiency is rare. Renal function should be monitored monthly

- Deferiprone (L1) currently is not approved for use in the United States, but is being used in Europe. Controversy exists at present about its potential toxicity, including idiosyncratic neutropenia/agranulocytosis, arthropathy and possible adverse redistribution of iron. Many studies find deferiprone clinically useful without unduly high risk of neutropenia. Preliminary data indicate that deferiprone may be particularly useful in reducing cardiac iron overload either as a single agent, or in combination with deferoxamine.

Splenectomy

1. Splenectomy reduces the transfusion requirements in patients with hypersplenism. It is usually performed in adolescents when transfusion requirements have increased secondary to hypersplenism. Splenectomy is avoided if possible due to the risk of infection, pulmonary hypertension and thromboembolism.
2. Indications for splenectomy include:
   - Persistent increase in blood transfusion requirements by 50% or more over initial needs for over 6 months
   - Annual packed cell transfusion requirements in excess of 250 ml/kg/year in the face of uncontrolled iron overload (ferritin greater than 1,500 ng/ml or increased hepatic iron concentration)
   - Evidence of severe leukopenia and/or thrombocytopenia.
3. At least 2 weeks prior to splenectomy, a polyvalent pneumococcal and meningococcal vaccine should be given. If the patient has not received a *Haemophilus influenzae* vaccine, this should also be given. Following splenectomy, prophylactic penicillin 250 mg bid is given to reduce the risk of overwhelming postsplenectomy infection. Management of the febrile splenectomized patient is detailed in Chapter 31.

Supportive Care

1. Folic acid is not necessary in hypertransfused patients; 1 mg daily orally is given to patients on low transfusion regimens.
2. Hepatitis A and B vaccination should be given to all patients.
3. Appropriate inotropic, antihypertensive and antiarrhythmic drugs should be administered when indicated for cardiac dysfunction.
4. Endocrine intervention (i.e., thyroxine, growth hormone, estrogen, testosterone) should be implemented when indicated.
5. Cholecystectomy should be performed if symptomatic gallstones are present.
6. Patients with high viral loads of hepatitis C that are not spontaneously decreasing, should be treated with PEG-interferon and ribavirin. Ribavirin increases hemolysis and transfusion requirements typically increase during therapy.
7. HIV-positive patients should be treated with the appropriate antiviral medications.
8. Genetic counseling and antenatal diagnosis (when indicated) should be carried out using chorionic villus sampling or amniocentesis.

9. Management of osteoporosis includes:
   - Periodic screening and prevention through early hormonal replacement
   - Yearly bone densitometry and gonadal hormone evaluation should be performed starting at age 10 years
   - Calcium and vitamin D intake should be monitored and supplements administered if poor intake or low vitamin D levels
   - Hormonal replacement therapy (estrogen/progesterone; testosterone) should be administered to those with gonadal insufficiency
   - Encourage physical activity. Discourage smoking
   - Agents used to treat osteoporosis include:
     - Calcitonin: prevents trabecular bone loss by inhibiting osteoclastic activity. Parenteral and intranasal preparations are available. Miacalcin is the intranasal preparation. The dose is 1 spray into alternating nostrils daily. Miacalcin should be taken with calcium carbonate 1,500 mg daily and vitamin D 400 units daily
     - Bisphosphonates: also inhibit osteoclast-mediated bone resorption. Alendronate, pamidronate and zolendronic acid have all been shown to have some efficacy in thalassemia.

Follow up of patients with thalassemia includes:

Monthly:

- Complete blood count
- Complete blood chemistry (including liver function tests, BUN, creatinine) if taking deferasirox
- Record transfusion volume.

Every 3 months:

- Measure height and weight
- Measure ferritin (trends in ferritin used to adjust chelation); perform complete blood chemistry, including liver function tests.

Every 6 months:

- Complete physical examination including Tanner staging, monitor growth and development, dental examination.

Every year:

- Cardiac function – echocardiograph, ECG, Holter monitor (as indicated)
- Endocrine function (TFTs, PTH, FSH/LH, fasting glucose, testosterone/estradiol, FSH, LH, IGF-1, Vitamin D levels)
- Opthalmological examination and auditory acuity

- Viral serologies (HAV, HBV panel, HCV (or if HCV+, quantitative HCV RNA PCR), HIV)
- Bone densitometry
- Ongoing psychosocial support.

Every 1–2 years:

- Evaluation of tissue iron burden
- Liver iron measurement – R2 MRI, SQUID, or biopsy
- T2* MRI measurement of cardiac iron (age >10 years).

Pharmacologic Enhancement of Fetal Hemoglobin Synthesis

High levels of fetal hemoglobin (Hb F) ameliorate the symptoms of β-thalassemia by increasing the hemoglobin concentration of the thalassemic red cells and decreasing the accumulation of unmatched α-chains, which cause ineffective erythropoiesis. Hydroxyurea has been demonstrated to increase Hb F production and mean hemoglobin levels in patients with thalassemia intermedia or Eβ-thalassemia, decreasing or eliminating need for transfusion. Additionally, there are reports of a few β-thalassemia major patients who became transfusion-free using hydroxyurea. Decitabine is another fetal hemoglobin-inducing agent that is currently being studied in thalassemia. Butyric acid analogs and erythropoietin as well as further testing with hydroxyurea are avenues of further investigation. Side effects of these agents include neutropenia, increased susceptibility to infection and possible oncogenicity.

Hematopoietic Stem Cell Transplantation

1. Stem cell transplantation is a curative mode of therapy.
2. Outcome is best for children <17 years with an HLA-identical sibling donor. Overall survival is greater than 90%.
3. The presence of hepatomegaly, liver fibrosis and/or history of poor adherence with chelation therapy has been associated with worse outcome; however, with the use of modified conditioning regimens for those with two or more of these risk factors, outcome is improved.
4. Although limited data are available, the outcome for matched unrelated donor transplantation with high-resolution molecular testing at HLA Class 1 and 2 loci appears to be comparable to matched sibling donor transplantation. Chronic graft versus host disease is seen in 18%.
5. Reduced intensity-conditioning regimens are under study.

Gene Therapy

Research is underway on methods of inserting a normal β-globin gene into mammalian cells. Ultimately, the aim is to insert the gene into stem cells and utilize these for stem cell transplantation.

Management of the Acutely Ill Thalassemic Patient

Acute illness requiring urgent treatment occurs secondary to:

- Sepsis, usually with encapsulated organisms. Iron-overload and chelation with deferoxamine also increase the risk of infection with *Yersinia entercolitica*
- Cardiomyopathy secondary to myocardial iron overload
- Endocrine crises such as diabetic ketoacidosis.

Prevention of these complications should be the primary treatment. Preventive measures include:

- Management of the splenectomized patient as outlined in Chapter 31
- Adequate chelation to prevent secondary hemochromatosis
- Routine monitoring of cardiac and endocrine function.

If a patient presents with signs of shock, the following measures should be instituted:

- Determine hemoglobin, electrolyte, calcium and glucose levels; perform urinalysis
- Obtain blood cultures
- Distinguish between cardiogenic shock and septic shock because the management of each differs. To distinguish between the two, obtain:
  - ECG
  - Echocardiograph, to determine left ventricular contractility
  - Central venous pressure (CVP).
- If the patient is in cardiogenic shock, management includes:
  - Diuretics
  - Inotropic support
  - Careful monitoring of CVP and cardiac output
  - Deferoxamine chelation as a continuous intravenous infusion at a dose of 50–60 mg/kg/day administered over 24 hours. If deferiprone is available, it may be added to deferoxamine to further increase iron excretion.
- If the patient is in septic shock, management consists of:
  - Blood cultures, at least two peripheral sites
  - Broad-spectrum antibiotics IV (e.g., third-generation cephalosporin and an aminoglycoside)
  - Discontinue deferoxamine until infection is under control
  - Fluid boluses of 10 cc/kg normal saline to restore blood pressure
  - Pressors such as dopamine, as indicated
  - Coagulation studies to evaluate for disseminated intravascular coagulation (DIC)
  - CVP monitoring to guide fluid management
  - Arterial blood gas and chest radiograph.

- If the patient is in diabetic ketoacidosis, manage the ketoacidosis in the usual manner with careful monitoring of cardiac function when the patient is being vigorously hydrated.

β-THALASSEMIA INTERMEDIA

Although patients are homozygous or doubly heterozygous, the resultant anemia is milder than in thalassemia major.

Clinical Features

1. Patients generally do not require transfusions and maintain a hemoglobin between 7 and 10 g/dl.
2. Marked medullary expansion, which may result in nerve compression, extramedullary hematopoiesis, hepatosplenomegaly, growth retardation and facial anomalies may occur in untransfused patients.
3. Pulmonary hypertension and increased risk of thrombosis, particularly in splenectomized patients.
4. Patients are most healthy if management is as vigorous as that for thalassemia major.

Management

1. Folic acid 1 mg/day PO should be administered.
2. Iron-fortified foods should be avoided. A cup of tea with every meal will reduce the absorption of nonheme iron.
3. Chelation therapy is required at an older age than in thalassemia major because patients have received fewer transfusions. Ferritin levels may not correlate well with total iron burden (usually lower than expected for the degree of iron loading). Indications for chelation include elevated transferrin saturation of 70% or ferritin >1,000 ng/ml. Liver iron quantitation may also be used to guide treatment.
4. Transfusions generally are not required except during periods of erythroblastopenia (aplastic crises) or during acute infection. If hemoglobin falls below 7 g/dl, growth is poor, or other complications develop, transfusion therapy should be initiated. Children should be monitored for facial bone changes, which can be prevented, but not reversed, by chronic transfusions.
5. Splenectomy may improve hemoglobin level. However, the risk of infection with encapsulated organisms, pulmonary hypertension and hypercoaguability are increased following splenectomy. The relative benefits and risks should be considered when making the decision to perform splenectomy.
6. Cardiac (including evaluation for pulmonary hypertension) and endocrine evaluation and bone densitometry should be performed as in thalassemia major.

β-THALASSEMIA MINOR OR TRAIT (HETEROZYGOUS β⁰ OR β+)

Clinical Features

1. Asymptomatic (physical examination is normal):
   - Discovered on routine blood test – slightly reduced hemoglobin, basophilic stippling, low MCV, normal RDW
   - Discovered in family investigation or family history of heterozygous or homozygous β-thalassemia
   - Confirmed with hemoglobin electrophoresis, demonstrating slightly decreased hemoglobin A (90–95% typically) increased hemoglobin A2 (>3.5%); hemoglobin F mildly elevated in 50% of cases.
2. Thalassemia trait of unusual severity. There are cases of β-thalassemia trait of unusual severity secondary to the coinheritance of α-gene duplication with increased α-globin synthesis, thereby increasing α- and β-chain imbalance, causing a β-thalassemia intermedia phenotype.

α-THALASSEMIAS

The major syndromes resulting from decreased α-chain synthesis are listed in Table 8-16. α-Thalassemia may present as silent carrier, thalassemia trait, hemoglobin H disease, or hydrops fetalis. Hemoglobin H disease is clinically milder than homozygous β-thalassemia and usually does not require regular red cell transfusions. Hemoglobin levels may fall with intercurrent illnesses and patients may require transfusion at such times. Hemoglobin H Constant Spring tends to produce a more severe phenotype; some patients may require chronic red cell transfusion therapy. Hydrops fetalis is not compatible with life and presents with intrauterine or neonatal death, though some babies have survived with fetal packed red blood cell transfusions when antenatal diagnosis was made. These patients should continue on hypertransfusion regimens and be treated like β-thalassemia major, or treated with allogeneic stem cell transplantation.

Differential Diagnosis

The differential diagnosis of the thalassemia syndromes and other microcytic anemias is listed in Table 3-10.

Suggested Reading

Adams RJ, Brambilla D. Discontinuing prophylactic transfusions used to prevent stroke in sickle cell disease. N Engl J Med. 2005;353:2769–2778.
Adams RJ, McKie VC, Hsu L, et al. Prevention of a first stroke by transfusions in children with sickle cell anemia and abnormal results on transcranial Doppler ultrasonography. N Engl J Med. 1998;339:5–11.

Aessopos A, Kati M, Tsironi M. Congestive heart failure and treatment in thalassemia major. Hemoglobin. 2008;32:63–73.

Bhatia M, Walters MC. Hematopoietic cell transplantation for thalassemia and sickle cell disease: past, present and future. Bone Marrow Transplantation. 2008;41:109–117.

Borgna-Pignatti C. Modern treatment of thalassaemia intermedia. Br J Haematol. 2007;138:291–304.

Chui DH, Fucharoen S, Chan V. Hemoglobin H disease: not necessarily a benign disorder. Blood. 2003; 101:791–800.

Cunningham MJ, Macklin EA, Neufeld EJ, Cohen AR. Complications of beta-thalassemia major in North America. Blood. 2004;104:34–39.

Fung EB, Harmatz P, Milet M, et al. Morbidity and mortality in chronically transfused subjects with thalassemia and sickle cell disease: A report from the multi-center study of iron overload. Am J Hematol. 2007;82:255–265.

Kato GJ, Gladwin MT, Steinberg MH. Deconstructing sickle cell disease: reappraisal of the role of hemolysis in the development of clinical subphenotypes. Blood Rev. 2007;21:37–47.

Koshy M, Weiner SJ, Miller ST, et al. Surgery and anesthesia in sickle cell disease. Cooperative Study of Sickle Cell Diseases. Blood. 1995;86:3676–3684.

Kwiatkowski JL. Oral iron chelators. Pediatr Clin North Am. 2008;55:461–482.

Ohene-Frempong K, Weiner SJ, Sleeper LA, et al. Cerebrovascular accidents in sickle cell disease: rates and risk factors. Blood. 1998;91:288–294.

Pegelow CH, Macklin EA, Moser FG, et al. Longitudinal changes in brain magnetic resonance imaging findings in children with sickle cell disease. Blood. 2002;99:3014–3018.

Pennell DJ. T2* magnetic resonance and myocardial iron in thalassemia. Ann N Y Acad Sci. 2005;1054: 373–378.

Platt OS, Thorington BD, Brambilla DJ, et al. Pain in sickle cell disease. Rates and risk factors. N Engl J Med. 1991;325:11–16.

Rosse WF, Gallagher D, Kinney TR, et al. Transfusion and alloimmunization in sickle cell disease. The Cooperative Study of Sickle Cell Disease. Blood. 1990;76:1431–1437.

Rund D, Rachmilewitz E. Beta-thalassemia. N Engl J Med. 2005;353:1135–1146.

Pierre TG, Clark PR, Chua-anusorn W, et al. Noninvasive measurement and imaging of liver iron concentrations using proton magnetic resonance. Blood. 2005;105:855–861.

Vichinsky E. Hemoglobin e syndromes. Hematology Am Soc Hematol Educ Program. 2007;79–83.

Vichinsky EP, Haberkern CM, Neumayr L, et al. A comparison of conservative and aggressive transfusion regimens in the perioperative management of sickle cell disease. The Preoperative Transfusion in Sickle Cell Disease Study Group. N Engl J Med. 1995;333:206–213.

Vichinsky EP, Styles LA, Colangelo LH, et al. Acute chest syndrome in sickle cell disease: clinical presentation and course. Cooperative Study of Sickle Cell Disease. Blood. 1997;89:1787–1792.

Voskaridou E, Terpos E. Pathogenesis and management of osteoporosis in thalassemia. Pediatr Endocrinol Rev. 2008;6(Suppl 1):86–93.

CHAPTER 9

Extracorpuscular Hemolytic Anemia

The causes of hemolytic anemia due to extracorpuscular defects are listed in Table 9-1; they may be immune or nonimmune.

IMMUNE HEMOLYTIC ANEMIA

Immune hemolytic anemia can be either isoimmune or autoimmune. Isoimmune hemolytic anemia results from a mismatched blood transfusion or from hemolytic disease in the newborn. In autoimmune hemolytic anemia (AIHA), shortened red cell survival is caused by the action of immunoglobulins, with or without the participation of complement on the red cell membrane. The red cell autoantibodies may be of the warm type, the cold type, or the cold–warm Donath–Landsteiner type.

Complement participation is usually confined to the IgM type of antibody; only rarely is it associated with IgG. AIHA may be idiopathic or secondary to a number of conditions listed in Table 9-1.

Warm Autoimmune Hemolytic Anemia

Antibodies of the IgG class are most commonly responsible for AIHA in children. The antigen to which the IgG antibody is directed is one of the Rh erythrocyte antigens in more than 70% of cases. This antibody usually has its maximal activity at 37°C and the resultant hemolysis is called warm antibody-induced hemolytic anemia.

Rarely, warm reacting IgA and IgM antibodies may be responsible for hemolytic anemia. As in all patients with AIHA, erythrocyte survival is generally proportional to the amount of antibody on the erythrocyte surface although rarely hemolysis can occur in patients with too few antibodies on the surface of the red cell to cause a positive direct antiglobulin test (DAT-negative hemolytic anemia).

Clinical Features
1. Severe, life-threatening condition.
2. Sudden onset of pallor, jaundice, dark urine.
3. Splenomegaly.

Manual of Pediatric Hematology and Oncology. DOI: 10.1016/B978-0-12-375154-6.00009-4

Table 9-1   Causes of Hemolytic Anemia due to Extracorpuscular Defects

I. Immune
   A. Isoimmune
      1. Hemolytic disease of the newborn
      2. Incompatible blood transfusion
   B. Autoimmune: IgG only; complement only; mixed IgG and complement, other antibody mediated mechanisms
      1. Idiopathic
         a. Warm antibody
         b. Cold antibody
         c. Cold–warm hemolysis (Donath–Landsteiner antibody)
      2. Secondary
         a. Infection, viral: infectious mononucleosis—Epstein–Barr virus (EBV), cytomegalovirus (CMV), hepatitis, herpes simplex, measles, varicella, influenza A, coxsackie virus B, human immunodeficiency virus (HIV); bacterial: streptococcal, typhoid fever, *Escherichia coli* septicemia, *Mycoplasma pneumoniae*(atypical pneumonia)
         b. Drugs and chemicals: quinine, quinidine, phenacetin, *p*-aminosalicylic acid, sodium cephalothin (Keflin), ceftriaxone, penicillin, tetracycline, rifampin, sulfonamides, chlorpromazine, pyradone, dipyrone, insulin; lead
         c. Hematologic disorders: leukemias, lymphomas, lymphoproliferative syndrome, paroxysmal cold hemoglobinuria, paroxysmal nocturnal hemoglobinuria
         d. Immunopathic disorders: systemic lupus erythematosus, periarteritis nodosa, scleroderma, dermatomyositis, rheumatoid arthritis, ulcerative colitis, agammaglobulinemia, Wiskott–Aldrich syndrome, dysgammaglobulinemia, IgA deficiency, thyroid disorders, giant cell hepatitis, Evans syndrome, (immune-mediated anemia associated with immune thrombocytopenia) autoimmune lymphoproliferative syndrome (ALPS), common variable immune deficiency
         e. Tumors: ovarian teratomata, dermoids, thymoma, carcinoma, lymphomas
II. Nonimmune
   A. Idiopathic
   B. Secondary
      1. Infection, viral: infectious mononucleosis, viral hepatitis; bacterial: streptococcal, *E. coli* septicemia, *Clostridium perfringens, Bartonella bacilliformis;* parasites: malaria, histoplasmosis
      2. Drugs and chemicals: phenylhydrazine, vitamin K, benzene, nitrobenzene, sulfones, phenacetin, acetinalimide; lead
      3. Hematologic disorders: leukemia, aplastic anemia, megaloblastic anemia, hypersplenism, pyknocytosis
      4. Microangiopathic hemolytic anemia: thrombotic thrombocytopenic purpura, hemolytic uremic syndrome, chronic relapsing schistocytic hemolytic anemia, burns, post cardiac surgery, march hemoglobinuria
      5. Miscellaneous: Wilson disease, erythropoietic porphyria, osteopetrosis, hypersplenism

Laboratory Findings

1. Hemoglobin level: very low in fulminant disease or normal in indolent disease.
2. Reticulocytosis: common although often the reticulocytes are destroyed by the antibody as well and reticulocytopenia may occur.

3. Smear: prominent spherocytes, polychromasia, macrocytes, autoagglutination (IgM), nucleated red blood cells, erythrophagocytosis.
4. Neutropenia and thrombocytopenia (occasionally).
5. Increased osmotic fragility and autohemolysis proportional to spherocytes.
6. Direct antiglobulin test positive (DAT) (Coomb's positive) established diagnosis of AIHA.
7. Hyperbilirubinemia and increased serum lactate dehydrogenase.
8. Haptoglobin level: markedly decreased.
9. Hemoglobinuria especially at first presentation, increased urinary urobilinogen.

Management

Because this is potentially a life-threatening condition, the following must be monitored carefully:

- Hemoglobin level (every 4 hours)
- Reticulocyte count (daily)
- Splenic size (daily)
- Hemoglobinuria (daily)
- Haptoglobin level (weekly)
- Direct antiglobulin test (DAT) (weekly).

Close attention should always be paid to supportive care issues such as folic acid supplementation, hydration status, urine output and cardiac status.

Treatment

Blood Transfusion

Transfusion should be avoided, where possible, because there will be no truly compatible blood available and the survival of transfused cells in this situation is quite limited and may fail to elevate the hemoglobin level significantly. Nonetheless, using the "least incompatible" blood may be required in properly selected situations in order to avoid cardiopulmonary compromise. The guidelines listed below should be followed:

- If a specific antibody is identified, a compatible donor may be selected. The antibody usually behaves as a panagglutinin and no totally compatible blood can be found
- Washed packed red cells should be used from donors whose erythrocytes show the least agglutination in the patient's serum
- The volume of transfused blood should only be of sufficient quantity to relieve any cardiopulmonary embarrassment from the anemia. Usually aliquots of 5 ml/kg are taken from a single unit and transfused at a rate of 2 ml/kg/h

- The use of such incompletely matched blood is made relatively safe by biologic cross-matching, transfusing of relatively small volumes of blood at any given time and concomitant use of high-dose corticosteroid therapy.

Corticosteroid Therapy

1. Prednisone 2–6 mg/kg/day orally or methylprednisone or 2–4 mg/kg/day IV for 3 days followed by oral prednisone.
2. High-dose corticosteroid therapy should be maintained for several days. Thereafter, corticosteroid therapy in the form of prednisone should be slowly tapered over a 3–4-week period.

The dose of prednisone should be tailored to maintain the hemoglobin at a reasonable level; when the hemoglobin stabilizes, the corticosteroids should be discontinued. The presence of a continued positive direct antiglobulin test does not preclude continuing to taper steroids as long as the hemoglobin is stable or rising and reticulocytosis continues to decrease or remain normal.

About 50% of patients respond within 4–7 days to corticosteroid therapy, but there are a number of patients who continue with profound hemolysis for the first week. For these patients and patients who appear dependent on steroids other alternatives need to be considered.

Intravenous Gammaglobulin

Doses in the range of 1–5 g/kg should be employed. Responses vary but are usually short-lived. It should be considered in patients with severe hemolysis who are requiring transfusion.

Rituximab

In patients with severe disease not responding early on or in patients exhibiting steroid dependence, Rituximab (anti-CD20) should be used in doses of 375 mg/m^2 once a week for 4 weeks. It has a very high rate of remission induction in autoimmune hemolytic anemia in children. The short-term allergic side effects such as itching, hives, hypotension and chest pain have been tolerable (most patients are pre-medicated and monitored carefully during each infusion). Expected increases in infection risk have not been apparent and intravenous gammaglobulin has been administered to offset potential losses of B cell function.

Splenectomy

Splenectomy is indicated if the hemolytic process is brisk despite the use of high-dose corticosteroid therapy, rituximab and transfusions and the patient cannot maintain a reasonable hemoglobin level safely or, if the patient enters a chronic phase.

The results of splenectomy are unpredictable, but it is beneficial in 60–75% of patients. Whenever possible, children should be over 5 years of age and the disease should be present for at least 6–12 months with no significant response to medical treatment.

Plasmapheresis

Plasmapheresis has been successful in slowing the rate of hemolysis in patients with severe IgG-induced immune hemolytic anemia. The effect is short-lived if antibody production is ongoing and success is limited, possibly because more than half of the IgG is extravascular and the plasma contains only small amounts of the antibody as most of the antibody is on the red cell surface. In IgG warm immune hemolytic anemia it should always be combined with an enhanced level of immunosuppression (e.g. rituximab) so that less antibody is being produced as the relatively inefficient plasmapheresis proceeds.

Immunomodulating Agents

1. Mycophenolate mofetil. This drug is showing promise in the treatment of a number of autoimmune diseases including autoimmune hemolytic anemia. It is also effective in Evans syndrome with or without markers for autoimmune lymphoproliferative syndrome. Agents such as this (and the antimetabolites below) often require 4–12 weeks for their effects to begin and are usually started as steroids are being weaned.
2. Cyclosporin has been frequently used in immune cytopenias and in Evans syndrome in patients non-responsive to steroids. However, it is used less often because of the availability of rituximab and mycophenolate mofetil.
3. Danazol. There has been some success with danazol (synthetic androgen), which has a masculinizing effect. Danazol's early effect appears to be due to decreased expression of macrophage Fc-receptor activity.

Antimetabolites

Azathioprine and 6-mercaptopurine: As with the immunomodulators they may take 4–12 weeks to provide a steroid-sparing effect.

Alkylating Agents

Cyclophosphamide, because of its known side effects should only be used in more severe situations which are unresponsive to steroids, Rituximab or immunomodulators.

Mitotic Inhibitors

Vincristine and vinblastine: These drugs are rarely used nowadays, but when given are used as a bridge to suppress hemolysis while waiting for an immunomodulator or cytotoxic agent to begin to work.

Giant Cell Hepatitis and Direct Antiglobulin Test-Positive Autoimmune Hemolytic Anemia

This is a specific rare entity of unknown etiology, although an autoimmune component has been suggested because of the association of direct antiglobulin test (DAT)-positive AIHA and response to immunosuppression.

Clinical Findings

1. Age: 6–24 months, occasionally older age.
2. Fever.
3. Pallor.
4. Jaundice (progressing to cirrhosis and liver failure).
5. Firm hepatomegaly and splenomegaly.
6. Associated convulsions.

Prognosis

1. Poor.

Laboratory Findings

1. Direct antiglobulin test: mixed (IgG and complement); no evidence of other autoimmunity.
2. Hemolytic anemia.
3. Liver function abnormality: high direct bilirubin, transaminase and serum globulin values; prolonged prothrombin time.
4. Liver histology: marked lobular fibrosis, extensive necrosis with central-portal bridging and giant cell transformation.

Treatment

The use of corticosteroids in combination with immunosuppressive therapy (e.g., azathioprine) has met with some success. Vincristine, α-interferon and intravenous immunoglobulin have also been used.

Cold Autoimmune Hemolytic Anemia

IgM antibodies are found less often in association with hemolysis in the pediatric age group. Most IgM autoantibodies that cause immune hemolytic anemia in humans are cold agglutinins and cold hemagglutinin disease is almost always caused by an IgM antibody. The destruction of red blood cells is usually triggered by cold exposure.

Cold hemagglutinin disease usually occurs during *Mycoplasma pneumoniae* infection. It may also occur with other infections, such as infectious mononucleosis, cytomegalovirus

and mumps. Cold hemagglutinin disease or IgM-induced hemolysis is usually due to reaction with antigens of the I/i system. Anti-I is characteristic of *M. pneumoniae*-associated hemolysis and anti-I cold agglutinins are usually found in infectious mononucleosis.
M. pneumoniae adherence to the red cell membrane appears to be mediated by sialic-acid-containing receptors, associated with terminal galactose residues of the I antigen. The association of the infecting organism with the red blood cell may alter the antigenic structure of red blood cell membrane antigen, rendering it immunogenic. In children IgM antibody is usually polyclonal and immunologically heterogeneous.

Clinical Features

This disease may be idiopathic but is more frequently seen in conjunction with infections such as *M. pneumoniae* (atypical pneumonia) and less commonly with lymphoproliferative disorders. The following are the clinical features:

- Hemoglobin is usually normal or mildly decreased. Reticulocyte count may be elevated
- The peripheral blood smear may show agglutination and polychromatophilia
- Spherocytosis is usually absent
- The direct antiglobulin test is positive for complement (polyspecific and anti C3 agents) only and is negative for anti-IgG
- Most blood banks do cold agglutinin testing only when the direct antiglobulin test is positive for complement.

Treatment

1. Control of the underlying disorder.
2. Transfusions may be necessary for patients with significant hemolysis who may be symptomatic. Identification of compatible blood may prove difficult and the blood bank may have to release least incompatible blood. Warming the blood to 37°C during administration by means of a heating coil or water bath is indicated to avoid further temperature activation of the antibody. Efficient in-line blood warmers (McGaw Water Bath; Fenwall Dry Heat Warmer) are designed to deliver blood at 37°C to the patient. Unmonitored or uncontrolled heating of blood is extremely dangerous and should not be attempted. Red cells heated too long are rapidly destroyed *in vivo* and can be lethal to the patient.
3. Warm the patient's room. Keeping a patient warm will help diminish hemolysis and peripheral agglutination.
4. Plasmapheresis is very efficient for the treatment of IgM disease as IgM is largely intravascular. Patients with severe hemolysis should undergo plasmapheresis. If the blood is obtained at 37°C, with the patient's arm warmed by hot pads, the warm unit can be separated quickly by centrifugation and the red cells returned to the patient through an efficient in-line blood warmer.

5. Drug therapy. If the anemia is severe, a drug trial is appropriate. Rituximab and cyclophosphamide have been used with plasmapheresis. Steroids are of marginal value in cold agglutinin disease.

Paroxysmal Cold Hemoglobinuria (PCH) Due to Donath–Landsteiner Cold Hemolysin

This is an unusual IgG antibody with anti-P specificity and a cold thermal amplitude, originally described in cases of syphilis. This antibody, although uncommon, is now most frequently found in young children with viral infections. Hemolysis is most commonly intravascular as a result of the unusual complement-activating efficiency of this IgG antibody.

Clinical Features

The most common clinical finding is a sudden bout of hemolysis with a drop in hemoglobin and hemoglobinuria. The hemoglobin drop is often serious enough to require transfusion (and sudden death from this disease has been reported). Children usually have a short-lived, explosive illness where the antibody is only produced for a short time. Although blood for transfusion will appear compatible, all red cells carry the P blood group specificity that the antibody is directed against.

Laboratory Findings

A positive complement test is present on antiglobulin testing and this should lead to testing for the Donath Landsteiner antibody (the IgG cold binding antibody) in the absence of an obvious IgM cold agglutinin.

Treatment

Keeping a patient warm is the mainstay of treatment and warming blood in a blood warmer prior to transfusion is important. Patients may respond to corticosteroids, unlike IgM-induced cold hemolysis. Plasmapheresis may also be effective in life-threatening PCH.

NONIMMUNE HEMOLYTIC ANEMIA

This group of conditions is due to extracorpuscular causes of hemolytic anemia in which the direct antiglobulin (Coombs) test is negative. The various causes are listed in Table 9-1. Conditions caused by various infections, drugs and underlying hematologic disease respond to treatment of the underlying condition, as well as the necessary acute supportive care including red cell transfusions as needed.

Microangiopathic Hemolytic Anemia

Microangiopathic hemolytic anemia (MAHA) is a result of diverse causes that have in common a relatively uniform hematologic picture and in general a common pathogenesis. Table 9-2 lists the various causes of MAHA.

Diagnosis

The blood smear is characterized by the presence of burr erythrocytes, schistocytes, helmet cells and microspherocytes. This occurs in association with evidence of hemolysis and usually, but not invariably, thrombocytopenia. The severity of both the anemia and the thrombocytopenia, as well as the degree of compensatory erythroid response, varies greatly. Intravascular hemolysis occurs in all forms; plasma hemoglobin levels may be elevated, haptoglobin absent, hemosiderinuria present and urinary iron excretion increased in the more chronic forms.

Elevated serum fibrin degradation products in some cases of MAHA may represent evidence of associated DIC. The thrombocytopenia is due to consumption of platelets in the microthrombi and is an example of excessive platelet destruction rather than a failure of

Table 9-2   Causes of Microangiopathic Hemolytic Anemia

Renal disease
  Hemolytic uremic syndrome
  Renal vein thrombosis
  Renal transplantation rejection
  Radiation nephritis
  Chronic renal failure
Cardiac conditions
  Malignant hypertension
  Coarctation of aorta
  Severe valvular heart disease
  Subacute bacterial endocarditis of aortic valve
  Intracardiac prosthesis
Liver disease
  Severe hepatocellular disease
Infections
  Disseminated herpes infection
  Meningococcal septicemia
  Cerebral falciparum malaria
Hematologic
  Thrombotic thrombocytopenic purpura (hereditary or secondary) (Chapter 12)
Miscellaneous
  Severe burns
  Giant hemangioma (Kasabach–Merritt syndrome)
  Disseminated intravascular coagulation of any causation; sometimes accompanied by consumption of circulating coagulation factors (consumption coagulopathy)

production. The bone marrow, therefore, shows normal numbers of megakaryocytes together with erythroid hyperplasia. Acute forms of MAHA are sometimes accompanied by disseminated intravascular coagulation (DIC).

Hypersplenism

Whether splenic enlargement is caused by infection or is secondary to such diseases as thalassemia, portal hypertension, or storage diseases, a shortened red cell survival with excessive sequestration can be demonstrated in many patients with clinical splenomegaly. Typically, hypersplenism is accompanied by moderate neutropenia and thrombocytopenia with active erythropoiesis, myelopoiesis and thrombopoiesis in the marrow. There may also be mild spherocytosis. Splenectomy is followed by the return to normal of the blood values.

Wilson Disease

Wilson disease is a rare inherited disease of copper metabolism that leads to copper deposition most prominently in the liver and central nervous system (CNS). It has an autosomal recessive inheritance pattern and usually presents with liver or CNS symptoms. Wilson disease rarely presents with anemia which is normochromic and normocytic without an intense reticulocytosis or indirect hyperbilirubinemia. Because of the lethal nature of the disease without treatment and the potential successful treatment if the disease is detected early, Wilson disease should be considered in any patient with unexplained hemolytic anemia that has no abnormal morphology and a negative direct antiglobulin test.

Suggested Reading

Dacie J. The Haemolytic Anaemias 3. The Auto-Immune Haemolytic Anaemias. 3rd ed. Edinburgh: Churchill Livingstone; 1992.

Gertz M. Management of cold haemolytic syndrome. British J. Haemat. 2007;138:422–429.

Giulino L, Bussel JB, Neufeld E. Treatment with Rituximab in benign and malignant hematologic disorders in children. J. Pediatrics. 2007;150:338–344.

Glader B. Autoimmune Hemolytic Anemia. In: Arceci R, Hann I, Smith O, eds. Pediatric Hematology. Boston: Blackwell Publishing Ltd.; 2006.

Lanzkowsky P. Hemolytic anemia. Pediatric Hematology Oncology: A Treatise for the Clinician. New York: McGraw-Hill; 1980.

CHAPTER 10

Polycythemia

The term polycythemia, particularly as it pertains to the newborn and children, should be more accurately termed erythrocytosis because it generally refers to conditions in which only erythrocytes are increased in number and volume usually as an appropriate response to various causes of hypoxia, the presence of high-oxygen-affinity hemoglobins (reduced P50 in whole blood) or increased production of erythropoietin or other circulating erythropoietic stimulating factors. True polycythemia, on the other hand, is due to congenital (germline) erythropoietin receptor or acquired (somatic) mutations that make erythroid progenitor cells exquisitely sensitive to circulating cytokines resulting in intrinsically hyperactive erythropoiesis.

POLYCYTHEMIA (ERYTHROCYTOSIS) IN THE NEWBORN

A venous hematocrit reading of more than 65% or a venous hemoglobin concentration in excess of 22.0 g/dl any time during the first week of life should be considered evidence of polycythemia. Capillary blood samples should not be relied on for the diagnosis of polycythemia because they are significantly higher than venous hemoglobin or venous hematocrit and vary with the temperature of the extremity from where the sample is taken. Hematocrit values determined on a microcentrifuge include a small amount of trapped plasma and have a higher value than hematocrit values determined from automated analyzers.

The incidence is 0.4–4.0% of all births and the incidence of neonatal polycythemia is higher at high altitudes than at sea level. Neonatal polycythemia is an appropriate physiological response to intrauterine hypoxia and due to increased oxygen affinity of fetal hemoglobin. This perinatal elevation of red cell mass is transient as it is associated with a rapid drop of erythropoietin (EPO) and ensuing decreased red cell production. The causes of neonatal polycythemia are listed in Table 10-1.

Symptoms

Symptoms are a consequence of the increase in blood viscosity. Hematocrit up to 65% has a linear correlation with viscosity and beyond 65% has an exponential relationship. Viscosity depends on a number of factors (Table 10-2). Table 10-3 lists the clinical and

Manual of Pediatric Hematology and Oncology. DOI: 10.1016/B978-0-12-375154-6.00010-0

Table 10-1   Causes of Neonatal Polycythemia

I. **Intrauterine hypoxia**
   A. Placental insufficiency
      1. Small-for-gestational age (SGA) (intrauterine growth factor)
      2. Dysmaturity
      3. Postmaturity
      4. Placenta previa
      5. Maternal hypertension syndromes (toxemia of pregnancy)
   B. Severe maternal cyanotic heart disease
   C. Maternal smoking
II. **Hypertransfusion**
   A. Twin-to-twin transfusion
   B. Maternal to fetal transfusion
   C. Placental cord transfusion (delayed cord clamping, cord stripping, third stage of labor underwater at body temperature, holding baby below mother with cord attached)
III. **Endocrine causes**
   A. Congenital adrenal hyperplasia
   B. Neonatal thyrotoxicosis
   C. Congenital hypothyroidism
   D. Maternal diabetes mellitus
IV. **Miscellaneous**
   A. Chromosomal abnormalities
      1. Trisomy 13
      2. Trisomy 18
      3. Trisomy 21 (Down syndrome)
   B. Beckwith-Wiedemann syndrome (hyperplastic visceromegaly)
   C. Oligohydramnios
   D. Maternal use of propranolol
   E. High altitude conditions
   F. High oxygen affinity hemoglobinopathies (Table 10-4)

laboratory findings and complications in neonatal polycythemia. Some of the symptoms may result from the underlying cause such as intrauterine hypoxia, maternal diabetes or placental insufficiency.

Laboratory Findings

When polycythemia is due to maternofetal transfusion, the following laboratory findings may be present:

- Increased quantities of immunoglobulin (IgA and IgM) in the infant's serum
- Reduction in fetal hemoglobin to less than 60%
- The presence of red cells bearing maternal blood group antigens in the baby's circulation and, if the infant is a male, the presence of XX cells of maternal origin in the baby's circulation.

Table 10-2 Factors Increasing Viscosity

1. Hematocrit >60%
2. Larger MCV (mean cell volume)
3. Decreased deformability of fetal erythrocytes
4. Plasma protein levels especially high fibrinogen
5. Decreased flow rate – Vessel diameter and endothelial integrity, e.g. increased levels of erythropoietin, in addition to inducing erythrocytosis, may induce other effects such as:
   a. a hematocrit-independent, vasoconstriction-dependent hypertension
   b. upregulation of tissue rennin
   c. increased endothelin production
   d. stimulation of endothelial and vascular smooth muscle proliferation
   e. change in vascular tissue prostaglandin production
   f. stimulation of angiogenesis

Table 10-3 Clinical and Laboratory Findings and Complications in Neonatal Polycythemia

Clinical Findings		
"Feeding problems" (20%)	Hypotonia (7%)	Hepatomegaly
Plethora (20%)	Tremulousness (7%)	Vomiting
Cyanosis (15%)	Difficult to arouse	Tachycardia
Lethargy (15%)	Weak suck	Cardiomegaly
Respiratory distress (9%)	Easily startled	Jaundice
Laboratory Findings		
Venous hemoglobin >22g/dl	Unconjugated	Chest radiograph
Venous hematocrit >65%	hyperbilirubinemia (22%)	• Increased vascularity
Thrombocytopenia	Hypoglycemia (12–40%)	• Pleural fluid
Reticulocytosis	Hypocalcemia (1–11%)	• Hyperaeration
Normoblastemia	Hypomagnesemia	• Alveolar infiltrates
Increased blood viscosity	EEG abnormal	• Cardiomegaly
(normal 12.1 cP ± 3.9)	ECG abnormal	
Presence of IgM or IgA in serum		
Complications		
Transient tachypnea of newborn	Intracranial hemorrhage (<1.0%)	Acute renal failure
Respiratory distress	Peripheral gangrene	Testicular infarction
Respiratory distress	Priapism	Disseminated intra-
Congestive heart failure	Necrotizing enterocolitis	vascular coagulation
Convulsions	Ileus	

(%) indicates the percentage of frequency of the symptoms.

When polycythemia is due to intrauterine hypoxia it is usually accompanied by an increase in the nucleated red blood cells (nRBC) in the blood during the early neonatal period. The mean value of nRBC in the first few hours of life in a healthy full term neonate is 500 nRBC /mm^3 or 0–10 nRBC/100 WBC. A value of greater than 1,000 nRBC/mm^3 or 10–20 nRBC/100 WBC is considered abnormal. The other hematologic indices of fetal hypoxia include higher absolute lymphocyte counts and lower platelet counts in comparison with normal full-term neonates without hypoxia during fetal life.

Table 10-4   Classification of Polycythemia

I. Relative polycythemia (hemoconcentration, dehydration)

II. Primary polycythemia (results from somatic or germline mutations of erythroid progenitor cells that make them exquisitely sensitive to erythropoietin or other cytokines)

Congenital: Erythropoietin receptor mutation (results from germline mutation)

Acquired: Polycythemia vera (results from somatic mutation)

III. Secondary polycythemia

A. Insufficient oxygen delivery (also known as *appropriate* polycythemia since it results from a normal response of erythron to hypoxia)

   1. Physiologic

      a. Fetal life

      b. Low environmental O_2 (high altitude)

   2. Pathologic

      a. Impaired ventilation: cardiopulmonary disease, obesity

      b. Pulmonary arteriovenous fistula

      c. Congenital heart disease with left-to-right shunt (e.g., tetralogy of Fallot, Eisenmenger syndrome)

      d. Abnormal hemoglobins (reduced P_{50} in whole blood)

         1. Methemoglobin (congenital and acquired)

         2. Carboxyhemoglobin

         3. Sulfhemoglobin

         4. High oxygen affinity hemoglobinopathies[a] (hemoglobin Chesapeake, Ranier, Yakima, Osler, Tsurumai, Kempsey and Ypsilanti)

         5. 2,3-DPG Mutase deficiency in red cells resulting in 2-3 bisphosphoglycerate (BPG) deficiency.

B. Increase in erythropoietin (also known as *inappropriate* polycythemia since it results from an aberrant production of erythropoietin or other growth factors)

   1. Endogenous

      a. Renal: Wilms' tumor[b], hypernephroma, renal ischemia, e.g. renal vascular disorder, congenital polycystic kidney, benign renal lesions (cysts, hydronephrosis), renal cell carcinoma. Post-renal transplantation erythrocytosis (occurs in 10–15% of renal graft recipients). Contributing factors include persistence of erythropoietin secretion from the recipients's diseased and ischemic kidney and secretion of angiotensin II androgen and insulin-like growth factor.

      b. Endocrine: pheochromocytoma, Cushing's syndrome, congenital adrenal hyperplasia, adrenal adenoma with primary aldosteronism

      c. Liver: hepatoma, focal nodular hyperplasia[c], hepatocellular carcinoma, hepatic hemangioma, Budd-Chiari syndrome (some of these patients may have overt or latent myeloproliferative disorder)

      d. Cerebellum: hemangioblastoma, hemangioma, meningioma

      e. Uterus: leiomyoma, leiomyosarcoma

      f. Ovaries: Dermoid cysts

   2. Exogenous

      a. Administration of testosterone and related steroids

      b. Administration of growth hormone

C. Polycythemia with characteristics of both primary and secondary polycythemias

      a. Chuvash polycythemia

      b. Non-Chuvash polycythemias

IV. Neonatal polythemia (Erythrocytosis) (see Table 10-1)

[a]Some of these are electrophoretically silent and require hemoglobin oxygen association kinetics for diagnosis.

[b]Associated with male gender, low clinical stage and usually >16 years of age. May occur with a normal serum erythropoietin.

[c]Histologically stains positively for erythropoietin by immunohistochemistry.

Treatment

Because instruments to measure viscosity are not clinically available, neonatal hyperviscosity is diagnosed by a combination of symptoms and an abnormally high hematocrit.

Treatment should be reserved for infants with respiratory, cardiac, or central nervous system (CNS) symptoms and a venous hematocrit of 65–69% or an asymptomatic infant with a venous hematocrit >70%. All polycythemic infants however, should be carefully monitored for evidence of hypoglycemia, hypocalcemia and hyperbilirubinemia. Treatment should be designed to reduce the venous hematocrit to approximately 50–55%. This can be accomplished by a partial exchange transfusion, using 5% human albumin, Ringer's lactate or normal saline. It is better to avoid the use of fresh frozen plasma because it may potentially transmit infectious agents. Normal saline or Ringer's lactate solutions have the advantages that they are easily available and equally effective. However, it is important to take into account the patient's renal status and serum sodium level when a decision is made to use albumin, Ringer's lactate or normal saline to avoid sodium overload. Partial exchange has a favorable effect on cerebral blood flow velocity in newborn infants with polycythemia because it reduces the hematocrit while maintaining blood volume.

The following formula is employed to approximate the volume of exchange required to reduce the hematocrit reading to the desired level:

$$\text{Volume of exchange (ml)} = \frac{\text{blood volume (ml)} \times (\text{observed Hct} - \text{desired Hct})}{\text{observed Hct}}$$

Partial exchange transfusion has been shown to increase capillary perfusion, cerebral blood flow and cardiac function and reduces the risk of tissue damage caused by ischemia in various organs, resulting from severe slowing in the microcirculation due to a high hematocrit and low shear rates. However, there is little evidence that the long-term outcome of infants is improved by the procedure.

POLYCYTHEMIA IN CHILDHOOD

The term polycythemia applies to an increase in circulating red cell mass to above the normal upper limits of 30 ml/kg body weight (excluding hemoconcentration due to dehydration). A hemoglobin level greater than the 99th percentile of method-specific reference range for age, sex and altitude of residence should be applied. For practical purposes, this means a hemoglobin level higher than 17 g/dl or a hematocrit level of 50% or more during childhood.

Table 10-4 classifies various causes of polycythemia.

Primary polycythemia results from congenital [(germline) erythropoietin receptor mutation] or acquired [(somatic) polycythemia vera] mutations that make erythroid progenitor cells proliferate independently or excessively in response to extrinsic regulators and are exquisitely sensitive to circulating cytokines resulting in increased red cell mass. Low serum erythropoietin (EPO) is their hallmark.

Secondary polycythemia on the other hand results from the action of an excessive amount of circulating cytokines on the normal responsive erythroid progenitor cells. The cytokine usually is erythropoietin. However, in some clinical conditions, non-erythropoietin growth factors (e.g. angiotensin II androgens and insulin-like growth factor I in recipients of renal graft during post-transplantation period) may also play a role in inducing erythrocytosis.

POLYCYTHEMIA VERA

Polycythemia vera (PV) is a clonal disorder of the multipotent hematopoietic stem cell that manifests as excess production of normal erythrocytes, with low EPO levels (EPO levels may be normal in the presence of Budd–Chiari syndrome, iron-deficiency anemia or post-phlebotomy) and variable overproduction of leukocytes and platelets. It is one of the *Philadelphia chromosome negative myeloproliferative disorders* and can usually be differentiated from them by the predominance of erythrocyte production.

Pathophysiology

The biology of PV is characterized by clonality and EPO independence. In PV, a single clonal population of erythrocytes, granulocytes, platelets and variable clonal B-cells arises when a hematopoietic stem cell gains a proliferative advantage over other non-mutated stem cells. EPO independence is the ability of erythroid progenitors (BFU-E) to form colonies without EPO.

Genome-wide scanning which compared clonal PV and nonclonal cells from the same individuals revealed a loss of heterozygosity in chromosome 9p, found in approximately 30% of patients with PV. This is not a classical chromosomal deletion, but rather a duplication of a portion of the chromosome and the loss of the corresponding parental region. This process is called *uniparental disomy*. The 9p region contains a gene encoding for the JAK2 tyrosine kinase. The JAK2, a member of family of kinases, transmits the activating signal in the EPO-EPO receptor (EPOR) signaling pathway. A mutation in an autoinhibitory JAK2 domain, known as *JAK2V617F* is a point mutation in exon 14 leading to a valine-to-phenylalanine mutation at codon 617 of the JAK2 gene, was discovered leading to a gain-of-function mutation affecting the kinase which at least partly explains EPO hypersensitivity/independence. The end result is constitutive phosphorylation of the JAK2 tyrosine kinase, thus the proliferative advantage seen in PV. Over 95% of patients with PV carry the *JAK2V617F* mutation, as well as approximately 50% of adults with essential thrombocythemia and idiopathic myelofibrosis. This mutation has also been identified in isolated case reports in adults in the following hematologic disorders:

- Chronic neutrophilic leukemia
- Myelodysplastic syndrome with ring sideroblasts and thrombocytosis
- Chronic eosinophilic leukemia (rare)
- Juvenile chronic myelomonocytic leukemia (rare).

There are compelling data against *JAK2V617F* being a disease-initiating mutation but rather that the *JAK2V617F* mutation plays a major role in behavior of the polycythemia vera clone. Most of the leukemic transformation, however, arises from *JAK2V617F* negative PV progenitor cells.

Clinical Features

PV in children is rare. The incidence in adults is approximately 10–20 per 100,000, of which 1% is present before the age of 25 and 0.1% present before the age of 20. They usually present with an elevated hemoglobin and/or hematocrit found on routine testing. Some patients are asymptomatic while others may have had various nonspecific symptoms recognized retrospectively to be consistent with PV. In adults, 33% will present with thrombosis or hemorrhage; thrombosis is about equally distributed between arterial and venous thromboses including cerebrovascular accidents, myocardial infarction, deep venous thrombosis and pulmonary embolism. Less frequent, but more specific for PV is Budd-Chiari syndrome (hepatic vein thrombosis). In children about 20–30% may present with Budd-Chiari syndrome.

Less than 5% will have erythromelalgia: erythema and warmth of the distal extremities and especially the hands and feet with a painful burning sensation that can progress to digital ischemia. Erythromelalgia is associated with augmented platelet aggregation and frequently responds within hours to low- or regular-dose aspirin therapy. Less commonly, PV presents with neurological symptoms due to spinal cord compression by extramedullary hematopoiesis and elevated uric acid with associated gout due to increased cell turnover. Hemorrhagic presentations are usually mild with gum bleeding and easy bruising although serious gastrointestinal hemorrhage can occur. About 40% of patients will experience life-altering pruritis. Typically the pruritis is worse after a warm shower or bath, known as aquagenic pruritis. This has been attributed to increased numbers of mast cells and elevated histamine levels. Potential physical findings include plethora and ruddiness of the face, erythromelalgia of the distal extremities, bruising and splenomegaly.

Diagnosis

The WHO criteria listed in Table 10-5 are used for diagnosis. While the presence of EPO-independent erythroid colony is specific for PV, this test is difficult and not generally available. *JAK2V617F* mutation may be expressed less commonly in children than in adults,

Table 10-5 Revised WHO Criteria

Diagnosis requires the presence of both major criteria and one minor criterion *or* the presence of the first major criterion together with two minor criteria.

Major criteria
1. Hemoglobin > 18.5 g/dl in men, 16.5 g/dl in women or other evidence of increased red cell volume (hemoglobin or hematocrit > 99th percentile of method-specific reference range for age, sex, altitude of residence or hemoglobin > 17 g/dl in men, 15 g/dl in women if associated with a documented and sustained increase of at least 2 g/dl from the individual's baseline value that cannot be attributed to correction of iron deficiency, or elevated red cell mass > 25% above mean normal value)
2. Presence of *JAK2V617F* or other functionally similar mutation such as *JAK2* exon 12 mutation

Minor criteria
1. Bone marrow biopsy showing hypercellularity for age with trilineage growth (panmyelosis) with prominent erythroid, granulocytic and megakaryocytic proliferation (not validated in prospective studies)
2. Serum erythropoietin level below the reference range for normal
3. Endogenous erythroid colony formation *in vitro*

which once again suggests a disease-initiating mutation not yet identified and raises the question of the applicability of the WHO criteria in children.

Treatment

Treatment of PV depends on whether the disease is in the plethoric (proliferative) or the spent phase when the bone marrow is transitioning into myelofibrosis. In the plethoric phase, the goal of treatment is controlling thrombotic episodes by restraining monoclonal proliferation with cytoreductive therapy.

- *Phlebotomy* is performed to maintain hemoglobin levels of 16–17 g/dl (i.e., less than 20 g/dl). Most patients can tolerate removal of 450–500 ml of blood every 2–4 days. As more blood is removed and the patient becomes iron-deficient, the hematocrit becomes easier to control and the phlebotomy schedule should be adjusted accordingly. However, in some patients the iron deficiency can become symptomatic and can cause neurocognitive impairment and decreased athletic abilities. Although phlebotomy is effective for controlling erythrocytosis, it does not affect the variable leukocytosis, thrombocytosis or thromboembolic events found in polycythemia vera. Nevertheless, phlebotomy is recommended therapy.
- *Low-dose aspirin* is employed to reduce the risk of thromboembolic events and results in a minor but statistically significant decreased risk of cardiovascular death, nonfatal myocardial infarction, nonfatal stroke, pulmonary embolism and major venous thrombosis without a significant increase in rates of hemorrhage.
- *Chemotherapeutic cytoreductive therapy*: Cytoreductive therapy or pegylated interferon is indicated as follows:

- History of thrombosis or transient ischaemic attacks (TIA)
- Leukocytosis, as it increases the risk for thrombosis
- A platelet count greater than 1.5 million/mm^3. Platelet counts at this level are a risk factor for bleeding due to an acquired von Willebrand disease. Acquired von Willebrand disease may be protective against thrombosis.

The following cytoreductive therapy is used:

- Hydroxyurea. Initial dose of 20–30 mg/kg daily. This dose is adjusted depending on the hematological response or signs of toxicity. The safety and efficacy is unclear in pediatric patients. Hydroxyurea reduces the risk of thrombosis compared to phlebotomy or phlebotomy and aspirin. While very effective in reducing cell counts its potential as a leukemogenic agent has to be considered. Unlike alkylating agents and radioactive phosphorus that lead to an increase in fatal PV leukemic transformation, such an association for hydroxyurea has not been proven
- Interferon (pegylated interferon-alfa-2a [Pegasys®]) achieves complete hematological response in a high percentage of cases and some patients achieving complete molecular response of the *JAK2V617F* that is durable. Most patients tolerate pegylated interferon well and no vascular events have been recorded. The acceptable tolerance, efficacy and extremely low leukemic risk may make interferon a first-line therapy in the future. Its role will be evaluated in a pending randomized trial comparing Pegasys® to hydroxyurea. The initial dose of interferon is 90 µg subcutaneously weekly for 2 weeks. The dose is then escalated every 2 weeks (and in the absence of toxicity) to 135 µg subcutaneously weekly and then to 180 µg subcutaneously weekly if no hematological response occurs at a lower level. Safety and efficacy is unclear in patients younger than 18 years old
- Anagrelide is also useful to decrease platelet counts. An induction dose of Anagrelide in children of 0.5 mg twice daily, followed by a maintenance dose of 0.5–1.0 mg twice a day adjusted to maintain a platelet count within the normal range is employed. The dosage is adjusted to the lowest effective dosage required to reduce and maintain the platelet count below 600,000/mm^3 and ideally to maintain it in the normal range.

Standard management of cardiovascular risk factors such as smoking, hypertension, diabetes and hyperlipidemia will also decrease the risk of vascular events.

PRIMARY FAMILIAL AND CONGENITAL POLYCYTHEMIA

Primary familial and congenital polycythemia (PFCP) is a primary polycythemia that is an autosomal dominant condition where the defect exists in the erythroid progenitor and thus presents with a low EPO level. In contrast to polycythemia vera, PFCP does not present with leukocytosis, thrombocytosis, or splenomegaly and does not progress to myelofibrosis or leukemia. Although PFCP is a rare disease, it is frequently misdiagnosed as PV. To date, 12 erythropoietin receptor (EPOR) mutations associated with PFCP have been described. These EPOR mutations lead to a hyperfunctional EPO receptor (by a gain-of-function mutation) involving deletion of the cytoplasmic negative regulatory subunit of EPOR. Those with phenotypic expression are generally asymptomatic, however there may be a predisposition of these families to cardiovascular disease and other thrombotic complications. Phlebotomy should be only used for those patients who have hyperviscosity symptoms.

CONGENITAL POLYCYTHEMIA DUE TO ALTERED HYPOXIA SENSING WITH NORMAL P50

Chuvash Polycythemia and Other Von Hippel Lindau (VHL) Mutations

Chuvash polycythemia is an endemic polycythemia found with high frequency on the west bank of the Volga River in the Chuvash Autonomous Republic in western Russia, the Italian island of Ischia and sporadically world-wide in all ethnic and racial groups. It is an autosomal recessive disorder characterized by a loss-of-function mutation of the VHL gene that delays ubiquitin degradation of HIF-1 (hypoxia inducible factor) and HIF-2 resulting in the upregulation of transcription in a number of target genes including EPO and vascular-endothelial growth factor (VEGF). Specific to the Chuvash polycythemia is a cytosine to thymine change at nucleotide 598 of the VHL gene. This mutation results in defective hypoxic sensing of kidney cells and increased production of EPO. Because EPO can be high, normal or increased, Chuvash polycythemia can be grouped with the secondary inappropriate polycythemias. However, because the erythroid progenitors in Chuvash polycythemia are hyper-responsive to EPO it also has features of primary polycythemia.

Clinical Manifestations

Patients with Chuvash polycythemia have normal arterial blood gases and normal P50. They often have a relatively low blood pressure, varicose veins and benign vascular abnormalities and increased risk of pulmonary hypertension. There is an increased risk for arterial and venous thrombotic and hemorrhagic complications and strokes, but no greater predisposition for developing malignancies typical of VHL.

Other congenital VHL mutations have been described in which there are simple heterozygous, compound heterozygous and even homozygous genotypes. Typically the congenital polycythemia is due to compound heterozygosity, the Chuvash mutation with another VHL mutation. These patients may present with isolated erythrocytosis, elevated EPO level and a normal P50. No cases have developed VHL syndrome-associated tumors.

Table 10-6   Clinical Manifestations of Polycythemia Vera (PV), Primary Familial and Congenital
Polycythemia (PFCP) and Chuvash Polycythemia (CP)

Clinical Entities	Polycythemia Vera (PV)	Primary Familial and Congenital Polycythemia (PFCP)	Chuvash Polycythemia (CP)
Frequency	Rare	Unknown	Unknown
Inheritance	None	Dominant	Recessive
Underlying Cause	None	Erythropoietin receptor mutation is found only in 12%	Functional deficiency of VHL
Symptoms of polycythemia (e.g. headache, dizziness lethargy, blurred vision)	Present	Usually diagnosed on routine blood count	Present
Signs	Plethora, Splenomegaly	Plethora, no splenomegaly	Plethora, no splenomegaly Varicosities of peripheral veins
Erythropoietin level	Undetectable	Normal or low	Increased but high or normal in Sporadic non-CP
Course	Thrombosis or hemorrhage	Benign	Thrombosis or hemorrhage
Diagnosis		Molecular analysis for truncation of cytosolic portion of ER and *in-vitro* hyper-sensitivity to EPO	Molecular analysis of VHL protein gene mutation. Increased levels of VEGF and PAI-1
Treatment	Phlebotomy, α-IFN, ASA, HU, Anagrelide	Phlebotomy	Phlebotomy

Abbreviations: VHL, Von Hippel-Lindau; VEGF, vascular endothelial growth factor; PAI-1, plasminogen activators inhibitor; α-IFN, α-interferon; ASA, aspirin; HU, hydroxyurea.

Table 10-6 compares the clinical manifestations of PV, PFCP and Chuvash polycythemia.

Hypoxia-Induced Factor 2α (HIF) Mutations

HIF mutations are a family of transcription factors with the important role of regulating EPO gene transcription. These mutations are composed of two subunits, HIFα and HIFβ. Although there are three forms of HIFα (HIF1α, HIF2α and HIF3α), HIF1α is the main transcription factor controlling renal EPO, while HIF2α is the main isoform of EPO regulation in liver and brain. Patients with a mutation of HIF2α will usually have isolated polycythemia and normal to elevated EPO levels.

PROLYL HYDROXYLASE DOMAIN (PHD) 2 MUTATIONS

PHD-containing enzymes hydroxylate HIF2α increasing the binding to VHL, thus leading to ubiquitin-mediated proteasome degradation. Patients with PHD mutations present with an isolated polycythemia and normal to elevated EPO levels. In one case a patient heterozygous for an 1121 A>G missense mutation was also described with recurrent paragangliomas. The paraganglioma demonstrated loss-of-heterozygosity of the PHD2 region thus suggesting that PHD2 is also a tumor suppressor gene.

CONGENITAL POLYCYTHEMIA DUE TO ALTERED HYPOXIA SENSING WITH DECREASED P50

High-Affinity Hemoglobinopathies

High-affinity hemoglobinopathies are autosomal dominant conditions. Most of these mutations occur within the β-globin chain where α1 and β2 chains contact. This change impairs intramolecular rotation or 2,3-biphosphoglycerate (BPG) binding making hemoglobin unable to transition from high-oxygen-affinity to low-oxygen-affinity states thus causing tissue hypoxia and compensatory polycythemia. Hemoglobin electrophoresis is insufficient to identify hemoglobin structural defects and some hemoglobin mutants will be missed. The only reliable screening test is P50 measurement either by the Hemox-Analyzer or calculated from venous blood gases. Often the mutation is identified by sequencing globin genes.

Treatment

Phlebotomy is usually not beneficial because it causes decrease in exercise tolerance.

2,3-BPG Deficiency

2,3-bisphosphoglycerate (BPG), also known as 2,3-DPG, promotes hemoglobin transition from a high-oxygen-affinity state to a low-oxygen-affinity state. 2,3-BPG binds to the central compartment of the hemoglobin tetramer changing its conformation and shifting the oxygen disassociation curve to the right. The deficiency is created by ineffective bisphosphoglyceratemutase (BPGM), a red cell enzyme of the early glycolytic pathway that converts 1,3-BPG to 2,3-BPG. Mutations of BPGM are extremely rare and are typically autosomal recessive. Diagnosis is confirmed by establishing a decreased P50 and excluding other hemoglobin mutants and by establishing a decreased 2,3-BPG quantity and BPGM enzyme activity.

Treatment

Phlebotomy is usually not beneficial because it causes decrease in exercise tolerance.

Methemoglobinemia

Methemoglobinemia is usually considered when the patient is cyanotic with low oxygen saturation by pulse oximetry, yet PaO_2 levels are normal. Methemoglobin levels are often included in blood gas measurements. Methemoglobin is generated when the oxygen-carrying ferrous iron (Fe^{2+}) has been oxidized to ferric iron (Fe^{3+}) thus unable to carry oxygen. In normal physiological conditions the body will reduce the methemoglobin to a level of <2%, via the enzymes cytochrome b5 and cytochrome b5 reductase (methemoglobin reductase or b5R). Congenital methemoglobinemia, an autosomal recessive disorder, is most commonly due to a cytochrome b5 reductase (b5R) deficiency. Methemoglobinemia caused by various mutations of globin genes, known as hemoglobins M, is autosomal dominant.

OTHER CAUSES OF POLYCYTHEMIA

Physiologically inappropriate polycythemia is often due to exogenous sources of EPO. Several malignancies, e.g., hepatoma, renal cell carcinoma, uterine myomas and cerebellar hemangiomas have been shown to produce EPO. Large bulky tumors produce erythrocytosis by mechanical interference with the blood supply to the kidneys resulting in false sensing of hypoxia and EPO production. Renal polycythemia is due to EPO produced by renal cysts, polycystic disease, or hydronephrosis.

Endocrine disorders such as pheochromocytomas, aldosterone-producing adenomas, Barter syndrome and dermoid cysts of the ovary can result in inappropriate EPO production through mechanical interference with renal blood supply or hypertensive damage to renal parenchyma resulting in false sensing of hypoxia by the kidneys.

Postrenal transplantation erythrocytosis occurs in some patients following kidney transplantation and is associated with dysregulation of angiotensin receptor. These patients respond to drugs that cause inactivation of the renin–angiotensin system, e.g. captopril, enalapril, losartan, lisinopril and fosinopril. Patients unable to tolerate angiotensin-converting enzyme inhibitors can be treated with an angiotensin II AT1 receptor antagonist, losartan. Androgens increase hematocrit by two mechanisms – stimulation of EPO production and an independent hyperproliferative effect on erythrocyte precursors.

DIAGNOSTIC APPROACH

Figure 10-1 shows a diagnostic algorithm for the diagnosis of polycythemia in children. The initial step in the diagnosis of polycythemia is to apply the appropriate age-specific reference range for confirmation and then repeat the laboratory study as the hemoglobin concentration may reflect transient decrease of plasma volume due to dehydration causing hemoconcentration (relative polycythemia). A determination has to be made as to whether

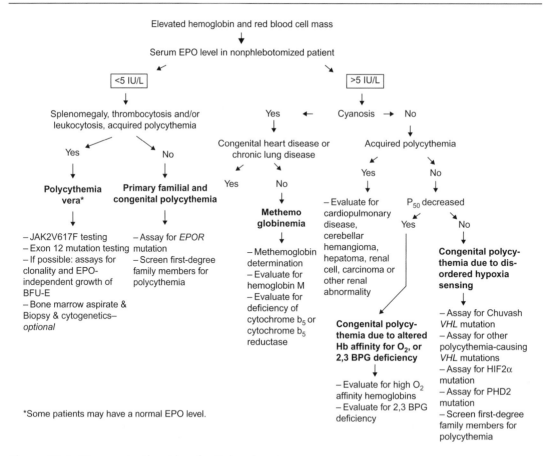

Figure 10-1 Diagnostic Algorithm for Polycythemia.
Abbreviation: EPO, erthropoietin; EPOR, erthropoietin receptor; VHL, vonHippel Lindau; 2,3 BPG, 2,3 biphosphiglycerate.

the increased hemoglobin level is acquired or congenital and whether it is familial. If the hemoglobin level is persistently elevated, hypoxia should be considered as the most common cause. An arterial oxygen saturation level (S_aO_2) of <92% suggests cardiac or pulmonary etiologies. The complete blood count (CBC), serum EPO level, the P50 level (partial pressure of oxygen in blood at which 50% of the hemoglobin is saturated with oxygen) and finally red cell and plasma volume studies to rule out spurious polycythemia due to chronic contraction of plasma volume (*Gaisbock syndrome*) should be done.

The CBC may reveal increased leukocytes, platelets and erythrocytes which often coexist in *polycythemia vera* (PV) along with splenomegaly. PV, an acquired clonal disorder, is a prototype of primary polycythemia. *Primary familial congenital polycythemia (PFCP)* is another primary polycythemia that presents with isolated elevated erythrocytes without leukocytosis or thrombocythemia and with a history of autosomal dominant inheritance (although *de novo* cases have been known to occur). An EPO level will be low in primary polycythemias; while in *secondary polycythemias* the EPO level will be inappropriately elevated or high normal that is inappropriate to the high hemoglobin level. The disorders resulting from high hemoglobin oxygen affinity such as high-affinity hemoglobin mutants or low 2,3-BPG concentrations are diagnosed with a decreased P50 from a Hemox-Analyzer, an instrument which records blood oxygen equilibrium curves. If a Hemox-Analyzer is not available, the P50 value can be calculated from freshly obtained venous blood gasses by applying the formula: $\log PO_2 (7.4) = \log PO_2 (observed) - [0.5(7.4 - pH(observed))]$. An abnormally decreased P50 pressure is <20 mmHg (Figure 10-1).

Suggested Reading

Cario H. Childhood polycythemias/erythrocytoses: classification, diagnosis, clinical presentation and treatment. Ann Hematol. 2005;84:137.

Cario H, McMullin MF, et al. Clinical and hematological presentation of children and adolescents with polycythemia vera. Ann Hematol. 2009;88:713.

Carobbio A, Finazzi G, et al. JAK2V617F allele burden and thrombosis: A direct comparison in essential thrombocythemia and polycythemia vera. Exp Hem. 2009;37:1016.

Carobbio A, Finazzi G, et al. Thrombocytosis and leukocytosis interaction in vascular complications of essential thrombocythemia. Blood. 2008;112:3135.

DiNisio M, Barbui T, et al. The haematocrit and platelet target in polycythemia vera. Br J Haem. 2007;136:249.

Gordeuk VR, Stockton DW, Prchal JT. Congenital polycythemias/erythrocytoses. Haematologica. 2005;90:110.

Kiladjian J, Cassinat B, et al. Pegylated interferon-alfa-2a induces complete hematologic and molecular responses with low toxicity in polycythemia vera. Blood. 2008;112:3065.

Landolfi R, Di Gennaro L, et al. Leukocytosis as a major thrombotic risk factor in patients with polycythemia vera. Blood. 2007;109:2446.

Landolfi R, Marchioli M, et al. Efficacy and safety of low-dose aspirin in polycythemia vera. New Eng J Med. 2004;350:114.

Najean Y, Rain J. Treatment of polycythemia vera: The use of hydroxyurea and pipobroman in 292 patients under the age of 65 years. Blood. 1997;90:3370.

Nussenzvieg R, Swierczek S, et al. Polycythemia vera is not initiated by JAK2V617F mutation. Exp Hem. 2007;35:32.

Patnaik MM, Tefferi A. The complete evaluation of erythrocytosis: congenital and acquired. Leukemia. 2009;23:834.

Prchal JT. Classification and molecular biology of polycythemias (erythrocytoses) and thrombocytosis. Hematol Oncol Clin N Am. 2003;17:1151–1158.

Prchal JT, Sokol L. "Benign erythrocytosis" and other familial and congenital polycythemias. Eur J Haematol. 1996;57:263–268.

Tefferi A, Thiele J, Vardiman JW. The 2008 World Health Organization classification system for myeloproliferative neoplasms: order out of chaos. Cancer. 2009;115(17):3842–3847.

Teofili L, Giona F, et al. Markers of myeloproliferative diseases in childhood polycythemia vera and essential thrombocythemia. J Clin Onc. 2007;25:1048.

Teofili L, Giona F, et al. The revised WHO diagnostic criteria for Ph-negative myeloproliferative diseases are not appropriate for the diagnostic screening of childhood polycythemia vera and essential thrombocythemia. Blood. 2007;110:3384.

CHAPTER 11

Disorders of White Blood Cells

QUANTITATIVE DISORDERS OF LEUKOCYTES

The total white blood cell count and the differential count are valuable guides in the diagnosis, treatment and prognosis of various childhood illnesses.

Leukocytosis

Table 11-1 lists the causes of leukocytosis. The normal leukocyte counts and the absolute counts of different classes of leukocytes vary with age in children and their ranges are listed in Appendix 1. Leukocytosis may be acute or chronic and may result from an increase in one or more specific classes of leukocytes.

Table 11-2 lists the causes of monocytosis and monocytopenia, Table 11-3 the causes of basophilia and Table 11-4 the causes of neutrophilia. Eosinophils and lymphocytes are discussed later in this chapter.

It is important to calculate the absolute count for each white blood cell (WBC) class rather than the relative percentage count for the purposes of quantitative interpretation. If nucleated red blood cells (NRBC) are present, the total WBC count includes the total nucleated cell count (TNCC). Under these circumstances, the true total WBC count is calculated by subtracting the absolute NRBC count from the TNCC. This correction is generally required in the hemolytic anemias and in newborns.

Blood smear examination of the white cell morphology is important in the diagnosis of various causes of leukocytosis. For example, in severe infections or other toxic states, the neutrophils may contain fine deeply basophilic granules (toxic granulations) or larger basophilic cytoplasmic masses (Döhle bodies). Vacuolization of neutrophils may also occur. Döhle bodies are also found in pregnancy, burns, cancer, May Hegglin anomaly and many other conditions.

Manual of Pediatric Hematology and Oncology. DOI: 10.1016/B978-0-12-375154-6.00011-2

Table 11-1   Causes of Leukocytosis

Physiologic	Poisoning
Newborn (maximal 38,000/mm^3)	Lead
Strenuous exercise	Mercury
Emotional disorders; fear, agitation	Camphor
Ovulation, labor, pregnancy	Acute hemorrhage
Acute infections	Malignant neoplasms
Bacterial, viral, fungal, protozoal, spirochetal	Carcinoma
Metabolic causes	Sarcoma
Diabetic coma	Lymphoma
Acidosis	Connective tissue diseases
Anoxia	Rheumatic fever
Azotemia	Rheumatoid arthritis
Thyroid storm	Inflammatory bowel disease
Acute gout	Hematologic diseases
Burns	Splenectomy, functional asplenia
Seizures	Leukemia and myeloproliferative disorders
Drugs	Hemolytic anemia
Steroids	Transfusion reaction
Epinephrine	Infectious mononucleosis
Endotoxin	Megaloblastic anemia during therapy
Lithium	Postagranulocytosis
Ranitidine	
Serotonin	
Histamine	
Heparin	
Acetylcholine	

In infants and children, there is a tendency to release immature granulocytes into the circulation and the white blood cell count may reach very high levels (>50,000/mm^3). This is called a leukemoid reaction. The shift to the left may be so marked as to suggest myeloid leukemia. Table 11-5 lists the distinguishing features of leukemoid reaction and true leukemia.

Leukopenias

Leukopenia exists when the total white blood cell count is less than 4,000/mm^3. Leukopenia may result from decrease in one or more specific classes of leukocytes. The causes of neutropenia are listed in Table 11-6, lymphopenia in Table 11-12 and monocytopenia in Table 11-2. Leukopenia can result from a number of conditions. However, isolated leukopenia resulting from a decrease in all classes of leukocytes is observed uncommonly.

Table 11-2   Causes of Monocytosis and Monocytopenia

Monocytosis
- Hematologic disorders
  - Leukemia
  - Acute myelogenous leukemia
  - Chronic myelogenous leukemia
  - Lymphoma (Hodgkin and non-Hodgkin)
  - Chronic neutropenia
  - Histiocytic medullary reticulosis
  - Recovery from myelosuppressive chemotherapy
- Connective tissue disorders
  - Systemic lupus erythematosus
  - Rheumatoid arthritis
  - Myositis
- Granulomatous diseases
  - Inflammatory bowel disease
  - Sarcoidosis
- Infections
  - Subacute bacterial endocarditis
  - Tuberculosis
  - Syphilis
  - Rocky Mountain spotted fever
  - Kala-azar
- Malignant disease (usually carcinomas)
- Miscellaneous disorders
  - Postsplenectomy state
  - Tetrachlorethane poisoning
  - Lipoidoses (e.g., Niemann–Pick disease)

Monocytopenia
- Glucocorticoid administration
- Infections associated with endotoxemia

Table 11-3   Causes of Basophilia

- Hypersensitivity reactions
  - Drug and food hypersensitivity
  - Urticaria
- Inflammation and infection
  - Ulcerative colitis
  - Rheumatoid arthritis
  - Influenza
  - Chickenpox
  - Smallpox
  - Tuberculosis
- Myeloproliferative diseases
  - Chronic myeloid leukemia
  - Myeloid metaplasia

Table 11-4   Causes of Neutrophilia

Increased production
   Clonal disease
      Myeloproliferative disorders
         Chronic myelogenous leukemia
         Chronic neutrophilic leukemia
      Juvenile myelomonocytic leukemia
      Transient myeloproliferative disorder of Down syndrome
   Hereditary
      Autosomal dominant form of hereditary neutrophilia
      Familial cold urticaria
   Reactive
      Chronic infection
      Chronic inflammation
         Juvenile rheumatoid arthritis
         Inflammatory bowel disease
         Kawasaki disease
      Hodgkin disease
      Drugs: Lithium, G-CSF, GM-CSF, chronic use of corticosteroids
      Leukemoid reaction
      Chronic idiopathic neutrophilia
Increased mobilization from marrow storage pool
   Drugs: Corticosteroids, G-CSF
   Stress
      Acute infection
      Hypoxia
Decreased Margination
   Exercise
   Epinephrine
Decreased egress from circulation
   Leukocyte adhesion deficiency (LAD)
      LAD type I: deficiency of CD 11/ CD 18 integrins on leukocytes
      LAD type II: absence of neutrophil sialyl Lewis X structures
Asplenia

Modified from: Dinaur MC The phagocyte system and disorders of granulopoiesis and granulocyte function: In Nathan and Oski's Hematology of Infancy and Childhood, 5th Edition, 1998, W.B. Saunders Company, Philadelphia, with permission.

Neutropenia

Neutropenia is defined as a decrease in the absolute neutrophil count (ANC). The ANC is calculated by multiplying the total WBC count by the percentage of segmented neutrophils and bands. In whites, neutropenia is defined as an ANC of less than $1,000/mm^3$ in infants between 2 weeks and 1 year of age and less than $1,500/mm^3$ beyond 1 year of age. African Americans have lower counts with ANC levels $200–600/mm^3$ less than in whites. Neutropenia can be transient or chronic. Neutropenia is considered "chronic" when it persists beyond 6 months.

Table 11-5   Features of Leukemoid Reaction and Leukemia

Feature	Leukemoid Reaction	Leukemia
Clinical	Evidence of infection	Hepatosplenomegaly Lymphadenopathy
Hematologic	No anemia No thrombocytopenia	Anemia Thrombocytopenia
Bone marrow	Normal, hypercellular	Blasts Decreased megakaryocytes Decreased erythroid precursors
Leukocyte alkaline phosphatase	High	Absent

Table 11-6   Causes of Neutropenia

I. **Decreased production**
   A. Congenital
      1. Neutropenia in various ethnic groups[a]
      2. Hereditary
         a. Severe congenital neutropenia: sporadic (most common) or autosomal dominant or Kostmann disease—autosomal recessive
         b. Familial benign chronic neutropenia—autosomal dominant
      3. Chronic benign neutropenia[b]
      4. Reticular dysgenesis
      5. Cyclic neutropenia
      6. Neutropenia associated with agammaglobulinemia and dysgammaglobulinemia
      7. Neutropenia associated with abnormal cellular immunity in cartilage–hair hypoplasia
      8. Neutropenia associated with pancreatic insufficiency (Shwachman–Diamond syndrome and Pearson syndrome); (Chapter 6)
      9. Neutropenia associated with hyperimmunoglobulin M syndrome
      10. Neutropenia associated with metabolic disease
         a. Glycogen storage disease (type IB)
         b. Idiopathic hyperglycinemia
         c. Isovaleric acidemia
         d. Methylmalonic acidemia
         e. Propionic acidemia
         f. Thiamine-responsive anemia in DIDMOAD syndrome (see Chapter 4)
         g. Barth Syndrome (see p.289)
      11. Bone marrow failure (Chapter 6)
         a. Fanconi anemia
         b. Familial congenital aplastic anemia without anomalies
         c. Dyskeratosis congenita
      12. Bone marrow infiltration: osteopetrosis, cystinosis, Gaucher disease, Niemann–Pick disease
   B. Acquired
      1. Acute
         a. Acute transient neutropenia
         b. Viral infection (e.g., HIV, EBV, hepatitis A and B, respiratory syncytial virus, measles, rubella, varicella, influenza)

(Continued)

Table 11-6   (Continued)

<blockquote>

      c. Bacterial infection (e.g., typhoid, paratyphoid, tuberculosis, brucellosis)

      d. Rickettsial infection (e.g., ehrlichiosis)

  2. Chronic

      a. Bone marrow aplasia

        (1) Idiopathic

        (2) Secondary: drugs, chemicals, irradiation, infection, immune reaction, malnutrition, copper deficiency, vitamin B_{12} deficiency, folate deficiency

      b. Bone marrow infiltration, neoplastic

        (1) Primary: leukemia

        (2) Secondary: neuroblastoma, lymphoma, rhabdomyosarcoma

II. Failure to release mature neutrophils from the bone marrow (myelokathexis and WHIM Syndrome [warts, hypogammaglobulinemia, infections, myelokathexis]) (ineffective myelopoiesis)

  A. Cortisone stimulation test (Table 11-8)

III. Increased margination of neutrophils (pseudoneutropenia)

  A. Epinephrine stimulation test (Table 11-8)

IV. Increased destruction

  A. Immune

    1. Drug induced (e.g., anticonvulsants)

    2. Alloimmune (Isoimmune)

      a. Maternofetal

      b. Multitransfusion

    3. Autoimmune neutropenia

      a. Idiopathic[a]

      b. Secondary: systemic lupus erythematosus, lymphoma, leukemia, rheumatoid arthritis, HIV infection (in 20–44% of AIDS patients), infectious mononucleosis, associated with autoimmune thrombocytopenia and/or autoimmune hemolytic anemia

  B. Nonimmune

    1. Infections

    2. Hypersplenism

</blockquote>

[a]Chronic, mild with a benign course.
[b]Probably same condition as idiopathic autoimmune neutropenia.

Severity and duration of neutropenia correlates with susceptibility to develop various types of bacterial infections. Severity of neutropenia is graded according to ANC as follows:

- Severe neutropenia: ANC less than $500/mm^3$
- Moderate neutropenia: ANC $500–1,000/mm^3$
- Mild neutropenia: ANC $1,000–1,500/mm^3$.

Neutropenic patients are usually infected with their own endogenous bacterial flora that resides in the mouth, oropharynx, gastrointestinal tract and skin. For this reason, the frequency of Gram-negative bacterial infections and *Staphylococcus aureus* infections is high in these patients. Neutropenia, alone per se, does not predispose them to parasitic, viral or fungal infections.

Benign ethnic neutropenia is observed in a variety of populations, including Africans, West Indians, Yemenite and Ethiopian Jews, Bedouin Arabs and Jordanians. In these groups, an ANC as low as 1,000/mm^3 may be considered as normal. There is no increased infection risk in these individuals.

Clinical Features

Severe neutropenia has the following common clinical manifestations:

- High fever, chills, severe prostration and irritability
- Extensive necrotic and ulcerative lesions: oropharyngeal and nasal tissues, skin, gastrointestinal (GI) tract, vagina and uterus
- Gram-negative septicemia.

The risk of infection is inversely proportional to the absolute neutrophil count (ANC). When the ANC falls below 1,000/mm^3, stomatitis, gingivitis and cellulitis dominate the clinical picture. More severe infections occur when the ANC is below 500/mm^3 with perirectal abscesses, pneumonia and sepsis being common.

Granulocyte colony-stimulating factor (G-CSF) produces a sustained neutrophil recovery in patients with severe chronic neutropenia, reduces the incidence and severity of infection and improves the quality of life. The drug is tolerated well and adverse effects are transient and mild.

Table 11-6 lists the causes of neutropenia and Figure 11-1 shows an approach to the diagnosis of neutropenia.

DECREASED POLYMORPHONUCLEAR LEUKOCYTE PRODUCTION

Table 11-7 summarizes the features of some of the congenital neutropenias.

Severe Congenital Neutropenia (SCN) and Kostmann Disease (KD)

Epidemiology

Severe congenital neutropenia (SCN) includes a heterogeneous group of disorders with different patterns of inheritance. Kostmann disease follows autosomal recessive pattern of inheritance. Its underlying genetic defect is thought to be due to homozygous mutations in the HAX 1 gene on chromosome 1. Other SCN may follow autosomal-dominant or sporadic patterns of inheritance and in this group of patients most of them ($\sim 60\%$) have diverse mutations in the neutrophil elastase gene (ELA-2). These mutations affect only one allele. The majority of patients present with a sporadic pattern, since autosomal dominant inheritance is relatively more lethal. In some patients, there may be germline mosaicism.

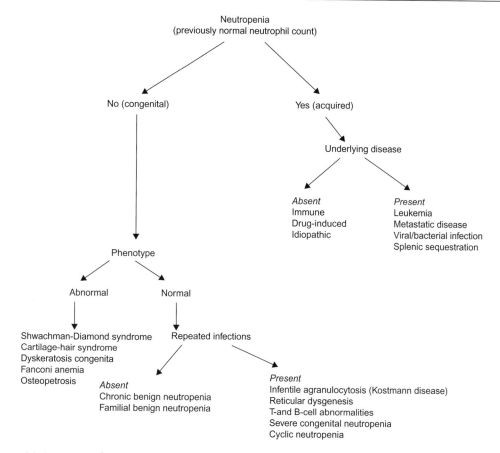

Figure 11-1 Approach to Diagnosis of Neutropenia.
From: Roskos RR, Boxer LA. Clinical Disorders of Neutropenia. Pediatr Rev 1991;12:202–218.

Incidence

The incidence of SCN is 2 per million population.

Clinical Manifestations

During the first year of life omphalitis, otitis media, upper respiratory tract infections, pneumonitis, skin abscesses, liver abscesses occur commonly with positive cultures for staphylococci, streptococci, pseudomonas, peptostreptococcus and fungi. Splenomegaly may be present. Other manifestations include the following:

- Blood counts reveal a normal WBC with an ANC less than $200/mm^3$ and a compensatory eosinophilia and monocytosis. Mild anemia and thrombocytosis may be present

Table 11-7   Clinical and Hematologic Features of Some Congenital Neutropenias

Disorders	Severe Congenital Neutropenia (SCN) Kostmann Disease (KD)	Familial benign neutropenia	Chronic benign neutropenia[a]	Reticular dysgenesis[b]
Inheritance	Autosomal recessive (KD) Autosomal dominant or sporadic (SCN)	Dominant	Not hereditary	Not hereditary
Severity	Severe illness Life-threatening pyogenic infections in first months of life	Variable; benign to severe infections	Benign	Severe, fatal thymic dysplasia Lymphoid hypoplasia
Clinical findings	Skin infection, Aphthous ulcers, Septicemia, Meningitis, Peritonitis, Lung abscess, Lymphadenopathy, Splenomegaly (20%)	Less troublesome to severe infections	Paronychia Gingivitis Impetigo: mild infections, localized	Severe bacterial and viral infection Neonatal death
Hematologic Findings	Anemia, Neutropenia $<200/mm^3$, Monocytosis, Eosinophilia, Risk of leukemia	Neutropenia, usually $<300/mm^3$ Monocytosis	No anemia Absent mature PMN Some band forms Monocytosis	Neutropenia Lymphopenia
Marrow findings	↑ Promyelocytes Absent MM, B, PMN ↑ Monocytes ↑ Eosinophils ↑ Plasma cells	↓ MM, B, PMN "Maturation arrest"	Absent PMN Normal myeloid cells to band stage; lymphocytes increased	Absent myeloid and absent lymphoid cells Normal thrombopoiesis and erythropoiesis
Treatment	Antibiotics Supportive measures G-CSF Stem cell transplantation	No therapy G-CSF, if indicated	Antibiotics, as indicated G-CSF, if indicated	Stem cell transplantation

[a]Most of these cases are autoimmune in origin and pathogenesis, even though antineutrophil antibodies may not always be demonstrated. This condition is probably the same as idiopathic autoimmune neutropenia.
[b]Failure of stem cells committed to myeloid and lymphoid development.
Abbreviations: B, bands; MM, metamyelocytes; PMN, polymorphoneuclear leukocytes; G-CSF, granulocyte colony-stimulating factor.

- Bone marrow: Bone marrow examination shows a maturation arrest of myelopoiesis at the promyelocyte or myelocyte stage with marked paucity of mature neutrophils. There is an increase in monocytes, eosinophils, macrophages and plasma cells.

Pathogenesis

In vitro bone marrow studies show reduced number of granulocyte-macrophage colonies in SCN patients. There is also a reduced number of CD34+/Kit+/G-CSFR+ myeloid progenitor cells in the bone marrow.

Neutrophil elastase gene (ELA-2) mutations: It has been hypothesized that mutations of ELA-2 in SCN result in a high rate of premature apoptosis in neutrophil precursors, which results in decreased myelopoiesis. Neutrophil elastase is a serine protease localized in the granules of neutrophils and monocytes. A mutant enzyme has a dominant negative effect on the normal wild-type elastase. This explains the defective proteolysis in the SCN neutrophils even though half of the normal amount of the elastase is present in the neutrophils of these patients.

The gene responsible for the autosomal recessive form of SCN in the original cases described by Kostmann has been recently identified as mutations in the HAX1 gene located on chromosome 1, causing deficiency of the protein. HAX1 is an inhibitor of apoptosis. In its absence, apoptosis proceeds unchecked. G-CSF can overcome this deficiency by activating other anti-apoptotic pathways.

Another gene defect recently reported in SCN is GFI-1 (growth factor-independent-1) on chromosome 1 q22.

Treatment

G-CSF

The initial dose of G-CSF employed is 5 µg/kg/day. The response occurs 7–10 days from the start of treatment. The goal of therapy is to achieve an ANC of approximately 1,000–1,500/mm^3 and maintain the patient free of infections. More than 95% of patients with SCN will respond to G-CSF. After beginning G-CSF therapy, the dose should be adjusted up or down at 1–2-week intervals until the lowest effective dose is reached.

Complications associated with the use of G-CSF include: bone pain, splenomegaly, hepatomegaly, thrombocytopenia, osteopenia/osteoporosis, Henoch-Schonlein purpura type of immune-complex-induced vasculitis of the skin and/or glomerulonephritis.

A baseline bone marrow cytogenetics should be obtained prior to G-CSF therapy. Initial cytogenetic studies of bone marrow at diagnosis usually are normal. However, during the course of the disease, clonal abnormalities may emerge, 50% of which are monosomy 7. Since 12% of patients with SCN develop myelodysplastic syndrome (MDS) and/or acute myelogenous leukemia (AML), it is important to perform periodic bone marrow examinations for morphology and cytogenetic studies in the follow-up of these patients. Patients who require higher doses of G-CSF (more than 8 µg/kg/d) are at higher risk to develop MDS/AML than those who are more G-CSF responsive (40% vs. 11% after 10 years of therapy).

G-CSF receptor is normal in patients with SCN. However, patients with SCN are predisposed to develop acquired (somatic) mutations of the cytoplasmic domain of G-CSF receptor. There is a good correlation between the development of leukemia/MDS and the acquisition of G-CSF receptor mutations in patients with SCN. The time interval between these two events varies considerably.

Hematopoietic Stem Cell Transplantation (HSCT)

The following are the indications for HSCT:

- Patients who require greater than 100 μg/kg/day dose of G-CSF
- Refractoriness to G-CSF treatment
- Emergence of MDS/AML.

However, some experts favor the treatment with HSCT for all patients with SCN who have a HLA-matched sibling donor available.

The leading causes of death in SCN are infection and MDS/leukemia.
Patients with severe chronic neutropenia should be registered with the Severe Chronic Neutropenia International Registry. This registry collects and maintains long-term outcome data as well as provides resources for physicians, patients and families.

Reticular Dysgenesis

Reticular dysgenesis is a disorder of stem cells in which maturation of both myeloid and lymphoid lineages is defective. Platelet and red cell production is normal. Affected individuals have severe neutropenia and moderate to severe lymphopenia. In addition, there is absence of peripheral lymphoid tissues, Peyer's patches, tonsils and splenic follicles. The mortality rate is high from infection at an early age. Treatment: HSCT can be curative.

Cyclic Neutropenia

1. Rare disorder; presents in infancy or childhood; may be benign, but 10% die from overwhelming infection. Pneumonia and sepsis due to peritonitis (especially from *Clostridium perfringens*) are the most common causes of death.
2. Autosomal-dominant mode of inheritance in some patients and sporadic in others.
3. Genetic and molecular studies by linkage analysis have localized the genetic defect to chromosome 19p13.3. Gene sequencing revealed mutations in the gene for neutrophil elastase (ELA-2), a serine protease synthesized during the promyelocyte/myelocyte stage. It has been speculated that neutropenia results from activation of apoptotic pathway by mutant forms of ELA-2.
4. Marked neutropenia (ANC less than 500/mm^3, usually less than 200/mm^3) at regular intervals, usually every 21 days, persisting for 3–6 days, ANC then increases to the

lower limit of normal, about 2,000/mm^3 and remains approximately at this level until the next neutropenic period. Cycles may vary from 14 to 40 days.

5. Coincident with neutropenia, the patient may manifest the following clinical signs:
   - Fever
   - Ulceration of oral, vaginal, or rectal mucous membranes
   - Gingivitis, stomatitis
   - Furunculosis, cellulitis
   - Perirectal abscess
   - Cervical lymphadenopathy
   - Fatal Clostridial bacteremia from gastrointestinal tract ulcers
   - Other infections (e.g., mastoiditis, pneumonia, adenitis).

6. Hematology
   - Reciprocal monocytosis during neutropenic nadir
   - Oscillations in reticulocyte and platelet counts are observed in most patients and fluctuations of eosinophil and lymphocyte counts are seen in some patients
   - Bone marrow shows absent late myeloid precursors before development of neutropenia, but during the neutropenic phases, the marrow shows myeloid hyperplasia
   - The diagnosis is usually established by obtaining a complete blood count 2–3 times per week (preferably 3) for a period of at least 6 weeks to document the cycling.

7. Management includes:
   - Antibiotic therapy (as indicated)
   - Diligent dental care is recommended
   - Prophylactic therapy with G-CSF may be recommended for patients with cyclic neutropenia to prevent the severe dental complications and also, to prevent development of life-threatening infections. G-CSF is started at a dose of 5 μg/kg/day. The dose is adjusted at 1–2-week intervals until the lowest effective dose is achieved. It may be administered daily or on alternate days as needed. G-CSF will not prevent the cycling but will decrease the length of the nadir.

Granulocyte-macrophage colony-stimulating factor (GM-CSF) is used less frequently for long-term treatment because of its side effects. It is less effective than G-CSF in elevating the ANC but is effective in reducing the nadir in cyclic neutropenia.

MDS/leukemia has not been reported to date in patients with cyclic neutropenia.

Neutropenia Associated with X-linked Agammaglobulinemia

1. Inherited as X-linked recessive trait, caused by mutations in the gene encoding a tyrosine kinase known as Bruton's or B-cell tyrosine kinase (BTK).

2. Severe decrease in serum IgG, IgA and IgM levels and marked decrease in B-lymphocytes, but normal T-cell function.
3. BTK gene is also expressed in myeloid cells and its product participates in signal transduction for myeloid maturation.
4. Patients with neutropenia are more prone to develop fungal or *Pneumocystis carinii* infections.
5. Neutropenia is observed only when rapid production of these cells is required.
6. A short course of G-CSF with antibiotic may be used, when needed, although no evidence-based data are available.

Neutropenia Associated with Autosomal Recessive Agammaglobulinemia

Autosomal recessive agammaglobulinemia due to mutations of the gene encoding for μ heavy chain is also associated with neutropenia and absent B cells. A short course of G-CSF may be used, when needed, although no evidence-based data are available.

Neutropenia Associated with Abnormal Cellular Immunity in Cartilage–Hair Hypoplasia

This condition is characterized by:

- Autosomal-recessive mode of inheritance; found in Amish and Finnish population
- Short-limbed dwarfism
- Fine hair
- Moderate to severe neutropenia and increased susceptibility to infection
- Lymphopenia, macrocytic anemia
- Impaired cellular immune function due to impaired T-cell function caused by mutations in the RMRP gene, which encodes the RNA component of a ribonuclear protein ribonuclease – diminished delayed skin hypersensitivity and rejection of skin allograft
- Treatment – allogeneic stem cell transplantation. There are no reports of the use of G-CSF in this disorder.

Neutropenia Associated with Common Variable Immunodeficiency (CVID)

CVID is used to describe patients with hypogammaglobulinemia of undetermined origin. IgG level is low, but IgM and IgA levels vary. *In vitro* response of B-cells to pokeweed mitogen to produce immunoglobulin is impaired in patients whose B-cells fail to differentiate into mature cells.

Clinical Manifestations

CVID usually becomes manifest in patients between 20–30 years of age and only occasionally in childhood. Most patients manifest recurrent infections of the sinopulmonary tract. They are also prone to develop noncaseating granulomas of the skin, gut and other organs.

These patients are predisposed to develop autoimmune disorders such as ITP, SLE and thyroiditis.

Treatment

Neutropenia in CVID is explained on an autoimmune basis. Patients respond to G-CSF treatment, IV immunoglobulin and antibiotics.

Myelokathexis and WHIM (Warts, Hypogammaglobulinemia, Infections, Myelokathexis) Syndrome

Inheritance

Autosomal dominant.

Hematologic Findings

1. Moderate to severe neutropenia. Neutrophils and eosinophils contain vacuoles, prominent granules, nuclear hypersegmentation with pyknotic nuclei connected to each other by thin filaments. Normal morphology of lymphocytes, monocytes and basophils. Neutrophil function is usually normal.
2. Bone marrow shows granulocytic hyperplasia and neutrophils with similar degenerative changes as in the blood, i.e. congenital dysmyelopoietic neutropenia. Ineffective myelopoiesis due to increased apoptosis of neutrophils in the bone marrow because of decreased expression of the antiapoptotic factor, *bcl-x*.

Treatment

Prompt response to G-CSF or GM-CSF. Immunoglobulin levels return to normal after treatment with G-CSF.

Selective IgA Deficiency and Neutropenia

Selective IgA deficiency may be associated with neutropenia that is autoimmune in nature. It is not known if the use of G-CSF is effective in this condition.

Dubowitz Syndrome

Dubowitz syndrome is characterized by: dysmorphic facies, mental retardation, microcephaly, growth retardation, eczema, associated with recurrent neutropenia and low IgG and IgA with elevated IgM levels. The mode of inheritance is autosomal recessive.

Neutropenia Associated with Pancreatic Insufficiency (Shwachman–Diamond syndrome) (SDS)

This condition is a rare multiorgan disease characterized by:

- Metaphyseal chondrodysplasia
  - Dwarfism
  - Impaired gait due to hip dysplasia.
- Dysmorphic features:
  - Short stature
  - Cutaneous syndactyly or clinodactyly, supernumerary metatarsals
  - Bifid uvula, short soft palate or cleft palate
  - Dental dysplasia
  - Hypertelorism, microcephaly
  - Retinitis pigmentosa.
- Abnormal hematopoiesis: Abnormal bone marrow stroma with its reduced ability to support hematopoiesis, increased expression of Fas on hematopoietic progenitor cells resulting in increased apoptosis and stem cell abnormality characterized by decreased growth potential of CFU-GM and CFU-E on culture
  - Neutropenia – 200–400 cells/mm^3; cyclic pattern (occasionally) without reciprocal monocytosis
  - Myeloid hypoplasia
  - Recurrent bacterial infection (e.g., otitis media, pneumonia)
  - May also have anemia and/or thrombocytopenia
- Polymorphonuclear motility constant defect due to an impairment of the cellular cytoskeleton or microtubular function
- Pancreatic exocrine insufficiency
  - Diarrhea, steatorrhea
  - Failure to thrive, growth failure.
- Increased frequency of myocardial necrosis.

Diagnosis

1. Fecal fat (72 hours).
2. Pancreatic function tests (duodenal intubation to demonstrate absence of trypsin, amylase and lipase).
3. Low serum trypsinogen levels in young patients, which may improve with age.
4. Congenital lipomatosis of the pancreas revealed on pancreatic CAT scan imaging to identify gross fatty changes, especially in the body of the pancreas and ductal ectasia and calcification.
5. Normal sweat electrolyte test.

Genetics

SDS is inherited as an autosomal recessive disorder. The gene has been identified as SBDS on chromosome 7. The function of the product is still unknown. It is postulated to be involved in RNA processing in the affected tissue based upon structural homology. The most common mutations result from gene conversions with a neighboring pseudogene, SBDSP. These are truncation deletions and are thought to result in a nonfunctional protein. Chromosome 7 abnormalities such as deletions of the long arm, isochromosome 7 and loss of chromosome 7 occur in the majority of patients with SDS who develop myelodysplasia. It is unclear if this somatic mutation in the marrow stem cells results because of the congenital mutations in the SBDS locus.

Hematologic Complications

1. Anemia.
2. Thrombocytopenia (9% have platelet counts less than $50,000/mm^3$); the development of thrombocytopenia may signal conversion to aplastic anemia or myelodysplasia.
3. Aplastic anemia.
4. Acute myelocytic anemia (AML); incidence increases with age, starting at about 10 years of age.
5. Myelodysplasia; incidence increases with age, starting at about 10 years of age
6. The overall incidence of AML and myelodysplasia is about 16%. Both are resistant to chemotherapy and only respond to allogeneic stem cell transplantation.

Treatment

1. Pancreatic enzyme replacement.
2. GCSF or GMCSF.
3. Monitor bone marrow annually for myelodysplasia and cytogenetic changes.
4. Allogeneic hematopoietic stem cell transplantation (HSCT): Indications and optimum timing for transplantation are unclear. Most of the patients have been transplanted for the treatment of myelodysplasia, AML or severe pancytopenia. The following contribute to unsatisfactory results of HSCT:
   - Hepatic and cardiac toxicity
   - Pre-existing infections
   - Poor nutritional status
   - Marrow stromal defect not corrected by HSCT
   - Presence of MDS/AML
   - Excessive sensitivity of critical organs to radiation therapy and/or chemotherapy.

Neutropenia Associated with Hyperimmunoglobulin M Syndrome

Immunodeficiency with hyperIgM may be caused by one of the following molecular genetic defects:

- X-linked recessive trait, caused by mutations in the gene for CD40 ligand, a molecule on T-cell that binds to its receptor CD40 on B-cell, to induce immunoglobulin class switching from production of IgM to IgG and IgA
- X-linked recessive form of the hyper IgM syndrome caused by a mutation in IκB kinase γ subunit/NF-κB essential modulator (NEMO) and associated with hypohidrotic ectodermal dysplasia (conical teeth, inadequate sweating and poor antibody production to polysaccharide antigens)
- An autosomal recessive form of hyper IgM syndrome caused by defect in the gene for activation-induced cytidine deaminase. This enzyme is needed for Ig class switch, recombination and somatic hypermutation.

The mechanism of neutropenia: A decreased interaction between T-cells and bone marrow stromal cells, resulting in reduced production of G-CSF.

Clinical Manifestations

1. Severe recurrent pyogenic bacterial infections.
2. Infections by opportunistic organisms including: *Pneumocystis jiroveci*, histoplasmosis, cryptosporidium and *Toxoplasma*.
3. Neutropenia, transient, cyclic (10%) or chronic (50% of cases).
4. DAT (direct antiglobulin test – Coombs)-positive autoimmune hemolytic anemia.
5. Lymphoid hyperplasia.
6. Low serum immunoglobulin A (IgA), IgE and IgG; elevated IgM.
7. Number of circulating B cells is normal or increased and T cells are normal.

Treatment

1. IV immunoglobulin and G-CSF, (mainstay of therapy).
2. Hematopoietic stem cell transplantation.

Neutropenia Associated with Metabolic Diseases

The presenting clinical features in neutropenia associated with metabolic diseases (Table 11-6) are lethargy, vomiting, ketosis and dehydration during the neonatal period, failure to thrive and growth retardation. The marrow is hypoplastic in these conditions, with decreased numbers of myeloid precursors. Idiopathic hyperglycinemia and methylmalonic acidemia also have associated thrombocytopenia. Individuals with isovaleric acidemia have a characteristic odor of "smelly feet."

In glycogen storage disease type IB, there is impairment of glucose-6-phosphate-translocase, an enzyme necessary for the transport of glucose-6-phosphate from the cytoplasm to the endoplasmic reticulum, the site where glucose-6-phosphate is hydrolyzed to glucose and inorganic phosphate by an enzyme glucose-6-phosphatase. As a result of low availability of glucose these patients develop hypoglycemia and defective chemotaxis and recurrent infections. The mechanism of neutropenia is not known in this disease. G-CSF can be used effectively in these patients. The bone marrow is hypercellular with abundant neutrophils.

Barth syndrome is characterized by cardiomyopathy, mild neutropenia, proximal skeletal myopathy, growth retardation, low creatinine levels, mitochondrial abnormalities and organic aciduria. It is an X-linked recessive disorder. It results from mutations in the gene G4.5, a member of the tafazzins family of proteins resulting in an inborn error of lipid metabolism. Despite the mild neutropenia, their cardiac defect places them at increased risk for morbidity when they develop infections.

Bone Marrow Disease

Bone marrow failure, whether congenital (Fanconi anemia, dyskeratosis congenita) or acquired (idiopathic or secondary), or due to bone marrow infiltration, whether non-neoplastic, for example, storage diseases (Gaucher disease, Niemann–Pick disease), osteopetrosis and cystinosis, or neoplastic (leukemia, neuroblastoma) may present with neutropenia as a component of the pancytopenia that may occur in these disorders.

INCREASED DESTRUCTION OR DISORDERS OF DISTRIBUTION OF POLYMORPHONUCLEAR LEUKOCYTES

Drug-Induced Neutropenia

Drug-induced neutropenia may be due to:

- Idiosyncratic suppression of myeloid production affecting a few exposed persons; for example, antibiotics (novobiocin, methicillin), sulfonamides, antidiabetics (tolbutamide, chlorpropamide), antithyroids (propylthiouracil, methimazole), antihistamines and antihypertensives (chlorothiazides, Aldomet)
- Regularly occurring dose-dependent myeloid suppression from cytotoxic drugs or antimetabolites; for example, 6-mercaptopurine, methotrexate and nitrogen mustard
- Destruction of white cells produced by the marrow due to differences in individual ability to metabolize a drug; for example, phenothiazine and thiouracil
- Drug – haptene disease, in which antibodies to the drug–leukocyte complex are produced, resulting in demonstrable *in vitro* leukoagglutinins; for example, amidopyrine-related drugs (dipyrone, phenylbutazone), sulfapyridine, mercurial diuretics and chlorpropamide.

Immune Neutropenia

Neonatal Immune Neutropenia

Immunoneutropenias may be alloimmune (isoimmune), with immunization to foreign antigens on the fetal leukocytes not present on maternal leukocytes analogous to Rh isoimmunization, or autoimmune, in infants born to mothers with neutropenia who have antileukocyte antibodies (e.g., a mother who has systemic lupus erythematosus [SLE]).

Alloimmune Neonatal Neutropenia

In alloimmune neonatal neutropenia, alloantibodies are frequently directed to the neutrophil-specific NA antigen system, of which there are two alleles – NA1 and NA2.

Clinical Features

1. Infants may be asymptomatic or they may have infections (e.g., pyoderma, omphalitis, pneumonia).
2. Neutropenia usually resolves by 2 months of age.
3. Bone marrow is hypercellular with an increase in neutrophil precursors and a paucity of mature neutrophils.

Treatment

1. Antibiotics, as indicated.
2. IVIgG.
3. G-CSF.

Autoimmune Neutropenia

Autoimmune neutropenia can be primary or secondary.

Primary Autoimmune Neutropenia[*]

Neutropenia may result from antibodies against leukocytes. It is analogous to autoimmune hemolytic anemia or autoimmune thrombocytopenia. Neutrophil antibodies may adversely affect the function of neutrophils, producing qualitative defects in the neutrophils and amplifying the risk of infection associated with neutropenia. Neutrophil autoantibodies may also affect myeloid precursor cells. When this occurs, it can produce profound neutropenia. The disease is characterized by the following:

- Age – 3–30 months; median age, 8 months; has occurred as early as 1 month of age. Occasionally, it may be congenital
- It is non-familial

---

[*]Probably the same entity as idiopathic chronic benign neutropenia of infancy and childhood.

- Physical examination – normal; occasionally, slight splenomegaly
- Benign infections of skin and upper respiratory tract, as well as otitis media in 90% of cases; not life-threatening; responsive to standard antibiotics.

Diagnosis

1. Neutrophil counts range from 0 to 1,000 cells/mm^3. Monocytosis is common.
2. The bone marrow may be normal or may show evidence of myeloid hyperplasia with marked reduction in segmented neutrophils due to their destruction by antibodies.
3. Epinephrine or hydrocortisone administration results in the rise of neutrophil counts.
4. Antineutrophil antibodies are not always present and screening has to be repeated for antibody detection. Immunoassay is more sensitive than leukoagglutination testing for diagnosing immune neutropenia. In most patients, the autoantibody is auto-anti-NA1 and some have auto-anti-NA2.

Table 11-8 lists the different types of antineutrophil assays:

- The *granulocyte immunofluorescence test (GIFT)* detects neutrophil-bound IgG by binding of glutaraldehyde-fixed patient neutrophils to fluorescent-labeled anti-human IgG
- The *granulocyte indirect immunofluorescence test (GIIFT)* uses the patient's serum with normal neutrophils or previously typed neutrophils and subsequent incubation with fluorescent-labeled anti-human IgG
- *Granulocyte agglutination test (GAT)* uses the patient's serum incubated with normal neutrophils followed by microscopic evaluation for leukoagglutination
- *Enzyme-linked immunoassay (ELISA)* uses microtiter plates with bound glutaraldehyde fixed normal neutrophils to detect antineutrophil antibodies in patient's sera. An anti-human IgG conjugated to a reporter enzyme is used for detection
- *Monoclonal antibody-specific immobilization of granulocyte antigens (MAIGA)* is the most specific. It involves incubation of type-specific neutrophils with both patient's serum and mouse monoclonal antibodies directed against another neutrophil surface antigen. The mixture is passed over an affinity column containing anti-mouse IgG antibodies and then is assayed for the presence of human IgG. This allows for detection of antibody and knowing its specificity in one assay.

GIFT and GAT are used most commonly to diagnose autoimmune neutropenia.

Prognosis

Spontaneous recovery usually occurs within a few months to a few years. The median age at recovery is 30 months (range, 7–73 months). In 95% of cases, recovery occurs by 4 years of age.

Table 11-8   Investigations of Patients with Neutropenia[a]

1. History of drug ingestion, toxin exposure, infectious history
2. Physical examination—nature of infectious lesions, growth and development, presence of anomalies, presence of enlarged lymph nodes or hepatosplenomegaly
3. Familial: absolute granulocyte count in family members
4. Blood count: CBC with differential and platelet count, absolute granulocyte count and reticulocyte count; CBC and differential three times per week for 6–8 weeks (to exclude cyclic neutropenia)
5. Bone marrow
   a. Maturation characteristics of myeloid series; ? reduction in mature granulocytes
   b. Maturation and number of megakaryocytes and erythroid precursors
   c. Karyotype (to identify myelodysplasia or acute myelocytic leukemia) and FISH studies for chromosome 7 and 5q
   d. Electron microscopy (subcellular morphology, congenital dysgranulopoiesis)
6. Detection of antineutrophil antibodies (see text for details)
   a. Granulocyte immunofluorescence test (GIFT)
   b. Granulocyte indirect immunofluorescence test (GIIFT)
   c. Granulocyte agglutination test (GAT)
   d. Enzyme linked immunoassay (ELISA)
   e. Monoclonal antibody specific immobization of granulocyte antigens (MAIGA)
7. Immunologic tests:
   a. Immune globulins (IgA, IgG, IgM, IgE)
   b. Cellular immunity (skin-test activity, purified protein derivative (PPD), lymphocyte subsets; suppressor T-cell assay)
   c. Antinuclear antibodies, C_3, C_4, CH_{50}
8. Evidence of metabolic disease
   a. Plasma and urine aminoacid screening
   b. Serum vitamin B_{12}, folic acid and copper
9. Evidence of pancreatic disease
   a. Exocrine pancreatic function: stool fat, pancreatic enzyme assays, CT scan of pancreas for pancreatic lipomatosis, serum levels of trypsinogen and isoamylase
10. Chromosomal breakage analysis (Fanconi anemia)
11. Radiographic bone survey (cartilage-hair hypoplasia, Shwachman–Diamond syndrome, Fanconi anemia)
12. Serum muramidase (ineffective myelopoiesis)
13. Flow cytometry for CD59 (or other GPI linked protein) (paroxysmal nocturnal hemoglobinuria) This study is much more specific and reliable than the Sucrose hemolysis test or HAM test that had been used in the past to make this diagnosis
14. Bone density studies (14% of patients with chronic neutropenia show nonclinical osteoporosis or osteopenia)
15. Many gene mutation analyses are commercially available including: Neutrophil elastase (ELA-2) (SCN and cyclic neutropenia), GFI-1 (SCN), WAS (X-linked neutropenia), SBDS (Shwachman-Diamond), HAX 1, TAZ (Barth syndrome), Fanconi family of genes, LYST (Chediak Higashi syndrome) and others that are continually being discovered. Molecular diagnostic studies have made diagnosing many of these entities more accurate. In the past, physicians had to rely on interpretation of colony-forming unit (CFU) assays and colony-stimulating activity (CSA) assays to try to distinguish between these different entities

(Continued)

Table 11-8   (Continued)

Other investigations that are rarely used today, but may prove useful in making a diagnosis in a particular patient are listed below:
1. Estimate of marginating granulocyte reserve pool
   Epinephrine stimulation tests (0.1 ml 1:1,000 epinephrine SC)
   (1) Absolute granulocyte counts at 5, 10, 15 and 30 minutes
   (2) Normal: double base count
2. Estimate of bone marrow granulocyte reserve pool
   a. Cortisone stimulation tests (5 mg/kg IV)
      (1) Absolute granulocyte counts hourly for 6 hours
      (2) Normal: Increase of more than 2,000 neutrophils/mm^3
   b. Typhoid stimulation tests (0.5-ml vaccine SC)
      (1) Absolute granulocyte count at 3, 6, 12 and 24 hours
      (2) Normal: threefold to fourfold increase
3. Rebuck skin window (to assess leukocyte migration and chemotaxis)
   Normal: at 3 hours, neutrophils; at 6 hours, mixed neutrophils and monocytes; at 24 hours, monocytes

[a]Absolute granulocyte count less than 1,500/mm^3.

Treatment

In patients with severe neutropenia and severe or recurrent infections the following treatment is given:

- Appropriate antibiotics for acute infections or prophylactic antibiotics such as trimethoprim/sulfamethoxazole
- Mouth care with oral rinses and good dental hygiene
- In patients with more serious infections, G-CSF or less commonly GM-CSF. G-CSF or GM-CSF is used to treat patients with active infections. There is an almost 100% response rate to G-CSF in a dose of 5 μg/kg/day subcutaneously. The injections are continued until the absolute neutrophil count is greater than 1,000–2,000/mm^3.

Secondary Autoimmune Neutropenia

This is more common in adults than in children.

The following diseases are associated with secondary autoimmune neutropenia:

- Evans syndrome
- Autoimmune hemolytic anemia
- Autoimmune thrombocytopenia
- Thyroiditis
- Insulin-dependent diabetes mellitus
- Common variable immune deficiency.

Autoantibody specificity: Pan-FcR γIIIb

Treatment

1. Treat the associated condition.
2. G-CSF.

Non-Immune Neutropenia

Pseudoneutropenia

In this condition, a normal neutrophil population may be shifted toward the marginating compartment, leaving fewer cells in the circulating compartment. The white blood cell count measures only the circulating cells and not the marginating pool; therefore, this represents a pseudoneutropenia. The bone marrow is normal in appearance. The neutrophils function normally and the leukocyte changes are usually found incidentally on blood count. Marginating neutrophils may be uncovered by the injection of epinephrine.

Ineffective Myelopoiesis

This condition is a chronic neutropenia with the ability to respond with a neutrophil leukocytosis at times of infection. Marrow examination shows a shift to the right among the granulocytic series, the predominant cells resembling degenerating polymorphs with dense pyknotic chromatin. The fundamental defect appears to be intramedullary death of the neutrophils (or ineffective myelopoiesis). Corticosteroids and splenectomy do not influence the course of the disease.

Infections

Viral infections and certain bacterial infections, such as typhoid fever, paratyphoid fever and rickettsial disease, may be associated with neutropenia. Staphylococcal or pneumococcal infections associated with neutropenia indicate a grave prognosis.

Hypersplenism

Hypersplenism causes peripheral sequestration not only of red cells and platelets but of granulocytes as well. The marrow in such cases shows myeloid hyperplasia with normal maturation to the polymorph stage. Splenomegaly from any cause (e.g., thalassemia, storage diseases, portal hypertension, lymphomas) may produce hypersplenism. Occasionally, splenomegaly and neutropenia of unknown cause (primary splenic neutropenia) occur in which the hematologic abnormality can be cured by splenectomy.

Investigations in Neutropenia

Table 11-8 lists the investigations to be considered when evaluating patients with neutropenia.

Management of Neutropenia

Table 11-9 lists the management required in the care of neutropenic patients.

Table 11-9   Management of the Neutropenic Patient[a]

1. Admit to hospital for persistent fever over 101°F and ANC <500 or patient is toxic appearing
2. Obtain appropriate cultures (blood, throat, urine, infected area) and sensitivity
3. Administer parenteral antibiotics (see Chapter 31)
   a. If an organism is isolated, 10–14 days intravenous treatment is required
   b. If no organism is isolated, antibiotic is continued until afebrile and neutropenia is resolved
4. Whenever possible, patient should be in a single-patient room. If not available, the patient sharing the room should be "infection-free"
5. Staff, family members and visitors should observe strict hand-washing procedures. Visitors should be free from colds or other infections
6. Wash skin carefully with a povodine or chlorhexidine-containing solution before all skin puncture procedures
7. Minimize manipulation of skin, oral mucosa, perineum and rectum; rectal temperatures and enemas are contraindicated
8. Treat mouth ulcerations and gingivitis with appropriate systemic antibiotics if secondary bacterial infection is found and 3% hydrogen peroxide–1% alum mouthwash which usually produces symptomatic relief. Use a soft toothbrush for brushing
9. Administer G-CSF[b] for treatment of Kostmann disease, Shwachman–Diamond syndrome, other congenital neutropenias and severe neutropenia following chemotherapy (the starting dose is 5 μg/kg SC with dose modification according to the patient's absolute neutrophil count)

[a]Neutrophil count less than 500 cells/mm^3.
[b]G-CSF specifically stimulates myeloid progenitor cells in the bone marrow and enhances neutrophil production and function.

EOSINOPHILS

Table 11-10 lists the nonclonal (reactive) causes of eosinophilia and Figure 11-2 shows the nonclonal (reactive) and clonal causes of eosinophilia.

Eosinophilia

Normal mean eosinophil count in the circulating blood is 400/mm^3. Normally, most of the eosinophils reside in the connective tissue located in the immediate proximity of the epithelial lining of the gut, respiratory tract and urogenital tract. Their number and activation increase as a response to antigens, especially when these antigens are deposited in the above tissues. A response is characterized by immediate hypersensitivity reaction, mediated by IgE or delayed hypersensitivity reaction, mediated by T-lymphocytes.

Severity of eosinophilia is graded according to the presence of their absolute number in the circulating blood as follows:

- Mild eosinophilia: 400–1,500/mm^3
- Moderate eosinophilia: 1,500–5,000/mm^3
- Severe eosinophilia: greater than 5,000/mm^3.

Table 11-10   Nonclonal (Reactive) Causes of Eosinophilia

Allergic disorders
  Asthma, hay fever, urticaria, drug hypersensitivity
Immunologic disorders
  Omenn syndrome (severe combined immunodeficiency and eosinophilia)
Skin disorders
  Eczema, scabies, erythema toxicum, dermatitis herpetiformis, angioneurotic edema, pemphigus
Parasitic infestations
  Helminthic: *Ascaris lumbricoides*,[a] trichinosis, echinococcosis, visceral larva migrans,[a,b] hookworm,[a]
    strongyloidiasis,[a] filariasis[a]
  Protozoal: malaria, pneumocystis, toxoplasmosis
Hematologic disorders
  Hodgkin disease, postsplenectomy state, eosinophilic leukemoid reaction, congenital immune deficiency
    syndromes, Fanconi anemia, thrombocytopenia with absent radii, Kostmann disease, infectious
    mononucleosis, familial reticuloendotheliosis
Familial eosinophilia
Irradiation
Pulmonary eosinophilia
  Eosinophilic pneumonitis (Loeffler syndrome), pulmonary eosinophilia with asthma, tropical
    eosinophilia
Miscellaneous
  Idiopathic hypereosinophilic syndrome,[b] periarteritis nodosa, metastatic neoplasm, cirrhosis, peritoneal
    dialysis, chronic renal disease, Goodpasture syndrome, sarcoidosis, thymic disorders, hypoxia
Gastrointestinal disorders
  Eosinophilic gastroenteritis, milk precipitin disease, ulcerative colitis, protein-losing enteropathy, regional
    enteritis, allergic granulomatosis
Idiopathic

[a]Helminth infestations associated with eosinophilia and pulmonary infiltrates.
[b]Conditions associated with striking eosinophilia. Leukocyte counts of $30,000-1,00,000/mm^3$ are characteristic, with
50–90% of leukocytes being eosinophils. In all other conditions, the white blood cell count is normal or only slightly
elevated and eosinophils make up 10–40% of the leukocyte count.

Figure 11-3 illustrates mechanisms of eosinophil production in bone marrow, release in circulation and migration in the tissues. Following activation, the eosinophil expresses its effector function, which includes the release of highly toxic granule proteins and other mediators of inflammation.

Finally, eosinophils undergo apoptosis or necrosis and are ingested by professional macrophages. Engulfment of apoptotic eosinophils prevents spillage of the eosinophil tissue-toxic contents. It also results in release of anti-inflammatory cytokines, such as transforming growth factor β, interleukin-10 and prostaglandin E2 by macrophages. In contrast to this, when eosinophils undergo necrosis, their tissue toxic contents, such as major basic protein, lipids, cationic proteins and neurotoxins, are released. Ingestion of necrotic eosinophils by macrophages results in release of proinflammatory cytokines, e.g. thromboxane B2 and granulocyte-macrophage colony-stimulating factor.

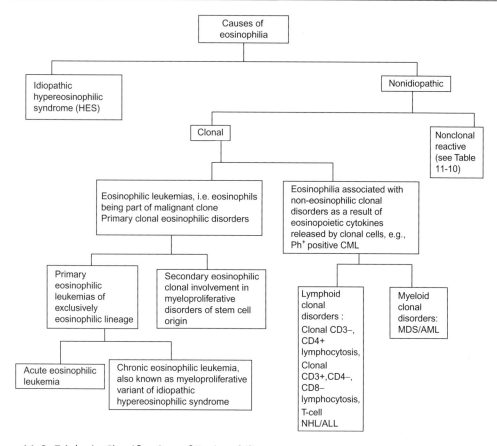

Figure 11-2 Etiologic Classification of Eosinophilia.
Abbreviations: NHL, Non-Hodgkin lymphoma; ALL, Acute lymphoblastic leukemia; MDS, Myelodysplatic syndrome; AML, Acute myelocytic leukemia; CML, Chronic myeloid leukemia.

For this reason, in treatment of eosinophilic diseases, the use of the drugs that induce eosinophil apoptosis is desirable, e.g. corticosteroids, cyclosporin and theophylline induce eosinophil apoptosis rather than the drugs that induce eosinophil necrosis.

Idiopathic Hypereosinophilic Syndrome (HES)

Definition

HES includes a heterogeneous group of disorders defined by:

- A persistent eosinophilia of $>1,500/mm^3$ for longer than 6 months
- Absence of evidence of known causes of eosinophilia despite a comprehensive work-up for such causes
- Signs and symptoms of organ involvement, directly attributable to eosinophilia, including hepatomegaly, splenomegaly, heart disease, diffuse or focal central nervous

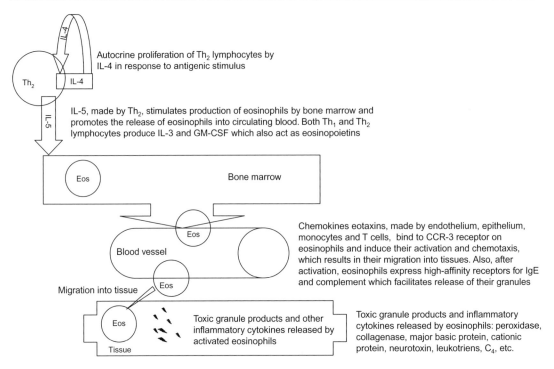

Figure 11-3 Mechanism of Eosinophil Production in Bone Marrow, Release in Circulation and Migration in Tissue.
Abbreviations: IL, Interleukin; Eos, Eosinophil; CCR, Chemokine receptor; Th, Helper T lymphocytes; GM-CSF, Granulocyte macrphage colony stimulating factor.

system (CNS) abnormalities, pulmonary fibrosis, fever, weight loss, or anemia (i.e. evidence of end organ damage with histologic demonstration of tissue infiltration by eosinophils or objective evidence of clinical pathology in any organ system associated with eosinophilia and not clearly attributable to another cause).

Epidemiology

HES most commonly occurs between the ages of 20 and 40 years with a male to female ratio of 4:1. Clinical manifestations of HES in pediatric age group are similar to HES in adult patients. In children, HES may be associated with trisomy 8 or trisomy 21.

Clinical Presentation

The disease generally has a gradual onset. The chief complaints include anorexia, fatigue, weight loss, recurrent abdominal pain, fever, night sweats, persistent non-productive cough, chest pain, pruritus, skin rash and congestive heart failure.

Organ Involvement

Cardiovascular Disease

HES-associated heart disease evolves through three stages:

- Early acute phase, associated with degranulating eosinophils in the heart muscle (5–6 weeks into eosinophilia)
- Subacute thrombotic stage (10 months into eosinophilia)
- Chronic stage of fibrosis (24 months into eosinophilia). Cardiac disease involves both ventricles and can cause incompetence of mitral and tricuspid valves.

Coagulation System

Eosinophilia can cause a hypercoagulable state, the etiology of which is unclear. Eosinophil major basic proteins inactivate thrombomodulin, thus, resulting in unavailability of activated protein C. Intracardiac thrombus, deep venous thrombosis, dural sinovenous thrombosis and/or arterial thrombosis can occur.

Nervous System Complications

1. Encephalopathy (altered behavior and cognitive function).
2. Thrombotic strokes.
3. Peripheral neuropathies including mononeuritis multiplex, symmetrical sensory-motor neuropathy and radiculopathy.
4. Retinal hemorrhages.

Gastrointestinal Complications

1. Hepatomegaly due to eosinophilic infiltration of the liver results in liver function abnormalities.
2. Enteropathy due to blunting of the villi and cellular infiltration in the lamina propria results in diarrhea and fat malabsorption.
3. Eosinophilic infiltration of colon results in colitis.

Spleen

Splenomegaly with disruption of its architecture can occur.

Dermatologic Manifestations

The most common lesions include pruritic papules and nodules, urticarial plaques and angioedema. Vesiculobullous lesions, generalized erythroderma and aquagenic pruritus occur in some patients. Digital necrosis may result from vasculitis and microthrombi.

Pulmonary Complications

Nocturnal cough, fever, diaphoresis can occur due to accumulation of eosinophils in the lungs. Pulmonary fibrosis can also occur.

Treatment

Table 11-11 lists the treatment in idiopathic HES.

Non-Idiopathic Eosinophilia

Eosinophilia, for which a cause is ascertained, can be clonal or reactive.

Primary Clonal Eosinophilic Disorders

These include eosinophilic leukemias of exclusively eosinophilic lineage, e.g. acute eosinophilic leukemia and chronic eosinophilic leukemia (also known as myeloproliferative

Table 11-11 Treatment of Eosinophilia

Treatment of reactive or non-clonal eosinophilia: Treat the underlying cause e.g. treatment of parasitic infections with appropriate anti-parasitic drugs
Treatment of clonal disease: Myeloid clonal disease: Treat with appropriate chemotherapy±hematopoietic stem cell transplantation. Lymphoid malignancies: Treat with appropriate chemotherapy CD3−, CD4+ Lymphoid clonal disease with high levels of IL-5, usually associated with dermatologic manifestations: Treat with cyclosporine A, glucocorticoids or 2CDA CD3+, CD3−, CD8− Lymphoid clonal induced eosinophilia with high levels of IL-5 and usually associated with dermatologic manifestations: Treat with glucocorticoids, cyclosporine A
Treatment of HES caused by interstitial deletion of 4q12 resulting in a fusion gene FIP1L1-PDGFRA: Imatinib mesylate, adult dose: 400 mg/day. Pediatric dose: not established
Treatment of idiopathic HES: Glucocorticoids, Hydroxyurea, α-interferon, vincristine, thioguanine, or etoposide.[a] Use these agents sequentially. If the response is unsatisfactory, then treat with imatinib mesylate at doses of 100–200 mg/day (of interest is that patients with normal serum Interleukin-5 values respond to imatinib, but not the ones with high values). During acute life threatening presentation of HES, high dose 10–20 mg/kg of solumedrol (methylprednisolone) may be required, but usually 1–2 mg/kg of prednisone may be sufficient. Treatment with allogeneic hematopoietic stem cell transplantation is reserved for patients with HES refractory to above mentioned therapies. Treatment of patients with idiopathic HES but without organ involvement: None. Treatment is not necessary, but continuous periodic monitoring for organ involvement and emergence of clonality is warranted. Also, continue search for rare reactive causes of eosinophilia.

The following eosinophilic disorders with single organ involvement may progress into HES:
     Eosinophilic gastroenteritis
     Gleich syndrome (episodic eosinophilia with angioedema)
     Loeffler syndrome
     Schulman syndrome (eosinophilic fascitis)
     Well syndrome (eosinophilic cellulitis)
     Parasitic infections with eosinophilia

[a]Doses of some of the drugs: Thioguanine, 40–60 mg/m^2/day orally, Vincristine, 1.5 mg/m^2/week IV, Etoposide, 60–100 mg/m^2/day for 3–5 days IV every 3–6 weeks, Hydroxyurea, 10–20 mg/kg/day orally, Cyclosporine 6 mg/kg/day orally (trough level 100–200 μg/L), α-Interferon 5×10^6 units/m^2/day subcutaneously or IM.

variant of idiopathic hypereosinophilic syndrome). The following karyotypic abnormalities associated with chronic eosinophilic leukemia have been reported: Majority of patients have t(5;12) (q33:p13). Sporadic patients have trisomy 8, i(17q), t(5;12) (q31:q13), t(1;5) (q23:q33), t(2;5) (p13:q35), t(5;9) (q32:q33), t(5;16)(q33:p13), trisomy 10, 17q+, 15q-, -7, t(7;12) (q11:p11) or t(4;16) (q11 or 12:p13).

HES with FIP1L1-PDGFRA Fusion Gene

A special variant of HES with a fusion gene FIP1L1-PDGFRA occurring as a result of interstitial deletion on chromosome 4q12 has recently been described. The fusion gene makes an activated tyrosine kinase, which results in a myeloproliferative variant of HES. It is characterized by increased levels of tryptase, increased atypical mast cells in bone marrow and tissue fibrosis (myelofibrosis, endomyocardial fibrosis, pulmonary fibrosis). They respond well to imatinib mesylate treatment, which targets the fusion tyrosine kinase. Some patients with FIP1L1-PDGFRA fusion gene may not have the classic characteristics of a myeloproliferative variant of HES and also respond well to imatinib. Some patients with HES may not have FIP1L1-PDGFRA fusion gene and still respond to imatinib.

Secondary clonal involvement of eosinophil lineage can occur in myeloproliferative disorders of stem cell origin, e.g. Ph[1]-positive chronic myeloid leukemia. Eosinophilia is observed less commonly in polycythemia vera, myelofibrosis and essential thrombocythemia.

Non-Clonal Eosinophilic Disorders

In non-clonal eosinophilic disorders, clonal cells release eosinopoietic cytokines and thus, are associated with eosinophilia. These noneosinophilic clonal disorders may be lymphoid or myeloid in their clonality. Lymphoid clonal disorders associated with eosinophilia include dermatologic patients with abnormal clones of T-cells producing interleukin-5, patients with acute lymphoblastic leukemia and T lymphoblastic lymphoma. Patients who present with T-cell lymphoblastic lymphoma and eosinophilia may be predisposed to developing secondary AML. Myeloid clonal disorders associated with eosinophilia include myelodysplastic syndromes, acute myeloid leukemia with chromosome 16 abnormality, the 8p myeloproliferative syndrome, myelodysplastic syndromes and systemic mastocytosis. The following cytogenetic abnormalities associated with acute myeloid leukemia with eosinophilia have been reported: inv (16) (p13:q22), t(16:16) (p13:q22), t(5;16) (q33:q22) and monosomy 7.

Reactive or nonclonal eosinophilic conditions are listed in Table 11-10.

Eosinophilia in Newborn Period

A mild eosinophilia with eosinophil count greater than 700/mm^3 is observed in 75% of growing preterm infants. It is present in the second or third week of life and persists for several days or sometimes for weeks. Eosinophilia of prematurity is considered to be benign, although it could be associated with a higher incidence of sepsis, especially, with Gram-negative bacteria.

A complete absence of eosinophils is observed in neonates who fare poorly and subsequently die.

Familial Eosinophilia

Familial eosinophilia is an autosomal dominant disorder. A genome wide search showed evidence of linkage on chromosome 5q31-33 between markers D55642 and D55816. Some of the affected members are found to have high white blood cell counts, lower red cell counts, intermittent thrombocytopenia, cellular infiltration with mast cells in the liver and bone marrow, or involvement of the heart and nervous system. The levels of IL-3, IL-5 and GM-CSF are normal.

Figure 11-4 shows a list of diagnostic studies for evaluation of eosinophilia.

Table 11-11 outlines the treatment of conditions that cause eosinophilia.

LYMPHOCYTES

Table 11-12 lists the causes of lymphocytosis and lymphopenia.

Acute Infectious Lymphocytosis

This disorder can be characterized by:

- Discovery on routine blood count. It is caused by a coxsackie virus
- Mild complaints – vomiting, diarrhea, upper respiratory tract infection, abdominal pain, slight or absent fever; symptoms of short duration
- No enlargement of liver, spleen, or lymph nodes
- Hematologic findings
  - White blood cell counts varying from 40,000 to 100,000/mm^3
  - Absolute increase in lymphocytes, with a lymphocyte predominance of more than 70%
  - No anemia or thrombocytopenia.
- No evidence of Epstein–Barr virus (EBV) infection
- Condition to be differentiated from acute leukemia, infectious mononucleosis and lymphocytosis accompanying certain infections, particularly pertussis
- Prognosis excellent and no treatment required.

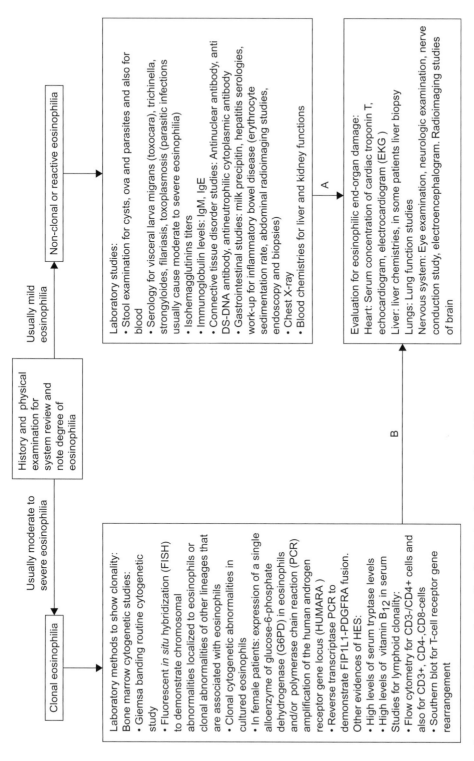

Figure 11-4 Diagnostic Studies for Evaluation of Eosinophilia.

A, in all patients.

B, in patients with prolonged eosinophilia.

Table 11-12   Causes of Lymphocytosis and Lymphopenia

  I. **Lymphocytosis**
  A. Physiologic: 4 months–4 years
  B. Infections
     1. Acute
        a. Moderate lymphocytosis: measles, rubella, varicella, mumps, roseola infantum, brucellosis, typhoid, paratyphoid, autoimmune diseases, granulomatous diseases, postimmunization states, drug reactions, graft rejection
        b. Marked lymphocytosis: acute infectious lymphocytosis,[a] infectious mononucleosis, cytomegalovirus infection, toxoplasmosis, pertussis
     2. Chronic
        a. Tuberculosis, syphilis
  C. Leukemia: acute lymphoblastic leukemia
 II. **Lymphopenia**
  A. X-linked agammaglobulinemia
  B. Reticular dysgenesis
  C. Severe combined immunodeficiency
III. **Alymphocytosis**
  A. Swiss-type agammaglobulinemia

[a]A case associated with Coxsackievirus B2 has been described.

Infectious Mononucleosis

Infectious mononucleosis is an acute infectious disease caused by EBV. This common disease occurs in epidemic form in children of all ages, but is rare in infants under 6 months of age. Infectious mononucleosis is usually a benign, self-limited disease; occasionally, it is associated with severe and fatal complications. In young children, the symptoms are often mild and infection may not be apparent. They often exhibit rashes, neutropenia and may have pneumonia. "Typical" infectious mononucleosis is more frequently diagnosed when the primary exposure occurs in adolescents of higher socioeconomic status in industrialized countries. Table 11-13 lists the frequency of various signs and symptoms of infectious mononucleosis and Table 11-14 lists the laboratory and diagnostic findings.

Differential Diagnosis

Mononucleosis-like syndromes are common in infants and children and are generally due to EBV infection. The EBV antibody-negative mononucleosis-like syndrome may be due to the following agents:

- Cytomegalovirus
- Toxoplasma gondii
- Drugs (para-aminosalicylic acid (PAS), Dilantin, sulfone)
- Other agents (adenovirus, herpes simplex, rubella).

Table 11-13   Approximate Frequency of Various Signs and Symptoms in Infectious Mononucleosis

Symptom or Sign	Percentage	Symptom or Sign	Percentage
Adenopathy[a]	100	Myalgia	12–30
Malaise and fatigue	90–100	Hepatomegaly	15–25
Fever	89–95	Rhinitis	10–25
Sweats	80–95	Ocular muscle pain	10–20
Sore throat, dysphagia	80–95	Chest pain	5–20
Pharyngitis[b]	65–85	Jaundice	5–10
Anorexia	50–80	Arthralgia	5–10
Nausea	50–70	Diarrhea or soft stools	5–10
Splenomegaly	50–60	Photophobia	5–10
Headache	40–70	Skin rash[c]	3–6
Chills	40–60	Conjunctivitis	<5
Bradycardia	35–50	Abdominal pain	<5
Cough	30–50	Gingivitis	<3
Periorbital edema	25–40	Pneumonitis	<3
Palatal enanthem	25–35	Epistaxis	<3
Liver or splenic tenderness	15–30		

[a]Chiefly cervical (posterior cervical); commonly axillary, epitrochlear and inguinal nodes, but any nodes may be involved. Rarely, mediastinal nodes are enlarged.
[b]Tonsils may be covered with membrane resembling diphtheria.
[c]Rash may be scarlatiniform, morbilliform, vesicular, maculopapular and salmon colored. When ampicillin is given, rash occurs in 75–100% of patients.

Leukemia and lymphomas must always be considered and can be excluded by a bone marrow examination, if sufficient doubt exists about the diagnosis. See Table 11-15 for a list of the causes of atypical lymphocytosis.

Complications

Numerous and bizarre complications have been described in infectious mononucleosis (Table 11-16).

Treatment

1. Acetaminophen, in a dose of 10–15 mg/kg, is usually sufficient as an analgesic and antipyretic agent.
2. Corticosteroids are indicated in the following clinical situations:
   * Severe pharyngitis with dysphagia and potential airway obstruction
   * Acute and severe hemolytic anemia
   * Severe life-threatening complications
   * Rapidly progressive thrombocytopenic purpura
   * Left upper quadrant abdominal pain with subcapsular splenic hemorrhage (not rupture). Prednisone 1–2 mg/kg is typically given under these circumstances. Corticosteroids are not without harmful side effects and should be used only for the treatment of disabling disease; the course of corticosteroids should be short.

Table 11-14   Laboratory Findings in Infectious Mononucleosis

1. Increase in atypical lymphocytes (seen in other conditions; see Table 11-15). 25% or more of leukocytes during second or third week are Downey type I, II and III cells[a]
2. Leukocyte count usually elevated (12,000–18,000, but counts of 30,000–50,000 are not uncommon), or may be normal or low. Neutropenia in 60–90% of cases. Thrombocytopenia in 50% of cases. Anemia, occasionally DAT-positive hemolysis can occur
3. EBV-specific antibodies[b]. The following indicates acute or recent primary infection:
   a. Presence of viral capsid antigen (VCA)-specific IgM antibodies
   b. High titers of VCA-specific IgG antibodies
   c. Detection of anti-early antigen (EA)
   d. Absence of anti-Epstein–Barr nuclear antigen (EBNA) (see Figure 11-5 for antibody responses during EBV-induced infectious mononucleosis)
4. Transient heterophile antibody test positive (4 or 5 days after onset) in up to 90% of patients at some point of the illness[b]
   a. Sheep red cell agglutinins also develop during serum sickness, in other disease entities (infectious hepatitis, rubella, tuberculosis, leukemia and Hodgkin disease) and in normal persons
   b. In the latter conditions, the antibodies exist in a low titer and can be differentiated from those in infectious mononucleosis by absorption tests with guinea pig kidney and beef red cells as follows:

	Titer After Absorption	
Sources	**Guinea Pig Kidney**	**Beef Red Cells**
Infectious mononucleosis	Present	Absent
Normal serum	Absent	Present
Serum sickness	Absent	Present

5. Monospot test
   a. 96–99% reliable in older children and adults, however, detects disease in only less than 50% of children under 4 years of age
   b. Can detect heterophile titers as low as 1:56
6. Occasionally false-positive Venereal Disease Research Laboratory (VDRL), antinuclear antibody (ANA) and febrile agglutination reactions
7. Evidence of hepatic dysfunction in 75% of cases
8. Abnormalities of immune system
9. EBV virus detection by PCR (polymerase chain reaction) on blood or other body fluid specimen

[a]These cells vary markedly in size and shape and are T lymphocytes. The cytoplasm is vacuolated, with a foamy appearance and the periphery of the cytoplasm is characteristically indented and deformed by the surrounding erythrocytes. The nucleus is round, bean shaped, or lobulated, with no nucleolus (Downey type I monocytoid cells). Some lymphocytes have deep blue-staining cytoplasm and nuclei that appear immature (Downey type II plasmacytoid cells); some may have nucleoli (Downey type III blastoid cells).
[b]Specific EBV antibody testing may prove more reliable, especially in the younger patient.

Chronic Active EBV

This term is used for patients who have the following clinical syndrome:

- Chronic and relapsing illness and fatigue, myalgia, adenopathy, hepatosplenomegaly, mild pharyngitis and intermittent fever that began as a primary EBV infection
- Absence of known predisposing cause or chronic illness

Table 11-15   Causes of Atypical Lymphocytosis

I. Less than 20%
  A. Infections
    1. Bacterial: brucellosis, tuberculosis
    2. Viral: mumps, varicella, rubeola, rubella, atypical pneumonia, herpes simplex, herpes zoster, roseola infantum, HIV
    3. Protozoal: toxoplasmosis
    4. Rickettsial: rickettsialpox
    5. Spirochetal: congenital syphilis, tertiary syphilis
  B. Radiation
  C. Miscellaneous
    1. Hematologic: Langerhans cell histiocytosis, leukemia, lymphoma, agranulocytosis
    2. Other: lead intoxication, stress
II. More than 20%
  A. Infectious mononucleosis
  B. Infectious hepatitis
  C. Post-transfusion syndrome
  D. Cytomegalovirus syndrome
  E. Drug hypersensitivity: *p*-aminosalicylic acid (PAS), phenytoin (Dilantin), mephenytoin (Mesantoin), organic arsenicals

Table 11-16   Complications of Infectious Mononucleosis

Neurologic[a]: Bell's palsy, cerebellar syndrome, meningoencephalitis, Guillain–Barré syndrome, myelitis, peripheral neuritis, radiculoneuritis, convulsions, coma

Cardiac: myocarditis, pericarditis

Respiratory: laryngeal obstruction, peritonsillar abscess, respiratory obstruction, pleural effusion, pneumonitis

Hematologic: acquired hemolytic anemia (usually DAT[b] positive), immunopathic thrombocytopenic purpura, neutropenia, pancytopenia, aplastic anemia, eosinophilia, splenic rupture

Oncologic: nasopharyngeal carcinoma, Burkitt's lymphoma, Hodgkin disease, lymphoproliferative disease, nasal T cell/natural killer cell lymphoma, lymphomatoid granulomatosis, angioimmunoblastic lymphadenopathy, central nervous system lymphoma in immunocompromised host, smooth muscle tumors in transplantation patients, gastric carcinoma, peripheral T cell lymphoma with virus-associated hemophagocytic syndrome (see Chapter 16)

In HIV patients with AIDS: oral hairy leukoplakia, lymphoid interstitial pneumonitis, non-Hodgkin lymphoma

Renal: nephritis, nephrotic syndrome, hematuria, hemoglobinuria, proteinuria

Hepatic: jaundice, hepatic dysfunction, hepatic necrosis, Reye syndrome

Gastrointestinal: protein-losing enteropathy, melena, pancreatitis

Genitourinary: orchitis, azoospermia, endocervicitis

Ocular: eyelid edema, conjunctivitis, papilledema, uveitis, nystagmus, diplopia, retro-orbital pain, scotomata

[a]May precede, follow, or occur simultaneously with acute phase of disease. Brain, meninges, spinal cord and cranial and peripheral nerves may be involved separately or in combination, producing bizarre neurologic signs and symptoms.
[b]DAT –, Direct antiglobulin test (Direct Coombs test).

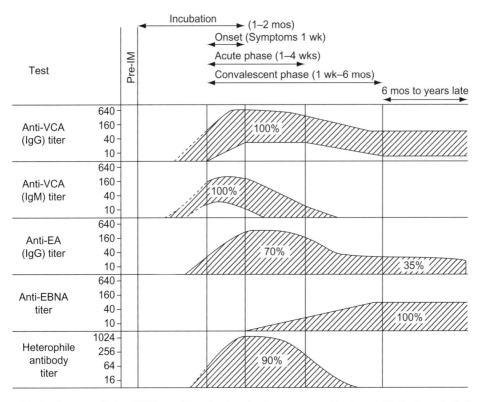

Figure 11-5 Characteristic EBV-specific Antibody Responses During EBV-induced Infectious Mononucleosis.
Adapted from: Henle G, Henle W, Horowitz CA. Epstein–Barr Virus Specific Diagnostic Tests in Infectious Mononucleosis. Hum Pathol 1974;5:551, with permission.

- Anemia, neutropenia, lymphocytopenia, or lymphocytosis, thrombocytopenia, polyclonal hypergammaglobulinemia may be present.

To diagnose this condition, the following criteria must be met and the symptoms must be present for more than 6 months:

- Elevated IgG antibody titers to viral capsid antigen (VCA) greater than 5120, VCA IgA positive. Early antigen diffuse (EAD) IgG greater than 640, EAD IgA positive. Early antigen restricted (EAR) IgG greater than 640 or detection of genomes in affected tissues, EBNA1 low or negative – an indicator of recent EBV infection
- Detection of EBV genome in affected tissues and high serum viral loads by PCR
- Oligoclonal or monoclonal EBV-infected lymphocyte population may be of T- or B-cell lineage. Occasionally, there may be hemophagocytic lymphohistiocytosis (HLH).

Therapy

Symptomatic patients with active EBV-infection are treated. Treatments are similar to those used for EBV-virus infected immunodeficient patients with:

- Antiviral drugs: acyclovir, ganciclovir, adenosine arabinoside
- Intravenous immunoglobulin (IVIgG)
- Chemotherapy: etoposide, dexamethasone for patients with EBV-associated hemophagocytic syndrome
- Anti-CD20 (Rituximab) antibody infusion in patients with B-cell proliferation.

DISORDERS OF LEUKOCYTE FUNCTION

Table 11-17 classifies the diseases of leukocyte dysfunction and lists the investigations to be carried out in patients with leukocyte dysfunction.

Leukocyte Adhesion Disorders

Leukocyte Adhesion Deficiency (LAD) Type I

This disorder is characterized by a deficiency of glycoprotein CD 11b which forms the α subunit of the Mac-1 β_2 integrin (CD 11b/CD18) on the cell surface of the neutrophils. Intercellular adhesion molecule-1 (ICAM-1) expressed on the vascular endothelium serves as a ligand to Mac-1 for the neutrophil adhesion. Because of the mutation and the absence of Mac-1, neutrophils in LAD type I are not able to attach to the endothelium and undergo transendothelial migration. Additionally, the ability for chemotaxis, phagocytosis, degranulation and respiratory burst activity is also impaired. Survival of neutrophils is prolonged in LAD type I.

Clinical Features

LAD type I is a rare autosomal-recessive disorder that causes lack of expression of the CD11/CD18 integrins. CD18 is located on chromosome 21. Most patients are compound heterozygotes with two different mutations in the CD18 gene. Clinical features include:

- Persistent neutrophilia and lack of pus formation. Frequent skin and periodontal infections
- Omphalitis, delayed umbilical cord separation (normal mean age of cord separation: 7–15 days)
- Perirectal abscess
- Sepsis
- Necrotizing enterocolitis

Table 11-17   Classification and Investigation of Diseases of Leukocyte Dysfunction

Function	Disease	Investigations[a]
Motility and migration	Leukocyte adhesion deficiency	All tests below may be abnormal
Chemotaxis	Lazy-leukocyte syndrome	Rebuck skin window: at 3 hours, neutrophils; at 6 hours, mixed neutrophils and monocytes; at 24 hours, monocytes
	Syndrome of elevated IgE, eczema and recurrent infection	
Opsonization	Complement deficiency	Serum complement levels
	Specific bacterial antibody deficiency	Immunoglobulin levels
		Specific bacterial antibody levels
Bacterial killing	Chédiak–Higashi syndrome	Morphologic tests
		Chédiak–Higashi giant granules
	Myeloperoxidase deficiency	Peroxidase stain
	Chronic granulomatous disease (CGD)	Bacterial test
		Killing of catalase-positive (*Staphylococcus aureus*) and catalase-negative (streptococcal) bacteria
	Job syndrome	
	Leukocyte glutathione peroxidase efficiency	Metabolic tests
		Nitroblue tetrazolium (NBT) reduction test[b]
	Glucose-6-phosphate dehydrogenase (G6PD) deficiency	Flow cytometry for DHR_{123} to R_{123}[c]
		Glucose $1-^{14}C$ oxidation with Phagocytosis
		Oxygen consumption during phagocytosis
		H_2O_2-dependent ^{14}C-formate oxidation with phagocytosis Iodine-125 fixation during phagocytosis[d]

[a]Gene testing is available for many of the diseases listed (see text).
[b]Impaired in CGD due to nicotinamide adenine dinucleotide phosphate (NADPH) oxidase deficiency, deficiency of glutathione reductase of G6PD. Normal in myeloperoxidase deficiency.
[c]Flow cytometric analysis to detect reduction of dihydroxyrhodamine 123 to rhodamine 123.
[d]Impaired in CGD and in myeloperoxidase deficiency.

- Pneumonia
- Sinusitis
- Infecting organisms: *Staphylococcus aureus*, *Pseudomonas aeruginosa*, *Proteus*, *Escherichia coli*, *Klebsiella* sp., *Candida albicans*, *Aspergillus* sp.

Diagnosis

1. Flow cytometry for assessment of expression of CD 11b or CD 18 on neutrophil cell surfaces with the use of specific monoclonal antibodies.
2. *In vitro* functional assays of neutrophil adhesion, chemotaxis and phagocytosis.

Treatment

For mild and moderate disease: Oral hygiene with antimicrobial mouthwash, e.g. chlorhexidine gluconate, prophylactic use of trimethoprim/sulfamethoxazole (co-trimoxazole). Aggressive treatment of infections. For severe cases: Hematopoietic stem cell transplantation.

Leukocyte Adhesion Deficiency (LAD) Type II

LAD type II is an extremely rare condition caused by the absence of neutrophil sialyl Lewis X structure. Patients have rare Bombay (hh) red cell phenotype. Compared with LAD type I, these patients suffer from less serious types of infections. They are unable to form pus in spite of leukocytosis of 30,000–150,000/mm^3.

Lazy-Leukocyte Syndrome

An altered membrane microfilamentous protein structure or function leads to rigidity and impaired mobility of the neutrophils, causing difficulty in their entering and exiting the circulation.

The disease is characterized by:

- Poor polymorphonuclear leukocyte response to bacterial pyrogen injection, indicating impaired marrow release of polymorphs
- Recurrent stomatitis, otitis, gingivitis and low-grade fever, frequently *Staphylococcus aureus* infections
- Low or normal absolute granulocyte counts with normal phagocytic and bactericidal activity in the polymorphs
- Normal bone marrow findings
- Defective recruitment of polymorphs to skin windows, recruitment of monocytes normal.

Syndrome of Elevated Immunoglobulin E, Eczema and Recurrent Infection (Job Syndrome)

Clinical Manifestations

1. Chronic eczema, delay in shedding primary teeth, hyperextensible joints, scoliosis, osteopenia and tendency to fractures, growth retardation and coarse facies.
2. Recurrent severe staphylococcal abscesses.
3. Recurrent cutaneous, pulmonary and joint abscesses.
4. Chronic candidiasis of mucosa and nails.

Laboratory Findings

1. Very high serum IgE level (greater than 2,500 IU/ml).
2. Defect in T-lymphocytes that results in reduced production of IFN-γ and tumor necrosis factor.
3. Molecular basis of hyper IgE syndrome is unknown.
4. Striking defect in neutrophil granulocyte chemotactic responsiveness; neutrophil migration, phagocytosis and bactericidal activity normal.

Treatment

1. Prophylactic antibiotics: Trimethoprim/sulfamethoxazole or dicloxacillin. rhIFN-γ: 50 µg/m^2 subcutaneously three times per week, however, evidence-based data for its efficacy are not available.

Localized Juvenile Periodontitis (LJP)/Localized Aggressive Periodontitis in Children

LJP is characterized by severe alveolar bone loss localized to the first molars and incisors. It is frequently associated with the presence of the bacterium *Actinobacillus actinomycetemcomitans*.

Age of Onset

1. Usually at puberty.

Etiology

1. Defective neutrophil chemotaxis in the majority of patients. This defect is attributed to neutrophil chemotactic inhibitors. Phagocytosis is also abnormal in many of these patients.
2. Unstimulated neutrophils from these patients show reduced Lewis X, sialyl Lewis X and L-selectin expression.

Treatment

1. Deep subgingival scaling and root planing.
2. Adjunctive systemic use of antibiotics: Simultaneous use of metronidazole and amoxicillin or amoxicillin with clavulanic acid (Augmentin).
3. Local delivery of antibiotics: Not much information is available. However, the use of doxycycline or minocycline in local delivery formulation, may be considered in eliminating infection with *A. actinomycetemcomitans*.

Chédiak–Higashi Syndrome

This syndrome is characterized by:

- Autosomal-recessive inheritance with repeated pyogenic infections
- Most cases are fatal with the mean age at death 6 years
- Most cases are caused by a defect in the CHS1 gene (also known as LYST gene), a lysosomal trafficking regulator protein, leading to abnormal granule formation. The granules are abnormally large and deficient in essential cidal proteins
- Photophobia, pale optic fundi, nystagmus, partial oculocutaneous albinism and excessive sweating
- Neurologic features: ataxia, muscle weakness, decreased deep tendon reflexes, sensory loss and abnormal electrical evaluations on electroencephalograms, visual- and auditory-evoked potentials

- Hematologic features
  - Anemia, neutropenia and thrombocytopenia
  - Giant refractile peroxidase-positive granules (1–4 μm) are present in the neutrophils, eosinophils, basophils and platelets. They stain greenish gray. Ring-shaped lysosomes are present in monocytes
  - Granules fail to discharge lysosomal enzymes into the phagocytic vacuole, leading to increased susceptibility to infection
  - A defect in chemotaxis along with the neutropenia contributes to further increase the risk of bacterial infection
  - Platelets are deficient in dense bodies and may contain the giant granules. Minor bleeding problems presenting with easy bruising and epistaxis.
- An accelerated phase occurs in 85% of cases. It is characterized by lymphoproliferative disease with infiltration of the central nervous system and peripheral nerves, liver, spleen and other organs by histiocytes and atypical lymphocytes. A lymphoma-like picture with fever, jaundice, hepatosplenomegaly, lymphadenopathy, bleeding tendency and pancytopenia develops. The accelerated phase may be triggered by an EBV infection.

Treatment

1. Ascorbic acid (20 mg/kg/day) may normalize the chemotactic defect and bactericidal function.
2. Antibiotics:
   - Prophylactic antibiotics: Trimethoprim/sulfamethoxazole
   - Therapeutic use of antibiotics to treat infections.
3. Vincristine, corticosteroids and antithymocyte globulin may induce temporary remissions when used in accelerated phase. Protocols for hemophagocytic lymphohistiocytosis consisting of corticosteroids, etoposide and cyclosporine have been employed in the accelerated phase.
4. Allogeneic hematopoietic stem cell transplantation is potentially curative.

Chronic Granulomatous Disease (CGD)

Pathogenesis

In CGD, neutrophils show normal phagocytosis but defective killing of microorganisms as a result of markedly deficient or absent superoxide production due to inherited mutations of polypeptides of reduced nicotinamide adenine dinucleotide phosphate (NADPH) oxidase (also known as respiratory burst oxidase). Superoxide is a precursor of microbicidal oxidants such as hydrogen peroxide and hypochlorous acid. Table 11-18 lists the reactions of respiratory burst pathway.

**Table 11-18   Reactions of Respiratory Burst Pathway in a Neutrophil
(Activated for Phagocytosis)**

1. Assembly of respiratory oxidase, also known as NADPH oxidase, a multisubunit enzyme complex consisting of four essential phagocyte oxidase (phox) polypeptides: $gp91^{phox}$, $p22^{phox}$, $p47^{phox}$ and $p67^{phox}$. NADPH oxidase catalyzes the transfer of an electron from NADPH to molecular oxygen as a result of which superoxide ($O_2 \cdot^-$) is formed $NADPH + 2O_2 \xrightarrow{NADPH\ oxidase} NADP + H^+ + O_2 \cdot^-$
2. Conversion of superoxide to hydrogen peroxide (H_2O_2) by superoxide dismutase or spontaneously and also, formation of hydroxyl radical ($OH \cdot^-$)
3. Hypochlorous acid (HOCL) formation: $H_2O_2 \xrightarrow{Myeloperoxidase} HOCl$
4. Conversion of hydrogen peroxide to water by glutathione peroxidase:
   GSH (reduced glutathione) $+ H_2O_2 \xrightarrow{glutathione\ peroxidase} GSSG$ (oxidized gluthathione) $+ H_2O$
5. Hydrogen peroxide is also converted to water by catalase
6. Restoration of GSH by conversion of GSSG by glutathione reductase:
   $NADPH + GSSG \xrightarrow{glutathione\ reductase} GSH + NADP$
7. Generation of NADPH through G6PD (glucose 6 phosphate dehydrogenase) reaction

Organisms that make their own catalase are more often responsible for severe infections in these patients. These organisms are able to convert their own hydrogen peroxide to water, thus making it unavailable to the phagocyte for the microbicidal purpose, As a consequence, the ingested bacteria or fungi remain viable in the phagocytes and are protected from host humoral immunity and from antibiotics, which fail to penetrate the cell.

The mobility of the phagocytes leads to generalized seeding of the reticuloendothelial system with live microorganisms. This results in the formation of generalized chronic granulomatous lesions, characterized by recurrent suppurative infection with bacteria of low virulence (e.g., *Staphylococcus aureus*, *Staphylococcus epidermidis*, *Aerobacter aerogenes*, *Serratia marcescens Burkholderia (Pseudomonas) cepacia* and *Salmonella* sp.) or infection with mycotic organisms (e.g., *Aspergillus*). *S. aureus* is the most frequently isolated organism. However, the most common causes of death are pneumonia and/or sepsis due to *Aspergillus* or *B. cepacia*.

Genetics

CGD results from mutations in any of the four genes encoding essential subunits of the NADPH oxidase. Table 11-19 shows the Genetic Classification of CGD according to the component of NADPH oxidase affected.

Clinical Features

1. Incidence, between one in 200,000 and one in 250,000 live births.
2. First symptoms may manifest in infancy or childhood.
3. Lymphadenopathy (nodes suppurate and drain pus) and hepatosplenomegaly.
4. Recurrent suppurative infections – pneumonitis, subcutaneous abscesses, impetiginous rashes and osteomyelitis (often small bones of the hands and feet).

Table 11-19   Genetic Classification of CGD

Component Affected	Gene Locus	Inheritance	Frequency (% of Cases)
gp91^{phox}	Xp21.1	X-linked	65
p22^{phox}	16q24	AR	7
p47^{phox}	7q11.23	AR	23
p67^{phox}	1q25	AR	5

Abbreviations: phox: phagocyte oxidase; AR, autosomal recessive.

5. Urologic problems (e.g., granulomatous ureteral or urethral strictures, bladder granulomas and urinary tract infections) in 38% of cases.
6. Gastrointestinal problems: colitis, enteritis, granulomatous obstruction of gastric outlet.
7. Hematologic features include:
    *   Appropriate neutrophil leukocytosis
    *   Anemia due to infection
    *   McLeod syndrome (i.e. mild hemolytic anemia, acanthocytosis, decreased expression of Kell antigen due to defect in K_x antigen on red cells): This occurs in rare patients with large deletions of Xp 21.1 gene.
    *   Hypergammaglobulinemia.
8. Noninfectious complications: Noninfectious granulomata resulting in colitis, granulomatous cystitis and urethritis, cutaneous granulomata, pericarditis and recurrent gastrointestinal strictures occur in patients with CGD, as a result of a failure of phagocytes to clear both exogenous and endogenous debris.

Diagnosis

A nitroblue tetrazolium (NBT) dye test can be used to establish a diagnosis. Newer, more precise testing for CGD is flow cytometric analysis to detect reduction of dihydroxy-rhodamine$_{123}$ to rhodamine$_{123}$, a fluorescent compound. This can also be used to detect carrier states in X-linked forms. As a quantitative assay, it can also distinguish between forms with no activity, such as gp91^{phox}, versus those with low activity, such as p47^{phox}.

Treatment

1. The use of prophylactic antibiotics and antifungal agents:
    *   Antibacterial: Trimethoprim/sulfamethoxazole (co-trimoxazole), or dicloxacillin for patients allergic to sulfa (Dose of co-trimoxazole: 5 mg/kg/day, once a day orally based on trimethoprim component) or a cephalosporin for those patients allergic to sulfa
    *   Antifungal antibiotic: Itraconazole: 3–5 mg/kg/day once a day orally.
2. Aggressive infection management:
    *   Meticulous wound care – prompt cleansing of wounds and abrasions with hydrogen peroxide

- Prompt treatment with broad-spectrum antibiotics to cover *B. cepacia* and *Staph.* species when serious infection is suspected
- Aggressive surgical intervention to drain abscesses when not responding to antibiotics appropriately.

3. The use of rhIFN-γ (interferon-γ) in a dose of 50 μg/m^2 subcutaneously three times a week: Beneficial effect of interferon-γ is probably related to increased synthesis of nitric oxide (NO) through nitric oxide synthase pathway. NO causes nitration of bacteria. Interferon-γ may also be stimulating other nonoxidative microbicidal pathways. It also increases superoxide production in phagocytes. Interferon-γ as prophylaxis has been recommended only in patients with significant infections despite appropriate oral agents, or as an adjunct to the treatment of deep-seated infections. Oral antibiotic prophylaxis appears to be adequate in most of the patients.
4. Granulocyte transfusions may provide short-term relief at times of crisis.
5. Steroids should be used cautiously to treat granulomatous disease in the gastrointestinal or urinary tract.
6. Allogeneic stem cell transplantation: Allogeneic stem cell transplantation is rarely used to treat CGD. It may be indicated in patients with severe disease.
7. Gene therapy is also a consideration for the future but no therapies are currently available.

Prognosis

With the use of prophylactic antibiotics and recombinant human interferon-γ (rhIFN-γ) the prognosis of CGD has improved remarkably in the past two decades.

Myeloperoxidase Deficiency

Myeloperoxidase participates in the halogenation reaction in the neutrophils and is useful in killing bacteria phagocytosed by the neutrophils. This condition has the following characteristics:

- Rare
- Autosomal-recessive inheritance
- Less severe than chronic granulomatous disease
- Normal NBT test and glucose metabolism
- Abnormal iodination, cytochemical peroxidase staining and bactericidal tests
- Infections generally mild, but with more susceptibility for *Candida* infections which need to be treated aggressively.

Glutathione Synthetase Deficiency

1. Autosomal recessive.
2. Relatively benign disorder.
3. Intermittent neutropenia.
4. Severe metabolic acidosis as a result of increased levels of 5-oxoproline, a metabolite formed during the biochemical steps of glutathione synthesis.
5. Hemolysis induced by oxidant stress.
6. Treatment:
   - Correction of acidosis
   - Use of vitamin E, 400 units per day, to treat hemolysis and infections
   - Antibiotic therapy for infections.
7. Respiratory burst is normal in neutrophils and phagocytosis abnormal only in patients with severe deficiency of glutathione synthetase.

Glutathione Reductase Deficiency

1. Autosomal recessive.
2. Accumulation of H_2O_2 in neutrophils.
3. Hemolysis induced by oxidant stress.
4. Frequency of infections not increased.
5. Avoid use of oxidant foods and drugs.

Glucose-6-Phosphate Dehydrogenase Deficiency in Leukocytes

In some cases of Caucasian glucose-6-phosphate dehydrogenase (G6PD) deficiency, the enzyme is severely depressed (less than 5% of normal) in the neutrophil, as a result the conversion of NADP to NADPH is decreased leading to decreased respiratory burst and abnormal NBT test. There is persistent and eventually fatal bacterial infection because of the inability of the leukocytes to generate H_2O_2. Clinical manifestations are similar to CGD. Hemolytic anemia may occur.

Treatment

1. Aggressive treatment of infections.
2. Avoidance of oxidant drugs and foods.
3. Red cell transfusions for anemia.

NEUTROPHIL PRODUCTION AND DESTRUCTION IN NEWBORN INFANTS

Neutrophil Production

Using soluble Fc receptor III (sFcR III) as a surrogate marker for estimation of total neutrophil mass, it has been shown that newborn infants born before 32 weeks of gestation have 20% of the adult neutrophil mass. Normal levels of neutrophil mass are attained at 4 weeks of age in premature neonates. However, full-term neonates have neutrophil stores within the normal adult range.

Neutrophil Function in the Newborn

Chemotaxis

It is decreased in newborn infants. Normal chemotactic ability is attained at 2 weeks of age in term and preterm neonates.

Vascular Rolling of Neutrophils

It is also decreased due to decreased expression of L-selectin on the neutrophil cell membrane.

Vascular Endothelial Adhesion

It is also decreased due to decreased expression of β_2 integrin Mac-1 on the neutrophil cell membrane in neonates.

Dynamics of Change of Shape

Newborn neutrophils are rigid because of impaired ability to form polymers of actin (P-actin) and reduced formation of microtubules.

Phagocytosis

Neutrophils of term neonates have a normal ability for phagocytosis of Gram-positive and Gram-negative bacteria. However, their ability for phagocytosis of candida is abnormal up to 2 weeks of age. Preterm neonates have abnormal bacterial phagocytosis. However, when treated with therapeutic doses of intravenous gamma globulin G (IVIg) the neutrophils are able to ingest bacteria normally.

Respiratory Burst

Respiratory burst in term neonates is normal under normal conditions. In contrast, it is less active under the conditions of stress and sepsis. For this reason, neonates are more susceptible to overwhelming infections with group B streptococci, *Staphylococcus epidermis*, *Staphylococcus aureus* and *E. coli*. Respiratory burst performance of neutrophils in preterm neonates remains abnormal for more than 2 months of age. In term neonates, the generation of superoxide and hydrogen peroxide is increased but the levels of lactoferrin and myeloperoxidase being low there is truncation of the later respiratory burst activity, resulting in abnormal bacterial killing.

Therapeutic Implications

G-CSF or GM-CSF may be helpful in term and preterm neonates during sepsis. The prophylactic use of G-CSF is more effective in preterm infants. The use of IVIgG to improve opsonization and phagocytosis has been disappointing.

Neonatal Pre-Eclampsia-Associated Neutropenia

It occurs in low-birth-weight neonates with a maternal history of pregnancy-induced hypertension.

Treatment

Prophylactic use of G-CSF.

Suggested Reading

Alter B. Bone Marrow Failure Syndromes. In: Nathan DG, Orkin SH, Ginsburg D, Look AT, eds. Nathan and Oski's Hematology of Infancy and Childhood. Sixth Edition Philadelphia: Saunders; 2003.

Berliner N. Acquired Neutropenia. In American Society of Hematology (ASH) Education Book. 2004:69.

Brito-Babapulle F. The eosinophilias, including the idiopathic hypereosinophilic syndrome. Br. J. Haematol. 2003;203–223.

Bux J, Mueller-Eckhardt C. Autoimmune neutropenia. Semin Hematol. 1992;29:45–53.

Carlsson G, Melin M, Dahl N, et al. Kostmann syndrome or infantile genetic agranulocytosis, part two: understanding the underlying genetic defects in severe congenital neutropenia. Acta Paediatrica. 2007;96:813.

Carr R. Neutrophil production and function in newborn infants. Br. J. Haematol. 2000;110:18–28.

Cartron J, Tchernia G, Cleton J, et al. Alloimmune neonatal neutropenia. Am J Pediatr Hematol/Oncol. 1991;13:21–25.

Dale, DC, Guest Editor: Severe Chronic Neutropenia, Seminars In Hematology 2002;39:73–133, Philadelphia, Saunders.

Dale DC, Cottle TE, Fier CJ, et al. Severe Chronic Neutropenia: Treatment and follow-up of the patients in the Severe Chronic Neutropenia International registry. Am J Hematol. 2003;72:82.

Dale DC, Link DC. The many causes of severe congenital neutropenia. N Engl J Med. 2009;360:3.

Dinauer MC. The phagocyte system and disorders of granulopoiesis and granulocyte function. In: Nathan DG, Orkin SH, Ginsburg D, Look AT, eds. Nathan and Oski's Hematology of Infancy and Childhood. Sixth Edition Philadelphia: Saunders; 2003.

Dror Y. Shwachman-Diamond syndrome. Pediatr Blood Cancer. 2005;45:892.

Horwitz M, Li F, et al. Leukemia in severe congenital neutropenia: Defective proteolysis suggests new pathways to malignancy and opportunities for therapy. Cancer Investgation. 2003;21:579–587.

Johannsen EC, Schooley RT, Kaye KM. Epstein-Barr Virus (Infectious Mononucleosis). In: Mandell Bennett, Dolin, eds. Principles and Practice of Infectious Diseases. Sixth Edition Philadelphia: Churchill Livingstone; 2005.

Karlsson S. Treatment of genetic defects in hematopoietic cell function by gene transfer. Blood. 1991;78:2481–2492.

Klion AO, Robyn J, et al. Molecular remission and reversal of myelofibrosis in response to imatinib mesylate treatment in patients with the myeloproliferative variant of hypereosinophilic syndrome. Blood. 2004;103:473–476.

Lalezari P, Khorshidi M, Petrosova M. Autoimmune neutropenia of infancy. J Pediatr. 1986;109:764.

Shastri KA, Logue GL. Autoimmune neutropenia. Blood. 1993;81:1984–1995.

Sullivan JL. Hematologic consequences of Epstein–Barr virus infection. Hematol/Oncol Clin N Am. 1987;1:397.

Vlachos A, Lipton JM. Hematopoietic stem cell transplantation for inherited bone marrow failure syndromes. In Pediatric Stem Cell Transplantation, Edited by Mehta, P. Jones and Bartlett Publishers: Boston; 2004.

Yang K, Hill H. Neutrophil function disorders: pathophysiology, prevention and therapy. J Pediatr. 1991;119:343–354.

CHAPTER 12

Disorders of Platelets

Platelets are an important component in the first phase of hemostasis known as platelet plug formation (see Chapter 13). Defects in platelet number or function may lead to bleeding. Bleeding due to platelet disorders usually involves skin and mucous membranes, including petechiae, purpura, ecchymosis, epistaxis, hematuria, menorrhagia and gastrointestinal and even intracranial hemorrhage.

Platelet characteristics include:

- Size: 1–4 μm (younger platelets are larger). Table 12-1 lists the causes of thrombocytopenia based on platelet size
- Mean platelet volume (MPV): $8.9 \pm 1.5\ \mu m^3$
- Distribution: one-third in the spleen, two-thirds in circulation
- Average lifespan: 9–10 days.

Table 12-2 lists the causes of thrombocytopenia according to pathophysiology and Table 12-3 lists the common and uncommon causes of thrombocytopenia in the neonate and child.

THROMBOCYTOPENIA IN THE NEWBORN

Neonatal thrombocytopenia is relatively common, occurring in 1–3% of healthy term infants and in 20–30% of sick neonates. Thrombocytopenia in sick neonates is often due to underlying illness such as sepsis, DIC or respiratory distress syndrome or due to maternal factors such as pregnancy-induced hypertension and gestational diabetes and intrauterine growth retardation (IUGR).

Table 12-4 lists the causes of neonatal thrombocytopenia and Figure 12-1 shows an approach to the diagnosis of thrombocytopenia in the infant.

Neonatal Alloimmune Thrombocytopenia (NAIT)

Neonatal alloimmune thrombocytopenia is the most common cause of severe thrombocytopenia in the newborn. It occurs in approximately 1 in 1,000 births. NAIT typically resolves in 2–4 weeks. First-born infants are often affected and subsequent affected pregnancies are more severe and therefore require antenatal treatment.

Manual of Pediatric Hematology and Oncology. DOI: 10.1016/B978-0-12-375154-6.00012-4

Table 12-1   Platelet Diseases Based on Platelet Size

Macrothrombocytes (MPV Raised)
ITP or any condition with increased platelet turnover (e.g., DIC)
Bernard–Soulier syndrome
May–Hegglin anomaly and other MYH-9 related diseases (Table 12-7)
Swiss cheese platelet syndrome
Montreal platelet syndrome
Gray platelet syndrome
Various mucopolysaccharidoses
Normal Size (MPV Normal)
Conditions in which marrow is hypocellular or infiltrated with malignant disease
Microthrombocytes (MPV Decreased)
Wiskott–Aldrich syndrome
TAR syndrome
Some storage pool diseases
Iron-deficiency anemia

Abbreviations: MPV, mean platelet volume (as determined by automated electronic counters); normal, $8.9 \pm 1.5 \, \mu m^3$; ITP, idiopathic thrombocytopenic purpura; DIC, disseminated intravascular coagulation; MYH-9, non-muscle myosin heavy chain 9 gene; TAR, thrombocytopenic absent radii.

Pathophysiology

Neonatal alloimmune thrombocytopenia can be thought of as a platelet analog of Rh incompatibility (i.e. hemolytic disease of the fetus and newborn). It differs from Rh incompatibility because 50% of cases are first-born infants, suggesting the antigenic exposure occurs early in pregnancy unlike in Rh which occurs primarily at the time of delivery. Neonatal alloimmune thrombocytopenia occurs when fetal platelets expressing platelet-specific antigens inherited from the father, are the target of maternal alloantibodies. Mothers who lack the platelet-specific surface antigen and who possess the immunologic predisposition to make antibodies to it, can become sensitized to the paternally derived antigens when they are expressed on fetal platelets. The most common antigen involved is $P1^{A1}$ (HPA-1a) which accounts for approximately 75% of cases. A further 10–20% of cases are due to maternal sensitization to HPA-5b. More than 13 other antigens are known to be involved in NAIT; HPA-4 is important in Asian populations which do not have the HPA-1A/B polymorphism. Specifically, mothers who possess the HLA-DR type DRB30101 have a high rate of sensitization to $P1^{A1}$. These IgG antibodies can cross the placenta and attach to the surface of fetal platelets, causing platelet destruction and perhaps inhibition of platelet production.

Clinical Features

1. Typically infants are otherwise-healthy full-term babies, who develop generalized petechiae within minutes of birth followed by ecchymosis and even cephalhematomata. Bleeding from the umbilicus, skin puncture site, or gastrointestinal or renal tract may also occur.

Table 12-2   Pathophysiological Classification of Thrombocytopenic States

I. Increased Platelet Destruction (normal or increased megakaryocytes in the marrow – megakaryocytic thrombocytopenia)

  A. Immune thrombocytopenias

    1. Idiopathic

      a. Immune (idiopathic) thrombocytopenic purpura

    2. Secondary

      a. Infection induced (e.g., viral – HIV, CMV, EBV, varicella, rubella, rubeola, mumps, measles, pertussis, hepatitis, parvovirus B19; bacterial – tuberculosis, typhoid)

      b. Drug-induced (see Table 12-10)

      c. Post-transfusion purpura

      d. Autoimmune hemolytic anemia (Evans syndrome)

      e. Systemic lupus erythematosus

      f. Hyperthyroidism

      g. Lymphoproliferative disorders

    3. Neonatal immune thrombocytopenias

      a. Neonatal autoimmune thrombocytopenia

      b. Neonatal alloimmune thrombocytopenia

      c. Erythroblastosis fetalis–Rh incompatibility

  B. Nonimmune thrombocytopenias

    1. Due to platelet consumption

      a. Microangiopathic hemolytic anemia: hemolytic-uremic syndrome (HUS), thrombotic thrombocytopenic purpura (TTP), hematopoietic stem cell transplantation (HSCT) associated microangiopathy

      b. Disseminated intravascular coagulation

      c. Virus-associated hemophagocytic syndrome

      d. Kasabach–Merritt syndrome (giant hemangioma)

      e. Cyanotic heart disease

    2. Due to platelet destruction

      a. Drugs (e.g., ristocetin, protamine sulfate, bleomycin)

      b. Left ventricular outflow obstruction

      c. Infections

      d. Cardiac (e.g., prosthetic heart valves, repair of intracardiac defects, left ventricular outflow obstruction)

      e. Malignant hypertension

II. Disorders of Platelet Distribution or Pooling

  A. Hypersplenism (e.g., portal hypertension, Gaucher disease, cyanotic congenital heart disease, neoplastic, infectious)

  B. Hypothermia

III. Decreased Platelet Production – Deficient Thrombopoiesis (decreased or absent megakaryocytes in the marrow – amegakaryocytic thrombocytopenia)

  A. Hypoplasia or suppression of megakaryocytes[a]

    1. Drugs (e.g., chlorthiazides, estrogenic hormones, ethanol, tolbutamide)

    2. Constitutional

      a. Thrombocytopenia absent radii – TAR syndrome

      b. Congenital amegakaryocytic thrombocytopenia

      c. Amegakaryocytic thrombocytopenia with radio-ulnar synostosis

      d. Thrombocytopenia agenesis of corpus callosum syndrome

      e. Paris-Troussea syndrome

(Continued)

Table 12-2 (Continued)

   f. Rubella syndrome
   g. Trisomy 13, 18
3. Ineffective thrombopoiesis
   a. Megaloblastic anemias (folate and vitamin B_{12} deficiencies)
   b. Severe iron-deficiency anemia
   c. Certain familial thrombocytopenias
   d. Paroxysmal nocturnal hemoglobinuria
4. Disorders of control mechanism
   a. Thrombopoietin deficiency
   b. Tidal platelet dysgenesis
   c. Cyclic thrombocytopenias
5. Metabolic disorders
   a. Methylmalonic acidemia
   b. Ketotic glycinemia
   c. Holocarboxylase synthetase deficiency
   d. Isovaleric acidemia
   e. Idiopathic hyperglycinemia
   f. Infants born to hypothyroid mothers
6. Hereditary platelet disorders[b]
   a. Bernard–Soulier syndrome
   b. May–Hegglin anomaly and other MYH-9 gene related disorders (Table 12-7)
   c. Wiskott–Aldrich syndrome
   d. Pure sex-linked thrombocytopenia
   e. Mediterranean thrombocytopenia
7. Acquired aplastic disorders
   a. Idiopathic
   b. Drug-induced (e.g., dose-related: antineoplastic agents; benzene, organic and in organic arsenicals, Mesantoin, Tridione, antithyroids, antidiabetics, antihistamines, phenylbutazone, insecticides, gold compounds; idiosyncrasy: chloramphenicol)
   c. Radiation-induced
   d. Viral infections (e.g., hepatitis, HIV, EBV)
   B. Marrow infiltrative processes
   1. Benign
      a. Osteopetrosis
      b. Storage diseases
   2. Malignant
      a. *De novo* – leukemias, myelofibrosis, Langerhans cell histiocytosis, histiocytic medullary reticulosis
      b. Secondary – lymphomas, neuroblastoma, other solid tumor metastases
IV. Pseudothrombocytopenia:
   A. Platelet activation during blood collection
   B. Undercounting of megathrombocytes
   C. *In-vitro* agglutination of platelets due to EDTA
   D. Monoclonal antibodies that bind to platelet glycoprotein receptors such as abciximab, eptifibatide, tirobifan

[a]A bone marrow biopsy, in addition to marrow aspiration, should always be carried out to avoid sampling errors and to establish the presence of a decreased number of megakaryocytes in the marrow.
[b]These conditions are associated with normal or increased bone marrow megakaryocytes.
Abbreviations: HUS, hemolytic-uremic syndrome; TTP, thrombotic thrombocytopenic purpura; HSCT, hematopoietic stem cell transplantation.

Table 12-3 Classification of Thrombocytopenic Purpura by Age and Frequency

	Common	Uncommon

Neonate	Sepsis
	Asphyxia
	Alloimmune
	 thrombocytopenia
	Necrotizing
	 enterocolitis
	Maternal ITP
		Cardiac (prosthetic heart valves, repair of intracardiac defects, left
		 ventricular outflow obstruction)
		Maternal hypertension
		Infections (rubella, CMV, HIV, Hepatitis B, syphilis)
		Amegakaryocytic thombocytopenias
		 Congenital amegakaryocytic thrombocytopenia without anomalies
		  (CAMT)
		 Congenital amegakaryocytic thrombocytopenia with bilateral
		  absence of radii (TAR syndrome)
		 Amegakaryocytic thrombocytopenia with radio-ulnar synostosis
		  (ATRUS)
		Wiskott Aldrich and X-linked thrombocytopenia
		Bernard Soulier syndrome
		MYH9 disorders
		Montreal Platelet syndrome
		Quebec syndrome
		Gray platelet syndrome
		Inborn errors of metablism
		Congenital leukemia (trisomies 13,18,21)

Child	ITP Drug induced	Type II von Willebrand disease Immunodeficiencies Autoimmune diseases such as SLE, JRA Infections Fanconi Anemia Leukemia and other malignancies with bone marrow involvement Autoimmune hemolytic anemia (Evans Syndrome) TTP/HUS Hyperthyroidism Megaloblastic anemias (folate and Vitamin B_{12} deficiencies) Severe iron-deficiency anemia Cyclic thrombocytopenias Lymphoproliferative disorders Aplastic Anemia Drug induced Radiation induced

2. Neonates may have higher rates of intracranial hemorrhage (ICH) (up to 10–20%) and when present, the ICH tends to be more severe because it is intraparenchymal. Intracranial hemorrhage can occur *in utero* and be detected on ultrasonography during apparently uncomplicated pregnancies. Death *in utero* may occur. This type of neonatal thrombocytopenia accounts for most of the cases of fetal morbidity and mortality. Hemorrhagic manifestations are variable but tend to be more severe than in the passive transfer of a platelet autoantibody across the placenta (NITP).

Table 12-4   Causes of Neonatal Thrombocytopenia

I. **Normal or Increased Megakaryocytes in the Marrow (Consumptive Thrombocytopenia)**
   A. Immune disorders
      1. Autoimmune (passive transfer of platelet antibody) (NITP)
         a. Maternal ITP
         b. Maternal drug-induced thrombocytopenia
         c. Maternal SLE
      2. Allo-immune (NAIT)
         a. Isolated platelet antigen incompatibility
   B. Infection
      1. Bacterial: Gram-negative and Gram-positive septicemia, listeriosis
      2. Viral: cytomegalovirus, rubella, herpes simplex, coxsackievirus, HIV
      3. Protozoal: toxoplasmosis
      4. Spirochetal: syphilis
   C. Drugs
      1. Immune: drug–hapten disease (e.g., quinine, quinidine, sedormid)
      2. Nonimmune: thiazide, tolbutamide (given to mother)
      3. Prolonged antibiotics or ganciclovir
      4. Chemotherapy
   D. Disseminated intravascular coagulation
      1. Antenatal causes
         a. Pre-eclampsia and eclampsia
         b. Abruptio placentae
         c. Dead twin fetus
         d. Amniotic fluid embolism
      2. Intranatal causes
         a. Breech delivery
         b. Fetal distress
      3. Postnatal causes
         a. Infections
         b. Hypoxia and acidosis
         c. Respiratory distress syndrome
         d. Renal vein thrombosis
         e. Indwelling catheters
         f. Giant hemangioma (usually in the months after birth)
   E. Inherited thrombocytopenia
      1. Sex-linked
         a. Gata-1
         b. Wiskott–Aldrich syndrome[a]
      2. Autosomal
         a. Bernard-Soulier syndrome
         b. MYH9-RD (was called May-Hegglin and other rarer types)
II. **Decreased or Absent Megakaryocytes in the Marrow (Amegakaryocytic Thrombocytopenia)**
   A. Isolated megakaryocytic hypoplasia
      1. Congenital amegakaryocytic thrombocytopenia associated with bilateral absent radii (TAR syndrome)
      2. Congenital amegakaryocytic thrombocytopenia associated with radio-ulnar synostosis (ATRUS)
      3. Congenital megakaryocytic hypoplasia without anomalies (CAMT)
      4. Congenital hypoplastic thrombocytopenia with microcephaly

(Continued)

Table 12-4   (Continued)

      5. Rubella syndrome[b]
      6. Congenital hypoplastic thrombocytopenia associated with trisomy syndromes
      7. Thrombocytopenia-agenesis of corpus callosum
      8. Fanconi anemia
      9. Hoyeraal-Hreidarsson syndrome
   B. Generalized bone marrow disorders
      1. Bone marrow aplasia
         a. Fanconi anemia
         b. Pancytopenia without congenital anomalies
         c. Osteopetrosis
      2. Bone marrow infiltration
         a. Congenital leukemia
         b. Langerhans cell histiocytosis
         c. Congenital neuroblastoma
   C. Metabolic causes
      1. Associated with acidosis and ketosis
         a. Hyperglycinemia
         b. Methylmalonic acidemia
         c. Isovaleric acidemia
         d. Propionic acidemia
      2. Other
         a. Maternal hyperthyroidism[c]

[a]Decreased production and poor survival of small defective platelets.
[b]Decreased megakaryocytes but not confirmed at autopsy. ? Error in sampling.
[c]Pathogenesis not determined.

3.   Early jaundice occurs in 20% of cases.
4.   Platelet count is very low at birth, usually <50,000/mm³.

Diagnosis

Neonatal alloimmune thrombocytopenia should be considered in all newborns with thrombocytopenia.

Ninety percent of cases of NAIT have a platelet count of <50,000/mm³, making it a reasonable screening tool for identifying NAIT. Two other reasons to suspect NAIT, even if the neonatal count is >50,000/mm³, include:

• The cause of thrombocytopenia is not clinically apparent
• There is a family history of transient neonatal thrombocytopenia in the family.

It is important to establish the diagnosis because treatment of subsequent pregnancies should be started as early as the 12th week of gestation. Laboratory evaluation should include screening for HPA-1, 3 and 5 antibodies, as well as HPA-4 antibodies in those of Asian descent. HPA-9 and 15 are the next most common antigen incompatibilities. To confirm the diagnosis of NAIT, testing must show both antigen incompatibility and antibodies to the antigen in question. If the initial post-partum testing does not detect an antibody to a platelet-specific antigen antibody, the optimal time for repeat antibody testing is in the 10–20-week period and every 10 weeks thereafter until or unless the diagnosis is confirmed.

Useful clinical criteria for the diagnosis include:

- Congenital severe thrombocytopenia
- Normal maternal platelet count and negative history of maternal ITP
- No evidence of systemic disease, infection, malignancy or hemangioma in the newborn
- Recovery of platelet count within 2–3 weeks
- Increased megakaryocytes in bone marrow aspiration; however, a reduced number has been noted in a few instances.

Treatment

The following treatment is recommended:

- Platelet transfusion 10–20 ml/kg. The mainstay of management of an affected neonate is random donor platelet transfusion. Maternal or matched platelets are not necessary for effective therapy since there is little if any difference in efficacy between transfusion of matched versus unmatched platelets in the setting of NAIT
- Matched donor platelets or washed maternal platelets may be used if available, especially if random donor platelet transfusions are ineffective
- IVIG 1 g/kg/day for 1–3 days depending on response with the goal platelet count being above 30–50,000/mm^3
- Methylprednisolone (1 mg IV) every 8 hours with IVIG until the IVIG is stopped. No tapering of the steroid is necessary
- Head ultrasound is mandatory for the thrombocytopenic neonate. If there are any abnormal neurological findings, a CT or MRI should be done. If ICH is present in NAIT on ultrasound, the target platelet count should be greater than 100,000/mm^3 and a head CT or MRI should be performed to better define the hemorrhage. Ultrasound should be repeated at one month to identify early hydrocephaly
- With or without treatment, follow-up is necessary until the platelet count is within the normal range to avoid missing other causes of thrombocytopenia.

Management of Subsequent Pregnancies

If a previous sibling has been identified, the likelihood of the next fetus being affected depends on the father's platelet typing. If the father is homozygous for the antigen responsible (as is the case in 75% of men with HPA-1A) then essentially all later fetuses will be affected. If the father is heterozygous, or if typing is unavailable or uncertain, amniocentesis or (if necessary) fetal blood sampling could determine fetal platelet type. However, due to the invasive nature and potential to cause fetal bleeding, it is recommended to avoid fetal blood sampling to evaluate platelet count whenever possible.

Instead mothers with a previously affected neonate *without* ICH should be started on a combination of weekly IVIG at 1 gram per kilogram per week and prednisone at 0.5 mg per kilogram per day starting at 20 weeks gestation until birth. The following management should be carried out in a mother who had a previously affected neonate *with* ICH.

Previous child with ICH in the second trimester:
Start treatment at 12 weeks gestation with IVIG 2 g/kg/week with or without prednisone 0.5–1 mg/kg/day (added at 20–26 weeks).

Previous child with ICH during third trimester:
Start treatment at 12 weeks gestation with IVIG 1 g/kg/week; with or without increasing dose of IVIG to 2 g/kg at week 20; with or without prednisone at 0.5–1 mg/kg/day (at 28 weeks).

The above schema of treatments and their timing could be modified based on knowledge of the fetal platelet count if fetal blood sampling is performed.

Neonatal Autoimmune Thrombocytopenia

Neonatal autoimmune thrombocytopenia is due to a passive transfer of autoantibodies from mothers with isolated ITP (it may be seen in association with other conditions such as maternal SLE, hypothyroidism and lymphoproliferative states).

Neonates born to mothers who have autoimmune thrombocytopenia are typically well after an uncomplicated delivery. To distinguish this entity from neonatal alloimmune thrombocytopenia, maternal history must be obtained and the maternal platelet count determined. A history of maternal thrombocytopenia during the pregnancy is not diagnostic of maternal ITP. Gestational thrombocytopenia (GTP) occurs in 5–10% of pregnancies, but maternal thrombocytopenia by definition is mild (70–100,000/mm^3) in GTP and GTP is not associated with neonatal thrombocytopenia. The platelet count in the mother swiftly returns to normal after delivery.

Neonatal thrombocytopenia in infants born to mothers with autoimmune thrombocytopenia is usually less severe than that seen in alloimmune thrombocytopenia (NAIT): only 10–15% of newborns have a platelet count less that 50,000/mm^3. Neonatal passive ITP is also associated with a lower bleeding rate than in NAIT and ICH only occurs rarely. The platelet count may often be near-normal at delivery, but then falls to a clinically significant nadir over the next 1–3 days.

Table 12-5 lists the pathogenesis and clinical differences between alloimmune thrombocytopenia (NAIT) and autoimmune thrombocytopenia.

Pathophysiology

Neonatal ITP occurs as a result of passive transfer of maternal antibodies across the placenta, as is seen in neonatal alloimmune thrombocytopenia. However, the target of the antibodies is

Table 12-5 Pathogenesis and Clinical Differences Between Alloimmune Thrombocytopenia (NAIT) and Autoimmune Thrombocytopenia

	Alloimmune Thrombocytopenia	Autoimmune Thrombocytopenia
Platelet Antigens	Antigens found on fetal platelets not present on maternal platelets (usually HPA-1A or HPA-5b)	Antigens common to both maternal and fetal platelets (usually GPIIB/IIIA and GPIb/IX complexes)
Platelet count	Often <20,000/mm^3	Birth counts often >50,000/mm^3
Time of presentation	Birth	Platelet count can be near normal at birth and then falls
Maternal History	Normal platelet count, no history of ITP, SLE, or hypothyroidism – may have GTP (unrelated)	Low platelet counts (unless mother is splenectomized) History of ITP, SLE, hypothyroidism
Intracranial Hemorrhage	10–20%	<1–2%
Treatment	Random donor platelets IVIG +/− Methylprednisolone Matched platelets	IVIG +/− Methylprednisolone Random platelets (if hemorrhage)
Resolution of thrombocytopenia	Usually in 2–4 weeks	Usually 3–12 weeks

Abbreviation: GTP, Gestational thrombocytopenic purpura.

an antigen present on maternal platelets that is also on fetal platelets (in contrast to alloimmune thrombocytopenia, where the antigen is not present on maternal platelets). The most frequently targeted antigens are the GPIIB/IIIA or GPIb/IX complexes. Differences in glycosylation of fetal and maternal platelets may explain the variability of this issue.

Diagnosis

Pregnant women with the following conditions may give birth to a neonate with autoimmune thrombocytopenia:

- History of previously affected infant
- Mother with ITP or previously splenectomized for ITP (mother may have platelet antibodies without being thrombocytopenic)
- Mother with thrombocytopenia (<100,000/mm^3) in current pregnancy
- Mother with systemic lupus erythematosis, hypoythyroidism, pre-eclampsia – HELLP syndrome (**H**emolytic anemia, **E**levated **L**iver enzymes and **L**ow **P**latelets)
- Maternal drug ingestion, e.g., thiazide.

Treatment

Blood counts, including platelet counts, should be performed daily (as the initial platelet count may be near normal). All infants with severe thrombocytopenia should have an ultrasound of the head to exclude ICH.

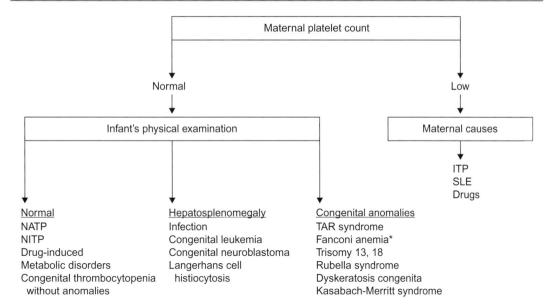

Figure 12-1 Diagnostic Approach to an Infant with Thrombocytopenia.
*Thrombocytopenia usually occurs at 4–6 years of age in Fanconi anemia.

Treatment is required when the infant's platelet count falls below $30,000/mm^3$ or if significant bleeding is present. The regimen is the same as for NAIT, using IVIG, IV methylprednisolone and random donor platelet transfusion if indicated. The duration of neonatal thrombocytopenia is usually about 3 weeks. Unlike NAIT, there is no benefit to the fetus from maternal treatment during pregnancy.

General Diagnostic Approach to a Newborn with Thrombocytopenia

Figure 12-1 depicts an algorithm for diagnosis of an infant with thrombocytopenia.

If the maternal platelet count is low, this suggests maternal ITP or another etiology for the maternal thrombocytopenia including inherited thrombocytopenias. If the maternal platelet count is normal but the mother has had a splenectomy for ITP, the neonatal ITP could still be due to the previous maternal ITP. Gestational thrombocytopenia, which complicates up to 10% of all normal deliveries, may have a platelet count as low as $50–70,000/mm^3$ but does not cause neonatal thrombocytopenia except in extremely rare cases.

If the platelet count is $<50,000/mm^3$ on the first day of life in an otherwise neonatal alloimmune thrombocytopenia (NAIT) should be considered and appropriate testing done as well as a head ultrasound. Platelet transfusion and possibly IVIG should be given. The next most common cause of severe neonatal thrombocytopenia is transplacental infection due to toxoplasmosis, rubella, cytomegalovirus or herpes (TORCH) which presents in a sick newborn occasionally with a blueberry muffin rash.

If improvement is not seen with treatment and testing does not confirm NAIT, congenital amegakaryocytic thrombocytopenia (CAMT) should be considered. If the baby is sick and the platelet count is $>50,000/mm^3$, it is likely that the underlying illness (e.g. RDS, sepsis) is the cause of the low platelets and the underlying condition should be treated.

Thrombocytopenia Associated with Erythroblastosis Fetalis or Exchange Transfusion

Severe cases of erythroblastosis frequently show petechiae and purpura in the first few hours after birth. This may be due to an isoimmune mechanism or, when associated with hyperbilirubinemia, it may be an effect of bilirubin toxicity on platelet survival.

Thrombocytopenia may also occur following exchange transfusion because of the paucity of platelets in stored blood or from the shorter survival time of transfused platelets. The thrombocytopenia is transient.

Thrombocytopenia Secondary to Maternal Diabetes, Pregnancy-Induced Hypertension, Intrauterine Growth Retardation (IUGR), or Hypoxia

Neonatal thrombocytopenia may be caused by maternal diabetes mellitus, pregnancy-induced hypertension, IUGR, perinatal asphyxia and/or placental insufficiency. These may be due in part to increased platelet destruction, but typically there is impaired mega-karyopoiesis and an elevated thrombopoietin level. Thrombocytopenia is usually not severe and is self-limited. The nadir tends to occur around day four with recovery by days 7–10. Often no treatment is required, but it may be appropriate to treat if there is sufficient asphyxia and thrombocytopenia so that the risk of ICH is increased in which case platelet transfusions are indicated.

Thrombocytopenia Secondary to Congenital Infections

Severe thrombocytopenia that occurs within 72 hours in a sick neonate is likely to be due to a perinatal infection, such as toxoplasmosis, rubella, cytomegalovirus, or herpes simplex (TORCH) infections; group B streptococcus; *Listeria monocytogenes*; *Escherichia coli*; or HIV. Of the TORCH infections, cytomegalovirus (CMV) infection most commonly causes severe thrombocytopenia. Infants with congenital infections have jaundice, pallor, hepatosplenomegaly and may have a classic "blueberry muffin" rash, which is not actually petechial or purpuric in nature, but rather represents sites of extramedullary hematopoiesis in the skin. These skin lesions may also be seen in congenital leukemia. Early-onset sepsis in a neonate causes thrombocytopenia because of:

- Platelet consumption associated with disseminated intravascular coagulation
- Bacteria, viruses, or immune complexes stick to platelets and are cleared by the mononuclear phagocyte system

- Impaired thrombopoiesis. Often there is insufficient compensation for platelet destruction or increased platelet clearance. In these infants, thrombocytopenia resolves with treatment and resolution of the underlying infection.

Late-Onset Thrombocytopenia Secondary to Late-Onset (Hospital-Acquired) Infections, Necrotizing Enterocolitis, Chronic Hepatitis, or Thrombosis

Thrombocytopenia occurring more than 5–7 days after birth is more likely to be related to late-onset sepsis, necrotizing enterocolitis, thrombosis, or liver disease. Hypogammaglobulinemia in very premature infants may contribute to these processes. However, IgG present in transfused plasma may be sufficient to maintain adequate IgG levels in the infant.

The causes of this type of thrombocytopenia are varied and include:

- Consumption related to infection (as in necrotizing enterocolitis)
- Deficient platelet production
- Disseminated intravascular coagulation (DIC)
- Thrombosis (e.g. renal vein thrombosis).

These causes may exist in combination or alone.

Thrombocytopenia Due to Aneuploidy

Thrombocytopenia in these settings may not become obvious for 3–4 days.

Thrombocytopenia is common in Down syndrome, occurring in up to 85% of cases. Some of these cases may represent a transient myeloproliferative syndrome. It is also seen in trisomy 13 and 18, Turner syndrome and triploidy, in which cases other congenital anomalies suggest the diagnosis. In these cases bleeding complications are not frequent, depending on the platelet count.

Rare Bone-Marrow Disease or Inborn Errors of Metabolism

The following marrow diseases, many relatively rare, may be associated with thrombocytopenia in the newborn:

- Osteopetrosis – generalized hyperostosis of bone and obliteration of bone marrow cavity resulting in extramedullary hematopoiesis and pancytopenia
- Metastatic neuroblastoma
- Gaucher disease often including hypersplenism and other coagulopathies
- Niemann–Pick disease

- Hemophagocytic lymphohistiocytosis – autosomal recessive – patients present with hepatosplenomegaly, fever, rash, jaundice, tachypnea, lymphadenopathy, afibrinogenemia, hypertriglyceridemia and absent cytotoxic activity
- Congenital leukemia – more commonly with AML, can be associated with skin infiltrates, hepatosplenomegaly and megakaryoblasts in the peripheral blood smear. Congenital leukemia carries a poor prognosis unless associated with Down syndrome.

Metabolic Causes

Hyperglycinemia with ketosis and the closely related metabolic disorder of *methylmalonic acidemia* may cause periodic thrombocytopenia, as well as neutropenia, during infancy. Infants with these metabolic disorders present with lethargy, vomiting and ketosis during the neonatal period. A similar disorder, *isovaleric acidemia,* is associated with a generalized marrow hypoplasia causing thrombocytopenia and neutropenia. Neonatal thrombocytopenic purpura has been reported in infants born to *hyperthyroid mothers*. The mechanism has not been defined.

Giant Hemangioma

Giant hemangioma coupled with thrombocytopenia probably represents a form of localized intravascular coagulation. The association of thrombocytopenic purpura with giant hemangioma is referred to as the Kasabach–Merritt syndrome. Platelet trapping has been demonstrated in giant hemangiomas, accompanied, in some instances, by evidence of the consumption of coagulation factors and an increase in fibrin degradation products. Most of the hemangiomas are cutaneous, however, visceral hemangiomas, e.g., liver, spleen, do occur and should be sought (using CT or MRI) in an infant with unexplained thrombocytopenia with or without DIC.

Treatment

1. Transfusions of platelets and other coagulation factors (e.g., fibrinigen concentrates, FFP, cryoprecipitate and antifibrinolytic drugs) have only transient effects. These modalities may be required because of active bleeding due to thrombocytopenia.
2. Corticosteroid therapy has little immediate effect but may bring about involution of the hemangioma, especially in very young infants. However, tumor regression is probably a consequence of vascular thrombosis and infarction.
3. External compression of the hemangioma by firm bandaging, when possible, may reduce blood flow and platelet trapping.
4. Surgical excision, when possible, corrects the thrombocytopenia.
5. Radiation therapy to reduce the size of the hemangioma should be considered when the hemorrhagic manifestations are severe.

6. Interferon α-2a has been shown to be effective. It inhibits angiogenesis, in part by inhibiting the proliferation of endothelial cells, smooth muscle cells and fibroblasts that have been stimulated by fibroblast growth factor (FGF), decreasing collagen production and increasing endothelial prostacyclin production.

INHERITED THROMBOCYTOPENIAS

Thrombocytopenias due to inherited causes are usually distinguished by characteristic clinical features including family history, platelet morphology, long-term stable low counts and lack of response to therapies of immune thrombocytopenia, e.g. IVIG. Most of the syndromes are associated with "giant" platelets (Table 12-1). In general, the thrombocytopenia is not severe and often there are characteristic features that define each syndrome other than the thrombocytopenia. In certain cases (i.e. Wiskott–Aldrich), bleeding symptoms can be more severe than would be expected by the platelet count alone, due to platelet dysfunction.

Table 12-6 lists the characteristics of congenital thrombocytopenic syndromes.

Bernard-Soulier Syndrome

Bernard-Soulier syndrome is a relatively rare autosomal recessive disorder. It is characterized by:

- Moderate thrombocytopenia (automated counting usually underestimates the true platelet count because of undercounting of very large platelets)
- Prolonged bleeding time and
- Characteristic platelet morphology.

Platelets are very large, equaling or exceeding the size of a red cell. These platelets contain two to four times the normal protein content and three times the usual number of dense granules. These dense granules can gather and give the appearance of a pseudonucleus. The findings are more severe in the rare homozygote (or compound heterozygote) than in the far more common heterozygote. Morphologically, the megakaryocytes have an abnormality in the demarcation membrane system, likely explaining the large platelets and the thrombocytopenia.

There can be complete absence of GPIb glycoprotein complex or a point mutation in the GPIb-alpha subunit known as the Bolzano variant; all of which leads to inability of von Willebrand factor (vWF) binding. Platelets fail to agglutinate in response to ristocetin, despite normal aggregation and secretion in response to ADP, epinephrine, thrombin and collagen. Because von Willebrand factor acts as a bridge in adhesion of platelets to exposed subendothelium, the absence of the vWF receptors inhibits normal platelet adhesion and

Table 12-6   Characteristics of Congenital Thrombocytopenic Syndromes

Clinical Features	TACC	TAR	ATRUS	FA	AMT	HH (DC)	WAS	BS	MH	Trisomy 13, 18, 21
Agenesis of corpus callosum	+	−	−	−	−	−	−	−	−	−
Hypoplasia of cerebellar vermis	+	−	−	−	−	+	−	−	−	+
Low birth weight	+	+	−	+	−	+	−	−	−	+
Growth delay	+	+	−	+	+	+	−	−	−	+
Dysmorphic face	+	+	−	+	+	+	−	−	−	+
Developmental delay	+	−	−	±	+	+	−	−	−	+
Thrombocytopenia	+	+	+	±	+	+	+	+	+	±
Platelet size	N	N	N	N	N	N	↓	↑	↑	N
Pancytopenia	−	−	+	+	−	+	−	−	−	−
Immunodeficiency	−	−	−	−	−	+	+	−	−	−
Megakaryocytes, bone marrow	N/↓	↓	↓	↓	↓	↓	N/↑	N/↑	N/↑	N
Skeletal deformities										
Radial X-ray	−	+	−	±	−	−	−	−	−	−
Clinodactyly/syndactyly	±	−	+	±	±	−	−	−	−	±
Enamel hypoplasia	±	−	−	−	−	−	−	−	−	−
Cardiac defects	±	±	−	±	±	−	−	−	−	±
Renal malformations	±	−	−	±	±	−	−	−	−	±
Cutaneous abnormalities	−	+	−	+	−	+*	+	−	−	−
Karyotypic abnormalities	−	−	−	−	−	−	−	−	−	+
Chromosome breaks	−	−	−	+	−	−	−	−	−	−

* Present at a mean age of 9 years.

Abbreviations: TACC, thrombocytopenia agenesis of corpus callosum; TAR, thrombocytopenia absent radii; ATRUS, amegakaryocytic thrombocytopenia with radio-ulnar synostosis; FA, Fanconi anemia; AMT, amegakaryocytic thrombocytopenia; HH, Hoyeraal-Hreidarsson syndrome; DC, dyskeratosis congenital; WAS, Wiskott-Aldrich syndrome; BS, Bernard-Soulier syndrome; MH, May-Hegglin anomaly; N, normal; ↑, increased; ↓, decreased; +, present; −, absent; ±, present occasionally.

causes a significant degree of bleeding (even in cases of mild thrombocytopenia), with mucocutaneous bleeding that can begin in early infancy, especially severe epistaxis.

Treatment

1. Antifibrinolytic therapy is the mainstay of treatment.
2. Platelet transfusion is the only reliable therapy, but is frequently reserved for managing severe bleeding episodes. There can be antibody formation with repeated platelet transfusion, because these patients lack the GPIb/IX complex. The alloantibody can inhibit normal platelet adhesion and obviate the benefit of transfusion.
3. Activated factor VIIa can be used in an emergency.

MYH9 Disorders

A mutation in the MYH9 gene encoding the non-muscle myosin heavy chain expressed only in neutrophils and platelets (myosin IIa) has been reported in the May–Hegglin anomaly, Sebastian, Fechtner, Epstein and Alport syndromes. These disorders are referred to as MYH9-related diseases. Currently, no clear genotype–phenotype correlation has been established for this group of disorders which may have other manifestations like high-tone hearing loss and glomerulonephritis. Primary hemostatic bleeding, as in Bernard–Soulier syndrome can be seen and local control measures are required. If bleeding is uncontrolled despite antifibrinolytic agents such as desmopressin acetate and aminocaproic acid (amicar) and platelet transfusions may be necessary. The newly licensed thrombopoietic agents are being explored.

Table 12-7 lists the clinical and morphological features of the MYH-9 macrothrombo-cytopenias.

May–Hegglin Anomaly

May–Hegglin anomaly is a MYH9-related syndrome associated with giant platelets and variable thrombocytopenia, although frequently of moderate degree. It has an autosomal dominant inheritance and its distinguishing features are the large bluish cytoplasmic inclusions in granulocytes and monocytes known as Döhle bodies. Bleeding is usually minor and is thought to be due to the sheer size of the platelets and the difficulty in shape change necessary for adhesion to damaged endothelium with collagen exposure in small vessels. Bleeding is mainly manifest with trauma, surgery, or dental extractions.

Sebastian Platelet Syndrome

Sebastian platelet syndrome is very similar to May–Hegglin, but even more rare. The Döhle body-like inclusions are smaller and are composed of ribosomes and dispersed filaments and lack an enclosing membrane when examined under the microscope, compared with

Table 12-7 Clinical and Morphological Features of the MYH9 Macrothrombocytopenias

| | CLINICAL FEATURE | | | | |
Disorder	Thrombocytopenia	Leukocyte Inclusions	Nephritis	High-tone Sensorineural Deafness	Cataracts
May Hegglin anomaly	+	+	−	−	−
Sebastian syndrome	+	+	−	−	−
Epstein syndrome	+	−	+	+	−
Fechtner syndrome	+	+	+	+	+
Alport syndrome	−	−	+	+	+

+, Present; −, Absent.

those of May–Hegglin. It is inherited in an autosomal dominant manner and bleeding tendency is considered to be mild to moderate.

Fechtner Syndrome

Fechtner syndrome is a variant of Sebastian syndrome, but also has specific additional features that are associated with Alport syndrome, including high-tone sensorineural deafness, cataracts and renal failure. White cell inclusions are also seen and resemble toxic Döhle bodies and May–Hegglin granulocyte inclusions. There is mild to moderate bleeding tendency with significant hemorrhage following trauma, dental extraction and surgery. Patients bleed more as renal failure progresses.

Epstein Syndrome

Epstein syndrome, like Fechtner syndrome, is another variant of Alport syndrome and is associated with sensorineural deafness and renal failure (but not cataracts). In addition to thrombocytopenia, it is characterized by bleeding secondary to defective platelet aggregation and secretion in response to ADP and collagen. Alport syndrome itself is not associated with thrombocytopenia.

Montreal Platelet Syndrome

Montreal platelet syndrome is a rare autosomal dominant disorder characterized by thrombocytopenia that tends to be severe, with large platelets seen on smear. Bleeding is typically delayed for several days. There may be spontaneous platelet aggregation and, unlike Bernard–Soulier syndrome, aggregation testing demonstrates normal response to ristocetin. There is a reduced response to thrombin-induced aggregation. These features also accompany deficient activity of platelet calpain activity, which is a calcium-dependent protease that is active on the platelet cytoskeleton. Deficiency of platelet calpain is thought to underlie this disorder.

Quebec Platelet Disorder

See page 367–368.

Gray Platelet Syndrome

See page 367.

Wiskott–Aldrich Syndrome (WAS)

This syndrome has X-linked inheritance and has the classic features of thrombocytopenia, eczema, recurrent bacterial and viral infections secondary to abnormalities in T cell function and a propensity to develop autoimmune disorders.

The molecular defect in this syndrome is an abnormal Wiskott–Aldrich syndrome protein (WASP) resulting from a mutation of the Wiskott–Aldrich syndrome gene located on band Xp11-12. WASP is known to be involved in signal transduction and it regulates actin filament assembly which explains the abnormalities in platelet and lymphocyte cytoskeleton and signaling.

Clinical Features

Infants present with thrombocytopenia in the first few months of life. Bleeding is frequently ushered in by melena during the neonatal period, later followed by purpura. Bleeding is often out of proportion to thrombocytopenia. The clinical course is punctuated by recurrent pyogenic infections, including otitis media, pneumonia and skin infections. There is also lowered resistance to nonbacterial infections, including herpes simplex and *Pneumocystis jiroveci* (formerly *carinii*) pneumonia. It is often fatal by early teens due to infection, lymphoproliferative malignancy, or bleeding.

Hematologic Findings

1. Thrombocytopenia (platelet count 10,000–100,000/mm^3); microthrombocytes; low mean platelet volume (MPV). This is not obvious in the newborn and the MPV is unreliable when the platelet count is low.
2. Platelets have abnormal aggregation in response to agonists such as ADP, epinephrine and collagen.
3. Platelets have reduced platelet survival to half normal.
4. Ineffective megakaryocytopoiesis reflected by a platelet turnover 25% that of normal megakaryocyte mass.
5. Anemia (due to blood loss).
6. Leukocytosis (due to infection).

7. Normal or increased megakaryocytes.
8. Absent isohemagglutinins, reduced IgM and normal or elevated IgG and IgA.
9. Defective cell-mediated immunity in some cases.

Treatment

Allogeneic stem cell transplantation is treatment of choice when there is a fully matched donor available. If no matched donor is available the patient should be managed as follows:

- Aggressive treatment of infections
- Platelet transfusions for hemorrhagic episodes
- Steroid cream for eczema
- Splenectomy. Reserved only for severe cases because of risk of overwhelming post-splenectomy infection. Usual pre- and post-splenectomy precautions pertain
- Thrombopoietic agents are being explored.

X-Linked Thrombocytopenia

A variant of WAS, known as *inherited X-linked thrombocytopenia (XLT),* has thrombocytopenia which is less severe than WAS and does not have the associated features of eczema or immunodeficiency. Like WAS, there is an abnormality in the Wiskott-Aldrich syndrome protein, but the mutations occur more commonly in exons 1 and 2, which gives rise to this milder phenotype, because only the megakaryocyte binding domain is affected. The distinction of the XLT form from full-blown WAS is not always clinically clear. Patients with XLT may be mistaken for chronic "refractory" ITP but have a higher incidence of post-splenectomy sepsis than patients with ITP.

Several families have been identified with variable anemia and thrombocytopenia with GATA-1 mutations recognized as the underlying cause. GATA-1 is a transcription factor important in erythrocyte and megakaryocyte development. Bone marrow shows many large megakaryocytes with nuclei pushed to the side and with unorganized granular content.

Thrombocytopenia with Absent Radii (TAR) Syndrome

Thrombocytopenia with absent radii syndrome is a rare disorder with autosomal recessive inheritance. Most cases are diagnosed either *in utero* or in the first day of life. There is bilateral absence of the radii manifesting as a shortening of the forearms and flexion at the elbows. Both thumbs are present, which helps to distinguish TAR from Fanconi anemia. Other defects of phalanges, humeri and lower limbs can be present, as well as cardiac anomalies. White blood cell (WBC) count elevation is frequently seen and anemia can be part of the clinical picture. Significant bleeding episodes, such as gastrointestinal and even intracerebral hemorrhage, occur in the first 6 months of life. Deaths can occur in the first year of life due to hemorrhage, but this can be mitigated by the use of prophylactic single-donor platelet transfusions when counts are very low. Typically, the thrombocytopenia improves with time for reasons which are not understood and the platelet count can be normal beyond the first year of life. There is a subgroup of patients who continue to have low counts and bleeding through adulthood. Neither the molecular basis nor the pathogenesis of TAR is known, although defective signaling through the thrombopoietin (TPO) receptor *c-mpl* has been demonstrated. Thrombopoietin levels studied in these patients are shown to be elevated, as is seen in all patients with reduced or absent numbers of megakaryocytes in the marrow. There have been several case reports of acute myeloid leukemia in early childhood in these patients, but there is no associated chromosomal fragility as seen in Fanconi anemia.

Treatment

1. Transfusion of red cells for anemia and single-donor platelet transfusions for severe bleeding.
2. Allogeneic stem cell transplantation may be required for symptomatic patients.
3. Corticosteroid and splenectomy are of no long-lasting benefit.

Congenital Amegakaryocytic Thrombocytopenia (CAMT)

Congenital amegakaryocytic thrombocytopenia is a bone marrow failure syndrome that presents with isolated thrombocytopenia in the neonatal period. Inheritance is autosomal recessive. The most common age at diagnosis of the thrombocytopenia is within the first month, because of petechiae and other bleeding symptoms. The diagnosis of CAMT, however, is not usually made until the infant is several weeks or months old when the bone marrow is examined. The thrombocytopenia is severe and, unlike NAIT does not resolve. Physical anomalies that may be seen in some of the patients are orthopedic, renal, or cardiac, but most have no physically distinguishable features. Bone marrow evaluation reveals absent or greatly reduced numbers of megakaryocytes with normal granulopoietic and erythroid elements. Thrombopoietin levels are very high as a result of the markedly decreased megakaryocytes and their progenitors.

CAMT is caused by mutations in the *c-mpl* gene (thrombopoietin receptor). Both frameshift and nonsense mutations have been described, which result in loss of *c-mpl* function. The type of mutation, such as frameshift versus missense mutation, may determine the degree of clinical severity and whether megakaryocytes are present in small numbers or are altogether absent. Due to *c-mpl* function in preventing stem cell apoptosis, many patients progress to total aplastic anemia which is thought to result from stem cell depletion. Recombinant thrombopoietin would likely play no role in therapy for these patients, as its receptor is unable to signal normally.

Current treatment for CAMT is supportive, using platelet transfusion. The only curative treatment is allogeneic stem cell transplantation. Gene therapy is being developed.

Amegakaryocytic Thrombocytopenia with Radio-Ulnar Synostosis (ATRUS)

Children with ATRUS present at birth with severe thrombocytopenia (similar to CAMT), but they have the following characteristic physical examination findings that can identify this specific syndrome:

• Proximal radio-ulnar synostosis (fusion of the radius and ulnar at the elbow)
• Clinodactyly (minor)
• Shallow acetabulae (minor).

An *HOXA11* mutation has been identified in two kindreds but not in others. The mutation appears to inhibit megakaryocytic differentiation. The disease is associated with aplastic anemia and possible leukemia.

Fanconi Anemia

The clinical manifestations of Fanconi anemia are fully discussed in Chapter 6. Thrombocytopenia is usually the first cytopenia to appear followed by granulocytopenia and ultimately severe aplastic anemia. The diagnosis of Fanconi anemia should always be considered in a child with an isolated cytopenia even when the classical somatic anomalies are absent as a significant number of these cases are physically normal.

Familial Platelet Syndrome with Predisposition to Acute Myelogenous Leukemia (FPS/AML)

FPS/AML involves a mild thrombocytopenia (usually a platelet count around $100,000/mm^3$) with dysfunctional platelets. There is an associated approximately 50% chance of malignancy, with two-thirds developing AML and one-third developing a solid tumor. The genetic defect has been traced to the transcription factor *Runx-1* (also known as AML1).

Paris-Trousseau Syndrome

This syndrome consists of mild thrombocytopenia with a subpopulation of platelets with giant α granules. There is an expansion of immature megakaryocyte progenitors in the bone marrow with normal erythroid and granulocytic maturation. A subset of these patients will have Jacobsen syndrome, which includes the same platelet defects as Paris-Trousseau with additional abnormalities such as trigonocephaly, facial dysmorphism, cardiac defects, syndactyly and psychomotor retardation.

Congenital Hypoplastic Thrombocytopenia with Microcephaly

Three infants have been reported with congenital hypoplastic thrombocytopenia with microcephaly and the persistence of thrombocytopenia beyond one year of life. Rubella syndrome, which may present in a similar manner, should be excluded.

Thrombocytopenia Agenesis of Corpus Callosum Syndrome (TACC)

Three female patients have been reported with thrombocytopenia, agenesis of the corpus callosum, low birth weight, growth delay and dysmorphic facial features. Megakaryocytes are absent and dysmorphic.

THROMBOCYTOPENIA IN CHILDHOOD

Immune Thrombocytopenic Purpura (ITP)

Immune thrombocytopenia is a disorder caused by antiplatelet antibodies which lead to an accelerated destruction of platelets and an inhibition of the production of platelets. ITP is the most common cause of thrombocytopenia in children. Peak occurrence is between 2 and 5 years of age. In most children the disease is self-limited, with resolution in 80% of patients within 6–12 months from diagnosis. North American studies report an incidence of 7.2–9.5/100,000 children between 1 and 14 years of age. There is a seasonal pattern to ITP with a peak in winter and early springtime, presumably mimicking the pattern of viral illnesses. There does not appear to be a race or sex predilection.

Pathophysiology

ITP is a heterogeneous disease with a complex pathogenesis. An acute infection often appears to be the initial trigger, but that may only potentiate an already-established immunologic disturbance.

Mechanisms:

- Antibody-mediated destruction:
    - Most of the identified autoantibodies are directed against GPIIb-GPIIIa, GPIb-GPIX and GPIa-IIa
    - Antibody-coated platelets are destroyed by activated Fc receptors on reticuloendothelial cells (mostly splenic) via recognition of the IgG Fc region of the antiplatelet antibody
    - Antiplatelet antibodies rarely have a significant effect on platelet function.
- Impaired megakaryopoiesis
    - Antibody and cellular cytotoxicity and immune-cell-derived cytokines have been implicated in impairment of megakaryocytes
    - Platelet kinetic studies have shown that autologous platelets labeled with indium have a 2–3-day half life (longer than expected for platelet count) indicative of reduced platelet production
    - Thrombopoietin levels are not substantially elevated.

- T cell activity
  - Glycoprotein-specific antibodies are absent in 20–40% of cases of ITP
  - There is an upregulation of genes involved in cell-mediated toxicity (e.g. granzyme, perforin) in CD3+CD8+ T cells in ITP patients
  - CD4+ T-helper cells regulate antiplatelet-antibody-secreting B cells
  - Th1-associated cytokines predominate in ITP.

Infections in ITP

Acute infections have been implicated in the initiation of ITP and are known to cause an acute decrease (or paradoxically an acute increase) in the platelet count in patients with ongoing ITP.

1. Virus-specific antibodies that cross-react with platelets have been demonstrated in several children with varicella (also HIV).
2. HIV
   - A clear relationship between platelet count and viral load has been demonstrated; suppressing the virus results in an increased platelet count in more than 80–90% of cases
   - The pathogenesis of HIV-related ITP may be different – First, severe T-cell depletion and immune dysregulation occurs but second, antiplatelet antibodies may cause intravascular platelet lysis
   - Megakaryocytes express receptors for HIV including CD4, CXCR4 and CCR5, suggesting that direct viral infection plays a role.
3. *H. pylori* – studies indicate that platelet counts may often be increased when *H. pylori* is treated in countries with a high background incidence of *H. pylori* such as Italy and Japan (but not the USA).
4. Hepatitis C and its therapy (interferon) are both associated with thrombocytopenia which may prevent successful eradication of the virus. Thrombopoietic agents may raise the platelet counts in these cases (e.g., Eltrombopag 50–75 mg orally daily). The effect of eradication of hepatitis C on thrombocytopenia is not clear.
5. ITP refractory to therapy may be exacerbated by CMV infection (either re-activation or *de novo*).

Clinical Manifestations

Typically patients are otherwise well and present with petechiae, purpura and nonpalpable ecchymoses 1–3 weeks after a viral infection. It may also occur after rubella, rubeola, chickenpox or live virus vaccination. Occasionally patients may present with mucosal bleeding (hematuria, hematochezia, menometrorrhagia, or epistaxis). Most often, bleeding symptoms are mild, but rarely patients may develop severe bleeding including intracranial hemorrhage, protracted epistaxis, hematuria, hemoptysis and gastrointestinal bleeding.

Table 12-8   Characteristics of ICH in ITP

Incidence:	0.2–0.8%
Age:	13 months–16 years
Platelet count:	<20,000 in 90% of cases
	<10,000 in 75% of cases
Interval between diagnosis of ITP and ICH:	At presentation (25% of cases)
	<1 week (45% of cases)
	week–6 months (25% of cases)
	Greater than 6 months (30%)

Identifiable risk factors for ICH include:
- Head injuries (33%) (versus 1% in ITP without ICH)
- Hematuria (22%) (versus 0% in ITP without ICH)
- Hemorrhage more than petechiae and bruises (63%) (versus 44% in ITP without ICH)
- AV malformation
- Aspirin treatment

Site of ICH
- Intracerebral (77% of cases) – 87% supratentorial; 13% posterior fossa
- Subdural hematoma (23% of cases)

Prior Treatment
- 70% had prior treatment

Survival
- 75% survive, but 1/3 have neurologic sequelae

Table 12-8 lists the clinical manifestations of intracranial hemorrhage in ITP.

With the exception of hemorrhagic manifestations, the physical examination is not significant. Pallor is usually absent unless there has been significant bleeding. The tip of the spleen is palpable in fewer than 10% of patients. The finding of splenomegaly suggests the probability of leukemia, systemic lupus erythematosus (SLE), infectious mononucleosis, or hypersplenism. Cervical lymphadenopathy is not present unless the precipitating factor is a viral illness.

Table 12-9 lists the features of newly diagnosed and chronic ITP.

Diagnosis

The diagnosis of ITP remains almost entirely one of exclusion. Demonstrating antiplatelet antibodies has not been shown to be of diagnostic or prognostic importance since antiplatelet antibodies are only present in approximately 60–80% of cases. There are three diagnostic criteria:

- Isolated thrombocytopenia with otherwise completely normal CBC and blood smear (particularly red cell and white cell morphology)
  - Smear often has large platelets (too many macrothrombocytes suggest other disorder, e.g. Bernard–Soulier, MYH9 disorders) (Table 12-1)

Table 12-9   Features of Newly Diagnosed and Chronic ITP

Feature	Newly Diagnosed	Chronic
Age	Children 2–6 years old	Adults
Sex distribution	Equal	Female:male = 2:1
Preceding infection	~80%	Unusual
Seasonal predilection	Springtime	None
Associated autoimmunity	Uncommon	More common
Onset	Acute	Insidious
Platelet count	$<20,000/mm^3$	$<20,000–80,000/mm^3$
Eosinophilia-lymphocytosis	Not uncommon	Rare
IgA/IgG levels	Normal	Infrequently low
Duration	2–8 weeks	1 to many years
Prognosis	Spontaneous remission in 70–80% of cases	Ongoing thrombocytopenia with occasional remission

- Artifactual low platelet count ("pseudothrombocytopenia") must be excluded
- Other possibilities such as TTP or Evans syndrome (anemia, high reticulocyte count) must be excluded.
- Absence of hepatosplenomegaly, lymphadenopathy and radial ray anomaly (oligodactyly, aplasia or hypoplasia of the thumb, aplasia or hypoplasia of the radius)
- Platelet response to classic ITP therapy (usually IVIG or anti-D, possibly steroids) is the only finding that helps to make the diagnosis of ITP (including secondary ITP).

Coagulation screening tests including PT, PTT and fibrinogen are all normal.

Other causes of thrombocytopenia must be excluded:

- Inherited thrombocytopenia by:
  - Family history
  - Presence of congenital defects (skeletal, cardiac, renal, neurologic).
- Pregnancy
- Chronic infection with HIV, Hepatitis C and CMV
- Immunodeficiency (e.g. CVID [hypogammaglobulinemia])
- Lymphoproliferative disease (benign or malignant)
- Medications, especially quinine, valproate, heparin, estrogen in any form and diphenylhydantoin and dietary supplements. Table 12-10 lists drugs proven or suspected to induce drug-dependent antibody-mediated immune thrombocytopenia.

Additional laboratory analyses may be required for nonresponsive, persistent, or chronic cases or for those cases which have *specific clinical indications*:

- Autoimmune screening – ANA and anti-double-stranded DNA antibodies, RA
- Thyroid screening (TSH, thyroid antibodies)
- Immune globulin measurements (IgG, IgA and IgM)

Table 12-10   Drugs Proven or Suspected to Induce Drug-dependent Antibody-mediated Immune Thrombocytopenia

Anti-inflammatory
- Acetaminophen
- Acetylsalicylic acid
- Diclofenac
- Ibuprofen
- Indomethacin
- Meclofenamate
- Mefenamic acid
- Naproxen
- Oxyphenbutazone
- Phenylbutazone
- Piroxicam
- Sodium-*p*-amino salicylic acid
- Sulfasalzine
- Sulindac
- Tolmetin

Antibiotics
- Antituberculous drugs
  - Ethambutol
  - Isoniazid
  - Para-aminosalicylic acid (PAS)
  - Rifampin
  - Streptomycin
- Penicillin group
  - Ampicillin
  - Methicillin
  - Penicillin
  - Mezlocillin
  - Piperacillin
- Cephalosporin
  - Cefamandole
  - Cefotetan
  - Ceftazidime
  - Cephalothin
- Sulfonamides
  - Sulfamethoxazole
  - Sulfamethoxypyridazine
  - Sulfisoxazole
- Other antibiotics
  - Amphotericin B
  - Ciprofloxacin
  - Clarithromycin
  - Fluconazole
  - Gentamicin
  - Indinavir
  - Nalidixic acid

Anticonvulsants, Sedatives and Antidepressants
- Amitriptyline
- Carbamazepine
- Desipramine
- Diazepam
- Doxepin
- Haloperidol
- Imipramine
- Lithium
- Mianserin
- Phenytoin
- Valproic acid

Cardiac and Antihypertensive Drugs
- Acetazolamide
- Amiodarone
- Alprenolol
- Captopril
- Chlorothiazide
- Chlorthalidone
- Digoxin
- Digitoxin
- Furosemide
- Hydrochlorothiazide
- α-methyldopa
- Oxprenolol
- Procainamide
- Spironolactone

H_2-antagonists
- Cimetidine
- Ranitidine

Cinchona Alkaloids
- Quinidine
- Quinine

Miscellaneous
- Antazoline
- Chlorpheniramine
- Chlorpropamide
- Danazol
- Desferrioxamine
- Diethylstilbestrol
- Etretinate
- Glibenclamide
- Gold salts
- Heparin
- Interferon-α

(Continued)

Table 12-10   (Continued)

Novobiocin	Iodinated contrast agents
Pentamidine	Isotretinoin
Sodium stibogluconate	Minoxidil
Stibophen	Levamisole
Suramin	Lidocaine
Vancomycin	Morphine
Antineoplastic	Papaverine
Actinomycin-D	Ticlopidine
Aminoglutethimide	**Foods**
Tamoxifen	Beans

- Liver function tests
- EBV, CMV, parvovirus and HIV testing by PCR
- *H. pylori* testing (in countries with a high background incidence)
- Bone marrow aspiration and biopsy. Patients with isolated thrombocytopenia, who fit the diagnostic criteria above, do not require bone marrow sampling to rule out leukemia. Failure to respond to ITP therapy e.g. anti-D and/or IVIG, requires a bone marrow aspirate and biopsy
- Antiphospholipid antibodies (present in up to 15% of cases) and lupus anticoagulant – in cases of prolonged PTT, persistent headache and/or thrombosis
- Antiplatelet antibodies – Not sensitive or specific and not required for routine clinical management.

Treatment

The goal of therapy in ITP is to increase the platelet count enough to prevent serious hemorrhage. Treatment decisions should be based on the potential for bleeding (e.g. activity level), patient's history of bleeding current platelet count, signs and symptoms and other factors as follows:

- Patients who have a platelet count greater than 30,000/mm^3 with no signs of bleeding generally do not need treatment, although certain risk factors (Table 12-8) and surgery may necessitate treatment
- Patients with a platelet count less that 20,000/mm^3 and moderate bleeding and those with counts less than 10,000/mm^3 should be treated
- Quality of life can be a major concern in ITP (in particular fatigue) and this may be a reason to treat even in the absence of bleeding.

Competitive contact sports should be avoided when the platelet count is known to be less than 30,000/mm^3. Depo-Provera or any other long-acting progesterone is useful in suspending menstruation for several months, to prevent excessive menorrhagia in menstruating females. Aspirin, nonsteroidal anti-inflammatory agents and any other drugs that interfere with platelet function should not be given.

The following drugs are employed in the treatment of ITP.

Corticosteroids

Prednisone 1–2 mg/kg/day in divided doses orally for 2–4 weeks with tapering after platelet response occurs. The initial response rate is 50–90% in adults (children tend to be at the higher end of the response range). Another approach is to use 4 mg/kg for a short period of time with rapid taper. As with any initial therapy of newly diagnosed ITP, many children go into long-term remission after a single course. Dexamethasone in a daily dose of 40 mg/kg/day (24 mg/m^2) orally for 4 days given every 14 days for three cycles has been shown to have an initial response rate of 85% in adults and children.

Prolonged use of steroids in ITP is undesirable. Large doses or prolonged steroid usage may perpetuate the thrombocytopenia and depress platelet production. It also leads to side effects including gastritis, ulcers, weight gain, Cushingoid facies, fluid retention, acne, hyperglycemia, hypertension, mood swings, pseudotumor cerebri, cataracts, growth retardation and avascular necrosis.

Mechanism of action of steroids:

- Inhibit the phagocytosis of antibody-coated platelets and thus prolong platelet survival
- Inhibit platelet antibody production by B lymphocytes
- Possibly improve capillary resistance and thereby reduce bleeding.

Immune Globulin (IVIG)

IVIG in a dose of 0.4–1 g/kg/day for 1–5 days for initial therapy or for relapsed disease. IVIG is preferred over steroids in children less than 2 years of age because they tend to have lower response rate to steroids. Meta-analysis of randomized controlled trials has shown a more rapid response to IVIG in children compared to corticosteroids. In addition, a large, retrospective study has suggested that there may be a lower rate of chronic disease in patients initially treated with IVIG compared to those treated with prednisone. However, IVIG as an alternative therapy to corticosteroid therapy is much more expensive and has significant side effects (see below).

Mechanism of action of IVIG: Early studies suggested that IVIG inhibits clearance of Ig-coated platelets. Recent studies of mouse models of ITP have lead to the hypothesis that the IgG molecule does this by upregulating FcγRIIb, the inhibitory FcγR, on phagocytes.

Adverse effects of IVIG are:

- Post-infusion headache in >50% of patients. It is transient but occasionally severe (in severe cases, administer high dose IV steroids, e.g., dexamethasone 0.15–0.3 mg/kg IV). Severe headache in ITP may suggest the presence of intracranial hemorrhage and if clinically indicated may require a CT scan, although most post-IVIG headaches occur with good platelet counts
- Fever and chills in 1–3% of patients. Prophylactic acetaminophen (10–15 mg/kg, 4 hourly, as required) and diphenhydramine (1 mg/kg, 6–8 hourly, as required) may be useful to reduce the incidence and severity
- Direct antiglobulin test (DAT)-positive hemolytic anemia because of the presence of blood group antibodies (anti-A, anti-B and anti-D) present in IVIG
- Anaphylaxis in IgA-deficient patients because of pre-existing IgA antibodies that react with small amounts of IgA present in commercially available gammaglobulin
- Aseptic meningitis (rare, very severe headache)
- Acute renal failure
- Pulmonary insufficiency
- Thrombosis
- Viral transmission (hepatitis C reported in the past, but not currently). No cases of HIV in the United States).

Anti-D Therapy

IV anti-D in a dose of 50–75 µg/kg for initial therapy or for recurrent disease. Approximately 70% of patients have an initial response to 75 µg/kg of anti-D therapy within 1 day (comparable to high-dose IVIG). The effect is more pronounced after 48–72 hours. Anti-D is plasma-derived, immune globulin with high titers against the Rhesus D antigen. It can be used only in Rh+ patients for the treatment of ITP. Hemolysis is expected when anti-D is used and hemoglobin levels usually decrease by 0.5–2 g/dl.

Mechanism of Action. Anti-D works by binding to Rhesus D antigen expressed on red blood cells, which leads to their recognition by Fc receptors on cells of the reticuloendothelial system. The coated red cells compete with the antiplatelet-antibody-coated platelets for the activated Fc receptors, thereby slowing platelet clearance.

Adverse Effects
1. Fever and chills.
2. Intravascular hemolysis.
3. Headache, vomiting.
4. Anaphylaxis (rare).

Splenectomy

Splenectomy is rarely indicated because of the increasing number of effective medical therapies that have been developed combined with the favorable natural history of ITP in children.

Splenectomy is only indicated for severe ITP with acute life-threatening bleeding nonresponsive to medical treatment or in patients with chronic ITP with bleeding and/or limitation of a patient's activities, e.g., contact sports (because of potential danger of ICH) and nonresponsive to medical treatment. Splenectomy can restore platelet counts in at least two-thirds of patients. Other modalities should be tried before splenectomy in pediatric patients because of:

- Long-term risk of infection with encapsulated organisms post-splenectomy
- Unpredictable response to splenectomy
- Unknown risk of late side effects of splenectomy.

Rituximab

Rituximab is a chimeric human–mouse monoclonal antibody directed against the transmembrane CD20 antigen present on B-cells licensed for the treatment of B-cell non-Hodgkin lymphoma. Rituximab has shown a substantial initial response rate (30–50%) in children with chronic ITP after a four-dose course, but response rates in children who have had chronic ITP for 3–5 years may be less. Rituximab has a long-term response rate in chronic severe ITP of 31 to 37%.

Dosage

The standard dosage of Rituximab is 375 mg/m^2 IV weekly for 4 weeks. Limited studies of lower doses have shown comparable efficacy. The optimal dose has not yet been determined.

Mechanism of Action

Rituximab works by attaching to CD20 antigen on B-cells causing:

- Apoptosis
- Antibody-dependent cellular cytotoxicity
- Complement modulation of auto-reactive T cells.

Adverse Effects

1. Fever and chills – common with first infusion.
2. Serum sickness – 5–10% in children with chronic ITP.
3. Headache, nausea and vomiting.
4. Hypotension (rare).
5. Tachycardia.
6. Mucocutaneous reactions including hives during first infusion, Stevens-Johnson syndrome, lichenoid dermatitis, vesiculobulbar dermatitis, toxic epidermal necrolysis.

7. Profound and prolonged peripheral B-cell depletion (hepatitis B reactivation has been seen so hepatitis B carriers should not receive rituximab).
8. Progressive multifocal leukoencephalopathy (PML) – one case has been reported in ITP.

Thrombopoietic Agents

These agents are very effective in adults. Efficacy in adults with refractory ITP is more than 75%. There is little experience with thrombopoietic agents in children with ITP, but one study of 22 children showed responses similar to those seen in adults. AMG531 (romiplostim, nplate) is given subcutaneously weekly at doses of 1–10 μg/kg. Eltrombopag (promacta) does not have an established pediatric dosage but in adults doses range from 25 to 75 mg orally daily (25 or 50 mg tablets). Toxicity includes risks of thromboembolism and reticulin fibrosis in the marrow; rebound thrombocytopenia may occur when treatment is discontinued. Eltrombopag infrequently results in transient abnormal liver tests.

Plasmapheresis

Plasmapheresis may rarely be useful to accelerate the effect of other therapies by the removal of previously synthesized platelet antibodies.

Platelet Transfusions

Platelet transfusions are required for rare emergency situations including intracranial hemorrhage, internal bleeding and emergency surgery. Platelet survival is short, but some hemostatic activity is usually obtained.

Drugs that may be effective but are infrequently used in pediatric ITP

1. Danazol.
2. Dapsone.
3. Azathioprine (or 6MP).
4. Cyclophosphamide.
5. Vinca alkaloids (Vincristine, Vinblastine).
6. Cyclosporine A.
7. Mycophenolate mofetil.

Emergency Therapy

Patients with profound mucosal bleeding or internal bleeding require immediate therapy. Combination therapy is optimal:

- IV methylprednisolone 30 mg/kg/day for 1–3 days
- IVIG 1 g/kg/day for 2–3 days with or without:
  - Anti-D 75 μg/kg (one dose)

- Platelet transfusion (bolus followed by continuous infusion if required)
- Recombinant human factor VIIa (rhuVIIa).

Emergency splenectomy in urgent, life-threatening bleeding is not usually recommended.

Chronic ITP

Chronic ITP is currently defined as thrombocytopenia that persists beyond 12 months from the diagnosis. Patients with ITP of 3–12 months duration are defined as persistent disease. It is more common in older children and those with altered immunity.

HIV-Associated Thrombocytopenia

Thrombocytopenia is common in patients with HIV and can be an early finding in the disease process. HIV-associated thrombocytopenia should be considered in a child with known HIV or a patient who presents with thrombocytopenia in conjunction with a compatible family or transfusion history and on physical examination has axillary or inguinal lymphadenopathy. The thrombocytopenia in this group of patients is multifactorial, but includes antiplatelet antibodies as a common cause. The antibodies are most commonly to GPIIb/IIIa as in ITP without HIV and have been shown to be directed to a specific peptide of GPIIIa leading to platelet intravascular lysis. The virus may also invade megakaryocytes and precursors, leading to altered thrombopoiesis.

Aggressive treatment of the underlying HIV infection with highly active antiretroviral therapy (HAART) has almost eliminated HIV-related thrombocytopenia as a clinical problem. Treatments used in non-HIV ITP are often effective. Steroids, however, should be avoided, as they contribute to the development of secondary infections.

Drug-Induced Thrombocytopenia

Table 12-10 lists the drugs proven or suspected to induce drug-dependent antibody-mediated immune thrombocytopenia. Selective serotonin reuptake inhibitor (SSRI) drugs used as antidepressants may also affect platelet function, thereby leading to bleeding symptoms without thrombocytopenia.

Heparin-Induced Thrombocytopenia

Heparin-induced thrombocytopenia (HIT) is a potentially severe cause of thrombocytopenia, more common in adults than children. HIT is caused by antibody formation to complexes of heparin and platelet factor 4. The antibodies activate FcγRIIa on the platelet surface and cause platelet activation. This in turn creates an increased risk for both venous and arterial thromboses.

Features of HIT include:

- Platelet count drops below 150,000/mm^3 or 50% drop from baseline
- Onset of thrombocytopenia following first heparin exposure occurs usually 5–10 days from start of treatment
- Occurs at a rate of 1–2% in pediatric patients who are in a pediatric ICU or who have recently undergone cardiac surgery
- To confirm diagnosis, Heparin-dependent antibodies must be demonstrated by:
  - Heparin-PF4 ELISA
  - Serotonin release assay for confirmation if needed.

When an infant is receiving heparin, no matter how small the dose (e.g. to keep a catheter patent or in total parenteral nutrition) a watchful eye should always be kept on the daily platelet count.

When HIT is suspected to be the likely cause of thrombocytopenia, heparin should be discontinued immediately and alternative anticoagulation (not coumadin) initiated (see Chapter 14 for details of management). Any suspicion of HIT should prompt evaluation for HIT-antibodies.

A separate entity called nonimmune heparin-associated thrombocytopenia can also occur. It is not immune-mediated, although its exact pathophysiology is not known. It is mild and occurs in the first day of heparin administration.

Thrombotic Thrombocytopenic Purpura

Thrombotic thrombocytopenic purpura (TTP) is a rare disease characterized by hemolytic anemia, thrombocytopenia, neurologic symptoms, renal impairment, fever and usually a high LDH. There is an acquired (idiopathic) and a congenital form.

Idiopathic (Acquired, Autoimmune) TTP

The incidence of idiopathic TTP in the United States is 4–11 cases per million population. One-third of patients successfully treated relapse and the mortality is approximately 10–20%. The disease is much more common in adults than children.

Risk factors are:

- Female sex (70% of patients are female)
- African descent
- Some cases are associated with pregnancy, autoimmune disease, infection or transplantation (both solid organ and bone marrow).

Pathology

The hallmark of TTP is the widespread presence of segmental hyaline microthrombi in the microvasculature. These hyaline deposits originate in the subendothelium, focally occlude vascular lumina and are composed of compacted platelet debris, para-aminosalicylic acid (PAS)-positive endothelial fibrils, VWF and variable amounts of fibrin. The hyaline occlusions of TTP are most consistently visible in lymph node biopsies and sections of spleen.

Pathophysiology

The majority of patients with idiopathic TTP have a severe deficiency in ADAMTS13 activity. ADAMTS13 (*a d*isintegrin *a*nd *m*etalloprotease with *t*hrombo*s*pondin type 1 repeats) is a protein responsible for cleaving unusually large multimers of VWF (UL-VWF) into biologically less-active multimers. Absence of this protein leaves large multimers present which cause "spontaneous" platelet adhesion and aggregation. In idiopathic TTP, the reduction in ADAMTS13 activity has been shown to be immune-mediated with antibodies to ADAMS13. These antibodies are not present in all cases.

Clinical and Laboratory Features

The spectrum of systemic involvement varies from patient to patient and from time to time and may be acute, chronic, or relapsing. It is most commonly seen in teenagers.

Clinical Features

1. Fever.
2. Malaise.
3. Nausea.
4. Vomiting.
5. Abdominal pain.
6. Chest pain.
7. Arthralgia.
8. Myalgia.
9. Fluctuating neurologic signs and symptoms.
10. Progressive renal failure (occurs in 25% of chronic patients).

Hematologic and Laboratory Findings

1. Microangiopathic hemolytic anemia (blood smear reveals polychromasia, basophilic stippling, schistocytes, microspherocytes and nucleated red blood cells).
2. Direct antiglobulin test (DAT) negative.
3. Thrombocytopenia.
4. DIC.

5. Purpura.
6. Pallor.
7. Jaundice.
8. Abnormally elevated factor VIII:vWF antigen in plasma.
9. Haptoglobin level reduced.
10. Hemoglobulinuria and hemosiderinuria usually present.
11. Unconjugated bilirbuin increased.
12. Lactate dehydrogenase levels increased (sensitive index of response to therapy or development of relapse).

Diagnosis

The histopathological diagnosis of TTP is often difficult. Skin and muscle biopsies yield a low percentage of positive findings (32%). Gingival biopsies are diagnostic in about 50% of cases.

Treatment

Exchange plasmapheresis provides remission in 50–80% of patients and is the mainstay of treatment. It removes antibodies to ADAMTS 13 and UL-VWF multimers and restores nonantibody-bound ADAMTS 13. Concentrates of ADAMTS 13 is under development.

Fresh frozen plasma (FFP) infusion provides the VWF-cleaving metalloproteinase.

Rituximab 375 mg/m^2 weekly for 2–8 weeks has been successful where other measures have failed.[*] Plasma exchange, if deemed clinically necessary, should be postponed as long as possible after Rituximab.

Immunosuppressive agents like glucocorticoids, cyclosporine, cyclophosphamide, vincristine and azathioprine have been used, but no systematic evaluations have been performed.

Congenital TTP

Congenital TTP occurs much less frequently than the acquired (idiopathic) form and is usually seen in the neonatal period or during childhood (acquired TTP in childhood is almost always seen in teenagers). Patients typically present with neonatal jaundice and thrombocytopenia. Some patients will not have episodes of overt TTP for several years until triggered by an infection. Patients with congenital TTP have low levels of ADAMTS13, but no antibodies. Several genetic defects on chromosome 9q34 have been identified as the underlying cause.

---

[*]Rituximab increases the ADAMTS 13 level and decreases the inhibitor level to ADAMTS 13 if it exists.

Because there are no circulating antibodies to ADAMTS13 and a small amount of the enzyme is needed for function; plasma infusion, usually administered at 2–3-week intervals is adequate therapy.

Disseminated Intravascular Coagulation

Thrombocytopenia is seen in several syndromes associated with disseminated intravascular coagulation (DIC), including purpura fulminans, overwhelming sepsis and giant hemangioma. In an unusual case of thrombocytopenia, PT, PTT, fibrinogen and fibrin split products should be determined to exclude DIC as the cause of the thrombocytopenia.

Autoimmune Disorders

Thrombocytopenia is often associated with a variety of autoimmune disorders.

1. Systemic lupus erythematosus (SLE): Thrombocytopenia occurs in 15–25% of patients with SLE as a result of peripheral destruction of platelets. The initial treatment is similar to that of ITP, including steroids, IVIG and a proclivity for immunosuppressive agents.
2. Autoimmune lymphoproliferative syndrome (ALPS): The syndrome is characterized by massive lymphadenopathy, hepatosplenomegaly, hypergammaglobulinemia and autoimmune cytopenias including ITP. The pathogenesis is inherited defects in *FAS* and other genes that regulate lymphocyte apoptosis. The initial treatment is steroids but long-term management is rituximab and/or mycophenolate mofetil (MMF).
3. Antiphospholipid antibody syndrome: Antiphospholipid antibodies enhance platelet activation. These patients have recurrent arterial or venous thrombi. The treatment is steroids and/or immunosuppressive agents.
4. Evans syndrome: Evans syndrome is the combination of autoimmune hemolytic anemia and thrombocytopenia and/or neutropenia. Patients have a poor response to steroids, IVIG, or splenectomy. A combination of these therapies is often required.
5. Other autoimmune processes: Hodgkin disease, non-Hodgkin lymphoma, juvenile rheumatoid arthritis, dermatomyositis, Graves disease, Hashimoto thyroiditis, myasthenia gravis, inflammatory bowel disease, sarcoidosis, protein-losing enteropathy may be associated with autoimmune thrombocytopenia.

Hemolytic-Uremic Syndrome

The triad of thrombocytopenia, hemolytic anemia and renal failure defines hemolytic-uremic syndrome (HUS). HUS is a common cause of renal failure in children and approximately 50% of children who develop the disorder will require dialysis. Typically patients develop bloody, sometimes painful, diarrhea followed by several days of apparent improvement and the development of the classic triad of findings.

Pathophysiology

There are several pathophysiologic mechanisms underlying HUS, but approximately 90% of cases are related to shiga toxin-producing *Escherichia coli* (usually O157:H7). The shiga toxin produces a prothrombotic state, particularly in the renal vasculature, leading to microangiopathic hemolytic anemia, platelet consumption and renal failure. Inherited and/or atypical cases may be caused by abnormalities of the alternative pathway of complement.

Treatment

HUS is treated with supportive care, e.g. dialysis. Red cell transfusions are commonly needed, but platelet transfusions should be used only when necessary, as they may worsen the prothrombotic state.

Type 2B von Willebrand Disease and Platelet-Type (Pseudo-von Willebrand) Disease

Type 2B von Willebrand and platelet-type von Willebrand both cause mucocutaneous bleeding out of proportion to any thrombocytopenia that may be present. Stress-induced worsening of thrombocytopenia is common. The management of these conditions is described in detail in Chapter 13.

Cyanotic Congenital Heart Disease

Thrombocytopenia frequently occurs in children with severe cyanotic congenital heart disease, usually when the hematocrit levels are more than 65% and when the arterial oxygen saturation is less than 65%. This may be due to margination of platelets in the small blood vessels, which may occur in the presence of a high hematocrit level. Cyanotic congenital heart disease may be associated with an impairment in platelet aggregation by adenosine diphosphate (ADP), norepinephrine and collagen. This impairment is correlated with the severity of hypoxia. Affected children usually experience little bleeding during corrective surgery, which, if successful, results in an improved platelet count.

Hypersplenism

A variety of conditions characterized by splenomegaly are associated with thrombocytopenia, presumably resulting from the sequestration or destruction of platelets by the enlarged spleen. This is usually associated with neutropenia and anemia. Megakaryocytes are plentiful in the marrow. Hypersplenism occurs in patients who have splenomegaly, irrespective of cause.

Thrombopoietin Deficiency

Thrombopoietin deficiency is a rare congenital cause of chronic thrombocytopenia with numerous immature megakaryocytes in the bone marrow. Infusion of normal plasma and plasma from patients with ITP is followed by a rise in the platelet count to normal, with apparent maturation of the megakaryocytes.

THROMBOCYTOSIS

Thrombocytosis is defined as a platelet count greater than two standard deviations above the mean. Severity of thrombocytosis can be classified into the following groups:

- mild ($500–700,000/mm^3$)
- moderate ($700–900,000/mm^3$)
- severe ($900,000–1$ million/mm^3)
- extreme (>1 million/mm^3).

Thrombocytosis is relatively common in young children, but is usually a transient, benign finding related to an infectious process. It may also occur secondary to iron deficiency, major trauma, surgery and post-splenectomy. Table 12-11 lists the conditions associated with thrombocytosis in infants and children.

Essential thrombocythemia (primary thrombocytosis) is rare in children, occurring in approximately 1/1,000,000 children. Table 12-12 lists the differentiation of essential thrombocythemia and reactive thrombocytosis.

Essential Thrombocythemia (Familial Thrombocytosis)

Essential thrombocythemia (ET) is rare in children. It is characterized by thrombocytosis and hyperplasia of megakaryocytes in the bone marrow. Other cell lines are not involved. ET is a clonal disorder that originates in a multipotent hematopoietic progenitor. *JAK2V617F* mutation has been identified in approximately half the patients with this disorder. (It occurs in more than 90% of patients with polycythemia vera and primary myelofibrosis.) Children with ET tend to have a more benign course compared to adults, but about one-third will experience thrombotic or mild bleeding complications.

Diagnosis of ET requires the following criteria:

- Persistent thrombocytosis ($>600,000/mm^3$)*
- Normal hemoglobin level (≤13 g/dl)

---

*Platelet counts greater than 1.5 million/mm^3 may be a risk factor for bleeding due to an acquired von Willebrand disease.

Table 12-11   Conditions Associated with Thrombocytosis in Infants and Children

Hereditary	**Auto-immune Diseases or Chronic Inflammation**
Asplenia	Kawasaki disease
Myeloproliferative disorder in Down syndrome	Inflammatory bowel disease
Nutritional	Rheumatoid arthritis
Iron deficiency (chronic blood loss)	Henoch-Schönlein purpura
Vitamin E deficiency	Polyarteritis nodosa
Megaloblastic anemia	**Myelodysplastic States**
Metabolic	5q-syndrome
Hyperadrenalism	Sideroblastic anemia
Immune	**Traumatic**
Graft versus host reaction	Surgery
Nephrotic syndrome	Fractures
Infectious	Hemorrhage
Viral (e.g., CMV)	Burns
Bacterial	**Miscellaneous**
Mycobacterial	Splenectomy
Fungal	Caffey disease
Drug-Induced	Pulmonary embolism
Vinca alkaloids	Thrombophlebitis
Citrovorum factor	Cerebrovascular accident
Corticosteroid therapy	Sarcoidosis
Epinephrine	Acute blood loss
Neoplastic	Hemolytic anemia
Chronic myeloid leukemia	Exercise
Polycythemia vera	Ankylosing spondylitis
Essential thrombocythemia	Spurious
Histiocytosis	Idiopathic
Lymphoma (Hodgkin and non-Hodgkin)	
Carcinoma of colon, lung	
Hepatoblastoma	
Wilms tumor	
Neuroblastoma	
Leukemia	

- Normal red cell mass (males <36 ml/kg; females <32 ml/kg)
- Normal stainable iron in bone marrow
- Absence of fibrosis in the bone marrow
- Absence of Philadelphia chromosome
- Abnormal platelet function tests
- No known cause for reactive thrombocytosis.

JAK2 V617F testing should be performed to confirm the diagnosis but absence does not rule out ET. PRV1 testing should also be performed.

Table 12-12 Differentiation of Essential Thrombocythemia and Reactive Thrombocytosis

	Reactive Thrombocytosis	Essential Thrombocythemia
Clinical and laboratory features		
Thromboembolism and hemorrhage	Uncommon	Common (in adults)[a]
Duration	Often transitory	Usually persistent
Splenomegaly	Usually absent	Present in 80% of cases
Platelet count	Usually <1 million/mm^3	Usually >1 million/mm^3
Bleeding time	Usually normal	Often prolonged
Platelet morphology and function	Usually normal	Often abnormal
Leukocyte count	Usually normal	Increased in 90% of cases
Thrombokinetic features		
Total megakaryocyte mass	Slightly increased	Greatly increased
Megakaryocyte number	Increased	Increased
Megakaryocyte volume	Decreased	Decreased
Platelet turnover or production rate	Increased	Increased
Total platelet mass	Increased	Increased
Platelet survival	Normal	Normal to slightly decreased

[a]Rare in younger children.
Modified from: Bithell TC. Thrombocytosis. In: Wintrobe's clinical hematology. 9th ed. Philadelphia: Lea & Febiger, 1993; 1391, with permission.

Treatment

Antiplatelet Agents

1. Acetylsalicylic acid (ASA): 80–160 mg orally once daily.
2. Dipyridamole: 3–6 mg/kg/day orally divided in three doses.
3. Plavix should be considered (dose is 75 mg daily in adults).

Platelet-Lowering Drugs

1. Hydroxyurea: 20–30 mg/kg orally daily to lower platelet count to between 300,000–600,000/mm^3. Maintenance therapy is necessary to sustain remission with dose adjustments as required.
2. Anegrelide hydrochloride: 1 mg PO daily to lower platelet count between 300,000–600,000/mm^3 with maintenance therapy necessary (dose adjustments may be required).
3. Alpha interferon (3 million units subcutaneously three times weekly).
4. Possible use of Jak2 inhibitors.

Hereditary Thrombocytosis

Hereditary thrombocytosis is an autosomal disorder resulting from mutations in either the thrombopoietin gene (*TPO*) or the thrombopoietin receptor gene. Mutations in the *TPO* gene lead to increased translation and subsequent thrombocytosis, whereas the *mpl mutation*

(S505N) leads to constitutive activation of the protein. The course of the disease is largely benign with rare thrombosis and no known leukemic potential.

QUALITATIVE PLATELET DISORDERS

Qualitative platelet disorders result from a variety of congenital or acquired conditions, but all demonstrate a bleeding tendency including petechiae, purpura and mucosal bleeding as well as bleeding following surgery or trauma. Various tests may demonstrate abnormalities depending upon the particular condition and the platelet count.

Table 12-13 provides a classification of congenital platelet function disorders, Table 12-14 lists the laboratory findings in inherited platelet function disorders and Table 12-15 lists the genetic transmission and recommended treatment for commonly known hereditary disorders of platelet function. In some of these disorders hemostasis will be improved by administration of DDAVP as in von Willebrand disease. Platelet transfusion is often necessary for hemostasis in those disorders not responsive to DDAVP. Leuko-depleted single donor platelets, HLA matched if available, are preferred in these patients to reduce the risk of platelet allo-sensitization. Iso-antibodies against the platelet proteins absent from the patient are common in these disorders and may render patients refractory to platelet transfusion. Administration of rFVIIa 90 µg/kg can effect hemostasis in some patients not responsive to platelet transfusion. For nonresponders, higher doses may be hemostatic. Antifibrinolytic agents are a useful adjunct in patients with these disorders.

Defects in Platelet Receptor–Agonist Interactions

Selective Impairments in Platelet Responsiveness to Adrenaline

The interaction of platelets with epinephrine *in vitro* in aggregation tests is mediated by α_2-adrenargic receptors and results in several responses, including the exposure of fibrinogen receptors, an increase in intracellular ionized calcium, the inhibition of adenylate cyclase activity and platelet aggregation. Some patients have impaired aggregation and secretion only to epinephrine, associated with a decrease in the number of platelet α_2-adrenargic receptors, with a history of easy bruising with minimally prolonged bleeding times. On the other hand, platelets in 10% of apparently normal people may fail to aggregate in response to epinephrine. The clinical significance of this finding is unknown.

Selective Impairment in Platelet Responsiveness to Collagen

There are two major collagen receptors, GPVI and $\alpha 2\beta 1$. Defects in GPVI structure and signaling have been identified in a total of nine patients, leading to mild–moderate bleeding symptoms. Very few patients have been shown to have bleeding due to abnormalities in $\alpha 2\beta 1$.

Table 12-13   Classification of Congenital Platelet Function Disorders

Defects in platelet–agonist interaction (receptor defects)
  Selective epinephrine defect
  Selective collagen defect
  Selective thromboxane A_2 defect
  Selective ADP defect
Defects in platelet–vessel wall interaction (disorders of adhesion)
  von Willebrand's disease (deficiency or defect in plasma VWF)
  Platelet type von Willebrand disease
  Bernard–Soulier syndrome (deficiency or defect in GP1b)
Defects in platelet–platelet interaction (disorders of aggregation)
  Congenital afibrinogenemia
  Glanzmann's thrombasthenia (deficiency or defect in GP11b/111a)
Disorders of platelet secretion
  Storage pool deficiency
    δ-storage pool deficiency
      Hermansky–Pudlak syndrome
      Chediak–Higashi syndrome
    α-storage pool deficiency (gray platelet syndrome)
  Abnormalities in arachidonic acid pathway
    Impaired liberation of arachidonic acid
    Cyclo-oxygenase deficiency
    Thromboxane synthetase deficiency
    Altered nucleotide metabolism
    Glycogen storage disease
    Fructose-1,6-diphosphanate ??? deficiency
  Primary secretion defect with normal granule stores and normal thromboxane synthesis
    Defects in calcium mobilization
    Defects in phosphatidylinositol metabolism
    Defects in myosin phosphorylation
  Disorders of platelet–coagulant protein interaction
    Defect in factor V^a–X^a interaction on platelets
Vascular or connective tissue defect
  Ehlers–Danlos syndrome
  Pseudoxanthoma elasticum
  Marfan syndrome
  Osteogenesis imperfecta
  Hereditary hemorrhagic telangiectasia (Osler–Weber–Rendu disease)

Defects in Thromboxane A_2 (TXA_2)

Defects in TXA_2 formation and TXA_2 receptor function have been identified as causes of mild but lifelong bleeding symptoms.

Selective Impairment in Platelet Response to ADP

A few patients have been identified with defects in P2Y12 (the ADP receptor responsible for macroscopic platelet aggregation), with associated bleeding symptoms.

Table 12-14   Laboratory Findings in Inherited Platelet Function Disorders

	Glanzmann's Thrombasthenia	Storage Pool Deficiency[a]	Collagen Receptor Defect	Release Defect[b]	Bernard-Soulier Syndrome	Von Willebrand Disease
Platelet count	Normal	Normal	Normal	Normal	Decreased	Normal
Platelet size	Normal	Microplatelets	Normal	Normal	"Giant platelets"	Normal[d]
Bleeding time	Prolonged	Variable	Variable	Variable	Prolonged	Prolonged
Platelet aggregation						
ADP	Absent	No second wave	Normal	No second wave	Normal	Normal
Arachidonic acid	Absent	Variable	Absent	Decreased	Normal	Normal
Collagen	Absent	Decreased	Normal	Absent	Normal	Normal
Ristocetin (1.5 mg/ml)	Normal	Normal	Normal	Normal	Absent	Decreased[c]
Ristocetin (0.5 mg/ml)	Absent	Variable	Absent	Absent	Normal	Decreased[c]
Storage nucleotide pool	Normal	Decreased	Normal	Normal	Normal	Normal
Others	HPA1[a] absent; GP IIb and III deficient	Dense bodies reduced; ATP/ADP ratio increased	GP Ia and IIa deficient	Cyclo-oxygenese; thromboxane synthetase or TxA$_2$ receptor deficient	GP Ib deficiency	Decreased FVIII, VWF antigen and ristocetin cofactor

[a]In Hermansky–Pudlak, Chediak–Higashi and Wiskott–Aldrich syndromes.
[b]Classically occurs with acetylsalicylic acid, omega3 ingestion and other drugs affecting arachidonic acid and prostaglandin pathway.
[c]Types I and II, decreased; Type IIb increased; Type III absent.
[d]Platelet in Type IIB will often be clumped together.

Table 12-15   Genetic Transmission and Recommended Treatment for Commonly Known Hereditary Disorders of Platelet Function

Disorder	Transmission/ Frequency	Defect	Primary Treatment for Platelets	Alternative Treatment
Disorders of receptors				
Glanzmann's thrombasthenia	Autosomal recessive, rare	GP IIb/IIIa complex	Platelet transfusion	rFVIIa in instances of platelet alloimmunization
Bernard-Soulier Syndrome	Autosomal recessive, rare	GP Ib/V/IX complex	Platelet transfusion	rFVIIa in instances of platelet alloimmunization
Defects in granule content, storage pool deficiency				
Grey platelet syndrome	Rare	Absent alpha granules (platelets look grey)	DDAVP, (Treatment rarely required)	Platelet transfusion for non-responders
Chediak-Higashi syndrome	Autosomal recessive, rare	Abnormal granules	DDAVP	Platelet transfusion for non-responders
Wiskott-Aldrich syndrome	X linked recessive	WAS protein. Primarily a quantitative defect, storage pool deficiency may also be present	Platelet Transfusion	Splenectomy TPO-R agonists HSCT
Hermansky-Pudlak	Autosomal recessive	Absent dense granules	DDAVP	Platelet transfusion
Storage pool release defects	Variable	Impaired secondary wave of aggregation	DDAVP	Platelet transfusion

Abbreviations: TPO-R, thrombopoietin receptor; HSCT, hematopoietic stem cell transplantation.

Defects in Platelet Vessel–Wall Interaction

Bernard–Soulier Syndrome (see page 335–337)

See discussion in "thrombocytopenia in the newborn".

Von Willebrand disease

See Chapter 13.

Defects in Platelet–Platelet Interaction

Glanzmann's Thrombasthenia

Glanzmann's thrombasthenia (GT) is an autosomal recessive bleeding disorder characterized by the following:

- Normal platelet count and morphology
- Prolonged bleeding time

- Mild to severe bleeding symptoms
- Reduced clot formation
- Defective platelet aggregation with all agonists except ristocetin.

GT represents a family of disorders of the platelet GPIIb/IIIa receptor complex. As this complex functions in platelet aggregation via fibrinogen, vWF and fibronectin, disruption of this pathway leads to impaired platelet aggregation. GT platelets attach normally to damaged subendothelium but fail to spread normally to form platelet aggregates. Platelets from patients with GT fail to aggregate in response to most physiologic agonists like ADP, thrombin and epinephrine; they do aggregate in response to ristocetin, however. The treatment of GT is platelet transfusions. Patients may become refractory to transfusions and in such cases activated recombinant human factor VIIa (rVIIa) may be required.

Disorders of Platelet Secretion

The category of storage pool disease includes patients with deficiencies of dense granules (δ-SPD), α-granules (α-SPD), or both types of granules (αδ-SPD).

δ-Storage Pool Deficiency

The total adenosine triphosphate (ATP) and adenosine diphosphate (ADP) platelet granule content in patients with δ-SPD is decreased as are other dense granule contents, including calcium, pyrophosphate and serotonin. Patients tend to have mild to moderate bleeding symptoms. Platelet aggregation studies show an absent second wave of aggregation with ADP and epinephrine. The aggregation response to collagen is impaired or absent. Response to arachidonic acid is variable. Disorders of δ-granules are seen in Hermansky–Pudlak syndrome (HPS), Chédiak–Higashi syndrome, Wiskott–Aldrich syndrome (WAS) and TAR syndrome.

Hermansky-Pudlak Syndrome (HPS)

HPS is an autosomal recessive disorder. The syndrome includes oculocutaneous albinism photophobia, rotatory nystagmus, loss of visual acuity and ceroid-like material accumulation in reticuloendothelial cells. Platelets have platelet dense-body and α-granule storage pool deficiency and the bleeding tendency is usually mild (related to storage pool defect and not thrombocytopenia which is not a feature of the syndrome). Patients generally present in childhood with mucosal bleeding symptoms, or prolonged bleeding after surgical procedures or dental extraction. HPS is a group of eight autosomal recessive disorders (HPS 1–8). The most common and severe form is HPS 1, which is a defect on chromosome 10q23. While rare worldwide, it is more common in northwest Puerto Rico, where 1 in

1800 individuals are affected. In this disorder, platelets contain very low levels of serotonin, adenine nucleotides and calcium. The most serious complication and most common cause of death is pulmonary fibrosis, which may be due to ceroid-like deposits in the lungs. Granulomatous colitis is also seen in this variant. HPS has been reported to be associated with nonfamilial hemophagocytic lymphohistiocytosis. Treatment is DDAVP for bleeding.

Chédiak–Higashi Syndrome (see Chapter 11)

This rare autosomal recessive syndrome includes partial oculo-cutaneous albinism, recurrent infections, mild coagulation defects and progressive neurologic dysfunction. Leukocytes, lymphocytes, monocytes and platelets have large intracytoplasmic granules. Patients also develop lymphoproliferative disease (pancytopenia, hepatosplenomegaly, lymphadenopathy and extensive tissue infiltration with lymphoid cells) and it has been associated with nonfamilial hemophagocytic lymphohistiocytosis.

DDAVP is used for bleeding and platelet transfusions are indicated for nonresponders. Stem cell transplantation may cure patients of immune and hemostatic defects, but does not improve the neurologic dysfunction.

α-Granule Storage Pool Deficiency (Gray Platelet Syndrome)

The term "gray platelet" describes the morphological appearance of platelets in Romanovsky-stained peripheral blood smears prepared from patients with a deficiency of α-granules. Electron microscopy reveals the virtual absence of α-granules and platelets appear gray because they are devoid of "purplish" cytoplasmic granulation on blood smear. Platelets from these patients contain absent or markedly reduced α-granule proteins including PF4, vWF, fibronectin and factor V. Affected patients have mild thrombocytopenia with a platelet count of around $100,000/mm^3$, prolonged bleeding times and a lifelong bleeding diathesis. These patients tend to develop myelofibrosis due to an inability of megakaryocytes to store newly synthesized, platelet-derived granule proteins such as TGF-beta, which "leaks out" in the marrow. The most consistent laboratory abnormality has been impairment in thrombin-mediated aggregation and secretion. Aggregation responses to collagen and ADP are variable.

Autosomal recessive, autosomal dominant and X-linked (GATA-1 mutation) inheritance have been observed.

Treatment is rarely required for this condition but DDAVP is used if bleeding occurs and platelet transfusions are indicated for nonresponders.

Quebec Platelet Disorder (Platelet Factor V Quebec)

Quebec syndrome is inherited as an autosomal dominant, with mild thrombocytopenia. However, it is associated with a significant degree of bleeding including mucosal and joint bleeding. In this disorder, there is an overproduction of the proteolytic enzyme urokinase-type plasminogen activator (uPA). This enzyme overexpression occurs in megakaryocytes and leads to degradation of several proteins, including factor V. There is a platelet aggregation deficiency, in the setting of epinephrine exposure in particular, for unclear reasons. The disorder is diagnosed by examination of platelet uPA levels and α-granule fibrinogen degradation products. The treatment of choice is antifibrinolytic agents. Patients do not respond adequately to platelet transfusions.

Arthogryposis–Renal Dysfunction–Cholestasis (ARC)

ARC syndrome is a multisystem disorder that includes platelet dysfunction and low granule content.

May Hegglin Anomaly

Some patients with this disorder may have platelet dysfunction – see page 337.

Miscellaneous

Platelet function abnormalities have been reported in Wiskott-Aldrich syndrome, TAR syndrome, hexokinase deficiency and glucose-6-phosphate deficiency.

Impaired Liberation of Arachidonic Acid Pathways

A major response of platelets during activation is the release of arachidonic acid from membrane-bound phospholipids and its subsequent oxygenation to thromboxane A2. Thromboxane A2 forms an important positive feedback that enhances platelet activation. Ingestion of omega-3 fatty acids in sufficient quantity over a several-week period can also reduce arachidonic release by the platelet membrane.

Cyclooxygenase and Thromboxane Synthetase Deficiency

Defects in thromboxane A2 due to deficiencies of cyclooxygenase and thromboxane synthetase have been reported.

Platelet Intracellular Signaling Defects

Kindlin-3

The intracellular signaling molecules, kindlin 2 and 3, are important in the activation of integrin αIIBβ3. Leukocyte adhesion deficiency (LAD) type III is a syndrome involving leukocytosis, chronic infections and a bleeding diathesis similar to Glanzmann's thrombasthenia. All patients identified with this disease have been shown to have mutations in the kindlin-3 gene. Cure can be achieved by stem cell transplantation.

Dysregulated Calcium Signaling

Calcium is an essential signaling molecule involved in platelet activation; several patients have been reported with defective calcium mobilization.

Deficiency of Platelet Procoagulant Activity

Scott Syndrome

Scott syndrome is a rare disorder involving mucosal and postsurgical bleeding. The disorder arises from impaired translocation of phosphatidylserine from the inner to the outer leaflet of the plasma membrane, resulting in impaired binding of factor Va and subsequent decrease in prothrombinase activity. In these patients, bleeding time, platelet aggregation and secretion studies are normal. Prothrombin time is shortened due to the reduced prothrombinase activity.

Isolated Defect in Membrane Vesiculation

Patients in four families have been reported with lifelong bleeding disorders who have impaired microvesicle secretion.

ACQUIRED PLATELET DISORDERS

Table 12-16 lists the acquired disorders causing defective platelet function.

Medications

Acetylsalicylic acid (ASA or aspirin) impairs platelet aggregation usually attributed to inhibition of cyclooxygenase-1(COX-1) and subsequent reduction in thromboxane A2 production, although other mechanisms have been implicated. A dose of 81 mg of aspirin daily can permanently affect platelets in circulation, such that the length of drug effect is related to platelet lifespan, not drug half life. Other NSAIDs impair thromboxane A2 production as well but the effects are related to drug half life.

Platelet dysfunction caused by aspirin is important in several clinical settings:

- Excessive post-tonsillectomy bleeding after ASA ingestion
- Development of purpura in children after ingesting aspirin
- Prolonged bleeding time in children with suspected hemostatic defect
- Misdiagnosing children with mild von Willebrand disease because of ASA ingestion prior to testing.

Platelet aggregation may be impaired in the newborn following maternal drug therapy. For this reason a history of drug ingestion is an essential part of the investigation of hemorrhagic states in the newborn period.

Table 12-16 Acquired Disorders Causing Defective Platelet Function

I. **Vascular or Connective Tissue Defects**
   A. Scurvy
   B. Amyloidosis
II. **Adhesion Defects**
   A. Acquired von Willebrand disease eg: high flow cardiac lesion
   B. Renal failure
   C. Drugs: dipyridamole
III. **Platelet Aggregation Defects**
   A. Fibrin or fibrinogen split products: DIC, liver disease
   B. Macromolecules: paraproteins, dextran
   C. Drugs: penicillin, semisynthetic penicillins, cephalosporins
IV. **Release Reaction Defects**
   A. Storage pool deficiency
      1. α-Granules: cardiopulmonary bypass
      2. Dense granules: ITP, SLE
      3. Drugs: reserpine, tricyclic antidepressants, phenothiazines
   B. Defective release
      1. Platelet dyspoiesis: myelodysplastic syndromes, acute leukemias, myeloproliferative syndromes
      2. Drugs: aspirin, other nonsteroidal anti-inflammatory agents, furosemide, nitrofurodantin
      3. Ethanol
   C. Altered nucleotide metabolism
      1. Drugs: phosphodiesterase inhibitors or stimulators of adenylcyclase
V. **Other Defects**
   A. Drugs: heparin, sympathetic blockers, clofibrate, antihistamines
   B. Infection: viral
   C. Hypothyroidism

Table 12-17 lists other commonly used drugs that have been implicated in platelet dysfunction.

Renal Failure

A generalized hemorrhagic state is known to occur in advanced renal failure. Thrombocytopenia is present in a minority of patients and reduced platelet adhesiveness to glass and defective platelet factor 3 occurs with platelet dysfunction being the primary hemostatic problem. Treatment with estrogen and DDAVP can be attempted but reversal of renal failure with dialysis may be required to control bleeding.

Liver Disease

In addition to a deficiency of coagulation factors of the prothrombin complex and others in liver disease, an abnormality of platelet function has been described. Platelet aggregation by ADP and thrombin is significantly impaired in patients with cirrhosis and prolonged bleeding time. This is due to known inhibition of platelet function by fibrinogen degradation products, resulting from excessive fibrinolysis seen in advanced liver disease.

Table 12-17 Drugs Implicated in Platelet Dysfunction

Antibiotics
  Ampicillin[a]
  Carbenicillin[a]
  Furadantin[a]
  Gentamycin[a]
  Keflin[a]
  Moxalactam[a]
  Nafacillin[a]
  Piperacillin[a]
  Quinacrin[b]
Anti-inflammatory Drugs
  Acetylsalicylic acid[b]
  Colchicine[c]
  Ibuprofen[b]
  Indomethacin[b]
  Naprosyn[b]
  Phenylbutazone[b]
Anesthetics
  Cocaine[a]
  Nupercaine[a]
  Procaine[a]
  Xylocaine[a]
Cardiovascular/Respiratory
  Aminophylline[d]
  Dicumarol[e]
  Dipyridamole[d]
  Heparin[e]
  Hydralazine[b]
  Nitroglycerin[f]
  Papaverine[e]
  Propranolol[a]
  Reserpine[a]
  Theophyllin[d]
  Verapamil[b]

Diuretics
  Acetazolamine[e]
  Ethacrynic acid[e]
  Furosemide[b]
Psychiatric Drugs
  Eventyl[a]
  Elavil[a]
  Norpramin[a]
  Sinequan[a]
  Stelazine[a]
  Tofranil[a]
Others
  Alcohol[a]
  Benedryl[a]
  Caffeine[d]
  Cyclosporine[a,b]
  Dextan[a]
  Glycerol guaiacolate[e]
  Hydrocortisone[b]
  Methylprednisolone[b]
  Phenergan[a]
  Tocopherol[b]
  Vinblastine[c]
  Vincristine[c]

[a]Interference with membrane receptors.
[b]Interference with prostaglandin synthesis.
[c]Interference with thrombostatin.
[d]Interference with phosphodiesterase.
[e]Unknown mechanism of action.
[f]Interference with platelet cyclic adenosine monophosphate (AMP).

Management of Defects in Platelet Function

The management of defects of platelet function may include:

- Removing any exogenous cause of platelet dysfunction (e.g., drugs)
- Treating the underlying disorder
- Using platelet transfusions for hemorrhagic episodes or surgery

- DDAVP or cryoprecipitate – this may shorten bleeding time in some patients with:
  - Renal failure
  - Inherited or acquired defects in release reaction
  - Inherited or acquired von Willebrand disease.
- EACA (ε-aminocaproic acid) in a dose of 50–100 mg/kg orally or IV every 6 hours – this may have some benefit in mucosal hemorrhage
- Activated recombinant factor VIIa (rVIIa).

INHERITED VASCULAR AND CONNECTIVE TISSUE DISORDERS

Disorders of the vascular system and connective tissues do not generally lead to clotting or platelet abnormalities; bleeding is instead caused by the fragility of the connective tissues of the skin, subcutaneous tissues and vessel wall.

Ehlers-Danlos Syndrome (EDS)

Ehlers-Danlos syndrome is an uncommon disorder of connective tissue characterized by skin hyperextensibility, delayed wound healing, joint hypermobility, bleeding tendency and connective tissue fragility. Patients may present with easy bruising, bleeding from the gums after dental extraction and prolonged menstruation. This condition may be associated with reduced aggregation with ADP, epinephrine and collagen. Epistaxis, petechiae, hematuria, hemoptysis and hemarthrosis are usually not seen. There are many subtypes of EDS. Type IV (vascular type) has an unusual sensitivity to aspirin and is particularly notable for a generalized vascular fragility which may manifest as arterial rupture and sudden death. This disorder is inherited as an autosomal dominant and is associated with decreased levels of type III collagen. Type IV EDS carries the worst prognosis due to the vascular complications.

Pseudoxanthoma Elasticum

Pseudoxanthoma elasticum is a rare disorder of elastic tissues that is inherited in an autosomal manner. Severely affected individuals may present with spontaneous hemorrhage resulting from defective vessels and bleeding may occur into the skin, eyes, kidney, joints, uterus and gastrointestinal tract. Patients may develop fatal subarachnoid or gastrointestinal hemorrhage. A similar disorder has been identified with deficiency of the vitamin-K-dependent clotting factors.

Marfan's Syndrome

Marfan's syndrome is characterized by skeletal abnormalities, cardiovascular abnormalities and dislocation of the lens. The syndrome is inherited in an autosomal dominant fashion. Affected individuals demonstrate easy bruising and may bleed excessively during surgery.

Osteogenesis Imperfecta

Osteogenesis imperfecta is an autosomal dominant disorder characterized by variable bone fragility, short stature and blue sclera and hearing impairment. Patients may present with a bleeding disorder characterized by bruising, epistaxis, hemoptysis and intracranial hemorrhage. The abnormality occurs as a result of mutations in COLA1 and COLA2 which code for type I collagen.

Hereditary Hemorrhagic Telangiectasia

Hereditary hemorrhagic telangiectasia (Osler-Weber-Rendu disease) is the most common of the inherited vascular disorders. The disease is inherited as an autosomal dominant trait and bleeding occurs from vascular lesions on the skin or mucous membranes. Lesions consist of dilated arterioles and capillaries lined by a thin endothelial layer. They are typical in appearance (1–3 mm in diameter, flat, round, red or violet in color) and they blanch on pressure. Histology of the abnormal vessels shows a deficiency of supporting elastic fibers. Typical lesions occur most often on the nasal mucosa, tongue, face, hands, gastrointestinal tract and, rarely, in the respiratory, gynecological and urinary tracts. Epistaxis is usually the most common symptomatology.

LABORATORY EVALUATION OF PLATELETS AND PLATELET FUNCTION

The following tests can be helpful in determining platelet number and function.

Examination of the Blood Smear

The blood smear should be evaluated in all cases where a platelet disorder is suspected. A review of the smear can quickly rule out pseudothrombocytopenia (the low platelet count by automatic electronic cell counter compared to an abundance of usually clumped platelets on blood smear). A disproportionate number of large platelets suggests increased platelet turnover. The large platelets represent a younger population. If the mean platelet volume using automatic electronic counters is raised (normal, 8.9 ± 1.5 fl) it indicates peripheral platelet destruction rather than impairment of platelet formation. Table 12-1 lists the various platelet diseases based on platelet size. If a true platelet abnormality is present, however, the morphology of the platelets including size and color may suggest one or more specific diagnoses.

Bleeding Time

The bleeding time test is an *in vivo* screening examination for the interaction between platelets and the blood vessel wall. Several techniques exist for measuring bleeding time.

The test has not been standardized for young children or infants (who have been shown to have shorter bleeding times). In general, bleeding time for children has been shown to be less than 9 minutes and is prolonged in cases where the platelet count is less than $100,000/mm^3$ or when disorders of platelet adhesion or aggregation are present (e.g. von Willebrand disease, acquired or congenital platelet function abnormalities).

This test is used infrequently because of difficulties with standardization.

Closure Time

The platelet function analyzer-100 (PFA-100; Dade-Behring, Deerfield, IL) is a rapid *in vitro* screening test for platelet function at high shear rates. Citrated whole blood is aspirated through a 150-μm diameter aperture in a membrane coated with collagen and either epinephrine or adenosine 5′-diphosphate (ADP). The machine measures closure time, which is the time it takes for a platelet plug to form and occlude the aperture. Closure time is useful in identifying most cases of von Willebrand disease, aspirin effect and for some cases of platelet dysfunction. There is a high negative predictive value for these disorders. Closure time can be affected by hematocrit and platelet count. Optimal results require a hematocrit between 25–50% and a platelet count of at least $50,000/mm^3$. Closure time is useful in attempting to identify a suspected bleeding disorder, but is a poor screening test in the general population (i.e. in pre-operative screening for bleeding disorders).

Platelet Aggregation in Platelet-Rich Plasma

Platelet-rich plasma (PRP) aggregometry is considered the gold standard for examining platelet function. Platelet aggregation is measured by a photo-optical instrument which compares platelet-rich plasma to platelet-poor plasma. As aggregating agents like ADP, thrombin and epinephrine are added, platelet aggregation is stimulated and platelet clumps form, causing the turbid platelet-rich plasma to become clear. The photo-optical device measures the increased light transmittance through the sample. The light received through the sample is converted through the electronic signals, amplified and recorded on chart paper. The following are normal wave responses to the addition of aggregating agents:

- A biphasic response (i.e. a primary and secondary wave of aggregation) is seen in response to:
  - ADP (low concentration)
  - Epinephrine
  - Thrombin
  - Ristocetin (note that aggregation with ristocetin may be reduced or absent in von Willebrand disease, ristocetin may be used in the quantitative assay of von Willebrand disease).

- A single wave of aggregation is seen with collagen
- Arachidonic acid causes rapid secondary aggregation. Nonsteroidal anti-inflammatory drugs (NSAIDs), such as indometacin and aspirin, are potent inhibitors of aggregation induced by arachidonic acid
- A platelet count in the platelet-rich plasma (PRP) used in the actual aggregation testing must exceed 100,000/mm^3 for accurate results.

Aspirin and other NSAIDs inhibit platelet aggregation by interfering with the release reaction of platelets, thereby reducing or eliminating the secondary wave of aggregation.

Several weaknesses are inherent in PRP aggregometry. Specifically, the centrifugation required can injure or activate platelets and may remove giant platelets. Centrifugation also removes erythrocytes and leukocytes, which are known to affect platelet function.

Platelet Aggregation in Whole Blood

Platelet aggregation may be measured with an impedance aggregometer (Lumi aggregometer). In this test, whole blood is placed in a cuvette containing two electrodes. An electric current is applied to the electrodes and agonist-stimulated platelets aggregate on the surface of the electrodes and impede the current flow in proportion to the degree of platelet aggregation. Because this method does not require centrifugation, it avoids some of the problems encountered in platelet-rich plasma (PRP) aggregometry.

Either method of testing platelet aggregation may identify platelet disorders like Bernard-Soulier syndrome, Glanzmann's thrombasthenia, cyclooxygenase deficiency, storage pool disease and von Willebrand disease.

NONTHROMBOCYTOPENIC PURPURA

Anaphylactoid Purpura

Henoch-Schönlein is a vasculitis which leads to nonthrombocytopenic, nonthrombopathic purpura. The purpuric eruptions are different to those seen in thrombocytopenia in that the lesions are maculopapular, initially resembling urticaria because of edema and perivascular infiltration and later becoming erythematous, with central areas of hemorrhage that finally fade to brown because of denaturation of the extravasated hemoglobin. The rash appears on the buttocks and on the extensor surfaces of the lower extremities. Accompanying the rash are joint or gastrointestinal symptoms, localized areas of edema and renal damage. Initial hematuria occurs in about one-third of patients. The platelet count and tests for hemostasis are normal. Treatment is symptomatic and steroids may be beneficial for severe abdominal pain. Follow-up for the patients with suspected renal disease to prevent irreversible renal damage is important.

Infections

Infections may present with nonthrombocytopenic purpura, including acute bacterial endocarditis, meningococcal septicemia, coxsackievirus and echovirus infections, rubella and atypical measles.

Drugs

Diffuse, benign and self-limited purpura has been described after exposure to certain drugs, particularly sulfonamides.

Purpura Factitia

Although rare, purpura factitia may present a diagnostic dilemma. It is self-inflicted purpura, usually linear and always on accessible parts of the body. It is more common in females than in males and frequently indicates deep-seated psychopathology. All hematologic tests are normal.

Gardner–Diamond Syndrome

Gardner–Diamond syndrome is a poorly characterized syndrome also known as psychogenic purpura which presents with bruising. It occurs most commonly in teenage girls. The etiology remains unknown. The "diagnostic test" involves injection of the patient's own red cells into the patient's skin and demonstrating an allergic reaction at the site.

Scurvy

Scurvy is a vascular disease occurring with profound deficiency of vitamin C secondary to a diet lacking in fresh fruits and vegetables. Platelet number and function are normal. It presents with bruising and oozing from the gums.

Suggested Reading

Althaus K, Greinacher A. MYH9-Related platelet disorders. Semin Thromb Hemost. 2009;35:213–223.

Bouw MC, Dors N, van Ommen H, et al. Thrombotic thrombocytopenic purpura in childhood. Pediatr Blood Cancer. 2009;53:537–542.

Bussel JB, Zacharoulis S, Kramer K, et al. Clinical and diagnostic comparison of neonatal alloimmune thrombocytopenia to non-immune cases of thrombocytopenia. Pediatr Blood Cancer. 2005;45:176–183.

Cines DB, Bussel JB, Leibman AL, et al. The ITP syndrome: pathologic and clinical diversity. Blood. 2009;113:6511–6521.

Dame C, Sutor HA. Primary and secondary thrombocytosis in childhood. Br J Haematol. 2005;165–177.

Isreals SJ. Diagnostic evaluation of platelet function disorders in neonates and children: an update. Semin Thromb Hemost. 2009;35:181–188.

Provan D, Stasi R, Newland A, et al. International consensus report on the investigation and management of primary immune thrombocytopenia. Blood doi:10.1182/blood-9009-06-225565.

Rodeghiero F, Statsi R, Gernsheimer T, et al. Standardization of terminology, definitions and outcome criteria in immune thrombocytopenic purpura in adults and children: report from an international working group. Blood. 2009;113:2386–2393.

Salles II, Feys HB, Iserbyt BF, et al. Inherited traits affecting platelet function. Blood Reviews. 2008;22:155–172.

Wei HW, Schoenwaelder SM, Andrews RK, et al. New insights into the haemostatic function of platelets. Br J Haematol. 2009;147:415–430.

CHAPTER 13

Hemostatic Disorders

PHYSIOLOGY OF HEMOSTASIS

As a result of injury to the blood vessel endothelium, three events take place concurrently:

- Vasoconstriction (vascular phase)
- Platelet plug formation (primary hemostatic mechanism – platelet phase)
- Fibrin thrombus formation (initiation, amplification and propagation phases).

Relevant Components of Hemostasis

Endothelial cells secrete substances that repel platelets (prostaglandin I2 (PGI2), adenosine diphosphate (ADP) and nitric oxide), initiate coagulation (collagen, fibronectin), promote platelet adhesion (von Willebrand factor [vWF]) and fibrin dissolution (tissue plasminogen activator), catalyze the inhibition of thrombin (heparin and thrombomodulin) and inhibit the initiation of fibrin dissolution (tissue plasminogen activator inhibitor).

Participation of **platelets** in hemostasis is a fundamental component of the physiologic process of coagulation. Platelet interactions in coagulation are initiated by adhesion to areas of vascular injury. Subsequent activation of platelets results in release of ADP, serotonin and calcium from "dense bodies" and fibrinogen, vWF, factor V, HMW kininogen, fibronectin, α-1-antitrypsin, α-thromboglobulin, platelet factor 4 (PF4) and platelet-derived growth factor from α granules. Platelets provide surfaces for the assembly of coagulation factors (e.g.,VIIIa/Ca^{2+}/IXa and Va/Ca^{2+}/Xa complexes). The platelets aggregate and increase the mass of the hemostatic plug. They also mediate blood vessel constriction (by releasing serotonin) and neutralize heparin.

All of the **plasma coagulation factors** are produced in the liver; factor VIII is also produced by endothelial cells. Table 13-1 lists the half-life and plasma levels of the coagulation factors. Factors II, VII, IX and X are vitamin K dependent and require vitamin K in order to undergo post-translational gamma carboxylation. These vitamin-K-dependent factors circulate in zymogen form, are activated on platelet phospholipid surfaces and upon activation have serine protease activity. The plasma coagulation factors work in an interdependent manner to generate thrombin (factor IIa) from prothrombin (factor II); thrombin then digests fibrinogen to form fibrin monomers. Fibrin monomers polymerize and establish a network. By incorporating into the hemostatic plug, thrombin becomes inactivated. Thrombin plays a central bioregulatory role, promoting platelet aggregation and release reactions and generating a biofeedback-positive loop to form more thrombin at a faster rate. Thrombin activates factor XIII, which in turn cross-links the fibrin network. Thrombin and thrombin complexed to thrombomodulin also activate thrombin activatable fibrinolysis inhibitor (TAFI), a procarboxypeptidase found in plasma, which attenuates fibrinolysis of the clot. The three components of hemostasis (blood vessels, platelets and plasma coagulation factors) do not function independently but are integrated and interdependent.

Primary Hemostatic Mechanism (Platelet Phase)

This mechanism leads to the formation of a reversible aggregate of platelets: a temporary hemostatic plug (Figure 13-1). Endothelial injury exposes von Willebrand factor and collagen from the subendothelial matrix to flowing blood and shear forces. Plasma von Willebrand factor binds to the exposed collagen, uncoils its structure and, in synergy with

Table 13-1 Half-life and Plasma Levels of Coagulation Factors

Factors	Common Name	Biologic Half-life (h)	Plasma Concentration (nM)	Plasma Levels (units/dl)
I	Fibrinogen	90	8800	200–400[b]
II	Prothrombin[a]	60	1400	50–150
III	Tissue thromboplastin	N/A		0
V	Proaccelerin, labile factor	12–36	20	50–150
VII	Proconvertin,[a] stable factor	6–8	10	50–150
VIII	Antihemophilic factor	8–12	0.7	50–150
IX	Christmas factor[a]	12–24	90	50–150
X	Stuart factor[a]	32–58	170	50–150
XI	Plasma thromboplastin antecedent	48–72	30	50–150
XII	Hageman factor	48–52	375	50–150
XIII	Fibrin-stabilizing factor	72–120	70	50–150
High-molecular-weight kininogen	Fitzgerald factor	136	6000	—
Prekallikrein	Flecher factor	N/A	450	—

[a]Vitamin-K-dependent.
[b]In mg/dl.

collagen supports the adhesion of platelets. Initially the von Willebrand factor interacts with the GPIb platelet receptor, tethering the platelets. As the platelet collagen receptors GPVI and α2β1 bind to collagen the platelets adhere and become activated with a resulting release of platelet alpha and dense granule contents. Platelet activation results in a conformational change in the αIIbβIII receptor, activating it and enhancing its avidity for von Willebrand factor, for vessel wall ligands and for fibrinogen. The enhanced avidity for von Willebrand factor and fibrinogen mediates platelet-to-platelet interactions which eventually lead to platelet plug formation.

Fibrin Thrombus Formation

The fibrin thrombus formation component of hemostasis occurs in three overlapping phases, initiation, amplification and propagation (Figure 13-2). The *initiation phase* begins with cell-based expression of tissue factor (TF) at the site of endothelial injury. Factor VII binds to the exposed TF and is rapidly activated. The factor VIIa/TF complex in turn generates factor Xa (FXa) and factor IXa (FIXa). FXa can activate

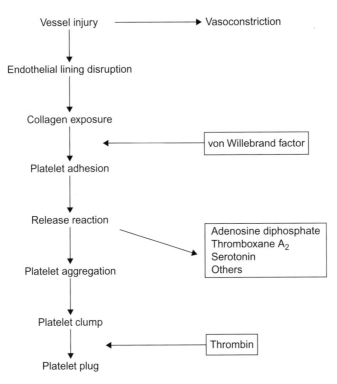

Figure 13-1 Primary Hemostatic Mechanism.

factor V (FV) which complexes with FXa and generates small amounts of thrombin. During the *amplification phase* the pro-coagulant stimulus is transferred to the surface of platelets at the site of injury. The small amounts of thrombin enhance platelet adhesion, fully activate the platelets and activate factors V, VIII and XI. In the *propagation phase* the "tenase" complex of FIXa–FVIIIa is assembled on the platelet surface and efficiently generates FXa. Similarly the "prothrombinase" complex of FXa–FVa is assembled on the platelet surface and efficiently generates thrombin. Unlike FXa generated from TF–FVIIa interactions, FXa complexed to FV is protected from inactivation by tissue factor pathway inhibitor, assuring adequate thrombin generation. The resulting pro-coagulant, thrombin, activates factor XIII and cleaves fibrinopeptides (FPs) A and B from fibrinogen. The residual peptide chains aggregate by means of loose

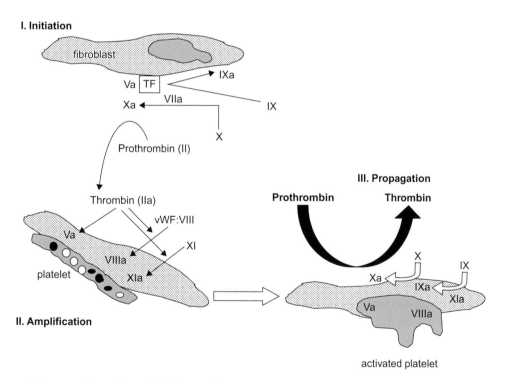

Figure 13-2 A Cell-based Model of Coagulation.
The three phases of coagulation occur on different cell surfaces: initiation on the tissue-factor bearing-cell; amplification on the platelet as it becomes activated; and propogation on the activated platelet surface.
Adapted from: Hoffman M, Monroe DM. A cell-based model of hemostasis. Thromb Haemost 2001;85:958–965, with permission.

hydrogen bonds to form fibrin monomers. Under the influence of FXIIIa, fibrin monomers are converted into fibrin polymers, forming a stable fibrin clot. In the presence of thrombin, the mass of loosely aggregated intact platelets is transformed into a densely packed mass that is bound together by strands of fibrin to form a definitive hemostatic barrier against the loss of blood.

Platelet–Vessel Interaction

Blood vessels form a circuit which maintains blood in a fluid state. With vascular injury, platelets and the coagulation system temporarily close the rent and prevent leakage of blood out of the blood vessel. Blood vessel wall characteristics exhibit properties that contribute to hemostasis or stop hemorrhage as well as prevent thrombosis. The media and adventitia of the vessel wall enable vessels to dilate or constrict. The subendothelial basement membranes contain adhesive proteins which provide binding sites for platelets and leukocytes. Remodeling which occurs after injury is enhanced by extracellular matrix metalloproteinases.

FIBRINOLYSIS

The fibrinolytic system provides a mechanism for removal of physiologically deposited fibrin. Clot lysis is brought about by the action of plasmin on fibrin. Fibrinolytic events are shown in Figure 13-3. Plasminogen from circulating plasma is laid down with fibrin during the formation of thrombin. Plasminogen is primarily synthesized in the liver and circulates in two forms, one with a NH_2-terminal glutamic acid residue (glu-plasminogen) and a second form with a NH_2-terminal lysine, valine, or methionine residue (lys-plasminogen). Glu-plasminogen can be converted to lys-plasminogen by limited proteolytic degradation. Lys-plasminogen has a higher affinity for fibrin and cellular receptors, it is also more readily activated to plasmin than glu-plasminogen. Both forms of plasminogen bind to fibrin through specific lysine-binding sites. These lysine-binding sites also mediate the interaction of plasminogen with its inhibitor, α_2-antiplasmin (α_2AP). Thrombin-activated fibrinolysis inhibitor (TAFI)-mediated removal of C-terminal lysine and arginine residues will prevent high-affinity plasminogen binding and will attenuate fibrinolysis. Plasminogen is converted to its enzymatically active form, plasmin, by several activators. These activators are widely distributed in body tissues and fluids. Tissue plasminogen activator (t-PA) is the principal intravascular activator of plasminogen. t-PA is a serine protease that binds to fibrin through lysine-binding sites. When t-PA is bound to fibrin its plasmin generation efficiency increases markedly. Urokinase type plasminogen activator (u-PA), a second physiological activator of plasminogen, is present in urine and activates plasminogen to plasmin equally well in the absence and

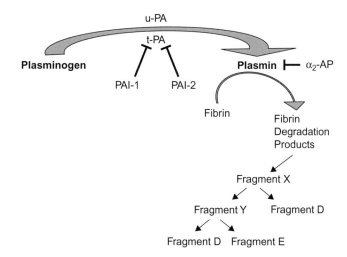

Figure 13-3 The Fibrinolytic Pathway.
Plasminogen is converted enzymatically to plasmin by t-PA or by u-PA. Plasmin cleaves fibrin and fibrinogen into fibrin degradation products. Major inhibitors of the fibrinolytic pathway are depicted. PAI-1 and PAI-2 Inhibit t-PA. Plasmin activity is inhibited by α_2-AP. t-PA: Tissue plasminogen activator, u-PA: urokinase, PAI: plasminogen activator inhibitor 1, PAI-2: plasminogen activator inhibitor 2, α_2-AP: alpha 2 anti-plasmin. ⊢, inhibition.

in the presence of fibrin. Plasmin splits fibrin and fibrinogen into fibrin-degradation products:

- Fragment X (molecular weight [MW], 270,000)
- Fragment Y (MW, 155,000)
- Fragment D (MW, 90,000)
- Fragment E (MW, 50,000).

Properties attributed to the various fibrin split products include heparin-like effects, inhibition of platelet adhesion and aggregation, potentiation of the hypotensive effect of bradykinin and chemotactic properties (monocytes and neutrophils). Increased fibrinolysis is usually a reaction to intravascular coagulation (secondary fibrinolysis) rather than the initial event (primary fibrinolysis).

The action of plasmin is negatively regulated by several inhibitors (shown in Table 13-2), these include α_2-antiplasmin and α_2-macroglobulin. Plasminogen activator is in turn regulated by two inhibitors, plasminogen activator inhibitor-1 (PAI-1) and plasminogen activator inhibitor-2 (PAI-2). PAI-1 is the more physiologically important of these inhibitors (Figure 13-3).

NATURAL INHIBITORS OF COAGULATION

In addition to the physiologic role of fibrinolysis, other inhibitors play critical roles in the control of hemostasis. Table 13-2 lists the plasma fibrinolytic compenents and hemostatic

Table 13-2   Plasma Fibrinolytic Components and Hemostatic Inhibitors

	Biologic Half-life	Proteases Inhibited	Concentration in Plasma (mg/dl)
Fibrinolytic Components			
Plasminogen[a]	48 h	—	10–15
Plasminogen activators			
Tissue	3–4 min	—	—
Urokinase	9–16 min	—	—
Plasminogen activator inhibitor[a]	—	Plasminogen activator, XII[a]	60–200[b]
Inhibitors			
Antithrombin[a] (heparin cofactor)	17–76 h	XII[a], XI[a], IX[a], X[a], thrombin, kallikrein, plasmin	10–14 104–121[b]
α_2-Plasmin inhibitor[a] (antiplasmin)	30 h	XII[a], XI[a], kallikrein, plasmin, thrombin	6–8 80–120[b]
α_2-Macroglobulin[a]	—	XII[a], XI[a], thrombin, kallikrein, plasmin	190–310
C1 Inhibitor[a]	—	XII[a], kallikrein	20–25
α_1-Antitrypsin[a]	—	Thrombin, XI[a], kallikrein	245–325
Thrombin activatable inhibitor of fibrinolysis (TAFI)			20–400[c]
Tissue Factor Pathway Inhibitor	—	Factor VII[a]/Tissue Factor complex	Endothelial bound
Protein C[a]	6 h	V[a],VIII[a], plaminogen activator inhibitor	0.4–0.6 71–109[b]
Protein S	60 h	V[a], VIII[a]	95–125[b]
Protein C inhibitor[a]	—	Protein C[a]	0.5

[a]Enzymatic activity: serine protease.
[b]Activity in plasma (%).
[c]Activity expressed as *nM/liter.*

inhibitors and their principal substrates. All members of this group have overlapping roles in the control of coagulation and fibrinolysis. Major antiproteases of this group of inhibitors include antithrombin α_2-antiplasmin, α_2-macroglobulin, the inhibitor of the activated first component of complement (C1 inhibitor) and α_1-antitrypsin.

Antithrombin (AT) neutralizes the procoagulants thrombin, factor IXa, Xa and XIa (Figure 13-4). When bound to circulating heparin or heparan sulfate on endothelial cells, AT undergoes a conformational change with a dramatic increase in this activity.

Tissue factor pathway inhibitor (TFP1) is responsible for inactivation of the FXa/FVIIa/ tissue factor complex.

The vitamin-K-dependent zymogen, protein C and its cofactor protein S, which is also a vitamin K protein, play an important role in the control of hemostasis by inhibiting activated factors V and VIII (Figure 13-4). Binding of thrombin to thrombomodulin on endothelial cells of small blood vessels neutralizes the procoagulant activities of thrombin

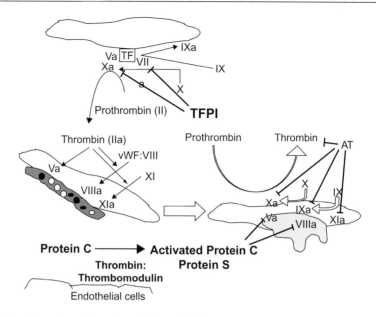

Figure 13-4 Major Inhibitory Proteins of Coagulation.
TFPI, AT, protein C and protein S are depicted with their target coagulation factor substrates.
TFPI, tissue factor pathway inhibitor; AT, antithrombin. ⊢, inhibition.

and activates protein C. Protein C binds to a specific receptor and the binding augments the activation of protein C by thrombin. Activated protein C inactivates factors Va and VIIIa in a reaction that is greatly accelerated by the presence of free protein S and phospholipids, thereby inhibiting the generation of thrombin. Free protein S itself has anticoagulant effects: it inhibits the prothrombinase complex (factor Xa, factor Va and phospholipid) which converts prothrombin to thrombin and inhibits the complex of factor IXa, factor VIIIa and phospholipid, which converts factor X to factor Xa.

HEMOSTASIS IN THE NEWBORN

Table 13-3 lists the hemostatic values in healthy preterm and term infants.

Plasma Factors

In comparison with hemostatic mechanisms in older children and adults, that of the newborn infants is not uniformly developed. In newborns plasminogen levels are only 50% of adult values and α_2AP levels are 80% of adult values, whereas plasminogen activator inhibitor-1 (PAI-1) and t-PA levels are significantly increased over adult values. The increased plasma levels of t-PA and PAI-1 in newborns on day 1 of life are in marked contrast to values from cord blood, in which concentrations of these two proteins are significantly lower than in adults. Newborns also have decreased activity of anticoagulant factors, especially

Table 13-3   Hemostatic Values in Healthy Preterm and Term Infants

	Normal Adults/Children	Preterm Infant (28–32 Weeks)	Preterm Infant (33–36 Weeks)	Term Infant
PT (s)	10.8–13.9	14.6–16.9	10.6–16.2	10.1–15.9
APTT (s)	26.6–40.3	80–168	27.5–79.4	31.3–54.3
Fibrinogen (mg/dl)	95–425	160–346	150–310	150–280
II (%)	100[a]	16–46	20–47	30–60
V (%)	100[a]	45–118	50–120	56–138
VII (%)	100[a]	24–50	26–55	40–73
VIII(%)	100[a]	75–105	130–150	154–180
vWF Ag (%)	100[a]	82–224	147–224	67–178
vWF (%)	100[a]	83–223	78–210	50–200
IX (%)	100[a]	17–27	10–30	20–38
X (%)	100[a]	20–56	24–60	30–54
XI (%)	100[a]	12–28	20–36	20–64
XII (%)	100[a]	9–35	10–36	16–72
XIII (%)	100[a]	–	35–127	30–122
PK (%)	100[a]	14–38	20–46	16–56
HMW-K (%)	100[a]	20–36	40–62	50–78

[a]Expressed as a percentage of activity in pooled control plasma.
Abbreviations: PK, prekallikrein; HMW-K, high-molecular-weight kininogen.

antithrombin, protein C and protein S. In addition the following other physiologic differences are present in normal newborn infants:

- Blood Vessels
  - Capillary fragility is increased
  - Prostacyclin production is increased.
- Platelets
  - Platelet adhesion is increased due to increased von Willebrand factor (vWF) and increased high-molecular-weight (HMW) vWF multimers
  - Platelet aggregation abnormalities
    a. Epinephrine-induced aggregation is decreased due to decreased platelet receptors for epinephrine
    b. Ristocetin-induced aggregation is increased due to increased vWF and increased HMW vWF multimers.
  - Platelet activation is increased: as evidenced by elevated levels of thromboxane A2, *B* thromboglobulin and PF4.
- Bleeding Time
  - The bleeding time is normal because of increased platelet–vessel wall interactions due to increased vWF and HMW vWF, a high hematocrit and the large red blood cell size.

DETECTION OF HEMOSTATIC DEFECTS

Evaluation of a patient for a hemostatic defect generally entails the following:

- Detailed history
  - Symptoms: epistaxis, gingival bleeding, easy bruising, menorrhagia, hematuria, neonatal bleeding, gastrointestinal bleeding, hemarthrosis, prolonged bleeding after lacerations
  - Response to hemostatic challenge: circumcision, surgery, phlebotomy, immunization/intramuscular injection, suture placement/removal, dental procedures
  - Underlying medical conditions: known associations with hemostatic defects (liver disease, renal failure, vitamin K deficiency)
  - Medications: antiplatelet drugs (nonsteroidal anti-inflammatory drugs), anti-coagulants (warfarin, heparin, low-molecular-weight heparin), antimetabolites (L-asparaginase), prolonged use of antibiotics causing vitamin K deficiency, long-term use of iron suggestive of ongoing blood loss causing iron-deficiency anemia
  - Family history: symptoms, response to hemostatic challenge (siblings, parents, aunts, uncles, grandparents), transfusions after surgeries.
- Complete physical examination: signs consistent with past coagulopathy: petechiae, echymoses, hematomas, synovitis/joint effusion, arthropathy, muscle atrophy, evidence of joint laxity or hyperextensibility
- Laboratory evaluation (Tables 13-1 to 13-4)

Initial screening tests:

- Complete blood count (CBC): quantitative assessment of platelets
- Bleeding time: prolonged with impaired platelet function, platelet counts reduced below 80–100,000/mm^3 or impaired vascular integrity. This test is no longer used because of difficulties in standardization and interpretation
- Platelet Function Analyzer (PFA 100): assesses flow through a membrane, membrane closure time is measured in response to ADP and to epinephrine. Often prolonged with impaired platelet function (see Chapter 12)
- Coagulation factor screening tests (from a laboratory perspective the coagulation system is divided into the intrinsic pathway, the extrinsic pathway and the common pathway. Such a division, though not relevant for *in vivo* hemostasis is useful for conceptualizing *in vitro* laboratory testing [Figure 13-5])
- Prothrombin time (PT) assay (assesses the extrinsic system): this test utilizes tissue thromboplastin and calcium chloride, to initiate the formation of thrombin via the extrinsic pathway

Table 13-4   Coagulation Tests and Normal Alues

Test	Normal Value
Platelet function	
Template bleeding time (min)	<9
Platelet retention (%)	40–90
Platelet aggregation	
Platelet factor 3 availability	
Prothrombin consumption(s)	>25
Clot retraction	Starts at hour 1; completes at hour 24
Clot solubility[a]	
Intrinsic system	
Activated partial thromboplastin time(s)	25–35
Extrinsic system	
Prothrombin time(s)	10–12
Stypven time[b]	
Factor assays	See Tables 11-1 and 11-3
Thrombin time(s)[c]	<24
Reptilase time(s)[c]	<25
Fibrinolytic system	
Staphylococcal clumping (μg/ml)[d]	<9
Thrombo-Wellcotest (μg/ml)[d]	<10
D-Dimer Wellcotest[d]	No latex agglutination

[a]Screening test for factor XIII. Normally clot remains intact after 30 minutes in 1% monochloroacetic acid.
[b]Russell's viper venom test: screening test for factor VII.
[c]Screening test for dysfibrinogenemia, hypofibrinogenemia and afibrinogenemia.
[d]Split products of fibrinogen.

- Partial thromboplastin time (PTT) assay (assesses the intrinsic system): this test utilizes a phospholipid reagent, a particulate activator (e.g., ellagic acid, kaolin, silica, soy extract) and calcium chloride to start the enzyme reaction that leads to the formation of thrombin via the intrinsic pathway.

Common confirmatory coagulation tests:

- Fibrinogen: quantitative measurement of fibrinogen, useful when both the PT and the PTT are prolonged
- Thrombin time: prolonged when fibrinogen is reduced or abnormal, in the presence of inhibitors (fibrin degradation products, D-dimers) and in the presence of thrombin inhibiting drugs. Useful test to diagnose dysfibrinogenemas (qualitatively abnormal fibrinogen). Useful when both the PT and a PTT are prolonged
- Mixing studies (performed to evaluate a prolonged PT or PTT): the respective assay is performed following addition of normal pooled plasma to patient plasma Normalization indicates a clotting factor deficiency that was corrected by addition of normal pooled

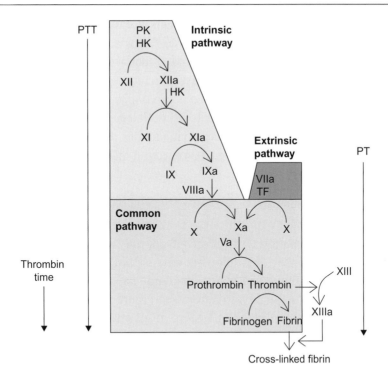

Figure 13-5 A Conceptualization of Commonly Used Screening Tests of Coagulation and the Coagulation Parameters they Measure.
Abbreviations: PTT, partial thromboplastin time; PT, prothrombin time.

plasma. Continued prolongation indicates presence of a coagulation inhibitor. Such inhibitors may be physiologically relevant or only of *in vitro* significance

- Clotting factor activity assays: Performed to identify clotting factor deficiencies if mixing studies normalize. FXII, FXI, FIX and FVIII assays are useful if the aPTT normalizes in mixing studies. The FVII assay is useful if the PT normalizes in mixing studies. FX, FV, FII and fibrinogen assays are useful if both PT and aPTT normalize in mixing studies. (Note: factor XIII deficiency does not result in prolongation of the PT or PTT)

- von Willebrand antigen: quantitative assay for von Willebrand factor, useful when bleeding time or PFA-100 closure time is prolonged or when von Willebrand disease is suspected

- von Willebrand factor (ristocetin cofactor activity): functional/qualitative assay for von Willebrand factor, useful when the bleeding time or PFA-100 closure time is prolonged or when von Willebrand disease is suspected

- von Willebrand factor multimers are useful when discrepancies between von Willebrand factor antigen (normal to low) and ristocetin co-factor activity (extremely low or ratio of ristocetin co-factor activity to von Willebrand factor antigen <0.6)
- Platelet aggregation studies: a qualitative assessment of platelet function, useful when the bleeding time or PFA-100 closure time is prolonged
- Urea Clot Lysis Assay: Useful screen for FXIII deficiency outside of the newborn period. Fetal fibrinogen interferes with this assay giving a false positive result. In the absence of fibrin cross linkage by FXIII, a clot will degrade with incubation in 5 M urea.

Pre-Operative Evaluation of Hemostasis

1. History (the most important element of the evaluation):
   - If negative: no coagulation tests are indicated; only a CBC
   - If positive or unreliable – the following tests should be performed: CBC, PT, aPTT, fibrinogen, thrombin time and von Willebrand panel.
2. Abnormal tests require further investigation (see Figure 13-6).
3. In a patient with a significant bleeding history, if all screening tests and von Willebrand panel are normal consider FXIII, PAI-1 activity, alpha 2 antiplasmin and platelet aggregation studies, vitamin C levels and rule out hyperextensibility (to diagnose Ehlers–Danlos syndrome) on clinical examination.

ACQUIRED COAGULATION FACTOR DISORDERS

Vitamin K Deficiency

The normal full-term infant is born with levels of factors II, VII, IX and X that are low by adult standards (Table 13-3). The coagulation factors fall even lower over the first few days of life, reaching their nadir on about the third day. This is due to the low body stores of vitamin K at birth. As little as 25 µg vitamin K can prevent this fall in activity of the vitamin K-dependent clotting factors. The vitamin K content of cow's milk is about 6 µg/dl and that of breast milk 1.5 µg/dl. It is a combination of low initial stores and subsequent poor intake of vitamin K that occasionally produces an aggravation of the coagulation defect causing primary hemorrhagic disease of the newborn. Vitamin K deficiency results in hemorrhagic disease between the second and fourth days of life and is manifested by gastrointestinal hemorrhage, hemorrhage from the umbilicus, or internal hemorrhage (classic hemorrhagic disease of the newborn). Bleeding attributable to this cause is responsive to parenteral vitamin K therapy; for this reason parenteral vitamin K is routinely administered to newborns.

In premature infants of low birth weight, both the vitamin K stores and the level of coagulation factors are even lower than in term infants. The response to vitamin K is slow and inconsistent, suggesting that the immature liver has reduced synthetic capability.

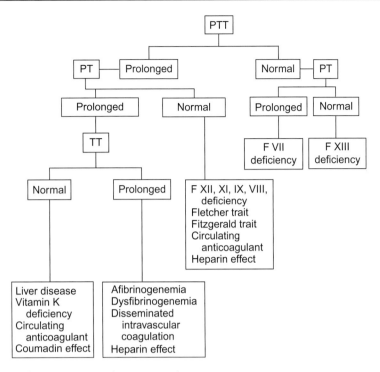

Figure 13-6 Coagulation Tests and Interpretation.
Abbreviations: PTT, activitated partial thromboplastin time; PT, prothrombin time; TT, thrombin time.

Maternal ingestion of certain drugs may cause early hemorrhagic disease of the newborn as a result of neonatal hypoprothrombinemia due to a reduction in factors VII, IX and X. These drugs include oral anticoagulants and anticonvulsants (phenytoin, primidone and phenobarbital).

Late hemorrhagic disease of the newborn occurs from one week after birth until 6 months of life due to the absence of adequate vitamin K prophylaxis at birth, malabsorption, liver disease, or idiopathically. Consider investigating for biliary atresia, alpha 1 antitrypsin deficiency if hemorrhagic disease of the newborn occurs after adequate vitamin K prophylaxis at birth.

Table 13-5 lists the conditions associated with deficiency of vitamin K-dependent factors in the newborn and in the older child and Table 13-6 lists the laboratory findings in vitamin K deficiency in relationship to the findings in liver disease and disseminated intravascular disease (DIC).

Hepatic Dysfunction

Any transient inability of the newborn's liver to synthesize necessary coagulation factors, even in the presence of vitamin K, can result in hemorrhagic disease that is nonresponsive

Table 13-5   Conditions Associated with Deficiency of Vitamin K-Dependent Factors

Normal newborn (normal by 3 months of age) Prematurity
  Vitamin K responsive
  Vitamin K nonresponsive (caused by immaturity, infection, hypoxia, hepatic under-perfusion)
Maternal medications: anticoagulants, antituberculosis drugs, valproate, carbamazepine
Dietary
  Cow's milk: 6 µg/l
  Human milk: 1.5 µg/l
Altered bacterial colonization
Vomiting
Severe diarrhea Malabsorption syndromes
  Celiac disease
  Cystic fibrosis
  Biliary atresia
  Obstruction of gastrointestinal tract
  Antibiotics (including antibiotics in breast milk)
Hepatocellular disease
  Acute
    Reye syndrome
    Acute hepatitis
  Chronic
    Cirrhosis
    Wilson disease
Drugs
  Coumarins

Table 13-6   Laboratory Findings in Vitamin K Deficiency, Liver Disease and Disseminated Intravascular Coagulation

Component	Vitamin K Deficiency	Liver Disease	DIC
Red cell morphology	Normal	Target cells	Fragmented cells, burr cells, helmet cells, schistocytes
aPTT	Prolonged	Prolonged	Prolonged
PT	Prolonged	Prolonged	Prolonged
Fibrin split products	Normal	Normal or slightly increased	Markedly increased
Platelets	Normal	Normal	Reduced
Factors decreased	II, VII, IX, X	I, II, V, VII, IX, X	Assays are of limited utility

Abbreviations: DIC, disseminated intravascular coagulation; PT, prothrombin time; PTT, partial thromboplastin time.

to vitamin K therapy. Hepatic dysfunction as a result of immaturity, infection, hypoxia, or underperfusion of the liver can all result in transient inability of the liver to synthesize coagulation factors. This is more prominent in small premature infants. The sites of bleeding in these cases are usually pulmonary and intracerebral with a high mortality. Other causes of hepatocellular dysfunction, affecting patients of all ages, are also listed in Table 13-5. In liver disease vitamin K-dependent factors and factor V and fibrinogen are

Table 13-7 Treatment of Disseminated Intravascular Coagulation and Purpura Fulminans

Treatment of the underlying disorder
  Treatment of infections with appropriate anti-infectives (antibiotics, anti-viral drugs, anti-fungal drugs)
  Correction of electrolyte imbalances, acidosis and shock
  Appropriate anti-neoplastic therapy
  Removal of triggering stimulus
Replacement therapy as indicated
  Platelet concentrates (1 unit/10 kg)
  Cryoprecipitate (50–100 mg/kg fibrinogen)[a]
  Fresh frozen plasma (10–15 ml/kg, initially; may need 5 ml/kg q6h)
Intravenous heparinization[b]
Intravenous direct thrombin inhibitors
Anti-platelet drugs
Antithrombin concentrate
Activated protein C concentrate

[a]One bag of cryoprecipitate contains about 200 mg fibrinogen.
[b]See Heparin therapy.

usually decreased, fibrin split products may be elevated due to impaired clearance (Table 13-6). In contrast, factor VIII levels are usually normal. There is no response to vitamin K. There is usually a clinical response to clotting factor replacement therapy, using fresh frozen plasma and cryoprecipitate (replacement guidelines are the same as those outlined in Table 13-7).

Disseminated Intravascular Coagulation

Disseminated intravascular coagulation is characterized by the intravascular consumption of platelets and plasma clotting factors. Widespread coagulation within the vasculature results in the deposition of fibrin thrombi and the production of a hemorrhagic state when the rapid utilization of platelets and clotting factors results in levels inadequate to maintain hemostasis. The accumulation of fibrin in the microcirculation leads to mechanical injury to the red cells, resulting in erythrocyte fragmentation and microangiopathic hemolytic anemia.

Widespread activation of the coagulation cascade rapidly results in the depletion of many clotting factors as fibrinogen is converted to fibrin throughout the body. Two important mechanisms take place:

- The generation of thrombin results in intravascular coagulation and rapidly falling platelet count, fibrinogen and FV, FVIII and FXIII levels. Paradoxically, *in vitro* bioassays for these factors may be elevated owing to generalized activation of the coagulation system

- Concurrently, plasminogen is converted to its enzymatic form (plasmin) by t-PA. Plasmin digests fibrinogen and fibrin (secondary fibrinolysis) into fibrin split products (FSPs), resulting in clot lysis.

Diagnosis of DIC relies on presence of a well-defined clinical situation associated with a thrombo-hemorrhagic disorder. The following test results are the most useful and reliable in the diagnostic and therapeutic evaluation of fulminant DIC:

- Increased PT and activated partial thromboplastin time (aPTT)
- Decreased fibrinogen – relative decrease in fibrinogen compared to the acute inflammatory state – a normal level in the presence of inflammation does not preclude a diagnosis of DIC
- Decreased platelet count
- Fibrin degradation (elevated fibrin degradation products, elevated D dimmers)
- Presence of fragmented red blood cells (e.g., schistocytes, triangle cells, helmet cells, burr cells)
- Increased PF4 (platelet factor 4)
- Increased FPA (fibrinopeptide A)
- Decreased FV, FVIII, FXIII (useful if levels are decreased).

The typical diagnostic findings of DIC are listed in Table 13-6. Disease states associated with DIC are listed in Table 13-8. Generally available treatment options for treatment of DIC are shown in Table 13-7.

In children the following conditions may be associated with low-grade DIC:

- Kasabach–Merritt syndrome
- Chronic inflammatory disorders
- Arteriovenous fistulae
- Vascular prosthesis
- Glomerulonephritis.

Low-grade DIC has the potential to accelerate into fulminant DIC. The following laboratory findings are useful in the diagnosis of low-grade DIC:

- Presence of fragmented red blood cells
- Usually normal or mildly decreased platelet count
- Normal or mildly increased PT/APTT
- Normal or mildly decreased fibrinogen
- Usually increased FSPs
- Usually increased PF4
- Usually increased FPA levels.

Table 13-8   Disease States Associated with Disseminated Intravascular Coagulation

Causative Factors	Clinical Situation
Tissue injury	Trauma/crush injuries
	Head injury
	Major surgery
	Heat stroke
	Burns
	Venoms
	Malignancy
	Obstetrical accidents
	Amniotic fluid embolism
	Placental abruption
	Stillborn fetus
	Abortion
	Fat embolism
Endothelial cell injury	Infection (bacterial, viral, protozoal)
AND/OR	Immune complexes
Abnormal vascular surfaces	Eclampsia
	Postpartum renal failure
	Oral contraceptives
	Cardiopulmonary bypass
	Giant hemangioma
	Vascular aneurysm
	Cirrhosis
	Malignancy
	Respiratory distress syndrome
Platelet, leukocyte, or red cell injury	Incompatible blood transfusion
	Infection
	Allograft rejection
	Hemolytic syndromes
	Drug hypersensitivity
	Malignancy

Treatment of low-grade DIC involves:

- Treatment of the underlying disease responsible for triggering low-grade DIC
- Antiplatelet therapy:
  - Aspirin (5–10 mg/kg/day)
  - Dipyridamole (3–5 mg/kg/day).

INHERITED COAGULATION FACTOR DISORDERS

Table 13-9 lists the genetics, prevalence, coagulation studies and symptoms of inherited coagulation factor disorders.

Table13-9   Genetics, Prevalence, Coagulation Studies and Symptoms of Inherited Coagulation Factor Deficiencies

Factor Deficiency	Genetics	Est. Prevalence	BT	APTT	PT	Associated with Bleeding Episodes
Afibrinogenemia	AR	1:1 million	N	P	P	++
Dysfibrinogenemia	AR		N	N/P	P	+/− thrombosis
II	AR	1:2 million	N	P	P	++
V (parahemophilia)	AR	1:1 million	N	P	P	++
VII	AR	1:500,000	N	N	P	+
VIII (hemophilia A)	XLR	1:10,000	N	P	N	+++
von Willebrand's disease		1:1000				
Type 1	AD		N/P	N/P	N	+
Type 2	AD		P	N/P	N	++
Type 3	AR		P	P	N	++
IX (hemophilia B)	XLR	1:60,000	N	P	N	+++
X	AR	1:1 million	N	P	N	++
XI (hemophilia C)	AR	1:1 million	N[a]	P	N	+
XII	AD		N	P	N	−
XIII	AR	1:1 million	N	N	N	+[c]
Prekallikrein (Fletcher trait)	AD		N	P[b]	N	−
HMW kininogen (Fitzgerald trait)	AR		N	P	N	−
Passovoy (?)	AR		N	P	N	+/−

Bleeding episodes occur in most individuals homozygous for the disorder. However, with FXI and FVII deficiency there is no correlation between factor levels and bleeding phenotype. Most studies consider factor levels <10% to be associated with bleeding symptoms; however, there is no consensus.
[a]Prolonged when associated with platelet dysfunction.
[b]Shortened with prolonged exposure to kaolin.
[c]Umbilical stump bleeding; clot soluble in 5 M urea or 1% monochloracetic acid.
Abbreviations: AD, autosomal dominant; APTT, activated partial thromboplastin time; AR, autosomal recessive; BT, bleeding time; N, normal; P, prolonged; PT, prothrombin time; XLR, X-linked recessive.

Hemophilia A and B

The most common coagulation disorders, following von Willebrand disease, are hemophilia A and B. Hemophilia A is an X-linked recessive bleeding disorder attributable to decreased blood levels of functional procoagulant factor VIII (FVIII, VIII:C, antihemophilic factor). Hemophilia B is also an X-linked recessive disorder and is indistinguishable from hemophilia A with respect to its clinical manifestations. In hemophilia B, the defect is a decreased level of functional procoagulant factor IX (FIX, IX:C, plasma thromboplastin component, or Christmas factor). The incidence of hemophilia is 1 per 5000 live male births. Factor VIII deficiency accounts for 80–85% of cases of hemophilia, with factor IX deficiency accounting for the remainder. Both types occur with similar incidence among all races and in all parts of the world.

Clinical Course of Hemophilia

Hemophilia should be suspected when unusual bleeding is encountered in a male patient. Clinical presentations of hemophilia A and hemophilia B are indistinguishable. The

Table 13-10   Relationship of Factor Levels to Severity of Clinical Manifestations of Hemophilia A and B

Type	Percentage Factor VIII/IX	Type of Hemorrhage
Severe	<1	Spontaneous; hemarthroses and deep soft tissue hemorrhages
Moderate	1–5	Gross bleeding following mild to moderate trauma; some hemarthrosis; seldom spontaneous hemorrhage
Mild	5–25	Severe hemorrhage only following moderate to severe trauma or surgery
High-risk carrier females	variable	Gynecologic and obstetric hemorrhage common, other symptoms depend on plasma factor level

Table 13-11   Common Sites of Hemorrhage in Hemophilia

Hemarthrosis
Intramuscular hematoma
Hematuria
Mucous membrane hemorrhage
  Mouth
  Dental
  Epistaxis
  Gastrointestinal
High-risk hemorrhage
  Central nervous system
    Intracranial
    Intraspinal
  Retropharyngeal
  Retroperitoneal
  Hemorrhage causing compartment syndrome/nerve compression
    Femoral (iliopsoas muscle)
    Sciatic (buttock)
    Tibial (calf muscle)
    Perineal (anterior compartment of leg)
    Median and ulnar nerve (flexor muscles of forearm)

frequency and severity of bleeding in hemophilia are usually related to the plasma levels of factor VIII or IX (Table 13-10), although some genetic modifiers of hemophilia severity have been identified. The median age for first joint bleed is 10 months, corresponding to the age at which the infant becomes mobile. Table 13-11 shows the common sites of hemorrhage in hemophilia. The incidence of severity and clinical manifestations of hemophilia are listed in Table 13-12.

For hemophilia B the Leyden phenotype (severe hemophilia as a child that becomes mild after puberty) has been described in families with defects in the androgen-sensitive promoter region of the gene.

Table13-12   Incidence of Severity and Clinical Manifestations of Hemophilia

Severity	Severe	Moderate	Mild
Incidence			
Hemophilia A	70%	15%	15%
Hemophilia B	50%	30%	20%
Bleeding manifestations			
Age of onset	≤1 year	1–2 years	2 years–adult
Neonatal hemorrhages			
Following circumcision	Common	Common	None
Intracranial	Occasionally	Rare	Rare
Muscle/joint hemorrhage	Spontaneous	Following minor trauma	Following major trauma
CNS hemorrhage	High risk	Moderate risk	Rare[a]
Postsurgical hemorrhage	Common	Common	Rare[a]
Oral hemorrhage[b]	Common	Common	Rare[a]

[a]FVIII, >25; FIX, >15.
[b]Following trauma or tooth extraction.

Hemophilia A Carrier Detection

Excessive lyonization may result in reduced FVIII levels in female carriers of hemophilia, hence a reduced FVIII level can have utility in diagnosing the carrier state.

Direct gene mutation analysis: The FVIII common intron 22 inversion, resulting from an intrachromosomal recombination, is identifiable in 45% of severe hemophilia A patients. For the remaining 55% of severe hemophilia A patients as well as all mild and moderate hemophilia A patients, the molecular defects can usually be detected by efficient screening of all 26 FVIII exons and splice junctions. Therefore, direct gene mutation analysis is the most accurate test for carrier detection and prenatal diagnosis for severe hemophilia A. For rare patients in whom a precise mutation cannot be identified, intragenetic and extragenetic linkage analysis of DNA polymorphisms can be useful with up to 99.9% precision (when an affected male patient and his related family members are available). Pre-implantation genetic diagnosis is another option available to carrier females.

When definitive diagnosis of the carrier state cannot be made, determination of the FVIII/vWF:Ag ratio (<1.0) can be used to detect 80% of hemophilia A carriers with 95% accuracy. Use of this methodology requires careful standardization of the laboratory performing the testing.

Hemophilia B Carrier Detection

Hemophilia B carriers have a wide range of FIX levels but, in a subset of cases, can be detected by the measurement of reduced plasma factor IX activity (in 60–70% of cases).

Direct gene mutation analysis: The factor IX gene is located centromeric to the factor VIII gene in the terminus of the long arm of the X chromosome. There is no linkage between the FVIII and FIX genes. The 34 kb FIX coding sequence comprises eight exons and encodes a

461 amino acid precursor protein that is approximately one-third the size of the factor VIII cDNA. Because of the smaller gene size, FIX mutations can be identified in nearly all patients. Direct FIX mutation testing is available through DNA diagnostic laboratories, with linkage analysis used in those cases where the responsible mutation cannot be identified.

Prenatal Diagnosis

Prenatal diagnosis of hemophilia can be performed by either chorionic villus sampling (CVS) at 10–12 weeks' gestation or by amniocentesis after 15 weeks' gestation. If DNA analysis is not available or if a woman's carrier status cannot be determined, fetal blood sampling can be performed at 18–20 weeks' gestation for direct fetal factor VIII plasma activity level measurement. The normal fetus at 18–20 weeks' gestation has a very low FIX level which an expert laboratory can distinguish from the virtual absence of FIX in a fetus with severe hemophilia B.

Maternal–fetal combined complication rates for amniocentesis and CVS are 0.5–1.0% and 1.0–2.0%, respectively. Fetal blood sampling is less available; the fetal loss rate for these procedures ranges from 1 to 6%.

Treatment (Factor Replacement Therapy)

Factor replacement therapy is the mainstay of hemophilia treatment. The degree of factor correction required to achieve hemostasis is largely determined by the site and nature of the particular bleeding episode. Commercially available products for replacement therapy are listed in Table 13-13.

- Commercially available FVIII products include high-purity recombinant preparations, highly purified plasma-derived concentrates (monoclonal/immunoaffinity purified) and intermediate-purity plasma-derived preparations
- Commerically available FIX products include recombinant FIX and plasma-derived high-purity FIX concentrate (coagulation FIX concentrate). Prothrombin complex concentrates, formerly a mainstay of hemophilia B treatment, are not utilized because of the risk of thrombotic complications associated with intensive treatment
- Factor concentrates are preferred over fresh frozen plasma or cryoprecipitate because the plasma-derived concentrates undergo viral inactivation treatment in addition to donor screening.

Source plasma for all plasma-derived factor concentrates undergoes donor screening and nucleic acid testing for a variety of viral pathogens. In addition, all plasma-derived and many recombinant factor concentrates undergo a viral inactivation treatment, typically with solvent detergent, wet or dry heat treatment, pasteurization or nanofiltration. Recombinant factor concentrates are widely accepted as the treatment of choice for previously untreated patients, minimally treated patients and patients who have not had transfusion-associated infections.

Table 13-13 Commercially Available Coagulation Factor Concentrates

Class	Product	Purity	Procedure	Primary Use
FVIII Plasma Derived	Koate-DVI (Bayer Corp)	Intermediate purity	SD/Dry Heat	Hemophilia A
	Humate P (Aventis Behring)	Intermediate purity	P	Hemophilia A/von Willebrand
	Alphanate SD-HT (Grifols Corp)	High purity	SD/Dry Heat	Hemophilia A/von Willebrand
	Hemofil M (Baxter Bioscience)	Ultra high purity	SD	Hemophilia A
	Monoclate P (Aventis Behring)	Ultra high purity	P	Hemophilia A
FVIII Recombinant	1st generation - Recombinate (Baxter Bioscience)	Ultra high purity	none	Hemophilia A
	2nd generation - Helixate (Aventis Behring)	Ultra high purity	SD	Hemophilia A
	2nd generation - ReFacto (Wyeth)	Ultra high purity	SD	Hemophilia A
	2nd generation - Kogenate FS (Bayer Corp)	Ultra high purity	SD	Hemophilia A
	3rd generation - Advate (Baxter Bioscience)	Ultra high purity	SD	Hemophilia A
	3rd generation - Xyntha (Wyeth)	Ultra high purity	SD	Hemophilia A
Prothrombin Complex	Profilnine SD (Grifols Corp)	Intermediate	SD	Hemophilia B
	Bebulin VH (Baxter Bioscience)	Intermediate	Vapor heat	Hemophilia B
FIX Coagulation Concentrate				
Plasma derived	Alphanine SD-VF (Grifols Corp)	High	SD/nanofiltration	Hemophilia B
Plasma derived	Mononine (Aventis Behring)	High	SD/ultrafiltration	Hemophilia B
Recombinant	BeneFIX (Wyeth)	High	nanofiltration	Hemophilia B
Activated Prothrombin Complex Concentrate	Autoplex T (NABI)	N/A	Dry heat	Inhibitor bypass therapy
	FEIBA VH (Baxter Bioscience)	N/A	Vapor heat	Inhibitor bypass therapy
Bypassing agent:	NovoSeven (NovoNordisk)	N/A	None	Inhibitor bypass therapy, FVII deficiency
Specialty items	Hyate C (Ipsen)	N/A	None	For FVIII inhibitor treatment
	Proplex T (Baxter Bioscience)	Intermediate	Dry heat	FVII replacement (3.5 U FVII/U FIX)

Notes: Purity (international units/mg protein before albumin is added); intermediate <100; high >100; ultra high >1,000.
Abbreviations: SD, solvent detergent; P, pasteurization; N/A, not applicable.

Table 13-14 provides generally accepted guidelines for treatment of most types of hemophiliac bleeding. When a bleeding episode is suspected hemostatic treatment should be rendered first then diagnostic evaluation(s) may be performed.

Strategies for hemophilia care include on-demand treatment of acute bleeding episodes or, for severe hemophilia patients, prophylactic administration of clotting factor concentrate to maintain trough factor levels >1% augmented with on-demand treatment of breakthrough bleeding episodes. The latter strategy pharmacologically converts the severe hemophilia phenotype to a moderate phenotype with an attendant reduction in frequency of bleeding episodes. One randomized multicenter United States national study suggested a markedly reduced incidence of hemophilic arthropathy when prophylaxis was instituted prior to onset of recurrent joint bleeds. This benefit must be balanced with the need for frequent prophylactic infusions (3–4 times/week for FVIII, 2 times/week for FIX), venous access considerations, potential requirements for central venous access devices, increased cost of treatment and the occasional patient who has a very mild clinical course.

Ancillary Therapy

DDAVP

In hemophilia A patients 1-Deamino-8-D-arginine vasopressin (DDAVP) increases plasma FVIII levels 2½–6-fold. It is commonly used to treat selected hemorrhagic episodes in mild hemophilia A patients. When used intravenously the dose is 0.3 µg/kg administered in 25–50 ml normal saline over 15–20 minutes. Its peak effect is observed in 30–60 minutes. Subcutaneous DDAVP is as effective as intravenous DDAVP, facilitating treatment of very young patients with limited venous access.

Concentrated intranasal DDAVP (Stimate), available as a 1.5 mg/ml preparation has approximately two-thirds of the effect of intravenous DDAVP. Care should be exercised to avoid inadvertent dispensing of the dilute intranasal DDAVP commonly used for treatment of diabetes insipidus. The peak effect of intranasal Stimate is observed 60–90 minutes after administration.

Recommended dosage for use of intranasal Stimate:

- Body weight <50 kg: 150 µg (one metered dose)
- Body weight >50 kg: 300 µg (two metered doses).

Recommendations during the administration of DDAVP are:

- Mild fluid restriction to two-third maintenance fluids and drinking of electrolyte-containing fluids (avoiding free water) to satisfy thirst
- Monitoring urinary output and daily weights may be useful to track fluid retention.

Table 13-14 Treatment of Bleeding Episodes

Type of Hemorrhage	Hemostatic Factor Level	Hemophilia A	Hemophilia B	Comment/Adjuncts
Hemarthrosis	30–50% minimum	FVIII 20–40 U/kg q12–24 h as needed; if joint still painful after 24 h, treat for further 2 days	FIX 30–40 U/kg q24h as needed; if joint still painful after 24 h, treat for further 2 days	Rest, immobilization, cold compress, elevation
Muscle	40–50% minimum, for iliopsoas or compartment syndrome 100% then 50–100% × 2–4 days	20–40 U/kg q12–24h as needed for iliopsoas or compartment syndrome initial dose is 50 U/kg	40–60 U/kg q24h as needed for iliopsoas or compartment syndrome initial dose is 60–80 U/kg	Calf/forearm bleeds can be limb threatening. Significant blood loss can occur with femoral-retroperitoneal bleed
Oral mucosa	Initially 50%, then EACA at 50 mg/kg q6hr × 7 days usually suffices	25 U/kg	50 U/kg	Antifibrinolytic therapy is critical. Do not use with PCC or APCC
Epistaxis	Initially 30–40%, use of EACA 50 mg/kg q6h until healing occurs may be helpful	15–20 U/kg	30–40 U/kg	Local measures: pressure, packing
Gastrointestinal	Initially 100% then 50% until healing occurs	FVIII 50 U/kg, then 25 U/kg q12h	FIX 100 U/kg, then 50 U/kg q day	Lesion is usually found, endoscopy is recommended, antifibrinolytic may be helpful
Hematuria	Painless hematuria can be treated with complete bed rest & vigorous hydration for 48 h. For pain or persistent hematuria 100%	FVIII 50 units/kg, if not resolved 30–40 u/kg q day until resolved	FIX 80–100 units/kg, if not resolved then 30–40 U/kg q day until resolved	Evaluate for stones or urinary tract infection. Lesion may not be found. Prednisone 1–2 mg/kg/d × 5–7 days may be helpful. Avoid antifibrinolytics
Central Nervous system	Initially 100% then 50–100% for 14 days	50 U/kg; then 25 U/kg q12h	80–100 U/kg; then 50 U/kg q24h	Treat presumptively before evaluating, hospitalize. Lumbar puncture requires prophylactic factor coverage
Retroperitoneal or retropharyngeal	Initially 80–100% then 50–100% until complete resolution	FVIII 50 units/kg; then 25 U/kg q12h until resolved	FIX 100 U/kg; then 50 U/kg q24h until resolved	Hospitalize
Trauma or surgery	Initially 100%; then 50% until would healing is complete	50 U/kg; then 25 U/kg q12h	100 U/kg; then 50 U/kg q24h	Evaluate for inhibitor prior to elective surgery

Abbreviations: Antifibrinolytic therapy, EACA; Epsilon-aminocaproic acid, AMICAR; syrup, 250 mg/5 ml; tablet, 500 mg, 1,000 mg; PCC, prothrombin complex concentrates; aPCC, activated PCC.

Responses to DDAVP vary from patient to patient, but in a given patient are reasonably consistent on different occasions. Therefore, a test dose of DDAVP should be administered at the time of diagnosis or in advance of an invasive procedure to assess the magnitude of the patient's response. DDAVP administration may be repeated at 24-hour intervals according to the severity and nature of the bleeding. Administration of DDAVP at shorter intervals results in a progressive tachyphylaxis over a period of 4–5 days. Depending on the response after a trial, DDAVP could be recommended for most minor procedures. For major procedures requiring maintenance of factor activity >80%, factor concentrates may be needed in mild hemophilia A patients.

Side effects of DDAVP include:

- Asymptomatic facial flushing
- Thrombosis (a rarely reported complication)
- Hyponatremia is more common in very young patients, in patients receiving repeated doses of DDAVP or large volumes of oral or intravenous fluid; hyponatremic seizures have been reported in children under 2 years of age. DDAVP is contraindicated in children under 2 years of age.

Antifibrinolytic Therapy

Antifibrinolytic drugs inhibit fibrinolysis by preventing activation of the proenzyme plasminogen to plasmin, mostly by activating the thrombin-activatable fibrinolysis inhibitor (TAFI). This intervention is useful for preventing clot degradation in areas rich in fibrinolytic activity including the oral cavity, the nasal cavity and the female reproductive tract. Approved antifibrinolytic drugs are:

- Epsilon aminocaproic acid (EACA; Amicar): Orally; dose is 50–100 mg/kg every 6 hours (maximum, 24 grams total dose per day). Gastrointestinal symptoms may occur at higher doses; therefore, the preferred starting dose is 50 mg/kg. The drug is available as 500 mg, 1000 mg tabs or as a flavored syrup (250 mg/ml)
- Tranexamic acid (Cyklokapron): 20–25 mg/kg (maximum, 1.5 g) orally or 10 mg/kg (maximum, 1.0 g) intravenously every 8 hours. At the time of writing, this drug is not currently available in the United States.

Note: Antifibrinolytic therapy should not be utilized in patients with urinary tract bleeding because of the potential for intrarenal clot formation.

To treat spontaneous oral hemorrhage or to prevent bleeding from dental procedures in pediatric patients with hemophilia, either drug is begun in conjunction with DDAVP or factor replacement therapy and continued for up to 7 days or until mucosal healing is complete. Antifibrinolytic drugs also have efficacy as an adjunct treatment for epistaxis and for menorrhagia. Antifibrinolytic drugs are safe to use in hemophilia B patients receiving coagulation factor IX concentrates but should not be used contemporaneously with prothrombin-complex concentrates (PCCs) because of the thrombotic potential of these concentrates. Initiation of oral antifibrinolytic drug therapy 4–6 hours after the last dose of a PCC appears to be well tolerated.

Management of Inhibitors in Hemophilia

Approximately 30% of patients with hemophilia A develop neutralizing allo-antibodies (inhibitors) directed against factor VIII. Inhibitors are a major cause of morbidity and mortality in hemophilia. Risk factors for inhibitors include:

* Early age at exposure
* Presence of the common inversion mutation
* Large deletions of the FVIII gene
* African-American ethnicity
* A hemophiliac sibling with an inhibitor.

Other factors not proven in randomized studies include recombinant factor treatment and continuous infusion of factor concentrates for treatment. Inhibitors are quantitated using the Bethesda assay and clinically are classified as low- and high-responder inhibitors.

Low Responders

Low-responder inhibitors have titers <5 Bethesda Units (BU) and do not exhibit anamnesis upon repeated exposure to FVIII; 1 BU neutralizes 50% of factor VIII/IX activity. Approximately half of patients with inhibitors will be low responders and, of these, approximately half will have transient inhibitors. Hemophilia A patients with low-responder inhibitors can generally be treated with factor VIII concentrate, albeit at an increased dosing intensity because of reduced *in vivo* recovery and shortened half-life of the FVIII.

For serious limb or life-threatening bleeding, a bolus infusion of 100 units/kg factor VIII is administered, repeat doses of 100 units/kg are administered at 12 hour intervals. Alternatively, the level is maintained with a continuous infusion rate based on the inhibitor titer and recovery and survival studies which estimate half-life of the factor. A factor VIII assay should be obtained 1 hour after the bolus infusion and trough or steady-state FVIII levels should be followed at least daily thereafter. For example, if a child has a 3 BU inhibitor titer, for treatment of an intracranial bleed a 100% FVIII level is needed, he needs to receive a bolus of 125 u/kg which will raise his level to 250% of which 150% will be neutralized because of his 3BU titer leaving him with 100% factor activity level. This should be followed by a continuous infusion of at least 8 u/kg/h if his half-life is 4 hours since 4 u/kg/h gives a 100% level. All these doses need to be adjusted based on FVIII activity levels drawn at least once a day on continuous infusions. Normal saline at 20 cc/h should be given distally to the infusion so as to avoid diluting the factor in order to reduce thrombophlebitis.

During prolonged treatment in FVIII *in vivo* recovery and half-life may transiently improve as inhibitor antibody is adsorbed by the FVIII.

High Responders

High-responder inhibitors have titers ≥5 BU and, although the titer may decay in the absence of FVIII exposure, these patients will display an anamnestic rise in titer upon rechallenge with FVIII. The clinical approach is different for high and low responders (Table 13-15).

Treatment of an acute bleed is achieved by use of bypassing agents (activated prothrombin complex concentrate [aPCC] and recombinant factor VIIa [rVIIa]) and inhibitor eradication is achieved through immune tolerance.

In high-responder inhibitor patients with limb or life-threatening bleeding, if the inhibitor titer is <20 BU, high-dose continuous infusion of factor VIII may saturate the antibody permitting a therapeutic FVIII level. The dose can be based on recovery and survival studies.

If a factor VIII level is not attainable or the antibody level is greater than 20 BU, then by-passing agents to initiate hemostasis independent of FVIII should be employed, such as treatment with aPCC or rFVIIa. There is no demonstrated difference in the efficacy of either product.

Table 13-15   Recommendations for Replacement Therapy for Treatment of Bleeding in Patients with Factor VIII Inhibitors

Type of Patient	Type of Bleed	Recommended Treatment
Low responder[a] (<5 BU)	Minor or major bleed	Factor VIII infusions using adequate amounts of factor VIII to achieve a circulating hemostatic level
High responder[b] with low inhibitor level (<5 BU)	Minor/major bleed	PCC, aPCC or rVIIa infusions
	Life threatening bleed	Factor VIII infusions until anamnestic response occurs, then aPCC or rVIIa infusions
High responder[c] with high inhibitor level (≥5 BU)	Minor bleed	PCC, aPCC or rVIIa infusions
	Major bleed	PCC, aPCC or rVIIa infusions

[a]Rise of inhibitor titer is slow to factor VIII challenge.
[b]Rise of inhibitor titer is rapid to factor VIII challenge.
[c]Hyate:C.
Abbreviations: BU, Bethesda units; PCC, prothrombin complex concentrates. Activated PCC, e.g., Autoplex (Hyland) and FEIBA (factor VIII inhibitor bypassing activity). Nonactivated PCC, e.g., Konyne (Cutter) and Proplex (Hyland).

Activated Prothrombin Complex Concentrate (aPCC) (Autoplex and FEIBA)

These products have increased amounts of activated factor VII (VIIa), activated factor X (Xa) and thrombin and are effective in patients even with high titer inhibitors (>50 BU). The initial dose is 75–100 units/kg, which can be repeated in 8–12 hours. Generally for a joint bleed 4–5 doses at 12-h intervals are employed after which risk of thrombogenicity is increased. Approximately 75% of patients with inhibitors respond to aPCC infusions. For some patients trace amounts of FVII in aPCC products may cause anamnesis of the inhibitor titer. If multiple doses are administered, the patient should be monitored for the development of DIC or even myocardial infarction. The simultaneous use of antifibrinolytic therapy (i.e., Amicar) should be avoided. Oral antifibrinolytic drug therapy 4–6 hours after the last dose of a PCC, however, appears to be well tolerated.

Recombinant Factor VIIa (NovoSeven)

Recombinant activated factor VII (rFVIIa) concentrate can be administered to achieve hemostasis in patients with high-titer inhibitors. The usual dose is 90 micrograms/kg rFVIIa repeated every 2 hours for 2–3 infusions. Higher doses up to 200 µg/kg for 2–3 doses have been used in pediatric patients in life-threatening situations to achieve better hemostasis. Recently, a single high dose of 270 µg/kg has been shown to be effective in the outpatient setting for the treatment of an acute bleed instead of using every 3–4-hour dosing. The subsequent frequency of infusion and duration of therapy must then be individualized, based on the clinical response and severity of bleeding. Early initiation of hemostatic treatment (within 8 h of a bleed) with rFVIIa produces response rates on the order of 90%. Treatment failures with conventional doses of rFVIIa may respond to higher doses. The incidence of thrombotic complications with this product has been low and anamnesis of the inhibitor does not occur.

Plasmapheresis with Immunoadsorption

When bleeding persists despite active treatment, extracorporeal plasmapheresis (4-liter exchange) over staphylococcal A columns may rapidly reduce the inhibitor titer by adsorbing out offending inhibitory IgG antibodies. This approach is cumbersome and may produce significant fever and hypotension due to the release of staphylococcal A protein from the solid-phase matrix of the chromatographic column. However, it can be life-saving in desperate situations and its efficacy can be enhanced by concomitant replacement therapy with FVIII-containing concentrates. Staphylococcal protein A columns for immunoadsorption is not readily available.

Immune-Tolerance Induction (ITI)

This intervention is instituted for inhibitor eradication in high-responder inhibitors and involves frequent administration of FVIII concentrate to induce immune tolerance to exogenous FVIII. Objectives of ITI are an undetectable inhibitor titer, restoring the ability to treat bleeds with FVIII concentrates and restoration of normal *in vivo* FVIII recovery and half life. A variety of regimens have been used for ITI (Table 13-16), including daily high-dose FVIII (100 U/kg twice daily or 200 U/kg daily regimen) with or without immunomodulatory therapy, daily intermediate dose FVIII (50–100 U/kg/d) and alternate day low-dose FVIII (25 IU/kg). Immune tolerance is eventually achieved in 60–75% of patients. Until tolerance is successfully attained, episodic bleeds require treatment with bypassing agents. A low historical peak inhibitor titer, a low inhibitor titer at initiation of ITI (<10 BU) and a low maximum inhibitor titer during ITI all favor success. Success rates may also be higher in young patients and in patients treated on higher-dose regimens. Cost and venous access are added obstacles to successfully completing immune tolerance. An ongoing International Immune Tolerance study randomizing patients to a high-dose (200 U/kg daily) or low-dose arm (50 U/kg three times a week) may provide better insight into ITI regimens since cost and venous access continue to remain challenging issues when conducting ITI.

Anti-CD20-antibody (Rituximab) has been used anecdotally with and without ITI with varying success rates. Larger studies are underway to determine efficacy and treatment toxicity.

Treatment of Factor IX Inhibitors

The frequency of inhibitory antibodies to factor IX is much lower (3–6%) than the frequency seen with factor VIII. The bypassing agents aPCC and rFVIIa have hemostatic efficacy in hemophilia B patients with inhibitors. Their use and dosages are the same as for the treatment of hemophilia A patients with inhibitors. Immune tolerance induction has had limited success in hemophilia B patients with inhibitors. However, many hemophilia B patients, typically those with a history of anaphylaxis to FIX concentrates, develop nephrotic syndrome in response to the frequent, high-dose, infusions of coagulation FIX concentrates, necessitating termination of the ITI effort. More recently, Rituximab (antiCD20 antibody) in combination with mycophenolate mofetil, dexamethasone and immunoglobulin has been shown to be successful in anecdotal cases along with ITI.

Spontaneously Acquired Inhibitory Antibodies to Coagulation Factors

Spontaneous auto-antibodies to factors VIII and IX and other coagulation factors may arise in nonhemophilia patients in a variety of clinical conditions.

Acquired Hemophilia A

Auto-antibodies (acquired inhibitors) may occur in nonhemophiliacs, resulting in acquired hemophilia A. These develop mainly in adults, more frequently in elderly subjects and in postpartum patients. Associated settings include pregnancy, immune diseases and hematologic neoplasms (CLL, lymphoma). Symptoms are hemorrhage, typically into soft tissues and mucous membranes with occasional cerebral hemorrhage. Diagnosis is based upon history (hemorrhage in the absence of prior patient or family history), prolonged PTT with failure to correct on mixing studies, reduced FVIII level and the presence of an anti-FVIII inhibitor on Bethesda assay. Affinity of acquired inhibitors for FVIII differs from that of inhibitors (allo-antibodies) associated with hemophilia A, resulting in complex kinetics and often incomplete inactivation of FVIII. Treatment options for minor bleeding episodes in patients with low-titer auto-antibodies include augmentation of endogenous FVIII using DDAVP or FVIII concentrates. For more severe bleeding associated with high-titer auto-antibodies hemostatic treatment options include porcine FVIII (if inhibitor titer against porcine FVIII is low), or bypassing agents as used for hemophilia A inhibitors (see above and Table 13-16). Plasmapheresis with or without a staphylococcal A column can be used to transiently lower the inhibitor titer, permitting effective FVIII replacement therapy. Immunosuppression-immunomodulation has been shown to be effective in eliminating auto-antibodies. Treatment modalities alone or in combination include high-dose IVIG infusion (400 mg/kg/day for 5 days), prednisone (1 mg/kg/day as a single agent), cyclophosphamide (2 mg/kg/day) or other agents (vincristine, azothioprine, etc.). Cyclosporine A and rituximab are other agents reported to be effective in elimination of auto-antibodies. Responses to any individual regimen are variable, requiring individualization of therapy. A thorough evaluation for autoimmune disease is warranted.

Acquired Antibodies to Other Coagulation Factors

Acquired auto-antibodies (inhibitors) to factor V are very rare. Some patients may have major hemorrhage (patients with antibodies that also bind to platelet factor V), others may not bleed, even at the time of major surgery. Available treatments for acute hemostasis include platelet transfusion (the factor V contained within the platelet α granule may be locally protected from the inhibitor) and the use of bypassing agents.

Auto-antibodies (inhibitors) to other factors (including prothrombin or factor XI) may occur, most commonly in the setting of SLE (either isolated or in addition to the lupus anticoagulant) or hematologic malignancy. Some patients, usually those with a profound decrease in prothrombin concentration, may have hemorrhagic symptoms. Acute hemostatic treatment strategies include infusion of fresh frozen plasma or prothrombin complex concentrates. Inhibitor elimination strategies, as for other acquired inhibitors, include the use of plasmapheresis, glucocorticoids and/or immunosuppression.

Auto-antibodies to factor XI are very rare in children. They have been reported following viral infections and are usually transient. Glucocorticoid therapy may be efficacious in their elimination.

Lupus Anticoagulant

See Chapter 14 on thrombotic disorders.

Table 13-16 Selected Immune Tolerance Induction Regimens for Hemophilia-Associated Inhibitors

Protocol	FVII Dose	Other Agents	% Success Rate (# Patients in Trial)	Inhibitor Elimination Time (Range in Months)
High dose:				
Bonn[a](Brackman et al.)	100–150 IU/kg twice daily	APCC as required	100 (60)	1–2
Malm[b](Berntorp/ Nilsson et al.)	Maintain FVIII>0.40 U/ml	Cyclophosphamide; dose is 12–15 mg/kg IV daily for 2 days followed by 2–3 mg/kg orally given daily for 8–10 days IV Ig; dose is 0.4 g/kg daily for 5 days	63 (16)	
Intermediate dose:				
Kasper/Ewing[c]	50–100 IU/kg daily	Oral prednisone PRN	79 (12)	1–10
Unuvar et al.[d]	50–100 IU/kg daily (induction phase) 25 IU/kg daily (reduction phase)		75 (14)	0–8
Low dose:				
Dutch[e] (Mauser-Bunschoten)	25 IU/kg alternate Days		87 (24)	2–28

References

[a] Oldenburg J, Schwaab R, Brackmann HH. Induction of immune tolerance in haemophilia A inhibitor patients by the "Bonn Protocol": predictive parameter for therapy duration and outcome. Vox Sanguinis 1999;77(suppl 1):49–54.

[b] Freiburghaus C, Berntorp E, Ekman M, et al. Tolerance induction using the Malmo treatment model 1982–1995. Haemophilia 1999;5:32–39.

[c] Ewing NP, Sanders NL, Dietrich SL, et al. Induction of immune tolerance to factor VIII in hemophilia A patients with inhibitors. JAMA 1988;259:65–68.

[d] Unuvar A, Warrier I, Lusher JM. Immune tolerance induction in the treatment of paediatric haemophilia A patients with factor VIII inhibitors. Haemophilia 2000;6:150–157.

[e] van Leeuwen EF, Mauser-Bunschoten EP, van Dijken PJ. Disappearance of factor VIII:C antibodies in patients with haemophilia A upon frequent administration of factor VIII in intermediate or low dose. Br J Haematol 1986;64:291–297.
Reproduced with permission of the Resource Group and Dannemiller Memorial Education Foundation.

Von Willebrand Disease

Von Willebrand disease (vWD) is an autosomally inherited congenital bleeding disorder caused by a deficiency (type 1), dysfunction (type 2) or complete absence (type 3) of von Willebrand factor (vWF). vWF has two functions:

* Plays an integral role in mediating adherence of platelets at sites of endothelial damage, promoting formation of the platelet plug (Figure 13-1) and
* Binds and transports FVIII, protecting it from degradation by plasma proteases.

vWF is a large multimeric glycoprotein that is synthesized in megakaryocytes and endothelial cells as pre-pro-vWF. Sequential cleavage releases mature vWF which undergoes multimerization and is stored in specific cellular storage granules such as the Weibel–Palade body in endothelial cells and the α granule in platelets. It is present in normal amounts in plasma and levels can be significantly increased by administering drugs such as desmopressin (DDAVP) that induce the release of vWF from storage sites into plasma. Deficiency of vWF results in mucocutaneous bleeding and prolonged oozing following trauma or surgery. vWD is the most common hereditary bleeding disorder, with biochemical evidence present in 1–2% of the population and biochemical evidence combined with a bleeding history of 0.1% of the population. Differences between vWD and hemophilia are displayed in Table 13-17. Table 13-18 shows the classification of vWD, Figure 13-7 shows the structure of vWF and indicates the location of mutations giving rise to variants of vWD.

Treatment of vWD

Recommended treatment of vWD is indicated in Table 13-19.

Type 1 vWD

This is the most common form of the disorder and is characterized by a mild to moderate decrease in the plasma levels of vWF. Plasma vWF from these individuals has a normal structure. Levels of ristocetin cofactor activity (R:Co or vWF activity) and vWF antigen

Table 13-17   Differences Between von Willebrand Disease and Hemophilia A

	von Willebrand Disease	Hemophilia A
Symptoms	Bruising and epistaxis	Joint bleeding
	Menorrhagia or mucosal bleeding	Muscle bleeding
Sexual distribution	Males = females	Males
Frequency	1:200 to 1:500	1:6000 males
Abnormal protein	vWF	Factor VIII
Molecular weight	$0.6–20 \times 10^6$ Da	280 kDa
Function	Platelet adhesion	Clotting cofactor
Site of synthesis	Endothelial cell or megakaryocytes	??
Chromosome	Chromosome 12	X chromosome
Inhibitor frequency	Rare	14–25% of patients
Laboratory tests		
History	Abnormal	Abnormal
aPTT	Normal or prolonged	Prolonged
Factor VIII	Borderline or decreased	Decreased or absent
vWF Ag	Decreased or absent	Normal or increased
vWF R:Co	Decreased or absent	Normal or increased
vWF multimers	Normal or abnormal	Normal

From: Montgomery RR, Gill JC, Scott JP. Hemophilia and von Willebrand disease. In: Nathan D, Orkin S, editors. Nathan and Oski's Hematology of Infancy and Childhood, 5th ed. Philadelphia: Saunders, 1998:1631–1675, with permission.

Table 13-18 Classification of von Willebrand Disease

	Type 1	Type 2A	Type 2B	Type 2N	Type 2M	Type 3	Platelet-type Pseudo-vWD
Genetic transmission	AD	AD	AD	AR	AD	AR	AD
Frequency (%)	70–80	10–12	3–5	1–2	1–2	1–3	0–3
Platelet count	N	N	N/decreased	N	N	N	decreased
FVIII	Decreased	N	Normal	Markedly decreased	Normal	Absent	N
VWF: Ag	Decreased	Decreased	N/decreased	Decreased	N	Absent	N/decreased
R:Co (VWF activity)	Decreased	Decreased	N/decreased	Decreased	Decreased	Absent	N/decreased
RIPA	N/decreased	Decreased	N	Normal	Decreased	Absent	N
RIPA-low dose	Absent	Absent	Increased	Absent	Absent	Absent	N
Multimeric structure	N	Absence of large multimers from plasma and platelets	Reduced large multimers from plasma	N	N	Absent	Reduced large multimers
Response to DDAVP							
FVIII, vWFAg, R:Co	Increases	Increases	Increases	Increases	Increases	NR	Increases
BT	N	Shortens	NR/shortens[a]	N	Shortens	NR	NR

[a]Causes platelet aggregation.

Abbreviations: AD, autosomal dominant; N, normal; vWF-Ag, von Willebrand's factor antigen (FVIII-related antigen); RIPA, ristocetin-induced platelet aggregation; RIPA-LD, low-dose RIPA; NR, no response.

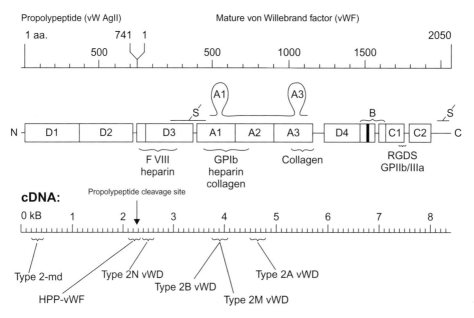

Figure 13-7 Protein Structure of vWF and its Large Propolypeptide, von Willebrand's Antigen II (vW AgII).

Domains of the vWF molecule have high degrees of similarity and various protein interactions have been localized to specific regions such as the interaction with platelet glycoprotein Ib or glycoprotein IIb/IIIa. This latter site may be related to an Arg-Gly-Asp-Ser (RGDS) sequence located in the C1 domain. The lower portion of this figure represents the clusters of complementary DNA mutations that cause some of the variants of von Willebrand's disease, including types 2A, 2B, 2M and 2N, as well as the less common variants that prevent propolypeptide cleavage (hereditary persistence of pro-vWF [HPP-vWF]) or a variant that prevents N-terminal multimerization (type 2-md). From: Montgomery RR, Gill JC, Scott JP. Hemophilia and von Willebrand Disease. In: Nathan DG, Orkin SH, editors. Nathan and Oski's Hematology of Infancy and Childhood. 5th ed. Philadelphia Saunders, 1998: 1631–1675, with permission.

tend to be decreased in parallel. DDAVP can be used to manage most hemostatic problems in patients with type I vWD. DDAVP treatment generally normalizes FVIII, vWF and the bleeding time. The standard dose may be repeated daily as necessary. Response to DDAVP should be assessed for each individual patient. Tachyphylaxis with repeated doses of DDAVP needs to be considered when prescribing DDAVP for a major surgical procedure when prolonged high levels of vWF may be needed to achieve hemostasis and promote healing. As described previously in the hemophilia treatment section a 3–5-fold rise above baseline levels for vWF antigen and ristocetin co-factor activity can be expected necessitating a DDAVP trial in an individual before a surgical procedure to document response.

Table 13-19 Recommended Treatment of von Willebrand Disease

	Type 1	Type 2A	Type 2B	Type 2N	Type 2M	Type 3	Platelet-type Pseudo-vWD
Severe hemorrhages; major surgical procedures	DDAVP	FVIII:vWF concentrates	FVIII:vWF concentrates	FVIII:vWF concentrates	FVIII:vWF concentrates	FVIII:vWF concentrates	Platelet transfusion
Mild hemorrhages; minor surgical procedures	DDAVP	FVIII:vWF concentrates (may respond to DDAVP)	FVIII:vWF concentrates	FVIII:vWF concentrates (may respond to DDAVP)	FVIII:vWF concentrates (may respond to DDAVP)	FVIII:vWF concentrates	Platelet transfusion
Oral surgical procedures	DDAVP and EACA	FVIII:vWF concentrates and EACA	FVIII:vWF concentrates and EACA	FVIII:vWF concentrates and EACA	FVIII:vWF concentrates and EACA	FVIII:vWF concentrates and EACA	Platelet transfusion and EACA

Abbreviations: DDAVP, desmopressin; EACA, amicar; Factor VIII: vWF concentrated; see Table 13-13.

Type 2A vWD

This variant is associated with decreased platelet dependent vWF function and a lack of large high-molecular-weight (HMW) multimers in plasma and platelets. In some patients with type 2A vWD, abnormalities have been localized to the A2 domain of vWF (Figure 13-7) and specific gene defects have been described. There is either abnormal synthesis or increased proteolysis of vWF due to the genetic defect. A much lower level of ristocetin cofactor activity as compared to the vWF antigen and FVIII level should raise suspicion for type 2 vWD. DDAVP treatment usually increases the FVIII level in type 2A, the bleeding time may improve in some patients. DDAVP may suffice to control some types of bleeding in patients with this variant. Other patients will require treatment with clotting factor concentrates containing both FVIII and vWF. Use of cryoprecipitate for vWF replacement therapy should not be utilized because this product does not undergo viral inactivation treatment.

Type 2B vWD

This variant is a rare form of vWD characterized by dominant gain in function mutations in the A1 domain of the vWF gene. The high-molecular-weight (HMW) multimers bind spontaneously to platelets and are continuously removed from circulation. There may be mild thrombocytopenia due to spontaneous agglutination of the platelets. *In vitro* this variant is characterized by a reduced threshold for aggregation of platelets upon exposure to the antibiotic ristocetin. A low-dose ristocetin-induced platelet aggregation (low-dose RIPA) in which platelet-rich plasma from patients with type 2B vWD will aggregate but not type 2A. DDAVP is contraindicated in type 2B vWD because of transient thrombocytopenia resulting from release and clearance of abnormal vWF. Clotting factor concentrates containing both FVIII and vWF are the mainstay of treatment for this vWD variant.

Type 2N vWD

Type 2N vWD is a result of an abnormal vWF molecule that does not bind factor VIII due to a genetic defect in the FVIII binding domain of the vWF molecule (Figure 13-7). The unbound factor VIII is rapidly cleared from circulation resulting in reduction of plasma factor VIII as compared with the level of vWF. Some of these patients may be misdiagnosed as having mild hemophilia A since FVIII levels tend to be in the 8–10% range. It appears that the phenotype is expressed only in the presence of a second allele that carries type 1 vWD (compound heterozygous) or in the rare situation of recessive inheritance. vWD FVIII levels increase following DDAVP infusion but the released FVIII circulates only for a short time because of impaired binding to vWF. For the same reason the half-life of infused high-purity factor VIII is markedly shortened. For major bleeding or surgery, the recommended treatment is a factor VIII concentrate containing high levels of vWF.

Type 2M vWD

Type 2M vWD results from of an abnormal binding site on vWF for platelet GP Ib, resulting in reduced ristocetin cofactor (R:Co) activity (functional defect). In this variant multimers of all sizes are present. The vWF protein is released by DDAVP so that patients heterozygous for this variant may respond clinically to standard doses of DDAVP. For homozygotes the vWF released by DDAVP is defective and a poor clinical response occurs. Major bleeding episodes or surgery should be managed with vWF replacement therapy.

Type 3 vWD

This vWD variant occurs in patients with homozygous or doubly heterozygous null mutations or deletions. The clinical phenotype is a severe bleeding disorder with major deficits in both primary and secondary hemostasis. The plasma level of FVIII and vWF is virtually undetectable. Patients are unresponsive to DDAVP (no releasable vWF stores) and require episodic treatment with vWF containing FVIII concentrates. Alloantibodies that inactivate von Willebrand factor develop in 10–15% of patients with type 3 disease who have received multiple transfusions. Administration of FVIII concentrates containing vWF to these patients is contraindicated since life-threatening anaphylactic reactions may result. In this setting, administration of recombinant FVIII, which is devoid of vWF, can raise FVIII levels to hemostatic levels. In the absence of vWF the half-life of the infused FVIII will be short (1–2 hours) and administration of high doses, at short intervals or by continuous infusion, will be required. Administration of rFVIIa at a dose of 90 μg/kg every 2 hours has produced hemostasis for some of these patients.

Platelet-type Pseudo–von Willebrand Disease

This platelet disorder has a phenotype similar to type 2B vWD. Platelet-type pseudo-vWD is due to a gain-of-function mutation defect in the *GPIBA* gene which causes the production of an abnormal GP1b-1X in the platelet GP1b receptor on platelets. Excessive binding of vWF to platelet GP Ib recepter causes platelet activation and vWF removal from the circulation, plasma concentrations of vWF are reduced and platelet aggregation is increased. Bleeding in this disorder should be treated with platelet transfusions.

Acquired von Willebrand Disease

Acquired von Willebrand disease may present as a marked reduction in levels of von Willebrand antigen in a person who does not have a lifelong bleeding disorder. The onset has been associated with a variety of conditions including Wilms tumor, polycythemia vera, other neoplasms, autoimmune diseases (e.g., systemic lupus erythematosus [SLE]), myeloproliferative disease, lymphoproliferative disorders, use of various drugs, as well as in individuals with angiodysplastic lesions. Proposed mechanisms include specific auto-antibodies, adsorption onto malignant cell clones and depletion in conditions of high vascular shear force. Therapeutic interventions are directed at controlling acute bleeding episodes, treating the underlying disorder in the hope of correcting the abnormalities of vWF and effecting antibody elimination. DDAVP infusion is the initial treatment of choice for achieving hemostasis, plasma-derived FVIII/vWF concentrates are the second choice. In either case clearance of the von Willebrand factor will be accelerated and levels must be monitored. IVIG, plasmapheresis and/or immunosuppressive drugs may be useful for eliminating antibody.

RARE INHERITED COAGULATION FACTOR DISORDERS (FACTOR II, V, VII, X, XIII, FIBRINOGEN DEFICIENCIES)

The exact prevalence of these disorders is unclear due to lack of epidemiological data. They seem to occur in a relatively low frequency and most of them are autosomal recessive in inheritance except for dysfibrinogenemias, which are autosomal dominant. They comprise 3–5% of inherited bleeding disorders. The natural history and spectrum of clinical features are not well established. Severe deficiencies are seen in populations with a high rate of consanguineous marriages (Iran, India, 10–20 times greater incidence in consanguinity). Clinical manifestations are usually seen in homozygous or compound heterozygous individuals and rarely in the heterozygous state. Patients generally present with mild to moderate bleeding tendency with a potential for severe bleeding. Bleeding manifestations tend to be less severe than hemophilia B with less frequent gastrointestinal (GI), genitourinary (GU), central nervous system (CNS) and musculoskeletal (MSK) bleeds. Severe bleeding and MSK bleeds occur with fibrinogen disorders, FX, FII and FXIII deficiencies. GI and CNS bleeds are commonly seen with FX deficiency while umbilical cord bleeds are seen with fibrinogen, FII, V and X deficiencies. Mucocutaneous bleeds are the most common bleeding sites with menorrhagia occurring in 50% of women with rare bleeding disorders. They are often diagnosed for the first time in adulthood after prolonged postoperative bleeding.

Diagnostically the screening PT and/or aPTT are abnormal, except in FXIII deficiency. Specific assays are required to measure activity and immunoassays for antigen to differentiate type 1 (hypoproteinemias) and type 2 (dysproteinemias) deficiencies. Genetic mutations have been identified in most deficiencies. These mutations are unique to each kindred. Deletions, null, nonsense and missense mutations have been identified and in 10–20% no mutation has been identified. A genotype–phenotype correlation is not well established. Homozygous individuals present with clinically significant bleeding while heterozygous individuals generally have no bleeding except with FVII and FXI deficiencies where severe bleeding may be observed in the heterozygous state.

Table 13-20   Treatment of Rare Coagulation Factor Deficiencies[a]

Factor Deficiency	Half-life (h)	Percentage Increase in Plasma Concentration after 1 unit/kg IV	Percentage Required for Hemostasis	Percentage Required for Minor Trauma	Percentage Required for Surgery and Major Trauma
Afibrinogenemia	56–82	1–1.5[a]	80[a]	150[a]	200[b]
II	45–60	1	20–30	30	50
V	36	1.5	10–15	10–15	25
VII	5	1	10–15	10–15	20
VWD[c]	24–48	1.5[d]	10	20–30	50
X	24–60	1	10–15	10–15	25
XI	48	1	30	30	>45
XIII	168–240	1	2–3	15	25

[a]See Table 13-14 for factor VIII and factor IX deficiency treatment guidelines.
[b]In mg/dl.
[c]vWD, von Willebrand's disease.
[d]Calculated in units of VWF:Rcof activity.

Purified concentrates are not readily available for treatment. Dosing is based on minimal hemostatic level and half-life of the particular factor. Fresh frozen plasma can be utilized for factor V deficiency, fibrinolytic factor deficiency and for most disorders when a diagnosis is not clearly established at first encounter in an emergency. Prothrombin complex concentrates can be used for FII and X deficiencies and cryoprecipitate can be used for FXIII and fibrinogen disorders when concentrates are not readily available. Purified concentrates are generally used for fibrinogen (recently licensed in the U.S.), FXIII (available as an Investigations New Drug [IND] status of the U.S. Food and Drug Administration) and in factor VII deficiency (recombinant products).

Antifibrinolytics such as epsilon aminocaproic acid and tranexamic acid may be used as adjuvant therapy for skin and mucocutaneous bleeds. Caution needs to be exercised when used concurrently with prothrombin complex concentrates. Oral contraceptive pills are used in women for menorrhagia in these bleeding disorders to control bleeding.

Table 13-20 lists the treatment of rare coagulation deficiencies.

HEREDITARY DISORDERS OF PLATELET FUNCTION

Qualitative disorders of platelet function usually manifest with mucocutaneous bleeding. Symptoms include petechiae, echymoses, epistaxis, menorrhagia, gastrointestinal bleeding and abnormal bleeding in association with injury or trauma. Physical examination may reveal associated abnormalities such as nystagmus and oculocutaneous albinism in the

Hermansky–Pudlack syndrome or eczema in the Wiskott–Aldrich syndrome (see Chapter 12). Laboratory evaluation includes morphologic inspection of the blood smear for platelet size, inclusions and presence of a gray or washed-out appearance. Assessment of platelet function by bleeding time, platelet aggregation studies, or PFA-100 closure time will usually be abnormal. Definitive diagnosis of suspected disorders will often require additional studies available only at reference laboratories (e.g. flow cytometry enumeration of receptors, electron microscopy, etc). Physiologically these disorders may be divided into those caused by abnormal platelet receptors and those with defects in granule content or storage pool release. See Chapter 12 for a detailed discussion on these conditions.

Suggested Reading

Acharya SS, Coughlin A, DiMichele DM. The North American Rare Bleeding Disorder Study Group. Rare Bleeding Disorders Registry Deficiencies of Factors II, V, VII, X, XIII, Fibrinogen and Dysfibrinogenemias. J Thromb Haemost. 2004;2:248–256.

Acharya, SS, DiMichele DM. Management of FVIII inhibitors. Francis CW, Blumberg N, Heale J, eds. Best Practice & Research. Clinical Haematology 2006;19(1):51–66.

Andrew M, Vegh P, Johnston M, et al. Maturation of the hemostatic system during childhood. Blood. 1992;80:1998.

Bick R. Disseminated intravascular coagulation: objective criteria for clinical and laboratory diagnosis and assessment of therapeutic response. Clin. Appl. Thrombosis/Hemostasis. 1995;1:3–23.

Hoffman M, Monroe III DM. A cell-based model of hemostasis. Thromb Haemost. 2001;85:958–965.

Hoyer LL. Hemophilia A. N Engl J Med. 1994;330:38–47.

Manco-Johnson MJ, et al. Prophylaxis versus episodic treatment to prevent joint disease in boys with severe hemophilia. N Engl J Med. 2007;357(6):535–544.

Mannucci P. Treatment of von Willebrand disease. N Engl J Med. 2004;351:683–694.

Mannucci PM. Desmopressin: a nontransfusional form of treatment for congenital and acquired bleeding disorders. Blood. 1988;72:1449–1455.

Nichols WL, et al. von Willebrand disease (VWD): evidence-based diagnosis and management guidelines, the National Heart, Lung and Blood Institute (NHLBI) Expert Panel report (USA) 1. Haemophilia. 2008;14(2): 171–232.

CHAPTER 14

Thrombotic Disorders

MECHANISMS OF THROMBOSIS IN INHERITED THROMBOPHILIA

In inherited thrombophilias, impaired neutralization of thrombin or a failure to control the generation of thrombin causes thrombosis. There is a malfunction in a system of natural anticoagulants that maintain the fluidity of the blood. Decreases in antithrombin III activity impair the neutralization of thrombin and reduced activity of protein C or protein S impairs or increases thrombin generation. Both of these mechanisms increase susceptibility to thrombosis (Figure 14-1).

The control of thrombin generation is also compromised by mutations in the gene for factor V or prothrombin. The Arg506Gln substitution in factor V Leiden involves the first of three sites on factor Va that are cleaved by activated protein C. This mutation slows down the proteolytic inactivation of factor Va, which in turn leads to the augmented generation of thrombin. Moreover, the mutant factor V has diminished cofactor activity in the inactivation of factor VIIIa by activated protein C. Both these abnormalities in factor V cause the *in vitro* phenomenon of resistance to activated protein C, resulting in the failure of activated protein C to prolong the activated partial thromboplastin time. For unknown reasons, the G20210A mutation in the 3 untranslated region of the prothombin gene is associated with increased levels of plasma prothrombin ($\approx$130% by either antigen or activity assays), an effect that promotes the generation of thrombin and impairs the inactivation of factor Va by activated protein C. The mechanisms by which increased levels of factor VIII, factor IX, factor XI, fibrinogen and homocysteine enhance venous thrombosis are incompletely understood.

Table 14-1 lists the clinical manifestations of a hypercoagulable state and laboratory findings in hypercoagulable states are listed in Table 14-2.

VENOUS THROMBOSIS

Venous thrombotic events develop under conditions of slow blood flow. It may occur by way of activation of the coagulation system with or without vascular damage. Venous thrombi are composed of large amounts of fibrin containing numerous erythrocytes,

Manual of Pediatric Hematology and Oncology. DOI: 10.1016/B978-0-12-375154-6.00014-8

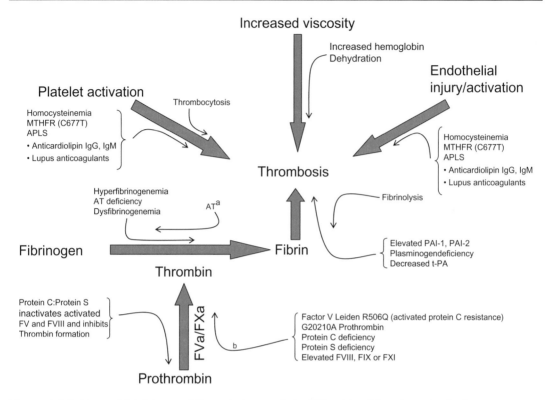

Figure 14-1 Natural Inhibitors of Coagulation and the Effect(s) of Pro-thrombotic States.
[a] Leads to neutralization of thrombin; [b] Decreased neutralization or increased generation of thrombin.

platelets and leukocytes (red thrombus). The incidence of a venous thrombotic event in children is estimated to be between 0.7 to 1.9 per 100,000 children. Venous thrombosis usually produces significant obstruction to blood flow. The most serious consequence is embolization from a deep vein thrombosis resulting in pulmonary embolism. Table 14-3 lists the predisposing causes of venous thrombosis. The detection of venous thrombosis and diagnosis of pulmonary embolism are given in Table 14-4 and Table 14-5, respectively.

Newborns are at a higher risk for a venous thrombotic event because of decreased activity of anticoagulant factors, specifically antithrombin, protein C and protein S. Fibrinolytic activity in the newborn period is also decreased by lower serum plasminogen levels. Renal, caval, portal and hepatic venous system thrombosis are well-known complications of peripartum asphyxia, sepsis, dehydration and maternal diabetes. The estimated annual incidence of venous thrombotic events is about 0.5 per 10,000 newborns. However, the majority of venous thrombotic events within the first year of life are associated with central venous access devices.

The treatment of venous thrombotic events is given in Table 14-6.

Table 14-1   Clinical Manifestations of Hypercoagulable States

Family history of thrombosis
Recurrent spontaneous thromboses
Thrombosis in unusual sites
Thrombosis at an early age — Myocardial infarcts, strokes, deep vein thromboses
Resistance to anticoagulant therapy
Warfarin-induced skin necrosis syndrome
Recurrent spontaneous abortions
Spinal surgeries associated with prolonged immobilization and thrombosis
Thrombosis associated with long airplane flights
Northern European ancestry
Thrombosis during pregnancy/oral contraceptives
Migratory superficial thrombophlebitis
Antiphospholipid syndrome
Autoimmune disorders
   Inflammatory bowel disease
   Lupus erythematosus (SLE)
   Behcet's disease
Malignancy
Nephrotic syndrome
Infections
   Varicella
   HIV
   Suppurative thrombophlebitis

ARTERIAL THROMBOSIS

Arterial thrombosis initially occurs under conditions of rapid blood flow and often is the result of a process that damages the vessel wall. The thrombus is composed of tightly coherent platelets that contain small amounts of fibrin and few erythrocytes and leukocytes (white thrombus). The most serious consequence of arterial thrombosis is vascular occlusion. Arterial thrombotic events occur in a number of congenital and acquired diseases (Table 14-7).

Arterial Catherization

The most common cause of an arterial thrombotic event in children is the use of arterial vascular catheters. Without prophylactic anticoagulation, the incidence of an arterial thrombotic event from femoral artery catherization is about 40%. Anticoagulation with heparin (100–150 U/kg) reduces the incidence of arterial thrombotic events to 8%. Incidence of an arterial thrombotic event from an umbilical arterial catheter is markedly reduced by a low-dose continuous heparin infusion (3–5 units/h).

Table 14-2 Laboratory Findings in Hypercoagulable States

Primary Hemostasis
Thrombocytosis
Platelet aggregation
  Hyperaggregation
  Spontaneous aggregation
  Circulating platelet aggregates
Increased platelet adhesiveness
Elevated levels of β-thromboglobulin, vWF and PF4
Short platelet life span
Secondary Hemostasis
Shortened PT/aPTT
Elevated coagulation factors (i.e., I, II, V, VII, VIII, IX, X, XI)
Reduced antithrombin, heparin cofactor II, plasminogen, tissue plasminogen activator, protein C, protein
  S, FXII, prekallikrein
Factor V Leiden/APC resistance
Prothrombin gene mutation (G20210A)
Elevated α-2-antiplasmin, α-2-macroglobulin, plasminogen activator inhibitor-1 (PAI-1)
Increased lipoprotein (a)
Thrombomodulin deficiency
Presence of fibrinopeptide A (FPA)
Increased fibrin degradation products (some measured as the D-dimer)
Short fibrinogen life span
Dysfibrinogenemia — diagnosed by prolonged thrombin and reptilase times
Homocysteinemia
Presence of lupus anticoagulants
Presence of anticardiolipin antibodies
Presence of beta 2 glycoprotein 1 antibodies

Cardiac Procedures

Blalock-Taussig Shunts

The short length and very high flow in these shunts may result in arterial thrombosis. The incidence varies from 1–17 percent. For anticoagulant management see p. 459.

Fontan Operation

In this procedure the incidence of an arterial thrombotic event ranges from 3–19%. Arterial thrombotic events may occur anytime following the surgery and it is the major cause of early and late morbidity and mortality. These children may need to be on life-long anticoagulation. For anticoagulant management see p. 458.

Endovascular Stents

These stents are used increasingly to manage some patients with congenital heart disease (e.g. pulmonary artery stenosis, pulmonary vein stenosis, coarctation of aorta). Therapeutic

Table 14-3   Predisposing Factors for Venous Thrombosis

Acquired[a]

Age

Trauma

Venipuncture

Indwelling venous access devices (intravenous catheters)

Surgery

Thermal injury

Immune complexes

Infections (HIV, varicella, supportive thrombophlebitis)

Severe dehydration

Shock

Prolonged immobilization

Cancer

L-Asparaginase therapy, Prednisone

Oral contraceptives or hormone replacement therapy

Cyanotic heart disease

Pregnancy and the puerperium

Nephrotic syndrome

Liver disease

Neonatal asphyxia

Infant of a diabetic mother

Vita,in B_{12}/Folic Acid deficiency (due to associated hyperhomocysteinemia)

Antiphospholipid syndrome

   Primary

   Secondary

     Systemic lupus erythematosus

     Rheumatoid arthritis

Resistance to activated protein C not due to factor V Leiden

Systemic lupus erythematosus (SLE)

   Anticardiolipin antibodies

   Lupus anticoagulants

Hyperviscosity syndromes

   Myeloproliferative disorders (Chapter 16)

   Polycythemia vera

   Chronic myelogenous leukemia

   Essential thrombocythemia (Chapter 12)

   Paroxysmal nocturnal hemoglobinuria (Chapter 7)

Hyperleukocytosis (acute leukemias)

Sickle cell disease, thalassemia intermedia not on chronic transfusion program

Infusion of concentrated vitamin K–dependent (II, VII, IX and X) factors (activated

   prothrombin complex concentrates) or fresh frozen plasma (FFP) frequently without close monitoring

Congenital/inherited[b]

R506Q mutation in the factor V gene ((factor V Leiden) – activated protein C resistance)

G20210A mutation in the prothrombin (factor II) gene

Antithrombin (AT) deficiency

Protein C deficiency

Protein S deficiency

(Continued)

Table 14-3 (Continued)

Plasminogen deficiency
Reduced levels of tissue thromboplastin activator (t-PA)
Dysfibrinogenemia
FXII (Hageman factor) deficiency
Prekallikrein deficiency
Heparin cofactor II (HC-II) deficiency
Hyperhomocysteinemia (due to C677T MTHFR or folate or vitamin B_{12} deficiency)
Homozygous homocystinuria
Increased levels of factor VIII, factor IX, factor XI or fibrinogen
Increased thrombin activatable fibrinolysis inhibitor (TAFI)

[a]These acquired factors increase the risk of venous thrombosis in inherited thrombophilia.
[b]The first thrombotic event may occur at an early age in patients who have more than one thrombophilia or who are homozygous for Factor V Leiden or the G20210A prothrombin gene mutation or asymptomatic heterozygotes who are relatives of patients with inherited thrombophilia who have had venous thrombosis. The combination of hyperhomocysteinemia with either factor V Leiden or G20210A prothrombin gene mutation significantly increase the risk of venous thrombosis. Recurrent thrombosis is more common in those with more than one thrombophilia.

Table 14-4 Detection of Venous Thrombosis

Clinical assessment
   Symptoms and signs of deep vein thrombosis
   Presence or absence of an alternative diagnosis
   Presence and number of predisposing factors for venous thrombotic event (Table 14-3)
Venous ultrasound
Impedance plethysmography
D-dimer blood testing
   High sensitivity and negative predictive value for PE more than deep vein thrombosis
Venogram
MRI/magnetic resonance venography (MRV)/CT venography especially for sinus venous thrombosis where
   CT scan may not be accurate

Table 14-5 Diagnosis of Pulmonary Embolism

Clinical assessment – typical signs and symptoms
High-resolution helical CT scan with good sensitivity and specificity
Ventilation–perfusion lung scanning
Pulmonary angiography
   The gold standard for the diagnosis of pulmonary emboli
Risk factors
   Recent surgery
   Immobilization (>3 days bed rest)
   Previous deep vein thrombosis or pulmonary emboli
   Lower extremity plaster cast
   Lower extremity paralysis
   Strong family history for deep vein thrombosis/pulmonary emboli
   Cancer
   Post-partum

Table 14-6   Treatment of Venous Thrombotic Events and Thromboembolism

Treatment	Indications
Anticoagulant therapy	Majority of patients with a venous thrombotic event (see anticoagulant therapy)
Thrombolytic therapy	Massive pulmonary embolism with hemodynamic compromise; patients with extensive iliofemoral thrombosis (see thrombolytic therapy)
Intracaval filter	Pulmonary embolism despite adequate anticoagulant therapy
Pulmonary embolectomy	Massive pulmonary embolism despite thrombolytic therapy

doses of heparin are given at the time of stent insertion, followed by aspirin therapy (5 mg/kg/day).

Renal Artery Thrombosis

Renal arterial thrombosis is commonly associated with kidney transplantation. The incidence is about 0.2–3.5% in children. Prophylactic administration of low-molecular-weight heparin (LMWH) 0.0.5 mg/kg twice daily for 21 days is effective anticoagulant therapy.

Hepatic Artery Thrombosis

Hepatic artery thrombosis is a very serious complication of liver transplantation. It usually occurs within 2 weeks. The reported incidence in children can be as high as 42%. The detection of an hepatic artery thrombosis includes Doppler, angiography and CT of the liver. The prophylactic use of LMWH and aspirin is controversial. The mortality from hepatic artery thrombosis may be as high as 70% in children. Anticoagulant therapy alone is not effective in hepatic artery thrombosis. Surgical intervention and retransplantation usually are necessary. Thrombolytic therapy has been successful in about 20% of children.

Kawasaki Disease

The acute phase of Kawasaki disease may be associated with arteritis, arterial aneurysms, valvulitis and myocarditis. The incidence of coronary artery aneurysm, stenosis, or thrombosis is about 25% without initial treatment. Early use of IV gamma-globulin and aspirin reduces the coronary artery involvement of Kawasaki disease.

Antiphospholipid Antibody Syndrome (APLS)

APLS may result in an arterial thrombotic event including stroke in children. The lupus anticoagulants and APLS are discussed on page 430–436.

Table 14-7   Predisposing Causes of Arterial Thrombosis

Acquired
  Catherization
    Cardiac
    Umbilical artery
    Renal artery–kidney transplantation
    Hepatic artery–liver transplantation
  Vascular
    Injury
    Infections
    Periarteritis nodosa
    Systemic lupus erythematosus (SLE)
    Kawasaki disease
    Hemolytic–uremic syndrome (HUS)
    Thrombotic thrombocytopenic purpura (TTP)
  Cardiac
    Blalock–Taussig shunts
    Fontan operation
    Endovascular stents
    Cyanotic congenital heart disease
    Primary endocardial fibroelastosis
    Enlarged left atrium with arterial fibrillation
    Hypertension
    Myocarditis
  Hematologic/Hemostatic
    Elevated levels of LP(a)
    Elevated PA1-1
    Reduced t-PA
    Elevated level of fibrinogen
    Antiphospholipid syndrome (APLS)
    Chronic disseminated intravascular coagulation (DIC)
    Elevated levels of fibrin degradation products (D-dimers)
    Hyperhomocysteinemia
    Hypereosinophilic syndrome
    Myeloproliferative disorders (MDS)
    Polycythemia vera (PV)
    Chronic myelogenous leukemia (CML)
    Essential thrombocythemia (ET)
    Paroxysmal nocturnal hemoglobinuria (PNH)
    Hyperleukocytosis (Acute leukemia)
    Sickle cell disease. Hemoglobin SC disease, thalassemia intermedia not on chronic transfusions
    Thrombotic thrombocytopenic purpura (TTP)
    Vitamin B_{12}/Folic acid deficiency (due to associated hyperhomocysteinemia)
    Activated prothrombin complex concentrate (aPCC)administration without monitoring
      for activation of coagulation
  Other
    Shock
    Nephrotic syndrome
    Diabetes mellitus

(Continued)

Table 14-7   (Continued)

Hyperlipidemia
Hypercholesterolemia
Cigarette smoking
Elevated CRP
Malignancy
Obesity
Physical inactivity

Congenital/inherited

R506Q mutation in the factor V gene [(factor V Leiden) – Activated Protein C Resistance]
G20210A mutation in the prothrombin (factor II) gene
Homozygous C677T mutation in the methylenetetrahydrofolate reductase gene (C677T MTHFR)
Antithrombin (AT) deficiency
Protein C deficiency
Protein S deficiency
Marfan syndrome
Familial hypercholesterolemia
Mitral valve prolapsed

Abbreviations: Lp(a), lipoprotein a; PAI-1, plasminogen activator inhibitor-1; t-PA, tissue thromboplasminogen activator; CRP, C-reactive protein.

Specific Risk Factors

Homocysteine

Homocysteine is a sulfur-containing amino acid formed during the metabolism of methionine, an amino acid derived from dietary protein. When excess methionine is present, homocysteine is linked with serine to form cystathionine in a reaction catalyzed by *cystathionine β-synthetase*, a vitamin-B_6-dependent enzyme. Otherwise homocysteine acquires a methyl group from N^5-methylhydrofolate in a reaction catalyzed by the vitamin B_{12}-dependent enzyme *methionine synthetase*. Hyperhomocysteinemia, a risk factor for venous thrombosis, can be caused by genetic disorders affecting the trans-sulfuration or remethylation pathways of homocysteine metabolism, or by folic acid deficiency, vitamin B_{12} deficiency, vitamin B_6 deficiency, renal failure, hypothyroidism, increasing age and smoking. A rare example of excessive hyperhomocysteinemia is homozygous homocystinuria due to cystathionine β-synthetase deficiency; 50% of affected patients present with venous or arterial thrombosis by the age of 29 years. Homozygosity for the C677T mutation in the methylenetetrahydrofolate reductase gene is a cause of mild hyperhomocysteinemia.

It has been shown that excess homocysteine has the following effects:

- Toxic effect on the endothelium
- Promotes thrombosis by platelet activation
- Causes oxidation of LDL cholesterol

- Increases levels of von Willebrand factor and thrombomodulin
- Increases smooth muscle proliferation.

The mean total homocysteine level in the newborn in various studies are 5.84 μmol/L, 7.4 μmol/L and 7.8 μmol/L. Daily folic acid (2.5 mg), vitamin B_6 (50 mg) and vitamin B_{12} (1 mg) lower plasma homocysteine levels. In countries with folate-fortification of various food items, hyperhomocystinemia is rarely found.

Lipoprotein (a) [Lp(a)]

Lp(a) is a distinct serum lipoprotein composed of a low-density lipid particle with disulfide links along a polypeptide chain (*apolipoprotein a*). Lp(a) regulates fibrinolysis by competing with plasminogen. Elevated levels of Lp(a) are associated with coronary artery disease and are an independent risk factor for spontaneous stroke and venous thrombosis in childhood. Measurement of Lp(a) levels needs to be performed in a reference laboratory for accurate interpretation of results and the sample should be drawn in a fasting state.

Fibrinogen

Higher levels of fibrinogen are associated with cardiovascular events in both men and women. Individuals with higher fibrinogen concentrations have a risk of cardiovascular disease about 2.3 times higher than individuals with lower levels. Cigarette smoking substantially increases fibrinogen levels, a relationship that is both dose-dependent and reversible.

Fibrin Degradation Products (D-Dimers)

The breakdown of crosslinked fibrin yields a number of degradation products, most notably D-dimers. Elevated levels of D-dimers are associated with risk of recurrence of thrombosis. Low or normal levels of D-dimers are useful for excluding a diagnosis of pulmonary embolus and acute thrombosis. D-dimers are a sensitive marker for fibrin turnover that allow recognition of covert coagulation. Alternately, it may indicate preclinical atherosclerosis, because levels are elevated for years before arterial thrombosis occurs.

Tissue Plasminogen Activator (t-PA)

This fibrinolytic protein is synthesized and secreted in the vascular endothelium and has a very short half-life. Individuals with reduced levels of t-PA are associated with higher risk for coronary artery disease. Low-dose aspirin therapy may reduce coronary artery disease in these individuals.

Plasminogen Activator Inhibitor-1 (PAI-1)

PAI-1 is an antifibrinolytic protein and is synthesized and secreted in the vascular endothelium. Individuals with elevated levels of PAI-1 antigen are prone to recurrent episodes of spontaneous thrombosis and coronary artery disease. Aspirin therapy may have a beneficial effect in individuals with elevated PAI-1 antigen.

C-Reactive Protein (CRP)

CRP is a typical acute phase reactant. In response to acute injury, infection or other stimuli, serum levels of CRP increase several fold. It is produced in the liver, regulated primarily via interleukin-6 production by activated leukocytes, fibroblasts and endothelial cells. Inflammation plays an essential role in plaque rupture and atherosclerosis. CRP may act as an indirect marker for preclinical atherosclerosis.

General Risk Factors

Cigarette Smoking

Smokers have a two-fold increased risk of developing coronary artery disease and a 50% increased risk of dying from it. Smoking interacts in synergistic fashion with hyper-cholesterolemia, hypertension and oral contraceptives.

Cholesterol

The precise mechanism by which LDL cholesterol increases the risk of coronary artery disease and stroke is not known. Lipid parameters other than total cholesterol and LDL cholesterol may be useful predictors of coronary artery disease or may aid in tailoring lipid-lowering therapy. Studies show that for every 1 mg/dl decrease in HDL cholesterol, there is a corresponding 3–4% increase in coronary artery disease.

Hypertension

Studies showed that a 7 mmHg increase from the baseline in diastolic blood pressure is associated with a 27% increase in risk of coronary artery disease and a 42% increase in stroke.

Pregnancy

Venous thrombosis occurs in 60% of women with antithrombin deficiency and in 20% with a deficiency of either protein C or protein S.

Diabetes

Among people with diabetes, arthrosclerotic complications are a major cause of morbidity and mortality. Coronary artery disease is the cause of death in 69% of individuals with diabetes.

Obesity

Obesity early in life is a strong predictor of later cardiovascular disease. Observational studies have demonstrated increased risk of coronary artery disease mortality and morbidity among obese individuals.

Physical Inactivity

Physical inactivity is second only to hypercholesterolemia as a contributing cause of coronary artery disease.

ANTIPHOSPHOLIPID SYNDROME

Antiphospholipid syndrome (APLS) includes:

- Lupus anticoagulant (LA) detected by prolonged aPTT (with no or partial correction on mixing studies). A greater than 3 s difference in a PTT value between control and patient after mixing studies suggests a lupus anticoagulant
- Anticardiolipin antibodies (ACLAs) detected by immunoassays
  - Anticardiolipin antibody IgA*
  - Anticardiolipin antibody IgG
  - Anticardiolipin antibody IgM
- Subgroups of antiphospholipid antibodies (APLA) detected by immunoassays
  - anti-beta-2-glycoprotein-I (β-2-GPI) (IgA, IgG, IgM)*
  - anti-phosphatidylserine (IgG IgA, IgM)*
  - anti-phosphatidylethenolamine (IgG, IgA, IgM)*
  - anti-phosphatidylcholine (IgG, IgA, IgM)*
  - anti-phosphatidylinositol (IgG, IgA, IgM)*
  - anti-annexin-V*

Antiphospholipid antibodies (APLA) are found in 1–5% of young healthy subjects and these increase in prevalence with increasing age. Among patients with systemic lupus erythematosus (SLE) the prevalence is 12–30% for ACLA and 15–34% for LA. Many patients have laboratory evidence of APLA without clinical consequences.

The APLAs may promote thrombosis by:

- Antibody-induced activation of endothelial cells and the secretion of cytokines
- Platelet activation
- Impaired activation of protein C, annexin V and tissue factors

---

*These APLAs are not included in the laboratory diagnosis of APLS (see Table 14-8).

- Inhibition of β2-GPI (thought to act as a natural anticoagulant)
- Impaired fibrinolytic activity
- Antibody against endothelial cells.

APLS is the most common acquired condition associated with venous and/or arterial thrombotic events. The thrombotic events associated with APLS include deep vein thrombosis, pulmonary embolus, coronary artery thrombosis, stroke, transient ischemic attacks, retinal vascular thrombosis and placental vascular thrombosis (leading to recurrent-miscarriage syndrome). Virtually any organ can be involved and the range of disorders observed within any one organ spans a diverse spectrum. The effects of APLS depend on two factors:

- Nature and size of the vessel effected
- Acuteness or chronicity of the thrombotic process.

Venous thrombosis, especially deep venous thrombosis of the legs, is the most common manifestation of the antiphospholipid syndrome. Up to half these patients have pulmonary emboli. Arterial thromboses are less common than venous thromboses and most frequently manifest with features consistent with ischemia or infarction. In arterial thromboses the severity of presentation relates to the acuteness and extent of occlusion. The brain is the most common site, with strokes and transient ischemic attacks accounting for almost 50% of arterial occlusions. Coronary occlusions account for an additional 23%; the remaining 27% involve diverse vessels, including subclavian, renal, retinal and pedal arteries.

In patients with APLS there is an increased rate of heart valve abnormalities, mainly valve masses or thickening affecting primarily the mitral valve. Up to 63% of patients with the antiphospholipid syndrome reveal at least one valvular abnormality on echocardiography. Emboli, especially from mitral-valve or aortic-valve vegetations, can lead to cerebral events. Not all arterial episodes of ischemia or infarction are thrombotic in origin.

Acute involvement at the level of the capillaries, arterioles, or venules may result in a clinical picture resembling hemolytic-uremic syndrome and thrombotic thrombocytopenia purpura as well as other thrombotic microangiopathies. Thrombotic microangiopathy may also occur as a more chronic process, resulting in slow, progressive loss of organ function, the underlying reason for which may only be determined by biopsy. Thus, organ involvement in patients with the antiphospholipid syndrome can present in a spectrum from rapidly progressive to clinically silent and indolent. Depending on the size of the vessels affected, organ failure has two predominant causes, thrombotic microangiopathy and ischemia secondary to thromboembolic events.

Other prominent manifestations of the antiphospholipid syndrome include thrombo-cytopenia, hemolytic anemia and livedo reticularis. Although renal manifestations are a very common feature of systemic lupus erythematosus, they were only recently recognized as part of the antiphospholipid syndrome.

Histopathologic features of APLS consist of:

• Thrombotic microangiopathy
• Ischemia secondary to arterial thromboses or emboli
• Peripheral embolization from venous, arterial or intracardiac sources.

Table 14-8 lists the criteria for diagnosis of APLS.

Lupus Anticoagulant and Thrombosis

The so-called "lupus anticoagulant antibody" (LA) was first described in patients with SLE who presented with prolonged aPTT. The term is actually a misnomer because only a few patients with SLE have LA and the inhibitor is not an anticoagulant. Despite the name LA antibodies are associated with thromboembolic events rather than clinical bleeding. Approximately 10% of patients with SLE have LA. LA is occasionally found in patients with immune thrombocytopenic purpura (ITP) in whom SLE subsequently develops. Human leukocyte antigen (HLA) HLA-DR antigens (DR7 and DRW35) may play a role in the development of LA in some individuals.

Thromboembolism occurs in about 10% of patients with SLE; however, in patients with SLE and LA the occurrence of thromboembolism is up to 50%. The LA is estimated to account for approximately 6–8% of thrombosis in otherwise healthy individuals. Primary LA thrombosis syndrome consists of patients with LA and thrombosis but who have no secondary underlying disease such as SLE or other autoimmune disorders, malignancy, infection; inflammation or ingestion of drugs inducing the LA. It is five-fold more common than the secondary type associated with an underlying disease.

Lupus anticoagulant may be associated with the following conditions:

• HIV infections
• Intercurrent viral infections
• Malignancy
• Systemic antibiotic therapy
• Certain medications (e.g., chlorpromazine, Dilantin, alpha-interferon, Fansidar, procainamide phenothiazines, phenytoin hydralazine, quinidine, quinine)
• Cocaine use
• SLE
• "Lupus-like" chronic autoimmune disorders
• Arterial or venous thrombotic events with no apparent underlying cause
• Women with recurrent fetal wastage.

Table 14-8   Criteria for Diagnosis of Antiphospholipid Antibody Syndrome (APS)[a]

APS is present if at least one of the clinical and one of the laboratory criteria are present.

Clinical criteria

1. Vascular thrombosis

One or more clinical episodes of arterial, venous, or small vessel thrombosis,[b] in any tissue or organ. Thrombosis must be confirmed by objective validated criteria (i.e. unequivocal findings of appropriate imaging studies or histopathology). For histopathologic confirmation, thrombosis should be present without significant evidence of inflammation in the vessel wall

2. Pregnancy morbidity
    (a) One or more unexplained deaths of a morphologically normal fetus at or beyond the 10th week of gestation, with normal fetal morphology documented by ultrasound or by direct examination of the fetus, or
    (b) One or more premature births of a morphologically normal neonate before the 34th week of gestation because of:
       (i) eclampsia or severe preeclampsia defined according to standard definitions, or
       (ii) recognized features of placental insufficiency[c]
    (c) Three or more unexplained consecutive spontaneous abortions before the 10th week of gestation, with no maternal anatomic or hormonal abnormalities and paternal and maternal chromosomal causes excluded

In studies of populations of patients who have more than one type of pregnancy morbidity, investigators stratify groups of subjects according to a, b, or c above

Laboratory criteria[d]

1. Lupus anticoagulant (LA) present in plasma, on two or more occasions at least 12 weeks apart, detected according to the guidelines of the International Society on Thrombosis and Haemostasis (Scientific Subcommittee on LAs/phospholipid-dependent antibodies)
2. Anticardiolipin (aCL) antibody of IgG and/or IgM isotype in serum or plasma, present in medium or high titer (i.e. >40 GPL or MPL, or greater than the 99th percentile), on two or more occasions, at least 12 weeks apart, measured by a standardized ELISA
3. Anti-b$_2$ glycoprotein-I antibody of IgG and/or IgM isotype in serum or plasma (in titer greater than the 99th percentile), present on two or more occasions, at least 12 weeks apart, measured by a standardized ELISA, according to recommended procedures

[a]Classification of APS should be avoided if less than 12 weeks or more than 5 years separate the positive aPL test and the clinical manifestation. Coexisting inherited or acquired factors for thrombosis are not reasons for excluding patients from APS trials. However, two subgroups of APS patients should be recognized, according to: (a) the presence or (b) the absence of additional risk factors for thrombosis, such as age (>55 in men and >65 in women) and the presence of any of the established risk factors for cardiovascular disease (hypertension, diabetes mellitus, elevated LDL or low HDL cholesterol, cigarette smoking, family history of premature cardiovascular disease, body mass index $\geq$30 kg/m^2, microalbuminuria, estimated GFR <60 mL/min/1.73 m^2), inherited thrombophilias, oral contraceptives, nephritic syndrome, malignancy, immobilization and surgery. Thus, patients who fulfill criteria should be stratified according to contributing causes of thrombosis. A thrombotic episode in the past could be considered as a clinical criterion, provided that thrombosis is proved by appropriate diagnostic means and that no alternative diagnosis or cause of thrombosis is found.
[b]Superficial venous thrombosis is not included in the clinical criteria.
[c]Generally accepted features of placental insufficiency include: (i) abnormal or non-reassuring fetal surveillance test(s), e.g. a non-reactive non-stress test, suggestive of fetal hypoxemia; (ii) abnormal Doppler flow velocimetry waveform analysis suggestive of fetal hypoxemia, e.g., absent end-diastolic flow in the umbilical artery; (iii) oligohydramnios, e.g., an amniotic fluid index of 5 cm or less; or (iv) a postnatal birth weight less than the 10th percentile for the gestational age.
[d]Investigators are strongly advised to classify APS patients in studies into one of the following categories: I, more than one laboratory criteria present (any combination); IIa, LA present alone; IIb, aCL antibody present alone; IIc, anti-b$_2$ glycoprotein 1 antibody present alone.
From: Miyakis S, Lockshin MD et al. J Thromb Haemost. 4:295–306:2006.

The LA is an acquired circulating immunoglobulin (usually IgG, sometimes IgM or IgA), which binds to negatively charged phospholipid components of factors Xa and Va, phospholipid and calcium. This circulating anticoagulant is detected by a prolongation of phospholipid-dependent clotting tests such as APTT that does not correct to normal when mixed with normal plasma. More specific tests include a dilute tissue thromboplastin inhibition test, platelet neutralization procedure, dilute Russell viper venom test (DRVVT), or antiphospholipid antibodies by enzyme-linked immunosorbent assay (ELISA). To exclude the presence of LA, two or more assays that are sensitive to these antibodies must be negative.

Because this antibody does not usually cause a bleeding disorder, the primary reason to diagnose it is to avoid intensive workup and therapy prior to surgical procedures. If an antibody is identified in a young child during an acute viral illness, testing should be repeated in several months in order to document its disappearance.

Despite the frequent concordance between LA and either anticardiolipin or anti B2-glycoprotein 1 antibodies, these antibodies are not identical. In general, LAs are more specific for APLS whereas ACLAs are more sensitive.

Anticardiolipin Antibodies (ACLAs)

Table 14-9 lists the syndromes of thrombosis associated with antiphospholipid antibodies.

Table 14-9 Syndromes of Thrombosis Associated with Antiphospholipid Antibodies

Type I Syndrome
Deep venous thrombosis with or without pulmonary embolus
Type II Syndrome
Coronary artery thrombosis
Peripheral artery thrombosis
Aortic thrombosis
Carotid artery thrombosis
Type III Syndrome
Retinal artery thrombosis
Retinal vein thrombosis
Cerebrovascular thrombosis
Transient ischemic attacks
Type IV Syndrome
Mixtures of types I, II and III
Type IV patients are rare
Type V (Fetal Wastage) Syndrome
Placental vascular thrombosis
Fetal wastage common in first trimester
Fetal wastage can occur in second and third trimesters
Maternal thrombocytopenia (uncommon)
Type VI Syndrome
Antiphospholipid antibody with no apparent clinical manifestations

Modified from: Bick RL. Antiphospholipid thrombosis syndromes. Hematol Oncol Clin North America 2003;17:115–147, with permission.

ACLAs are generally IgG, IgA and IgM anticardiolipin idiotypes. These antibodies are found in both SLE and non-SLE patients. The ACLAs are associated with venous and arterial thrombotic events. ACLA IgG, IgA and IgM and antibodies β-2-GP-1, phosphatidylserine, phosphatidylinositol, phosphatidylcholine, phosphatidylethanolamine, phosphatidylglycerol, or annexin-V are independent risk factors for thrombosis of all types (types I–V). Although there is an association between the LA and ACLAs and an association between LAs and the aforementioned syndromes, LA, ACLAs and antibodies to β-2-GP-I, phosphatidylserine, phosphatidylinositol, phosphatidylcholine, phosphatidylethanolamine, phosphatidylglycerol, or annexin-V are separate entities. Most individuals with ACLAs do not have a LA and most with the LA do not have ACLAs.

ACLAs and Venous or Arterial Thrombosis

ACLAs are associated with many types of venous thrombosis, including DVT of the upper and lower extremities, intracranial veins, inferior and superior vena cava, hepatic vein (Budd–Chiari syndrome), portal vein, renal vein and retinal veins. Arterial thrombotic sites associated with ACLAs include coronary arteries, carotid arteries, cerebral arteries, retinal arteries, subclavian or axillary artery (aortic arch syndrome), brachial arteries, mesenteric arteries, peripheral (extremity) arteries and both proximal and distal aorta.

ACLAs and Cardiac Disease

In patients with high ACLA levels who have coronary artery bypass surgery, there is about a 33% chance of late graft occlusion. More than 20% of young survivors of myocardial infarction have high ACLAs. In those surviving, 60% having these antibodies may experience a later thromboembolic event. ACLAs are also associated with cardiac valvular abnormalities.

ACLAs and Cutaneous Manifestations

ACLAs are associated with livido reticularis. This cutaneous finding is associated with recurrent arterial and venous thromboses, valvular abnormalities and cerebrovascular thromboses with concomitant essential hypertension. Other cutaneous manifestations include a syndrome of recurrent deep venous thrombosis, necrotizing purpura and stasis ulcers of the ankles. Other common cutaneous manifestations include livido vasculitis or reticularis, unfading acral microlivido, peripheral gangrene, necrotizing purpura, hemorrhage (ecchymosis and hematoma formation) and crusted ulcers about the nail beds.

ACLAs and Central Nervous System (CNS) Manifestations

ACLAs associated with CNS manifestations include TIAs, stroke, arterial and venous retinal occlusive disease, cerebral arterial and venous thrombosis, migraine headaches, Guillain–Barre syndrome, chorea, seizures and optic neuritis. The CNS manifestations of systemic lupus erythematosus are commonly, but not always, associated with positive antiphospholipid antibodies. Although it is clear that patients with lupus with antiphospholipid antibodies may experience cerebrovascular thromboses, cerebral ischemia and infarction, these events occur more commonly in patients with the primary anticardiolipin antibody thrombosis syndrome with no underlying autoimmune disease. Multiple cerebral infarctions in patients with antiphospholipid antibodies may result in dementia.

ACLAs and Fetal Wastage Syndrome

These antibodies are associated with a high incidence of recurrent miscarriage. Characteristics of this syndrome are:

• Frequent abortions in the first trimester due to placental thrombosis or vasculitis
• Recurrent fetal loss in second and third trimesters, due to thrombosis or vasculitis
• Premature delivery due to pregnancy associated hypertensive disease and placental insufficiency
• Maternal thrombocytopenia.

In these individuals there is the presence of high-level IgG ACLAs. This syndrome has been treated successfully by the institution of heparin plus low-dose aspirin. Low-molecular-weight heparin may be substituted for standard heparin. Women with ACLAs have approximately a 50–75% chance of fetal loss and successful anticoagulant therapy can increase the chances of normal-term delivery to approximately 97%.

Treatment

Recommended antithrombotic regimens for syndromes of thrombosis associated with APLAs are listed on Table 14-10.

HEREDITARY THROMBOTIC DISORDERS

Inherited thrombophilia[*] should be suspected when:

• Patient has recurrent or life-threatening venous thromboembolism
• Family history of venous thrombosis
• Patient is younger than 45 years of age
• No apparent risk factors for thrombosis
• History of multiple abortions, stillbirths, or both.

---

[*]In thrombophilia multiple gene defects often coexist with the clinical penetrance of the syndrome being the end result of the number of gene defects present in an individual. This has been shown for patients with inherited deficiencies of antithrombin, protein C or protein S whose risk of developing thrombotic manifestations is enhanced when there is coexistence of factor V Leiden.

Table 14-10   Recommended Antithrombotic Regimens for Syndromes of Thrombosis-associated with Antiphospholipid Antibodies

Type I Syndrome
Acute treatment with heparin or LMWH followed by long-term[a] administration of LMWH
Type II Syndrome
Acute treatment with heparin or LMWH followed by long-term administration of LMWH
Type III Syndrome
Cerebrovascular
Long-term administration of subcutaneous LMWH
Retinal
Long-term[a] self-administration of subcutaneous porcine heparin/LMWH
Type IV Syndrome
Therapy depends on types and sites of thrombosis
Type V (fetal wastage) syndrome
Low-dose aspirin before conception and add fixed, low-dose LMWH every 12 hours
immediately after conception[b]
Type VI Syndrome
No indications for antithrombotic therapy

[a]Antithrombotic therapy should not be stopped unless the ACLA has been absent for 4–6 months.
[b]In women with prior thromboembolism higher doses of heparin are recommended for full anticoagulation.
Abbreviations: LMWH, low-molecular-weight heparin;
Modified from: Bick RL; Antiphospholipid thrombosis syndromes. Hematol Oncol Clin North America 2003;17:115–117, with permission.

Factor V Leiden (Activated Protein C (APC) Resistance)

In Caucasians, factor V Leiden is the single most common inherited disorder predisposing to thrombosis. It results from a single G to A point mutation at nucleotide 1691 within the factor V gene, arginine is replaced by glutamine at position (R506Q), rendering activated factor V relatively resistant to inactivation by protein C. Approximately 95% of patients with activated protein C resistance test positively for factor V Leiden mutation. The remaining 5% of cases are attributable to oral contraceptive usage, presence of a lupus anticoagulant, or other rare mutations in the factor V gene.

The HR_2 haplotype of the factor V gene causes resistance to activated protein C and increases the risk of venous thrombosis when co-inherited with factor V Leiden. There is also a rare mutation in the second of the three sites in factor Va that activated protein C cleaves (Arg306Thr). Additional causes of resistance to activated protein C, probably genetic but as yet unidentified, also increase the risk of venous thrombosis.

Three to 8% of Caucasians carry the factor V Leiden mutation and approximately 0.1% are homozygous. In contrast this mutation is relatively uncommon in African and Asian populations with a prevalence of <1.0% (heterozygous). Children with homozygous and heterozygous factor V Leiden usually have their first thrombotic event following puberty with an estimated annual incidence of 0.28%. Heterozygous factor V Leiden increases the risk of thrombosis 5–10-fold, whereas homozygous individuals have an 80-fold increased risk. APC resistance even in heterozygotes is a significant risk factor for thrombosis because >20% of patients presenting with a thrombotic event exhibit APC resistance. Forty-two percent of patients who sustain a first venous thrombotic event before 25 years of age and 21% of those between 54 and 70 years of age have APC resistance.

Both heterozygous and homozygous cases of factor V Leiden mutation have increased risk for either venous or arterial thrombosis throughout life but usually are asymptomatic in youth unless associated with other acquired or genetic prothrombotic conditions, including central venous catheters, trauma, surgery, cancer, pregnancy, oral contraceptives, deficient protein C, deficient protein S, or homocysteinemia.

Factor V Leiden mutation has been associated with cerebral infarction and with venous thrombosis in children.

Diagnosis

aPTT-based assay (a screening test) can be used to obtain a diagnosis of APC resistance. Patients positive for APC resistance in the clotting assay should undergo genetic testing for factor V Leiden mutation by analyzing genomic DNA in peripheral blood mononuclear cells. This can be readily accomplished by amplifying a DNA fragment containing the factor V mutation site by polymerase chain reaction (PCR).

Prothrombin G20210A Mutation (FII G 20210 A)

The genetic defect is at nucleotide position 20210 in the prothrombin gene. The mutation results in abnormally high prothrombin levels and contributes to thrombotic risk by promoting increased thrombin generation. Prothrombin gene mutation is the second most common inherited thrombotic defect. The frequency of this mutation is about 2–3% in Caucasian population and 4–5% in Mediterranean populations. Homozygous state is extremely rare. Homozygotes for the prothrombin G200210A mutation have less severe clinical presentations than homozygotes for antithrombin, protein C and protein S deficiencies. The individual with this genetic defect usually presents with thrombotic episodes in adulthood. The diagnosis of this defect can be made by PCR testing.

Antithrombin (AT) Deficiency

AT functions by forming a complex with the activated clotting factors thrombin, Xa, IXa and XIa. The relatively slow formation of this complex is greatly accelerated in the presence of heparin or cell-surface heparin sulfate. The incidence of AT deficiency is about 0.2–0.5%. Congenital AT deficiency is a heterozygous disorder; homozygous deficiency, probably incompatible with life, has not been reported.

Two major types of AT deficiency have been described. Type I AT deficiency (low activity and low antigen level) is a quantitative defect caused by a mutation resulting in both decreased synthesis and functional activity of AT whereas type II AT deficiency is a qualitative defect characterized by decreased AT activity and normal antigenic levels.

Thrombotic complications of AT deficiency usually occur in the second decade of life. However, few pediatric cases with venous thrombotic events have been reported with heterozygous AT deficiency. In affected individuals AT levels are about 40–60% (normal range, 80–120%). The risk of developing thrombotic complications depends on the particular subtype of AT deficiency and on other coexisting inherited or acquired risk factors. In children with heterozygous AT deficiency the relative risk of developing thrombotic episodes is increased ten-fold.

Treatment

AT deficiency is usually treated with oral warfarin-type anticoagulants (Coumadin) to decrease the level of vitamin-K-dependent procoagulants so that they are in balance with the level of AT. Heparin requires binding with AT to anticoagulate blood and, for this reason, heparin administration is usually ineffective in AT deficiency states. Because the administration of warfarin may take several days to decrease the vitamin-K-dependent factors, acute thrombotic episodes are usually managed with AT replacement therapy, using either fresh frozen plasma or recombinant AT concentrates. Patients are initially treated with approximately 50 units/kg AT, which will increase the baseline level of AT by approximately 50%. AT levels should be maintained at 80% or higher. In patients with severe recurrent thrombosis who cannot be managed with oral anticoagulants, therapeutic or prophylactic replacement therapy can be monitored with AT concentrates.

5,10-Methylenetetrahydrofolate Reductase (MTHFR) Mutation

In this condition there is a cytosine to thymine mutation at nucleotide 677 (C677T) of *MTHFR* gene. This mutation may be a risk factor for stroke in children, venous thrombosis in the young and coronary artery disease in adults. In children *MTHFR* mutation has not been shown to be a big risk factor for thrombosis. *MTHFR* is essential for the re-methylation of homocysteine to methionine and homozygosity for mutation of *MTHFR* is associated with hyperhomocysteinemia. Hyperhomocysteinemia is a common risk factor for deep-vein thrombosis and increases the risk for deep-vein thrombosis in patients with factor V Leiden. The mechanisms whereby excess homocysteine may be a risk factor in thrombosis are listed on page 427.

Protein C Deficiency

Protein C (PC) is a vitamin-K-dependent plasma glycoprotein which when activated functions as an anticoagulant by inactivating factors Va and VIIIa. PC activity is enhanced by another vitamin-K-dependent inhibitory cofactor, protein S.

PC deficiency is inherited in an autosomal-dominant manner and is subdivided into two types. Type I PC deficiency (low activity and low antigen level) is a quantitative deficiency with decreased plasma concentration and functional activity to approximately 50% of normal. Type II PC deficiency (low activity and normal antigen level) is less common and is characterized by a qualitative decrease in functional activity, despite normal levels of PC antigen. Plasma levels of PC below 50% (normal range, 70–110%) are associated with the risk of thrombotic complications.

The population prevalence of heterozygous PC deficiency is estimated at 0.2%. The clinical manifestation of heterozygous PC deficiency is primarily venous thrombotic episodes during the second decade of life or young adulthood. Heterozygous deficiency of PC is not a significant problem during the neonatal period, but homozygous deficiency or compound heterozygous individuals with protein C deficiency may cause a fatal thrombotic disorder. It may present with neonatal purpura fulminans, disseminated intravascular coagulation, progressive skin necrosis with microvascular thrombosis and thrombosis of the renal veins, mesenteric veins and dural venous sinuses. Several of the neonates with homozygous deficiency have been subsequently found to be blind.

Treatment

Initially, affected neonates are treated for several months with replacement therapy with fresh frozen plasma (5–10 ml/kg q12 h); later, they can usually be managed with long-term oral warfarin or protein C replacement (fresh frozen plasma or prothrombin complex concentrate). Recently a purified plasma-derived protein concentrate has been approved for use in children with protein C deficiency.

Heterozygous PC deficiency is also one of the major causes of warfarin-induced skin necrosis. Because the half-life of PC is extremely short (2–8 hours), warfarin decreases PC more rapidly than factors IX, X and prothrombin, resulting in a hemostatic imbalance with resultant microvascular thrombosis. Warfarin-induced skin necrosis is most likely to occur in these patients when loading doses of warfarin are used. This complication can usually be avoided by overlapping heparin and warfarin usage until the desired INR is attained and then discontinuing the heparin.

Protein S Deficiency

Protein S deficiency is inherited as an autosomal-dominant manner and has a prevalence of 1:33,000. Protein S (PS) is also a vitamin-K-dependent anticoagulant that circulates in the plasma in two forms: a free active form (40%) and an inactive form bound to C4b-free binding protein (60%). PS functions as a cofactor to PC by enhancing its activity against factors Va and VIIIa. Free PS levels (normal range, 44–92%) correlate clinically with thrombotic episodes. PS deficiency is classified into three subtypes according to a quantitative defect (type I) (low activity and low antigen level) or a qualitative defect (type II) (low activity and normal antigen level) in which PS activity is reduced as follows:

	Total PS	Free PS	PS Function
Type 1	↓	↓	↓
Type II a	N	↓	↓
Type II b	N	N	↓

The clinical presentation of protein S deficiency is indistinguishable from that of protein C deficiency. Heterozygous PS deficiency manifests in adulthood as either venous or arterial thrombotic events. Only a few cases with heterozygous protein S deficiency have been reported with thrombotic episodes during childhood. A small number of newborns with either homozygous or compound heterozygous PS deficiency have been reported. These infants, like those with homozygous PC deficiency, may present with purpura fulminans.

Treatment

Asymptomatic patients are not usually treated but need prophylaxis for high-risk procedures, e.g. surgery. Pregnancy may be accompanied with a reduction in free PS. Oral contraceptives will also result in a reduction of PS and may precipitate thrombosis in a patient with heterozygous PS deficiency.

The acute episodes are usually treated with standard anticoagulation therapy with heparin, followed by oral warfarin administration for 3–6 months.

First thrombotic events in heterozygous thrombophilia in the presence of a triggering factor (e.g. infection or inflammation) may not warrant life-long anticoagulation. In these cases careful monitoring of serological markers of inflammation (ESR, CRP) and recurrence for thrombosis (FVIII, fibrinogen, d-dimer, LA, ACA, anti-beta 2 glycoprotein 1 antibodies) is required before a decision is made to discontinue anticoagulation therapy.

Recurrent thrombosis or a life-threatening thromboembolic event in a patient with PS deficiency is usually managed with long-term oral anticoagulation.

Dysfibrinogenemia

Approximately 45 different dysfibrinogenemias that predispose to thrombotic events have been described, with the majority due to single-point mutations. Dysfibrinogenemias are usually inherited as an autosomal-recessive condition. Dysfibrinogenemia is a very rare condition, it is estimated to occur in 0.8% of adults presenting with thrombotic episodes. Impaired binding of thrombin to an abnormal fibrin and defective fibrinolysis occurs. Plasminogen activator (t-PA) and plasminogen activation on the abnormal fibrin have been implicated in the development of thrombosis. Congenital dysfibrinogenemias may also be associated with abnormal interactions with platelets and defective calcium binding. Although dysfibrinogenemias are more often associated with bleeding, thrombosis occurs in 20% of cases. The clinical presentation is variable, consisting of venous thrombosis, pulmonary embolism and arterial occlusion. Some cases of dysfibrinogenemia may have an associated bleeding diathesis. Although it is very rare, intracranial hemorrhage is the most common cause of death in these patients. Homozygosity for dysfibrinogenemia is very rare and associated with juvenile arterial stroke, thrombotic abdominal aortic occlusions and postoperative thrombotic episodes. Fibrinogen infusions may precipitate the thrombotic events. A normal fibrinogen antigen with low functional fibrinogen together with a prolonged thrombin time (TT) or reptilase time is indicative of dysfibrinogenemia.

Heparin Cofactor II (HC-II) Deficiency

HC-II deficiency is inherited as an autosomal dominant trait. Clinical manifestations of HC-II deficiency are arterial and venous thrombotic events. HC-II deficiency seems to be a rare cause of unexplained thrombotic events.

Hereditary Defects of Fibrinolytic System

Type I Dysplasminogenemia

Homozygous deficiency of this condition clinically manifests as pseudomembranous conjunctivitis, hydrocephalus, obstructive airway disorder and abnormal wound healing secondary to failure of removing of fibrin deposits in various organs. Replacement therapy with plasminogen corrects these defects by allowing the lysis of fibrin deposits. Infants with homozygous deficiency do not seem to have a higher incidence of thrombotic events.

Type II Dysplasminogenemia

This type is inherited as a mutation at various loci in the plasminogen molecule leading to functional abnormalities and failure of plasminogen activation. Type II defect is also associated with increased incidence of thrombotic events.

Tissue Plasminogen Activator (t-PA) Deficiency

There are several families reported with this defect with failure to release fibrinolytic activity following venous thrombotic events.

Plasminogen Activator Inhibitor (PAI-1) Deficiency

These polymorphic variations of human PAI-1 gene have been reported where specific alleles may be associated with decreased PAI-1 levels. PAI-1 genotype abnormality may represent a risk factor for venous thrombotic events in the setting of PS deficiency.

Prophylaxis in Relatives of Patients with Thrombophilia

First-degree relatives of index patients who are asymptomatic but who are affected by a thrombophilia should be advised of the risk of venous thrombosis. Primary prophylaxis in these persons includes the administration of low-molecular-weight heparin in high-risk conditions, such as during surgery, trauma, immobilization and the 6-week postpartum period; maintenance of normal weight and homocysteine levels; and avoidance of contraceptives and hormone-replacement therapy. Women with antithrombin deficiency, combined thrombophilia, or homozygosity for factor V Leiden or the G20210A mutation in the prothrombin gene should be treated throughout pregnancy and for 6 weeks postpartum with low-molecular-weight heparin.

THROMBOTIC DISORDERS IN NEWBORNS

Congenital

Newborns comprise the largest group of children developing thrombotic events. Stroke in newborns appears to be a diverse condition with many potential risk factors. Inherited thrombophilia is a multigene disorder in which the likelihood to develop thrombotic episodes increases with the number of genetic risk factors present in a subject. The "thrombophilia burden"[*] is found to be increased in newborn and children with stroke.

Acquired

Systemic Venous Thromboembolic Disorders

Incidence of symptomatic venous thrombotic events in newborns (exclusive of CNS) is 0.24 per 10,000 newborn and child births. Neonatal thrombotic events including CNS events is

---

[*]Concomitant presence or any combination of APLA, Factor V Leiden, FIIG20210A, AT deficiency, PC deficiency, PS deficiency, Lp(a). A full thrombophilia work-up should be carried out in newborn infants with unexplained prenatal or neonatal cerebral infarction especially if there is a family history of venous thrombosis, early stroke or heart disease.

0.51 per 10,000 births with 50% being venous thrombotic events and the other 50% arterial thrombotic events.

The incidence of *sinovenous thrombosis* is 0.7 per 100,000 children per year. The following factors predispose a newborn to sinovenous thrombosis:

- During birth, the normal molding and overlapping of cranial sutures may damage the cerebral sinus structures and provoke sino-venous thrombosis
- Asphyxia
- Dehydration
- Sepsis and meningitis.

MRI scanning with venography is the most sensitive and specific test for the diagnosis of sinovenous thrombosis. CT scan alone could miss the diagnosis. Doppler flow ultrasound may demonstrate absent or decreased flow within sinovenous channels.

The use of anticoagulants in neonates is controversial and probably not indicated.

Central Venous Catheter-Related Thrombosis

More than 80% of venous thrombotic events are secondary to central venous catheters. Thrombotic events result from:

- Damage to vessel walls
- Disrupted blood flow
- Infusion of substances in total parenteral nutrition which damage endothelial cells
- Thrombogenic catheter materials.

Right atrial thrombosis commonly occurs in newborns with central venous catheters. Right atrial thrombosis may result in cardiac failure, persistent sepsis and fatal pulmonary emboli.

Umbilical Venous Catheters

Major clinical symptoms of umbilical arterial catheter-related thrombotic events occur in about 3% of infants. Loss of patency due to umbilical arterial catheter-related thrombotic events occur in 13–73% without unfractionated heparin being administered and in 0–13% with unfractionated heparin.

Renal Vein Thrombosis

Renal vein thrombosis is the most common non-catheter-related venous thrombotic event in newborns. The incidence is about 10% of all venous thrombotic events and is bilateral in

24% of cases. Almost 80% present within the first week of life. The risks of renal vein thrombosis include:

- Perinatal asphyxia
- Shock
- Polycythemia
- Cyanotic congenital heart disease
- Maternal diabetes
- Sepsis which results in reduced renal flow, hyperviscosity, hyperosmolality and hypercoagulability.

Diagnostic ultrasonography is the radiographic test of choice.

Systemic Arterial Thromboembolic Disorders

Arterial Ischemic Stroke

The incidence of arterial ischemic stroke in newborns is 93 per 100,000 live births. A primary risk factor is identifiable in about 70% of affected newborns. The risk factors for perinatal stroke are listed on Table 14-11. More than one risk factor is identified in many cases.

Diagnostic radiographic studies include CT scan, MRI, MRA and less frequently conventional angiogram. Cranial ultrasound has a limited role in arterial ischemic stroke.

Use of anticoagulants is controversial and thrombolytic therapy is rarely indicated.

Diagnosis and treatment of thrombotic events are shown on Table 14-12.

ANTITHROMBOTIC AGENTS

Heparin Therapy

Heparin mediates its activities through catalysis of the natural inhibitor AT. The activities of heparin are considered anticoagulant and antithrombotic. The antithrombotic activities of heparin are influenced by plasma concentration of AT. In the presence of heparin, thrombin generation is decreased in young children compared to adults. Clearance of heparin is also faster in the young. For this reason the optimal dosing of heparin is different in children from adults. In pediatric patients activated partial thromboplastin time (aPTT) values correctly predict heparin therapy only 70% of the time. The following dosage schedule is utilized for heparinization:

Table 14-11   Risk Factors for Perinatal Stroke

Cardiac Disorders
  Congenital heart disease
  Patent ductus arteriosus
  Pulmonary valve atresia
Hematologic Disorders
  Polycythemia
  Disseminated intravascular coagulopathy
  Factor-V Leiden mutation
  Protein-S deficiency
  Protein-C deficiency
  Prothrombin G20210a mutation
  High homocysteine level
  High lipoprotein (a) level
  MTHFR C677T mutation
  Factor VII
Infectious Disorders
  CNS infection
  Systemic infection
Maternal Disorders
  Autoimmune disorders
  Coagulation disorders
  Anticardiolipin antibodies
  Twin to twin transfusion syndrome
  In utero cocaine exposure
  Infection
  Diabetes mellitus – Type I, type 2, or gestational
Placental Disorders
  Placental thrombosis
  Placental abruption
  Placental infection (chorioamnionitis)
  Fetomaternal hemorrhage
Vasculopathy
  Vascular maldevelopment
Trauma and Catheterization
  Central venous catheters, umbilical venous catheters
Birth Asphyxia
Dehydration
Extracorporeal Membrane Oxygenation

- Loading dose: 75 units/kg IV over 10 minutes
- Initial maintenance dose:
  - ≤1 year of age: 28 units/kg/h
  - >1 year of age: 20 units/kg/h.
- Repeat aPTT 4 hours after administration of the heparin loading dose and monitor heparin dose to maintain an APTT at 2½ times normal for the reference laboratory as follows:

Table 14-12 Diagnosis and Treatment of Neonatal Thromboembolism

Thrombosis	Diagnosis	Treatment
Systemic venous thrombotic events	VG/US	LMWH/UFH/TT[a]
Pulmonary emboli	V/Q Scan	LMWH/UFH
Central venous catheter-related thromboembolic event	VG/US	LMWH/UFH/TT[a]
Right atrial thrombosis	ECHO	LMWH/UFH
Umbilical venous catheter	VG	LMWH/UFH
Renal vein thrombosis	US	LMWH/UFH indicated if thrombosis extends to IVC or renal failure
Umbilical arterial catheter	AG	LMWH/UFH prophylaxis: UFH 0.25 unit/ml
Peripheral arterial thrombotic event	AG	LMWH/UFH prophylaxis: UFH 1 unit/ml

[a]Recommended only if potential loss of life, organ or limb.
Abbreviations: ECHO, echocardiogram; V/Q Scan, ventilation perfusion scan; IVC, inferior vena cava; AG, arteriogram; VG venogram; US, ultrasonogram; LMWH, low-molecular-weight heparin; UFH, unfractionated heparin; TT, thrombolytic therapy.

aPTT Times of Normal	Bolus (Units/kg)	Dose Change (Units/kg/hour) (Increase or Decrease of Bolus Dose)	Intervals of Testing aPTT (Hours)
<2	50	20% increase	4
2	–	10% increase	4
>2	–	No change	4
$2\frac{1}{2}$	–	No change	24
$>2\frac{1}{2}$[a]	–	10% decrease	4
>3[b]	–	20% decrease	4

[a]Hold the dose 30 minutes.
[b]Hold the dose 1 hour.
Heparin solution at concentrations of 80 units/ml for children 10 kg or less or 40 units/ml for children greater than 10 kg.

- When the patient's aPTT achieves a therapeutic level, repeat aPTT, CBC with platelet count daily
- The anti-Xa level should be monitored after a therapeutic aPTT level is attained to ensure a level between 0.35–0.70 units/ml by an *anti-factor* Xa assay
- When heparin is interrupted for more than 1 hour, re-establish the heparin maintenance infusion at the previous rate until the aPTT result is available. After that, administer heparin in accordance with the aPTT results
- When the platelet count is 100,000/mm^3 or less, consider discontinuing heparin therapy and instituting alternative therapy, because the risk of heparin-induced thrombocytopenia is greater after 5 days of therapy.

Duration of Heparin Therapy

1. Deep-vein thrombosis: A minimum of 5–7 days; maintenance warfarin therapy can be instituted on day 1 or 2 of heparin therapy.

2. Pulmonary embolus: 7–14 days; start warfarin therapy on day 5.
3. Neonates may be treated for 10–14 days without warfarin.
    *   *Avoid acetylsalicylic acid-ASA (aspirin) or other antiplatelet drugs, arterial sticks and intramuscular injections during heparin therapy when possible.*

Heparin Antidote

1. If heparin needs to be discontinued, termination of the heparin infusion will be sufficient (because of the rapid clearance of heparin).
2. If an immediate effect is required, protamine sulfate administration may be indicated.
3. Following administration of IV protamine sulfate, neutralization occurs within 5 minutes.

The dose of protamine sulfate required to neutralize heparin is as follows:

Last Dose of Heparin	Protamine Dose[a] (per 100 mg Heparin)
<30 minutes	1 mg
30 minutes	0.5 mg
1 hour	0.75 mg
>1 hour	0.375 mg
>2 hours	0.25 mg

[a]Maximum dose of protamine sulfate, 50 mg, to be administered in a concentration of 10 mg/ml at a rate not to exceed 5 mg/min. If administered faster, it may cause cardiovascular collapse; patients with known hypersensitivity reactions to fish and those who have received protamine-containing insulin or previous protamine therapy may be at risk for hypersensitivity reactions to protamine sulfate. Repeat APTT 15 minutes after the administration of protamine sulfate.

Low-Molecular-Weight Heparin

The anticoagulant activities of low-molecular-weight heparin (LMWH) are also mediated by catalysis of the natural inhibitor AT. LMWHs preferentially inhibit factor Xa over thrombin due to the decreased capacity to bind both AT and thrombin simultaneously for thrombin inhibition. Preparations used for children are enoxaparin (Lovenox, Rhone-Poulenc) and reviparin (Clivarine, Knoll Pharma). Young infants have an accelerated clearance of LMWH compared to older children.

Indications for Low-Molecular-Weight Heparin Therapy

1. Neonates.
2. Patients requiring anticoagulation and deemed to be at increased risk for hemorrhage.
3. Patients in whom venous access for administration and monitoring of standard heparin therapy is difficult.

Dose (Enoxaparin) [Lovenox, Aventis]

Age (Months)	Treatment[a] (Dose)	Prophylactic (Dose)
<2	1.5 mg/kg every 12 h SC	0.75 mg/kg every 12 h SC
>2	1.0 mg/kg every12 h SC	0.5 mg/kg every 12 h SC

[a]Maximum dose, 2.0 mg/kg q12h.
Abbreviation: SC, Subcutaneously.

Monitoring of Low-Molecular-Weight Heparin Therapy

1. Prior to the initiation of LMWH therapy, a complete blood count, including platelet count, PT and aPTT, should be determined.
2. Aspirin or other antiplatelet drugs should be avoided during the therapy.
3. Intramuscular injections and arterial punctures should be avoided during the therapy.
3. If the platelet count drops to $100,000/mm^3$ or less, heparin-induced thrombocytopenia must be ruled out; it rarely occurs with LMWH therapy.
4. The post-treatment anti-factor Xa level should be determined after three doses; thereafter, weekly monitoring should be sufficient 3–4 hours after the SC administration of LMWH. The therapeutic anti-factor Xa level is 0.5–1.0 units/ml and the prophylactic anti-factor Xa level is 0.1–0.3 units/ml.
5. For long-term LMWH therapy, bone densitometry studies should be performed at baseline and then at 6-month intervals to assess for possible osteoporosis.

Duration of Low-Molecular-Weight Heparin Therapy

1. LMWH is usually administered up to 3 months without warfarin.
2. Extensive thrombus or pulmonary embolus. LMWH for 7–14 days should be administered and warfarin should be started on day 5.
3. Newborns may be treated for 10–14 days with LMWH alone.

Adjusting Low-Molecular-Weight Heparin Dose

The dose of LMWH can be adjusted according to the anti-factor Xa level achieved:

Anti-FXa Level (Units/ml)[a]	Dose
<0.35	25% increase
<0.5	10% increase
0.5–1.0	No change
>1.0	20% decrease
>1.5	30% decrease
>2.0	Hold for 24 hours

[a]Repeat anti-FXa level 4 hours post next dose until 0.5–1.0 units/ml, then once weekly at 4 hours post dose.

Antidote for Low-Molecular-Weight Heparin

1. Termination of LMWH is usually sufficient.
2. When an immediate effect is required, 1 mg protamine per 100 units (1 mg) of LMWH may be given. It is generally not as effective as when used for unfractionated heparin.
3. The protamine should be administered intravenously over a 10-minute period. A rapid infusion of protamine may cause hypotension.

HEPARIN-INDUCED THROMBOCYTOPENIA (SEE CHAPTER 12)

Heparin-induced thrombocytopenia is an uncommon complication of heparin therapy and can cause significant morbidity and mortality. It usually begins 5–15 days after commencing heparin therapy (median 10 days) but can occur earlier in patients with prior exposure to heparin. An abrupt decrease of platelet count or a decrease of platelet count by half in 1–2 days should raise suspicion of heparin-induced thrombocytopenia. Any type of heparin including LMWH, unfractionated heparin used to flush lines should be discontinued. Discontinuation of heparin will result in the platelet count returning to normal. In the setting of heparin-induced thrombocytopenia associated with thrombotic events heparin must be discontinued and alternative anticoagulation therapy must be initiated until the antibody complex causing heparin-induced thrombocytopenia is cleared. Alternative intravenous anticoagulation therapy may include the following.

Argatroban

1. No loading dose.
2. Maintenance dose: Begin at 2 μg/kg/min continuous IV infusion (maximum 10 μg/kg/min).
3. Monitor with aPTT beginning 2 hours after the start of therapy; target aPTT is 1.5–3-fold greater than the pre-treatment value and <100 sec.
4. Excretion occurs via the liver and the dose should be reduced in patients with hepatic insufficiency.
5. Clinical data for use in patients under age 18 are extremely limited.

Lepirudin (Recombinant Hirudin)

1. Loading dose 0.4 mg/kg (up to 44 mg) as an IV bolus.
2. Maintenance dose: 0.15 mg/kg/hour (up to 16.5 mg/hour).
3. Monitor with aPTT beginning 4 hours after start of infusion. Adjust infusion for a target aPTT between 1.5- and 2.5-fold greater than the pre-infusion value. Recheck aPTT 4 hours after any dosage changes:
   - If aPTT is greater than 2.5 $\times$ original value: the infusion should be stopped for 2 hours and resumed thereafter at 5% of prior dose.

- If aPTT is less than 1.5 × original value the infusion rate should be increased by 20%
- Antibodies may develop in up to 40% of patients affecting pharmacokinetics, aPTT monitoring must be ongoing
- Excretion occurs via the kidneys and the dose should be reduced in patients with renal insufficiency
- Clinical data for use in patients under age 18 are extremely limited. Dosing requirements in neonates and newborns may be markedly reduced.

Orgaran (Heparan Sulfate) (Danaproid Sodium)

1. Loading dose: 30 units/kg.
2. Maintenance dose: 1.2–2.0 units/kg/h.
3. Antifactor Xa activity can be monitored immediately following the bolus dose and every 4 hours until a steady state is reached and then daily to maintain a therapeutic range of 0.4–0.8 units/ml.
4. Orgaran is contraindicated in patients with severely impaired renal function.

Ancrod (Russell Pit Viper Venom)

1. Ancrod cleaves fibrinopeptide A from fibrinogen.
2. Loading dose: 1–2 units/kg intravenous or subcutaneous infusion over 6–24 hours.
3. Maintenance dose: 1–2 units/kg every 24 hours (intravenous or subcutaneous infusion) with the dose adjusted to achieve plasma fibrinogen levels of 0.2–1 mg/ml.
4. Fresh frozen plasma or anti-ancrod antibody can be used to reverse ancrod's effect.

Warfarin Therapy

Warfarin (4-hydroxycumarin) (Coumadin [Bristol-Myers, Squibb]) competitively inhibits vitamin K, an essential cofactor for the post-translational carboxylation of gamma glutamic acid residues on factors II, VII, IX and X. Coumadin is the trade name for the most commonly available warfarin preparation.

The following is a detailed description of warfarin management and duration of treatment in special clinical conditions:

- A loading dose of 0.2 mg/kg by mouth as a single daily dose (maximum initial dose, 10 mg) should be employed when the INR is less than 1.3. When it is greater than 1.3 the loading dose should be reduced to 0.1 mg/kg. For a patient with Fontan procedure or liver dysfunction, the daily dose should be reduced by 50%
- The subsequent dose is age-dependent (infants [0.32 mg/kg] and teenagers [0.09 mg/kg]) and based on the international normalized ratio (INR) response

(see the following table). Before the INR was adopted, the prothrombin time test was used for many years to control oral anticoagulant therapy. The variability of different thromboplastin reagents and instruments used in the performance of the PT test made the transferability of results among laboratories difficult and impeded the development of universal guidelines for patient therapy

- Generally warfarin is started after the patient is on a regular diet and should be taken on an empty stomach or 2 h after meals, without any other medications in order to facilitate absorption. Green leafy vegetables which contain large amounts of vitamin K are discouraged but a moderate daily intake helps prevent fluctuations in INR and the need for frequent monitoring

Warfarin Daily Loading Doses (Approximately 3–5 Days)

INR[a]	Warfarin Loading Doses
1.1–1.3	Repeat initial loading dose
1.4–1.9	50% of initial loading dose
2.0–3.0	50% of initial loading dose
3.1–3.5	25% of initial loading dose
>3.5	Hold until INR<3.5, then restart at 50% less than previous dose

[a]The international normalized ratio (INR) provides a standardized scale for monitoring patients who are receiving oral anticoagulant therapy. The INR is effectively the PT ratio upon which the patient would have been measured had the test been made using the primary World Health Organization international reference preparation (IRP). The INR is calculated by use of the international sensitivity index (ISI), which is established by the manufacturer for each lot of thromboplastin reagent using a specific instrument: $INR = (Patient\ PT/Normal\ PT)^{ISI}$.

Warfarin Maintenance Doses for Long-term Therapy

INR	Warfarin Dose
1.1–1.4	Increase dose by 20%
1.5–1.9	Increase dose by 10%
2.0–3.0	No change
3.1–3.5	Decrease dose by 10%
>3.5	Hold until INR <3.5, then restart at 20% less than previous dose

- The warfarin loading period is approximately 3–5 days for most patients before a stable maintenance phase is achieved
- Warfarin should be started on day 1 or day 2 of heparin therapy. Heparin should be continued for a minimum of 5 days' duration. When the target INR is greater than 2.0 for 2 consecutive days and at least 5 days of heparin are completed, heparin can be discontinued
- For extensive deep vein thrombosis with or without pulmonary emboli, warfarin should be started on day 5 of heparin therapy

- The INR should be maintained between 2.0 and 3.0 for the vast majority of patients. Children with mechanical heart valves require an INR between 2.5 and 3.5
- Once the patient has two INRs between 2.0–3.0 (or 2.5–3.5 for mechanical valves) obtained weekly, the INR determinations could be carried out every 2 weeks. If INR remains stable, then the INR could be determined once monthly
- The INR should be monitored a minimum of once monthly
- If the patient is receiving total parenteral nutrition, vitamin K should be removed from the amino acid solution before warfarin therapy begins
- Both diet and co-administration of certain drugs can have a marked effect on the magnitude of Warfarin action (see Table 14-13) and the loading dose may require adjustment
- The INR should be obtained 5–7 days after initiating a new dose. The maintenance guidelines should be employed for making changes in dosage.

Table 14-13   Effect of Drugs on Warfarin Response

Medications that Potentiate the Effect of Warfarin	
Acetaminophen	Oxyphenbutazone
Acetohexamide	p-Aminosalicyclic acid
Allopurinol	Paromomycin
Androgenic and anabolic steroids	Phenylbutazone
α-Methyldopa	Phenytoin
Antibiotics that disrupt intestinal	Phenyramidol
flora (tetracyclines, streptomycin	Propylthiouracil
erythromycin, kanamycin, nalidixic	Quinidine
acid, neomycin)	Salicylate
Cephaloridine	Sulfinpyrazone
Chloramphenicol	Sulfonamides
Chlorpromazine	Thyroid hormone
Chlorpropamide	Tolbutamide
Chloral hydrate	**Medications that Reduce the Effect of Warfarin**
Cimetidine	Antipyrine
Clofibrate	Barbiturates
Diazoxide	Carbamazepine
Disulfiram	Chlorthalidone
Ethacrynic acid	Cholestyramine
Glucagon	Digitalis
Guanethidine indometacin	Ethanol
Isoniazid	Ethchlorvynol
Mefenamic acid	Glutethimide
Methimazole	Griseofulvin
Methotrexate	Haloperidol
Methylphenidate	Phenobarbital
Nalidixic acid	Prednisone
Nortriptyline	All vitamin preparations
	containing vitamin K

- Vitamin K is the antidote for warfarin. The patient may also be given fresh frozen plasma or prothrombin complex concentrate infusions to reverse the effects of warfarin:
  - If there is no active bleeding, vitamin K_1 0.5–2 mg SC can be given
  - If there is active bleeding, vitamin K_1 0.5–2 mg SC (not IM) plus FFP 20 cc/kg IV can be given
  - If bleeding is significant and life-threatening, vitamin K_1 5 mg IV (by slow infusion over 10–20 minutes because of the risk of anaphylactic shock) and prothrombin complex concentrates 50 units/kg IV can be given.
- Children with mechanical heart valves or repeated thromboembolic complications should receive warfarin indefinitely
- Children with uncomplicated deep vein thrombosis and pulmonary emboli should receive warfarin for a minimum of 3 months
- Children with a thrombotic event and a persistent, significant, underlying predisposing factor (e.g., continued presence of a central venous catheter, persistence of an antiphospholipid antibody) may be placed on low-dose warfarin (0.1 mg/kg) target INR −1.5–2.0 following 3 months of treatment with full-dose warfarin until the predisposing factor is no longer present
- If low-risk procedures have to be carried out, INR should be reduced to 1.5 or less prior to the procedures. Warfarin should be discontinued 72 hours prior to any high-risk surgery
- Where surgical procedures have to be carried out and where the risk of thrombosis is high and anticoagulant therapy cannot be reversed for even a short period of time, the following may be considered:
  - Discontinue warfarin 72 hours prior to surgery
  - Initiate heparin infusion therapy without a bolus at appropriate dose for age
  - If the INR is more than 1.5 at 12 hours prior to surgery, low-dose vitamin K_1 0.5 mg SC should be given and the INR should be determined approximately 6 hours later
  - Intravenous heparin infusion should be discontinued approximately 6 hours prior to surgery. Preoperative PT and aPTT should be within normal limits
  - Heparin should be resumed 8 hours following surgery
  - If the patient develops any signs of bleeding, heparin infusion should be immediately discontinued
  - Oral warfarin should be resumed on the second day post surgery
  - Heparin infusion should be discontinued when a therapeutic INR is reached.
- Where the risk of thrombosis is low and there is no thrombotic event for several weeks:
  - Warfarin should be discontinued 72 hours prior to surgery
  - Warfarin maintenance should be initiated on the day following surgery.

Antiplatelet Therapy

The two most commonly used antiplatelet agents for children are aspirin and dipyridamole. Aspirin acetylates the enzyme cyclo-oxygenase and thereby interferes with the production of thromboxane A2 and platelet aggregation. Dipyridamole interferes with platelet function by increasing the cellular concentration of adenosine 3,5-monophosphate (cyclic AMP). This latter effect is mediated by inhibition of cyclic nucleotide phosphodiesterase and/or by blockade of uptake of adenosine, which acts at A2 receptors for adenosine to stimulate platelet adenylcyclase.

Antiplatelet agents are used in the following conditions:

- Cardiac disorders
  - Mechanical prosthetic heart valves
  - Blalock-Tausig shunts
- Endovascular shunts
- Cardiovascular events
- Kawasaki disease.

Dosage

1. Aspirin: 5–10 mg/kg/day.
2. Dipyridamole: 3–5 mg/kg/day.

THROMBOLYTIC THERAPY

In contrast to the anticoagulants heparin and warfarin, which function to prevent fibrin clot formation, the thrombolytic agents act to dissolve established thrombus by converting endogenous plasminogen to plasmin, which can lyse existing thrombus. Thrombolytic therapy should be considered for arterial thrombi using tissue plasminogen activator, urokinase or streptokinase.

Tissue Plasminogen Activator

1. Tissue plasminogen activator (t-PA) infusion should be given at a rate of 0.5 mg/kg/h IV for 6 hours (standard dose t-PA). However, because of the risk of bleeding with standard dose t-PA, low-dose t-PA (0.03–0.06 mg/kg/h) up to 96 hours has been used with equivalent efficacy and lower bleeding risk. Heparin in a dose of 20 units/kg/hours should be continued with the t-PA.
2. Heparin should be given (20 units/kg/h) during t-PA infusion if the patient is not already on heparin.

3. Following 6 hours of t-PA infusion and if there is no response to treatment, the plasminogen level should be determined. If the plasminogen level is low, FFP 20 cc/kg IV q8h should be administered. A repeat infusion of t-PA may be considered

4. No arterial sticks or intramuscular injections should be given during t-PA administration.

Urokinase

Urokinase (UK) should be administered in a loading dose of 4,000 units/kg over 10 minutes, followed by 4,000 units/kg/h for 6 hours:

- Heparin should be given (20 units/kg/h) during UK infusion if the patient is not already on heparin
- Following 6 hours of UK infusion and if there is no response to treatment, the plasminogen level should be determined. If the plasminogen level is low, FPP 20 cc/kg IV q8h should be administered. A repeat infusion of UK may be considered
- Streptokinase (SK) can be administered when t-PA and UK are not available.

Streptokinase

Streptokinase (SK) should not be administered if it was used previously or if the plasminogen level is low (i.e., newborns). The loading dose consists of 4,000 units/kg (maximum 250,000 units) over 10 minutes, followed by 2,000 units/kg/h for 6 hours:

- Heparin should be given (20 units/kg/h) during SK infusion if the patient is not already on heparin
- If there is no response to treatment, the plasminogen level should be determined. If the plasminogen level is low, FPP 20 cc/kg IV q8h should be given. UK or t-PA should be utilized (not SK) for further thrombolytic therapy
- Patients must be premedicated with Tylenol and Benadryl before SK (repeat every 4–6 hours)
- Patients should not receive more than one course of SK because of the potential for allergic reactions. An anaphylactic reaction can occur in 1–2% of patients receiving SK. In the event of an anaphylactic reaction discontinue SK immediately and administer epinephrine, steroids and antihistamines.

Monitoring Response of Thrombolytic Therapy

1. Obtain a PT and aPTT every 6 hours.
2. Determine the fibrinogen level and/or fibrin/fibrinogen degradation products (FDPs) or D-dimer, every 6 hours.

3. Determine the plasminogen level at the end of the 6-hour infusion if there is no response or prior to proceeding to another course of therapy.
4. The fibrinogen concentration may decrease by at least 20–50%; and the fibrinogen concentration must be maintained at approximately 100 mg/dl by cryoprecipitate infusions (1 unit/5 kg).
5. When the fibrinogen concentration is less than 100 mg/dl and the patient is still receiving infusion of SK, UK, or t-PA, the dose of the thrombolytic agent should be decreased by 25%.
6. The platelet count should be maintained at 100,000/mm^3 or higher. Every urine specimen should be examined for blood and every stool should have a guaiac test.
7. Six hours following thrombolytic therapy, heparin therapy may be sufficient for 24 hours before re-instituting thrombolytic therapy. There may be ongoing thrombolysis even in the absence of continued administration of the thrombolytic agent.

Complications of Thrombolytic Therapy

Minor bleeding may occur in up to 50% of patients (i.e., oozing from a wound or puncture site). Supportive care and application of local pressure may be sufficient.

When severe bleeding occurs the infusion of the thrombolytic agent should be terminated and cryoprecipitate should be administered (usual dose of 1 unit/5 kg).

When life-threatening bleeding occurs the infusion of the thrombolytic agent should be terminated.

The fibrinolytic process can be reversed by infusing Amicar 100 mg/kg (maximum, 5 g) bolus, then 30 mg/kg/h (maximum, 1.25 g/h) until bleeding stops (maximum, 18 g/m^2/day). Protamine sulfate may be required to reverse the heparin effect.

The table below lists the management of blocked central venous catheters (CVC) using tissue plasminogen activator (t-PA).

Weight	Single Lumen-CVC	Double Lumen-CVC	Subcutaneous CVC (Mediport)
<10 kg	0.5 mg t-PA diluted in 0.9% NaCl to volume required to fill line	0.5 mg t-PA diluted in 0.9% NaCl per lumen to fill volume of line. Treat one lumen at a time	0.5 mg t-PA diluted with 0.9% NaCl to 3 ml
>10 kg	1.0 mg t-PA in 1.0 ml 0.9% NaCl. Use amount required to fill volume of line, to maximum of 2 ml (2 mg t-PA)	1.0 mg t-PA in 1.0 ml 0.9% NaCl. Use amount required to fill volume of line, to a maximum of 2 ml (2 mg t-PA per lumen). Treat one lumen at a time	2.0 mg t-PA diluted with 0.9% NaCl to 3 ml

ANTITHROMBOTIC THERAPY IN SPECIAL CONDITIONS

Mechanical Valve Replacement

Heparin

1. Heparinization starting at 48 hours postoperatively.
2. See heparin dose (p. 446).
3. Heparin should be stopped for 2 hours before intracardiac lines are removed.
4. Chest tube removal or insertion is not a contraindication for heparin.

Warfarin

1. Warfarin begins with oral intake and is continued indefinitely.
2. See warfarin dose schedule (p. 451).
3. Aim for an INR of 2.5–3.5.

Aortic Valve Replacement

1. Low-dose aspirin begins with oral intake (3–5 mg/kg/day) and is continued for 3 months.

Mitral and Tricuspid Valve Replacement

1. Atrial fibrillation or a proven intra-atrial thrombus, treatment as per mechanical valve replacement (INR of 2.5–3.5).
2. If normal sinus rhythm, anticoagulation will be with warfarin for 3 months (INR of 2.0–3.0) and low-dose aspirin (ASA) (3–5 mg/kg/day), indefinitely.

Pulmonary Valve Replacement

1. Currently no treatment.

Fontan Procedure

Heparin

1. No heparin loading dose.
2. Low-dose heparin infusion of 10 units/kg/h starting 48 hours postoperatively.
3. Heparin is stopped for 2 hours before intracardiac lines are removed.
4. Chest tube removal or insertion is not a contraindication for heparin.

Warfarin

1. Warfarin begins with oral intake and is continued for 3 months.
2. Patients after a Fontan procedure are very sensitive to oral anticoagulants. Loading doses of warfarin should be decreased.
3. Warfarin dosing should be adjusted according to the warfarin protocol (INR of dose 2.0–3.0).

Aspirin

1. Low-dose ASA (3–5 mg/kg/day) starts with oral intake and continues indefinitely.

Blalock-Taussig Shunts

Heparin

1. Loading dose of heparin: 75 units/kg over 10 min starts immediately postoperative.
2. Maintenance dose: 28 units/kg/h (≤1 year old): 20 units/kg/h (>1 year old).

Aspirin

1. Low-dose ASA (3–5 mg/kg/day) starts with oral intake and continues indefinitely.

Acute Arterial Infarct with Evidence of Dissection in Cerebral or Carotid Arteries

Heparin

1. Loading dose of heparin followed by maintenance dose (see heparin dose, p. 446).

Warfarin

1. Warfarin therapy for a minimum of 3 months (see warfarin dose, p. 451).

Aspirin

1. Low-dose ASA (3–5 mg/kg/day) begins with oral intake and is continued for 3 months.

Idiopathic Arterial Infarct Without Evidence of Dissection

1. Low-dose ASA (3–5 mg/kg/day) for a minimum of 3 months.

Arterial Infarct with Moyamoya Syndrome

1. Anticoagulation with heparin or LMWH for 5 days.
2. Low-dose ASA (3–5 mg/kg/day) starts at diagnosis and continues indefinitely.

Transient Ischemic Attacks

1. Anticoagulation with heparin or LMWH for 5 days.
2. Low-dose ASA (3.5 mg/kg/day) starts at diagnosis and continues indefinitely.

Acute Stroke without Hemorrhage

1. Anticoagulation with heparin or LMWH for 5 days.
2. Subsequent anticoagulation will be according to the underlying etiology and depends on the presence or absence of dissection

Sinovenous Thrombosis

1. Anticoagulation with heparin or LMWH 10–14 days in newborns. In infants or older children heparin or LMWH for 5 days followed by warfarin for 3 months (for dosage see heparin, nomogram, LMWH dose adjustment schedule and warfarin dose adjustment schedule).

Suggested Reading

Baker Jr. WF. Thrombolytic therapy clinical application. Hematol Oncol Clin N Am. 2003;17:283–311.

Barnard T, Goldenberg NA. Pediatric arterial ischemic stroke in children. Pediatr Clin North Am. 2008;55:323–338.

Bick RL. Antiphospholipid thrombosis syndrome. Hematol Oncol Clin N Am. 2003;17:115–147.

Bick RL. Prothrombin G 20210A mutation, heparin Cofactor II, protein C and protein S defects. Hematol Oncol Clin N Am. 2003;17:9–36.

Crowther MA, Kelton JG. Congenital thrombophilic states associated with venous thrombosis. A qualitative overview and proposed classification system. Ann Intern Med. 2003;138:128–134.

Hirsch J, et al. Antithrombotic and thrombolytic therapy. Chest. 2008;133(6 suppl1):110s–112s.

Kwann HC, Nabhan C. Hereditary and acquired defects in the fibrinolytic system associated with thrombosis. Hematol Oncol Clin N Am. 2003;17:103–114.

Levine JS, Branch W, Rauch J. Antiphospholipid syndrome. New Engl J Med. 2002;346:752–763.

Meinardi JR, Middeldorp S, De Kam PJ, et al. The incidence of recurrent venous thromboembolism in carriers of factor V Leiden is related to concomitant thrombophilic disorders. Brit J Hematol. 2002;116:625–633.

Monagle P, Chalmers E, Chan A, et al. Antithrombotic therapy in neonates and children. Chest 2008;133:887S–968S.

Roach ES, Golomb MR, Adama R, et al. Management of stroke in infants and children. A Scientific Statement from a Special Writing Group of the American Heart Association Stroke Council and the Council on Cardiovascular Disease in the Young. Stroke 2008;39:2544.

Selighsohn U, Lubertsky A. Genetic susceptibility to venous thrombosis. N Eng J Med. 2001;344:1222–1231.

Wang M, et al. Low-dose tissue plasminogen activator thrombolysis in children. J Pediatr Hematolol Oncolol. 2003;25(5):379–386.

CHAPTER 15

Lymphadenopathy and Splenomegaly

Lymphadenopathy and splenomegaly are common findings in children. Both benign and malignant processes can produce these findings and it is important to distinguish between the two so that appropriate management can be undertaken.

LYMPHADENOPATHY

Enlarged lymph nodes are commonly found in children. Lymphadenopathy might be caused by proliferation of cells intrinsic to the node, such as lymphocytes, plasma cells, monocytes or histiocytes or by infiltration of cells extrinsic to the node, such as neutrophils and malignant cells. In most instances, lymphadenopathy represents transient proliferative responses to local or generalized infections. Reactive hyperplasia, defined as a polyclonal proliferation of one or more cell types, is the most frequent diagnosis in pediatric lymph node biopsies.

Lymphadenopathy is also a presenting sign of malignancies such as leukemia, lymphoma, or neuroblastoma and it is important to be able to differentiate benign from malignant lymphadenopathy.

Lymphadenopathy in the head and neck region must be differentiated from several other masses due to congenital malformations (Table 15-1).

Systematic palpation of the lymph nodes is important and should include examination of the occipital, posterior auricular, preauricular, tonsillar, submandibular, submental, upper anterior cervical, lower anterior cervical, posterior upper and lower cervical, supraclavicular, infraclavicular, axillary, epitrochlear and popliteal lymph nodes. Many children have small palpable nodes in the cervical, axillary and inguinal regions which are usually benign in nature. However, adenopathy in the supraclavicular regions is usually pathologic.

When a child presents with lymphadenopathy, management is based on the following factors.

History

This involves the duration of the lymphadenopathy; fever; recent upper respiratory tract infection; sore throat; skin lesions or abrasions, or other infections in the lymphatic region

Manual of Pediatric Hematology and Oncology. DOI: 10.1016/B978-0-12-375154-6.00015-X

Table 15-1   Differential Diagnosis of Non-lymph Node Masses in Neck

- Cystic hygroma
- Branchial cleft anomalies, branchial cysts
- Thyroglossal duct cysts
- Epidermoid cysts
- Neonatal torticollis
- Lateral process of lower cervical vertebra may be misdiagnosed as supraclavicular node

drained by the enlarged lymph nodes; immunizations; medications; previous cat scratches, rodent bites, or tick bites; arthralgia; sexual history; transfusion history; travel history and consumption of unpasteurized milk. Significant weight loss, night sweats, or other systemic symptoms should also be recorded as part of the patient's history.

Age

In children younger than 6 years, the most common cancers of the head and neck are neuroblastoma, rhabdomyosarcoma, leukemia and non-Hodgkin lymphoma. In children 7–13 years of age, non-Hodgkin lymphoma and Hodgkin lymphoma are equally common followed by thyroid carcinoma and rhabdomyosarcoma; and for those older than 13 years Hodgkin disease is the more common cancer encountered.

Location

Enlargement of tonsillar and inguinal lymph nodes is most likely secondary to localized infection; enlargement of supraclavicular and axillary lymph nodes is more likely to be of a serious nature. Enlargement of the left supraclavicular node, in particular, should suggest a malignant disease (e.g., malignant lymphoma or rhabdomyosarcoma) arising in the abdomen and spreading via the thoracic duct to the left supraclavicular area. Enlargement of the right supraclavicular node indicates intrathoracic lesions because this node drains the superior areas of the lungs and mediastinum. Palpable supraclavicular nodes are an indication for a thorough search for intrathoracic or intra-abdominal pathology.

Lymphadenopathy is either localized (one region affected) or generalized (two or more noncontiguous lymph node regions involved). Although localized lymphadenopathy is generally due to local infection in the region drained by the particular lymph nodes, it may also be due to malignant disease, such as Hodgkin disease or neuroblastoma. Generalized lymphadenopathy is caused by many disease processes. Lymphadenopathy may initially be localized and subsequently become generalized.

Size

Nodes in excess of 2.5 cm should be regarded as pathologic. In addition, nodes that increase in size over time are significant.

Character

Malignant nodes are generally firm, rubbery and matted. They are usually not tender or erythematous. Occasionally, a rapidly growing malignant node may be tender. Nodes due to infection or inflammation are generally warm, tender and fluctuant. If infection is considered to be the cause of the adenopathy, it is reasonable to perform a 2-week trial of antibiotic therapy. Failure to produce a reduction in the size of the lymph node within this period is an indication for careful observation. If the size, location and character of the node suggest malignant disease, the node should be biopsied.

Diagnosis of Lymphadenopathy

Table 15-2 outlines the differential diagnosis of lymphadenopathy.

Figure 15-1 provides a diagnostics algorithm for evaluation of mononucleosis-like illness and Figure 15-2 for diagnostic evaluation of cervical lymphadenitis.

The following investigations should be carried out to elucidate the cause of either localized or generalized lymphadenopathy:

- Thorough history of infection, contact with rodents or cats and systemic complaints
- Careful examination of the lymphadenopathy including size, consistency, mobility, warmth, tenderness, erythema, fluctuation and location. All the lymph-node-bearing areas as outlined above should be carefully examined
- Physical examination for evidence of hematologic disease, such as hepatosplenomegaly and petechiae
- Blood count and erythrocyte sedimentation rate (ESR)
- Skin testing for tuberculosis
- Bacteriologic culture of regional lesions (e.g., throat)
- Specific serologic tests for Epstein–Barr virus (EBV), *Bartonella henselae* (IFA), syphilis (VDRL) toxoplasmosis, cytomegalovirus (CMV), human immunodeficiency virus (HIV), tularemia, brucellosis, histoplasmosis, coccidiomycosis
- Chest radiograph and CT scan (if necessary); abdominal sonogram and CT, if indicated
- Ultrasonography is useful in an acute setting in assessing whether a swelling is nodal in origin, an infected cyst or other soft tissue mass. It may detect an abscess requiring drainage
- EKG and echocardiogram if Kawasaki disease is suspected
- Lymph node aspiration and culture; helpful in isolating the causative organism and deciding on an appropriate antibiotic when infection is the cause of the lymphadenopathy
- Fine needle aspiration; may yield a definite or preliminary cytologic diagnosis and occasionally obviate the need for lymph node biopsy; it provides limited material in the

Table 15-2   Differential Diagnosis of Lymphadenopathy

I. **Nonspecific reactive hyperplasia (polyclonal)**
II. **Infection**
    A. Bacterial:
        Staphylococcus, streptococcus, anaerobes, tuberculosis, atypical mycobacteria, *Bartonella henselae* (cat scratch disease, brucellosis, *Salmonella typhi*, diphtheria, *C. trachomatis* lymphogranuloma venereum), calymmatobacterium granulomatis, francisella tularensis
    B. Viral:
        Epstein–Barr virus, cytomegalovirus, adenovirus, rhinovirus, coronavirus, respiratory syncytial virus, influenza, coxsackie virus, rubella, rubeola, varicella, HIV, herpes simplex virus, human herpes virus 6 (HHV-6)
    C. Protozoal:
        Toxoplasmosis, malaria, trypanosomiasis
    D. Spirochetal:
        Syphilis, rickettsia typhi (murine typhus)
    E. Fungal:
        Coccidioidomycosis (valley fever), histoplasmosis, cryptococcus, aspergillosis
    F. Postvaccination:
        Smallpox, live attenuated measles, DPT, Salk vaccine, typhoid fever
III. **Connective tissue disorders**
    A. Rheumatoid arthritis
    B. Systemic lupus erythematosus
IV. **Hypersensitivity states**
    A. Serum sickness
    B. Drug reaction (e.g., Dilantin, mephenytoin, pyrimethamine, phenylbutazone, allopurinol, isoniazid, antileprosy and antithyroid medications)
V. **Lymphoproliferative disorders** (Chapter 16)
    A. Angioimmunoblastic lymphadenopathy with dysproteinemia
    B. X-linked lymphoproliferative syndrome
    C. Lymphomatoid granulomatosis
    D. Sinus histiocytosis with massive lymphadenopathy (Rosai–Dorfman disease)
    E. Castleman disease benign (giant lymph node hyperplasia, angiofollicular lymph node hyperplasia)
    F. Autoimmune lymphoproliferative syndrome (ALPS) (Canale-Smith syndrome)
    G. Post-transplantation lymphoproliferative disorder (PTLD)
VI. **Neoplastic diseases**
    A. Hodgkin and non-Hodgkin lymphomas
    B. Leukemia
    C. Metastatic disease from solid tumors: neuroblastoma, nasopharyngeal carcinoma, rhabdomyosarcoma, thyroid cancer
    D. Histiocytosis
        1. Langerhans cell histiocytosis
        2. Familial hemophagocytic lymphohistiocytosis
        3. Macrophage activation syndrome
        4. Malignant histiocytosis
VII. **Storage diseases**
    A. Niemann–Pick disease
    B. Gaucher disease
    C. Cystinosis

(Continued)

Table 15-2   (Continued)

VIII. **Immunodeficiency states**
   A. Chronic granulomatous disease
   B. Leukocyte adhesion deficiency
   C. Primary dysgammaglobulinemia with lymphadenopathy
  IX. **Miscellaneous causes**
   A. Kawasaki disease (mucocutaneous lymph node syndrome)
   B. Kikuchi–Fujimoto disease (self limiting histiocytic necrotizing lymphadenitis)
   C. Sarcoidosis
   D. Beryllium exposure
   E. Hyperthyroidism
   F. Periodic fever, aphthous stomatitis, pharyngitis and cervical adenitis syndrome (PFAPA syndrome)

---

event flow cytometry is required and negative results cannot rule out a malignancy because the sample may be inadequate

- Bone marrow examination if leukemia or lymphoma is suspected
- Lymph node biopsy is indicated if:
  - Initial physical examination and history suggest malignancy
  - Lymph node size is greater than 2.5 cm in absence of signs of infection
  - Lymph node persists or enlarges
  - Appropriate antibiotics fail to shrink node within 2 weeks
  - Supraclavicular adenopathy.

Close communication between surgeon, oncologist and pathologist is critical to maximize results from lymph node biopsy. In addition the following precautions should be observed:

- Upper cervical and inguinal areas should be avoided; lower cervical and axillary nodes are more likely to give reliable information
- The largest node should be biopsied, not the most accessible one. The oncologist should select the node to be biopsied in consultation with the surgeon
- The node should be removed intact with the capsule, not piecemeal
- The lymph node should be immediately submitted to the pathologist fresh or in sufficient tissue culture medium to prevent the tissue from drying out. The node must not be left in strong light, where it will be subject to heat and it should not be wrapped in dry gauze, which may produce a drying artifact. Fresh and frozen samples should be set aside for additional studies, as noted below.

Intraoperative frozen section and cytologic smears should be performed. These findings, together with the clinical data, will determine which of the following additional studies may be required:

- Gram stain and culture (bacterial including mycobacterial, viral and/or fungal) if clinically warranted or if intraoperative frozen section suggests an infection

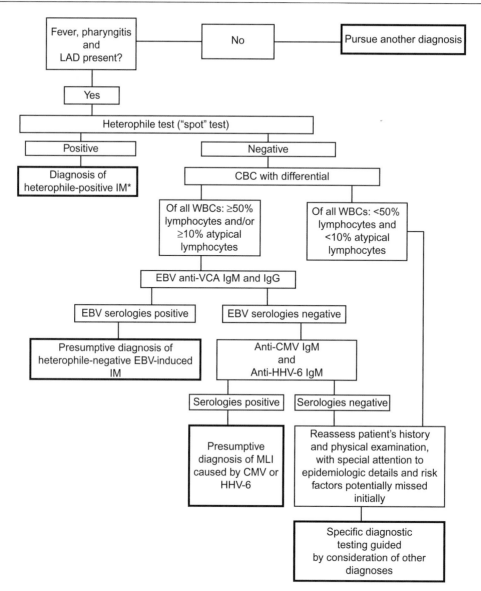

Figure 15-1 Diagnostic Algorithm for Evaluation of Mononucleosis-like Illness (MLI).
Abbreviations: CMV, cytomegalovirus; EBV, Epstein-Barr virus; HHV-6, human herpes virus 6; IM, infectious mononucleosis; LAD, lymphadenopathy; VCA, viral capsid antigen; WBC, white blood cell.
*Consider possibility of false-positive heterophile test due to HIV-1 before finalizing diagnosis.
Adapted from: Hurt C, Tammaro D. Diagnostic evaluation of mononucleosis-like illnesses. Am J Med 2007;120(10):911e1–911e8, with permission.

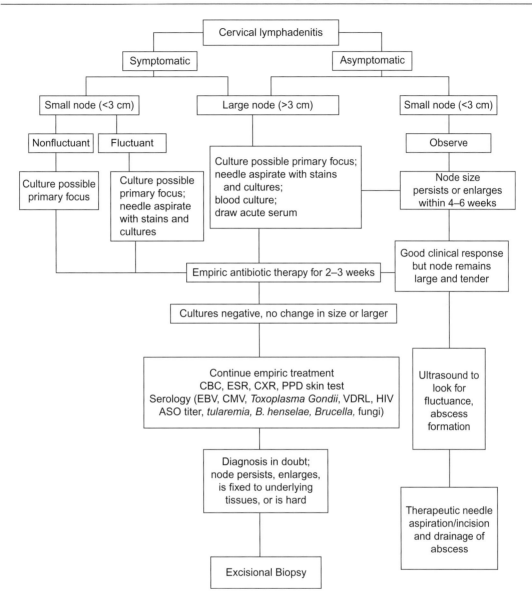

Figure 15-2 Diagnostic Evaluation of Cervical Lymphadenitis.

Abbreviations: ASO, anti-streptolysin titer; CXR, chest radiography; CBC, complete blood cell count; CMV, cytomegalovirus; EBV, Epstein-Barr virus; ESR, erythrocyte sedimentation rate; HIV, human immunodeficiency virus; PPD, purified protein derivative; VDRL, venereal disease research laboratories.

Adapted from: Gosche JR, Vick L. Acute, subacute and chronic cervical lymphadenitis in children. Semin Pediatr Surg 2006;15:99–106, with permission.

- Tissue in tissue culture medium for cytogenetic analysis in cases of suspected malignancy. Smears or touch preparations of the node on slides can be air dried for fluorescent *in situ* hybridization studies to confirm certain malignancies
- Tissue frozen immediately for molecular studies
- Immunohistochemical stains to help differentiate and classify tumor types
- Flow cytometry for classifying and subtyping leukemias and lymphomas
- Gene rearrangement studies for the T-cell receptor and the immunoglobulin gene may be required to determine monoclonality in leukemia or lymphoma
- These can be performed on fresh frozen tissue, or less optimally in formalin-fixed paraffin-embedded tissue
- Formalin fixation for light microscopic analysis.

Once the cause of the lymphadenopathy is ascertained, appropriate management can be undertaken.

SPLENOMEGALY

The tip of the spleen is frequently palpable in otherwise normal infants and young children. It is usually palpable in premature infants and in about 30% of full-term infants. It may normally be felt in children up to 3 or 4 years of age. At an older age, the spleen tip is generally not palpable below the costal margin and a palpable spleen usually indicates splenic enlargement two to three times its normal size.

Splenoptosis

In children, a palpable spleen may occasionally be due to visceroptosis rather than true splenomegaly. This distinction is important to make so that extensive investigations for the cause of splenomegaly are not undertaken unnecessarily. Visceroptosis may result from congenital or acquired defects in the supporting mechanism responsible for maintaining the spleen in the correct position. The visceroptosed spleen may be felt anywhere from the upper abdomen to the pelvis and may undergo torsion. When the spleen is felt in the upper abdomen, it can easily be pushed under the left costal margin. This finding is helpful in diagnosing visceroptosis and in differentiating it from true splenomegaly.

In addition to this finding, an abdominal radiograph in the upright position may reveal intestinal gas bubbles between the left dome of the diaphragm and the spleen. This sign may be helpful in suggesting the diagnosis.

Splenomegaly

The significance of splenomegaly depends on the underlying disease. Splenomegaly can be caused by diseases that result in hyperplasia of the lymphoid and reticuloendothelial systems (e.g., infections, connective tissue disorders), infiltrative disorders (e.g., Gaucher disease, leukemia, lymphoma), hematologic disorders (e.g., thalassemia, hereditary sphero-cytosis) and conditions that cause distention of the sinusoids whenever there is increased pressure in the portal or splenic veins (portal hypertension). Table 15-3 lists the various causes of splenomegaly.

Diagnostic Approach to Splenomegaly

Detailed History

1. Fever or rigors indicative of infection (e.g., subacute bacterial endocarditis [SBE], infectious mononucleosis, malaria).
2. Neonatal omphalitis, umbilical venous catheterization (inferior vena cava or portal vein thrombosis).
3. Jaundice (evidence of liver disease).
4. Abnormal bleeding or bruising (hematologic malignancy).
5. Family history of hemolytic anemia (e.g., hereditary spherocytosis or thalassemia major).
6. Travel to endemic areas (e.g., malaria).
7. Trauma (splenic hematoma).

Physical Examination

1. Size of spleen (measured in centimeters below costal margin); consistency, tenderness, audible rub.
2. Hepatomegaly.
3. Lymphadenopathy.
4. Fever.
5. Ecchymoses, purpura, petechiae.
6. Stigmata of liver disease such as jaundice, spider angiomata, or caput medusa.
7. Stigmata of rheumatoid arthritis or systemic lupus erythematosus (SLE).
8. Cardiac murmurs, Osler nodes, Janeway lesions, splinter hemorrhages, fundal hemorrhages (SBE).

Laboratory Investigations

The extent to which the following investigations are undertaken must be guided by clinical judgment. It is not necessary to perform all the evaluations. If the child appears well and the index of suspicion is low, it is reasonable to do no further investigations and re-examine the child in 1–2 weeks. If the splenomegaly persists, the following investigations should be done:

• *Blood count*: Red cell indices, reticulocyte count, platelet count, differential white blood cell count and blood film (which may demonstrate evidence of hematologic malignancy, hemolytic disorders, viral and protozoal infections)

Table 15-3   Causes of Splenomegaly

I. **Infectious splenomegaly** (due to antigenic stimulation with hyperplasia of the reticuloendothelial and lymphoid systems)
  A. Bacterial: acute and chronic systemic infection, subacute bacterial endocarditis, abscesses, typhoid fever, miliary tuberculosis, tularemia, plague
  B. Viral: infectious mononucleosis (Epstein–Barr virus), cytomegalovirus, HIV, hepatitis A, B, C
  C. Spirochetal: Syphilis, lyme disease, leptospirosis
  D. Rickettsial: Rocky Mountain spotted fever, Q fever, typhus
  E. Protozoal: malaria, babesiosis, toxoplasmosis, toxocara canis, toxocara catis, leishmaniasis, schistosomiasis, trypanosomiasis
  F. Fungal: Disseminated candidiasis, histoplasmosis, coccidioidomycosis, South American blastomycosis
II. **Hematologic disorders**
  A. Hemolytic anemias, such as thalassemia, splenic sequestration crisis in sickle cell disease, hereditary spherocytosis
  B. Extramedullary hematopoiesis as in osteopetrosis and myelofibrosis
  C. Myeloproliferative disorders (e.g., polycythemia vera, essential thrombocythemia)
III. **Infiltrative splenomegaly**
  A. Nonmalignant
    1. Langerhans cell histiocytosis
    2. Storage diseases such as Gaucher disease, Niemann–Pick disease, GM-1 gangliosidosis, glycogen storage disease type IV, Tangier disease, Wolman disease, mucopolysaccharidoses, hyperchylomicronemia types I and IV, amyloidosis and sarcoidosis
  B. Malignant
    1. Leukemia
    2. Lymphoma: Hodgkin and non-Hodgkin
IV. **Congestive splenomegaly**
  A. Intrahepatic (portal hypertension): cirrhosis of the liver (e.g., neonatal hepatitis, α_1-antitrypsin deficiency, Wilson's disease, cystic fibrosis)
  B. Prehepatosplenic or portal vein obstruction (e.g., thrombosis, vascular malformations)
V. **Immunologic diseases**
  A. Serum sickness, graft-versus-host disease
  B. Connective tissue disorders (e.g., systemic lupus erythematosus, rheumatoid arthritis—Felty syndrome, mixed connective tissue disorder, Sjogren syndrome, macrophage activation syndrome, systemic mastocytosis)
  C. Common variable immunodeficiency
  D. Autoimmune lymphoproliferative syndrome (ALPS) (Canale-Smith syndrome)
VI. **Primary splenic disorders**
  A. Cysts
  B. Benign tumors (e.g., hemangioma, lymphangioma)
  C. Hemorrhage in spleen (e.g., subcapsular hematoma)
  D. Partial torsion of splenic pedicle leading to congestive splenomegaly, cyst and abscess formation

---

- *Evaluation for infection*: Blood culture and viral studies (CMV, EBV panel, HIV, toxoplasmosis, smear for malaria, tuberculin test)
- *Evaluation for evidence of hemolytic disease*: Blood count, reticulocyte count, blood smear, haptoglobin level, serum bilirubin, urinary urobilinogen, direct antiglobulin test (Coombs test), osmotic fragility, autohemolysis and red cell enzyme assays, if indicated

- *Evaluation for liver disease*: Liver function tests, α_1-antitrypsin deficiency, serum copper, ceruloplasmin (to exclude Wilson disease) and liver biopsy (if indicated)
- *Evaluation for portal hypertension*: Ultrasound and Doppler of portal venous system and endoscopy (if indicated to exclude esophageal varices)
- *Evaluation for connective tissue disease*: ESR, C3, C4, CH_{50}, antinuclear antibody (ANA), rheumatoid factor, urinalysis, blood urea nitrogen (BUN) and serum creatinine
- *Evaluation for infiltrative disease (benign and malignant)*:
  - Bone marrow aspiration and biopsy, looking for blasts, Langerhans cell histiocytes, or storage cells
  - Enzyme assay for Gaucher disease
- *Lymph node biopsy*: If there is significant lymphadenopathy, lymph node biopsy may provide the diagnosis
- *Imaging studies*:
  - CT scan
  - Magnetic resonance imaging (MRI)
  - Liver – spleen scans with ^{99m}Tc-sulfur colloid
- *Biopsy*: If less invasive studies have failed to provide the diagnosis, it may be necessary to perform a splenectomy or a partial splenectomy. Biopsy and splenic aspiration are rarely performed because there is a significant risk of bleeding. Biopsy material must be processed for cultures and Gram stain, as well as for histology, flow cytometry, histochemical stains, electron microscopy and gene rearrangement studies. Once the etiology of the splenomegaly is ascertained, further management for the underlying disorder can be instituted.

Suggested Reading

Behrman R, Kliegman RM, Jenson HB. Nelson Textbook of Pediatrics. 17th ed. Elsevier; 2004.

Gosche JR, Vick L. Acute, subacute and chronic cervical lymphadenitis in children. Semin Ped Surg. 2006; 15(2):99–106.

Hoffman R, Benz EJ, Shattil SJ, et al. Hematology: Basic Principles and Practice. 3rd ed. Churchill Livingstone; 2000.

Hurt C, Tammaro D. Diagnostic evaluation of mononucleosis-like illnesses. Am J Med. 2007;120(10): 911.e1–911.e8.

Kim DS. Kawasaki disease. Yonsei Med J. 2006;47(6):759–772.

La Barge 3rd DV, Salzmam KL, Harnsberger HR, et al. Sinus histiocytosis with massive lymphadenopathy (Rosai-Dorfman disease): imaging manifestations in the head and neck. Am J Roentgenol. 2008;191(6): W299–W306.

Leung AKC, Robson WLM. Childhood cervical lymphadenopathy. J Pediatr Health Care. 2004;18:3–7.

Nathan D, Orkin S. Nathan and Oski's Hematology of Infancy and Childhood. 6th edn. Philadelphia: Saunders; 2003.

Paradela S, Lorenzo J, Martinez-Gomez W, et al. Interface dermatitis in skin lesions of Kikuchi-Fujimoto's disease: a histopathological marker of evolution into systemic lupus erythematosus? Lupus. 2008;17 (12):1127–1135.

Tracy Jr TF, Muratore CS. Management of common head and neck masses. Semin Pediatr Surg. 2007;16(1):3–13.

Twist CJ, Link MP. Assessment of lymphadenopathy in children. Pediatr Clin N Am. 2002;49:1009–1025.

CHAPTER 16

Lymphoproliferative Disorders, Myelodysplastic Syndromes and Myeloproliferative Disorders

LYMPHOPROLIFERATIVE DISORDERS

Lymphoproliferative disorders manifest with uncontrolled hyperplasia of lymphoid tissues (lymph nodes, spleen, bone marrow, liver). They are a heterogeneous group of diseases that range from reactive polyclonal hyperplasia (immunologic disorders) to true monoclonal (malignant) diseases.

Angioimmunoblastic Lymphadenopathy with Dysproteinemia (AILD)

Manifestations of this clinicopathological syndrome include:

- Generalized lymphadenopathy (80%)
- Hepatosplenomegaly (70%)
- Fever (70%), malaise, weight loss, polyarthralgia
- Quantitative changes in serum proteins (polyclonal hypergammaglobulinemia, 70%); hypocomplementemia
- Autoantibodies; circulating immune complexes, anti-smooth muscle antibody
- Rashes
- Pulmonary infiltrates, pleural effusions
- Thrombocytopenia
- Hemolytic anemia (often direct antiglobulin test [Coombs'] positive).

Diagnosis

The lymph node shows architectural effacement, absence of germinal center, arborization of postcapillary venules and a polymorphous infiltrate that includes immunoblasts and plasma cells. Immunoblasts are CD4 positive. Lymph node cytogenetic studies have shown nonrandom abnormalities, including +3, 14q+ and del(8) (p21). AILD was previously considered to be a benign lymphoproliferative disorder with the potential to transform

Manual of Pediatric Hematology and Oncology. DOI: 10.1016/B978-0-12-375154-6.00016-1

into lymphoma. Since it is a monoclonal disorder, it is now recognized as a type of angioimmunoblastic peripheral T-cell lymphoma.

Prognosis

Prognosis is poor. Death usually results from overwhelming infections. The median overall survival is 1.5 years.

Treatment of AILD-type Lymphoma

Chemotherapy: Adriamycin-containing regimens such as CHOP (see Table 16-7 later) have been used. Fludarabine has been used successfully in some patients with AILD-type lymphoma.

Small Lymphocytic Infiltrates of the Orbit and Conjunctiva (Ocular Adnexal Lymphoid Proliferation, Pseudolymphoma, Benign Lymphoma, Atypical Lymphocytic Infiltrates)

Lymphocytic infiltrates of the orbit and conjunctiva may be divided into three histologic groups:

*   Monomorphous infiltrates of clearly atypical lymphocytes
*   Infiltrates composed of small lymphocytes with minimal or no cytologic atypia
*   Benign inflammatory pseudotumor or reactive follicular hyperplasia.

On the basis of immunophenotypic criteria, they can be divided into two classes:

*   Infiltrates with monotypic immunoglobulin expression
*   Infiltrates with polytypic immunoglobulin expression.

For the localized small lymphocytic infiltrates, monotypic (monoclonal) immunoglobulin expression confers a 50% risk of dissemination. The initial immunophenotypic (mono- or polyclonal) and molecular studies of various histologic groups fail to correlate with the eventual outcome of these cases because the initial polyclonal tumors may become monoclonal.

All patients presenting with small lymphocyte infiltrates of the orbit and conjunctiva should have a systemic evaluation with serum chemistries, blood counts and appropriate imaging studies at initial diagnosis and every 6 months for 5 years thereafter.

For localized disease, local radiotherapy is commonly used, regardless of histologic grading. DNA from *Chlamydia psittaci* has been isolated in a large percentage of biopsies and treatment with doxycyline may be efficacious.

Angiocentric Immunolymphoproliferative Disorders

These are a collection of entities classified as peripheral T-cell disorders and include lymphomatoid granulomatosis, midline granuloma and post malignancy angiocentric immunolymphoproliferative lymphoma.

There are three grades of angiocentric immunolymphoproliferative (AIL) disorders:

- *Grade I* – polymorphic infiltrates with minimal necrosis, few large atypical lymphoid cells and small lymphocytes lacking nuclear irregularities
- *Grade II* – cytologic atypia of small lymphocytes, scattered large atypical lymphoid cells and intermediate amount of necrosis
- *Grade III* – lymphoma, either diffuse, mixed, large cell, or immunoblastic, with prominent necrosis.

The cellular origin of AIL lymphoma remains uncertain because of the following findings:

- Immunophenotype of T cells (CD2+, CD3+, CD4$\pm$, CD5$\pm$, CD7$\pm$)
- Absence of clonal rearrangements of T-cell receptors
- Expression of natural killer (NK) cell antigens (CD16+, CD56+, CD57+).

AIL lymphoma has been postulated to be a clonal process induced by Epstein–Barr virus (EBV) infection of T lymphocytes.

Clinical Features

Lymphomatoid Granulomatosis

This is a systemic disease in which the lungs are typically involved. Patients may present with cough, dyspnea and chest pain, or they may be asymptomatic and discovered incidentally on a chest radiograph. Chest radiographs may show bilateral nodules, consolidation, diffuse bilateral reticulonodular infiltrates, lymphadenopathy (mediastinal and/or hilar) and/or pleural effusion. Purplish skin nodules occur and may undergo spontaneous central necrosis and ulceration. The kidneys, central nervous system (CNS), skeletal muscles, nasopharynx, or peripheral nerves may also be involved.

Midline Lethal Granuloma

This presents with a progressive necrotizing and destructive process involving the upper airways. There may or may not be a tumor mass. The most common sites of the disease are the nasal fossa, nasal septum, nasopharynx, palate and adjacent soft tissue or bony structures. There is often a history of longstanding sinusitis with purulent and foul-smelling nasal discharge.

Postmalignancy Angiocentric Immunolymphoproliferative Lymphoma

This rarely occurs in children. Five of the seven childhood cases had acute lymphoblastic leukemia (ALL) previously. The interval between the remission of ALL to the diagnosis of AIL lymphoma ranges from 1 month to 4–5 years. The prognosis is poor.

Treatment

There is no standard treatment; the majority of patients are treated with chemotherapy based on adult non-Hodgkin lymphoma protocols and rituximab. The most commonly used agents are cyclophosphamide, prednisone, adriamycin and vincristine.

Castleman Disease (Angiofollicular Lymph Node Hyperplasia, Benign Giant Lymph Node Hyperplasia, Angiomatous Lymphoid Hamartoma)

Castleman disease is characterized by an accumulation of non-malignant lymphoid tissue interspersed with plasma cells and blood vessels.

Vascular hyperplasia has been attributed to a humoral vasoproliferative factor.

Table 16-1 shows the relationship between histologic types and clinical features in Castleman disease.

The production of interleukin-6 (IL-6) by B cells in the germinal centers of hyperplastic lymph nodes in Castleman disease plays a central role in inducing the variety of symptoms in this disease. In localized disease, human herpes virus 8 (HHV-8) DNA sequences have

Table 16-1 Relationship between Histologic Types and Clinical Features in Castleman Disease

Histologic Type (Frequency)	Disease Sites	Symptoms
Hyaline vascular type (80%)	Solitary lymph node 1.5–16 cm or chain of lymph nodes; two-thirds in mediastinum; sometimes other sites such as peripheral lymph nodes, abdomen and Pelvis	>90% Asymptomatic; pressure effects referable to the location of the mass may be present; 5–10% patients have constitutional symptoms
Plasma cell variety (20%)	Two types: Unicentric (single node or chain) Multicentric (multiple nodal groups) splenomegaly may be present	Associated with systemic symptoms such as fever, sweats, arthralgia, rashes, growth retardation, peripheral neuropathy, nephrotic syndrome Laboratory data: Hypergammaglobulinemia, microcytic anemia, raised ESR, amyloidosis; plasma cell proliferation usually polyclonal; increased IL-6 production

been detected in CD19 B cells. In multicentric disease, HHV-8 sequences have been detected in CD19 B cells and CD2 T cells.

Clinical Features

Unicentric hyaline vascular variant:
* Single lymph node or chain: cervical and mediastinal nodes most common
* 5–10% of patients have constitutional symptoms, but most are asymptomatic
* May have reactive diffuse shotty non-pathologic lymphadenopathy
* Associated with thrombotic thrombocytopenic purpura.

Unicentric plasma cell variant:
* Single lymph node or chain: abdominal nodes most common
* Associated with increased IL-6
* May have reactive diffuse shotty non-pathologic lymphadenopathy and splenomegaly
* Frequently have constitutional symptoms, hematologic abnormalities (anemia, thrombocytopenia, lymphocytosis), hypoalbuminemia and hypergammaglobulinemia.

Multicentric plasma cell variant:
* Similar clinical presentation as unicentric plasma cell variant; however, it involves multiple lymph nodes and chains. Hepatosplenomegaly is more common
* May also have neurologic manifestations, peripheral edema, ascites and pleural effusions
* Associated with HIV and HHV8 infections.

Prognosis

Localized disease: Excellent.

Multicentric disease: Poor. A minority with multicentric disease progress rapidly and develop hemolytic anemia and fatal intercurrent infection. Some develop non-Hodgkin lymphoma or Kaposi sarcoma.

Treatment

Localized disease: Surgical resection is curative. If complete resection is not possible, prednisone may be used to reduce size to enable complete surgical resection. Local radiation may be used for nonresectable disease.

Multicentric disease:
* Glucocorticoids alone or with vincristine
* Monoclonal antibodies: anti interleukin-6, anti interleukin-6 receptor, anti interleukin-1 receptor antibodies and anti-CD20 antibodes (rituximab)
* Multiagent chemotherapy (Table 16-7 later) with rituximab is considered the standard of care in many centers.

Epstein–Barr Virus (EBV)-Associated Lymphoproliferative Disorders (LPD) in Immunocompromised Individuals

EBV is a gamma herpes virus, a subfamily distinguished by its limited tissue tropism and latency. EBV enters via the oropharyngeal route and infects resting B lymphocytes through the interaction between the EBV viral envelope glycoprotein (gp350/220) and the C3d complement receptor (CD21). Infected B cells induce immunologic response of both virus-specific and -nonspecific T cells during primary infection which leads to regression of the majority of infected B cells. However, the virus persists in its latent state lifelong, by its continued presence in a small number of B cells, which express latent membrane protein 2A (LMP2 A) and small EBV encoded RNA 1 and 2 (EBERs 1 and 2). Table 16-2 lists the functions of EBV latent antigens.

EBV Antigens Associated with the Lytic Cycle

- Early antigen (EA)
- Viral capsid antigen (VCA).

During acute infectious mononucleosis (AIM) humoral responses are directed against both lytic and latent proteins. IgM antibodies to cellular proteins and heterophile antigens are also present during AIM.

Detectable levels of IgG antibodies to VCA and EBNA1 persist through lifetime.

In conjunction with humoral response, the CD4 and CD8 T-cell-mediated responses also play an important role in controlling the EBV infection by identifying and destroying latently infected cells.

Table 16-2   Functions of EBV Latent Antigens

EBV Antigen	Function
LMP1	Prevention of apoptosis (through NFκB and MAPK, induction of antiapoptotic proteins AP-1, bcl-2) and immortalization of B cells, increase in tumor invasiveness (by induction of matrix metalloproteinase)
LMP 2A	Inhibits lysis of B cells
Lytic antigens	vIL-10 and, BHRF1 facilitate lysis of cells and induce survival and proliferation of nearby, latently infected B cells
EBNA 1	Maintenance and replication of the EBV episome in latency
EBNA 2	Has an immortalizing function

Abbreviations: MAPK, mitogen activated protein kinase; NFκB, nuclear factor κ B; vIL-10, viral homolog of cellular interleukin-10; BHRF1, viral homolog of cellular bcl-2;.bcl-2, B cell lymphoma protein.

Cellular Responses in the Control of EBV Infection

The lymphocytosis during AIM consists of activated CD4 and CD8 T lymphocytes. They are specific for EBV lytic and latent proteins.

Cell	Reactivity (Function)
CD8+ & CD4+ T cells	Directed towards latent phase proteins: EBNA3, EBNA4, LMP1, EBNA6
NK (natural killer cells)	Lysis of virus containing cells during lytic phase

Reasons for persistence of latency in EBV infection:

- Paucity of production of neutralizing antibodies to the gp 350/220 component of the viral glycoprotein, a ligand for interaction with C3d on B cells
- Absence of cytotoxic T lymphocytes (CTL) response to EBNA1 which is primarily responsible for replication of and maintenance of viral genome during host cell division.

A state of balance between the control of EBV infection in latent state and host's immune response is achieved in a healthy host after the control of AIM phase. However, this state of balance is offset if the host is immunocompromised. Under this situation, the EBV can proliferate and can cause chronic active EBV infection (CAEBV), hemophagocytic lymphohistiocytosis (HLH), LPD, lymphomas and other rare complications.

Figure 16-1 shows latency types and corresponding patterns of gene expression observed in different diseases.

The following immunodeficiencies are associated with the development of lympho-proliferative disorders (LPDs):

- Inherited immunodeficiencies
  - Ataxia telangiectasia
  - Wiscott–Aldrich syndrome
  - Common variable immunodeficiency
  - Severe combined immunodeficiency (SCID)
  - Antibody deficiency syndromes such as hypergammaglobulinemia, hyper IgM syndrome, X-linked agammaglobulinemia, IgA and IgG subclass deficiency
  - Bloom syndrome
  - Chédiak–Higashi syndrome.
- Biological factors of significance involved in the pathogenesis of LPDs in this population are:
  - EBV, which causes B-cell proliferation
  - Imbalanced production of cytokines (e.g., predominance of interleukin-4 (IL-4) and interleukin-6 (IL-6) activates B-cells, i.e., humoral arm of immune system)

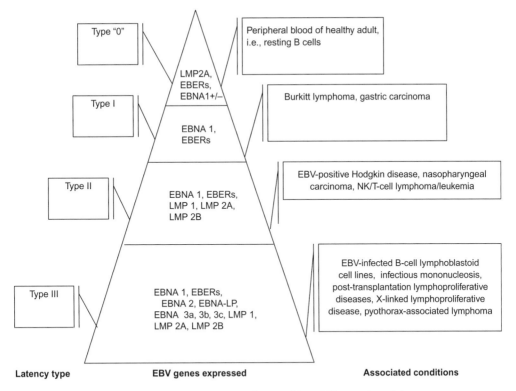

Figure 16-1 Patterns of EBV Gene Expression Observed in Different Conditions.
Abbreviations: LMP, Latent Membrane Protein. EBNA, Epstein-Barr Virus Nuclear Antigen. EBNA-LP, EBNA-leader Protein. EBETs, Epstein-Barr-encoded RNAs.
Modified from: Swaminathan, S: Molecular Biology of Epstein-Barr Virus and Kaposi's Sarcoma Associated Herpesvirus. Semin Hematol. 2003;40:107–115 and Tsuchiya S: Diagnosis of Epstein-Barr-Associated Diseases. Crit Rev Oncol/Hematol. 2002;44:227–238.

- Genetic defects resulting in ineffective or aberrant rearrangement of T-cell receptor and immunoglobulin genes.
- Iatrogenically induced immunodeficiencies
  - *Organ transplantation recipients*: Long-term treatment with immunosuppressive drugs, including the calcineurin inhibitors cyclosporine and tacrolimus, increases the risk of infections with EBV. EBV status of the organ transplantation recipient plays a significant role. If the patient is EBV seronegative and the donor is seropositive, then the risk of developing clinical infectious mononucleosis followed frequently by LPD increases remarkably. The incidence is six times higher in children and it is more likely to present with clinical features of fulminating infectious mononucleosis. Many children have not been previously exposed to EBV and, thus, are more likely to develop a primary EBV infection. Post transplantation-LPD (PT-LPDs) after solid

organ transplantation are of B-cell origin and are associated with EBV infection in most patients. A few cases of T-cell PT-LPDs have been reported

- *Hematopoietic stem cell transplantation (HSCT) recipients*: HSCT differs in the following respects:

  The patient receives a new immune system from a donor who is either a full histocompability locus antigen (HLA) match or a partial HLA match. The patient's immune system is ablated prior to HSCT. HSCT may require T-cell depletion of the donor's bone marrow.

  The complete recovery of the donor's immune system is delayed by the use of cyclosporine, methotrexate, prednisone and calcineurin inhibitors, which are used to prevent rejection of graft and graft versus host disease (GVHD). It takes 6 months to 2 years for immunologic recovery to occur. LPD after HSCT is also associated with EBV-infected B cells. The incidence of LPD in standard matched HSCT is low, but it is 6–12% in HLA-mismatched HSCT due to an increased use of T-cell depletion of the donor's marrow (relative risk factor = 12.3). Also, the use of antithymocyte globulin (ATG) in a preparative regimen increases the risk of PT-LPD (relative risk factor = 3.1). Depletion of both T and B cells from the graft are associated with a lower incidence of PT-LPD (relative risk factor = 2), since, by removing B cells, it depletes the graft of latent virus creating a balance between latent virus and cellular T cell immunity. PT-LPD has also been reported after autologous transplantation after CD34+ selection, resulting in complete depletion of lymphocytes.

  If HSCT is performed for immunodeficiency, a relative risk factor for post HSCT LPD is 2.5–3.8.

Table 16-3 shows a classification of post-transplantation lymphoproliferative disorders.

The frequency of organ involvement with PT-LPDs varies with the type of transplantation. For example, liver involvement is 50% in HSCT recipients; lung involvement is 54% in heart transplantation recipients and 38% in HSCT recipients; CNS 24% in renal transplantation recipients; and kidneys 24% in HSCT recipients and 33% in liver transplantation recipients.

Table 16-4 lists the World Health Organization (WHO) classification of post-transplantation lymphoproliferative disorders (PT-LPD).

Diagnosis of Post-Transplantation Lymphoproliferative Disorders

1. Physical examination: Table 16-5 shows the common sites of involvement in B cell post-transplantation LPDs.
2. Biopsy of the appropriate presenting site for histologic studies of immune phenotyping, cytogenetics, molecular analysis and the examination of EBV genomes in the lymphoid cells (i.e., *in situ* hybridization, Southern blot, polymerase chain reaction (PCR),

Table 16-3   Classification of Post-Transplantation Lymphoproliferative Disorders

1. Plasmacytic hyperplasia
      Commonly arise in the oropharynx or lymph nodes
      Nearly always polyclonal
      Usually contains multiple clones of EBV
      No oncogenes and tumor suppressor gene alterations
2. Polymorphic B-cell hyperplasia and polymorphic B-cell lymphoma
      Can involve lymph nodes or extranodal sites
      Nearly always monoclonal
      Usually contain a single clone of EBV
      No oncogenes or tumor suppressor gene alterations
3. Immunoblastic lymphoma or multiple myeloma
      Present with widely disseminated disease
      Always monoclonal
      Contains a single clone of EBV
      Contains alterations of one or more oncogenes or tumor suppressor genes (N-*ras*, p53, c-*myc*)

From: Magrath IT, Shad AT, Sandlund JT. Lymphoproliferative disorders in immunocompromised individuals. In: Magrath IT, editor. The non-Hodgkin lymphoma. 2nd ed. London: Arnold, 1997:955–974, with permission.

Table 16-4   WHO Classification of Post-Transplantation Lymphoproliferative Disorders (PT-LPDS)

Early lesions
   Reactive plasmacytic hyperplasia
   Infectious mononucleosis type
PT-LPD, polymorphic
   Polyclonal (rare)
   Monoclonal
PT-LPD, monomorphic (classify according to lymphoma classification)
   B-cell lymphomas
      Diffuse large cell lymphoma (immunoblastic, centroblastic, anaplastic)
      Burkitt/Burkitt-like lymphoma
      Plasma cell myeloma
   T-cell lymphomas
      Peripheral T-cell lymphoma, not otherwise categorized
      Other types (hepatosplenic, gamma/delta, T/NK)
Other types, rare
   Hodgkin disease-like lesions (associated with methotrexate therapy)
   Plasmacytoma-like lesions

From: Harris NL, Jaffe ES, Diebold J, Flandrin,G, et al: World Health Organization classification of neoplastic Disease of the hematopoietic and lymphoid tissues: Report of the Clinical Advisory Committee Meeting-Airlie House, Virginia, November 1997; J Clin Oncol 1999;17:3835–3849.
Abbreviation: T/NK, T-cell natural killer.

Table 16-5   Common Sites of Involvement in B-Cell Post-Transplantation
Lymphoproliferative Disorders

Site	Percentage of Patients
Lymph nodes	59
Liver	31
Lung	29
Kidney	25
Bone marrow	25
Small intestine	22
Spleen	21
Central nervous system	19
Large intestine	14
Tonsils	10
Adrenals	9
Skin/soft tissue	7
Blood	6
Heart	5
Salivary glands	4

From: Magrath IT, Shad AT, Sandlund JT. Lymphoproliferative disorders in immunocompromised individuals. In: Magrath IT, editor. The non-Hodgkin lymphoma, 2nd ed. London: Arnold, 1997:955–974, with permission.

terminal repeat probes for clonality). Also, immunostaining of tissues for EBV-RNAs (EBERs) and EBV antigens (proteins) are required for a diagnosis of EBV LPD.

3. Complete blood count.
4. Kidney and liver chemistries, lactic dehydrogenase (LDH).
5. EBV serology: PT-LPD patients have very high anti-VCA titers but lack anti-EBNA antibodies; however, this is not always a consistent finding.
6. A real-time quantitative polymerase chain reaction (qPCR) for EBV-DNA in serum or plasma.
7. A real-time qPCR for EBV-DNA copies in serum or plasma can be used for pre-emptive diagnosis of post-hematopoietic stem cell PTLD and post solid organ transplantation PTLD.

Table 16-6 shows the comparison of characteristics of lymphoproliferative diseases in post solid organ transplantation (SOT) patients and post hematopoietic stem cell transplantation (HSCT) patients.

Treatment

The indications for treatment of LPD with chemotherapy are:

• Patient in whom all other measures to control LPD have failed and patient has widespread LPD
• Patient who has developed monoclonal LPD.

Table 16-6   Comparison of Lymphoproliferative Disorders (LPD): Post-Solid Organ Transplantation and Post-Hematopoietic Stem Cell Transplantation (HSCT)

Characteristic	LPD Post-Solid Organ Transplantation	LPD Post-Hematopoietic Stem Cell Transplantation
Immune system	Patient's own i.e. organ recipient's own	Replaced by donor's
Immunosuppressive therapy	Life-long	About 6 months post transplantation, or as long as graft-versus-host disease (GVHD) persists.
Onset of LPD	Highest incidence: 6–12 months after transplantation	Highest incidence: 6–12 months after transplantation
EBV association	Early-onset LPD: always associated with EBV, Late onset LPD: multifactorial, many patients being EBV negative	Almost always EBV associated
Cellular origin of LPD	LPD arises in patient's, i.e. organ recipient's own lymphocytes	LPD arises in HSCT donor's lymphocytes, however, sometimes it may arise from blood transfusions
Incidence	EBV status of recipient: Seronegative 15.8% Seropositive 5.4% Organ transplantation: Kidney 1.2–9% Liver 6.8–13% Thoracic organ 3.8–11.7% Heart 5–15% Lung 10–20% Intestinal/multivisceral 31% Type of immunosuppression Antithymocyte globulin (ATG) 11.4%	Cumulative incidence 1% at 10 years Mismatched 1% Mismatched and T-cell depletion 1–8% Unrelated 1.5% Unrelated and T-cell depletion 5–29% Unrelated and Campath depletion 1.3%
Treatment	Pre-emptive therapy Intravenous immunoglobulin infusion Acyclovir or gancyclovir Infusion of patient's (organ recepient's) own EBV-specific cytotoxic T-cells (CTL) Infusion of anti-CD20 antibody (Rituximab) Therapeutic Same as pre-emptive plus chemotherapy	Pre-emptive therapy Intravenous immunoglobulin infusion Acyclovir or gancyclovir Infusion of donor's EBV-specific cytotoxic T-cells Infusion of anti-CD20 antibody (Rituximab) Therapeutic Same as pre-emptive plus chemotherapy

Treatment of B-Cell Lymphoproliferative Disease in Immunosuppressed Patients

General Treatment

The following treatment strategy is employed in all patients:

- Reduction or withdrawal of immunosuppressive therapy (including prednisone and other drugs such as cyclosporine A, azathioprine and tacrolimus) is attempted as first-line treatment. However, this can aggravate graft versus host disease (GVHD) in HSCT patients and cause graft rejection in organ transplantation patients. In addition, transition of immunosuppressive therapy to mTOR (mammalian target of rapamycin) inhibitors may be effective in treating PT-LPD

- High-dose acyclovir (500 mg/m^2 every 8 hours with standard adjustment for impaired renal function) and intravenous immunoglobulin (IVIgG): 500 mg/kg day once a week for 2–4 weeks. Antiviral agents, acyclovir or ganciclovir, may be effective as a pre-emptive measure.

These measures are usually not effective in HSCT patients because the regenerating donor's immune system cannot provide enough immunity to eradicate EBV-infected B-cells. In solid organ transplantation patients, regression of LPD has been observed in 23–86% of patients with these measures.

For localized LPD, surgery or local radiation may be helpful.

Specific Treatment

1. Inherited immunodeficiency syndromes
   Polyclonal or monoclonal with no lymphoma-specific genetic abnormalities
   a. Localized LPD
      Radiation (low-dose) or surgery[*]
   b. Generalized LPD
      Interferon (see below)
      Anti-B-cell monoclonal antibodies (see below)
      Chemotherapy (Table 16-7) if no response or if LPD is rapidly progressive or recurrent
   Monoclonal with a lymphoma-specific genetic abnormality[**]
      Chemotherapy[***]
2. Post-solid organ transplantation
   Polyclonal or monoclonal with no lymphoma-specific genetic abnormalities
   Reduce dose of immunosuppressive drugs if possible
   a. Localized LPD
      Radiation or surgery
   b. Generalized LPD
      i. Interferon:
         Interferon-α (IFN-α) Dose: 3 million units/m^2 once a day subcutaneously. IFN-α stimulates natural killer (NK) cells. It also inhibits proliferation of EBV-infected B cells.

---

[*]Avoid radiation therapy in patients with chromosomal instability/DNA repairs defect diseases e.g. ataxia telangiectasia.
[**]Any B-cell malignancy.
[***]Use smaller doses of chemotherapy agents for patients with chromosomal instability/DNA repair defect diseases e.g. ataxia telangiectasia (Table 16-7).

Table 16-7   Treatment of Post-transplantation Lymphoproliferative Disorder with Cyclophosphamide, Doxorubicin, Vincristine, Prenisone (CHOP)[a]

Cyclophosphamide	750 mg/m^2 Intravenous on Day 1 of each cycle
Doxorubicin	50 mg/m^2 intravenous on Day 1 of each cycle
Vincristine	1.4 mg/m^2 intravenous (with maximum dose 2 mg) on day 1 of each cycle
Prednisone	100 mg/m^2 orally (with maximum dose 100 mg) per day on days 1 through 5

[a]This same regimen (CHOP) can also be utilized for treatment of angioimmunoblastic lymphadenopathy with dysproteinemia (AILD)-type lymphoma and Castleman disease.
Treatment cycles are repeated every 21 days or, following hematopoietic recovery. Patients who experience complete remission after four cycles are treated with two additional cycles of chemotherapy. Patients are maintained on reduced cyclosporine levels indefinitely.

    ii.   Anti-B-cell monoclonal antibodies:
- Humanized anti-CD21 and anti-CD24, although effective in treatment of postorgan transplantation LPD, are not commercially available
- Humanized anti-CD20 monoclonal antibody (Rituximab) is commercially available. It is administered at a dose of 375 mg/m^2 once a week for 4 weeks after reducing the dose of immunosuppressive drugs. Complete remission rates of 50–65% have been reported. However, it should be used cautiously since it may increase the risk of graft rejection, especially, when a reduction in immunosuppressive therapy is attempted at the same time. A second course of anti-CD20 may be given for the relapsed patients
- Pre-emptive therapy with one infusion of 375 mg/m^2 of Rituximab in recipients of T-cell-depleted allo-HSCT who develop reactivation of EBV (1,000 or more genome-equivalent (geq)/mL) results in a significant reduction of LPD and prevention of LPD-related-mortality. Reduction in EBV DNA to less than 50 geq/mL is considered effective therapy.

   iii.   Adoptive immunotherapy
Infusion of patient's own EBV-specific cytotoxic T-cells (CTL) has been tried in few patients with an 80% rate of normalization of EBV DNA viral load.

   iv.   Other therapies:
- Anti-Interleukin-6 (IL-6) antibody treatment: IL-6 induces proliferation and maturation of EBV-infected B-cells. Anti-IL-6 antibody has been tried in few patients with some success
- Vaccination of EBV-seronegative recipient before undergoing transplantation
- Chemotherapy: If there is no response or rapidly progressive, or recurrent disease chemotherapy should be utilized.

Table 16-7 lists the treatment regimen with cyclophosphamide, doxorubicin, vincristine and prednisone (CHOP Regimen) used for post-transplantation lymphoproliferative disorder.

3. Post–HSCT

Interferon (see above)

Anti-B-cell monoclonal antibodies (see above)

Donor leukocyte infusions:

- Use of unmanipulated donor's T-cells
  - Without transduction of donor's T-cells with a suicide gene that can be turned on in case of worsening of GVHD
  - Transduced donor's T-cells with a suicide gene, herpes simplex virus thymidine kinase, (HSV-tk), which can be activated by ganciclovir therapy in the event of worsening of GVHD.
- Use of selectively *ex vivo* expanded EBV-specific donor's cytotoxic T-cells (CTL)

Chemotherapy (Table 16-7) should be employed if no response or recurrence occurs.

Monoclonal with a lymphoma specific genetic abnormality: Chemotherapy (Table 16-7).

X-Linked Lymphoproliferative Syndrome

X-linked lymphoproliferative (XLP) syndrome is a rare disorder characterized by dysregulation of T-cell-mediated immune response, which is induced by EBV.

Clinical manifestations of XLP such as dysgammaglobulinemia, aplastic anemia and lymphoproliferative disease have been described in the absence of EBV infection.

Pathophysiology

Familial XLP results from mutations in *SH2D1A* gene that makes *SAP* (SLAM-associated protein) protein. Normally, in activated T cells and NK cells SAP levels increase. SAP interacts with SLAM (signaling lymphocyte activation molecule family), a molecule expressed on T cell surface, B cell surface and dendritic cell surface. In T cells, SAP regulates T cell receptor (TCR) induced IFN-γ. In NK cells, SAP binds to 2B4 and NTB-A (which also belong to SLAM family receptors) and activates NK-cell-induced cytotoxicity.

In XLP, various types of *SAP (SH2D1A)* gene mutations have been described including deletion, nonsense, missense and splice site mutations. As a result of this, SAP protein may be absent or truncated, or may contain altered amino acid residue at highly conserved sites.

No correlation has been found between genotypes and phenotypes and outcomes of these patients. It has been postulated that the mutation of SAP protein in XLP causes defective helper and cytotoxic T cell function. In some patients, XLP is related to a mutation in XIAP. In contrast to familial XLP, no mutations of SAP occur in sporadic XLP.

Clinical Manifestations

1. Fulminant infectious mononucleosis (frequency 58%, survival 4%): It is characterized by infiltration of various organs with polyclonal B and T cells, production of inflammatory cytokines and necrosis of liver, bone marrow, lymph nodes and spleen caused by the invading cytotoxic T-cell and uncontrolled killer cell activity. Death is generally attributable to liver failure with hepatic encephalopathy or bone marrow failure with fatal hemorrhage in the lungs, brain or gastrointestinal tract and occurs within 1 month of onset of symptoms.
2. Secondary dysgammaglobulinemia (frequency 30%, survival 55%).
3. B-cell lymphoproliferative disease including malignant lymphoma (extranodal nonHodgkin lymphoma) (frequency 25%, survival 35%).
4. Aplastic anemia.
5. Virus-associated hemophagocytic syndrome. Patients usually die within one month of onset of symptoms.
6. Vasculitis and pulmonary lymphomatoid granulomatosis.
7. In the same patient, several sequential phenotypes of the disease may manifest over time. This phenotypic variation most often includes dysgammaglobulinemia, malignant lymphoma and marrow aplasia.

Laboratory Manifestations

Patients with the XLP syndrome reveal many humoral and cellular immunologic defects, which include the following:

- Selective impaired immunity to EBV but normal immune responses to other herpes viruses
- Uncontrolled T_H1 responses (with high levels of IFN-γ in some patients)
- Inverted CD4/CD8 ratio (due to increase in CD8+ cells)
- Dysgammaglobulinemia: low IgG, high IgM and high IgA
- Defective NK cell activity
- Decreased T-cell regression assay
- Failure to switch from IgM to IgG class response after secondary challenge with OX174 and diminished mitogen-induced transformation of lymphocytes
- Unchanged lymphocyte-mediated antibody-dependent cellular cytotoxicity.

Diagnosis is established by sequencing analysis of *SH2D1A* and *XIAP* genes and SAP protein quantification.

Treatment

1. Prevention of EBV infection: Prophylactic use of intravenous immunoglobulin (IVIG) has been used without much success. High-dose IVIG, 600 mg/kg per month however, has been effective in some patients.

2. Treatment of acute EBV infection in XLP patients: High-dose IVIG and/or acyclovir are ineffective; however, rituximab may be helpful.
3. Treatment of virus associated hemophagocytosis: Corticosteroids, Etoposide and cyclosporine A are effective agents. Etoposide decreases macrophage activity and cyclosporine A decreases T cell activity. HLH (2004) protocol of the Histiocytic Society may be used (see Chapter 18, Histiocytosis Syndromes, p. 567–596).
4. Treatment of aplastic anemia: Etoposide and cyclosporine A are effective.
5. Treatment of lymphoma: Standard therapy for lymphoma is used, i.e. chemotherapy and radiation therapy where indicated.

Allogeneic hematopoietic stem cell transplantation (HSCT) is the only curative therapy for XLP syndrome.

Prognosis

Seventy percent of patients with XLP die before 10 years of age.

Autoimmune Lymphoproliferative Syndrome (ALPS) (Canale-Smith Syndrome)

Autoimmune lymphoproliferative syndrome (ALPS) is a disorder of disrupted lymphocyte homeostasis caused by defective fas-mediated apoptosis. Normally, as part of the down-regulation of the immune response, activated B and T lymphocytes up-regulate fas expression and activated T lymphocytes up-regulate expression of fas ligand. Fas and fas ligand interact triggering the caspase cascade, leading to cellular apoptosis. Patients with ALPS have a defect in this apoptotic pathway leading to chronic lymphoproliferation, autoimmunity and secondary malignancies.

Clinical Manifestations

1. Chronic (>6 months) non-malignant lymphoproliferation manifesting as lymphadeno-pathy and/or hepatomegaly and/or splenomegaly. Lymphoproliferation may be massive and tends to wax and wane. The majority of patients present with lymphoproliferation prior to 2 years of age; however, lymphoproliferation may not occur until adult years in rare cases.
2. Autoimmune disease. Greater than 80% of patients develop autoimmune disease. The most common manifestation is autoimmune destruction of blood cells (autoimmune cytopenias) including immune thrombocytopenia, autoimmune hemolytic anemia and autoimmune neutropenia. Patients may also develop autoimmune disease of almost any organ system, including autoimmune hepatitis, nephritis, gastritis and colitis. Patients may develop urticaria, alopecia and bronchiolitis obliterans.
3. Secondary malignancy. Occurs in greater than 10% of ALPS patients, most commonly lymphoma.

4. Common variable immunodeficiency (CVID). A subset of ALPS patients develops secondary CVID.

Laboratory Manifestations

1. Elevated DNTs (double-negative T cells; cell phenotype CD3+/TCRα/β+/CD4−/ CD8−). DNTs are a rare population representing <1% of circulating T lymphocytes in normal individuals. When markedly elevated, DNTs are virtually pathognomonic for ALPS. Mild elevations can be found in systemic lupus erythematosis and other autoimmune diseases.
2. *In vitro* evidence of defective fas-mediated apoptosis. Only performed in a few specialized laboratories.
3. Elevated gamma/delta DNTs, CD57+ T cells, CD5+ B cells, CD8+ T cells and HLA-DR+ T cells.
4. Elevated serum interleukin-10.
5. Increased soluble fas-ligand.
6. Increased vitamin B_{12}.
7. Hypergammaglobulinemia (or hypogammaglobulinemia).
8. Autoantibodies (DAT, anti-platelet, anti-neutrophil, ANA, Rf, anti-phospholipid).
9. Eosinophilia.

Pathophysiology

ALPS has been attributed to defective apoptosis (programmed cell death) of lymphocytes, most often arising as a result of mutations in the gene encoding the lymphocyte apoptosis receptor FAS/APO-1/CD95. Because of the failure of the affected lymphocytes to die after their response to antigen has been completed, there is an accumulation and build up of an excessive number of polyclonal lymphocytes, which leads to hepatosplenomegaly and lymphadenopathy.

There are subtypes of ALPS on the basis of phenotypic and genotypic differences:

- *ALPS 0*: This is caused by complete deficiency of Fas, resulting from homozygous null mutations of *fas (TNFRSF-6, tumor necrosis factor receptor super family member 6)* gene. Its mouse mutant model counterpart is *lpr/lpr,* Fas ko (knock out)
- *ALPS Ia*: This is the most common type of ALPS affecting 60–70% of patients. It is caused by heterozygous mutations of *fas (TNFRSF-6)* gene. Mutant Fas exerts a transdominant effect on wild-type Fas. Its mouse mutant model counterpart is *lpr^{cg}/lpr^{cg}*
- *ALPS Ib*: This is caused by a dominant mutation of Fas L (Fas ligand). Clinically, it manifests with features of systemic lupus erythematosus and chronic lympho-proliferation. However, it lacks the classical features of ALPS i.e. expansion of DNT cells and splenomegaly. Its mouse mutant counterpart is gld/gld

- *ALPS Is*: This is caused by somatic mutations of Fas in DNTs. These patients have a normal *in vitro* fas-mediated apoptosis assay, because DNTs do not survive in culture and apoptosis is tested in "normal" T-lymphocytes. Diagnosis is made by genetic analysis of sorted DNTs
- *ALPS II*: This is caused by caspase 10 deficiency. Caspase 8 deficiency was formerly considered a subset of ALPS; however, it is now considered a separate disease as these patients have defective apoptosis of B, T and NK cells and have difficulty with mucocutaneous herpes virus infections
- *ALPS III*: Molecular defect to account for this subtype of ALPS is unknown; 20–30% of ALPS patients are type III
- *ALPS IV*: NRAS mutations.

Diagnosis

The following criteria have been recommended by the National Institute of Health (NIH) ALPS Group to diagnose patients with ALPS.

Required Criteria (ALL three Mandatory for Diagnosis)

1. Chronic nonmalignant lymphoproliferation.
2. Defective lymphocyte apoptosis *in vitro*.
3. Greater than or equal to 1% TCR $\alpha/\beta+$ CD4$-$ CD8$-$ T cells ($\alpha/\beta+$-DNT cells) in peripheral blood and/or presence of DNT cells in lymphoid tissue.

Supporting Criteria

1. Autoimmunity/autoantibodies.
2. Mutations in TNFRSF6, FasL, or caspase10 gene.

There are problems with the NIH classification system, including dependence on a test only performed in a research setting (*in vitro* apoptosis assay) that is normal in patients with type 1b and 1s ALPS. In addition, marked elevation of DNTs is pathognomonic for ALPS, whereas lesser elevations of DNTs can be found in other disease entities. Finally, the NIH criteria do not take into account other clinical and laboratory manifestations associated with ALPS. For these reasons the following criteria have been recommended by the Children's Hospital of Philadelphia:

- Major criteria:
  - Chronic non-malignant lymphoproliferation
  - Elevated peripheral blood DNTs greater than or equal to 5%
  - Defective *in vitro* fas mediated apoptosis
  - Germ-line or somatic mutation in fas, fas-ligand, caspase-10, or NRAS.

- Minor criteria:
  - Hypergammaglobulinemia
  - Elevated serum IL-10
  - Increased DNTs on histopathology and/or elevated peripheral blood DNTs >2 S.D. above mean of reference lab but less <5%
  - Elevated serum vitamin B-12 level
  - Elevated serum fas-ligand
  - Autoimmune cytopenias
    - Neutropenia, hemolytic anemia and/or thrombocytopenia AND
    - Proven autoimmunity by either autoantibody and/or clinical response to immunosuppression.

Three major or two major plus two minor of the above criteria are required for diagnosis. One of the criteria (major or minor) must be elevated peripheral blood DNTs and/or increased DNTs on histopathology.

Mechanism of autoimmunity: IL-10 is a likely cytokine playing a role in autoimmune manifestations in ALPS. IL-10 induces T cell differentiation towards the Th2 cell type and these cells in turn stimulate autoreactive B cells to make autoantibodies. IL-10 also causes increased expression of BCL-2 proteins which contributes to inhibition of mitochondrial pathway of apoptosis.

The origin of $\alpha/\beta-$ DNT cells: these cells in ALPS are either cytotoxic CD8+ T cells that have lost CD8 expression or Tregs.

Treatment

Lymphoproliferation rarely requires treatment unless patients develop hypersplenism and/or organ compression, in which case it should be treated with immunosuppressive agents similarly to autoimmune disease. Splenectomy should be avoided because ALPS patients have a high risk of post-splenectomy sepsis even with vaccination and antimicrobial prophylaxis. Splenectomy is only indicated in patients with clinically significant hypersplenism who do not respond to immunosuppression. Other treatments include:

- Corticosteroids. Most commonly used. Typically very effective; however, many patients have chronic problems and develop steroid side-effects
- Second-line therapies: Mycophenolate mofetil and sirolimus. Patients who are refractory to or intolerant of corticosteroids may be treated with alternative immuno-suppressants. Mycophenolate mofetil (MMF) has the longest published track record; however, many patients are refractory to MMF. Sirolimus has been demonstrated to result in complete responses in MMF and steroid refractory patients. Other agents that may be effective include azathioprine, tacrolimus, mercaptopurine, methotrexate, vincristine and cyclosporine. Rituximab may be effective; however, a large percentage

of ALPS patients treated with rituximab have developed common variable immunodeficiency (CVID) and it should be avoided if possible. Early data suggested fansidar (pyrimethamine/sulphadoxine) may be effective; however, it was found ineffective in a larger trial.

Allogeneic stem cell transplantation is occasionally necessary.

Prognosis

ALPS may improve with age.

The severity of ALPS varies from mild to severe within the same family. This may be because other pathways of apoptosis are compensating for the FAS pathway.

The following malignancies have been reported in ALPS families:

• Burkitt lymphoma, T-cell rich B cell lymphoma and atypical lymphoma
• Nodular lymphocyte predominance Hodgkin disease
• Breast cancer, lung cancer, basal cell carcinoma of the skin, squamous cell carcinoma of the tongue and colon cancer.

Most of the cases of lymphoma occur in families with intracellular mutations of TNFRSF-6.

Dianzani Autoimmune Lymphoproliferative Disease (DALD)

This ALPS-like syndrome is characterized by:

• Defective function of the Fas receptor
• Autoimmune conditions, predominantly involving blood cells; ITP, autoimmune hemolytic anemia and autoimmune neutropenia
• Polyclonal accumulation of lymphocytes in the spleen and lymph nodes.

DALD differs from ALPS in that it lacks expansion of DNT cells. DALD patients have a high level of osteopontin (OPN) due to an increase in the frequency of B and C haplotypes of OPN. These haplotypes are responsible for the increased levels of OPN. *In vitro*, observations show that high levels of OPN decrease activation-induced T cell apoptosis. For this reason, high levels of OPN have been implicated in the apoptotic defect of DALD.

Lymphomatoid Papulosis in Children

Lymphomatoid papulosis (LyP) is a clinically benign skin disorder characterized by chronic and recurrent self-healing papulonodular lesions. This disorder occurs rarely in children. The sites of lesions include the limbs and/or trunk. The eruption recurs episodically, often ulcerates and heals spontaneously in 3–8 weeks, occasionally leaving an atrophic scar. LyP results from a clonal T-cell proliferation, which may explain its evolution or coexistence with Hodgkin disease, mycosis fungoidis, or anaplastic large cell lymphoma (ALCL).

Histology

LyP is characterized by superficial and deep infiltrates consisting of atypical lymphocytes with a hyperchromatic, pleomorphic nucleus that can obscure the dermoepidermal junction. Similar cellular infiltrates can also be found in the epidermis along with spongiosis, parakeratosis and neutrophils. Two types of activated T-cell infiltrates are found:

- LyP type A lesions: LyP lesions of this cell type predominantly contain Ki-1+ (CD30+), Reed–Sternberg-type cells with a pale-staining, convoluted nucleus, prominent nucleoli and moderately basophilic cytoplasm
- LyP type B lesions: LyP lesions of this cell type contain the Sezary-type cell with a cerebriform, hyperchromatic nucleus and scant cytoplasm. Only some of these cells are CD30+.

The atypical cells of LyP have been shown to be of T-cell origin, in most cases expressing CD2, CD3 and CD4. These cells lack the expression of some of the usual pan-T-cell antigens, such as CD5, CD7, or both. Clonal rearrangement of the T-cell antigen receptor (TCR- and/or clone genes) has been demonstrated in most of the cases studied. Thus, LyP serves as an example of a benign lymphoproliferative disorder in spite of the evidence of clonal T-cell receptor gene rearrangement.

A lymphoma evolving from lymphomatoid papulosis can be recognized by enlarging or persistent skin lesions, peripheral lymphadenopathy, or circulating atypical lymphocytes. A biopsy of suspicious skin lesions or enlarged lymph nodes for histologic, immunologic, cytogenetic, or gene rearrangement studies is indicated. It has also been suggested that a thorough physical examination every 6 months in children with lymphomatoid papulosis with attention to growth and development (as a tool in assessing occult malignancies in children), skin lesions and lymph nodes should be carried out. Additional studies such as bone marrow examination and radioimaging studies should be performed as needed.

Table 16-8 shows the clinical and histologic differences among cutaneous CD30+ NHL (including ALCL), borderline cases and LyP.

Treatment

Treatment involves oral antibiotics, systemic corticosteroids, low-dose methotrexate, psoralen with ultraviolet radiation (PUVA), topical steroids, or ultraviolet beam.

Table 16-8   Clinical and Histologic Differences Among Cutaneous CD30 + NHL (Including ALCLs), Borderline Cases and LyP

Variable	Cutaneous CD30 + NHL (Including ALCLs)	Borderline Cases	LyP
Clinical criteria			
Extent of skin lesion	Solitory > regional > generalized	Regional	Regional or generalized
Type of skin lesion	Nodule, tumor	Nodule	Papules or papulonodular
Spontoneous regression	Rare	Frequent	Always
Extracutaneous disease	Possible	Rare	Rare
Histology			
Wedge shape of infiltrates	Never	Sometimes	Regular
Infiltration of subcutis	Regular	Rare	Rare
Inflammatory infiltrate (with neutophils)	Infrequent	Regular, admixed	Regular, admixed
Epidermotropism	Infrequent	Epidermis infiltrated by small cells and large atypical CD30+ cells	Epidermis infiltrated by small cells, occasionally by large CD30+ cells
CD30 expression	>80% of tumor cells	Small clusters of CD30+ cells	Scattered CD30+cells

Abbreviations: NHL, nonHodgkin lymphoma; ALCL, anaplastic large cell lymphoma.
Modified from: Paulli M, Berti E. Rosso R, Boveri E, et al. CD30/Ki-1 positive lymphoproliferative disorders of the skin-clinicopathologcal correlation and statistical analysis of 86 cases: a multicentric study of the European Organization of Research and treatment of cancer cutaneous lymphoma project groups. J Clin Oncol 1995;13:1343–1354, with permission.

Prognosis

The duration of LyP ranges from 1 to 40 years. Approximately 23 children have been described and evolution to lymphomas has occurred in two of the 23 cases.

MYELODYSPLASTIC SYNDROME

Myelodysplastic syndrome (MDS) results from acquired clonal disorders of the pluripotent (i.e. a stem cell involved in myelopoiesis, erythropoiesis, megakaryopoiesis and lymphopoiesis) or multipotent (i.e. a stem cell restricted to myelopoiesis, erythropoiesis and megakaryopoiesis) hematopoietic progenitor cell. It is characterized by absolute and/or functional cytopenias with usually a hypercellular or normocellular bone marrow containing less than 20% blasts. It has the propensity to evolve into acute myeloid leukemia.

Table 16-9 provides minimal diagnostic criteria for MDS and Table 16-10 provides a list of investigations for the diagnosis and pertinent differential diagnosis of myelodysplasia.

Table 16-9 Minimal Diagnostic Criteria for Myelodysplastic Syndrome

At least two of the following:
• Sustained unexplained cytopenia (neutropenia, thrombocytopenia, or anemia)
• At least bilineage morphologic myelodysplasia
• Acquired conal cytogenetic abnormality in hematopoietic cell
• Increased blasts (5%)

From: Hasle H, et al. A pediatric approach to the WHO classification of myelodysplastic and myeloproliferative diseases. Leukemia 2003;17:277–282, with permission.

Table 16-10 Investigation for the Diagnosis of Myelodysplasia (MDS) and Pertinent Differential Diagnosis (D/D)

Blood
Hemoglobin level and red cell indices:
Macrocytosis D/D – Drugs, folate or B_{12} deficiency, congenital bone marrow failure syndromes, MDS, JMML
Microcytic and ring sideroblasts: unlikely to be MDS, exclude mitochondrial diseases
Normocytic red cells
White cell count, differential count, platelet count
Blood film
Fetal hemoglobin concentration[a]: D/D: MDS, JMML, congenital bone marrow failure syndromes
Immunodeficiencies: Occasionally, immunodeficiency may be associated with MDS or vice versa
Pertinent tests to be performed on peripheral blood in the context of above D/D:
Fetal hemoglobin (HbF)
Cytogenetics (mitomycin C or di-epoxybutane [DEB] study for excessive chromosomal breakage)[b]
Erythrocyte adenosine deaminase (ADA) level
Leukocyte alkaline phosphatase
Vitamin B_{12} and folate levels
Quantitative immunoglobulin levels and T and B lymphocyte quantitation
Bone marrow
Aspirate and trephine biopsy[c]
Cytochemistry and iron stain
Cytogenetics: Conventional and fluorescent *in-situ* hybridization for chromosome 7, 8 and BCR/ABL (Ph chromosome)
Molecular analysis: RT-PCR for BCR/ABL and FLT3 ITD
Additional tests
Neutrophil function
Platelet function
Colony-forming unit assay for various lineages on bone marrow cells
Assays for expression of CD 55 and CD 59 antigens on erythrocytes

[a]Draw blood for Hemoglobin electrophoresis, CD55 and CD59 antigen tests and adenosine deaminase (if macrocytosis) before transfusing patient with red blood cells.
[b]Patients with Fanconi anemia may present with myelodysplasia. The significance of this observation is that all patients with myelodysplasia should have chromosomal breakage analysis performed to exclude Fanconi anemia because Fanconi anemia is a recessive disorder and warrants genetic counseling. Additionally, preparative regimen for HSCT is different for patients with Fanconi anemia.
[c]Bone marrow trephine biopsy: Two additional types of MDS have now been recognized. A) Hypoplastic MDS and B) MDS with myelofibrosis. Also, in some reports emphasis is placed on recognizing abnormal location of immature precursor cells (ALIP) i.e. presence of blasts in intertrabecular areas in bone marrow biopsy specimens, since it may have prognostic significance. However, no significance has been found thus far, in the major reports of Pediatric MDS series.
Note: If bone marrow is hypocellular aplastic anemia and/or paroxysmal nocturnal hemoglobinuria should be considered. Hepatomegaly or splenomegaly or hepatosplenomegaly favors diagnosis of JMML or AML.

Table 16-11 shows the French–American–British (FAB) classification of MDS and the 2001 and 2008 World Health Organization (WHO) classification and criteria of MDS. These two classifications are poorly suited for children with MDS. A classification of MDS in children that conforms to the WHO suggestions and also allows for the special problems of MDS in children has not yet been developed. Table 16-12 shows diagnostic categories of MDS and myeloproliferative diseases in children.

The concept of monosomy 7 as a distinct syndrome has been abandoned. Myeloid leukemia in children with Down syndrome (DS) is distinct from the disease in non-DS children. MDS often precedes acute myeloid leukemia (AML) in DS. In some children with DS, AML is preceded by a transient myeloproliferative disorder. Nonclonal disorders with dysplastic morphology should not be considered to be MDS.

Epidemiology

1. Incidence: 1.8 per million children per year in age group 0–14 years.
2. Constitutes 4% of all hematological malignancies.
3. Constitutional abnormalities are present in 30% of children with MDS. The most common is Down syndrome (DS).

Table 16-13 lists constitutional abnormalities associated with JMML and MDS in children. Familial MDS is observed in 10% of children with MDS. It is associated with 7- or 7q-chromosome abnormalities.

Chemotherapy-related MDS

1. Alkylating agents-induced MDS is characterized by deletions or loss of whole chromosome. Latency period: 3–5 years.
2. Epipodophyllotoxins-induced MDS is characterized by translocations involving chromosome band 11q23. Latency period: 1–3 years.
3. Incidence of therapy-related MDS; Represents 5% of all childhood MDS and occurs in 13% of children treated for malignancies.

Children with chemotherapy-induced MDS or chemotherapy-induced AML (t-MDS/t-AML) compared to children with *de novo* AML or MDS, have the following characteristics:

- Older at presentation, have lower WBC counts and are less likely to have hepatomegaly or splenomegaly or hepatosplenomegaly
- More likely to have trisomy 8 and less likely to have classic AML translocations. The therapy-related form of the disease is associated with an increased frequency of abnormal karyotypes, complex cytogenetic abnormalities and deletions involving either chromosome 5 or chromosome 7 or both

Table 16-11 Classifications of MDS

FAB	WHO 2001	WHO 2008	WHO 2008 Diagnostic Criteria
RA	RA	RCUD	
(n/a)	RA	RA	Anemia (Hb <10 g/dl); ± neutropenia or thrombocytopenia; $<1\%$ circulating blasts; $<5\%$ medullary blasts; unequivocal dyserythropoiesis in $\geq10\%$ erythroid precursors; dysgranulopoiesis and dysmegakaryopoiesis, if present, in $<10\%$ nucleated cells; $<15\%$ RS; no Auer rods
(n/a)	(n/a)	RN	Neutropenia (absolute neutrophil count $<1.8\times10^9/L$); ± anemia or thrombocytopenia; $<1\%$ circulating blasts; $<5\%$ medullary blasts; $\geq10\%$ dysplastic neutrophils; $<15\%$ RS; no Auer rods
(n/a)	(n/a)	RT	Thrombocytopenia (platelet count $<100\times10^9/L$); ± anemia or neutropenia; $<1\%$ circulating blasts; $<5\%$ medullary blasts; $\geq10\%$ dysplastic megakaryocytes of ≥30 megakaryocytes; $<10\%$ dyserythropoiesis and dysgranulopoiesis; $<15\%$ RS; no Auer rods
RARS	RARS	RARS	Anemia; no circulating blasts; $<5\%$ medullary blasts; dyserythropoiesis only, with RS $\geq15\%$ of 100 erythroid precursors; no Auer rods
(n/a)	RCMD and RS	RCMD	Cytopenia(s); $<1\times10^9/L$ circulating monocytes; $<1\%$ circulating blasts; $<5\%$ medullary blasts; dysplasia among $>10\%$ cells of ≥2 lineages; no Auer rods
		(n/a)	(n/a)
RAEB	RAEB-1	RAEB-1	Cytopenia(s); $<1\times10^9/L$ circulating monocytes ; $<5\%$ circulating blasts; 5–9% medullary blasts; dysplasia involving ≥1 lineage(s); no Auer rods
(n/a)	RAEB-2	RAEB-2	Cytopenia(s); $<1\times10^9/L$ circulating monocytes; 5–19% circulating blasts; 10–19% medullary blasts; dysplasia involving ≥1 lineage(s); ± Auer rods
(n/a)	(n/a)	RAEB-F	Similar to RAEB-1 or RAEB-2, with at least bilineage dysplasia and with diffuse coarse reticulin fibrosis, with or without collagenous fibrosis
RAEB-T	(AML)	(AML)	(n/a)
(n/a)	MDS, U	MDS, U	Pancytopenia; $<1\times10^9/L$ circulating monocytes; $\leq1\%$ circulating blasts; $<5\%$ circulating blasts; dysplasia in $<10\%$ cells of ≥1 lineage(s); demonstration of MDS-associated chromosomal abnormality(ies), exclusive of trisomy 8, del(20q) and -Y
	MDS, isolated del(5q)	MDS, isolated del(5q)	Anemia; platelet count may be normal or increased; $<1\%$ circulating blasts; $<5\%$ medullary blasts; megakaryocytes with characteristic nuclear hypolobulation; isolated del(5q) cytogenetic abnormality involving bands q31-q33
CMML	(MDS/MPN)	(MDS/MPN)	(n/a)
(n/a)	(n/a)	RCC	Thrombocytopenia, anemia and/or neutropenia; $<2\%$ circulating blasts; $<5\%$ medullary blasts; unequivocal dysplasia in ≥2 lineages, or in $>10\%$ cells of one lineage; no RS

Abbreviations: RA, refractory anemia; RCUD, refractory cytopenia with unilineage dysplasia; CMML, chronic myelomonocytic leukemia; MDS, U, myelodysplastic syndrome, unclassifiable; MPN, myeloproliferative neoplasm; n/a, not applicable; RAEB-F, refractory anemia with excess blasts with fibrosis; RAEB-T, refractory anemia with excess blasts in transformation; RCC, refractory cytopenia of childhood; RN, refractory neutropenia; RS, ring sideroblasts; RT, refractory thrombocytopenia; RARS, refractory anemia with ringed sideroblasts; RCMD, refractory cytopenia with multilineage dysplasia.

From: Nguyen, P. The Myelodysplastic Syndromes. Hematol Oncol Clin North Am 2009;23:680-681 with permission.

Note: The main distinguishing feature of the subgroups is the proportion of myeloblasts in the bone marrow: less than 5% in refractory anemia and RARS, 5 to 20% in RAEB, 21 to 30% in RAEB-T and 0 to 20% in CMML. In addition, RARS is notable for more than 15% ring sideroblasts in the erythroid precursor population and CMML is notable for monocytosis ($>1.0\times10^9$ cells per liter). Ring sideroblasts, a defining feature of RARS, can also be seen in disease variants and probably represent defective iron transport between mitochondria and cytoplasm.

Table 16-12   Diagnostic Categories of Myelodysplastic and Myeloproliferative Diseases in Children

I. **Myelodysplastic/myeloproliferative disease**
   • Juvenile myelomonocytic leukemia (JMML)
   • Chronic myelomonocytic leukemia (CMML) (secondary only)
   • BCR/ABL negative chronic myeloid leukemia (Ph CML)
II. **Down syndrome (DS) disease**
   • Transient abnormal myelopoiesis (TAM)
   • Myeloid leukemia of DS
III. **Myelodysplastic syndrome (MDS)**
   • Refractory cytopenia (RC) (peripheral blood blasts less than 2% and bone marrow blasts less than 5%)
   • Refractory anemia with excess blasts (RAEB) (peripheral blood blasts 2–19% and bone marrow blasts 5–19%)

From: Hasle, H, et al.: A pediatric approach to the WHO classification of myelodysplastic and myeloproliferative diseases. Leukemia (2003);17:277–262, with permission.
Abbreviation: RAEB, Refractory Anemia with Excess Blasts.

Table 16-13   Constitutional Abnormalities Associated with JMML and MDS in Children

A. **Associated with JMML**
  Constitutional conditions
    Neurofibromatosis type 1 (NF-1)
    Noonan syndrome
    Trisomy 8 mosaicism
B. **Associated with MDS**
  Constitutional conditions
    Congenital bone marrow failure
      Fanconi anemia
      Kostmann syndrome
      Shwachman–Diamond syndrome
      Diamond–Blackfan syndrome
      Dyskeratosis Congenita
      Severe congenital neutropenia
    Trisomy 8 mosaicism
    Familial MDS (at least one first-degree relative with MDSAML)
  Acquired conditions
    Prior chemotherapy/radiation
    Aplastic anemia

From: Hasle H, et al. A pediatric approach to the WHO classification of myelodysplastic and myeloproliferative diseases. Leukemia 2003;17:177–282, with permission.

• Less likely to attain remission after induction therapy (50% versus 72%) and less likely to have a longer overall survival (26% versus 47%) and event free survival (21% versus 39%).

Their disease-free survival after attaining remission is similar to children with *de novo* AML or MDS (45% vs 53%).

Pathophysiology

MDS is a clonal disease arising either in a pluripotent or multipotent hematopoietic stem cell. It is a heterogeneous disease with different pathophysiologic mechanisms playing roles in its initiation and progression. It has been postulated that initially apoptosis dominates the process and with time, as more genetic abnormalities accumulate in the MDS cells, arrest of maturation and proliferation occur, resulting in transformation to acute myeloid leukemia (AML). There is not much information regarding the pathophysiologic mechanisms of MDS in childhood.

Clinical Features

Clinical features are related to anemia and other cytopenias (e.g. pallor, bruises, petechiae, infections). Usually, there is no lymphadenopathy or hepatosplenomegaly. Children with refractory anemia may not present with low hemoglobin levels, but they have macrocytosis and elevated fetal hemoglobin levels. Blood smears frequently have a dimorphic red-cell population including oval macrocytes, nuclear hyposegmentation of neutrophils (pseudo-Pelger-Huët cells) and hypogranulation of neutrophils and platelets.

Bone marrow dysplasia involving at least 10% of the cells of a specific myeloid lineage is the cardinal feature of the myelodysplastic syndromes. Ring sideroblasts, which reflect abnormal accumulation of iron in mitochondria, sometimes accompany signs of dyserythropoiesis.

Cytogenetics

Monosomy 7 is the most common cytogenetic abnormality in childhood MDS. Trisomy 8 and 21 are the second most common abnormalities. Other cytogenetic abnormalities are del(5q), del(20q) and loss of the Y chromosome, which is considered an age-related phenomenon and not always indicative of a clonal disorder.

Differential Diagnosis

1. Clinical course and the response to therapy for M6 (erythroleukemia), M7 (acute megakaryoblastic leukemia) and AML with monosomy 7 are more similar to MDS than the other types of AML. For this reason, it is important for these conditions to be diagnosed accurately. AML with monosomy 7 has a poor prognosis, whereas it is not an unfavorable prognostic factor in MDS.
2. Patients with 20% or greater bone marrow blasts are classified as having AML. The MDS subtype RAEB-T (refractory anemia with excess blasts in transformation [Table 16-11]) has been abandoned.
3. It is difficult to distinguish hypoplastic MDS (refractory anemia [RA]) from aplastic anemia in the absence of laboratory abnormalities. The majority of children who

develop MDS following a diagnosis of aplastic anemia, present with MDS within the first 3 years from the diagnosis of aplastic anemia. Patients with mild to moderate aplastic anemia may be more likely to develop a clonal disease than a patient with severe aplastic anemia. Repeated evaluation for both conditions including bone marrow examinations may become necessary to reach a diagnosis.

4. In a case of refractory anemia with ringed sideroblasts (RARS) a search for mitochondrial disease is warranted since this subtype of MDS is rare in children.

5. Erythroid dysplasia can be a secondary change in a variety of disorders that must be excluded before the diagnosis of a myelodysplastic syndrome is made. These disorders include deficiency of vitamin B_{12}, folate and copper; viral infections, including infection with the human immunodeficiency virus; treatment with hydroxyurea or other chemotherapeutic agents; lead or arsenic poisoning; and inherited disorders such as congenital dyserythropoietic anemias.

Prognosis

The International Prognostic Scoring System for MDS is listed on Table 16-14 but its applicability to MDS in children remains to be determined.

The FPC (hemoglobin F, platelet counts, cytogenetics)-based scoring system for prognostic classification of childhood MDS is listed in Table 16-15.

Table 16-16 shows International Prognostic System (IPSS) for childhood MDS and JMML.

Also, in an analysis reported by the European Working Group on MDS in childhood, preliminary data on childhood MDS revealed poor prognosis in patients with two to three lineage cytopenia and a blast count greater than 5% in bone marrow.

Table 16-14   International Prognostic Scoring System (IPSS) for Myelodysplastic Syndrome (MDS)

Prognostic Variable	Score Value				
	0	0.5	1.0	1.5	2.0
BM Blasts (%)	<5	5–10	–	11–20	21–30
Karyotype[a]	Good	Intermediate	Poor		
Cytopenias	0/1	2/3			

[a]Good, normal, -Y, del(5q), del(20q); Poor, complex (≥3 abnormalties) or chromosome 7 anomalies; Intermediate, other abnormalities.
Scores for risk groups are as follows: Low, 0; INT-1, 0.5–1.0; INT-2, 1.5–2.0; and High ≥2.5.
From: Greenberg P, et al. International scoring system for evaluating prognosis in myelodysplastic syndromes. Blood 1997;89:2085.

Table 16-15 FPC (Hemoglobin F, Platelet Count, Cytogenetics)-based Scoring System for Prognostic Classification of Childhood Myelodysplastic Syndrome[a]

Criteria (Before Treatment)	Points
Hemoglobin F (HbF) >10%	1
Platelet count < or = 40,000/mm^3	1
Cytogenetic studies	
No detectable clonal abnormalities	0
Simple abnormalities arising from one event, i.e., single translocation, other structural abnormalities, or gain or loss of one chromosome	1
Complex abnormalities arising from 2 or more structural or numerical events	2

Conclusion

Score	5-year Survival
0	61%
1	20%
2 or 3	0%

[a]Patients included: MDS as per FAB classification, Juvenile chronic myeloid leukemia, Infantile monosomy 7 syndrome, MDS with eosinophilia (data tabulated from information available from: Passmore SJ, et al. Blood 1995;85:1742–1750.

Table 16-16 International Prognostic Scoring System (IPSS)[a] for Childhood MDS and JMML (European Working Group on MDS in Childhood)

IPSS Risk Group	5-Year Survival (%)
Low	100
Intermediate 1	64
Intermediate 2	55
High	54

[a]Refer to Table 16-14 for definitions of IPSS risk groups for MDS which was generated on the basis of series of adult patients. Comment: Cytopenia and Bone Marrow blasts >5% correlated with poor survival, however, cytogenetic studies were not informative. Data tabulated from information available from: Hasle H, et al. Leukemia 2004;18:2010.

Treatment

Refractory anemia (RA) or refractory cytopenia: Children with RA may have a long and stable clinical course without treatment, the median time to refractory anemia with excess blasts (RAEB) being 47 months. For this reason, some investigators recommend that until the children with RA become susceptible to infections, usually due to neutropenia, or become transfusion-dependent they be observed with supportive therapy. Once these events set in, then, as soon as possible they should be treated with a matched related stem cell donor transplantation. If a matched related donor is not available then a matched unrelated donor transplantation should be considered. Chemotherapy prior to stem cell transplantation is not necessary. Other investigators recommend performing a hematopoietic stem cell transplantation (HSCT) as soon as the diagnosis is confirmed and a donor is available. Some reports have suggested immunosuppressive therapy, including corticosteroids, ATG and cyclosporine may be effective in a subset of patients with RA-MDS.

Refractory anemia with excess blasts (RAEB): Many investigators recommend AML-like therapy (see Chapter 17 on acute leukemias) before HSCT in children with RAEB. Other investigators, on the other hand, recommend proceeding directly to HSCT with the best available donor. A suitable matched related donor is preferred for HSCT. If a suitable matched related donor is not available then a matched unrelated donor should be considered.

Allogeneic HSCT is the only curative treatment for MDS. Other therapies are ineffective. Newer agents including thalidomide analogs and demethylating agents are being used more frequently in adult MDS; however, they have not be studied in childhood MDS.

Results of HSCT – Disease-free survival:

*   In *de novo* pediatric MDS patients
    *   HLA-matched family donor: 50%
    *   Matched unrelated donor: 35%
*   Children with secondary MDS: 20–30%.

MDS in Children with Down Syndrome (DS)

Incidence

It has been suggested that approximately 1 in 150 children with DS develops MDS/AML by the age of three. DS children are 10–20 times more at risk of developing MDS/AML compared with non-DS children. The most common form of AML in DS children is acute megakaryoblastic leukemia (M7).

Ten percent of DS infants develop transient myeloproliferative disorder (TMD) at birth. It is characterized by accumulation of immature megakaryoblasts in blood and bone marrow. The majority of patients attain spontaneous remission in 3 months. However, 30% of TMD patient develop M7 AML by age 3 years.

Biology

Acquired mutations in GATA1 (a hematopoietic transcription factor gene) are associated with M7 AML and transient myeloproliferative disorder of DS. A subset of patients with M7 AML and DS have mutations in Jak3.

Treatment

DS patients with MDS/AML have improved survival rates (disease-free survival [DFS] at 4 years is 88%) compared to non-DS patients. Nevertheless, DS patients do not tolerate (nor need) as intensive therapy as non-DS patients with AML/MDS. DS patients are treated with similar agents as non-DS patients but with a less-intensive schedule (see Chapter 17 on acute leukemias).

Treatment of Transient Myeloproliferative Disorder (TMD)

Generally, supportive treatment is given. In the case of severe organ dysfunction exchange transfusion, leukapheresis and chemotherapy are used. Chemotherapies of various intensities with Ara-C have been used. Uncommon complications of TMD include hydrops fetalis, renal failure, organ infiltration, pleural effusion, respiratory failure, hepatic fibrosis, disseminated intravascular coagulopathy.

JUVENILE MYELOMONOCYTIC LEUKEMIA (JMML)

Epidemiology

1. Incidence: 1.2 per million children per year.
2. Median age at diagnosis: 1.8 years; 35% below 1 year of age and only 4% above 5 years of age.
3. Male:female ratio of 2:1.
4. Association with neurofibromatosis type 1 (NF-1): Children with NF-1 have more than a 200-fold increased risk of JMML. Fifteen percent of children with JMML have NF-1.

Children with Noonan syndrome or trisomy 8 mosaicism are at increased risk of developing JMML. Most cases of neonates with Noonan syndrome with JMML-like presentation resolve spontaneously.

Clinical Features

1. Age: usually before 2 years of age.
   Physical findings:
   - Skin: eczema, xanthoma, café-au-lait spots, macular-papular rash
   - Lymphadenopathy
   - Hepatosplenomegaly
   - Bleeding
   - Respiratory symptoms: chronic tachypnea, cough, wheezing
   - Fever, infection.

Laboratory Features

Blood

1. White blood cell (WBC) count: The majority of patients have >25,000/mm^3 WBC counts.
2. Monocytosis may be present for months before overt symptoms of JMML.
3. Thrombocytopenia.
4. Immature myeloid cells with orderly maturation in peripheral blood.
5. Nucleated red blood cells.

Bone Marrow

- Increased cellularity, increased myeloid series, increased monocytes, less than 20% blasts.

Cytogenetics

1. Monosomy 7: 25–30%, patients with monosomy 7 show lower WBC counts, higher percent monocytes in the blood, decreased myeloid: erythroid ratio in the bone marrow, high mean corpuscular volume and normal to moderately high fetal hemoglobin levels. Monosomy 7 occurs with the same frequency in JMML, whether it is associated with NF-1 or not.
2. 7q–: 5%.
3. Normal karyotype: 60%.
4. Other cytogenetic abnormalities: 5%.

Table 16-17 shows diagnostic guidelines for juvenile myelomonocytic leukemia (JMML).

Differential Diagnosis

It is important to distinguish JMML from chronic active viral infections such as Epstein–Barr virus, cytomegalovirus and human herpes virus-6 infections. All of these infections can present with hepatosplenomegaly, lymphadenopathy,

Table 16-17   Diagnostic Guidelines for Juvenile Myelomonocytic Leukemia (JMML)

All of the Following	At Least Two of the Following
1. Absence of the t(9;22) BCR/ABL fusion gene 2. Absolute monocyte count >1,000/mm^3 3. <20% blasts in the bone marrow	1. Circulating myeloid precursors 2. White blood cell count >10,000/mm3 3. Increased fetal hemoglobin (HgF) 4. GM-CSF hypersensitivity

From: Chan et al. Juvenile myelomonocytic leukemia: A report from the 2nd international JMML symposium. Leukemia Res 2009;33:355–362.

leukocytosis, thrombocytopenia and elevated hemoglobin F levels. The following tests are useful:

- Bone marrow examination to see if hemophagocytosis is present. If present, then it favors viral illness
- Viral antibody titers
- Polymerase chain reaction (PCR) for viruses
- Bone marrow colony-forming units granulocyte macrophage (CFU-GM) studies for granulocyte macrophage colony-stimulating factor (GM-CSF) sensitivity.

Biology

JMML is a clonal disorder that arises from a pluripotent hematopoietic stem cell. The leukemic progenitor cell of JMML is capable of producing erythroid, myeloid, monocytic, megakaryocytic and possibly lymphoid lineages. It is probably a heterogeneous disease in the context of clonal involvement of different lineages.

The distinct characteristic of JMML mononuclear cells, of both blood and bone marrow, is that they yield excessive numbers of granulocyte-macrophage colonies when cultured in a semisolid system, even without the addition of exogenous growth factors. The endogenous production of interleukin-1 (IL-1), granulocyte-macrophage colony stimulating factor (GM-CSF) and tumor necrosis factor-α (TNF-α) by monocytes accounts for the exuberant spontaneous growth of colony forming units-granulocyte-macrophage (CFU-GM) growth.

TNF-α exerts bidirectional actions:

- It inhibits normal hematopoiesis
- It induces proliferation of the JMML clone-derived monocytes-macrophage elements.

IL-1 stimulates accessory cells to produce more GM-CSF.

The inhibitory effect of TNF-α on normal hematopoiesis causes bone marrow suppression and results in anemia and thrombocytopenia. Splenomegaly also contributes to the development of anemia and thrombocytopenia.

Molecular Events

The constitutive activation of the Ras signaling pathway plays a central role in the proliferative responses to growth factors in JMML. Over 75% of patients with JMML have an identifiable genetic mutation that can activitate Ras signaling, including mutations in NF1, NRas, KRas, PTPN11 and CBL. Ras signaling proteins regulate cellular proliferation by switching between an active guanosine triphosphate state (Ras-GTP) and an inactive guanosine diphosphate state (Ras-GDP). In JMML, there may be mutations of Ras gene or

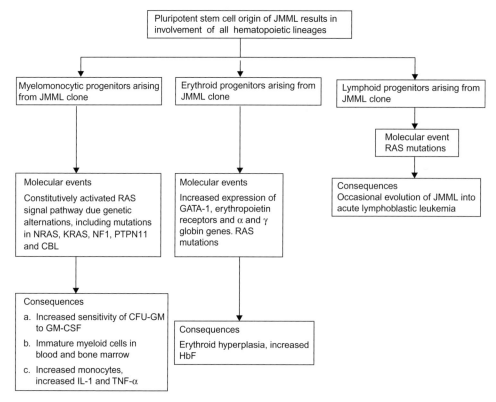

Figure 16-2 Biology of JMML and Its Correlation with Hematologic Findings.
JMML, Juvenile myelomonocytic leukemia; IL-1, Interleukin-1 which stimulates accessory cells to produce GM-CSF; CFU-GM, colony forming unit granulocute macrophage; GM-CSF, granulocute marcrophage colony stimulating factor; TNF-α, tumor necrosis factor-α which inhibits normal hematopoiesis; HbF, fetal hemoglobin.

there may be defective regulation of Ras gene which results in the aberrant transmission of proliferative signals from GM-CSF to the nucleus. The conversion of Ras-GTP to the Ras-GDP is induced by GTP-ase-activating proteins (GAPS); thus, GAPS acts as a negative regulator of Ras. There are two GAPS: Ras-GAP and neurofibromin, the protein made by nf-1 gene. An understanding of the crucial role of Ras in the pathogenesis of JMML has led to trials with molecularly targeted therapies.

Figure 16-2 shows the biology of JMML and its correlation with hematologic findings in JMML.

Natural History

1. Spontaneous remission has been reported in a very few cases. The majority of patients progress, if untreated.

2. Blast transformation occurs in 15% of patients and is associated with additional cytogenetic abnormalities in some patients.
3. Most often patients die of respiratory failure, due to leukemic infiltrates and/or infections.
4. Erythroleukemia-like phase with anemia, erythroid hyperplasia and megaloblastoid cells may develop in some patients.
5. Occasionally, pre-B acute lymphoblastic leukemia (ALL) develops.

Prognosis

Adverse Factors

1. Age at diagnosis: 2 years or more.
2. High fetal hemoglobin level at diagnosis.
3. Platelet count below 33,000/mm^3 at diagnosis (considered to be the strongest indicator of prognosis).

Treatment

JMML responds poorly to chemotherapy.

13-Cis Retinoic Acid (CRA)

1. 13-Cis retinoic acid (isotretinoin, accutane) is used to induce durable responses. 13-Cis retinoic acid reduces spontaneous growth of JMML cells *in vitro*. Additionally, it induces maturation of normal and JMML hematopoietic cells.
2. *Dose*: 100 mg/m^2/day or 3 mg/kg/day for children less than 1 year of age.
3. *Note*: Each dose of 13-cis retinoic acid (CRA) must be drawn from the caplet into a tuberculin syringe and given immediately to the patient.

Hematopoietic Stem Cell Transplantation (HSCT)

HSCT is the only curative treatment available for children with JMML. If a suitable family member donor is available the patient should receive a HSCT. Splenectomy is often recommended before HSCT. Allogeneic HSCT cures 30% of the patients with JMML.

If a suitable family member donor is not available an unrelated donor should be employed, if available. Treatment with CRA should be given in the interim.

Chemotherapy

If there is progression of disease treatment with intensive chemotherapy may be required. However, the responses are usually short-lived.

The following chemotherapy is suggested:

- Ara-C 100 mg/m^2/day by continuous infusion (days 0 to 4)
- Etoposide 100 mg/m^2/day intravenously (days 0 to 4)
- Low dose Ara-C 15 mg/m^2/day subcutaneously (days 6 to 15).

It is to be noted there are no standard chemotherapy regimens available for treatment of JMML. Ara-C based regimens appear to be used most often.

Other Investigational Therapies:

- GM-CSF analogs
- Farnesyl protein transferase (FPTase) inhibitors
- GM-CSF/diphtheria fusion molecule
- Clofarabine
- mTOR inhibitors
- RAF kinase inhibitors
- MEK inhibitors.

CHRONIC MYELOPROLIFERATIVE DISEASES

Table 16-18 shows the WHO classification of chronic myeloproliferative diseases.

CHRONIC MYELOID LEUKEMIA (CML)

Chronic myeloid leukemia (CML) is a clonal myeloproliferative disorder of the primitive hematopoietic stem cell and is characterized by the presence of the Philadelphia (Ph^1) chromosome. This abnormal chromosome results from reciprocal translocation involving the long arms of chromosome 9 and 22, t(9;22)(q34;q11).

Incidence

One per 100,000 for persons younger than 20 years and 1–3% of all childhood leukemia. One hundred cases of CML are diagnosed in the United States per year.

Table 16-18   WHO Classification of Chronic Myeloproliferative Diseases

Chronic myeloid leukemia (Ph chromosome, t(9;22) (q34;q11), BCR/ABL positive)
Chronic neutrophilic leukemia
Chronic eosinophilic leukemia (and the hypereosinophilic syndrome)
Polycythemia vera
Chronic idiopathic myelofibrosis (with extramedullary hematopoiesis)
Essential thrombocythemia
Chronic myeloproliferative disease, unclassifiable

From: Vardiman JW, Harris NL, Brunning RD. The World Health Organization (WHO) Classification of the myeloid neoplasms. Blood 2002;100:2292–2302, with permission.

Chronic Phase

Signs and Symptoms

1. Clinically stable for several years.
2. Nonspecific complaints: fever, night sweats, abdominal pain, bone pain.
3. Symptoms resulting from hyperviscosity:
   - Neurologic dysfunction: headache, strokes
   - Visual disturbances: retinal hemorrhages, papilledema
   - Priapism.
4. Hepatomegaly, splenomegaly.
5. Pallor.

Laboratory Findings

Hematologic findings:

- Mild normocytic, normochromic anemia
- Leukocytosis with shift to left characterized by sequential orderly maturation of myeloid series in all stages starting from myeloblasts to segmented neutrophils
- Increased absolute eosinophil and basophil counts
- Thrombocytosis
- Decreased leukocyte alkaline phosphatase score
- Deterioration of neutrophil function progressively
- Bone marrow examination: hypercellularity, myeloid bulge with sequential orderly maturation, increase eosinophilic and basophilic series, increased megakaryocytes. Blood or bone marrow contain less than 10% leukemic blasts. There may be myelofibrosis. Gaucher like cells or sea-blue histiocytes may be present
- Cytogenetics: presence of Ph[1] chromosome.

Blood Chemistry

1. Elevation of lactic dehydrogenase, uric acid, vitamin B_{12} and transcobalamin 1 levels.

Accelerated Phase

1. Poorly defined intermediate phase characterized by increasing blood and bone marrow leukemic blasts (10–19%) and cytopenias.
2. Dyspoiesis, rising basophil count, progressive myelofibrosis, or refractoriness to therapy
3. Clonal karyotypic evolution.
4. Signs and symptoms: fever, night sweats, weight loss.

Blast Crisis

1. Blasts greater than 20% in blood or bone marrow; or extramedullary or intramedullary clusters of blasts.
2. *Myeloid blast crisis*: most common type of blast crisis (80%), may be myeloblastic or myelomonocytic. Associated karyotypic evolution: duplication of Ph^1 chromosome, trisomies of 8, 19 or 21 chromosomes, i(17q), t(7;11), acute myeloid leukemia specific rearrangements, for example, t(15;17).
3. *Lymphoid blast crisis*: less common type of blast crisis (15–20%), more often B lineage than T. Associated karyotypic evolution: duplication of Ph^1 chromosome, trisomy 21, inv(7), t(14;14).
4. *Multilineage blast crisis*: mixed myelocytic, erythrocytic, megakaryoblastic and lymphocytic. Blasts may co-express antigens of different lineage or two distinct populations of blasts may be present (uniclonal with bilineages).
5. *Erythrocytic blast crisis*: rare in pure form.
6. *Megakaryoblastic blast crisis*: rare in pure form, associated with inv(3)(q21;q26) or t(3)(q21;q26).
7. *Mast cell crisis*: extremely rare.

Signs and Symptoms

Pallor, easy bruisability, pruritis, urticaria, bone pain.

Biology of CML

Molecularly targeted therapies have revolutionized cancer therapeutics and are much more selective in their actions than traditional chemotherapy agents. CML is a disease that exemplifies how the discoveries in molecular biology have helped design molecularly targeted therapies.

Figure 16-3 depicts the Ph^1 chromosome, the molecular genetic studies of which have revealed that the ABL (Abelson) segment from chromosome 9 is translocated to chromosome 22 at its major breakpoint cluster region (BCR) and thus forming a novel fusion gene termed the BCR-ABL. The BCR-ABL fusion gene usually makes an oncoprotein of 210 kd molecular weight. The BCR/ABL oncoprotein activates directly or indirectly many downstream signal transduction pathways e.g. Ras, P13K, JAK-STAT, Bcl-2, BCLx, F-actin and ROS, which induce transformation, proliferation, inhibition of apoptosis and oxidative damage to DNA. Additionally, cytoskeletal proteins are tyrosine phosphorylated and this results in alteration of cytoskeletal function. CML cells are less adherent and egress from the marrow into circulating blood prematurely. Figure 16-3 depicts various cellular effects of BCR/ABL oncoprotein through downstream transduction pathways. Table 16-19 lists some of the functional domains of BCR-ABL oncoprotein.

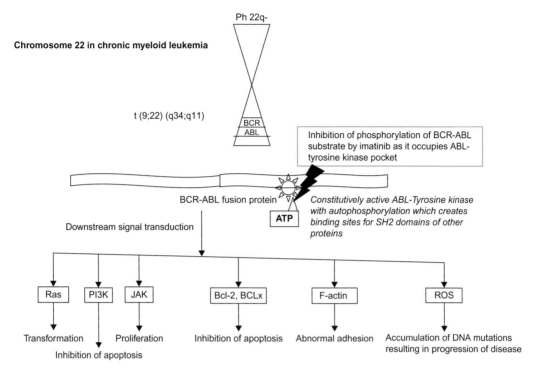

Figure 16-3 t(9;22), Fusion Gene BCR-ABL, Fusion Protein BCR-ABL, Cellular Effects of BCR-ABL and Inhibition of BCR/ABL by Imatinib in Chronic Myeloid Leukemia

Several mechanisms play roles in disease progression which lead eventually to blast crisis of CML. They include:

- Drug resistance mechanisms:
  - BCR/ABL gene amplification

Table 16-19   Functional Domains of BCR-ABL Oncoprotein

Domain	Location	Function
Oligomerization domain	BCR	Activation of ABL kinase
Y^{177}	BCR	Y^{177} regulates the Ras and P13K pathways through recruitment of GRB2-SOS and GAB2-P13K
		These pathways are important for transformation and proliferation
Serine-threonine kinase	BCR	Activation of signal transduction proteins
Y kinase (SH1)	ABL	Constitutive tyrosine kinase activity which is responsible for phosphorylation of signal and adaptor proteins and plays a central role in leukemogenesis
Actin binding domain	ABL	Interference with adhesion

- Overexpression of BCR/ABL transcripts
- Mutations in the ABL tyrosine kinase domain at ATP binding site
- Clonal evolution
- Drug efflux.
- Genomic instability
- Impaired DNA repair
- Tumor suppressor gene inactivation
- Differentiation block
- BCR/ABL-independent activation of more oncogenes and thus, participation of more oncoproteins resulting in redundant pathways for proliferation, decreased apoptosis and differentiation block.

Treatment

The treatment of CML has changed dramatically over the past few years. In the past, HSCT was the only known curative therapy and was the recommended treatment of choice. Over the past few years, complete responses have been documented with tyrosine kinase inhibitors (TKIs) directed against BCR-ABL and front-line use of a BCR-ABL targeting TKI is now the standard of care for most patients with CML.

Treatment of Chronic Phase of CML

Imatinib is used as the first line of treatment.

Mechanism of Action of Imatinib.
Imatinib inhibits ABL-tyrosine kinase by occupying the adenosine triphosphate (ATP)-binding site in the kinase domain of the ABL component of the BCR/ABL oncoprotein (see Figure 16-3). As a result of this, phosphorylation of ABL-tyrosine kinase-substrates fails to occur and activation of downstream leukemogenic signal transduction is prevented.

Dose

1. In adults:
   Chronic phase of CML 400 mg/day orally
   Accelerated phase of CML 600 mg/day orally
   Blastic transformation of CML 600 mg/day orally
2. In children: Chronic phase of CML 340 mg/m^2/day orally.

Frequency of Monitoring Response to Imatinib

1. Hematologic: Monitor every 2 weeks until CHR then every 3 months.
2. Cytogenetic: Monitor every 6 months until CCR then annually.
3. Molecular: Monitor every 3 months.

Table 16-20   Response Criteria for Treatment with Imatinib

Category	Suboptimal Response	Treatment Failure
Hematologic	No CHR after 3 months	No response after 3 months
Cytogenetic	No MCR after 6 months; No CCR after 12 months	No Response after 6 months; No MCR after 12 months; No CCR after 18 months
Molecular	No MMR after 18 months	

Any loss of a hematologic or cytogenetic complete response that is confirmed on repeat testing is a treatment failure. Any loss of a major molecular response is considered a suboptimal response but not a treatment failure.
Abbreviations: CHR, complete hematologic response; MCR, major cytogenetic response; MMR, major molecular response (see text for definitions).

Definitions: Response Criteria

1. Complete hematologic response (CHR): Platelet count $<450,000/mm^3$; WBC $<10,000/mm^3$; differential white cell count with no immature granulocytes and with less than 5% basophils; nonpalpable spleen.
2. Major cytogentic response (MCR): $<35\%$ Ph+ cells in bone marrow
3. Complete cytogentetic response (CCR): No detectable Ph+ cells in bone marrow
4. Major molecular response (MMR): BCR-ABL ratio $<0.1\%$ by RT-PCR
5. Complete molecular response (CMR): No detectable BCR-ABL by RT-PCR.

Table 16-20 lists the definitions of a suboptimal response and treatment failure.

If there is failure or resistance to imatinib, test for highly resistant mutations. If mutation is found and mildly resistant (IC50<5-fold unmutated BCR-ABL), increase imatinib dose. If mutation is highly resistant (IC50≥5-fold unmutated BCR-ABL) change to second-generation TKI (dasatinib or nilotinib). If fails (or intolerant) to second-generation TKI, consider experimental TKIs, HSCT, or hydroxyurea. If with imatinib there is a suboptimal response, increase dose.

Side Effects of Imatinib

1. Myelosuppression.
2. Gastrointestinal toxicity: Nausea, vomiting, diarrhea.
3. Edema and fluid retention: Use of diuretics may be necessary.
4. Muscle cramps, bone pain and arthralgia: Use of calcium and magnesium supplements is effective in controlling muscle cramps, despite normal serum levels of ionized calcium and magnesium.
5. Skin rashes: Use of antihistamines and/or topical steroids may control rashes. Occasionally, Stevens–Johnson syndrome may develop. In that case, imatinib should be discontinued and systemic steroids started.
6. Hepatotoxicity: Transaminitis.

7. Drug interactions: Imatinib is metabolized in the liver by the CYP3A4/5 cytochrome P450 enzyme system. Main inducers of CYP3A4/5 activity, which causes a decrease in the level of imatinib, include carbamazepine, dexamethasone, phenytoin, phenobarbital, progesterone, rifampin and St. John's wort. Drugs that inhibit CYP3A4/5 activity, which causes an increased level of imatinib, include cimetidine, erythromycin, fluoxetine, ketoconazole, ritonavir, itraconazole, verapamil, cyclosporine, clotrimazole, clarithromycin, azithromycin, isoniazid and metronidazole. Grapefruit juice should also be avoided.

Metabolic Management

1. Tumor lysis syndrome: Hydration, recombinant urate oxidase (rasburicase) and allopurinol are employed (see Chapter 31). Tumor lysis is rare in CML and usually only occurs in blast crisis or advanced accelerated phase.
2. Hyperleukocytosis: Consider hydroxyurea (50–75 mg/kg/day) and/or leukopheresis. Leukopheresis is rarely indicated. Some investigators recommend leukapheresis when WBC $>600,000/mm^3$.

Treatment of Advanced Phases of CML

Treatment of CML in Accelerated Phase.
Patients with accelerated phase are treated similarily to patients with chronic phase; however, they are more likely to fail TKIs and more likely to require HSCT.

Treatment of CML in Blastic Phase (Blast Crisis):
Patients with blastic phase who are newly diagnosed may respond to TKIs and these agents should be tried first. A majority of patients will need HSCT and cytoreductive chemotherapy resembling AML therapy for patients with myeloid blasts or ALL therapy for lymphoid blasts can be used (see Chapter 17).

The remission rates in blast cell phase of CML are low and duration of remission is usually short.

Allogeneic Stem Cell Transplantation for Chronic Myeloid Leukemia (CML)

For patients who fail TKIs, HSCT may be indicated. HSCT is very successful in patients with CML partially because of a strong graft versus leukemia effect. The graft versus leukemia effect (GVL) results from genetic disparity between donor and recipient. Therefore, greatest GVL effect occurs when the donor is unrelated to the recipient, because of a greater number of incompatible minor histocompatibility antigens and also, disparities between more major histocompatibility antigens.

Strongest GVL occurs when the transplantation is performed during the chronic phase in the presence of minimal residual disease, detectable by molecular testing. Weakest GVL

Table 16-21 Specific Antigens Implicated in the GVL Response in CML

Antigen	Protein	Peptide
Minor histocompatibility antigens	–	HA-1
	–	HA-2
	SMYC	HY-1
Tissue-restricted	Proteinase 3	PR-1
		PR-7
Leukemia specific	BCR/ABL	KQSSKALQR

effect occurs when the transplantation is performed in a blastic phase and/or T-cell depletion is used for prevention of graft versus host disease (GVHD). The specific antigens implicated in the GVL response in CML are listed in Table 16-21.

The immune response also generates CD4+ Th_1 and CD4+ Th_2 cells. Th_1 cells make interleukin-2 (IL-2), interferon-y (IFN-γ) and tumor necrosis factor (TNF) which induce proliferation of T-cells (induced by IL-2) and apoptosis of CML cells (induced by IFN-γ and TNF). BCR/ABL-specific Th_1 cells exist in circulating blood, but they do not recognize the CML cells, indicating involvement of antigens other than BCR/ABL, in GVL effect induced by Th_1 cells. In this regard, Th_1 cells exhibit leukemia-specific cytotoxicity through tissue-restricted antigens. Th_2 cells make interleukin-4 (IL-4), interleukin-5 (IL-5) and interleukin-10 (IL-10) and are implicated in humoral responses.

Treatment of Post-Transplantation Relapse

1. Donor leukocyte infusion (DLI).
2. TKIs.
3. DLI plus TKIs.
4. Retransplantation.

If the patient relapses with accelerated or blast phase, intensive chemotherapy followed by a stem cell transplantation may be performed.

Essential Thrombocythemia

See Chapter 12.

Polycythemia Vera

See Chapter 10.

Agnogenic Myeloid Metaplasia with Myelofibrosis (AMM)

AMM is a clonal myeloproliferative disorder characterized by leukoerythroblastosis, myeloid metaplasia and varying degrees of myelofibrosis. It is a rare disease in children and they have a better prognosis than adults. Its etiology is unknown.

Clinical Features

1. Malaise, night sweats, weight loss and discomfort from splenomegaly.

Hematologic Findings

1. Blood smear: tear drop cells, leukoerythroblastosis, Pelger-Hüet cells.
2. Bone marrow is difficult to aspirate. Bone marrow biopsies show fibrosis.
3. Cytogenetic abnormalities may be present in bone marrow cells.

Differential Diagnosis

1. Myelofibrosis secondary to metastatic disease to bone marrow. This diagnosis is more common in children than AMM.

Complications

1. Thrombocytosis, bleeding.
2. Splenomegaly – Hypersplenism, refractory thrombocytopenia, hemolytic anemia, some children may not have hepatosplenomegaly.
3. Portal hypertension.
4. Transformation into leukemia.

Treatment

1. Hydroxyurea.
2. Interferon-α.
3. Splenectomy: for hypersplenism and/or severe discomfort and mechanical problems.
4. Hematopoietic stem cell transplantation should be considered.

Suggested Reading

Lymphoproliferative Disorders

Bassiri H, Janice WC, et al. X-linked lymphoproliferative disease (XLP): a model of impaired anti-viral, anti-tumor and humoral immune response. Immunol Res. 2008;42:145–159.

Dham A, Peterson BA. Castleman disease. Curr Opin Hematol. 2007;14:354–359.

Frey NV, Tasi DE. The management of posttransplantation lymphoproliferative disorder. Med Oncol. 2007;24:125–136.

Rao VK, Straus SE. Causes and consequences of the autoimmune lymphoproliferative syndrome. Hematology. 2006;11:15–23.

MDS and JMML

Chan RJ, Cooper T, et al. Juvenile myelomonocytic leukemia: a report from the 2nd international JMML symposium. Leuk Res. 2009;33:355–362.

Germing U, Aul C, et al. Epidemiology, classification and prognosis of adults and children with myelodysplastic syndromes. Ann Hematol. 2008;87:691–699.

Hasle H, Fenu S, et al. International prognostic scoring system for childhood MDS and JMML. Leukemia. 2000;14:968.

Luna-Fineman S, Shannon KM, et al. Myelodysplastic and myeloproliferative disorders of childhood: a study of 167 patients. Blood 1999;93:459–466.

Passamore SJ, Hann CA, et al. Pediatric myelodysplasia: a study of 68 children and a new prognostic scoring system. Blood 1995;85:1742–1750.

Woods WG, Barnard DR, et al. Prospective study of 90 children requiring treatment for juvenile myelomonocytic leukemia or myelodysplastic syndrome. J Clin Oncol. 2002;20:434–440.

Myeloproliferative Disorders

Deininger MWN, O'Brien SG, et al. Practical management of patients with chronic myeloid leukemia receiving imatinib. J Clin Oncol. 2003;21:1637–1647.

Jabbour E, Cortes J, et al. Treatment selection after imatinib resistance in chronic myeloid leukemia. Target Oncol. 2009;4:3–10.

Mauro MJ. Appopriate sequencing of tyrosine kinase inhibitors in chronic myelogenous leukemia: when to change? A perspective in 2009. Curr Opin Hematol. 2009;16:135–139.

Sattler M, Griffin JD. Molecular mechanisms of translocation by the BCR-ABL oncogene. Semin Hematol. 2003;40:4–10.

CHAPTER 17

Leukemias

Acute leukemias represent a clonal expansion and arrest at a specific stage of normal myeloid or lymphoid hematopoiesis. They constitute 97% of all childhood leukemias and consist of the following types:

- Acute lymphoblastic leukemia (ALL) – 75%
- Acute myeloblastic leukemia (AML), also known as acute nonlymphocytic leukemia (ANLL) – 20%[*]
- Acute undifferentiated leukemia (AUL) – <0.5%
- Acute mixed-lineage leukemia (AMLL).

Chronic myeloid leukemias constitute 3% of all childhood leukemias (Chapter 16) and consist of:

- Philadelphia chromosome positive (Ph[1] positive) myeloid leukemia
- Juvenile myelomonocytic leukemia (JMML).

INCIDENCE

1. ALL: 3–4 cases per 100,000 white children, 2,500–3,000 children diagnosed in the United States per year; AML: 500 new cases in the United States per year
2. Peak incidence between 2 and 5 years of age
3. Accounts for 25–30% of all childhood cancers.

ETIOLOGY

The etiology of acute leukemia is unknown. The following factors are important in the pathogenesis of leukemia:

- Ionizing radiation
- Chemicals (e.g., benzene in AML)
- Drugs (e.g., use of alkylating agents either alone or in combination with radiation therapy increases the risk of AML)

---

[*]AML and ANLL are used interchangeably in this chapter and refer to the same disorder.

Manual of Pediatric Hematology and Oncology. DOI: 10.1016/B978-0-12-375154-6.00017-3

- Genetic considerations:
  - Identical twins – If one twin develops leukemia during the first 5 years of life the risk of the second twin developing leukemia is 20%
  - Incidence of leukemia in siblings of leukemia patient is four times greater than that of the general population
  - Chromosomal abnormalities:

Group	Risk	Time Interval
Trisomy 21 (Down syndrome)	1 in 95	<10 years of age
Bloom syndrome	1 in 8	<30 years of age
Fanconi anemia	1 in 12	<16 years of age

  - Increased incidence with the following genetically determined conditions:
    - Congenital agammaglobulinemia
    - Poland syndrome[*]
    - Shwachman–Diamond syndrome
    - Ataxia telangiectasia
    - Li–Fraumeni syndrome (germ line p53 mutation) – the familial syndrome of multiple cancers in which acute leukemia is a component malignancy
    - Neurofibromatosis
    - Diamond–Blackfan anemia
    - Kostmann disease
    - Bloom syndrome[**]

Most cases of leukemia do not stem from an inherited genetic predisposition but from somatic genetic alterations.

CLINICAL FEATURES OF ACUTE LYMPHOBLASTIC LEUKEMIA (ALL)

Table 17-1 shows the common clinical and laboratory presenting features in ALL.

---

[*]Poland syndrome is unilateral absence of pectoralis major and minor muscles and ipsilateral abnormally short, webbed fingers of the hand. It is more common in males and may have variable associated other features.

[**]Bloom syndrome (congenital telangiectatic erythema) is a rare autosomal recessive disorder characterized by telangiectases and photosensitivity, growth deficiency of prenatal onset, variable degrees of immunodeficiency and increased susceptibility to neoplasms of many sites and types. It is caused by a mutation in the gene designated *BLM* traced to band 15q 26.1. The presence of genetic variants of *BLM*, as well as proteins that form complexes with *BLM* (e.g., TOP3A, RMI1), also increases cancer risk. Twenty percent develop malignancies (acute leukemia, lymphoma, gastrointestinal adenocarcinoma). It is more common in European Ashkenazi Jews.

General Systemic Effects

1. Fever (60%).
2. Lassitude (50%).
3. Pallor (40%).

Hematologic Effects Arising from Bone Marrow Invasion

1. Anemia – causing pallor, fatigability, tachycardia, dyspnea and sometimes congestive heart failure.
2. Neutropenia – causing fever, ulceration of buccal mucosa and infection.
3. Thrombocytopenia – causing petechiae, purpura, easy bruisability, bleeding from mucous membrane and sometimes internal bleeding (e.g., intracranial hemorrhage).
4. One to 2% of patients present initially with pancytopenia and may be erroneously diagnosed as having aplastic anemia or bone marrow failure (represents 5% of acquired

Table 17-1   Clinical and Laboratory Presenting Features in Acute Lymphoblastic Leukemia

Clinical and Laboratory Presenting Features	Percentage of Patients
Symptoms and physical findings	
Fever	61
Bleeding (e.g., petechiae or purpura)	48
Bone pain	23
Lymphadenopathy	50
Splenomegaly	63
Hepatosplenomegaly	68
Laboratory features	
Leukocyte count (mm^3)	
<10,000	53
10,000–49,000	30
>50,000	17
Hemoglobin (g/dl)	
<7.0	43
7.0–11.0	45
>11.0	12
Platelet count (mm^3)	
<20,000	28
20,000–99,000	47
>100,000	25
Lymphoblast morphology	
L1	84
L2	15
L3	1

From: Pizzo PA, Poplack DG. Principles and Practice of Pediatric Oncology. 5th ed. Philadelphia: Lippincott-Williams and Wilkins, 2006, with permission.

aplastic anemia) and ultimately develop acute leukemia. In these cases the illness is characterized by:

- Pancytopenia or single cytopenia
- Hypocellular bone marrow
- No hepatosplenomegaly
- Diagnosis of leukemia 1–9 months after onset of symptoms.

The treatment consists of supportive transfusions initially and specific antileukemic chemotherapy, when leukemia is diagnosed.

Clinical Manifestations Arising from Lymphoid System Infiltration

1. Lymphadenopathy – sometimes presents with bulky mediastinal lymphadenopathy causing superior vena cava syndrome. More common in T-cell leukemia in adolescents.
2. Splenomegaly.
3. Hepatomegaly.

Clinical Manifestations of Extramedullary Invasion

Central Nervous System (CNS) Involvement

This occurs in less than 5% of children with ALL at initial diagnosis. It may present with the following:

- Signs and symptoms of raised intracranial pressure (e.g., headache, morning vomiting, papilledema, bilateral sixth-nerve palsy)
- Signs and symptoms of parenchymal involvement (e.g., focal neurologic signs such as hemi paresis, cranial nerve palsies, convulsions, cerebellar involvement – ataxia, dysmetria, hypotonia, hyperflexia)
- Hypothalamic syndrome (polyphagia with excessive weight gain, hirsutism and behavioral disturbances)
- Diabetes insipidus (posterior pituitary involvement)
- Chloromas of the spinal cord (very infrequent in ALL) – may present with back pain, leg pain, numbness, weakness, Brown–Séquard syndrome and bladder and bowel sphincter problems
- CNS hemorrhage – complication that occurs more frequently in patients with AML than in ALL. It is caused by:
  - Leukostasis in cerebral blood vessels, leading to leukothrombi, infarcts and hemorrhage
  - Thrombocytopenia and coagulopathy, contributing to CNS hemorrhage.

Genitourinary Tract Involvement

Testicular Involvement

1. Usually presents with painless enlargement of the testis.
2. Occurs in 10–23% of boys during the course of the disease at a median time of 13 months from diagnosis.
3. Occult testicular involvement is recognized in 10–33% of boys undergoing bilateral wedge biopsies.
4. Risk factors for the development of testicular involvement include:
   - T-cell ALL
   - Leukocytosis at diagnosis ($>20,000/mm^3$)
   - Presence of a mediastinal mass
   - Moderate to severe hepatosplenomegaly and lymphadenopathy
   - Thrombocytopenia ($<30,000/mm^3$).

Ovarian Involvement

1. Occurs very rarely.

Priapism

Occurs rarely. It is due to involvement of sacral nerve roots or mechanical obstruction of the corpora cavernosa and dorsal veins by leukemic infiltrates or by the coagulation of the platelet-rich leukemic blood in the corpora cavernosa.

Renal Involvement

1. Occasionally may present with hematuria, hypertension and renal failure.
2. Evaluated in many patients by ultrasonography; renal involvement more common in T-cell ALL or mature B-cell ALL.

Gastrointestinal Involvement

1. The gastrointestinal (GI) tract may be involved in ALL. The most common manifestation is bleeding.
2. Leukemic infiltrates in the GI tract are usually clinically silent until terminal stages when necrotizing enteropathy might occur. The most common site for this is the cecum, giving rise to a syndrome known as typhlitis (Chapter 30).

Bone and Joint Involvement

Bone pain is one of the initial symptoms in 25% of patients. It may result from direct leukemic infiltration of the periosteum, bone infarction, or expansion of marrow cavity by leukemic cells. Radiologic changes occasionally seen include:

- Osteolytic lesions involving medullary cavity and cortex
- Transverse metaphyseal radiolucent bands

- Transverse metaphyseal lines of increased density (growth arrest lines)
- Subperiosteal new bone formation.

Skin Involvement

Skin involvement occurs occasionally in neonatal leukemia or AML.

Cardiac Involvement

One-half to two-thirds of patients have demonstrated cardiac involvement at autopsy, although symptomatic heart disease occurs in less than 5% of cases. Pathologic findings may include leukemic infiltrates and hemorrhage of the myocardium or the pericardium.

Lung Involvement

Lung involvement is uncommon and may be due to leukemic infiltrates or hemorrhage.

Diagnosis

Laboratory Studies

1. *Blood count*
   a. *Hemoglobin*: Moderate to marked reduction. Normocytic; normochromic red cell morphology. Low hemoglobin indicates longer duration of leukemia; higher hemoglobin indicates a more rapidly proliferating leukemia
   b. *White blood cell count*: Low, normal, or increased
   c. *Blood smear*: Blasts are present on blood smear. Very few to none (in patients with leukopenia). When the white blood cell (WBC) count is greater than 10,000/mm^3, blasts are usually abundant. Eosinophilia is seen uncommonly in children with ALL; 20% of patients with AML have an increased number of basophils
   d. *Thrombocytopenia*: 92% of patients have platelet counts below normal. Serious hemorrhage (GI or intracranial) occurs at platelet counts less than 20,000/mm^3.
2. *Bone marrow*: Bone marrow is usually replaced by 80–100% blasts. Megakaryocytes are usually absent. Leukemia must be suspected when the bone marrow contains more than 5% blasts. The hallmark of the diagnosis of acute leukemia is the blast cell, a relatively undifferentiated cell with diffusely distributed nuclear chromatin, one or more nucleoli and basophilic cytoplasm. Special bone marrow studies, which help in detailed cell classification, include the following:
   - Histochemistry
   - Immunophenotyping
   - Cytogenetics.

3. *Chest radiograph*: Mediastinal mass in T-cell leukemia.
4. *Blood chemistry*: Electrolytes, blood urea, uric acid, liver function tests, immuno-globulin levels.
5. *Cerebrospinal fluid*: Chemistry and cells. Cerebrospinal fluid findings for the diagnosis of CNS leukemia require:
   - Presence of more than 5 WBCs/mm^3
   - Identification of blast cells on cytocentrifuge examination. CNS involvement in leukemia is classified as follows:
     - CNS1 <5 WBCs/mm^3, no blasts on cytocentrifuge slide
     - CNS2 <5 WBCs/mm^3, blasts on cytocentrifuge slide
     - CNS3 >5 WBCs/mm^3, blasts on cytocentrifuge slide
   - TdT stain for suspicious cells
   - If a lumbar puncture is traumatic in a patient with peripheral blasts the following formula can be helpful in defining the presence of CNS leukemia. CNS disease is present if:

$$\frac{CSF\ WBC}{CSF\ RBC} > \frac{Blood\ WBC}{Blood\ RBC}$$

6. *Coagulation profile*: Decreased coagulation factors that frequently occur with AML are: hypofibrinogenemia, factors V, IX and X.
7. *Cardiac function*: Electrocardiogram (ECG) and echocardiogram.
8. *Infectious disease profile*: Varicella antibody titer, cytomegalovirus (CMV) antibody titer, herpes simplex antibody, hepatitis antibody screening.
9. *Immunologic screening*: Serum for immunoglobulin levels, C3 and C4.

Classification

Acute leukemia can be classified based on morphologic characteristics (Tables 17-2 and 17-3), cytochemical features (Table 17-4), immunologic characteristics (Fig. 17-1) and cytogenetic and molecular characteristics (Fig. 17-2). The World Health Organization has developed a new classification of ALL based on cytogenetic and molecular characteristics (Table 17-5).

Light microscopy, cytochemistry, immunophenotyping and cytogenetics are necessary studies to characterize leukemic subtypes.

Morphology

Light Microscopy

Certain morphologic criteria have been established to differentiate lymphoblasts from myeloblasts (Table 17-2). Acute lymphoblastic leukemia can be further subclassified

Table 17-2 Morphologic Characteristic of Lymphoblasts and Myeloblasts

Characteristic	Lymphoblasts	Myeloblasts
Size	10–20 μm	14–20 μm
Nucleus		
Shape	Round or oval	Round or oval
Chromatin	Smooth, homogeneous	Spongy, loose, finely developed meshwork
Nucleoli	0–2 and indistinct	2–5 and distinct "punched-out"
Nuclear membrane	Smooth, round	Irregular
Nuclear–cytoplasmic ratio	High	Low
Cytoplasm		
Color	Blue	Blue-gray
Amount	Thin rim	More abundant
Granules	Absent	Present
Auer rods	Absent	Present

Table 17-3 Cytologic Features of the Morphologic Types of Acute Lymphoblastic Leukemias According to French–American–British (FAB) Classification

Cytologic Features[a]	L1	L2	L3[b]
Cell size	Small cells predominate	Large, heterogeneous in size	Large and heterogeneous
Nuclear chromatin	Homogeneous	Variable, heterogeneous	Finely stippled and homogeneous
Nuclear shape	Regular, occasional clefting or indentation	Irregular, clefting and indentation common	Regular, oval to round
Nucleoli	Not visible, or small and inconspicuous	One or more present, often large	Prominent, one or more vesicular
Amount of cytoplasm	Scanty	Variable, often moderately abundant	Moderately abundant
Basophilia of cytoplasm	Slight or moderate, rarely intense	Variable, deep in some	Very deep
Cytoplasmic vacuolation	Variable	Variable	Often prominent

[a]For each of the features considered, up to 10% of the cells may depart from the characteristic of the type.
[b]The only immunologically pure type of ALL that can be consistently recognized morphologically and invariably carries IgM surface receptor on its membrane.
From: Bennett JM, Catovsky D, Daniel MT, et al. Proposals for the classification of the acute leukemias. Br J Haematol 1976;33:451, with permission.

Table 17-4 Cytochemical Characteristics of the Various Types of Acute Leukemias

Staining Reaction	ALL	Acute Nonlymphoblastic Leukemia			
		AML	Acute Myelomonocytic Leukemia	Erythroleukemia	Megakaryoblastic Leukemia
Nonenzymatic					
PAS	Present as coarse granules or blocks in a variable number of cells	Negative or diffusely positive	Negative or fine granulation	Strongly positive granular	Positive or negative
Sudan black	Negative	Positive	Positive	Positive	Negative
Enzymatic					
Peroxidase	Negative	Positive[a]	Usually negative	Positive	Negative
Alkaline phosphatase	Normal	Low	High	Normal or high	–
Esterases					
Naphthol AS-D Chloroacetate	Negative	Positive	Negative	Negative	–
Naphthol AS-D acetate	Negative or weakly positive	Positive (not inhibited by fluoride)	Strongly positive (inhibited by fluoride)	Weakly positive	Positive or negative
α-Naphthyl acetate	Negative	Negative	Strongly positive	Strongly positive	Positive or negative
Acid phosphatase	Positive in T-ALL	Negative	Negative	Negative	Positive (localized pattern)

[a]The peroxidase reaction, when positive, is considered to indicate the presence of myeloblastic rather than lymphoblastic elements. Unfortunately, myeloblasts that contain no specific granules will be peroxidase negative. It is these cells that cause the greatest difficulty in morphologic classification. Generally, when morphologic classification is difficult, these histochemical tests are of little help. However, demonstration of MPO (myeloperoxidase) by immunologic techniques or expression of MPO by molecular methods can be performed in specialized research laboratories.

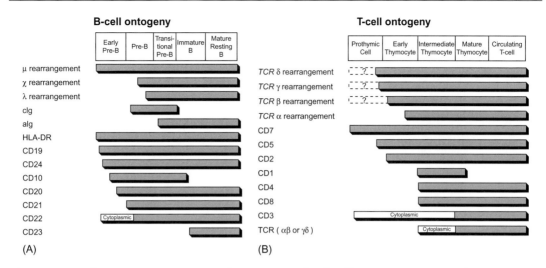

Figure 17-1 Schematic Representation of Human Lymphoid Differentiation. (A) Hypothetical schema of marker expression and gene rearrangement during normal B-cell ontogeny. (B) Hypothetical schema of marker expression and gene rearrangement during normal T-cell ontogeny.
From: Pui CH, Behm FG, et al. Clinical and biologic relevance of immunologic marker studies in childhood acute lymphoblastic leukemia. Blood 1993;82:343, with permission.

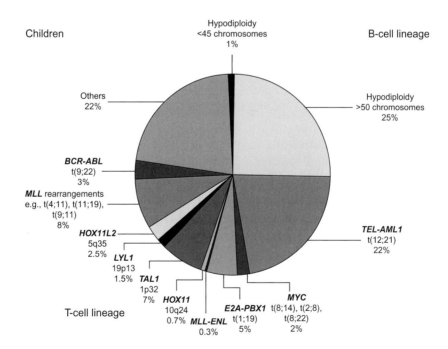

Figure 17-2 Estimated Frequency of Specific Genotypes of ALL in Children.
Modified from: Pui CH, Relling MV, Downing JR. Acute lymphoblastic leukemia. New Engl J Med 2004;350:1535–1548.

Table 17-5   World Health Organization Classification of ALL

B lymphoblastic leukemia/lymphoma
B lymphoblastic leukemia/lymphoma, not otherwise specified (NOS)
B lymphoblastic leukemia/lymphoma with recurrent genetic abnormalities
B lymphoblastic leukemia/lymphoma with t(9;22)(q34;q11.2);*BCR-ABL 1*
B lymphoblastic leukemia/lymphoma with t(v;11q23); *MLL* rearranged
B lymphoblastic leukemia/lymphoma with t(12;21)(p13;q22) *TEL-AML1 (ETV6-RUNX1)*
B lymphoblastic leukemia/lymphoma with hyperdiploidy
B lymphoblastic leukemia/lymphoma with hypodiploidy
B lymphoblastic leukemia/lymphoma with t(5;14)(q31;q32) *IL3-IGH*
B lymphoblastic leukemia/lymphoma with t(1;19)(q23;p13.3);*TCF3-PBX1*
T lymphoblastic leukemia/lymphoma

according to the French–American–British (FAB) classification as L1, L2 and L3 morphologic types (Table 17-3).

Uncommon ALL subtypes:

- *B-cell ALL with L1 morphology*: ALL patients have been described with lymphoblasts having L1 morphology but displaying B-cell ALL immunophenotype. The blast cells are TdT+ in some cases. L3 lymphoblasts generally express immunoglobulin on their cell surface, whereas L1 do not

- *Transitional Pre-B ALL*: This is characterized by:
  - Blasts cells that may express cytoplasmic and surface heavy chains but not Ig κ or λ light chains, indicating that these cells are in transition between pre-B and B stages of differentiation
  - Blast cells that lack FAB L3 morphology or chromosomal translocations associated with B-cell ALL
  - Very good clinical outcome.

Cytochemistry

The cytochemical characteristics of the various types of leukemia are listed in Table 17-4.

Immunology

The putative immunologic classification and cellular characteristics of B-lineage ALL and T-cell ALL are shown in Figure 17-1.

A panel of antibodies is used to establish the diagnosis of leukemia and to distinguish among the immunologic subclones. The panel should include at least one marker that is highly lineage-specific, for example, CD19 for B lineage, CD7 for T lineage and CD13 or

CD33 for myeloid cells. In addition, the use of cytoplasmic CD79a for early pre-B lineage, cytoplasmic CD3 for T lineage and cytoplasmic myeloperoxidase for myeloid cells can be helpful in differentiating unclear immunophenotypes.

Immunophenotype Distribution of Acute Lymphoblastic Leukemia

Pre-B cell accounts for 80% of ALL cases and is subdivided on the basis of cytoplasmic immunoglobulin into transitional pre-B or common ALL antigen (CALLA positive).

Mature B cell accounts for 1–2% of ALL cases. These are surface immunoglobulin positive and are treated as Burkitt's lymphoma. The prognosis has improved and is now similar to other subtypes of high-risk ALL.

T cell accounts for 15–20% of ALL cases. This subtype is associated with:

- Older age at presentation
- High initial white cell count
- Presence of extramedullary disease
- Poor prognosis (treatment on high-risk intensive therapies has improved the prognosis).

Cytogenetics and Molecular Characteristics

The cytogenetic abnormalities observed in leukemia have biologic and prognostic significance. The realization of the biologic significance of leukemia cytogenetics has resulted in the application of molecular methods to understand various mechanisms of leukemogenesis. Specific cytogenetic abnormalities have been shown to have prognostic significance (Table 17-6).

Table 17-6   Prognostic Significance of Chromosomal Abnormalities in ALL

Chromosomal Abnormalities	Event-Free Survival
Hyperploidy	
Trisomies of chromosomes 4,10,17	89% 8 yr EFS
Tel /AML1 Fusion positive	86% 5 yr EFS
Hypodiploidy, <45 chromosomes	39% 8 yr EFS
MLL-rearranged	
Infants (0–12 months)	37% 4 yr EFS
Non infants (>12 months)	50% 8 yr EFS
t(9;22)	39% 8 yr EFS[*]

[*]Studies have shown a 3 year EFS of 80% using intensive chemotherapy with Imatinib.

Molecular Genetics of Leukemia of ALL

Figure 17-2 shows the distribution of molecular rearrangements in childhood acute lymphoblastic leukemia.

Seventy-five percent of childhood ALL cases have evidence of chromosomal trans-locations. The following translocations result in activation of protein kinases by oncogenes and activation of transcription factors:

- Tel–AML1 fusion gene t(12;21) (p13q22). t(12;21) is detected by standard cytogenetics in less than 1 in 1,000 cases; whereas, using molecular techniques, it is detected in 22% of pre-B cases. This translocation is associated with an excellent prognosis
- BCR–ABL fusion gene t(9;22) (q34q11). t(9;22) occurs in only 3% of pediatric ALL cases. This is in contrast to adult ALL where this translocation is present in 25% of cases and 95% of adult chronic myelogenous leukemia (CML) cases. In pediatric BCR–ABL-positive ALL, the BCR breakpoint produces a 190-kb protein (p190) in contrast to CML where a different protein (p210) is usually produced. The t(9;22) in pediatric ALL is usually associated with older age, higher WBC count and frequent CNS involvement at diagnosis
- E2A-PBX1 fusion gene t(1;19) (q23p13.3). t(1;19) is frequently associated with an elevated WBC at diagnosis and occurs in 25% of cases with a pre-B cytoplasmic immunoglobulin-positive phenotype. Intensive therapy is necessary in this subtype
- MLL gene rearrangement at chromosome band 11q23 affects 80% of ALL cases in infants, 3% of ALL cases in older children and 85% of secondary AML involving the topoisomerase II inhibitor. This translocation carries a very poor prognosis with a survival of less than 20%, despite intensive therapy
- B-cell ALL translocations involve MYC genes on chromosome 8q24. Eighty percent of B-ALL cases contain t(8;14) (q24q32); the remaining cases have t(2;8) (p12q24) or t(8;22) (q24q11). All of these translocations deregulate MYC expression and need to be treated intensively with the same agents used to treat Burkitt lymphoma
- More than 50% of the cases of T cell ALL have activating mutations that involve *NOTCH1*, a gene encoding a transmembrane receptor that regulates normal T cell development. When *NOTCH* receptors become activated there is a cascade of proteolytic cleavages which causes intracellular *NOTCH1* translocation to the nucleus which then regulates by transcription a set of oncogenes including *MYC*. It is believed that aberrant *NOTCH* signaling causes constitutive expressing of *MYC* and activation of *RAS*. Interference with *NOTCH* signaling by small molecule inhibition has the potential for induction of remission in T cell ALL.

Prognostic Factors

Table 17-7 lists the prognostic factors in childhood ALL.

Table 17-7   **Prognostic Factors in Childhood Acute Lymphoblastic Leukemia**

Factor	Favorable	Intermediate	Unfavorable
Age (years)	1–9	≥10[a]	<1 and *MLL*+
White blood cell count (×10^9/L)	<50	≥50[a]	
Immunophenotype	Precursor B cell	T cell[a]	
Genetics	Hyperdiploidy >50 chromosomes or DNA index >1.16	Diploid t(1;19)/*TCF3-PBX1*[a]	t(9;22)/*BCR-ABL1* t(4;11)/*MLL-AF4*
	Trisomies 4,10 and 17 t(12;21)/*ETV6-CBFA2*		Hypodiploid <44 chromosomes
CNS status	CNS1	CNS2[a] Traumatic spinal tap with blasts	CNS3
MRD (end of induction)	<0.01%	0.01% to 0.99%	≥1%

[a]These factors used to carry an unfavorable prognosis; however, outcome has improved with risk-directed contemporary therapy.
Abbreviation: MRD, minimal residual disease.
Adapted from: Jeha S, Pui CH. Risk adapted treament of pediatric acute lymphoblastic leukemia. Hematol Oncol Clin N Am. 2009;23:973–990.

Patients who are between age 1–9 with an initial WBC <50,000/mm^3 (standard risk), which includes two-thirds of pre-B ALL patients, have a 4-year event-free survival of 80%. The remaining patients (high risk) have a 4-year event-free survival of 65%. Factors that should be included in risk classification are:

- Age – Patients under 1 year of age and greater than 10 years of age have a worse prognosis than children >1 year and <10 years of age. Infants under 1 year of age have the worst prognosis
- White cell count – Children with the highest WBC tend to have a poor prognosis
- Immunophenotype – Early pre-B cell ALL has the best prognosis. Mature T cell ALL has a worse survival due to its association with older age and with higher WBC at diagnosis. Mature B cell ALL previously had a poor prognosis with early relapses and CNS involvement but recent aggressive therapies have improved prognosis
- DNA index >1.16 hyperdiploid ALL with greater than 50 chromosomes has been associated with good outcome due to increased apoptosis and increased sensitivity to chemotherapeutic agents
- Cytogenetics – The combinations of trisomies of chromosomes 4, 10 and 17 have been associated with good outcome. Translocations involving the MLL rearrangement on 11q23 have been associated with a worse prognosis. The Philadelphia chromosome t (9;22)(q34;q11) ALL is the most difficult translocation to treat and has a bad prognosis. Hypodyploid ALL is also associated with a poor prognosis
- Central nervous system disease – The presence of CNS disease at diagnosis is an adverse prognostic factor despite intensification of therapy with CNS irradiation and

Table 17-8   Classification of Bone Marrow Remission Status in ALL

Classification	% Blasts in Bone Marrow
M_1 Bone Marrow	$\leq 5\%$
M_2 Bone Marrow	$\leq 25\%$
M_3 Bone Marrow	$>25\%$

Based on 200 cell count, true remission requires M_1 marrow status with normal marrow cellularity and presence of all cell lines.

additional intrathecal therapy. The presence of blasts on cytospin without an increased WBC (CNS2 status) is also associated with poor outcome
- Early response to induction therapy. Patients who are not in remission at the end of induction therapy have a very poor prognosis. Bone marrow results on day 7 and day 14 of induction therapy have been used to estimate response to therapy. Table 17-8 lists the classification of bone marrow remission status in ALL
- Minimal residual disease (MRD) after induction therapy is an important prognostic factor for outcome. The worse outcome is observed in patients who have MRD present at the end of consolidation therapy. MRD as measured by flow cytometry on day 8 induction blood samples and day 28 of induction bone marrow aspirate samples are associated with shorter event-free survival. Patients with MRD even at levels of $>0.01\%$ to 0.1% at day 29 of induction have poorer outcomes compared to patients with no MRD at those time points. Day 29 induction MRD values of greater than 0.01% indicate a poorer outcome and the need for more intensification of therapy.

Future Directions in ALL Classifications

1. Gene expression profiling using DNA microassay technology can define biologically and prognostically distinctive ALL subsets, can identify genes which may be responsible for leukemogenesis and may identify genes for which targeted therapy could be developed.
2. Host pharmacogenomics
   Polymorphisms for genes that encode drug-metabolizing enzymes can influence the efficacy and toxicity of chemotherapy.
   a. Gene polymorphisms for thiopurine methyltransferase (TPMT) is a gene that catalyzes the inactivation of mercaptopurine. Ten percent of the population carry at least one variant allele. This results in high levels of active metabolites of mercaptopurine. These patients, especially homozygous patients, have an increased risk of side effects and require marked reduction of 6-mercaptopurine doses
   b. Gene polymorphism for thymidylate synthetase (target of methotrexate) has been associated with increased enzyme expression and poor outcome in ALL.
3. Identification of DNA copy number abnormalities in ALL: Investigators have identified abnormalities of genes that encode regulators of B cell development. The alteration of *IKZF1*, a gene that encodes the lymphoid transcription factor *IKAROS*,

has been associated with poor outcome. A deletion or a mutation of *IKZF1* is strongly associated with elevated levels of minimal residual disease. Future studies will use this genetic abnormality in assigning patients to high-risk therapies. Another mutation associated with poor prognosis and in combination with *IKAROS* has activated mutations in the Janus kinase genes *JAK1, JAK2* and *JAK3*. The Janus genes *JAK1* and *JAK2* have previously been shown to be mutated in Down syndrome patients with ALL and T cell ALL.

Treatment

General Care

More detailed information on supportive medical care is given in Chapter 31.

At presentation, patients may be dehydrated, infected, bleeding and anemic and may have impaired renal and hepatic functions due to leukemic infiltration. Blood urea, creatinine, electrolytes, calcium, phosphorous, serum glutamic pyruvic transaminase (SGPT), serum protein and uric acid levels should be determined, as well as cultures of the blood, cerebrospinal fluid (CSF) and urine.

Supportive care including the use of packed red cells, platelet transfusions and human recombinant granulocyte colony-stimulating factor (G-CSF), are required fairly frequently, especially when aggressive multiagent chemotherapy is employed, resulting in temporary bone marrow failure. When high fever and possible septicemia occur in the presence of neutropenia, antibiotic therapy should be started after taking appropriate blood cultures and a chest radiographs. See Chapter 31 for the treatment of the febrile neutropenic patient. Platelet transfusions (1 unit of random donor platelets per 10 kilogram body weight up to 6 units) should be administered to patients with overt bleeding or when the platelet count is below 10,000/mm^3.

A total of 10 mg/kg/day of allopurinol in divided doses is given in all cases before the commencement of antileukemic drugs. Because allopurinol interferes with the metabolism of 6-mercaptopurine, these two drugs must not be given together. When the blast cell count is more than 50,000/mm^3 or there are large tumor masses, allopurinol is obligatory, together with a fluid intake of 2–3 L/m^2/day.*

These measures are aimed at preventing hyperuricemic acidosis and uric acid nephropathy, which could be fatal. Lactic acidosis, hyperkalemia and hypocalcemia may also occur during induction requiring specific measures to correct the abnormalities.

Recombinant urate oxidase (rasburicase) is used to prevent tumor lysis syndrome and uric acid nephropathy in patients who have a high WBC count, high uric acid and high tumor

---

*Bicarbonate is no longer used to alkalinize the urine.

burden. The decrease in uric acid is very rapid with the use of 0.2 mg/kg daily as a single infusion for up to 5 days. The drug should not be used in patients with G6PD deficiency and allopurinol should be withheld during urate oxidase treatment. All uric acid levels drawn during rasburicase therapy should be sent to the laboratory on ice.

In cases involving hyperleukocytosis (WBC >100,000/mm^3), leukapheresis can be performed to rapidly reduce the level of circulating blasts within 2–4 hours. Leukapheresis, however, does not correct the metabolic problems (e.g., hyperphosphatemia, hypocalcemia and hyperuricemia) or correct the anemia (see Chapter 30, Table 30.2). In patients who weigh less than 10 kilograms partial exchange transfusion may have to be performed.

Live vaccines are contraindicated. There is evidence that the altered immune mechanism existing in the leukemic child may impair the child's ability to handle such antigenic stimuli; therefore, inactivated vaccine should be employed. Eighty percent of leukemic children who received intensive chemotherapy at our institution have at least one defect in cellular or humoral immunity. Fifty percent have two or more unprotective specific antibody responses to measles, mumps, rubella, polio, *Haemophilus influenzae* B, or pneumococcus. Therefore, children with ALL in long-term continuous remission may be at risk for developing serious life-threatening infections and should be evaluated for immune competence at the end of therapy. When protective specific antibody responses are absent they should be revaccinated. This is a "Swiss cheese-like" humoral immune defect caused by the disease and/or chemotherapy.

Psychosocial support of child and family is discussed in Chapter 33.

Minimal Residual Disease and its Implication in the Management of Leukemia

Remission status in leukemia is usually determined on the basis of morphologic findings alone. However, methods with increased sensitivities are being utilized for the detection of minimal residual disease (MRD) and application of these techniques may have a profound influence on the management of leukemia for the following reasons:

- A more precise definition of remission can be obtained based on morphologic, cytogenetic, immunologic and molecular findings
- Treatments may be modified, depending on the level of MRD
- Recent studies have attempted to define the prognostic significance of positive MRD. Patients who have a molecular or immunologic remission (defined as leukemic cells less than 0.01% of nucleated bone marrow cells) are predicted to have a better outcome. These levels of MRD detection are now used as prognostic markers after induction. Patients with >0.01% leukemic cells after the end of induction have a worse prognosis and may require more intensive therapy.

Table 17-9 shows the levels of detection of MRD in leukemia by various methods.

Table 17-9   Levels of Detection of Minimal Residual Disease in Leukemia

Method	Lowest Levels of Detection[a]
Morphology	5 per 100
Conventional cytogenetics	2 per 100
Karyotyping by flow cytometry	1 per 100
Southern blot for receptor gene rearrangement	1 per 100
Karyotyping by fluorescent *in situ* hybridization (FISH)	1 per 1,000
Double immunological marker analysis (leukemia-associated phenotype)	1 per 100,000
Polymerase chain reaction (PCR)	1 per 1,000,000
Combining the use of above methods (in theory)	Unknown (theoretically may increase sensitivity)

[a]Number of leukemic cells per number of normal bone marrow cells.

Treatment of Newly Diagnosed Acute Lymphoblastic Leukemia

There are many successful treatment regimens for ALL. All ALL regimens include certain treatment elements: remission induction, consolidation (intensification of remission), prevention of central nervous system leukemia and maintenance therapy. Remission induction rates with three or four drugs (i.e., vincristine, prednisone, L-asparaginase, intrathecal chemotherapy with or with anthracyclines, depending on the risk classification) produce a 95% remission induction rate.

The aim of therapy in acute leukemia is to cure the patient and includes the following:

*   To induce a clinical and hematologic remission
*   To maintain remission by systemic chemotherapy and prophylactic CNS therapy
*   To treat the complications of therapy and of the disease.

A complete remission is defined as:

*   No symptoms attributable to the disease (e.g., fever and bone pain)
*   No hepatosplenomegaly, lymphadenopathy, or other clinical evidence of residual leukemic tissue infiltration; normal CSF examination (including cytocentrifugation)
*   A normal blood count, with minimal levels of $500/mm^3$ granulocytes, $75,000/mm^3$ platelets and 12 g/dl hemoglobin with no blast cells seen on the blood smear
*   A moderately cellular bone marrow with a moderate number of normal granulocytic and erythroid precursors, together with adequate megakaryocytes and less than 5% blast cells, none of which possess frank leukemic features.

Relapse is defined by the appearance of any of the following:

*   More than 50% blasts in a single bone marrow aspirate
*   Progressive repopulation of blasts in excess of 5%, culminating in more than 25% in two or more bone marrow samples separated by 1 week or more

- More than 25% blasts in the bone marrow and 2% or more circulating lympho-blasts
- Leukemic cell infiltration in extramedullary organs, for example, CNS or gonads (biopsy proven) (for the diagnosis of isolated extramedullary relapse, the bone marrow should contain less than 5% blasts)
- Blasts in the CSF with a cell count greater than 5 WBC/mm^3.

Acute lymphoblastic leukemia (ALL) trials of the Children's Oncology Group (COG) have determined the following therapeutic principles:

- Initial response to therapy is an important predictor of outcome
- Delayed intensification improves outcome for standard- and higher-risk patients
- Standard-risk patients who receive delayed intensification (DI) only require dexamethasone, vincristine and L-asparaginase and intrathecal methotrexate for induction
- Improved systemic therapy eliminates cranial irradiation for standard-risk patients
- Substitution of dexamethasone for prednisone improves outcome for standard-risk patients
- Standard-risk patients who receive the Capizzi methotrexate regimen during interim maintenance (see Table 17-10) have improved survival. Table 17-10 lists the protocol used for standard-risk pre-B cell ALL. Pre-B cell ALL event-free survival at 5 years is over 85%
- The high-risk pre-B cell ALL protocol with the augmented regimen (Table 17-11) rescues higher-risk patients with >25% blasts on day 7.

Table 17-11 lists a protocol for the treatment of high-risk pre-B cell ALL, which includes an augmented therapy protocol for delayed early responders (i.e., greater than 25% blasts at day 7 of therapy after drug induction).

Treatment of Lesser-Risk B-Lineage ALL

Table 17-12 shows treatment of lesser-risk B-lineage ALL. Patients eligible for this protocol are children who have all of the following characteristics:

- Age: between 1 and 9 years of age
- WBC count less than 50,000/mm^3
- Cytogenetic findings of trisomy of chromosomes 4 and 10 or DNA index >1.16
- No evidence of CNS leukemia.

Treatment duration is 2.5 years. The 6-year event-free survival for this protocol was 86.6% with an overall survival of 97.2%.

Table 17-10   Therapy for Standard-Risk ALL

Induction (1 month)	Oral dexamethasone for 28 days (6 mg/m^2/day in three divided doses), IV vincristine (1.5 mg/m^2 on days 0, 7, 14 and 21), intramuscular pegylated L-asparaginase (2,500 units/m^2, on day 4), (pegylated asparaginase can also be given intravenously using the same dose over 2 hours) and at time of initial lumbar puncture age-adjusted intrathecal cytarabine (age 1 to less than 2 years 30 mg; age 2 to less than 3 years 50 mg; age 3 years and older 70 mg) and intrathecal methotrexate (age 1 to less than 2 years, 8 mg; age 2 to less than 3 years, 10 mg; older than 3 years, 12 mg on day 14)
Consolidation (1 month)	Oral 6-mercaptopurine (75 mg/m^2/d on days 1–28 of consolidation), IV vincristine (1.5 mg/m^2 on day 1) and age-adjusted (see above) intrathecal methotrexate on days 1, 8 and 15 for patients without CNS disease at diagnosis.
Interim maintenance	IV vincristine. 1.5 mg/m^2 (max dose 2 mg) on days 1, 11, 21, 31 and 41, IV methotrexate starting dose of 100 mg/m^2/dose over 10–15 minutes on day 1 thereafter escalate by 50 mg/m^2/dose on days 11, 21, 31 and 41 (discontinue escalation and resume at 80% of last dose if there is a delay because of myelosurpression or mucositis). Age-adjusted intrathecal methotrexate (see Induction) on day 31
Delayed intensification (2 months)	Oral dexamethasone in all patients (10 mg/m^2/d on days 1–7 and 14–21 days), IV vincristine (1.5 mg/m^2 on days 0, 7 and 14), intramuscular or intravenous pegylated L-asparaginase (2,500 u/m^2 on day 4), doxorubicin (25 mg/m^2, IV push, on days 0, 7 and 14), IV cyclophosphamide (1,000 mg/m^2 over 30 minutes on day 28), oral 6-thioguanine (60 mg/m^2/day on days 28 to 41), cytarabine (75 mg/m^2/day, IV push, on days 29–32 and 36–39) and age-adjusted intrathecal methotrexate (see Induction) on day 28
Maintenance (girls, 20 months; boys, 32 months)	Dexamethasone (6 mg/m^2/day on days 0 to 4, 28 to 32 and 56 to 60), oral mercaptopurine (75 mg/m^2/day on days 0 to 83), IV vincristine (1.5 mg/m^2 on days 0, 28 and 56), weekly oral methotrexate (20 mg/m^2 beginning on day 7 of each course) and age-adjusted intrathecal methotrexate (see Induction) on day 0 of each course

Treatment of B-Cell ALL

Table 17-13 lists a protocol for treatment of B-cell ALL (LMB 89 protocol), which has an EFS of 84% at 5 years using this protocol.

The Children's Oncology Group is investigating whether the addition of Rituximab (an anti-CD20 monoclonal antibody) to this therapy improves the prognosis.

Treatment of T-Cell ALL

Table 17-14 lists the protocol for treatment of T-cell ALL.

T-cell ALL has an EFS of approximately 75% at 5 years using this protocol.

Table 17-11   High-risk Pre-B Cell ALL Protocol for Rapid Early Responders (RER)[*] and for Slow Early Responders (SER)[**]

Phase	Treatment	Dose
Induction	Prednisone	60 mg/m^2/day PO for 28 days
	Vincristine	1.5 mg/m^2/week IV, days 1, 8, 15, 22
	Daunomycin	25 mg/m^2/week IV, days 1, 8, 15, 22
	PEG Asparaginase	2,500 units/m^2/day IM, day 4
	Cytarabine[a]	Age-adjusted IT, day 0
	Methotrexate[a]	Age-adjusted IT, day 8
Consolidation (9 weeks)	Cyclophosphamide	1,000 mg/m^2/day IV, days 0, 28
	Cytarabine	75 mg/m^2/day IV, days 1–4, 8–11, 29–32, 36–39
	Mercaptopurine	60 mg/m^2/day PO, days 0–13 and days 28–41
	Vincristine	1.5 mg/m^2/day IV, days 14, 21, 42, 49
	PEG asparaginase	2,500 units/m^2 IM days 14, 42
	Methotrexate[a]	Age adjusted IT, days 1, 8, 15, 22
Interim Maintenance 1 (7 Weeks)	Vincristine	1.5 mg/m^2 per day IV days 0, 10, 20, 30, 40 Interim Maintenance I
	Methotrexate	100 mg/m^2/day IV days 0, 10, 20, 30, 40, (7 Weeks)
		Escalate by 50 mg/m^2 per dose
	PEG Asparaginase	2,500 units/m^2 IM days 1, 21
	Methotrexate	Age-adjusted IT days 0, 30
Delayed Intensification I (8 weeks)		
Reinduction (4 weeks)	Dexamethasone	10 mg/m^2/day PO, days 0–7, 14–20
	Vincristine	1.5 mg/m^2/day IV, days 0, 7, 14
	Doxorubicin	25 mg/m^2/day IV, days 0, 7, 14
	PEG Asparaginase	2,500 units/m^2/day IM, day 3,
	Methotrexate	Age-adjusted IT day 0

Reconsolidation (4 weeks)		
	Cyclophosphamide	1,000 mg/m^2/day IV day 28
	Thioguanine	60 mg/m^2/day PO days 28 to 41
	Cytarabine	75 mg/m^2/day SC or IV days 29 to 32, 36 to 39
	Methotrexate	Age-adjusted IT days 28, 35
	Vincristine	1.5 mg/m^2 IV days 42, 49
	PEG Asparaginase	2,500 units/m^2 IM day 42

For SER only give a second Interim Maintenance (for 7 weeks same as interim maintenance 1) and a second delayed intensification (for 8 weeks same as adjusted intensification 1).

Maintenance (12 weeks)		
	Vincristine	1.5 mg/m^2/day IV days 0, 28, 56
	Prednisone	40 mg/m^2/day PO days 0 to 4, 28–32, 56–60
	Mercaptopurine	75 mg/m^2/day PO days 0–83
	Methotrexate	20 mg/m^2/day PO days 7, 14, 21, 28 (hold cycles 1–4 when receiving IT methotrexate), 35, 42, 49, 56, 63, 70, 77
	Methotrexate	Age-adjusted IT day 0 in addition on day 28 cycles 1–4 in RER ONLY

* RER (Day 7 Bone Marrow <25% Blasts).

** SER-Augmented Regimen (Day 7 Bone Marrow >25% Blasts).

[a] The doses are age adjusted: IT methotrexate age 1 to 1.9 yrs 8 mg; age 2 to 2.9 years 10 mg; age>3 years 12 mg, IT cytarbine age 1 to 1.9 yrs 30 mg; age 2 to 2.9 years 50 mg, age >3 years 70 mg.

The cycles of maintenance are repeated until the duration of therapy beginning with the first interim maintenance period reaches 2 years for girls and 3 years for boys.

During the first 2 weeks of consolidation therapy patients with central nervous system disease at diagnosis receive 2,400 cGy to the cranium in 12 fractions and 600 cGy to the spinal cord in 3 fractions. Patients with testicular disease at diagnosis receive 2,400 cGy bilateral testicular radiation in 8 fractions during consolidation therapy.

Patients with central nervous system disease at diagnosis do not receive intrathecal methotrexate on days 15 and 22 of consolidation therapy.

Abbreviations: IV, intravenously; PO, orally; IT, intrathecally; SQ, subcutaneously; IM, intramuscularly.

Modified from: Siebel NL Stenherz PG, Harland HN, et al. Early post induction intensification therapy improves survival for children and adolescents with high risk acute lymphoblastic leukemia: a report from the Children's Oncology Group Blood 2008;111:2548–2555.

Table 17-12   Treatment of Lesser Risk B-Lineage ALL

Induction (weeks 1–4)

Vincristine 15 mg/m^2 IV on days 1, 8, 15 and 22

Prednisone 40 mg/m^2 in three divided doses on days 1 to 28

L-asparaginase 6,000 units/m^2 IM on days 2, 5, 8, 12, 15, 19

Intrathecal methotrexate therapy is age-adjusted (age 1 to 1.9 years 8 mg; 2 to 2.9 years 10 mg and >3 years 12 mg) on days 1 and 22. Additional doses are given on days 8 and 15 when CNS 2 status is present

Consolidation (weeks 5–24)

Methotrexate IV 1 gram/m^2 as 24 hour infusion on weeks 7, 10, 13, 16, 19 and 22 with delayed leukovorin rescue (10 mg/m^2) orally or IV every 6 hours for five doses beginning 48 hours after start of methotrexate infusion

6 mercaptopurine 50 mg/m^2 orally daily on weeks 5–24

Intrathecal methotrexate (age-adjusted as above) on weeks 10, 13, 16, 19 and 22

Vincristine 1.5 mg/m^2 IV on weeks 8 and 9 and weeks 16 and 17

Prednisone 40 mg/m^2 orally in three divided doses for 7 days on weeks 8 and 16

Maintenance (weeks 25–130)

6 mercaptopurine 75 mg/m^2 orally daily on weeks 25–130

Methotrexate 20 mg/m^2 IM weekly on weeks 25–130 (half dose on days of intrathecal methotrexate)

Vincristine 1.5 mg/m^2 IV weekly on weeks 25 and 26, 41 and 42, 57 and 58, 73 and 74, 89 and 90, 105 and 106

Prednisone 40 mg/m^2 for 7 days orally on weeks 25 to 26, 41 to 42, 57 to 58, 73 to 74, 89 to 90 and 105 to 106

Intrathecal methotrexate (age-adjusted as above) is given every 12 weeks (weeks 25, 33, 41, 49, 57, 65, 73, 81, 89, 97 and 105)

Abbreviations: IV, intravenously; IM, intramuscularly.
Adapted from: Chauvenet AR, Martin PL, Devidas M, et al. Antimetabolite therapy for lesser-risk B-lineage acute lymphoblastic leukemia of childhood: a report from the Children's Oncology Group. Blood 2007;110:1105–1111.

Infant Leukemia

Infant ALL accounts for 2–5% of childhood leukemias. Infants under 12 months of age with ALL have an extremely poor prognosis, worse than any other age group. Patients go into remission but have a high incidence of early bone marrow or extramedullary relapse. These patients have a high incidence of the following poor prognostic features:

* High initial WBC
* Massive organomegaly
* Thrombocytopenia
* CNS leukemia
* Failure to achieve complete remission by day 14.

Infant ALL is biologically unique. The leukemia arises from a very early stage of commitment to B-cell differentiation and has the following characteristics:

Table 17-13 Protocol for B-Cell ALL and B-cell NHL with Marrow Involvement with or without CNS Involvement (LMB 89 Protocol)

Reduction Phase
COP
  Cyclophosphamide — 300 mg/m^2 IV, day 1
  Vincristine — 1 mg/m^2 (maximum dose, 2 mg) IV, day 1
  Prednisone — 60 mg/m^2/day PO or IV, days 1–7
  Methotrexate (MTX) and hydrocortisone (HC) — 15 mg IT, days 1, 3 and 5 (dose adjusted for patients <3 years of age)[a]
  Ara-C — 30 mg IT, days 1, 3 and 5 (dose adjusted for patients <3 years of age)[a]
Folinic acid — 15 mg/m^2 q6h PO, days 2 and 4

Induction: 2 courses, COPADM1 and COPADM2 started on day 8 after first day of reduction (COP) phase
COPADM1:
  Vincristine — 2 mg/m^2 (maximum dose, 2 mg) IV, day 1
  MTX high dose (HD) — 8 g/m^2 (over 4 hours) IV, day 1
    Folinic acid — 15 mg/m^2 PO every 6 h×12 doses, days 2–4 MTX level at 72 hours, <0.1 μmol/l
  MTX and HC — 15 mg IT days 2, 4 and 6 (dose adjusted for patients <3 years of age)[a]
  Ara-C — 30 mg IT, days 2, 4 and 6 (dose adjusted for patients <3 years of age)
  Cyclophosphamide — 500 mg/m^2/day IV, days, 2–4 (in 2 injections per day every 12 h and hydration)
  Adriamycin — 60 mg/m^2 IV, day 2
  Prednisone — 60 mg/m^2 PO or IV, days 1–6
COPADM2 (after hematologic recovery)
Same as COPADM1 except
  Cyclophosphamide — Doses doubled (1 g/m^2/day, always in 2 injections per day at 12 h intervals)
  Second vincristine — Day 6

Consolidation: 2 courses of CYVE (after hematologic recovery)
CYVE
  Cytarabine (Ara-C) — 50 mg/m^2/12-hour continuous infusions (hours 1–12), days 1–5
  Cytarabine HD — 3 g/m^2/day (in 3 hours) IV (hours 12–15), days 1–4
  VP16 — 200 mg/m^2/day (in 2 hours) IV (hours 15–17), days 1–4

Maintenance 4 courses monthly in succession
Course 1
  Prednisone — 60 mg/m^2 PO in 2 divided doses, days 1–5
  MTX HD — 8 g/m^2 (in 4 hours) IV, day 1
    Folinic acid — 15 mg/m^2 every 6 hours IV, days 2–4
  MTX IT and HC — 15 mg IT, day 2 (doses adjusted for patients <3 years of age)[a]
  Ara-C — 30 mg IT, day 2 (doses adjusted for patients <3 years of age)[a]
  Cyclophosphamide — 500 mg/m^2/day IV, days 1 and 2 (in 2 divided doses every 12 h per day)
  Adriamycin — 60 mg/m^2 IV, day 2
  Vincristine — 2 mg/m^2 (maximum dose, 2 mg) IV, day 1
  Cranial irradiation — 2,400 cGy to commence on day 8 in case of Initial meningeal involvement (except in case of isolated compression of spinal cord)

(Continued)

Table 17-13   (Continued)

Course 2 (after hematologic recovery)	
Cytarabine	100 mg/m^2/day SC, days 1–5 (in 2 injections q12h)
VP16	150 mg/m^2/day IV, days 1–3 (over 2 hours)
Course 3 (after hematologic recovery)	
Same as course I except without MTX HD and without IT	
Course 4 (after hematologic recovery)	
Same as course 2	

^{a}Age-adjusted doses for intrathecal use (triple intrathecal).

Age	MTX	HC	Ara-C
<1 year	8 mg	8 mg	15 mg
1 year	10 mg	10 mg	20 mg
2 years	12 mg	12 mg	25 mg
>3 years	15 mg	15 mg	30 mg

Abbreviations: MTX and HC, methotrexate and hydrocortisone; Ara-C, cytarabine, cytosine arabinoside; IT, intrathecal; HD, high dose; SC, subcutaneous.
Modified from: Patte C, Auperin A, Michon J, et al. The Societe Francaise d' Oncologie Pediatrique LMB89 protocol: highly effective multiagent chemotherapy tailored to the tumor burden and initial response in 561 unselected children with B-cell lymphomas and L3 leukemia. Blood 2001;97:3370–3379.

- The leukemic cells are usually CALLA (CD10)-negative
- A chromosomal abnormality on chromosome 11, particularly band 11q23 where the MLL/ALL1 gene is located, is commonly found in infant ALL and is associated with a poor prognosis
- The cells frequently express myeloid antigens
- Blasts have fetal characteristics and have a greater resistance to therapy.

Independent adverse prognostic features in this group are:

- Age <3 months
- WBC >75,000/mm^3
- CD10-negative
- Slow response to induction
- Presence of translocation at 11q23.

Infant ALL requires very intensive therapy with advanced supportive care. Most centers advocate that after remission is attained, an allogeneic stem cell transplant be performed for infant ALL with an 11q23 cytogenetic abnormality or a molecular MLL abnormality. Intensive chemotherapy protocols have an event-free survival of 40%. The Children's

Table 17-14   Protocol for T-cell ALL

Induction	
4 Weeks	Vincristine 1.5 mg/m^2 weekly
	Prednisone 40 mg/m^2/day for 28 days
	Doxorubicin 30 mg/m^2 on days 5 and 6
	IT Ara-C (dosed by age) on days 5 and 19
	MTX 4 gm/m^2 on day 5 eight hours after 2nd dose of doxorubicin with leukovorin rescue beginning at hour 36
CNS treatment 3 Weeks	Cranial XRT 1,800 cGy with IT medications 4 doses during cranial radiation
Intensification 6–9 months	Prednisone 120 mg/m^2/day for 5 days every 3 weeks
	Doxorubicin 30 mg/m^2 every 3 wk until cumulative dose of Doxorubicin reaches 360 mg/m^2 and then substitute methotrexate as in continuation
	6-mercaptopurine 50 mg/m^2/d po for 14 days
	L-asparaginase 25,000 IU/m^2 every wk for 20 wks
	Vincristine 2.0 mg/m^2 IV every 3 weeks
	IT Ara-C and methotrexate every 18 weeks
Continuation until 2 yrs continual complete remission	Prednisone 120 mg/m^2/d for 5 days ⎫
	6-mercaptopurine 50 mg/m^2/d po for 14 d ⎪
	Vincristine 2.0 mg/m^2 IV every 3 wk ⎬ Every 3 weeks
	Methotrexate 30 mg/m^2 IV weekly ⎪
	IT Ara-C and methotrexate every 18 weeks ⎭

IT doses: IT Ara-C 15 mg <1 year, 1–2 years 20 mg, 2–3 years 30 mg, ≥3 years 40 mg. Methotrexate 8 mg ages 1–2, 10 mg ages 2–3, 12 mg ages >3 years.
Modified from: Goldberg JM, Silverman LB, Levy DE, et al. Childhood T cell acute lymphoblastic leukemia. The Dana-Farber Cancer Institute. Acute Lymphoblastic Leukemia Consortium experience. J Clin Oncol 21;2003:3616–3622.

Oncology Group has studied an intensive therapy which may make stem cell transplantation unnecessary.

Pediatric ALL under 1 year of age has been treated initially with 7 days of prednisone followed by a protocol which contains elements of both ALL and AML therapy. Infants receiving this therapy have a 4-year event-free survival of 47%.

Philadelphia-Positive ALL

Philadelphia-positive ALL is present in only 3–5% of children with ALL. Previously less that 40% of these patients were cured with intensive chemotherapy. The use of imatinib (340 mg/m^2/day) added to an intensive chemotherapy regimen has improved the outcome in this population at 3 years to an EFS of 80%.

Bone Marrow Relapse in Children with ALL

Despite current intensive front-line treatments, 20% of children with ALL experience bone marrow relapse. Relapse may be an isolated event in the bone marrow or may be combined with relapse in other sites. Reinduction remission rates for patients with first relapse range from 71 to 93%, depending on the timing and site of relapse. Survival of patients experiencing relapse can be predicted by the site of relapse and the length of first complete remission. Children who have a bone marrow relapse during therapy or within 6 months after completing therapy have a very poor long-term survival (early relapse). Allogeneic stem cell transplantation should be considered for these patients.

Children whose bone marrow relapse occurs 6 months after completion of therapy have a significantly longer second remission. Although clinical remission can be achieved in most relapses, long-term survival rates range from 40–50%.

Reinduction of patients with relapsed ALL commonly includes conventional agents largely identical to those used at initial diagnosis.

Studies have shown the impact of MRD on outcome for patients with relapsed ALL. Patients who are MRD-negative at the end of the first block of chemotherapy have improved survival compared with those who are MRD-positive. MRD positivity also correlates strongly with the duration of initial remission. Patients experiencing relapse less that 18 months from initial diagnosis have the highest proportion of MRD positivity.

The authors have used the Memorial Sloan Kettering NY2 protocol (Table 17-15) to treat ALL bone marrow relapses with satisfactory results. Table 17-16 lists an alternative induction regimen for recurrent ALL. This regimen is now being studied with the addition of epratuzumab, an anti-CD22 monoclonal antibody.

Clofarabine

Clofarabine is indicated for pediatric patients with relapsed refractory lymphoblastic or myeloblastic leukemia after second relapse. It inhibits both DNA polymerase and ribonucleotide reductase leading to impaired DNA synthesis and repair and directly induces apoptosis.

Table 17-15   Memorial Sloan-Kettering-New York 2 Protocol for Acute Lymphoblastic Leukemia Relapse

Induction

Day 0: cytosine arabinoside 20, 30, 50, or 70 mg IT for ages <1, 1-2, 2-3, >3 years, respectively

Days 0, 1: daunorubicin 60 mg/m^2/day × 2 or daunorubicin 120 mg/m^2 over 48 hours continuous infusion

Day 2: cyclophosphamide 1200 mg/m^2 IV Days 2,9, 16, 23: vincristine 1.5 mg/m^2 IV Days 2-23: prednisone 60 mg/m^2/day PO and 9-day tapering dose

Day 4 and every Monday, Wednesday and Friday thereafter: L-asparaginase 6,000 units/m^2/day IM

Days 15, 22: methotrexate 6, 8, 10, or 12 mg IT for ages <1, 1-2, 2-3 and >3 years, respectively

Consolidation (begins on the nearest Monday to day 28 or when ANC >500/mm^3 and platelet count >100,000/mm^3)

Days 28, 29: cytosine arabinoside 3,000 mg/m^2 over 3 hours IV

Days 28, 30, 32, 37, 39, 42, 44, 46, etc. until the beginning of maintenance: L-asparaginase 6,000 units/m^2/day IM

Day 31: methotrexate 150 mg/m^2 IV over 4 hours

Days 32 and 39: vincristine 1.5 mg/m^2 IV

Days 32-39: prednisone 180 mg/m^2/day PO

First maintenance (day 56 of consolidation is day 0, begins with ANC >1000/mm^3 and platelet count >100,000/mm^3)

Days 0, 7, 15, 22 methotrexate IT

Days 0-3: 6-mercaptopurine 300 mg/m^2/day PO

Day 4: cyclophosphamide 600 mg/m^2 IV days 4, 11, 18, 25, 32, 39, 46, 53, 60: L-asparaginase 25,000 units IM

Days 11, 18, 25: vincristine 1.5 mg/m^2 IV

Days 18 to 25: prednisone 180 mg/m^2/day PO

Day 25: methotrexate 150 mg/m^2 IV over 4 hours

Days 40, 41: daunorubicin 20 mg/m^2/day IV or daunorubicin 40 mg/m^2 over 48 hours infusion intravenous

Days 42-44: cytosine arabinoside 100 mg/m^2/day, 72-hour continuous infusion

Days 42-44: thioguanine 40 mg/m^2/day PO every 12 hours × 6

Subsequent maintenance cycles

Day 0: methotrexate IT

Days 0-3: 6-mercaptopurine 300 mg/m^2/day PO cyclophosphamide 1,200 mg/m^2 IV

Days 11, 18, 25: vincristine 1.5 mg/m^2 IV

Days 18-25: prednisone 180 mg/m^2/day PO

Day 25: methotrexate 200 mg/m^2 IV over 4 hours, escalate dose by 50 mg/m^2 each subsequent cycle until mucositis or ANC <500/mm^3 occurs

Days 40, 41: daunorubicin 20 mg/m^2/day IV or daunorubicin 40 mg/m^2 over 48 hour Infusion intravensously or dactinomycin 750 ug/m^2 IV

Day 41 (1 day only instead of 2-day Daunorubicin) once the limit of anthracycline has been reached

Days 42-44: cytosine arabinoside 100 mg/m^2/day, 72-hour continuous infusion

Days 42-44: thioguanine 40 mg/m^2/day PO every 12 hours × 6

Duration of maintenance therapy: 2 years

Abbreviations: IT, intrathecal; IV, intravenous; PO, oral; IM, intramuscular; CNS, central nervous system; ANC, absolute neutrophil count.

From: Steinherz PG, Redner A, Steinherz L, et al. Development of a new intensive therapy for acute lymphoblastic leukemia in children at increased risk of early relapse. Cancer 1993;72:3120-3130, with permission.

Table 17-16   Alternate Induction for Recurrent Acute Lymphoblastic Leukemia

Drug and Dosage	Day Number
Block 1	
Vincristine, 1.5 mg/m^2 IV	1, 8, 15 and 22
Prednisone, 40 mg/m^2/d PO	1–29
PEG-asparaginase, 2,500 U/m^2 IM	2, 9, 16 and 23
Doxorubicin, 60 mg/m^2 IV	1
Intrathecal cytarabine	1
Intrathecal methotrexate	8 and 29 (CNS–)
Triple intrathecal therapy	8, 15, 22 and 29 (CNS+)
Block 2	
Cyclophosphamide, 440 mg/m^2 IV	1 to 5
Etoposide, 100 mg/m^2 IV	1 to 5
Methotrexate, 5 g/m^2 IV	22 (pending blood count recovery)
Intrathecal methotrexate	1 and 22 (CNS–)
Triple intrathecal therapy	1 and 22 (CNS+)
G-CSF, 5 μg/kg SQ	6 until ANC >1,500/μl × 2 days
Block 3	
Cytarabine, 3 g/m^2 IV every 12 hours	1, 2, 8 and 9
L-asparaginase, 6,000 U/m^2 IM	2 and 9 at hour 42 after cytarabine
G-CSF, 5 μg/kg SQ	10 until ANC >1,500/μl × 2 days

Abbreviations: IV, intravenous; PO, oral; IM, intramuscular; SQ, subcutaneous; Ph, Philadelphia chromosome; G-CSF, granulocyte colony-stimulating factor; LP, lumbar puncture. Triple intrathecal therapy (methotrexate, cytarabine and hydrocortisone) is continued weekly beyond four doses until two successive lumbar puncture CSF are free of blasts. All intrathecal medications are dosed based on age.
Modified from: Raetz EA, Borowitz MJ, Devidas M, et al. J Clin Oncol 2008;28:3971–3978.

The recommended dose and schedule of clofarabine as a single agent in pediatric patients is 52 mg/m^2 administered by intravenous infusion over 2 hours daily for 5 consecutive days. Treatment cycles are repeated following blood count recovery or return to baseline organ function approximately every 2–6 weeks. To prevent drug incompatibilities, no other medication should be administered through the same intravenous line. Clofarabine should be administered along with IV fluids to reduce the effects of tumor lysis. The use of prophylactic steroids (100 mg/m^2 hydrocortisone on days 1 through 3) may be of benefit in preventing signs or symptoms of systemic inflammatory response syndrome (SIRS) or capillary leak syndrome. If signs or symptoms of SIRS or capillary leak occur (e.g., hypotension) clofarabine should be discontinued and the use of steroids, diuretics and albumin administered. Clofarabine can be re-instituted when the patient is stable and organ function has returned to baseline with a 25% dose reduction. New trials are exploring the use of clofarabine in combination with cyclophosphosmide and etoposide and clofarabine in combinatation with cytarabine.

Central Nervous System Relapse

Treatment

1. Intrathecal chemotherapy alone fails to cure CNS leukemia. However, temporary remission can be achieved with intrathecal chemotherapy alone.
2. Bone marrow relapse occurs in more than 50% of patients achieving CNS remission, regardless of the use of intensified chemotherapy at the time of CNS relapse.

One of the recommended regimens in children relapsing only in the CNS who have not previously received cranial irradiation is listed in Table 17-17. This protocol should be employed in CNS relapses occurring after 18 months of first remission. However, early CNS relapses (<18 months first remission) should be treated with chemotherapy followed by allogeneic stem cell transplantation. In patients who had previously received cranial irradiation and then develop a CNS relapse the author favors a reinduction regimen followed by allogeneic bone marrow transplantation.

Toxicity of CNS Treatment

1. Significant decline in mean values on global IQ scale.
2. Possibility of increased risk of leukoencephalopathy.

For these reasons recent protocols have further delayed radiation therapy to permit more extensive chemotherapy to be delivered prior to radiation therapy and have decreased cranial radiation therapy to 1,800 cGy.

The 4-year event-free survival after CNS relapse is 70%. For patients with a first complete remission ≥18 months is 78%, while for those with a first complete remission less than 18 months it is 52%.

Testicular Relapse

The treatment for isolated testicular relapse includes:

- Local radiotherapy to both testes
- Reinduction and continuation of systemic chemotherapy and CNS chemoprophylaxis (Table 17-16). However, trials have been using the same protocol as for CNS relapse with the elimination of cranial irradiation.

Future trials will investigate the use of high-dose methotrexate to intensfty therapy in an attempt to eliminate testicular radiation therapy.

Table 17-17 Treatment Schema for ALL Patients with Isolated CNS Relapse

Induction (weeks 1–4)
  DEX: 10 mg/m^2 orally daily for 28 weeks
  VCR: 1.5 mg/m^2 IV weekly for 4 weeks
  DNR: 25 mg/m^2 weekly for 3 weeks
  TIT: MTX/HC/Ara-C (age-adjusted dose)* IT weekly for 4 weeks
Consolidation (weeks 5–10)
  Ara-C: 3 g/m^2 IV every 12 hours for 4 doses on each of weeks 5 and 8 (a total of 8 doses)
  L-asp: 10,000 IU/m^2 IM weekly on weeks 5, 6, 8, 9
  TIT: MTX/HC/Ara-C (age-adjusted dose) IT week 10
Intensification 1(weeks 11–22)
  MTX: 1,000 mg/m^2 IV over 24 hours, followed by
  MP: 1,000 mg/m^2 IV over 8 hours weeks 11, 14, 17 and 20
  VP-16: 300 mg/m^2 IV, followed by
  CYC: 500 mg/m^2 IV weeks 12, 15, 18 and 21
  TIT: MTX/HC/Ara-C (age-adjusted dose)* IT weeks 13, 16, 19 and 22
Reinduction (weeks 23–26)
  DEX: 10 mg/m^2 orally daily for 28 days
  VCR: 1.5 mg/m^2 IV weekly for 4 weeks
  DNR: 25 mg/m^2 IV weekly for 3 weeks
Intensification 2 (weeks 27–50)
  Ara-C : 3 g/m^2 IV every 12 hours for 4 doses in weeks 27, 33, 39, 45
  L-asp: 10,000 IU/m^2 IM on weeks 27, 28, 33, 34, 39, 40, 45, 46
  TIT : MTX/HC/Ara-C (age-adjusted dose)* IT weekly on weeks 30, 36, 42, 48
  MTX: 1,000 mg/m^2 IV over 24 hours on weeks 31, 37, 43, 49
  6MP: 100 mg/m^2 IV over 8 hours on weeks 31, 37, 43, 49
  VP16: 300 mg/m^2 IV on weeks 32, 38, 44, 50
  CYC: 500 mg/m^2 IV on weeks 32, 38, 44, 50
Irradiation (weeks 51–54)
CR1<18 months Craniospinal: 24 Gy cranial and 15 Gy spinal
CR1>or = 18 months Cranial 18 Gy ONLY (not spinal)
  Dex 10 mg/m^2/ orally daily for 21 days
  VCR 1.5 mg/m^2 IV weekly for 3 weeks 51, 52, 53
  L-asp:10,000 IU/m^2 IM x9 on weeks 51–54
Maintenance (weeks 55–104)
  MP: 75 mg/m^2 orally daily $\times$ 42
  MTX: 20 mg/m^2 IM weekly $\times$ 6
  Alternate with
  VCR: 1.5 mg/m^2 IV weekly $\times$ 4
  CYC: 300 mg/m^2 IV weekly $\times$ 4

*See Table 17-13 for age-adjusted doses for intrathecal therapy.
Abbreviations: DEX, dexamethasone; VCR, vincristine; DNR, daunomycin; TIT, triple intrathecal therapy; HC, hydrocortisone;
Ara-C, Cytarabine; L-asp, L-asparaginase; MP, 6-mercaptopurine; VP-16, Etoposide; CYC, cyclophosphamide; IV,
intravenous, IT, intrathecally; IM, intramuscularly; CR1, first complete remission.
The authors have substituted 6-mercaptopurine 50 mg/m^2/day orally for 7 days instead of IV 6-mercaptopurine.
Adapted from: Barredo JC, Devidas M, Lauer SJ, et al. Isolated CNS relapse of Acute Lymphoblastic Leukemia treated with
intensive systemic chemotherapy and delayed CNS radiation: a pediatric oncology study. J Clin Oncol 2006;24:3142–3149.

Future Drugs in ALL Therapy

1. Imatinib mesylate inhibits the BCR-ABL tyrosine kinase and is being used in conjunction with chemotherapy in newly diagnosed Ph+ ALL.
2. Nelarabine (Arabinosylguanine) (506 U) is effective in T cell ALL and is being studied in conjunction with other chemotherapy in newly diagnosed T cell ALL.
3. Monoclonal antibodies targeting CD 22 and other antigens are being studied in addition to chemotherapy in relapsed ALL patients.
4. Small molecule inhibitors of FLT3 are being studied in infant ALL.

Table 17-18 describes the mechanism of action of new drugs being studies in pediatric ALL.

Table 17-18   Selected Antileukaemic Drugs being Tested in Clinical Trials

Drug	Mechanism of Action	Subtype of Leukemia Targeted
Clofarabine	Inhibits DNA polymerase and ribonucleotide reductase; disrupts mitochondria membrane	All
Nelarabine	Inhibits ribonucleotide reductase and DNA synthesis	T-cell
Forodesine	Inhibits purine nucleoside phosphorylase	T-cell
γ-secretase inhibitors	Inhibits γ secretase, an enzyme required for NOTCH signaling	T-cell
Rituximab	Anti CD20 chimeric murine-human monoclonal antibody	CD20-positive
Epratuzumab	Anti-CD22 humanized monoclonal antibody	CD22-positive
Imatinib mesilate	ABLkinase inhibitor	BCR-ABL positive
Nilotinib	ABL kinase inhibition	BCR-ABL positive
Dasatinib	BCR-ABL kinase inhibition	BCR-ABL positive
MK-0457	Aurora kinase inhibition	BCR-ABL positive
Lestaurtinib	FMS-like tyrosine kinsase 3 inhibition	MLL-rearranged
Tipifamib	Farnesyltransferase inhibition	All
Azacytidine, decitabine	DNA metyltransferase inhibition	All
Romidepsin, vorinostat,	Histone deacetylase inhibition	All
Valproic acid, MD-27,	Histone deacetylase inhibition	
AN-9	Histone deacetylase inhibition	
Sirolimos,temsirolimus	Mammalian target of rapamycin inhibition	All
Bortzezomib	Inhibition of ubiquitin proteasome pathway	All
Flavopiridol	Serine-threonine-cyclin-dependent kinase Inhibition	All
Oblimersen	Downregulation of BCL2	All
17-AAG	Heat shock protein-90 inhibition	BCR-ABL positive ZAP-70 positive

Modified from: Pui Ch, Robinson LL, Look AT. Acute lymphoblastic leukemia. Lancet 2008;371:1030–1043.

ACUTE MYELOID LEUKEMIA

The age incidence of AML in childhood is constant except for a peak of incidence in the neonatal period and a slight increase in incidence during adolescence. The following inherited disorders predispose a patient to AML:

- Down syndrome
- Fanconi anemia
- Kostmann syndrome
- Bloom syndrome
- Diamond–Blackfan anemia.

Secondary AML can evolve from:

- Myelodysplastic syndromes and myeloproliferative syndromes (Chapter 16)
- Exposure to ionizing radiation treatment with chemotherapy agents. The following chemotherapy agents are associated with secondary AML:
  - Nitrogen mustard
  - Cyclophosphamide
  - Ifosfamide
  - Chlorambucil
  - Melphalan
  - Etoposide.

Clinical Features of AML

Many of the clinical features of AML are similar to those described in ALL in this chapter. The morphologic features of myeloblasts and the cytochemical features of AML are tabulated in Tables 17-2 and 17-4, respectively.

Classification of AML

The World Health Organization (WHO) classification of AML defines that 20% blasts are required for the diagnosis of AML. In addition, patients with clonal cytogenetic abnormalities t(8;21) (q22:q22), inv(16) (p13q22) or t(16:16) (p13;q22) and t(15;17) (q22; q12) are considered to have AML regardless of the blast percentage.

Table 17-19 lists the WHO classification of acute myeloid leukemia and related neoplasms and acute leukemias of ambiguous lineage.

Table 17-20 lists the French–American–British (FAB) classification of AML.

FAB M5 and M7 are more common in early childhood while older children are more likely to have FAB M_0, M_1, M_2 and M_3.

Table 17-19 WHO Classification of Myeloid Neoplasms and Acute Leukemia

Myeloproliferative neoplasms (MPN)
  Chronic myelogenous leukemia, *BCR-ABL1*–positive
  Chronic neutrophilic leukemia
  Polycythemia vera
  Primary myelofibrosis
  Essential thrombocythemia
  Chronic eosinophilic leukemia, not otherwise specified
  Mastocytosis
  Myeloproliferative neoplasms, unclassifiable
Myeloid and lymphoid neoplasms associated with eosinophilia and abnormalities of
  PDGFRA,PDGFRB, or FGFR1
  Myeloid and lymphoid neoplasms associated with *PDGFRA* rearrangement
  Myeloid neoplasms associated with *PDGFRB* rearrangement
  Myeloid and lymphoid neoplasms associated with *FGFR1* abnormalities
Myelodysplastic/myeloproliferative neoplasms (MDS/MPN) (see Chapter 16)
  Chronic myelomonocytic leukemia
  Atypical chronic myeloid leukemia, *BCR-ABL1*–negative
  Juvenile myelomonocytic leukemia
  Myelodysplastic/myeloproliferative neoplasm, unclassifiable
      Provisional entity: refractory anemia with ring sideroblasts and thrombocytosis
Myelodysplastic syndrome (MDS) (see Chapter 16)
  Refractory cytopenia with unilineage dysplasia
      Refractory anemia
      Refractory neutropenia
      Refractory thrombocytopenia
  Refractory anemia with ring sideroblasts
  Refractory cytopenia with multilineage dysplasia
  Refractory anemia with excess blasts
  Myelodysplastic syndrome with isolated del(5q)
  Myelodysplastic syndrome, unclassifiable
  Childhood myelodysplastic syndrome
      Provisional entity: refractory cytopenia of childhood
Acute myeloid leukemia and related neoplasms
  Acute myeloid leukemia with recurrent genetic abnormalities
      AML with t(8;21)(q22;q22); *RUNX1-RUNX1T1*
      AML with inv(16)(p13.1q22) or t(16;16)(p13.1;q22); *CBFB-MYH11*
      APL with t(15;17)(q22;q12); *PML-RARA*
      AML with t(9;11)(p22;q23); *MLLT3-MLL*
      AML with t(6;9)(p23;q34); *DEK-NUP214*
      AML with inv(3)(q21q26.2) or t(3;3)(q21;q26.2); *RPN1-EVI1*
      AML (megakaryoblastic) with t(1;22)(p13;q13); *RBM15-MKL1*
      Provisional entity: AML with mutated NPM1
      Provisional entity: AML with mutated CEBPA
  Acute myeloid leukemia with myelodysplasia-related changes
  Therapy-related myeloid neoplasms
  Acute myeloid leukemia, not otherwise specified
      AML with minimal differentiation
      AML without maturation

(Continued)

Table 17-19   (Continued)

AML with maturation
Acute myelomonocytic leukemia
Acute monoblastic/monocytic leukemia
Acute erythroid leukemia
Pure erythroid leukemia
Erythroleukemia, erythroid/myeloid
Acute megakaryoblastic leukemia
Acute basophilic leukemia
Acute panmyelosis with myelofibrosis
Myeloid sarcoma
Myeloid proliferations related to Down syndrome
Transient abnormal myelopoiesis
Myeloid leukemia associated with Down syndrome
Blastic plasmacytoid dendritic cell neoplasm
Acute leukemias of ambiguous lineage
Acute undifferentiated leukemia
Mixed phenotype acute leukemia with t(9;22)(q34;q11.2); *BCR-ABL1*
Mixed phenotype acute leukemia with t(v;11q23); *MLL* rearranged
Mixed phenotype acute leukemia, B-myeloid, NOS
Mixed phenotype acute leukemia, T-myeloid, NOS
Provisional entity: natural killer (NK) cell lymphoblastic leukemia/lymphoma

Adapted from: Vardiman JW, Thiele J, Arber DA, Brunning RD, et al. The 2008 revision of the World Health Organization (WHO) Classification of myeloid and acute leukemia: rationale and important changes. Blood 2009;114:937–951.

Table 17-20   FAB Classification of AML

Type M0 – acute undifferentiated leukemia
Type M1 – myeloblastic leukemia without maturation; morphologically indistinguishable from L2 morphology (Table 17-3)
Type M2 – myeloblastic leukemia with differentiation
Type M3 – acute promyelocytic leukemia (APML); most cells abnormal hypergranular promyelocytes; cytoplasm contains multiple Auer rods
Type M3V – microgranular variant of APML; cells with deeply notched nucleus; typical hypergranular promyelocytes less frequent
Type M4 – both myelocytic and monocytic differentiation present in varying proportions
Type M4EOS – associated with prominent proliferation of eosinophils
Type M5 – monocytic leukemia containing poorly differentiated and/or well-differentiated monocytoid cells (the M4 and M5 subtypes are particularly common in children under 2 years of age)
Type M6 – ertyoleukemia (Diguglielmo disease)
Type M7 – megakaryoblastic leukemia; associated with myelofibrosis; frequently observed in children with trisomy 21. M7 leukemia has the following characteristics:
  • The blast morphology is heterogeneous in appearance, resembling L1 or L2 cells with or without granules and having one to three nucleoli; the cytoplasm has blebs
  • Immunophenotype is CD41, CD42, CD61 positive – in addition to CD13 and CD33 positivity
  • Electron microscopy demonstrates positive platelet peroxidase (PPO) reaction localized exclusively on the nuclear membrane and the endoplasmic reticulum

Table 17-21 Quantitative Bone Marrow Criteria for the Diagnosis of Acute Meloblastic Leukemia Subtypes[a]

Bone Marrow Cells	M1 (%)	M2 (%)	M4 (%)	M5 (%)	M6 (%)
Blasts					
All nucleated cells	–	>30	>30	–	<30 or >30
Non-erythroid cells	90	>30	>30	>80[e]	>30
Erythroblasts, all nucleated cells	–	<50	<50	–	>50
Granulocytic component,[b] Non-erythroid cells	<10	>10	>20[d]	<20	Variable
Monocytic component,[c] non-erythroid cells	<10	<20	>20	>80[e]	Variable

[a]Lysozyme estimations and cytochemical tests are required if the peripheral blood monocyte count is 5×10^9/L or more, but the marrow suggests M2 and marrow suggests M4, but the peripheral blood monocyte count is less than 5×10^9/L.
[b]Promyelocytes, myelocytes, metamyelocytes and neutrophils.
[c]Promyelocytes and monocytes.
[d]May include myeloblasts.
[e]Monoblasts in M5a; in M5b; the predominant cells are promonocytes and monocytes.
From: Bennett JM, Catovsky D, Daniel MT, et al. Proposed revised criteria for the classification of acute myeloid leukemia. Ann Intern Med 1985;103:626, with permission.

The quantitative bone marrow criteria for the diagnosis of acute myeloblastic leukemia are summarized in Table 17-21.

Uncommon Subtypes of AML Leukemia

1. *AML with erythrophagocytosis:* M4 or M5 morphology with prominent erythrophagocytosis; fibrinolysis and consistent karyotype t(8;16) (p11;p13); associated with a poor prognosis.

2. *M0:* Lymphoblastic (L2) morphology; negative myeloperoxidase (MPO) and Sudan black by conventional histochemical methods, despite the positivity for myeloid antigens (CD13 and/or CD33). MPO detected by the monoclonal antibody method (which detects the α chain as well as the inactive proenzyme form) or by electron microscopy may be positive. Lymphoid markers other than TdT and CD7 are negative.

3. *M3V-microgranular variant of acute promyelocytic leukemia (APML):* M3V is characterized by bilobed cells, multilobular cells, or cells with reniform nucleus and cytoplasm with minimal or no granulation; generally associated with a few typical M3 cells. It presents with hyperleukocytosis and severe coagulopathy. Prognosis is poor as a result of fatal hemorrhages in vital organs during induction therapy. It displays the same cytochemical (MPO+), cytogenetic t(15;17) phenotypic (HLA-DR, CD13+, CD15+, CD33+) and molecular (RAR and PML rearrangements) features as the hypergranular (M3) APML. Morphologic diagnosis can be difficult.

Immunophenotype of AML

Antibodies to cell surface proteins are useful in the diagnosis of AML and can be correlated with the FAB subtypes. Table 17-22 lists the relationship among immunologic surface markers with FAB subtypes of AML.

Table 17-22   Relationship Among Immunologic Surface Markers with FAB Subtypes of Acute Myeloblastic Leukemia

FAB Subtype of AML	Immunologic Surface Marker										
	HLA-DR	CD11b	CD13	CD14	CD15	CD33	CD34	Glycophorin	CD41	CD42	CD61
M1/M2	+				+	+	+				
M3/M3V		+	+		+	+	+				
M4/M5	+	+	+	+	+	+	+				
M6	+		+			+	+	+			
M7	+		+			+	+		+	+	+
M0			+			+	+				

Molecular Genetics of AML

1. Fifteen percent of AML have a t(8;21) (q22q22), which is associated with FAB M2. The translocation creates an AML1-ETO fusion gene. The translocation seems to interfere with the expression of a myeloid-specific gene.
2. A related myeloid transcription factor is also altered by the cytogenetic inv(16) and t(16;16), which occurs in 15% of AML cases. These cases are associated with myelomonocytic differentiation with abnormal bone marrow eosinophils and a favorable progression. They result in a chimeric protein (CBFB-MYH11).
3. Acute promyelocytic leukemia is associated with a balanced translocation of the retinoic acid receptor-alpha (RARα gene at 17q21) and the PML gene at 15q21. RARα is a transcription factor that binds retinoids and interacts directly with DNA.

These molecular abnormalities have clinical importance. APML can be treated with trans-retinoic acid due to its binding to the RARα receptor. Detection of AML-ETO or CBFB-MYH11 is associated with a high rate of long-term remission after treatment with high-dose cytosine arabinoside.

Table 17-23 lists the common cytogenetic abnormalities in pediatric AML, the French–American–British (FAB) subtype associations, the affected genes, the frequency of clinical presentations of these subtypes. Figure 17-3 identifies the distribution of translocations in pediatric AML.

Gene expression profiling by microassay technology has been able to define distinct expression profiles of leukemias based on their genetic mutations. These profiles may in the future be able to uncover linkages between molecular subclass and clinical outcome which cannot be identified by standard cytogenetic analysis and clinical variables at present.

Table 17-23 Cytogenetic Abnormalities in Childhood Acute Myelogenous Leukemia (AML)[a]

Chromosome Abnormality	AML French–American–British type	Affected Genes	Functions	Frequency	Comments
t(8;21)(q22;q22)	M1, M2	ETO-AML1	Transcription factors	5–15%	Auer rods common; chloromas
t(15;17)(q22;q12)	M3, M3v	PML-RARA	Transcription factor; hormone receptor	6–15%	Coagulopathy; ATRA responsiveness
t(11;17)(q23;q12)	M3	PLZF-RARA	Transcription factor; hormone receptor	Rare	Coagulopathy; ATRA unresponsiveness
inv (16)(p13q22);t (16;16)	M4Eo	MYH11-CBFB	Muscle protein; transcription factor	2–11%	CNS leukemia; eosinophilia with basophilic granules
t(8;16)	M5b	MOZ-CBP	Transcription factors	1%	Infants or young adults; high WBC; chloromas; erythrophagocytosis; secondary leukemia after epipodophyllotoxins
t(9;11)(p22;q23)	M4, M5a	AF9-MLL	Homeodomain proteins	5–13%	Infants or young adults; high WBC; chloromas; erythrophagocytosis; secondary leukemia after epipodophyllotoxins
t(10;11)(p12;q23)	M5	AF10-MLL	Homeodomain proteins	Rare	Infants or young adults; high WBC; chloromas; erythrophagocytosis; secondary leukemia after epipodophyllotoxins
t(11;17)(q23;q21)	M5	MLL-AF17	Homeodomain proteins	Rare	Infants or young adults; high WBC; chloromas; erythrophagocytosis; secondary leukemia after epipodophyllotoxins
t(11q23)[b]	M4, M5	MLL, other partners	Homeodomain proteins	2–10%	Infants; high WBC, CNS and skin involvement; poor prognosis often associated with these, especially t(4;11)
t(1;22)	M7	?	?	2–3%	M7 AML in infants with Down syndrome; myelofibrosis
t(6;9)	M2, M4, MDS	DEK-CAN	Transcription factor	<1%	Basophilia nuclear protein
inv(3)(q21q26) t(3;3)(q21q26)	M2, M4, MDS	EVI1	Transcription factor	<1%	Prior MDS; thrombocytosis and abnormal platelets
-7/del(7)(q22-q36)	All subtypes, MDS	?	?	2–7%	Toxic exposure; prior MDS; more common in other adults, bacterial infections common
-5/del(5)(q11-q35)	All subtypes, MDS	?	?	Rare	Toxic exposure; prior MDS; more common in older adults
+8	All subtypes	?	?	5–13%	Prior MDS; older patients

[a]*Note:* This table is not a complete list of all chromosomal abnormalities associated with childhood AML, but represents some of the more common abnormalities or those with important phenotypic characteristics.

[b]11q23 translocations involving the MLL gene have been shown to have many different fusion partners.

Abbreviations: ATRA, all-*trans*-retinoic acid; CNS, central nervous system; MDS, myelodysplastic syndrome; MLL, malignant lymphoma, lymphoblastic; WBC, white blood cell count.

Adapted from: Pizzo P, Poplack D. Principles and practice of pediatric oncology, vol 4, 2002. Lippincott Williams and Wilkins, Philadelphia Pa.

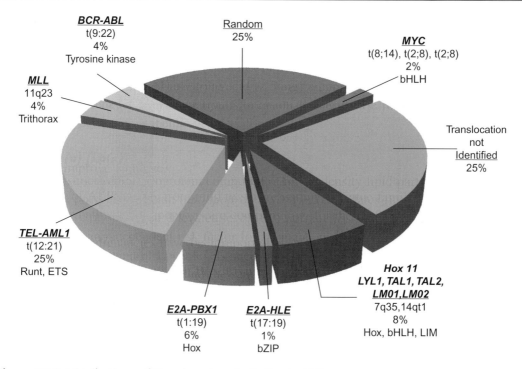

Figure 17-3 Distributions of Translocations in Pediatric AML.
From: Rubnitz JE, Look AT. Molecular genetics of childhood leukemia. J Pediatr Hematol/Oncol 1998;20:1–11, with permission.

Receptor tyrosine kinase mutations (FLT 3 mutations) have been described in pediatric AML. FLT-3 is a receptor tyrosine kinase that is highly expressed on myeloid blasts. FLT 3 internal tandem duplications (FLT 3 ITD) have been identified in up to 16.5% of pediatric AML patients. Patients with FLT 3-ITD mutations have a poorer prognosis.

Chemotherapy trials are using this mutation to identify poorer risk subgroups. FLT-3 tyrosine kinase inhibitors are being tested in AML.

Treatment of Newly Diagnosed AML

The general supportive treatment described under ALL should be employed in the management of AML. The intensification of the induction therapy in AML has led to an increased risk of sepsis and fungal infection. Rapid initiation of broad-spectrum intravenous antibiotics and possibly antifungal therapy is necessary for any febrile episodes while neutropenic. Newly diagnosed AML patients should remain hospitalized during induction therapy.

Table 17-24 lists therapy for newly diagnosed AML.

Table 17-24   Therapy for Newly Diagnosed AML-protocol (Modified from MRC 12)

Course 1	ADE	Daunorubicin 50 mg/m^2 IV days 1, 3, 5
		Cytosine arabinoside 100 mg/m^2 IV bolus every 12 hours days 1 to 10 (20 doses)
		Etoposide 100 mg/m^2 IV 1 hour infusion days 1 to 5
		Intrathecal cytarabine age adjusted doses at time of diagnostic LP
Course 2	ADE	Daunorubicin 50 mg/m^2 IV daily on days 1, 3, 5
		Cytosine arabinoside 100 mg/m^2 IV bolus every 12 hours on days 1 to 8 (16 doses)
		Etoposide 100 mg/m^2 IV daily (1 hour infusion) days 1 to 5
		Intrathecal cytarabine age adjusted dosing on day 1
Course 3	MACE	Amsacrine[a] 100 mg/m^2 IV daily (1 hour infusion) days 1 to 5
		Cytosine arabinoside 200 mg/m^2/d IV (continuous infusion) days 1 to 5
		Etoposide 100 mg/m^2 IV daily (1 hour infusion) days 1 to 5
		Intrathecal cytarabine age adjusted dosing on day 1
Course 4	MidAC	Mitoxantrone 10 mg/m^2 IV daily (short infusion) days 1 to 5
		Cytosine arabinoside 1.0 gram/m^2 12-hourly IV (2 hour infusion) days 1 to 3 (6 doses)
		Intrathecal cytarabine age adjusted dosing on day 1
Course 5	CLASP	Cytosine arabinoside 3.0 grams/m^2 IV every 12 hours for 4 doses on days 1, 2, 8 and 9
		L-asparaginase 6,000 IU/m^2 IM days 2 and 9 (3 hours after completion of Ara-c)

All doses are reduced by 25% for children less than 1 year of age.

Age-adjusted intrathecal chemotherapy with cytosine arabinoside:

   Age 0 to 1 year 20 mg; 1 to 2 years 30 mg; 2 to 3 years 50 mg; 3 year or older 60 mg.

For patients with CNS disease at diagnosis IT therapy with cyatarabine is given twice per week until CSF is clear with two additional doses after clearing of CSF with a minimum of 4 doses of intrathecal therapy.

[a]Not available in the United States.

Abbreviations: LP, lumbar puncture.

Adapted from: Gibson BES, Wheatley K, Hann IM, Stevens RF, Webb D, Hills RK, De Graaf SSN, Harrison CJ. Treatment strategy and long-term results in paediatric patients treated in consecutive UK AML trails. Leukemia 2005;19:2130–2138.

The MRC AML protocol results in a 92% remission rate and a 56% 5-year event-free survival. Patients receiving HLA-matched sibling donor allogeneic stem cell transplant have a 70% long-term survival.

The authors recommend that if a suitable family HLA-matched donor is available, all patients with AML, with the exception of the M3 subtype (acute promyelocytic leukemia), t(8,21) or inv16 karyotypes and Down syndrome patients, receive an allogeneic stem cell transplant in the first remission. In the absence of a suitable matched donor, patients should be treated with intensive chemotherapy. Patients who fail to go into remission after induction or who have FTL3-ITD or monosomy 5 or 7 or secondary AML should be considered for unrelated bone marrow transplantation in first remission.

Prognosis

Factors associated with poor prognosis in AML:

- WBC >100,000/mm^3
- Monosomy 7

- Secondary AML (see Chapter 16)
- FLT 3-ITD
- Minimal residual disease present after induction.

The Children's Oncology Group has modified the MRC protocol due to the unavailability in the United States of amsacrine. They have substituted course 3 with the following chemotherapy: cytosine arabinoside 1,000 mg/m^2/dose IV every 12 hours for a total of 10 doses on days 1–5 and etoposide 150 mg/m^2/dose IV daily on days 1–5 with age-adjusted intrathecal cytosine arabinoside on day 1 of course 3.

Factors associated with prognosis when the MRC protocol is employed can be subdivided into:

- Favorable karyotypes: t(8;21), t(15;17), inv 16 have a 5-year disease-free survival of 75%
- Intermediate karyotype: normal karyotype or karyotype which is abnormal but not in favorable or unfavorable group has a 5-year disease-free survival of 62%
- Unfavorable karyotype: monosomy 5, monosomy 7, del (5q), abn (3q) or other complex karyotypes or >15% blasts after course 1 have a 5-year disease-free survival of 41%.

The Children's Oncology Group is investigating the use of the above MRC regimen with the addition of the anti CD33 monoclonal antibody gemtuzumab ozogamicin (mylotarg) in course 1 and course 4 at a dose of 3 mg/m^2.

Gemtuzumab ozogamicin (Mylotarg) is a new drug being used to treat AML. The drug is a humanized anti-CD33 monoclonal antibody conjugated to an antitumor antibiotic calicheamicin. Side effects are fever, chills, prolonged pancytopenia, abnormalities in liver function tests and raised bilirubin. Veno-occlusive disease (VOD) of the liver has been seen in some patients treated with gemtuzumab ozogamicin. Care should be taken in allogeneic stem cell transplant with these patients since they have an increased risk of VOD.

Down Syndrome and AML

Down syndrome children with AML, despite having an increased frequency of M7 leukemia, have a markedly superior outcome compared to children with non-Down syndrome AML. Down syndrome AML patients usually present with a lower initial WBC, no central nervous system involvement and fewer cytogenetic abnormalities. These patients should be treated with less-intensive chemotherapy regimens using high-dose cytarabine, reduced anthracycline and reduced intrathecal therapy. Maintenance therapy, cranial radiation and stem cell transplantation are not required in the Down syndrome AML patients. The remission rates are approximately 90% with event-free survival approximating 70–80%. This may be partially explained by the increased sensitivity of Down syndrome blasts to cytarabine because of the GATA1 mutation seen in these patients (see myelodysplastic syndrome in children with Down syndrome (Chapter 16, p. 502).

Treatment of Acute Promyelocytic Leukemia (APML) (type M3) with Trans-Retinoic Acid

The induction of remission of APML by trans-retinoic acid (ATRA) is associated with the differentiation of immature promyelocytes into mature granulocytes, followed by the restoration of normal hematopoiesis as the patient enters remission. Data in support of this mechanism include the:

*   Absence of bone marrow hypoplasia during induction with ATRA
*   Persistence of Auer rods in morphologically mature granulocytes
*   Persistence of t(15;17) in morphologically mature granulocytes
*   Appearance of immunophenotypically unique intermediate cells that express both mature and immature cell surface antigens (CD33+, CD16+ cells), indicating their origin from leukemic promyelocytes.

Dose of Trans-Retinoic Acid (ATRA)

1. Dose range: 10–100 mg/m^2/day.
2. Most commonly employed dose: 45 mg/m^2/day in a single or two equally divided doses.
3. Time required to induce remission: 1–3 months (median time to remission, 44 days); resolution of APML-associated coagulopathy is frequently the first sign of response to ATRA.
4. Range of duration of remission: 1–23 months (median duration of remission, 3–5 months).

The unique pharmacologic feature of ATRA leads to the resistance of APML cells to ATRA and, for this reason, ATRA is not useful as a maintenance agent in the therapy of APML. ATRA does not cross the blood–brain barrier. Therefore, ATRA is ineffective in the treatment of CNS involvement with APML.

Side Effects

The side effects of ATRA include skin dryness, itching, peeling, angular stomatitis, headache, pseudotumor cerebri, hypertriglyceridemia and hypercholesterolemia.

The following side effects occur only in patients with APML:

- Fever
- Hematologic
    a. Marked leukocytosis
    b. Thrombosis
- Gastro-intestinal
    a. Hepatic liver enzymes and bilirubin level elevations
- Cardiovascular
    a. Congestive heart failure
    b. Fluid overload
    c. Pericardial effusion
- Pulmonary
    a. Retinoic acid syndrome[*]

Leukocytosis Following the use of ATRA

ATRA treatment is associated with leukocytosis involving a count of $20,000/mm^3$ or more leukocytes in 50% of patients with APML. The peripheral white blood cells comprise myeloblasts, promyelocytes, intermediate cells (CD33+, CD16+ cells) and neutrophils. Nuclear shrinkage, nuclear elongation and marked nuclear and cytoplasmic vacuolation characterize the granulocytic cells.

Management of Acute Promyelocytic Leukemia

1. Treat with ATRA as described previously ($45 mg/m^2$/day in a single or equally divided doses).
2. Perform surveillance coagulation studies periodically and treat as follows:
    - *Disseminated intravascular coagulation (DIC).* Treat with platelet and fresh frozen plasma transfusion to maintain a platelet count of $50,000/mm^2$ or more and a fibrinogen level of at least 100 mg/dl. Heparin is indicated for patients with marked or persistent elevation of fibrin degradation products
    - *Fibrinolysis.* The use of epsilon-aminocaproic acid is reserved for patients with life-threatening hemorrhages.
3. After the induction of remission with ATRA, the patient should be treated with three cycles of the cytosine arabinoside and anthracycline regimen commonly employed in the treatment of AML as follows:
    - Daunorubicin $60 mg/m^2$/day IV bolus for 3 days and
    - Cytarabine $200 mg/m^2$/day by continuous infusion IV for 7 days.

---

[*]This syndrome is characterized by respiratory distress, fever, pulmonary infiltrates radiographically, pleural effusion, weight gain, death due to progressive hypoxemia and multiple organ failure. Prompt treatment with dexamethasone (12 mg every 12 hours for 3 days) stops the progression of this syndrome.

After hematologic recovery then the second course is followed by:

- Daunorubicin 45 mg/m^2/day IV bolus for 3 days with
- Cytarabine 1 g/m^2 every 12 hours for 4 days.

Maintenance therapy consists of:

- ATRA 45 mg/m^2/day for 15 days every 3 months and
- 6 mercaptopurine 90 mg/m^2/day and
- Methotrexate 15 mg/m^2/wk orally for 2 years.

Assessment of MRD: The use of a RT-PCR (reverse transcriptase polymerase chain reaction) for detecting PML–RAR∝ fusion transcripts may be helpful for the identification of patients with MRD and it is possible that they may benefit from further antileukemic therapy.

Use of arsenic trioxide in acute promyelocytic leukemia: Arsenic trioxide induces both differentiation and apoptosis of APML blasts and also degrades the PML-RARA fusion protein. Using arsenic trioxide as a single agent in APML has resulted in remission rates of 65–85%. Studies have shown promising results in newly diagnosed APML. ATRA and arsenic trioxide have been used safely together during induction in APML.

Arsenic trioxide is usually well tolerated at doses of 0.15 mg/kg per day orally 5 days per week for 20 doses.

Side Effects

Arsenic trioxide can produce a prolonged QTc interval. It can also cause fluid retention and pleural and pericardial effusions. Polyneuropathy can occur with repeated courses. Other side effects include rashes, keratosis, hyperpigmentation, transient elevations of liver function tests, nausea, vomiting, abdominal pain, constipation, electrolyte abnormalities, hyperglycemia and headache.

Patients with Refractory or Recurrent Acute Myeloblastic Leukemia

Treatment of these patients is difficult. However, induction of remission may be attempted with one of the following regimens:

- Intermediate-dose Ara-C and Mitoxantrone (Table 17-25). This is the preferred regimen, or
- High-dose Ara-C and L-asparaginase (Table 17-26) or
- Clofarabine has also been used in refractory or recurrent acute myeloblastic leukemia (see p. 546).

The author is presently using clofarabine and Ara-C.

The reinduction protocol outlined in Table 17-25 achieves a remission rate of 76% in relapsed or refractory AML.

Table 17-25   Reinduction Protocol for Relapsed AML

Cytarabine (Ara-C) 1,000 mg/m^2 as a 2 hour infusion every 12 hours for 4 days (8 doses) starting hour 0 day 0
Mitoxantrone 12 mg/m^2 as 1 hour infusion every 24 hours for 4 days beginning day 2 hour 11
Cytarabine (Ara-C) intrathecally on day 0 with age adjusted doses
Decadron eye drops every 4 hours during and for 48 hours after completion of cytosine arabinoside
Granulocyte colony stimulating factor 5 micrograms per kilogram subcutaneously starting 24 hours after last dose of mitoxantrone

Adapted from: Wells RJ, Adams MT, Alonzo TA, et al. Mitoxantrone ad Cytarabine Induction, High Dose Cytarabine and Etoposide Intensification for pediatric patients with relapsed or refractory Acute Myeloid Leukemia: Children's Cancer Group study 2951. J Clin Oncol 2003;21:2940–2947.

Table 17-26   High-dose Ara-C Regimen (The Capizzi Regimen)

Course 1
Ara-C 3 g/m^2 IV over 3 hours every 12 hours for four doses followed by L-asparaginase 6,000 units/m^2 IM 3 hours after the completion of the fourth dose of Ara-C. Decadron eyedrops[a] every 4 hours beginning before, during and 1 day after the Ara-C treatment
Course 2
Same as Course 1, beginning on day 8. Decadron eyedrops every 4 hours as above

[a]To prevent severe conjunctivitis, which can cause pain and photophobia.

We have used gemtuzumab ozogamicin 6–9 mg/m^2 to induce remission in relapsed AML prior to allogeneic stem cell transplantation.

Allogeneic stem cell transplantation should be carried out once a remission is attained. In the absence of a suitable matched donor, a search for a matched unrelated donor should be made. Using a mismatched stem cell transplant may also need to be considered.

The prognosis of children with AML who fail to enter remission or relapse is very poor. Chemotherapy alone results in less than 10% 1-year disease-free survival, whereas autologous or allogeneic stem cell transplant in second remission leads to a 30–50% long-term survival.

Acute Undifferentiated Leukemia Subtypes (M0)

In some patients with acute leukemia, assignment to specific lineage is not possible because of the lack of lineage-associated antigens on the cell surface. These leukemias arise from clonal expansion of poorly differentiated hematopoietic cells and are referred to as acute undifferentiated leukemia (AUL). The prognosis for AUL is generally poor.

Acute Mixed-Lineage Leukemia (AMLL)

Acute mixed lineage leukemia consists of:

- Acute lymphoblastic leukemia expressing two or more myeloid-associated antigens (My^+ALL) – 6% of ALL cases
- Acute myeloid leukemia expressing two or more lymphoid-associated antigens (Ly^+AML) – 17% of AML cases.

True AMLL displays the unequivocal features of more than one lineage on the same blast (as in acute biphenotypic leukemia) or separately on two populations of blasts (as in acute bilineal and acute biclonal leukemia). Bilineal leukemia may exist synchronously or develop metachronously. To distinguish bilineal metachronous leukemia as a result of spontaneous lineage switch, from therapy-induced second primary leukemia, the reappearance of the original clone must be demonstrated by the presence of identical cytogenetic abnormalities or identical gene rearrangements. Lineage switch from ALL to AML is observed more commonly than vice versa.

Therapy of AMLL should be initially based on the predominant cell population, either myeloid or lymphoid (see AML or ALL therapy earlier in this chapter) and then followed by therapy for the second lineage.

Leukemias that express a single marker of another lineage do not qualify for the mixed-lineage category.

Acute biphenotypic lymphoblastic leukemia exhibiting features of both T and B lineages has also been described (e.g., CD2+, CD19+ ALL, putatively originating from its normal [CD2+, CD19+] counterpart).

Congenital Leukemias

Leukemia diagnosed from birth to 6 weeks of age is defined as congenital leukemia. This is a rare disease. Its etiology is unknown. Congenital leukemia has been associated with:

- Trisomy 21
- Turner syndrome
- Mosaic trisomy 9
- Mosaic monosomy 7.

A few examples of congenital JMML have been reported.

Clinical Features

1. Nodular skin infiltrates (bluish, fibroma-like tumors, leukemia cutis).
2. Hepatosplenomegaly.

3. Lethargy, poor feeding, pallor.
4. Purpura/petechiae.
5. Respiratory distress.

Laboratory Studies

1. Usually monocytic subtype of ANLL.
2. Occasionally ALL (pre-B immunophenotype).

Management

1. Congenital leukemia in Down Syndrome
   a. Therapy should be withheld as long as possible, because spontaneous remission may occasionally occur
   b. If disease progresses and the hematological or clinical condition deteriorates, appropriate AML chemotherapy or low-dose cytarabine should be administered, as mentioned earlier in this chapter.
2. Congenital AML leukemia with chromosomal anomalies in blast cells: This leukemia progresses clinically and hematologically and requires institution of therapy (Table 17-24).
3. Congenital ALL presents with a high WBC and a high incidence of mll gene rearrangement and CD10-negative B lineage blasts. Congenital ALL has been treated on the infant ALL protocols but have a 2-year event-free survival of only 20%.

Suggested Reading

Advani AS, Hunger SP, Burnett AK. Acute leukemia in adolescents and young adults. Semin Oncol. 2009;36:213–226.

Baredo JC, Devidas M, Lauer SJ, et al. Isolated CNS relapse of acute lymphoblastic leukemia treated with intensive systemic chemotherapy and delayed CNS radiation: a pediatric oncology group study. J Clin Oncol. 2006;19:3142–3149.

Borowitz MJ, Devidas M, Hunger SP, et al. Clinical significance of minimal residual disease in childhood acute lymphoblastic leukemia and its relationship to other prognostic factors: a Children's Oncology Group study. Blood 2008;111:5477–5485.

Chauvenet AR, Martin PL, Devidas M, et al. Antimetabolite therapy for lesser risk B-lineage acute lymphoblastic leukemia of childhood: a report from Children's Oncology Group Study P9201. Blood 2007;110:1105–1111.

deBoton S, Coiteux V, Chevret S, et al. Outcome of childhood acute promyelocytic leukemia with all-trans-retinoic acid and chemotherapy. J Clin Oncol. 15. 2004;22(8):1404–1412.

Fox E, Razzouk BI, Widemann BC, et al. Phase 1 trial and pharmacokinetic study of arsenic trioxide in children and adolescents with refractory or relapsed acute leukemia including acute promyelocytic leukemia or lymphoma. Blood 2008;111:566–573.

Gamis AS. Acute myeloid leukemia and Down syndrome evolution of modern therapy – State of the art review. Ped Blood Cancer 2005;44:13–20.

Gibson BES, Wheatley K, Hann IM, et al. Treatment strategy and long-term results in paeditric patients treated in consecutive UK AML trials. Leukemia 2005;19:2130–2138.

Goldberg JM, Silverman LB, Levy DE, et al. Childhood T-cell acute lymphoblastic leukemia: The Dana Farber Cancer Institute. Acute lymphoblastic leukemia consortium experience. J Clin Oncol. 2003;21:3616–3636.

Gregory J, Kim H, Alonzo T, et al. Treatment of children with promyelocytic leukemia: Results of the first North American Intergroup trial INT0129. Pediatr Blood Cancer 2009;53:1005–1010.

Harnod TM, Gaynon PS. Treating refractory leukemia in childhood, role of clofarabine. Ther Clin Risk Manage. 2008;4:327–336.

Jeha S. New therapeutic strategies in acute lymphoblastic leukemia. Semin Hematol. 2009;46:76–88.

Jeha S, Razzouk B, Rytting M, et al. Phase 2 study of clofarabine in pediatric patients with refractory or relapsed acute myeloid leukemia. J Clin Oncol. 27:4392–4397.

Ko RH, Ji L, Barnette P, et al. Outcome of patients treated for relapsed or refractory acute lymphoblastic leukemia: a therapeutic advances in childhood leukemia consortium study. J Clin Oncol. 2009;22:2950.

Meshinchi S, Arceci RJ. Prognostic factors and risk-based therapy in pediatric acute myeloid leukemia. The Oncologist 2007;12:341–355.

Mullighan CG, Su X, Zhang J, et al. Deletion of IKZF1 and prognosis in acute lymphoblastic leukemia. N Engl J Med. 2009;360:470–480.

Nguyen K, Devidas M, Cheng SC, et al. Factors influencing survival after relapse from acute lymphoblastic leukemia: a Children's Oncology Group study. Leukemia 2008;22:2142–2150.

Patte C, Auperin A, Michon J, et al. The Societe Francaise d'Oncologie Pediatrique LMB89 protocol: highly effective multiagent chemotherapy tailored to the tumor burden and initial response in 561 unselected children with B cell lymphomas and L3 leukemia. Blood 2001;97:3370–3379.

Pieters R, Schrappe M, Delorenzo P, et al. A treatment protocol for infants younger than 1 year with acute lymphoblastic leukemia (Interfant –99); an observational study and multicentre randomized trial. Lancet 2007;370:240–250.

Pui CH, Evans WE. Treament of acute lymphoblastic leukemia. New Engl J Med. 2006;354:166–178.

Pui CH, Howard S. Current management and challenges of malignant disease in the CNS in paediatric leukemia. Lancet Oncol. 2008;9:257–268.

Pui CH, Robison LI, Look AT. Acute lymphoblastic leukemia. Lancet 2008;371:1030–1043.

Rabin KR, Whitlock JA. Malignancy in children with trisomy 21. The Oncologist 2009;14:164–173.

Raetz EA, Cairo MS, Borowitz MJ, et al. Chemoimmunotherapy reinduction with epratuzumzb in children with acute lymphoblastic leukemia in marrow relapse: a children's oncology group pilot study. J Clin Oncol. 2008;26:3756–3762.

Raetz EA, Borowitz MJ, Devidas M, et al. Reinduction platform for children with first marrow relapse of acute lymphoblastic leukemia: a children's oncology group study. J Clin Oncol 26: 3971–3978.

Schultz KH, Bowman P, Aledo A, et al. Improved early event-free survival with imatinib in Philadelphia chromosome-positive acute lymphoblastic leukemia: a children's oncology group study. J Clin Oncol. 2009;27:5175–5181.

Siebel NL, Stenherz PG, Sather HN, et al. Early postinduction intensification therapy improves survival for children and adolescents with high-risk acute lymphoblastic leukemia: a report from the Children's Oncology Group. Blood 2008;111:2528–2556.

Smith M, Arthur D, Camitta B, et al. Uniform approach to risk classification and treatment assignment for children with acute lymphoblastic leukemia. J Clin Oncol. 1996;14:18–24.

Stanulla M, Schrappe M. Treatment of childhood acute lymphoblastic leukemia. Semin Hematol. 2009;46: 52–63.

Vardiman JW, Thiele J, Arber DA, et al. The 2008 revision of the World Health Organization (WHO) classification of myeloid and acute leukemia: rationale and important changes. Blood 2009;114:937–951.

Vrooman LM, Silverman LB. Childhood acute lymphoblastic leukemia: update on prognostic factors. Curr Opin Pediatrics 2009;21:1–8.

Wells RJ, Adams MT, Alonzo TA, et al. Mitoxanthrone cytarabine induction, high-dose cytarabine and etoposide intensification for pediatric patients with relapsed or refractory acute myeloid leukemia. Children's Oncology Study Group 2951. J Clin Oncol. 2003;21:2940–2947.

CHAPTER 18

Histiocytosis Syndromes

The term histiocytosis identifies a group of disorders that have in common the proliferation of cells of the mononuclear phagocyte and the dendritic cell systems. In the context of histiocytosis syndromes, histiocytes are defined as a group of immune cells, including macrophages and dendritic cells.

Macrophages have several key functions, including tissue remodeling, tumor repressing and promoting activities and in antigen-processing and immune stimulation; dendritic cells function primarily as antigen-presenting cells and immune system regulators. Although the principal pathological cells of these syndromes are histiocytes, the term histiocytosis is a very selective term in the sense that it does not include storage diseases, hyperlipidemic xanthomatosis, or granulomatous reaction in chronic infections, such as tuberculosis or foreign body granuloma. Figure 18-1 shows the ontogeny of histiocytes and Table 18-1 lists the immunophenotypic markers of non-neoplastic macrophages and dendritic cells of the histiocyte system. Table 18-2 shows classification of histiocytic disorders.

LANGERHANS CELL HISTIOCYTOSIS

Incidence

The ratio of male to female incidence is 2:1. Most cases occur in children between 1 and 15 years of age. The prevalence is 1 in 50,000 and the incidence is 1.08 in 200,000 per year.

Pathology and Classification

The Histiocyte Society established the criteria required for the histological, immunohisto-chemical and electron microscopic diagnosis of Langerhans cell histiocytosis (LCH) (Table 18-3).

The characteristic histopathology required for a **presumptive diagnosis** is well described and usually shows a granulomatous-like lesion with a predominance of dendritic cells (Langerhans cells) that have characteristic bean-shaped, folded nuclei and pale cytoplasm. A **definitive diagnosis** of LCH requires the immunohistochemical identification of the presence of Langerhans cells by cell surface CD1a or by the presence of cells with Birbeck

Manual of Pediatric Hematology and Oncology. DOI: 10.1016/B978-0-12-375154-6.00018-5

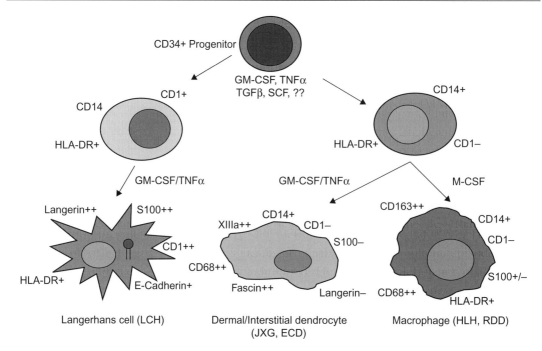

Figure 18-1 Much of the Basis for this Scheme of Ontogeny Derives from *In vitro* Work in Developing Systems for the Production of Dendritic Cells for Immunotherapy.

Cytokines (GM-CSF, α-TNF, IL-4, etc.) play important roles in these pathways and GM-CSF is fundamental to the differentiation of the Langerhans cells. Langerhans cells, indeterminate cells of the dermis and interdigitating dendritic cells are considered to be in a cytological continuum. Indeterminate cells are precursors to Langerhans cells.

Abbreviations: JXG, juvenile xanthogranulomatous disease; ECD, Erdheim-Chester disease; LCH, Langerhans cell histiocytosis; RDD, Rosai-Dorfman disease; HLH, hemophagocytic lymphohistiocytosis.

Adapted from Weitzman S, Jaffe R. Uncommon histiocytic disorders: The non-Langerhans cell histiocytoses, Pediatric Blood Cancer 2005;45:256–264; and Bechan GI, Egeler RM, Arceci RJ. Biology of Langerhans cells and Langerhans cell histiocytosis. Int Rev Cytol 2006;254:1–43.

granules by electron microscopy or immunochemical detection of Langerin, a protein representing the light microscopic evidence of Birbeck granule. The CD1a surface antigen can now be identified routinely from paraffin-embedded specimens.

Most data would support an abnormal Langerhans cell as the key pathological cell in LCH. A characteristic feature of Langerhans cells is Birbeck granules, which are derived from the cytoplasmic membrane and appear as racquet-shaped structures in the cytoplasm on electron microscopy or can be identified by staining for langerin by regular immunohistochemistry. They are involved in receptor CD1a-mediated endocytosis, as well as in nonreceptor-mediated endocytosis. The LCH lesion is an often granulomatous collection of dendritic cells, eosinophils and lymphocytes; multinucleated giant cells are often present. Early in the

Table 18-1 Immunophenotypic Markers of Non-neoplastic Macrophages and Dendritic Cells[*]

Marker	LC	IDC	FDC	PDC	Mφ	DIDC
MHC Class II	+C	++S	–	+	+	+/–
Fc-receptors	–	–	+	–	=	–
CD1a	++	–	–	–	–	–
CD4	+	+	+	+	+	+/–
CD21	–	–	++	–	–	–
CD35	–	–	++	–	–	–
CD68	+/–	+/–	–	++	++	+
CD123	–	–	–	++	–	–
CD163	–	–	–	–	++	–
Factor XIIIa	–	–	+/–	–	–	++
Fascin	–	++	+/++	–	–/+	+
Langerin	++	–	–	–	–	–
Lysozyme	+/–	–	–	–	+	–
S100	++	++	+/–	–	+/–	+/–
TCL1	–	–	–	+	–	–

Abbreviations: FCR, Fc IgG receptors include CD16, CD32, CD64 on some cells; LC, Langerhans cell; IDC, interdigitating dendritic cell; FDC, follicular dendritic cell; PDC, plasmacytoid dendritic cell; Mφ, macrophage; DIDC, dermal/interstitial dendritic cell; c, cytoplasmic; s, surface; Expression is semiquantitatively graded 0 or - through ++, + present, ++ high, +/– low or varies with cell activity.
[*]From Jaffe, R, Pileri SA, Facchetti, F Jones DM, Jaffe ES, Histiocyytic and dendritic cell neoplasms: Introduction, pp. 354–367; WHO Classification of Tumours of haematopoietic and Lymphoid Tissues, ed. By Steven H. Swerdlow, Elias Campo, Nancy Lee Harris, Elaine S. Jaffe, Stefano A. Pileri, Harald Stein, Jurgen Thiele, James W. Vardiman, WHO Press, Lyon, 2008, with permission.

course of the disease the lesions are usually proliferative and locally destructive. Later, the lesions undergo necrosis.

Several reports have shown that LCH is a clonal disorder of the Langerhans cell with immature features, decreased telomere length, some evidence of chromosome instability and abnormal oncogene expression. Such data have suggested that some presentations of LCH represent neoplastic disorders, possibly similar to the transient myeloproliferative disease described in Down syndrome. Other data have stressed the aberrant cytokine and chemokine expression as well as the spontaneous remission of LCH lesions to suggest a primary immune etiology.

The following organs are commonly involved in LCH: bone (pelvis, femur, ribs, skull and orbit), skin, lymph nodes, bone marrow, lungs, hypothalamic pituitary axis, spleen and liver. The damaged tissue is cleared by macrophages, which eventually undergo xanthomatous changes followed by fibrosis.

Clinical Features

Clinical manifestations depend on the site of lesions, number of involved sites and extent to which the function of key involved organs is compromised.

Table 18-2   Classification of Histiocytic Disorders

Disorders of varied biological behavior
   Dendritic cell and dermal dendrocyte related
      Langerhans cell histiocytosis
      Secondary dendritic cell processes
      Juvenile xanthogranuloma and related disorders
      Solitary histiocytomas of various dendritic cell phenotypes
   Macrophage related
Hemophagocytic syndromes
      Primary hemophagocytic lymphohistiocytosis
        (Familial and sporadic; commonly elicited by viral infections)
      Rosai–Dorfman disease (sinus histiocytosis with massive lymphadenopathy)
      Solitary histiocytoma with macrophage phenotype
   Secondary hemophagocytic syndromes
      Infection associated
      Malignancy associated
      Other
Malignant disorders
   Monocyte related
      Leukemias (FAB and revised FAB classifications)
        Monocytic leukemia M5A and B
        Acute myelomonocytic leukemia M4
        Chronic myelomonocytic leukemia
      Extramedullary monocytic tumor or sarcoma (monocytic counterpart of granulocytic sarcoma)
   Dendritic cell-related histiocytic sarcoma (localized or disseminated)
      Specify phenotype: follicular dendritic cell, interdigitating dendritic cell, etc.
   Macrophage-related histiocytic sarcoma (localized or disseminated)

From: Favara BE, Feller AC, with members of the WHO Committee on Histiocyte/Reticulum Cell Proliferations: Pauli M, et al. Contemporary classification of histiocyte disorders. Med Pediatr Oncol 1997;29:157–166, with permission.

Table 18-3   Histological, Histochemical and Electron Microscopic Diagnosis of Langerhans Cell Histiocytosis

1. Presumptive diagnosis: light morphologic characteristics
2. Designated diagnosis
   a. Light morphologic features plus
   b. Two or more supplemental positive stains for
      (1) Adenosine triphosphatase
      (2) S-100 protein
      (3) α-D-Mannosidase
      (4) Peanut lectin
3. Definitive diagnosis
   a. Light morphologic characteristics plus
   b. Birbeck granules in the lesional cell with electron microscopy and/or
   c. Staining positive for CD1a antigen on the lesional cell

From: Writing Group of the Histiocyte Society (Chu T, D'Angio GJ, Favara B, Ladisch S, Nesbit M, Pritchard J). Lancet 1987;1:208–209, with permission.

Although replaced by more prognostically reliable systems the classic designations, eosinophilic granuloma, Hand-Schuller-Christian disease and Abt-Letterer-Siwe disease are useful descriptions of the various clinical manifestations of LCH.

Eosinophilic granuloma, solitary (SEG) or multifocal (MEG), are found predominantly in older children, as well as in young adults, usually within the first three decades of life with the incidence peaking between 5 and 10 years of age. SEG and MEG represent approximately 60–80% of all instances of LCH. Patients with systemic involvement frequently have similar bone lesions in addition to other manifestations of disease.

Hand-Schüller-Christian disease (multisystem disease) was historically described as the clinical triad of lytic lesions of bone, exophthalmos and diabetes insipidus, although this form of LCH is now usually considered to be more typical multisystem LCH without key organ dysfunction. It is most commonly described in younger children aged 2–5 years and represents 15–40% of such patients, although this type of involvement can be observed in all ages. Signs and symptoms include bone defects with exophthalmos due to tumor mass in the orbital cavity. This usually occurs from involvement of the roof and lateral wall of the orbital bones. Orbital involvement may rarely result in vision loss or strabismus due to optic nerve or orbital muscle involvement, respectively. The most frequent sites of skeletal involvement include the flat bones of the skull, ribs, pelvis and scapulae. There may be extensive involvement of the skull, with irregularly shaped, lucent lesions, giving rise to the so-called geographic skull. Somewhat less frequently, long bones and lumbosacral vertebrae, usually the anterior portion of the vertebral body, are involved. Oral involvement commonly affects the gums and/or palate. Erosion of the lamina dura gives rise to the characteristic floating tooth seen on dental radiographs. The entire mandible may be involved, with loss of bone leading to diminished height of the mandibular rami. Erosion of gingival tissue causes premature eruption, decay and tooth loss. Parents of affected children, particularly infants, frequently report precocious eruption of teeth due to receding, diseased gums. Chronic otitis media, due to involvement of the mastoid and petrous portion of the temporal bone and otitis externa are not uncommon.

Abt-Letterer-Siwe disease is the rarest (10% of cases) and most severe manifestation of LCH. Typically, patients are less than 2 years of age and present with a scaly seborrheic, eczematoid, sometimes purpuric rash that involves the scalp, ear canals, abdomen and intertriginous areas of the neck and face. The rash may be maculopapular or nodulopapular. Ulceration may result, especially in intertriginous areas. Ulcerated and denuded skin may serve as a portal of entry for microorganisms, leading to sepsis. Draining ears, lymphadenopathy, hepatosplenomegaly and, in severe cases, hepatic dysfunction with hypoproteinemia and diminished synthesis of clotting factors can occur. Anorexia, irritability, failure to thrive and significant pulmonary symptoms such as cough, tachypnea and pneumothorax may occur as well. One of the most significant areas of involvement is that of the hemato-poietic system, with bone marrow infiltration resulting in pancytopenia.

Other presentations of LCH are commonly seen. LCH can have a strictly *nodal presentation*, not to be confused with sinus histiocytosis with massive lymphadenopathy (Rosai-Dorfman disease). This presentation is characterized by significant enlargement of multiple lymph node groups, with little or no other signs of disease. In the *pulmonary syndrome* there is often exclusive involvement of the lungs. This condition is usually seen in young adults in their third or fourth decade (occasionally in adolescents) and may follow a severe and often chronic, debilitating course; patients may present with pneuomothorax. This type of involve-ment is often associated with cigarette smoking. Pulmonary disease may also occur as part of multisystem disease. *Cutaneous disease* with no evidence of dissemination has been described in infants, children and adults.

Involvement by Site of Disease

Skeleton

Painful bone lesions affecting hematopoietically active bones are common. Radiographi-cally, the lesions are lytic. They occur commonly in the skull as punched-out lytic lesions, without evidence of marginal sclerosis or periosteal reaction. Bone involvement of the mandible and maxilla and soft-tissue involvement of the gingivae may result in loss of teeth. Involvement of vertebra can result in vertebral collapse (vertebra plana) and lesions of long bones could result in fractures. There is often an inability to bear weight and tender, sometimes warm swelling due to soft-tissue infiltrations overlying the bone lesions occur. Radionuclide bone scan (^{99m}Tc-polyphosphate) may show localized increased uptake at the site of involvement. Magnetic resonance imaging (MRI) shows bone lesions not identifiable by either radiographic or radionuclide scans. The differential diagnoses include osteomyelitis, malignant bone tumors and bony cysts. Only rarely are the wrists, hands, bones of the feet, or cervical vertebrae involved. Table 18-4 lists the distribution of the sites of bone lesions.

Skin

Cutaneous eruptions consist of:

- Diffuse papular scaling lesions, resembling seborrheic eczema (most common)
- Petechiae and purpura
- Granulomatous ulcerative lesions
- Xanthomatous lesions.

Table 18-4 Distribution of Bone Lesions in Langerhans Cell Histiocytosis

Site	Incidence (%)
Skull	49
Innominate bone	23
Femur	17
Orbit	11
Ribs	8
Humerus	7
Mandible	7
Tibia	7
Vertebra	7
Clavicle	5
Scapula	3
Fibula	2
Sternum	1
Radius	1
Metacarpal	1

Lungs

Lung involvement may result in pulmonary dysfunction with tachypnea and/or dyspnea, cyanosis, cough, pneumothorax, or pleural effusion. Radiographic densities or infiltrates consisting of diffuse cystic changes, nodular infiltrations, or extensive fibrosis can occur. The radiographic appearance may resemble miliary tuberculosis.

Liver

The liver may be enlarged. Liver dysfunction may consist of the following: hypoproteinemia (total protein less than 5.5 g/dl and/or albumin less than 2.5 g/dl), edema, ascites and/or hyperbilirubinemia (bilirubin level greater than 1.5 mg/dl, not attributable to hemolysis).

Pretreatment liver biopsy more often reveals portal triditis and less often fibrohistiocytic infiltrates or bile duct proliferation. Sclerosing cholangitis, fibrosis and liver failure may lead to the need for liver transplantation.

Hematopoietic System

The pathophysiology of hematopoietic dysfunction can be due to hypersplenism as well as direct involvement of the bone marrow by Langerhans cells and/or reactive macrophages.

Hematopoietic system dysfunction may consist of the following:

- Anemia (hemoglobin level less than 10 g/dl, not due to iron deficiency or superimposed infection), leukopenia (neutrophils less than 1,500/mm^3), or thrombocytopenia (less than 100,000/mm^3)

• An excessive number of histiocytes in the marrow aspirate is not considered evidence of dysfunction.

Lymph Nodes

Occasionally, massive lymph node enlargement of cervical nodes occurs without other evidence of histiocytosis.

Endocrine System

Short stature has been found in up to 40% of children with systemic LCH. Chronic illness and steroid therapy play an important role in its causation. However, short stature may also be a consequent of anterior pituitary involvement and growth hormone deficiency, which may occur in at least half of the patients with initial anterior pituitary dysfunction. Posterior pituitary involvement with diabetes insipidus is characterstic of systemic LCH. Other endocrine manifestations include hyperprolactinemia and hypogonadism due to hypothalamic infiltration. Pancreatic and thyroid involvement have also been reported.

Gastrointestinal System

Gastrointestinal tract disease can occur. Two to 13% of patients have biopsy-proven gastrointestinal involvement and/or digestive tract symptoms, although this may be an underestimate.

Central Nervous System

Clinically, four groups of patients can be distinguished:

• Patients who present with a disorder of the hypothalamic pituitary system
• Patients who present with site-dependent symptoms of space-occupying lesions such as headache and seizures
• Patients who exhibit neurologic dysfunction characterized by a neurodegenerative picture of reflex abnormalities, ataxia, intellectual impairment, sometimes hydrocephalus, tremor and dysarthria with variable progression to severe CNS deterioration
• Patients who present with an overlap of the aforementioned symptoms.

Patients who develop CNS disease are more likely to have multisystem disease and skull lesions.

Table 18-5 shows the clinical characteristics of patients with LCH who developed CNS disease compared to those who did not develop CNS disease. It reveals that patients who developed CNS disease are more likely to have multisystem disease with skull and temporal bone lesions, orbital involvement, diabetes insipidus and endocrinopathies. Table 18-6 shows the classification of CNS lesions according to MRI morphology.

Table 18-5   Extent of Langerhans Cell Histiocytosis and Organs Involved at Diagnosis in Patients Who Develop CNS Disease Compared to Those Who Do Not Develop CNS Disease

	Percentage in 38 CNS Patients	Percentage in 275 LCH Patients
Multisystem disease	72	40
Single-system bone disease	18	53
Single-system skin, lymph node	0	7
Primary CNS disease	10	0
Bone	84	79
Skull	74	40
Temporal bone	34	8
Skin	58	25
Diabetes insipidus	31	6
Orbits	24	2
Endocrinopathies	18	3
Lungs	16	6
Gastrointestinal tract	10	5
Liver	10	11
Spleen	10	9

From: Grois NG, Favara BE, Mostbeck GH, Prayer D. Central nervous system disease in Langerhans cell histiocytosis. Hematol/Oncol Clin N Am 1998;12:287–305, with permission.

Table 18-6   Classification of Central Nervous System Lesions According to Magnetic Resonance Imaging Morphology[a]

Type		Number	Percentage
Ia	White-matter lesions without enhancement	21	55
Ib	White-matter lesions with enhancement	9	24
IIa	Gray-matter lesions without enhancement	19	50
IIb	Gray-matter lesions with enhancement	3	8
IIIa	Extraparenchymal dural based	12	32
IIIb	Extraparenchymal arachnoidal based	2	5
IIIc	Extraparenchymal choroid plexus based	3	8
IVa	Infundibular thickening	8	21
IVb	(Partial) empty sella	14	37
IVc	Hypothalamic mass lesions	4	10
Va	Atrophy: diffuse	10	26
Vb	Atrophy: localized	6	16
VI	Therapy-related with enhancement	6	15

[a]LCH-CNS Study, $n = 38$.
From: Grois NG, Favara BE, Mostbeck GH, Prayer D. Central nervous system disease in Langerhans cell histiocytosis. Hematol/Oncol Clin N Am 1998;12:287–305, with permission.

Histopathology

There are likely four types of LCH involvement of the CNS as follows:

- Hyperplastic proliferative
- Granulomatous
- Xanthomatous
- Fibrosis.

Lesions in the hyperplastic proliferative phase are most likely to contain the diagnostic LCH cells. In the cerebellum and cerebrum, white matter may show demyelination and there may also be destruction of Purkinje cells in the absence of histiocytes. Extensive gliosis with lymphocytic infiltrates may also be found.

Hypothalamic Pituitary Involvement

Signs and Symptoms

1. *Hypothalamic involvement*: disturbances in social behavior, appetite, temperature regulation, sleep patterns.
2. *Posterior pituitary involvement*: diabetes insipidus (DI), polyuria, polydypsia.
3. *Anterior pituitary involvement*: growth failure, precocious or delayed puberty, amenorrhea, hypothyroidism.

Of all these, DI is the most common manifestation. The incidence of this complication ranges from 5 to 35% depending upon the extent and location of disease. Most children present within 4 years of diagnosis. DI is due to infiltration by Langerhans cells into the hypothalamus with or without involvement of the posterior pituitary gland. Local tissue damage may be a consequence of IL-1 and prostaglandin E2 production. Polydypsia and polyuria may develop at presentation, during active disease (even when there is improvement in other areas), or after therapy is discontinued and there is no other apparent active disease.

Diagnosis

Water deprivation test with at least 3 hours of no intake and with serum and urine electrolytes and osmolalities before and after deprivation plus arginine vasopression (ADH) levels A fasting morning urinalysis for specific gravity is also helpful as a screen.

The inability to concentrate the urine, low ADH levels and a therapeutic response to a dose of vasopressin are all useful in making the diagnosis of DI. These tests discriminate partial from complete DI. Partial DI fluctuates spontaneously. However, partial DI is very rare in LCH.

Gadolinium-enhanced MRI studies show thickening of the hypothalamic pituitary stalk (>2.5 mm) and absence of a posterior pituitary bright signal in T1-weighted images. These lesions are caused by the infiltration of LCH cells.

Management

There is no convincing evidence that established DI can be reversed by any treatment modality but new-onset DI is considered to represent active disease and rapid initiation of treatment, usually LCH-directed chemotherapy, is recommended, although there are few data about the true response rate.

Replacement therapy with desmopressin (DDAVP) is recommended for patients with DI. The rapid institution of effective systemic chemotherapy for disseminated disease may prevent the occurrence of DI and might be responsible for the low frequency of DI, although this has not been definitively proven. A small pituitary or empty sella indicates combined anterior and posterior pituitary insufficiency. This may be a result of disease or may be observed following cranial radiotherapy for DI.

Patients presenting with isolated idiopathic DI and morphologic changes in the suprasellar area should be observed very closely. Stereotactic biopsy performed because of an enlarged pituitary stalk can distinguish a variety of conditions such as sarcoidosis, granulomatosis, tuberculosis, nonspecific lymphocytic hypophysitis and LCH. However the biopsy is risky and should be avoided if possible. Thus, patients without a definitive diagnosis and with DI are often followed with serial contrast MRIs, once germinoma or other disorders are investigated. If the CNS lesion enlarges and no definitive diagnosis has been made, then a biopsy is indicated. Treatment is determined by the histologic diagnosis.

Space-Occupying Central Nervous System Lesions

These lesions most often arise from adjacent bone lesions, brain meninges, or choroid plexus. They usually give rise to signs and symptoms of increased intracranial pressure. They are also site-specific and size-dependent. Such symptoms include headaches, vomiting, papilledema, optic atrophy, seizures and other focal symptoms. Even diffuse meningitis-like manifestations can occur. These lesions may occur without any other evidence of LCH.

Mass lesions respond well to treatment, leaving minimal or no residual defects. They do not necessarily require radiation therapy and chemotherapy is often sufficient treatment.

Cerebellar Syndrome/Neurologic Degeneration

The cerebellum is the second most common site of LCH CNS involvement, the first being the hypothalamic–hypophyseal axis. Neurologic symptoms occasionally predate the diagnosis of LCH. The symptoms mainly follow the pontine–cerebellar pattern, beginning as a discrete reflex abnormality or gait disturbance and/or nystagmus. Sometimes patients may also present with hydrocephalus. They can progress to disabling ataxia. Pontine symptoms include dysarthria, dysphagia and other cranial nerve deficits, ultimately leading to fatal neurodegeneration. On MRI, often enhancing lesions involve the pons, basal ganglion and cerebellar peduncles show. MRI of the cerebrum may show white-matter lesions in the periventricular area.

There is no established effective treatment available and prognosis is poor. Aggressive early treatment may reduce the incidence of this syndrome, but this too is unproven. Biopsy shows a primarily inflammatory, lymphocyte response associated with gliosis, demyelination and neuronal cell death. The etiology of this neurodegenerative process is unknown, but is believed to be an immune-mediated paraneoplastic response. In some cases, the onset of CNS symptoms occurs many years after the initial diagnosis of LCH.

Clinical and Laboratory Evaluation

Table 18-7 lists the laboratory and radiographic evaluation of new patients with LCH and their follow-up and Table 18-8 lists the evaluation of patients upon specific clinical

Table 18-7   Required Laboratory and Radiographic Evaluation of New Patients with Langerhans Cell Histiocytosis

Test	Follow-up Test Interval When Organ System Is:		
	Involved	Not Involved	Single-bone Lesion
Hemoglobin and/or hematocrit	Monthly	6 months	None
White blood cell count and differential count	Monthly	6 months	None
Platelet count	Monthly	6 months	None
Ferritin, iron, transferrin ESR	Monthly	6 months	None
Liver function tests (SGOT, SGPT, alkaline phosphatase, bilirubin, total proteins, albumin)	Monthly	6 months	None
Coagulation studies (PT, PTT, fibrinogen)	Monthly	6 months	None
Chest radiograph (PA and lateral)	Monthly	6 months	None
Skeletal radiograph survey[a]	6 months	None	Once, at 6 months
Urine osmolality measurement after overnight water deprivation	6 months	6 months	None
Bone marrow aspirate and biopsy[b]			
HLA-typing[c]			

[a]Radionuclide bone scan is not as sensitive as the skeletal radiograph survey in most patients. It may be performed optionally but should not replace the skeletal survey. If suspicion of lesion exists (e.g., pain) and both radiographic and radionuclide tests are negative, an MRI should be performed.
[b]From patients with multisystem disease.
[c]From patients with high risk disease (multisystem with major organ dysfunction).
Abbreviations: SGOT, serum glutamic-oxaloacetic transaminase; SGPT, serum glutamic pyruvate transaminase; PT, prothrombin time; PTT, partial thromboplastin time; PA, posteroanterior.
Modified from: Broadbent V, Gadner H, Komp D, Ladish S. Histiocytosis in children II: approach to the clinical and laboratory evaluation of children with Langerhans cell histiocytosis. Med Pediatr Oncol 1989;17:492.

Table 18-8 Evaluation Required Upon Specific Clinical Indication

Indication	Tests	Follow-up Tests
Anemia, leukopenia or thrombocytopenia	Bone marrow aspirate and trephine biopsy	6 months
Abnormal chest radiograph, tachypnea, intercostal retractions	Bone marrow aspirate and trephine biopsy, pulmonary function tests, Chest CT	Every 6 months
Patients with abnormal chest radiographs in whom chemotherapy is being considered, to exclude opportunistic infection	Bronchoalveolar lavage; if not helpful, lung biopsy necessary	None
Unexplained chronic diarrhea or failure to thrive, evidence of malabsorption	Small-bowel series and biopsy 72-hour stool fat	None
Liver dysfunction, including hypoproteinemia not due to protein-losing enteropathy, to differentiate active LCH of the liver from cirrhosis	Liver biopsy	To be performed when all disease resolved but liver dysfunction persists to distinguish cirrhosis from continuing LCH
Hormonal, visual, or neurologic abnormalities	CT of brain/hypothalamic pituitary axis, with IV contrast enhancement or MRI preferable	Every 6 months
Oral involvement	Panoramic dental radiograph of mandible and maxilla; oral surgery consultation	Every 6 months
Short stature, growth failure, diabetes insipidus, hypothalamic syndromes, galactorrhea, precocious or delayed puberty; CT or MRI abnormality of hypothalamus/pituitary	Endocrine evaluation	None
Annual discharge, deafness	Otolaryngology consultation and audiogram	Every 6 months

Abbreviations: CT, computed tomography; MRI, magnetic resonance imaging.
From: Broadbent V, Gadner H, Komp D, Ladish S. Histiocytosis in children II: Approach to the clinical and laboratory evaluation of children Langerhans' cell histiocytosis. Med Pediatr Oncol 1989;17:492.

indications and their follow-up. Every patient should have a thorough physical examination including temperature, height, weight and head circumference, pubertal status, skin and scalp for rash, pallor or jaundice, external and middle ear, face and orbits, oropharynx, dentition, chest and lungs, abdomen for organomegaly, extremity and spine. There are few evidenced-based guidelines, but the ones listed represent general best practice recommendations.

Diagnosis

A chest radiograph and skeletal survey should usually be performed.

Follow-up Radiograph

A chest radiograph should be performed monthly if the lungs are involved and every 6 months if the lungs are not involved. A follow-up chest film is not required when a monostotic lesion is found at presentation; however, a one-time skeletal survey should be obtained at 6 months in this situation. When multiple bones are affected, skeletal surveys are obtained every 6 months. No follow-up skeletal survey is required if bones are not involved at presentation.

Special Situations

1. Presence of malabsorption, unexplained chronic diarrhea, or failure to thrive – endoscopic biopsy, upper gastrointestinal study with small-bowel follow-up; 72-hour stool fat.
2. Patients with hormonal, visual, or neurologic abnormalities – MRI scan or contrast-enhanced computed tomography (CT) scan of the brain and hypothalamic pituitary axis.
3. Patients with oral involvement – panoramic dental radiographs of the mandible and maxilla every 6 months.
4. Patients with suspected spinal cord compression – MRI of the spine.
5. Patients with pulmonary symptoms or significant mediastinal widening of chest film – high-resolution CT of the lungs, pulmonary function tests.
6. Superior vena cava syndrome – CT with contrast.
7. Significant cervical lymphadenopathy – CT or MRI of the neck.
8. Ear involvement – CT of temporal bones.
9. Hepatosplenomegaly – ultrasound of the abdomen.
10. Soft-tissue tumors – MRI of involved tissue.

Technetium-99m labeled with methylene diphosphonate (MDP) scintography for bone lesions and routine radiographic skeletal examinations are complementary to each other, because radiography is more likely to detect older and quiescent lesions and scintography may detect early aggressive lesions. The ability of scintography to determine the activity of a lesion can be useful in the evaluation of a persistent radiographic abnormality.

Treatment

A generally accepted standard for the initial treatment of patients with LCH is to use an appropriate amount of the least toxic therapy to treat the disease. In patients with potentially life-threatening disease at presentation, or in those developing life-threatening disease during the course of treatment, alternative and sometimes more aggressive treatment should be considered, including hematopoietic stem cell transplantation. Although recurrence rates have been shown to be significantly reduced in patients with multisystem low-risk disease when patients are treated for 12 months instead of 6 months with vinblastine and prednisone, the role of more intensive or prolonged treatment to prevent diabetes insipidus, central nervous system degeneration, sclerosing cholangitis is less clear.

Specific Sites

Solitary Bone Lesions

1. Curettage (radical surgery is not indicated).
2. Intralesional steroids (e.g., methylprednisolone acetate 40 mg/ml, 1–4 ml, depending on size of lesion).
3. Radiotherapy (600–900 cGy for most lesions although some lesions, especially in adults, may require 1,500 cGy) is reserved for isolated lesions inaccessible to intralesional steroid treatment or lesions that have potential to compromise vital structures (e.g., optic nerve, spinal cord and occasionally other sites, such as mastoids).
4. Systemic therapy for disease involving multiple bones or organ involvement.

Localized Skin Involvement

1. Topical steroid application.
2. Systemic steroids for patients with skin and multisystem disease.
3. Thalidomide has also been used.

Severe or Refractory Skin Disease

1. Local application of 20% nitrogen mustard only to involved skin, avoiding the surrounding normal skin.
2. Psoralen and ultraviolet A irradiation.
3. Topical tacrolimus.
4. Electron beam radiotherapy.
5. Systemic therapy with chemotherapy vinblastine/prednisone or thalidomide.

Solitary Lymph Node

1. Excisional biopsy should be performed and observation of the patient.

Regional Lymph Node Involvement

1. Systemic therapy, usually with vinblastine plus prednisone should be administered.

Meningeal Disease and Brain Parenchyma Disease

1. Systemic therapy should be administered.

Multisystem Disease

1. For appropriate risk-adapted therapy patients with multisystem disease are grouped as shown in Table 18-9.

Table 18-9   Risk Groups According to the Histiocyte Society LCH-III Trial

Group 1 – Multisystem "risk" patients
Multisystem patients with involvement of one or more Risk organs (i.e. hematopoietic system, liver, spleen or lungs)
Group2 – Multisystem "low-risk" patients
Multisystem patients with multiple organs involved but without involvement of Risk organs
Group 3 – Single system "multifocal bone disease" or localized "special site" involvement
Patients with multifocal bone disease, i.e., lesions in two or more different bones or patients with localized special site involvement, such as lesions with intracranial soft tissue extension or vertebral lesions with intraspinal soft tissue extension

Certain principles of treatment were derived from the Histiocyte-Society-sponsored randomized studies, LCH-I, II and III.

- LCH-1 showed that the rapidity of the initial response correlates with prognosis
- Results from the LCH-II study suggest that there is no significant advantage from adding etoposide in terms of survival or the frequency of disease recurrence.
  A retrospective analysis of a subset of patients on this trial, however, has suggested an increased response in patients with multisystem, risk-organ-involved disease (hematopoietic system, liver, spleen, or lungs) when treated with vinblastine, etoposide and prednisone
- In an attempt to improve the response rate and possibly reduce the frequency of recurring disease, the LCH-III trial addressed whether the initial response rate is improved by the addition of intermediate-dose methotrexate to prednisone and vinblastine and whether overall outcome is improved using 6 or 12 months of continuation therapy. This study has demonstrated no advantage to adding methotrexate and a significantly lower recurrence rate for patients treated for 12 months compared to 6 months. Table 18-10 lists systemic therapy for patients with LCH and summarizes the LCH III study.

Recurrent or Refractory Disease

For patients with recurrent and/or refractory disease, alternative treatment has not been standardized. Patients with recurrent disease, i.e., disease that reappears after a period of remission, often respond well to the drugs with which they were initially treated. Several studies have demonstrated significant activity to 2-chlorodeoxyadenosine (2-CdA) in recurrent and refractory LCH. In addition, the combination of 2-CdA and high-dose cystosine arabinoside (AraC) has been used in highly refractory patients. Patients with relapsed, refractory disease can receive two courses of 2-CdA (5 mg/m^2/day in 50 ml normal saline over 2 hours for

Table 18-10 Systemic Therapy for LCH*

Risk Group	Initial Treatment	Continuation Treatment	Duration
Group 1	Prednisone[a] Vinblastine[b] ± Methotrexate[c]	6-MP[d] Prednisone[e] Vinblastine[f] ± Methotrexate[g]	12 months
Group 2	Same as Group 1	Prednisone[e] Vinblastine[f]	6 vs 12 months
Group 3	Same as Group 1	Same as Group 2	6 months only

*Patients on systemic chemotherapy should receive standard supportive care including sulfamethoxazole/trimethoprim (5 mg/kg/day of trimethoprim) in two divided doses per day for 3 days per week. Sulfamethoxazole/trimethoprim should not be administered during methotrexate administration.

[a]Oral prednisone, 40 mg/m^2 daily in three divided doses as a 4-week course, followed by a tapering doses over a period of 2 weeks. Poor responders should receive a further 6-week course of prednisone 40 mg/m^2 daily on days 1–3 of each week for an additional 6 weeks.

[b]Vinblastine 6 mg/m^2 i.v. bolus on day 1 of week 1, 2, 3, 4, 5 and 6. This 6-week course of vinblastine is repeated for poor responders.

[c]Methotrexate 500 mg/m^2 24-hour infusion with folinic acid (leucovorin) rescue day 1 of weeks 1, 3 and 5. Ten percent of the dose is given as i.v. bolus over 30 minutes, followed by 90% of the dose as a 23.5 hour infusion with 2,000 ml/m^2 hydration. Folinic acid 12 mg/m^2 is given 24 hours and 30 hours after methotrexate infusion is completed (at 48 and 54 hours after methotrexate therapy is started). It has not been established whether the addition of methotrexate improves results.

[d]Oral 6-mercaptopurine (6-MP) 50 mg/m^2 daily until the end of the 12th month from commencement of therapy.

[e]Pulses of oral prednisone 40 mg/m^2 daily in three doses on day 1–5 every 3 weeks, starting on day 1 of week 7 in patients who have no active disease after course 1 or on day 1 of week 13 in patients who have no active disease or the active disease is better after course 2, continued until the end of month 12.

[f]Vinblastine 6 mg/m^2/day IV bolus once weekly for 3 weeks, starting on day 1 of week 7 in patients with no active disease after course 1 or on day 1 of week 13 in patients with no active disease or the active disease is better after course 2, continued until the end of month 12.

[g]Methrotrexate 20 mg/m^2 orally, once weekly, until the end of month 12. It has not been established whether the addition of methotrexate improves results.

5 consecutive days every 3–4 weeks). If there is a good response then patients receive 2–4 additional courses of 2-CdA for not more than six total courses. If there is a poor initial response after two courses of CdA or in some patients with extensive disease, the combination of 2CdA (9 mg/m^2/day as a 2-hour iv infusion given daily for 5 days. 2-CdA is started on the second day of the course and is given at hours 23, 47, 71, 95 and 119) and cystosine arabinoside (AraC) (500 mg/kg in 250 ml/m^2 twice a day, i.e., every 12 hours, for 5 days as a 2-hour iv infusion can be used). This latter regimen is that of the stem cell transplantation for high-risk or refractory patients warrants further evaluation. A regimen from The Netherlands has used vincristine, low-dose cytosine arabinoside and prednisone for both patients with newly diagnosed and recurrent disease. There is anecdotal experience using TNF-inhibitors in recurrent or refractory disease as well as bisphosphanates for skeletal lesions.

Prognosis

Historic prognostic factors used to stratify therapy include:

* Response to initial therapy
* Age at diagnosis (<24 months, 55–60% mortality) although response to initial 6–12 weeks of therapy outweighs this factor
* Number of organs involved at diagnosis:

Number of Organs	Mortality (%)
1–2	0
3–4	35
5–6	60
7–8	100

* Organ dysfunction (e.g., lung, liver, bone marrow) at diagnosis:

Organ Dysfunction	Mortality (%)
Present	66
Absent	4

* Natural history on treatment:

Group	Description	Mortality (%)
A	No disease progression over 6–12 months	0
B	Progressive disease without organ dysfunction	20
C	Development of organ dysfunction during course of disease	100

* Congenital self-healing histiocytosis: condition that manifests in neonates with skin lesion pulmonary nodules with or without bone lesions (these lesions regress spontaneously and require no treatment).

Sequelae and Complications

The risk factors for developing residual adverse sequelae include:

* Generalized disease with skeletal disease and especially lesions of the orbit, sphenoid, mastoid or temporal bones with associated dural involvement
* Smoldering disease.

Long-term complications include:

- Pulmonary: Progressive fibrosis, pulmonary cyst formation and chronic pneumothoraces. There is no effective therapy for these complications and progression to cor pulmonale and respiratory failure commonly occurs
- Hepatic: Sclerosing cholangitis has been reported and may lead to secondary biliary cirrhosis, portal hypertension and liver failure. The etiology is not understood. The only successful treatment has been liver transplantation
- Neuropsychiatric: CNS manifestations can occur without any relationship to radiotherapy or other treatments. This may manifest with learning disability, ataxia, pyramidal signs and behavioral changes. MRI studies with gadolinium contrast are helpful in localizing structural changes in the brain
- Endocrine: Diabetes insipidus and growth retardation are the most frequent complications. They result from histiocytic infiltration of the pituitary and hypothalamus. Such lesions that initially present with diabetes insipidus will result in panhypopituitism in approximately 50–60% of patients. A hyperphagic syndrome associated with extreme weight gain may also occur as a result of hypothalamic lesions
- Orthopedic: Deformities of the spine can result in long-term disabilities
- Dental: Loss of teeth and jaw abnormalities may occur
- Hearing: Patients with mastoid and middle ear involvement may develop permanent hearing loss
- Malignancies: Second primary malignancies associated with radiation therapy include astrocytoma, medulloblastoma, meningioma, hepatoma, osteosarcoma of the skull and thyroid carcinoma. Secondary AML has been associated with the use of chemotherapy, particularly with 2CdA.

OTHER HISTIOCYTIC DISORDERS

Secondary Dendritic Cell Processes

The accumulation of dendritic cells and Langerhans cells occurs in the lymph nodes in Hodgkin disease, lymphoma and tumor of the lung, thyroid and other sites. The secondary dendritic cell processes involute with control of the primary disease. This is a histopathologic finding of no clinical significance in terms of the diagnosis of a primary histiocytic disorder.

DERMAL DENDROCYTE DISORDERS

Juvenile Xanthogranuloma

Clinically, juvenile xanthogranuloma (JXG) is characterized by multiple cutaneous nodules consisting of dermal dendrocytes. These lesions sometimes involute spontaneously and no treatment is required. Occasionally there is systemic involvement. When this occurs therapy similar to that for LCH is used. When the disease is disseminated, it can be referred to as xanthoma disseminatum.

Erdheim–Chester Disease

Erdheim–Chester disease (ECD) occurs primarily in adults but can occur rarely in young individuals. ECD is characterized by accumulations of xanthomatous macrophages particularly in the retroperitoneum which often can lead to renal failure. In addition, ECD commonly affects the lungs and heart, along with bilateral long bone involvement leading to severe and chronic pain. In addition, ECD can lead to diabetes insipidus and other CNS signs and symptoms similar to LCH. Treatment for patients with ECD usually involves alpha-interferon or treatments similar to those used to treat patients with LCH.

Solitary Histiocytomas with Dendritic Cell Phenotypes

These tumors are composed of dendritic cells without malignant features. They have variable phenotypes identified by various immunological and cytochemical markers (Table 18-1), for example, indeterminate cell or interdigitating dendritic cell phenotype. They occur in the cutaneous tissue and less often in the central nervous system. Surgical resection when possible is a reasonable therapeutic approach; on occasions, radiation therapy or chemotherapy such as that used in patients with LCH is used.

Sinus Histiocytosis with Massive Lymphadenopathy (Rosai–Dorfman Disease)

Manifestations of sinus histiocytosis with massive lymphadenopathy (SHML) include the following:

* Worldwide occurrence with higher incidence among blacks
* Onset usually within the first two decades of life
* Massive painless bilateral cervical lymphadenopathy with involvement of other groups of lymph nodes
* Snoring, when there is involvement of retropharyngeal lymphoid tissue; possibility of sleep apnea
* Extranodal infiltration in approximately 25% of patients (skin, orbit, eyelid, liver, spleen, testes, CNS, salivary glands, bone, respiratory tract)
* Immunologic abnormalities with manifestations of autoimmune disorders (e.g., hematologic antibodies, glomerulonephritis, amyloidosis, or joint disease) in 10% of patients
* Fever
* Leukocytosis with neutropenia, mild anemia, elevated ESR
* Polyclonal hypergammaglobulinemia.

Complications
1. Retropharyngeal involvement causing respiratory compromise.
2. Epidural involvement causing spinal cord compression.
3. Visual loss due to optic nerve compression or corneal involvement.

Diagnosis
Lymph node shows marked dilatation of sinuses by proliferation of benign histiocytes with prominent phagocytosis of lymphocytes, plasma cells and erythrocytes by sinus histiocytes. Plasma cell infiltrates in the medullary cords and capsular fibrosis may also be evident.

Prognosis
Twenty percent of patients have spontaneous resolution or improvement within 3–9 months. The majority of patients have stable and persistent disease lasting up to several years. Seven percent of patients have a fatal outcome, especially if immunologic abnormalities and extra-nodal involvement are present.

Treatment
The following treatment has been employed:

- No treatment is generally warranted because this is a self-limited disease
- Prednisone 2 mg/kg PO may be administered for life-threatening complications
- If prednisone fails, the combination of vinblastine plus steroid can be used with anecdotal information suggesting that dexamethasone is the more effective steroid
- 6-mercaptopurine 50–75 mg/m^2/day PO in combination with methotrexate 10–20 mg/m^2/week PO have also been used with evidence of response
- Intermediate- and high-dose methotrexate with leukovorin rescue have been reported in small numbers of patients to result in good, partial responses
- 2CdA has been reported to result in good partial response
- α-Interferon has also been reported to be effective in some cases.

Hemophagocytic Lymphohistiocytosis (Hemophagocytic Syndromes)

Hemophagocytic lymphohistiocytosis (HLH) falls into two categories:

- *Familial hemophagocytic lymphohistiocytosis (FHLH)* (familial or sporadic): This is an autosomal-recessive disease that affects immune regulation. Although it is commonly termed familial HLH, because the disease has an autosomal recessive inheritance, sporadic cases with no obvious family inheritance occur. The signs and symptoms of FHLH may be triggered by infections
- *Non-familial HLH*: A lymphohistiocytic proliferation with hemophagocytosis may also develop from marked immunological activation during viral, bacterial and parasitic infections (infection-associated hemophagocytic lymphohistiocytosis). This

may also be associated with malignancies, prolonged administration of lipids, rheumatoid arthritis (macrophage activation syndrome), immune deficiencies associated with cytotoxic T and/or NK cell dysfunction such as DiGeorge syndrome (del 22q11.2), Chédiak–Higashi syndrome, Griscelli syndrome[*], X-linked lymphoproliferative disease (XLP) and lysinuric protein intolerance (LPI). Hermansky-Pudlak syndrome has been reported to be associated with HLH.

Table 18-11 lists the diagnostic guidelines for HLH. It should be noted that there is no specific diagnostic feature for primary HLH (FHLH). For this reason, when the index of

Table 18-11   Diagnostic Guidelines for Hemophagocytic Lymphohistiocytosis

The diagnosis of HLH can be established if one of either 1 or 2 below is fulfilled:
  (1) A molecular diagnosis consistent with HLH
  (2) Diagnostic criteria for HLH fulfilled (five out of the eight criteria below)
      (A) Initial diagnostic criteria (*to be evaluated in all patients with HLH*)
          Fever
          Splenomegaly
          Cytopenias (affecting ≥2 of 3 lineages in the peripheral blood):
              Hemoglobin <9 g/dl (in infants <4 weeks: hemoglobin <10.0 g/dl)
              Platelets <100,000/mm^3
              Neutrophils <1,000/mm^3
          Hypertriglyceridemia and/or hypofibrinogenemia:
              Fasting triglycerides ≥3.0 mmol/L (i.e., ≥265 mg/dl)
              Fibrinogen ≤1.5 g/L
          Hemophagocytosis in bone marrow or spleen or lymph nodes
          No evidence of malignancy
      (B) New diagnostic criteria
          Low or absent NK-cell activity (according to local laboratory reference)
          Ferritin ≥500 μg/L
          Soluble CD25 (i.e., soluble IL-2 receptor) ≥2,400 U/ml

Comments:
  (1) If hemophagocytic activity is not proven at the time of presentation, further search for hemophagocytic activity is encouraged. If the bone marrow specimen is not conclusive, material may be obtained from other organs. Serial bone marrow aspirates over time may also be helpful
  (2) The following findings may provide strong supportive evidence for the diagnosis: (a) spinal fluid pleocytosis (mononuclear cells) and/or elevated spinal fluid protein, (b) histological picture in the liver resembling chronic persistent hepatitis (biopsy)
  (3) Other abnormal clinical and laboratory findings consistent with the diagnosis are: cerebromeningeal symptoms, lymph node enlargement, jaundice, edema, skin rash. Hepatic enzyme abnormalities, hypoproteinemia, hyponatremia, VLDL ↑, HDL ↓.

Abbreviations: VLDL, very-low density-lipoprotein; HDL, high-density lipoprotein; ↑, increased; ↓, decreased.
From: Henter JI, Horne A, Arico M, et al. Pediatric Blood and Cancer 2007:48;124–131, with permission.

---

[*]Type 2; mutation in RAB27A (15q21), decreased T- and NK-cell function, hypogammaglobulinemia and partial albinism.

suspicion is strong for primary HLH, treatment may be started before extensive disease activity causes irreversible organ damage and the likelihood of a response to therapy decreases.

Familial Hemophagocytic Lymphohistiocytosis (FHLH)

Pathophysiology, Immunology and Genetics

In the absence of perforin activity the resulting inability to kill infected target cells results in sustained NK and cytolytic T cell (CTL) activity. This, in turn, results in the overexpression of inflammatory cytokines (soluble IL-2 receptor, IL-6, TNF-α, IL 10 and IL-12) leading to excessive macrophage activation, dissemination and organ infiltration and the signs, symptoms and laboratory abnormalities that characterize HLH.

Impaired NK cell activity is key to the diagnosis. FHLH is inherited as an autosomal-recessive disorder. The absence of intracytoplasmic perforin may be used as a reliable marker in the 20–40% of patients with familial HLH type 2 associated with the 10q21-22 mutations in the gene at chromosome 10q22 resulting in perforin gene (PRF1) mutations. Perforin functions by perforating the cytolytic target cell membrane allowing for the entry of cytolytic granules that in turn initiate the apoptotic cell death pathway. The pathways leading to the synthesis of perforin, subcellular compartmentalization as well as directional targeting and release of cytolytic granules all represent potential points that could be mutated and contribute to different genetic causes of inherited HLH syndromes. One such is the description of mutations in the Munc 13-4 gene, which is essential for cytolytic granule fusion. Inactivating mutations in this gene, located at chromosome 17q25, have now been shown to cause familial HLH, termed FHLH type 3.

Those patients with absent perforin expression should undergo PRF1 and the MUNC 13-4 (FHLH type 3) mutation analysis. FHLH-4 is characterized by mutations in the *STX11* gene encoding Syntaxin 11 on chromosome 6(q24). Syntaxin is believed to also be involved in the intracellular movement of cytolytic granules. FHLH type 5 has been shown to be characterized by mutations in the MUNC 18-2 that appears to be involved in a multiprotein complex with Syntaxin 11. FHLH type 1, linked to the 9q21.3 locus, represents approximately 10% of the cases and can be recognized by impaired NK cell function not associated with the absence of perforin expression.

Patients with infection-associated HLH may have transiently impaired NK activity, thus mutation analysis will distinguish these cases from familial HLH. In the absence of a positive mutation analysis, re-evaluation of NK function after successful treatment should be undertaken.

In summary, a wide array of immune dysfunction may result in defective NK and CTL target cell killing resulting in the failure to eliminate infected cells permitting a sustained inflammatory response complete with excessive cytokine production and sustained systemic macrophage activation that characterizes a final pathway of HLH.

Clinical Features

1. The age of onset is less than 1 year of age in 70% of cases. There is no known upper age limit for the onset of disease.
2. Signs and symptoms of FHLH include:
   - Fever (91%), splenomegaly (98%) and hepatomegaly (94%) are the most common early findings
   - Lymph node enlargement (17%), skin rash (6%) and neurologic abnormalities (20%) may also occur. Neurologic findings including irritability, bulging fontanel, neck stiffness, hypotonia, hypertonia, convulsions, cranial nerve palsies, ataxia, hemiplegia, blindness and unconsciousness
   - Multisystem involvement includes lungs, bone marrow and leptomeninges. Occasionally, ocular, heart, skeletal muscles and kidney involvement have been noted.

Treatment

Patients should be treated as per modern protocols (e.g., HLH 2004 of the Histiocyte Society.[*] Non-familial disease is initially treated in a similar manner. The following treatment protocol has been utilized:

- Dexamethasone, 10 mg/m^2/day for 2 weeks followed by a decrease every 2 weeks to 5 mg/m^2, 2.5 mg/m^2 and 1.25 mg/m^2 for a total of 6 weeks
- Etoposide IV, 150 mg/m^2 IV 2-hour infusions daily, twice weekly for 2 weeks, then weekly
- Cyclosporine A, 3–5 mg/kg/day by continuous IV infusion starting week 8 to reach a blood trough level of 150–200 ng/ml and switching to oral administration of 6–10 mg/kg/day in two divided doses
- Intrathecal methotrexate (IT MTX), age-adjusted doses of intrathecal methotrexate weekly for 3–6 weeks as follows if there are progressive neurological symptoms or if abnormal cells persist in the CSF:

Age	IT MTX Dose (mg)
<1 year	6
1–2 years	8
2–3 years	10
>3 years	12

[*]For details, contact the local chapter of the Histiocyte Society in various countries or the Histiocyte Society, 302 North Broadway, Pitman, NJ 08071. Fax: 1-609-589-6614; Telephone: 1-609-589-6606.

5. Allogeneic hemopoietic stem cell transplantation (HSCT) after cytotoxic chemotherapy for patients with familial disease or those with persistent or recurrent non-familial disease.

Without treatment, FHLH is usually rapidly fatal, with a median survival of about 2 months. Chemotherapy and immunosuppressive therapy may prolong survival in FHLH but only stem cell transplantation may be curative. Patients with known familial disease or severe or persistent acquired disease should then receive hematopoietic stem cell transplantation. The 3-year actuarial survival in familial HLH with this approach has been reported to be approximately 50–55% overall but 64% following HSCT.

Non-familial Hemophagocytic Lymphohistiocytosis

A number of inherited disorders associated with impaired cytotoxic T or NK cell function result in HLH. HLH may also be associated with a number of infectious agents and malignancies.

Infection-associated Hemophagocytic Lymphohistiocytosis

The findings in children with infection-associated HLH (IAHLH) are similar to those in FHLH. However, decreased or absent NK cells are found more often in FHLH. NK cell activity in IAHLH patients is reconstituted as soon as the infection is cleared.

Table 18-12 lists the triggering organisms and clinical outcomes for IAHLH. Viruses include human herpes virus-6 (HHV-6), cytomegalovirus (CMV) (most common of the viruses), adenovirus, parvovirus, varicella zoster, herpes simplex virus (HSV), Q-fever virus and measles.

Table 18-12   Infection-associated Hemophagocytic Syndrome in Children: Associated Organisms and Clinical Outcome

Organism	Number of Patients	Clinical Outcome		
		Dead	Alive	No Data
Epstein–Barr virus	121	72	27	22
Other viruses	28	11	13	4
Bacteria	11	2	9	0
Fungi	2	1	1	0
Protozoae	1	0	1	0
No organism	56	13	34	11

From: Janka G, Imashuku S, Elinder G, Schneider M, Henter JI. Infection and malignancy associated hemophagocytic syndromes. Hematol/Oncol Clin N Am 1998;12:435–443, with permission.

Treatment

Epstein–Barr virus (EBV)-related IAHLH: In addition to HLH-directed treatment, the addition of rituximab, a monoclonal antibody against the B cell antigen, CD20, has been used.

Other infections: antibiotics for bacterial infections, antiviral drugs for viruses, in addition to corticosteroids and/or etoposide. Patients with persistent HLH may require FHLH treatment and hematopoietic stem cell transplant. Patients with resolved disease may discontinue therapy at 8 weeks. If they recur therapy should be restarted and HSCT should be employed.

Table 18-13 shows IAHLH in children by age and clinical outcome prior to the use of effective protocols.

Malignancy-associated Hemophagocytic Syndrome

Table 18-14 shows the malignancies associated with the development of hemophagocytic syndromes.

Table 18-13   Infection-associated Hemophagocytic Syndrome in Children by Age and Clinical Outcome

Age	Number of Patients	Clinical Outcome		
		Dead	Alive	No Data
<3 years	77	40	26	11
>3 years	82	29	47	6
Children[a]	60	34	22	4
Totals	219	103/198	95/198	

[a]Age unknown from records reviewed.
From: Janka G, Imashuku S, Elinder G, Schneider M, Henter JJ. Infection and malignancy associated hemophagocytic syndromes. Hematol/Oncol Clin N Am 1998;12:435–443.

Table 18-14   Malignancies Associated with the Development of Hemophagocytic Syndromes

1. Development of a hemophagocytic syndrome before and/or during the treatment for malignancy, such as
     Acute lymphoblastic leukemia
     Acute myeloid leukemia
     Multiple myeloma
     Germ cell tumor
     Thymoma
     Carcinoma
2. Development of a hemophagocytic syndrome with a masked hematolymphoid malignancy in the background, such as
     T/NK-cell leukemia
     Lymphomas
         Angiocentric immunoproliferative lesion-like
         Large cell anaplastic lymphoma
         Adult B-cell lymphoma

From: Janka G, Imashuku S, Elinder G, Schneider M, Henter JJ. Infection and malignancy associated hemophagocytic syndromes. Hematol/Oncol Clin N Am 1998;12:435–443.

Treatment

If malignancy-associated hemophagocytic syndrome (MAHS) occurs in an immunocompromised host before treatment, therapy of malignancy and infection is suggested. If it develops in association with an infection during chemotherapy, cessation of chemotherapy may be considered if the malignancy is under control. Rapidly progressive MAHS may also need to be treated according to the HLH-2004 protocol (see p. 591).

Macrophage Activation Syndrome in Systemic Juvenile Rheumatoid Arthritis and Other Chronic Conditions (Reactive Hemophagocytic Lymphohistiocytosis)

Macrophage activation syndrome (MAS) is caused by an excessive activation and proliferation of mature macrophages. It is observed in a number of conditions, including infections, neoplasms and rheumatologic diseases. Triggers for MAS in juvenile rheumatoid arthritis (JRA) include gold therapy, aspirin, other nonsteroidal anti-inflammatory drugs and viral infections. Typically, patients with such chronic conditions present with the following features of acute illness:

- Persistent fever
- Hepatosplenomegaly
- Pancytopenia
- Easy bruisability and mucosal bleeding
- Low erythrocyte sedimentation rate (ESR)
- Elevated liver enzymes.

Prolonged prothrombin time, prolonged partial thromboplastin time, hypofibrinogenemia, low levels of vitamin-K-dependent clotting factors may occur. Fibrin degradation products and serum liver enzyme values may be elevated. Easy bruisability and mucosal bleeding may occur. Numerous, well-differentiated hemophagocytic histiocytes, a pathognomonic feature of this condition, are found in various organs, e.g., bone marrow and lymph nodes.

In patients with systemic juvenile rheumatoid arthritis (JRA), the acute deterioration is preceded by either a viral infection or major changes in therapy (e.g., administration of gold therapy or nonsteroidal anti-inflammatory drugs).

Macrophage activation syndrome as a complication of JRA is associated with considerable morbidity and death. Table 18-15 shows the clinical differentiation of macrophage activation syndrome associated with JRA from a typical exacerbation of systemic JRA.

Treatment

Because of its seriousness, this syndrome should be recognized promptly on the basis of the clinical findings. Early treatment, consisting of the use of corticosteroids (after performing a bone marrow examination) along with intravenous gamma-globulin, should be considered. This treatment usually results in a rapid resolution of symptoms. The dose of steroids should be tapered slowly to prevent relapse. If this type of treatment fails to rapidly result in a

Table 18-15   Features That Differentiate Acute Exacerbation of Juvenile Rheumatoid Arthritis from Macrophage Activation Syndrome Complicating Juvenile Rheumatoid Arthritis

	Acute Exacerbation of JRA	Macrophage Activation Syndrome in JRA[a]
Fever	One spike or two spikes daily	Persistent
Generalized lymphadenopathy	Present	Not present
Hepatosplenomegaly	Present	Present
Laboratory findings		
Blood count	Marked polymorphonuclear leukocytosis and thrombocytosis	Pancytopenia
Clotting studies	Hyperfibrinogenemia	Hypofibrinogenemia, prolonged PT, prolonged PTT, increased D-dimers
ESR	Increased ESR	Decreased ESR
Liver enzymes	Mildly elevated	Moderately elevated

[a]Must be clearly differentiated from malignant histiocytic conditions.

significantly improved condition, then treatment with HLH-directed regimens should be initiated.

Malignant Histiocytic Disorders in Children

Table 18-16 shows a contemporary classification of malignant histiocytic disorders, based on recently available phenotypic and genotypic data.

Nosology

Figure 18-2 shows the nosology for malignant histiocytic disorders in children.

The previously mentioned malignant histiocytic disorders are categorized as monocyte–macrophage histiocytic sarcomas (MMHs). The term sarcoma, although not an ideal name, is used instead of lymphoma because lymphomas are malignant tumors of different lymphoid cell types and, thus, they cannot be histiocytic.

Table 18-17 lists the criteria for true malignant monocyte–macrophage-related disorders.

Treatment

The following treatments are recommended for MMHs:

- Monocyte-related leukemias and extramedullary monocytic sarcoma: Same as the treatment for acute nonlymphocytic leukemias (see Chapter 17)
- Macrophage-related histiocytic sarcomas (localized and disseminated): The BFM-NHL protocol for Ki-1+ anaplastic large cell lymphoma (ALCL) (see Table 20-9 and Figure 20-4, Chapter 20)
- Dendritic-cell-related histiocytic sarcoma

Table 18-16   Classification of Malignant Histiocytic Disorders

Monocyte related
   Leukemias (FAB and revised FAB classifications)
      Monocytic leukemia M5A and B
      Acute myelomonocytic leukemia M4
      Chronic myelomonocytic leukemia
   Extramedullary monocytic tumor or sarcoma (monocytic counterpart of granulocytic sarcoma)
   Macrophage-related histiocytic sarcoma (localized or disseminated)
   Dendritic-cell-related histiocytic sarcoma (localized or disseminated)
      Follicular dendritic cell
      Dermal dendrocyte
      Interdigitating dendritic cell
      Indeterminate cell
      Langerhans cells

From: Favara BE, Feller AC, with members of the WHO Committee on Histiocytic/Reticulum Cell Proliferations: Pauli M, et al. Contemporary classification of histiocytic disorders. Med Pediatr Oncol 1997;29:157–164.

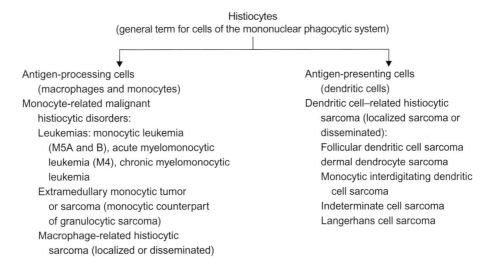

Figure 18-2 Nosology for Malignant Histiocytic Disorders in Children.

    a. Localized disease: Wide excision only and observe as these tumors are usually only locally aggressive
    b. Disseminated disease: The BFM-NHL protocol for Ki-1+ ALCL.

These recommendations are based on the observation of a small number of patients because these disorders are rare.

Table 18-17   Criteria for True Malignant Monocyte–Macrophage-related Disorders

Immunophenotyping
Cells are negative for T- and B-cell-associated antigens and express antigens associated with monocyte–macrophage origin including:
Myelomonocytic antigen: Cd11b, CD11c, CD13, CD15, CD33, lysozyme
Monocyte–macrophage antigens: CD36, CD68, MAC-387, α-1-antitrypsin, α-1-antichymotrypsin
Antibodies associated with other lineages that may show reactivity cells of monocyte–macrophage lineage: HLA-DR, CD41, CD43, CD45RO, CD45 (LCA), CD74
Gene rearrangement studies
Negative B immunoglobulin and T-cell antigen receptor gene rearrangement studies[a]

[a]It is generally assumed that Ig and TCR gene rearrangements are functional only in B and T cells, respectively. Such rearrangements have been described to occur occasionally in cells from other lineages as well, known as cross-lineage rearrangements and particularly V-D-J rearrangements in the malignant cells of such disorders arising following the diagnosis of a T or B cell lymphoma or leukemia.

Data from: Egeler RM, Schmitz L, Sonneveld P, et al. Malignant histiocytosis: a reassessment of cases formerly classified as histiocytic neoplasms and a review of the literature. Med Pediatr Oncol 1995;25:1; and from Bucsky P, Egeler RM. Malignant histiocytic disorders in children. Hematol/Oncol Clin N Am 1998;12:465–471.

The treatment for MHS is the same as the BFM-NHL 90 protocol or equivalent for ALCL. (See Table 20-9 and Figure 20-4, in Chapter 20; the table and the figure must be read in conjunction with each other.)

Treatment is usually stratified into three categories, depending on the stage of the disease.

- Category I: Stages I and II (Murphy staging system of NHL) with completely resected disease – three courses of A–B–A
- Category II: Stage II not resected and stage III (Murphy staging system of NHL) – six courses of A–B–A–B–A–B
- Category III: Stage IV (Murphy staging system of NHL) – six intensified courses with higher doses of methotrexate and cytarabine AA–BB–CC–AA–BB–CC.

Suggested Reading

Allen CE, McClain KL. Interleukin-17A is not expressed by CD207(+) cells in Langerhans cell histiocytosis lesions. Nat Med. 2009;15:483–485.

Allen CE, McClain KL. Langerhans cell histiocytosis: a review of past, current and future therapies. Drugs Today (Barc) 2007;43:627–643.

Allen CE, Flores R, Rauch R, et al. Neurodegenerative central nervous system Langerhans cell histiocytosis and coincident hydrocephalus treated with vincristine/cytosine arabinoside. Pediatr Blood Cancer 2009.

Arceci RJ. When T cells and macrophages do not talk: the hemophagocytic syndromes. Curr Opin Hematol. 2008;15:359–367.

Arceci RJ, Longley BJ, Emanuel PD. Atypical cellular disorders. Hematology Am Soc Hematol Educ Program 2002:297–314.

Bechan GI, Egeler RM, Arceci RJ. Biology of Langerhans cells and Langerhans cell histiocytosis. Int Rev Cytol. 2006;254:1–43.

Bechan GI, Meeker AK, De Marzo AM, et al. Telomere length shortening in Langerhans cell histiocytosis. Br J Haematol. 2008;140:420–428.

Beverley PC, Egeler RM, Arceci RJ, Pritchard J. The Nikolas Symposia and histiocytosis. Nat Rev Cancer 2005;5:488–494.

Bhatia S, Nesbit Jr ME, Egeler RM, Buckley JD, Mertens A, Robison LL. Epidemiologic study of Langerhans cell histiocytosis in children. J Pediatr. 1997;130:774–784.

Caselli D, Arico M, Party EPW. The role of BMT in childhood histiocytoses. Bone Marrow Transplant. 2008;41 (Suppl 2):S8–S13.

Coury F, Annels N, Rivollier A, et al. Langerhans cell histiocytosis reveals a new IL-17A-dependent pathway of dendritic cell fusion. Nat Med. 2008;14:81–87.

da Costa CE, Egeler RM, Hoogeboom M, et al. Differences in telomerase expression by the CD1a+ cells in Langerhans cell histiocytosis reflect the diverse clinical presentation of the disease. J Pathol. 2007;212:188–197.

Dhall G, Finlay JL, Dunkel IJ, et al. Analysis of outcome for patients with mass lesions of the central nervous system due to Langerhans cell histiocytosis treated with 2-chlorodeoxyadenosine. Pediatr Blood Cancer 2008;50:72–79.

Egeler RM, de Kraker J, Voute PA. Cytosine-arabinoside vincristine and prednisolone in the treatment of children with disseminated Langerhans cell histiocytosis with organ dysfunction: experience at a single institution. Med Pediatr Oncol. 1993;21:265–270.

Egeler RM, Neglia JP, Arico M, et al. The relation of Langerhans cell histiocytosis to acute leukemia, lymphomas and other solid tumors. The LCH-Malignancy Study Group of the Histiocyte Society. Hematol Oncol Clin North Am. 1998;12:369–378.

Favara B, Feller AC. with members of the WHO Committee on Histiocytic/Reticulum Cell Proliferations: Pauli M, et al. Contemporary classification of histiocyte disorders. Med Pediatr Oncol. 1997;29:157–166.

Gadner H, Grois N, Arico M, et al. A randomized trial of treatment for multisystem Langerhans' cell histiocytosis. J Pediatr. 2001;138:728–734.

Gadner H, Grois N, Potschger U, et al. Improved outcome in multisystem Langerhans cell histiocytosis is associated with therapy intensification. Blood 2008;111:2556–2562.

Grois N, Prayer D, Prosch H, Minkov M, Potschger U, Gadner H. Course and clinical impact of magnetic resonance imaging findings in diabetes insipidus associated with Langerhans cell histiocytosis. Pediatr Blood Cancer 2004;43:59–65.

Grois N, Potschger U, Prosch H, et al. Risk factors for diabetes insipidus in langerhans cell histiocytosis. Pediatr Blood Cancer 2006;46:228–233.

Haupt R, Nanduri V, Calevo MG, et al. Permanent consequences in Langerhans cell histiocytosis patients: a pilot study from the Histiocyte Society-Late Effects Study Group. Pediatr Blood Cancer 2004;42:438–444.

Henter JI, Samuelsson-Horne A, Arico M, et al. Treatment of hemophagocytic lymphohistiocytosis with HLH-94 immunochemotherapy and bone marrow transplantation. Blood 2002;100:2367–2373.

Henter JI, Horne A, Arico M, et al. HLH-2004: Diagnostic and therapeutic guidelines for hemophagocytic lymphohistiocytosis. Pediatr Blood Cancer 2007;48:124–131.

Horne A, Trottestam H, Arico M, et al. Frequency and spectrum of central nervous system involvement in 193 children with haemophagocytic lymphohistiocytosis. Br J Haematol. 2008;140:327–335.

Janka G. Hemophagocytic syndromes. Blood Rev. 2007;21:245–253.

Janka G. Hemophagocytic lymphohistiocytosis: when the immune system runs amok. Klin Padiatr. 2009;221:278–285.

Maarten Egeler R. Cognitive deficits in patients with multisystem Langerhans cell histiocytosis – a commentary. Eur J Cancer 2004;40:922–923.

McClain KL, Kozinetz CA. A phase II trial using thalidomide for Langerhans cell histiocytosis. Pediatr Blood Cancer 2007;48:44–49.

Minkov M, Grois N, Heitger A, et al. Response to initial treatment of multisystem Langerhans cell histiocytosis: an important prognostic indicator. Med Pediatr Oncol. 2002;39:581–585.

Minkov M, Prosch H, Steiner M, et al. Langerhans cell histiocytosis in neonates. Pediatr Blood Cancer 2005;45:802–807.

Minkov M, Steiner M, Potschger U, et al. Reactivations in multisystem Langerhans cell histiocytosis: data of the international LCH registry. J Pediatr. 2008;153:700–705.

Prayer D, Grois N, Prosch H, Gadner H, Barkovich AJ. MR imaging presentation of intracranial disease associated with Langerhans cell histiocytosis. AJNR Am J Neuroradiol. 2004;25:880–891.

Weitzman S, Jaffe R. Uncommon histiocytic disorders: the non-Langerhans cell histiocytoses. Pediatr Blood Cancer 2005;45:256–264.

Weitzman S, Braier J, Donadieu J, et al. 2'-Chlorodeoxyadenosine (2-CdA) as salvage therapy for Langerhans cell histiocytosis (LCH) results of the LCH-S-98 protocol of the Histiocyte Society. Pediatr Blood Cancer 2009;53:1271–1276.

Zur Stadt U, Beutel K, Kolberg S, et al. Mutation spectrum in children with primary hemophagocytic lymphohistiocytosis: molecular and functional analyses of PRF1, UNC13D, STX11 and RAB27A. Hum Mutat. 2006;27:62–68.

zur Stadt U, Rohr J, Seifert W, et al. Familial hemophagocytic lymphohistiocytosis type 5 (FHL-5) is caused by mutations in Munc18-2 and impaired binding to syntaxin 11. Am J Hum Genet. 2009;85:482–492.

CHAPTER 19

Hodgkin Lymphoma

Hodgkin lymphoma (HL) is characterized by progressive enlargement of the lymph nodes. It is considered unicentric in origin and has a predictable pattern of spread by extension to contiguous nodes.

ETIOLOGY AND EPIDEMIOLOGY

1. Etiology is unknown.
2. Comprises 8.8% of all childhood cancers under the age of 20.
3. Overall annual incidence rate in the United States is 12.1 per million for children under 20 years.
4. Incidence increases to 32 per million for adolescents 15–19 years.
5. Bimodal age – incidence curve with one peak at 15–35 years of age and the other above 50 years of age (incidence is highest among 15–19-year-olds).
6. Association with Epstein–Barr virus (EBV).
7. Incidence increased among consanguineous family members and among siblings of patients with HL.
8. In United States, increased risk of HL in families with higher incomes and higher education levels.

RISK FACTORS

Several factors have been associated with an increased risk of developing HL. These include family history of HL, EBV infections and socioeconomic status.

Familial Hodgkin Disease

Familial HL represents 4.5% of all HL cases. For adolescents and young adults there is a 99-fold increased risk among monozygotic twins and a 7-fold increased risk among siblings.

EBV-Associated Hodgkin Disease

EBV incorporated into the tumor genome has been most commonly reported with the mixed cellularity histologic subtype. In turn, this subtype is most common in children from

Manual of Pediatric Hematology and Oncology. DOI: 10.1016/B978-0-12-375154-6.00019-7

underdeveloped countries, in males under age 10 years and in those with other immunodeficiencies.

Socioeconomic Status and Hodgkin Disease

There is an association between HL and socioeconomic status. In children less than 10 years of age and in underdeveloped nations, HL is associated with lower socioeconomic status and in households with more children. However, in young adult patients and in developed nations, HL incidence increases with higher socioeconomic status and with smaller households with fewer children. These findings may be related to an association with infections; increased infections in early childhood may decrease the risk of HL in young adults.

Lymphomatoid papulosis in children may evolve into or co-exist with HL (Chapter 16, p. 492).

BIOLOGY

Reed-Sternberg (R-S) cells and its variants most commonly are derived from a neoplastic clone originating from B lymphocytes in lymph node germinal centers. Most cases of HL arise from preapoptotic germinal center B cells that cannot synthesize immunoglobulin. The R-S cell appears to be resistant to apoptotic stimuli. Deregulation of the nuclear transcription factor nuclear factor kappa B (NFκB) in the R-S cells may explain this resistance to apoptosis.

PATHOLOGY

Macroscopic Features

The spread of HL occurs mostly by contiguity from one chain of lymph nodes to another. Involvement of the left supraclavicular nodes often follows abdominal para-aortic node involvement with spread up the thoracic duct, whereas involvement of the right supraclavicular nodes tends to be associated with mediastinal adenopathy. Para-aortic node involvement commonly occurs in association with involvement of the spleen, which in turn is commonly followed by liver or bone marrow involvement, or both. Nodular sclerosis classical Hodgkin lymphoma shows the greatest propensity to spread by contiguity, whereas noncontiguous dissemination, when it occurs, is more than twice as frequent in the mixed cellularity and lymphocyte-depleted classical Hodgkin lymphoma histologic types.

The R-S cell is the hallmark of classical HL and is characterized by a binucleated or a multinucleated giant cell that is often characterized by a bilobed nucleus, with two large

Table 19-1 Histologic Variants of Hodgkin Lymphoma

"CLASSICAL" HODGKIN LYMPHOMA

Nodular Sclerosis (NS) Classical Hodgkin Lymphoma (NSCHL)

The nodular sclerosis subtype is characterized by collagen bands and lacunar variants of RS cells. The presence of one or more sclerotic bands is the defining feature. These bands usually radiate from a thickened lymph node capsule often following the course of a penetrating artery. These bands are composed of mature, laminated, relatively acellular collagen. They are birefringent in polarized light. In most cases, several broad collagenous bands can be identified, or fibrosis can be so extensive that isolated nodules of lymphoid tissue remain.

The collagenous bands of nodular sclerosis enclose nodules of lymphoid tissue containing variable numbers of Hodgkin cells and reactive infiltrates. Lacunar cells are a common type of RS cell present and may be found in large numbers or in sheets. They tend to aggregate at the center of nodules, sometimes forming a rim around central areas of necrosis. Diagnostic Reed–Sternberg cells are present in variable numbers and may be difficult to identify in small biopsy specimens. Eosinophils, histiocytes and sometimes even neutrophils are often numerous. Plasma cells are usually less conspicuous.

Mixed Cellularity (MC) Classical Hodgkin Lymphoma (MCCHL)

This intermediate subtype falls between lymphocyte-rich classical HL and lymphocyte-depleted classical HL. The capsule is usually intact and of normal thickness. A vague nodularity may be present at low magnification, but the presence of any definite fibrous bands would warrant classification as nodular sclerosis rather than mixed cellularity. At high magnification, a heterogeneous mixture of Hodgkin cells (HC), small lymphocytes, eosinophils, neutrophils, epithelioid and non-epithelioid, histiocytes, plasma cells and fibroblasts are present. Diagnostic RS cells and mononuclear variants are usually easy to find. Small foci of necrosis may be present, but the extent is much less than that seen in nodular sclerosis.

Lymphocyte-depleted Classical Hodgkin Lymphoma (LDCHL)

Lymphocyte-depleted HL encompasses two variants: Diffuse fibrosis and reticular. The most characteristic features are a marked degree of reticulin fibrosis surrounding single cells along with lymphocyte depletion. In contrast to nodular sclerosis, this subtype is not characterized by the presence of thick fibrous bands and the fibrosis envelops individual cells, not nodules of cells. HC are usually easily identified, but increased numbers of HC are not essential to the diagnosis. In the reticular variant, sheets of HC, often showing pleomorphic features, are found.

Lymphocyte-rich Classical Hodgkin Lymphoma (LRCHL)

Many cases of lymphocyte-rich classical HD have a resemblance to mixed cellularity HL, with vaguely nodular and less often diffuse pattern at low magnification. Hodgkin and Reed–Sternberg cells are relatively rare and the background is dominated by small mature lymphocytes. Eosinophils and neutrophils are usually absent and if present are scanty and usually within the interfollicular areas. Reed–Sternberg cells and variants are not easy to find but when encountered have identical features to the Hodgkin cells of mixed cellularity. Some cases of lymphocyte-rich HD may show a distinctly nodular appearance that may closely mimic nodular lymphocyte predominance HL and often contain relatively small germinal centers, with Hodgkin and Reed–Sternberg cells present in and near the mantle zone, a pattern that has been called *follicular* HL.

Unclassified Cases (UC)

UC is defined as histopathology that does not fit into a definite or provisional category (they may be T cell, B cell, or undefined; they may be borderline between HL and NHL).

Nodular Lymphocyte Predominant Hodgkin Lymphoma (NLPHL)

NLPHL is a B-cell neoplasm with a nodular or nodular and diffuse proliferation of scattered large neoplastic cells termed "popcorn" cells [formerly called L&H cells (for lymphocytic and/or histiocytic Reed–Sternberg cell variants)]. These large cells resemble centroblasts but are larger and have folded

(Continued)

Table 19-1   (Continued)

or multilobulated nuclei and multiple small basophilic nucleoli are often present adjacent to the nuclear membrane. The cytoplasm is broad and only slightly basophilic. These large cells are present within spherical nodules with numerous dendritic cells, histiocytes and small lymphocytes. Ultrastructural studies demonstrate that popcorn cells have the appearance of centroblasts of germinal center. Epithelioid histiocytes are preferentially found in the outer rim of nodules. They are arranged in small groups or clusters and well-formed granulomas may be present in rare cases. Eosinophils and neutrophils are rare. Plasma cells are not common and are seen only between follicles. In diffuse areas, the popcorn cells are still often arranged in a vaguely nodular pattern. Classic Hodgkin and Reed–Sternberg cells are completely lacking or are few in number. In some cases, popcorn cells may resemble lacunar cells because both cell types show irregularly shaped or lobulated nuclei, small nucleoli and broad pale to slight basophilic cytoplasm. The popcorn cells are often surrounded by rosettes of CD3 (+), CD57(+) T-lymphocytes.

nucleoli, giving the classically described "owl's eye" appearance. The R-S cells are embedded within a benign-appearing reactive infiltrate of lymphocytes, macrophages, granulocytes and eosinophils. Important exceptions to this histologic picture apply to the categories of nodular sclerosis type and nodular variant lymphocyte predominance type in which peculiar variant forms of the tumor giant cells can be used to establish the diagnosis (lacunar cell variants).

Histology

Hodgkin lymphomas (HL) consist of two disease groups, as was proposed in the Revised European-American Classification of Lymphoid Neoplasms (REAL classification) and incorporated into the 2008 4th edition of the WHO Classification of Tumors of Haematopoietic and Lymphoid Tissues. These two groups are:

- "Classical" HL includes the nodular sclerosis classical Hodgkin lymphoma (NSCHL), mixed cellularity HL (MCHL), lymphocyte-depleted (LDHL) and lymphocyte-rich classical HL (LRCHL) subtypes
- Nodular lymphocyte-predominant HL (NLPHL).

Table 19-1 describes the histologic variants in Hodgkin lymphoma and Table 19-2 a histopathological classification of Hodgkin lymphoma by age.

Immunophenotypic Features

Tumor cells in classic HL usually do not express B-cell antigens such as CD20 and CD79a, but virtually all express CD30 and most express CD15. In comparison, the tumor cells of NLPHL always express B-cell antigens such as CD20, CD79a and are almost always negative for CD30 and CD15. The majority of RS cells and popcorn cells express

Table 19-2   Histopathological Classification of Hodgkin Lymphoma by Age

Age Group	LRCHL (%)	NSHL (%)	MCHL (%)	LDHL (%)	UC/IF (%)
<10 years	14	45	32	0	9
11–16 years	7	78	11	1	3
≥17 years	5	72	17	1	5

Abbreviations: LP, lymphocyte predominance; NS, nodular sclerosis; MC, mixed cellularity; LD, lymphocyte depletion; UC, unclassifiable Hodgkin disease; IF, interfollicular Hodgkin's disease.
From: Donaldson SS, Link MP. Pediatr Clin N Amer. 1991;38:457–473, with permission.

Table 19-3   Immunophenotypic Markers in Classical HL and NLPHL

	CD15	CD30	CD45	CD20	CD79a	Pax5
Classical HL	+/−	+	−	−/+	−/+	+
NLPHL	−	−	+	+	+	+

+, all cases are positive; +/−, majority of cases positive; −/+, minority of cases positive; −, all cases are negative.

B-cell-specific activator protein PSX-5 (BSAP). Table 19-3 shows the common immuno-phenotypic markers in HL.

CLINICAL PRESENTATION

Table 19-4 lists the frequency of the presenting signs and symptoms in children with Hodgkin lymphoma.

Lymphadenopathy (90% of All Cases)

1. Painless swelling of one or more groups of superficial lymph nodes; rarely painful.
2. Cervical nodes involved in 60–80% of cases; associated with mediastinal involvement in 60% of cases.
3. Axillary, inguinal, mediastinal and retroperitoneal nodes also frequently involved.
4. Involved nodes are discrete, elastic, painless and usually "rubbery"; tenderness is rare.
5. The bulk of palpable lymph nodes is defined by the single largest dimension (in centimeters) of the single largest lymph node or conglomerate node mass in each region of involvement. A node or nodal mass ≥10 cm is defined as *bulky*. Abdominal nodal bulk is generally determined using CT.
6. Alcohol-induced pain: This occurs in areas of nodal enlargement and has been reported to occur in fewer than 5% of cases. The pain may begin within minutes of alcohol ingestion and may occur in the chest with radiation to the arms, back and legs. The onset of alcohol-induced pain may precede the diagnosis of HL. The pain diminishes with treatment of HL and may recur before other signs or symptoms of relapse. The

Table 19-4   Frequency of Presenting Signs and Symptoms in Children

Presenting Symptoms	Percentage
Lymphadenopathy	90%
Mediastinal adenopathy	
Adolescents and young adults	75%
Children less than 10 years of age	25%
Splenomegaly	25%
B symptoms	20%
Weight loss	
Night sweats	
Fever	
Puritis	15%
Pulmonary symptoms	10%
Cough	
Dyspnea	
Alcohol induced pain	5%

mechanism of alcohol-induced pain is unknown, but edema of lesions, vasodilation and release of histamine have been proposed as causes.

Mediastinal Adenopathy (60% of All Cases)

Approximately 20% of patients have *bulky mediastinal disease* defined as a large mediastinal mass (>10 cm in maximum dimension) or which, on a posterior–anterior chest radiograph has a maximum width equal to or greater than one-third of the internal transverse diameter of the thorax at the level of the T5–6 interspace (mass to thoracic ratio ≥ 0.33). The chest radiograph should be taken with maximal inspiration in the upright position at a source-skin distance of 2 meters. Relapse occurs more frequently in bulky mediastinal disease than do those with small masses. The principal sites of relapse are the margins of the radiation ports and adjacent organs.

Adolescents and young adults present with a mediastinal mass in 75% of cases, as opposed to children less than 10 years of age where mediastinal disease is present in only 35% of cases. This is thought to be due to these younger patients having either mixed cellularity or lymphocyte predominant histology, where peripheral adenopathy is more common.

Clinical Presentation

1. Persistent nonproductive cough; however, this site is often asymptomatic.
2. Superior vena cava syndrome (enlargement of the vessels of the neck, hoarseness, dyspnea and dysphagia).

Management

See Chapter 30.

Splenomegaly

The spleen is commonly enlarged on physical examination or by ^{99m}Tc-sulfur colloid scan. However, its size is not indicative of splenic involvement with Hodgkin lymphoma.

In 13% of cases, the spleen is the only site of subdiaphragmatic disease. Spleen is involved in 26% of spleens resected in staging laparotomy. It is very important to determine splenic disease. On CT the accuracy of "Splenic Index"[*] is only 59%. Fluorodeoxyglucose-PET (FDG-PET) scan appears to be more sensitive (about 92% accuracy) to identify splenic HL.

Systemic Symptoms (30% of All Cases)

1. Intermittent fever (Pel–Ebstein), anorexia, fatigue, weakness, nausea, night sweats and weight loss.
2. Mild itching may be seen in 15–25% of patients with HL, although severe itching is less common. It is more common in patients with advanced-stage disease and may be severe enough that excoriations are produced from scratching. It may be the first manifestation of Hodgkin lymphoma. Pruritus does not correlate with prognosis and is no longer considered one of the B symptoms.

Pulmonary Disease (17% of All Cases)

The mediastinal and hilar nodes are the pathway to the lungs for the spread of HL. Pulmonary HL does not occur without mediastinal or hilar disease. About two-thirds of patients with HD have intrathoracic involvement. There is a striking predilection of the nodular sclerosis type for mediastinal involvement. There are four forms of pulmonary involvement:

- Contiguous lesions
- Peribronchovascular disease
- Subpleural spread with plaque-like pleural thickening or effusion
- Intraparenchymal involvement, either nodular or alveolar in nature.

Symptoms: cough and dyspnea may occur with pulmonary involvement.

Radiographically: uncommonly Hodgkin lymphoma may have single or multiple cavitating lesions. The radiologic differential diagnosis of these cavitating lesions is tuberculosis and fungal disease.

---

[*]Splenic index is defined as the measurement of the spleen on CT scan. It is the product of the length $\times$ width $\times$ thickness of the spleen. If splenic index in cm^3 exceeds $500 + 20 \times$ age (in years) it is considered positive for splenic enlargement.

Pleural effusion occurs because of lymphatic obstruction in the mediastinum or hilum. Pleural invasion occurs but is relatively rare in comparison with its incidence in non-Hodgkin lymphoma.

Neurologic Manifestations

Neurologic dysfunction is usually a late manifestation. The disease can reach the nervous system in the following ways:

* Spread from paravertebral lymph nodes along nerve roots, blood vessels, or perineural lymphatics into the spinal canal by way of the intervertebral foramina and similarly into the intracranial region
* Hematogenous dissemination.

Involvement of the central neuroaxis is epidural and produces compression symptoms, which vary according to the area of the nervous system involved.

Intracranial disease may present with signs and symptoms of increased intracranial pressure or with hemiparesis, hemisensory defects, focal seizures, or occasionally papilledema. When the base of the brain is involved, there may be palsies of the cranial nerves.

Bone Disease (2% of All Cases)

Osseous HL typically presents with bone pain and the majority of patients have concurrent nonosseous lesions detected at staging. The radiographic features of osseous Hodgkin lymphoma vary but indicate an aggressive malignant process. The histologic diagnosis may be problematic and immunohistochemical stains aid in establishing the diagnosis of HL in bone.

The major complication of vertebral involvement is spinal compression resulting from collapse of a vertebral body or from extension of the tumor into the epidural area.

Bone Marrow Involvement (5% of All Cases)

Symptomatic patients, particularly those with anemia, leukopenia, or thrombocytopenia, may have bone marrow involvement. Multiple biopsies are indicated, because HL tends to involve the marrow in a focal fashion.

Liver Disease (2% of All Cases)

Mild hepatomegaly, nonspecific abnormalities on liver scan and abnormal liver function tests do not correlate with actual histologic involvement of the liver. Liver biopsy is the only method for accurate diagnosis of liver involvement.

To establish liver involvement, the following findings should be present:

* Evidence of discrete lesions that features one of the histologic subtypes of HL
* Conclusive evidence of a Reed–Sternberg cell or one of its variants.

Nonspecific histologic findings are common and do not constitute evidence of invasion of the liver.

The presence of jaundice in patients is an ominous prognostic sign and may be terminal or preterminal. However, other possible causes of jaundice should be excluded (e.g., hemolytic anemia, viral hepatitis, toxic hepatitis, toxoplasmosis and cholestasis of unknown etiology).

Renal Manifestations

HL may directly involve the kidney in 13% of cases. Renal involvement may be unilateral or bilateral and may be present as diffuse involvement, discrete nodules, or microscopic disease. Renal involvement with HL may result from ureteral obstruction and patients with HL may also have renal dysfunction related to renal vein thrombosis, hypercalcemia and hyperuricemia.

Hematologic Manifestations

1. Blood count: Anemia – normocytic and normochromic, neutrophilia (50% cases), eosinophilia (50% cases), lymphocytopenia.
2. Myelosuppression in HL may be caused by hypersplenism or bone marrow infiltration.
3. A positive direct antiglobulin test (DAT) may or may not be associated with overt hemolysis. Patients with a positive DAT test frequently have advanced-stage disease and systemic symptoms.
4. Immune thrombocytopenia may accompany HL in 1–2% of cases. Thrombocytopenia may develop before, concurrently with, or after the diagnosis of HL.
5. Autoimmune neutropenia has also been described in patients with HL. Neutropenia may occur before the diagnosis.
6. Eosinophilia is frequently associated with HL. This occurs in 15% of patients. The presence of eosinophilia is not related to stage or histology. IL-5 production by Reed-Sternberg cells, may explain the etiology of eosinophilia accompanying HL.

Constitutional B symptoms

Approximately 20% of patients have associated B symptoms. B symptoms are defined as:

- Unexplained weight loss or more than 10% of body weight in the 6 months preceding the diagnosis
- Unexplained fever with temperatures $>38°C$ for more than 3 days
- Drenching night sweats.

DIAGNOSTIC EVALUATION AND STAGING

The diagnostic workup for HL includes a detailed history and a thorough physical examination. Staging is based on a combination of physical examination and imaging

Table 19-5   Diagnostic Investigations and Staging for Hodgkin Lymphoma

Surgical
   Excisional lymph node biopsy
   Bilateral bone marrow biopsies
Imaging studies
   CT scan of neck, chest, abdomen and pelvis
   FDG-PET
   Technetium-99 bone scintography
Laboratory studies
   Complete blood count
   Blood chemistries for renal and hepatic function
   Erythrocyte sedimentation rate
   Ferritin

Abbreviations: CT, computed tomography; FDG-PET, fluorodeoxyglucose positron emission tomography.
Adapted from: Friedman D, Schwartz C, Hodgkin Lymphoma in The Lymphoid Neoplasms, 3rd edition, 2008, with permission.

studies. Staging and evaluation of disease status is done initially at the diagnosis and performed again early in the course of treatment and at the end of treatment.

Diagnostic investigations and staging of HD are summarized in Table 19-5.

It is important for the diagnostic evaluation to begin with a detailed history of evidence of B symptoms and a physical examination that documents the sites and size of all adenopathy as well as splenic involvement and organ dysfunction.

Laboratory studies include a complete blood count, chemistries to evaluate liver and renal function and acute phase reactants such as ferritin and erythrocyte sedimentation rate, which may be seen as nonspecific markers of tumor activity.

Computed tomography (CT) scans of the neck, thorax, abdomen and pelvis should be performed to help determine the status of involvement of all lymph node groups as well as extranodal disease by extension or metastasis. In addition, an upright chest radiograph (CXR) should be performed with posteroanterior (PA) and lateral views to document any mediastinal disease. The CXR is important for establishing the criterion of bulky mediastinal lymphadenopathy, although this is likely to change with newer technologies.

In addition to CT, nuclear imaging is important for staging HL and for monitoring response to therapy. Functional imaging such as fluorodeoxyglucose positron emission tomography (FDG-PET) is now an essential component of the Harmonization Consensus Criteria for diagnosis, staging and response. It is also a reliable method to assess bone involvement and may be useful in demonstrating unsuspected lesions in the spleen. FDG-PET has been recommended as the functional imaging procedure for initial staging. In addition, with newer imaging techniques and equipment, combined CT-PET scans can evaluate disease simultaneously with both modalities. Technetium-99 bone scintography should be considered in patients with bone pain or elevated alkaline phosphatase.

Surgical procedures for staging of HL include bone marrow biopsies and lymph node biopsies. (Care should be taken when administering anesthesia. Patients with large mediastinal masses are at risk for cardiac or respiratory compromise during anesthesia or sedation – see Chapter 30.) Bilateral bone marrow biopsies are recommended for patients with B symptoms or those with higher stage (stage III or IV).

After a careful and thorough physical examination and imaging has been undertaken, the least-invasive procedure should be used to establish the diagnosis of HL. If possible, the diagnosis should be established by lymph node biopsy. Fine-needle biopsy is not recommended because the small number of cells present in the specimen may be inadequate for immunophenotyping needed to classify HL into one of the subtypes.

Staging

The staging classification currently used for HL is the Ann Arbor Classification as shown in Table 19-6. The Ann Arbor Classification was originally adopted in 1971 and was updated in 1989; it has also been called the Cotswald Classification.

PROGNOSTIC FACTORS

Several pretreatment factors have been associated with an adverse outcome in one or more studies and these include:

- Advanced stage of disease (Stage IIB, IIIB, or IV disease)
- The presence of B symptoms
- The presence of bulk disease
- Extranodal extension
- Male sex
- Elevated erythrocyte sedimentation rate
- White blood cell count 11,500/mm^3 or higher
- Hemoglobin less than 11.0 g/dl
- The combination of B symptoms and bulky disease are inferior prognostic indicators.

Histology has been noted to be an important prognostic indicator as well. Patients with nodular lymphocyte predominance HL (NLPHL) appear to have an overall better prognosis than those with classical HL (CHL).

Table 19-6   Modified Ann Arbor Classification of Hodgkin lymphoma

Stage	Description
I:	Involvement of single lymph node region (I) or localized involvement of a single extralymphatic organ or site (1E)
II:	Involvement of two or more lymph node regions on the same side of the diaphragm (II) or localized contiguous involvement or a single extralymphatic organ or site and its regional lymph node(s) with involvement of one or more lymph node regions on the same side of the diaphragm (IIE)
III:	Involvement of lymph node regions on both sides of the diaphragm (III), which may also be accompanied by localized contiguous involvement of an extralymphatic organ or site (IIIE), by involvement of the spleen (IIIS), or both (IIIS+E)
IV:	Disseminated (multifocal) involvement of one or more extralymphatic organs or tissue, with or without associated lymph node involvement, or isolated extralymphatic organ involvement with distant (non-regional) nodal involvement

Symptoms and Presentation

A symptoms:	Lack of "B" symptoms
B symptoms:	At least one of the following:
	Unexplained weight loss >10% in the preceding 6 months
	Unexplained recurrent fever >38°C
	Drenching night sweats
X:	Bulky disease: >10 cm in maximum dimension or involves >1/3 of the chest diameter (seen on chest X-ray)
S:	Splenic involvement
E:	Involvement of a single extranodal site that is contiguous or proximal to the known nodal site

From: Lister TA, Crowther D, Sutcliffe SB, et al. Report of a committee convened to discuss the evaluation and staging of patients with Hodgkin's disease: Cotswalds meeting. J Clin Oncol 7 (11);1630–1636:1989.

Serum markers are currently being studied as prognostic indicators; those associated with poor prognosis include: tumor necrosis factor, soluble CD30, beta 2 macroglobulin, transferring and serum IL-10 level. Favorable outcomes have been noted with high levels of caspase 3 in Reed–Sternberg cells.

Other prognostic indicators include initial response to chemotherapy. Early response assessment is based on volume reduction of disease, functional imaging status or both. Significant reduction in disease volume and PET negativity after two cycles of chemotherapy is associated with a favorable outcome in children and adolescents and failure to achieve this has been associated with an unfavorable outcome in adults. This therapeutic response may permit titration of therapy based on early response and this paradigm is being studied in cooperative group and consortium trials.

TREATMENT

In general, the use of combined chemotherapy and low-dose involved field radiation therapy (LD-IFRT) is considered standard of care. This allows for maximal therapy with reduction of individual drug-related or radiation-related toxicities. The volume of radiation and the intensity/duration of chemotherapy are determined by prognostic factors at presentation and include:

- The presence of B symptoms
- Initial stage of disease
- Presence of bulky disease.

Due to the large number of severe long-term effects associated with HL therapy, cooperative groups have tailored HL therapy for young children to include reduction of alkylating agents, anthracycline agents and radiation dose and volume. Pediatric oncologists agree that standard dose radiation therapy, particularly applied to large volumes including critical organs normally in the mantle field, have unacceptable toxicities which include growth dysfunction, increased risk for breast cancer in young females and cardiovascular complications. Thus, all children and adolescents generally receive combination chemotherapy as initial treatment based on prognostic factors. The general treatment strategy for children with HL is to use chemotherapy for all patients with low-dose radiation therapy. The number of cycles and intensity of chemotherapy is determined by extent of disease and may be further modified by rapidity and degree of response, as is the radiation dose and volume. Table 19-7 gives some of the commonly accepted standard strategies for newly diagnosed children and adolescent patients with HL. Current COG studies continue to focus on decreasing long-term effects by minimizing radiation and chemotherapy. Studies focus on titrating therapy based on risk group and early response to chemotherapy, with the goal to reduce exposure to alkylating agents, anthracyclines and radiotherapy.

Treatment for Nodular Lymphocyte Predominant HL (NLPHL)

Children treated for NLPHL have a favorable outcome, particularly if disease is stage I (localized). Current standard therapy for children is chemotherapy and LD-IFRT. There are reports of patients who have been treated with chemotherapy alone or with complete resection. COG is currently evaluating patients with NLPHL. Those patients with stage I, completely resected disease are observed without further treatment. Patients with stage I or IIA disease with no bulk disease and residual disease following initial biopsy receive three cycles of vincristine, doxorubicin and cyclophosphamide chemotherapy. Patients showing a complete response receive no further therapy; those patients with partial response receive LD-IFRT.

Table 19-7 Accepted Standard Treatment Strategies for Newly Diagnosed Children and Adolescent Patients with Classical Hodgkin Lymphoma

Low Risk Group (IA, IIA, no bulk disease without any B symptoms)			
VAMP X 4 courses + LD-IFRT			
Vincristine	6 mg/m^2/day	IV	Days 1 and 15
Adriamycin	25 mg/m^2/day	IV	Days 1 and 15
Methotrexate	40 mg/m^2/day	IV	Days 1 and 15
Prednisone	40 mg/m^2/day	PO	Days 1–14
Repeat every	28 days		
		OR	
COPP/ABV hybrid courses ×4 + LD-IFRT			
Cyclophosphamide	600 mg/m^2/day	IV	Day 1
Vincristine	1.4 mg/m^2/day	IV	Day 1
Prednisone	40 mg/m^2/day	PO	Days 1–14
Procarbazine	100 mg/m^2/day	PO	Days 1–7
Adriamycin	35 mg/m^2/day	IV	Day 7
Bleomycin	10 units/m^2/day	IV	Day 7
Vinblastine	6 mg/m^2/day	IV	Day 7
Repeat every	28 days		
		OR	
DBVE ×4 courses + LD-IFRT			
Doxorubicin	25 mg/m^2/day	IV	Days 1 and 15
Bleomycin	10 units/m^2/day	IV	Days 1 and 15
Vincristine	1.5 mg/m^2/day	IV	Days 1 and 15
Etoposide	100 mg/m^2/day	IV	Days 1–5
Repeat every	28 days		
		OR	
OEPA(males) or OPPA(females) ×2 courses + LD-IFRT			
Vincristine	1.5 mg/m^2/day	IV	Days 1, 8, 15
Etoposide	125 mg/m^2/day	IV	Days 3–6
Prednisone	60 mg/m^2/day	PO	Days 1–15
Doxorubicin	40 mg/m^2/day	IV	Days 1 and 15
Repeat after	28 days		
OR			
Vincristine	1.5 mg/m^2/day	IV	Days 1, 8, 15
Procarbazine	100 mg/m^2/day	PO	Days 1–15
Prednisone	60 mg/m^2/day	PO	Days 1–15
Doxorubicin	40 mg/m^2/day	IV	Days 1 and 15
Repeat after	28 days		
Intermediate Risk Group (All Stages I and II not classified as early stages; Stage IIIA, Stage IVA)			
COPP/ABV ×6 courses + LD-IFRT			
Cyclophosphamide	600 mg/m^2/day	IV	Day 1
Vincristine	1.4 mg/m^2/day	IV	Day 1
Prednisone	40 mg/m^2/day	PO	Days 1–14
Procarbazine	100 mg/m^2/day	PO	Days 1–7
Adriamycin	35 mg/m^2/day	IV	Day 7

(Continued)

Table 19-7   (Continued)

Intermediate Risk Group (All Stages I and II not classified as early stages; Stage IIIA, Stage IVA)			
Bleomycin	10 units/m^2/day	IV	Day 7
Vinblastine	6 mg/m^2/day	IV	Day 7
Repeat every	28 days		
OR			
DBVE-PC ×3–5 courses + LD-IFRT			
Doxorubicin	25 mg/m^2/day	IV	Days 1 and 2
Bleomycin	5 units/m^2/day	SQ	Day 1
	10 units/m^2/day	SQ	Day 8
Vincristine	1.4 mg/m^2/day	IV	Days 1 and 4
Etoposide	125 mg/m^2/day	IV	Days 1, 2, 3
Prednisone	40 mg/m^2/day	PO	Day 1–7
Cyclophosphamide	800 mg/m^2/day	IV	Day 1
Repeat every	21 days		
OR			
OEPA (males) or OPPA (females) ×2 courses + COPP ×2 courses + LD-IFRT			
Vincristine	1.5 mg/m^2/day	IV	Days 1, 8, 15
Etoposide	125 mg/m^2/day	IV	Days 3–6
Prednisone	60 mg/m^2/day	PO	Days 1–15
Doxorubicin	40 mg/m^2/day	IV	Days 1 and 15
OR			
Vincristine	1.5 mg/m^2/day	IV	Days 1, 8, 15
Procarbazine	100 mg/m^2/day	PO	Days 1–15
Prednisone	60 mg/m^2/day	PO	Days 1–15
Doxorubicin	40 mg/m^2/day	IV	Days 1 and 15
Repeat after	28 days		
PLUS			
Cyclophosphamide	500–600 mg/m^2/day	IV	Day 1, 8
Vincristine	1.5 mg/m^2/day	IV	Days 1, 8
Prednisone	40 mg/m^2/day	PO	Days 1–14
Procarbazine	100 mg/m^2/day	PO	Days 1–14
Repeat after	28 days		
High Risk Group (Stages IIIB and IVB)			
DBVE-PC × 3–5 courses + LD-IFRT			
Doxorubicin	30 mg/m^2/day	IV	Days 1 and 2
Bleomycin	5 units/m^2/day	SQ	Day 1
	10 units/m^2/day	SQ	Day 8
Vincristine	1.4 mg/m^2/day	IV	Days 1 and 8
Etoposide	75 mg/m^2/day	IV	Days 1–5
Prednisone	40 mg/m^2/day	PO	Day 1–7
Cyclophosphamide	800 mg/m^2/day	IV	Day 1
Repeat every	21 days		
OR			
BEACOPP ×8 courses + LD-IFRT			
Bleomycin	10 units/m^2/day	IV	Day 1
Etoposide	100 mg/m^2/day	IV	Days 1–3

(Continued)

Table 19-7   (Continued)

High Risk Group (Stages IIIB and IVB)			
Adriamycin	35 mg/m^2/day	IV	Day 1
Cyclophosphamide	1,200 mg/m^2/day	IV	Day 1
Vincristine	2 mg/m^2/day	IV	Day 1
Prednisone	40 mg/m^2/day	PO	Days 1–14
Procarbazine	100 mg/m^2/day	PO	Days 1–7
Repeat every	21 days		
OR			
BEACOPP ×4 courses + ABVD ×2 courses + LD-IFRT			
Bleomycin	10 units/m^2/day	IV	Day 1
Etoposide	100 mg/m^2/day	IV	Days 1–3
Adriamycin	35 mg/m^2/day	IV	Day 1
Cyclophosphamide	1,200 mg/m^2/day	IV	Day 1
Vincristine	2 mg/m^2/day	IV	Day 1
Prednisone	40 mg/m^2/day	PO	Days 1–14
Procarbazine	100 mg/m^2/day	PO	Days 1–7
Repeat every	21 days		
PLUS			
Adriamycin	25 mg/m^2/day	IV	Days 1 and 15
Bleomycin	10 units/m^2/day	IV	Days 1 and 15
Vinblastine	6 mg/m^2/day	IV	Days 1 and 15
Dacarbazine	375 mg/m^2/day	IV	Days 1 and 15
Repeat every	28 days		
OR			
OEPA (males) or OPPA (females) ×2 courses + COPP × 4 courses + LD-IFRT			
Vincristine	1.5 mg/m^2/day	IV	Days 1, 8, 15
Etoposide	125 mg/m^2/day	IV	Days 3–6
Prednisone	60 mg/m^2/day	PO	Days 1–15
Doxorubicin	40 mg/m^2/day	IV	Days 1 and 15
Repeat after	28 days		
OR			
Vincristine	1.5 mg/m^2/day	IV	Days 1, 8, 15
Procarbazine	100 mg/m^2/day	PO	Days 1–15
Prednisone	60 mg/m^2/day	PO	Days 1–15
Doxorubicin	40 mg/m^2/day	IV	Days 1 and 15
Repeat after	28 days		
PLUS			
Cyclophosphamide	500–600 mg/m^2/day	IV	Days 1, 8
Vincristine	1.5 mg/m^2/day	IV	Days 1, 8
Prednisone	40 mg/m^2/day	PO	Days 1–14
Procarbazine	100 mg/m^2/day	PO	Days 1–14
Repeat every	28 days		

Abbreviations: LD-IFRT, low-dose involved field radiation therapy (15–25 Gy).
From: Friedman D, Schwartz C, Hodgkin Lymphoma in The Lymphoid Neoplasms, 3rd edition, 2008, with permission.

Radiation

Hodgkin lymphoma is a radiosensitive disease. Treatment for HL in children and adolescents employs radiation therapy in the context of multimodality therapy. Most newly diagnosed children and adolescents are treated with risk-adapted chemotherapy alone or in combination with LD-IFRT. In general, doses of 15–25 Gy are used with modification based on patient's age, the presence of bulk disease, normal tissue concerns and potential acute and long-term effects.

Megavoltage energies are now the treatment of choice for pediatric HL. A 4–6 mV linear accelerator should be used to treat the supradiaphragmatic fields, whereas 8–15 mV machines may be appropriate for treating para-aortic nodes. A linear accelerator with beam energy of 6 mV is desirable due to its penetration, well-defined edge and homogeneity throughout an irregular treatment field. Immobilization techniques (i.e. sedation) may be necessary for young children to ensure accuracy and reproducibility. Treatment of involved supradiaphragmatic fields or a mantle field requires precision because of the distribution of lymph nodes and critical adjacent tissues.

Volume considerations for treatment generally include the initial involved lymph node regions. Additional considerations relate to the location of the disease (pericardium and chest wall). When it is necessary to treat the pelvis, special attention must be given to the ovaries and testes. If possible, the ovaries should be relocated (laterally along the iliac wings or centrally behind the uterus), marked with surgical clips to permit appropriate shielding. Ideally the ovaries should be exposed to less than 3 Gy to preserve fertility. The testes may be exposed to 5–10% of the administered pelvic dose; this low dose can cause azoospermia depending on the total pelvic dose. For males, the greatest shielding to the testes can be done by using the frog-leg position together with an individually fitted shield. It is important to restrict radiation therapy to areas of bulky disease while still protecting normal tissue. This can often be accomplished by careful positioning of the patient and by

Table 19-8   Guidelines for Involved Field Radiation

Involved Nodes	Radiation Therapy Field
Cervical	Neck, supraclavicular/infraclavicular
Supraclavicular	Neck and infraclavicular/supraclavicular, axilla
Axilla	Axilla ± infraclavicular/supraclavicular
Mediastinum	Mediastinum, hila, infraclavicular/supraclavicular
Hila	Hila, mediastinum
Spleen	Spleen ± paraaortic
Para-aortics	Para-aortics ± spleen
Iliac	Ipsilateral iliacs, inguinal + femoral
Inguinal	Inguinal + femoral ± iliac
Femoral	Inguinal + femoral ± iliac

customizing shield blocks. Large-volume radiation therapy can compromise organ function and may limit the intensity of retrieval therapy if relapse occurs. Guidelines for involved field radiation are shown in Table 19-8.

Daily radiation therapy is delivered using equally weighted anterior and posterior fields with fractional doses of 1.5–1.8 Gy daily, five times a week.

Treatment for Relapsed and Progressive Disease

Treatment failure in children and adolescents with HL can be divided into:

- Relapsed disease, or
- Primary progressive disease.

Relapsed patients can still be cured even if relapse occurs after initial treatment. Relapse usually occurs within the first 4 years following treatment. The presence of B symptoms and extranodal disease at the time of relapse are poor prognostic features. A study conducted by the Society for Pediatric Oncology and Hematology (GPOH) showed the following results:

- Patients with an early relapse (defined as occurring between 3–13 months from the end of therapy) had a 10-year event-free survival (EFS) of 55% and overall survival (OS) of 78%
- Patients who had a late relapse (defined as occurring more than 12 months from the end of therapy) had a 10-year EFS and OS of 86% and 90%, respectively
- Patients with favorable disease at diagnosis (i.e. stage IA or stage IIA; no bulk symptoms; no B symptoms), with relapse confined to an area of initial involvement after chemotherapy and no radiation, can generally be salvaged with further chemotherapy and LD-IFRT.

Currently there is no standard reinduction protocol for HL though there is a spectrum of treatment options that exists. Potentially curative options include

- Conventional chemotherapy and LD-IFRT radiation therapy protocols for those with low-risk relapses (Table 19-7)
- Reinduction chemotherapy followed by autologous hematopoietic stem cell transplantation for those with relapses that follow dose-dense therapies or are otherwise higher risk.

Autologous stem cells are used as opposed to allogeneic transplantation due to higher transplant-related morbidity and mortality with allogeneic transplantation. Following autologous stem cell transplantation, the projected survival rate is 45–70% and progression-free survival is 30–65%.

There are cases where a myeloablative or reduced-intensity allogeneic transplantation may be the treatment of choice. For those patients who fail autologous stem cell transplantation or for patients who cannot mobilize sufficient numbers of autologous stem cells, allogeneic stem cells have been used with varying degrees of success. LD-IFRT should be given to sites of relapsed disease even if these sites had not received previous radiation. LD-IFRT is usually administered after high-dose chemotherapy and stem cell rescue.

Patients treated with stem cell transplant may experience relapse as late as 5 years after their transplant; they should be monitored closely for relapse along with long-term effects. Patients with progressive or primary refractory HL have poorer salvage rates even with stem cell transplant and radiation therapy.

LONG-TERM COMPLICATIONS

Survivors of HL are at risk for numerous late complications of chemotherapy and radiation therapy. Late effects include secondary malignancies, organ dysfunction and psychosocial sequelae. See Chapter 32 for more information on long-term complications.

Secondary Malignancies

The incidence of secondary malignancies in survivors is 7–18 times higher than the general population. Many of these patients had received high-dose alkylating agents and radiation. Secondary hematological malignancies (most commonly acute myeloid leukemia and myeloid dysplasia) are related to the use of alkylating agents, anthracycline and etoposide. The risk of leukemia appears to plateau at 10–15 years post therapy while the risk of second solid malignancies increases.

Female survivors of HL have an increased risk for breast cancer, directly related to the dose of radiation therapy received. There is a 3.2-fold increase in the risk of developing breast cancer for females who received 4 Gy and an 8-fold increase in risk for females who received 40 Gy. Female patients treated with both radiation therapy and alkylating agent chemotherapy have a lower relative risk of developing breast cancer than women who received radiation therapy alone. Second solid tumors are consistently noted in patients receiving at least 35 Gy radiation.

Cardiac Toxicity

HL survivors exposed to doxorubicin or thoracic radiation therapy are at an increased risk for long-term toxicity. The risks to the heart are related to:

- Cumulative anthracycline dose and method of administration
- Amount of radiation delivered to different depths of the heart, volume and specific areas of the heart irradiated, total or fractional irradiation doses
- Age at exposure

- Latency periods
- Gender.

Increased risk of doxorubicin-related cardiomyopathy is associated with:

- Female sex
- Cumulative doses higher than 200–300 mg/m^2
- Younger age at time of exposure
- Increased time from exposure.

Late effects to the heart include:

- Delayed pericarditis
- Pancarditis (which includes pericardial and myocardial fibrosis)
- Cardiomyopathy
- Coronary artery disease
- Functional valve injury
- Conduction injuries.

Nontherapeutic risk factors for coronary heart disease including family history, obesity, smoking, hypertension, diabetes and hypercholesterolemia are likely to impact the frequency of disease.

Pulmonary Dysfunction

Pulmonary fibrosis is a late complication of HL following radiation therapy. Bleomycin-associated pulmonary fibrosis with decreased diffusion capacity is most commonly seen following doses greater than 200–400 units/m^2. Acute pneumonitis as evidenced by fever, congestion, cough and dyspnea can follow radiation therapy alone at doses greater than 40 Gy to focal lung volumes or after lower doses (15–20 Gy) when combined with chemotherapy and including generous or whole lung volumes.

Thyroid Dysfunction

Primary hypothyroidism is the most common thyroid dysfunction noted following radiation therapy to the neck for HL; though hyperthyroidism, goiter, or nodules have also been noted. The incidence of thyroid dysfunction varies with the dose of radiation and the length of follow-up. For hypothyroidism, there is a clear dose response with a 20-year risk of 20% for patients who had received less than 35 Gy, 30% following 35–44.9 Gy and 50% following greater than 45 Gy to the thyroid gland. The risk for both hypo- and hyperthyroidism increases in the first 3–5 years after diagnosis. The incidence of nodules increases 10 years from the time of diagnosis.

Gonadal Dysfunction and Infertility

Male Gonadal Toxicity

Gonadal toxicity in males is likely to manifest as:

- Infertility
- Lack of sexual development
- Small atrophic testicles
- Sexual dysfunction.

Infertility caused by azoopsermia is the most common manifestation of gonadal toxicity. Alkylating agents and radiation therapy are common agents associated with male infertility. Males who receive less than 4 g/m^2 of cyclophosphamide without testicular radiation or any other alkylating agents are likely to retain their fertility. Cumulative doses of cyclo-phosphamide greater than 9 g/m^2 are likely to result in infertility. Radiation doses less than 30 Gy are unlikely to affect endocrine function and boys usually progress through puberty normally. There can be temporary oligospermia following low doses of radiation though permanent azoopsermia may occur.

Female Gonadal Toxicity

The risk of female gonadal toxicity increases with age at treatment. Menstrual irregularities including amenorrhea and premature ovarian failure occur more commonly in women treated with cyclophosphamide and other alkylating agents than in adolescent and prepubertal females. Several studies have shown that the incidence of early menopause in young female survivors of HL may be as high as 37%.

Growth Dysfunction

Long-term effects on bone and soft-tissue growth among survivors of childhood HL are related to the dose and volume of radiation received, as well as the age of the child when the radiation was received. In one study of 124 children with HL, prepubertal children who received >33 Gy radiation to the entire spine had an average height "loss" of 13 cm (2 SD of the United States population mean), whereas prepubertal children who received <33 Gy and pubertal and postpubertal children who received >33 Gy had little height impairment. High-dose radiation can also cause narrowing of the thoracic apex with symmetric narrowing of the clavicles and atrophy of the soft tissues of the neck.

Pyschosocial Problems

Six percent of HL survivors have symptoms of depression and 15% have somatic distress higher than that for other malignancies or when compared to sibling cohorts. Risk factors associated with increased risk for somatic distress include female gender, lower household income, less than a high school education and lack of current employment (see Chapter 33).

Monitoring for Adverse Long-term Outcomes

Specific long-term follow-up guidelines after treatment of childhood cancer have been published by the Children's Oncology Group and are available at www. survivorshipguidelines.org. Table 19-9 reviews general guidelines for radiation and

Table 19-9   General Guidelines for Radiation and Chemotherapy Late Effects Assessment and Management

System	Agents	Potential Effects	Monitoring Recommendations[*]
Cardiac	Anthracyclines Radiotherapy	Cardiomyopathy Pericarditis Coronary artery disease Valvular disease	Electrocardiogram and echocardiogram
Pulmonary	Bleomycin Radiotherapy	Pulmonary fibrosis	Pulmonary function tests (including DLCO and spirometry)
Thyroid	Radiotherapy	Hypothyroidism Thyroid nodules, cancer Hyperthyroidism	Free T4 and TSH Thyroid examination
Gonadal(F)	Alkylating agents	Delayed/arrested puberty Early menopause Ovarian failure	FSH, LH, estradiol
Gonadal(M)	Alkylating agents	Germ cell failure Infertility/azoopsermia Leydig cell dysfunction Hypogonadism Delayed/arrested puberty	FSH, LH, testosterone, semen analysis
Bone	Corticosteroids	Osteopenia Osteoporosis Osteonecrosis	Bone density evaluation (DEXA) MRI to evaluate for AVN
SMN	Alkylating agents Topoisomerase II Inhibitors Anthracyclines Radiotherapy	Leukemia, MDS Solid organ tumors Skin cancer (non-melanoma far more frequent than melanoma)	Cancer Screening per COG Guidelines

[*]Monitoring recommendations all include a risk adapted health history and physical examination. Studies and laboratory tests should be performed based on dose of radiation therapy, chemotherapy exposures, age at exposure and other clinical parameters. Detailed follow-up guidelines developed by the Children's Oncology Group can be found at www. survivorshipguidelines.org.

Abbreviations: DLCO, carbon monoxide diffusing capacity; F, females; M, males; T4, thyroxine; TSH, thyroid stimulating hormone; FSH, follicle stimulating hormone; LH, luteinizing hormone; MRI, magnetic resonance imaging; AVN, avascular necrosis; SD, standard deviation; CNS, central nervous system; RT, radiation therapy; SMN, second malignant neoplasm.

chemotherapy late effects assessment and management (see Chapter 32 for more information concerning late effects of cancer therapy).

Although to date, there are no fully agreed-upon and evidence-based long-term follow-up plans for survivors of childhood and adolescent HL, the following is suggested follow-up in HL during and after completion of therapy. Recommendations depend on the specific therapy received and the patient's response to same.

FOLLOW-UP EVALUATIONS
During Therapy

History and Physical Examination
1. Special reference to lymph nodes, liver, spleen and indication of response.

Laboratory Studies
1. *Complete blood count (CBC)*: Generally performed weekly with chemotherapy and radiotherapy.
2. *Erythrocyte sedimentation rate (ESR)*: Generally performed prior to each cycle of chemotherapy and radiotherapy until normalized.
3. *Liver function tests, blood chemistries*: Generally performed prior to each cycle of chemotherapy and radiotherapy.

Imaging Studies
1. *CT scan*: A CT of the neck, chest, abdomen and pelvis is performed at diagnosis and upon completion of chemotherapy, prior to radiotherapy. This may be repeated at intervals during therapy which is generally protocol-specific and may include only more limited scans of the neck and chest.
2. *FDG-PET scan*: This is the nuclear medicine imaging study of choice and should be used in place of gallium scan. It is performed at diagnosis and upon completion of chemotherapy and prior to radiotherapy. It is repeated at intervals during therapy dependent generally on the protocol being used.

Organ Toxicity Monitoring
1. *Cardiac function tests*: An echocardiogram and ECG are performed at diagnosis and the echocardiogram is generally repeated after a cumulative doxorubicin dose of 200–300 mg/m^2 and/or prior to radiation therapy.
2. *Pulmonary function tests*: These are performed at diagnosis and generally repeated after a cumulative bleomycin dose of 45 mg/m^2 and/or prior to radiation therapy.

After Completion of Therapy

Following the completion of therapy, monitoring is required for disease recurrence as well as for long-term sequelae of therapy.

Disease Monitoring

Imaging studies for disease recurrence generally include a CT of the neck and chest and of other primary site (abdomen/pelvis). In the first 18 months, this is generally obtained every 3 months. Some protocols recommend alternating CT with chest radiographs (for chest primary) during the first 18 months. Thereafter, CT screening generally continues every 6 months until 48 months and then annually through 5 years. After 5 years no routine imaging for disease recurrence is required. At these same time points, all patients should have a careful history and physical examination, a CBC and ESR.

Long-term Sequelae Monitoring

This monitoring is highly dependent on the treatment received and guidelines from the Children's Oncology Group should be used to guide follow-up (http://www. survivorshipguidelines.org). Generally, all patients should be followed lifelong for health promotion with history and physical and routine dental care. Anthropomorphic measures and Tanner stage should be monitored until puberty is complete.

Laboratory studies include thyroid function tests for patients who received neck or mantle radiotherapy, sex hormones for those exposed to alkylating agents or pelvic radiotherapy, cardiac studies for those exposed to anthracyclines or thoracic radiotherapy and pulmonary function tests for those exposed to bleomycin or thoracic radiotherapy. The frequency of monitoring is dependent on many factors such as the therapy received and the age and sex of the patient.

Suggested Reading

Cheson BD, et al. Revised response criteria for malignant lymphoma. J Clin Oncol. 2007;25(5):579–586.

Friedman DL, Meadows AT. Late effects of lymphoma treatment. In: Weinstein HJ, Link M, Hudson M, eds. Pediatric Lymphomas. London: Springer Verlag; 2008.

Friedmann AM, et al. Treatment of unfavorable childhood Hodgkin's disease with VEPA and low-dose, involved-field radiation. J Clin Oncol. 2002;20(14):3088–3094.

Gallamini A, et al. Early interim 2-[18F]fluoro-2-deoxy-D-glucose positron emission tomography is prognostically superior to international prognostic score in advanced-stage Hodgkin's lymphoma: a report from a joint Italian–Danish study. J Clin Oncol. 2007;25(24):3746–3752.

Gleeson HK, Darzy K, Shalet SM. Late endocrine, metabolic and skeletal sequelae following treatment of childhood cancer. Best Pract Res Clin Endocrinol Metab. 2002;16(2):335–348.

Grufferman S, Delzell E. Epidemiology of Hodgkin's disease. Epidemiol Rev. 1984;6:76–106.

Hudson MM, Constine L. Hodgkins' disease. Pediatric Radiation Oncology. In: Halperin EC, Tarbell NJ, et al., eds. Philadelphia: Lippincott Williams and Wilkins; 2005:223–260.

Hudson MM, et al. Risk-adapted, combined-modality therapy with VAMP/COP and response-based, involved-field radiation for unfavorable pediatric Hodgkin's disease. J Clin Oncol. 2004;22(22):4541–4550.

Hutchinson RJ, et al. MOPP or radiation in addition to ABVD in the treatment of pathologically staged advanced Hodgkin's disease in children: results of the Children's Cancer Group Phase III Trial. J Clin Oncol. 1998;16(3):897–906.

Kaste SC, et al. 18F-FDG-avid sites mimicking active disease in pediatric Hodgkin's. Pediatr Radiol. 2005;35(2):141–154.

Landier W, Bhatia S, Eshelman DA, et al. Development of risk-based guidelines for pediatric cancer survivors: the Children's Oncology Group Long-Term Follow-Up Guidelines from the Children's Oncology Group Late Effects Committee and Nursing Discipline. J Clin Oncol. 2004;22:4979–4990.

Nachman JB, et al. Randomized comparison of low-dose involved-field radiotherapy and no radiotherapy for children with Hodgkin's disease who achieve a complete response to chemotherapy. J Clin Oncol. 2002;20(18):3765–3771.

Neglia JP, et al. Second malignant neoplasms in five-year survivors of childhood cancer: childhood cancer survivor study. J Natl Cancer Inst. 2001;93(8):618–629.

Percy CL, Linet M, et al. Lymphomas and reticuolendothelial neoplasms. Bethesda: National Cancer Institute, SEER Program, 1999.

Schwartz CL. Prognostic factors in pediatric Hodgkin disease. Curr Oncol Rep. 2003;5(6):498–504.

Schwartz CL. Special issues in pediatric Hodgkin's disease. Eur J Haematol. 2005;(Suppl):55–62.

Schwartz C, London W. Intensive response based therapy with or without dexrazoxane for intermediate/high stage (I/HS) pediatric Hodgkin's disease (HL): a Pediatric Oncology Group (POG) Study. In: Polliack A, Diehl V, et al., eds. Fifth International Symposium on Hodgkin's lymphoma. Koln, Germany: Harwood Academic, 2001.

Schwartz CL, et al. A risk-adapted, response-based approach using ABVE-PC for children and adolescents with intermediate- and high-risk Hodgkin lymphoma: the results of P9425. Blood 2009;114(10):2051–2059.

Tebbi CK, et al. Treatment of stage I, IIA, IIIA1 pediatric Hodgkin disease with doxorubicin, bleomycin, vincristine and etoposide (DBVE) and radiation: a Pediatric Oncology Group (POG) study. Pediatr Blood Cancer, 2006;46(2):198–202.

Weiner MA, et al. Randomized study of intensive MOPP-ABVD with or without low-dose total-nodal radiation therapy in the treatment of stages IIB, IIIA2, IIIB and IV Hodgkin's disease in pediatric patients: a Pediatric Oncology Group study. J Clin Oncol. 1997;15(8):2769–2779.

CHAPTER 20

Non-Hodgkin Lymphoma

INTRODUCTION

Non-Hodgkin lymphoma (NHL) results from malignant proliferation of cells of lymphocytic lineage. Although malignant lymphomas are generally restricted to lymphoid tissue such as lymph nodes, Peyer's patches and spleen, it is not uncommon to find bone marrow involvement in children. Bone and primary CNS lymphomas are rare presentations.

INCIDENCE AND EPIDEMIOLOGY

Incidence

1. NHL represents 8–10% of all malignancies in children between 5–19 years of age.
2. It accounts for approximately 60% of all lymphomas in children and adolescents.
3. It is the third most common childhood malignancy with an annual incidence of 750–800 cases per year in children <19 years of age in the USA.
4. There has been a stable incidence for children less than 15 years of age over the last two decades, but in adolescents 15–19 years of age there has been an increase of 50%.

Epidemiology

Sex: male:female is 2.5:1 overall and over 3:1 in ages 5–14.
Age: relatively similar across all ages, but peaks at ages 15–19 years, largely because of a significant increase in the incidence of diffuse large cell lymphoma at this age.
Risk factors:
  - Genetic: Immunological defects (Bruton type of sex-linked agammaglobulinemia, common variable agammaglobulinemia, severe combined immunodeficiency ataxia-telangiectasia, Bloom syndrome Wiskott–Aldrich syndrome, autoimmune lymphoproliferative syndrome [ALPS])
  - Post-transplant immunosuppression, e.g. post bone-marrow transplantation (especially with use of T-cell depleted marrow); post-solid organ transplantation
  - Lymphomatoid papulosis in children may evolve into or co-exist with anaplastic large cell lymphoma (ALCL)

Manual of Pediatric Hematology and Oncology. DOI: 10.1016/B978-0-12-375154-6.00020-3

- • *Drugs*: Diphenylhydantoin, Infliximab and other immunosuppressive agents
- • *Radiation*: Children treated with chemotherapy and radiotherapy for Hodgkin lymphoma
- • *Viral*: Epstein–Barr virus (EBV), human immune deficiency virus (HIV) and possible link to human T-lymphotropic virus (HTLV).

PATHOLOGIC CLASSIFICATION

Table 20-1 presents the World Health Organization classification for lymphoid neo-plasms from the International Lymphoma Study Group. The preferred classification for NHL is based on currently recognized histologic (morphologic), immunophenotypic and genetic features, their clinical presentation and course. Table 20-2 lists the morpho-logic, immunologic and genetic features in childhood NHL. Table 20-3 lists the epide-miologic, immunologic and molecular features of endemic and sporadic Burkitt lymphoma (BL).

CLINICAL FEATURES

The clinical manifestations of childhood NHL depend primarily on pathological subtype and sites of involvement. Tumors which grow rapidly can cause symptoms based on size and location. Approximately 70% of children present with advanced-stage disease including extranodal disease with gastrointestinal, bone marrow and central nervous system (CNS) involvement.

Abdomen

Patients with Burkitt's lymphoma (BL) and diffuse large B-cell lymphoma (DLBCL) com-monly have primary tumors in the abdomen. The primary site in 35% of childhood NHL is abdominal, involving the ileocecal region, appendix, ascending colon, or some combination of these sites. These patients may present with abdominal pain and distention, nausea and vomiting, bowel changes, palpable masses, intussusception, hepatosplenomegaly, obstruc-tive jaundice, ascites, or peritonitis. Bleeding and perforation of the intestine, although rare, may occur. Lymphoma is the most frequent cause of intussusception in children over 6 years of age.

Head and Neck

In 13% of cases, the head and neck are involved, causing enlargement of the cervical node(s) and parotid gland, jaw swelling and unilateral tonsillar hypertrophy. The disease may present with nasal obstruction, rhinorrhea, hearing difficulty and cranial nerve palsies. The endemic

Table 20-1   WHO Classification of Lymphoid Neoplasms (2008)

Precursor Lymphoid Neoplasms
  B lymphoblastic leukemia/lymphoma NOS
  B lymphoblastic leukemia/lymphoma with recurrent genetic abnormalities
  B lymphoblastic leukemia/lymphoma with t(9;22); bcr-abl1
  B lymphoblastic leukemia/lymphoma with t(v;11q23); MLL rearranged
  B lymphoblastic leukemia/lymphoma with t(12;21); TEL-AML1 & ETV6-RUNX1
  B lymphoblastic leukemia/lymphoma with hyperploidy
  B lymphoblastic leukemia/lymphoma with hypodiploidy
  B lymphoblastic leukemia/lymphoma with t(5;14); IL3-IGH
  B lymphoblastic leukemia/lymphoma with t(1;19); E2A-PBX1 & TCF3-PBX1
  T lymphoblastic leukemia/lymphoma
Mature B-cell Neoplasms
  Chronic lymphocytic leukemia/small lymphocytic lymphoma
  B-cell prolymphocytic leukemia
  Splenic marginal zone lymphoma
  Hairy cell leukemia
  Lymphoplasmacytic lymphoma/Waldenstrom macroglobulinemia
  Heavy chain disease
  Plasma cell myeloma
  Solitary plasmacytoma of bone
  Extraosseous plasmacytoma
  Extranodal marginal zone B-cell lymphoma of mucosa-associated lymphoid tissue (MALT) type
  Nodal marginal zone lymphoma
  Follicular lymphoma
  Primary cutaneous follicular lymphoma
  Mantle cell lymphoma
  Diffuse large B-cell lymphoma, NOS (T-cell/histiocyte-rich type; primary CNS type; primary leg skin type &
    EBV+ elderly type)
  Diffuse large B-cell lymphoma with chronic inflammation
  Lymphomatoid granulomatosis
  Primary mediastinal large B-cell lymphoma
  Intravascular large B-cell lymphoma
  ALK+ large B-cell lymphoma
  Plasmablastic lymphoma
  Large B-cell lymphoma associated with human herpes virus 8 (HHV8) + Castleman disease
  Primary effusion lymphoma
  Burkitt lymphoma
  B cell lymphoma, unclassifiable, Burkitt-like
  B cell lymphoma, unclassifiable, Hodgkin lymphoma-like
Mature T-cell & NK-cell Neoplasms
  T-cell prolymphocytic leukemia
  T-cell large granular lymphocytic leukemia
  Chronic lymphoproliferative disorder of NK-cells

  Aggressive NK-cell leukemia
  Systemic EBV+ T-cell lymphoproliferative disorder of childhood
  Hydroa vacciniforme-like lymphoma

(Continued)

Table 20-1   (Continued)

Adult T-cell lymphoma/leukemia
Extranodal T-cell/NK-cell lymphoma, nasal type
Enteropathy-associated T-cell lymphoma
Hepato-splenic T-cell lymphoma
Subcutaneous panniculitis-like T-cell lymphoma
Mycosis fungoides
Sézary syndrome
Primary cutaneous CD30+ T-cell lymphoproliferative disorder
Primary cutaneous gamma-delta T-cell lymphoma
Peripheral T-cell lymphoma, NOS
Angioimmunoblastic T-cell lymphoma
Anaplastic large cell lymphoma, ALK+ type
Anaplastic large cell lymphoma, ALK- type
Hodgkin Lymphoma (Hodgkin Disease)
Nodular lymphocyte-predominant Hodgkin lymphoma
Classic Hodgkin lymphoma
Nodular sclerosis Hodgkin lymphoma
Lymphocyte-rich classic Hodgkin lymphoma
Mixed cellularity Hodgkin lymphoma
Lymphocyte depletion Hodgkin lymphoma
Post-transplant Lymphoproliferative Disorders (PTLD)
Plasmacytic hyperplasia
Infectious mononucleosis like PTLD
Polymorphic PTLD
Monomorphic PTLD (B & T/NK cell types)
Classic HD type PTLD
Histiocytic and Dendritic Cell Neoplasms
Histiocytic sarcoma
Langerhans cell histiocytosis
Langerhans cell sarcoma
Interdigitating dendritic cell sarcoma
Follicular dendritic cell sarcoma
Fibroblastic reticular cell tumor
Indeterminate dendritic cell sarcoma
Disseminated juvenile xanthogranuloma

Abbreviation: NOS, not otherwise specified.
Adapted from research originally published by Jaffe, ES, Harris, NL, Stein, H and Isaacson, PG. Classification of lymphoid neoplasms: the microscope as a tool for disease discovery. Blood 2008;112:4384–4399.

form of BL has a much higher incidence of facial involvement, with around 50–60% at presentation.

Mediastinum

The frequency of involvement of the mediastinum is 26%. Patients with large mediastinal masses may have superior mediastinal syndrome (SMS) and present with distended neck

Table 20-2 Morphologic, Immunologic and Genetic Features in Childhood NHL

	DLBCL	Burkitt	Lymphoblastic	ALCL
Morphology				
Cell size	Large	Intermediate	Small to intermediate	Small to large
Nuclear chromatin	Clumped, vesicular	Coarse	Fine, blastic	Clumped, vesicular
Nucleoli	Variable	Variable	Absent	Variable
Cytoplasm	Moderate to abundant	Moderate to scanty with prominent vacuoles	Scanty	Moderate to abundant
Nodal pattern	Diffuse	Diffuse, starry-sky pattern	Diffuse, starry sky pattern	Sinusoidal or diffuse
Immunophenotype	CD20+, CD22+	CD20+, CD22+	CD2 (80%), CD19 (20%)	T-cell, null cell, CD30+, ALK+
Genetic Features	t(8;14)(q24;q32), 12 gain, 21 gain, 15 loss, 3q27 rearrangements, 6q deletion, 7q gain, 11q gain	t(8;14)(q24;q32), t(8;22)(q24;q11), t(2;8)(q11;q24), 1q duplication, 13q abnormality, 6q deletion, 17p abnormality, 11q abnormality	T-cell receptor rearrangements (Tα, Tβ, Tδ, Tγ), TCL-1 t(7;14)(q35;q32) t(14;14) (q11;q32), TCL-2 t(11;13)(p13;q11), TCL-3 t(10;14)(q24;q11), TAL-1 t(1;14)(q32;q11)	t(2;5)(p23;q25), ALK translocation with chromosome 1, 2, 3, 17

Abbreviations: DLBCL, diffuse large B-cell lymphoma; ALCL, anaplastic large cell lymphoma.
Adapted from: Carroll WL, Finlay JL. Cancer in Children and Adolescents, Chapter 14, copyright 2010: Jones and Bartlett Publishers, Sudbury, MA. www.jbpub.com. Reprinted with permission.

Table 20-3 Epidemiologic, Immunologic and Molecular Features of Endemic and Sporadic Burkitt Lymphoma

Feature	Endemic	Sporadic
Epidemiologic		
Population affected	African	Worldwide
Age affected	Children (peak age 7 years)	Young adults (peak age: 11 years)
Organ involvement	Jaw, orbit, paraspinal, abdomen, ovary	Abdomen, bone marrow, nasopharynx, Ovary
Epstein-Barr Virus	Present in >97% of cases	Present in 15–20% of cases
CNS involvement	More frequent than bone marrow involvement	Less frequent than bone marrow involvement
Immunologic		
IgM secretion	Little or none	Prominent
Fc and C3	+	−
CALLA	±	+
IgH gene rearrangement	DH or JH	IgH switch region
Molecular		
Breakpoints on chromosome 8	Upstream of C-myc	Within C-myc
Site of cell origin	Germinal center of lymph node	Bone marrow

veins, edema of the neck and face, marked dyspnea, orthopnea, dizziness, headache, dysphagia, epistaxis, altered mental status and syncope.

Lymphoblastic lymphoma (LL) commonly presents with an intrathoracic or mediastinal mass along with spread to other sites, e.g., bone marrow, meninges and gonads. BL and anaplastic large cell lymphoma (ALCL) involve the mediastinum less commonly. Primary mediastinal B-cell lymphoma (PMBL) is currently classified by the WHO as a distinct subtype of DLBCL although there are increasing data to suggest that it is a separate entity that shares clinical, morphologic and genetic features with both NHL and Hodgkin lymphoma. PMBL is typically a malignancy of adolescents and young adults. Outcomes in children with PMBL are worse than in pediatric patients with other NHL subtypes. Differences in tumor biology may account for this discrepancy in outcomes.

Mediastinal tumors may lead to pleural effusion either through direct pleural involvement or may result from the compression of lymphatics by the mediastinal mass. The presence of significant pericardial effusion may cause cardiac tamponade.

Bone Marrow

Both BL and LL have a predilection for spreading to the bone marrow. Rarely, ALCL and DLBCL will present with bone marrow involvement. In both BL and LL, bone marrow involvement of ≥25% is consistent with acute leukemia. Patients with BL have a significantly worse prognosis if there is bone marrow involvement, however, this difference in prognosis is not as pronounced in LL.

Central Nervous System

As with bone marrow involvement, BL and LL also present more commonly with CNS disease. In DLBCL or ALCL this is rare. Involvement of the CNS upstages the risk classification at diagnosis to advanced-stage lymphoma and confers a worse overall survival, often requiring intensification of treatment.

Other Primary Sites

Other sites involved (11%) include skin and subcutaneous tissue, orbit, thyroid, bone (with or without hypercalcemia), kidney, epidural space, breast and gonads. The peripheral nodes are affected in 14% of cases.

Constitutional Symptoms

Fever and weight loss are relatively uncommon except in ALCL. Weight loss may also occur secondary to mechanical bowel obstruction.

DIAGNOSIS

Table 20-4 lists the diagnostic tests required to evaluate NHL. Recommended laboratory and radiologic testing includes: complete blood count, electrolytes, uric acid, calcium, phosphorus, creatinine; bone marrow aspiration and biopsy; lumbar puncture with CSF cytology, cell count and protein; chest radiograph and neck, chest, abdominal and pelvic CT scans. PET scan should also be performed to help in initial staging as well as to measure tumor response during the course of therapy in addition to traditional RECIST (response evaluation criteria in solid tumors) criteria. Tumor tissue (i.e., biopsy, bone marrow, CSF, or pleural/paracentesis fluid) should be tested by flow cytometry for immunophenotypic origin as well as cytogenetics.

STAGING

Staging of NHL requires an expeditious investigation to determine the clinical extent of the disease, the degree of organ impairment and biochemical disturbance present. Correct staging is critically important at the time of diagnosis in childhood NHL due to the high number of patients who present with advanced disease. Prior staging systems for childhood NHL entailed a modification of the Ann Arbor Staging System for Hodgkin lymphoma while taking into consideration the common presentations of childhood NHL such as extranodal involvement and metastatic spread to the bone marrow and CNS. A new French, American and British (FAB) childhood NHL staging classification has been developed (Table 20-5) that more clearly defines the staging of childhood NHL. This classification recognizes the

Table 20-4   Evaluation of Patient with Non-Hodgkin Lymphoma

Detailed History and Physical Examination
Laboratory Tests
  Complete blood count
  Serum electrolytes (calcium, phosphorous, magnesium)
  Evaluation of renal function (urinalysis, BUN, uric acid, creatinine)
  Evaluation of hepatic function (bilirubin, alkaline phosphatase, ALT, AST)
  Lactic dehydrogenase level
  Soluble interleukin-2 receptor levels (if possible)
  Viral studies: HIV antibody, Hepatitis A, B, C serology, CMV antibody, varicella antibody, HSV antibody

Tissue Tests
  Adequate surgical biopsy for cytochemical, immunologic, cytogenetic and molecular studies
  Bone marrow aspiration and biopsy (bilateral)
  Spinal fluid, peritoneal, pericardial or pleural fluid examination – cytochemical, immunologic, cytogenetic

Radiologic Tests
  Chest radiograph
  CT scan of the chest
  CT scan of the abdomen and pelvis with contrast
  PET scan
  Magnetic resonance imaging (MRI) when clinically indicated, especially for bone involvement
    (e.g., vertebrae)

Table 20-5   FAB* Staging System for Childhood DLBCL, BL and BLL

Group A
  Completely resected stage I (Murphy)
  Completely resected abdominal stage II (Murphy)
Group B
  All patients not eligible for Group A or Group C
Group C
  Any CNS involvement** and/or bone marrow involvement (≥25% blasts)

*FAB: French–American–British.
**CNS: Any L3 blast, cranial nerve palsy or compression, intracerebral mass and/or parameningeal compression.

common occurrence of dissemination to either the bone marrow or the central nervous system, or both and the fact that this is indicative of a poor prognosis.

MANAGEMENT

Emergency Treatment

Three clinical manifestations that require immediate attention include:

- Superior vena cava (SVC) syndrome secondary to a large mediastinal mass obstructing blood flow

- Respiratory airway compression and
- Tumor lysis syndrome (TLS) secondary to severe metabolic abnormalities from massive lysis of tumor cells.

The management of these critical conditions is fully described in Chapter 30.

Chemotherapy

Proliferation of lymphoblasts may be extremely rapid, with doubling times as short as 24 hours in some histologic types, such as BL. Successful therapy of NHL requires multiagent systemic chemotherapy and intrathecal chemotherapy.

Treatment groups are divided into the following categories:

- Lymphoblastic lymphoma (LL)
  a. Localized and advanced disease treated the same
- B-lineage NHL (Burkitt lymphoma [BL], Burkitt-like lymphoma [BLL] and diffuse large cell lymphoma [DLCL])
  a. Low risk
  b. Intermediate risk
  c. High risk (CNS or bone marrow involvement)
- ALCL
  a. Low stage
  b. Advanced stage.

LYMPHOBLASTIC LYMPHOMA

Patient Eligibility

1. B or T lineage LL.
2. Localized and advanced disease.

Treatment

Table 20-6 and Figure 20-1 show the treatment strategy used (Berlin–Frankfurt–Munster (BFM) approach) for both localized and advanced lymphoblastic lymphoma. Patients with localized LL receive Induction protocol I, Consolidation protocol M and then proceed directly to maintenance therapy. They do not receive prophylactic cranial radiation or a re-induction phase as do patients with advanced LL.

Results

1. Localized T-lymphoblastic lymphoma: 5-year event-free survival is 100%.
2. Advanced T-lymphoblastic lymphoma: 6-year event-free survival is 90% + 3% for stage III and 95% + 5% for stage IV (Figure 20-2).

B LINEAGE NHL (BC, BLL AND DLCL)

Low Risk (Localized BL,BLL and DLCL)

Patient Eligibility

1. Completely resected stage I or abdominal stage II disease.

Treatment

1. Two cycles of COPAD, second cycle given when ANC $> 1,000/mm^3$ and platelet count $>100,000/mm^3$:

C	Cyclophosphamide	$250 \ mg/m^2$/dose every 12 hours as a 15 minute IV infusion for 6 doses on days 1, 2 and 3. First dose prior to doxorubicin infusion on day 1. Hydration at $125 \ ml/m^2$/hour until 12 hours after last cyclophosphamide.
O	Vincristine	$2 \ mg/m^2$ (maximum 2 mg) IV on days 1 and 6.
P	Prednisone	$60 \ mg/m^2$ divided twice daily PO/IV days 1–5 inclusive then taper over 3 days.
AD	Doxorubicin	$60 \ mg/m^2$ as a 6-hour infusion on day 1.
G-CSF		5 micrograms/kg/day subcutaneously from day 7 until ANC $>3,000/\mu l$.

Results

Excellent event-free survival and overall survival rates of 98.3 and 99.2%, respectively, occur in pediatric patients with localized B-cell NHL, treated with two cycles of COPAD.

Intermediate Risk: BL, BLL and DLCL

Patient Eligibility

1. Incompletely resected stage I.
2. All other stage II, (except completely resected abdominal stage II).
3. Stage III.
4. Stage IV (CNS negative, bone marrow with less than 25% blasts).

Treatment

See Table 20-7 and Figure 20-3 for details.

Results

1. All stages: 90.2% 4-year event-free survival.
2. Stage I and II non-resected: 98.4% 4-year event-free survival.
3. Stage III: 89.8% 4-year event-free survival.

Table 20-6　Therapy Non-B Lymphoma (BFM-NHL Protocol)

(Read in Conjunction with Figure 20-1)

Drug	Dose	Days when administered[a]
Induction protocol I[b]		
Prednisone (PO)	60 mg/m^2	1 to 28, then taper over 9 days
Vincristine (IV)	1.5 mg/m^2 (maximum dose, 2 mg)	8, 15, 22, 29
Daunorubicin (IV over 1 h)	30 mg/m^2	8, 15, 22, 29
Asparaginase (IV over 1 h)	10,000 units/m^2	12, 15, 18, 21, 24, 27, 30, 33
Cyclophosphamide (IV over 1 h)[c]	1,000 mg/m^2	36, 64
Cytarabine (IV)	75 mg/m^2	38–41, 45–48, 52–55, 59–62
Mercaptopurine (PO)	60 mg/m^2	36 to 63
MTX (IT)[d,e]	12 mg	1,15,29,45,59
Consolidation protocol M[f]		
Mercaptopurine (PO)	25 mg/m^2	1 to 56
MTX (24-h infusion)[g]	5 g/m^2	8, 22, 36, 50
MTX (IT)[e]	12 mg	8, 22, 36, 50
Reinduction protocol II		
Dexamethasone (PO)	10 mg/m^2	1 to 21, then taper over 9 days
Vincristine (IV)	1.5 mg/m^2 (maximum dose, 2 mg)	8, 15, 22, 29
Doxorubicin (IV)	30 mg/m^2	8, 15, 22, 29
Asparaginase (IV)	10,000 units/m^2	8, 11, 15, 18
Cyclophosphamide (IV over 1 h)[c]	1,000 mg/m^2	36
Cytarabine (IV)	75 mg/m^2	38 to 41, 45 to 48
Thioguanine (PO)	60 mg/m^2	36 to 49
MTX (IT)[e]	12 mg	38, 45
Maintenance		
Mercaptopurine (PO)	50 mg/m^2/day	Daily[h]
Methrotrexate (PO)	20 mg/m^2/day	Weekly[h]

[a]Adjustments of time schedule are made for clinical condition and marrow recovery.

[b]Patients with less than 70% response at day 33 $\frac{\text{Initial tumor volume} - \text{tumor volume day 33}}{\text{Initial tumor volume}} \times 100$ were treated as high risk ALL.

[c]with Mesna.

[d]Additional doses at days 8 and 22 for CNS-positive patients.

[e]Doses adjusted for children less than 3 years of age as follows:

Methotrexate intrathecal

Age (years)	Dose (mg)
>3	12
2–3	10
1–2	8
Up to 1	6

[f]If residual tumor present after induction, it should be resected. It viable tumor present, treat with high-risk ALL therapy.

[g]Methotrexate with leucovorin (CF) rescue.

　Methotrexate (MTX): 5 g/m^2 is administered IV over 24 hours as follows: 10% of the MTX dose of 5 g/m^2 (500 mg/m^2) is administered IV over 30 minutes and 90% of the dose (4,500 mg/m^2) is administered over 23.5 hours as a continuous infusion.

CF Rescue: 30 mg/m2 IV at hour 42 followed by 6 doses of 15 mg/ m2 every 6 hours IV/PO until serum MTX level is <0.4 µmolar.

[h]Total duration of therapy: 24 months.

Cranial radiation (CRT): Postpone CRT for patients <1 year of age with overt CNS involvement until second year of life and give a total dose of 18 Gy. Older children receive 24 Gy. Prophylactic CRT is given to all patients without CNS involvement using 12 Gy. Infants under 1 year are not irradiated prophylactically. Timing: CRT is delivered during the second phase of the re-induction protocol, ie., 27–29 weeks from the time of initial diagnosis. Note that CRT is given only for patients with stage III and IV disease.

Testicular involvement: In males with biopsy proven persistent testicular involvement after protocol M deliver 20 Gy to the testes.

From: Reiter A, Schrappe M, Ludwig W-D, et al. On behalf of the Berlin-Frankfurt-Münster Group. Intensive ALL-type therapy without local radiotherapy provides a 90% event-free survival for children with T-cell lymphoblastic lymphoma: a BFM Group report. Blood 2000;95 416–421.

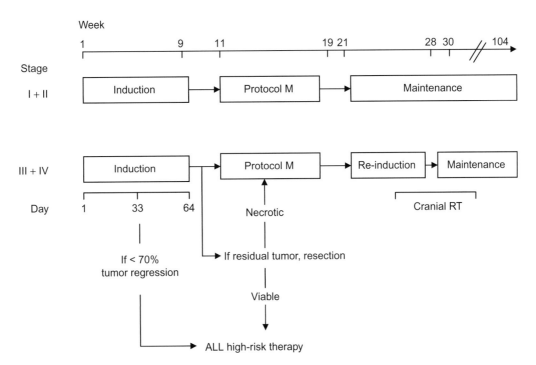

Figure 20-1 Berlin-Frankfurt-Munster (BFM) Treatment Strategy for Lymphoblastic Lymphoma (read in conjunction with Table 20-6).

Abbreviation: RT, radiotherapy.

From: Reiter A, Schrappe M, Ludwig W-D, et al. On behalf of the Berlin-Frankfurt-Münster Group. Intensive ALL-type therapy without local radiotherapy provides a 90% event-free survival for children with T-cell lymphoblastic lymphoma: a BFM Group report. Blood 2000;95:416–421.

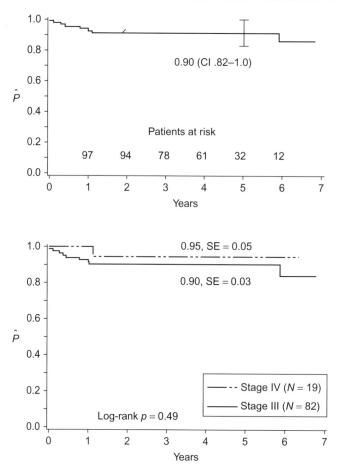

Figure 20-2 Probability of Duration of Event-free Survival NHL-BFM 90 Trial.
Adapted from: Reiter A, Schrappe M, Ludwig W-D, et al. On behalf of the Berlin-Frankfurt-Münster Group. Intensive ALL-type therapy without local radiotherapy provides a 90% event-free survival for children with T-cell lymphoblastic lymphoma: a BFM Group report. Blood 2000;95:416–421.

4. Stage IV: 85.6% 4-year event-free survival.
5. LDH less than twice normal: 95.8% 4-year event-free survival.
6. LDH greater than twice normal: 85.7% 4-year event-free survival.

High Risk: (Advanced Stage BL, BLL and DLCL Including B-Cell ALL and B-Cell NHL with Marrow Involvement with or without CNS Involvement)

Patient Eligibility

1. CNS involvement demonstrated by any of the following criteria:
   - Presence of any L3 lymphoblasts in the CSF

- Cranial nerve palsy without other explanation
- Spinal cord compression
- Parameningeal disease
- Intracerebral mass.

2. Greater than 25% lymphoblasts present in bone marrow.

Treatment

1. See Table 20-8 and Figure 20-3 for details.

Table 20-7 Protocol for Intermediate Risk BL, BLL and DLCL (LMB-96 Protocol)

Reduction phase		
COP		
Cyclophosphamide	300 mg/m^2 IV	Day 1
Vincristine	1 mg/m^2 (maximum 2 mg) IV	Day 1
Prednisolone	60/mg/m^2/day PO or IV	Days 1–7
Methotrexate (MTX) and Hydrocortisone (HC)	15 mg IT[a] (dose adjusted for patients <3 years of age)	Day 1
Induction: 2 courses of COPADM started on day 8 after first day of reduction (COP) phase		
COPADM 1 & 2		
Vincristine	2 mg/m^2 (maximum 2 mg) IV	Day 1
MTX high dose (HD)	3 g/m^2 (over 3 hours) IV	Day 1
Folinic acid (FA)	15 mg/m^2 PO q 6 hr × 12 doses[b]	Days 2–4
MTX and HC	15 mg IT[a]	Day 2, Day 6
Cyclophosphamide	250 mg/m^2/dose IV q 12 hr	Days 2–4 (6 doses total)
Doxorubicin	60 mg/m^2 IV 6 hour infusion	Day 2 (after cyclophosphamide)
Prednisolone	60/mg/m^2/day PO or IV	Days 1–5
G-CSF	5 micrograms/kg/day	Day 7 until ANC >3,000/mm^3
CYM 1 & 2		
MTX high dose (HD)	3 g/m^2 (over 3 hours) IV	Day 1
Folinic acid (FA)	15 mg/m^2 PO q6hr × 12 doses[b]	Days 2–4
Cytarabine	100 mg/m^2/day 24 hour continuous IV infusion	Days 2–6, inclusive
MTX	15 mg IT[a]	Day 2
HC	15 mg IT[a]	Day 2, Day 7
Cytarabine	30 mg IT[a]	Day 7
G-CSF	5 micrograms/kg/day	Day 7 until ANC >3,000/mm^3

[a]Age-adjusted doses for intrathecal use

Age	MTX	HC	Cytarabine
<1 year	8 mg	8 mg	15 mg
1 year	10 mg	10 mg	20 mg
2 years	12 mg	12 mg	25 mg
>3 years	15 mg	15 mg	30 mg

[b]Do a serum methotrexate level at 72 hours and if level is >0.1 μM/L continue FA until serum methotrexate level is <0.1 μM/L.

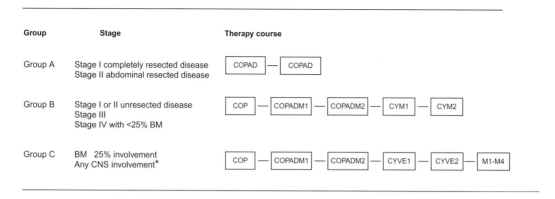

Group	Stage	Therapy course

Figure 20-3 Stratification, Staging and Therapy Criteria for B-NHL in Children.
(Should be read in conjunction with Table 20-7 for intermediate risk or with Table 20-8 for advanced BL, BLL and DLCL disease). *Patients with CNS disease receive additional high-dose methotrexate 8 g/m^2 as well as three additional doses of IT therapy.

2. The addition of Rituximab to the induction and consolidation phases in children with advanced B-cell leukemia/lymphoma being treated with LMB/FAB therapy is currently being investigated.

Results

1. Patients randomized to receive full course therapy had 90% 4-year event-free survival. Reducing the number of maintenance courses resulted in significantly worse outcome with 80% 4-year event-free survival, therefore, the full therapy outlined in Table 20-7 remains the standard
   a. Bone marrow involvement only: 91% 4-year event-free survival
   b. CNS involvement only: 85% 4-year event-free survival
   c. Bone marrow and CNS involvement: 66% 4-year event-free survival
   d. CNS patients who received high-dose methotrexate (without cranial RT) had similar overall survival to those on prior studies who received cranial RT.

Ki-1+ Anaplastic Large cell Lymphoma (ALCL)

Diagnosis

Must carefully distinguish Hodgkin lymphoma from the Hodgkin-like variant of ALCL and differentiate from malignant histiocytosis. Lymphomatoid papulosis may coexist with ALCL (see Chapter 16, Table 16-8).

Treatment

1. See Table 20-9 and Figure 20-4 for details

Table 20-8   Protocol for Advanced BL, BLL and DLCL (LMB-96 Protocol)

Reduction Phase:

COP:

Cyclophosphamide	300 mg/m^2 IV (over 15 minutes)	Day 1
Vincristine	1 mg/m^2 (maximum 2 mg) IV	Day 1
Prednisolone	60/mg/m^2/day PO or IV	Days 1-7 (then 3 day taper)
Methotrexate (MTX)	15 mg IT[a] (age adjusted)	Days 1, 3, 5
Hydrocortisone (HC)	15 mg IT[a] (age adjusted)	Days 1, 3, 5
Cytarabine	30 mg IT[a] (age adjusted)	Days 1, 3, 5
Folinic acid	15 mg/m^2 PO q12h	Days 2 and 4

Induction: 2 courses of COPADM started on day 8 after first day of reduction (COP) phase

COPADM 1

Vincristine	2 mg/m^2 (maximum 2 mg) IV	Day 1
MTX high dose (HD)	8 g/m^2 (over 4 hours) IV	Day 1
Folinic acid (FA)	15 mg/m^2 PO q6hr × 12 doses[b]	Days 2-4
MTX and HC	15 mg IT[a] (age adjusted)	Days 2, 4, 6
Cytarabine	30 mg IT[a] (age adjusted)	Days 2, 4, 6
Cyclophosphamide	250 mg/m^2/dose IV q12hr	Days 2-4
Doxorubicin	60 mg/m^2 IV 6 hour infusion	Day 2 (after cyclophosphamide)
Prednisolone	60/mg/m^2/day PO or IV	Days 1-5 (then 3 day taper)
G-CSF	5 micrograms/kg/day SQ	Day 7 until ANC >3,000/mm^3

COPADM 2 Begin when ANC >1,000/cm^3 and platelet count >100,000/cm^{3c,d}

CYVE 1 & 2 Begin when ANC >1,000/cm^3 and platelet count >100,000 cm^{3c}

Cytarabine	50 mg/m^2/day as 12 h infusion (hours 1-12)	Days 1-5, inclusive
Cytarabine (HD)	3,000 mg/m^2/day over 3 hours (hours 13-15)	Days 2-5, inclusive
Etoposide	200 mg/m^2/day IV over 2 hours (hours 17-19)	Days 2-5, inclusive
G-CSF	5 micrograms/kg/day SQ	Day 7 until ANC >3,000/mm^3

Maintenance 1: Begin when ANC > 1,000/μl and platelets >100,000/μl[b]

Vincristine	2 mg/m^2 (maximum 2 mg) IV	Day 1
MTX high dose (HD)	8 g/m^2 (over 4 hours) IV	Day 1
Folinic acid (FA)	15 mg/m^2 PO q6hr × 12 doses[b]	Days 2-4
MTX and HC	15 mg IT[a] (age adjusted)	Day 2
Cytarabine	30 mg IT[a] (age adjusted)	Day 2
Cyclophosphamide	500 mg/m^2 IV over 30 minutes	Days 2 and 3
Doxorubicin	60 mg/m^2 IV 6 hour infusion	Day 2 (after cyclophosphamide)
Prednisolone	60/mg/m^2/day PO or IV	Days 1-5 (then 3 day taper)

Maintenance 2: Begin when ANC >1,000/cm^3 and platelet count >100,000/cm^3

Cytarabine	50 mg/m^2/dose SQ injection every 12 hours	Days 1-5, inclusive
Etoposide	150 mg/m^2 IV over 90 minutes	Days 1-3, inclusive

(Continued)

Table 20-8   (Continued)

Maintenance 3: Begin when ANC >1,000/cm^3 and platelet count >100,000/cm^3

Vincristine	2 mg/m^2 (maximum 2 mg) IV	Day 1
Cyclophosphamide	500 mg/m^2 IV over 30 minutes	Days 1 and 2
Doxorubicin	60 mg/m^2 IV 6 hour infusion	Day 1 (after cyclophosphamide)
Prednisolone	60/mg/m^2/day PO or IV	Days 1–5 (then 3 day taper)

Maintenance 4: Begin when ANC >1,000/cm^3 and platelet count >100,000/cm^3

Cytarabine	50 mg/m^2/dose SQ injection every 12 hours	Days 1–5, inclusive
Etoposide	150 mg/m^2 IV over 90 minutes	Days 1–3, inclusive

[a]Age-adjusted doses for intrathecal use.

Age	MTX	HC	Cytarabine
<1 year	8 mg	8 mg	15 mg
1 year	10 mg	10 mg	20 mg
2 years	12 mg	12 mg	25 mg
>3 years	15 mg	15 mg	30 mg

[b]Do a serum methotrexate level at 72 hours and if level is >0.1 µM/L then continue FA until serum methotrexate level is <0.1 µM/L.
[c]G-CSF should not be administered for 48 hours prior to beginning of next course.
[d]If no hematologic recovery by Day 25, perform bone marrow aspirate. If blasts present, begin COPADM 2.

2. Stage I and II resected disease – treatment is Strategy 1, which consists of a cytoreductive chemotherapy phase followed by three courses of alternating chemotherapy courses A, B, A.

3. Stage II unresected disease and stage III – treatment is Strategy 2 consisting of a cytoreductive chemotherapy phase followed by six alternating chemotherapy courses A and B.

4. Stage IV (or multiple bone involvement) Ki-1+ ALCL – treatment is Strategy 3 consisting of a cytoreductive chemotherapy phase followed by six alternating chemotherapy courses AA, BB and CC.

Results

1. Patients stratified to Strategy 1: 100% 5-year event-free survival.
2. Patients stratified to Strategy 2: 73% 5-year event-free survival.
3. Patients stratified to Strategy 3: 79% 5-year event-free survival.
4. Composite data from BFM, French SFOP and UKCCSG series for systemic ALCL is: 69% 5-year event-free survival and 81% overall survival. Prognostic factors determined

Table 20-9 **Treatment of Ki-1+ Anaplastic Large Cell Lymphoma (BFM-NHL 90 Protocol)**
(Read in Conjunction with Figure 20-4)

Drug	Dose	Days Administered
Cytoreductive Phase (course V)		
Prednisone (orally/IV)	30 mg/m^2	1–5
Cyclophosphamide (IV)	200 mg/m^2	1–5
Course A		
Dexamethasone (orally)	10 mg/m^2	1–5
Ifosfamide (1 hour infusion)	800 mg/m^2	1–5
Methotrexate (24 hour infusion)[a]	500 mg/m^2	1
Methotrexate (IT)[b]	12 mg[b]	1
Cytarabine (IT)[b]	30 mg[b]	1
Prednisolone (IT)[b]	10 mg[b]	1
Cytarabine (1 hour infusion)	150 mg/m^2 q12h	4, 5
Etoposide (1 hour infusion)	100 mg/m^2	4, 5
Course B		
Dexamethasone (orally)	10 mg/m^2	1–5
Cyclophosphamide (1 hour infusion)	200 mg/m^2	1–5
Methotrexate (24 hour infusion)[a]	500 mg/m^2	1
Methotrexate (IT)[b]	12 mg[b]	1
Cytarabine (IT)[b]	30 mg[b]	1
Prednisolone (IT)[b]	10 mg[b]	1
Doxorubicin (1 hour infusion)	25 mg/m^2	4, 5
Course AA		
Dexamethasone (orally)	10 mg/m^2	1–5
Ifosfamide (1 hour infusion)	800 mg/m^2	1–5
Vincristine (IVP)	1.5 mg/m^2 (maximum, 2 mg)	1
Methotrexate (24 hour infusion)[a]	5 g/m^2	1
Methotrexate (IT)[b]	6 mg[b]	1, 5
Cytarabine (IT)[b]	15 mg[b]	1, 5
Prednisolone (IT)[b]	5 mg[b]	1, 5
Etoposide (1 hour infusion)	100 mg/m^2	4, 5
Course BB		
Dexamethasone (orally)	10 mg/m^2	1–5
Cyclophosphamide (1 hour infusion)	200 mg/m^2	1–5
Vincristine (IV)	1.5 mg/m^2 (maximum, 2 mg)	1
Methotrexate (24 hour infusion)[a]	5 g/m^2	1
Methotrexate (IT)[b]	6 mg[b]	1, 5
Cytarabine (IT)[b]	15 mg[b]	1, 5
Prednisolone (IT)[b]	5 mg[b]	1, 5
Doxorubicin (1 hour infusion)	25 mg/m^2	4, 5
Course CC		
Dexamethasone (orally)	20 mg/m^2	1–5
Vindesine (IV)[c]	3 mg/m^2 (maximum, 5 mg)	1
Cytarabine (3 hour infusion)	2 gram/m^2 q12h	1, 2
Methotrexate (IT)[b]	12 mg[b]	5
Cytarabine (IT)[b]	30 mg[b]	5
Prednisolone (IT)[b]	10 mg[b]	5

(Continued)

Table 20-9   (Continued)

Drug	Dose	Days Administered
Etoposide (1 hour infusion)	150 mg/m^2	3–5

[a]IV infusion of MTX and Leucovorin (CF) rescue: In courses A and B, 10% of the MTX dose of 500 mg/m^2 (=50 mg/m^2) is administered IV over 30 minutes and 90% (=450 mg/m^2) as a 23.5 hour continuous IV infusion. Leucovorin 15 mg/m^2 is administered as a bolus at 48, 51 and 54 hours.

In courses AA and BB, the MTX dose is increased to 5 g/m^2. The infusion schedule and leucovorin rescue are the same as in Non-B Lymphoma protocol (see Table 20-6 footnote).

[b]Doses adjusted for children <3 years of age; given 2 hours after beginning of Methotrexate infusion.

[c]May be replaced with Vinblastine 6 mg/m^2 if Vindesine is not commercially available.

Age-adjusted doses of IT Methotrexate (MTX), Ara-C and Prednisolone in courses AA, BB:

Age	MTX	Ara-C	Prednisolone
<1 year	3	8	2 mg
1–2 years	4	10	3 mg
2–3 years	5	13	4 mg
≥3 years	5	15	5 mg

Age adjusted dose of IT Methotrexate (MTX), Ara-C and Prednisolone in courses A, B, CC:

Age	MTX	Ara-C	Prednisolone
<1 year	6	16	4 mg
1–2 years	8	20	6 mg
2–3 years	10	26	8 mg
≥3 years	12	30	10 mg

From research originally published in: Seidemann K, Tiemann M, Schrappe M, et al. Short-pulse B-non Hodgkin lymphoma-type chemotherapy is efficacious treatment for pediatric anaplastic large cell lymphoma: a report of the Berlin-Frankfurt-Munster Group Trial NHL-BFM 90. Blood 2001;97:3699–3706.

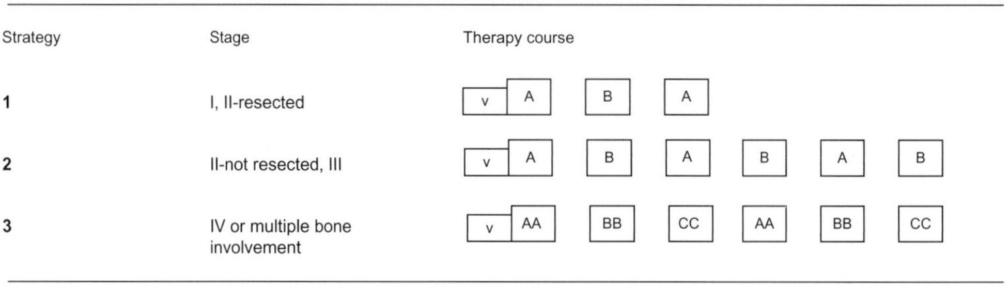

Figure 20-4 Therapy, Strategy and Stratification Criteria for ALCL (NHL-BFM 90 Protocol) (Read in Conjunction with Table 20-9).

v-cytoreductive phase.

From: Seidemann K, Tiemann M, Schrappe M, et al. Short-pulse B-non Hodgkin lymphoma-type chemotherapy is efficacious treatment for pediatric anaplastic large cell lymphoma: a report of the Berlin-Frankfurt-Munster Group Trial NHL-BFM 90. Blood 2001;97:3699–3706.

to have increased risk of failure were found to be mediastinal involvement, visceral involvement and skin lesions.

5. Table 20-10 provides a comparison of outcomes in advanced ALCL in various trials.

Radiation Therapy

Generally, radiotherapy is not indicated on an elective basis. It is only indicated for acute life-threatening complications refractory to initial chemotherapy, such as superior mediastinal syndrome. Doses and fields should be limited. It is used in many of the protocols to treat advanced-stage lymphoblastic lymphoma with CNS involvement. In B-NHL with CNS disease, cranial irradiation should not be used nor should it be used for CNS prophylaxis. Radiation should be used for the treatment of acute paraplegia.

Surgical Therapy

Although surgical biopsy is critical in the original diagnosis, as well as for evaluation of residual masses and response to therapy, the role of surgery in NHL as a treatment is limited. It should be performed on patients in whom there is good reason to believe that total resection can be achieved (e.g., localized bowel disease) without a mutilating procedure (e.g., amputation, extensive maxillofacial surgery, exenterations) or an excessively risky procedure. Patients with widespread lymphoma are not eligible for surgical resection.

Table 20-10   Advanced Anaplastic Large Cell Lymphoma in Children

	BFM Seidemann et al.	POG Laver et al.	SFOP Brugieres et al.	St. Jude Sandlund et al.	MSKCC Mora et al.	CCG Lowe et al.	EICNHL Deley et al.	EICNHL Brugieres et al.
Patients (N)	89	67	82	18	19	86	225	352
Protocol	NHL-BFM 90	POG 9315	HM89-91	CHOP-based	LSA_2L_2, LSA_4	CCG-5941	BFM, SFOP, UKCCSG	Modified NHL-BFM 90
Duration (months)	2–5	12	7–8	6–18	14–36	12	2–8	4–12
EFS (Est) 2–5 yr	76%	73%	66%	57%	56%	68%	69%	75%
OS 2–5 yr	NR	93%	83%	84%	84%	80%	81%	94%

Abbreviations: EFS, event-free survival; Est, estimate; OS, overall survival.
Adapted from: Carroll WL, Finlay JL. Cancer in Children and Adolescents, Chapter 14, copyright 2010: Jones and Bartlett Publishers, Sudbury, MA. www.jbpub.com. Reprinted with permission.

Prognosis

Prognosis depends on the appropriateness of protocols. Expected event-free survival has been mentioned with each category of treatment in the above discussion. Table 20-11 lists clinical and biological factors associated with favorable and unfavorable prognosis.

Management of Relapse

Relapse indicates a poor prognosis, regardless of the site of relapse, tumor histology, other original prognostic factors, prior therapy, or time from diagnosis to relapse. For this reason, the selection of most effective front-line treatment is critical. Relapsed patients are first treated to induce remission. For lymphoblastic lymphoma and B-cell lymphomas (see Table 20-12), chemotherapy with ifosfamide, carboplatin and etoposide (VP-16) is a useful salvage regimen following which autologous stem cell harvest is generally easily achieved. The addition of Rituximab in patients with B-lineage lymphomas that are CD20+ may help induce remission.

Table 20-11   Factors Associated with Favorable and Unfavorable Prognosis*

Favorable
Localized (stage I and II)
Low LDH (less than 500 or twice the upper normal level)
Nonvisceral ALCL (no mediastinal, hepatic/splenic or skin)
Unfavorable
High LDH (greater than 500 or twice the upper normal level)
Combined bone marrow and central nervous system B-NHL
Poor response to cyclophosphamide, vincristine, prednisone (COP) reduction therapy in B-NHL
Primary mediastinal B cell lymphoma as primary B-NHL
Visceral anaplastic large cell lymphoma (mediastinal, hepatic/splenic or skin)
Minimal disease in bone marrow at diagnosis

*All the above variables essentially reflect tumor burden – the higher the burden the worse the prognosis.
IL-2R is a more specific marker of lymphoid malignancy than LDH and can be used as a marker to differentiate small round cell tumors of lymphoid origin from nonlymphoid origin.
Abbreviations: ALCL, anaplastic large cell lymphoma; LDH, lactate dehydrogenase.

Table 20-12   Treatment for Relapsed NHL

• Rituximab 375 mg/m^2 IV on Days 1 and 3 of Course 1 and 2 and on Day 1 of Course 3, if given
• Ifosfamide 3,000 mg/m^2/day as a 2 hour IV infusion daily for 3 days (Days 3,4,5) with Mesna 600 mg/m^2 IV immediately with ifosfamide and as a bolus at hours 3, 6, 9 and 12 post-ifosfamide)
• Carboplatin 635 mg/m^2 as 1 hour IV infusion on Day 3 only
• VP-16 100 mg/m^2/day as 1 hour IV infusion daily for 3 days (Days 3,4,5)
• Repeat second cycle 3 weeks from the first cycle, then reevaluate radiographically
– If in complete remission or very-good partial remission, proceed to consolidation with transplantation
– If responsive, may give 3rd cycle prior to consolidation with transplantation
– If resistant, alternate therapies indicated

After induction of complete or very good partial remission, consolidation with stem cell transplantation is usually indicated. Patients who are chemotherapy-resistant are unlikely to be cured using autologous stem cell transplantation. ALCL may require less aggressive salvage therapy. In this case, autologous stem cell transplant does not provide added benefit. Patients who have a second relapse may benefit from alternative therapies such as vinblastine (6 mg/m^2/week for 12–18 months) or Cis-retinoic acid (1 mg/kg/day orally in 2–3 divided doses for 8 weeks), with or without interferon alpha (4.5 mega units (MU) subcutaneously every other day for 8 weeks). Both of these approaches have been reported to maintain prolonged remissions in some patients.

Adoptive immunotherapy using allogeneic stem cell transplantation is theoretically preferable as numerous studies demonstrate a substantially lower relapse rate as a result of graft-versus-lymphoma effects. There is a tendency toward improved survival with allogeneic compared to autologous transplantation; however, the increased risk of allogeneic transplant related mortality reduces this advantage somewhat. Disease-free survival for allogeneic transplants ranges from 24–68% and in autologous transplants from 22–46%. Myeloablative chemotherapy regimens such as CBV (Chapter 29, Table 29-9) or BEAM (Chapter 29, Table 29–10) have been utilized followed by stem cell transplantation. Other myeloablative regimens that have been utilized consist of total body irradiation, etoposide and cyclophosphamide. Post-transplantation interleukin-2 may be used to reduce recurrence rate. Current strategies addressing the increased risk of allogeneic transplant utilize reduced-intensity (nonablative) conditioning with Busulfan and Fludarabine (see Chapter 29, p. 818) which significantly reduces transplant-related mortality for improved overall long-term outcomes.

Suggested Reading

Bradley MB, Cairo MS. Stem cell transplantation for pediatric lymphoma: past, present and future. Bone Marrow Transplant 2008;41(2):149–158.

Brugieres L, Le Deley M-C, Rosolen A, et al. Impact of the methotrexate administration dose on the need for intrathecal treatment in children and adolescents with anaplastic large-cell lymphoma: results of a randomized trial of the EICNHL group. J Clinical Oncol. 2009;27(6):897–903.

Cairo MS. Non-Hodgkin's Lymphoma and Lymphoproliferative Disorders in Children. In: Carroll WL, Finlay JL, eds. Cancer in Children and Adolescents. Jones & Bartlett Publishers; 2009.

Cairo MS, Gerrard M, Sposto R, et al. Results of a randomized international study of high-risk central nervous system B non-Hodgkin lymphoma and B acute lymphoblastic leukemia in children and adolescents. Blood 2007;109:2736–2743.

Cairo MS, Sposto R, Perkins S, et al. Burkitt's and Burkitt-like lymphoma in children and adolescents: a review of the Children's Cancer Group Experience. Br J Haematol. 2003;120:660–670.

Gerrard M, Cairo MS, Weston C, et al. Excellent survival following two courses of COPAD chemotherapy in children and adolescents with resected localized B-cell non-Hodgkin's lymphoma: results of the FAB/LMB 96 international study. Br J Haematol. 2008.

Goldman SC, Holcenberg JS, Finklestein JZ, et al. A randomized comparison between rasburicase and allopurinol in children with lymphoma or leukemia at high risk for tumor lysis. Blood 2001;97(10):2998–3003.

Griffin TC, Weitzman S, Weinstein H, et al. A study of rituximab and ifosfamide, carboplatin and etoposide chemotherapy in children with recurrent/refractory B-cell (CD20+) non-Hodgkin lymphoma and mature B-cell acute lymphoblastic leukemia: A report from the Children's Oncology Group. Pediatr Blood Cancer. 2009;52(2):177–181.

Hochberg J, Cairo MS. Tumor lysis syndrome: current perspective. Haematologica 2008;93(1):9–13.

Hochberg J, Waxman I, Kelly K, et al. Adolescent non-Hodgkin lymphoma and Hodgkin lymphoma: state of the science. Br J Haematol. 2009;144(1):24–40.

Le Deley MC, Reiter A, Williams D, et al. Prognostic factors in childhood anaplastic large cell lymphoma: results of a large European intergroup study. Blood 2008;111:1560–1566.

Link MP, Shuster JJ, Donaldson SS, et al. Treatment of children and young adults with early stage non-Hodgkin's Lymphoma. New Engl J Med. 1997;337:1259–1266.

Lowe EJ, Sposto R, Perkins SL, et al. Intensive chemotherapy for systemic anaplastic large cell lymphoma in children and adolescents: final results of Children's Cancer Group Study 5941. Pediatr Blood Cancer 2009;52(3):335–339.

Magrath IT. Malignant Non-Hodgkin's Lymphomas in Children. In: Pizzo PA, Poplack DG, eds. Principles and Practice of Pediatric Oncology. 4th ed. Philadelphia: Lippincott, Williams & Wilkins; 2002.

Neth O, Seidemann K, Jansen P, et al. Precursor B-cell lymphoblastic lymphoma in childhood and adolescence: Clinical features, treatment and results in trials NHL-BFM 86 and 90. Med Pediatr Oncol. 2000;35:20–27.

Patte C, Auperin A, Gerrard M, et al. Results of the randomized international FAB/LMB96 trial for intermediate risk B-cell non-Hodgkin lymphoma in children and adolescents: it is possible to reduce treatment for the early responding patients. Blood 2007;109:2773–2780.

Reiter A, Schrappe M, Ludwig WD, et al. Intensive ALL-type therapy without local radiotherapy provides a 90% event-free survival for children with T-cell lymphoblastic lymphoma: a BFM Group report. Blood 2000;95(2):416–421.

Seidemann K, Tiemann M, Schrappe M, et al. Short-pulse B-non-Hodgkin lymphoma-type chemotherapy is efficacious treatment for pediatric anaplastic large cell lymphoma: a report of the Berlin-Frankfurt-Munster Group Trial NHL-BFM 90. Blood 2001;97(12):3699–3706.

CHAPTER 21

Central Nervous System Malignancies

Tumors of the central nervous system (CNS) are frequently encountered tumors in children; approximately 35 cases per million in children under 15 years of age. Brain tumors are the second most common group of malignant tumors in childhood, accounting for 20% of all childhood malignancies.

Most childhood brain tumors (60–70%) arise from glial cells and tend not to metastasize outside the CNS unless there is operative intervention. The relative frequency in regard to specific tumor types can be seen in Table 21-1. Most are infratentorial in location.

There is an association between primary CNS tumors and the following conditions:

- Neurofibromatosis type 1 (NF-1)
- Tuberous sclerosis
- Von Hippel–Lindau syndrome
- Children of families with Li–Fraumeni syndrome (families with an excess occurrence of breast, brain and lung cancer, sarcomas and adrenocortical carcinoma associated with a germ line mutation of p53, a suppressor oncogene)
- Turcot's syndrome (polyposis of colon and CNS malignancy)
- Gorlin's syndrome (nevus basal cell carcinoma syndrome [NBCCS]), which is characterized by multiple basal cell carcinomas, keratocysts of the jaw, palmar and plantar pits and skeletal abnormalities.

PATHOLOGY

Supratentorial Lesions

1. Cerebral hemisphere: low- and high-grade glioma, ependymoma, meningioma, primitive neuroectodermal tumor (PNET).
2. Sella or chiasm: craniopharyngioma, pituitary adenoma, optic nerve glioma.
3. Pineal region: Pineoblastoma, pineocytoma, germ cell tumors, astrocytoma.

Manual of Pediatric Hematology and Oncology. DOI: 10.1016/B978-0-12-375154-6.00021-5

Table 21-1   Relative Frequency of Common Childhood Brain Tumors

Location and Histology	Frequency (%)
Supratentorial	
Low-grade glioma	16
Ependymoma	3
High-grade glioma	5
Primitive neuroectodermal	3
Tumor (PNET)	2
Other	
Sella/chiasm	
Craniopharyngioma	6
Optic nerve glioma	5
Total	40
Infratentorial	
Cerebellum	
Medulloblastoma	20
Astrocytoma	18
Ependymoma	6
Brain stem	
Glioma	14
Other	2
Total	60

Infratentorial Lesions

1. Posterior fossa: medulloblastoma, glioma (low more frequent than high grade), ependymoma, meningioma.
2. Brain stem tumors: low- and high-grade glioma, PNET.

Ventricular Lesions

1. Choroid plexus papilloma, choroid plexus carcinoma, neurocytoma.

The most useful pathologic classification for brain tumors is based on embryonic derivation and histologic cell of origin. Tumors are further classified by grading the degree of malignancy within a particular tumor type. This grading is useful in astrocytoma and ependymoma. Criteria that are useful microscopically in grading the degree of malignancy include:

- Cellular pleomorphism
- Mitotic index
- Anaplasia and necrosis
- MIB-1 index.

Table 21-2   Cytogenetic Loci Implicated in Malignant Brain Tumors

Tumor	Chromosome	Disorder
PNET/MB	5q21-22	Turcot Syndrome[a]
	9q22.3	NBCCS
	17p13.3	
	17q	
Astrocytoma grade III–IV	7p12	
	17p13.1	Li Fraumeni syndrome
	17q11.2	Neurofibromatosis-1
	3p21, 7p22	Turcot syndrome[a]
Subependymal giant cell tumor	9q34	Tuberous sclerosis
	16p13	
Meningioma/Ependymoma	22q12	Neurofibromatosis-2
AT/RT	22q11.2	
Hemangioblastoma	3p25	Von-Hippel Lindau

[a]A rare and distinctive form of multiple intestinal polyposis associated with brain tumors. Autosomal-recessive inheritance caused by mutation in one of the mismatch repair genes MLH1 or chromosome 3p, MPS2 on chromosome 7p or the adenomatous polyposis gene (APC) on chromosome 5q.
Abbreviations: PNET/MB, primitive neuroectodermal tumor, medulloblastoma; AT/RT, atypical teratoid rhabdoid; NBCCS, nevoid basal cell carcinoma syndrome.
Adapted from: Biegel JA. Cytogenetics and molecular genetics of childhood brain tumors. Neuro-Oncology 1999;1:139–151.

Molecular Pathology of CNS Neoplasms

The majority of CNS neoplasms are sporadic. Only a small percentage is associated with the inherited genetic disorders listed above.

Chromosomal abnormalities have been identified in many pediatric brain tumors. These chromosomal abnormalities are helpful in the pathologic classification especially in the differentiation of PNET from atypical rhabdoid tumors. Identification of the genes involved and how they contribute to tumor genesis is ongoing.

Table 21-2 describes common cytogenetic abnormalities identified in brain tumors.

CLINICAL MANIFESTATIONS

Intracranial Tumors

Symptoms and signs are related to the location, size and growth rate of tumor:

* Slow-growing tumors produce massive shifts of normal structures and may become quite large by the time they first become symptomatic
* Rapidly growing tumors produce symptoms early and present when they are relatively small.

The most common presenting signs and symptoms of an intracranial neoplasm are increased intracranial pressure (ICP) and focal neurologic deficits.

If symptoms and signs of increased ICP precede the onset of localized neurologic dysfunction, a tumor of the ventricles or deep midline structures is most likely. If localizing signs (seizures, ataxia, visual field defects, cranial neuropathies, or corticospinal tract dysfunction) are predominant in the absence of increased ICP it is more probable that the tumor originates in parenchymal structures (cerebral hemispheres, brain stem, or cerebellum).

General Signs and Symptoms of Intracranial Tumors

1. Headache. In young children headache can present as irritability. Often worse in the morning, improving throughout the day.
2. Vomiting.
3. Disturbances of gait and balance.
4. Hemiparesis.
5. Cranial nerve abnormalities.
6. Impaired vision
   a. Diplopia (6th nerve palsy): In young children diplopia may present as frequent blinking or intermittent strabismus
   b. Papilledema from increased ICP may present as intermittent blurred vision
   c. Parinaud syndrome (failure of upward gaze and setting-sun sign, large pupils and decreased constriction to light).
7. Mental disturbances – somnolence, irritability, personality or behavioral change, or change in school performance.
8. Seizures – usually focal.
9. Endocrine abnormalities: Midline supratentorial tumors may cause endocrine abnormalities due to effects on the hypothalamus or pituitary and visual field disturbances as a result of optic pathway involvement.
10. Cranial enlargement in infants (characteristic of increased ICP).
11. Diencephalic syndrome can be seen in patients aged 6 months to 3 years with brain tumors who present with sudden failure to thrive and emaciation. The syndrome is caused by a hypothalamic tumor in the anterior portion of the hypothalamus or the anterior floor of the third ventricle.

Spinal Tumors

Spinal tumors of children may be found anywhere along the vertebral column. They cause symptoms by compression of the contents of the spinal canal. Localized back pain in a child or adolescent should raise suspicion of a spinal tumor, especially if the back pain is

Table 21-3   Major Signs and Symptoms of Spinal Tumors

Back pain (in 50%, increased in supine position or with Valsalva maneuver)
Resistance to trunk flexion
Paraspinal muscle spasm
Spinal deformity (especially progressive scoliosis)
Gait disturbance
Weakness, flaccid or spastic
Reflex changes (especially decreased in arms and increased in legs)
Sensory impairment below level of tumor (30% of cases)
Decreased perspiration below level of tumor
Extensor plantar responses (Babinski signs)
Sphincter impairment (urinary or anal)
Midline closure defects of skin or vertebral arches
Nystagmus (with lesions of upper cervical cord)

worse in the recumbent position and relieved when sitting up. The major signs and symptoms of spinal tumors are listed in Table 21-3. Most spinal tumors have associated muscle weakness and the muscle group affected corresponds to the spinal level of the lesions.

Spinal tumors can be divided into three distinct groups:

* *Intramedullary.* These tumors tend to be glial in origin and are usually gliomas or ependymomas
* *Extramedullary–Intradural.* These tumors are likely to be neurofibromas often associated with neurofibromatosis. If they arise in adolescent females, meningiomas are more likely
* *Extramedullary–Extradural.* These tumors are most often of mesenchymal origin and may be due to direct extension of a neuroblastoma through the intervertebral foramina or due to a lymphoma. Tumors of the vertebra may also encroach upon the spinal cord, leading to epidural compression of the cord and paraplegia (e.g. PNET or Langerhans cell histiocytosis occurring in a thoracic or cervical vertebral body).

DIAGNOSTIC EVALUATION

Computed Tomography

The computed tomography (CT) scan is an important procedure in the detection of CNS malignancies. Scans performed both with and without iodinated contrast agents detect 95% of brain tumors. However, tumors of the posterior fossa, which are common in children, are better evaluated with MRI. CT scans should be performed using thin sections (usually 5 mm). Sedation is often necessary to avoid motion artifacts. CT is more useful than MRI in:

* Evaluating bony lesions
* Detection of calcification in tumor
* Investigating unstable patients because of the shorter imaging time.

Magnetic Resonance Imaging

Magnetic resonance imaging (MRI) provides the following additional advantages:

* No ionizing radiation exposure (especially important in multiple follow-up examinations)
* Greater sensitivity in detection of brain tumors especially in the temporal lobe and posterior fossa (these lesions are obscured by bony artifact on CT)
* Ability to directly image in multiple planes (multiplanar), which is of value to neurosurgical planning (CT is usually only in axial planes, though reconstructions may be performed)
* Ability to apply different pulse sequences which is useful in depicting anatomy (T1-weighted images) and pathology (T2-weighted images)
* Ability to map motor areas with functional MRI.

MRI specificity is enhanced with the contrast agent gadolinium diethylenetriaminepentaacetic acid dimeglumine (Gd-DTPA), which should be used in the evaluation of childhood CNS tumors and has the following advantages:

* Highlights areas of blood–brain barrier breakdown that occur in tumors
* Useful in identifying areas of tumor within an area of surrounding edema
* Improves the delineation of cystic from solid tumor elements
* Helps to differentiate residual tumor from gliosis (scarring).

The major difficulty with MRI in infants and children is the long time required to complete imaging and for this reason adequate sedation is required.

Magnetic Resonance Angiography

Magnetic resonance angiography (MRA) has been utilized in the pre-operative evaluation of the normal anatomic vasculature (e.g. dural sinus occlusion) but has not been particularly useful in the assessment of tumor vascularity.

Magnetic Resonance Spectroscopy

Magnetic resonance spectroscopy may be helpful in both diagnosing pediatric brain tumors and during follow-up investigations. This technique has been shown to be able to distinguish between malignant tumors and areas of necrosis by comparing creatine/choline ratios with n-acetyl aspartate/choline ratios. This technique, in combination with tumor characteristics as identified by MRI, tumor site and other patient characteristics, may be able to more accurately predict the tumor type preoperatively. In addition, it may be helpful in identifying postoperative residual tumor from postoperative changes.

Positron Emission Tomography

Positron emission tomography (PET) is a potentially useful technique for evaluating CNS tumors. The fluorine-18-labeled analog of 2-deoxyglucose (FDG) is used to image the metabolic differences between normal and malignant cells. Astrocytomas and oligo-dendrogliomas are generally hypometabolic whereas anaplastic astrocytomas and glioblastoma multiforme are hypermetabolic.

PET is useful to determine:

- Degree of malignancy of a tumor
- Prognosis of brain tumor patients
- Appropriate biopsy site in patients with multiple lesions, large homogeneous and heterogeneous lesions
- Recurrent tumor from necrosis, scar and edema in patients who have undergone radiation therapy and chemotherapy
- Recurrent tumor from post-surgical change.

Evaluation of the Spinal Cord

MRI and Gd-DTPA has replaced myelography in the evaluation of meningeal spread of brain tumors in the spinal column and delineating spinal cord tumors.

Cerebrospinal Fluid Examination

The cerebrospinal fluid (CSF) should have the following studies performed:

- Cell count with cytocentrifuge slide examination for cytology of tumor cells
- Glucose and protein content
- CSF α-fetoprotein (AFP)
- CSF human chorionic gonadotropin (hCG).

Polyamine assays in the CSF are of value in the evaluation of tumors that are in close proximity to the circulating CSF (medulloblastoma, ependymoma, brain stem glioma). The assay is not useful in glioblastoma multiforme and not predictive in astrocytomas. AFP and HCG of the CSF may be elevated in CNS germ cell tumors.

Bone Marrow Aspiration and Bone Scan

These studies are indicated in medulloblastoma and high-grade ependymomas with evidence of cytopenias on the blood count because a small percentage of these patients have systemic metastases at the time of diagnosis.

TREATMENT

Surgery

The purpose of neurosurgical intervention is threefold.

1. To provide a tissue biopsy for purposes of histopathology and cytogenetics.
2. To attain maximum tumor removal with fewest neurologic sequelae.
3. To relieve associated increased ICP due to CSF obstruction.

Use of pre-operative dexamethasone can significantly decrease peritumoral edema, thus decreasing focal symptoms and often eliminating the need for emergency surgery. For patients with increased hydrocephalus that is moderate to severe, endoscopic or standard ventriculostomy can decrease ICP. Tumor resection is safer when performed 1–2 days following reduction in edema and ICP by these means.

Technical advances in neurosurgery that have improved the success of surgery include ultrasonic aspirators; image guidance (allows 3-D mapping of tumor); functional mapping and electrocorticography, which allow pre- and intraoperative differentiation of normal and tumor tissue; and neuroendoscopy. Intraoperative MRI scanning is useful in limited circumstances, when visualization by the neurosurgeon is compromised. Stereotactic biopsies allow biopsy of deep-seated midline tumors. The ideal goal is a gross-total resection of tumor (likely leaving microscopic residual) as removal of a margin of normal tissue would cause devastating neurologic sequelae.

Radiotherapy

Most patients with high-grade brain tumors require radiotherapy to achieve local control of microscopic or macroscopic residual. Radiation therapy for intracranial tumors consists of external beam irradiation using conventional fields or three-dimensional conformal radiotherapy. The latter decreases radiation to normal brain tissue by up to 30%.

Intensity-modulation radiation therapy (IMRT) uses more complex computerized planning and intensity modulation of the radiation to further decrease radiation to normal tissue under certain circumstances.

The wider use of ionizing radiation in pediatric brain tumors has resulted in improved long-term survival. However, the significant long-term effects on cognition and growth, especially in patients requiring craniospinal irradiation (e.g. PNET) can be devastating.

The total dose of radiotherapy depends on:

- Tumor type (which also influences volume of treatment)
- Age of the child
- Volume of the brain or spinal cord to be treated.

Current efforts seek to decrease radiation by conforming better to the tumor and by using chemotherapy in addition. Children 3 years of age are most drastically affected; in this group newer strategies that avoid or delay radiation therapy, by initial treatment with chemotherapy, have promising preliminary results. In medulloblastoma therapy, this approach allows up to 50% cure without ionizing radiation.

Brachytherapy, stereotactic radiosurgery and fractionated stereotactic radiosurgery are alternatives to conventional radiation therapy presently under continuing study and may prove useful in relapsed patients.

Chemotherapy

Chemotherapy plays an expanding role in the management of recurrent disease and in many newly diagnosed patients.

Two anatomic features of the CNS make it unique with respect to the delivery of chemotherapeutic agents.

1. The tight junction of the endothelial cells of the cerebral capillaries – the blood–brain barrier.
2. The ventricular and subarachnoid CSF.

The blood–brain barrier inhibits the equilibration of large polar lipid-insoluble compounds between the blood and brain tissue, while small nonpolar lipid-soluble drugs rapidly equilibrate across the blood–brain barrier. The blood–brain barrier is probably not crucial in determining the efficacy of a particular chemotherapeutic agent, since in many brain tumors the normal blood–brain barrier is impaired. Factors such as tumor heterogeneity, cell kinetics and drug administration, distribution and excretion play a more significant role in determining the chemotherapeutic sensitivity of a particular tumor than the blood–brain barrier. Tumors with a low mitotic index and small growth fraction are less sensitive to chemotherapy; tumors with a high mitotic index and larger growth fraction are more sensitive to chemotherapy.

The CSF circulates over a large surface area of the brain and provides an alternate route of drug delivery; it can function as a reservoir for intrathecal administration or as a sink after systemic administration of chemotherapeutic agents. The rationale of instillation of chemotherapy into the CSF compartment is that significantly higher drug concentrations can be attained in the CSF and surrounding brain tissue. This mode of administration is most applicable in cases of meningeal spread or in those tumors in which the risk of spread through the CSF is high.

Adjuvant chemotherapy is used in select primary brain tumors in addition to recurrent disease. Table 21-4 lists chemotherapy regimens commonly employed in brain tumors. In certain cases chemotherapy allows for decreased radiation doses with equal or improved cure rates. In others adjuvant chemotherapy improves outcome with standard radiation

Table 21-4   Chemotherapy Regimens for Brain Tumors

Regimen	Dosage/Route	Frequency
CCNU	100 mg/m^2 PO, day 1	
Vincristine[a]	1.5 mg/m^2 IV, days 1, 8	Every 6 weeks for 8 cycles
Prednisone	40 mg/m^2 PO, days 1–14	
OR		
CCNU	100 mg/m^2 PO, day 1	
Vincristine[a]	1.5 mg/m^2 IV, days 1, 8	Every 6 weeks for 8 cycles
Procarbazine	100 mg/m^2 PO, days 1–14	
OR		
CCNU	75 mg/m^2 PO, day 1	
Vincristine[a]	1.5 mg/m^2 IV, days 1, 8, 14	Every 6 weeks for 8 cycles
Cisplatin	75 mg/m^2 IV, day 1 over 6 hours	
OR		
Temozolomide	90 mg/m^2/day PO for 42 days	During radiation therapy
Temozolomide	200 mg/m^2/day PO for 5 days	After radiation therapy every 4 weeks for 10 cycles
OR		
Vincristine[b]	0.065 mg/kg IV, days 1, 8, 29, 36	Every 12 weeks for 4 cycles for children ages 24–36 months;
Cyclophosphamide	65 mg/kg IV days 1, 29	For children less than 24 months of
Cis-platinum	4 mg/kg IV over 6 hours day 57	age at diagnosis 8 cycles should
Etoposide	6.5 mg/kg IV over 1 hour days 59, 60	be given[c]

[a]Maximum dose 2 mg.
[b]Maximum dose 1.5 mg.
[c]Children less than 24 months of age at diagnosis are given a longer period of chemotherapy so that if radiotherapy is required they will have reached an age at which they will have less complications from the radiotherapy.

therapy. Trials of new agents, combinations of agents and standard drugs as radio-sensitizing agents are ongoing.

ASTROCYTOMAS

Astrocytomas account for approximately 50% of the CNS tumors with peaks between ages 5–6 and 12–13 years. The WHO grades these tumors I–IV in increasing malignancy. Low-grade tumors (WHO I and II) are distinguished histologically from high-grade astrocytomas (WHO III and IV) by the absence of cellular pleomorphism, high cell density, mitotic activity and necrosis. The following are the histologic subtypes:

- *Pilocytic astrocytoma* has a fibrillary background, rare mitoses and classically, Rosenthal fibers. It usually behaves in a benign fashion. These tumors are well circumscribed and grow slowly (WHO grade I)
- *Diffuse or fibrillary astrocytoma* is more cellular and infiltrative and more likely to undergo anaplastic change (WHO grade II)
- *Anaplastic astrocytoma* is highly cellular with significant cellular atypia. It is locally invasive and aggressive (WHO grade III)
- *Glioblastoma multiforme* demonstrates increased nuclear anaplasia, pseudopalisading and multinucleate giant cells (WHO grade IV).

Pleiomorphic xanthoastrocytomas are usually classified as WHO grade II subtype, but often behave more aggressively.

Pilomyxoid astrocytomas have variable predictability and may present with diffuse disease. They generally respond as WHO grade II tumors.

The majority of cerebellar astrocytomas remain confined to the cerebellum. Very rarely do they have neuraxis dissemination.

Management of Astrocytomas

Low-Grade Astrocytomas (WHO Grades I and II)
Low-grade astrocytomas present with hydrocephalus, focal signs, or seizures.

Surgery
Surgical excision is the initial treatment. Gross total resection is desirable. Pilocytic astrocytomas are slow-growing and well-circumscribed with a distinct margin. These features permit complete resection in 90% of patients with posterior fossa tumors and a majority with hemispheric tumors. By contrast, diffuse low-grade astrocytomas are

infiltrative and are less often completely resected. Diencephalic tumors are amenable to gross total resection in less than 40% of cases. If removal is complete, no further treatment is recommended. Pilomyxoid astrocytomas may be localized or have diffuse leptomeningeal spread. In the latter case, surgical resection of the primary lesion with adjuvant chemotherapy is warranted. Patients with significant residual tumor postoperatively may require further therapy if the risk of subsequent surgery to remove progressive tumor is too great.

Radiotherapy

The ability to re-resect hemispheric lesions usually allows postponement of adjuvant therapy. Radiotherapy is used when chemotherapy has failed in unresectable symptomatic tumors, especially in older patients. Dosing is 5,000–5,500 cGy, depending on age, to the original tumor bed with a 2 cm margin.

In deep midline locations, chemotherapy tends to be used to avoid long-term effects of radiation to vital areas, including the pituitary.

Chemotherapy

Carboplatin and vincristine is recommended in patients with newly diagnosed, progressive low-grade astrocytoma in patients younger than 5 years of age (Table 21-5). Responses occur in 56% of patients. In patients with recurrent disease there is a response rate of 52%.

In patients older than 5 years of age a regimen consisting of eight 6-week cycles of 6-thioguanine, procarbazine, CCNU and vincristine (TPCV) is recommended (Table 21-5). This regimen has an event-free survival of 48.9%.

Use of single-agent vinblastine demonstrated a 42% response rate in phase II trial for recurrent disease and, at the time of writing, is in trials for newly diagnosed patients.

Prognosis

The 5- and 10-year overall survival rates for completely resected low-grade supratentorial astrocytomas treated with surgery alone are 76–100% and 69–100%, respectively. In the posterior fossa these rates approach 100%. In patients with partially resected low-grade astrocytomas who are observed without treatment or who are treated with postoperative radiotherapy the 5- and 10-year survival rates are 67–87% and 67–94%, respectively.

Recurrence

Recurrent low-grade astrocytomas should be approached surgically when possible. If not completely resectable, chemotherapy and/or radiation should be given, depending on prior therapy received.

Table 21-5 Chemotherapy for Low-Grade Astrocytomas and Optic Gliomas

A. Patients Younger than 5 Years of Age

Induction: (One 12-week Cycle)

Week	1	2	3	4	5	6	7	8	9	10	11–12
	V	V	V	V	V	V	V	V	V	V	
	C	C	C	C			C	C	C	C	rest

Maintenance (Eight 6-week Cycles)

Week	1	2	3	4	5–6
	V	V	V		
	C	C	C	C	rest

(C) Carboplatin	175 mg/m^2 IV over 1 hour
(V) Vincristine	1.5 mg/m^2 IV (maximum dose 2 mg)

B. Patients older than 5 years of age

TPCV (Eight 6-week cycles)

Thioguanine	30 mg/m^2/dose orally every 6 hours for 12 doses, Days 1–3, hours 0–66
Procarbazine	50 mg/m^2/dose orally every 6 hours for 4 doses, Days 3–4, hours 60–78
CCNU (Lomustine)	110 mg/m^2/dose once on Day 3, hour 72
Vincristine	1.5 mg/m^2 (0.05 mg/kg in children <12 kg) (maximum dose 2 mg) intravenous push on Days 14 and 28

High-Grade Astrocytomas (WHO Grades III and IV)

Surgery

Complete surgical removal of these tumors is rarely accomplished because of their infiltrative nature. In about 20–25% of cases, the contralateral hemisphere may be involved, due to spread through the corpus callosum. The main purpose of surgery in these cases is to reduce the tumor burden. Patients who have total or near-total resection have longer survival than do patients who have partial resection or simple biopsy. Maximal surgical debulking is recommended unless the neurologic sequelae may be too devastating.

Radiotherapy

Postoperative irradiation increases survival in high-grade astrocytomas. The port should include the tumor bed and a 2-cm margin of normal surrounding tissue. The dose is 5,000–6,000 cGy in children over 5 years of age. The use of stereotactic radiotherapy is being investigated as a technique to increase local tumor doses.

Chemotherapy

In children, multi-agent adjuvant chemotherapy added to postoperative radiation results in a significant (but modest) improvement in disease-free survival compared with postoperative radiation alone. Procarbazine (or prednisone), CCNU and vincristine (Table 21-4) is the current standard, but without a reasonable resection, chemotherapy is ineffective. Progression-free survival is approximately 33% at 5 years. Temozolomide significantly improved survival in adults when given for six cycles after radiotherapy (Table 21-4) but did not change outcome in pediatric patients.

Relapsed High-Grade Astrocytoma

While salvage of patients with relapsed disease is unusual, high-dose chemotherapy (Table 21-6) with autologous stem cell rescue (aSCR) improves outcome. About 20% of patients treated in this manner survive event-free for 8 or more years compared to 0% of patient treated with conventional therapy. Patients who are surgically debulked (<3 cm of residual tumor diameter) and treated with aSCR have a 46% survival compared to 0% for conventional chemotherapy.

MEDULLOBLASTOMA

Medulloblastoma (posterior fossa PNET) is the most common CNS tumor in children, representing approximately 20% of all childhood brain tumors with 80% of the cases presenting before the age of 15 years. The tumor presents in the posterior fossa and widespread seeding of the subarachnoid space may occur. The frequency of CNS spread outside the primary tumor can be as high as 40% at the time of diagnosis. Extraneural spread to bone, bone marrow, lungs, liver, or lymph nodes occurs in approximately 4% of patients.

Staging studies should include MRI of the spine (preferable preoperatively), lumbar CSF cytology, liver function tests and, if clinically symptomatic, bone scan and/or bone marrow examination. Histology and cytogenetics of the original tumor are essential to evaluate for large cell anaplastic subtype and for monosomy 22, which is characteristic of atypical teratoid/rhabdoid tumor. These more aggressive tumors fare poorly when treated as typical medulloblastoma. The Chang staging system (Table 21-7), which evaluates tumor size, local extension and metastases is used to assess prognosis. Patients are divided into

Table 21-6 Conditioning Regimen for Autologous Stem Cell Transplantation

Carboplatin	500 mg/m^2/day IV*	days –8 to –6
Thiotepa	300 mg/m^2/day IV	days –5 to –3
Etoposide	250 mg/m^2/day IV	days –5 to –3
Rest		days –2 to –1
Stem Cell Infusion		day 0
G-CSF	5 µg/kg/day	day 1 until neutrophil engraftment

*Modified to AUC of 7 mg/ml/min by pediatric Calvert formula.

Table 21-7   Chang System for Posterior Fossa Medulloblastoma (PNET)

Classification	Description
T1	Tumor less than 3 cm in diameter and limited to the classic midline position in the vermis, the roof of the fourth ventricle and less frequently to the cerebellar hemispheres
T2	Tumor 3 cm or greater in diameter, further invading one adjacent structure or partially filling the fourth ventricle
T3	Divided into T3a and T3b:
T3a	Tumor further invading two adjacent structures or completely filling the fourth ventricle with extension into the aqueduct of Sylvius, foramen of Magendie, or foramen of Luschka, producing marked internal hydrocephalus
T3b	Tumor arising from the floor of the fourth ventricle or brain stem and filling the fourth ventricle
T4	Tumor further spreading through the aqueduct of Sylvius to involve the third ventricle or midbrain or tumor extending to the upper cervical cord
M0	No evidence of gross subarachnoid or hematogenous metastasis
M1	Microscopic tumor cells formed in cerebrospinal fluid
M2	Gross nodular seeding demonstrated in cerebellar, cerebral subarachnoid space, or in the third or lateral ventricles
M3	Gross nodular seeding in spinal subarachnoid space
M4	Metastasis outside the cerebrospinal axis

From: Laurent J. Classification system for primitive neuroectodermal tumors (medulloblastoma) of the posterior fossa. Cancer 1985;56:1807, with permission.

Table 21-8   Medulloblastoma Risk Categories

	Average Risk	High Risk
Extent of disease	Negative CSF cytology	Positive CSF cytology
	Normal MRI of spine	Positive MRI of spine with Gd-DTPA
Extent of residual tumor[*]	≤ 1.5 cm^2 residual	>1.5 cm^2 residual
Histology	Undifferentiated	Large cell anaplastic
Age at diagnosis	>3 years	<3 years

[*]On a 2-dimensional measurement.

average- and high-risk categories based on extent of disease (i.e. CSF cytology or spinal cord involvement), volume of residual tumor, histology and age at diagnosis (Table 21-8).

Surgery

Surgical excision is employed as the initial therapy with an objective of gross total resection or near total resection with less than 1.5 cm^2 of residual tumor. Disease-free survival improves for patients who have had a radical resection of the tumor. Occasionally a child presents with life-threatening raised intracranial pressure due to obstruction of the fourth ventricle. In these cases, preoperative endoscopic third ventriculostomy or shunting should be performed to reduce the intraoperative mortality of definitive tumor resection in the presence of increased ICP.

Radiotherapy

Radiation therapy plays a critical role in treatment. Medulloblastomas are one of the most radiosensitive primary CNS tumors of childhood. The standard dose of radiotherapy is 5,400–5,580 cGy to the area of the primary tumor, with 2,340 cGy given to the neuroaxis. Use of intensity modulated radiation therapy (IMRT) to spare the cochlea is important in decreasing therapy-induced hearing loss.

Radiation therapy to children under 3 years of age with medulloblastoma is controversial. Chemotherapy should be utilized so that radiation can either be omitted or postponed in this young age group. At the time of writing, the Children's Oncology Group is evaluating the use of chemotherapy and focal radiation in patients 8–36 months. Preliminary results demonstrate a 50% 3-year disease-free survival. Other approaches include consolidation with high-dose chemotherapy followed by autologous stem cell rescue (Table 21-6).

Chemotherapy

The relatively rapid rate of growth and high mitotic index of medulloblastoma result in this tumor's responsiveness to a number of chemotherapeutic agents.

1. In patients who have high-risk disease (Table 21-8) adjuvant chemotherapy in combination with 36 Gy craniospinal radiation improves the disease-free survival for medulloblastoma. The use of high-dose chemotherapy with autologous stem cell rescue (Table 21-6) remains experimental.
2. Patients with average-risk disease (Table 21-8) have a greater than 60% 5-year disease-free survival with surgery and radiotherapy. The Children's Oncology Group study utilizing craniospinal irradiation following surgery and adjuvant CCNU, vincristine and cisplatinum (Table 21-4) chemotherapy has demonstrated an 89% progression-free survival rate in nondisseminated posterior fossa medulloblastoma with decreased craniospinal irradiation. Trials are now underway comparing chemotherapy regimens in average-risk medulloblastoma patients in an attempt to decrease local recurrences and to attempt to further decrease radiation therapy dosages.

Prognosis

Survival for patients with average-risk medulloblastoma treated with adjuvant radiation and chemotherapy is 82% at 5 years. In high-risk patients using craniospinal radiation with chemotherapy, 45–50% of patients are disease-free at 5 years. When chemotherapy dose intensity is increased with more aggressive regimens and high-dose chemotherapy with autologous stem cell rescue (Table 21-6), 2-year disease-free survival data are encouraging at 73–78%.

Relapsed Medulloblastoma

Surveillance scanning is very important after therapy since it can detect most recurrences prior to the onset of symptoms. Studies have shown some improvement of survival in relapsed patients using high-dose carboplatin, thiotepa and etoposide followed by autologous stem cell rescue (Table 21-6), with salvage rates up to 30%.

BRAIN STEM TUMORS

Brain stem gliomas comprise 15–20% of all childhood CNS tumors. The median age at presentation is 6–7 years of age. Fifty percent of the patients present with cranial nerve and long tract signs.

The majority of patients have diffuse, infiltrating pontine tumors, sometimes with an exophytic component. These are typically WHO grades II–IV gliomas with 1-year and 2-year survival rates of <10–20%. A subset of patients have focal lesions that are usually grade I tumors and have 2-year survival rates of about 60% or greater. Localized brainstem tumors may also be PNETs. While sensitivity to chemotherapy as well as radiation is common, prognosis remains poor, at around 10% 2-year survival.

Surgery

Surgical resection is not usually possible because of the proximity to vital structures, limited room for expansion and swelling and possible damage to medullary structures. There is no apparent benefit from a surgical biopsy when the imaging and clinical picture are indicative of a diffuse infiltrating pontine glioma. For focal tumors (nontectal), complete resection may be safe and may not require any further therapy. Partial resection of exophytic tumors will establish the diagnosis and reduce the obstructing mass within the fourth ventricle. If hydrocephalus is present, a CSF diversion should be performed.

Radiotherapy

Limited field irradiation is standard palliative care in patients with infiltrative pontine gliomas. A tumor dose of approximately 5,400 cGy is standard. The treatment field should include the extent of the defined tumor and a 2-cm margin around the tumor. In irradiating these tumors, precaution should be taken to minimize brain stem edema, especially in patients who have not been shunted. The use of high-dose steroids is valuable during treatment and may be required throughout the treatment period. Children with diffusely infiltrating pontine gliomas often initially respond to radiation therapy but progressive disease is usually seen within 8–12 months. Timing of radiation for partially resected focal lesions is unclear. It is not known whether immediate adjuvant therapy is superior to observation and therapy only at the time of progression.

Chemotherapy

The use of combination chemotherapy after radiotherapy has not improved the disease-free survival in brain stem tumors. Until some chemotherapeutic regimen confirms a survival advantage in newly diagnosed patients, chemotherapy currently plays a palliative role.

Prognosis

The overall prognosis for brain stem tumors is poor. Most centers report a 5–20% 2-year survival rate when high doses of irradiation are employed. Children with diffusely infiltrating pontine gliomas often respond initially to radiation therapy but progressive disease is usually seen within 8–12 months. Improved survival is seen in focal low-grade brain stem astrocytomas especially those seen in patients with NF-1 and those in the tectal area.

EPENDYMOMAS

Fifty percent of all ependymomas occur during childhood and adolescence and they constitute approximately 9% of all primary childhood CNS tumors. The tumors occur either infratentorially or supratentorially. The fourth ventricle is the most common location. Ependymomas can also occur in the spinal cord and account for 25% of spinal cord tumors. Obstructive hydrocephalus is the most common presenting condition.

Surgery

Total removal of these tumors is difficult to accomplish, especially in tumors originating from the fourth ventricle. However, since gross total tumor resection predicts a greatly improved outcome, this should be attempted. In the posterior fossa, the use of evoked potentials helps to make this process safer, but patients may still have multiple cranial nerve palsies and may require tracheotomy and gastrostomy for months until normal swallowing recovers. Patients with a subtotal resection should receive two courses of chemotherapy before potential second-look surgery.

Radiotherapy

The recurrence rate with surgery alone is extremely high and postoperative radiotherapy results in a significant increase in survival. Local disease without evidence of subarachnoid spread should receive local irradiation with a margin (that should include extension to lateral ventricles or superior cervical spine when appropriate). Use of 3D-conformal radiation or IMRT is appropriate and is currently under investigation. The dose is 5,400–6,000 cGy for intracranial lesions.

Chemotherapy

No advantage has been demonstrated from the use of adjuvant chemotherapy and although platinum agents are the most active they are not currently part of standard therapy. In infants, chemotherapy is used to postpone radiotherapy. There is some evidence that low dose metronomic chemotherapy (i.e. oral etoposide at 50 mg/m^2/day for 21 or 28 days) can be effective in treating relapsed disease. Optimal length of therapy is not yet determined.

Prognosis

The overall prognosis for ependymomas is dependent on the initial extent of resection, presence or absence of anaplasia and gender. Patients with total resection who receive radiation have a 5-year survival of 67–86% compared to 22–47% 5-year event-free survival for patients who receive radiation after subtotal resection. Patients with anaplasia or male gender have significantly worse event-free survival.

OPTIC GLIOMA

Optic gliomas constitute 5% of the primary CNS tumors in childhood and the majority is diagnosed by 5 years of age with almost all diagnosed by age 20. Involvement of the optic chiasm is usually seen in older children. Neurofibromatosis is present in up to 70% of patients with tumors of the optic nerve or chiasma tumors; isolated optic nerve tumors are more commonly in patients with NF-1. Patients may present with decreased vision, proptosis, optic atrophy and papilledema. In young children asymmetric nystagmus may be the presenting sign of a chiasmatic tumor. Tumors that extend beyond the optic pathway may be associated with hypothalamic dysfunction. Histologically approximately 90% are low-grade astrocytomas.

Surgery

Biopsy may compromise vision, MRI and CT should be used to make a clinical diagnosis and biopsy should be used for unusual circumstances. Surgery should be reserved for tumor extension into the optic canal or increasing visual compromise.

Radiotherapy

The role of radiotherapy in optic gliomas has not been completely established but it may have a beneficial role in preserving vision. Radiation may be indicated in patients with progressive disease instead of surgery, especially for chiasmatic–hypothalamic lesions. Patients with NF-1 may have exacerbation of moya moya, which one must consider before recommending radiation. A local field to the tumor of 4,500–5,000 cGy over 5–6 weeks is usually recommended.

Chemotherapy

Chemotherapy has been used to treat progressive disease. Regimens used include:

- Actinomycin D and vincristine. Actinomycin D 0.015 mg/kg/d for 5 days (0.5 mg maximum per dose) IV every 12 weeks for six cycles. Vincristine 1.5 mg/m^2 (2 mg maximum dose) IV weekly for 8 weeks with a 4-week rest between cycles. Four-year progression-free survival (PFS) is 62.5% but 7-year PFS is only about 33%, or
- Carboplatin 560 mg/m^2 IV every 4 weeks for up to 18 months with an approximately 30% response rate, or
- Carboplatin 175 mg/m^2 IV weekly for 4 weeks followed by a 2-week rest and then four more weekly doses of carboplatin. Vincristine 1.5 mg/m^2 (maximum dose 2 mg/weekly) for 10 weeks is given concurrently with the carboplatin course. If a response or stabilization is obtained, maintenance therapy is given consisting of courses of carboplatin 175 mg/m^2 weekly for 4 weeks with vincristine 1.5 mg/m^2 (maximum dose 2 mg) weekly for the first 3 weeks of each 4-week cycle. There is a 3-week rest between each maintenance cycle. In children with stable or improved disease the regimen is continued for 8–12 cycles (Table 21-9).

Recommendations

Since optic glioma can run an indolent course, the decision for therapeutic intervention should be based on radiographic findings and visual assessment of visual acuity, visual fields and visual-evoked response. The following treatment plan is recommended:

- *Evidence of optic nerve tumor and normal visual assessment*: No therapeutic intervention is recommended. CT/MRI scan and visual assessments are performed every 6 months to monitor progression of disease.

Table 21-9   Chemotherapy for Optic Glioma

Induction: (One 12-week cycle)										
Week 1	2	3	4	5	6	7	8	9	10	11–12
V	V	V	V	V	V	V	V	V	V	REST
C	C	C	C			C	C	C	C	

Maintenance (Eight 6-week cycles)				
Week 1	2	3	4	5–6
V	V	V		REST
C	C	C	C	

(C) Carboplatin	175 mg/m^2 IV
(V) Vincristine	1.5 mg/m^2 IV (maximum dose 2 mg)

- *Evidence of progression on visual assessment, with or without tumor extending posteriorly into the optic canal*: A trial of chemotherapy is recommended in an attempt to preserve useful vision.

Prognosis

Progression-free survival rates with carboplatin and vincristine are 68% at 3 years. Current studies seek to determine the benefit of adding temozolomide to vincristine and carboplatin.

CRANIOPHARYNGIOMAS

Craniopharyngiomas may involve the pituitary gland. They account for 6–9% of all CNS tumors in children. The peak incidence is 5–10 years of age. Patients present with symptoms of increased ICP, visual loss and endocrine deficiencies. They typically require replacement therapy with cortisone, thyroxine, growth hormone and/or sex hormones. These tumors are slow-growing benign lesions amidst vital anatomic structures.

Surgery

Craniopharyngiomas should be completely removed if possible without producing untoward neurologic sequelae. Complete excision is possible in 60–90% of cases. If radical excision can be accomplished without significant postoperative morbidity, a primary surgical approach is warranted. However, controversy exists over whether subtotal resection followed by radiation will produce less morbidity and be an equally effective strategy. Complete tumor removal is most easily accomplished in cystic tumors and is most difficult in solid or mixed tumors larger than 3 cm in size.

Radiotherapy

In tumors in which conservative surgery consisting only of drainage or subtotal removal is performed, the addition of radiotherapy reduces the local recurrence rate and improves long-term survival. For children older than 5 years, 5,000–5,500 cGy are given. In children less than 5 years of age the dose may be reduced to 4,000–4,500 cGy.

For patients with complete resections, radiation may be reserved for those who relapse.

Chemotherapy

At present there is no established role for systemic chemotherapy in craniopharyngioma. Interferon-alpha 2a systemically shrank the lesion in 25% of patients treated, primarily those with cystic lesions.

Prognosis

The long-term survival of patients treated with radical and total removal is 80–90% at 5 years and 81% at 10 years. The 5-year recurrence-free survival after subtotal removal is approximately 50%. Survival after partial removal and radiation therapy is 62–84%.

INTRACRANIAL GERM CELL TUMORS

Primary intracranial germ cell tumors (GCT) comprise 1–3% of primary pediatric brain tumors. The peak age is between 10 and 21 years of age. Multiple tumor types can be seen: the majority are germinomas ($\sim$55%), teratomas and mixed germ cell tumors ($\sim$33%) and the remaining 10% are malignant endodermal sinus tumors, embryonal cell carcinomas, choriocarcinomas and teratocarcinomas. In all but the germinomas, serum and cerebrospinal fluid alpha-fetoprotein and βHCG may be elevated. MRI of the spine with Gd-DTPA should be performed, as leptomeningeal spread is relatively common.

The outcome of patients with pure germinoma is considerably better than with mixed germ cell tumors.

Surgery

Surgical biopsy is indicated in all germ cell tumors to make an appropriate diagnosis unless elevated serum or CSF tumor markers establish the diagnosis. For patients with benign tumors such as teratomas or dermoids, surgery can be curative.

Radiotherapy

Conventional radiotherapy for a CNS germinoma includes doses to the primary tumor of 5,000 cGy with 3,000 cGy craniospinal therapy. However, there are several studies which demonstrate the ability to use chemotherapy to reduce the dose of radiation to between 3,060 cGy and 5,040 cGy to gross tumor only (in nondisseminated disease) depending on response while maintaining outcomes of >90% event-free survival. Nongerminoma germ cell tumors (NGGCT) respond poorly to radiation therapy but the use of chemotherapy with radiation appears to improve survival substantially in preliminary studies. These data need to be replicated in larger studies.

Chemotherapy

In both germinomas and nongerminoma germ cell tumors, cycles of cisplatin 20 mg/m^2/day IV days 1–5 with etoposide 100 mg/m^2/day IV days 1–5 alone or alternating with cyclophosphamide 1 g/m^2/day for 2 days with vincristine 1.5 mg/m^2/day weekly for 3 days has been used. For high-risk disease (nongerminomatous disease, HCG >50 mIu/ml or elevated AFP) doses of cisplatin and cyclophosphamide have been doubled or ifosfamide has been added to the cisplatin/etoposide above at 1.5 g/m^2/day for days 1–5. These approaches are generally combined with graded doses of radiation depending on initial risk and response.

Prognosis

The germinomas have the best overall survival rate followed by teratomas and pineal parenchymal tumors. Nongerminoma germ cell tumors such as embryonal carcinoma, endodermal and choriocarcinoma have a worse survival. However, preliminary data from combined chemotherapy/radiation approaches demonstrate encouraging data indicating that survival may now approach 60–80%.

MALIGNANT BRAIN TUMORS IN INFANTS AND CHILDREN LESS THAN 3 YEARS OF AGE

Infants and very young children with brain tumors have a worse prognosis than older children. They are also at higher risk for neurotoxicity including mental retardation, growth failure and leukoencephalopathy. Due to these factors there is a reluctance to treat infants and young children with radiation therapy. Recent studies have been designed to use chemotherapy and to either withhold radiation therapy or postpone its use to a time when the patient is older. Postoperative chemotherapy with cyclophosphamide, vincristine, cisplatin and etoposide in children under 36 months of age at diagnosis are utilized (Table 21-4) followed by delayed radiation. Chemotherapy is administered for 48–96 weeks, depending on age at diagnosis, to delay radiation until as close to age 3 as possible or beyond. Ongoing studies are evaluating adding high-dose methotrexate to higher-risk cases. Newer approaches using high-dose chemotherapy with autologous stem cell rescue (Table 21-6) to intensify the chemotherapeutic regimen have been employed in an attempt to avoid radiation altogether.

Average-risk medulloblastoma and other PNETs, which are generally chemosensitive tumors do well using this approach with 2-year PFS of about 70%, in those who received chemotherapy and delayed, reduced craniospinal irradiation. The results are worse when radiation is not employed, with 5-year EFS and OS of 32 and 43% for medulloblastoma. Gliomas do not do as well with this therapy. One-year EFS is generally respectable, between 34 and 77%, however, there is a drop-off to 20% for the entire group. However, with localized disease that is completely resected, ependymoma 4-year PFS is approximately 58%.

Atypical teratoid/rhabdoid tumors are very aggressive tumors of infancy that present similarly to medulloblastoma and are at times indistinguishable pathologically except for monosomy 22 (lack of expression of INI-1). While uniformly fatal if treated with conventional medulloblastoma therapy, these patients have a survival approaching 50% if treated with high-dose chemotherapy followed by autologous stem cell rescue.

Trials of intensive chemotherapy followed by stem cell rescue have also been attempted in young children with malignant brain tumors with promising results with avoidance of radiation therapy. The current COG study uses three tandem auto stem cell rescues after three cycles of chemotherapy. These approaches require further research. Additionally, with 3D-conformal radiation and IMRT, the timing and use of radiation therapy in subsets of disease that do worse (i.e. subtotal resections) is being studied in children as young as 8 months with focal radiotherapy alone. The use of second-look surgery is also being evaluated in current trials.

Suggested Reading

Allen J, Finlay J. Intensive chemotherapy and bone marrow rescue for young children with newly diagnosed malignant brain tumors. J Clin Oncol. 1998;16:210–221.

Balmaceda C, Heller G, Rosenblum M, et al. Chemotherapy without irradiation – a novel approach for newly diagnosed CNS germ cell tumors: Results of an international cooperative trial. J Clin Oncol. 1996;14:2908–2915.

Biegel JA. Cytogenetics and molecular genetics of childhood brain tumors. Neuro Oncol. 1999;1(2):139–151.

Blaney SM, Kun LE, Hunter J, et al. Tumors of the Central Nervous System in Principles and Practice of Pediatric Oncology. In: Pizzo PA, Poplack DG, eds. Lippincott-Wilkins & Williams. 4th ed. 2006.

Duffner PC, Horowitz ME, Krischer JP, et al. Myeloablative chemotherapy with autologous bone marrow rescue in children and adolescents with recurrent malignant astrocytoma: outcome compared with conventional chemotherapy: a report from the Children's Oncology Group. Pediatr Blood Cancer. 2008;51(6):806–811.

Geyer JR, Sposto R, et al. Multiagent chemotherapy and deferred radiotherapy in infants with malignant brain tumors: a report from the Children's Cancer Group. J Clin Oncol. 2005;23(30):7621–7631.

Halperin EC, Constine LS, Tarbell NJ, Kun LE. Pediatric Radiation Oncology. 4th ed., Philadelphia: Lippincott Williams and Wilkins, 2005.

MacDonald TJ, Rood BR, Santi MR, et al. Advances in the diagnosis, molecular genetics and treatment of pediatric embryonal CNS tumors. The Oncologist. 2003;8:174–186.

Mason WP, Growas A, Halpern S, et al. Carboplatin and vincristine chemotherapy for children with newly diagnosed progressive low-grade glioma. J Neurosurg. 1997;86:747–754.

Thompson D, Phipps K, et al. Craniopharyngioma in childhood: our evidence-based approach to management. Childs Nerv Syst. 2005;21(8–9):660–668.

CHAPTER 22

Neuroblastoma

Neuroblastoma originates from primordial neural crest cells that normally give rise to adrenal medulla and the sympathetic ganglia.

EPIDEMIOLOGY

- Neuroblastoma is the most common extracranial solid tumor in children, accounting for 7% of all childhood malignancies. It accounts for 15% of childhood cancer mortality
- Neuroblastoma is the most common malignancy in infants with an annual incidence of 10 per million live births
- At diagnosis, 50% of patients are under age 2, 75% under age 4 and 90% under age 10. Peak age of incidence is 2 years
- *In situ* neuroblastoma is detected in 1 in 259 autopsies among infants less than 3 months of age. This 400-fold increase in the autopsy incidence of neuroblastoma compared with the clinical incidence of the tumor indicates that involution or maturation of the tumor occurs spontaneously in a moderate number of infants

PREDISPOSITION

- A family history of neuroblastoma is identified in 1–2% of cases and the tumors that arise in patients with familial disease are heterogeneous. The study of neuroblastoma predisposition has yielded remarkable insights into the biology of this disease
- Disorders involving the autonomic nervous system are occasionally seen in patients with neuroblastoma. Conditions associated with neuroblastoma are listed in Table 22-1
- Mutations in PHOX2B (a key regulator of normal development of the autonomic nervous system) have been noted in a subset of patients with hereditary neuroblastoma
- Heritable mutations in the anaplastic lymphoma kinase (ALK) gene are an important cause of familial neuroblastoma

PATHOLOGY AND BIOLOGY

The diagnosis of neuroblastoma is based on the presence of characteristic histopathologic features of tumor tissue or the presence of tumor cells in a bone marrow aspirate or biopsy

Manual of Pediatric Hematology and Oncology. DOI: 10.1016/B978-0-12-375154-6.00022-7

Table 22-1   Conditions Associated with Neuroblastoma

Neurofibromatosis type I
Hirschsprung disease with aganglionic colon
Congenital central hypoventilation syndrome
Pheochromocytoma
Heterochromia
Turner syndrome
Noonan syndrome
Cardiac malformations

accompanied by elevated levels of urinary catecholamines. Neuroblastoma is one of the small round blue cell tumors of childhood and tumors are classified as histologically favorable or unfavorable. The prognostic grouping of neuroblastic tumors according to histological classification systems as characterized by the International Neuroblastoma Pathology Classification System (INPC) and the Shimada classification is shown in Table 22-2.

- Biologic features of neuroblastoma tumors are of critical importance for risk assessment. Genomic amplification of *MYCN* is detected in approximately 20% of primary tumors and is strongly correlated with advanced-stage disease and treatment failure. Its association with poor outcome in patients with otherwise favorable disease patterns highlights its importance as a prognostic factor
- Alterations involving chromosomes 1p, 11q and 17q also have prognostic significance
- Diploid tumor cell DNA content appears to be associated with a less favorable outcome in subsets of children with neuroblastoma while hyperdiploid DNA content appears to be associated with more favorable outcome.

CLINICAL FEATURES

Neuroblastoma usually presents with an adrenal mass or less commonly as a tumor arising anywhere along the sympathetic neural chain. The most common presenting feature is an asymptomatic abdominal mass, with distant metastases detected at the time of diagnosis in 50% of cases.

Anatomic Site

Clinical findings related to the anatomic site of the primary tumor are listed below:

- Head and neck
  - Unilateral palpable neck mass
  - Horner's syndrome (myosis, ptosis, enophthalmos, anhydrosis).

Table 22-2 Prognostic Grouping of Neuroblastic Tumors According to Histologic Classification Systems: International Neuroblastoma Pathology Classification System (INPC) and Shimada Classification

International Neuroblastoma Pathology Classification		Original Shimada Classification	Prognostic Group
Neuroblastoma	(Schwannian Poor)[a]	Stroma-Poor	
Favorable		Favorable	Favorable
<1.5 years	Poorly differentiated or differentiating & low or intermediate MKI		
1.5–5 years	Differentiating & low MKI		
Unfavorable		Unfavorable	Unfavorable
<1.5 years	a) undifferentiated tumor[b]		
	b) high MKI		
1.5–5 years	a) undifferentiated or poorly differentiated tumor		
	b) intermediate or high MKI tumor		
≥5 years	All tumors		
Ganglioneuroblastoma, intermixed	(Schwannian stroma-rich)	Stroma-rich Intermixed (favorable)	Favorable
Ganglioneuroma	(Schwannian stroma-dominant)		
Maturing		Well differentiated (favorable)	Favorable[c]
Mature		Ganglioneuroma	
Ganglioneuroblastoma, nodular	(composite Schwannian stroma-rich/ stroma dominant and stroma poor)	Stroma-rich nodular (unfavorable)	Unfavorable[c]

[a]Neuroblastoma subtypes detailed in Shimada H, Ambros IM, Dehner LP, et al. The International Neuroblastoma Pathology Classification (the Shimada system). Cancer 1999;86:364–372.
[b]Rare subtype, especially diagnosed in this age group. Further investigation and analysis required.
[c]Prognostic grouping not related to age.
Abbreviations: MKI, mitosis-karyorrhexis index.
Adapted from: Shimada H, Ambros IM, Dehner LP, et al. The International Neuroblastoma Pathology Classification (the Shimada system). Cancer 1999;86:364–372.

- Orbit and eyes
  - Orbital metastases with periorbital hemorrhage ("raccoon eyes")
  - Exophthalmos, palpable supraorbital masses, ecchymosis, edema of eyelids and conjunctiva, ptosis
  - Opsoclonus ("dancing eyes syndrome")
  - Heterochromia iridis, anisocoria.
- Chest
  - Upper thoracic tumors: dyspnea, pulmonary infections, dysphagia, lymphatic compression, Horner syndrome
  - Lower thoracic tumors: usually no symptoms.

- Abdomen
  - Anorexia, vomiting
  - Abdominal pain
  - Palpable mass
  - Massive involvement of the liver with metastatic disease; may result in respiratory insufficiency in the newborn.
- Pelvis
  - Constipation
  - Urinary retention
  - Presacral mass palpable on rectal examination.
- *Paraspinal area* (dumbbell or hourglass-shaped neuroblastoma).
  - Localized back pain and tenderness
  - Limp
  - Weakness of lower extremities
  - Hypotonia, muscle atrophy, areflexia, hyperreflexia
  - Paraplegia
  - Scoliosis
  - Bladder and anal sphincter dysfunction.
- Lymph nodes: enlarged
- Bone: pain, limping and irritability in the young child associated with bone and bone marrow metastasis
- Lung: Pulmonary parenchyma or pleura may rarely be involved. Lung involvement should be proven by biopsy or MIBG (I^{131}-meta-iodobenzylguanidine scan) avidity due to its rarity in this disease
- Brain: Metastatic involvement of the brain has been described, though it is more common in the setting of recurrent rather than newly diagnosed disease.

Other Symptoms

1. Nonspecific constitutional symptoms include lethargy, anorexia, pallor, weight loss, abdominal pain, weakness and irritability. These symptoms are more commonly seen in patients with metastatic disease at diagnosis.
2. Intermittent attacks of sweating, flushing, pallor, headaches, palpitation and hypertension related to catecholamine release can occur, but are relatively rare.
3. Hypertension in children with neuroblastoma is usually due to reno-vascular compromise rather than catecholamine secretion.

Paraneoplastic Syndromes

1. Vasoactive intestinal peptide (VIP) secretion (Kerner–Morrison syndrome):
   - Signs and symptoms include intractable watery diarrhea resulting in failure to thrive, associated with abdominal distention and hypokalemia

- Symptoms are due to the secretion of an enterohormone, VIP by the neuroblastoma cells
- VIP-secreting tumors typically have favorable histology.

2. Opsoclonus myoclonus ataxia syndrome (OMAS):
   - Signs and symptoms include bursts of rapid involuntary random eye movements in all directions of gaze (opsoclonus); and motor incoordination due to frequent, irregular, jerking of muscles of the limbs and trunk (myoclonic jerking)
   - Symptoms may or may not resolve after the tumor is removed
   - Prognosis for survival is favorable because OMAS is typically seen in patients with low-stage, biologically favorable tumors
   - There is a high incidence of chronic neurologic deficits, including cognitive and motor delays, language deficits and behavioral abnormalities
   - All children presenting with this syndrome should be evaluated for the presence of neuroblastoma.

NEUROBLASTOMA IN THE NEONATE

Stage 4S (S=special) disease occurs in about 5% of cases. 4S neuroblastoma is restricted to infants under 1 year of age who have disseminated disease. These children typically have small primary tumors with metastases in liver, skin and/or bone marrow (<10% marrow involvement) that almost always spontaneously regresses. Unlike older infants, those under 2 months of age may require treatment for 4S neuroblastoma due to extensive and rapidly progressive intrahepatic disease that may result in respiratory compromise.

DIAGNOSIS AND STAGING

When neuroblastoma is suspected, urinary levels of the catecholamines homovanillic acid (HVA) and vanillylmandelic acid (VMA) should be analyzed to facilitate noninvasive monitoring of disease status. Collection of the sample prior to biopsy or resection of the primary tumor or metastases is recommended. Tumor biopsies should contain enough tissue to permit assessment of tumor histology and sufficient material for the molecular genetic analyses that are central to correct risk stratification and treatment planning. At present, open biopsies are preferred rather than needle biopsies if the risks of open biopsies can be minimized. Criteria for neuroblastoma diagnosis are described in Table 22-3.

Once the diagnosis of neuroblastoma has been confirmed, a complete staging workup should be performed consisting of the following:

- Computed tomography (CT) for evaluation of tumors in the abdomen, pelvis or mediastinum

Table 22-3   Recommended Criteria for Neuroblastoma Diagnosis

Established if:
(1) Unequivocal pathologic diagnosis* is made from tumor tissue by light microscopy (with or without immunohistology, electron microscopy) and/or increased urine or serum catecholamines or metabolites**
OR
(2) Bone marrow aspirate or trephine biopsy contains unequivocal tumor cells* (e.g., syncytia or immunocytologically positive clumps of cells) and increased urine or serum catecholamines or metabolites**

*If histology is equivocal, karyotypic abnormalities in tumor cells characteristic of other tumors (e.g. t(11;22)) excludes a diagnosis of neuroblastoma, whereas genetic features characteristic of neuroblastoma (1p deletion, *MYCN* amplification) support this diagnosis.
**Includes dopamine, HVA and/or VMA (levels must be >3.0 SD above the mean per milligram creatinine for age to be considered increased and at least two of these must be measured).
From: Brodeur GM, Pritchard J, Bethold F, et al. Revisions of the International Criteria for Neuroblastoma Diagnosis, Staging and Response to Treatment. J Clin Oncol 1993;11:1466–1477.

- Magnetic resonance imaging is helpful for imaging of paraspinal lesions and is essential when evaluating intraforaminal extension with the potential for cord compression
- I^{123}-Meta-iodobenzylguanidine (MIBG) scintigraphy is recommended for detection of bony metastases and occult soft tissue disease. MIBG is taken up by approximately 90% of neuroblastoma
- ^{99}Tc-diphosphonate bone scan is only recommended for evaluation of bone involvement when tumors do not take up MIBG or MIBG scan is unavailable
- ^{18}F-fluorodeoxyglucose positron emission tomography (FDG-PET) may be useful to visualize primary and metastatic sites of disease, especially soft tissue disease, in patients whose tumors are not MIBG avid
- Bone marrow disease should be assessed by performing bilateral bone marrow aspirates and biopsies. Standard histological analyses should be performed on the specimens
- Tumor detection in blood, bone marrow or peripheral blood stem cell apheresis product (see Treatment section) can be performed using immunocytochemical (ICC) or polymerase chain reaction (PCR) methodologies. Confirmation of lack of tumor detection in the stem cell product by ICC techniques is standard. Further investigations are underway to assess whether low level detection of tumor cells using PCR techniques will have prognostic significance
- Table 22-4 shows the recommended tests for the assessment of extent of disease.

Staging Systems

Table 22-5 shows the International Neuroblastoma Staging System (INSS) and the recently developed International Neuroblastoma Risk Group system (INRGS).

Table 22-4 Tests Recommended for Assessment of Extent of Disease

Tumor Site	Recommended Tests
Primary Tumor	CT and/or MRI scan* with 3D measurements; MIBG scan**
Metastatic Sites	
Bone marrow	Bilateral posterior iliac crest marrow aspirates and trephine (core) bone marrow biopsies required to exclude marrow involvement. A single positive site documents marrow involvement. Core biopsies must contain at least 1 cm of marrow (excluding cartilage) to be considered adequate
Bone	MIBG scan; if tumor does not take up MIBG consider ^{99}Tc- diphosphonate scan or ^{18}F-FDG-PET
Lymph nodes	Clinical examination (palpable nodes), confirmed histologically. CT scan for nonpalpable nodes (3D measurements)
Abdomen/liver	CT and/or MRI scan* with 3D measurements
Chest	CT/MRI necessary if abdominal mass/nodes extend into chest, or evidence of soft tissue MIBG uptake. Consider inclusion if patient has non-thoracic primary with disseminated disease to allow evaluation for distant nodal metastases

*Ultrasound considered suboptimal for accurate 3D measurements.
**The MIBG scan is applicable to all sites of disease. I-123 MIBG is the recommended tracer.
Abbreviation: AP, anteroposterior.
From: Brodeur GM, Pritchard J, Bethold F, et al. Revisions of the International Criteria for Neuroblastoma Diagnosis, Staging and Response to Treatment. J Clin Oncol 1993;11:1466–1477.
Sharp SE, Shulkin BL, Gelfand MJ, et al. ^{123}I-MIBG scintigraphy and ^{18}F-FDG PET in neuroblastoma. J Nucl Med 2009;50:1237–1243.

- The INSS utilizes extent of surgical resection, presence of lymph node involvement and metastatic disease to determine disease stage
- The INRGS uses presurgical assessments of extent of disease based upon results of imaging studies and bone marrow morphology. Radiologic features are used to distinguish locoregional tumors that do not involve local structures (INRGS L1) from locally invasive tumors (INRGS L2, exhibiting image-defined risk factors). Table 22-6 shows neuroblastoma image-defined risk factors (IDRF). Stages M and MS refer to tumors that are widely metastatic or have an INSS 4S pattern of disease, respectively.

PROGNOSIS AND RISK STRATIFICATION

Factors of prognostic significance are listed in Table 22-7.

- Clinical variables utilized for risk stratification include stage and age
- Although previous risk stratification schemes made use of a cut-off point of 12 months of age in designating higher or lower risk, data indicate that a higher age cut off of 18 months has greater prognostic significance.

Table 22-5 **Neuroblastoma Staging: International Neuroblastoma Staging System (INSS) and International Neuroblastoma Risk Group System (INRGS)**

INSS	INRGS
Stage 1. Localized tumor with complete gross excision, with or without microscopic residual disease; representative ipsilateral lymph nodes negative for tumor microscopically (nodes attached to and removed with the primary tumor may be positive)	*Stage L1.* Localized tumor not involving vital structures (as defined by the list of image-defined risk factors, Table 22-6) and confined to one body compartment
Stage 2A. Localized tumor with incomplete gross excision; representative ipsilateral nonadherent lymph nodes negative for tumor microscopically.	*Stage L2.* Locoregional tumor with presence of one or more image-defined risk factors
Stage 2B. Localized tumor with or without complete gross excision, with ipsilateral nonadherent lymph nodes positive for tumor. Enlarged contralateral lymph nodes must be negative microscopically	
Stage 3. Unresectable unilateral tumor infiltrating across the midline,[a] with or without regional lymph node involvement; or localized unilateral tumor with contra-lateral regional lymph node involvement; or midline tumor with bilateral extension by infiltration (unresectable) or by lymph node involvement	
Stage 4. Any primary tumor with dissemination to distant lymph nodes, bone, bone marrow, liver, skin and/or other organs (except as defined for stage 4S)	*Stage M.* Distant metastatic disease (except Stage MS)
Stage 4S. Localized primary tumor (as defined for stage 1, 2A or 2B), with dissemination limited to skin, liver and/or bone marrow[b] (limited to infants <1 yr of age)	*Stage MS.* Metastatic disease in children younger than 18 months with metastases confined to skin, liver and/or bone marrow

Multifocal primary tumors (e.g., bilateral adrenal primary tumors) are staged according to the greatest extent of disease in both systems.

[a]The midline is defined as the vertebral column. Tumors originating on one side and crossing the midline must infiltrate to or beyond the opposite side of the vertebral column. A grossly resectable tumor arising in the midline from pelvic ganglia or the organ of Zuckerkandl would be considered stage 1. A midline tumor that extended beyond one side of the vertebral column and was unresectable would be considered stage 2A. Ipsilateral lymph node involvement would make it stage 2B, whereas bilateral lymph node involvement would make it stage 3. A midline primary tumor with bilateral infiltration that was not resectable would be considered stage 3. A tumor of any size with malignant ascites or peritoneal implants would be stage 3 (but a thoracic tumor with malignant pleural effusion unilaterally would be stage 2B).

[b]Marrow involvement in stage 4S should be minimal i.e., <10% of total nucleated cells identified as malignant on bone marrow biopsy or on marrow aspirate. More extensive marrow involvement would be considered to be stage 4. The MIBG scan should be negative in the marrow.

From: Brodeur GM, Pritchard J, Bethold F, et al. Revisions of the International Criteria for Neuroblastoma Diagnosis, Staging and Response to Treatment. J Clin Oncol 1993;11:1466–1477, with permission.

Monclair T, Brodeur GM, Ambros PF, et al. The International Neuroblastoma Risk Group (INRG) staging system: an INRG Task Force report. J Clin Oncol 2009;10;27:298–303, with permission.

Table 22-6   Neuroblastoma Image-Defined Risk Factors (IDRF) By Location

Neck:
  Tumor encasing carotid and/or vertebral artery and/or internal jugular vein
  Tumor extending to base of skull
Cervico-thoracic junction:
  Tumor encasing brachial plexus roots
  Tumor encasing subclavian vessels and/or vertebral and/or carotid artery
  Tumor compressing the trachea
Thorax:
  Tumor encasing the aorta and/or major branches
  Tumor compressing the trachea and/or principal bronchi
  Lower mediastinal tumor, infiltrating the costovertebral junction between T9 and T12
  Significant pleural effusion with or without presence of malignant cells
Thoraco-abdominal:
  Tumor encasing the aorta and/or vena cava
Abdomen/pelvis:
  Tumor infiltrating the porta hepatis
  Tumor infiltrating the branches of the superior mesenteric artery at the mesenteric root
  Tumor encasing the origin of the celiac axis and/or of the superior mesenteric artery
  Tumor invading one or both renal pedicles
  Tumor encasing the aorta and/or vena cava
  Tumor encasing the iliac vessels
  Pelvic tumor crossing the sciatic notch
  Ascites with or without presence of malignant cells
Dumbbell tumours with symptoms of spinal cord compression:
  Any localization
Involvement/infiltration of adjacent organs/structures:
  Pericardium, diaphragm, kidney, liver, duodeno-pancreatic block, mesentery and others

From: Monclair T, Brodeur GM, Ambros PF, et al. The International Neuroblastoma Risk Group (INRG) staging system: an INRG Task Force report. J Clin Oncol 2009;27:298–303, with permission.

- Risk stratification schemes also rely on variables including tumor histology (favorable vs. unfavorable according to INPC) and biologic features of neuroblastoma tumors (*MYCN* status, DNA index and the presence or absence of chromosomal alterations)
- Table 22-8 shows the approach to risk group assignment for neuroblastoma.

TREATMENT

The treatment modalities used in neuroblastoma are surgery, chemotherapy and radiotherapy. The role of each is determined by clinical and molecular characteristics determined at neuroblastoma diagnosis such as age, stage and biological features.

Surgery

The role of surgery is to establish diagnosis, provide tissue for biologic studies, stage the disease and excise the tumor (when feasible).

Table 22-7   Factors of Prognostic Significance in Neuroblastoma

Prognostic factor	Comparison[a]	5-year Overall Survival
Pathology	Differentiated > Undifferentiated	89% vs. 72%
	Low MKI > High MKI	82% vs. 44%
Age at diagnosis	<18 months > 18 months or older	89% vs. 49%
Ferritin level (ng/ml)	<92 ng/ml > 92 ng/ml or greater	87% vs. 52%
LDH	Low (<1,500 μ/ml) > high (>1,500 μ/ml)	85% vs. 58%
Stage	1, 2, 3, 4s > 4	91% vs. 42%
MYCN gene amplification	Normal > amplified copy number	82% vs. 34%
DNA index (DI)	Hyperdiploid (DI >1) > Diploid (DI = 1)	82% vs. 60%
Segmental Chromosomal Aberrations	Absent > Present	
Chromosome 1p	Normal > ch1p aberration	83% vs. 48%
Chromosome 11q	Normal > ch11q aberration	79% vs. 57%
Chromosome 17q	Normal > Gain ch17q	74% vs. 55%

[a]>, better prognosis.
Abbreviations: MKI, mitosis-karyorrhexis index; ch, chromosome.
From: Cohn SL, Pearson AD, London WB, et al. The International Neuroblastoma Risk Group (INRG) classification system: an INRG Task Force report. J Clin Oncol 2009;27:289–297, with permission.

Table 22-8   Children's Oncology Group Risk Group Assignment for Neuroblastoma

Risk Group	INSS	Age	MYCN	Ploidy	Shimada
Low risk	1	any	any	any	any
Low risk	2a/2b	any	not amp	any	any
High risk	2a/2b	any	amp	any	any
Intermediate risk	3	<547d	not amp	any	any
Intermediate risk	3	≥547d	not amp	any	FH
High risk	3	any	amp	any	any
High risk	3	≥547d	not amp	any	UH
High risk	4	<365d	amp	any	any
Intermediate risk	4	<365d	not amp	any	any
High risk	4	365–<547d	amp	any	any
High risk	4	365–<547d	any	DI=1	any
High risk	4	365–<547d	any	any	UH
Intermediate risk	4	365–<547d	not amp	DI>1	FH
High risk	4	≥547d	any	any	any
Low risk	4s	<365d	not amp	DI>1	FH
Intermediate risk	4s	<365d	not amp	DI=1	any
Intermediate risk	4s	<365d	not amp	any	UH
High risk	4s	<365d	amp	any	any

Abbreviations: INSS, International Neuroblastoma Staging System; amp, amplified; DI, DNA index; FH, favorable histology; UH, unfavorable histology.
Adapted from: Park JR, Eggert A, Caron H. Neuroblastoma: biology, prognosis and treatment. Pediatr Clin North Am 2008;55(1):97–120, with permission.

- Timing of surgical resection is dependent upon radiographic features of response to treatment and upon a patient's risk stratification
- Residual tumor does not adversely affect prognosis in low- or intermediate-risk disease, while an attempt at complete surgical resection is recommended for high-risk disease.
- Attempted surgical resection of localized tumors is the mainstay of therapy
- Surgical resection that may result in loss or damage to vital organs should be avoided, especially in the setting of low- and intermediate-risk neuroblastoma. In such cases, chemotherapy should be administered prior to attempted surgical resection
- Delayed surgical resection of primary disease should be performed in patients with evidence of metastatic disease.

Radiation Therapy

Although neuroblastoma is an extremely radiosensitive tumor, the long-term risks of radiotherapy are generally not warranted in patients with low- and intermediate-risk disease except in specific settings. Radiotherapy is, however, an important component of therapy for patients with high-risk disease.

Radiotherapy may be warranted in the following clinical scenarios involving patients with low- or intermediate-risk neuroblastoma:

- Neonates with INSS stage 4S neuroblastoma who develop respiratory distress secondary to hepatomegaly or hepatic failure and for whom treatment with chemotherapy is ineffective
- Children with neuroblastoma-associated spinal cord compression and neurologic compromise for whom there is a contraindication to the use of emergent chemotherapy. Laminectomy has also been used to rapidly reduce cord compression. Both radiation and surgery are associated with a significant incidence of vertebral body damage and scoliosis.

DISEASE STRATIFICATION AND THERAPY

Table 22-8 displays the currently utilized risk stratification schema for patients with neuroblastoma.

1. Clinical features of age and stage, tumor histology and tumor molecular characteristics including *MYCN* amplification and DNA index (ploidy) are of critical prognostic importance.
2. Additional features of segmental chromosomal aberrations, including loss of heterozygosity at chromosome 11q, 1 p or gain of chromosome 17q, have proven to be of prognostic importance and assist in individualizing treatment within each risk group, particularly for patients with loco-regional disease.

3. Therapy varies greatly across risk groups.
4. Response criteria are based upon disease status at primary and metastatic sites. Neuroblastoma-specific definitions of response to treatment are listed in Table 22-9.

Low-Risk Group

The low-risk group consists of (see Table 22-8):

- All patients with INSS stage 1
- All patients with INSS stage 2A, 2B without *MYCN* amplification
- Patients with INSS stage 4S disease, favorable Shimada histology, hyperdiploid and without *MYCN* amplification
  a. Two sequential multi-institutional clinical trials in North America (CCG3881 and COG P9641) have demonstrated that surgical resection is the mainstay of treatment of patients with INSS Stage 1 or 2 neuroblastoma without *MYCN* amplification
  b. Chemotherapy is reserved for those whose tumor can not be surgically removed without damaging vital structures or who present with life-threatening symptoms or neurologic compromise
  c. Incomplete surgical resection does not compromise relapse-free survival in the setting of localized disease and favorable biologic features
  d. Chemotherapy is recommended if less than 50% of the tumor is resected at diagnosis. Current clinical trials in Children's Oncology Group are investigating whether excellent outcomes can be maintained while decreasing exposure to chemotherapy for those patients with localized tumors but less than 50% tumor resection at diagnosis

Table 22-9   Modified INRC Definitions of Response to Treatment

Response	Primary Tumor*	Metastatic Sites*
CR	No tumor	No tumor; catecholamines normal; MIBG normal
VGPR	Decreased by 90–99%	No tumor; catecholamines normal; residual ^{99}Tc bone changes allowed, MIBG normal
PR	Decreased by >50%	All measurable sites decreased by >50%. Bones and bone marrow: number of positive bone sites decreased by >50%; no more than 1 positive bone marrow site allowed**
MR	No new lesions; >50% reduction of any measurable lesion (primary or metastases) with <50% reduction in any other; <25% increase in any existing lesion	
SD	No new lesions; <25% increase in any existing lesion; exclude bone marrow evaluation	
PD	Any new lesion; increase of any measurable lesion by >25%; previous negative marrow positive for tumor	

*Evaluation of primary and metastatic disease as outlined in Table 22-4.
**One positive marrow aspirate or biopsy allowed for PR if this represents a decrease from the number of positive sites at diagnosis.
Abbreviations: CR, complete response; VGPR, very good partial response; PR, partial response; MR, mixed response; SD, stable disease; PD, progressive disease.
From: Brodeur GM, Pritchard J, Bethold F, et al. Revisions of the International Criteria for Neuroblastoma Diagnosis, Staging and Response to Treatment. J Clin Oncol 1993;11:1466–1477.

    e. Reduction in chemotherapy is not recommended in patients with tumors exhibiting loss of heterozygosity of chromosome 11q or 1p, as these patients have increased risk for recurrence and lower overall survival than patients without these chromosomal aberrations

    f. Disease recurrence is rare in low-risk patients regardless of extent of surgical resection. When recurrent disease is seen in this population, it typically occurs at the primary site or within regional lymph nodes. Treatment of recurrent disease is patient-specific and may include repeat surgical resection or chemotherapy. Decision-making is based upon site of disease recurrence and associated symptoms.

Treatment of Low-Risk Group

All patients INSS stage 1

Treatment:

1. Surgical removal of the primary tumor with close observation for recurrent disease. Three-year event-free survival (EFS): 94%; overall survival (OS) 99%.

All patients with INSS stage 2A, Stage 2B without MYCN amplification

Treatment:

1. Surgical removal of the primary tumor without damage to vital organs. Post-surgical observation is only required in patients with >50% resection of primary tumor mass.
2. For patients with <50% resection: four cycles of moderate-dose chemotherapy utilizing carboplatin, etoposide, cyclophosphamide, carboplatin and doxorubicin. Chemotherapy agents and dosing (cycles 1–4) are according to the intermediate-risk chemotherapy regimen detailed in Table 22-10. The current COG protocol ANBL0531 is investigating whether therapy can be reduced to two cycles of chemotherapy. Three-year EFS 85%; OS 99%.

Patients with INSS stage 4S disease

The majority of patients with INSS stage 4S disease fall into the low-risk category with EFS 86% and OS 92%.

1. The majority of 4S tumors will undergo spontaneous regression, although patients less than 2 months of age have a higher incidence of respiratory compromise and liver dysfunction due to diffuse infiltration of the liver with tumor.
2. In the absence of life-threatening complications, no therapy is indicated.
3. Surgical resection of the primary tumor is not usually necessary, although biopsy of primary or metastatic site of disease should be undertaken to ascertain biologic characteristics.
4. Chemotherapy is utilized in those patients with life-threatening complications including respiratory compromise and severe liver dysfunction. Studies have demonstrated that a

Table 22-10   Intermediate-Risk Chemotherapy Regimen (COG A3961 SCHEMA)

Course #	Cycle #	Day	Agents
I	1	0	Carboplatin and Etoposide
		1	Etoposide
		1	Etoposide
	2	21	Carboplatin, Cyclophosphamide and Doxorubicin
	3	42	Cyclophosphamide and Etoposide
		43	Etoposide
		44	Etoposide
	4	63	Carboplatin, Etoposide and Doxorubicin
		64	Etoposide
		65	Etoposide
II	5	84–86	Repeat chemotherapy drugs in cycle 3
	6	105	Repeat cycle 2
	7	126–128	Repeat cycle 1
	8	147	Cyclophosphamide and Doxorubicin

For children <1 year of age or who are ≤12 kg in weight, the doses of chemotherapy will be adjusted and given as mg/kg. Each cycle is given at 3 weeks (21 days) interval
Dose of chemotherapeutic agents:
Carboplatin 560 mg/m^2 or 18.6 mg/kg IV over 1 hour for 1 day
Etoposide 120 mg/m^2 or 4 mg/kg IV over 2 hours daily for 3 days
Cyclophosphamide 1,000 mg/m^2 or 33.3 mg/kg over 1 hour daily for 1 day
Doxorubicin 30 mg/m^2 or 1 mg/kg IV over 60 minutes daily for 1 day
Myeloid growth factors (G-CSF or GM-CSF) should be given for infants less than 60 days of age. G-CSF 5 μg/kg/day or GM-CSF 250 μg/m^2/day should start 24 hours post chemotherapy and continue until ANC ≥1,000/mm^3 for 2 consecutive days.

short course of low-dose oral cyclophosphamide (5 mg/kg/day for 5 days every 2–3 weeks as needed) or up to four cycles of intermediate-risk chemotherapy (cycles 1–4, Table 22-10) frequently induces remission. Chemotherapy should be discontinued if remission is reached prior to completing four cycles of chemotherapy. Emergent low-dose radiotherapy can also be utilized (150 cGy two or three times to the anterior two-thirds of the liver through lateral oblique ports). This dose is generally sufficient to halt the progression of the tumor and induce regression.

5. The rare stage 4S patient with unfavorable biological features may be a candidate for more intensive treatment.

Intermediate-Risk Group

The intermediate-risk group (Table 22-8) consists of the following:

*   Patients less than 18 months of age with INSS stage 3, without *MYCN* amplification
*   Patients 18 months of age or older with INSS stage 3, with favorable histology, without *MYCN* amplification

- Patients less than 12 months of age with INSS Stage 4, without *MYCN* amplification
- Patients with Stage 4S disease less than 1 year of age with either unfavorable histology and/or diploid tumor without *MYCN* amplification
- Patients between 12 months to 18 months of age with INSS Stage 4, with favorable histology, hyperdiploid tumor without *MYCN* amplification.

Treatment

Chemotherapy

1. Chemotherapy regimens utilized include cyclophosphamide, carboplatin, etoposide and doxorubicin. The now-completed clinical trial COG A3961 outlined in Table 22-10, utilizes four cycles of chemotherapy (course I) for the following patients:
   - Intermediate-risk stage-3 patients (groups 1–2 above)
   - Patients less than 12 months of age with stage 4 neuroblastoma exhibiting favorable biology characteristics (group 3 above).

   Eight cycles of the same chemotherapy agents (Course I and II) are administered to the following patients:

   - Patients less than 12 months with stage 4 or with stage 4s without *MCYN* amplification but with unfavorable tumor histology or diploid tumor cell DNA content at time of diagnosis (groups 3–4 above)
   - Patients with residual metastatic disease or less than 90% reduction in primary tumor (very good partial response, VGPR).
2. The current clinical trial in Children's Oncology Group (ANBL0531) is investigating whether the excellent outcome for intermediate-risk neuroblastoma can be maintained despite further reduction in chemotherapy (as few as two cycles for patients with favorable biologic characteristics, e.g. cycles 1 and 2 of intermediate-risk chemotherapy per Table 22-10) and for all patients ending chemotherapy once the patient has achieved at least 50% reduction in primary tumor (partial response, PR).
3. Chemotherapy should not be reduced for patients with tumors exhibiting loss of heterozygosity of chromosome 11q or 1p as these patients have increased risk for recurrence and lower overall survival than those without these chromosomal aberrations.
4. Data suggest that patients with stage 3 and 4 neuroblastoma between the age of 12 and 18 months with favorable histology and hyperdiploid tumors without *MYCN* amplification have an excellent prognosis similar to patients less than 12 months of age (group 5 above). Historically these patients have been treated using high-risk therapy algorithms.
5. Based on these data, COG ANBL0531 will also assess whether a significant dose reduction in therapy can be administered while maintaining excellent survival in patients with stage 3 and 4 neuroblastoma between the age of 12 and 18 months with

favorable histology and hyperdiploid tumors without *MYCN* amplification have an excellent prognosis similar to patients less than 12 months of age (group 5 above). In lieu of high-risk therapy, patients receive eight cycles of intermediate-risk chemotherapy followed by six cycles of 13-cis retinoic acid therapy (80 mg/m^2/dose for patients >12 kg and 2.67 mg/kg/dose for patients ≤12 kg given twice per day for 14 consecutive days in each 28-day cycle for a total of six cycles) using a goal endpoint of at least 90% reduction in tumor.

Surgery
1. Surgical resection following pre-operative chemotherapy is frequently performed to improve primary tumor response although extent of response necessary remains controversial.
2. Clinical trials performed in Germany reveal excellent overall survival with limited surgical intervention for patients with stage 3 intermediate-risk disease.
3. Mass screening studies suggest that tumors occurring in infancy can regress without chemotherapy or surgical resection.
4. Based on these data, the ANBL0531 will investigate whether a further reduction in therapy can be achieved for patients with stage 3 disease and favorable biology or patients less than 18 months of age with stage 4 disease and favorable biology by using an endpoint of partial response (PR; at least 50% reduction in primary tumor) rather than very good partial response (VGPR, at least 90% reduction in primary tumor). Residual tumor will be monitored with appropriate imaging studies.

Radiation Therapy
Radiation therapy has a limited role in the treatment of intermediate-risk neuroblastoma.

1. Radiotherapy is to be administered when there are life-threatening symptoms or a risk of organ impairment/neurologic compromise due to tumor bulk not responding to initial chemotherapy.
2. Intraspinal neuroblastoma with neurologic deficits will be managed with chemotherapy alone in an effort to avoid neurosurgical laminectomy or radiation as primary therapy. Laminectomy or radiation will be used if there is progression of neurological signs despite chemotherapy.

High-Risk Group

The high-risk group (Table 22-8) consists of the following:

- INSS stage 4 greater than 18 months of age
- INSS stage 4 with *MYCN* amplification, regardless of age
- INSS stage 4 between ages of 12 and 18 months with unfavorable histology and/or diploid tumor DNA content

- INSS stage 3 with amplified *MYCN*, regardless of age
- INSS stage 3 and unfavorable histology, greater than 18 months of age
- INSS stage 2 with *MYCN* amplification, regardless of age
- INSS stage 4S with *MYCN* amplification.

The probability of long-term survival in this group of patients is approximately 40% following comprehensive treatment approaches that include:

- Induction therapy that incorporates alternating cycles of dose-intensive multiagent chemotherapy and surgical resection of primary disease
- Consolidation therapy that includes high-dose consolidation therapy with autologous hematopoietic stem cell rescue and external beam radiotherapy to the primary tumor site
- Continuation/maintenance therapy for minimal residual disease.

Treatment

Induction Therapy

Neuroblastoma is usually responsive to chemotherapy even in the presence of *MYCN* amplification. The goal of induction therapy is to induce maximum reduction in the tumor bulk and metastatic sites. The efficacy of induction chemotherapy is assessed by the response rate.

- The quality of remission at the end of induction therapy is associated with long-term survival
- Retrospective analyses have determined that persisting bony lesions and bone marrow involvement were independent adverse prognostic factors
- Delayed resolution of MIBG avid disease during induction therapy correlates with decreased event-free survival (EFS) and overall survival (OS)
- There are many treatment protocols available for induction chemotherapy with a 5–7-month duration of induction therapy. Table 22-11 describes recent regimens being investigated in North America and Europe. The induction regimen currently being used by the Children's Oncology Group is shown in Table 22-12
- Despite marked increases in the dose intensity of active agents such as cisplatin, cyclosphosphamide and etoposide, an end induction complete remission is achieved in less than 75% of patients. Approximately 10% of patients will have disease progression during induction.

Consolidation Therapy

Patients without progressive disease during induction therapy should proceed with consolidation therapy.

Table 22-11 Induction Chemotherapy Regimens and Response Rates from Multicenter Clinical Trials

Study	Regimen[a]	Response Rate (CR + VGPR + PR)[b]
CCG 3891	CDDP, DOX, ETOP, CPM	76%
	Repeat every 28 days × 5 cycles	
POG 9341	CDDP + ETOP: VCR + DOX + CPM;	78%
	IFOS + ETOP; CARBO + ETOP;	
	CDDP + ETOP	
	Five sequential cycles every 28 days	
COG A3973	VCR + DOX + CPM cycles 1, 2, 4, 6	78%
	CDDP + ETOP cycles 3, 5	
	Cycles given every 21 days	
ENSG-5	VCR + CARBO + ETOP cycle 1, 3, 5, 7	85%
	VCR + CDDP cycles 2, 4, 6, 8	
	Cycles given every 10 days	

[a]All drugs are given intravenously.

[b]Responses all include surgery and chemotherapy.

Abbreviations: CARBO, carboplatin, CDDP, Cisplatin; CPM, cyclophosphamide; CR, complete response; DOX, doxorubicin; PR, partial response; VCR, vincristine; ETOP, Etoposide; IFOS, Ifosfamide.

CCG 3891

Cisplatin 60 mg/m^2/dose on day 1, doxorubicin 30 mg/m^2/dose on day 3, etoposide 100 mg/m^2/dose on days 3 and 6 and cyclophosphamide 1,000 mg/m^2/dose on days 4 and 5.

POG 9341

Cycles 1 and 5: Cisplatin 40 mg/m^2/day on days 1 to 5, Etoposide 100 mg/m^2/dose twice daily on days 1 to 3.

Cycle 2: Cyclophosphamide 1,000 mg/m^2/dose on days 1 and 2, Doxorubicin 60 mg/m^2/dose on day 1, Vincristine 1.5 mg/m^2/dose on days 1, 8 and 15.

Cycle 3: Ifosfamide 2 g/m^2/dose on days 1 to 5, Etoposide 75 mg/m^2/dose twice per day on days 1 to 3.

Cycle 4: Carboplatin 500 mg/m^2/dose on days 1 and 2, Etoposide 75 mg/m^2/dose twice per day on days 1 to 3.

COG A3973

Vincristine 0.67 mg/m^2/24 hours or 0.022 mg/kg/24 hours (whichever is lower) by continuous infusion over 72 hours on days 1, 2 and 3 (total dose 2 mg/m^2/over 72 hours or 0.067 mg/kg/72 hours with MAXIMUM total per cycle of 2 mg). For patients ≤12 kg, use 0.022 mg/kg/24 hours by continuous infusion for 72 hours (total dose 0.066 mg/kg/3 days). For infants <12 months, use 0.017 mg/kg/24 hours by continuous infusion for 72 hours, total dose 0.051 mg/kg/3 days. Cyclophosphamide 2.1 g/m^2/day on days 1 and 2. For patients ≤12 kg 70 mg/kg/day. Doxorubicin 25 mg/m^2/24 hours by continuous infusion over 72 hours on days 1, 2 and 3. (Total dose 75 mg/m^2/72 hours). For patients ≤12 kg, use 0.83 mg/kg/24 hours (total dose 2.49 mg/kg/72 hours). Etoposide 200 mg/m^2/day (for patients ≤12 kg use 6.67 mg/kg/day) on days 1, 2 and 3. (Total dose 600 mg/m^2/3 days or 20 mg/kg/3 days if ≤12 kg). Cisplatin 50 mg/m^2/day (for patients ≤12 kg use 1.66 mg/kg/day) on days 1, 2, 3 and 4 (Total dose 200 mg/m^2/4 days or 6.64 mg/kg/4 days).

ENSG-5

Cycles 1 and 5: Vincristine 1.5 mg/m^2 on day 1, Carboplatin 750 mg/m^2 on day 1, Etoposide 175 mg/m^2 on days 1 and 2.

Cycles 2, 4, 6 and 8: Vincristine 1.5 mg/m^2 on day 1, Cisplatin 80 mg/m^2 over 24 hours beginning on day 1.

Cycles 3 and 7: Vincristine 1.5 mg/m^2 on day 1, Etoposide 175 mg/m^2 on days 1 and 2, Cyclophosphamide 1,050 mg/m^2 on days 1 and 2.

Adapted with permission from: Matthay KK, Reynolds CP, Seeger RC, et al. Long-term results for children with high-risk neuroblastoma treated on a randomized trial of myeloablative therapy followed by 13-cis-retinoic acid: A Children's Oncology Group study. J Clin Oncol 2009;27:1007–1013, with permission.

Zage PE, Kletzel M, Murray K. et al. Outcomes of the POG 9340/9341/9342 trials for children with high-risk neuroblastoma: a report from the Children's Oncology Group. Pediatric Blood & Cancer 2008;51:747–753, with permission.

Pearson AD, Pinkerton CR, Lewis IJ, et al. High-dose rapid and standard induction chemotherapy for patients aged over 1 year with stage 4 neuroblastoma: a randomised trial. Lancet Oncol 2008;9:247–256.

**Table 22-12   Induction Therapy for High-risk Group Neuroblastoma
(Children Oncology Group)**

Cycle 1	Topotecan	1.2 mg/m^2/dose IV once daily for 5 doses
	Cyclophosphamide	>12 kg 400 mg/m^2/dose IV once daily for 5 doses
		≤12 kg 13.3 mg/kg/dose IV once daily for 5 doses
Cycle 2	Same as Cycle 1	
Peripheral Stem Cell Harvest		
Cycle 3	Cisplatin	>12 kg 50 mg/m^2/dose IV once daily for 4 doses
		≤12 kg 1.66 mg/m^2/dose IV once daily for 4 doses
	Etoposide	>12 kg 200 mg/m^2/dose IV once daily for 3 days
		≤12 kg 6.67 mg/kg/dose IV once daily for 3 days
Cycle 4	Cyclophosphamide	>12 kg 2,100 mg/m^2/dose IV once daily for 2 doses
		≤12 kg 70 mg/kg/dose IV once daily for 2 doses
	Doxorubin	>12 kg 25 mg/m^2/dose IV once daily for 3 doses
		≤12 kg 0.83 mg/kg/dose IV once daily for 3 doses
	Vincristine	>12 kg & ≥12 months 0.67 mg/m^2/dose IV once daily for 3 days
		≤12 kg & ≥12 months 0.022 mg/kg/dose IV once daily for 3 days
		<12 months 0.017 mg/kg/dose IV once daily for 3 days
Cycle 5	Same as Cycle 3	
Surgery		
Cycle 6	Same as Cycle 4	

- The goal of consolidation therapy is to achieve a state of minimal residual disease. The efficacy of myeloablative chemotherapy with autologous hematopoietic stem cell rescue as consolidation therapy has been demonstrated in several multicenter randomized clinical trials
- Myeloablative therapy with purged autologous bone marrow transplant (AMBT) improves outcome for patients with high-risk neuroblastoma compared to maintenance chemotherapy or observation
- Sequentially administered (tandem) myeloablative chemotherapy regimens may provide a further decrement in risk of neuroblastoma relapse, forming the basis for an ongoing COG randomized trial comparing single versus tandem transplant consolidation for high-risk neuroblastoma.

Consolidation regimens differ in each national trial. Trial CG3891 utilizes a regimen that combines myeloablative doses of carboplatin, etoposide and melphalan (CEM) with total body irradiation (TBI). In patients with creatinine clearance or glomerular filtration rate ≥100 ml/min/1.73 m^2 regimen includes:

- carboplatin (425 mg/m^2/dose IV for patients >12 kg and 14.2 mg/kg/dose IV for patients ≤12 kg once daily for four doses)
- melphalan (70 mg/m^2/dose IV for patients >12 kg and 2.3 mg/kg/dose IV for patients ≤12 kg once daily for three doses)
- etoposide (338 mg/m^2/dose IV for patients >12 kg and 11.3 mg/kg/dose IV for patients ≤12 kg once daily for four doses).

Doses for patients with altered renal function (creatinine clearance or glomerular filtration rate ≥ 60 and <100 ml/min/1.73 m^2) should be carefully modified. Given the toxicity of delivering TBI to young children, including risk for secondary malignancies, more recent treatment regimens have incorporated non-TBI-containing myeloablative regimens.

Retrospective analyses suggest that regimens containing only chemotherapy are as efficacious as those containing TBI and many centers now use CEM rather than a TBI-containing preparative regimen. A current clinical trial is comparing efficacy of a busulfan plus melphalan regimen to a CEM-based regimen.

Peripheral blood stem cells (PBSC) are the recommended source for hematopoietic stem cell rescue following consolidation myeloablative chemotherapy.

- PBSC are less likely than bone marrow to contain neuroblastoma tumor cells as measured by immunohistochemical techniques and PBSC permits earlier engraftment
- No benefit has been demonstrated in terms of risk for tumor recurrence following *ex vivo* antineuroblastoma antibody PBSC purging and that effective "*in vivo*" purging of stem cell product is achieved.

Continuation/Maintenance Therapy
The majority of patients with high-risk neuroblastoma experience relapse, suggesting the existence of minimal residual disease at completion of consolidation therapy. The aim of maintenance therapy is to eradicate any residual tumor cells using noncytotoxic agents which are active against chemoresistant tumor cells that may have survived intensive induction and consolidation regimens. The following drugs are used:

- **13-*cis*retinoic acid (cisRA)**: The administration of cisRA (80 mg/m^2/dose for patients >12 kg and 2.67 mg/kg/dose for patients ≤ 12 kg given twice per day for 14 consecutive days in each 28-day cycle for a total of six cycles) following consolidation therapy decreases risk of developing disease recurrence and it has been incorporated into standard post-consolidation therapy for high-risk neuroblastoma
- **Anti-ganglioside (GD2) antibody therapy**: The addition of anti-GD2 immunotherapy (chimeric 14.18 antibody) combined with cytokines granulocyte macrophage colony stimulating factor (GMCSF) and interleukin-2 (IL2) to standard cisRA significantly decreases risk for recurrence and improves overall survival for high-risk neuroblastoma. There is currently no clinical source of anti-GD2 antibody therapy and it must be obtained in the context of clinical trials in North America and Europe.

POST-THERAPY MONITORING

Routine monitoring of patients during and following completion of therapy is required given the risk for recurrent disease. Table 22-13 lists the recommended disease status

Table 22-13   Recommended Disease Status Monitoring for High-Risk Neuroblastoma

Diagnostic Test	Frequency Post-Therapy		
	1st Year	2nd Year	After 2nd Year
CT and/or MRI	3 months or as clinically indicated	3–6 months or as clinically indicated	6–12 months if clinically indicated
^{123}I-MIBG Scintography if MIBG avid at diagnosis	3 months or as clinically indicated	3–6 months or as clinically indicated	6–12 months if clinically indicated
^{18}F-FDG PET or ^{99m}Tc-scintography if not MIBG avid	3 months or as clinically indicated	3–6 months or as clinically indicated	6–12 months if clinically indicated
Bone marrow biopsy and aspirate (bilateral)	3 months or as clinically indicated	3–6 months or as clinically indicated	6–12 months if clinically indicated
VMA/HVA (if abnormal at diagnosis)	if abnormal at end of therapy or as clinically indicated	if abnormal at end of therapy or as clinically indicated	if abnormal at end of therapy or as clinically indicated

monitoring for high-risk neuroblastoma. Patients must also be monitored for long-term complications of therapy including anthracycline-induced cardiomyopathy, platinum-induced ototoxicity and renal dysfunction, chemotherapy and radiation-induced risk for secondary malignancies, infertility, growth and development deficiencies.

- The risk for disease recurrence is highest in the initial 2 years following completion of therapy but can occur as long as 4 years from completion of therapy
- Disease recurrence primarily occurs in metastatic sites with bone and bone marrow most frequent followed by nodal recurrence
- Primary site recurrence is less frequent, occurring in 10–15% of children following aggressive surgical resection and local radiotherapy.

Post-Therapy Surveillance Recommendations

1. I^{123}-MIBG scan (if MIBG avid at diagnosis), bone marrow assessment and CT/MRI scans of primary site of disease are warranted during the initial 5 years from completion of therapy.
2. Frequency of surveillance (MIBG scans, CT/MRI scans and bone marrow aspirations) should be carried out every 3–6 months for the initial 2 years as the risk for recurrence is greatest in the initial 2 years from diagnosis. For those patients who remain in complete remission after 2 years the frequency of surveillance scans can be extended to every 6–12 months until 5 years from diagnosis.
3. Patients who survive long term from high-risk neuroblastoma should be monitored in a comprehensive survivorship program skilled at facilitating management of therapy-related morbidities.

SPECIAL TREATMENT CONSIDERATIONS

Dumbbell Neuroblastoma and Spinal Cord Compression

Spinal cord decompression is a neurological emergency and treatment can be accomplished by: chemotherapy, surgery and radiation. Complete neurological recovery is observed in 30–40% of patients. Neurological improvement is observed in 65–70% of patients.

1. **Decompression by chemotherapy combined with the use of corticosteroids (e.g. decadron)**: Emergent initiation of decadron should be instituted upon diagnosis of neuroblastoma-related spinal cord compression. Diagnosis and staging should be performed rapidly to avoid delay in initiation of chemotherapy. Intensity of chemotherapy is determined according to stage and age risk criteria as previously described. The majority of patients (>85%) will have rapid tumor shrinkage and prevention or improvement in neurologic compromise. The patient's neurologic status must be monitored closely with consideration of using radiotherapy or surgical laminectomy should the patient exhibit new onset or worsening neurologic compromise following initiation of decadron and chemotherapy.
2. **Decompression by laminectomy**: Surgery is reserved for the uncommon patient given the likely initial response to chemotherapy and the potential development of kyphoscoliosis and spinal column instability following a laminectomy procedure.
3. **Decompression by radiotherapy**: No longer indicated in the initial management of neuroblastoma but should be considered if the patient is not responding to initial chemotherapy.

Opsoclonus Myoclonus Ataxia Syndrome (OMA)

Clinical features of the OMA syndrome are detailed above. Clinical observations suggest an immune-mediated mechanism for OMA syndrome.

• Patients with OMA syndrome have serum and cerebrospinal fluid IgG and IgM auto-antibodies that bind to the cytoplasm of cerebellar Purkinje cells, central and peripheral axons and neural proteins of the neurofilaments
• Diffuse and extensive lymphocytic infiltration with lymphoid follicles is a characteristic histologic feature of neuroblastic tumors with associated opsoclonus myoclonus
• The administration of chemotherapy may alleviate symptoms.

Treatment

Definitive treatment remains controversial; suppression of the immune system is the mainstay of therapy.

- Chemotherapy should be utilized as appropriate based upon documented clinical risk group at the time of diagnosis
- Immunomodulatory therapies such as ACTH, prednisone, dexamethasone and low-dose cyclophosphamide have been utilized for treatment
- High-dose intravenous immunoglobulin (IVIG 150–500 mg/kg/day for 4–5 days) may benefit patients or provide a steroid-sparing effect
- A COG randomized clinical trial (ANBL00P2) is investigating whether the addition of IVIG to steroids and low-dose chemotherapy will improve neurologic outcome for patients with neuroblastoma-associated OMA syndrome.

NEUROBLASTOMA IN THE ADOLESCENT AND YOUNG ADULT

1. The distribution of primary neuroblastoma sites in this group of patients is similar to that seen in pediatric age group patients.
2. The tumor growth rates of neuroblastoma in young adults and adults are slower than in pediatric age group patients.
3. Neuroblastoma in this group of patients is relatively more resistant to chemotherapy than is neuroblastoma in pediatric patients.
4. The best treatment approach remains to be determined. However, intensive chemotherapy with autologous bone marrow transplantation, following surgery with or without radiation may be considered, especially for those patients with advanced regional or metastatic disease.

Suggested Reading

Adkins ES, Sawin R, Gerbing RB, et al. Efficacy of complete resection for high-risk neuroblastoma: a Children's Cancer Group study. J Pediatr Surg. 2004;39:931–936.

Ambros PF, Ambros IM, Brodeur GM, et al. International consensus for neuroblastoma molecular diagnostics: report from the International Neuroblastoma Risk Group (INRG) Biology Committee. Br J Cancer 2009;100:1471–1482.

Berthold F, Boos J, Burdach S, et al. Myeloablative megatherapy with autologous stem-cell rescue versus oral maintenance chemotherapy as consolidation treatment in patients with high-risk neuroblastoma: a randomised controlled trial. Lancet Oncol. 2005;6:649–658.

Brodeur GM. Neuroblastoma: biological insights into a clinical enigma. Nat Rev Cancer. 2003;3:203–216.

Brodeur GM, Pritchard J, Berthold F, et al. Revisions of the International Criteria for Neuroblastoma Diagnosis, Staging and Response to Treatment. J Clin Oncol. 1993;11:1466–1477.

Cohn SL, Pearson AD, London WB, et al. INRG Task Force. The International Neuroblastoma Risk Group (INRG) classification system: an INRG Task Force report. J Clin Oncol. 2009;27:289–297.

George RE, Li S, Medeiros-Nancarrow C, et al. High-risk neuroblastoma treated with tandem autologous peripheral-blood stem cell-supported transplantation: long-term survival update. J Clin Oncol. 2006; 24(18):2891–2896.

Guglielmi M, et al. Resection of primary tumor at diagnosis in stage IV-S neuroblastoma: does it affect the clinical course? J Clin Oncol. 1996;14:1537–1544.

Kushner BH, Kramer K, LaQuaglia MP, et al. Reduction from seven to five cycles of intensive induction chemotherapy in children with high-risk neuroblastoma. J Clin Oncol. 2004;22:4888–4892.

Maris JM, Hogarty MD, Bagatell R, Cohn SL. Neuroblastoma. Lancet 2007;369:2106–2120.

Maris JM, Woods WG. Screening for neuroblastoma: a resurrected idea? Lancet 2008;371:1142–1143.

Matthay KK, Blaes F, Hero B, et al. Opsoclonus myoclonus syndrome in neuroblastoma a report from a workshop on the dancing eyes syndrome at the advances in neuroblastoma meeting in Genoa, Italy, 2004. Cancer Lett. 2005;228:275–282.

Matthay KK, Reynolds CP, Seeger RC, et al. Long-term results for children with high-risk neuroblastoma treated on a randomized trial of myeloablative therapy followed by 13-cis-retinoic acid: a Children's Oncology Group study. J Clin Oncol. 2009;27:1007–1013.

Modak S, Cheung NK. Disialoganglioside directed immunotherapy of neuroblastoma. Cancer Invest. 2007;25:67–77.

Monclair T, Brodeur GM, Ambros PF, et al. INRG Task Force. The International Neuroblastoma Risk Group (INRG) staging system: an INRG Task Force report. J Clin Oncol. 2009;27:298–303.

Mossé YP, Wood A, Maris JM. Inhibition of ALK signaling for cancer therapy. Clin Cancer Res. 2009; 15:5609–5614.

Nickerson HJ, Matthay KK, Seeger RC, et al. Favorable biology and outcome of stage IV-S neuroblastoma with supportive care or minimal therapy: a Children's Cancer Group study. J Clin Oncol. 2000;18:477–486.

Park JR, Eggert A, Caron H. Neuroblastoma: biology, prognosis and treatment. Pediatr Clin North Am. 2008;55:97–120.

Pearson AD, Pinkerton CR, Lewis IJ, et al. High-dose rapid and standard induction chemotherapy for patients aged over 1 year with stage 4 neuroblastoma: a randomised trial. Lancet Oncol. 2008;9:247–256.

Pritchard J, Cotterill SJ, Germond SM, et al. High dose melphalan in the treatment of advanced neuroblastoma: results of a randomised trial (ENSG-1) by the European Neuroblastoma Study Group. Pediatr Blood Cancer 2005;44(4):348–357.

Shimada H, Umehara S, Monobe Y, et al. International neuroblastoma pathology classification for prognostic evaluation of patients with peripheral neuroblastic tumors: a report from the Children's Cancer Group. Cancer 2001;92:2451–2461.

CHAPTER 23

Renal Tumors

WILMS' TUMOR

Clinical Features

Incidence

1. Renal tumors comprise approximately 6% of all childhood cancers and nearly 10% of all malignancies among children aged 1–4 years.
2. Wilms' tumor (nephroblastoma) is the most common primary renal tumor of childhood and the sixth most common childhood malignancy in the United States.
3. Male to female ratio 0.92:1.00 in unilateral Wilms' and 0.6:1.00 in bilateral Wilms' tumor.
4. Wilms' tumor, clear cell sarcoma of the kidney and malignant rhabdoid tumor of the kidney occur predominantly in younger children and are rarely seen after 10 years of age. The majority of renal tumors occurring in adolescents are renal cell carcinomas.
5. Seventy-eight percent of children with Wilms' tumor are diagnosed at 1–5 years of age, with a peak incidence occurring between 3 and 4 years of age. Median age of presentation is 44 months in unilateral disease and 32 months in bilateral disease.
6. Wilms' tumor is usually sporadic, but 1% of cases are familial.

Associated Congenital Anomalies

Congenital anomalies occur in 12–15% of cases. The most frequently diagnosed congenital anomalies are aniridia, genitourinary anomalies and hemihypertrophy. Table 23-1 lists the incidence of congenital anomalies in patients with Wilms' tumor.

Three syndromes associated with Wilms' tumor include:

* WAGR syndrome
* Denys–Drash syndrome and
* Beckwith–Wiedemann syndrome.

Manual of Pediatric Hematology and Oncology. DOI: 10.1016/B978-0-12-375154-6.00023-9

Table 23-1   Incidence of Congenital Anomalies in Patients with Wilms' Tumor

Anomaly	Incidence (%)
Genitourinary anomalies: horseshoe kidney, dysplasia of kidney, cystic disease of kidney, hypospadias, cryptorchidism, duplication of collecting system	4.4
Congenital aniridia	1.1
Congenital hemihypertrophy	2.9
Musculoskeletal anomalies: clubfoot, rib fusion, distal phocomelia, hip dislocation	2.9
Hamartomas: hemangiomas, "birthmarks," multiple nevi, café-au-lait spots	7.9

WAGR Syndrome

1. The syndrome consists of Wilms' tumor, aniridia, genitourinary malformations and mental retardation, but other associated anomalies are seen (Table 23-2).
2. A chromosomal deletion has been consistently found in the short arm of chromosome 11 (11p13 deletion). This area encompasses both the *WT1* and *PAX6* genes; *WT1* is implicated in Wilms' tumor and *PAX6* is implicated in aniridia.
3. Aniridia is present in 1.2% of children with Wilms' tumor.
4. A child with sporadic aniridia has a 5% chance of developing Wilms' tumor; most cases of sporadic aniridia have *PAX6* mutations/deletions but not *WT1* deletions. A child with *WT1* deletion has a 30–40% risk of Wilms' tumor.
5. WAGR syndrome is associated with focal segmental glomerular sclerosis (FSGS) and renal failure in 30–40% of individuals 20 years following Wilms' tumor diagnosis.

Denys–Drash Syndrome

1. The syndrome consists of Wilms' tumor, early renal failure with mesangial sclerosis and pseudohermaphroditism.
2. Associated with point mutations of the *WT1* gene.
3. Associated with 90% risk of Wilms' tumor.

Beckwith–Wiedemann Syndrome

1. This syndrome consists of:
   - Hyperplastic fetal visceromegaly involving the kidney, adrenal cortex, pancreas, gonads and liver, hemihypertrophy, macroglossia, abdominal wall defects (omphalocele, umbilical hernia, diastasis recti), ear pits or creases, microcephaly, mental retardation, hypoglycemia and postnatal somatic gigantism

Table 23-2 Associated Anomalies in Patients with WAGR Syndrome

Clinical Finding	% patients
Ocular	
Cataracts	67
Glaucoma	44
Nystagmus	41
Optic nerve hypoplasia	15
Macular/foveal hypoplasia	13
Retinal detachment	9
Strabismus	7
Genitourinary/renal	
Cryptorchidism	35
Ambiguous genitalia	7
Hypospadias	7
Neurologic	
Hypotonia/hypertonia	11
Epilepsy	7
Renal	
Proteinuria	26
FSGS	11
Cardiorespiratory	
Asthma	15
Recurrent pneumonia	11
Ears/Nose/Throat	
Tonsillectomy/adenoidectomy	41
Tympanostomy tube	35
Recurrent sinusitis	28
Obstructive sleep apnea	20
Recurrent otitis media	19
Severe dental malocclusion	17
Behavioral	
Attention deficit/hyperactivity	22
Autism spectrum	19
Obsessive compulsive	9
Anxiety disorder	7
Musculoskeletal	
Hypertense Achilles	17
Scoliosis	15
Metabolic	
Obesity	19

Abbreviations: FSGS, focal segmental glomerular sclerosis.
Modified from: Fischbach BV, Trout KL, Lewis J, et al. WAGR syndrome: A clinical review of 54 cases. Pediatrics 2005;116:984–988.

- Predisposition to embryonal tumors (Wilms' tumor, hepatoblastoma, neuroblastoma and rhabdomyosarcoma) and adrenocortical carcinoma. Risk of Wilms' tumor is approximately 5%.

2. Associated with imprinting defects at several genes at chromosome 11p15.5 (*IGF2*, *H19*, *p57*). This locus is often referred to as WT2.

Screening of Children with WAGR, Denys–Drash and Beckwith–Wiedemann Syndromes

Children with Beckwith–Wiedemann syndrome should have an abdominal ultrasound every 3 months until 7 years of age to evaluate for abdominal malignancies. Serum α-fetoprotein level should be screened during infancy and early childhood because of the association with hepatoblastoma. Genetic testing should be considered because new data indicate that certain Beckwith–Wiedemann mutations are more tumor-prone than others. Children with WAGR and Denys–Drash syndromes should have screening abdominal ultrasounds every 3 months until 5 years of age. Children with sporadic aniridia should be referred for FISH (fluorescence *in situ* hybridization) of the *WT1* gene because if *WT1* is not deleted, the risk of Wilms' tumor is similar to the population risk.

Signs and Symptoms

1. Initial signs and symptoms, in order of frequency, are listed in Table 23-3.
2. Abdominal mass is the most common presenting symptom and sign. Occasionally, there is abdominal pain, especially when hemorrhage occurs in the tumor following trauma.
3. Hematuria is not common but is more often seen microscopically than on gross examination.
4. Hypertension is seen in approximately 25% of patients due to elaboration of renin by tumor cells or, less commonly, due to distortion or compression of renal vasculature.
5. Polycythemia is occasionally present. Erythropoietin levels are usually increased but can also be normal. Polycythemia is usually associated with males, older age and low clinical stage. All children with unexplained polycythemia should be investigated for Wilms' tumor.

Table 23-3   Initial Signs and Symptoms of Wilms' Tumor in Order of Frequency

Sign/Symptom	Frequency (%)
Palpable mass in abdomen	60
Hypertension	25
Hematuria	15
Obstipation	4
Weight loss	4
Urinary tract infection	3
Diarrhea	3
Previous trauma	3
Other signs/symptoms: nausea, vomiting, abdominal pain, inguinal hernia, cardiac insufficiency[a], acute surgical abdomen, pleural effusion, polycythemia, hydrocephalus[b]	8

[a]Due to propagation of tumor clot from inferior vena cava into right atrium.
[b]Intracerebral secondary, associated congenital anomaly or associated central nervous system tumor.

6. Bleeding diathesis can occur due to the presence of acquired von Willebrand disease (reduced von Willebrand factor antigen level, prolonged bleeding time, decreased factor VIII and factor VIII ristocetin cofactor activity). The frequency of this complication is unknown.

Diagnostic Studies

Table 23-4 lists the standard clinical, laboratory and radiographic investigations required to evaluate a patient with a suspected renal tumor. Since the most common initial surgical approach is nephrectomy the following information is required prior to surgery:

- Presence/absence of a functioning kidney on the contralateral side
- Presence/absence of lung or liver metastases
- Presence/absence of bone or brain metastases if indicated by history or physical examination
- Presence/absence of tumor thrombus in the renal vein or inferior vena cava.

Table 23-4   Investigations in Wilms' Tumor

History: family history of cancer, congenital defects and benign tumors
Physical examination: congenital anomalies (aniridia, hemihypertrophy, genitourinary anomalies), blood pressure, liver enlargement
Complete blood count: presence/absence of polycythemia, anemia, thrombocytopenia
Urinalysis: presence/absence of hematuria
Serum chemistries: blood urea nitrogen, creatinine, uric acid, serum glutamic-oxaloacetic transaminases (SGOT), serum glutamic pyruvic transaminases (SGPT), lactic dehydrogenase (LDH), alkaline phosphatase
Assessment of coagulation factors: prothrombin time, partial thromboplastin time, fibrinogen level, PFA-100 closure time, factor VIII level, von Willebrand factor antigen level, factor VIII ristocetin cofactor activity should be considered to screen for acquired von Willebrand disease
Assessment of cardiac status: electrocardiogram and echocardiogram in all patients who will receive doxorubicin; echocardiogram may also be useful in detecting the presence of tumor in the IVC or right atrium
Abdominal ultrasound including Doppler imaging of renal veins and IVC
Abdominal CT scan with special attention to:
  Presence and function of the opposite kidney
  Evidence of bilateral lesions
  Evidence of involvement of renal vein or IVC with tumor
  Lymph node involvement
  Liver metastasis
Chest CT scan
Bone scintigraphy: only in cases of clear cell sarcoma, renal cell carcinoma, or rhabdoid tumor of kidney
Magnetic resonance imaging of brain: only in cases of clear cell sarcoma, renal cell carcinoma, or rhabdoid tumor of kidney
Peripheral blood for chromosomal analysis: in cases of congenital anomalies, such as aniridia, Beckwith–Wiedemann syndrome, hemihypertrophy

Staging System

Table 23-5 describes the current staging system for renal tumors utilized by the Renal Tumors Committee of the Children's Oncology Group (COG).

Table 23-5   Staging System for Renal Tumors (COG Renal Tumors Committee)

Stage I
Completely resected tumor limited to kidney with intact capsule
No biopsy or rupture of tumor prior to removal
No involvement of vessels or renal sinuses
No tumor at or beyond margins of resection
Regional lymph nodes negative for tumor
Stage II
Completely resected tumor
No tumor at or beyond margins of resection
Regional lymph nodes negative for tumor
One or more of the following findings:
   Penetration of the renal capsule
   Invasion of vasculature extending beyond renal parenchyma
Stage III
Residual tumor present after surgery, confined to abdomen, with one or more of the following present:
   One or more regional lymph nodes positive for tumor
   Tumor implanted on or penetrating through peritoneum
   Presence of gross unresected tumor or tumor at margin of resection
   Any tumor spillage occurring before or during surgery, including biopsy
   Tumor removed in more than one piece
Stage IV
Presence of hematogenous metastasis (e.g., lung, liver, bone, brain)
Presence of lymph node metastasis outside the abdomen and pelvis
Stage V
Wilms' tumor in both kidneys

Pathology

Wilms' tumor is derived from primitive metanephric blastema and is characterized by histopathologic diversity. The classic Wilms' tumor is composed of persistent blastema, dysplastic tubules (epithelial) and supporting mesenchyme or stroma. The coexistence of epithelial, blastemal and stromal cells has led to the term *triphasic* to characterize the classic Wilms' tumor. Each of the cell types may exhibit a spectrum of differentiation, generally replicating various stages of renal embryogenesis. The proportion of each cell type may also vary significantly from tumor to tumor. Some Wilms' tumors may be biphasic or even monomorphous in appearance. The absence of anaplastic features identifies Wilms' tumors as having favorable histology.

Clear cell sarcoma of the kidney, rhabdoid tumor of the kidney and renal cell carcinomas have unique histologies and are not Wilms' tumor variants.

Anaplastic Wilms' Tumor

Histiologic Findings

Anaplastic histology is identified by the presence of cells with nuclear enlargement, nuclear atypia and irregular mitotic figures. The presence of anaplasia is the only criterion for "unfavorable" histology in a Wilms' tumor.

Focal anaplasia is distinguished from diffuse anaplasia by the distribution of anaplastic elements. Anaplasia well-contained within a single region of the tumor is considered focal anaplasia and anaplasia found in multiple regions of the tumor or outside the renal parenchyma is considered diffuse anaplasia.

Favorable histology, as the name implies, is associated with the best prognosis, followed by focal anaplasia, followed by diffuse anaplasia having significantly worse outcomes.

A diagnosis of diffuse anaplasia requires the following characteristics:

- Anaplasia in any extrarenal site, including vessels of the renal sinus, extracapsular infiltrates, or nodal or distant metastases
- Anaplasia in a random biopsy specimen
- Anaplasia unequivocally expressed in one region of the tumor, but with extreme nuclear pleomorphism approaching the criteria of anaplasia (extreme nuclear unrest) elsewhere in the lesion
- Anaplasia in more than one tumor slide, unless: (1) it is known that every slide showing anaplasia came from the same focused region of the tumor; or (2) anaplastic foci on the various slides are minute and surrounded on all sides by nonanaplastic tumor.

Loss of Heterozygosity as a Prognostic Factor for Wilms' Tumor

Loss of heterozygosity (LOH) for polymorphic DNA markers on chromosomes 1p and 16q in favorable histology Wilms' tumor cells has been shown to be associated with inferior relapse-free survival (RFS) and overall survival (OS) in patients with favorable histology Wilms' tumor. More recently, LOH for 1p and 16q was studied prospectively as a biologic component of NWTS-5. Table 23-6 summarizes the prognostic significance of LOH for 1p and 16q in favorable-histology Wilms' tumor. Although LOH 1p/16q is currently being utilized for risk stratification and therapy determination in ongoing COG studies, it is not known at this time whether alteration of upfront therapy will improve the inferior RFS and OS conferred by LOH of 1p and 16q.

Table 23-6 **Prognostic Significance of Loss of Heterozygosity (LOH) for 1p and 16q in Favorable-Histology Wilms' Tumor**

Stage	LOH Status	4-Yr RFS	RR	P	4-Yr OS	RR	P
I or II	Neither	91.2%	—	—	98.4%	—	—
I or II	Both	74.9%	2.88	0.001	90.5%	4.25	0.01
III or IV	Neither	83.0%	—	—	91.9%	—	—
III or IV	Both	65.9%	2.41	0.01	77.5%	2.66	0.04

Abbreviations: LOH, loss of heterozygosity; RFS, relapse-free survival; RR, relapse risk; OS, overall survival.
From: Grundy PE, et al, J Clin Oncol 2005, with permission.

Table 23-7 **Treatment recommendations for previously untreated Wilms' tumor**

Stage	Histology	Surgery	Radiation Therapy	Chemotherapy Agents	Dose and Schedules	Duration (Weeks)
I and II	FH	YES	NO	AMD + VCR	Table 23-8	18
III and IV	FH	YES	YES	AMD + VCR + DOX	Table 23-9	24
I, II and III	FA DA	YES	YES	AMD + VCR + DOX	Table 23-9	24
IV	FA	YES	YES	VCR + DOX + CTX + ETOP	Table 23-10	24
II, III and IV	DA					

Abbreviations: FH, Favorable Histology; DA, Diffuse Anaplasia; FA, Focal Anaplasia; AMD, Dactinomycin (Actinomycin D); VCR, Vincristine; DOX, Doxorubicin (Adriamycin); CTX, Cyclophosphamide; ETOP, Etoposide.
Infants less than 12 months of age should receive one-half the recommended dose of all drugs, as full doses lead to prohibitive hematologic toxicity in this age group. Full doses of chemotherapeutic agents should be administered to children more than 12 months of age.

Treatment

Standard treatment of Wilms' tumor in North America consists of surgery and chemotherapy as shown in Tables 23-7 to 23-10. Treatment of Wilms' tumor in other parts of the world generally involves prenephrectomy chemotherapy. The chemotherapy regimens use the same agents at slightly different doses and durations of therapy.

Surgery

1. Complete exploration of the involved kidney should be undertaken.
2. Manual exploration of the liver and contralateral kidney is not required if high-quality preoperative CT imaging is available.

3. In all cases, lymph nodes should be sampled from the perihilar and para-aortic/ paracaval regions at minimum, in addition to all visible abnormal-appearing nodes. All lymph nodes removed should be identified and the site marked.

4. In unilateral renal tumor cases, radical excision, often nephrectomy, is advised so that all tumor tissue can be completely removed. The junction of suspicious abnormal areas with normal kidney should be removed to facilitate the accurate diagnosis of small lesions.

5. In cases involving bilateral renal tumors, a renal-sparing approach is warranted.

Table 23-8 Chemotherapy for Previously Untreated Stage I–II Favorable Histology Wilms' Tumor

Week 0		AMD
Week 1	VCR1	
Week 2	VCR1	
Week 3	VCR1	AMD
Week 4	VCR1	
Week 5	VCR1	
Week 6	VCR1	AMD
Week 7	VCR1	
Week 8	VCR1	
Week 9	VCR1	AMD
Week 10	VCR1	
Week 11		
Week 12	VCR2	AMD
Week 13		
Week 14		
Week 15	VCR2	AMD
Week 16		
Week 17		
Week 18	VCR2	AMD

Drug dosing information		
VCR1	Vincristine	0.025 mg/kg/day for 1 day IV for infants <12 months 0.05 mg/kg/day for 1 day IV for children ≥12 months & <30 kg 1.5 mg/m^2/day for 1 day IV for children >30 kg (Maximum dose 2 mg)
VCR2	Vincristine	0.034 mg/kg/day for 1 day IV for infants <12 months 0.067 mg/kg/day for 1 day IV for children ≥12 months & <30 kg 2 mg/m^2/day/day for 1 day IV for children >30 kg (Maximum dose 2 mg)
AMD	Actinomycin-D	0.023 mg/kg/day for 1 day IV for infants <12 months 0.045 mg/kg/day for 1 day IV for children ≥12 months & <30 kg 1.35 mg/m^2/day for 1 day IV for children >30 kg (Maximum dose 2.3 mg)
G-CSF (filgrastim) is not recommended with this regimen Doses of AMD reduced 50% if given within 6 weeks of whole lung or whole abdominal radiation therapy		

Table 23-9 Chemotherapy for Previously Untreated Stage III–IV Favorable Histology, Stage I–III Focal Anaplastic and Stage I Diffuse Anaplastic Wilms' Tumor

Week 0		AMD	
Week 1	VCR1		
Week 2	VCR1		
Week 3	VCR1		DOX1
Week 4	VCR1		
Week 5	VCR1		
Week 6	VCR1	AMD	
Week 7	VCR1		
Week 8	VCR1		
Week 9	VCR1		DOX1
Week 10	VCR1		
Week 11			
Week 12	VCR2	AMD	
Week 13			
Week 14			
Week 15	VCR2		DOX2
Week 16			
Week 17			
Week 18	VCR2	AMD	
Week 19			
Week 20			
Week 21	VCR2		DOX2
Week 22			
Week 23			
Week 24	VCR2	AMD	

Drug dosing information		
VCR1	Vincristine	0.025 mg/kg/day for 1 day IV for infants <12 months 0.05 mg/kg/day for 1 day IV for children ≥12 months & ≤30 kg 1.5 mg/m^2/day for 1 day IV for children >30 kg (Maximum dose 2 mg)
VCR2	Vincristine	0.034mg/kg/day for 1 day IV for infants <12 months 0.067 mg/kg/day for 1 day IV for children ≥12 months & ≤30 kg 2 mg/m^2/day for 1 day IV for children >30 kg (Maximum dose 2 mg)
AMD	Actinomycin-D	0.023mg/kg/day for 1 day IV for infants <12 months 0.045 mg/kg/day for 1 day IV for children ≥12 months & ≤30 kg 1.35 mg/m^2/day for 1 day IV for children >30 kg (Maximum dose 2.3 mg)
DOX1	Doxorubicin	0.75 mg/kg/day for 1 day IV for infants <12 months 1.5 mg/kg/day for 1 day IV for children ≥12 months & ≤30 kg 45 mg/m^2/day for 1 day IV for children >30 kg
DOX2	Doxorubicin	0.5 mg/kg/day for 1 day IV for infants <12 months 1 mg/kg/day for 1 day IV for children ≥12 months & ≤30 kg 30 mg/m^2/day for 1 day IV for children >30 kg

G-CSF (filgrastim) is not recommended with this regimen. Doses of DOX and AMD reduced 50% if given within 6 weeks of whole lung or whole abdominal radiation therapy

Table 23-10   Chemotherapy for Previously Untreated Stage IV Focal Anaplastic and Stage II–IV Diffuse Anaplastic Wilms' Tumor

Week 1	DOX			
Week 2		VCR1		
Week 3		VCR1		
Week 4			CTX5	ETOP
Week 5		VCR1		
Week 6		VCR1		
Week 7	DOX	VCR1	CTX3	
Week 8		VCR1		
Week 9		VCR1		
Week 10			CTX5	ETOP
Week 11		VCR1		
Week 12		VCR1		
Week 13	DOX	VCR2	CTX3	
Week 14		VCR2		
Week 15				
Week 16			CTX5	ETOP
Week 17				
Week 18				
Week 19	DOX	VCR2	CTX3	
Week 20				
Week 21				
Week 22			CTX5	ETOP
Week 23				
Week 24				
Week 25	DOX	VCR2	CTX3	

Drug dosing information		
DOX	Doxorubicin	0.75 mg/kg/day for 1 day IV for infants <12 months 1.5 mg/kg/day for 1 day IV for children ≥12 months & ≤30 kg 45 mg/m^2/day for 1 day IV for children >30 kg
VCR1	Vincristine	0.025 mg/kg/day for 1 day IV for infants <12 months 0.05 mg/kg/day for 1 day IV for children ≥12 months & ≤30 kg 1.5 mg/m^2/day for 1 day IV for children >30 kg (Maximum dose 2 mg)
VCR2	Vincristine	0.034 mg/kg/day for 1 day IV for infants <12 months 0.067 mg/kg/day for 1 day IV for children ≥12 months & ≤30 kg 2 mg/m^2/day for 1 day IV for children >30 kg (Maximum dose 2 mg)
CTX5	Cyclophosphamide	7.35 mg/kg/day IV for 5 days for infants <12 months 14.7 mg/kg/day IV for 5 days for children ≥12 months & ≤30 kg 440 mg/m^2/day IV for 5 days for children >30 kg
CTX3	Cyclophosphamide	7.35 mg/kg/day IV for 3 days for infants <12 months 14.7 mg/kg/day IV for 3 days for children ≥12 months & ≤30 kg 440 mg/m^2/day IV for 3 days for children >30 kg
ETOP	Etoposide	1.65 mg/kg/day for 5 days for infants <12 months 3.3 mg/kg/day IV for 5 days for children 1 ≥12 months & ≤30 kg 100 mg/m^2/day IV for 5 days for children >30 kg

G-CSF (filgrastim) 5 mg/kg/day subcutaneous starting 24 hours after cyclophosphamide until the ANC is >2,000/μL after the expected nadir
Doses of DOX reduced 50% if given within 6 weeks of whole lung or whole abdominal radiation therapy

Treatment of Tumors Considered Inoperable

Preoperative therapy should be used only in carefully selected cases after pathological diagnosis is obtained. Indications for preoperative therapy include bilateral Wilms' tumor, tumor thrombus in the inferior vena cava above the level of the hepatic veins, tumors that invade other organs if resection of the tumor would involve resection of the other organ (excluding the adrenal gland), tumors that, in the surgeon's judgment, would result in excessive morbidity if removed before chemotherapy. The following management plan is recommended:

- The chemotherapy regimen shown in Table 23-9 (vincristine, doxorubicin, dactinomycin [actinomycin-D]) should be administered
- An abdominal computed tomography (CT) scan should be performed after approximately 6 weeks of therapy
- Shrinkage usually occurs in 6 weeks. Surgery is planned according to tumor shrinkage and when a radiographic assessment suggests that the tumor can be totally excised and should be accomplished no later than week 12 of therapy
- The primary tumor should be considered stage III, regardless of the findings at surgery. Postoperative radiation therapy is required
- Pathology at definitive surgery should dictate the chemotherapy approach utilized postoperatively.

Radiation Therapy

Table 23-7 indicates which patients should receive radiation.

1. Radiation therapy (RT) is usually begun shortly after surgery (within 10–14 days), with the aim of eradicating tumor cells that may have spilled during surgery. The size of the field depends on the findings at surgery, but in all cases, the liver, spleen and opposite kidney should be carefully shielded.
2. Abdominal stage III favorable histology (FH) and abdominal stages I–III anaplastic Wilms' tumor, clear cell sarcoma and rhabdoid tumor require irradiation to the tumor bed at a dose of 1,080 cGy as determined by the preoperative radiographic findings. The tumor bed is defined as the outline of the kidney and any associated tumor. The portals should be extended to include areas of more diffuse involvement (e.g., the para-aortic chains) when those nodes are found to be involved.
3. When there is peritoneal seeding or gross tumor spillage, whole abdominal irradiation should be given at same doses as flank RT through anterior and posterior portals from the diaphragmatic domes to the inferior margin of the obturator foramen, sparing the femoral heads. The contralateral kidney should be shielded.
4. Unresected abdominal lymph node metastasis requires RT at a dose of to 1,980 cGy.

5. Actinomycin-D and doxorubicin doses should be decreased 50% if given within 6 weeks following whole-lung or whole-abdomen RT.
6. Whole abdominal radiotherapy is unnecessary for patients with tumor spills confined to the flank or for those who had prior biopsy of the neoplasm.
7. Metastatic disease to lungs requires whole lung RT 1,200 cGy (1,050 cGy if <12 months age).
8. *Pneumocystis jiroveci* prophylaxis with trimethoprim/sulfamethoxazole or pentamidine should be instituted in patients receiving lung RT.

Bilateral Wilms' Tumor

Synchronous bilateral Wilms' tumor occurs in about 5% of all Wilms' tumor patients. A significant concern for patients with bilateral Wilms' tumor is a vast increase in risk for end-stage renal disease, 11.5% compared to 0.6% for unilateral Wilms' tumor patients. Historically, patients with bilateral Wilms' tumor have had poorer outcomes than patients of similar histology, age and local stage with unilateral lesions.

Treatment Approach

The main objective in the treatment of bilateral Wilms' tumor is eradicating the tumor with maximum preservation of renal tissue. The goal is for patients to undergo unilateral or bilateral partial nephrectomies, if feasible.

Current therapy recommendations for bilateral Wilms' tumor include:

- Open or percutaneous biopsy of each lesion to determine that the tumor is indeed Wilms' tumor. Some physicians advocate not doing a biopsy because the overwhelming majority of bilateral renal tumors are Wilms' tumor. Moreover, biopsies are insensitive in the detection of anaplasia
- Six weeks of preoperative chemotherapy appropriate for the lesion with poorest prognosis stage and histology (Tables 23-8 to 23-10)
- Open wedge biopsy or nephrectomy should be performed at week 6 of therapy for patients whose disease has shrunk <50% during initial therapy, as the likelihood of histopathologic findings other than favorable histology are increased in the setting of disease that is poorly responsive to 6 weeks of therapy
- For patients who have experienced significant response but renal-sparing definitive surgery is still not possible, 6 additional weeks of chemotherapy may be given
- Definitive surgery should not be delayed beyond week 12, as further delays may result in chronic undertreatment of resistant disease and increased risk for relapse
- Chemotherapy should be adjusted accordingly if anaplastic histology is identified at the time of definitive surgery. Due to increased risk of injury to the remaining renal parenchyma, radiation therapy is reserved for those patients with positive tumor margins present following definitive surgery.

Post-Therapy Follow-Up

1. In patients with favorable-histology Wilms' tumor without metastatic disease, chest radiographs and abdominal ultrasounds should be performed alternately with chest and abdominal/pelvic CT scans every 3 months during the first 3 years following completion of therapy. Chest radiographs and abdominal ultrasounds should be performed every 6 months for 2 years in the fourth and fifth year following completion of therapy.

2. In patients with metastatic or anaplastic disease at diagnosis, chest and abdominal/pelvic CT scans should be performed every 3 months during the first 2 years following completion of therapy. Chest radiographs and abdominal ultrasounds should be performed every 6 months in the third through fifth years following completion of therapy.

Prognosis

Table 23-11 lists the percentage of relapse-free survival (RFS) and overall survival (OS) 4 years from diagnosis in relationship to the stage of disease, histology and chemotherapy treatment.

Table 23-11   Four-Year Outcomes in National Wilms' Tumor Study 5

Stage/Histology	Drug Treatment	4-Year RFS (%)	4-Year OS (%)
I FH ($n = 75$) (age <24 m, tumor <550 g)	Nephrectomy alone	86.5	98.7
I FH ($n = 97$) (age <24 m, tumor <550 g)	VCR, AMD	97.9	99.0
I FH ($n = 243$) (age ≥24 m, tumor ≥550 g)	VCR, AMD	93.8	97.0
II FH ($n = 555$)	VCR, AMD	84.4	97.2
III FH ($n = 488$)	VCR, AMD, DOX	86.5	94.4
IV FH ($n = 198$)	VCR, AMD, DOX	75.1	85.2
V FH ($n = 68$)	VCR, AMD, DOX	66.4	87.7
I FA + DA ($n = 29$)	VCR, AMD	69.5	82.6
II DA ($n = 23$)	VCR, DOX, CTX, ETOP	82.6	81.5
III DA ($n = 43$)	VCR, DOX, CTX, ETOP	64.7	66.7
IV DA ($n = 15$)	VCR, DOX, CTX, ETOP	33.3	33.3
V DA ($n = 20$)	VCR, DOX, CTX, ETOP	25.1	41.6

Abbreviations: FH, favorable histology; FA, focal anaplasia; DA, diffuse anaplasia, RFS, relapse-free survival; OS, overall survival; VCR, vincristine; AMD, actinomycin-D; DOX, doxorubicin; CTX, cyclophosphamide; ETOP, etoposide.
From: Gratias EJ, Dome JS. Pediatric Drugs 2008;10(2):122, with modification, with permission.

Treatment in Relapse

Patients with relapsed Wilms' tumor may be divided into three risk categories: standard risk, higher risk and very high risk.

Standard Risk Relapse

Patients with all the following factors are considered "standard risk" relapse:

- Favorable-histology Wilms' tumor at diagnosis
- Stage I or II at time of diagnosis
- Treatment initially with only dactinomycin and vincristine
- No radiotherapy given during initial therapy.

This group may be treated with the chemotherapy regimen shown in Table 23-10 and have 70% relapse-free survival and 80% overall survival post-relapse.

Higher Risk Relapse

Patients with favorable histology Wilms' tumor who do not meet standard risk criteria are considered "higher risk" relapse.

This group may appropriately be treated with one of the following regimens:

- Cyclophosphamide/etoposide alternating with carboplatin/etoposide (Table 23-12). With this regimen, there is a 42% event-free survival and 48% overall survival
- ICE protocol. The details of chemotherapy in the ICE protocol are shown in Table 23-13
- Topotecan has been demonstrated to have activity in recurrent Wilms' tumor; and it may be added to the more conventional agents
- High-dose chemotherapy (carboplatin, etoposide, melphalan [CEM]) with autologous stem cell rescue (HDC/ASCR) shown in Table 23-14, has been successfully used for higher risk relapsed Wilms' tumor in patients whose disease remains chemosensitive. It is not clear whether HDC/ASCR provides a benefit compared to conventional dose chemotherapy.

Very High Risk Relapse

These relapsed renal tumors include the following: anaplastic Wilms' tumor, clear cell sarcoma, rhabdoid tumor and renal cell carcinoma, regardless of other factors.

For very high risk relapsed patients, conventional chemotherapy offers very little hope. Intensive study of the biology of these lesions and development of novel therapeutic agents offers the best hope for improved prognosis for this group of patients. These patients should be treated with promising investigational new agent protocols.

Local Control for Relapsed Patients

Treatment for local control of relapsed patients is as follows:

- *Pulmonary relapse*: radiation to both lungs, in addition to aforementioned chemotherapy

- *Liver relapse*: surgery with partial hepatectomy if the tumor is localized to one lobe with or without radiotherapy, in addition to aforementioned chemotherapy
- *Local relapse in tumor bed*: surgical excision and local radiotherapy, in addition to aforementioned chemotherapy.

Table 23-12   Chemotherapy for Higher Risk Relapsed Favorable Histology Wilms' Tumor

Week 1	CTX5	ETOP5		
Week 4	CTX5	ETOP5		
Week 7			CBDCA	ETOP3
Week 10			CBDCA	ETOP3
Week 13	SURGERY			
Week 14	RADIATION THERAPY		CTX5 (50%)	ETOP5
Week 17			CBDCA	ETOP3
Week 20	CTX5	ETOP5		
Week 23	CTX5	ETOP5		
Week 26			CBDCA	ETOP3
Week 29			CBDCA	ETOP3
Week 32	CTX5	ETOP5		
Week 35	CTX5	ETOP5		
Week 38			CBDCA	ETOP3
Week 41			CBDCA	ETOP3
Week 44	CTX5	ETOP5		
Week 47	CTX5	ETOP5		
Week 50			CBDCA	ETOP3
Week 53			CBDCA	ETOP3
Week 56	CTX5	ETOP5		
Week 59	CTX5	ETOP5		
Week 62			CBDCA	ETOP3
Week 65			CBDCA	ETOP3
Week 68	CTX5	ETOP5		
Week 71	CTX5	ETOP5		
Week 74			CBDCA	ETOP3
Week 77			CBDCA	ETOP3
Week 80	CTX5	ETOP5		
Week 83	CTX5	ETOP5		
Week 86			CBDCA	ETOP3
Week 89			CBDCA	ETOP3

Drug dosing information		
CTX5	Cyclophosphamide	7.35 mg/kg/day IV for 5 days for infants <12 months 14.7 mg/kg/day IV for 5 days for children ≥12 months & ≤30 kg 440 mg/m^2/day IV for 5 days for children >30 kg
ETOP5	Etoposide	1.65 mg/kg/day for 5 days for infants <12 months 3.3 mg/kg/day IV for 5 days for children ≥12 months & ≤30 kg 100 mg/m^2/day IV for 5 days for children >30 kg
CBDCA	Carboplatin	8.35 mg/kg/day IV for 2 days for infants <12 months 16.7 mg/kg/day IV for 2 days for children ≥12 months & ≤30 kg 500 mg/m^2/day IV for 2 days for children >30 kg
ETOP3	Etoposide	1.65 mg/kg/day IV for 3 days for infants <12 months 3.3 mg/kg/day IV for 3 days for children ≥12 months & ≤30 kg 100 mg/m^2/day IV for 3 days for children >30 kg

Chemotherapy should not be initiated if ANC <1,000/μl or platelets <100,000/μl
G-CSF (filgrastim) 5 mcg/kg/day subcutaneous starting 24 hours after completion of chemotherapy until the ANC is ≥10,000/μL after the expected nadir
MESNA administration is recommended with all cyclophosphamide cycles
Doses of CTX reduced 50% if given within 6 weeks of radiation therapy

Table 23-13 ICE Protocol for Higher Risk Relapse Wilms' Tumor

Drug	Route	Dose		Day(s)
Ifosfamide (I)	IV over 1 hour	66.7 mg/kg/day for infants < 12 months 2,000 mg/m² for children ≥ 12 months		1 to 3
Carboplatin (CARBO)	IV over 1 hour	**GFR**	**Dose**	1
		>150 ml/min/1.73 m²	560 mg/m² (18.7 mg/kg for infants)	
		100–150 ml/min/1.73 m²	500 mg/m² (16.6 mg/kg for infants)	
		75–99 ml/min/1.73 m²	370 mg/m² (12.3 mg/kg for infants)	
		50–74 ml/min/1.73 m²	290 mg/m² (9.7 mg/kg for infants)	
		30–49 ml/min/1.73 m²	200 mg/m² (6.7 mg/kg for infants)	
		<30 ml/min/1.73 m²	Hold carboplatin	
Etoposide (ETOP)	IV over 1 hour	3.3 μg/kg/day for infants <12 months 100 mg/m²/day for children ≥12 months		1 to 3

G-CSF (following ICE): 5 μg/kg/day subcutaneously from Day 4 until ANC >5,000/mm³ for 1 day after 7 or more days of G-CSF

MESNA 400 mg/m²/dose IV as a loading dose with Ifosfamide over 1 hour; then 400 mg/m²/dose as a 3 hour infusion followed by 400 mg/m²/dose boluses every 3 hours for three doses IV over 15–30 minutes on Days 1 through 3 (this would be 100% of Ifosfamide dose in our heavily pretreated patients)

*Hydration: Pre-hydrate with D5½NS at 125 ml/m²/hr for 2 hr. Then continue at 125 ml/m²/h for 2 h following completion of chemotherapy

NEPHROBLASTOMATOSIS

Nephrogenic rests are regions of persistent embryonal tissue in the kidney that represent potential precursors to Wilms' tumor. Nephroblastomatosis is the presence of multiple nephrogenic rests or diffusely distributed nephrogenic rests, giving a rind-like appearance to the kidneys. Management of nephroblastomatosis involves:

- Close surveillance with ultrasounds, CT, or MRI scans (preferred) to assess the pattern of nephroblastomatosis and growth
- Conservative renal tissue-preserving surgery for growing lesions that could be representative of Wilms' tumor
- Chemotherapy may be beneficial in shrinking nephroblastomatosis. Vincristine and dactinomycin are used but the optimal duration of therapy is unknown.

CONGENITAL MESOBLASTIC NEPHROMA

Congenital mesoblastic nephroma (CMN) is a low-grade fibroblastic sarcoma that usually presents in the first 3 months of life. Two main histologic types have been described: (1) classic CMN, which is histologically identical to infantile fibromatosis;

Table 23-14 CEM Preparative Regimen for HDC/ASCR in Higher Risk Relapsed Wilms' Tumor

Carboplatin	Per dose calculation specified below*, IV over 1 h, Days -7 through -3
Etoposide	200 mg/m^2 IV over 1 h, Days -7 through -3
Melphalan	180 mg/m^2 IV over 1 h, Day -2

During the preparative regimen, patient will receive hydration at 3,000 ml/m^2/day

*CARBOPLATIN dose is calculated according to a target area under the curve (AUC) of 4 mg-min/ml/day. The Calvert formula is used to calculate the actual dose. Actual dose = 4 × [GFR + (15 × body surface area)], where GFR is the raw clearance in ml/min. Total carboplatin dose is AUC of 20

Example of CARBOPLATIN dosage calculation:
Body surface area = 0.5 m^2
GFR (raw, uncorrected #) = 20 ml/min
Total dose in mg/day
= 4 × [GFR + (15 × body surface area)
= 4 × [20 + (15 × 0.5)]
= 4 × [20 + 7.5]
= 4 × 27.5
= 110 mg/day

Day 0: Stem cell infusion
G-CSF infusion (5 μg/kg/day subcutaneously) begins 24 hours after stem cell infusion and continues daily until ANC is greater for 2,000/mm^3 for three consecutive days.

and (2) cellular CMN, which is histologically identical to infantile fibrosarcoma and carries the characteristic t(12;15). Some tumors have mixed classic and cellular elements. Complete nephrectomy without adjuvant therapy is usually adequate treatment, though local recurrence and/or metastasis have been associated with cellular histology and stage III disease. Patients who experience a local recurrence or metastases may require additional surgery or treatment with chemotherapy (vincristine, dactinomycin and doxorubicin or vincristine, dactinomycin and cyclophosphamide or vincristine, doxorubicin and cyclophosphamide have been employed).

CLEAR CELL SARCOMA OF THE KIDNEY

1. Second most common pediatric renal neoplasm.
2. Peak incidence between age 3 and 5 years.
3. Bone, lung and liver are most common sites of metastasis at presentation, but the brain is the most common site of metastasis at recurrence.
4. Treatment consists of nephrectomy, radiation therapy and chemotherapy with cyclophosphamide, etoposide, vincristine and doxorubicin for 24 weeks (see Table 23-10).
5. Event-free survival is >80% for disease stages I–III. Event-free survival is about 35% for stage IV disease.

RHABDOID TUMOR OF THE KIDNEY

1. Represents 2% of renal tumors, median age of presentation is 1 year old.
2. Chromosomal deletion or mutations of the *INI1* gene at chromosome 22q11-12 are present in the tumors of most patients.
3. Some patients have constitutional *INI1* mutations and multifocal disease in the brain (atypical teratoid/rhabdoid tumors) or rhabdoid tumors at other sites.
4. Metastases common to lung, liver and brain.
5. Poor prognosis: overall mortality is 80% for all stages of disease. The tumor responds suboptimally to current chemotherapy and radiotherapy with high rates of recurrence. Multiple case series have shown some response to an aggressive approach using intensive chemotherapy (vincristine-doxorubicin-cyclophosphamide alternating with ICE plus radiation therapy and surgical resection of all possible sites of disease). The patients with the best chance of cure are those with localized disease (stage I/II) and age over 3 years.

RENAL CELL CARCINOMA (RCC)

1. Five to six percent of primary renal tumors under 21 years of age.
2. Most common renal neoplasm in adults, but pediatric subtypes of RCC are distinct from the clear cell subtype commonly seen in adults.
3. "Translocation" histologic type (involving TFE3 gene on X chromosome) is the most common type of pediatric RCC.
4. Survival for stage I disease after complete resection >90%.
5. Survival for stage II and III 70–80%.
6. Stage IV has poor prognosis – <15% survival.
7. Treatment: Surgery is mainstay of therapy. Anecdotal responses have occurred with cytotoxic chemotherapy. Tumors not as responsive to antiangiogenic therapies or immunotherapy as clear cell renal cell carcinoma.

Suggested Reading

Abu-Gosh AM, Krailo MD, Goldman SC, et al. Ifosfamide, carboplatin and etoposide in children with poor risk relapsed Wilms' tumor: a Children's Cancer Group report. Ann Oncol. 2002;13:460–469.

Beckwith JB, Zuppan CE, Browning NG, et al. Histological analysis of aggressiveness and responsiveness in Wilms' tumor. Med Pediatr Oncol. 1996;27:422–428.

D'Angio GJ. Pre or post-operative treatment for Wilms' tumor? Who, what, when, where, how, why and which. Med Pediatr Oncol. 2003;41:545–549.

Dome JS, Coppes MJ. Recent advances in Wilms' tumor genetics. Curr Opin Pediatr. 2002;14:5–11.

Dome JS, Cotton CA, Perlman EJ, et al. Treatment of anaplastic histology Wilms' tumor: results from the Fifth National Wilms' Tumor Study. J Clin Oncol. 2006;20:2352–2358.

Dome JS, Roberts CM, Argani P. In: Orkin SH, Ginsburg D, Nathan DG, et al., eds. Oncology of Infancy and Childhood. Elsevier Academic Publishers; 2009:541–573.

Geller JI, Dome JS. Local lymph node involvement does not predict poor outcome in pediatric renal cell carcinoma. Cancer 2004;101:1575–1583.

Green DM, Breslow NE, Beckwith JB, et al. Comparison between single dose and divided dose administration of dactinomycin and doxorubicin for patients with Wilms' tumor: a report from the National Wilms' Tumor Study Group. J Clin Oncol. 1998;16:237–248.

Green DM, Cotton CA, Malogolowkin M, et al. Treatment of Wilms' tumor relapsing after initial treatment with vincristine and actinomycin D: a report from the National Wilms' Tumor Study Group. Pediatr Blood Cancer 2007;48:493–499.

Grundy PE, Breslow NE, Li S, et al. Loss of heterozygosity for chromosomes 1p and 16q is an adverse prognostic factor in favorable-histology Wilms' tumor: a report from the National Wilms' Tumor Study Group. J Clin Oncol. 2005;23:7312–7321.

Kremens B, Gruhn B, Klingebiel T, et al. High dose chemotherapy with autologous stem cell rescue in children with nephroblastoma. Bone Marrow Transplantation 2002;30:893–898.

Malogolowkin M, Cotton CA, Green DM, et al. Treatment of Wilms' tumor relapsing after initial treatment with vincristine, actinomycin D and doxorubicin. A report from the National Wilms' Tumor Study Group. Pediatr Blood Cancer 2007.

Metzger ML, Stewart CF, Freeman BB, et al. Topotecan is active against Wilms' tumor: results of a multi-institutional phase II study. J Clin Oncol. 2007;21:3130–3136.

Pein F, Michon J, Valteau-Couanet D, et al. High dose melphalan, etoposide and carboplatin followed by autologous stem cell rescue in pediatric high risk recurrent Wilms' tumor: a French Society of Pediatric Oncology study. J Clin Oncol. 1998;16:3295–3301.

Seibel NL, Li S, Breslow NE, et al. Effect of duration of treatment on treatment outcome for patients with clear-cell sarcoma of the kidney: a report from the National Wilms' Tumor Study Group. J Clin Oncol. 2004;22:468–473.

Tomlinson GE, Breslow NE, Dome J, et al. Age at diagnosis as a prognostic factor for rhabdoid tumor of the kidney: a report from the National Wilms' Tumor Study Group. J Clin Oncol. 2005;23:7641–7645.

Rhabdomyosarcoma and Other Soft-Tissue Sarcomas

Soft-tissue sarcomas (STS) constitute a heterogeneous group of malignant tumors that are derived from primitive mesenchymal cells. These tumors arise from muscle, connective tissue, supportive tissue and vascular tissue. As a group, they are locally highly invasive and have a high propensity for local recurrence. When they metastasize, it is usually via the bloodstream and, less commonly, via the lymphatics. The STS can be divided into two groups: rhabdomyosarcoma (RMS) and non-rhabdomyosarcoma soft-tissue sarcomas (NRSTS). Collaborative studies of rhabdomyosarcoma (RMS) have dramatically improved cure rates. NRSTS in children frequently have a better prognosis than in adults. In infants and young children, NRSTS often behave in a benign manner and have an excellent prognosis with surgery alone. There are various chromosomal translocations that are characteristic of NRSTS which have led to the refinement of the histopathological classification of pediatric STS. Table 24-1 summarizes the clinical and biological features of the non-rhabdomyosarcoma soft-tissue sarcomas.

INCIDENCE AND EPIDEMIOLOGY

Pediatric STS account for 7% of childhood malignancies. RMS is the most common pediatric STS, with an incidence of 4–5 per million in children less than 15 years old.

- RMS is the third most common solid extracranial tumor, following neuroblastoma and Wilms' tumor
- It accounts for 3% of all malignant neoplasms in children, with approximately 350 new cases diagnosed in the United States each year in children up to 19 years of age
- There is a slight male predominance with a male:female ratio of 1.4:1
- There are two age peaks: 2–6 years and 15–19 years. The adolescent peak has primarily been noted in males
- Incidence in African-American females is half that of Caucasian females. Incidence is lower in Asian populations residing in Asia or the West.

Manual of Pediatric Hematology and Oncology. DOI: 10.1016/B978-0-12-375154-6.00024-0

Table 24-1   Clinical and Biological Features of the Non-Rhabdomyosarcoma Soft Tissue Sarcomas (NRSTs)

Tumor[a]	Cell Origin/ Cytogenetics	Common Sites	Common Ages	Good Prognostic Factors[b]	Outcome	Therapy
Synovial sarcoma	Mesenchymal cells/t(X;18)	Extremities (lower>upper)	Adolescence/young adulthood	Age <14 years Size <5 cm, calcification	Stages I & II, 70% Stages III & IV, poor	WLE with/without RT Chemo: Ifosfamide/ Doxorubicin
Malignant fibrous histiocytoma	Fibroblast/19p+	Lower extremity, trunk, head and neck	40–60 years In children, 10–20 years	Extremity site	5 year survival, 27–53%	WLE Chemo: No established role; responsive to VAC +/− Doxorubicin
Angiomatoid form		Extremity	Young children		Excellent with surgery alone	WLE
Malignant peripheral nerve sheath tumor	Schwann cell or fibroblast/ 17q, 22q loss or rearrangement	Extremity, retroperitoneum trunk	Younger in patients with Neurofibromatosis (NF)	Size <5 cm, No NF	53% survival without NF, 16% with NF	WLE with/without RT Chemo: Neo-adjuvant role, Ifosfamide and etoposide or Ifosfamide and doxorubicin
Fibrosarcoma	Fibroblast				Excellent with surgery alone	WLE, avoid amputation RT/chemo if WLE not possible Neo-adjuvant VA useful
Congenital	t(12;15)	Extremity, trunk	Most <2 years	<5 years	5-year survival 84%	
Adult form	t(X;18), t(2;5), t(7;22)	Extremity (thigh, knee)	Adolescence		5-year survival 34–60%	WLE with/without RT Chemo: No established role
Leiomyosarcoma	Smooth muscle t(12;14) Common in HIV+ children	Retroperitoneum GI tract, vascular tissue	40–70 years, when in children, any age	<5 cm	33% disease-free survival at 1–5 years	WLE Chemo: doxorubicin, ifosfamide and dacarbazine

	Cell of origin / cytogenetics	Site	Age	Good prognostic factors	Prognosis	Treatment
Epitheliod form[c]		Stomach			Excellent with surgery alone	WLE
Alveolar soft part sarcoma	Paraganglionic mesoderm (?), neuroepithelial (?), muscle (?) t(X;17)	Lower extremity, head and neck	15–35 years	Young age, orbital site, <5 cm	5-year survival 27–59% (indolent; death from disease after 10–20 years	WLE Chemo or RT only after recurrence Chemo: no clear role. Possible role of Vascular endothelial growth factor inhibitors being explored
Hemangiopericytoma	Pericytes t(12;19), t(13;22)	Extremity, retroperitoneum head and neck	20–70 years, when in children, 10–20 years	Low stage, <5 cm	Stage I & II, 30–70% 5-year survival with adjuvant therapy Stages III & IV, poor	WLE, with/without RT Chemo: No established role
Infantile form		Extremity, trunk	Rare, but typically <1 year		Excellent with surgery alone	WLE
Liposarcoma	Primitive mesenchyme t(12;16)	Extremity, retroperitoneum	0–2 years and second decade; sixth decade most common	Child, myxoid type	Very good with WLE, rarely metastasizes	WLE, with/without RT RT important in retroperitoneal lesions Chemo: No established role
Clear Cell Sarcoma (also known as malignant melanoma of soft parts)	Mesoderm, melanin deposits t(12;22)(p13; q12) EWS-ATF1	Tendons and aponeuroses of lower extremity	Young adults	< 5 cm, no necrosis, non-metastatic	Adverse prognosis; 5-year survival rates of 60–70%. However, only 30–40% are long-term survivors due to late recurrences	WLE ± sentinel node biopsy No clear role for adjuvant chemotherapy. Potential role for immunotherapy (e.g., translocation-targeted vaccines, interferon, GM-CSF secreting vaccine)

[a]Listed in order of decreasing incidence. [b]Low histologic grade and low stage are good prognostic factors. [c]Carney triad: A condition consisting of gastric epithelioid leiomyosarcoma, pulmonary chondroma, functioning extra-adrenal paraganglioma.

Abbreviations: WLE, wide local excision; RT, radiation therapy; VAC, Vincristine, actinomycin D, cyclophosphadmide; VA, Vincristine; Actinomycin D.

Most cases are sporadic but some reports estimate that up to 30% of cases may have an underlying genetic risk factor which may include:

* Germline mutations of the p53 suppressor gene, as in Li-Fraumeni familial cancer syndrome. There is an association between early onset breast cancer, sarcomas, brain tumors and adrenocortical tumors in family members
* Ionizing radiation
* Neurofibromatosis with NF1
* Associated with anomalies of the central nervous system (CNS), genitourinary system, gastrointestinal system and cardiovascular system (i.e. Beckwidth–Weidemann syndrome and Costello syndrome[*]).
* Associated with maternal and paternal use of marijuana and cocaine and first-trimester pre-natal X-ray exposure, possibly as an environmental interaction with a genetic trigger.

PATHOLOGIC CLASSIFICATION

Specific assays may be necessary to aid in the differential diagnosis of rhabdomyosarcoma from the other small round blue cell tumors of childhood (i.e., lymphoma, Ewing sarcoma, neuroblastoma). These assays include electron microscopy, immunocytochemistry and cytogenetics. Table 24-2 lists the histologic subtypes with reference to their morphology, site of origin and age distribution.

GENETICS

1. Alveolar rhabdomyosarcoma (ARMS) has a characteristic translocation of the FKHR gene at 13q14 with PAX3 at 2q35 or less commonly PAX7 at 1p36. The fusion protein functions as transcription factors that activate transcription from PAX binding sites that are 10–100 times more active than wild-type PAX7 and PAX3. This alteration in growth, differentiation and apoptosis results in tumorigenic behavior.
2. Embryonal rhabdomyosarcoma (ERMS) has a loss of heterozygosity (LOH) at the 11p15.5 locus. This LOH involves loss of the imprinted maternal genetic information and all that remains is expression of the paternal genetic material. This LOH includes loss of tumor suppressor genes that have been implicated in oncogenesis which results in the overproduction of insulin-related growth factor-2

---

[*]This is a genetic disorder characterized by delayed development and mental retardation, unusually flexible joints, hypertrophic cardiomyopathy and short stature. Increased risk of developing cancerous and noncancerous tumors, the most frequent being rhabdomyosarcoma.

Table 24-2 Histologic Subtypes of Rhabdomyosarcoma (RMS)

Pathologic Subtype	Morphology	Usual Site of Origin	Usual Age (Years) Distribution
Embryonal (ERMS) solid	Resembles skeletal muscle in 7–10-week fetus. Moderately cellular with loose myxoid stroma. Actin and Desmin positive	Head and neck, orbit genitourinary tract	3–12
Botryoid variant	Only one microscopic field of cambium layer necessary to diagnose as botryoid. Grossly presents with grapelike configuration	Bladder, vagina, naso-pharynx, bile duct	0–8
Spindle cell variant	Spindle-shaped cells with elongated nuclei and prominent nucleoli. Low cellularity. Collagen-rich and -poor variants	Paratesticular	2–12
Alveolar (ARMS)	Resembles skeletal muscle in 10–21-week fetus. Basic cell is round with scanty eosinophilic cytoplasm; alveolar pattern may be lost if densely packed; cross striations more common than embryonal variety. To be defined as ARMS requires >50% alveolar component	Extremities, trunk, perineum (adolescents)	6–21
RMS, NOS	Heterogeneous unable to subtype due to paucity of tissue	Extremities, trunk	6–21

Abbreviations: NOS, not otherwise specified.

(IGF-2). IGF-2 stimulates the growth of RMS and blockade of the IGF-2 receptor inhibits RMS growth *in vivo*.

3. The allelic loss of 11p15.5 is also seen in translocation-negative alveolar rhabdomyosarcoma (ARMS); this observation, combined with recent microarray clustering analyses and global gene profiling data support the notion that most cases of "histologically" alveolar, translocation-negative ARMS are, in fact, genetically similar if not identical to ERMS.

4. The DNA content or ploidy has been implicated in prognosis, particularly in tumors of embryonal histology. Hyperdiploid tumors with 51 or more chromosomes have a better prognosis.

CLINICAL FEATURES

Primary Sites

Rhabdomyosarcoma may occur in any anatomic location of the body where there is skeletal muscle, as well as in sites where no skeletal muscle is found (e.g., urinary bladder, common bile duct). Rhabdomyosarcoma in children under 10 years of age generally involves the head and neck or genitourinary areas. Adolescents more commonly develop extremity, truncal, or paratesticular lesions. Table 24-3 lists the relative frequency of the various primary sites and sites of regional spread and distant metastases.

Table 24-3   Frequency of Primary Sites, Sites of Regional Spread and Distant Metastatic Sites in Rhabdomyosarcoma

Primary Site	Relative Frequency (%)	Regional Spread and Distant Metastatic Sites
Head and neck	40	
Orbit	8	Nodes rarely involved; rare lung metastasis
Parameningeal[a]	25	Regional spread to bone, meninges, brain; lung and bone metastases
Other[b]	7	Nodes rarely involved; lung metastases
Genitourinary tract	29	
Bladder, prostate	10	Nodes rarely involved; metastases to lung, bone and bone marrow (primarily prostate primaries)
Vagina, uterus	5	Nodes rarely involved; metastases to retroperitoneal nodes (mainly from uterus)
Paratesticular	14	Retroperitoneal nodes in 50% of boys ten or older; metastases to lung and bone
Extremities	14	Nodes involved in up to 50% of cases; metastases to lung, bone marrow, bone, CNS
Trunk	12	Nodes rarely involved; metastases to lung and bone
Other	5	Nodal involvement site dependent (increased in perineal/perianal primaries); metastases to lung, bone and liver

[a]Parameningeal sites are adjacent to the meninges at the base of the skull; they consist of nasopharynx, middle ear, paranasal sinuses and infratemporal and pterygopalatine fossae.
[b]Nonorbital, nonparameningeal sites consist of larynx, oropharynx, oral cavity, parotid, cheek and scalp.

Signs and Symptoms

The most frequent presenting sign is a painless mass.

Specific clinical manifestations vary with the site of origin of the primary lesion and are outlined in Table 24-4. An important clinical feature of parameningeal head and neck primary tumors (nasopharynx, sinuses, middle ear) is that approximately 25–50% have infiltration of tumor through the skull base with intracranial extension of disease that may manifest as cranial nerve palsies, meningeal symptoms, or respiratory difficulty due to brain stem infiltration.

Rare primary sites for rhabdomyosarcoma include the gastrointestinal–hepatobiliary tract (3%), where it presents with obstructive jaundice and a large abdominal mass. These tumors arise in the common bile duct and may extend into both lobes of the liver. Other rare primary sites are the intrathoracic region (2%) and the perineal–perianal area (2%).

Approximately 20% of RMS have metastatic disease at diagnosis and the most common site is the lung followed by the bone marrow, lymph nodes and bone.

Table 24-4   Clinical Manifestations of Rhabdomyosarcoma in Various Anatomic Locations

Location	Signs and Symptoms
Head and Neck[a]	
Neck	Soft-tissue mass
	Hoarseness
	Dysphagia
Nasopharynx	Sinusitis
	Local pain and swelling
	Epistaxis
Paranasal Sinus	Sinus obstruction/sinusitis
	Unilateral nasal discharge
	Local pain and swelling
	Epistaxis
Middle ear/mastoid	Chronic otitis media – purulent blood stained discharge
	Polypoid mass in external canal
	Peripheral facial nerve palsy
Orbit	Proptosis
	Ocular palsies
	Conjunctival mass
Genitourinary	
Vagina and uterus	Vaginal bleeding
	Grapelike clustered mass protruding through vaginal or cervical opening (i.e., sarcoma botryoides)
Prostate	Hematuria
	Constipation
	Urinary obstruction
Bladder	Urinary obstruction
	Hematuria
	Tumor extrusion
	Recurrent urinary tract infections
Paratesticular	Painless scrotal or inguinal mass
Retroperitoneum	Abdominal pain
	Abdominal mass
	Intestinal obstruction
Biliary tract	Obstructive jaundice
Pelvic	Constipation
	Genitourinary obstruction
Extremity/trunk	Asymptomatic or painful mass

[a]All can extend through multiple foramina and fissures into the epidural space and infiltrate the central nervous system with cranial nerve palsies, meningeal symptoms and brain stem signs.

DIAGNOSTIC EVALUATION

The diagnostic evaluation should delineate the extent of the primary tumor and the extent of metastatic disease and should consist of the following:

- Complete history and physical examination including measurements of the primary tumor and assessment of regional lymph nodes

- Laboratory tests including: Complete blood count, comprehensive metabolic panel including liver functions and urinalysis. In cases of advanced-stage ARMS, a coagulation profile should be performed as these patients may present with tumor-induced DIC
- Primary tumor imaging
  - Magnetic resonance imaging (MRI) for extremities, body wall and head/neck
  - Computed tomography (CT) for intrathoracic, retroperitoneal, paraspinal, skull base
  - Ultrasound for paratesticular, bladder/prostate or biliary tree.
- Metastatic workup
  - CT scan: chest CT. For lower extremity and/or pelvic primaries: abdominal and pelvic CT are indicated. In paratesticular primaries: Thin-cut (3 mm) slices through the retroperitoneum are recommended for optimal evaluation of regional adenopathy
  - Bone scan
  - MRI/CT of draining lymph nodes
  - Indications for surgery include:
    - Bilateral bone marrow aspiration and biopsy
    - Sampling of suspicious lymph nodes
    - Required sentinel node evaluation for extremity primaries
    - Required ipsilateral retroperitoneal lymph node dissection for paratesticular RMS for boys 10 years and older.
- PET scan for determining initial extent of disease and to monitor response to therapy
- Additional diagnostic evaluation depending on primary site
  - Ear, nose and throat examination
  - Ophthalmologic examination
  - Dental evaluation with Panorex radiography
  - Lumbar puncture for parameningeal head and neck
  - Cystoscopy or vaginoscopy
  - Radiation oncology consultation
  - Surgical consultation for percutaneous, incisional, or excisional biopsy
  - Physical occupation, speech therapy.

STAGING

It is essential to fully stage RMS. The prognosis, selection of systemic therapy and the design of optimal local therapy depend on the following:

- Primary site
- Histologic type
- Tumor size

- Degree of regional spread
- Nodal involvement
- Distant metastatic disease and
- Extent of prechemotherapy tumor resection.

RMS staging systems consist of:

Pretreatment clinical stage: The Intergroup Rhabdomyosarcoma Study (IRS) Group incorporated the most significant prognostic variables (primary site invasiveness, size, lymph node involvement) into a TNM (tumor–node–metastasis) staging system. This soft-tissue sarcoma – Children's Oncology Group (STS-COG) pretreatment TNM staging system is listed in Table 24-5
Postoperative clinical group: A unique aspect of the RMS staging is the postoperative clinical grouping system, which is based on the extent of the surgical resection and takes

Table 24-5   Soft Tissue Sarcoma – Children's Oncology Group (STS-COG) Pretreatment TNM Staging of RMS

T=Tumor	N=Regional nodes	M=Metastasis[a]
T1=Confined to anatomic site of origin	N0=Not clinically involved	M0=No distant metastasis
T2=Infiltration into surrounding tissue	N1=Clinically involved	M1=Distant metastasis present
a=≤5 cm in diameter	NX= Clinical status unknown	
b=>5 cm in diameter		

TNM pretreatment staging classification for intergroup rhabdomyosarcoma study IV

Stage	Sites	T Invasiveness	T Size	N	M
I	**Favorable Sites** Orbit, Head and neck (excluding parameningeal), Genitourinary (nonbladder/nonprostate)	T1 or T2	a or b	N0 or N1 or NX	M0
II	**Unfavorable Sites** Bladder/prostate, Extremity Cranial parameningeal Other (includes trunk, retroperitoneum, etc.)	T1 or T2	a	N0 or NX	M0
III	**Unfavorable Sites** Bladder/prostate, Extremity Cranial parameningeal Other (includes trunk retroperitoneum, etc.)	T1 or T2	a b	N1 N0 or N1 or NX	M0
IV	**Any Sites**	T1 or T2	a or b	N0 or N1	M1

[a]Distant metastatic disease consists of lung, liver, bones, bone marrow, brain and distant muscle and nodes.
The presence of positive cytology in CSF, pleural, or abdominal fluids, as well as implants on pleural or peritoneal surfaces, also constitutes stage IV disease.
Modified from: Mandell LR. Ongoing progress in the treatment of childhood rhabdomyosarcoma. Oncology 1993;7:71-84, with permission.

into account the lymph node evaluation (Table 24-6). In some locations, when complete resection is possible with negative margins overall survival is improved.

Risk group classification (Table 24-7): The risk group classification and treatment is based on:

- Pretreatment stage
- Postoperative clinical group and
- Histology.

The *low-risk group* includes all nonmetastatic favorable-site ERMS (stage 1, regardless of degree of initial surgical resection) and completely resected (groups I and II) nonmetastatic unfavorable site ERMS.

The *intermediate group* includes all nonmetastatic alveolar tumors and embryonal tumors in unfavorable primary sites (stages 2 or 3) that have been incompletely resected (group III).

Table 24-6   Postoperative Clinical Grouping System (Intergroup RMS Study [IRS])

Group	Definition (Incidence)
I.	No residual disease (16%)
	A. Localized, completely resected, confined to site of origin
	B. Localized, completely resected, infiltrated beyond site of origin
II.	Microscopic residual disease (20%)
	A. margins positive, lymph nodes negative
	B. margins negative, lymph nodes positive
	C. margins positive, lymph nodes positive
III.	Gross residual disease (48%)
	A. Biopsy only
	B. Grossly visible disease after ≥50% resection of primary tumor
IV.	Distant metastasis present at diagnosis (16%)

Table 24-7   Rhabdomyosarcoma Risk Group Classification from Soft Tissue Sarcoma – Children's Oncology Group (STS-COG)

Risk Group	Histology	Pretreatment Stage	Postoperative Clinical Group
Low	Embryonal	1 (All favorable sites)	I or II
Subset A		1 (orbit only)	III
		2 (unfavorable sites ≤5 cm)	I or II
Low	Embryonal	1 (non-orbit only)	III
Subset B		3 (unfavorable sites >5 cm, or unfavorable site, any size, with regional nodal involvement)	I or II
Intermediate	Embryonal	2, 3	III
Intermediate	Alveolar	1, 2, 3	I, II, III
High	Embryonal	4	IV
High	Alveolar	4	IV

The ***high-risk group*** includes all metastatic tumors both alveolar and embryonal.

Since prognosis is determined by the risk group, the risk group classification determines treatment.

PROGNOSIS

Prognosis for RMS

Overall, RMS is curable in the majority of children receiving optimal therapy (>70% survival 5 years after diagnosis).

- Extent of disease is the most important prognostic factors. The 3-year failure-free survival (FFS) by pretreatment clinical stage and postoperative group is summarized in Table 24-8. Children with localized, completely resected disease do better than those with widespread or disseminated disease
- Patients with gross residual disease after surgery (group III) have a statistically significantly lower 3-year FFS of 73% compared with 83 and 86% FFS for groups I and II, respectively
- Patients with smaller tumors (<5 cm) have improved survival compared to children with tumors >5 cm
- Those with metastatic disease at diagnosis have the worst prognosis. Amongst those with metastatic disease, two or fewer metastases is significantly better than three or more
- Patients with an alveolar histology have a worse prognosis. The alveolar subtype is frequently associated with an extremity site, which is a poor prognostic factor. Table 24-9 shows the 3-year failure-free survival according to histologic subtype and presence of metastatic disease

Table 24-8   Failure-free Survival (%) of Children at 3 Years by Stage and Group on IRS-IV

	Clinical Stage (Pretreatment)	Clinical Group (Post-operative)
I	86	83
II	80	86
III	68	73
IV	25	25

Table 24-9   Failure-free Survival (%) of Children at 3 Years According to Histologic Subtype and Presence of Metastatic Disease

Histologic Type	Non-metastatic	Metastatic, ≤2 Sites
Embryonal	83	40
Alveolar/unspecified	66	21

Table 24-10   Risk Stratification and Survival Data from IRS-III AND IRS-IV

Characteristics	Risk Group	Number of Patients	3 Year FFS	OS
Embryonal Histology, Favorable site	Low	279	88	94
Embryonal Histology, Unfavorable site, Group I/II	Low	63	88	93
Embryonal Histology, Unfavorable site, Group III	Intermediate	275	76	83
Embryonal Histology, Any site, Group IV, <10 years old	Intermediate*	98	55 (5-year)	59
Alveolar Histology, Any site, Group I-III	Intermediate	165	55 (5-year)	–

*Group IV ERMS under age 10 now classified as high risk on current front-line COG treatment protocols.
Abbreviations: FFS, failure free survival; OS, overall survival.

- Nonparameningeal head and neck sites, nonbladder/nonprostate male and female genitourinary tract sites and biliary tract are considered "favorable sites"; all other sites are considered "unfavorable."
- Table 24-10 shows the 3-year FFS and overall survival (OS) by risk stratification.
- Children between 1 and 9 years of age have a better prognosis than those older than 9 years of age (83 vs. 68% 3-year FFS). This finding might be due to the higher incidence of advanced disease and alveolar type in older children.

Prognosis for NRSTS

NRSTS are broadly divided into low-grade and high-grade tumors based upon their propensity for local invasion and distant metastases. The single most important prognostic variable for nonmetastatic tumors remains initial surgical resection.

- If tumors are completely resected the overall survival approaches 90%; most completely resected tumors are low grade
- In unresected tumors (most of which are high grade and invasive) 5-year survival is 55%; however, event-free survival is only 33%
- Metastatic tumors have a 5-year survival of 15%.

Factors for local recurrence overlap with those for metastases. Local recurrence and metastatic spread are substantially more likely to occur in the following conditions:

- Microscopically positive margins for high-grade tumors
- An intra-abdominal primary tumor
- Tumor size greater than 5 cm are more likely to metastasize
- Tumors that have a high histologic grade are more likely to be locally invasive
- High-grade tumors that recur locally following adequate local therapy will eventually metastasize in two-thirds of patients and curative treatment for such patients rarely occurs.

At present, molecular determinants of outcome for patients with high-grade NRSTS do not exist.

TREATMENT

General Principles

Successful treatment of RMS requires achievement of both local and systemic control of disease.

Local Control

For all patients with ARMS and for the vast majority of patients with ERMS with microscopic or gross residual disease following initial surgery, local control is accomplished using radiotherapy.

Systemic Control

This is accomplished with chemotherapy. Patients with localized disease require systemic therapy to eradicate micrometastatic disease typically present at diagnosis.

Surgery

Surgery is most effective if the STS can be completely excised with an adequate margin of uninvolved tissue. If the STS cannot be resected because of proximity to blood vessels and nerves or because this would produce major functional or cosmetic sequelae, an incisional or debulking biopsy is performed. This is followed by induction chemotherapy and radiation to cytoreduce the tumor. Second-look surgery is limited to situations where complete surgical resection of the postinduction chemotherapy mass may result in a meaningful reduction in the dose of postoperative radiation. For example (particularly important in infants and toddlers and very young children) nonmutilating surgical removal of postchemotherapy gross residual disease during second-look surgery can result in a radiation dose reduction from 50.4 Gy to 36 Gy if negative margins or to 41.4 Gy if positive margins (and, in selected instances, to no postoperative radiation being given). In other instances, for example, in patients with exceptionally large tumors in the thorax or retroperitoneum, surgical removal of postchemotherapy gross residual disease may permit the use of a lower (potentially more tolerable) tumoricidal dose of radiation. Surgical resection of postradiation residual masses is generally not indicated unless imaging studies (including PET scan) strongly suggest the presence of residual viable tumor.

Additionally prior to starting chemotherapy there may be a role for primary re-excision under the following conditions:

* If patient had a noncancer operation initially or
* If gross or microscopic residual disease is amenable to wide excision or
* If there is uncertainty about residual disease and or margins.

For primary tumors arising in the orbit, head and neck and certain extremity locations, aggressive surgical debulking, such as enucleation, head and neck dissection and amputation, are not indicated. Instead, chemotherapy and radiotherapy should be relied on to control the tumor at the primary site. For tumors arising in the bladder and prostate the goal is to preserve function. Radical surgical debulking is reserved for cases with residual disease after chemotherapy and radiation therapy.

The role of lymph node biopsy or dissection as part of the primary approach depends on the tumor site. Sites with a high incidence of regional node involvement include the extremity (30–50%), genitourinary (20%), perirectal (33%) and paratesticular (40%). Consideration should be given to lymph node biopsy, when management may be affected by the results of the biopsy or excision of clinically suspicious nodes at any site. Regional nodal exploration, possibly with sentinel node mapping, is required in extremity lesions and ipsilateral retroperitoneal lymph node dissection is mandatory for boys with paratesticular tumors who are 10 years of age and older.

Radiotherapy

STS are only moderately radiation-sensitive. In general NRSTS require higher radiation doses than do RMS to achieve local control. The treatment volume is determined by the extent of tumor at diagnosis prior to surgical resection with an appropriate margin which may be influenced by the surrounding normal tissue structures. Radiation dose guidelines for RMS and NRSTS are outlined in Table 24-11. Use of intensity-modulated radiation therapy (IMRT) should be considered in order to spare normal tissues. There is no added benefit to hyperfractionation and the current recommendation is fractionated daily doses of 180 cGy for most sites.

Chemotherapy

Treatment of RMS

Patients can either receive "standard" chemotherapy or a treatment regimen determined by risk classification.

Standard Treatment of RMS

All patients with RMS regardless of their initial stage or group receive combination chemotherapy as "standard therapy" consisting of VAC:

- Vincristine (V) age-adjusted (0.025 mg/kg for infants under age 1 year; 0.05 mg/kg for toddlers 1–3 years of age; 1.5 mg/m^2, maximum dose 2 mg, for children 3 years of age and older; all vincristine doses are given as an IV push over 1 minute)

Table 24-11   Guidelines for Radiotherapy for Rhabdomyosarcoma (RMS) and Nonrhabdomyosarcoma Soft-Tissue Sarcomas (NRSTS)

RMS	
Clinical Group I Embryonal	0 Gy
Clinical Group I Alveolar	36 Gy
Clinical Group IIA (nodes negative)	36 Gy
Clinical Group IIB/C (nodal sites)	41.4 Gy
Clinical Group III (orbit only)	45 Gy
Clinical Group III (other sites)	50.4 GY
Second look surgery	
Negative margins	36 Gy
Positive margins	41.4 Gy
Clinical group IV (metastases)	50.4 Gy
Second look surgery	
Negative margins	36 Gy
Positive margins	41.4 Gy
NRSTS	
Low-grade Tumors	
Grossly resected (any margins-observation only)	0 Gy
Unresectable tumors	
Preoperative	45 Gy
Postoperative	
Negative Margins	0 Gy
Positive Margins	10.8 Gy
Macroscopic Margins	19.8 Gy
High-grade Tumors	
Grossly resected ≤5 cm	
Negative margins (observation only)	0 Gy
Positive margins	55.8 Gy
Grossly resected >5 cm	
Negative or positive margins	55.8 Gy
Unresectable tumors	
Preoperative RT	45 Gy
Postoperative	
Negative margins	0 Gy
Positive margins	10.8 Gy
Macroscopic margins	19.8 Gy
Inoperable tumors	63 Gy

- Actinomycin D (A) age-adjusted (0.025 mg/kg for infants under age 1 year; 0.045 mg/kg for children 1 year of age and older, maximum dose 2.5 mg; actinomycin D doses are given as an IV push over 1 minute and this agent is omitted during radiotherapy)
- Cyclophosphamide (C) age-adjusted (36 mg/kg for infants under age 1 year; 73 mg/kg for toddlers 1–3 years of age; 2.2 g/m^2 for children 3 years of age and older; all cyclophosphamide doses should be administered IV over 30–60 minutes with appropriate pre- and postinfusion IV hydration and Mesna to protect the bladder

from bleeding. Each dose of Mesna is 20% of the cyclophosphamide dose given IV every 3 hours for 3–5 doses beginning just prior to the cyclophosphamide infusion).

Dactinomycin (actinomycin-D) and cyclophosphamide are given every 3 weeks and vincristine is given weekly for 9–12 weeks; doses of all agents are age-adjusted because of the higher risk of hepatic veno-occlusive disease (VOD) in children under 3 years of age. Dactinomycin is omitted during radiation therapy; vincristine may be given weekly or every 3 weeks in combination with cyclophosphamide during radiation therapy. This regimen is typically given for 12–14 cycles lasting 8–10 months. Patients with low-risk rhabdomyosarcoma may receive VA with or without C, depending on their specific risk factors.

Other effective agents in RMS include ifosfamide, etoposide, doxorubicin, irinotecan, topotecan, carboplatin and vinorelbine.

Treatment Regimens by Risk-Adapted Classification (See Table 24-7)

Low-Risk Group

Low-risk group patients (patients with ERMS that has been completely resected [Group I or II] either in a favorable site or incompletely resected localized ERMS). Low-risk patients represent approximately 30% of patients diagnosed with RMS. Amongst patients with low-risk tumors there are two subsets:

- Subset A – those with especially favorable outcome (including those who do not require radiation therapy). This subset has a 5-year FFS of 83% with an OS of 95% with only VA and the addition of cyclophosphamide improving the FFS to 93% and OS to 98%
- Subset B – those with favorable outcome who require longer therapy. Survival for this group of patients with favorable site nonmetastatic ERMS and gross totally resected unfavorable site ERMS have an expectation for cure of greater than or equal to 85%. Most treatment failures in low-risk patients are at the primary site.

The goal of the low-risk protocol is to maintain the excellent prognosis of patients with low-risk tumors but to improve the overall quality of survival by shortening the duration of treatment and reducing treatment-related costs – including the short- and long-term toxicities of therapy, particularly myelosuppression and infertility.

Radiation Therapy
1. Group I embryonal patients do not receive radiation therapy.
2. Patients with microscopic, locoregional, or gross residual tumor receive radiation therapy at week 13 of chemotherapy.

Chemotherapy

All low-risk patients receive the following chemotherapy:

- Vinristine IV one dose weekly for 9 weeks
- Dactinomycin IV and cyclophosphamide IV one dose each on the first week and then on weeks 4, 7 and 10. Cyclophosphamide dose should be reduced from a maximum of 2.2 to 1.2 g/m^2/course.

Low-risk subset A patients receive the additional following treatment:

- Vincristine IV one dose weekly from week 13 to week 21
- Dactinomycin IV one dose on weeks 13, 16, 19 and 22.

In patients where radiation is indicated it is started on week 13 (and dactinomycin is omitted during radiation).

Low-risk subset B patients receive the additional following treatment:

- Vincristine IV one dose weekly on weeks 13 to 21; 25 to 33 and 37 to 45
- Dactinomycin IV one dose on every 3 weeks starting on week 13 until week 46.

Some may have additional surgery at week 13 and in patients where radiation is indicated it is started on week 13 (and dactinomycin is omitted during radiation).

Intermediate-Risk Group

Intermediate-risk group patients account for the majority of the patients with RMS (55%) with an overall survival of 65%. This group includes unfavorable-site ERMS that have gross residual disease and nonmetastatic ARMS tumors. There has been no improvement in outcome for this group of patients over the past two decades. For patients with embryonal histology the risk of locoregional failure remains the dominant risk of treatment failure, whereas for patients with ARMS the risk of distant treatment failure is greater.

Figure 24-1 shows a chemotherapy regimen for intermediate-risk RMS.

- The current intermediate-risk COG study (ARST 0531) is investigating early radiotherapy (week 4) to improve local control and survival outcomes and to further assess whether the addition of irinotecan will improve outcome by comparing VAC versus VAC with reduced cyclophosphamide maximum dose from 2.2 to 1.2 g/m^2 alternating with vincristine-irinotecan (VI). Irinotecan is given in a dose of 50 mg/m^2/day for 5 days. This study compares 14 cycles of VAC with reduced-dose cyclophosphamide versus seven cycles of VAC with seven cycles VI. Cycles are given every 21 days as tolerated

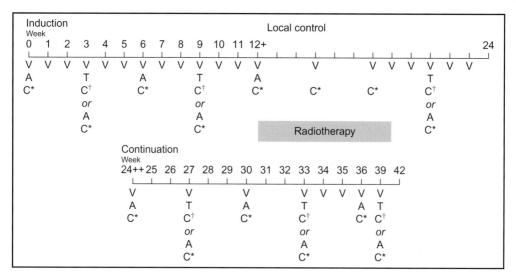

Figure 24-1 Chemotherapy for Intermediate-risk RMS (D9803).
Treatment schema: randomized schedule of either vincristine, dactinomycin and cyclophosphamide (VAC) or VAC alternating with vincristine, topotecan and cyclophosphamide (VAC/VTC) at weeks 3, 9, 21, 27, 33 and 39. Doses of drugs for children age 3 years and older: vincristine (V) 1.5 mg/m^2 (maximum dose, 2 mg) × 1 day; dactinomycin (A) 0.045 mg/kg × 1 day (maximum dose, 2.5 mg); topotecan (T) 0.75 mg/m^2 × 5 days; cyclophosphamide (C*) 2.2 g/m^2 × 1 dose (with mesna uroprotection); cyclophosphamide (C†) 250 mg/m^2/dose for 5 days.
Adapted from: Arndt CAS, et al. J Clin Oncol 2009;27:5182–5188, with permission.

- Although several small pilot series have reported excellent results with more intensive, 5-drug "Ewing's-type" therapy (vincristine/doxorubicin/cyclophosphamide and ifosfamide/etoposide), there remain no randomized clinical trial results demonstrating improved outcome of any non-VAC or VAC-plus regimen.

High-Risk Group

High-risk group patients have metastatic disease and represent 20% of patients with RMS. These patients have an overall survival of less than 30%. The dominant risk of treatment failure is from the inability to control systemic disease, although local treatment failure also appears to be higher in this group of patients compared to those without metastases. The following are associated with severely reduced event-free survival (EFS):

- Infants less than 1 year of age and children 10 years or older
- Unfavorable primary site
- Alveolar histology
- More than one metastatic site.

The standard therapy for high-risk patients is still VAC. No improvement in outcome has been demonstrated for this group of patients with single or tandem high-dose chemotherapy plus autologous stem-cell transplant regimens.

Treatment of NRSTS

The mainstay of therapy for this histologically and biologically diverse group of tumors is surgical resection. Tumors are broadly divided into two main categories:

- Low-grade lesions which may recur locally but are unlikely to metastasize and
- High-grade lesions, which have a substantially greater risk of both local invasiveness and distant spread.

The role of chemotherapy remains unclear for many patients. Preoperative chemotherapy may improve resectability and some studies suggest an improvement in metastasis-free survival with chemotherapy. However, the majority of randomized trials do not show improved overall survival. Chemotherapy appears to be of more use in synovial cell sarcoma and in extremity tumors. Ifosfamide and doxorubicin or etoposide seem to be the most active chemotherapy agents. The standard dose of ifosfamide is 1,800 mg/m^2/day for 5 days every 3 weeks for 5–7 cycles with 75 mg/m^2 per cycle of doxorubicin either as a continuous infusion or in a divided daily dose every 3 weeks for five cycles. During RT doxorubin is omitted and some investigators recommend replacing the doxorubicin with etoposide 100 mg/m^2/day for 5 days. For unresectable tumors the length of therapy depends upon response. In general, tumors that have not become resectable after five cycles of chemotherapy should receive radiation therapy and resectability should be re-evaluated after 45 Gy.

Low-grade tumors are, for the most part, observed following total surgical resection. The approach to high-grade tumors is based on their size, respectability and status of the margins. Small tumors (5 cm or less) that are completely resected with negative margins are generally observed without further treatment. Postoperative radiation is generally recommended for small tumors with positive margins and for all tumors more than 5 cm, even if margins are negative, because of the high risk of local recurrence. Neoadjuvant chemotherapy (i.e., chemotherapy before surgery or radiation therapy) is generally reserved for those patients with unresectable tumors and those with metastatic disease; adjuvant chemotherapy is typically recommended for patients with large tumors more than 5 cm, even if completely resected. The aim is to reduce the size or extent of the cancer before receiving surgery, thus making procedures easier and more likely to be successful and reducing the consequences of a more extensive surgery that would have to be done if the tumor was not reduced in size or extent.

FOLLOW-UP AFTER COMPLETION OF THERAPY

First Year After Completion of Therapy

1. Physical examination including complete blood count every 3 months.
2. Chest radiograph every 3 months.
3. Appropriate imaging studies (i.e., CT or MRI of the involved region) every 3 months.
4. Appropriate studies outlined below.

Second and Third Years After Completion of Therapy

1. Physical examination including complete blood count every 4 months.
2. Chest radiograph every 4 months.
3. Appropriate imaging studies every 4 months.

Fourth Year

1. Above studies every 6 months.

Fifth to Tenth Years After Completion of Therapy

1. Annual visit for physical examination and studies outlined below.
2. Referral to a survivorship program strongly recommended.

Ten Years After Completion of Therapy

1. Maintain yearly visit or phone contact.
2. Record attainment of puberty and pregnancy.
3. Long-term follow-up in a survivorship program strongly recommended.

General Studies

1. Height and weight at 6-month intervals.
2. Annual BP measurements.
3. Tanner staging annually until stage V.
4. Testicular size annually.
5. Onset of menses in girls.
6. Patients who receive cyclophosphamide: evaluation of gonadal function (FSH, LH and testosterone/estradiol).
7. Monitor for secondary malignancies, especially in cases with known familial risk factors.

Studies for Specific Sites

Head and Neck

1. Annual growth measurements.
2. Annual ophthalmologic examination.
3. Annual dental examination.

Trunk

1. Radiographs of irradiated bones every 1–2 years.
2. Kidney function, urinalysis, creatinine annually.
3. Sitting height/standing height ratio.
4. If patient received radiotherapy to chest or lungs, history of exercise tolerance or shortness of breath.

Genitourinary Sites

1. Kidney function evaluation every 1–2 years.
2. Girls – follow-up of sexual maturation and ovarian function.
3. Boys – history regarding ejaculatory function, semen analysis.

Extremities

1. Annual bilateral limb length measurements.
2. Radiographs of primary sites for bone growth abnormalities.

Because of the widespread need for radiotherapy to achieve local control and the young age of the patients, late-onset radiation-associated complications are the dominant source of treatment-related morbidities after 5 years.

RECURRENT DISEASE

The overwhelming majority of recurrences of RMS occur within the first 3 years of completing therapy. Although some children attain durable remissions with secondary therapy, the long-term prognosis for children with progressive or recurrent disease is extremely poor. The 3-year survival following relapse for patients is less than 15%. When recurrent disease is suspected, a biopsy should be done and an extent-of-disease workup should be undertaken. In formulating a treatment plan, the following factors should be considered:

- Timing of the recurrence
- Extent of disease at diagnosis and recurrence
- Type of chemotherapy and radiation therapy previously given.

Table 24-12   Alternative Chemotherapy for Recurrent Soft-Tissue Sarcomas

Regimen 1[*]		
Irinotecan IV	50 mg/m^2/day	Days 1–5
Temozolomide po	100 mg/m^2/day	Days 1–5
Regimen 2[*]		
Vinorelbine IV	20–30 mg/m^2/day	Days 1, 8
Irinotecan IV	20 mg/m^2/day	Days 1–5 and Days 8–12
Regimen 3[*]		
Vinorelbine IV	25 mg/m^2/day	Days 1, 8, 15
Cyclophosphamide po	25 mg/m^2/day	Days 1–28, 28

[*]Personal communication with Leonard H. Wexler, MD.
Regimens can be given for six cycles up to 24 cycles if tolerated.

Durable remission is most difficult to attain in patients who have unresectable or widespread disease at diagnosis and in those who progressed on therapy or relapsed early.

Treatment

Complete surgical resection, when feasible, improves outcome. Surgery should be followed by adjuvant chemotherapy and radiation (if further radiation is feasible). Durable remissions in paratesticular and vaginal recurrences can be attained with this salvage approach. Disseminated recurrence is essentially incurable.

While surgical resection of metastatic lesions is not curative, it may be palliative. Even when followed with intensive multiagent therapy with vincristine/doxorubicin/cyclophosphamide and ifosfamide/etoposide (plus irinotecan for responding patients), the median postrelapse survival time is 1.1 years and fewer than 20% of patients are alive 3 years postrelapse. Multiple trials have used intensive regimens followed by autologous stem cell rescue, but no large studies have demonstrated any significant salvage rate. Active agents in relapsed RMS include combinations such as ifosfamide/carboplatin/etoposide, docetaxel/gemcitabine and irinotecan in combination with vinorelbine or temozolomide. Chemotherapy regimens in use for recurrent soft-tissue sarcomas are outlined in Table 24-12.

FUTURE PERSPECTIVES

Survival has dramatically increased over the past 30 years; however, improvements in disease control are still needed, especially for the majority of patients with gross residual after resection or metastatic disease at the time of diagnosis. Risk stratification into low-, intermediate- and high-risk strata will hopefully help to reduce the short- and long-term toxicities of therapy in lower-risk patients and to allow for novel approaches with the high-risk patients. Ongoing investigations include antibodies directed at the IGF-1 receptor to overcome the IGF-1R signaling that activates the anti-apoptotic pathways and suppresses apoptosis protecting against chemotherapy and radiation as well as agents to inhibit mTOR and its analogs and induce apoptosis in cells with dysfunction p53 in the absence of exogenous IGF. Additional trials are using angiogenesis inhibitors such as bevacizumab or pazopanib.

Suggested Reading

Arndt C, Hawkins DS, Meyer WH, et al. Comparison of a pilot study of alternating vincristine/doxorubicin/cyclophosphamide and etoposide/ifosfamide with IRS-IV in intermediate risk rhabdomyosarcoma: a report from the Children's Oncology Group. Pediatr Blood Cancer 2008;50:33–36.

Arndt C, Stoner JA, Hawkins DS, et al. Vincristine, actinomycin and cyclophosphamide compared with vincristine, actinomycin and cyclophosphamide alternating with vincristine, topotecan and cyclophosphamide for intermediate-risk rhabdomyosarcoma: children's Oncology Group Study D9803. J Clin Oncol. 2009;27:5182–5188.

Blakely M, Spurbeck W, Pappo A, et al. The impact of margin of resection on outcome in pediatric nonrhabdomyosarcoma soft tissue sarcoma. J Pediatr Surg. 1999;34:672.

Casanova M, Ferrari A, Bisogno G, et al. Vinorelbine and low-dose cyclophosphamide in the treatment of pediatric sarcomas. Pilot study for the upcoming European rhabdomyosarcoma protocol. Cancer 2004;101:1664–1671.

Crist W, Anderson J, Meza J, et al. Intergroup Rhabdomyosarcoma Study – IV: results for patients with nonmetastatic disease. J Clin Oncol. 2001;19:3091–3102.

Dantonello TM, Int-Veen C, Winkler P, et al. Initial patient characteristics can predict pattern and risk of relapse in localized rhabdomyosarcoma. J Clin Oncol. 2008;26:406–413.

Davicioni E, Anderson MJ, Finckenstein FG, et al. Molecular classification of rhabdomyosarcoma – genotypic and phenotypic determinants of diagnosis. A report from the Children's Oncology Group. Am J Pathol. 2009;174:550–564.

Ferrari A, Grosso F, Stacchiotti S, et al. Response to vinorelbine and low-dose cyclophosphamide chemotherapy in two patients with desmoplastic small round cell tumor. Pediatr Blood Cancer 2007;49(6):864–866.

Klem ML, Grewal RK, Wexler LH, et al. PET for staging in rhabdomyosarcoma: An evaluation of PET as an adjunct to current staging tools. J Pediatr Hematol Oncol. 2007;29:9–14.

Klingebiel T, Boos J, Beske F, et al. Treatment of children with metastatic soft tissue sarcoma with oral maintenance compared to high dose chemotherapy: report from HD CWS-96 trial. Pediatr Blood Cancer 2008;50:739–745.

Lawrence W, Gehan E, Hays D, et al. Prognostic significance of staging factors of the UICC staging system in childhood rhabdomyosarcoma: a report from the Intergroup Rhabdomyosarcoma Study (IRS-II). J Clin Oncol. 1987;5:46–54.

Mattke AC, Bailey EJ, Schuck A, et al. Does the time-point of relapse influence outcome in pediatric rhabdomyosarcomas? Pediatr Blood Cancer 2009;52:772–776.

Meza JL, Anderson J, Pappo AS, et al. Analysis of prognostic factors in patients with nonmetastatic rhabdomyosarcoma treated on the intergroup rhabdomyosarcoma studies III and IV: the Children's Oncology Group. J Clin Oncol. 2006;24:3844–3851.

Nathan PC, Tsokos M, Long L, et al. Adjuvant chemotherapy for the treatment of advanced pediatric nonrhabdomyosarcoma soft tissue sarcoma: the National Cancer Institute experience. Pediatr Blood Cancer. 2005;44(5):449–454.

Oberlin O, Rey A, Lyden E, et al. Prognostic factors in metastatic rhabdomyosarcoma: results of a pooled analysis from the United States and European cooperative groups. J Clin Oncol. 2008;26(14):2384–2389.

Pappo, Devidas M, Jenkins J, et al. Phase II trial of neoadjuvant vincristine, ifosfamide, doxorubicin with GCSF support in children and adolescents with advanced stage NRSTSs: a Pediatric Oncology Group study. J Clin Oncol. 2005;23:18.

Puri DR, Wexler LH, Meyers PA, et al. The challenging role of radiation therapy for very young children with rhabdomyosarcoma. Int J Radiation Oncol Biol Phys. 2006;65:1177–1184.

Rodeberg DA, Stoner JA, Hayes-Jordan A, et al. Prognostic significance of tumor response at the end of therapy in Group III rhabdomyosarcoma. A report from the Children's Oncology Group. J Clin Oncol. 2009;27:3705–3710.

Spunt SL, Hill DA, Motosue AM, et al. Clinical features and outcome of initially unresected nonmetastatic pediatric nonrhabdomyosarcoma soft tissue sarcoma. J Clin Oncol. 2002;20:3225–3235.

Walterhouse D, Meza JL, et al. Dactinomycin (A) and vincristine (V) with or without cyclophosphamide (C) and radiation therapy (RT) for newly diagnosed patients with low risk embryonal/botryoid rhabdomyosarcoma (RMS). An IRS-V report from the Soft Tissue Sarcoma Committee of the Children's Oncology Group (STS COG). J Clin Oncol Suppl. 2006;24(18S):502s.

CHAPTER 25

Malignant Bone Tumors

Malignant bone tumors constitute approximately 6% of all childhood malignancies. In the United States, the annual incidence in children under 20 years of age is 8.7 per million. Osteosarcoma (56%) and Ewing sarcoma (34%) comprise the most frequently encountered malignant bone tumors in children and adolescents. In the United States 650–700 children are diagnosed with malignant bone tumors each year. Two-thirds of these are osteosarcoma and one-third Ewing sarcoma.

Figure 25-1 shows age-specific incidence rate of bone cancers for all races and both sexes.

The peak incidence of bone cancer is age 15, coincident with the adolescent growth spurt, after which rates show a decline. Osteosarcoma incidence has a bimodal age distribution with peaks in early adolescence and in adults over 60 years of age. Ewing sarcoma is a disease of children and young adults; it is very rare in older adults.

OSTEOSARCOMA

Epidemiology

The incidence is higher in males overall and in each age group except in those under 15 years of age. The incidence in African-American children is slightly higher than in whites and in those in the other race categories but varies by age. Although the etiology of osteosarcoma is unknown, certain factors appear to correlate with its occurrence:

- *Bone growth*: The development of osteosarcoma is correlated with linear bone growth as suggested by:
  - The peak incidence occurs during the pubescent growth spurt. The age peak in girls is 12 years and 16 years in boys, correlating with the different average age for pubertal development
  - Most studies suggest patients with osteosarcoma are taller than average
  - The most common sites are metaphyses of the most rapidly growing bones (distal femur, proximal humerus, proximal tibia)
  - Large-breed and not small-breed canines develop osteosarcoma which is genetically, histologically and clinically analogous to the human disease

Manual of Pediatric Hematology and Oncology. DOI: 10.1016/B978-0-12-375154-6.00025-2

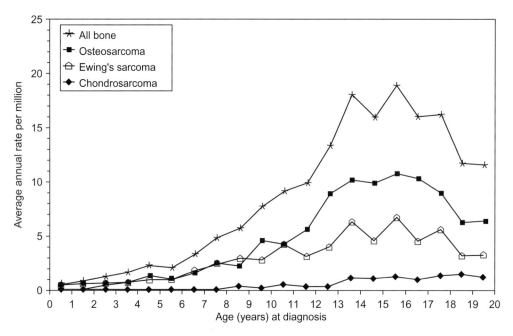

Figure 25-1 Bone Cancer Age-specific Incidence Rate by Histology for All Races and Both Sexes. SEER, 1976–94 and 1986–94 Combined.
From: Gurney JG, Swensen AR, Bukterys M. Malignant bone tumors. Cancer incident and survival among children and adolescents: United States SEER Program, 1975–1995. NHI Pub No 99-4649. Bethesda, MD: National Cancer Institute. SSER Program, 1999:99–110, with permission.

- *Genetic factors*: Osteosarcoma is a genetically complex tumor with bizarre karyotypes, numerous genetic abnormalities and lack of recurrent chromosomal translocations. Osteosarcoma can be hereditary in some rare cases. Human predisposition syndromes, murine models, genetic analyses of osteosarcoma tumor specimens and environmental factors may all help to understand the pathogenesis of this disease
  - Human predisposition syndromes
    - Retinoblastoma – In patients with early onset bilateral retinoblastoma, who are likely to have germline alteration in the Rb gene, the development of a secondary osteosarcoma is independent of the therapeutic modalities or radiation field used in the treatment of the retinoblastoma. Approximately 40% of patients will develop this secondary malignancy by 40 years of age
    - P53 – Osteosarcoma is seen frequently in families with Li–Fraumeni syndrome, in which members of the affected families have breast cancer, brain tumors, soft tissue sarcomas, leukemia, adrenocortical carcinoma and osteosarcoma, associated with a germ-line mutation of p53, a suppressor oncogene
    - Other syndromes such as Rothmund–Thomson syndrome (Rothmund–Thomson syndrome is an inherited autosomal recessive disorder consisting of skeletal

abnormalities, short stature, underdeveloped or missing forearm bones or thumbs, cataracts, abnormalities of teeth and nails, sparse hair and red, swollen patches on the skin) and Werner syndrome (Werner syndrome [also known as progeria adultorum] is a genetic disorder consisting of premature aging consisting of scleroderma-like, thin, wrinkled skin, loss of subcutaneous fat, graying and loss of hair, bilateral cateracts, hypogonadism and premature menopause) are associated with increased incidences of osteosarcoma among other malignancies.

- Murine models that produce osteosarcoma include those with alterations in the p53 pathway, Rb pathway, MYC, FOS as well as chronic parathyroid hormone exposure
- Genetic analyses of osteosarcoma tumor specimens
  - Osteosarcomas are genetically complex with numerous alterations in each tumor sample. These include a multitude of genes involved in cell cycle regulation, differentiation, oncogenic transformation, cell signaling, signal transduction, among many pathways. Consistent chromosomal gains and losses occur at genetic sites in which the gene(s) of interest have not been clearly indentified
  - Deletions and rearrangements of the Rb gene, altered expression of the Rb transcripts, or altered Rb protein have been described in osteosarcoma tumors
  - Inactivation of the p53 gene product with the loss of a growth regulator that inhibits cellular proliferation is important in the growth of osteosarcoma. The p53 pathway is altered in the vast majority of osteosarcoma specimens. Despite this, only 3% of patients with spontaneous osteosarcoma are found to have a germ-line alteration in p53 or Li–Fraumeni syndrome if it is screened for
  - There are nongerm-line mutations of both p53 and Rb in osteosarcoma tumors. Twenty-five percent have point mutation of p53, 10–20% have p53 rearrangements, 75% have loss of one 17q allele and 60% of OS tumors have a loss of heterozygosity at 13q of the site of the Rb gene.
- *Environmental factors*: The environmental factor most consistently associated with the development of osteosarcoma is radiation therapy. Bone turnover also may be related to osteosarcoma development. The evidence for these is:
  - Osteosarcoma may arise in any previously irradiated bone
  - The interval between irradiation and the appearance of osteosarcoma has ranged from 4 to 40 years but is generally 10–20 years, suggesting it is not a major etiologic factor in the younger patients
  - Bone diseases: Paget disease and other disorders associated with bone turnover are associated with a higher incidence of osteosarcoma

Table 25-1   Classification of Osteosarcoma

I. Classic
  A. Osteoblastic
  B. Chondroblastic
  C. Fibroblastic
II. Telangiectatic
III. Small cell
IV. Periosteal
V. Parosteal

Pathology

The histologic diagnosis of osteosarcoma is dependent on the demonstration of malignant osteoid in association with a malignant spindle cell. Table 25-1 lists the classification of osteosarcoma. Although several of the rarer variants may be confused radiographically or histologically with other entities the pathologic hallmark of osteosarcoma is the production of osteoid.

Clinical Manifestations

Symptoms are usually present for several months before the diagnosis is made, with an average duration of approximately 3 months. Patients commonly present with the following symptoms:

- Local pain (90%)
- Local swelling (50%)
- Decreased range of motion (45%)
- Pathologic fracture (8%)
- Joint effusion can occur rarely. Its presence may suggest an extension of the tumor into the joint space.

It is important to be aware of the sites and ages at which osteosarcoma presents as this helps distinguish it from other entities which occur more commonly in this population such as growing pains and osteomyelitis. Osteosarcoma can arise in any bone but has a propensity for the metaphyseal portions of long bones, particularly in the lower extremity around the knee (>50% involving distal femur or proximal tibia). A small number of patients can present with multifocal disease (lesions at multiple sites even in the absence of lung metastases). Additionally, osteosarcoma can metastasize to other bones. The most common presenting symptom is persistent pain and swelling at the initial primary site. Systemic manifestations of osteosarcoma are exceedingly rare at initial presentation and when it occurs it is usually associated with significant metastatic disease.

Figure 25-2 shows skeletal distribution of patients with osteosarcoma. The most common primary sites in individuals less than 25 years of age are:

- Lower long bones (74.5%)
- Upper long bones (11.2%)

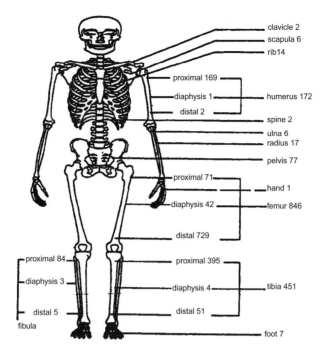

clavicle 2
scapula 6
rib14
proximal 169
diaphysis 1 — humerus 172
distal 2
spine 2
ulna 6
radius 17
pelvis 77
proximal 71
hand 1
diaphysis 42 — femur 846
distal 729
proximal 84
proximal 395
diaphysis 3
diaphysis 4 — tibia 451
distal 5
distal 51
fibula
foot 7

Figure 25-2 Skeletal Distribution of 1702 Patients with Osteosarcoma.
From: Bielack SS, Kampf-Bielack B, Delling G, et al. Prognostic factors in high grade osteosarcoma of the extremities or trunk: an analysis of 1,702 patients treated on neoadjuvant cooperative osteosarcoma study group protocols. J Clin Oncol 2002;20[3]:776–790, with permission.

- Pelvic region (3.6%)
- Face or skull (3.2%)
- Mandible (1.9%)
- Chest region (1.8%)
- Vertebral column (1.2%)
- Lower short bones (1.1%)
- Upper short bones (0.3%).

Diagnostic Evaluation

The diagnostic evaluation should delineate the extent of the primary tumor and the presence or absence of metastatic disease. All patients should have the following evaluation:

- History
- Physical examination
- Blood count
- Urinalysis
- Blood urea nitrogen (BUN), creatinine, liver enzymes, alkaline phosphatase (approximately 50% of patients have elevated alkaline phosphatase), total bilirubin
- Radiographs of the affected bone

- Computed tomography (CT) or magnetic resonance imaging (MRI) of affected bone
- Bone scan
- Chest CT
- Assessment of renal function (institutional practice varies with some obtaining a creatinine clearance or radionuclide GFR study to assess renal function prior to chemotherapy whereas others rely solely on the serum creatinine)
- Echocardiogram to assess cardiac function prior to chemotherapy
- Baseline hearing test prior to chemotherapy
- Fertility preservation – Sperm banking should be offered whenever feasible. Cryopreservation of ovaries has not become a clinical standard at present but interest in this approach as a means of preserving fertility continues to increase
- The clinical utility of PET imaging at diagnosis is debated.

Radiographically, osteosarcoma produces dense sclerosis in the metaphysis of the affected long bone along with the following findings:

- Soft-tissue extension (75%)
- Radiating calcification, or sunburst (60%)
- Osteosclerotic lesion (45%)
- Lytic lesion (30%)
- Mixed lesion (25%).

MRI gives valuable information in planning limb salvage surgery, such as intraosseous extent of tumor and involvement of muscle groups, subcutaneous fat, joint and neurovascular structures around the tumor. Approximately 5% of patients will have evidence of bone metastases on bone scan. The use of chest CT is essential for staging patients, because it detects metastases not seen on chest radiographs. If a lesion detected on chest CT cannot be diagnosed as metastatic, histologic confirmation is indicated.

Treatment

The two therapeutic modalities currently used in the primary treatment of osteosarcoma are surgery and chemotherapy. The tumor is generally viewed as radioresistant and radiation therapy only rarely plays a role in initial management.

Surgery

The surgical approach to osteosarcoma includes surgical biopsy, followed by either amputation or limb salvage surgery.

A biopsy is required to confirm the diagnosis. An open biopsy is the preferred method. The biopsy should be performed with surgery in mind, by an orthopedic surgeon experienced in the management of malignant bone tumors. A poorly placed biopsy may jeopardize future limb salvage surgery. Since the biopsy tract and skin are contaminated with tumor cells, the biopsy site needs to be excised en bloc with the tumor during definitive surgery.

Surgery should remove all gross and microscopic tumor with a margin of normal tissue to prevent local recurrence. Appropriate surgical procedures are limb salvage surgery or amputation. The rate of local recurrence and survival after amputation and limb salvage surgery are comparable.

Definitive surgery is generally performed following preoperative chemotherapy. Delaying definitive surgery until after preoperative chemotherapy has become the standard treatment approach because neoadjuvant (preoperative) chemotherapy offers several potential and theoretical advantages:

- Decrease in tumor-related edema and, sometimes, shrinkage of the tumor at the primary site so as to facilitate less radical surgery such as limb salvage
- Initial therapy directed towards the micrometastases which are present in 80–90% of patients and their responsiveness is the major determinant of patient survival
- Assessment of sensitivity of primary tumor to chemotherapy which has prognostic significance.

The type of surgery is determined by tumor location, size, extramedullary extent, presence of metastatic disease, age, skeletal development and patient's lifestyle choices. If complete excision cannot be accomplished with limb-sparing surgery, amputation is indicated.

Limb salvage surgery should be performed by orthopedic surgeons experienced with the technique. Guidelines for the use of limb-sparing surgery include:

- No major neurovascular involvement by tumor
- Ability to have a wide resection of affected bone with a normal muscle cuff in all directions
- Ability to perform en bloc removal of all previous biopsy sites and all potentially contaminated tissue
- Adequate motor reconstruction.

Limb salvage is feasible in the vast majority of patients.

Chemotherapy

The use of chemotherapy in a neoadjuvant (preoperative) or adjuvant (postoperative) fashion results in significant improvement in disease-free survival in patients with osteosarcoma. More than 80–90% of patients treated with surgery alone will develop metastatic disease. With the addition of multiagent chemotherapy, 60–70% of nonmetastatic extremity osteosarcoma patients will survive without evidence of recurrence. The standard components of multiagent cytotoxic chemotherapy for osteosarcoma are high-dose methotrexate (HDMTX),

(a) Schema of typical administration scheduled for MAP therapy for osteosarcoma

Week	1 2 3 4 5	6 7 8 9 10	11	12 13 14 15 16	17 18 19 20 21	22 23 24 25	26 27 28 29
Chemo	A M M	A M M	surgery	A M M	A M M	A M M	A M M
	P	P		P	P		

A = Doxorubicin 75 mg/m^2 continuous infusion over 48 hours
P = Cisplatin 60 mg/m^2/day × 2 days
M = Methotrexate 12 gm/m^2 × 1 dose, maximum dose 20 gm*

(b) Schema of MAP chemotherapy plus ifosfamide and etoposide

Week	1 4 5	6 9 10	11	12 15 16	19 20	23 24	27 28	31 32	35 36	39 40
Chemo	A M M	A M M	surgery	A M	I M	A M	I M	A M	I M	A M M
	P	P		P	E	i	E	P	E	i

A = Doxorubicin 75 mg/m^2 continuous infusion over 48 hours
P = Cisplatin 60 mg/m^2/day × 2 days
M = Methotrexate 12 gm/m^2 × 1 dose, maximum dose 20 g
I = Ifosfamide 2.8 gm/m^2/day × 5 days
i = Ifosfamide 3 gm/m^2/day × 3 days
E = Etoposide 100 mg/m^2/day × 5 days

* Methotrexate is administered in 1L of D5W ¼ NS with 50m Eq of sodium bicarbonate per liter over 4 hours
Hydration is administered to achieve a urine output of 1,400–1,600 ml/m^2 for each 24-hour period and a urine pH greater than 7.0

Figure 25-3 (a) and (b) Protocol Schema for Ongoing EURAMOS Clinical Trial, Including Standard Arm With Methotrexate, Doxorubicin and Cisplatin (MAP) and Experimental Arm for Patients with Poor Histologic Necrosis with MAP Plus Ifosfamide and Etoposide (MAPIE). From: Protocol document and with permission of Mark Bernstein.

doxorubicin, or adriamycin (ADM) and cisplatin (CDP), a regimen often referred to as MAP. Figure 25-3a shows the schedule and doses of MAP regimens.

Ifosfamide (IFO) and etoposide (VP16) are other agents with demonstrated antiosteosarcoma activity. Whether the addition of ifosfamide and etoposide to MAP chemotherapy results in improved overall and event-free survival is the subject of ongoing investigation. Figure 25-3b also shows how IFO and VP16 can be incorporated into the MAP regimen (MAPIE).

Specific HDMTX administration guidelines must be followed to prevent renal injury and life-threatening toxicity. Normal renal function should be documented prior to administration. Alkalinizing intravenous fluids should be administered with HDMTX and continued until a sufficient decrease in MTX blood levels has occurred. MTX levels should be monitored starting 24 hours after the beginning of MTX administration. Leucovorin should be initiated 24 hours after HDMTX administration in dose of 10 mg every 6 hours until the MTX level is less than 1.0×10^7 mol/L. Patients whose methotrexate excretion is delayed require increased dose or duration of leucovorin. Figure 25-4 is a nomogram of plasma methotrexate concentration over time, providing indication for increased leucovorin rescue.

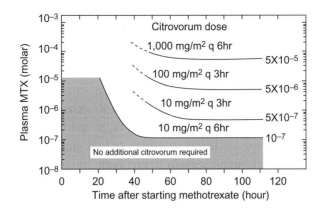

Figure 25-4 Nomogram of Plasma Methotrexate Concentration Over Time, Providing Indication for Increased Leucovorin Rescue.
From: Pediatric Oncology Group Protocol 9351, with permission.

Table 25-2   Histologic Response of Osteosarcoma to Neoadjuvant (Preoperative) Chemotherapy

Grade	Description
I	Little or no effect of chemotherapy
II	Partial response to chemotherapy with more than 50% tumor necrosis but with areas of viable tumor still demonstrable
III	Near-complete response to chemotherapy with more than 90% tumor necrosis but with areas of viable tumor still demonstrable
IV	Complete response to chemotherapy with no viable tumor cells demonstrable

A study conducted by the Pediatric Oncology Group and Children's Cancer Group demonstrated that the addition of muramyl-tripeptide phosphatidylethanolamine (MTP-PE) to standard chemotherapy significantly improved 6-year overall survival, from 70 to 78%, but did not significantly improve event-free survival. MTP-PE, a liposomal encapsulated bacille Calmette-Guerin (BCG) cell wall component, is approved for use in Europe but not in the United States.

The degree of tumor necrosis following neoadjuvant chemotherapy is an independent predictor of event-free and overall survival, presumably reflecting tumor resistance to chemotherapy. Table 25-2 shows histologic response of osteosarcoma to neoadjuvant (preoperative) chemotherapy. Patients whose tumors exhibit a good response (>90% necrosis or grades III and IV) have superior event-free and overall survival compared with poor responders. In patients whose response is suboptimal, attempts to improve outcome by modifying chemotherapy are continuing.

Treatment of Metastatic Osteosarcoma at Presentation

Approximately 15–20% of patients with osteosarcoma present with metastatic disease. This group of patients has a worse prognosis. Because of the poor prognosis, metastatic osteosarcoma is often treated with MAP chemotherapy plus IFO and VP16 (Figure 25-3b). Since it is not clear that the addition of IFO and VP16 to MAP chemotherapy improves outcome for these patients MAP alone is an acceptable regimen for the treatment of metastatic osteosarcoma at presentation. Surgical management of the primary tumor is performed as it would be for nonmetastatic osteosarcoma. When the tumor arises in a site that is not resectable, radiation may be used for local control even though the tumor is usually not radiosensitive.

Complete surgical removal of all metastatic sites, if feasible, is an important component of the treatment of metastatic osteosarcoma. About one-third of patients with pulmonary metastases from osteosarcoma have pulmonary nodules that are not visible on CT but can be identified by intraoperative palpation. Because it allows the surgeon to identify these nodules by palpation, open thoracotomy is the recommended approach for resection of pulmonary metastases. Patients with unilateral lung metastases should have an open thoracotomy on the contralateral side to permit identification and resection of palpable, nonvisible pulmonary metastases.

Treatment of Relapsed Osteosarcoma

More than 85% of osteosarcoma recurrences are in the lung and in a significant proportion of patients the lungs are the only site of recurrence. Patients with recurrent osteosarcoma have a poor prognosis. In retrospective studies, patients who have complete resection of all sites of recurrent disease have a better prognosis than patients who do not have complete resection. Therefore, surgical resection of all sites of recurrent disease, if feasible, is recommended. Whether chemotherapy improves outcome in recurrent osteosarcoma is not clear. When chemotherapy is administered, the choice of chemotherapeutic agents depends on the patient's prior therapy.

Newer Agents Under Investigation

Several institutions have intensified the postoperative treatment to improve the outcome of patients who have a poor response (<90% necrosis) to neoadjuvant chemotherapy. Most of these studies show that the intensified treatment does not change outcome. In addition, several institutions have intensified their neoadjuvant treatments to attempt to influence this outcome. Although the percentage of good responders increases, the event-free and overall survival rates are unchanged. For this reason, the evaluation of targeted or biologically based therapies such as bisphosphonates, denosumab, interferon, trastuzumab, insulin-like growth factor receptor-1 antibody and granulocyte macrophage stimulating factor are being investigated.

Follow-up Studies

1. CT of the chest every 3 months for 2 years after completion of therapy; then every 6 months for 1 year, then every year for 2 years.
2. Chest radiograph yearly for 5 years starting after the last CT of the chest.
3. Evaluation of the primary site every 3 months for 2 years after completion of therapy; then every 6 months for 3 years, then every year for 5 years.

Prognosis

The actuarial disease-free survival for patients with nonmetastatic osteosarcoma treated with adjuvant chemotherapy is 60–80%. Patients who present with metastatic disease have a significantly worse prognosis with 2-year survival of 10–30%. Key prognostic factors are surgical remission and response to chemotherapy (Figure 25-5). For newly diagnosed osteosarcoma, the following factors have been associated with a poor prognosis:

- Detectable metastatic disease at diagnosis
- Incomplete resection
- Poor response to chemotherapy
- Axial skeletal primaries
- Larger tumor size
- Proximal location within the limb.

For recurrent osteosarcoma, the following factors have been associated with a poor prognosis:

- Incomplete surgical resection
- Early (less than 2 years post-treatment) relapse
- More than one or two pulmonary nodules
- Bilateral pulmonary involvement
- Pleural disruption by metastatic disease.

EWING SARCOMA

Ewing sarcoma is a malignant tumor of bone or soft tissue in which the predominant histologic feature is densely packed small cells with round nuclei without prominent nucleoli. It is the second most common malignant primary bone tumor of childhood. Historically, a range of terms (Ewing sarcoma, Askin tumor and primitive neuroectodermal tumor) have been used to describe this entity. Ewing sarcoma can demonstrate a range of neural differentiation. Tumors previously referred to as classic Ewing sarcoma showed little neural differentiation while tumors previously referred to as peripheral primitive neuroectodermal tumors showed more prominent neural features. Based on an improved understanding of the

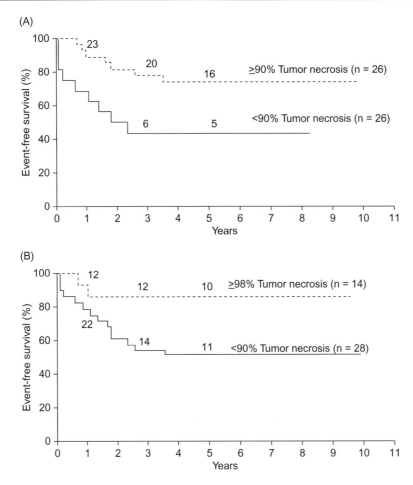

Figure 25-5 Event-free Survival Comparing Pathologic Response at (A) Less Than 10% Residual Viable Tumor Versus ≥ Residual Viable Tumor and (B) Less Than 2% Residual Viable Tumor Versus ≥ 2% Residual Viable Tumor. Numbers on the Curves Represent Patients Still at Risk.
From: Goorin AM, Schwartzentruber DJ, Devidas M, et al. Presurgical chemotherapy compared with immediate surgery and adjuvant chemotherapy for non-metastatic osteosarcoma: Pediatric Oncology Group Study POG-8651. J Clin Oncol 2003;21(8):1574–1580, with permission.

molecular pathogenesis of these tumors, they are now all recognized to be biologically equivalent. In the light of this, these tumors are therefore referred to simply as Ewing sarcoma throughout the remainder of this chapter.

Epidemiology

The etiology of Ewing sarcoma is unknown. Certain epidemiologic association studies have indicated higher rates of Ewing sarcoma in children with a history of inguinal hernia and in children of farm workers. Eighty percent of patients with Ewing sarcoma are younger than

20 years at diagnosis. Ewing sarcoma has a striking difference in racial incidence with an extremely low incidence in African–American and Asian children. It is not usually associated with familial cancer syndromes. Unlike osteosarcoma, the risk of Ewing sarcoma has not been shown to increase following radiation exposure.

Pathology

Ewing sarcoma is one of the small round blue cell tumors of childhood, with unknown histogenesis.

The pathologic features of Ewing sarcoma are:

* Highly cellular aggregates of small round cells compartmentalized by strands of fibrous tissue
* Tumor cells have round or oval nuclei, occasionally with indistinct nucleoli
* Nuclei containing finely dispersed chromatin (ground-glass appearance)
* Less than two mitotic figures per high-power field.

In order to differentiate Ewing sarcoma from other small round blue cell tumors, immunohistochemical studies are necessary. Table 25-3 shows the immunohistochemistry of Ewing sarcoma and other small round blue cell tumors of childhood. Glycoprotein p30/32 mic2

Table 25-3   Immunohistochemistry of Ewing Sarcoma and Other Small Round Blue Cell Tumors of Childhood

Marker	Ewing Sarcoma	Neuroblastoma	Rhabdomyosarcoma	Lymphoma[a]
NSE	+/−	+	+/−	−
S-100	+/−	+	+/−	−
NFTP	+/−	+	+/−	−
Desmin	+/−	−	+	−
Actin	+/−	−	+	−
Vimentin	+/−	−	+	+
Cytokeratin	+/−	−	+/−	−
LCA	−	−	−	+
HNK-1	+/−	+	+	−
β2-Microglobulin	+	−[b]	+/−	+
CD99	+	−	+[c]	+/−[d]

[a]Nonhistiocytic.
[b]Only in ganglion cells, Schwann cells and stage IVS neuroblastoma.
[c]Only well-differentiated rhabdomyoblasts.
[d]Positive in T-cell lymphoblastic lymphoma.
Abbreviations: −, negative; +/−, positive in greater than 10% of cases or variably positive; +, positive.
Adapted from: Ginsberg JP, Woo SY, Johnson ME, et al. Ewing sarcoma family of tumors: Ewing sarcoma of bone and soft tissue and the peripheral primitive neuroectodermal tumors. In: Pizzo PA, Poplack DG, editors. Principles and Practice of Pediatric Oncology. 4th ed. Philadelphia: Lippincott–Raven, 2003, with permission.

(CD99) has proven to be one of the most useful in supporting the diagnosis of Ewing sarcoma. CD99 reacts with the cytoplasmic membrane of Ewing sarcoma cells, giving a honeycomb staining pattern. It is also expressed by other tumors and normal tissues including nonblastematous portion of Wilms' tumor, clear cell sarcoma of the kidney, T-cell lymphoma and leukemia, pancreatic islet cell tumors, immature teratomas, testicular embryonal carcinoma, ependymomas and rhabdomyosarcoma. The pattern of CD99 staining in other tissues is often cytoplasmic, rather than the membranous pattern seen in Ewing sarcoma.

Molecular Genetics

Ewing sarcoma is characterized by the presence of one of several chromosomal translocations (Table 25-4). The t(11;22) translocation is the most common and has been found to involve the juxtaposition of two normally distinct genes, FL11 on chromosome 11 and EWS on chromosome 22. The fused gene results in an EWS/FL11 chimeric protein. Other variant translocations have been described that fuse EWS to other ETS family transcription factors, such as t(21;22) and EWS/ERG fusion. While the exact role of the fusion proteins remains unknown, they are felt to act primarily as aberrant transcription factors that play an important role in growth deregulation.

Table 25-4   Frequency of Chromosomal Translocations in Ewing Sarcoma

Chromosomal Abnormality	Frequency
t(11;22)(q24;q12)	90–95%
t(21;22)(q2;q12)	5–10%
t(7;22)(p22;q12)	<1%
t(17;22)(q12;q12)	<1%
t(2;22)(q33;q12)	<1%

These translocations can be easily detected by reverse transcription polymerase chain reaction (RT-PCR) and fluorescent *in-situ* hybridization (FISH). Recent studies have utilized the RT-PCR method to detect metastatic disease in the blood and bone marrow.

Clinical Features

Patients commonly present with the following symptoms:

*   Local pain (96%)
*   Local swelling (61%)
*   Fever (21%)
*   Pathologic fracture (16%).

Table 25-5   Distribution of Primary Site for 303 Patients with Ewing Sarcoma of Bone

Primary Site	Percentage
Pelvic	20.0
Ilium	12.5
Sacrum	3.3
Ischium	3.3
Pubis	1.7
Lower extremity	45.6
Femur	20.8
Fibula	12.2
Tibia	10.6
Feet	2.0
Upper extremity	12.9
Humerus	10.6
Forearm	2.0
Hand	0.3
Axial skeleton/ribs	12.9
Face	2.3

From: Grier HE. The Ewing Family of tumors: Ewing sarcoma and primitive neuroectodermal tumors. Pediatr Clin N Am 1997;64:991–1004, with permission.

It is important to realize that, because fever is a frequent symptom of Ewing sarcoma, it may be initially mistaken for osteomyelitis. Patients may have symptomatology for 3–6 months before a diagnosis is established. A small proportion of patients may have symptoms for longer than 6 months before the tumor is identified. The length of symptoms does not necessarily correlate with the presence of metastases, since these appear to be related to the biological behavior of the tumor rather than to the length of symptoms. Interestingly, the pain may be intermittent in nature.

Ewing sarcoma most commonly involves the lower extremity, with the pelvis being the next most common site. Table 25-5 lists the distribution of the primary sites of origin in Ewing sarcoma of bone. In long bones, Ewing sarcoma usually begins in the diaphysis. Ewing sarcoma of the thoraco-pulmonary region involving the chest wall has been referred to as Askin tumor.

Diagnostic Evaluation

The diagnostic evaluation typically begins with a history, physical examination and plain radiographs of the affected bone. Radiographically, Ewing sarcoma produces changes in the diaphysis of long bones with extension toward the metaphysis. Table 25-6 lists the radiographic findings in Ewing sarcoma. Patients with soft-tissue primary tumors may have normal plain radiographs.

Table 25-6   Radiographic Findings in Ewing Sarcoma

Finding	Percentage
Bone destruction	75
Soft-tissue extension	64
Reactive bone formation	25
Lamellated periosteal reaction (onion skin)	23
Radiating spicules (sunburst)	20
Periosteal thickening	19
Sclerosis	16
Fracture	13

From: Green DM. Diagnosis and management of solid tumors in infants and children. Boston: Nijhoff, 1985, with permission.

The remainder of the staging studies for Ewing sarcoma should consist of the following:

- Blood count
- Erythrocyte sedimentation rate (ESR)
- Urinalysis
- BUN, creatinine, liver enzymes, alkaline phosphatase, serum LDH
- MRI of primary site
- Chest CT scan
- Bone marrow aspirate and biopsy
- Bone scan.

The role of fluorodeoxyglucose-positron emission tomography (PET) scans in the management of Ewing sarcoma remains unclear. Some early studies have suggested that these scans may be useful for evaluating the presence of metastatic disease and for monitoring response to therapy. However, their role in the staging of Ewing sarcoma remains controversial.

Approximately 20–30% of patients have detectable metastases at the time of diagnosis. The most common metastatic sites are:

- Lung (38%)
- Bone (31%)
- Bone marrow (11%).

The site of the primary tumor is related to the incidence of metastases at diagnosis:

- Central primary – 40% incidence of metastases
- Proximal primary – 30% incidence of metastases
- Distal primary – 15% incidence of metastases.

Treatment

The treatment of Ewing sarcoma involves a multimodal approach, using surgery and/or radiation therapy for local control of the primary lesion and chemotherapy for eradication of subclinical micrometastases. Most current treatment protocols utilize neoadjuvant chemotherapy, followed by local control and then adjuvant chemotherapy.

Surgery

Biopsy should ideally be performed at the center at which ultimate surgical treatment will be rendered. The biopsy site should be chosen carefully and placed inline with potential resection site or radiation portals. The biopsy should be preferably taken from the extraosseous component to prevent pathologic fracture. Frozen section analysis of all biopsy specimens is essential to ensure adequate material has been obtained for light microscopy, immunohistochemistry and cytogenetic or molecular diagnostic studies. The biopsy site must be included en bloc with the resection or in the planned radiation field.

The preservation of function and the reduction of long-term sequelae should be taken into consideration in determining the mode of local control selected for an individual patient. For those tumors felt to be resectable with negative margins, definitive surgical local control is generally preferred to radiation as local control. This preference is largely due to the increased risk of second malignancy following radiation as well as to the improved surgical techniques. With advances in orthopedic oncology techniques, more patients are now candidates for complete surgical resection with salvage procedures. Definitive surgical resection typically takes place after several cycles of chemotherapy. Tumors that initially appear unresectable may become resectable with chemotherapy. Given that Ewing sarcoma is radioresponsive, amputation and debulking surgical procedures are used less commonly than in osteosarcoma.

Radiation

Radiation plays an important role in the treatment of Ewing sarcoma. Indications for radiation therapy include:

- Presence of an unresectable tumor
- Postoperative radiation for tumors resected with positive margins
- Postoperative radiation for resected tumors with poor necrosis (favored in Europe)
- Preoperative radiation for tumors that may become resectable (favored in Europe)
- Hemithorax radiation for patients with chest wall tumors and malignant pleural effusions
- Emergent radiation for patients presenting with paraspinal tumors and neurologic compromise
- Whole lung radiation for patients with lung metastases at initial diagnosis
- Palliation of recurrent disease.

Current recommendations for definitive radiation of the primary site are doses ranging from 55 to 60 Gy. In the postoperative setting, gross disease requires 55.8 Gy and microscopic disease requires 45 Gy. The appropriate radiation volume is an involved field of the pre-treatment tumor plus 2–2.5 cm margin, followed by a boost to the postinduction tumor volume with margin.

Chemotherapy

The major impact on improved long-term survival in Ewing sarcoma has been systemic chemotherapy. Prior to the routine use of chemotherapy, almost all patients with Ewing sarcoma died of distant metastatic disease. The current standard chemotherapy approach utilized in localized Ewing sarcoma consists of:

- vincristine 2 mg/m^2 (maximal dose 2 mg)
- doxorubicin 37.5 mg/m^2/day for 2 days and cyclophosphamide 1,200 mg/m^2 alternating with ifosfamide 1,800 mg/m^2/day for 5 days and etoposide 100 mg/m^2/day for 5 days. The courses are administered every 2 weeks for a total of 14 courses. Re-evaluation and local control occurs after six cycles of chemotherapy.

Treatment of Metastatic Ewing Sarcoma

Patients with metastatic Ewing sarcoma have a poor outcome. These patients appear to have more chemotherapy-resistant disease. The addition of ifosfamide and etoposide to vincristine, doxorubicin and cyclophosphamide does not result in an improved outcome for these patients. The impact of interval compression of chemotherapy cycles has not yet been evaluated in this subgroup of patients.

Patients with metastatic disease should be offered participation in clinical trials evaluating novel approaches to therapy. For patients with lung metastases at initial diagnosis, whole lung radiation appears to improve outcomes. An ongoing study is evaluating the role of a myeloablative regimen with autologous stem cell rescue for these patients, though this approach has not improved the outcome for these patients in prior single-arm studies.

Treatment of Relapsed Ewing Sarcoma

The prognosis for patients with recurrent disease is poor. The time from diagnosis to relapse appears to be one of the strongest determinants of outcome following relapse. Therapy for these patients must be individualized. Common salvage chemotherapy regimens include topotecan/cyclophosphamide and irinotecan/temozolomide. These patients are also commonly treated on phase I and II clinical trials of investigational agents. In recent years, insulin-like growth factor receptor monoclonal antibodies have been evaluated in this population with some early indications of activity. Retrospective studies have evaluated the role of myeloablative therapy and autologous stem cell rescue with equivocal results.

Prognosis

The overall 5-year disease-free survival for localized Ewing sarcoma treated with surgery, radiation and multiagent chemotherapy is approximately 60–70%. Patients who present with metastases at diagnosis have a 5-year survival rate of 20–30%, though there are some indications that patients with lung metastases only may have an improved outcome compared to patients with more widespread metastatic disease.

While the presence of metastatic disease appears to be the most important prognostic factor, other commonly reported adverse prognostic factors include:

- Older age
- Axial tumor location
- Larger tumor size
- Elevated serum LDH level.

Other prognostic factors that have been reported include fever, anemia, type of *EWS* translocation, ploidy, proliferative index and chemotherapy-induced necrosis.

Suggested Reading

Osteogenic Sarcoma

Bielack SS, Kampf-Bielack B, Delling G, et al. Prognostic factors in high grade osteosarcoma of the extremities or trunk: an analysis of 1,702 patients treated on neoadjuvant cooperative osteosarcoma study group protocols. J Clin Oncol. 2002;20(3):776–790.

Bielack SS, Kampf-Bielack B, Branscheid D, et al. Second and subsequent recurrences of osteosarcoma: presentation, treatment and outcomes of 249 consecutive cooperative osteosarcoma study group patients. J Clin Oncol. 2009;27(4):557–565.

Bielack SS, Marina N, Ferrari S, et al. Osteosarcoma: the same old drugs or more? J Clin Oncol. 2008;26 (18):3102–3103.

Briccoli A, Rocca M, Salone M, et al. Resection of recurrent pulmonary metastases in patients with osteosarcoma. Cancer. 2005;104(8):1721–1725.

Ferrari S, Smeland S, Mercuri M, et al. Neoadjuvant chemotherapy with high-dose Ifosfamide, high-dose methotrexate, cisplatin and doxorubicin for patients with localized osteosarcoma of the extremity: a joint study by the Italian and Scandinavian Sarcoma Groups. J Clin Oncol. 2005;23(34):8845–8852.

Goorin AM, Schwartzentruber DJ, Devidas M, et al. Presurgical chemotherapy compared with immediate surgery and adjuvant chemotherapy for non-metastatic osteosarcoma: Pediatric Oncology Group Study POG-8651. J Clin Oncol. 2003;21(8):1574–1580.

Goorin AM, Harris MB, Bernstein M, et al. Phase II/III trial of etoposide and high dose ifosfamide in newly diagnosed metastatic osteosarcoma: a Pediatric Oncology Group trial. J Clin Oncol. 2002;20(2):426–433.

Gorlick R, Meyers PA. Osteosarcoma necrosis following chemotherapy: innate biology versus treatment-specific. J Pediatr Hematol Oncol. 2003;25(11):840–841.

Hunsberger S, Freidlin B, Smith MA. Complexities in interpretation of osteosarcoma clinical trial results. J Clin Oncol. 2008;26(18):3103–3104.

Kager L, Zoubek A, Potschger U, et al. Primary metastatic osteosarcoma: presentation and outcome of patients treated on neoadjuvant Cooperative Osteosarcoma Study Group protocols. J Clin Oncol. 2003;21 (10):2011–2018.

Kayton ML, Huvos AG, Casher J, et al. Computed tomographic scan of the chest underestimates the number of metastatic lesions in osteosarcoma. J Pediatr Surg. 2006;41(1):200–206.

Kempf-Bielack B, Bielack SS, Jurgens H, et al. Osteosarcoma relapse after combined modality therapy: an analysis of unselected patients in the Cooperative Osteosarcoma Study Group (COSS). J Clin Oncol. 2005;23(3):559–568.

Lewis IJ, Nooij MA, Whelan J, et al. Improvement in histologic response but not survival in osteosarcoma patients treated with intensified chemotherapy: a randomized phase III trial of the European Osteosarcoma Intergroup. J Natl Cancer Inst. 2007;99(2):112–128.

Link MP, Gebhardt MC, Meyers PA. Osteosarcoma. In: Pizzo PA, Poplack DG, eds. Principles and Practice of Pediatric Oncology. 4th ed. Philadelphia: Lippincott–Raven; 2003.

Marina N, Gebhardt M, Teot L, Gorlick R. Biology and therapeutic advances for pediatric osteosarcoma. Oncologist. 2004;9(4):422–441.

Meyers PA, Schwartz CL, Krailo MD, et al. Osteosarcoma: the addition of muramyl tripeptide to chemotherapy improves overall survival – a report from the Children's Oncology Group. J Clin Oncol. 2008;26(4):633–638.

Ewing Sarcoma

Cangir A, Vietti TJ, Gehan EA, et al. Ewing's sarcoma metastatic at diagnosis. Cancer. 1990;66(5):887–893.

Donaldson SS. Ewing sarcoma: radiation dose and target volume. Pediatr Blood Cancer. 2004;42:471–476.

Gaetano B, Ferrari S, Bertoni F, et al. Prognostic factors in nonmetastatic Ewing Sarcoma of bone treated with adjuvant chemotherapy: analysis of 359 patients at the Istituto Ortopedico Rizzoli. J Clin Oncol. 2000;18 (1):4–11.

Ginsberg JP, Woo SY, Johnson ME, Hicks MJ, Horowitz ME. Ewing sarcoma family of tumors: Ewing sarcoma of bone and soft tissue and the peripheral primitive neuroectodermal tumors. In: Pizzo PA, Poplack DG, eds. Principles and Practice of Pediatric Oncology. 4th ed. Philadelphia: Lippincott–Raven; 2003.

Granowetter L, Womer R, Devidas M, et al. Dose intensified compared with standard chemotherapy for nonmetastatic Ewing sarcoma family of tumors: A Children's Oncology Group Study. J Clin Oncol. 2009;27(15):2536–2541.

Grier HE, Krailo MD, Tarbell NJ, et al. Addition of ifosfamide and etoposide to standard chemotherapy for Ewing sarcoma and primitive neuroectodermal tumor of bone. N Engl J Med. 2003;348:694–701.

Gurney JG, Swensen AR, Bukterys M. Malignant bone tumors. Cancer incidence and survival among children and adolescents: United States SEER Program 1975–1995. Bethesda, MD: National Cancer Institute, SEER Program. NIH Pub No 99-4649. 1999: 99–110.

Kushner BH, Meyers PA. How effective is dose intensive/myeloablative therapy against Ewing sarcoma/primitive neuroectodermal tumor metastatic to bone or bone marrow? The Memorial Sloan-Kettering experience and a literature review. J Clin Oncol. 2001;19(3):870–880.

Leavey P, Mascarenhas L, Marina N, et al. Prognostic factors for patients with Ewing sarcoma at first recurrence following multi-modality therapy. A report from the Children's Oncology Group. Ped Blood Cancer. 2008;51(3):334–338.

Meyers PA, Krailo MD, Ladanyi M, et al. High-dose melphalan, etoposide, total-body irradiation and autologous stem-cell reconstitution as consolidation therapy does not improve prognosis. J Clin Oncol. 2001;19(11):2812–2820.

Paulussen M, Ahrens S, Burdach S, et al. Primary metastatic (stage IV) Ewing tumor: Survival analysis of 171 patients from the EICESS studies. Ann Oncol. 1998;9(3):275–281.

Womer R, West DC, Krailo MD, et al. Randomized comparison of every 2-week vs. every 3-week chemotherapy in Ewing sarcoma family tumors. J Clin Oncol. 2008;26(Suppl): A10504.

CHAPTER 26

Retinoblastoma

Retinoblastoma is a malignant tumor of the embryonic neural retina. It affects young children under the age of 5 years. Tumors may be unilateral or bilateral, unifocal or multifocal. There are hereditary and nonhereditary forms of the disease and the disease can be sporadic or familial. Intraocular growth occurs first, prior to invasion of structures within the globe or spread to metastatic sites. In developed nations, presentation with metastatic disease is unusual. However, metastatic disease is not uncommon in developing nations, where it is a significant cause of morbidity and mortality. Retinoblastoma is the paradigm for a genetically inherited cancer and provides the basis for Knudson's two-hit hypothesis of carcinogenesis.

INCIDENCE

1. Retinoblastoma is the most common intraocular malignancy of childhood, occurring at a rate of 1 in 20,000 live births.
2. The annual incidence is 10–14 per million under 5 years of age.
3. Approximately 200–300 new cases of retinoblastoma occur each year in the United States.
4. Retinoblastoma accounts for 11% of cancers developing in the first year of life, but for only 3% of all cancers diagnosed in children younger than 15 years of age.
5. The average age at diagnosis is 18 months.
6. Approximately 40% of cases are hereditary. The majority of these patients present with bilateral disease with an average of three tumors per eye and present in the first year of life. Of these, only 15% have an established family history of retinoblastoma.
7. The remaining 60% of cases are nonhereditary, most often presenting with unilateral and unifocal disease in the second and third years of life.

CLASSIFICATION

There are three overlapping methods for classifying retinoblastoma:

- *Laterality*: Tumors may be unilateral or bilateral
- *Focality*: Tumors may be unifocal or multifocal
- *Genetics*: Tumors may be hereditary or non-hereditary.

Manual of Pediatric Hematology and Oncology. DOI: 10.1016/B978-0-12-375154-6.00026-4

Family history: The disease may be familial or sporadic.

Laterality

Bilateral Tumors

Patients have one or more tumors in both eyes. Tumor development can be synchronous or metachronous. It is presumed that these patients have the hereditary form of the disease, even in the absence of a positive family history. These patients almost always present prior to age 2 years and most present in the first year of life. The severity of tumors may be variable in the two eyes and this becomes important in weighing potential treatment options.

Unilateral Tumors

Patients have one or more tumors in one eye. Median age of presentation is 23 months, with few presenting under a year of age. Tumor development within the affected eye can be synchronous or metachronous. For those with unifocal unilateral disease, it is presumed that these patients have the nonhereditary form of the disease. Patients with multifocal unilateral disease or those that present at a younger age are more likely to have the hereditary form of the disease.

Focality

Unifocal Tumors

A single tumor focus exists. These tumors are more likely to be nonhereditary.

Multifocal Tumors

Multiple tumor foci are noted in one or both eyes. These patients are more likely to harbor a germline mutation in the *RB1* gene.

Genetics

The Two-Hit Hypothesis

Knudson's two-hit hypothesis proposes that as few as two stochastic events are required for tumor initiation. The first can be either germline or somatic and the second occurs somatically in the individual retinoblast cells.

The *RB1* Gene

Retinoblastoma occurs as a result of such mutations of the *RB1* gene located on chromosome 13q14. This is a tumor suppressor gene that spans 183-kilo bases of genomic DNA, consisting of 27 exons and coding for a 110-kd protein p110, with 928 amino acids. Positive and negative regulation of transcription and thus, cell proliferation are linked to the phosphorylation of the RB protein. Involved in this process are E2F1, a transcription factor that regulates cell cycle during G1, histone deacetylase 1 and downstream cell cycle-specific kinases. Loss of function is the initiating event for retinoblastoma. Mutations may be germline or somatic and there are a broad array of types and locations of mutations, ranging from single-base changes to large deletions. There have not been identified hot spots of common mutation.

Hereditary Retinoblastoma

A "two hit" model has been proposed to explain the different clinical features of hereditary and nonhereditary cases of retinoblastoma. These patients inherit a germline mutation in the *RB1* gene that is present in every cell of the body. Eighty-five percent of these are new spontaneous mutations (sporadic), while the remaining 15% have a positive family history (familial). For the sporadic cases, the paternal allele is affected in approximately 94% of cases. Postconception, a second somatic mutation occurs in the remaining *RB1* gene in one or more retinoblasts and tumor results. These cases are at risk of multifocal and bilateral tumors.

Nonhereditary Retinoblastoma

These patients inherit two normal copies of the *RB1* gene. Postconception, two somatic mutations occur, one in each copy of the *RB1* gene in a retinoblast and tumor results. These cases are usually unifocal or unilateral tumors.

Genetic Counseling

The *RB1* gene is inherited in an autosomal dominant fashion. Penetrance is high, approximately 90–95%. Genetic counseling should be an integral part of the therapy for a patient with retinoblastoma, whether unilateral or bilateral. Genetic counseling, however, is not always straightforward. Families with retinoblastoma may have a founder with embryonic mutagenesis causing genetic mosaicism of gametes. A significant proportion (10–18%) of children with retinoblastoma have somatic genetic mosaicism, making the genetic story more complex and contributing to the difficulty of genetic counseling.

As a guide, the following schema can be applied to retinoblastoma genetic counseling, although this is far from absolute and exceptions exist.

Unilateral or Bilateral Disease with a Positive Family History of Retinoblastoma

1. Risk for parents to have another child with retinoblastoma is 40%.
2. Risk for the affected patient to have offspring with retinoblastoma is 40%.
3. Risk for a normal sibling of the affected patient to have offspring with retinoblastoma is 7%.

Bilateral Disease Without a Positive Family History of Retinoblastoma

1. Risk for parents to have another child with retinoblastoma is 6%.
2. Risk for the affected patient to have offspring with retinoblastoma is 40%.
3. Risk for a normal sibling of the affected patient to have offspring with retinoblastoma is 1%.

Unilateral Disease Without a Positive Family History of Retinoblastoma

1. Risk for parents to have another child with retinoblastoma is 1%.
2. Risk for the affected patient to have offspring with retinoblastoma is 8%.
3. Risk for a normal sibling of the affected patient to have offspring with retinoblastoma is 1%.

Prenatal Diagnosis and Further Genetic Counseling

With sequencing of the *RB1* gene available, prenatal diagnosis can be undertaken, particularly if there is a family history and the mutation has been identified. However, a negative result cannot categorically rule out disease, as 100% of mutations are not picked up by current screening methods. The positive predictive value of screening however is rapidly improving and this is now offered in some centers as a clinical service. This will also assist in the process of genetic counseling and remove some uncertainty in our predictions for risk for other cases in families.

OTHER EPIDEMIOLOGIC DATA

1. There is an identified 13q deletion syndrome associated with an increased risk of retinoblastoma. This led to the identification of the *RB1* gene. This occurs in <0.05% of patients with retinoblastoma and is notable for the following features:
   - Microcephaly
   - Broad nasofrontal bones
   - Hypertelorism
   - Micro-ophthalmia
   - Epicanthic folds
   - Ptosis
   - Micrognathia
   - Hypoplastic or absent thumbs.
2. No known associations exist for race or gender.
3. There is no eye predilection.
4. Environmental and demographic risk factors that have been identified in some studies, but are inconsistent or limited include:
   - Increased paternal age
   - Paternal employment in the military, metal manufacturing, welder machinist
   - Maternal use of steroid hormones
   - High birthweight.

RISK FOR SECOND MALIGNANT NEOPLASMS (SMN)

1. As the *RB1* gene is a tumor suppressor gene, individuals with heritable retinoblastoma are at high risk for the development of second and subsequent malignancies.
2. The most common second malignancies reported have been osteosarcoma, followed by soft-tissue sarcomas and melanoma. Leukemia, lymphoma and breast cancer are also reported in excess of that expected.
3. The risk is highest for those children with the heritable form of the disease who are treated with full-dose, external-beam radiotherapy delivered without use of conformal fields under the age of 12 months.
4. The 50-year risk is approximately 50% for those treated with radiotherapy and 28% for those treated without radiotherapy. Those who received radiotherapy at less than 1 year of age are at highest risk.
5. The risk for those with patients with unilateral disease is approximately 5%, likely representative of the small fraction of genetic cases with only a single eye affected.
6. For those who survive an SMN, the risk of developing yet another primary malignancy is about 2% per year.
7. Approximately 60% of SMNs occur within the radiotherapy field, but the remainder occur outside the field.
8. Patients with retinoblastoma should be counseled carefully regarding their increased risk of SMN and should be followed by routine clinical evaluation. Any signs or symptoms potentially referable to a SMN should be promptly evaluated.

PATHOLOGY

Retinoblastoma

The tumor is composed mainly of undifferentiated anaplastic cells that arise from the nuclear layers of the retina. Histology shows similarity to other embryonal tumors of childhood, such as neuroblastoma and medulloblastoma, including features such as aggregation around blood vessels, necrosis, calcification and Flexner-Wintersteiner rosettes. Retinoblastomas are characterized by marked cell proliferation as evidenced by high mitosis counts and extremely high MIB-1-labeling indices.

Retinocytoma

This is a benign variant of retinoblastoma and may be referred to as either retinoma or retinocytoma. These tumors are composed of benign-appearing cells with a high degree of photoreceptor differentiation. Eyes with such tumors have normal vision. These tumors behave in a benign manner. The main issue is that these tumors will not regress in the same way as retinoblastoma, when treated with chemotherapy or local ophthalmic therapy and thus differentiating these benign variants from the malignant form of the tumor is essential, so unnecessary treatment is not employed.

CLINICAL FEATURES

Presenting Signs and Symptoms

1. Leukocoria.
2. Strabismus.
3. Decreased visual acuity.
4. Inflammatory changes.
5. Hyphema.
6. Vitreous hemorrhage, resulting in a black pupil.

Patterns of Spread

The most common routes of metastatic spread are intraocular and extraocular. Intraocular spread occurs with direct infiltration via the optic nerve to the central nervous system, or spread via the choroid to the orbit. Extraocular spread occurs via a variety of ways: dispersion of the tumor cells through the subarachnoid space to the contralateral optic nerve or through the cerebrospinal fluid to the central nervous system; hematogenous dissemination to the lung, bone, or brain; and lymphatic dissemination if the tumor spreads anteriorly into the conjunctivae and eyelids, or extends into extraocular tissue.

Intraocular

1. With endophytic growth, there is a white hazy mass.
2. With exophytic growth, there is retinal detachment.
3. Most tumors have combined growth.
4. Retinal cells frequently break off from the main mass and seed the vitreous, or new locations on the retina.
5. Glaucoma may result from occlusion of the trabecular network on from iris neovascularization.

Extraocular

1. Retinoblastoma spreads first to soft tissues surrounding the eye or invades the optic nerve. From there it can spread directly along the axons to the brain, or may cross into the subarachnoid space and spread via the cerebrospinal fluid to the brain.
2. Hematogenous spread leads to metastatic disease, most commonly to brain, lungs, bone marrow, or bone.

3. Lymphatic spread is rare since there is minimal lymphatic drainage of the orbit. Occasionally retinoblastoma spreads lymphatically to the preauricular and submandibular nodes.

In patients with the genetic form of retinoblastoma, central nervous system (CNS) disease is less likely the result of metastatic or regional spread than another primary intracranial focus, such as a pineoblastoma, associated with the trilateral retinoblastoma syndrome.

Trilateral Retinoblastoma

Trilateral retinoblastoma is a well-recognized syndrome that occurs in children under the age of 5 years. It usually consists of bilateral hereditary retinoblastoma associated with an intracranial neuroblastic tumor of the pineal gland and occurs in approximately 5–15% of children with familial, multifocal, or bilateral retinoblastoma. With the onset of systemic neoadjuvant chemotherapy, the incidence of this syndrome appears to be decreasing.

DIAGNOSTIC PROCEDURES

Screening

All children should have screening performed as part of well-child check-ups, primarily by eliciting red reflexes in the eye. However, most cases of retinoblastoma are diagnosed after a parent or other relative notices an abnormality of the eye and this prompts further evaluation.

Siblings of children with retinoblastoma should be screened by ophthalmology at regular intervals at least through age 2–3 years. If the proband has an identifiable mutation in the *RB1* gene, siblings can be screened for this mutation, although a negative result does not categorically rule out risk of disease.

Diagnosis of Intraocular Retinoblastoma

Diagnosis is made by ophthalmologists, retinal specialists, or ocular oncologists using a combination of an ophthalmologic examination generally performed under sedation or anesthesia, together with retinal camera (RetCam) imaging, ultrasound, CT, or MRI. Due to concern about rupturing the tumor and causing both intraocular and extraocular spread, surgical biopsies are not performed for confirmation.

Defining Extent of Disease

The ophthalmologic examination will determine the extent of intraocular tumor and presence or absence of orbital extension. It is crucial that intraocular examination with a

Table 26-1   Investigations for Diagnosis of Retinoblastoma (RB)

Examinations	Imaging Studies	Laboratory Evaluations	Diagnostic Studies
Examination under anesthesia by pediatric ophthalmologist Examination and consultation with pediatric oncologist Audiology evaluation if systemic carboplatin is considered	CT or MRI scan of brain and orbits (should include the pineal gland to exclude trilateral RB)	Complete blood count with differential white cell count Blood chemistries, electrolytes and urinalysis if systemic chemotherapy is considered Creatinine clearance if systemic carboplatin is considered	Lumbar puncture only when there is radiographic or clinical suspicion of CNS disease Bone scan only when bone pain or other extraocular disease Bone marrow biopsy only when there is abnormal blood counts (without alternative explanation) or other extraocular disease Pathologic evaluation if enucleation is performed

binocular indirect ophthalmoscope be performed on both eyes with the pupils maximally dilated. The ophthalmologist should use diagrams of the retina to show the number, size and location of all tumors. These diagrams are now being supplemented by the use of RetCam images, which are helpful in determining not only the extent of disease, but response to treatment. When the binocular indirect ophthalmoscope is used, the location of the tumor(s) should be related to specific landmarks such as the optic nerve head, fovea and ora serrata. Size of the lesion is estimated by comparison with the optic nerve head diameter.

Extraocular Extent of the Disease

All children with retinoblastoma should be referred to a pediatric oncologist for evaluation of extraocular disease.

Table 26-1 summarizes the investigations to be performed for diagnosis of retinoblastoma.

STAGING

The standard staging system for intraocular retinoblastoma is not truly a staging system, but rather a grouping system and is the Reese–Ellsworth classification, which is shown in Table 26-2. This system was designed to predict outcome when disease was treated with external beam radiotherapy and the prognosis associated with each group refers to the probability of retaining useful vision rather than survival.

With the advent of the use of systemic chemotherapy together with local ophthalmic therapies in the 1990s, it became evident that this grouping system was not useful in stratifying patients with respect to outcome following these newer treatment modalities. Therefore, newer treatment protocols use the International Classification System for Intraocular Retinoblastoma, which is based on the extent and location of intraocular retinoblastoma (Table 26-3).

Table 26-2   Reese–Ellsworth Classification of Intraocular Retinoblastoma

Group I: very favorable for maintenance of sight
    A. Solitary tumor, less than 4 disc diameters in size at or behind the equator
    B. Multiple tumors, none over 4 disc diameters in size at or behind the equator
Group II: favorable for maintenance of sight
    A. Solitary tumor, 4–10 disc diameters in size at or behind the equator
    B. Multiple tumors, 4–10 disc diameters in size behind the equator
Group III: possible for maintenance of sight
    A. Any lesion anterior to the equator
    B. Solitary tumors larger than 10 disc diameters behind the equator
Group IV: unfavorable for maintenance of sight
    A. Multiple tumors, some larger than 10 disc diameters
    B. Any lesion extending anterior to the ora serrata
Group V: very unfavorable for maintenance of sight
    A. Massive tumors involving more than one-half the retina
    B. Vitreous seeding

Table 26-3   International Classification System for Intraocular Retinoblastoma

Group A: Small intraretinal tumors away from foveola and disc
    • All tumors are 3 mm or smaller in greatest dimension, confined to the retina *and*
    • All tumors are located further than 3 mm from the foveola and 1.5 mm from the optic disk
Group B: All remaining discrete tumors confined to the retinal
    • All other tumors confined to the retina not in Group A
    • Tumor associated subretinal fluid less than 3 mm form the tumor with no subretinal seeding
Group C: Discrete local disease with minimal subretinal or vitreous seeding
    • Tumors are discrete
    • Subretinal fluid, present or past, without seeding involving one fourth of the retina
    • Local fine vitreous seeding may be present close to discrete tumor
    • Local subretinal seeding less than 3 mm (2 disk diameters) from the tumor
Group D: Diffuse disease with significant vitreous or subretinal seeding
    • Tumors may be massive or diffuse
    • Subretinal fluid present or past without seeding, involving up to total retinal detachment
    • Diffuse or massive vitreous disease may include "greasy" seeds or avascular tumor masses
    • Diffuse subretinal seeding may include plaques or tumor nodules
Group E: Presence of any one or more of these poor prognostic features
    • Tumor touching the lens
    • Tumor anterior to the anterior vitreous face involving ciliary body or anterior segment
    • Diffuse infiltrating retinoblastoma
    • Neovascular glaucoma
    • Opaque media from hemorrhage
    • Tumor necrosis with aseptic orbital cellulites
    • Phthisis bulbi

TREATMENT

In order to maximize the preservation of useful vision, treatment should be undertaken in specialized centers, where there is close collaboration between pediatric oncology and pediatric ophthalmology. The initial therapy for retinoblastoma is dependent on both the intraocular and extraocular extent of the disease. Therapeutic modalities include the following and a combined modality approach is not uncommon:

- Systemic chemotherapy
- External beam radiotherapy
- Local ophthalmic therapy, including local administration of chemotherapy
- Intra-arterial chemotherapy, using an interventional neuro-radiology approach
- Enucleation.

Treatment of Intraocular Retinoblastoma

Systemic Chemotherapy

For intraocular retinoblastoma, chemotherapy is used in a neoadjuvant setting, often in combination with local ophthalmic therapies, where its purpose is to decrease tumor size and volume to allow the successful utilization of local ophthalmic therapies. In this setting, it is often referred to as chemoreduction. Systemic chemotherapy can also be in combination with enucleation, where one eye is enucleated and chemotherapy is delivered to treat tumors in the remaining eye. Indications for chemotherapy include:

- Vision salvage
- Delay or avoidance of radiotherapy in patients with bilateral disease
- Tumor shrinkage in patients with unilateral disease
- Good vision
- A tumor that is too large for isolated local therapy
- Metastatic disease
- Risk factors identified after enucleation (i.e., massive choroid invasion or post-lamina involvement of the optic nerve)
- Trilateral retinoblastoma.

Chemotherapy regimens generally include vincristine, carboplatin and etoposide (Table 26-4).

The Children's Oncology Group (COG) Retinoblastoma Committee has developed a number of trials examining the efficacy of chemotherapy on retinoblastoma.

- A study of unilateral retinoblastoma with and without histopathologic high-risk features and the role of adjuvant chemotherapy using a standard regimen of six cycles of carboplatin, etoposide and vincristine (Histopathologic risk factor protocol, COG ARET 0332)

Table 26-4   Adjuvant Chemotherapy Protocol for Intraocular Retinoblastoma*

Six cycles of the following are given every 28 days:
   Vincristine 0.05 mg/kg (1.5 mg/m^2 if >age 3 years) Day 1
   Carboplatin 18.6 mg/kg (360 mg/m^2 if >age 3 years) Day 1
   Etoposide 5 mg/kg (150 mg/m^2 >age 3 years) Days 1 and 2

*There are ongoing protocols evaluating the utility of higher doses of carboplatin and etoposide for patients with Group C or D disease. There are ongoing protocols for patients with Group B disease, evaluating the elimination of etoposide from this regimen. There are other protocols that include cyclosporine A aimed at decreasing drug resistance.

- Trial of systemic neoadjuvant chemotherapy (use of vincristine and carboplatin chemoreduction combined with local ophthalmic therapies, without the use of etoposide) for group B intraocular retinoblastoma (Group B, COG ARET 0331)
- A single-arm trial of systemic (higher doses of systemic carboplatin combined with subconjunctival carboplatin) and subtenon chemotherapy for groups C and D intraocular retinoblastoma (Group C/D, COG ARET 0231).

External Beam Radiotherapy

Standard of care for intraocular retinoblastoma for many years has included external beam radiotherapy with doses of 40–45 Gy. This resulted in a number of late effects and increased risk of second malignancies both in and out of the field of radiotherapy. Newer methods of delivering radiotherapy with more conformal fields are being used at present, so that the normal structures receive less scatter and thus the risk for adverse long-term outcomes will be less. Conformal radiotherapy, stereotactic radiotherapy, proton-beam radiotherapy as well as intensity-modulated radiation therapy all use technology that minimizes doses to nontarget structures. In addition, there are current protocols testing doses of 23–36 Gy with encouraging results. Lastly, increased risk of secondary malignancy is reduced when radiation is given after 1 year of age to children with retinoblastoma. The use of systemic neoadjuvant chemotherapy may allow delay of external beam radiotherapy to post 1 year of age.

Local Ophthalmic Therapies

Local therapy is used to eradicate local disease after reduction of the tumor volume by chemotherapy and may include cryotherapy, green laser, infrared laser and/or radioactive plaque. The goal of local therapy is to achieve type I regression pattern with calcification or type IV with flat chorioretinal scars, or avascular, linear, white gliosis.

Cryoablation

Cryotherapy can be utilized for ablating tumor remnants/recurrences after chemotherapy up to 3 mm in height and 6 mm in diameter that are located at or anterior to the equator. It is recommended that no more than four different sites be frozen in one eye at a single session. There is less likelihood of creating vitreous seeds with cryotherapy if a tumor has been previously treated with chemotherapy. Extensive cryotherapy has been associated with significant persistent retinal detachment, particularly if the retina was originally detached prior to chemotherapy. Retinal breaks can be caused by cryotherapy.

Green Laser Photoablation

Green laser (argon or 532 nm frequency-doubled YAG) can be used to directly coagulate tumors up to 8 mm in thickness, especially posterior to the equator following chemotherapy. The tumor is outlined with burns half on and half off the retina. There should be 30% spot overlap. After outlining the tumor, the entire tumor should be covered with 30% overlapping spots. Complete coverage is considered "one laser treatment." Each numbered lesion should receive a minimum of three complete "laser treatments" with only one "complete laser treatment" given at each session. Laser-induced hemorrhage has been associated with vitreous seeding. Inadequate dilation of the pupil can cause laser burns to the iris.

Infrared Laser Photoablation

Infrared laser can be selected to treat tumors up to 8 mm thickness that have an intact retinal pigment epithelium following chemoreduction. The use of excessive power may result in hemorrhage or vitreous dissemination of tumor, but starting with low power and gradually increasing avoids these problems.

Episcleral Plaque Radiotherapy

Plaque radiotherapy (I-125 bracytherapy) may be used to treat local recurrences up to 8 mm in thickness and 15 mm in base. Plaque radiotherapy involves securing a radioactive plaque to the eye wall at the apex of the tumor. It may be used as a primary treatment or as an adjunct to surgery. Iodine-125 or ruthenium-106 can be utilized for plaques. The tumor dose is prescribed at the apex of the tumor. The total dose to the tumor apex is 30–35 Gy. Proliferative retinopathy secondary to plaque can occur.

Subtenon (Subconjuctival) Chemotherapy

For patients with more advanced intraocular disease (Reese Ellsworth Group V, International classification groups C and D), pilot studies have been conducted using subtenon carboplatin, in addition to systemic chemotherapy and other local ophthalmic therapies. The ophthalmologist makes a 3-mm incision in the conjunctiva and anterior tenons. A 5-cc syringe containing the carboplatin is fitted with an olive-tip irrigating cannula. The olive-tip cannula is placed through the conjunctival incision and is gently and bluntly passed through posterior tenon's capsule while maintaining constant contact with the globe. Once the irrigating cannula has been passed posteriorly to its full extent, the opening in the conjunctiva and tenons is pulled tightly over the shank of the needle with a forceps and the carboplatin is slowly delivered into the retrobulbar space by gentle pressure on the syringe plunger. The cannula is withdrawn once the retrobulbar injection is complete. No sutures are necessary on the conjunctival incision but that is the surgeon's choice. A combination antibiotic ointment should be instilled into the conjunctival sac upon completion of the injection. While the early results of such studies are promising, larger trials are required to fully evaluate efficacy.

Intra-Arterial Chemotherapy

There are several trials of intra-arterial melphalan for unilateral disease ongoing. While early results are promising, further study is warranted before this can be considered standard of care.

Enucleation

Enucleation of the eye is recommended when there is no chance for useful vision even if the entire tumor is destroyed. It is also indicated when high-risk features for the development of extraocular or metastatic disease are present. These include:

- Anterior chamber seeding
- Choroidal involvement
- Tumor beyond the lamina cribrosa
- Intraocular hemorrhage or
- Scleral and extrascleral extension.

Careful examination of the enucleated specimen by an experienced pathologist is necessary to determine if high-risk features for metastatic disease are present.

External beam radiotherapy or systemic adjuvant therapy is generally required in patients with certain high-risk features assessed by pathologic review after enucleation to prevent the development of metastatic disease. Adjuvant therapy with external beam radiotherapy and chemotherapy is required if there is tumor at the cut end of the optic nerve. An orbital implant (typically hydroxyapatite) is placed at the time of enucleation.

Treatment of Extraocular Retinoblastoma

There is no clearly proven effective therapy for the treatment of extraocular retinoblastoma. Clinical trials are underway to improve the overall dismal outcome (survival of approximately 10%) for this group of patients. Those with CNS metastases appear to do worse than those with other forms of extraocular disease. Chemotherapy has included the following regimens:

- Conventional doses of vincristine, cyclophosphamide and doxorubicin and although they produce an initial response, relapse is common
- Carboplatin, ifosfamide and etoposide have shown more promise for remission
- Other induction regimens include the use of cisplatin or carboplatin together with etoposide, cyclophosphamide and vincristine.

There is no recommended standard regimen and it is largely based on institutional preference. Generally, induction chemotherapy is given for four cycles and this is followed by high-dose chemotherapy followed by stem cell rescue. Following recovery from stem cell transplantation, radiotherapy is generally given to sites of initial bulky disease. As with induction regimens, there is no standard stem cell transplantation regimen. A regimen using etoposide, carboplatin and melphalan is used by some centers. Others use regimens of carboplatin, etoposide and cyclophosphamide or carboplatin, thiotepa and etoposide (Table 26-5). There are a few trials of topotecan in upfront therapy of unilateral or bilateral disease as well.

Treatment of Recurrent Retinoblastoma

The prognosis for a patient with recurrent or progressive retinoblastoma depends on the site and extent of the recurrence or progression. Recurrence is not uncommon for those individuals treated with systemic chemotherapy without radiation therapy or enucleation. For these individuals recurrence typically occurs in the first 6 months post chemotherapy. Responses as high as 85% have been reported following treatment with etoposide and

Table 26-5 Potential Preparative Chemotherapy Regimens for Stem Cell Transplantation for Metastatic Retinoblastoma (all are Followed by Infusion of Autologous Stem Cells)*

Carboplatin–Etoposide–Cyclophosphamide Carboplatin 250–350 mg/m^2/day Days 1–5 Etoposide 350 mg/m^2 Days 1–5 Cyclophosphamide 1.6 gm/m^2/day Days 2–5
Carboplatin–Thiotepa–Etoposide Carboplatin 500 mg/m^2/day or dosed to attain an area under the curve (AUC) of 7 mg/min/ml   using the Calvert formula Days 1–3 Thiotepa 300 mg/m^2/day Days 1–3 Etoposide 250 mg/m^2 Days 1–3

*There are many alternative preparative regimens and no current data to support any one as standard of care. All are followed by autologous stem cell infusion.

carboplatin. If the recurrence or progression of retinoblastoma is confined to the eye and is small, the prognosis for sight and survival may be excellent with local ophthalmic therapy only. If the recurrence or progression is confined to the eye but is extensive, the prognosis for sight is poor; however, the survival remains excellent. If the recurrence or progression is extraocular, the prognosis is more guarded and the treatment depends on many factors and individual patient considerations.

POST-TREATMENT MANAGEMENT

Disease-Related Follow-Up

Patients should be monitored for follow-up of the primary tumor. The majority of recurrences appear within 3 years of diagnosis and recurrences are extremely rare after 5 years of age. The follow-up schedule is largely dependent on the therapy used and the extent of original disease. For example, children who are treated with neoadjuvant chemotherapy and local ophthalmic therapies, who retain their eye and are not treated with external beam radiotherapy, require closer follow-up than those who have had an enucleation and pathology failed to show any high-risk features.

For patients treated with chemotherapy for intraocular or extraocular disease, an evaluation under anesthesia (EUA) is recommended to be performed every 3–4 weeks until there is no active tumor seen on a minimum of three EUAs, then every 6–8 weeks until age 3 years and then every 4–6 months to age 10 years. Patients should be examined without anesthesia when old enough to cooperate.

For patients who have undergone an enucleation and/or received external beam radiotherapy, an evaluation should be performed 4–6 weeks post treatment, then every 2–3 months for the first year post therapy, every 3–4 months the following year, every 6 months until age 5 years and then annually. For patients with extraocular disease, assessment for recurrent metastatic disease should also be performed, which should include CT or MRI of the brain and orbits, CSF evaluation, bone marrow evaluation and bone scan.

To screen for trilateral retinoblastoma, patients diagnosed with bilateral retinoblastoma, especially those diagnosed at less than 1 year of age, or those with positive family history are recommended to have a head CT or MRI every 6 months from the end of therapy until 5 years of age.

Toxicity-Related Follow-Up

This is largely dependent on the therapy received. For those treated with chemotherapy or radiation therapy, a history and physical examination should be performed at least every 3 months until 2 years off therapy, then every 6 months until 5 years off therapy and yearly thereafter. A visual acuity assessment should be conducted at least yearly. Attention should be paid to late effects of chemotherapy and radiotherapy, growth and development and surveillance for second malignant neoplasms (SMN). Parents (and patients as they get older) should be counseled regarding their risk of SMN and should have a comprehensive risk-directed evaluation annually lifelong.

If systemic chemotherapy has been used, the Children's Oncology Group Long-Term Follow-Up Guidelines (www.survivorshipguidelines.org) have the most comprehensive guidelines for follow-up based on therapeutic exposures.

There is no clear screening known to be effective for SMNs. Patients who have received chemotherapy associated with the occurrence of secondary leukemia (etoposide, alkylating agents, doxorubicin, nitrosoureas) should have annual blood counts with differential white cell count performed for 10–15 years from exposure. Patients should be made aware of signs and symptoms of the most common second malignancies following retinoblastoma, osteosarcoma and soft-tissue sarcomas, as well as skin cancers and breast cancer. No routine imaging studies are recommended for surveillance for skin and solid organ tumors unless there is a clinical indication. However, careful physical examinations by the patient and physician are clearly indicated for early detection and successful treatment.

FUTURE PERSPECTIVES

Until recently, retinoblastoma was largely a disease managed by ocular oncologists, pediatric ophthalmologists and radiation oncologists. With the advent of neoadjuvant chemotherapy for intraocular disease and dose-intensive myeloablative protocols for extraocular disease, the role of the pediatric oncologist in the management of this disease is growing. Pediatric oncologists are also well versed in the late effects of therapy and surveillance for second malignant neoplasms. Having a nationwide cooperative trials group committee for retinoblastoma will allow for accrual of sufficient numbers of patients across multiple institutions to rigorously test clinical and scientific hypotheses. Standardized databases regarding diagnosis, group and stage determination, treatment and late effects could be established for the studies. Uniform response criteria to local and systemic chemotherapy could be established that will be relevant to clinical outcomes. Long-term outcomes can be studied in a more systematic fashion. The goal of neoadjuvant systemic chemotherapy is to avoid high-dose external beam radiotherapy and enucleation and the acute and long-term effects associated with both. It is only with these cooperative group studies that long-term data can be systematically collected to demonstrate that changes in approach towards eye salvage provides an acceptable acute toxicity profile, while decreasing risks for adverse late outcomes. Biologic samples could be obtained to better understand retinoblastoma genetics and to test new treatment strategies.

Suggested Reading

Abramson DH, Frank CM, Dunkel IJ. A phase I/II study of subconjunctival carboplatin for intraocular retinoblastoma. Ophthalmology 1999;106(10):1947–1950.

Bunin GR, Meadows AT, Emanuel BS, et al. Pre- and postconception factors associated with sporadic heritable and nonheritable retinoblastoma. Cancer Res. 1989;49(20):5730–5735.

Chintagumpala M, Chevez-Barrios P, Paysse EA, et al. Retinoblastima: review of current management. Oncologist 2007;12(10):1237–1246.

Dunkel IJ, Aledo A, Kernan NA, et al. Successful treatment of metastatic retinoblastoma. Cancer 2000;89 (10):2117–2121.

Friedman DL, Himelstein B, Shields CL, et al. Chemoreduction and local ophthalmic therapy for intraocular retinoblastoma. J Clin Oncol. 2000;18(1):12–17.

Gallie BL, Budning A, DeBoer G, et al. Chemotherapy with focal therapy can cure intraocular retinoblastoma without radiotherapy. Arch Ophthalmol. 1996;114(11):1321–1328.

Hurwitz RL, Shields CL, Shields JA, et al. Retinoblastoma. In: Pizzo PA, Poplack DG, eds. Principles and Practice of Pediatric Oncology. 4th ed. Philadelphia: Lippincott, Williams and Wilkins; 2002:825–846.

Knudson AG. Cancer genetics. Am J Med Genet. 2002;111(1):96–102.

Moll AC, Imhof SM, Schouten-Van Meeteren AY, et al. Second primary tumors in hereditary retinoblastoma: a register-based study, 1945–1997: is there an age effect on radiation-related risk? Ophthalmology 2001;108(6):1109–1114.

Murphree AL, Villablanca JG, Deegan 3rd WF, et al. Chemotherapy plus local treatment in the management of intraocular retinoblastoma. Arch Ophthalmol. 1996;114(11):1348–1356.

Namouni F, Doz F, Tanguy ML, et al. High-dose chemotherapy with carboplatin, etoposide and cyclo-phosphamide followed by a haematopoietic stem cell rescue in patients with high-risk retinoblastoma: a SFOP and SFGM study. Eur J Cancer 1997;33(14):2368–2375.

Rodriguez-Galindo C, Wilson MW, Haik BG, et al. Treatment of metastatic retinoblastoma. Ophthalmology 2003;110(6):1237–1240.

Shields CL, Mashayekhi A, Au AK, et al. The International Classification of Retinoblastoma predicts chemoreduction success. Ophtalmology 2006;113(12):2276–2280.

Wong FL, Boice Jr JD, Abramson DH, et al. Cancer incidence after retinoblastoma. Radiation dose and sarcoma risk. JAMA 1997;278(15):1262–1267.

CHAPTER 27

Germ Cell Tumors

Germ cell tumors are neoplasms that develop from primordial germ cells of the human embryo, which are normally destined to produce sperm or ova. Primordial germ cells appear to originate in the yolk sac endoderm and migrate around the hindgut to the genital ridge on the posterior abdominal wall where they become part of the developing gonad. Figure 27-1 depicts the histogenesis of tumors of germ cell origin. Viable germ cells arrested along this path of migration may form neoplasia in midline sites, such as the pineal region (6%), mediastinum (7%), retroperitoneum (4%), sacrococcygeal region (42%) or in the ovary (24%), testis (9%) and other sites (8%).

INCIDENCE

Tumors of germ cell origin account for approximately 2–3% of childhood malignancies. The incidence of germ cell tumors is 2.5 per million in white children and 3.0 per million in African-American children under 15 years of age with a male:female ratio of 1.0:1.1. Germ cell tumors are more common in the ovaries and testes than in extragonadal sites.

PATHOLOGY

The germ cells are the precursors of sperm and ova and retain the potential to produce all the somatic (embryonic) and supporting (extraembryonic) structures of a developing embryo. *Yolk sac tumors (endodermal sinus tumors)* are derived from a totipotential germ cell that differentiate to extraembryonic structures. *Teratomas* are embryonal neoplasms that contain tissues from all three germ layers (ectoderm, endoderm and mesoderm). Teratomas are mature or immature and may occur with or without malignant germ cell elements (yolk sac tumor, choriocarcinomas, embryonal carcinomas, or germinoma) or rarely malignant somatic elements (such as primitive neuroectodermal tumors).

The malignant histologic variants of germ cell tumors are:

- Germinoma
  - Dysgerminoma (ovary)
  - Seminoma (testis).

Manual of Pediatric Hematology and Oncology. DOI: 10.1016/B978-0-12-375154-6.00027-6

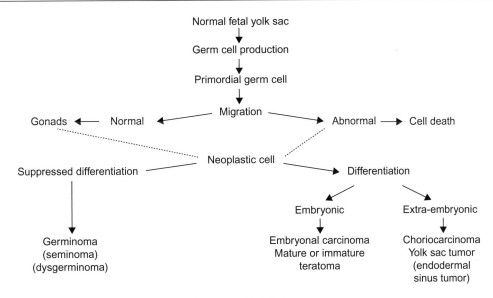

Figure 27-1 Histogenesis of Tumors of Germ Cell Origin.

- Immature teratoma
- Embryonal carcinoma
- Yolk sac tumor (endodermal sinus tumor)
- Choriocarcinoma.

Table 27-1 describes the pathology of germ cell tumors and associated biochemical markers. Cytogenetic analysis has identified isochromosome 12 p (i12p)) as a specific abnormality in more than 80% of germ cell tumors in postpubertal patients.

CLINICAL FEATURES

The signs and symptoms of germ cell tumors are dependent on the site of origin. Table 27-2 lists the clinical features of germ cell tumors at different sites. In addition, certain histologic variants may have associated clinical findings, as listed in Table 27-3. Although germ cell tumors are a diverse group histologically, all originate from primordial germ cells and have a common pattern of spread, irrespective of the primary site as follows:

- Lungs
- Liver
- Regional nodes
- Central nervous system (CNS)
- Bone and bone marrow (less commonly).

Most recurrences develop within 2 years from diagnosis.

Table 27-1   Pathology of Germ Cell Tumors and Associated Markers

Histologic Variant	Morphology	Common Sites of Origin	Markers* AFP	β-hCG	PLAP
Germinoma	Cells are round with discrete membranes, abundant clear cytoplasm. Round nucleus with one to several prominent nucleoli. Cells arranged in nests or lobules separated by fibrous stroma. Areas of necrosis with granulomatous reaction Multinucleated giant cells may be present	Ovary: dysgerminoma Testis: Seminoma Anterior mediastinum	−	−	+
Mature teratoma	Mature tissue derived from Ectoderm: squamous epithelium, neuronal tissue Mesoderm: muscle, teeth, cartilage, bone Endoderm: mucous glands, gastrointestinal and respiratory tract lining No mitoses seen	Sacrococcygeal/ presacral Gonads Mediastinum	−	−	−
Immature teratoma	Immature tissue derived from the three germinal layers. Lesions are graded histologically: Grade 1: Some immaturity with neuroepithelium absent or limited to a rare low magnification (×40) field and not more than one such field in any slide Grade 2: Immaturity and neuroepithelium present but does not exceed three low-power microscopic fields in any slide Grade 3: Immaturity and neuroepithelium prominent; occupying four or more low-magnification microscopic fields	Sacrococcygeal/ presacral Gonads Mediastinum	−	−	−

Tumor	Histology	Location	AFP	β-hCG	PLAP
Embryonal carcinoma	Tumor cells polygonal with abundant pink vacuolated cytoplasm with ill-defined cellular membranes. Nucleus irregular and pleomorphic with prominent nucleoli. Cells arranged in solid sheets with scanty stroma or in tubules, acini, or papillary structures. Typical and atypical mitoses present.	Testis (young adult)	–	±	±
Yolk sac tumor (Endodermal sinus)	Five characteristics are seen microscopically: 1. Aggregates of small undifferentiated embryonal cells 2. Areas of stellate mesodermal cells 3. Areas of perivascular formation consisting of a mesodermal core with a capillary in the center lined by columnar cells (Schuller-Duvall body) 4. Cystic structures which form small cavities lined by flat cells continuous with the parietal lining of the endodermal sinuses 5. Intra- and extracellular hyaline globules that are PAS positive and that contain α-fetoprotein, α₁-antitrypsin, albumin, or transferrin as demonstrated by immunohistochemical staining	Testis (Infant) Ovary Presacral	+	–	–
Choriocarcinoma	Presence of 1. Syncytiotrophoblastic cells that have multiple hyperchromatic nuclei and abundant eosinophilic cytoplasm AND 2. Cytotrophoblasts that have a single nucleus with abundant clear cytoplasm and well-defined cell borders. The two cell populations are often arranged in biphasic plexiform pattern	Ovary Mediastinum Pineal region	–	+	–

Abbreviations: AFP, Alpha-fetoprotein; β-hCG, human chorionic gonadotropin; PLAP, placental alkaline phosphatase.

Table 27-2 Clinical Features of Germ Cell Tumors

Tumor Type	Median Age (yr)	Relative frequency (%)	Features
Pediatric Ovarian Tumors			
Dysgerminoma	16	24	Rapidly developing; 14-25% with other germ cell elements; very radiosensitive
Yolk sac tumor (endodermal sinus tumor)	18	16	↑ AFP; 75% stage I; all patients require chemotherapy because of high risk of relapse even in low-stage disease
Teratoma			
Mature (solid, cystic)	10–15	31	Neuroglial implants may occur with cystic or solid teratomas, but do not affect prognosis; surgery is mainstay of treatment
Immature	11–14	10	Grading system based on amount of neuroepithelium present; prognosis inversely related to stage and grade; 30% with ↑ AFP
Embryonal carcinoma	14	6	47% prepubertal; ↑β-hCG and precocious puberty common; chemotherapy indicated
Malignant mixed germ cell tumor	16	11	40% premenarchal; 30% sexually precocious; AFP or β-hCG may be increased
Gonadoblastoma	8–10	1	Associated with dysgenetic gonads and sexual maldevelopment; removal of both gonads is treatment of choice
Other (polyembryoma, choriocarcinoma)	NA	<1	Rare in children
Pediatric Testicular Tumors			
Yolk sac tumor (endodermal sinus tumor)	2	26	Most common of malignant germ cell tumors of the testes; ↑AFP; compared to adult cases, pediatric tumors are pure histologically; 85% stage I; chemotherapy reserved for higher stage or recurrent disease
Teratoma	3	24	Poorly differentiated histologic features do no impart a malignant course in children. Surgery alone is usually sufficient treatment
Embryonal carcinoma	Late teens	20	Uncommon in young children; ↑AFP ± β-hCG; managed as for adults, with retroperitoneal lymphadenectomy ± chemotherapy ± irradiation based on stage
Teratocarcinoma	Late teens	13	80% stage I with 75% survival after surgery alone; more advanced disease requires multimodality therapy
Gonadoblastoma	5–10	<1	Associated with sexual maldevelopment syndromes; bilateral involvement in 30%; bilateral removal of gonads is treatment of choice
Other (polyembryoma, choriocarcinoma)	NA	16	Rare in children

Abbreviations: AFP, α-fetoprotein; β-hCG, human chorionic gonadotropin.
From: Pizzo PA, Poplack DG. Principals and Practice of Pediatric Oncology, 4th ed. Philadelphia, 2002, Lippincott-Raven, with permission.

Table 27-3   Clinical Association of Different Histologic Variants of Germ Cell Tumors

Histology	Clinical Association
Teratoma	Musculoskeletal anomalies
	Rectal stenosis
	Congenital heart disease
	Microcephaly
Ovarian dysgerminoma (postpubertal)	Amenorrhea
	Menorrhagia
	46XY (male pseudohermaphrodite)
Ovarian embryonal carcinoma	Precocious puberty
	Amenorrhea
	Hirsutism

DIAGNOSTIC EVALUATION

The following evaluations should be carried out:

- History
- Physical examination
- Complete blood count (CBC)
- Liver function tests, electrolytes, blood urea nitrogen (BUN), creatinine
- α-Fetoprotein (AFP)
- Human chorionic gonadotropin (β-hCG)
- Lactic dehydrogenase (LDH) isoenzyme 1. LDH may also correlate with disease activity such as:
  - Tumor bulk
  - Residual tumor after surgery
  - Response to chemotherapy and radiotherapy
  - Tumor recurrence.
- Radiographic evaluation of primary site and regional disease
  - Mediastinum – chest and upper abdominal computed tomography (CT)
  - Ovary – ultrasound, pelvic and abdominal CT
  - Sacrococcygeum – pelvic and abdominal CT
  - Testis – ultrasound, pelvic and abdominal CT.
- Radiographic evaluation for distant metastases
  - Chest radiographs (posteroanterior and lateral)
  - Chest CT
  - Bone scan.

TUMOR MARKERS

Certain histologic variants of germ cell tumors secrete the tumor markers AFP and β-hCG. The production of these markers can be assessed by immunohistologic staining of tissue sections or measurement of blood levels and has important diagnostic value.

AFP is a major serum protein of the human fetus. It is produced in the embryonic liver, in the yolk sac and, in smaller amounts, in the gastrointestinal tract. In general, the highest AFP levels are seen in yolk sac tumor with embryonal carcinoma exhibiting intermediate levels.

The beta subunit of human chorionic gonadotropin (β-hCG) can serve as a tumor marker when positive. It may be positive when syncytiotrophoblasts are present in the tumor and it is found to be elevated in embryonal carcinomas and choriocarcinomas. In choriocarcinoma, there is generally a marked elevation of β-hCG.

Tissue staining and/or measurements of the serum levels of AFP and β-hCG are extremely useful in evaluating teratomas. Pure teratomas are not associated with AFP or β-hCG production; elevation of either marker in association with teratoma indicates the presence of more malignant germ cell elements and requires review of the histologic material or study of more histologic sections. There are infants whose malignant germ cell tumors do not produce AFP. AFP is difficult to evaluate as an indicator of residual or recurrent germ cell tumors in infants less than 8 months of age. The half-life of AFP also varies with age, being 5.5 days at birth to 2 weeks of age, 11 days from 2 weeks to 2 months of age, 33 days from 2 to 4 months of age. The half-life of AFP (beyond 8 months of age) and β-hCG is 5 days and 16 hours, respectively.

AFP and β-hCG levels are useful for clinical evaluation and for assessing disease activity.

Table 27-4 lists the mean and standard deviation of normal serum AFP levels in infants at various ages. Increasing levels of serum AFP are not necessarily indicative of tumor

Table 27-4 Mean and Standard Deviation of Normal Serum α-Fetoprotein in Infants at Various Ages

Age	Mean ± SD (ng/ml)	
Premature	134,734	± 41,444
Newborn	48,406	± 34,718
Newborn–2 weeks	33,113	± 32,503
2 weeks–1 month	9,452	± 12,610
2 months	323	± 278
3 months	88	± 87
4 months	74	± 56
5 months	46.5	± 19
6 months	12.5	± 9.8
7 months	9.7	± 7.1
8 months	8.5	± 5.5

Abbreviation: SD, standard deviation.
From: Wu JT, Book L, Sudar K. Serum α-fetoprotein (AFP) levels in normal infants. Pediatr Res 1981;15:50–52, with permission.

progression because abrupt escalation in serum AFP can occur after chemotherapy-induced tumor lysis. Elevations of serum AFP could also be caused by alterations in hepatic function such as viral hepatitis, cirrhosis, hepatoblastoma, pancreatic and gastrointestinal malignancies and lung cancers. β-hCG can also experience sudden increases secondary to cell lysis during chemotherapy or can be increased with malignancies of the liver, pancreas, GI tract, breast, lung and bladder. The utility of CA-125 in monitoring ovarian germ cell tumors in children has not been thoroughly evaluated.

STAGING

The staging systems for germ cell tumors differ depending on the site of origin and are listed in Table 27-5.

Table 27-5   Staging Systems for Germ Cell Tumors

Stage	Extent of Disease
Extragonadal	
I	Complete resection at any site, including coccygectomy for sacrococcygeal site; negative tumor margins; tumor markers positive but fall to normal or markers negative at diagnosis; lymphadenectomy negative for tumor
II	Microscopic residual disease; lymph nodes negative; tumor markers positive or negative
III	Gross residual or biopsy only; retroperitoneal nodes negative or positive; tumor markers positive or negative
IV	Distant metastases, including liver
Ovarian	
I	Limited to ovary or ovaries; peritoneal washings negative for malignant cells; no disease beyond ovaries; presence of gliomatosis peritonei[a] does not result in changing stage I disease to a higher stage; tumor markers normal after appropriate half-life decline (AFP, 5 days; β-hCG, 16 hours)
II	Microscopic residual or positive lymph nodes (≤2 cm); peritoneal washings negative for malignant cells; presence of gliomatosis peritonei does not result in changing stage II disease to a higher stage; tumor markers positive or negative
III	Lymph node involvement (metastatic nodule) >2 cm; gross residual or biopsy only; contiguous visceral involvement (omentum, intestine, bladder); peritoneal washings positive for malignant cells; tumor markers positive or negative
IV	Distant metastases, including liver
Testicular	
I	Limited to testes; complete resection by high inguinal orchiectomy or transcrotal orchiectomy with no spill at surgery; no clinical, radiographic, or histologic evidence of disease beyond the testes; tumor markers normal after appropriate half-life decline; patients with normal or unknown tumor markers at diagnosis must have a negative ipsilateral retroperitoneal node dissection to confirm to stage I disease
II	Transcrotal orchiectomy with gross spill of tumor; microscopic disease in scrotum or high in spermatic cord (≤5 cm from proximal end); retroperitoneal lymph node involvement (≤2 cm); increased tumor markers after appropriate half-life decline
III	Retroperitoneal lymph node involvement (>2 cm) but no visceral or extra-abdominal involvement
IV	Distant metastases, including liver

[a]Peritoneal implants that contain only mature glial elements with no malignant elements.

TREATMENT

The treatment strategies for germ cell tumors depend on:

- Histologic subtype
- Site of origin
- Stage of disease.

Surgical resection is essential for treatment of most germ cells. It is the treatment of choice for benign teratomas, immature teratomas and low-stage malignant germ cell tumors. Chemotherapy has significantly improved outcome for children with malignant germ cell tumors.

Germinoma

Germinoma is the most common pure malignant germ cell tumor. When it arises in extrago-nadal sites such as the pineal region, anterior mediastinum and retroperitoneum, it is termed germinoma. It has been designated seminoma when it occurs in the testes and dysgermino-ma when it involves the ovary. Germinomas comprise 10% of ovarian tumors in children and 15% of all germ cell tumors. Seminomas are the most common malignancy found in undescended testes. Ovarian dysgerminomas are sometimes associated with precocious sexual development, but the majority of patients are developmentally normal.

Ovarian Dysgerminoma

Rapid development of signs and symptoms of an abdominal mass is the usual presentation. Abdominal pain is not common unless torsion is present. Seventy-five percent of patients have stage I disease at presentation. Patterns of spread include contiguous extension, metastasis to regional lymph nodes and rarely to liver and lungs.

Surgery

Dysgerminoma is the only germ cell tumor of the ovary in which there is a high incidence of bilateral ovarian involvement (5–10%). Bilateral involvement is particularly common in women with a Y chromosome and gonadal streaks.

Conservative surgery consisting of unilateral salpingo-oophorectomy and wedge biopsy of the contralateral ovary and sampling of regional lymph nodes is recommended for patients with the following criteria:

- Stage I unilateral encapsulated tumor less than 10 cm in diameter
- Normal contralateral ovary
- No evidence of retroperitoneal lymph node metastases
- No ascites; negative cytology of peritoneal washing
- Normal female 46XX karyotype.

Patients with stage I disease require no further postsurgical therapy.

In the past, the standard surgical management of patients with stage II and III disease had been total abdominal hysterectomy (TAH) and bilateral salpingo-oophorectomy (BSO) followed by postoperative radiotherapy. However, because the tumor most commonly affects children and young women, a program employing limited surgery, chemotherapy and/or radiotherapy may be more appropriate. Such an approach preserves as much endocrine and reproductive function as possible without compromising survival, compared to TAH and BSO plus radiotherapy.

Chemotherapy

Dysgerminoma and seminoma (testicular tumor analogous to dysgerminoma) are responsive to combination chemotherapy. Tables 27-6 and 27-7 list the therapy regimens used for the treatment of germ cell tumors.

Table 27-6   Various Chemotherapy Regimens for Germ Cell Tumors

Regimen 1 (PEB)	<12 Months of Age (Every 3 Weeks)	≥12 Months of Age (Every 3 Weeks)
Cisplatin (P)	0.7 mg/kg/IV Days 1, 2, 3, 4 and 5	20 mg/m^2/IV on Days 1, 2, 3, 4 and 5
Etoposide (E)	3 mg/kg/IV Days 1, 2, 3, 4 and 5	100 mg/m^2/IV over Days 1, 2, 3, 4 and 5
Bleomycin (B)	0.5 units/kg/IV Day 1	15 units/m^2/IV on Days 1
	OR	
Regimen 2 (JEB)		
Carboplatin	600 mg/m^2/IV Day 1	
Etoposide	120 mg/m^2/IV Days 1, 2 and 3	
Bleomycin	15 units/m^2 Day 3	

Disease is evaluated in week 12 after four cycles. If complete remission then discontinue chemotherapy. If partial remission then surgery (week 12) followed by two more cycles of chemotherapy as above (weeks 13 and 16).

Table 27-7   Chemotherapy Regimens for Malignant Germ Cell Tumors

Site	Histology	Therapy
Testicular	Stage I	Radical inguinal orchiectomy only
Testicular	Stage II	PEB or JEB every 3 weeks for four cycles
Testicular	Stages III and IV	PEB or JEB every 3 weeks, for four cycles followed by surgery
		If no disease (CR), discontinue therapy; if PR, two more cycles and repeat surgery
Ovarian	Stages I and II	PEB or JEB every 3 weeks for four cycles
Ovarian	Stages III and IV	PEB every 3 weeks for four cycles followed by surgery
		If CR, discontinue therapy; if PR, 2 more cycles and repeat surgery
Malignant Extragonadal germcell tumor	Stages I–IV	PEB every 3 weeks for 4 cycles followed by surgery
		If CR, discontinue therapy; if PR, 2 more cycles and repeat surgery
Immature teratoma[a]		Surgery alone

[a]If AFP is increased, chemotherapy as per Table 27-6.
Abbreviations: CR, complete remission; PR, partial remission.

The primary chemotherapy approach in patients with stage II and III disease, combined with limited surgery, has been effective in children. This approach has the advantage of preserving as much reproductive and endocrine function as possible without compromising long-term survival. It is recommended for pediatric and adolescent patients with 46XX karyotype.

Radiotherapy

Dysgerminoma is the most radiosensitive of the ovarian germ cell tumors. However, radiation therapy is reserved for patients with persistent disease after chemotherapy.

Occasionally dysgerminomas occur in combination with other malignant germ cell elements such as malignant teratomas, yolk sac tumor or embryonal carcinoma. They are called mixed germ cell tumors. Management should be based on the most malignant component present.

Prognosis

The prognosis for ovarian dysgerminoma correlates with extent of disease at diagnosis as shown in following table:

Extent of Disease	5-year Survival (%)
Stage I	95
Stage II	75
Stage III	60
Stage IV	33

Extragonadal Germinoma

Incomplete migration of germ cells seems to be responsible for the origin of extragonadal germ cell tumors. Most extragonadal germinomas occur in the mediastinum. In contrast to other germ cell tumors in the mediastinum, hematogenous metastases are rare and the primary therapeutic strategy is aimed at local control.

Surgery

Although excellent cure rates of 70–80% have been reported with mediastinal germinomas treated by surgical excision, most tumors are not resectable because of their size and/or proximity to great vessels and vital structures. In most cases, biopsy or minimal debulking is the only feasible surgical approach.

Radiotherapy

These tumors are radiosensitive and radiation to the mediastinum in the range of 4,500–5,000 cGy cures approximately 50–60% of patients.

Chemotherapy

Combination chemotherapy has been employed in advanced extragonadal germinomas. Cisplatin is particularly effective with excellent results. The chemotherapy regimens employed are shown in Table 27-6. In patients with large bulky mediastinal masses, an approach using combination chemotherapy with or without radiotherapy (depending on the response to two to three courses of chemotherapy) may improve the results compared to patients treated with radiotherapy alone.

Teratoma

Pathologically, teratomas can be divided into three main groups for the purpose of therapeutic strategies.

Mature Teratoma

These tumors contain mature tissue and the treatment is surgical excision, irrespective of the site. It is critically important that multiple histologic sections be examined to exclude the presence of focal immature tissue and malignant germ cell elements.

Immature Teratoma

These tumors contain elements of immature tissue and are graded histologically (see Table 27-1). The treatment of choice is surgical resection. Most immature teratomas in prepubertal children do not respond to chemotherapy. However, if AFP is increased, a malignant germ cell component is present and chemotherapy should be used, in addition to surgery.

Teratoma with Malignant Germ Cell Elements

These tumors contain foci of frankly malignant tissue – usually embryonal carcinoma, yolk sac tumor, or choriocarcinoma. These lesions are treated in an aggressive multimodal fashion. Chemotherapy improves local and metastatic control. The benign teratomatous portion will not respond to chemotherapy or radiation therapy and must be treated surgically. Table 27-7 lists the chemotherapy regimen utilized in malignant germ cell tumors.

Sacrococcygeal Teratoma

Sacrococcygeal teratomas are the most common germ cell tumors of childhood and account for 44% of all germ cell tumors and 78% of extragonadal tumors. These tumors are relatively rare with an incidence of 1 in 40,000 live births. Females are more frequently affected. Because most of these neoplasms are exophytic and are visible externally, approximately 80% are diagnosed within the first month of life. Seventeen percent are malignant at diagnosis and 5% have distant metastases. The average age at diagnosis in the patients with

**Table 27-8   Anatomic Location of Sacrococcygeal Teratoma and Percentage
with Malignant Histology**

Type	Location	Malignant Histology (%)
I	Tumor predominantly external (sacrococcygeal) with only a minimal presacral component (the most common type)	8
II	Tumor presenting externally but with a significant intrapelvic extension (second in frequency to type I)	21
III	Tumor minimal external component but with the predominant mass extending into the pelvis and abdomen	34
IV	Tumor internalized presacral tumor without external presentation	38

Table 27-9   Histologic Grading of Sacrococcygeal Teratoma

Grade	Description
0	All tissue mature; no embryonal tissue
1	Rare foci of embryonal tissue not exceeding 1 low-power field per slide
2	Moderate quantity of embryonal tissue and some atypia but not exceeding 3 lower-power fields per slide
3	Large quantity of immature tissue exceeding 3 lower-power fields per slide, with abundant mitoses and cellular atypia

metastatic disease is 22 months of age. Congenital anomalies, musculoskeletal and central nervous system defects are seen in up to 18–20% of patients.

Sacrococcygeal tumors can be diagnosed antenatally on fetal ultrasound. Large tumors can be associated with congestive cardiac failure and hydrops fetalis due to arteriovenous shunting within the tumor.

Tables 27-8 and 27-9 describe, respectively, the anatomic location and histologic grading of sacrococcygeal teratomas.

The histology of the malignant component of sacrococcygeal teratoma is almost always yolk sac tumor. Approximately 17% of sacrococcygeal teratomas are malignant in nature. The incidence of a malignant component of the tumor is related to the type of the tumor, 38% in type IV versus 8% in type I.

Treatment

Surgery
Complete surgical excision should be performed, which necessitates removal of the entire coccyx. In almost all cases, the tumor is attached to the coccyx and failure to remove the coccyx results in local tumor recurrence in 30–40% of cases.

In patients with complete removal with grade 0 or 1 histology, no further therapy is required. However, continuous monitoring of these patients is required because malignant germ cell tumor can recur either from missed malignant tissue in the original tumor or malignant conversion of residual tumor. In patients with histologic grade 3, malignant foci, or elevated tumor markers (for age), postsurgical chemotherapy is required.

Radiotherapy

Immature teratomas respond to radiotherapy but the dose required is in the range of 4,500–5,000 cGy. Because most patients are under 1 year of age, such high-dose radiation therapy can have severe long-term sequelae. For this reason, radiotherapy should not be used routinely and may be used with caution for disease that persists after chemotherapy and attempted re-excision.

Chemotherapy

Sacrococcygeal teratoma grade 3 or with malignant foci of yolk sac tumor is responsive to combination chemotherapy. The regimens used are outlined in Table 27-6.

Prognosis

The prognosis of sacrococcygeal teratoma depends on the following factors:

- Age
- Surgical resectability
- Histological grading.

Patients under 2 months of age have a favorable outlook, because only 7% of tumors in females and 10% in males are malignant at that age. Patients over 2 months of age have a less favorable outlook; the incidence of malignancy is 48% in females and 67% in males.

Patients whose tumors are initially resectable have a more favorable outcome. Patients whose tumor can not be resected do poorly, even though these lesions are initially responsive to chemotherapy.

The histologic grading is extremely important. Patients with grade 0, 1 and 2 lesions have a 90–95% cure rate with complete surgical excision. Patients with grade 3 or malignant elements have an approximately 45–50% 2-year disease-free survival when treated with surgery and chemotherapy.

Ovarian Teratoma

These tumors comprise 40–50% of the ovarian tumors seen in childhood and adolescence. Many are classified as immature teratomas. The clinical behavior of immature ovarian teratoma correlates well with the histologic grading but poorly with the stage.

Surgery

Surgery is essential, especially for immature teratomas that do not have malignant elements. Pure teratomas and immature teratomas do not usually respond to chemotherapy.

The surgical approach is similar to that for other ovarian tumors. The incidence of bilateral involvement is unusual, but the contralateral ovary should be inspected and biopsied if abnormalities are found on inspection and palpation.

Prognosis

Evidence of overt malignant elements (yolk sac tumor, embryonal carcinoma, choriocarcinoma) requires intensive chemotherapy to be administered. Subsequent surgery may be needed to remove residual nonmalignant elements.

Mediastinal Teratoma

Mediastinal teratomas are located in the anterior mediastinum. The average age of the pediatric patient is 3 years. They are more common in males. Teratoma subtypes (mature, immature or malignant) comprise the bulk of the tumor but yolk sac tumor and choriocarcinoma may occasionally be seen. Mediastinal teratomas occasionally have sarcomatous foci resembling rhabdomyosarcoma or undifferentiated sarcoma. These foci are extremely aggressive therefore appropriate metastatic survey including serum markers (AFP and β-hCG and LDH) is recommended. Most of these tumors are benign lesions (grade 0).

Treatment

Surgery
Patients with benign lesions are treated successfully with surgical excision. Patients who have malignant lesions (grade 2 or 3) are not generally amenable to complete surgical excision because of infiltration of surrounding vital structures. Surgical debulking must be attempted after reduction with chemotherapy.

Chemotherapy
Aggressive chemotherapy is used for those with foci of malignant elements. The combinations appropriate for patients with malignant germ cell elements are shown in Table 27-6. If other elements are present, such as rhabdomyosarcoma, therapy must be adjusted to treat these more malignant elements.

Radiotherapy
Germ cell tumors of the mediastinum are not especially radiosensitive.

Prognosis

The two critical prognostic factors in mediastinal germ cell tumors are age and histology. In patients under 15 years of age, mediastinal teratomas tend not to have malignant elements or metastasize. In older patients, these tumors often contain yolk sac tumor, embryonal carcinoma, or choriocarcinoma and have a worse prognosis, with 50% of patients developing metastases within 1 year.

Yolk Sac Tumor (Endodermal Sinus Tumor)

Yolk sac tumor is the most common malignant histology in children. It is often the only malignant element and is frequently found in immature teratomas. When present in an extragonadal site, yolk sac tumor behaves in a highly malignant fashion. It is important to recognize that foci of yolk sac tumor present in immature teratoma require aggressive chemotherapy similar to that for pure yolk sac tumor. The sacrococcygeal area is the major site of involvement in the newborn and infant; the ovary is the most common location for yolk sac tumors in older children and adolescents. Testicular yolk sac tumors have two peaks of incidence in infancy and adolescence. They are the most frequent malignant testicular tumors in young boys.

Testicular Yolk Sac Tumor

This tumor is localized (stage I) in 85% of cases. Overall survival rate is higher than 85% which seems to correlate with age. The management of this tumor is considered separately in two age groups – prepubertal and postpubertal. Tumors in these two age groups behave differently clinically and are different cytogenetically (postpubertal tumors have i12p).

Prepubertal Males

Surgery

The definitive surgical treatment is inguinal orchiectomy with high ligation of the spermatic cord. Retroperitoneal lymph node dissection is not indicated in young patients with disease limited to the testes because less than 5–6% will have positive retroperitoneal lymph nodes. Patients should be staged with CT of chest and abdomen. Any abdominal nodal or metastatic disease can be treated successfully with chemotherapy alone. Infants with a negative metastatic workup and normal AFP and β-hCG require no further therapy following orchiectomy. Close observation with serum AFP and radiologic evaluation of chest and abdomen are indicated because even with normalization of AFP in stage I disease 20–40% false-negative results have been seen and retroperitoneal disease has occurred before elevation of AFP.

Prognosis

In infants with normal AFP and β-hCG following orchiectomy, the cure rate is 85–90% with surgery alone. In the small number of patients with recurrence, combination chemotherapy can produce subsequent cure. In higher-stage disease treated with chemotherapy, overall survival at five years approaches 100%.

Postpubertal Males

Surgery

Radical orchiectomy should be combined with appropriate staging. In patients with no evidence of retroperitoneal adenopathy, retroperitoneal lymphadenectomy (RPLND) has been recommended. If on exploration the ipsilateral nodes are clinically involved, the contralateral nodes should be examined. Contralateral nodal involvement is seen in 15–30% of cases when the ipsilateral nodes are involved. In patients with stage II or III disease, initial chemotherapy followed by surgical debulking and possibly retroperitoneal lymphadenectomy should be done. There is some controversy concerning the need for RPLND which has been well documented in adult literature. Some institutions observe CT-negative patients, while others administer two cycles of chemotherapy. An additional problem with adolescent testicular germ cell tumors is residual benign lesions after chemotherapy. These lesions should be removed as they pose a risk for recurrence as malignant germ cell tumor.

Chemotherapy
In patients with stage I disease, no further therapy is given. In patients with stage II, III, or IV disease, combination chemotherapy is administered (see Tables 27-6 and 27-7).

Ovarian Yolk Sac Tumor
Ovarian yolk sac tumor occurs in both prepubertal and postpubertal females and is characterized by rapid growth with intrapelvic and intra-abdominal spread and widespread involvement of the peritoneal surfaces. Yolk sac tumors spread rapidly to lymphatic and peritoneal structures with a short duration of symptoms, high frequency of abdominal pain and high frequency of high-stage disease at diagnosis. Distant metastases are seen in liver, lungs, lymph nodes and rarely in bones. Most patients have elevated AFP levels.

Surgery
Unilateral salpingo-oophorectomy is performed for patients with stage I disease. In patients with more extensive disease, debulking should be performed, including omentectomy and retroperitoneal lymph node sampling. For unilateral disease, unilateral oophorectomy with biopsy of uninvolved ovary should be performed. Bilateral disease evaluation includes bilateral oophorectomy. Peritoneal washings should be obtained in all cases.

Chemotherapy
Chemotherapy is currently indicated in all cases postoperatively (Tables 27-6 and 27-7). In 85% of patients with stage I disease treated with surgery alone, recurrence will develop. There is some debate as to whether an appropriately staged, stage I ovarian yolk sac tumor can be observed. However, in all pediatric trials staging of ovarian germ cell tumors was not precise.
In patients who have unresectable gross disease at initial surgery, four cycles of chemotherapy should be followed by second-look surgery to confirm eradication of disease and to remove residual disease if present.

Radiotherapy
Radiotherapy does not play a major role in management.

Prognosis

The prognosis depends on the stage and the chemotherapeutic regimen employed, but with platinum-based combination therapy disease-free survival may be as high as 80%.

Embryonal Carcinoma

Embryonal carcinoma occurs either in pure form or in combination with a mixed germ cell tumor (malignant teratoma). In either case, the management is the same.

Testicular Embryonal Carcinoma

Embryonal carcinoma occurs more commonly in late adolescence or in early adulthood. The presenting symptoms include enlarging scrotal mass, metastatic abdominal or mediastinal disease, or localized peripheral lymphadenopathy. Serum AFP or β-hCG may be elevated. The therapy required for cure depends on stage of disease and quantitation of AFP.

Surgery

The standard surgical approach for patients with clinical stage I and II disease is radical orchiectomy with high ligation of the spermatic cord and retroperitoneal lymphadenectomy as described previously for postpubertal males.

Prognosis

The cure rate for patients with testicular embryonal carcinoma is excellent, provided that the appropriate chemotherapeutic regimens are employed and that patients who are treated with surgery alone have careful follow-up evaluation, including radiographic evaluation and serial AFP and β-hCG determinations to detect recurrence.

Ovarian Embryonal Carcinoma

Ovarian embryonal carcinoma is distinct from yolk sac tumor of the ovary. This rare tumor resembles embryonal carcinoma of the testis histologically and resembles it biochemically (with secretion of AFP and β-hCG). It occurs predominantly in adolescent females with a mean age of 15 years. The treatment approach is similar to that of ovarian yolk sac tumor.

RELAPSED AND RESISTANT GERM CELL TUMORS

The prognosis for children with malignant germ cell tumors is excellent. However, there are patients who do not respond or relapse after therapy. They are usually found to have persistent marker elevation or significant marker elevation after treatment.

Table 27-10   High-Dose Chemotherapy Regimen Followed by Autologous Stem Cell Transplant

Regimen 1
  Thiotepa 300 mg/m^2 IV over 3 hours on Days -5, -4 and -3.
  Etoposide 500 mg/m^2 IV over 3 hours Day -5, -4 and -3
  Rest period Day -2 and -1
  Autologous stem cell infusion Day 0

OR

Regimen 2
  Etoposide 1,500 mg/m^2 IV continuous infusion over 24 hours on Day -9.
  Thiotepa 300 mg/m^2 IV over 2 hours Day -8, -7, -6
  Cyclophosphamide 50 mg/kg IV over 2 hours Day -5, -4, -3, -2
  Rest Day -1
  Autologous stem cell infusion Day 0

Recurrent germ cell tumors often respond to various chemotherapy regimens such as, carboplatin, ifosfamide and etoposide or placlitaxel, ifosfamide and carboplatin. However, surgical resection is often essential. Though there are minimal data from pediatric trials some patients whose recurrent disease is responsive to chemotherapy may be candidates for ablative chemotherapy and autologous hematopoietic stem cell transplantation (Table 27-10).

Suggested Reading

Balmaceda C, Heller G, Rosenblum M, et al. Chemotherapy without irradiation a novel approach for newly diagnosed CNS germ cell tumors: results of an international cooperative trial. J Clin Oncol. 1996;12:2908–2915.

Cushing B, Perlman E, Marina N, Castleberry RP. Germ cell tumors. In: Pizzo PA, Poplack DG, eds. Principles and practice of pediatric oncology. 4th ed. Philadelphia: Lippincott–Raven; 2002.

Cushing B, Giller R, Cullen J, et al. Randomized of combination chemotherapy with etoposide, bleomycin and either high-dose or standard-dose Cisplatin in children and adolescents with high-risk malignant germ cell tumors: a pediatric intergroup study – Pediatric Oncology Group 9049 and Children's Cancer Group 8882. J Clin Oncol. 2004;22(13):2691–2700.

De Backer A, Madern GC, Oosterhuis JW, et al. Ovarian germ cell tumors in children: a clinical study of 66 patients. Pediatr Blood Cancer 2006;46(4):459–464.

Einhorn LH, Williams SD, Chamness A, et al. High-dose chemotherapy and stem cell rescue for metastatic germ-cell tumors. N Engl J Med. 2007;357(45):340–348.

Hawkins EP, Perlman E. Germ cell tumors in childhood: morphology and biology. In: Parham DM, ed. Pediatric neoplasia: morphology and biology. New York: Raven Press; p. 297.

Mann JR, Gray ES, Thornton C, et al. Mature and immature extracranial teratomas in children: the UK Children's Cancer Study Group experience. J Clin Oncol. 2008;26(21):3590–3597.

Mann JR, Raafat F, Robinson K, et al. The United Kingdom's Cancer Study Group's second germ cell study: carboplatin, etoposide and bleomycin are effective treatment for children with malignant extracranial germ cell tumors, with acceptable toxicity. J Clin Oncol. 2000;18(22):3809–3818.

Marina NM, Cushing B, Giller R, et al. Completer surgical excision is effective treatment for children with immature teratomas with or without malignant elements: a Pediatric Oncology Group/Children's Cancer Group intergroup study. J Clin Oncol. 1999;17:2137–2143.

Marina N, London WB, Frazier AL, et al. Prognostic factors in children with extragonadal germ cell tumors: a pediatric intergroup study. J Clin Oncol. 2006;24(16):2544–2548.

Motzer RJ, Nichols CJ, Margolin KA, et al. Phase III randomized trial of conventional-dose chemotherapy with or without high-dose chemotherapy with autologous hematopoietic stem-cell rescue as first-line treatment for patients with poor-prognosis metastatic germ cell tumors. J Clin Oncol. 2007;25(3):247–256.

Nichols CG, Fox FP. Extragonadal and pediatric germ cell tumors in children. Hematol/Oncol Clin. 1990;5:1189–1209.

Rogers PC, Olson TA, Cullen JW, et al. Treatment of children and adolescents with stage II testicular and stages I and II ovarian malignant germ cell tumors: A pediatric intergroup study–Pediatric Oncology Group 9048 and Children's Cancer Group 8891. J Clin Oncol. 2004;22(17):3563–3569.

Schlatter M, Rescorla F, Giller R, et al. Excellent outcome in patients with stage I germ cell tumor of the testes: a study of the Children's Cancer Group/Pediatric Oncology Group. J Pediatr Surg. 2000;38(3):319–324.

CHAPTER 28

Hepatic Tumors

Primary hepatic neoplasms are rare and account for only 1–2% of all childhood cancers. Hepatoblastoma accounts for approximately two-thirds of liver tumors. Hepatoblastoma and hepatocellular carcinoma (HCC) constitute the two most common malignancies that arise *de novo* in the liver. Table 28-1 lists the most prevalent malignant and benign liver tumors in children.

INCIDENCE

Table 28-2 lists the demographic features of hepatoblastoma and hepatocellular carcinoma. Hepatoblastoma occurs primarily in young children, with 80% of the cases reported before 3 years of age. The incidence of hepatoblastoma has increased over the last 25 years and that increase may be associated with increased survival of very-low-birth-weight infants, who are at increased risk for development of hepatoblastoma.

EPIDEMIOLOGY

The etiology of hepatoblastoma and hepatocellular carcinoma is unknown. Certain disorders increase the risk of liver cancer (see Table 28-3). The following are congenital anomalies which have been reported with hepatoblastoma:

- Hemihypertrophy
- Beckwith–Wiedemann syndrome
- Meckel's diverticulum
- Congenital absence of adrenal gland
- Congenital absence of kidney
- Umbilical hernia.

Hepatic tumors have a wide geographic variation in incidence:

- It is the third most common abdominal cancer in Japan
- It is seen more frequently in Asian and African children.

The geographic variation is thought to reflect the etiologic role of environmental conditions.

Manual of Pediatric Hematology and Oncology. DOI: 10.1016/B978-0-12-375154-6.00028-8

Table 28-1   Malignant and Benign Liver Tumors in Children

Malignant	Benign
Hepatoblastoma	Hemangioma
Hepatocellular carcinoma	Hemangioendothelioma
Rhabdomyosarcoma	Angiomyolipoma
Undifferentiated embryonal	Hamartoma
Angiosarcoma	Biliary cyst
Mesenchymal (mixed)	Adenoma
Sarcoma	Teratoma
Rhabdoid tumor	Myofibroblastic tumor
Yolk sac tumor	
Leiomyosarcoma	

Table 28-2   Demographic Features of Hepatoblastoma and Hepatocellular Carcinoma in Children

Host Factor	Hepatoblastoma	Hepatocellular Carcinoma
Incidence		
White males	1.4 per million	0.5 per million
African-American males	0.9 per million	0.0 per million
White females	0.5 per million	0.9 per million
African-American females	0.0 per million	0.0 per million
Median age	1 year	12 years
Male:female ratio	1.7:1.0	1.0:1.1

PATHOLOGY

Hepatic tumors are divided into two major histologic types: hepatoblastoma and hepato-cellular carcinoma. The pathologic classification is as follows:

- Hepatoblastoma
    - a. Epithelial type
        - (1) Embryonal pattern
        - (2) Fetal pattern
        - (3) Macrotrabecular type
        - (4) Small cell undifferentiated type or anaplastic
    - b. Mixed epithelial and mesenchymal type
- Hepatocellular carcinoma
- Fibrolamellar carcinoma (a histologic variant of hepatocellular carcinoma and has a more favorable prognosis).

Table 28-3   Disorders Associated with Increased Risk of Hepatoblastoma and Hepatocellular Carcinoma

Hepatoblastoma	Hepatocellular Carcinoma
Low-birth-weight infant	Familial cholestatic cirrhosis of childhood
Von Gierke disease	Biliary cirrhosis due to bile duct atresia
Congenital cystathioninuria and hemihypertrophy	Cirrhosis following giant cell hepatitis
Maternal use of hormonal therapy	Chronic carrier of hepatitis B virus
Exposure to metals such as in welding and soldering fumes	Hereditary tyrosinemia
	Androgen therapy
Beckwith-Wiedeman syndrome	Methotrexate therapy
Li-Fraumeni syndrome	α_1-Antitrypsin deficiency
Trisomy 18	Fanconi anemia
Fetal alcohol syndrome	Type 1 glycogen storage disease
Familial adenomatous polyposis	Neurofibromatosis
Gardner syndrome[a]	Soto syndrome[b]
Type I glycogen storage disease	Hepatoblastoma
Prader–Willi syndrome	Alagille syndrome
	Familial adenomatous polyposis
	Ataxia-telangiectasia
	Wilms' tumor with liver metastases treated with hepatic radiation

[a]Gardner syndrome, a variant of familial adenomatous polyposis, is an autosomal syndrome associated with deletions on the long arm of chromosome 5. It is characterized by colonic polyps that undergo malignant change and benign and malignant extracolonic lesions. Tumors frequently associated with Gardner syndrome include carcinoma of the ampulla of Vater, papillary carcinoma of the thyroid and, in children, hepatoblastoma. The childhood malignancies often precede the appearance of other manifestations by several years.
[b]A rare genetic disorder characterized by excessive physical growth during first 2–3 years of life. They tend to be larger at birth. May have mild mental retardation, delayed development and hypotonia.

CLINICAL FEATURES

The most common symptom of primary liver malignancies is an upper abdominal mass or generalized abdominal enlargement. Table 28-4 outlines the clinical features of the various pathologic types of hepatic tumors. Table 28-5 shows the frequency of signs and symptoms at diagnosis in children with hepatoblastoma and hepatocellular carcinoma. The clinical manifestations are similar in both.

DIAGNOSTIC EVALUATION

Patients should have the following evaluations:

- History
- Physical examination

Table 28-4   Clinical Features of the Various Pathological Types of Hepatic Tumors

Feature	Hepatoblastoma	Hepatocellular Carcinoma	Fibrolamellar Variant
Usual age of presentation	0–3 yr	5–18 yr	10–20 yr
Associated congenital anomalies	Dysmorphic features Hemihypertrophy Beckwith–Wiedemann syndrome	Metabolic	None
Advanced disease at presentation	40%	70%	10%
Usual site of origin	Right lobe	Right lobe, multifocal	Right lobe
Abnormal liver function tests	15%–30%	30–50%	Rare
Jaundice	5%	25%	Absent
Elevated AFP	60–70%	50%	10%
Positive hepatitis B serology	Absent	Present in some	Absent
Abnormal B12-binding protein	Absent	Absent	Present
Chromosomal abnormalities	11p15.5, 18, 17p13 occasional loss of heterozygosity	17q11.2, 20p12, 11q22-23 rare presence of TP53 mutation	Not reported
Distinctive radiographic appearance	None	None	None
Pathology	Fetal and/or embryonal cells + mesenchymal component	Large pleomorphic tumor cells and tumor giant cells	Eosinophilic hepatocytes with dense fibrous stroma

From Pizzo PA, Poplack DG (eds): Principles and Practice of Pediatric Oncology, 4th Ed JB Lippincott, Philadelphia, 2002, with permission.

- Complete blood count (anemia with moderate leukocytosis is commonly seen; thrombocytosis with platelet counts greater than $500,000/mm^3$ is the most frequent abnormality)
- Urinalysis
- Liver profile and electrolytes
- Fibrinogen, partial thromboplastin time (PTT) and prothrombin time (PT)
- Hepatitis B surface antigen (HB_sAg), core antigen (HB_cAg) and core antibody
- α-fetoprotein (AFP)
- β-hCG
- Carcinoembryonic antigen (CEA)
- Radiographic evaluation of intrahepatic disease determine:
  a. Sonogram
  b. Abdominal CT

Table 28-5   Frequency of Signs and Symptoms in Children with Hepatoblastoma and Hepatocellular Carcinoma

Sign/Symptom	Hepatoblastoma (%)	Hepatocellular Carcinoma (%)
Abdominal mass	80	60
Abdominal distention	27	34
Anorexia	20	20
Weight loss	19	19
Abdominal pain	15	21
Vomiting	10	10
Pallor	7	Rare
Jaundice	5	10
Fever	4	8
Diarrhea	2	Rare
Constipation	1	Rare
Pseudoprecocious puberty	Occasional	Not reported

    c.   Magnetic resonance imaging (MRI)
    d.   MRI angiogram
    e.   MRI cholangiogram
- Radiographic evaluation of extrahepatic disease
    a.   Chest radiographs (PA and lateral)
    b.   Chest CT
    c.   Bone scan
- Bone marrow aspirate/biopsy.

The diagnostic evaluation should determine:

- Extent of intrahepatic disease
- Potential for hepatic resectability
- Presence or absence of extrahepatic disease.

Approximately 90% of children with hepatoblastomas and 50% of children with hepatocellular carcinoma have elevated AFP. AFP is a valuable marker for monitoring residual or metastatic disease following resection of the primary tumor or for monitoring the response of an unresectable primary tumor to therapy. AFP is normally elevated in the newborn period and then declines (Chapter 27, Table 27-4). Concentrations of AFP greater than 500,000 ng/ml are not unusual in hepatoblastoma. Failure of AFP to return to normal after surgery is an indication of incomplete tumor removal or metastases. Serial determinations of AFP are the most precise measurement of the effectiveness of hepatoblastoma therapy.

β-hCG is measured in the occasional patient with hepatic tumor who has associated precocious puberty. Unlike AFP, β-hCG levels do not necessarily reflect the clinical course of the tumor.

Table 28-6   Staging of Hepatic Tumors (Based on Postsurgical Findings) in Children and Percentage of Cases

Stage	Description	Percentage of Cases by Stage
I	Complete resection of tumor by wedge resection lobectomy or by extended lobectomy as initial treatment	25
II	Tumors rendered completely resectable by initial radiotherapy or chemotherapy A. Residual disease confined to one lobe	4
III	A. Gross residual tumor involving both lobes of liver B. Regional lymph node involvement	48
IV	Metastatic disease irrespective of extent of liver involvement	23

MRI angiography and MRI cholangiogram are useful complements to CT for evaluation of intrahepatic disease because they aid in determining surgical resectability. In hepatoblastoma, the right lobe of the liver is more commonly involved than the left, but the tumor involves both lobes in 30% of patients.

Approximately 10% of patients with hepatic tumors have demonstrable pulmonary metastases on chest radiographs. The widespread use of chest CT has increased the number of patients detected with lung metastases.

STAGING

The following two staging systems are employed for liver tumors:

- System based on post-surgical findings – on the degree of resectability of the primary lesion and the presence or absence of metastatic disease. This staging system for hepatic tumors, used in the United States, is described in Table 28-6
- System based on pre-surgical findings – extent of disease (PRETEXT) at diagnosis, used in European protocols, is described in Table 28-7. This system was designed for international treatment programs in which only patients with PRETEXT stage 1 would undergo an initial attempt at resection. All other patients would receive chemotherapy prior to surgery. There is considerable inter-observer variability and some patients with potential postsurgical (United States) stage 1 would be upstaged in PRETEXT system and have surgery after chemotherapy. Despite the approach used, the outcomes are similar.

TREATMENT

Hepatoblastoma

Only patients in whom complete resection can be achieved have a reasonable chance of cure. This is possible in 40–50% of patients. Chemotherapy plays an ancillary role in eradicating

Table 28-7   PRETEXT (Based on Pre-surgical Findings) Staging for Hepatoblastoma

PRETEXT Stage	Description
PRETEXT stage 1	Tumor involves one quadrant. Three adjoining quadrants are free of disease
PRETEXT stage 2	Tumor involves two adjoining quadrants with remaining two, free of disease
PRETEXT stage 3	Tumor involves three adjoining quadrants or two non-adjoining quadrants. One quadrant or two non-adjoining quadrants are free of disease
PRETEXT stage 4	Tumor involves all four quadrants. No quadrant is free of disease

subclinical metastases in completely resected disease and chemotherapy plus radiation therapy may be effective in permitting initially unresectable disease to become resectable.

Surgery

The optimal management of hepatoblastoma is complete resection. Patients who are suitable candidates for complete resection include those with:

- Tumors confined to the right lobe
- Tumors originating in the right lobe that do not extend beyond the medial segment of the left lobe
- Tumors confined to the left lobe.

Patients who have tumor involvement of both lobes are not candidates for curative surgical resection. In these cases, biopsy should be performed. In the PRETEXT (European) staging model, chemotherapy is administered prior to surgery to all patients except PRETEXT stage 1.

Radical hepatic resection results in many potential postoperative complications, including the following:

- Hypovolemia
- Hypoglycemia
- Hypoalbuminemia
- Hypofibrinogenemia and deficiency of coagulation proteins
- Hyperbilirubinemia persists for 2–4 weeks after resection. Hepatic regeneration is complete by 1–3 months postsurgery.

Chemotherapy

1. Stage I and II – completely resected with pure fetal histology require no chemotherapy. Other histologic patterns require combination chemotherapy.
2. Stage III – chemotherapy followed by resection (including orthotopic liver transplantation if not resectable).
3. Stage IV – as for stage III and resection of pulmonary metastases.

Table 28-8   Various Combination Chemotherapy Regimens for Liver Tumors

Regimen 1 (every 3–4 weeks)
   Doxorubicin 25 mg/m^2/day $\times$ 3 continuous infusion by central line;
   Cisplatin 20 mg/m^2/day $\times$ 5 days continuous infusion, days 0–4 (6 cycles of chemotherapy)
Patients less than 10 kg
   Doxorubicin 0.83 mg/kg $\times$ 3 days; cisplatin 0.66 mg/kg/day for 5 days (6 cycles of chemotherapy)
     OR
Regimen 2 (every 3–4 weeks)
   Cisplatin 100 mg/m^2 IV infused over 6 hours, day 1;
   Vincristine 1.5 mg/m^2 IV (max 2 mg), day 3, 10, 17;
   Fluorouracil 600 mg/m^2 IV, day 3
under 1 year of age:
   Cisplatin 3.3 mg/kg and vincristine 0.05 mg/kg
After 4 cycles, attempt surgical resection, followed by 4 more cycles
     OR
Regimen 3 (every 3–4 weeks)
   Cisplatin 90 mg/m^2 IV infused over 6 hours (hours 0–6), day 0;
   Doxorubicin 20 mg/m^2/day continuous infusion $\times$ 4 days, day 0–3
Under 1 year of age:
   Cisplatin 3 mg/kg;
   Doxorubicin 0.66 mg/kg
After four cycles, attempt surgical resection followed by 4 more cycles

Combination chemotherapy plays several roles in the management of hepatoblastoma:

*   Adjuvant therapy for patients who have undergone complete resection, because its use improves disease-free survival
*   Preoperative therapy for patients who have initial unresectable disease to shrink the primary tumor
*   Palliative therapy for patients with metastatic disease at diagnosis.

Table 28-8 lists three commonly used chemotherapy regimens for liver tumors. Regimen 2 has been shown to be as effective as regimen 1 with an equivalent survival. Regimen 2 eliminates the use of doxorubicin and has minimal toxicity. However, controversy still remains as to the most effective regimen since many successful regimens include doxorubicin.

Radiation

Radiotherapy is not curative for intrahepatic disease because the effective tumor dose exceeds hepatic tolerance. However, radiotherapy is useful in shrinking unresectable disease or microscopic residual disease. Radiation dosages used to treat hepatic tumors range from 1,200 to 2,000 cGy. Occasionally, higher doses to localized areas of tumor have been associated with tumor regression. Radiotherapy, immediately after hepatic resection, will limit hepatic regeneration.

Liver Transplantation

Liver transplantation has been used successfully in selected cases of unresectable hepato-blastoma. Patients should be seen in consultation at centers with liver transplantation services, soon after diagnosis.

Hepatocellular Carcinoma

Surgery

Complete surgical resection is essential for survival in hepatocellular carcinoma. Chemotherapy (cisplatin and doxorubicin) may be given in some cases to make surgical removal possible.

Chemotherapy

1. Stage I and II – adjuvant chemotherapy as described in Table 28-8.
2. Stage III – chemotherapy followed by resection (including orthotopic liver transplantation if not respectable).
3. Stage IV – no chemotherapy combination has been effective, though cisplatin and doxorubicin have been used.

PROGNOSIS

The factors that determine prognosis include staging and histology. The overall survival rate is 70% for hepatoblastoma and 25% for hepatocellular carcinoma. Using the staging system shown in Table 28-6, patients with hepatocellular carcinoma stage I have a good outcome and stage IV are usually fatal. The 5-year survival rates for hepatoblastoma are as follows:

Stage	Percentage
I/II	90
III	60
IV	20

Histology also influences prognosis, as shown below:

Pathology	Prognosis
Hepatoblastoma – fetal pattern Fibrolamellar carcinoma	Favorable histology
Hepatoblastoma – embryonal pattern Hepatocellular carcinoma	Unfavorable histology

The reduction in AFP level is a good predictor of outcome in patients with unresected tumors who undergo chemotherapy initially because of unresectability. A report from Children's Cancer Group demonstrated that patients who show a decline in AFP by 2 logs after the initial four courses of chemotherapy have a 75% survival.

Studies of DNA content have shown that diploid tumors with low proliferation index have a better prognosis than aneuploid tumors and high proliferative index.

Suggested Reading

Bilk P, Superina R. Transplantation for unresectable liver tumors in children. Transplan Proc. 1997;29: 2834–2836.

Buckley JD, Sather H, Ruccione K, et al. A case control study of risk factors for hepatoblastoma: a report from the Children's Cancer Study Group. Cancer 1989;64:1169–1176.

Douglass EC, Reynolds M, Finegold M, et al. Cisplatin, vincristine and fluorouracil therapy for hepatoblastoma: a Pediatric Oncology Group study. J Clin Oncol. 1993;11:96–99.

Katzentein HM, Krailo MD, Malogolowkin MH, et al. Hepatocellular carcinoma in children and adolescents: results from the Pediatric Oncology Group and the Children's Cancer Group intergroup study. J Clin Oncol. 2002;20(12):2789–2797.

Litten JB, Tomlinson GE. Liver tumors in children. Oncologist. 2008;13(7):812–820.

Ortega JA, Douglass E, Feusner J, et al. Randomized comparison of cisplatin/vincristine/5-flourouracil and cisplatin/continuous infusion doxorubicin for treatment of pediatric hepatoblastoma. A report from the Children's Cancer Group and the Pediatric Oncology Group. J Clin Oncol. 2000;18(14):2665–2675.

Ortega JA, Krailo MD, Haas JE, et al. Effective treatment of unresectable or metastatic hepatoblastoma with cisplatin and continuous infusion doxorubicin chemotherapy: a report from the Children's Cancer Study Group. J Clin Oncol. 1991;9:2167–2176.

Otte JB, Pritchard J, Aronson DC, et al. Liver transplantation for hepatoblastoma: results from the International Society of Paediatric Oncology (SIOP) study – SIOPEL 1 and review of world literature. Pediatr Blood Cancer 2004;42(1):74–83.

Perilongo G, Shafford E, Mailbach R, et al. Risk-adapted treatment for childhood hepatoblastoma. Final report of the second study of the International Society of Paediatric Oncology-SIOPEL 2. Eur J Cancer 2004; 40(3):411–421.

Raney B. Hepatoblastoma in children: a review. J Pediatr Hematol/Oncol. 1997;19:418–422.

Tomilinson GE, Finegold MJ. Tumors of the liver. In: Pizzo PA, Poplack DG, eds. Principles and Practice of Pediatric Oncology. 4th ed. Philadelphia: Lippincott-Raven; 2002.

CHAPTER 29

Hematopoietic Stem Cell Transplantation

Hematopoietic stem cell transplantation (HSCT) has become an accepted therapeutic modality for a wide variety of diseases and is increasingly utilized for the treatment of malignant and nonmalignant disorders. Tables 29-1 and 29-2 list the indications for allogeneic and autologous hematopoietic stem cell transplantation. Preparation for HSCT involves delivery of high-dose chemotherapy with or without radiation to ablate hematopoiesis and provide sufficient immunosuppression to allow donor cell engraftment.

The rationale for high-dose chemotherapy involves the theory of the steep dose–response curve for many chemotherapeutic agents. Most drugs exhibit a log-linear relationship between tumor cell kill and dose over a certain range, followed by flattening of the curve in the upper dose ranges. For this reason, small changes in dose can produce significant changes in response for chemotherapeutic agents whose major dose-limiting toxicity is myelosuppression. Hematopoietic growth factor support offers the potential to maximize the dose response of high-dose chemotherapy. A 3–10-fold increase in drug dose may result in a multiple log increase in tumor cell killing. Multidrug therapy, compared to a single agent, is necessary to overcome tumor heterogeneity and drug resistance.

The most common type of stem cell transplantation to treat hematological malignancies and nonmalignant disorder is allogeneic, using a human leukocyte antigen (HLA)-matched histocompatible donor, usually a sibling. Solid tumors have been treated with high-dose chemotherapy followed by autologous peripheral blood stem cell transplantation (PBSCT) or bone marrow transplantation (purged or unpurged depending on whether the bone marrow is actually or potentially invaded with malignant cells) with concomitant use of hematopoietic growth factors, such as, granulocyte colony-stimulating factor (G-CSF), or granulocytic-macrophage colony-stimulating factor (GM-CSF), to reduce the effects of neutropenia caused by escalating doses of chemotherapy.[*]

Table 29-3 lists the different sources of hematopoietic stem cells for transplantation. Tables 29-4 and 29-5 list the advantages and disadvantages of allogeneic and autologous hematopoietic stem cell transplantation.

---

[*]G-CSF and GM-CSF stimulate the production, maturation and function of the myeloid and monocytic cell lineages.

Manual of Pediatric Hematology and Oncology. DOI: 10.1016/B978-0-12-375154-6.00029-X

Table 29-1   Indications for Allogeneic Hematopoietic Stem Cell Transplantation

Malignant Disorders

This group includes patients under 21 years of age with any of the following:

Acute Leukemia

Acute myelogenous leukemia (AML)[a]

1. High-risk first complete remission (CR1), defined as:
   a. Having preceding myelodysplasia (MDS) or
   b. High risk cytogenetics: del (5q) −5, −7, abn (3q),t (6;9) complex karyotype (≥5 abnormalities) or
   c. Requiring >1 cycle chemotherapy to obtain CR; or
   d. FAB classification: M6
2. Second or greater CR

Acute lymphocytic leukemia (ALL)

1. High-risk first remission, defined as:
   a. Ph+ ALL;
   b. MLL (myeloid/lymphoid or mixed lineage leukemia) rearrangement with slow early response [defined as having M2 (5–25% blasts) or M3 (>25% blasts on bone marrow examination on Day 14 of induction therapy)]; or
   c. Hypodiploidy (<44 chromosomes or DNA index <0.81); or,
   d. M3 bone marrow at end of induction, or
   e. M2 bone marrow at end of induction with M2-3 at Day 42.
2. High-risk second remission, defined as:
   a. Ph+ ALL or
   b. Bone marrow relapse <36 months from induction or
   c. T-lineage relapse at any time or
   d. Very early isolated CNS relapse (6 months from diagnosis); or
   e. Slow re-induction (M2-3 bone marrow at Day 28) after relapse at any time.
3. Any third or subsequent CR
4. Bi-phenotypic or undifferentiated leukemia in any CR

Chronic Myelogenous Leukemia Ph+ (CML)

1. Chronic stable phase[b]
2. Early accelerated phase or
3. Chronic phase after treatment for blast crisis

Juvenile Chronic myeloid Leukemia (JCML)

Juvenile Myelo-monocytic Leukemia (JMML)

Lymphomas (Second or subsequent Complete Remission or Partial remission)

1. Hodgkin
2. Non-Hodgkin

Myelodysplasia Including Monosomy 7 Syndrome

Myelofibrosis

Familial Hemophagocytic Lymphohistiocytosis

Non-malignant Disorders

Congenital

1. Immunodeficiency syndromes
   a. Severe combined immunodeficiency syndromes
   b. Chronic mucocutaneous candidiasis
   c. Wiskott–Aldrich syndrome

(Continued)

Table 29-1   (Continued)

2. Hematologic disorders
   a. Hemoglobinopathies:
      Sickle cell anemia (selected cases)
      Thalassemia
   b. Fanconi anemia (progressive)
   c. Shwachman-Diamond syndrome
   d. Kostmann agranulocytosis[c]
   e. Diamond-Blackfan Anemia
   f. Dyskeratosis congenita
   g. Thrombocytopenia absent radii syndrome (TAR)
   h. Chronic granulomatous disease
   i. Chediak–Higashi syndrome
   j. CD11/19 deficiency (leukocyte adhesion deficiency)
   k. Neutrophil actin defects
3. Lysosomal storage diseases[d]
4. Infantile osteopetrosis

Acquired
1. Severe aplastic anemia
2. Paroxysmal nocturnal hemoglobinuria

[a]After initial remission in AML, investigators differ regarding the optimum treatment strategy. Comparison of HSCT for HLA-matched sibling donors with patients without such donors who receive chemotherapy demonstrates a significantly better disease-free survival for HSCT (50–57%) compared with those receiving chemotherapy (36–37%). Beyond first remission, HSCT from unrelated donor or mismatched family member is recommended.
[b]Patients without HLA matched sibling should receive Gleevac, if no molecular remission is achieved then should receive unrelated donor transplant.
[c]G-CSF may be successful in treatment and avoiding HSCT.
[d]Enzyme replacement may avoid HSCT.
Abbreviation: CR, complete remission.

Table 29-2   Indications for Autologous Hematopoietic Stem Cell Transplantation

Hematologic malignancies
   1. Non-Hodgkin lymphoma
   2. Hodgkin disease
   3. ANLL[a] (investigational)
   4. CML[a] (investigational)
Solid tumors
   1. Neuroblastoma stage IV
   2. Ewing's sarcoma, primitive neuroectodermal tumor[b]
   3. Rhabdomyosarcoma[b]
   4. Germ cell tumor[b]
   5. Brain tumor
   6. Testicular cancer[b]
   7. Wilms' tumor[b]

[a]When HLA-compatible donor is not available.
[b]Metastatic disease at presentation but achieved partial remission with conventional chemotherapy or relapsed but has chemo-sensitive disease.

Table 29-3   Sources of Hematopoietic Stem Cells for Transplantation

1. Allogeneic bone marrow or peripheral blood stem cell (PBSC) using HLA-matched sibling
2. Allogeneic bone marrow or PBSC using 5 out of 6 HLA matched family members if available
3. Syngeneic bone marrow or PBSC (an identical twin)
4. Allogeneic bone marrow or PBSC using HLA-matched unrelated donors
5. Autologous bone marrow: Patient's own marrow is cryopreserved and reinfused after patient has received aggressive chemotherapy and/or radiation therapy to treat the underlying malignancy *in vivo* purging or *ex vivo* purging[a]
6. Autologous peripheral blood stem cells: The number of circulating stem cells can be increased by the use of recombinant growth factors, which can increase stem cell numbers by 100-fold prior to apheresis. This induces a more rapid hematopoietic recovery after transplantation[b]
7. Umbilical cord blood stem cells (always allogeneic). using HLA-matched sibling or 4 or 5 out of 6 HLA matched unrelated cord blood. *Ex vivo* expansion of cells in culture to increase the number of CD34+ cells available for transplant has been employed experimentally
8. Fetal liver stem cells administered to patients *in utero* (26–30-week gestation) with severe immunodeficiency and inborn errors of metabolism (experimental)

[a]*Ex vivo* purging for removing malignant cells from peripheral blood stem cells or marrow rely upon physical (density or velocity sedimentation, filtration), pharmacologic (e.g., 4-hydroperoxycyclophosphamide [4HC]), or immunologic (monoclonal antibodies, magnetic immunobeads) principles.
[b]This is a useful method in patients who have received previous pelvic radiation therapy or because of tumor in the marrow. G-CSF 10 μg/kg/day or GM-CSF 500 mg/m^2 for 5-7 days can be used to mobilize progenitor cells. Blood progenitor cells can be collected for 2–5 days beginning on day 4 or 5 after initiation of growth factor.

Table 29-4   Advantages and Disadvantages of Allogeneic Hematopoietic Stem Cell Transplantation

Advantages
1. Lower relapse rate
2. Graft versus leukemia effect[a]

Disadvantages
1. Compatible donor availability limited (approximately 25–30% of siblings are compatible)
2. Graft versus host disease (GVHD), except in syngeneic (identical twins)
3. Increased risk of viral infections such as cytomegalovirus or adenovirus and interstitial pneumonia (this risk is 2.5 times greater in unrelated matched donors compared to matched siblings)
4. Risk of veno-occlusive disease greater in unrelated donors
5. Immune supression due to GVHD prophylaxis or treatment drugs

[a]The relapse rate is 2.5 times lower in allogeneic recipients who have grade II–IV acute GVHD compared to recipients without GVHD. Leukemic cells have been reported to disappear during episodes of acute GVHD.

SOLID TUMORS

Patients with solid tumors with less than a 30% chance of long-term survival with conventional chemotherapy, only if a dose–response relationship to chemotherapy is established, should be considered for autologous or allogeneic (if bone marrow is invaded) stem cell transplantation in remission or for minimal residual disease.

Table 29-1 also lists other indications for stem cell transplantation.

Table 29-5   Advantages and Disadvantages of Autologous Hematopoietic Stem Cell Transplantation

Advantages
1. No need for an allogeneic donor
2. No GVHD
3. Lower risk of opportunistic infection (e.g., CMV, *Pneumocystis jiroveci* [formerly *Pneumocystis carinii*])
Disadvantages
1. Increased risk of relapse (approximately 50%)
2. No graft versus leukemia effect
3. Graft failure due to:
    a. Effect of *in vitro* purging
    b. Effect of previous chemotherapy and radiotherapy

ALLOGENEIC STEM CELL TRANSPLANTATION

Donor Selection

Any HLA-identical sibling should be considered a donor. ABO mismatch is not a contraindication, but if multiple donors are available, donors with the same ABO type should be used. Major ABO mismatch (A or B or O) requires removal of the red cells and plasma from the marrow prior to infusion. The HLA match can be phenotypic or genotypic. Finding the right donor necessitates matching of HLA antigens or histocompatibility genes by serology or molecular methodology. The major histocompatibility complex (MHC) is divided into two groups. The class I molecules are HLA-A, HLA-B and HLA-C and the class II molecules are HLA-DR, HLA-DQ and HLA-DP. Currently the matching is generally confined to major class I and II loci. However, it is increasingly appreciated that minor histocompatibility antigens are also likely to play important roles in the outcome of allogeneic HSCT. One of the major obstacles in allogeneic bone marrow transplantation is that 70–75% of patients have no acceptable donors within their families. In these cases, volunteer unrelated donors who are HLA, A, B and DR compatible with the recipients may be available as bone marrow donors worldwide through the U.S. National Marrow Donor Program (NMDP). Partial HLA-matched donors are sometimes used. Immunosuppressive drugs are given after transplantation to inhibit or modify the graft versus host immune reaction. Table 29-6 lists the blood bank support required for hematopoietic stem cell transplantation.

Histocompatibility Testing

The major histocompatibility complex (MHC) genes are mapped within a region called HLA (human leukocyte antigen system A) located on the short arm of chromosome 6.

Table 29-6   Blood Bank Support for Hematopoietic Stem Cell Transplantation

Problem	Solution
Transfusion associated graft versus host disease	Standard: gamma radiation (2,500–3,000 cGy) Investigational: a. Leukocyte depletion b. Selective T-cell depletion c. Ultraviolet radiation
CMV transmission (CMV negative donor and recipient)	Standard: use of only CMV-negative blood products or through an in-line filter to remove lymphocytes and Leukocytes that may harbor latent viruses Alternative: leukocyte-depleted blood products
Alloimmunization	
1. Graft failure or rejection (especially aplastic anemia)	Avoidance or minimizing of pretransplant transfusions; avoidance of transfusions from family members, especially potential donors
2. Refractory thrombocytopenia[a]	Most effective: leukocyte-depleted blood products a. Single-donor platelets b. Ultraviolet irradiation c. HLA-matched donors d. Cross-matched platelets
ABO incompatibility	
1. Major Patient O, donor A-IHT >1:16 Patient O, donor B-IHT >1:16	Removal of red blood cells from donor marrow
2. Minor Patient A, donor O-IHT >1:128 Patient B, donor O-IHT >1:128	Remove plasma from donor marrow
Donor red cell transfusion requirements	Autologous red blood cell salvage Predeposit autologous red blood cell units

[a]Non-alloimmunization causes of refractory thrombocytopenia include drugs, hepatic veno-occlusive disease, hypersplenism, or sepsis. Another more unusual cause of refractory thrombocytopenia is a syndrome resembling thrombotic thrombocytopenic purpura, often associated with the use of cyclosporine, total-body irradiation, or the development of acute GVHD.
Abbreviations: CMV, cytomegalovirus; IHT, isohemagglutinin titer.
From: Rowe JM, et al. Recommended guidelines for the management of autologous and allogeneic bone marrow transplantation: a report from the Eastern Cooperative Oncology Group ECOG. Ann Int Med. 1994;120:143–158.

Disparity at major loci requires vigorous immunologic intervention to avoid rejection, or, should engraftment occur, subsequent graft versus host disease (GVHD).

The class I molecules are composed of an α chain and $\beta2$ microglobulin, are highly polymorphic and are expressed on most nucleated cells and on platelets. The HLA class II loci are also highly polymorphic and are involved in exogenous antigen processing. The HLA class II antigens are heterodimeric cell surface molecules formed by the α and β chains, each of which are polymorphic. HLA terminology is designated by the World Health Organization (WHO) nomenclature committee for factors of the HLA system and is updated at regular intervals. The broadest designation is on serologic typing and highest resolution is based on actual DNA sequence.

Typing of HLA class II alleles can be determined using nucleotide sequence polymorphism. The following different techniques have been used for identification of class I and II polymorphism for determination of bone marrow donor and recipient selection by:

- Sequence-specific oligonucleotide probe (SSOP)
- Sequence specific primer (SSP).

Medical evaluation of bone marrow donors includes:

- Physical examination
- Complete blood count
- Biochemistry profile
- Cytomegalovirus (CMV) antibody, EBV profile, herpes antibody titers, HTLV I and II
- Human immunodeficiency virus (HIV)
- Hepatitis screen
- VDRL for syphilis.

Further evaluation is required in 2–20% of donor candidates.

Bone Marrow Harvesting and Transplantation

Bone marrow harvesting is carried out using general or epidural anesthesia under sterile conditions. The posterior and/or anterior iliac crests of the donor are prepared and draped. Approximately 2 ml of bone marrow are aspirated from each site, avoiding dilution with blood by taking multiple aspirates from the anterior and posterior iliac crests. The marrow is collected in a heparinized saline container. The concentration of heparin (preservative free) is 3–5 units/ml of bone marrow. The quantity of nucleated bone marrow cells required to ensure engraftment is $2–5 \times 10^8$ cells per kilogram of recipient body weight. The usual volume of marrow required to achieve this cell yield is 10–20 ml/kg of recipient body weight. Marrow from children, especially infants, has a higher proportion of marrow-repopulating cells than marrow from older donors. The marrow is filtered through a filtering apparatus to remove bone and tissue fragments and placed in a blood transfer pack. In allogeneic transplants, the pack containing the bone marrow is then given to the recipient as an intravenous infusion over a period of a few hours.

For autologous bone marrow transplantation, some form of processing to remove the erythrocytes is necessary prior to the collected marrow being cryopreserved at either $-80°C$ or $-139°C$. At this temperature, the marrow can be preserved for periods of up to 10 years or possibly longer. The cryopreserved bone marrow is thawed at the bedside in a

37°C water bath and given intravenously to the recipient over a period of a few minutes to an hour, depending on the volume infused.[*]

Peripheral blood stem cells (PBSC) have been collected from autologous donors after stimulation with G-CSF or GM-CSF and have been successfully used to reconstitute hematopoiesis in recipients of autologous, syngeneic and allogeneic grafts who underwent a myeloablative preparative regimen. Patients who receive more than 5×10^6 CD34+ cells/kg engraft satisfactorily.

Umbilical cord blood (UCB) has been collected, cryopreserved and used as the source of pluripotential hematopoietic stem cells when a related or unrelated stem cell donor is not available. UCB cells have increased proliferative capacity and decreased allo-reactivity with a lower incidence of GVHD. This property, coupled with an absence of donor risk, is an advantage in the use of this source of hematopoietic cells but a disadvantage in the inability to evaluate the genetic history of these donors.

ABO incompatibility between donor and recipient is encountered in 25–30% of allogeneic transplants. A major incompatibility occurs when the recipient plasma contains isohemagglutinins directed against the donor red blood cell (RBC) antigens (e.g., group O recipient, group A donor) and minor incompatibility occurs when the donor plasma contains isohemagglutinins directed against recipient RBC antigens (e.g., group A recipient, group O donor). In both instances, appropriate anti-A or anti-B isohemagglutinin titers must be determined before infusion of the cells. After the transplant procedure, all patients should have immunohematologic testing for the appearance of donor-derived red blood cells and changes in recipient isohemagglutinin titers.

Several techniques have been developed for special treatment of marrow before infusion or cryopreservation. This includes removal of T lymphocytes from allogeneic donor marrow in an attempt to decrease the risk of GVHD or to remove potentially or known malignant cells from autologous marrow (e.g., use of monoclonal antibodies combined with complement, monoclonal antibodies linked to toxins and various chemotherapy drugs).

Table 29-7 lists the pretransplantation evaluation of the HSCT recipient.

---

[*]Generation of a "buffy coat" or sedimentation at unit gravity in hetastarch both produce an erythrocyte-poor product that contains most of the nucleated cells, including mature granulocytes, while Ficoll–Hypaque density separation provides mononuclear cells depleted of erythrocytes and mature myeloid cells. Most investigators freeze cells in tissue culture medium containing 10% dimethyl sulfoxide (DMSO) and a colloid (autologous plasma or human serum albumin are common); use of bags for marrow cryopreservation simplifies thawing and allows direct infusion afterwards. Because of the possibility of bag breakage, it is recommended that at least two bags (preferably more) be used for each patient.

Table 29-7   Pre-Transplantation Evaluation of the Recipient

Physical examination
Complete blood count
Biochemistry profile
24-hour urine for creatinine clearance
Blood type including red cell antigen panel
HIV, HTLV I and II
EBV, CMV, hepatitis, herpes titers
VDRL for syphilis
Echocardiogram and EKG
Pulmonary function test (PFT)
Chest and sinus radiographs
Dental evaluation
Bone marrow aspiration/biopsy if applicable
Psychosocial evaluation

Table 29-8   Commonly Used Stem Cell Transplantation Preparative Regimens

Drug	Total Dose	Cyclophosphamide Total Dose	Total Body Irradiation (cGy)
Cyclophosphamide	120 mg/kg	–	1,200–1,350
Cytosine arabinoside	36 g/m^2	–	1,200
Etoposide	60 mg/kg	–	1,200
Etoposide	30 mg/kg	120 mg/kg	1,200
Busulfan	16 mg/kg	200 mg/kg	–
Melphalan	140 mg/m^2		1,350

Pretransplantation Preparative Regimens

Pretransplantation conditioning is used both for the purpose of eradicating disease and as a means of immunosuppressing the recipient sufficiently to allow the acceptance of a new immunologically disparate hematopoietic system. Table 29-8 shows various preparative regimens. The current mainstay preparatory regimen for HSCT consists of ablative doses of either total body irradiation (TBI) or busulfan (Bu) together with either one or two chemotherapy agents. On completion of the preparative regimen, the donor marrow, PBSC, or UCB is infused.

Preconditioning Regimens

Leukemia (Chapter 17)

Two preconditioning regimens, which have antitumor as well as immunosuppressive properties, are commonly used prior to stem cell transplantation. Total body irradiation (fractionated doses) and cyclophosphamide are utilized in ALL. A combination of TBI and cyclophosphamide or busulfan and cyclophosphamide in AML, CML, myelodysplasia, JCML and JMML as follows:

---

1. Total body irradiation (TBI) and cyclophosphamide:

Day	Schema I
−8	TBI 150 cGy twice daily
−7	TBI 150 cGy twice daily
−6	TBI 150 cGy twice daily
−5	TBI 150 cGy twice daily
−4	TBI 150 cGy once daily
−3	Cyclophosphamide 60 mg/kg/day IV
−2	Cyclophosphamide 60 mg/kg/day IV
−1	Rest day
0	Stem cells harvested and given intravenously over a period of 3–4 hours

2. Busulfan* and cyclophosphamide:

Day	Schema II
−9	Busulfan 0.8–1 mg/kg IV, 6 hourly for four doses
−8	Busulfan 0.8–1 mg/kg IV, 6 hourly for four doses
−7	Busulfan 0.8–1 mg/kg IV, 6 hourly for four doses
−6	Busulfan 0.8–1 mg/kg IV, 6 hourly for four doses
−5	Cyclophosphamide 50 mg/kg/day IV
−4	Cyclophosphamide 50 mg/kg/day IV
−3	Cyclophosphamide 50 mg/kg/day IV
−2	Cyclophosphamide 50 mg/kg/day IV
−1	Rest
0	Stem cell harvested and given intravenously over a period of 3–4 hours

---

*IV Busulfan is preferred if available because of the difficulty taking oral Busulfan and the unreliable absorption. It is recommended to measure the plasma levels of Busulfan and adjust the dose as per area under the curve (AUC 900–1,300 ng/ml) or as per concentration at steady state (CSS 600–900 ng/ml).

Non-Hodgkin Lymphoma (NHL) (Chapter 20)

Myeloablative chemotherapy regimens such as cyclophosphamide, BCNU, Etoposide (VP 16) (CBV) (Table 29-9) or BCNU, Etoposide, cytosine arabinoside and Melphalan (BEAM) (Table 29-10) are preconditioning regimens utilized prior to stem cell transplantation in relapsed NHL.

Wilms' Tumor (Chapter 23)

High-dose chemotherapy with autologous stem cell rescue (HDC/ASCR) has been successfully used for higher-risk relapsed Wilms' tumor in patients whose disease remains chemosensitive. The preconditioning regimen employed can be found in Chapter 23, Table 23-14.

Table 29-9   CBV Regimen for Stem Cell Transplantation in Non-Hodgkin Lymphoma

Days	Chemotherapy
−8, −7, −6	BCNU: 100 mg/m^2/day IV over 3 hours (total dose 300 mg/m^2) Etoposide 800 mg/m^2/day IV as a 72-hour continuous infusion (total dose 2,400 mg/m^2)
−5, −4,−3, −2	Cyclophosphamide: 1,500 mg/m^2/day IV over 1 hour. Use MESNA (total dose 2,200 mg/m^2 IV every 24 hours)
−1	No treatment
0	Stem cell infusion

Methylprednisolone 1 mg/kg/day in divided doses given every 6 hours for pulmonary protection from BCNU toxicity on days –9 to –2. Wean by 20% every 2 days for an 8-day taper.

Table 29-10   BEAM Regimen for Stem Cell Transplantation in Non-Hodgkin Lymphoma

Day	Chemotherapy
−6	BCNU: 300 mg/m^2 IV over 3 hours
−5, −4, −3, −2	Etoposide 200 mg/m^2/day IV over 1 hour (total dose 800 mg/m^2)
−5, −4, −3, −2	Cytosine arabinoside: 400 mg/m^2/day IV over 1 hour (total dose 1,600 mg/m^2)
−1	Melphalan 140 mg/m^2 IV over 30 minutes
0	Stem cell infusion

Methylprednisolone 1 mg/kg/day in divided doses given every 6 hours for pulmonary protection from BCNU toxicity on days –7 to –2. Wean by 20% every 2 days for an 8-day taper.

Aplastic Anemia (Chapter 6)

Patients who have received less than five blood transfusions usually receive cyclophosphamide 50 mg/kg/day for 4 days, followed by 1 day of rest and stem cell transplantation on the following day.

In multiply transfused patients, rejection of the bone marrow is a major problem (can be as high as 20–25%). To improve long-term survival in these patients, pretransplantation conditioning requires more intensive immunosuppression to be employed.

As graft rejection is the major cause of morbidity and mortality in HSCT for severe aplastic anemia (SAA) the pretransplantation preparative regimen of cyclophosphamide (Cytoxan) and antithymocyte globulin, ATG (ATGAM/thymoglobulin)[*] with the inclusion of cyclosporin A, (CSA [Sandimmune]) in the graft-versus-host-disease (GvHD) prophylaxis regimen is designed to be highly immunosuppressive. Table 29-11 describes a typical immunosuppressive HSCT preparative regimen for SAA. Even when an identical twin donor is used a similar preparatory regimen is recommended. Long-term survival in the range of >90% can be expected with HSCT using histocompatible related donors. The overall risk of unrelated or mismatched related donor transplantation precludes its use as front-line therapy

---

[*]Throughtout this chapter ATG refers to horse ATG.

Table 29-11   Hematopoietic Stem Cell Transplantation Preparative Regimen for Severe Aplastic Anemia

Day −5	Morning: Cyclophosphamide, 50 mg/kg IV over 1 hour
	Afternoon: ATG, 30 mg/kg IV. First dose of ATG is given over 8 hours; subsequent doses are given over 4 hours
This is repeated on Day −4 and Day −3	
Day −2	Morning: Cyclophosphamide, 50 mg/kg IV over 1 hour
Day −1	Rest, cyclosporine A, 10 mg/kg/d PO daily adjusted for serum levels
Day 0	Marrow infusion

Abbreviation: ATG, antithymocyte globulin.
Adapted from: Vlachos A, Lipton JM. In: Conn's Current Therapy, W.B. Saunders Company, 2002, with permission.

for SAA at this time. As improved HLA typing, preparative regimens and GVHD prophylaxis are utilized HSCT is becoming available to a wider group of patients with SAA.

The following regimens could also be employed:

- Cyclophosphamide and total nodal irradiation or
- Cyclophosphamide and TBI (according to schedule on page 815) or
- Busulfan and cyclophosphamide (according to schedule on page 815).

Fanconi Anemia (Chapter 6, p. 138)

The immunosupressive regimen recommended consists of one of the following:

- Fludarabine 30 mg/m^2/day for 4 days
- Cyclophosphamide 300 mg/m^2/day for 4 days
- Anti-thymocyte globulin 15 mg/kg/day for 3 days

OR

- Fludarabine 35 mg/m^2/day for 5 days
- Cyclophosphamide 5 mg/kg/day for 4 days
- Anti-thymocyte globulin 30 mg/kg/day for 5 days.

Miscellaneous Conditions

In hemoglobinopathies, selected immunodeficiency and other inherited disorders the conditioning regimen used is busulfan and cyclophosphamide with or without ATG.

Nonmyeloablative (Reduced-Intensity) Regimens

These regimens are based on the ability of donor T-cells to ablate the host hematopoiesis or malignancy or both. The object is to use the donor lymphoid cells to eradicate the disease in the host. This allows the use of decreased doses of myeloablative chemotherapy with less toxicity. However, aggressive immunosuppressive therapy is still required to eliminate the potential host rejection of the donor cells. This approach is referred to as nonmyeloablative or reduced-intensity transplantation.

At present this approach is still in the experimental stages and is being used in both selected cases of malignant and non-malignant diseases such as:

- **Malignant diseases**
  - CML
  - AML
  - lymphoma
  - Hodgkin disease
- **Non-malignant diseases:**
  - Immune deficiency disorders
  - Metabolic disorders
  - Bone marrow failure syndromes such as:
    - Severe aplastic anemia
    - Fanconi anemia
    - Congenital neutropenia,
    - Congenital thrombocytopenia,
    - Dyskeratosis congenita,
    - Chronic granulocytopenia
  - Hemophagocytic lymphohistiocytosis
  - Langerhans cell histiocytosis
  - Hemoglobinopathies.

Second Transplantation Regimens

The following regimens are being used for second transplantation following graft failure or relapse of disease:

- Fludarabine: 30 mg/m^2 IV for 4 days
- Melphalan: 70 mg/m^2 IV for 2 days
- Anti-thymocyte globulin 30 mg/kg IV for 3 days

OR

- Fludarabine: 30 mg/m^2 IV for 4 days
- Busulfan: 4 mg/kg/day IV for 2 days
- Anti thymocyte globulin 15 mg/kg for 3 days

OR

- Cyclophosphamide: 50 mg/kg IV for 1 day

- Fludarabine: 35 mg/m^2 IV for 4 days
- Anti-thymocyte globulin 30 mg/kg IV for 3 days

OR

- Low-dose melphalan and fludarabine
- Low-dose busulfan, melphalan and fludarabine
- Low-dose melphalan, fludarabine and campath
- Low-dose cyclophosphamide and limited-field irradiation
- Low-dose total body irradiation.

ENGRAFTMENT

Evidence of Engraftment

Evidence of engraftment is usually seen within 3 weeks following stem cell transplantation and is identified as follows:

- Direct Evidence
  - Fluorescence in situ hybridization (FISH) in sex-mismatched transplantation
  - Variable number of tandem repeats (VNTR) in same sex-matched transplantation
  - HLA typing: mismatched transplantation
  - Donor-type red cell antigens
  - Immunoglobulin allotypes of donor
  - In severe combined immunodeficiency disease (SCID): the presence of adenosine deaminase (ADA) in white blood cells.
- Indirect Evidence
  - Aplastic anemia and leukemia:
    - Adequate bone marrow cellularity
    - Increase in white blood cells with normal morphology and increase in platelet counts.
  - SCID: improved immunologic parameters.

Causes of Failure to Engraft

- Increased genetic disparity
- T-cell depletion or GVHD prophylaxis
- Inadequate stem cell dose
- History of multiple transfusions in aplastic anemia
- Disordered microenvironment in storage disorders and osteopetrosis.

GRAFT VERSUS HOST DISEASE (GVHD)

Prophylaxis

Methotrexate, cyclosporine and methylprednisolone are the most common drugs used singularly or in combination for GVHD prophylaxis beginning immediately following stem cell infusion. The most commonly used combinations are cyclosporine plus methotrexate or cyclosporine plus methylprednisolone. The drugs are used in the following dosages.

Methotrexate (Short Course)

Methotrexate 15 mg/m^2 IV on day following transplantation and 10 mg/m^2 IV on days 3, 6 and 11 post transplantation.

Cyclosporine

Cyclosporine 1.5 mg/kg twice daily IV starting on the day before transplantation and given until patient can take oral medication. At that time, a dose of 10 mg/kg/day in two divided doses is given orally. Serum levels should be kept between 150 and 300 µg/ml. Cyclosporine is slowly tapered by 10% weekly to start from 60 to 180 days post transplantation.

Complications

Complications of cyclosporine are fully discussed in Chapter 6, p. 134.

Methylprednisolone

Methylprednisolone is administered in a dose of 0.5 mg/kg orally every 12 hours from day 5 until day 28 post transplantation and thereafter it is tapered slowly if there is no evidence of GVHD.

The other major technique to prevent GVHD is purging of the donor marrow of T lymphocytes.[*] Use of this technique has the following disadvantages:

- Increase in early and late engraftment failures
- Disease relapse
- Secondary tumors.

Results

Table 29-12 lists the 5-year disease-free survival in different disorders following allogeneic stem cell transplantation using HLA-identical matched siblings. The disease-free survival varies among different published series.

---

[*]T-cell purging can be done by monoclonal antibodies accompanied by complement-mediated lysis, immunotoxins, or immunomagnetic beads. Physical methods of T-cell depletion include counter elutriation and soybean lectin agglutination plus E-rosette depletion.

Table 29-12 **HLA-Identical Matched Sibling Allogeneic Stem Cell Transplantation 5-year Disease-Free Survival**

Disease	Disease-Free Survival (DFS) (%)
ANLL, first remission	55–70
ALL, first or second remission	30–50
ANLL, transplant in relapse	10–30
CML, chronic phase[a]	60–80
CML, accelerated or blastic phase[a]	10–30
Lymphoma, Hodgkin's disease, after failure of first-line therapy	15–40
Aplastc anemia, untransfused	85–90
Aplastic anemia, transfused	50–70
Thalassemia major	70–85

[a]Age dependent; DFS decreases with increasing age.

Most of the hematopoietic stem cell transplantation in patients with ALL and ANLL have been performed in second or subsequent remissions using the regimen of cyclophosphamide and total body irradiation or cyclophosphamide and busulfan. The long-term disease survival is 20–50%.

AUTOLOGOUS STEM CELL TRANSPLANTATION

Autologous stem cell transplantation offers a viable alternative when a matched sibling is not available for transplantation. Normal appearing bone marrow is removed from the patient and may be treated *ex vivo* with cytotoxic drugs or monoclonal antibodies directed at the malignant cells. A concentrated selection of pluripotent hematopoietic cells can be obtained using CD34+ (stem cell antigen) as a marker for stem cells to reduce the volume of stem cells required for reinfusion. This is performed *in vitro* using CD34+-labeled columns. The stem cells are then cryopreserved. The patient then receives intensive chemotherapy with hematopoietic stem cell rescue and hematopoietic growth factor. Table 29-13 lists the high-risk solid tumors that may respond to intensive therapy followed by autologous stem cell transplantation.

Indications

1. Intermediate- and high-grade lymphomas in first relapse or partial remission.
2. Hodgkin disease in first relapse or partial remission.
3. High-risk tumors at the time of diagnosis where long-term survival is less than 30% and there is a chemotherapy dose response demonstrated (most consistent indicator of a poor prognosis is disseminated disease).
4. Failure or partial response to initial therapy or recurrence of disease (i.e., progressive disease).

Table 29-13   High-Risk Solid Tumors that may Respond to Intensive Therapy and Autologous Stem Cell Transplantation[a]

Tumor	Indication
Supratentorial astrocytoma	Recurrence
Medulloblastoma	Recurrence
Ependymoma	Recurrence
Ewing sarcoma/PNET	Primary metastatic
	Recurrence
Rhabdomyosarcoma	Primary stage IV
	Recurrence
Neuroblastoma	Stage IV *n-myc* positive
	Recurrence
Wilms tumor	Recurrence
Retinoblastoma	Primary extraocular
	Recurrence

[a]All tumors must have demonstrated some response to initial chemotherapy prior to autologous SCT.
Abbreviation: PNET, primitive neuroectodermal tumor.

Drugs appropriate for high-dose chemotherapy should have the following properties:

- The drug should be active against the tumor type
- The drug should be tolerable at 3–10 times non-marrow-ablative dose
- The drug should generate a steep and straight line dose–response curve over many logs of tumor cell kills
- The combination of drugs should have non-cross-resistance and no overlapping toxicities
- The dose-limiting toxicity should be marrow ablation only with tolerable toxicity to other organs.

The following factors contribute to the outcome:

- Timing of intensive chemotherapy
- Chemotherapy schedule (dose and drugs) and expected toxicity
- Supportive care – appropriate use of antibiotics and other drugs for managing side effects
- Appropriate use of peripheral stem cell transplantation and growth factors, which can reduce the period of neutropenia and hospitalization.

Prognosis

Solid Tumors

See the relevant chapters for specific tumors.

Table 29-14 presents the results of autologous stem cell transplantation disease-free survival for various malignancies.

Table 29-14   Autologous Stem Cell Transplantation

Disease	Disease-free Survival (%)
Lymphoma	30–60
Neuroblastoma	45–50
Ewing's sarcoma	25–30
Rhabdomyosarcoma	30
Wilms' tumor	50
Brain tumor	50

COMPLICATIONS OF HEMATOPOIETIC STEM CELL TRANSPLANTATION

Despite quantitative myeloid (neutrophil) recovery after bone marrow transplantation, the functional recovery of humoral and cellular immunity may take a year or longer. The type of graft used (autologous or allogeneic), the type and method of administration of immunosuppressive therapy after transplant (cyclosporine, corticosteroids, ATG) and whether GVHD has occurred (especially chronic GVHD) all influence the rate of lymphoimmunologic reconstitution.

Immunodeficiency

After stem cell transplantation, a transient state of combined immunodeficiency develops in all patients. The natural history of immune reconstitution is similar for autologous and allogeneic transplants but often is altered in allogeneic transplants by GVHD as well as by immunosuppressive therapy. Although natural killer cells reconstitute to normal levels usually within the first month after transplantation, other immunologic defects may persist.

The importance of the recognition of the delayed overall immunologic reconstitution relates to the clinically observed incidence of recurrent bacterial (*Streptococcus pneumoniae*) and opportunistic infections (*Pneumocystis jiroveci*, fungal, herpes zoster, CMV) that can occur many months after transplantation. In addition to the use of prophylaxis against *Pneumocystis jiroveci* infection, many centers add penicillin as post transplant prophylaxis against *Streptococcus pneumoniae* infection, particularly if the patient has chronic GVHD. The suppressed immunity also has major practical implications for consideration of the timing of vaccinations after bone marrow transplantation. The following approach for the timing of vaccinations should be considered:

- At the end of 1 year, individuals who do not have chronic GVHD should receive influenza (yearly) and pneumococcal polysaccharide vaccines and should also have vaccinations with inactivated poliovirus and diphtheria–pertussis–tetanus (DPT). It is probably also safe to administer hepatitis B and *Haemophilus influenzae* conjugate vaccines

Table 29-15   Reimmunization Schedule After Stem Cell Transplantation

Vaccine	Time to Begin Reimmunization
Diphtheria–tetanus toxoid	6–12 months
Inactivated poliovirus	6–12 months
Measles–mumps–rubella	24 months

- Immunosuppressed patients with chronic GVHD are less likely to develop an adequate antibody response after vaccinations. If vaccines are administered, antibody titers should be checked after vaccination to determine the efficacy of the response
- For patients who are 2 years post-stem cell transplantation, have no evidence of GVHD and are not receiving immunosuppressive therapy, measles, mumps and rubella (MMR) live vaccines should be given. Table 29-15 lists a reimmunization schedule after transplantation.

The risks of HSCT are related to the underlying disease, the pretransplantation conditioning with cytotoxic drugs or irradiation or both, post-transplantation immunosuppression (infections) and graft versus host disease. The early mortality can be as high as 20%, depending on the conditioning regimens, the disease and the patient's status.

The potential complications post transplantation of conditioning regimens include the following:

- Toxic vasculitis
- Severe mucositis
- Veno-occlusive disease of the liver
- Acute alveolitis
- Obliterative bronchiolitis
- Hemolytic–uremic syndrome
- Multiple organ failure.

Infections

Patients undergoing stem cell transplantation have increased risks of infection related to their disease and its status. Patients with more than two relapses who undergo stem cell transplantation develop more infectious complications than those who receive transplants during the first remission. The adequacy of treatment of any infection in the patient prior to marrow ablation influences the risk of subsequent complications. Prolonged duration of neutropenia, use of central venous catheters and slow speed of marrow engraftment are important risk factors.

Table 29-16 lists prophylaxis and supportive care for stem cell transplantation. Table 29-17 lists the bacterial and fungal infections most frequently seen at different times post transplantation.

Table 29-16   Prophylaxis and Supportive Care for Stem Cell Transplantation

1. Mouth care: Peridex or biotin solution and mycostatin every 3–4 hours daily
2. Recombinant hematopoietic growth factors: G-CSF or GM-CSF (Chapter 31)[a]
3. Prophylaxis against *Pneumocystis jiroveci:* trimethoprim/sulfamethoxazole 5 mg/kg in two divided doses (three times per week) starting after engraftment; continue if chronic GVHD is present[b]
4. Prophylaxis against *Candida* and other fungal infections: fluconazole 4–6 mg/kg once a day (see Chapter 31)
5. Prophylaxis for herpes simplex and CMV: IV acyclovir 250 mg/m^2 every 8 hours for first 3–6 months
6. Additional prophylaxis for CMV:
   CMV-negative blood products should be used if patient and donor are CMV-negative; if CMV negative blood is not available, leukocyte filters to remove leukocytes should be employed
7. IVIG (only in recipient of allogeneic stem cell transplant): 500 mg/kg every 2–3 weeks for first 3 months after transplant[d] and then check IgG levels periodically if less than within normal limits then infuse accordingly
8. Nutrition: total parenteral nutrition given if enteral nutrition is not possible or inadequate

[a]Does not increase relapse and GVHD rates. Stimulates hematopoietic recovery.
[b]If patients are allergic to trimethoprim/sulfamethoxazole or if severe cytopenia is present, aerosolized pentamidine or intravenously 4 mg/kg once a month can be used.
[c]Drug-related neutropenia is a major side effect.
[d]For its immunomodulatory (decreased incidence of GVHD) effect and antimicrobial activity, especially CMV. It also decreases Gram-negative sepsis.

During the first 30 days, the most frequently documented infection is coagulase-negative *Staphylococcus epidermidis* infection associated with the use of indwelling central venous catheters.

During the second and third months, infections with CMV and interstitial pneumonia occur. CMV infection occurs within 20–100 days after transplantation, with the highest infection rate in patients who are CMV-seropositive before transplantation or seronegative recipients receiving a seropositive graft. CMV infection arises from exogenous introduction of virus in blood products or from reactivation of endogenous virus.

The clinical manifestations of CMV infection are variable, ranging from asymptomatic viral excretion to fever, arthralgia, arthritis, hepatitis, secondary bone marrow hypoplasia with thrombocytopenia and leukopenia, retinitis, esophagitis, gastroenteritis and pneumonia with up to 80% mortality.

Early treatment with the antiviral drug ganciclovir plus high-titer CMV intravenous gammaglobulin has reduced morbidity to less than 20%. Prophylactic or preemptive therapy in seropositive patients or receiving a seropositive graft is an effective way to prevent the development of CMV infection.

The following guidelines for the prophylaxis and treatment of CMV infections should be carried out (as recommended by the U.S. Center for Disease Control along with American Society of Blood and Marrow Transplantation):

Table 29-17   Infections Most Frequently Seen at Different Times Post-Transplantation

I. Infections in first 30 days post-transplantation
   A. Bacteremia
       Gram-positive organisms: *Staphylococcus epidermidis*
       Gram-negative aerobes and anaerobes
   B. Invasive fungal infections: *Aspergillus, Candida*
   C. Reactivation of herpes simplex I
II. Infections 30–120 days post transplantation
   A. Protozoal infections
       Pneumocystis jiroveci
       Toxoplasma
   B. Viral infections
       Cytomegalovirus (CMV)
       Adenovirus
       Epstein–Barr virus (EBV)
       Human herpes virus 6 (HHV-6)
   C. Fungal infections
       Candida (C. albicans and *C. tropicalis)*
       Aspergillus
       Trichosporon
       Fusarium
       Candida kruseii
III. Infections after 120 days post transplantation
   A. Sinopulmonary infections with encapsulated organisms
   B. Viral infections
       Cutaneous herpes zoster

- Allogeneic transplantation patients who are seronegative at the time of transplant and have a seronegative donor do not require additional CMV prophylaxis other than the strict use of CMV-negative blood products
- Allogeneic transplantation patients who are CMV positive at transplantation or those who received transplantation from a CMV-positive donor should receive CMV safe, leukocyte-filtered blood products
- Allogeneic transplantation patients should have surveillance for CMV antigen-emia assay or a polymerase chain reaction (PCR) assay and blood, urine and throat cultures obtained weekly for the first 120 days post transplantation. Patients with CMV-positive cultures from blood, bronchial washings, throat, or gastrointestinal tract should be treated with ganciclovir and gammaglobulin at least until day 100 after transplantation or for 2–3 weeks post the latest date of positive cultures
- Autologous transplantation patients who are seronegative at the time of transplantation should preferably receive CMV-negative blood products but may, as an alternative, receive leukocyte-filtered blood products.

Interstitial Pneumonitis

Interstitial pneumonitis is an important complication and may be due to any of the following causes:

- CMV
- *Pneumocystis jiroveci*
- Undiagnosed infections (e.g., fungal)
- Drugs
- Radiation toxicity
- Immune reactions involving the lung
- Idiopathic (16%).

The mortality rate for interstitial pneumonitis is 60%. The use of prophylactic bactrim has reduced *Pneumocystis jiroveci* infection to less than 2%.

After 120 days post-bone marrow transplantation, the most common infections are sino-pulmonary infections with encapsulated organisms and cutaneous infections with herpes zoster.

Other localized or disseminated viral infections that occur include adenovirus, herpes simplex virus, varicella virus, respiratory syncytial virus, para-influenza and influenza A or B.

Table 29-18 outlines the management of infection during stem cell transplantation and during the first 30 days post-transplantation.

Pancytopenia

In addition to the corrections of anemia and thrombocytopenia by the use of irradiated leukocyte-depleted packed red cells and platelets, recombinant hematopoietic growth factor plays the following roles:

- It is safe and effective to administer growth factors to allogeneic transplantation patients. All allogeneic and autologous transplantation patients should receive G-CSF therapy following stem cell transplantation
- With the use of ganciclovir to treat CMV infections and other myelosuppressive therapies, growth factors may play an added role in the prevention or treatment of myelosuppression caused by these agents after stem cell transplantation.

Graft Versus Host Disease

The incidence of acute GVHD varies among different transplantation centers and depends on the primary diagnosis. In patients with primary immunodeficiency and aplastic anemia,

Table 29-18   Management of Infections During First 30 Days

Gram-negative infections
  Empiric combination therapy (Chapter 31)
Gram-positive infections
  Vancomycin
Fungal infections
  Amphotericin B
  Liposomal amphotericin
  Voriconazol
  Caspofungin
  Posaconazole
Removal of indwelling catheter
Management of interstitial pneumonia

1. Bronchoalveolar lavage (BAL)
2. Open lung biopsy
3. Treatment depends on etiology
  a. *Pneumocystis:* bactrim, pentamidine
  b. CMV: ganciclovir plus CMV high-titer gammaglobulin
  c. Idiopathic: supportive care, steroids
  d. Fungal: antifungals
Management of herpes zoster
  Acyclovir

the incidence of GVHD may vary from 10–35%, whereas in patients with acute leukemia, it may be approximately 50%. The probability of acute GVHD grade II–IV is approximately 70% in unrelated matched donors compared to 30% in patients with HLA-A, -B and -DR matched sibling transplantation. Approximately 10–30% of all patients with sustained allogeneic engraftment die of acute GVHD or its complications.

GVHD is caused by donor alloreactivity in an immunocompromised host and results from T-lymphocytes contained in the stem cell graft that proliferates and differentiates in the host. These T-cells recognize host allo-antigens as foreign and through both direct effector mechanisms and by inflammatory mediators released by T-cells, monocytes and production of cytokines especially IL-1 and tumor necrosis factor in the host may cause tissue damage.

Graft versus host disease is divided into two stages:

- Acute GVHD usually manifests within 30–40 days
- Chronic GVHD manifests 100 days after stem cell transplantation.

Acute Graft Versus Host Disease

The clinical manifestations of acute GVHD range from a mild maculopapular eruption to generalized erythroderma, hepatic dysfunction, gastroenteritis, stomatitis and lymphocytic bronchitis. Thrombocytopenia and anemia have been reported in GVHD. Ocular symptoms, including photophobia, hemorrhagic conjunctivitis and pseudomembrane formation, have also been observed. It is occasionally difficult to separate the clinical manifestations of GVHD from other disorders in the post-transplant patient and, in these cases, a biopsy of the skin or liver may be required. Table 29-19 lists the clinical manifestations and grades of acute GVHD and Table 29-20 describes the histologic grades of acute GVHD. Table 29-21 presents a comparison between acute and chronic GVHD.

Treatment

Prophylaxis

Complete matching of the donor and recipient at the level of HLA class I and II by DNA typing are the most important factors in preventing GVHD. Consideration of the sex and parity of the donor are advisable if there is a choice of donor. If the patient is CMV seronegative, use of a CMV negative donor appears to reduce the risk of CMV infection as well as risk of GVHD. Table 29-22 lists additional measures that can be employed in the prophylaxis of GVHD, as mentioned earlier in this chapter. Prophylactic drugs have been used in an attempt to inhibit T-cell response by relying on *in vivo* immunosuppression. Currently all agents are used in a combination that targets different molecular intermediates of T-cell signals.

Therapy

Grade II, III and IV GVHD require therapeutic intervention with high-dose methylprednisolone 3–5 mg/kg/day for 7 days, after which the dose is decreased to 2 mg/kg/day. The progression of GVHD despite steroid therapy in these doses requires an increase in the dose of methylprednisolone to 20 mg/kg/day for 3 days, 10 mg/kg/day for 3 days, 5 mg/kg/day for 3 days, 3 mg/kg/day with tapering doses of steroids, depending on the response or the use of ATG. ATG is usually given in doses of 10–15 mg/kg every other day for 7–14 days. Newer immunosuppressive agents like tacrolimus, sirolimus, mycophenolate mofetil and methods targeting CD5, CD3, the interleukin-2 receptor such as daclizumab and TNFα and its receptor have been used with some success.

Chronic Graft Versus Host Disease

Chronic GVHD occurs 100 days after transplantation. It may:

* Follow a progressive extension of acute GVHD, or
* Follow a quiescent period after acute GVHD has resolved, or
* occur in a patient who did not have acute GVHD.

Table 29-19   Clinical Manifestations and Grades of Acute Graft versus Host Disease (GVHD)

Clincal Grade	Skin[a]	Liver (Serum Total Bilirubin, mg/dl)[b]	Gut (Diarrhea)[c]	
			Children (ml/kg/day)	Adults (ml/day)
I (mild)	Maculopapular rash, <25% body surface. May be pruritic or painful	1.5–3.0	10–15	500–1,000 Nausea and vomiting
II (moderate)	Maculopapular rash, 25–50% body surface	3.0–6.0	16–20	1,000–1,500 Nausea and vomiting
III (severe)	Maculopapular rash, >50% body, or generalized erythroderma	6.0–15.0	21–25	>1,500 Nausea and vomiting
IV (life threatening)	Desquamation and bullae	>15	>25 Pain or ileus	>2,500 Pain or ileus

[a]Differential diagnosis includes chemoradiotherapy-induced rash, drug allergy and viral exanthema. Skin biopsies are informative for patients more than 21 days after bone marrow transplantation.
[b]The degree of hyperbilirubinemia does not correlate well with clinical outcome. Differential diagnosis includes hepatotoxic drug reactions and viral infections. Liver biopsy is helpful in establishing the diagnosis.
[c]Diarrhea may be severe and associated with crampy abdominal pain. The diarrhea is green, mucoid, watery and mixed with exfoliated cells that may form fecal casts. It may progress to bleeding and ileus. Intestinal radiographs show mucosal and submucosal edema with rapid barium transit time and loss of haustral folds. A variant of enteric GVHD in 13% of patients has presenting features of anorexia and dyspepsia. Patients with upper gastrointestinal disease may not manifest lower tract involvement. Gastrointestinal endoscopy and biopsy are mandatory for the diagnosis of upper gastrointestinal tract disease. Lower tract disease may be diagnosed with rectal biopsy, which shows necrosis of individual cells in crypts that may progress to dropout of entire crypts and loss of epithelium.

Table 29-20   Histologic Grades of Acute Graft Versus Host Disease

Grade	Skin	Liver	Gut
I	Epidermal base cell vacuolar degeneration	Fewer than 25% small inter lobular bile ducts abnormal (degeneration or necrosis)	Single-cell necrosis of epithelial cells
II	Grade I changes plus "eosinophilic bodies"	25–50% bile ducts abnormal	Necrosis
III	Grade II changes plus separation of the dermal–epidermal junction	50–75% bile ducts abnormal	Focal microscopic mucosal denudation
IV	Frank epidermal denudation	More than 75% bile ducts abnormal	Diffuse mucosal denudation

Table 29-21 Comparison of Acute and Chronic Graft Versus Host Disease

Characteristic	Acute	Chronic
Incidence	40–60%	20–40%
Onset, days	7–60 (up to 100)	>100
Clinical manifestations		
Skin	Erythematous rash	Sclerodermatous changes
Gut	Secretory diarrhea	Dry mouth, esophagitis, malabsorption
Liver	Hepatitis	Cholestasis
Lung	Diffuse alveolar hemorrhage	Pulmonary dysfunction
Other	Fever	Contractures (due to sclerodermatous skin damage)
		Alopecia
		Thrombocytopenia
Target cells	Epidermal	Mesenchymal
Basis of clinical grading	Severity of disease	Extent of disease

Table 29-22 Prophylaxis of Graft Versus Host Disease

1. Major histocompatibility complex (MHC) matching
2. Irradiation of blood products
3. Prevention of infection
4. *In vivo* treatment of recipient[a]
   a. Cyclosporine
   b. Methotrexate
   c. Corticosteroids
   d. Tacrolimus
   e. Mycophenolate mofetil
   f. ATG
   g. Daclizumab
5. Depletion of donor T cells from marrow graft such as:
   a. Purging of marrow with monoclonal antibodies OK T3, T12
   b. Soybean lectin and E-rosette depletion
   c. Pretreatment of marrow with various antibody-ricin conjugates
   d. Magnetic beads

[a]Current prevention in many centers consists of methotrexate and cyclosporine.

In matched-sibling grafts, the incidence of chronic GVHD is 13% for children under 10 years of age and 28% for children 10–19 years of age. In unrelated-donor or mismatched-family-member donor, the incidence of chronic GVHD ranges from 42–56%.

The predictors for chronic GVHD are:

- Donor–recipient HLA disparity
- Increasing patient age
- Latent donor viral infection
- Female donor with a history of parity (sensitized)

Table 29-23   Classification of Chronic Graft Versus Host Disease

Limited
   Either or both
      Localized skin involvement
      Hepatic dysfunction
Extensive
   Either
      Generalized skin involvement
      Hepatic dysfunction
               PLUS
      Ocular involvement (Schirmer's test, <5 mm wetting)
      Oral mucosal involvement (lip biopsy positive)
      Any other target organ

- Source of allogeneic cells, number of T cells collected during pharesis is greater than that collected during bone marrow harvest.

One-third of patients who develop chronic GVHD are at risk for bacterial and opportunistic infections. The duration of chronic GVHD varies and the morbidity and mortality range from 10–15%. Chronic GVHD can be divided into two groups (limited or extensive) (Table 29-23).

Limited (Usually Involving Only One Organ)
- Localized skin involvement: erythema, hyperkeratosis, patchy scleroderma-like lesions, reticular hyperpigmentation, desquamation, alopecia, nail loss; rash initially involving palms and soles, back of neck and ears and later the trunk and extremities
- Hepatic dysfunction: predominantly cholestatic abnormalities.

Extensive
- Generalized skin involvement: resembling lichen planus or scleroderma; atrophy of skin with ulceration; fibrosis of skin with or without limitation of joint movement
- Hepatic dysfunction: liver histology showing chronic aggressive hepatitis, bridging necrosis, or cirrhosis
- Ocular symptoms: dry eyes, Schirmer's test with less than 5 mm wetting; keratoconjunctivitis sicca including burning, irritation, photophobia and pain
- Buccal cavity: dry mouth, sensitivity to acidic or spicy foods and pain; oral atrophy with depapillation of tongue; erythema and lichenoid lesions of buccal and labial mucosa; involvement of minor salivary glands or oral mucosa demonstrated on labial biopsy specimen
- Gastrointestinal tract: malabsorption due to submucosal fibrosis of gastrointestinal tract; dysphagia, pain and weight loss
- Genitourinary tract: vaginitis and vaginal strictures

- Polyarthritis, myositis and fascitis leading to contractures
- Pulmonary complications: interstitial pneumonia and bronchiolitis obliterans.

Prognosis

Morbidity and mortality are highest in patients with progressive-onset chronic GVHD that directly follows acute GVHD, are intermediate in patients after resolution of acute GVHD and are lowest in those with *de novo* onset. The median day of diagnosis of chronic GVHD in recipients of HLA-identical sibling BMT is 200 days and the onset of the disease in recipients of bone marrow from an HLA-non-identical or unrelated donor is approximately 145 days. Screening studies at day 100 with skin examination with biopsy, oral examination with lip biopsy, ocular examination with Schirmer's test and liver function studies are useful in predicting patients at risk for the development of chronic GVHD. Even in the absence of signs or symptoms, a positive random skin biopsy or history of previous acute GVHD independently predicts a threefold increase in the relative risk of subsequent chronic GVHD.

Treatment

Methylprednisolone 1–2 mg/kg alone or in combination with PO cyclosporine 10–12 mg/kg/day or tacrolimus 0.15–0.3 mg/kg/day in two divided doses or mycophenolate mofetil 600 mg/m^2/dose twice a day. Azathioprine 1.5 mg/kg once a day may also be utilized. The azathioprine dose is adjusted to keep the white blood cell count between 2,500 and 3,500/mm^2. This can have a steroid-sparing effect if unable to taper steroids.

Extracorporeal photophoresis (ECP) has also been found to be effective in controlling chronic GVHD and in reducing the need for prednisone in patients with refractory GVHD. The ECP procedure involves collecting the patient's donor-derived white cells and treating the cells with the drug Uvadex and exposing them to ultraviolet light. The treated cells are then given back to the patient.

Thalidomide is also effective in chronic GVHD.

The following supportive care should be given in addition:

- Intravenous gammaglobulin 500 mg/kg every 2–4 weeks (as per IgG levels)
- Bactrim prophylaxis for *Pneumocystis jiroveci* infection
- Antibiotics for chronic sinuopulmonary infection.

Therapy should be continued for a minimum of 9 months or until all clinical and pathologic evidence of chronic GVHD has cleared and when biopsies of the skin and lip have returned to normal.

Veno-Occlusive Disease

Veno-occlusive disease (VOD) occurs as a complication of chemotherapy and/or radiation in allogeneic stem cell transplant. It is a common life-threatening complication of preparative-regimen-related toxicity of stem cell transplantation. It occurs after approximately 20% of allogeneic stem cell transplantations and after about 10% of autologous stem cell transplantations. Veno-occlusive disease is characterized by fibrous obliteration of small hepatic vessels. It usually occurs within the first 30 days.

Clinical Manifestations

1. Jaundice.
2. Weight gain.
3. Ascites.
4. Hepatomegaly.
5. Right upper quadrant pain.

Because not all patients exhibit the full spectrum of the syndrome, a common clinical definition requires the presence of any two of the above listed features with the onset occurring between 7–21 days from stem cell infusion. Because these clinical manifestations are not specific to veno-occlusive disease, all other causes of hepatic dysfunction must be excluded. Another condition that may mimic hepatic veno-occlusive disease is nodular regenerative hyperplasia of the liver, a diffuse nonfibrotic nodulation of the liver with areas of regenerative activity alternating with areas of atrophy. This syndrome affects the same transplantation population at risk for hepatic veno-occlusive disease and both conditions are frequently associated with the development of ascites. Although no specific treatments are available for either entity, the mortality rate is considerably higher for patients who develop hepatic veno-occlusive disease than for those who develop nodular regenerative hyperplasia.

Predisposing Factors

The cause of veno-occlusive disease remains unknown; however, patients with the following conditions have an increased risk:

- Pre-existing hepatitis (one of the most strongly predictive risk factors)
- Elevated liver function tests
- Refractory leukemia
- Age over 15 years
- Underlying metastatic tumor to liver
- Positive CMV serologic status
- Conditioning agents (such as TBI, busulfan, cytosine arabinoside)
- Mismatched or unrelated allogeneic transplants

- Certain antimicrobial agents (especially vancomycin and acyclovir when given before transplantation) have also been implicated.

Prophylaxis

The following measures have been employed as prophylaxis in VOD:

- Lower or fractionated doses of TBI shielding the liver during TBI
- Targeted busulfan levels
- Use of T-cell depleted marrow
- Use of ursodeoxycholic acid
- Continuous infusion of low-dose heparin has been used to reduce toxicity
- Defibrotide (DF) has shown promising results.

Treatment

1. Supportive care.
2. Sodium restriction.
3. Spironolactone therapy.
4. Defibrotide.
5. Recombinant tissue plasminogen activator (rt-PA).
6. Thrombolytic therapy (continuous infusion of heparin during pretransplant conditioning and through the first 2 weeks post-transplant in a dose of 100–150 units/kg/day or recombinant human tissue plasminogen activator (t-PA) and heparin or antithrombin III (AT III) infusion if AT III levels are low).[*]
7. Prostaglandin E (may be a promising agent in prophylaxis).
8. Portocaval shunting.
9. Liver transplantation.

LATE SEQUELAE OF STEM CELL TRANSPLANTATION

Table 29-24 lists the late sequelae of allogeneic stem cell transplantation in children.

POST-STEM CELL TRANSPLANTATION THERAPY

Post-stem cell transplantation patients do not require any chemotherapy with the exception of certain patients with lymphoma, who may require localized radiotherapy to the primary

---

[*]These approaches are limited by the risk of fatal intracerebral and pulmonary bleeding. Heparin has an effect on the overall incidence of VOD and may lessen the severity of the clinical illness. A dose of 100 units/kg/day of heparin has been shown to be safe.

Table 29-24   Late Sequelae of Allogeneic Stem Cell Transplantation in Children

1. Chronic GVHD
2. Multiple endocrine disorders (hypothyroidism, delayed pubescence, gonadal failure, growth hormone deficiency)[a]
3. Second malignancies (incidence of 0.6 per 100 person-years)[b]
4. Sterility due to gonadal failure
5. Cataracts (secondary to radiation therapy); 20–50% incidence at 5–6 years
6. Renal insufficiency (nephrotoxic antibiotics and cyclosporine)
7. Obstructive and restrictive pulmonary disease (occurs in 10–15% of patients with chronic GVHD)
8. Cardiomyopathy
9. Aseptic necrosis of bone
10. Leukoencephalopathy (especially with IT MTX)
11. Immunologic dysfunction
12. Disturbance in dental development (from TBI) particularly if HSCT takes place in patients less than 6 years of age

[a]Associated with the use of TBI.
[b]At 15 years after HSCT, 6% develop second malignancies if not prepared with TBI and 20% for those prepared with TBI regimens.
Note: The frequency and severity of these sequelae vary considerably with the conditioning required preceding SCT.

site. Because the relapse rate is about 50%, a number of investigators have considered the following post transplantation therapy:

- Use of double autologous HSCT to eradicate the few remaining logs of tumor cells. Sufficient stem cells can be cryopreserved from a single harvest for two transplants. Non-cross-resistant regimens, each with activity, if not causing considerable toxicity, would be advantageous
- Use of cytokines to enhance hematopoietic (shortening the period of neutropenia) and immunologic recovery. Cytokines are also used for their antitumor effects, which include the following:
  a. Stimulating host antitumor effector cells
  b. Direct cytotoxic effect
  c. Cytostatic for tumor cells
  d. Interfering with nutrient and oxygen supply to tumor by acting on tumor vasculature.
- Use of differentiation inducers. These agents can arrest cell proliferation and induce terminal differentiation and may prove effective against minimal disease after autologous HSCT (e.g., 13-cis-retinoic acid). It does not have hematopoietic toxicity and should be well tolerated in autologous HSCT.
- Use of antimyeloid monoclonal antibody in myeloid leukemia such as gemtuzumab ozogamicin (mylotarg).

- Use of low-dose cyclosporine to induce GVHD, which also has a beneficial effect on graft versus leukemia reaction and thus might reduce the relapse rate.

Suggested Reading

Antin J. Stem cell transplantation-harnessing of graft-versus malignancy. Curr Opin Hematol. 2003;10:440–444.

Coppell JA, Richardson PG, Soiffer R, et al. Veno-occlusive disease following stem cell transplantation: Incidence, clinical course and outcome. Biology of blood and marrow transplantation. J Am Soc Blood Marrow Transplant 2009.

Eapen M, Rubinstein P, Zhang MJ, et al. Comparable long-term survival after unrelated and HLA matched sibling donor hematopoietic stem cell transplantations for acute leukemia in children younger than 18 months. J Clin Oncol. 2006;24(1):145–151.

Feinstein L, Sandmaier B, Maloney D, et al. Nonmyeloablative hematopoietic cell transplantation. Replacing high-dose cytotoxic therapy by the graft-versus-tumor effect. Ann N Y Acad Sci. 2001;938:328–337.

Finke J, Bethge W, Schmoor C, et al. for the ATG-Fresenius Trial Group. Standard graft-versus-host disease prophylaxis with or without anti-T-cell globulin in haematopoietic cell transplantation from matched unrelated donors: a randomised, open-label multicentre phase 3 trial. The Lancet Oncol. 2009;10 (9):855–864.

Gea-Banacloche JC, Wade JC. Improvements in the prevention and management of infectious complications after hematopoietic stem cell transplantation. Cancer Treat Res. 2009;144:1–35.

Grosskreutz C, Ross V, Scigliano E, et al. Low-dose total body irradiation, fludarabine and antithymocyte globulin conditioning for nonmyeloablative allogeneic transplantation. Biol Blood Marrow Transplant. 2003;9(7):453–459.

Kosaka Y, Koh K, Kinukawa N, et al. Infant acute lymphoblastic leukemia with MLL gene rearrangements: outcome following intensive chemotherapy and hematopoietic stem cell transplantation. Blood 2004;104 (12):3527–3534.

Ladenstein R, Pötschger U, Hartman O, et al. EBMT Paediatric Working Party. 28 Years of high-dose therapy and SCT for neuroblastoma in Europe: lessons from more than 4000 procedures. Bone Marrow Transplant. 2008;41(Suppl 2):S118–27.

Li M, Sun K, Welniak LA, Murphy WJ. Immunomodulation and pharmacological strategies in the treatment of graft-versus-host disease. Expert Opin Pharmacother. 2008;9(13):2305–2316.

Marsh SG, Albert ED, Bodmer WF, et al. Nomenclature for factors of the HLA system, 2004. Tissue Antigens 2005;65(4):301–369.

Pawson R, Potter MN, Theocharous P, et al. Treatment of relapse after allogeneic bone marrow transplantation with reduced intensity conditioning (FLAG +/− Ida) and second allogeneic stem cell transplant. Br J Haematol. 2001;115(3):622–629.

Rocha V, Cornish J, Sievers E, et al. Comparison of outcomes of unrelated bone marrow and umbilical cord blood transplants in children with acute leukemia. Blood 2001;97(10):2962–2971.

Sawczyn KK, Quinones R, Malcolm J, et al. Cord blood transplant in childhood ALL. Pediatr Blood Cancer 2005;45(7):964–970.

Seebach Jörg D, Stussi Georg, Horowitz Mary M, Olle Ringdén AJohn. Barrett and GVHD Working Committee of the Center for International Blood and Marrow Transplant Research. ABO Blood Group Barrier in Allogeneic Bone Marrow Transplantation Revisited. Biology of Blood and Marrow Transplantation 2005;11(12):1006–1013.

Slavin S, Nagler A, Naparstek E, et al. Nonmyeloablative stem cell transplantation and cell therapy as an alternative to conventional bone marrow transplantation with lethal cytoreduction for the treatment of malignant and nonmalignant hematologic diseases. Blood 1998;91(3):756–763.

Trigg ME, Bond R. Preparative therapies used for those with Fanconi's anemia undergoing stem cell transplants and subsequent engraftment rates. Hematology 2003;8(6):397–402.

Walters MC, Storb R, Patience M, et al. Impact of bone marrow transplantation for symptomatic sickle cell disease. Blood 2000;95(6):1918–1924.

Woods WG, Neudorf S, Gold S, et al. A comparison of allogeneic bone marrow transplantation, autologous bone marrow transplantation and aggressive chemotherapy in children with acute myeloid leukemia in remission. Blood 2001; 97: 56–62.

CHAPTER 30

Management of Oncologic Emergencies

Survival in children with cancer has increased dramatically during the past five decades. This progress is due not only to advances in specific oncologic therapies, but also to advances in supportive care and an improved ability to manage life-threatening complications.

Cancer therapy requires the management of the following conditions:

- Hyperleukocytosis and tumor lysis syndrome
- Hypercalcemia
- Syndrome of inappropriate antidiuretic hormone secretion/hyponatremia
- Superior vena cava syndrome and mediastinal masses
- Acute abdominal processes
- Neurologic emergencies.

Oncologic emergencies can occur as an initial manifestation of cancer, as a side effect of therapy, or at the time of progression or recurrence of the disease and arise as a result of the following:

- Metabolic and endocrine disturbances that result from malignancy or are secondary to therapy
- Space-occupying lesions that can press on or obstruct vital organs and cause mechanical or surgical emergencies
- Pancytopenia secondary to bone marrow replacement or chemotherapy that may result in hemorrhage, anemia and susceptibility to overwhelming infection.

METABOLIC AND ENDOCRINE EMERGENCIES

Hyperleukocytosis

A total white cell count greater than $100,000/mm^3$ is considered hyperleukocytosis. Hyperleukocytosis is seen at presentation in 9–13% of children with acute lymphocytic leukemia (ALL) and 5–22% of children with acute myeloid leukemia (AML). It occurs in an even higher percentage of patients with chronic myeloid leukemia. Hyperleukocytosis is associated with an adhesive reaction between abnormal endothelium and leukemic blasts leading to leukostasis, thrombosis and secondary hemorrhage.

Manual of Pediatric Hematology and Oncology. DOI: 10.1016/B978-0-12-375154-6.00030-6

The presence of blasts in the microcirculation leads to sludging which interferes with oxygenation of local tissue, ultimately leading to tissue ischemia. The higher metabolic rate of the blasts and the local production of cytokines also contribute to tissue hypoxia. Thrombi in the pulmonary circulation may lead to vascular damage and parenchymal ischemia manifested as pulmonary hemorrhage and edema. The presence of blasts in the cerebral circulation increases the risk of cerebral hemorrhage and cerebrovascular ischemia. Myeloblasts are generally larger, less deformable and more adherent to vasculature than lymphoblasts. Due to these intrinsic properties, leukostasis and thrombosis are far more prevalent in AML than in ALL. Patients with AML can have intracranial hemorrhage or thrombosis or pulmonary hemorrhage and leukostasis whereas in ALL hyperleukocytosis is more likely to lead to metabolic disturbance from tumor lysis syndrome (TLS).

The definition of leukostasis varies based on the disease process. Leukostasis is defined as clinical symptoms and signs in the presence of an absolute myeloblast count greater than $50,000/mm^3$ in AML, an absolute lymphoblast count of greater than $100,000/mm^3$ in ALL or lymphocyte count greater than $300,000/mm^3$ in CLL. Hyperleukocytosis is associated with early morbidity and mortality.

Clinical Features

1. Central nervous system (CNS): blurred vision, confusion, somnolence, delirium, stupor, coma and papilledema; computed tomography (CT) may reveal hemorrhage or leukemic plaques.
2. Pulmonary: tachypnea, dyspnea, hypoxia; chest radiograph may reveal varying degree of diffuse interstitial or alveolar infiltrates, leukemic emboli
3. Genitourinary: oliguria, anuria, priapism.
4. Vascular symptoms which include disseminated intravascular coagulation, retinal hemorrhage, myocardial infarction and renal vein thrombosis.

Risk Factors

1. Age <1 year.
2. M4, M5 AML (higher lysozyme activity).
3. Cytogenetic abnormalities 11q23, t(4:11), Philadelphia chromosome.

Tumor Lysis Syndrome (TLS)

Tumor lysis syndrome arises due to the rapid release of intracellular metabolites (such as phosphorous, potassium and uric acid) from necrotic tumor cells in quantities that exceed the excretory capacity of the kidneys.

In patients with high tumor burden, especially with Burkitt or Burkitt-like lymphoma, B-cell acute lymphoblastic leukemia and T-cell leukemia or lymphoma, uric acid nephropathy may commence prior to initiating therapy. Extremely rapid cell proliferation is accompanied by significant cell death and release of intracellular ions and may result in the following metabolic complications before starting chemotherapy:

*   Hyperuricemia
*   Hyperkalemia
*   Hyperphosphatemia
*   Hypocalcemia
*   Hypercalcemia
*   Renal failure.

If not successfully treated, tumor lysis syndrome can result in cardiac arrhythmias, renal failure, seizures, coma, disseminated intravascular coagulation (DIC) and death.

Diagnostic Criteria for Laboratory TLS (LTLS) and Clinical TLS (CTLS)

LTLS: The presence of two or more abnormal serum values at presentation (i.e. uric acid ≥ 8 mg/dl, potassium ≥ 6 mg/dl, phosphate ≥ 2.1 mmol/L, calcium ≤ 1.75 mmol/L) or change in serum values by 25% within 3 days before or 7 days after initiation of therapy.

CTLS: presence of LTLS and one or more of the following clinical complications: renal insufficiency, cardiac arrhythmias, seizures, or sudden death.

Recognition of risk factors, close monitoring and appropriate preventative intervention are vital in managing TLS.

Table 30-1 lists patient stratification by risk for tumor lysis syndrome for various types of cancer.

Treatment of Tumor Lysis Syndrome

1. Close laboratory monitoring:
    *   Patients at risk for TLS: serum lactate dehydrogenase, uric acid, sodium, potassium, creatinine, BUN, phosphorus, calcium, urine pH, specific gravity should be carried out every 8–12 hours for first 3 days then less frequently
    *   Patients with evidence of TLS: above laboratory investigations every 6 hours and then less frequently; venous blood gases, EKG, body weight every 24 hours.
2. Prevention:
    *   Fluids and hydration
        a. Promote excretion of uric acid and phosphorus; increase GFR and renal blood flow
        b. Hydration at a rate of >2 liters/m^2/day should start 24 hours before chemotherapy

Table 30-1   Patient Stratification by Risk for Tumor Lysis Syndrome for Various Types of Cancer

| Type of Cancer | Risk | | |
	High	Intermediate	Low
NHL	Burkitt, lymphoblastic, B-ALL	DLBCL	Indolent NHL
ALL	WBC $\geq$ 100,000/mm^3	WBC 50,000–100,000/mm^3	WBC $\leq$ 50,000/mm^3
AML	WBC $\geq$ 50,000/mm^3	WBC 10,000–50,000/mm^3	WBC $\leq$ 10,000/mm^3
CLL		WBC 10,000–100,000/mm^3 Tx w/fludarabine	WBC $\leq$ 10,000/mm^3
Other hematologic malignancies (including CML and multiple myeloma) and solid tumors		Rapid proliferation with expected rapid response to therapy	Remainder of patients

Abbreviations: NHL, non-Hodgkin lymphoma; B-ALL, Burkitt acute lymphoblastic leukemia; DLBCL, diffuse large B-cell lymphoma; ALL, acute lymphoblastic leukemia; AML, acute myeloid leukemia, CLL, chronic lymphocytic leukemia; Tx, treatment; CML, chronic myeloid leukemia.
From: Coiffer B, Altman A, Pui CH, Younes A, Guidelines for Management of Pediatric and Adult Tumor Lysis Syndrome.

    c. Urine output goal of 3 ml/kg/hour should be maintained. Diuretics may be needed to maintain this urine output
- Allopurinol
  a. A xanthine analog which blocks conversion of xanthine and hypoxanthine to uric acid; works as competitive inhibitor of xanthine oxidase
  b. Does not remove preformed uric acid
  c. May take days to reduce uric acid levels
  d. Decreases risk of uric acid crystallization in kidney tubules
  e. Adverse effects include hypersensitivity reaction and a build up of xanthine and hypoxanthine which may lead to renal insufficiency as well
- Rasburicase (only to be used in patients in the high risk group)
3. Active management:
- Table 30-2 outlines the management of hyperleukocytosis and tumor lysis syndrome.

Current Management of Various Metabolic Abnormalities in Tumor Lysis Syndrome

Hyperuricemia (>8 mg/dl)

1. Most common complication.
2. Caused by breakdown of purines into uric acid.
3. Results in precipitation of uric acid in distal renal tubules causing a decrease in glomerular filtration rate and ultimately acute renal failure.

Table 30-2   Management of hyperleukocytosis and tumor lysis syndrome

Management/ objective	Guidelines
Aggressive hydration[a]	Usually 2–4 times maintenance
Diuresis	Furosemide 0.5–1 mg/kg
	Mannitol 0.5 g/kg can be used if patient has oliguria unresponsive to increased hydration and furosemide
Uric acid reduction	Rasburicase (recombinant urate oxidase)[b] 0.15–0.2 mg/kg/day IV should be considered the first line agent in patients with hyperleukocytosis and TLS. Allopurinol 300 mg/m^2/day or 10 mg/kg/day PO (maximum dose 800 mg/day) or 200 mg/m^2/day IV (maximum dose 600 mg/day) may be used as an alternative agent, if rasburicase is unavailable
Leukocyte reduction	Leukopheresis or exchange transfusion (for infants) can be used if the patient is symptomatic or if the initial white cell count is greater than 100,000/mm^3. Leukopheresis may be difficult in younger children
Transfusion	Platelet transfusion to keep platelet count over 20,000/mm^3 to decrease the risk of intracranial hemorrhage. Avoid packed RBC transfusions, because it increases blood viscosity
	Fresh frozen plasma (FFP) transfusion and administration of vitamin K, if coagulopathy is present
Chemotherapy	Chemotherapy should be started when patient is stabilized and has adequate urine output
Dialysis	Dialysis is indicated for progressive renal failure with potassium >6 mEq/L, phosphate >10 mg/dl, oliguria, anuria and volume overload unresponsive to the above measures
Monitor	Electrolytes; calcium, phosphorus, uric acid, BUN, creatinine every 6 hours
	Complete blood counts every 12 hours
	Urine output, pH and specific gravity every 6 hours
	Respiratory, CNS and cardiac monitoring if hyperkalemia or hypocalcemia are present
Imaging	Brain CT with contrast if no renal insufficiency in the presence of neurologic symptoms or signs. MRI, magnetic resonance angiography or venography should be carried out if thrombosis is suspected

[a]Alkalinization of urine is no longer recommended. Evidence has shown that alkalinization has no effect on solubility of xanthine and hypoxanthine. In addition, alkalinization in combination with allopurinol may promote the crystallization of xanthine and calcium phosphate in the urine.
[b]Rasburicase converts uric acid to allantoin which is 5–10 times more soluble in urine compared to uric acid. It should not be used in patients with methemoglobinemia or G6PD deficiency (due to excessive hydrogen peroxide accumulation predisposing to oxidation of hemoglobin) and allopurinol should be withheld when Rasburicase is used. All uric acid levels drawn during Rasburicase therapy should be sent to the laboratory on ice.

4. Pharmacologic options for prevention of hyperuricemia include allopurinol and recombinant urate oxidase (rasburicase).

5. Pharmacologic treatment should be based on risk stratification as outlined in Figure 30-1 which shows a treatment algorithm for the prevention and management of hyperuricemia.

- For low-risk patients: Observation and frequent laboratory monitoring are required. Allopurinol should be initiated as first-line therapy at diagnosis
- For intermediate risk: Allopurinol should be initiated as first-line therapy at diagnosis. Rasburicase should be considered if laboratory or clinical evidence of TLS is present despite allopurinol
- For high risk: Rasburicase should be initiated at diagnosis for tumors with high proliferative index, large tumor burden, or tumors that are highly chemosensitive. Rasburicase should also be used as first-line therapy for patients with *clinical evidence* of TLS despite risk stratification.

6. Hydration >2 liters/m^2/day should start at admission and urine output goal of >100 ml/m^2/h should be maintained during chemotherapy. Strict measurement of intake and output should be carried out every 2–4 hours. Diuretics may be needed to maintain this urine output (i.e. furosemide 0.5–1 mg/kg/dose or mannitol 0.5 g/kg of the 25% solution over 5–10 minutes repeated every 6 hours as needed). If patient is unresponsive to these

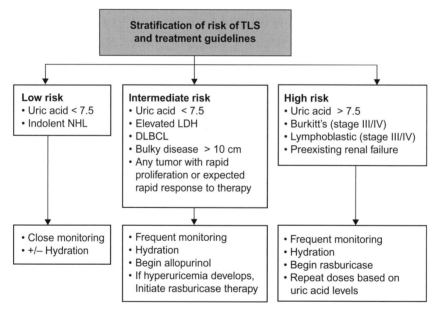

Figure 30-1 Treatment algorithm for the prevention and management of hyperuricemia. The treatment approach for low-risk patients is close observation with or without hydration. For those with intermediate risk, rasburicase is recommended if hyperuricemia develops despite prophylactic treatment with allopurinol. Vigorous hydration is recommended for all patients in the intermediate-to-high-risk groups, or those with diagnosed TLS. The use of rasburicase is recommended for the treatment of patients with hyperuricemia associated with diagnosed TLS, or in the initial management of patients considered to be at high risk of developing TLS.
Abbreviations: TLS, tumor lysis syndrome; NHL, non-Hodgkin lymphoma; LDH, lactate dehydrogenase; DLBCL, diffuse large B-cell lymphoma.
Adapted from: Hochberg J, Cairo MS. Rasburicase: future directions in tumor lysis management. Expert Opin Biol Therapy 2008;8(10):1595–1604, with permission.

interventions, continuous infusion of furosemide, beginning at 0.05 mg/kg/hour and titrated to desired effect should be considered.

Hyperkalemia (>5.0 mEq/L)

1. Caused by release of intracellular potassium during cell lysis coupled with the kidney's inability to clear the large amount of potassium effectively.
2. Potassium should not be administered until tumor lysis is controlled.
3. Mild (<6 mEq/L) and asymptomatic
   - Hydration and diuretics as described in hyperuricemia
   - Sodium polysterene sulfonate (Kayexelate) at dose of 1 g/kg every 6 hours with sorbitol 50–150 ml will remove 1 mEq of potassium per liter per gram of resin over 24 hours.
4. Moderate to severe hyperkalemia (>6 mEq/L)
   - Requires EKG
   - Rapid insulin (0.1 units/kg/h) plus glucose (dextrose 0.5 g/kg/h); in emergency cases, 50% dextrose can be used at 1 ml/kg through a central line. Monitor serum glucose closely
   - For life-threatening arrythmias IV calcium gluconate (100–200 mg/kg) or calcium chloride (10 mg/kg) slow IV infusion rate. The onset of action is within minutes and duration of activity lasts about 30 minutes (cannot be administered in the same line as $NaHCO_3$)
   - $NaHCO_3$ to stabilize myocardial cell membrane and to reverse acidosis; 1–2 mEq/kg IV. For every increase of 0.1 pH unit, potassium is decreased about 1 mEq/L. Onset of action is within 30 minutes and duration of action lasts several hours
   - Dialysis should be considered while these rescuing approaches are underway.

Hyperphosphatemia (>6.5 mg/dl)

1. Caused by release of intracellular phosphate during cell lysis.
2. Lymphoblasts contain four times the amount of phosphate present in normal lymphocytes; a low-phosphate diet should be prescribed.
3. Aluminum hydroxide 150 mg/kg/day divided every 4–6 hours should be administered. This will prevent absorption of oral phosphate intake, but has little direct effect on lowering serum phosphate. If serum calcium is not elevated use oral calcium carbonate (45–65 mg/kg/d PO divided qid) to avoid aluminum exposure. Sevelamer hydrochloride (RenaGel), a noncalcium phosphate binder (400 mg twice daily for older children) can also be used.
4. Hydration necessary to maintain a urine output of ≥3 ml/kg/hour.
5. Dialysis may be necessary to lower serum phosphate and prevent metastatic calcium deposition.

Hypocalcemia (Ionized Calcium <1.5 mEq/L)

1. As the product of serum calcium and phosphate increases over 60 due to hyperphosphatemia, a secondary hypocalcemia may occur due to calcium phosphate binding. At a calcium X phosphate product of >60 renal calcification can occur and exacerbate renal damage.
2. For symptomatic hypocalcemia (e.g., tetany), 10 mg/kg of elemental calcium (i.e. 0.5–1.0 ml/kg of 10% calcium gluconate) should be given. Calcium administration should be discontinued when symptoms resolve. Dialysis should be carried out if hyperphosphatemia persists.[*]

Hypercalcemia

1. Hypercalcemia has four main causes:
   * Osteolytic bone lesions (particularly in T-cell leukemia and lymphoma)
   * Bone demineralization secondary to parathyroid-like hormone (PTHrP) produced by tumors (paraneoplastic syndrome)
   * Immobilization
   * Defect in renal excretion.
2. Patients become symptomatic when serum calcium exceeds 12 mg/dl. Symptoms are nonspecific and include:
   * Anorexia, nausea, vomiting and constipation
   * Weakness
   * Coma
   * Pruritis
   * Bone pain
   * Polyuria, polydypsia, nephrogenic diabetes insipidus
   * Bradycardia, arrythmias
   * Dehydration, impaired renal function
   * DIC.
3. Treatment
   * Dehydration and electrolyte disturbances should be corrected. Stop calcium-containing medications
   * Renal calcium excretion should be increased by inducing diuresis with normal saline at two to threefold maintenance and furosemide 1–2 mg/kg/dose every 6 hours
   * Calcium mobilization from bone should be decreased by:
     a. Bisphosphonates such as pamidronate 0.5–1 mg/kg IV over 4–6 hours with very close monitoring of serum calcium, phosphate, magnesium for 2 weeks
     b. Prednisone 1.5–2.0 mg/kg daily (in lymphoproliferative disorders).

---

[*]Caution: Do not administer calcium in the same line as $NaHCO_3$.

Renal Failure

Mechanisms of renal failure include:

- Precipitation of urates in the acid environment of urine
- Precipitation of hypoxanthine when the urine pH exceeds 7.5
- Increase in the hypoxanthine levels after starting treatment with allopurinol
- Precipitation of calcium phosphate in renal microvasculature and renal tubules when the product of serum calcium and phosphate values exceeds 60.

Indications for dialysis include the following:

- Presence of hyperphosphatemia (>6 mg/dL) and hypercalcemia which promotes deposition in renal interstitium and tubular system, exacerbating kidney damage
- An estimated glomerular filtration rate (GFR) less than 50%
- Persistent hyperkalemia with QRS interval widening and/or level exceeding 6 mEq/L
- Severe metabolic acidosis
- Volume overload unresponsive to diuretic therapy
- Anuria and overt uremic symptoms (i.e. encephalopathy)
- Severe symptomatic hypocalcemia
- Hypertension (BP >150/90) and inadequate urine output at 10 hours from start of treatment
- Congestive heart failure.

Hemodialysis or hemofiltration should be used when renal failure occurs. Continous renal replacement therapy (CRRT) may be used for hemodynamically unstable patients because it is less likely to exacerbate hypotension. CRRT is not as effective for the treatment of hyperphosphatemia. Peritoneal dialysis should not be used. Dialyzable chemotherapy such as cyclophosphamide is given immediately after dialysis and not before. Renal dialysis usually needs to be repeated every 12 hours while there is continuous rapid tumor lysis.

Syndrome of Inappropriate Antidiuretic Hormone Secretion (SIADH)

1. Involves continuous pituitary release of antidiuretic hormone (ADH), irrespective of plasma osmolality.
2. Leads to significant hyponatremia, serum hyposmolality and water intoxication.
3. Results from physiologic stress, pain, surgery, mechanical ventilation, infections, CNS and pulmonary lesions, lymphomas and leukemias.
4. Occurs as a side effect of vincristine, vinblastine, cyclophosphamide, ifosfamide, cisplatin and melphalan.
5. Occurs in overhydration with hypotonic fluids, diabetes insipidus with free water replacement and cerebral salt wasting.

Clinical Features

*   Oliguria
*   Weight gain
*   Often asymptomatic
*   Early symptoms can include fatigue, headache and nausea
*   Late manifestations include lethargy, confusion, hallucinations, seizures and coma.

Laboratory features of SIADH:

*   Low serum osmolality (<280 mOsm/L)
*   High urine osmolality (>500 mOsm/L)
*   Urine to serum osmolality ratio >1
*   Hyponatremia (sodium, <130 mEq/L)
*   Increased urine specific gravity
*   Treatment
    a. Fluid restriction
    b. Furosemide 1 mg/kg should be administered to increase diuresis of free water and reduce the impact of the excess ADH
    c. Hydration with normal saline limited to insensible losses (500 mL/m^2/24 hours) plus ongoing losses
    d. In case of severe neurologic involvement (seizures or coma), hydrate carefully with hypertonic saline 3%. The rate of sodium correction should be limited to 2 mEq/L/h over the first two hours. Subsequent to that the rate of correction should target a change of serum sodium by 10 meq over the first twenty four hours and 18 meq by 48 hours of treatment. Too rapid correction may lead to further permanent neurologic sequelae.

ONCOLOGIC EMERGENCIES BY ANATOMIC REGION

Thoracic Emergencies: Superior Vena Cava Syndrome and Superior Mediastinal Syndrome

Superior vena cava syndrome (SVCS) consists of the signs and symptoms of superior vena cava (SVC) obstruction due to compression or thrombosis. This condition is frequently due to a large anterior mediastinal mass compressing the SVC. Rapid growth of the mediastinal mass does not permit the development of effective collateral circulation to compensate and results in the signs and symptoms of compression of the SVC.

Superior mediastinal syndrome (SMS) consists of SVCS with tracheal compression.

The trachea and main stem bronchus are more compressible in children making them more susceptible to SMS. In pediatrics, the term SVCS and SMS are usually used synonymously.

Etiology

1. Intrinsic causes: vascular thrombosis, e.g., following the introduction of a catheter.
2. Extrinsic causes: malignant anterior mediastinal tumors:
   - Hodgkin lymphoma
   - Non-Hodgkin lymphoma
   - Teratoma or other germ cell tumor
   - Thyroid cancer
   - Thymoma.

Clinical Features

1. Superior vena cava syndrome:
   - Swelling, plethora and cyanosis of the face, neck and upper extremities
   - Suffusion of the conjunctiva
   - Engorgement of collateral veins
   - Altered mental status.
2. Superior mediastinal syndrome:
   - Respiratory symptoms: cough, hoarseness, dyspnea, orthopnea, wheezing and stridor. Supine position worsens symptoms
   - Dysphagia
   - Chest pain
   - Altered mental status and syncope.

Management

1. Extreme care is required in handling the patient. The following may precipitate respiratory arrest:
   - Supine position (as for CT or operative procedures)
   - Stress
   - Sedation (conscious sedation, anxiolytics, or general anesthesia). The patient may have to be intubated. Extubation may not be possible until the anterior mediastinal mass has significantly decreased in size. Extra corporeal membrane oxygenation (ECMO) may be required if intubation is not possible.
2. Diagnosis should be made quickly in the least invasive manner:
   - Radiograph of the chest and CT (if tolerated)
   - Screening blood work such as CBC, LDH, uric acid, α-fetoprotein and β-hCG (to diagnose teratoma and rule out lymphoma)
   - Echocardiogram, if no evidence of mass on chest radiograph
   - Determine anesthesia risk. If high risk, perform the least invasive technique with local anesthesia (bone marrow, pleurocentesis, pericardiocentesis, lymph node

biopsy, or fine needle aspirate). If low risk use sedation or anesthesia and monitor closely.

3. Therapy:
   - Establishing a tissue diagnosis may not be possible and patients may need empiric treatment as a life-saving measure. Treatment options are radiotherapy and steroids even though both treatments may confound the diagnosis. Prednisolone 60 mg/m^2/day (2 mg/kg/day) or methylprednisolone 48 mg/m^2/day (1.6 mg/kg/day) divided into two daily doses should be employed. This will treat hematologic malignancies and decrease airway edema. Patient should undergo biopsy as soon as the mass shrinks and the patient is stable
   - Specific chemotherapy should be instituted after a biopsy has been obtained
   - For symptomatic venous thrombosis, with no evidence of hemorrhage, thrombolysis can be initiated using systemic or low-molecular-weight heparin (LMWH). (See Chapter 13 for more details)
     a. Unfractionated heparin can be started with a 75 U/kg bolus followed by 18 U/kg/ hour (for children) to 28 U/kg/hour (infants) continuous infusion. Titrate to a goal-activated partial thromboplastin time of 60–85 seconds or anti-Xa level of 0.3–0.7 U/ml
     b. Low-molecular-weight heparin 1 mg/kg every 12 hours. Titrate to a goal anti-Xa level of 0.5–1 U/ml. Convert to LMWH for outpatient therapy.

Abdominal Emergencies

1. Esophagitis: the most common gastrointestinal (GI) problem in oncology patients.
2. Gastric hemorrhage: especially in patients on corticosteroid therapy.
3. Typhlitis: seen only in neutropenic patients.
4. Perirectal abscess: in prolonged neutropenia.
5. Hemorrhagic pancreatitis: especially in patients on L-asparaginase therapy.
6. Massive hepatic enlargement from tumor: especially in infants with stage IVS neuroblastoma.

Table 30-3 provides an evaluation and management of common causes of abdominal pain.

Evaluation and Diagnosis

1. History regarding onset, timing, location and radiation of pain.
2. Observation and gentle examination including mouth and perirectal area in particular. If rectal examination is deemed necessary it should be performed very gently (the classic signs of an acute abdomen may be muted in a neutropenic patient or a patient on steroids).

Table 30-3   Evaluation and Management of Common Causes of Abdominal Pain

Diagnosis	Signs & Symptoms	Clinical Set-up	Evaluation	Management
Bowel obstruction	Pain Decreased bowel sounds	Tumor Adhesions	Abdominal Radiograph	Bowel rest/NG Surgical consultation
Constipation/ileus	Hard stools Pain	Narcotics Post-operative Vincristine	Abdominal Radiograph	Stool motility agents Stool softeners Suppository/enema
Gastritis/esophagitis	Gastric/throat pain Metallic taste	Chemotherapy, steroids Candida	Oral examination	Oral or IV antacid Anti-fungal
Hepatic enlargement	RUQ mass	Neuroblastoma 4S	CT, urine VMA/HVA	Chemotherapy
Pancreatitis	RUQ pain Emesis	L-Asparaginase, steroids Increased intracranial pressure	Amylase, lipase US/CT for pseudocyst	Bowel rest/NG
Perirectal abscess	Erythema Induration Pain with defecation	Severe myelosuppression	Peri-rectal examination	Broad-spectrum antibiotics (gram neg & anaerobes) Sitz baths
Typhilitis/colitis	Acute abdomen (RLQ pain) Bloody diarrhea Hypotension	Severe myelosuppression Acute leukemia	Abdominal Radiograph or CT (free air or thickened bowel wall) Stool cultures	Nothing by mouth Bowel rest/NG Surgical consultation Broad-spectrum antibiotics (gram neg, anaerobes, fungus)
Veno-occulsive disease	Hepatomegaly, RUQ mass, Edema, weight gain, Jaundice	Post-transplantation 6-thioguanine, Actinomycin Myelotarg	US for reversal of flow Bilirubin level Weight checks	Stop 6-thioguanine, Actinomycin Defibrotide

Abbreviations: RUQ, right upper quadrant of abdomen; RLQ, right lower quadrant of abdomen; US, ultrasound; NG, naso-gastric tube suctioning.

3. Serial blood counts to evaluate for hemorrhage, inflammation, infection.
4. Blood, stool and urine cultures as indicated.
5. Laboratory tests, liver enzymes, bilirubin, amylase, lipase, electrolytes.
6. Vital signs monitoring.
7. Abdominal radiography: sonography, CT and magnetic resonance imaging (MRI) as indicated.

Typhlitis

1. Typhlitis, a necrotizing colitis localized in the cecum, occurs in the setting of severe neutropenia, particularly in patients with leukemia and in stem cell transplant recipients.
2. It should be strongly suspected in patients with right lower quadrant pain or the development of a partially obstructive right lower quadrant mass.
3. It is a result of bacterial or fungal invasion of the mucosa and can quickly progress from inflammation to full-thickness infarction to perforation, peritonitis and septic shock.
4. The responsible pathogens include *Pseudomonas* species, *Escherichia coli*, other Gram-negative bacteria, *Staphylococcus aureus*, α-hemolytic *Streptococcus*, *Clostridium*, *Aspergillus*, and *Candida*.
5. Typhilitis in patients receiving chemotherapy is linked to mucosal injury caused by cytotoxic chemotherapeutic agents.
6. Diagnosis.
    *   Typhlitis is usually diagnosed clinically when a neutropenic patient presents with right lower quadrant pain
    *   Physical examination may reveal an absence of bowel sounds, bowel distention, tenderness on palpation maximal in right lower quadrant, or a palpable mass in right lower quadrant. Serial abdominal examinations are required
    *   Imaging studies may aid in the diagnosis of typhlitis:
        a. Radiograph of the abdomen may reveal pneumatosis intestinalis, free air in the peritoneum or bowel wall thickening
        b. Ultrasonography may reveal thickening of the bowel wall in the region of the cecum and is becoming a more commonly used non-radiation modality to image for typhlitis
        c. CT scan is the definitive imaging study and may demonstrate diffuse thickening of the cecal wall
        d. Barium enema may show severe mucosal irregularity, rigidity, loss of haustral markings and occasional fistula formation.
7. Treatment.
    *   Medical management is the initial treatment, consisting of:
        *   Discontinuation of oral intake
        *   Nasogastric tube suctioning

- Broad-spectrum antibiotics (anaerobic and Gram-negative coverage)
- Intravenous fluid and electrolytes
- Packed red cell and platelet transfusions, as indicated
- Vasopressors, as needed (hypotension is associated with a poor outcome).
- Indications for surgical intervention:
  - Persistent GI bleeding despite resolution of neutropenia and thrombocytopenia
  - Evidence of free air in the abdomen on abdominal radiograph (indicating perforation)
  - Clinical deterioration requiring fluid and pressor support, indicating uncontrolled sepsis from bowel infarction.

  Surgery consists of removing necrotic positions of the bowel and diversion via colostomy. Healing can occur with fibrosis and stricture formation

  Mortality is related to bowel perforation, bowel necrosis and sepsis.

Perirectal Abscess

Inflammation and infection of the rectum and perirectal tissue occur commonly in patients receiving chemotherapy or radiation therapy, especially in patients with prolonged neutropenia. Most abscesses are caused not by a single organism but rather by a combination of aerobic organisms, such as staphylococci, streptococci, *E. coli*, *Pseudomonas* and anaerobic Gram-positive and -negative organisms.

Presentation includes anorectal pain, tenderness and painful bowel movements. An abscess or draining fistula may be present; however, in the neutropenic patient, pus will be absent and the patient will present with a brawny edema and dense cellulitis.

Management

1. Initial therapy with intravenous antibiotics to cover Gram-negative organisms and anaerobes.
2. Granulocyte colony-stimulating factor (G-CSF) to shorten period of neutropenia.
3. Sitz baths four times a day and meticulous attention to perirectal hygiene.
4. Surgical incision and drainage of obviously fluctuant areas or draining fistulas that do not resolve with medical management.

Neurologic Emergencies

Neurologic emergencies, such as seizure, alterations in mental status (AMS), cerebrovascular accidents (CVA), spinal cord compression and increased intracranial pressure (ICP) occur in over 10% of pediatric oncology patients. Initial intervention for acute neurologic deterioration requires immediate stabilization of the patient.

Evaluation

1. Detailed history including current medications, recent chemotherapy and radiation therapy, prior events.
2. Thorough physical and neurologic examination, pulse oximetry, vital signs.
3. Laboratory tests – blood count, electrolytes, ammonia, liver and renal function tests.
4. Head CT acutely, MRI/MRA.
5. Spinal tap.
6. Electroencephalogram.

Management

1. Stabilize the patient – oxygen and hydration as needed.
2. Stop any intravenous infusions, especially chemotherapy, narcotics.
3. Transfuse platelets if thrombocytopenic.
4. Start broad-spectrum antibiotics after blood culture is obtained, a spinal tap can be performed once the CT reveals normal pressure.
5. If actively having seizures administer ativan or dilantin load.
6. Emergent head CT to look for CNS bleed, hydrocephalus, herniation, or mass.
7. If nonhemorrhagic CVA is found on MRI, thrombolysis and anticoagulation should be considered.

Differential Diagnosis

1. Seizures – CNS tumors, intrathecal chemotherapy, metabolic derangements.
2. Raised intracranial pressure – CNS tumors, blocked shunt, pseudotumor cerebri, infection.
3. Cerebrovascular accident (CVA) – L-asparaginase, hyperleukocytosis, coagulopathy/ hemorrhage, radiation-induced vasculopathy.
4. Alterations in mental status (AMS) – CNS tumor, opiates, benzodiazepines, steroids, intrathecals, high-dose cytarabine or methotrexate, ifosfamide, nelarabine, post-ictal, CVA, CNS infection, metabolic derangements, postradiation somnolence.

Spinal Cord Compression

Incidence and Etiology

1. Three to five percent of children with cancer develop acute spinal cord or cauda equina compression. Sarcomas account for about half of the cases of spinal cord involvement in childhood; the remainder are caused by neuroblastoma, germ cell tumor, lymphoma, leukemia and drop metastasis of CNS tumors.
2. The spinal cord can be compressed by tumor in the epidural or subarachnoid space or by metastases within the cord parenchyma.

The mechanisms of tumor spread to the spinal cord include:

* Direct extension of the tumor
* Metastatic spread to the vertebrae with secondary cord compression
* Spread to the epidural space via infiltration of the vertebral foramina
* Subarachnoid spread down the spinal cord from primary CNS tumor (such as medulloblastoma).

Clinical Presentation

1. Back pain with localized tenderness occurs in 80% of patients.
2. Incontinence, urinary retention and other abnormalities of bowel or bladder function are not frequent if spinal compression is diagnosed early.
3. Loss of strength and sensory deficits with a sensory level may also occur.
4. Any child with cancer and back pain should be presumed to have spinal cord involvement until further workup indicates otherwise.

Evaluation

1. A thorough history and neurologic examination should be included in the evaluation.
2. Spinal radiographs are useful if the compression is due to vertebral metastases but they will miss epidural disease in 50% of cases.
3. MRI with and without gadolinium is necessary to detect the presence and extent of epidural involvement.
4. Lumbar puncture myelography is an older procedure not necessary since MRI is now readily available.
5. Cerebrospinal fluid (CSF) analysis is important in the evaluation of subarachnoid disease, but it is not helpful in localizing epidural disease.

Treatment

Because the potential for permanent neurologic damage is high, it is crucial to initiate treatment immediately.

* Dexamethasone is initiated to decrease local edema, prior to diagnostic
  * In the presence of neurologic abnormalities, immediately start dexamethasone 1–2 mg/kg/day loading dose. Follow with 1.5 mg/kg/day divided every 6 hours and obtain emergent MRI
  * With back pain and the absence of neurologic symptoms, start dexamethasone 0.25–1 mg/kg/dose every 6 hours and perform MRI within 24 hours
* If an epidural mass is identified, treatment is aimed at rapid decompression Chemotherapy, radiation therapy, or surgical decompression may be used
  * If tumor is known to be radiosensitive, give local radiation including the full volume of the tumor plus one vertebra above and below the lesion

- Three daily doses of 300 cGy for a total of 3,000–4,000 cGy, depending on the responsiveness of the tumor
- Surgical emergent laminotomy or laminectomy may be indicated for rapid decompression especially in lesions expected to be less radioresponsive (e.g. sarcoma)
- Specific chemotherapy can be instituted in addition to the use of radiation and dexamethasone in lymphoma, leukemia and neuroblastoma.

Suggested Reading

Albanese A, Hindmarsh P, Stanhope R. Management of hyponatraemia in patients with acute cerebral insults. Arch Dis Child 2001;85(3):246–251.

Anghelescu DL, et al. Clinical and diagnostic imaging findings predict anesthetic complications in children presenting with malignant mediastinal masses. Paediatr Anaesth. 2007;17(11):1090–1098.

Antunes NL, DeAngelis LM. Neurologic consultations in children with systemic cancer. Pediatr Neurol. 1999;20(2):121–124.

Bowers DC, et al. Late-occurring stroke among long-term survivors of childhood leukemia and brain tumors: a report from the Childhood Cancer Survivor Study. J Clin Oncol. 2006;24(33):5277–5282.

Coiffier B, et al. Guidelines for the management of pediatric and adult tumor lysis syndrome: an evidence-based review. J Clin Oncol. 2008;26(16):2767–2778.

Cole JS, Patchell RA. Metastatic epidural spinal cord compression. Lancet Neurol. 2008;7(5):459–466.

DiMario Jr FJ, Packer RJ. Acute mental status changes in children with systemic cancer. Pediatrics 1990;85(3):353–360.

Fisher MJ, et al. Diffusion-weighted MR imaging of early methotrexate-related neurotoxicity in children. AJNR Am J Neuroradiol. 2005;26(7):1686–1689.

Ho VT, et al. Hepatic veno-occlusive disease after hematopoietic stem cell transplantation: review and update on the use of defibrotide. Semin Thromb Hemost. 2007;33(4):373–388.

Kaste SC, Rodriguez-Galindo C, Furman WL. Imaging pediatric oncologic emergencies of the abdomen. AJR Am J Roentgenol. 1999;173(3):729–736.

Kearney SL, et al. Clinical course and outcome in children with acute lymphoblastic leukemia and asparaginase-associated pancreatitis. Pediatr Blood Cancer 2009;53(2):162–167.

Kerdudo C, et al. Hypercalcemia and childhood cancer: a 7-year experience. J Pediatr Hematol Oncol. 2005;27(1):23–27.

Lowe EJ, et al. Early complications in children with acute lymphoblastic leukemia presenting with hyper-leukocytosis. Pediatr Blood Cancer 2005;45(1):10–15.

McCarville MB, et al. Typhlitis in childhood cancer. Cancer 2005;104(2):380–387.

Monagle P, et al. Antithrombotic therapy in neonates and children: American College of Chest Physicians Evidence-Based Clinical Practice Guidelines. 8th ed. Chest 2008;133(6 Suppl):887S–968S.

Perger L, Lee EY, Shamberger RC. Management of children and adolescents with a critical airway due to compression by an anterior mediastinal mass. J Pediatr Surg. 2008;43(11):1990–1997.

CHAPTER 31

Supportive Care of Patients with Cancer

MANAGEMENT OF INFECTIOUS COMPLICATIONS

Children being treated for cancer are rendered significantly immunocompromised because of chemotherapy. For this reason, it is extremely important to expeditiously evaluate and treat the febrile oncology patient. Any delay in antibiotic therapy while waiting culture results may lead to uncontrolled progression of infection in a neutropenic patient.

Factors leading to the susceptibility to infections in oncology patients are:

- Underlying disease: Patients with leukemia, advanced-stage lymphoma and uncontrolled tumors are more prone to infections
- Type of therapy: Dose-intensive therapies, high-dose cytosine arabinoside and stem cell transplantation render patients more susceptible to infections
- Degree and duration of neutropenia: The most important determinant of susceptibility to bacterial and fungal infections is the number of circulating neutrophils. Patients who are neutropenic (absolute neutrophil count [ANC] $<500/mm^3$) are more susceptible to infection. Neutropenia can be secondary to disease (leukemia, aplastic anemia) or chemotherapy. Neutrophil function may also be impaired by disease and by chemotherapy
- Disruption of normal barriers: The normal mechanical barriers to infection in the skin, respiratory, gastrointestinal and genitourinary systems are disrupted
- Nutritional status: Malnutrition affects the function of lymphocytes, neutrophils, mononuclear cells and the complement system
- Humoral immunity: Defects in humoral immunity result in susceptibility to encapsulated bacteria including *Streptococcus pneumoniae*, *Haemophilus influenzae* type b and *Neisseria meningitidis*
- Cell-mediated immunity: Defects in cellular immunity produce susceptibility to viruses, fungi and intracellularly multiplying bacteria (e.g., *Listeria*, *Salmonella* and *Mycobacterium tuberculosis*). Patients with Hodgkin disease and non-Hodgkin

lymphoma have impaired cell-mediated immunity. Chemotherapy, radiation and corticosteroids induce defects in lymphocyte function. Lymphopenia may persist after completion of chemotherapy
- Colonizing microbial flora: Most bacterial infections arise from endogenous microflora
- Foreign bodies: Indwelling central vascular catheters, ventriculoperitoneal shunts

Table 31-1 lists the predominant pathogens in cancer patients.

Table 31-1   Predominant Pathogens in Cancer Patients

Gram-positive Bacteria
  Staphylococci (coagulase-negative, *Staphylococcus aureus* including methicillin-resistant *S. aureus*)
  Streptococci (α-hemolytic) especially *Streptococcus mitis*
  Enterococci
  Corynebacteria
  Listeria sp.
Gram-negative Bacteria
  Enterobacteriaceae (*Escherichia coli, Klebsiella, Enterobacter, Serratia*)
  Pseudomonas aeruginosa, Stenotrophomonas maltophilia (and other oxidase-positive multi-resistant
    Gram-negative bacteria)
Anaerobic Bacteria
  Clostridium difficile
  Bacteroides sp.
  Proprionobacterium acnes
Fungi
  Candida sp.
  Aspergillus sp.
  Zygomycetes
  Cryptococci
  Pneumocystis jiroveci (formerly *P. carinii*)
Other
  Toxoplasma gondii
  Strongyloides stercoralis
  Cryptosporidium
Viruses
  Herpes simplex virus
  Varicella–zoster virus
  Cytomegalovirus
  Epstein–Barr virus
  Respiratory syncytial virus
  Adenovirus
  Influenza virus
  Parainfluenza virus
  Papovaviruses BK and JC

Modified from: Alexander SW, Walsh TJ, Freifield AG, Pizzo PA. Infectious complications in pediatric cancer patients. In: Pizzo PA, Poplack DG, editors. Principles and Practice of Pediatric Oncology. 4th ed. Philadelphia: Lippincott Williams & Wilkins, 2002; 1239–1284, with permission.

Fever is defined as a temperature greater than or equal to 38.3°C (101°F) occurring once or a temperature greater than or equal to 38°C (100.4°F) occurring twice during a 24-hour period. To appropriately evaluate and manage fever in oncology patients it is helpful to classify febrile patients in one or more of the following four categories:

- Non-neutropenic patients
- Patients with an indwelling catheter
- Neutropenic patients
- Splenectomized patients.

Febrile Non-Neutropenic Patients

The non-neutropenic patient remains susceptible to infections secondary to lymphocyte dysfunction. Evaluation should include a careful history and physical examination. Blood and urine cultures should be obtained. In addition, patients with localized findings should undergo the appropriate diagnostic procedures (e.g., throat culture, stool culture, chest radiograph). If the patient does not have an indwelling catheter and appears well, empiric antibiotic therapy is rarely indicated. They should be followed closely for clinical or microbiological evidence of infection. If infection is documented, they should be treated with appropriate antibiotics.

Febrile Patients with Indwelling Venous Catheters

If the patient has a subcutaneously tunneled catheter such as a Broviac or Hickman, or a totally implantable venous access device or "port" (e.g., Mediport) and is neutropenic, the guidelines for the neutropenic patient should be followed. If not neutropenic, the patient should be managed as follows:

- Examine the site for evidence of inflammation at the exit site or on the skin overlying the subcutaneously tunneled portion of the catheter. If there is pus at the exit site, collect and send for Gram-stained smear and culture
- Obtain blood for culture from each lumen of the catheter and from a peripheral venous site
- A broad-spectrum, third-generation cephalosporin (ceftriaxone) is appropriate if there is no sign of tunnel infection and the patient does not appear ill. If there is tunnel infection, the catheter should be removed and the patient treated with intravenous antibiotics following catheter removal. If the patient appears seriously ill, broad-spectrum antimicrobial therapy that includes vancomycin should be administered and catheter removal should be strongly considered. The patient's medical record should be checked for prior catheter infections or blood culture isolates and the antimicrobial

susceptibility of previously isolated microorganisms should be considered in the selection of antibotics

• Cultures should be followed for 48–72 hours; if they are negative, antibiotics should be discontinued.

Febrile Neutropenic Patients

Patients with an ANC of less than 500/mm^3 and those with an ANC of less than 1,000/mm^3 and decreasing are considered neutropenic. Because of the high risk of morbidity and mortality in the febrile neutropenic patient, the workup should be thorough and expeditious and empiric broad-spectrum antibiotics should be started as soon as possible.

Management

1. Management includes careful history and physical examination with special attention to important sites of infection in the neutropenic patient:
   • Oral mucosa and periodontium
   • Pharynx
   • Lower esophagus
   • Lungs
   • Skin including sites of vascular access, bone marrow aspiration sites, tissue around the nails
   • Perineum and anus.
2. In the neutropenic patient, signs of inflammation may be remarkably minimal but pain is usually preserved. Accordingly, even subtle signs of inflammation or a complaint of localized pain without physical findings must be considered as a potential source of infection and the site should be evaluated and cultured as appropriate.
3. Obtain a complete blood count, serum chemistries, urinalysis and urine culture.
4. Obtain blood cultures from all lumens of indwelling catheters and from a peripheral vein.
5. Obtain a chest radiograph.
6. Start broad-spectrum antibiotic therapy. Broad-spectrum therapy is required in the neutropenic patient even if a single source or pathogen is isolated, because the occurrence of infection with a second pathogen during continuing neutropenia is not infrequent.
7. These patients should be reassessed daily including an assessment of new symptoms, a meticulous daily physical examination, review of previous culture results, examination of the status of vascular catheters, additional cultures and diagnostic imaging of any organ suspected of having infection.

Empiric Antibiotic Therapy

The prompt initiation of antibiotic therapy in febrile neutropenic patients has reduced the mortality rate for Gram-negative infections significantly. Every empiric initial antibiotic regimen should include drugs with antipseudomonal activity. Because Gram-positive, Gram-negative and mixed infections occur, broad-spectrum coverage must be provided. The choice of antibiotics should be influenced by the predominant organisms and antibiotic susceptibility patterns at the local hospital. Acceptable initial regimens include:

*   Monotherapy with cefepime, ceftazidime, imipenem, or meropenem (generally, imipenem or meropenem should not be prescribed routinely as empiric therapy for neutropenic patients in order to minimize the development of antimicrobial resistance to these broad-spectrum agents)
*   Two-drug regimen: ticarcillin/clavulanate, piperacillin/tazobactam, cefepime, or ceftazidime AND gentamicin (or tobramycin or amikacin).

Table 31-2 shows the dose, route of administration and schedule of frequently used antibiotics. Initial therapy exclusively with oral antibiotics is not recommended in children. Figure 31-1 provides an algorithm for the management of the febrile neutropenic patient.

Many febrile neutropenic patients have no identifiable cause for their fever. A number of risk factors have been identified in febrile neutropenic children that correlate with their risk of serious or life-threatening infection. Risk assessment of patients with fever of unexplained origin helps further define treatment options. Patients may be divided into two groups, low risk and high risk, based on the granulocyte count, anticipated duration of neutropenia, degree of fever, remission status and initial physical examination.

1. Low-risk patients: The definition of low risk is variable but has included patients with an ANC >100 neutrophils/mm^3, an anticipated length of neutropenia less than 10 days, a nontoxic appearance, no identifiable source of fever, a normal chest radiograph, no comorbidities and a malignancy in remission status. In these patients, antibiotics may be stopped when the absolute neutrophil count reaches 500/mm^3. For patients with the availability of close ambulatory follow-up, physicians may consider discontinuing antibiotics when the absolute neutrophil count rises for two consecutive days.
2. High-risk patients: Patients with neutropenia that is expected to persist for more than 10 days and there is no evidence of bone marrow recovery. If the patient is afebrile but remains granulocytopenic, antibiotics should be continued until absolute neutrophil count reaches 500/mm^3. If the patient is continuously febrile for 5 days, antifungal treatment should be added.

Table 31-2   Frequently Used Antibiotics in Oncology and Stem Cell Transplantation Patients

Drug	Dose	Route	Schedule
Aminoglycosides			
Amikacin	22 mg/kg/day	IV	Divided q8h
Gentamicin	6.0–7.5 mg/kg/day	IV	Divided q8h
Tobramycin	6.0–7.5 gm/kg/day	IV	Divided q8h
Penicillin-related Drugs			
Nafcillin	100–200 mg/kg/day	IV	Divided q4–8h
Piperacillin/tazobactam	300 mg/kg/day of piperacillin component	IV	Divided q6h
Ticarcillin/clavulanate	300 mg/kg/day (maximum 24 g/day)	IV	Divided q4–6h
Cephalosporins			
Ceftazidime	100–150 mg/kg/day	IV	Divided q8h
Cefazolin	50–100 mg/kg/day (maximum 6 g/day)	IV	Divided q8h
Cefepime	150 mg/kg/day (maximum 2 g)	IV	Divided q8h
Carbapenems*			
Imipenem/cilastatin	50 mg/kg/day (maximum 4 g/day)	IV	Divided q6–8h
Meropenem	60–120 mg/kg/day (maximum dose 6 g/day)	IV	Divided q8h
Other			
Vancomycin	45 mg/kg/day	IV/PO	Divided q8h
Linezolid**	30 mg/kg/day (<2 years of age)	IV/PO	Divided q8h
	20 mg/kg/day (≥2 years of age) maximum dose 1.2 g/day)	IV/PO	Divided q 12h
Anaerobic Drugs			
Clindamycin	40 mg/kg/day	IV	Divided q6–8h
Metronidazole	30 mg/kg/day (loading dose initially 15 mg/kg)	IV	Divided q6h

*Generally reserved for patients with infections resistant to other agents.
**Prescribe with caution because of myelosuppression associated with this drug.
Modified from: Wolff LJ, Altman AJ, Berkow RL, Johnson FL. The management of fever and neutropenia. In: Altman AJ, editor. Supportive care of children with cancer. Current therapy and guidelines from the Children's Oncology Group. 3rd ed. Baltimore: The Johns Hopkins University Press, 2004; 200–220, with permission.

In a select group of low-risk patients oral therapy with amoxicillin/clavulante plus ciprofloxacin has demonstrated equal efficacy to therapy with IV cephalosporin with or without aminoglycoside coverage. Early discharge of low-risk neutropenic patients with oral antibiotic coverage or with observation alone may be appropriate in a select group of low-risk patients. Amoxicillin/clavulanate may be valuable for the patient in palliative or end-of-life situations.

Antibiotic therapy may need to be modified according to the patient's clinical status, focus for infection, or results of cultures and susceptibility testing. Empiric antifungal coverage should be initiated for patients with neutropenia and fever that persists for 3–7 days despite broad-spectrum antibacterial therapy (see "Fungal Infections" section below). Institution of empiric antifungal therapy has been found to decrease the risk for mortality significantly.

Table 31-3 shows modification of initial antimicrobial regimens for febrile neutropenic cancer patients under various circumstances.

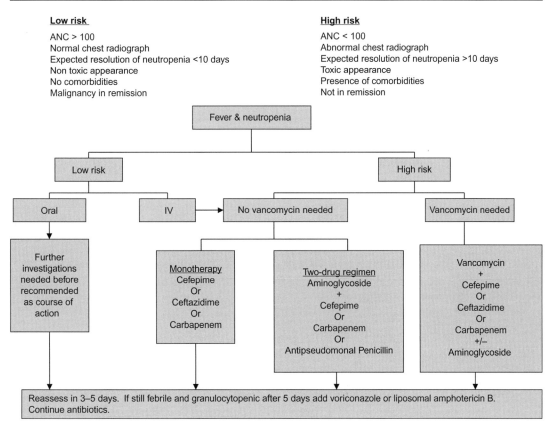

Figure 31-1 Algorithm for the Management of the Febrile Neutropenic Patient. Adapted from: Hughes et al. Clinical Infect. Dis 2002;34:730–751.

Febrile Splenectomized Patients

The spleen acts both as a mechanical filter and as an immune effector organ. Immune functions of the spleen are as follows:

- Production of antibodies to polysaccharide antigens
- Removal of damaged cells and opsonin-coated organisms from the circulation.

Splenectomized patients are immunocompromised in the following ways:

- They are deficient in antibody production when challenged with particulate antigens
- They have decreased levels of immunoglobulin M (IgM) and properdin
- They are deficient in the phagocytosis-promoting protein tuftsin. They are at risk for fulminant and rapidly fatal septicemia due to encapsulated bacteria, including *S. pneumoniae, H. influenzae* type b and *N. meningitidis*. Other pathogens which may

Table 31-3 **Modifications of Initial Antimicrobial Regimen for Febrile Neutropenic Cancer Patients**

Status of Symptom	Modifications of Primary Regimen
Fever	
Persistent for >5 d	Add empiric antifungal therapy
Recurrence after ≥5 d in patient with persistent neutopenia	Add empiric antifungal therapy. Consider broadening spectrum of antibiotics
Persistent or recurrent fever at time of recovery from neutropenia	Image liver and spleen for hepatosplenic candidiasis using computed tomography, ultrasonography, or magnetic resonance imaging and evaluate need for antifungal therapy
Bloodstream Infection (Obtain Cultures *Before* Antibiotic Therapy)	
Gram-positive organism	Add vancomycin pending further identification
Gram-negative organism	Maintain regimen if patient is stable and isolate is susceptible
	If *Pseudomonas aeruginosa, Enterobacter, Serratia,* or *Citrobacter* is isolated, add an aminoglycoside or if resistant to cephalosporin, add a carbapenem if patient is stable and isolate is susceptible
Organism Isolated *During* Antibiotic Therapy	
Gram-positive organism	Add vancomycin
Gram-negative organism	Change to new combination regimen (e.g., imipenem plus gentamicin or piperacillin/tazobactam plus amikacin)
Head, Eyes, Ears, Nose, Throat Infection	
Necrotizing or marginal gingivitis	Add specific antianaerobic agent (clindamycin or metronidazole) to empiric therapy or change Gram-negative coverage to carbapenem
Vesicular or ulcerative lesions	Suspect herpes simplex infection. Culture and initiate acyclovir therapy
Sinus tenderness or nasal ulcerative lesions	Suspect fungal infection with aspergillus or mucor. Consider endoscopy for biopsy and culture
Gastrointestinal Tract Infection	
Retrosternal burning pain	Suspect esophagitis due to *Candida,* herpes simplex, or both. Add antifungal therapy and, if no response, acyclovir. Bacterial esophagitis is also a possibility. For patients who do not respond within 48 hours, endoscopy should be considered
Acute abdominal pain	Suspect typhlitis, as well as appendicitis, if pain in right lower quadrant. Stool for *C. difficile.* Add specific antianaerobic coverage to empirical regimen and monitor closely for need for surgical intervention, especially when patient is recovering from neutropenia
Perianal tenderness	Add drug with anaerobic activity to empirical regimen and monitor need for surgical intervention, especially when patient is recovering from neutronpenia
Lung Infection	
New focal lesion in patient recovering from neutropenia	Observe carefully, because this may be a consequence of inflammatory response in concert with neutrophil recovery
New focal lesion in patient with continuing neutropenia	*Aspergillus* is the chief concern. Perform appropriate cultures and consider biopsy. If patient is not a candidate for a procedure,

(Continued)

Table 31-3   (Continued)

Status of Symptom	Modifications of Primary Regimen
New interstitial pneumonitis	consider therapy with voriconazole with or without a second agent such as caspofungin or liposomal amphotericin Attempt diagnosis by examination of induced sputum or bronchoalveolar lavage. If not feasible, begin empiric treatment with trimethoprim-sulfamethoxazole (alternative: pentamidine). Consider infectious causes and the need for open lung biopsy if condition has not improved after 4 days of therapy

Adapted from: Alexander SW, Walsh TJ, Freifeld AG, Pizzo PA. Infectious complications in pediatric cancer patients. In: Pizzo PA, Poplack DG, eds. Principles and Practice of Pediatric Oncology. 4th ed. Philadelphia: Lippincott Williams & Wilkins, 2002; 1239–1284, with permission.

cause septicemia are *Salmonella* species, *Capnocytophaga canimorsus* and intraerythrocytic parasites such as *Babesia microti* and *Plasmodium falciparum*.

Overwhelming postsplenectomy infection is a serious complication of asplenia. Infection may rapidly progress to bacteremic septic shock, accompanied by hypotension, anuria and disseminated intravascular coagulation.

Management

To decrease the risk of overwhelming postsplenectomy infection, the following measures are recommended:

- Vaccination: Two weeks or longer before splenectomy, *Haemophilus influenzae* type b vaccine, the 23-valent pneumococcal polysaccharide vaccine and quadrivalent meningococcal conjugate vaccine should be administered. If pneumococcal conjugate vaccine has not been given as part of early childhood immunization then the age-related appropriate number of doses should be given followed by the 23-valent pneumococeal polysaccharide vaccine. If the splenectomy is elective, administration of indicated vaccines should be completed 2 weeks before surgery to increase the likelihood of eliciting an optimal antibody response. Unimmunized patients who have had splenectomy should be immunized as soon as possible. A second dose of pneumococcal polysaccharide vaccine is recommended 5 years after the initial dose

- Antibiotic prophylaxis: Oral penicillin (125 mg twice daily for children under 3 years; 250 mg twice daily for children 3 years of age and older) should be prescribed at least until the child is 5 years of age for children with functional asplenia due to sickle cell disease. For asplenic children the appropriate duration of prophylaxis has not been established. For patients 5 years of age or older who have undergone splenectomy due to trauma, discontinuation of prophylaxis can be considered at least 1 year after splenectomy. For children postsplenectomy who have underlying disorders (e.g., thalassemia) prophylaxis should be continued long term and perhaps lifelong. Amoxicillin is an alternative to penicillin

- If the splenectomized patient becomes febrile:
  a. Obtain two blood cultures
  b. Start broad-spectrum IV antibiotic therapy (the currently recommended drug is ceftriaxone 2 g IV every 24 hours for adults or 75 mg/kg IV every 24 hours plus vancomycin until infection with ceftriaxone-resistant pneumococcus is ruled out) Diagnostic evaluation should not delay the initiation of antibiotic therapy
  c. Continue antibiotic therapy until the blood cultures are negative for 72 hours and the patient is afebrile for 24–48 hours
  d. If the patient is febrile and neutropenic, manage as for the febrile neutropenic patient, but include an antibiotic with excellent activity against pneumococci.

BACTERIAL INFECTIONS

Table 31-4 shows the common pathogens and treatment of various infections in children with cancer. Table 31-2 shows the dose, route of administration and schedule of frequently used antibiotics. Both the penicillin/beta-lactamase inhibitor drugs, ticarcillin/clavulanate and piperacillin/tazobactam and the carbapenems, imipenem and meropenem, provide broad-spectrum activity for Gram-positive pathogens, Gram-negative pathogens and anaerobic bacteria and are good choices for empiric therapy of patients with suspected intra-abdominal sepsis, with or without administration of an aminoglycoside. However, neither imipenem nor meropenem should be prescribed routinely in order to minimize the development of antimicrobial resistant to these very broad-spectrum agents.

Vancomycin is not recommended *routinely* for empiric therapy because its use promotes the emergence of resistant strains such as vancomycin-resistant enterococci. Vancomycin should only be added to the initial regimen if any of the following is present:

- Suspected central venous catheter-related infection
- Patients with acute myeloblastic leukemia because of the increased risk of infection with alpha hemolytic streptococcus such as *Streptococcus mitis*
- Prior history of alpha hemolytic streptococcus bacteremia
- Patients colonized with resistant organisms that are treatable only with vancomycin
- Patients with a recent history of bacteremia or venous catheter-related infection requiring treatment with vancomycin
- Intensive chemotherapy causing mucositis (such as high-dose cytosine arabinoside)
- Hypotension
- Patients who have developed fever despite quinolone prophylaxis.

If vancomycin is prescribed and cultures do not grow a pathogen requiring vancomycin, it should be discontinued within 48–72 hours.

Table 31-4   Common Pathogens and Treatment of Various Infections in Children with Cancer

Infection	Common Pathogens	Treatment
Bacteremia	*S. aureus, S. epidermidis* or other coagulase-negative staphylococci, *Streptococcus* sp., *Enterococcus* sp., *E. coli, K. pneumoniae, Enterobacter*	Cefepime or ceftazidime and gentamicin if Gram-negative and vancomycin if Gram-positive followed by appropriate antibiotics based on antibiotic susceptibility of the isolate
Catheter-associated bacteremia	*S. epidermidis,* other Gram-positive and Gram-negative bacteria	Vancomycin or appropriate antibiotics for 7–10 days. Antibiotic infusion should be rotated through each lumen if multilumen catheter. Use of antimicrobial lock solution should be strongly considered
	Candida	Fluconazole, amphotericin B, caspofungin. If infected with Candida (or if multiple organisms, cultures are persistently positive despite antibiotics, or there is recurrent bacteremia with the same organism), remove the catheter
Local catheter infections: exit site infection, tunnel infection	Gram-positive cocci, gram-negative bacilli	Vancomycin or appropriate antibiotic for 7–10 days. If there is tunnel infection, remove the catheter
Otitis media	*S. pneumoniae, H. influenzae,* other Gram-positive and Gram-negative bacteria	Amoxicillin-clavulanate for 10–14 days
Sinusitis	*S. pneumoniae, H. influenzae, Moraxella catarrhalis, S. aureus,* gram-negative aerobes. anaerobes. fungi (*Aspergillus, Candida, mucor*)	If patient not neutropenic, amoxicillin, clavulanate If patient neutropenic, ticarcillin/ clavulanate or piperacillin/ tazobactam. If no Improvement in 72 hours, perform aspirate or biopsy to rule out fungal infection
Localized pulmonary infiltrate	Gram-positive and gram-negative bacteria, Nocardia *Aspergillus sp., P. boydii, Fusarium sp., Trichosporon beigelii, Zygomycetes, Candida sp., H. capsulatum*	Cefepime or ceftazidime AND azithromycin. If there is no improvement within 72 hours, a broncho-alveolar lavage or open lung biopsy should be considered
Diffuse pulmonary infiltrate	PCP	TMP/SMX (alternative: pentamidine) and short course of steroids
	CMV	Ganciclovir and IVIG
	HSV or VZV	Acyclovir IV
	RSV	Ribavirin
	Influenza	Osteltamivir and/or amantadine
	Adenovirus, parainfluenza, HHV-6	
Infections in oral cavity:		
Thrush	*C. albicans*	Topical antifungal agents, fluconazole
Mucositis	HSV	Acyclovir

(Continued)

Table 31-4 (Continued)

Infection	Common Pathogens	Treatment
Periodontal disease	Anaerobic bacteria, mixed aerobic	Clindamycin, metronidazole, or imipenem
Esophagitis	Candida HSV	Fluconazole, amphotericin B Acyclovir

Typhlitis	Anaerobes, Gram-negative bacilli,
	  C. difficile
		Metronidazole AND ticarcillin/
		  clavulanate or piperacillin/
		  tazobactam +/− gentamicin

Intra-abdominal sepsis	Gram negative bacilli, *Bacteroides fragilis*,
	  enterococcus
		Piperacillin/tazobactam OR ticarcillin/
		  clavulanate +/− gentamicin

Colitis	*C. difficile*	Oral metronidazole (alternative: oral vancomycin)
Hepatitis	Hepatitis A, B, C HSV CMV EBV, coxsackie B, adenovirus	 Acyclovir Ganciclovir, foscarnet

Perianal cellulitis	Aerobic gram-negative bacilli
	  (*P. aeruginosa, K. pneumoniae, E. coli*),
	  enterococci, bowel anaerobes,
	  Staphylococcus aureus
		Cefepime or ceftazidime AND
		  gentamicin AND clindamycin/
		  metronidazole

Intraventricular shunt and positive Ommaya reservoir	Coagulase-negative and staphylococci,
	  S. aureus, Corynebacterium sp.,
	  Propionibacterium acnes, gram-negative
	  bacilli
		Vancomycin followed by appropriate
		  antimicrobials for the isolate

Meningitis	*S. epidermidis*, a-hemolytic streptococci,
	  Listeria, *Enterococcus, E. coli*,
	  K. pneumoniae, P. aeruginosa,
	  C. albicans, Aspergillus
		Ceftriaxone and vancomycin followed by
		  appropriate antimicrobials for the
		  isolate

Urinary tract infections	*E. coli, Klebsiella sp., Proteus sp.*,
	  P. aeruginosa, enterococci, *C. albicans*,
	  Candida tropicalis, Candida glabrata,
	  adenovirus, polyomavirus (BK virus)
		Cefepime or ceftazidime and gentamicin
		  followed by appropriate
		  antimicrobials for the isolate

Skin infections	*S. aureus, Streptococcus pyogenes*,
	  P. aeruginosa, Aeromonas hydrophilia,
	  S. marcenses, Aspergillus, Candida,
	  HSV, VZV
		Appropriate antimicrobials for the
		  organism
		
		Acyclovir

FUNGAL INFECTIONS

The risk of fungal infections is related to the cytotoxicity of the chemotherapeutic regimen as well as the duration of neutropenia. The major causative fungi are *Aspergillus* and *Candida* with a mortality approaching 30–60% with documented invasive fungal infections (IFIs) and is even higher among post-hematopoietic stem cell transplantation patients. Prompt diagnosis and treatment is paramount to improving outcomes in these patients. Difficulty in diagnosing fungal infections is caused by the frequent absence of localizing signs and symptoms and high likelihood of negative blood cultures. Notwithstanding these limitations, *empiric antifungal coverage must be initiated for patients with neutropenia and fever that persists for 3–7 days despite broad-spectrum antibacterial therapy.* Institution of empiric antifungal therapy has been found to decrease the risk for mortality significantly.

In certain situations, e.g., stem cell transplantation recipients, antifungal prophylaxis results in a significant reduction in the risk for fungal-related death and documented IFIs and a larger relative risk reduction when used in conjunction with antibacterial prophylaxis in contrast to antibacterial prophylaxis alone.

Currently available antifungal therapy belongs to three classes of agents: polyenes, azoles and echinocandins.

1. Polyenes
   * Amphotericin B deoxycholate
   * Liposomal amphotericin B (Ambisome®)
   * Amphotericin B lipid complex (Abelcet®)

   Polyenes are fungicidal by increasing fungal cell membrane permeability. Traditionally, amphotericin B deoxycholate has been the standard drug for antifungal therapy since it offers broad-spectrum coverage and is inexpensive. However, adverse reactions such as nephrotoxicity and hypokalemia and infusion-related reactions such as fever and chills limit its use. Lipid preparations of amphotericin, e.g., liposomal amphotericin, are now more frequently prescribed because they are less nephrotoxic and some result in fewer infusion-related adverse effects.

2. Azoles: fluconazole, itraconazole, voriconazole and posaconazole
   * Fluconazole is a first-line drug for candida infections but lacks activity against *Aspergillus*
   * Voriconazole is the drug of choice for *Aspergillus* infections because it has been shown to be more effective than amphotericin B for *Aspergillus* pulmonary infections. Voriconazole is available in intravenous and oral formulations and has excellent oral bioavailability. Oral voriconazole is an effective and relatively inexpensive agent for empiric antifungal therapy in patients with persisting fever and neutropenia despite broad-spectrum antibacterial therapy. In addition, voriconazole may also play a role in treating central nervous system disease
   * Posaconazole is a broad-spectrum azole effective against many yeasts and fungi including histoplasma and zygomycetes. It is available only for enteral administration. Food substantially enhances the bioavailability of posaconazole and the rate and extent of absorption, thereby limiting its utility in cases where oral intake is not feasible.

3. Echinocandins: caspofungin and micafungin. Echinocandins inhibit beta-1,3-D-glucan synthetase in the fungal cell wall. They are fungicidal for yeasts and fungistatic for molds such as *Aspergillus*. They are generally well-tolerated and are not nephrotoxic.

- Caspofungin has been found to be equally efficacious as amphotericin B against systemic candida infection and demonstrates activity against *Aspergillus*. Hepatotoxicity with concomitant use of cyclosporine has been reported, which may limit its use in some HSCT patients
- Micafungin has a similar spectrum of activity to micafungin but experience in the pediatric age group is limited and it is not currently FDA-approved for use in children.

Table 31-5 shows a list of fungal infections and their treatment and Table 31-6 gives doses and routes of administration of antifungal agents.

Antifungal Agents for Empiric Therapy of Patients with Prolonged Fever and Neutropenia

First-line antifungal agents include:

- Liposomal amphotericin (Ambisome®)
- Voriconazole
- Caspofungin
- Amphotericin B deoxycholate.

Second-line agents include:

- Fluconazole
- Posaconazole
- Itraconazole
- Micafungin (micafungin may be considered a first-line agent when FDA-approved for children).

Antifungal Prophylaxis in Hematopoietic Stem Cell Transplantation Patients

1. Autologous HSCT
   a. Micafungin/caspofungin
   b. Fluconazole
2. Allogeneic HSCT
   a. Micafungin/caspofungin
   b. Liposomal amphotericin
   c. Voriconazole
   d. Fluconazole
3. Postengraftment, GvHD, AML, MDS
   a. Posaconzole if eating well and tolerating oral fluids
   b. Voriconazole
   c. Fluconazole if no longer neutropenic and liver function tests normal.

Table 31-5   Drugs for Invasive and Other Serious Fungal Infections

Disease	Intravenous		Oral	Intravenous of Oral		
	Amphotericin B	Caspofungin,[a] Micafungin,[a,b] or Anidulafungin[a,b]	Flucytosine	Itraconazole[a]	Fluconazole[a]	Voriconazole
Aspergillosis	A	A	—	M	—	P
Blastomycosis	P	—	—	M	M	—
Candidiasis:						
Chronic, mucocutaneous	A	A	—	A	P	A
Oropharyngeal, Esophageal (severe cases)	P	P	—	A	P	A
Systemic	P[d]	P	S	—	P,M	A
Coccidiodomycosis	P	—	S	P,M	P,M	—
Cryptococcosis	P,S	—	S	—	A,M	—
Fusariosis	P	—	—	—	—	P
Histoplasmosis	P	—	—	P,A	A,M	—
Mucormycosis (zygomycosis)	P	—	—	—	—	—
Paracoccidioidomycosis	P[c]	—	—	P,M	—	—
Pseudallescheriasis	—	—	—	M	—	P
Sporotrichosis	P	—	—	M	—	—

[a]Efficacy has not been established for children.

[b]Approved by the Food and Drug Administration for adults.

[c]Usually in combination with itraconazole or a sulfonamide.

[d]Preferred treatment in neonates; alternate treatment for children and adults.

P, indicates preferred treatment in most cases; A, efficacy less well established or alternative drug; M, for mild and moderately severe cases; S, combination recommended if infection is severe or central nervous system is involved.

From: Pickering LK, ed. Red Book: 2009 Report of the Committee on Infectious Diseases, 28th ed., Elk Grove Village, IL: American Academy of Pediatrics; 2009, 772, with permission.

Table 31-6   Doses and Routes of Administration of Antifungal Agents

Drug	Route	Dose
Amphotericin B colloidal dispersion (Amphotec)	IV	3–5 mg/kg/day
Amphotericin B deoxycholate (Fungizone)	IV	Test dose 0.1 mg/kg to a maximum dose of 1 mg over 1 hour; if tolerated, may use 0.4 mg/kg and increase to 1 mg/kg/day to maximum dose of 1.5 mg/kg/day
Amphotericin B lipid complex (Abelcet)	IV	5 mg/kg/day
Amphotericin B liposomal (Ambisone)	IV	3–5 mg/kg/day
Caspofungin	IV	70 mg/m^2 loading dose, then 50 mg/m^2 once daily.
Fluconazole	IV, PO	3–12 mg/kg/day (maximum dose 800 mg/day)
Flucytosine	PO	50–150 mg/kg/day divided every 6 hours
Itraconazole	IV, PO	5–10 mg/kg/day every 12 hours
Micafungin	IV	Data in children limited. Adult dose: 50–150 mg daily. Children: 4–12 mg/kg once daily (higher doses needed for patients <8 years of age)
Posaconazole	PO	Children >13 years of age *Prophylactic dose:* 200 mg 3 times a day with meals *Treatment dose:* 400 mg twice a day with meals
Voriconazole	IV	Adults: 6 mg/kg every 12 hours, IV for first 24 hours and then 4 mg/kg every 12 hours. Children: 7 mg/kg every 12 hours IV for first 24 hours, then 7 mg/kg every 12 hours IV or PO.
	PO	Oral dose: <40 kg, 100 mg every 12 hours (may increase to 150 mg every 12 hours); >40 kg, 200 mg every 12 hours (may increase to 300 mg every 12 hours if needed)

Modified from: Wolff LJ, Altman AJ, Berkow RL, Johnson FL. The management of fever and neutropenia. In: Autumn AJ, ed. Supportive care of children with cancer. Current therapy and guidelines from the Children's Oncology Group, 3rd ed. Baltimore: The Johns Hopkins University Press, 2004:25–38, with permission. Reference used: Red Book 2009, pp 768–771.

Antifungal Prophylaxis of *Candida* Infection

1. Nystatin oral swish and swallow; 2 ml twice daily in infants, 5 ml twice daily in children and adults.
2. Clotrimazole troche one bid.
3. Fluconazole 2 mg/kg/day; maximum 200 mg.

Pneumocystis jiroveci (*Formerly* P. carinii)

Pneumocystis jiroveci has been reclassified as a fungus and for this reason it has been included in the section of this chapter dealing with fungal infections. It is a ubiquitous organism that causes severe or fatal pneumonitis in immunocompromised patients.

Clinical Features

1. Rapid onset of symptoms with fever and tachypnea.
2. Progressive respiratory distress with nasal flaring and intercostal retractions.
3. Absence of rales on physical examination.
4. Hypoxemia on arterial blood gas.
5. Development of bilateral pneumonitis on chest radiograph.
6. Rapid progression of respiratory distress/respiratory failure over a few days; usually fatal if untreated.

Diagnosis

1. Typical clinical syndrome.
2. Demonstration of the organism in sputum or material from endobronchial washing or percutaneous needle biopsy or open lung biopsy.
3. Identification of organism by Gomori methenamine silver stain.

Treatment

1. Start treatment as soon as the patient develops tachypnea and hypoxemia, even if there is no laboratory confirmation of the organism.
2. Trimethoprim/sulfamethoxazole (TMP/SMX) 20 mg/kg/day of TMP component IV divided into four doses is the treatment of choice. Treatment should continue for 21 days. TMP/SMX causes myelosuppression.
3. If TMP/SMX is not tolerated, treat with pentamidine 4 mg/kg/day IV. TMP 5 mg/kg PO every 6 hours and dapsone 100 mg/day PO for 21 days is another alternative.
4. Monitor carefully for hypotension. Pentamidine is recommended if patients fail to respond to TMP/SMX in 72 hours or if they develop pneumocystis in spite of TMP/SMX prophylaxis.
5. Concommitant administration of corticosteroids, e.g., 1 mg/kg methylprednisolone or prednisone given 2–4 times per day for 5–7 days followed by a taper over the next 1–2 weeks, is indicated for patients with moderate or severe infection proven to be caused by PCP.

Prophylaxis

1. TMP/SMX 5 mg/kg/day (150 mg/m^2/day), maximum dose 320 mg/day (based on TMP) divided into two daily doses for three consecutive days per week. TMP/SMX may cause rash, neutropenia and GI symptoms.
2. Alternative drugs for pneumocystis PCP prophylaxis for use in patients 1–2 months old, allergic to TMP/SMX, with G6PD deficiency, or excessive myelosupppression with TMP/SMX are dapsone, pentamidine, or atovaquone.

3. Dapsone 2 mg/kg/day PO once daily, maximum dose 100 mg/day; may be preferred agent in children under the age of 2 months who may have liver immaturity and be at risk for methemoglobinemia if TMP/SMX treatment is used.
4. Pentamidine aerosolized formulation for children old enough to cooperate: Infant dose: 2.27 mg/kg×nebulizer output (L/minute)×wt (kg) divided by alveolar ventilation (L/minute); Children <5 years: a dose of 8 mg/kg is recommended, children ≥5 years: 300 mg/dose.
5. Pentamidine intravenous. For children unable to cooperate with inhaled formulation, may receive 4 mg/kg administered intravenously every 4 weeks.
6. Atovaquone 30 mg/kg/day once daily in children 1–3 months and those older than 2 years of age. The recommended dose is 45 mg/kg/day for children between 3 and 24 months of age. Atovaquone is used for pentamidine breakthrough or intolerance to other agents.

VIRAL INFECTIONS

Children on chemotherapy can tolerate many common viral infections, but defects in cellular immunity predispose them to unusually severe infections with certain viruses, particularly of the herpes virus group. Primary varicella infection, in particular, can produce serious morbidity, including encephalitis and pneumonitis and mortality in 7% of patients. In stem cell transplantation recipients the herpes virus cytomegalovirus (CMV) is an important cause of interstitial pneumonitis, bone marrow aplasia and other infections. Infection with an influenza virus can cause significant morbidity in children on chemotherapy. Stem cell transplant recipients can develop life-threatening pneumonitis due to RSV, parainfluenza and rhinoviruses. Adenoviruses and papovaviruses BC are important causes of hemorrhagic cystitis.

Table 31-7 lists the indications, route of administration and usually recommended dosage of the antiviral drugs available for viral infections in children with cancer.

Antiviral drugs are indicated only if there is clinical or laboratory evidence of viral infection or for prophylaxis of herpes simplex infection or for prophylaxis or pre-emptive therapy of cytomegalovirus infection in stem cell transplant recipients.

Antiviral Prophylaxis

Acyclovir prophylaxis 750–1,500 mg/m^2/day IV divided every 8 hours or 600–1,000 mg/day PO divided every 8 hours during the risk period. It is prescribed to prevent reactivations of HSV infection in seropositive stem cell transplant or high-dose chemotherapy patients.

Table 31-7 Antiviral Drugs for Viral Infections in Children with Cancer

Generic (Trade Name)	Indication	Route	Usually Recommended Dosage
Acyclovir[1,2] (Zovirax)	Varicella in immuno-compromised host	IV	For infants <1 year of age 30 mg/kg/ day in three divided doses for 7-10 days; some experts also recommend this dose for children ≥1 year of age.
		IV	For children ≥1 year of age: 1,500 mg/m^2 of body surface area per day in three divided doses for 7-10 days
	Herpes simplex	IV	Children <12 years of age, 30 mg/kg/day in three divided doses for 7-14 days
		IV	Children 12 years and older, 15 mg/kg/day in three divided doses for 7-14 days
		IV	CNS infection: 30 mg/kg/day in three divided doses for 14-21 days
Foscarnet[1] (Foscavir)	CMV retinitis in patients with acquired immunodeficiency syndrome	IV	Adult dose: 180 mg/kg per day in three divided doses for 14-21 days, then 90-120 mg/kg once a day as maintenance dose
	HSV infection resistant to acyclovir in immunocompromised host	IV	80-120 mg/kg per day in two or three divided doses until infection resolves
Ganciclovir[1]	Acquired CMV retinitis in immunocompromised host[3]	IV	10 mg/kg per day in two divided doses for 14-21 days; for long-term suppression, 5 mg/kg per day for 5-7 days/wk
	Prophylaxis of CMV in high-risk host	IV	10 mg/kg per day in two divided doses for 1 wk, then 5 mg/kg per day in one dose for 100 days
Lamivudine	Treatment of chronic hepatitis B	Oral	For children ≥2 years of age, 3 mg/kg per day (maximum 100 mg/day)

[1]Dose should be decreased in patients with impaired renal function.
[2]Oral dosage of acyclovir in children should not exceed 80 mg/kg per day.
[3]Some experts use ganciclovir in immunocompromised host with CMV gastrointestinal tract disease and CMV pneumonitis (with or without CMV immune globulin intravenous).
Abbreviations: IV indicates intravenous; IO, intraocular.
For more information on individual drug see Physician's Desk Reference (Greenwood Village, CO: Thomson Micromedex) or http://pdrel.thomsonhc.com/pdrel/librarian/action/command.command).
Modified from: Pickering, LK, ed. Red Book 2009. Report of the Committee on Infectious Diseases, 28th Ed. Elk Grove Village, IL: American Academy of Pediatrics 2009, 777-782, with permission.

PROTOZOAN INFECTIONS

Toxoplasma gondii is the one protozoan infection that occurs relatively frequently. HIV-infected and other immunocompromised patients may develop re-activation of latent infection that most frequently presents as a focal encephalitis. The first-line therapy is pyrimethimine (with folinic acid) and sulfadiazine. Clindamycin can be substituted for sulfadiazine in patients intolerant of sulfadiazine. Prophylaxis for PCP with trimethoprim-sulfamethoxazole is effective for prevention of toxoplasmosis.

G-CSF AND GRANULOCYTE INFUSIONS

Therapy with G-CSF at a dose of 5 μg/kg/day subcutaneously is recommended under certain conditions in which worsening of the clinical course is predicted and there is an expected long delay in the recovery of the marrow. It should also be considered for patients who remain severely neutropenic and have documented infections that do not respond to appropriate antibiotic treatment. There are no specific indications for standard use of granulocyte transfusions. Granulocyte transfusions may be useful in patients with profound neutropenia, in whom the infection progresses despite optimal antibiotic treatment and G-CSF and in cases of severe uncontrollable fungal infection (see p. 868–872 for the indications for granulocyte transfusions). When granulocyte transfusions are given concomitantly with amphotericin B preparations, respiratory insufficiency may occur as a side effect.

IMMUNIZATIONS

Current CDC recommendations for immunizations are as follows:

- Immunocompromised patients should not be administered live vaccines
- Oral poliovaccine should not be administered to any household contact of a severely immunocompromised person. Measles–mumps–rubella (MMR) vaccine is not contraindicated for the close contacts (including health-care providers) of immunocompromised persons and should be given to susceptible household members
- Patients with leukemia in remission status who have not received chemotherapy for at least 3 months are not considered severely immunosuppressed and can receive live-virus vaccines
- When chemotherapy or immunosuppressive therapy is being considered vaccination ideally should precede the initiation of chemotherapy or immunosuppression by greater than or equal to 2 weeks
- Vaccination during chemotherapy or radiation therapy should be avoided
- Patients vaccinated while on immunosuppressive therapy or within 2 weeks before starting therapy, should be considered unimmunized and should be revaccinated at least 3 months after discontinuation of therapy
- When exposed to a vaccine-preventable disease such as measles, severely immunocompromised children should be considered susceptible regardless of their history of vaccination.

Measles

Severely immunocompromised patients who are exposed to measles should receive immune globulin (IG), regardless of prior vaccination status. The recommended dose of immune globulin for measles prophylaxis of immunocompromised patients is 0.5 ml/kg of body weight intramuscularly (maximum dose, 15 ml). The immunogenicity of measles vaccine is decreased if vaccine is administered within 6 months of immunoglobulin. For immuno-compromised patients receiving immunoglobulin for measles prophylaxis, measles vaccination should be delayed for 6 months following immunoglobulin administration. For patients receiving immunoglobulin for replacement of humoral immune deficiencies (320 mg/kg intravenously), measles vaccination should be delayed until 8 months following immuno-globulin administration.

Varicella

Leukemic children who are in remission status and do not have evidence of immunity to varicella should be vaccinated with antiviral therapy being available should complications arise. Patients with leukemia, lymphoma, or other malignancies whose disease is in remission and whose chemotherapy has been terminated for at least 3 months can receive live-virus vaccines. When immunizing persons in whom some degree of immunodeficiency might be present, only single-antigen varicella vaccine should be used rather than MMRV. Because varicella vaccine is recommended for all healthy children and adults without evidence of immunity, household contacts of immunocompromised persons should be vaccinated routinely. Vaccination of household contacts provides protection for immuno-compromised patients by decreasing the likelihood that wild-type VZV will be introduced into the household. Vaccination of household contacts of immunocompromised persons theoretically poses a risk for transmission of vaccine virus to immunocompromised patients. However, no cases have been reported of transmission of vaccine virus to immunocompromised persons. Vaccine recipients in whom vaccine-related rash occurs, particularly house-hold contacts of immunocompromised persons, should avoid contact with susceptible persons who are at high risk for severe complications. If the susceptible, immunocompromised person is inadvertently exposed to a person who has a vaccine-related rash, post-exposure prophylaxis with varicella-zoster immunoglobulin is not needed because disease associated with this type of virus is expected to be mild.

Varicella Exposure

Patients who had household exposure to active varicella, or have been in the same room with an individual in the contagious state of varicella (1–2 days before and 5 days after eruption of vesicles) for at least 1 hour should receive varicella-zoster immune globulin (available in the United States as investigational Varizig) 1 vial/10 kg IM (maximum dose, 5 vials) within 96 hours of exposure. Children who have been vaccinated and have positive titers do not need to receive varicella-zoster immune globulin after exposures.

Pneumococcal Vaccine

In addition to pneumococcal conjugate vaccine that is routinely administered to all children with doses at 2, 4, 6 and 12–15 months of age, 23-valent pneumococcal polysaccharide vac-cine is recommended for use in children 2 years of age or older with chronic illnesses associated with increased risk of pneumococcal disease or its complications (e.g., anatomic or functional asplenia [sickle cell disease], nephrotic syndrome, cerebrospinal fluid leaks and conditions associated with immunosuppression). Vaccination in these high-risk individuals includes two doses of pneumococcal conjugate vaccine if not previously completed and the pneumococcal polysaccharide vaccine should be given at least 2 months after the completion of the PCV-7 vaccines. A second dose of pneumococcal polysaccharide vaccine should be administered 5 years later for children at highest risk of infection, i.e., those with asplenia or sickle cell disease.

Meningococcal Vaccine

Routine immunization with the quadrivalent meningococcal conjugate vaccine is recommended for all children starting at 11 years of age and for certain high-risk children at age 2 through 10 years including children with terminal complement component deficiencies and those with anatomic or functional asplenia.

Influenza Vaccine

Because influenza may result in serious illness and complications for immunocompromised persons, annual vaccination with inactivated vaccine is recommended. However, response to this vaccine may be suboptimal in immunocompromised patients. In an attempt to decrease the exposure of such persons to influenza, all household members and all health-care personnel should be immunized with influenza vaccine (inactivated or live, attenuated nasal vaccine) annually.

BLOOD COMPONENT THERAPY

The two main causes of pancytopenia in children with malignancies are:

*   Invasion or replacement of the bone marrow by malignant cells
*   Myelosuppression due to chemotherapeutic agents.

Blood component administration is an integral component of management. The risks associated with transfusions (Table 31-8) are:

*   Transfusion reactions (febrile and hemolytic reactions)
*   Development of alloimmunization and refractoriness
*   Transmission of infectious agents
*   Development of graft versus host disease (GVHD).

Transfusion reactions include:

*   Acute hemolytic transfusion reactions
*   Delayed hemolytic transfusion reactions

Table 31-8   Relative Risks of Transfusion Complications per Unit

Risk	Risk Ratio
Hepatitis C	1:1,600,000
HIV	1:1,900,000
Wrong blood in tube	1:1,000
Wrong recipient of auto unit	1:16,000
TRALI	1:5,000
Hemolysis from ABO-incompatible plasma in an apheresis platelet	1:10,000–46,000

Abbreviation: TRALI, transfusion-related acute lung injury.

- Febrile nonhemolytic transfusion reactions
- Allergic reactions
- Transfusion-related acute lung injury (TRALI).

Hemolytic Reactions

Most hemolytic reactions are acute hemolytic transfusion reactions and result from blood group incompatibility (both major and minor red cell antigens) between the donor and the recipient. They can be prevented by thorough typing and cross-matching of all packed red cell transfusions. Major acute hemolytic transfusion reactions present with fever, chills, back or abdominal pain, dark urine, pallor, bleeding, or shock during the transfusion.

Laboratory investigations may show positive direct antiglobulin test for complement, spherocytes, decreased haptoglobin, hyperbilirubinemia and hemoglobinuria. The transfusion should be stopped immediately. The unit and a sample of the patient's blood should be sent to the blood bank for investigation. The patient should be vigorously hydrated and the use of diuretics (furosemide or mannitol) may be indicated to maintain urine flow.

Delayed hemolytic transfusion reactions occur 2–14 days after red cell transfusion. Patients present with low-grade fever, decreased hemoglobin and jaundice. The direct antiglobulin test is moderately positive for IgG and microscopic examination of the resuspended red cells reveals mixed field agglutination (clusters of agglutinated cells in a field of nonagglutinated cells). This phenomenon occurs because delayed transfusion reaction is the result of an anamnestic response to a donor antigen to which the patient was previously sensitized (and where screening of the patient failed to detect the sensitization). Antibody is directed against the transfused cells only, while the patient's cells remain nonagglutinated. Antibody specificity is usually easily detected and future transfusions should avoid that antigen in the donor.

Febrile Reactions

Febrile reactions are almost always caused by sensitization to leukocyte antigens. They can be largely prevented by two measures:

- Leukocyte-depleted products (usually through the use of leukocyte filters). Prestorage leukocyte filtration offers the most effective approach. Inflammatory cytokines (interleukin (IL)-1, IL-6 and tumor necrosis factor α [TNF-α]) secreted from leukocytes cannot be filtered by poststorage filtration and may still lead to febrile reactions
- Premedication of the recipient with Tylenol 10–15 mg/kg.

Allergic Reactions

Allergic reactions are due to plasma proteins. Mild cutaneous hypersensitivity reactions (itching, rash, redness) respond to antihistamines. Accordingly, it is common to premedicate patients with diphenhydramine 1 mg/kg (max. 50 mg) orally or IV prior to transfusion. Severe allergic reactions are seen and are often due to development of anti-IgA in IgA-deficient patients. Patients may require epinephrine and parenteral steroids. If reactions continue premedication with hydrocortisone may help. For those allergic to plasma products, frozen washed red cells may be necessary to avoid serious reactions.

Transfusion-Related Acute Lung Injury (TRALI)

Transfusion-related acute lung injury has become the leading cause of transfusion-related morbidity and mortality. Transfusion-related acute lung injury occurs within 4–6 hours of transfusions. Leukocyte antibodies in the donor blood interact with recipient white cells and lead to activation and aggregation of the white cell complexes in the pulmonary vasculature. This in turn results in increased vascular permeability and exudation of fluid and protein into the alveoli. The clinical picture of TRALI resembles adult respiratory distress syndrome. Patients present with dyspnea, hypoxia, hypotension, fever, tachycardia and noncardiogenic pulmonary edema. Radiograph of the chest reveals bilateral fluffy infiltrates. Treatment is supportive. Unlike ARDS, patients usually recover with resolution of pulmonary findings within 96 hours. Mortality rate has been reported to be between 5–10%. Since no diagnostic tests are available for diagnosis, TRALI remains a diagnosis of exclusion. Current measures to prevent TRALI are not to use donors previously implicated in a case of TRALI. Implicated donors typically have been found to be females with a history of pregnancy or transfusion. One strategy that is under consideration is using only male donor plasma for transfusion and limiting use of female donor plasma for fractionation and plasma derived products.

Transfusion-Associated Circulatory Overload (TACO)

Transfusion-associated circulatory overload is the result of cardiogenic pulmonary edema following a large volume or rapid infusion of blood product. Primary symptoms of TACO are dyspnea, orthopnea, hypertension and peripheral edema. Chest radiograph may reveal pulmonary edema and cardiomegaly. Patients at increased risk for TACO include infants, patients over the age of 60 years, patients with severe anemia and patients with pulmonary, renal, or cardiac failure. Mortality rate ranges from 1–3%. Differentiation between TACO and TRALI is aided with the use of brain natriuretic peptide (BNP). In transfusion-related circulatory overload, the BNP will be elevated, whereas in patients with TRALI, there will be no elevation of BNP. Treatment for TACO is supportive. Prevention of TACO requires a controlled rate of infusion.

Table 31-9 lists the clinical benefits from using leuko-reduced blood products.

Alloimmunization and Refractoriness

To some extent, all transfusion recipients become sensitized to foreign leukocyte antigens. However, the use of leukocyte filters helps to prevent this problem.

The problem of refractoriness applies particularly to platelet transfusions. Patients who are refractory show no rise or significantly less than the expected rise in platelet count following a platelet transfusion. In some patients, refractoriness occurs much sooner than in others. Specific factors affecting platelet survival will be discussed in the section on platelet transfusions.

Transmission of Infectious Agents

The agents that can be transmitted by transfusions are viruses, including cytomegalovirus (CMV), hepatitis and human immunodeficiency virus (HIV), bacteria and protozoa. Table 31-10 lists the risk estimates of viral transmission by transfusion. Careful donor selection and improvements in screening and blood product testing have significantly ameliorated this problem. The other infectious risk is bacterial contamination of the donor units of blood. Infection is most commonly due to *Yersinia enterocolitica* and other Gram-negative organisms. Platelet units are most commonly contaminated with *S. aureus*, *Klebsiella pneumoniae*, *Serratia marcenses* and *Staphylococcus epidermidis*. Parasitic infections that can be transmitted through blood transfusion are malaria, Chagas disease and *Babesia*.

Table 31-9   Clinical Benefits from Using Leuko-reduced (LR) Blood Products

Prevention of recurrent febrile nonhemolytic transfusion reactions
Prevention of primary HLA-alloimmunization and resulting complications, including platelet refractoriness
Prevention of transfusion-transmitted cytomegalovirus infection in at-risk patients
Reduction of prion (vCJD) transmission
Reduction of transmission of other leukotropic viruses (HTLV I-II and EBV)
Reduction of viral reactivation
Reduction of parasitic or bacterial infections
Reduction of transfusion-related acute lung injury

Abbreviation: CJD, Creutzfeldt-Jakob disease.

Table 31-10   Risk Estimates of Viral Transmission by Transfusion

Virus/Infection	Estimated Risk in 2002
Human immunodeficiency virus (HIV)	1:2,135,000
Hepatitis C (HCV)	1:1,935,000
Hepatitis B (HBV)	1:1,205,000
Human T-cell lymphotropic virus (HTLV-I/II)	1:641,000

From: Jamali F, Ness PM. Infectious complications. In: Hillyer CD, Strauss RG, Luban NLC, eds. Handbook of Pediatric Transfusion Medicine. San Diego: Elsevier Academic Press, 2004;329–339.

Donated blood is screened for hepatitis B and C and HIV. Positive units are discarded. CMV may be transmitted via CMV antibody-positive blood. Eighty-five percent of the population is positive for CMV antibodies. It is, therefore, impossible to discard all CMV-positive units. Because of the high risk of CMV interstitial pneumonitis in severely immunocompromised patients, many institutions use CMV serology-negative donors for neonates and bone marrow transplant patients. There are a number of reports, however, that suggest that the current leukodepletion filter technology may suffice for bone marrow transplant patients. Patients who will not be undergoing transplant or transplant patients who are already CMV positive might receive CMV-positive, leukodepleted products. Leukodepletion reduces the risk of CMV and, in theory, Creutzfeldt–Jakob disease.

Transfusion-Related Graft Versus Host Disease

GVHD has been observed to arise 4–30 days after the administration of nonirradiated blood products to immunocompromised patients. It results from viable donor precursor cells or stem cells engrafting in the immunocompromised host's marrow. The clinical manifestations are fever, erythematous, maculopapular skin rash, anorexia, nausea, vomiting, diarrhea, elevated liver enzymes and hyperbilirubinemia. A very high morbidity and mortality rate is associated with transfusion-acquired GVHD. Leukodepletion is not sufficient to protect patients from this condition. Patients at risk for transfusion-associated graft versus host disease must have all of their blood products irradiated with 2,500 cGy.

General Guidelines

All oncology patients should receive leukocyte-filtered, irradiated blood products. CMV-negative products should be used according to the previously mentioned guidelines.

Red Cell Transfusions

Causes of anemia in cancer patients in children include:

- Replacement of the bone marrow with malignant cells
- Chemotherapy-induced myelosuppression

- Anemia of inflammation
- Blood loss due to thrombocytopenia
- Iatrogenic blood loss
- Hemolysis.

Packed red cell transfusions are indicated for patients whose hemoglobin is less than 8 g/dl (with no expectation of spontaneous rise in the hemoglobin) or those whose hemoglobin is greater than 8 g/dl but who are cardiovascularly unstable and have respiratory failure or bleeding.

Patients should receive 10–15 ml/kg of packed red cells over 3 hours. Each transfusion should be completed within 4 hours after the unit or aliquot has been started.

Patients who are profoundly anemic (Hgb, <5 g/dl) should be transfused at a slower rate (2 ml packed red cells/kg/h) with careful monitoring of vital signs. Repeat transfusions can be given.

The expected increment can be estimated as follows: 1 ml/kg of packed red cells increases the hematocrit by 1%.

Platelet Transfusions

The causes of thrombocytopenia in children who have cancer are:

- Decreased production
- Increased destruction
- Hypersplenism
- Consumption due to brisk bleeding or massive transfusion.

Children begin to ooze colonic blood at platelet counts less than 10,000/mm^3. Since young children are physically active and do not avoid trauma, platelet transfusions are considered when:

- The platelet count is less than 15,000–20,000/mm^3
- The platelet count is greater than 20,000/mm^3, but the patient is bleeding (e.g., prolonged epistaxis, GI bleeding, or CNS bleeding)
- The patient is scheduled for an invasive procedure and the platelet count is less than 50,000/mm^3.

Platelets are collected in two ways:

- Platelets are collected from units of routinely donated whole blood and pooled (random-donor) platelets; 1 unit of random-donor platelets has 5–10×10^{10} platelets

- Platelet concentrates can be collected from a single donor by platelet apheresis; 1 unit of single-donor platelets has $3–6\times10^{11}$ platelets (roughly equivalent to 6 units of random donor platelets).

The dose of platelets given per transfusion is 1 unit random-donor platelets/10 kg or 1 unit single-donor platelets/50 kg. Apheresis units may be divided for smaller patients. Use ABO-type-specific platelets. The expected platelet increment can be calculated by corrected platelet count index (CCI) as follows:

$$CCI = \frac{\text{posttransfusion platelet count} - \text{pretransfusion platelet count} \times \text{body surface area}}{\text{Number of platelets transfused}(\times10^{11})}$$

One unit of random-donor platelets per 10 kg of body weight should increase the platelet count by 40,000–50,000/mm^3 within 1 hour after the infusion. In the average-sized adolescent or adult, 6 units of platelet concentrates or 1 single-donor apheresis unit will increase the platelet count by 50,000/mm^3 or greater. If the post-transfusion platelet count is below the expected count, the patient is refractory. Table 31-11 lists the factors leading to decreased platelet transfusion response and approaches to platelet refractoriness. Once the patient is determined to be platelet refractory, cross-matched or HLA-matched platelets may be required for adequate response.

Table 31-11   Factors Leading to Decreased Platelet Transfusion Response and Approaches to Management of Platelet Refractoriness

Factors in platelet transfusion response
Age of transfused platelets
Washing platelets
Fever, infection
Disseminated intravascular coagulation
Graft versus host disease
Hepatosplenomegaly, hypersplenism
Alloantibodies/autoantibodies
Massive blood loss
Hemolytic uremic syndrome/thrombotic thrombocytopenic purpura (platelet transfusions discouraged/ contraindicated)
Necrotizing enterocolitis
Medications (amphotericin, vancomycin, ciprofloxacin)
Approaches to Platelet Refractoriness
Transfuse ABO identical platelets
Use fresh platelets
Use HLA-matched platelets
Use cross-matched platelets
Intravenous gammaglobulin 1 g/kg prior to transfusion
Plasmapheresis
Corticosteroids

Other measures to prevent bleeding in patients with severe thrombocytopenia include:

- Avoid invasive procedures: nasogastric tube or urinary catheter insertion, rectal examination, intramuscular injections, deep venipunctures
- Apply local pressure for wounds and epistaxis
- Avoid aspirin and other drugs that interfere in platelet function (Chapter 12)
- Epsilon-aminocaproic acid or prednisone can be used as adjunct treatment in bleeding patients
- Estrogen or high-dose birth control pills may be used for menometrorrhagia.

Granulocyte Transfusions

Neutropenia (ANC $<1,000/mm^3$) and severe neutropenia (ANC $<500/mm^3$) commonly occur in patients on chemotherapy. Severe neutropenia greatly increases the risk for overwhelming sepsis and invasive fungal infections. Most neutropenic patients are treated with G-CSF to shorten the period of neutropenia (see later section "Hematopoietic Growth Factors"). Occasionally, however, patients are treated with granulocyte transfusions. Granulocyte transfusions are becoming more readily available.

Indications

1. Serious bacterial (particularly Gram-negative) or fungal infection with persistent (48 hours or more) positive cultures despite appropriate antibiotic coverage in severely neutropenic patients (ANC $<200/mm^3$).
2. The ANC is not expected to increase to more than $500/mm^3$ for 5–7 days.
3. Prolonged survival of the patient is reasonably expected if the infection is controlled.
4. In addition, patients with severe granulocyte dysfunction (e.g., chronic granulomatous disease) who have severe infection.
5. There is no role for prophylactic administration of granulocytes.

Risks

1. CMV infection.
2. Graft versus host disease.
3. Respiratory distress with pulmonary infiltrates (particularly in patients concurrently receiving amphotericin).
4. Alloimmunization and platelet refractoriness.
5. Hemolytic reactions.

Precautions

1. Use granulocytes from an ABO-Rh compatible donor.
2. Use CMV-seronegative donor, if available, for CMV sero-negative recipients.

3. If the patient is on amphotericin, wait 4–6 hours between amphotericin and granulocyte transfusion.
4. Do not administer with a leukocyte-depleting filter.
5. Administer granulocytes as soon as possible after collection to maximize effectiveness. If not used immediately, store at room temperature.
6. Premedicate with diphenhydramine and hydrocortisone.
7. Granulocytes must always be irradiated with 2,500 cGy to prevent transfusion-associated GVHD.

Dose and Duration

1. The dose of granulocytes should be as close to 1×10^{11} as possible.
2. The transfusion is administered through a 170-μm filter at a rate of 150 ml/m^2/hour.
3. Granulocytes are usually administered for 5–7 days or until the ANC has risen to more than 500/mm^3.[*]

HEMATOPOIETIC GROWTH FACTORS: BASIC BIOLOGY OF GROWTH FACTORS

Hematopoietic cells are derived from self-renewing, pluripotent stem cells. Pluripotent stem cells are able to differentiate into committed progenitor cells. The most primitive form of committed progenitor is termed a colony-forming unit granulocyte, erythrocyte, macrophage and megakaryocyte (CFU-GEMM). This cell is capable of producing colonies containing neutrophils, macrophages, erythrocytes, megakaryocytes, eosinophils and basophils. CFU-GEMMs appear to have a limited capacity for self-renewal and therefore, are not true "stem cells." The committed hematopoietic progenitors (erythroid burst-forming unit [BFU-E], erythroid colony-forming unit [CFU-E], granulocyte macrophage colony-forming unit [CFU-GM], granulocyte colony-forming unit [CFU-G], macrophage colony-forming unit [CFU-M], eosinophil colony-forming unit [CFU-Eo], basophil colony-forming unit [CFU-Baso] and megakaryocyte colony-forming unit [CFU-Meg]) are capable of giving rise to colonies containing cells of only one or two types.

The hematopoietic growth factors and interleukins are cytokines that regulate the growth, differentiation and functional activities of progenitor cells in peripheral blood, bone marrow and placental-cord blood. A second important action of many of these factors is augmentation of the function of mature cells. Some growth factors are specific for one type of progenitor, whereas others affect many types of progenitors. Figure 31-2 illustrates the hierarchy of hematopoietic progenitor cells and the sites of responsiveness of growth factors and interleukins. Table 31-12 lists the uses of hematopoietic growth factors.

---

[*]Transfused granulocytes have a half-life of 6–10 hours and do not increase the ANC.

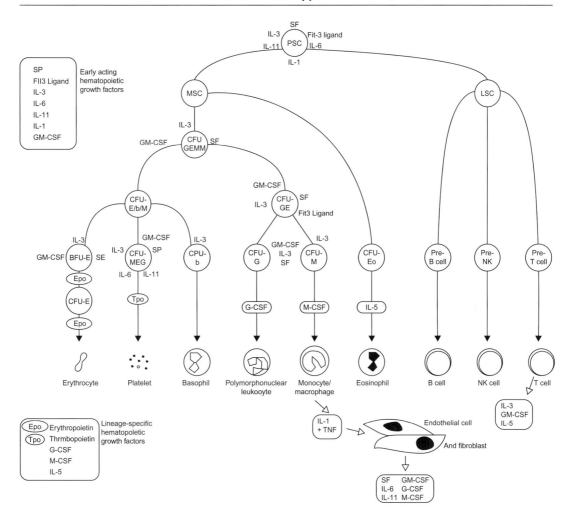

Figure 31-2 Major Cytokine Sources and Actions.

Cells of the bone marrow microenvironment such as macrophages (ma), endothelial cells (ec) and reticular fibroblastoid cells (fb) produce machrophage colony-stimulating factor (M-CSF), granulocyte-macrophage colony-stimulating factor (GM-CSF), granulocyte colony-stimulating factor (G-CSF), interleukin (IL) 6 and probably Steel factor (SF: cellular sources not yet precisely determined) after induction with endotoxin (ma) or IL-1/tumor necrosis factor (TNF) (ec, fb). T-cells produce IL-3, GM-CSF and IL-5 in response to antigenic and IL-1 stimulation. These cytokines have overlapping actions during hematopoietic differentiation, as indicated and for all lineages optimal development requires a combination of early- and late-acting factors.

Abbreviations: PSC, pluripotent stem cells; MSC, myeloid stem cells; CFU-E/b/M, erythrocyte, basophil and megakaryocyte colony-forming unit; CFU-E, erythroid colony-forming unit; CFU-G, granulocyte colony-forming unit; CFU-M macrophage colony-forming unit; CFU-Eo, eosinophil colony forming unit; NK, natural killer.

Adapted from: Clark SC, Nathan DG, Sieff CA. The anatomy and physiology of hematopoiesis. In: Nathan DG, Gingburg ND, Orkin SH, Look AS, eds. Hematology of Infancy and Childhood. 6th ed. Philadelphia: Saunders, 2003; 180, with permission.

Table 31-12   Uses of Hematopoietic Growth Factors

Enhances hematopoietic recovery, lowers hospital cost, permits increased cytotoxic drug dose
Early hematopoietic recovery reduces nonhematologic toxicity (e.g., infection, mucositis, pneumonia)
Mobilizes peripheral-blood progenitor cells
Augments transplantation using smaller number of hematopoietic cells
Expands hematopoietic cells

CLINICAL USE OF G-CSF, GM-CSF, ERYTHROPOIETIN AND IL-11

Recombinant Human G-CSF

Indications for Use

1. Prophylactic use in patients in whom the expected incidence of neutropenia following chemotherapy is greater than or equal to 40%. It accelerates myeloid recovery when the expected duration of severe neutropenia is 7 or more days.
2. After a prior episode of febrile neutropenia.
3. After high-dose chemotherapy with autologous progenitor stem cell support to accelerate myeloid recovery.
4. Mobilization of peripheral blood progenitor cells for harvesting prior to autologous stem cell transplant.
5. To increase granulocyte count in patients with aplastic anemia, myelodysplastic syndromes, congenital neutropenia and other congenital neutropenic disorders.

Contraindication

1. Hypersensitivity to G-CSF or to *E. coli*-derived proteins.

Dose and Duration

1. Start with a dose of 5 μg/kg/day SC or IV. When given intravenously, it is diluted with albumin and given over 15–30 minutes. The dose is 10 μg/kg/day SC or IV for peripheral blood stem cell mobilization in preparation for autologous transplant. A PEGylated form of G-CSF (pegfilgrastim) is used at a dose of 6 mg SC once per chemotherapy cycle in patients over 45 kg. There is no experience with pegfilgrastim in pediatrics.
2. Continue until the ANC is greater than $10,000/mm^3$.
3. Start G-CSF 24 hours after the last dose of chemotherapy.
4. Do not resume chemotherapy until 24–48 hours after discontinuing G-CSF.

Adverse Effects

1. Common: bone pain, elevation of uric acid, lactate dehydrogenase and alkaline phosphatase.

2. Uncommon: fever, nausea, vomiting, rash, diarrhea, splenomegaly, erythema at injection site, hypotension, exacerbation of psoriasis, allergic reaction, acute respiratory distress syndrome, splenic rupture.

Recombinant Human GM-CSF

Indications

1. To accelerate myeloid recovery following stem cell transplantation.
2. To accelerate myeloid recovery in a neutropenic patient with fungal infection, because it also accelerates macrophage recovery.

Contraindications

1. Excessive myeloid blasts in bone marrow or peripheral blood (>10%).
2. Juvenile chronic myeloid leukemia or monosomy 7 syndrome.
3. Hypersensitivity to GM-CSF or yeast-derived proteins.

Dose and Duration

1. The starting dose is 250 μg/m^2/day SC or IV over 2 hours.
2. When given to prevent engraftment delay, the first dose should be given 2–4 hours after stem cell transplantation and continued for 21 days or until the ANC is greater than 10,000 cells/mm^3.
3. Do not give within 24 hours of last chemotherapy or 12 hours of last radiation dose.

Adverse Effects

1. Common: bone pain, reaction at injection site.
2. Uncommon: fluid retention (peripheral edema, pleural effusion, pericardial effusion), leukocytosis, diarrhea, rash, malaise, fever, asthenia, headache, chills, arthralgias, chest pain, dyspnea, thrombocytopenia, thrombophlebitis, thrombosis, eosinophilia, weight gain, respiratory distress syndrome, bundle branch block, supraventricular arrhythmias. The first dose of GM-CSF may be followed within hours by a characteristic reaction involving skin flushing, tachycardia, hypotension, musculoskeletal pain, dyspnea, nausea and vomiting and arterial oxygen desaturation.

Use of G-CSF or GM-CSF and Risk of Leukemia

1. Numerous clinical studies have not shown detectable adverse effects of G-CSF on stimulation of myeloid leukemic cell growth.
2. In aplastic anemia, the risk of myelodysplastic syndrome (MDS) or leukemia is increased in children treated with immunosuppressive therapy and G-CSF.

3. Seven percent of patients with severe chronic neutropenia treated for more than 8 years with G-CSF develop AML or myelodysplastic syndrome (MDS). Patients with cyclic neutropenia treated with G-CSF have not developed a secondary malignancy.

Recombinant Human Erythropoietin

When planning to use recombinant human erythropoietin (rHuEPO), the following points must be considered:

1. If the degree of anemia is disproportionate to the degree of thrombocytopenia and neutropenia, this may be due to other causes of low hemoglobin such as hemorrhage, iron deficiency, or hemolysis and these will not be resolved by rHuEPO.
2. Not all patients respond to rHuEPO because of "end-organ" problems, such as myelodysplastic syndrome and some anemias of chronic illness.
3. Patients on cisplatin-containing regimens who suffer renal damage may respond well to rHuEPO.

Indications

Thrombotic events have been associated with the use of the various forms of recombinant erythropoietin. In addition, especially in adults, there are data showing a shortening of progression to relapse in patients with various cancers including lymphoma. Any decision to use these drugs should take these factors into account. There are no proven indications for the use of recombinant human erythropoietin in pediatric oncology and concerns exist about appropriate dosing schedule, cost and risk of thrombovascular events. Possible indications in pediatrics include:

*   Anemia secondary to renal failure
*   Anemia of chronic illness, such as human immunodeficiency virus and juvenile rheumatic arthritis
*   Anemia secondary to chemotherapy (to decrease the need for blood transfusions, particularly in children with solid tumors)
*   Anemia associated with radiation therapy
*   Anemia after allogeneic stem cell transplantation
*   Anemia secondary to myelodysplastic syndromes when the serum erythropoietin is low.

Contraindications

1. Anemia secondary to nutritional deficiencies, hemorrhage, or hemolysis.
2. Uncontrollable hypertension.
3. Hypersensitivity to mammalian-cell-derived products or to human albumin.

Dose and Dose Modification Schedule

1. Obtain baseline serum erythropoietin and ferritin levels prior to starting therapy.
2. If ferritin is low, prescribe ferrous sulfate 6 mg elemental iron/kg/day divided into three daily doses.
3. Start rHuEPO at a dose of 50–100 units/kg/day subcutaneously (SC) three times a week. Darbepoetin alpha is an analog of erythropoietin with a longer plasma half-life and can be used once every 2–4 weeks. The starting dose is 2.25 µg/kg/dose.
4. If there is no response within 2–4 weeks, increase dose up to 300 Units/kg/day three times a week or the darbepoetin dose to 4.5 µg/kg/week.
5. If the hemoglobin reaches 13 g/dl, discontinue rHuEPO until the hemoglobin is 12 g/dl; then resume at 25% dose.
6. If the hemoglobin increases very rapidly (>1 g/dl in 2 weeks) or hemoglobin reaches 12 g/dl, reduce the dose by 25%.
7. Give rHuEPO concurrently with chemotherapy.
8. Continue rHuEPO until the patient no longer requires red cell support.

Early initiation (prior to the need for transfusions), a low observed or predicted erythropoietin level (<0.825) and evidence of early response (more than 1 g/dl rise in hemoglobin within the first 4 weeks) are all associated with increased likelihood of response. Patients meeting all three of these criteria show an 85% chance of significant response (more than 2 g/dl rise in hemoglobin). Almost all of these patients have renal-failure-associated anemia and were treated with subcutaneous erythropoietin.

Adverse Reactions

1. Common: hypertension, pain at the injection site, headache, fever, diarrhea.
2. Uncommon: nausea, malaise, seizures, thrombosis. There are reports of pure red cell aplasia due to neutralizing antibodies during erythropoietin treatment.

Interleukin-11

IL-11 acts synergistically with IL-3 and thrombopoietin (TPO) to stimulate various stages of megakaryocytopoiesis and thrombopoiesis. IL-11 has been evaluated in several human clinical trials and is currently approved for the prevention of severe chemotherapy-induced thrombocytopenia in patients with nonmyeloid malignancies. Low dose IL-11 has also been shown to be effective in patients with bone marrow failure.

Indications

1. Prevention of severe thrombocytopenia after myelosuppressive chemotherapy.

Contraindications

1. Hypersensitivity.
2. Children younger than 12 years of age.
3. Myeloid malignancies.
4. In patients with heart failure, atrial arrhythmias, thromboembolic disorders, major organ dysfunction, papilledema, or CNS tumors, IL-11 should be used with caution.

Dose and Dose Modification Schedule

The dose is 50 µg/kg /day in adults. A dose of 50–75 µg a day has been suggested for children. IL-11 has not been commonly employed because of the side effects, especially capillary leak. No large pediatric trials are available.

Adverse Reactions

1. Common: edema, tachycardia, headache, fatigue, dizziness, anorexia, nausea, conjunctivitis and papilledema in children, dyspnea, skin rashes, arthralgia, myalgia.
2. Uncommon: atrial arrhythmia, thrombosis, cerebral infarction, syncope.

Thrombopoietin

Thrombopoietin is a hematopoietic factor that controls platelet production and levels of thrombopoietin rise 24 hours after the onset of thrombocytopenia. Studies using recombinant thrombopoetin have led to the development of antithrombopoietin antibodies in normal subjects which raises the issue of safety of its use in treating chemotherapy-related thrombocytopenia.

Interest has been increasing for the thrombopoietin mimetic agents or thrombopoietin receptor agonists that have proven their value in clinical trials of ITP. Trials are underway to assess their usefulness in the prevention of thrombocytopenia as well as in reversal of thrombocytopenia in myelodysplastic syndromes and bone marrow failure states.

PREVENTION OF ORGAN TOXICITY

Prevention of Cardiac Toxicity

A baseline EKG, echocardiogram and/or multigated acquisition scan (MUGA) scan should be obtained before anthracycline treatment or thoracic radiotherapy. Cardiac toxicity is seen in patients receiving chemotherapy with anthracyclines (doxorubicin, daunorubicin, idarubicin) as well as thoracic radiotherapy. Toxicity is anticipated in patients receiving a cumulative dose exceeding 450 mg/m^2 for doxorubicin and daunorubicin and exceeding 125 mg/m^2 for idarubicin. Toxicity may also occur in patients receiving radiation to the heart with doses exceeding 1,000 cGy.

During therapy, echocardiogram or radionuclide cardiac cineangiocardiography (RNA) should be performed regularly. If the planned cumulative dose of daunomycin/doxorubicin is less than 300 mg/m^2, testing is repeated after half the cumulative dose has been given. After the cumulative dose reaches more than 360 mg/m^2, testing is repeated before each cycle. If an abnormality is detected in either the echocardiogram or MUGA is appreciated, the other test should be performed as confirmation of the abnormality.

The following are criteria for progressive deteriorating cardiac function:

- A decrease in the fractional shortening (SF) by an absolute value of 10% from the previous test
- SF less than 29%
- A decrease in the RNA–left ventricular ejection fraction (RNA LVEF) by an absolute value of 10% from the previous test
- RNA LVEF less than 5%
- A decrease in the RNA LVEF with stress
- Development of arrhythmia.

If there is a documented infection or development of arrhythmia during therapy, anthracyclines should be discontinued and a cardiology consultation should be obtained.

Modification of Anthracycline Therapy

1. If both the SF and the RNA LVEF are abnormal, anthracycline therapy should be discontinued unless there is a recovery of SF and RNA LVEF to normal on two serial tests taken 1 month apart.
2. If one of the preceding test modalities is abnormal, while the other is normal, anthracyclines should be temporarily discontinued. Both tests should be repeated after 1 month. If one remains normal and the other becomes normal or does not deteriorate further, therapy should be resumed. Therapy should be discontinued if further deterioration occurs in either tests. Patients who receive anthracyclines and/or thoracic radiotherapy should be evaluated by EKG and echocardiogram every year for 5 years.

Low-Dose Radiotherapy with Combination Chemotherapy

1. Appropriately lowered cumulative dose of anthracycline should be employed when radiotherapy is administered over the heart.
2. Use of less than 450 mg/m^2/cumulative dose for both doxorubicin and daunomycin and 125 mg/m^2 for idarubicin and 160 mg/m^2 for mitoxantrone. Patients who have received a cumulative dose of doxorubicin of more than 450 mg/m^2 should not receive mitoxantrone. The recommended cumulative dose of mitoxantrone for patients who have received prior treatment with doxorubicin is 120 mg/m^2.

3. Use of cardioprotective agents such as dexrazoxane (ICRF-187, Zinecard, DZR). The dose of dexrazoxane is 300 mg/m^2 (10 times the doxorubicin dose) IV bolus over 15 minutes followed by doxorubicin or daunomycin IV over 15 minutes, that is, before the elapsed time of 30 minutes (from the beginning of the dexrazoxane). This drug is not standard of care and is still investigational.
4. Cardiac tissue damage serum marker (e.g., determination of cardiac troponin levels (cTnT), creatinine kinase (CK)-MB isozyme). cTnT is a thin-filament contractile protein that is released from damaged cardiomyocytes. Unlike CK-MB, cTnT is not found in the serum of a healthy person.
5. ECHO/MUGA surveillance studies as recommended.
6. Use of new anthracyclines with decreased potential for cardiotoxicity is being investigated.

See Chapter 32 for guidelines on long-term follow-up of patients who receive anthracyclines.

Prevention of Renal Toxicity

1. Maintain hydration at 125 ml/m^2/h (i.e., twice maintenance) for a minimum of 2 hours prior to start of chemotherapy with agents such as high-dose methotrexate, moderate to high dose of cyclophosphamide, cisplatin, or ifosfamide to increase urine output to more than 3 ml/kg/h and decrease urinary specific gravity below 1.010.
2. During infusion of these agents, maintain hydration at 125 ml/m^2/h and urinary output at 5 ml/kg/h.
3. After infusion, continue hydration at 90–125 ml/m^2/h.
4. Strict monitoring of intake, output and daily weights is critical. Maintain isovolemic fluid balance. When needed, use furosemide 0.5–1 mg/kg IV or mannitol 6 g/m^2 (200 mg/kg) in at least 25 ml of fluid over 15–60 minutes.
5. Use mesna for its uroprotective effect to prevent hemorrhagic cystitis caused by high-dose cyclophosphamide and ifosfamide. A total mesna dose equal to the total dose of cyclophosphamide (0.8 mg of mesna per 1 mg of cyclophosphamide) is recommended.
6. Adjust the doses of chemotherapy according to the glomerular filtration rate (GFR) (Table 31-13).
7. During infusion of cisplatin, add mannitol 15 g/m^2 (10–24 g/m^2/L) and MgSO$_4$ in a dose of 20 mEq/L to the hydrating solution to prevent hypomagnesemia.
8. Continue postchemotherapy hydration with 5% dextrose/0.45% NaCl+20 mEq/L KCl+20 mEq/L MgSO$_4$+20 g/L mannitol.
9. Measurement of renal function and electrolytes prior to and during each cycle. Correct metabolic derangements in a timely manner.

Table 31-13   Suggested Percentage Dose of Chemotherapeutic Agents Adjusted for Glomerular Filtration Rate

>60 ml/min:	30-60 ml/min (50-75% Baseline)[a]		10-30 ml/min (25-49% Baseline)		< 10 ml/min (<25% Baseline)	
100% Dose	75%	Omit	75%	Omit	50%	Omit
Adriamycin[b]	Bleomycin[c]	Cisplatin	Carboplatin[d]	Cisplatin	Cyclophosphamide	Cisplatin
Bleomycin		Methotrexate	Ifosfamide[d]	Methotrexate		Methotrexate
Cisplatin		Nitrosureas	Nitrosureas	Nitrosureas		
Cyclophosphamide		Bleomycin[a]				
Cytarabine[b]		Cisplatin				
5-fluorouracil[b]						
Ifosfamide						
Melphalan[b]						
Methotrexate						
Nitrosureas						
Vinblastine[b]						
Vincristine						

[a]Percentage of baseline may not be suitable if high urine output renal dysfunction present.
[b]No dose modification for decreased glomerular filtration rate.
[c]Recommendations vary per protocol.
[d]Reduce by 50%.
From: Barnard DR, Friedman DL, Landier W, et al. Monitoring and management of drug toxicity. In: Altman AJ, ed. Supportive care of children with cancer. Current therapy and guidelines from the Children's Oncology Group, 3rd ed. Baltimore: The Johns Hopkins University Press, 2004: 100–148, with permission.

Prevention of Neurotoxicity

Neurotoxicity ranges from peripheral neuropathy following vincristine and cisplation to acute encephalopathy following intrathecal chemotherapy or high-dose methotrexate. Following administration of vincristine, patients may complain of jaw pain, constipation, foot drop, lid lag and/or tingling and numbness in the distal extremities. Supportive care includes frequent neurologic examinations, pain management and oral laxatives to prevent constipation. If neurologic deficits persist, dose modifications or discontinuation of medications may be necessary.

PAIN MANAGEMENT

Adequate pain management is crucial to the quality of life of a child with cancer. At diagnosis, pediatric oncology patients have tumor-related pain. As the treatment evolves, treatment-related pain predominates. The causes of pain in a child with cancer are:

- Cancer-related pain: bone pain, compression of central or peripheral nervous system structures
- Procedure-related pain: venipuncture, bone marrow aspiration and biopsy, lumbar puncture
- Treatment-related pain:
  a. Chemotherapy: mucositis, peripheral neuropathy, aseptic necrosis of bone, steroid-induced myopathy
  b. Radiation: mucositis, radionecrosis, myelopathy, brachial/lumbar plexopathies
  c. Postsurgical: acute postoperative pain, phantom pain.

Pain management can be attained through an integration of drugs and other modalities. Pharmacologic analgesia is the mainstay of pain management. Nonpharmacologic methods of pain control are:

- Preparation of the child prior to a procedure
- Physical: massage, heat and cold, electrical nerve stimulation, acupuncture
- Behavioral: art and play therapy, exercise, relaxation, biofeedback, desensitization, operant conditioning
- Cognitive: distraction, imagery, attention, hypnosis, music therapy, psychotherapy
- Complementary and alternative medicine.

The recommended starting doses, route of administration and schedules for analgesic medications are listed in Table 31-14.

The dosage and frequency of the analgesics are adjusted to the patient's pain and are gradually increased as needed to maintain the patient free of pain. The next dose should be given before the previous dose has worn off fully to erase the memory and fear of pain.

Table 31-14   Recommended Starting Doses for Analgesic Medications

Medication	Dose (mg/kg)	Route	Schedule
Nonsteroidal anti-inflammatory drugs			
Acetaminophen	10–15	PO	q4h
	15–20	PR	q4h
Choline-magnesium Salicylate	10–15	PO	q6–8h
Ibuprofen	5–10	PO	q6h
Rofecoxib	0.6 mg/kg (adult dose 25 mg)	PO	qd
Ketorolac	0.5	IV	q6h
Opioids			
Codeine	0.5–1	PO	q4h
Fentanyl	0.001–0.002 (1–2 µg/kg)	IV	q1–2h
	0.002–0.004 (2–4 µg/kg)	IV	qh continuous infusion
Hydromorphone	0.010–0.015	IV	q3–4h
	0.1	PO	q4h
Methadone	0.1	IV	q4h×2–3 doses; then q8–12h
	0.3–0.7/day	PO	Divide into three equal doses
Morphine	0.08–0.1	IV	q2–3h
	0.03–0.05	IV	qh continuous infusion
	0.2–0.3	PO	q4h
Morphine (MS contin)	0.3–0.6	PO	q12h long acting
Oxycodone	0.15	PO	q4h
Oxycodone (Oxycontin)	0.15	PO	q12h long acting
Adjuvants			
Amitriptyline	0.1–0.2	PO	qday at bedtime; advance to 0.5–2.0 mg/kg/day
Gabapentin	Initial 5 mg/kg or 300 mg	PO	qday at bedtime; advance dose to TID
Methylphenidate	0.1–0.2	PO	qdose; slowly advance dose as tolerated
Dextroamphetamine	0.1–0.2	PO	qdose; slowly advance dose as tolerated

Note: For all medications, dosages should be modified based on individual circumstances. Many of these agents have not yet received specific approval for infants and younger children. Certain doses are based on extrapolation from adult doses or from unpublished experience. For nonintubated infants aged 4 months or less, reduce intial opioid doses to one-third to one-quarter of the recommended doses. Administer opioids with the patient in a location that permits close observation and immediate intervention. For the management of severe ongoing acute or chronic pain, increase opioid doses until comfort is achieved or until side effects prohibit further dose escalation.
From: Sacks N, Ringwals-Smith K, Hale G. Nutritional support. In: Autumn AJ, ed. Supportive Care of Children with Cancer. Current therapy and guidelines from the Children's Oncology Group, 3rd ed. Baltimore: The Johns Hopkins University Press, 2004: 243–261, with permission.

Simplest dosage schedule and least-invasive pain management modality should be used first. Medications for persistent cancer-related pain should be administered on an around-the-clock basis, with additional "as needed" doses because regularly scheduled dosing maintains a constant level of drug in body. Table 31-15 lists the opioid analgesic initial dosage guidelines. With ongoing administration of narcotic analgesics, tolerance builds up and

Table 31-15   Opioid Analgesic Initial Dosage Guidelines

Drug	Equianalgesic Doses		Usual Starting IV or SC Doses and Intervals		Parenteral-Oral Dose Ratio	Usual Starting Oral Doses and Intervals	
	Parenteral	Oral	Child <50 kg	Child >50 kg		Child <50 kg	Child >50 kg
Codeine	120 mg	200 mg	NR	NR	1:2	0.5–1.0 mg/kg q3–4h	30–60 mg q3–4h
Morphine	10 mg	30 mg (chronic) 60 mg (single dose)	Bolus: 0.1 mg/kg q2–4h Infusion: 0.03 mg/kg/h	Bolus: 5–8 mg q2–4h Infusion: 1.5 mg/h	1:3 (chronic) 1:6 (single dose)	Immediate release 0.3 mg/kg q3–4h Sustained release: 20–35 kg: 10–15 mg q8–12h 35–50 kg: 15–30 mg q8–12h	Immediate release 15–20 mg q3–4h Sustained release 30–45 mg q8–12h
Oxycodone	NA	15–20 mg	NA	NA	NA	0.1–0.2 mg/kg q3–4h	5–10 mg q3–4h
Methadone[a]	10 mg	10–20 mg	0.1 mg/kg q4–8h	1:2	1:1.5–1:2	0.15–0.2 mg/kg q4–8h	7–10 mg q4–8h
Fentanyl	100 µg (0.1 mg)	NA	Bolus: 0.5–1.0 µg/kg q1–2h Infusion: 0.5–2.0 µg/kg/h	Bolus: 25–50 µg q1–2h Infusion: 25–100 µg/h	NA	NA	NA
Hydromorphone	1.5–2.0 mg	6–8 mg	Bolus: 0.02 mg q2–4h Infusion: 0.006 mg/kg/h	Bolus: 1 mg q2–4h Infusion: 0.3 mg/h	1:4	0.04–0.08 mg/kg q3–4h	2–4 mg q3–4h
Meperidine[b] (pethidine)	75–100 mg	300 mg	Bolus: 0.8–1.0 mg/kg q2–3h	Bolus: 50–75 mg q2–3h	1:4	2–3 mg/kg q3–4h	100–150 mg q3–4h

[a]Methadone requires additional vigilance, because it can accumulate and produce delayed sedation. If sedation occurs, doses should be withheld until sedation resolves. Thereafter, doses should be substantially reduced or the dosing interval should be extended to 8–12 hours (or both).

[b]Meperidine should generally be avoided if other opioids are available, especially with chronic use, because its metabolite can cause seizures.

Abbreviations: NA, not available; NR, not recommended.

Note: Doses refer to patients older than 6 months. In infants younger than 6 months, initial doses per kg should begin at roughly 25% of the doses per kg recommended here. All doses are approximate and should be adjusted according to clinical circumstances.

From: Berde CB, Sethna NF. Analgesics in the treatment of pain in children. N Engl J Med 2002: in press; and from Cancer Pain Relief and Palliative Care in Children. Geneva: World Health Organization Press, 1998, with permission.

increasing doses must be given to maintain the same level of analgesia. Tolerance and physical dependence are expected with long-term use and should not be confused with psychological dependence ("addiction") Combining a narcotic with tricyclic antidepressants, anticonvulsants, amphetamines, steroids, or topical agents may potentiate its analgesic activity. Guidelines for opioid therapy for persistent cancer pain are listed in Table 31-16 and Table 31-17 lists examples and dosages of adjuvant analgesics. Constipation, nausea and somnolence are common side effects of opioid treatment. Management aspects of these opioid side effects are listed in Table 31-18.

Analgesic medications include:

- Nonopioids
  a. Acetaminophen: Inhibits prostaglandin synthesis in the brain. The dose is 15 mg/kg/dose every 4–6 hours
  b. Aspirin and nonsteroidal anti-inflammatory drugs (NSAIDs): Contraindicated in oncology patients because of adverse effects on platelet function. Choline magnesium salicylate has minimal effects on platelet function; however, its safety has not been established in severely thrombocytopenic patients. COX-2 inhibitors do not cause platelet dysfunction, but have not been tried in severely thrombocytopenic patients
- Opioids:
  a. Weak opioids:
     (1) Codeine: 0.5–1 mg/kg/dose orally every 4 hours. Codeine is a relatively weak analgesic
     (2) Oxycodone: 0.05–0.15 mg/kg/dose orally every 4–6 hours.
  b. Strong opioids
     (1) Used for treatment of moderate to severe pain
     (2) Important to use on a regular schedule rather than as needed
     (3) Oral route is preferred route – most convenient and cost-effective. If this is not available or appropriate, rectal or transdermal routes should be attempted, if not IV or subcutaneous route should be utilized
     (4) Adjuvant drugs may be used to counteract side effects
        - Hydroxyzine for nausea and/or anxiety
        - Caffeine to counteract sedation and provide additional analgesia
        - Phenothiazines for nausea
     (5) Documentation of failure of maximal doses or opioids and coanalgesics administered through other routes should precede use of intraspinal analgesia
     (6) Side effect profile
        - Sedation
        - Nausea and vomiting

Table 31-16 Guidelines for Opioid Therapy for Persistent Cancer Pain

Comprehensive assessment
• Define pain etiology, pathophysiology and syndrome
• Clarify status of disease
• Determine impact of the pain and comorbid physical and psychosocial disturbances
Drug selection
• Consider age and whether major organ failure is present, especially renal, hepatic, or respiratory
  Consider drug-selective differences in side effect or toxicity profile
• Consider the effects of concurrent drugs with possible pharmacokinetic and pharmacodynamic
  interactions
• Consider individual differences (note prior treatment outcomes) and patient preference
• Be aware of available preparations for route (e.g., oral, IV, subcutaneous injection, topical) and
  formulation (e.g., immediate or controlled-release)
• Be aware of cost differences
Route selection
• Use least invasive route possible
• Consider patient convenience and compliance
Dosing
• Consider previous dosing requirements and relative analgesic potencies when initiating therapy
• Start with low dose and increase until adequate analgesia occurs or dose-limiting side effects are
  encountered
• Consider dosing schedule (e.g., around-the-clock or as needed) depending on the anticipated time
  course of pain
• Consider "rescue medication" for breakthrough pain
• Recognize that tolerance is rarely the driving force for dose escalation; consider disease progression
  when increasing dose requirements occur
Treat side effects
• Give laxatives prophylactically
• Be prepared to treat nausea, itch and somnolence
• Trial of alternative opioids
Monitoring
• Monitor treatment efficacy and pain status over time and consider modification if necessary

From: Portenoy RK, Lesage P. Pain management—the series. Part 4: Cancer Pain and End-of-Life Care. American Medical Association, with permission.

Table 31-17 Examples and Dosages of Adjuvant Analgesics

Dexamethasone	0.08–0.3 mg/kg/day PO, IV divided every 6–12 hours
Amitriptyline	0.1 mg/kg/dose PO at bedtime; increase as needed and tolerated to 0.5–2 mg/kg/dose
Gabapentin	300 mg PO three times a day (for children over 12 years old and adults)
Methylphenidate	0.3 mg/kg/dose before breakfast and lunch; may gradually increase to 1 mg/kg/day

• Respiratory depression – naloxone titration, when used to reverse opioid-induced respiratory depression, should be given at doses that improve respiratory function without reversing analgesia
• Corticosteroids – pain secondary to visceral distention, bony destruction, cerebral edema.

Table 31-18   Approaches in the Management of Opioid Side Effects

Opioid Side Effect	Treatment
Constipation	*General approach* • Increase fluid intake and dietary fiber • Encourage mobility and ambulation if appropriate • Ensure comfort and convenience for defecation • Rule out or treat impaction if present *Pharmacologic approach* • Contact laxative plus stool softener (e.g., senna plus docusate) • Osmotic laxative (e.g., milk of magnesia) • Lavage agent (e.g., oral propylethylene glycol) • Prokinetic agent (metoclopramide) • Oral naloxone
Nausea	*General approach* • Hydrate as appropriate • Progressive alimentation • Good mouth care • Correct contributory factors • Adjust medication *Pharmacological approach* • If associated with vertiginous feelings, antihistamine (e.g., scopolamine, meclizine) • If associated with early satiety, prokinetic agent (e.g., metoclopramide) • In other cases, dopamine antagonist drugs (e.g., prochlorperazine, metoclopramide)
Somnolence or Cognitive Impairment	*General approach* • Reassurance • Education • Treatment of potential etiologies *Pharmacologic approach* • If analgesia is satisfactory, reduce opioid dose by 25–50% • If analgesia is satisfactory and the toxicity is somnolence, consider a trial of a psychostimulant (e.g., amphetamine or methylphenidate)
Itching	• Begin with an antihistamine at low doses, especially in patients who receive premedication (daily dose should not to exceed 5 mg/kg/day) • Low dose nalaxone in doses of 0.25–2.0 μg/kg/h by continuous infusion may dramatically reduce narcotic-induced itching without interference in pain control

Adapted from: Portenoy RK, Lesage P. Pain management – the series. Part 4: Cancer Pain and End-of-Life Care. American Medical Association, with permission.

Neuropathic Pain

Neuropathic pain may result from compression or infiltration of neural tissue by tumor. Several nonanalgesic mediations are available for the treatment of neuropathic pain:

• Tricyclic antidepressants
• Anticonvulsants (valproate, carbamazepine, topiramate)

- Gabapentin: first-line therapy
- Amitriptyline: for pain and insomnia.

GENERAL GUIDELINES FOR MANAGEMENT OF PAIN IN ONCOLOGY PATIENTS

Determine the Class of Drug Based on Severity of Pain

1. Mild to moderate: NSAID or acetaminophen, unless contraindicated.
2. Moderate at outset or persistent mild pain: opioid should be added (not substituted), i.e. codeine, oxycodone, hydrocodone. May be used in fixed-dose combinations but if pain persists, recommend separate dosages to avoid excessive doses of NSAIDs or acetaminophen.
3. Moderate to severe pain at outset or persistent moderate pain: increase opioid potency or dose, i.e. morphine, hydromorphone, methadone, fentanyl.
4. Severe, chronic pain: best managed with immediate-release and long-acting morphine.

Patient-Controlled Analgesia

Patient-controlled analgesia (PCA) is a method of opioid administration using a computer-controlled pump that enables the patient to deliver small boluses as needed up to a preset maximum. It can be used with a baseline continuous infusion. The advantages of PCA are:

- It permits titrated dosing to compensate for individual variation in pharmacokinetics and pain intensity
- It permits the patient to exercise control and diminishes anxiety
- It allows the patient to balance analgesia against its side effects such as sedation
- It helps determine conversion to oral regimen once steady level of appropriate analgesia is attained
- It is safe for home and hospital use.

The usual starting dose is 0.01–0.02 mg/kg of morphine every 6–10 minutes with or without a basal infusion of 0.01–0.02 mg/kg/h.

Anesthesia for Painful Procedures

Painful or invasive procedures are a major source of anxiety for the child undergoing therapy for cancer. These procedures include insertion of intravenous lines, accessing of mediports, spinal taps, bone marrow aspirations and pleural and peritoneal taps. Adequate local or general anesthesia, if indicated, can significantly reduce the child's fear.

Local Anesthesia

Local anesthesia is adequate for inserting intravenous lines, for accessing mediports and for performing spinal taps and bone marrow aspirations on some children. Local anesthesia can be provided via:

- EMLA (Eutectic Mixture of Local Anesthetics) cream (lidocaine 2.5%+prilocaine 2.5%) applied topically to the skin and covered with an occlusive dressing 1 hour prior to the procedure
- Lidocaine 2% injected superficially into the area of the procedure
- Skin cooling with ice or fluorocarbon cooling sprays.

General Sedation/Anesthesia

General sedation/anesthesia is required for more painful procedures, such as bone marrow biopsies and for lesser procedures in the young child who is extremely fearful. The following agents may used:

- Propofol:
  a. Sedative hypnotic, not analgesic
  b. May produce respiratory depression and hypotension
  c. Supplied in an emulsion with egg lecithin; avoid use in patients with egg allergies
  d. Short-acting anesthetic; patients awaken when infusion is terminated
  e. Dosage: 2 mg/kg IV bolus followed by continuous infusion of 40–200 μg/kg/minute; reduce dose by half for patients with neurologic, cardiac, or respiratory disease
  f. Some patients develop tolerance and require higher doses
- Brevital (methohexital):
  a. Very short-acting anesthetic, barbiturate
  b. May produce respiratory depression, hypotension
  c. Dosage: 1–2 mg/kg/dose IV or 5–10 mg/kg/dose IM
- Ketamine:
  a. Rapid-acting anesthetic with profound analgesia
  b. Produces fugue-like state, disorientation, delirium, bizarre dreams
  c. Can be given along with midazolam dose to cause amnesia of the dreams
  d. Dosage: 2 mg/kg IV; increments of one half to the full dose may be repeated as needed
- DPT (this combination is rarely used and should be avoided in very young children because of reports of life-threatening toxicity from phenergan):
  a. Demerol 2 mg/kg IV, phenergan 1 mg/kg IV, thorazine 1 mg/kg IV
  b. Sedative, analgesic and anesthetic
  c. Produces sedation for 3–4 hours

The current American Academy of Pediatrics Committee guidelines for pediatric patients undergoing conscious sedation require the presence of medical personnel trained in the administration of medications used for procedural sedation and management of complications associated with these medications. Management requires the medical staff to identify when the level of sedation has exceeded the intended effect and be skilled in rescue methods to reverse such effects. Anesthiologists and other physicians trained in the skilled management of airway problems are the preferred practitioners for procedural sedation. When an anesthesiologist is present, propofol is preferred because it is extremely short acting and the dose can be titrated as needed.

MANAGEMENT OF NAUSEA AND VOMITING

In most cases, nausea and vomiting can be prevented. Every effort should be made to provide adequate antiemetic coverage. The receptors for nausea and vomiting should be blocked before treatment starts and should continue as long as the effect is likely to occur. Scheduled doses should be administered regardless of symptoms.

Vomiting is mediated by the vomiting center located in the medullary lateral reticular formation. This center gets input from five sources:

1. Chemoreceptor trigger zone.
2. Vagal and sympathetic afferents from viscera.
3. Midbrain receptors that detect changes in intracranial pressure.
4. Labyrinthine apparatus.
5. Limbic system.

Chemotherapeutic agents may stimulate vomiting by direct effect on the vomiting center or through the chemoreceptor trigger zone. Serotonin (5-HT) receptors and substance P play a role in mediating emesis. The emetogenic potential of chemotherapeutic agents is listed in Table 31-19.

The etiology of vomiting should be identified before therapy is initiated. Unusual severity, timing, or duration should alert the physician to a diagnosis other than chemotherapy-induced nausea and vomiting.

Chemotherapy-induced nausea and vomiting (CINV) is divided into:

*   Anticipatory: prior to chemotherapy administration – results from learned or conditioned response
*   Acute: within 24 hours following chemotherapy
*   Delayed: 24 hours after chemotherapy
*   Breakthrough: despite the administration of antiemetics.

Table 31-19   Emetogenic Potential of Chemotherapeutic Agents[a]

Acute Symptoms
  Highly emetogenic
    Actinomycin-D
    Cisplatin ($\geq$40 mg/m^2)
    Cyclophosphamide ($\geq$1 g/m^2)
    Cytarabine ($\geq$1 g/m^2)
    Dacarbazine
    Ifosfamide
    Mechlorethamine
  Moderately emetogenic
    Anthracyclines (daunorubicin, doxorubicin, idarubicin)
    Carboplatin
    Cisplatin (<40 mg/m^2)
    Cyclophosphamide (<1 g/m^2)
    Cytarabine (IV <1 g/m^2, or IT)
    Mercaptopurine (IV)
    Methotrexate (IV >1 g/m^2)
    Nitrosoureas (carmustine, lomustine)
  Mildly emetogenic
    Bleomycin
    Epipodophyllotoxins (etoposide, teniposide)
    Paclitaxel
    Procarbazine
    Topotecan
    Vinblastine
  Nonemetogenic
    Asparaginase
    Mercaptopurine (PO)
    Methotrexate (low-dose IV, IM, PO, IT)
    Steroids
    Thioguanine
    Vincristine
Delayed Symptoms
  Severe
    Cisplatin
  Moderate
    Cyclophosphamide

[a]Independent of route or dose unless specific route or dose is specified.
From: Barnard DR, Friedman DL, Landier W, et al. Monitoring and management of drug toxicity. In: Altman AJ, ed. Supportive Care of Children with Cancer. Current therapy and guidelines from the Children's Oncology Group, 3rd ed. Baltimore: The Johns Hopkins University Press, 2004: 100–148, with permission.

Chemotherapy agents are classified into four categories of emesis risk without use of antiemetics:

- High (>90%)
- Moderate (30–90%)

- Low (10–30%)
- Minimal (<10%).

Table 31-20 lists dosage and side effects of commonly used antiemetic agents.

Table 31-21 lists the antiemetic regimens used for the initial cycle of chemotherapy or radiotherapy.

There are five classes of antiemetic available:

- Serotonin antagonists
- Neurokinin antagonists
- Cannabinoids
- Dopamine antagonists
- Corticosteroids

Serotonin (5-HT3) Receptor Antagonists

1. Ondansetron (zofran).
2. Ganisetron (kytril).
3. Palonosetron (Aloxi) novel 5-HT3 receptor antagonist has 100-fold greater binding affinity for the serotonin receptor and a longer half-life (≈40 hours longer).
   a. Used for prevention of acute and delayed CINV with moderately and highly emetogenic chemotherapy for patients ≥18yrs
   b. Palonsetron in combination with dexamethasone provides complete response of 40.7% compared to ondansetron plus dexamtheasone which has a 25.2% response rate.

Neurokinin-1 Antagonists Aprepitant (Emend)

1. Aprepitant blocks neurokinin receptor and enhances activity of 5-HT3 receptor antagonists. It is metablolized by CYP34A enzyme
   a. Used in patients >40 kg and ≥12 years
   b. When combined with ondansetron and dexamethasone provides complete response rate of 72.7% on days 1–5 however, aprepitant alone or in combination with dexamethasone does not control acute emesis as well as ondansetron and dexamethasone
   c. It is an inducer of CYP2CP and can therefore react with many drugs. The dose of steroid given concomitantly as antiemetics should be decreased by 25% if given IV or 50% if given orally. Doses of benzodiazepine may also need to be decreased. Aprepitant should be used with caution when administering with etoposide, ifosfamide, imatinib, irinotecan, paclitaxel, vincristine or vinblastine.

Table 31-20   Dosage and Side Effects of Commonly Utilized Antiemetic

Agents dosing should routinely be scheduled and not as needed. Initial dose should precede initial chemotherapy by 30 minutes IV or 1 hour if given PO

Medication	Dosage	Side Effects	Comments
5-HT₃ Serotonin antagonist Ondansetron (Zofran)	PO: *Dose based on body surface area* BSA: <0.3 m²: 1 mg 3 times/day BSA 0.3–0.6 m²: 2 mg 3 times/day BSA 0.6–1 m²: 3 mg 3 times/day BSA >1 m²: 4–8 mg 3 times/day IV: 0.15 mg/kg/dose IV q 4 hours ×3 doses; single daily dosing: 0.45 mg/kg/dose IV once in 24 hours; maximum single daily dose IV is 32 mg PO: *dose based on age:* Age <4 yrs use dose based on body surface area from above Child 4–11 years: 4 mg before chemotherapy and every 4 hrs for 2 doses thereafter 4 mg 3 times/day Alternative therapy: single 12 mg dose before chemotherapy Child >11 yrs and adults: 8 mg before chemotherapy and every 4 hrs for 2 doses thereafter 8 mg 3 times/day Alternative therapy: single 24 mg dose before chemotherapy or 24 mg once daily *Intravenous* Child >3 yrs: 0.15 mg/kg/dose before chemotherapy and every 4 hours for 2 doses	Headache, asthenia, lightheadedness, dizziness, asymptomatic prolongation of electrocardiographic interval, constipation, diarrhea, abdominal pain, ataxia, fever, tremor, transient serum transaminase elevations, sedation, fatigue, warm sensation on IV administration	Oral therapy is considered as efficacious as IV therapy and less costly. Single dose regimens preferred over multiple-dose regimens in adult patients. Antiemetic of choice for prophylactic therapy in pediatric and adult patients receiving chemotherapy with emetic potential of levels 3 through 5 (in combination with a corticosteroid). At equipotent doses, ondansetron, dolasetron and granisetron are equally efficacious. Use in combination with corticosteroids for delayed emesis

(Continued)

Table 31-20 (Continued)

Medication	Dosage	Side Effects	Comments
Granisetron (Kytril)	*Oral/Intravenously* Child >2 years: 10–20 μg/kg/dose every 12 hours (maximum single dose 1,000 μg) and 20–40 μg/kg/dose every 24 hours (maximum single dose 2,000 μg)		
Dolasetron (Anzemet)	*Oral/intravenously* Child >2 years: 1.8 mg/kg before chemotherapy and every 24 hours (maximum single dose: 100 mg)		See text for cautions about the use of this drug
NK-1 Receptor Antagonist Aprepitant (Emende)	>40 kg and ≥12 years 125 mg orally once a day, then 80 mg orally on subsequent 2 days		
Corticosteroids Dexamethasone (Decadron)	*Oral/intravenously* Standard dose: 10 mg/m²/dose as loading dose (maximum 20 mg) before chemotherapy and 5 mg/m²/dose (maximum 120 mg) every 6–12 hours as needed Low dose: 2.5–3 mg/m²/dose (maximum dose 5 mg every 12 hours)	Anxiety, insomnia, dyspepsia, hyperglycemia, euphoria, behavioral changes, agitation, perineal burning after rapid IV infusion	Can be used as a single agent in prophylactic therapy for adult and pediatric patients receiving chemotherapy regimens with level 2 emetic potential Should be added to $5-HT_3$ receptor antagonist for adult and pediatric patients receiving chemotherapy with potential emetic levels 3 through 5
Methylprednisolone (Solumedrol)	*Oral/intravenously* 0.5–1 mg/kg/dose before chemotherapy and every 4 hours for 2 doses		Oral therapy as safe and effective as IV therapy Use for breakthrough nausea and vomiting in adult and pediatric patients Use with metoclopramide or $5-HT_3$ receptor antagonist for delayed emesis in adult patients Use with lorazepam $5-HT_3$ receptor

Drug	Route/Dose	Side effects	Comments
Benzodiazepine Lorazepam (Ativan)	*Oral/intravenously* Child: 0.02-0.05 mg/kg/dose every 6-8 hours as needed (maximum 2 mg/dose) Adult: 1-2 mg/dose every 6 hours as needed	Sedation, amnesia, behavior changes	antagonist or chlorpromazine for delayed emesis in pediatric patients May be effective to reduce anticipatory vomiting Use for breakthrough nausea and vomiting in adult and pediatric patients Use in combination with dexamethasone for delayed emesis in pediatric patients
Phenothiazine Chlorpromazine	*Oral* Child >6 months: 0.6 mg/kg/dose every 4-6 hours as needed Maximum dose 40 mg/day if <5 years; 75 mg/day if 5-12 years *Intravenously* Child >6 months: 0.5 mg/kg/dose every 6-8 hours as needed. Maximum dose 40 mg/day if >5 years 75 mg/day if 5-12 years	Sedation, extrapyramidal effects	Use for breakthrough nausea and vomiting in adult and pediatric patients Use in combination with dexamethasone for delayed emesis in pediatric patients
Promethazine (Phenergan)	*Oral/intravenously* Child: 0.25-1 mg/kg/dose every 4-6 hours as needed. Maximum 25 mg/dose Adult: 12.5-25 mg/dose every 4 hours as needed	Sedation, extrapyramidal effects	Use for breakthrough nausea and vomiting in adult and pediatrics patients Use in combination with dexamethasone for delayed emesis in pediatric patients
Benzamide Metoclopramide (Reglan)	*Oral/intravenously* Acute emesis: 1-2 mg/kg/dose every 2-6 hours for 2-5 days. Maximum single dose 50 mg Delayed emesis: 0.5 mg/kg/dose or 30 mg IV every 4-6 hours for 3-5 days Premedicate with diphenhydramine to prevent extrapyramidal reactions	Sedation, diarrhea, extrapyramidal effects	Use for breakthrough nausea and vomiting in adult patients Use in combination with dexamethasone in delayed emesis in adult patients
Butryophenone Haloperidol	*Oral/intravenously* Adult: 1-4 mg every 6 hours	Sedation, hypotension, tachycardia, extrapyramidal effects	Use for breakthrough nausea and vomiting in adult patients

(Continued)

Table 31-20   (Continued)

Medication	Dosage	Side Effects	Comments
Cannabinoid Dronabinol (Marinol)	*Oral* 5 mg/m^2/dose every 4–6 hours. May increase dose by 2.5 mg/m^2/dose increments to maximum of 15 mg/m^2/dose	Euphoria, dysphoria, drowsiness, dry mouth, anxiety, irritability, confusion	Use for breakthrough nausea and vomiting in adult patients
Nabilone (Cesamet)	Standard adult dose: 1–2 mg orally twice a day. Maximum daily dose 2 mg three times a day		
Diphenhydramine (Benadryl)	1 mg/kg/dose IV/PO every 4–6 hours Maximum single dose 50 mg		Antihistimines have been used and should be viewed as adjunctive agents when dopamine antagonists are used; they are not recommended as single agents for emesis control
Hydroxyzine (Vistaril, Atarax)	0.5–1 mg/kg/dose IV/PO every 4–6 hours. Maximum 50 mg. IV administration is not recommended unless given by slow IV infusion via central line		

Abbreviations: BSA, body surface area; PO, orally; IV, intravenously; NK-1, Neurokinin-1.

Table 31-21 Antiemetic Regimen for Initial Cycle of Chemotherapy or Radiotherapy[a]

Goal	Emetogenicity of the Chemotherapy or Radiotherapy	Treatment
Prophylaxis to prevent acute symptoms	High	5-HT$_3$ receptor antogonist, steroid
	Moderate	5-HT$_3$ receptor antagonist
	Mild	5-HT$_3$ receptor antagonist
Rescue (treatment of breakthrough of acute symptoms): advance up ladder, starting after agents already given for prophylaxis	ANY	1. 5-HT$_3$ receptor antagonist 2. Steroid 3. Lorazepam 4. Dronabinol (if >6 yr old) 5. Metoclopramide (with diphen-hydramine)
Prophylaxis to prevent delayed symptoms	High	Steroid (consider metoclopramide or 5-HT$_3$ receptor antogonist if break-through emesis in the first 24 h)
	Moderate	None (consider metoclopramide or 5-HT$_3$ receptor antagonist if break-through emesis in the first 24 h)
	Mild	None
Rescue (treatment of breakthrough delayed symptoms): advance up ladder, starting after agents already given for prophylaxis	Any	1. Dexamethasone 2. Metoclopramide 3. 5-HT$_3$ receptor antagonist 4. Lorazepam, dronabinol

[a]Data from the Dana-Farber Cancer Institute and Children's Hospital, Boston.
Abbreviations: 5-HT$_3$, 5-hydroxytryptamine subtype 3.
From: Barnard DR, Friedman DL, Landier W, et al. Monitoring and management of drug toxicity. In: Altman AJ, ed. Supportive Care of Children with Cancer. Current therapy and guidelines from the Children's Oncology Group, 3rd ed. Baltimore: The Johns Hopkins University Press, 2004:100–148, with permission.

Cannabinoids

1. Physiology: tetrahydrocannobinol (THC) activates:
   a. CB-1 receptor found in large quantities in cerebellum, basal ganglia, brainstem and hypothalamus
   b. CB-2 on cells of the immune function
2. Drugs available.
   a. dronabinol (Marinol)
   b. nabilone (Cesamet) – side effects include drowsiness, sedation, euphoria, heightened sensation, dizziness.

Dopamine antagonists

1. Inhibits dopaminergic function.
2. Metoclopramide (Reglan).

Corticosteroids

1. Dexamethasone preferred agent.[*]
2. Methylprednisone.

Antiemetic agents can be given in combination to increase their efficacy. For patients receiving chemotherapy with high emetic risk, there is no role for using antiemetic drugs of lower antiemetic potential and attempting to escalate the antiemetic drugs once emesis has begun. Agents of lower antiemetic potential should only be used in high-emetic-risk situations when patients are intolerant of or refractory to 5-HT3 serotonin receptor antagonists, NK_1 antagonists and dexamethasone.

Commonly used effective regimens include: 5-HT3 receptor antagonist in combination with dexamethasone and Emend on day 1, followed by dexamethasone on days 2-4 and Emend on days 2 and 3.

For *moderate-emetic-risk* agents: Two-drug regimen consisting of a 5-HT3 receptor antagonist in combination with dexamethasone on day 1 followed by either dexamethasone alone on days 2–4 or in combination with 5-HT3 antagonist for delayed CINV.

For *low-emetic-risk* agents:

*   Dexamethasone alone is recommended when appropriate or 5-HT3 antagonist alone on Day 1
*   No routine preventive use of antiemetics for delayed emesis is suggested.

For *minimal-emetic-risk* agents: Prescribe only as required.

If vomiting and nausea continue despite recommended prophylaxis:

*   Conduct a careful re-evaluation of emetic risk
*   Review disease status
*   Consider concurrent illness and medications
*   Confirm that the best regimen is being given based on emetic risk
*   Consider adding lorazepam or alprazolam to regimen.

The following drugs could be used on an ambulatory basis when breakthrough occurs:

*   Prochloraperazine (Compazine) 0.1 mg/kg PO every 6 hours
*   Metoclopramide (Reglan) 1–2 mg/kg PO every 4 hours to be given with diphenhydramine
*   Dronabinol (Marinol) 5–10 mg PO q3–6h

---

[*]For anthracyclines: 5HT3 antagonists, dexamethasone and aprepitant, followed by aprepitant alone for days 2 and 3. Otherwise the use of aprepitant in this category is optional and based on individual response

- Promethazine (Phenergan) 0.25–1 mg/kg PO every 6 hours (do not use in children less than 2 years of age)
- Scopolamine patch
  - ½ patch every 3 days in infants 8–15 kg body weight
  - 1 patch every 3 days if >15 kg body weight
- Hydroxyzine (Vistaril, Atarax) 0.02–0.05 mg/kg PO every 4–6 hours
- Lorazepam (Ativan) 0.02–0.05 mg/kg PO every 6 hours.

Radiation-Induced Emesis

1. High emetic risk: (total body irradiation): 5-HT3 antagonist alone or in combination with corticosteroid before each fraction and for at least 24 hours after.
2. Moderate emetic risk (upper abdomen, hemibody irradiation, abdominal-pelvic, mantle, craniospinal irradiation, cranial radiosurgery): 5-HT3 antagonist before each fraction.
3. Low emetic risk (lower thorax, cranium and craniospinal): 5-HT3 antagonist before each fraction.
4. Minimal emetic risk (radiation of breast, head and neck, cranium and extremities): medication as required.

Anticipatory Emesis

1. Use of the most appropriate antiemetic regimens based on chemotherapy being administered to prevent acute or delayed emeis.
2. Ativan 0.025 mg/kg PO in morning before chemotherapy.
3. Cognitive therapy and behavioral modification.
4. Hypnosis with desensitization procedures.

NUTRITIONAL SUPPORT

Progressive weight loss, protein energy malnutrition (PEM) and cachexia occur in advanced malignancies in children. Cachexia leads to decreased strength, impaired immune function, decreased pulmonary function, increased disability and death. The clinical features of cachexia include:

- Wasting
- Anorexia
- Weakness
- Anemia
- Hypoalbuminemia
- Hypoglycemia
- Lactic acidosis
- Hyperlipidemia

- Impaired liver function
- Glucose intolerance
- Skeletal muscle atrophy
- Anergy.

The following factors interact to produce cachexia in pediatric oncology patients:

- Decreased intake secondary to mucositis
- Anorexia and altered taste perception (dysgeusia)
- Nausea and vomiting
- Decreased absorption secondary to the effects of chemotherapy, radiation therapy, or surgery
- Protein-losing nephropathy
- Steroid-induced diabetes
- Increased metabolic expenditure secondary to rapid tumor growth and metastases
- Endogenously produced cytokines such as TNF, IL-1, IL-6 and γ-interferon
- Hepatotoxicity with impaired liver synthetic capacity and altered protein, carbohydrate and lipid metabolism
- Psychosocial factors such as depression and anticipatory vomiting.

The net result of these factors is failure of nutritional intake to meet increased metabolic demands as a result of which, tissue wasting occurs.

Treatment

It is crucial to increase nutritional intake both to improve quality of life and to increase survival from the underlying malignancy. Methods of nutritional support include:

- Oral feeding with high-calorie supplements
- Nasogastric tube feeding
- Total parenteral nutrition (intravenous hyperalimentation).

Oral Feeding

If the child tolerates oral intake and can absorb nutrients, oral feeding is the treatment of choice. Supplemental diet formulas and elemental diets are helpful in these children. Table 31-22 lists the characteristics of various enteral products.

Tube Feeding

Oral feeding may not be adequate or may not be tolerated. Adequate nutrition may have to be supplied via nasogastric intubation if the patient has adequate blood counts and does not have mucositis. In cases where chronic tube feeding will be required, a gastrostomy tube

Table 31-22 Characteristics of Enteral Products

Product Description	1–10 Years of Age	>10 Years of Age	
	Tube Feeding and/or Oral	Tube Feeding	Oral
1. Standard or polymeric	Pediasure	Isocal	Ensure
Intact macronutrients intended as meal replacements	Kindercal	Osmolite	Boost
Requires normal digestive and absorptive capacity	Nutren Junior	Nutren	NuBasics
Usually lactose free unless otherwise indicated	Resources just for Kids TM	Compleat Modified	Meritene
	Compleat Pediatric	Promote	Resource Fortashake Scandishake Carnation Instant Breakfast
2. High nitrogen		Isocal HN	Ensure HN
Intact macronutrients with >15% total calories as protein		Osmolite HN	Boost HN
		Criticare HN	TwoCal HN
Useful for patients with increased protein need (i.e., poor wound healing, radiation)			
3. Concentrated or high calorie	Nutren 1.5	Nutren 1.5	NuBasic Plus
Contain higher calorie per milliliter than standard formulas	Nutren 2.0	Nutren 2.0	Ensure Plus
		TwoCal HN	Resource Plus
		Deliver 2.0	Boost Plus
May be used with fluid restriction		Comply	TwoCal HN
		Isosource	Jevity Plus
		Jevity Plus	
4. Predigested/elemental	Peptamen Junior	Peptamen	Vital HN
Predigested or partially hydrolyzed peptide-based diet that may be beneficial for child with impaired gastrointestinal function (diarrhea, mucositis, intestinal villous atrophy)	Neocate One Plus	Peptamen VHP	
	VivonexPeds TM	Reabillin	
	Pediasure with Fiber	Tolerex	
	EleCare	Vivonex TEN	
Many contain medium-chain triglycerides to minimize fat tolerance	L-Emental Pediatric	Vivonex Plus	
		Criticare HN	
Can have high osmolality		Travasorb MCT	

(Continued)

Table 31-22 (Continued)

Product Description	1–10 Years of Age Tube Feeding and/or Oral	>10 Years of Age Tube Feeding	Oral

5. Fiber containing
   Containing fiber from
     natural sources or added
     soy polysaccharides to
     aid in bowel function
	Kindercal
	Pediasure with fiber
	Nutren Junior with fiber
		Jevity
		Ultracal
		Nutren 1.0 fiber
		FiberSource
			Jevity
			Ultracal
			Nutren 1.0 with fiber

6. Disease specific
   Macro- and micronutrients
     modified for disease state
		Diabetes: choice
		   Dm, Glucerna
		Renal: Nepro, Suplena
		Pulmonary: Pulmocare,
		   Nutrivent
		Liver: NutriHep
		Metabolically stressed:
		   Replete, Impact,
		Perative
			Diabetes: choice
			   DM, Glucerna
			Renal: Nepro, Suplena
			Pulmonary: Pulmocare
			   Nutrivent
			
			Metabolically stressed:
			   Impact, Replete

7. Infant formulas (variety of
   formulas available for
   premature infants and
   for infants with poor
   tolerance).
  Many infants are lactose
   intolerant after chemo-
   therapy and benefit from
   lactose-free formulas.
  Human breast milk should be
   used when possible
   (contains lactose and may
   not be tolerated well after
   chemotherapy)
	Similac
	Enfamil
	Lactofree (lactose free)
	Gerber
	Carnation Good Start
	Similac PM 60/40
		Isomil
		Prosobee
			Pregestimil
			Alimentum
			Nutramigen

8. Modular components	Protein
	Casec
	Pro-mix
	ProMod
	NutriSource Protein
	Elementra
		Carbohydrate
		Polycose
		Moducal
		Liquid
		   Carbohydrate
		   (LC)
		Sumacal
		NutriSource
		Carbohydrate
		NutriSource (long- or
		   medium-chain
		   Triglycerides)
			Fat
			Vegetable oil (long-
			   chain triglycerides)
			Microlipid (long-chain
			   triglycerides)
			Medium-chain
			   triglycerides

9. Oral electrolyte solutions
   Provides electrolytes,
   calories and water during
   mild to moderate
   dehydration
	Pedialyte
	Infalyte
	Equalyte		

From: Sacks N, Ringwals-Smith K, hale G. Nutritional support. In: Altman AJ, ed. Supportive Care of Children with Cancer. Current Therapy and Guidelines from the Children Oncology Group, 3rd ed. Baltimore: The Johns Hopkins University, 2004: 243–261, with permission.

may have to be placed. If bolus feeds are not tolerated, continuous low-volume infusions can be utilized to maximize caloric intake.

Intravenous Hyperalimentation or Total Parenteral Nutrition

Total parenteral nutrition (TPN) is indicated when enteral feeds cannot be tolerated. It is the most intensive modality of nutritional supplementation and requires a central intravenous device. Full nutritional requirements can usually be met. TPN should be continued until oral feedings can be resumed. Complications of TPN are hypoglycemia, hyperglycemia, fatty liver and cholestatic jaundice.

Indwelling Intravenous Catheters

With ongoing therapy, it is difficult to find adequate peripheral venous access in pediatric oncology patients. In addition, chemotherapeutic agents produce venous sclerosis, further aggravating the problem of finding peripheral venous access.

To alleviate this problem, permanently indwelling right atrial catheters are used. These catheters are recommended for:

- All young children and infants receiving chemotherapy
- Most older children and adults receiving chemotherapy if the protocol is prolonged and of moderate to high intensity
- Patients requiring TPN.

Types of Catheters and Methods of Insertion

Two major types of long-term indwelling tunneled catheters are in use:

- External tunneled catheters that lead to tubing that exits the skin. They can be single, double and occasionally triple lumen. The most frequently used types are the Broviac catheter and the Hickman catheter
- Total implantable venous access devices or "ports" are right atrial catheters that lead to reservoirs totally embedded under the skin, replacing the external portion of the catheter The device is accessed using a special non-boring needle that tranverses the skin and enters the device. Such devices are more cosmetically satisfactory to older children and adolescents and have a lower infection rate than external tunneled catheters.

Catheters are inserted via the subclavian, internal, or external jugular vein with a subcutaneous tunnel to the anterior or lateral chest wall. Alternatively, catheters can be placed via the femoral vein to the inferior vena cava with a tunnel to the abdominal wall.

Maintenance

1. Broviacs (or Hickmans) require daily or every-other-day heparin flushes, in addition to dressing changes three times a week.

2. Mediports require heparin flushes only once every 3–4 weeks and do not require dressing changes.
3. In both cases, completely sterile technique must be used in accessing the catheters.

Complications

The major complications are:

- Infection: Episodes of catheter-related bacteremia occur and are most commonly due to coagulase-negative staphylococci. Management consists of appropriate intravenous antibiotic therapy. Antibiotic therapy without catheter removal is successful in approximately 75% of episodes and higher if antibiotic lock therapy is used. If the bacteremia fails to resolve, the patient develops uncontrolled clinical sepsis, or if there is infection overlying the subcutaneous tunnel, then the catheter must be removed. Occasionally, catheter-related fungal infection occurs and, in these cases catheter removal is necessary
- Thrombosis: Thrombosis may be cleared by instilling tPA 0.5–1 mg and leaving it in for 30–60 minutes. If the catheter remains occluded after two doses, a continuous infusion of tPA is indicated. The dose is 0.01 mg/kg/hour for 6 hours. The dose may be increased to 0.02 mg/kg/hour for 6 hours, then to 0.03 mg/kg/hour for 6 hours if there is no resolution. The fibrinogen level should be monitored every 6 hours. If the line is still occluded, it should be removed.

Most patients develop the preceding complications from time to time. In the majority of cases, however, the morbidity associated with these complications is outweighed by the advantages of reliable, adequate venous access.

PSYCHOSOCIAL SUPPORT AND END-OF-LIFE CARE

Throughout the course of the disease, children with cancer and their families will face a number of psychological issues. In addition to the oncologist's skills in this area, specially trained psychologists, psychiatric social workers, psychiatrists and pastoral care practitioners can be helpful in dealing with these matters as they arise (see Chapter 33).

Suggested Reading

Agency of Health Care Policy and Research clinical guideline #9 publication 94-0592. Management of Cancer Pain, March 1994.

Alexander SW, Walsh TJ, Freifeld AG, Pizzo PA. Infectious complications in pediatric cancer patients. In: Pizzo PA, Poplack DG, eds. Principles and Practice of Pediatric Oncology. 4th ed. Philadelphia: Lippincott, Williams & Wilkins; 2002:1239–1284.

Barnard DR, Friedman DL, Landier W, et al. Monitoring and management of drug toxicity. In: Altman AJ, ed. Supportive Care of Children with Cancer. Current Therapy and Guidelines from the Children's Oncology Group, 3rd ed. Baltimore: The Johns Hopkins University Press; 2004:100–148.

Bechard LJ, Adiv OE, Jaksic T, Duggan C. Nutritional supportive care. In: Pizzo PA, Poplack DG, eds. Principles and Practice of Pediatric Oncology, 4th ed. Philadelphia: Lippincott, Williams & Wilkins; 2002:1285–1300.

Berde CB, Billett AL, Collins JJ. Symptom management in supportive care. In: Pizzo PA, Poplack DG, eds. Principles and Practice of Pediatric Oncology, 4th ed. Philadelphia: Lippincott, Williams & Wilkins; 2002:1301–1332.

Brook I, Frazier E. The aerobic and anaerobic bacteriology of perirectal abscesses. J Clin Microbiol. 1997; 35:2974–2976.

Feusner J, Hastings C. Recombinant human erythropoietin in pediatric oncology: a review. Med Pediat Oncol. 2002;39:463–468.

Freifeld A, et al. A double-blind comparison of empirical oral and intravenous antibiotic therapy for low risk febrile with neutropenia during cancer chemotherapy. New Engl J Med. 1999;341:305–311.

Goldman S, et al. Feasibility study of IL-11 and granulocyte-stimulating factor after myelosuppressive chemotherapy to mobilize peripheral blood stem cells from heavily pretreated patients. J Pediatr Hematol Oncol. 2001;23(5):300–305.

Gorlin JB. Noninfectious complications of pediatric transfusion. In: Hillyer CD, Strauss RG, Luban NLC, eds. Handbook of Pediatric Transfusion Medicine. San Diego: Elsevier Academic Press; 2004; 317–325.

Guidelines for Monitoring and Management of Pediatric Patients During and After Sedation for Diagnostic and Therapeutic Procedures: An Update. Pediatrics 2006;118(6):2587–2602.

Hughes WT, Armstrong D, Bodey GP, et al. Guidelines for the use of antimicrobial agents in neutropenic patients with cancer. Clin Infect Disease 2002;34:730–751.

Jamali F, Ness PM. Infectious complications. In: Hillyer CD, Strauss RG, Luban NLC, eds. Handbook of Pediatric Transfusion Medicine. San Diego: Elsevier Academic Press; 2004:329–339.

Kern WV, et al. Oral versus intravenous empirical antimicrobial therapy for fever in patients with granulocytopenia who are receiving chemotherapy. New Engl J Med. 1999;341:312–318.

Koo et al. Efficacy and safety of caspofungin for the empiric management of fever in neutropenic children. Pediatr Infect Dis. DOL:10.1097 ISSN 0891-3668 Lippincott Williams & Wilkins, 2007.

Logdberg LE. Leukoreduced products: prevention of leukocyte-related transfusion-associated adverse effects. In: Hillyer CD, Strauss RG, Luban NLC, eds. Handbook of Pediatric Transfusion Medicine. San Diego: Elsevier Academic Press; 2004:85–90.

Michallet M, Ito J. Approaches to management of invasive fungal infections in hematological malignancy and hematopoietic stem cell transplantation. J Clin Oncol. 2009;27(20):10.

Oppenheimer BA. Management of febrile neutropenia in low risk cancer patients. Thorax. 2000;55:563–569.

Orudjev E, Lange BJ. Evolving concepts of management of febrile neutropenic in children with cancer. Med Pediatr Oncol. 2002;39:77–85.

Ozer H, Armitage JO, Bennett CL, et al. Update of recommendations for the use of hematopoietic colony-stimulating factors: evidence-based, clinical practice guidelines. J Clin Oncol. 2000;18(20):3558–3585.

Pizzo P. Principles and Practice of Pediatric Oncology, 5th ed. Philadelphia, PA: Lippincott, Williams & Wilkins; 2006:1202–1230

Robertson N. Shilkofski N. The Harriet Lane Handbook, 18th ed. The Johns Hopkins Hospital. 2008:783.

Rizzo JD, Lichtin AE, Woolf SH, et al. Use of epoetin patients with cancer: evidence-based clinical practice guidelines of the American Society of Clinical Oncology and the American Society of Hematology. Blood 2002;100(7):2303–2320.

Rogers ZR, Aquino VM, Buchana GR. Hematologic supportive care and hematopoietic cytokines. In: Pizzo PA, Poplack DG, eds. Principles and Practice of Pediatric Oncology, 4th ed. Philadelphia: Lippincott, Williams & Wilkins; 2002:1205–1238.

Sacks N, Ringwals-Smith K, Hale G. Nutritional support. In: Altman AJ, ed. Supportive Care of Children with Cancer. Current Therapy and Guidelines from the Children's Oncology Group, 3rd ed. Baltimore: The Johns Hopkins University Press; 2004:243–261.

Schwartzberg LS. Chemotherapy-induced nausea and vomiting: which antiemetic for which therapy? Ncbi.nlm. nih.gov 2007 July;21(8):946–953.

Talcott J, et al. Risk assessment in cancer patients with fever and neutropenia: A prospective, two-center validation of a prediction rule. J Clin Oncol. 1992;10(2):316–322.

Triulzi D. Transfusion-Related Acute Lung Injury: an Update. American Society of Hematology. 2008:497–501

Wolff LJ, Altman AJ, Berkow RL, Johnson FL. The management of fever and neutropenia. In: Altman AJ, ed. Supportive Care of Children with Cancer. Current Therapy and Guidelines from the Children's Oncology Group, 3rd ed. Baltimore: The Johns Hopkins University Press; 2004:200–220.

Zempsky WT, Schechter NL, Altman AJ, Weisman SJ. The management of pain. In: Altman AJ, ed. Supportive Care of Children with Cancer. Current Therapy and Guidelines from the Children's Oncology Group, 3rd ed. Baltimore: The Johns Hopkins University Press; 2004:200–220.

CHAPTER 32

Evaluation, Investigations and Management of Late Effects of Childhood Cancer

Since the 1970s, outcomes for childhood cancer have shown remarkable and steady improvements. Five-year overall survival from childhood cancer now exceeds 75% and as many as 1 in 450 American adults will be survivors of childhood cancer. Despite these successes, over 60% of survivors will have at least one chronic health condition, over 25% will have a severe or life-threatening condition and almost 20% will die within 30 years of diagnosis. The complications faced by survivors and the care they require, are multisystem and complex. The most common complications include cardiac, endocrine, musculoskeletal and pulmonary complications, as well as treatment-related secondary neoplasms. Survivors also often face societal, emotional and psychological barriers – such as learning challenges, school difficulties and problems obtaining insurance – and almost one in five suffer from stress-related mental disorders such as post-traumatic stress disorder. Extensive, intricate, long-term medical and psychological follow-up is required to maintain survivors' health and quality of life.

Chemotherapy, radiation therapy and surgery may all cause "late effects" involving any organ system. A definition of "*late effects*" includes any physical or psychological outcome that develops or persists beyond 5 years from the diagnosis of cancer. Some late effects of therapy identified during childhood and adolescence resolve without consequence, while other late effects become chronic and may progress to become adult medical problems.

As the study of late effects has grown as a scientific discipline, our ability to predict and ameliorate late effects based on exposures has resulted in the development of screening guidelines. Several organizations have produced such guidelines. These include:

- Scottish Intercollegiate Guideline Network
  http://www.sign.ac.uk/pdf/sign76.pdf
- Late Effects Group of the United Kingdom Children's Cancer Study Group
  http://www.ukccsg.org/public/followup/PracticeStatement/index.html
- Children's Oncology Group
  http://www.survivorshipguidelines.org

Manual of Pediatric Hematology and Oncology. DOI: 10.1016/B978-0-12-375154-6.00032-X

These guidelines provide evidence-based screening recommendations that can be individually tailored based on exposure. In order to effectively utilize the guidelines it is critical to be familiar with the patient's exposures and to be able to elicit relevant information from the survivor. Important information to be obtained in follow-up of childhood cancer survivors is listed in Table 32-1. Selected late effects associated with different childhood cancers are listed in Table 32-2 and the evaluation for suspected late effects can be found in Table 32-3. Table 32-4 lists the radiosensitivity of various cell types and Table 32-5 lists the radiation dose toxicity levels for subacute and late visceral effects.

The remainder of this chapter reviews the specific complications by organ system related to the treatment of childhood cancer.

MUSCULOSKELETAL DYSFUNCTION

Surgery

Surgery remains the primary therapy for many musculoskeletal tumors and the most visibly obvious late effect associated with therapy for childhood cancer remains amputation. As internal prostheses have become more refined, the ability to perform limb-salvage procedures has dramatically improved. Although there are many advantages to a limb-salvage approach to extremity tumors, there are significant disadvantages as well. In particular, the prostheses generally cannot tolerate extreme stress, so sport participation is

Table 32-1   Information to be Elicited in Follow-up of Survivors of Childhood Cancer

History of previous cancer treatment:
A. Cumulative doses of chemotherapeutic agents
B. Doses and sites of radiotherapy
C. Surgeries
History of intercurrent illnesses
Review of systems
Development of any benign tumors or other cancers
Medications (including prophylactic antibiotics and hormone replacement)
Educational status
A. Highest grade completed and grade point average
B. Results of neurocognitive evaluations
C. Is there an Individualized Education Plan in place at the school?
Employment status
Insurance coverage
A. Does the patient have an individual policy, or coverage through his/her parents?
B. Has the patient had difficulties obtaining insurance?
Marital history
Menses, libido, sexual activity
Pregnancy outcome (patient or spouse)
High-risk behaviors (smoking, alcohol or drug use, multiple sexual partners)

Adapted from: Pizzo and Poplock Ed. *Principles and Practice of Pediatric Oncology*, 4th Ed. 2002 Lippincott Williams and Wilkens, Philadelphia, with permission.

Table 32-2   Selected Potential Late Effects Associated with Childhood Cancer

Cancer	Potential Late Effects	
Leukemias	• Neurocognitive effects • Second malignant neoplasms • Abnormal growth and maturation • Cardiomyopathy • Blood-borne infection	• Dental problems • Obesity • Reduced bone mineral density • Avascular necrosis of bone
Brain cancer	• Neurocognitive effects • Second malignant neoplasms • Abnormal growth and maturation • Hearing loss	• Kidney damage • Blood born infection • Infertility • Vision problems • Endocrine abnormalities
Hodgkin lymphoma	• Second malignant neoplasms (especially breast cancer) • Cardiomyopathy • Infertility • Lung damage	• Blood-born infection • Reduced bone mineral density • Hypothyroidism (secondary to mantle or neck radiation) • Abnormal growth and maturation
Non-Hodgkin lymphoma	• Cardiomyopathy • Blood-borne infection • Neurocognitive effects	• Infertility • Reduced bone mineral density
Bone tumor	• Amputation/disfigurement • Functional activity limitations • Damage to soft tissues and underlying bones (radiation may cause scarring, swelling, or inhibit growth)	• Hearing loss • Cardiomyopathy • Kidney damage • Second malignant neoplasms • Blood-born infection • Infertility
Wilms' tumor	• Cardiomyopathy • Kidney damage • Damage to soft tissues and underlying bones (radiation may cause scarring, swelling, or inhibit growth)	• Second malignant neoplasms • Infertility • Scoliosis
Neuroblastoma	• Cardiomyopathy • Damage to soft tissues and underlying bones (radiation may cause scarring, swelling, or inhibit growth)	• Neurocognitive effects • Hearing loss • Blood-born infection • Second malignant neoplasms • Kidney damage
Soft tissue sarcoma	• Amputation/disfigurement • Functional activity limitations • Cardiomyopathy • Damage to soft tissues and underlying bones (radiation may cause scarring, swelling, or inhibit growth)	• Second malignant neoplasms • Hepatitis C • Kidney damage • Cataracts • Infertility • Neurocognitive effects

Adapted from: Hewitt M, Weiner SL, Simone JV Editors. Childhood Cancer Survivorship. Improving Care and Quality of Life. 2003, National Academies Press, Washington, DC, with permission.

Table 32-3   Evaluation for Suspected Late Effects

Late Effect	Screening Test	Recommendations if Screening Results Abnormal
Short stature	Growth curve	Bone-age, growth hormone stimulation test
	Sitting height	Thyroid function tests[a]
	Parental heights	Endocrine consultation
Obesity or weight loss	Growth curve	Thyroid function tests[a]
	Diet history	Nutritionist, endocrinologist, consultation
Scoliosis	Physical examination	Spine radiography; evaluate again during adolescent growth spurt
		Orthopedist consultation
Bone asymmetries (hypoplasia, atrophy)	Bone lengths, circumference	Orthopedist consultation; bone radiography; plastic surgeon consultation
Avascular necrosis or osteoporosis	History of pain, fractures	MRI/bone scan
	Bone radiography	Serum estradiol level; Ca, P Orthopedist consultation; physical therapist consultation
Soft tissue hypoplasia, contractures, edema	Physical examination	Plastic surgeon consultation
Dental abnormalities	Physical examination	Dentist, oral surgeon consultation
Learning disabilities	Communication with school, family; psychological testing	CT or MRI scan of head; special education classes
Leukoencephalopathy	CT or MRI (see also Learning disabilities, above)	Cerebrospinal fluid basic myelin protein; neurologist consultation
Neuropathy	Physical examination	Neurologist consultation
Hearing loss	Audiogram	Otorhinolaryngologist consultation; audiologist consultation
Infertility	History (primary versus secondary dysfunction)	Endocrinologist consultation
	Gonadal function testing[b]	Obstetrician or gynecologist consultation
Thyroid dysfunction	Thyroid function testing[a]	Endocrinologist consultation
Cardiomyopathy or pericarditis	Electrocardiogram; echocardiogram; radio nuclide angiography	Cardiologist consultation
Vaso-occlusive disease	Angiography; Doppler pulses	Vascular surgeon
Pneumonitis or pulmonary fibrosis	Chest radiography	Lung biopsy
	Pulmonary function tests	Pulmonologist consultation
Chronic enteritis	Growth curves	Serum folate, carotene
	Nutritional assessment	Small-bowel studies; barium enema; gastroenterologist consultation
Hepatitis or cirrhosis	Liver function tests	Liver biopsy, hepatitis screen; liver scan; gastroenterologist consultation

(Continued)

Table 32-3   (Continued)

Late Effect	Screening Test	Recommendations if Screening Results Abnormal
Nephritis, rickets (tubular defects)	Urinalysis; BUN, creatinine, serum electrolytes, CO_2, Ca, P, alkaline phosphatase; wrist radiographs	24-h creatinine clearance, glomerular filtration rate, intravenous urogram or sonogram; nephrologist consultation
Hemorrhagic cystitis	Urinalysis	Cytoscopy; urologist consultation
Thrombotic thrombocytopenic purpura	CBC, platelets, BUN, creatinine; peripheral blood smear	
Sepsis	Compliance with prophylactic antibiotics	
Second malignancy	Studies on an individual basis	Oncologist consultation

[a]Thyroid function tests include thyroxine (T_4), TSH.
[b]Gonadal function tests: Tanner staging for boys older than 14 years at a the time of evaluation or girls not yet menstruating by age 12 years or if menses become irregular; follicle-stimulating hormone, luteinizing hormone and testosterone (semen analysis) or estradiol, as appropriate.
Abbreviations: BUN, blood urea nitrogen; CBC, complete blood cell count; CT, computed tomography; MRI, magnetic resonance imaging.
From Pizzo PA, Poplack DG, editors. Principles and Practice of Pediatric Oncology. 4th ed. Philadelphia; Luppincott, Williams and Wilkins, 2002, with permission.

Table 32-4   Cell Types in Order of Radiosensitivity

Radiosensitivity	Examples	Cell Types
High	Hematopoietic stem cells; intestinal crypt cells; basal cells of the epidermis; lymphocytes[a]	Stem cells
Intermediate to high	Differentiating hematopoietic cells; spermatogonia and spermatocytes	Differentiating cells
Intermediate	Endothelial cells; fibroblasts	Connective tissue cells
Low to intermediate	Duct cells of the salivary glands, kidney cells, liver cells, epithelial cells; smooth muscle	Differentiated cells with some ability to divide
Low	Neurons; muscle cells; most mature hematopoietic cells	Differentiated, nondividing cells

[a]Differentiated, nondividing cells, but highly radiosensitive.
From: Kun LE. General Principles of Radiation Therapy. In: P.A. Pizzo and D.G. Poplack, editors. Principles and Practice of Pediatric Oncology. Philadelphia: Lippincott–Raven, 1997:289–321, with permission.

highly restricted, whereas the only limits following amputation are those of physics. In order to maintain the optimal quality of life, the choice of amputation versus limb-salvage must be tailored to the individual.

Chemotherapy

The most common chemotherapy-induced musculoskeletal late effects are steroid-induced avascular necrosis (AVN) and reduced bone mineral density. The increased use of

Table 32-5 Radiation Dose Toxicity Levels for Subacute and Late Visceral Effects

Organ	Toxicity	Whole Organ Irradiation			Partial Organ Irradiation				Chemotherapy Effect
		5–10% Incidence Dose[a] (Gy)	RT±CTx	>25–50% Incidence (Gy)	Volume	<5–10% Incidence Dose[a] (Gy)	RT±CTx	>25–50% Incidence (Gy)	
Lung	Subacute pneumonitis or late fibrosis	18–20	(CTx–)	24–25	<30% of one lung	25–30	(CTx–)	45–50	++
		15–18	(CTx+)	21–24		<20	(CTx+)	>35	
Heart	Subacute pericarditis	35–40		>50	<50%	40–50		60–70	+/–
	Late cardiomyopathy	45–50	(CTx–)	>60	<25%	<40	(CTx–)	>70	+++
		30–40	(CTx+)	>50			(CTx+)	>50	
Live	Subacute hepatopahy	25–30	(CTx–)	>40	<50%	30–40	NA	40	++
		20–25	(CTx+)	>35					
Kidney	Subacute nephropathy	18–20	(CTx–)	24–26	>50	20–24	NA	>30	+
	Late nephropathy	16–18	(CTx+)	22–26					
	Late hypertension	Similar to subacute nephropathy			>50	20–25	NA	>35	+/–
Small bowel	Subacute enteropahy	15–25	NA	>40	<50%	20–25	NA	>60	+
Brain	Late necrosis	54–60	NA	>65	<50%	Equal to whole brain	NA		+/– +++
Spinal cord	Subacute–late necrosis	40–45	NA	>50	<15%	45–50	NA	>55	–

[a] Dose assumes conventional fractionation (150–200 cGy once daily, 5 days per week). If ≥40% of liver excluded from dose ≥18 Gy, remainder of organ can receive doses up to 40+ Gy.

Abbreviations: (CTx–), without prior, concurrent, or subsequent chemotherapy; (CTx+), interation with one or more chemotherapeutic agents; NA, not applicable.

Adapted from: Kum LE. General Principles of Radiation Therapy. In: Pizzo PA, Poplack DG, editors. Principles and Practice of Pediatric Oncology. Philadelphia: Lippincott-Rave, 1997:289–321, with permission.

dexamethasone in ALL therapy has resulted in an increased number of cases of avascular necrosis, with adolescents appearing to be at highest risk. Protocols have attempted to decrease this incidence by decreasing the number of continuous weeks of exposure to dexamethasone and by utilizing more prednisone than dexamethasone. The incidence of reduced bone mineral density in survivors of childhood cancer is markedly elevated. Routine recommendations include adequate daily intake of calcium (1000–1500 mg) and vitamin D (200 IU daily) and weight-bearing exercises. The use of bisphosphonates and other absorption-reducing therapies in childhood cancer survivors remain investigational.

Radiation

For many musculoskeletal tumors, especially those not surgically resectable, radiation remains a keystone of therapy. Despite its effectiveness, radiation therapy carries with it the risk of significant consequences which include:

- Scoliosis
- Atrophy or hypoplasia of muscles
- Avascular necrosis
- Reduced bone mineral density
- Poor enamel and teeth formation with increased risk of cavities, malocclusion, hypoplasia of mandible
- Discrepancy in lengths of extremities
- Alteration in sitting-to-standing height ratio
- Second malignant neoplasms.

These late consequences are related to dose, site, volume and age of the child at the time of radiation. The higher the dose and the younger the child, the more pronounced the late effects. Some protocols have been designed to reduce potential radiation-associated late effects through the use of:

- Low-volume, low-dose radiation combined with effective systemic therapy
- Improvements in the delivery of radiation with advances in radiation technology
- Dental prophylaxis prior to radiotherapy in maxillofacial sites.

Screening and Management

Regular physical examinations with appropriate imaging (e.g. MRI for suspected AVN) can identify musculoskeletal problems early and potentially reduce their impact on a survivor's quality of life. Survivors who have been exposed to high doses of steroids, especially in conjunction with radiation or high-dose methotrexate, may benefit from an evaluation of bone mineral density at entry into long-term follow-up. Survivors should be counseled to engage in routine weight-bearing exercise and to consume the recommended daily allowance of vitamin D (200 IU daily) and calcium (1,000–1,500 mg daily).

CARDIOVASCULAR DISORDERS

Chemotherapy, radiotherapy and especially their combined use, have the potential to cause both early and late cardiac complications. The primary cardiac late effect is the development of cardiomyopathy and congestive heart failure. Other late cardiac complications include valvular heart disease and the onset of adult coronary artery disease at a younger age.

Chemotherapy

Anthracyclines

Anthracycline-induced myocyte death results in hypertrophy of existing myocytes, reduced thickness of the wall of the heart and interstitial fibrosis. The heart is unable to compensate adequately to meet the demands of growth, pregnancy, or other cardiac stress, which results in late-onset anthracycline-induced cardiac failure. The incidence of cardiomyopathy is related to the cumulative dose of anthracyclines, occurring in $\sim 10\%$ of survivors after a cumulative dose of less than 400 mg/m^2, $\sim 20\%$ after 400–599 mg/m^2, $\sim 50\%$ after 600–799 mg/m^2 and almost 100% after 800 mg/m^2.

Anthracycline-induced cardiomyopathy is a progressive disorder. It manifests with signs of congestive heart failure, including exercise intolerance, dyspnea, peripheral edema, pulmonary rales, S3 and S4 and hepatomegaly. Rapid progression of symptoms may occur with pregnancy, anesthesia, isometric exercise, the use of illicit drugs (e.g., cocaine), prescription drugs, or alcohol.

Cyclophosphamide

Cyclophosphamide-induced cardiac effects occur primarily with high-dose preparatory regimens for stem cell transplantation. Cyclophosphamide causes intramyocardial edema and hemorrhage, often in association with serosanguineous pericardial effusion and fibrous pericarditis. This cardiotoxicity is usually reversible.

Radiation

In addition to acting in concert with anthracyclines to increase the risk of cardiomyopathy, radiation to the heart (e.g. mantle radiation for Hodgkin lymphoma or total body irradiation as conditioning for bone marrow transplantation) can also induce valvular damage, pericarditis, or coronary vessel damage increasing the risk of ischemic heart disease. Cardiac doses over 20 Gy confer the highest risk. Radiation to the neck can damage the carotid vessels, increasing the risk of stroke.

Screening and Management

In addition to a good interval history and physical examination, important screening tools for evaluating cardiac function following exposure to anthracyclines or radiation include:

- EKG
- ECHO
- Radionuclide angiocardiography.

Electrocardiogram (EKG) findings include prolonged QT_c (≥ 0.45 or longer), second-degree atrio-ventricular (A–V) block, complete heart block, ventricular ectopy, ST elevation or depression and T-wave changes. An association between prolongation of QT_c interval and anthracycline dose over 300 mg/m^2 has been described in childhood cancer survivors.

Echocardiogram (ECHO) findings include fractional shortening (SF) and velocity of circumferential fiber shortening (VCF) for the measurement of left ventricular contractility.

Radionuclide cardiac cineangiocardiography (RNA or MUGA) determines the ejection fraction and is useful for those patients in whom a good ECHO cannot be obtained. A left ventricular ejection fraction (LVEF) of 55% or more indicates normal systolic function.

The following criteria define progressively deteriorating cardiac function:

- A decrease in the SF by an absolute value of 10% from the previous test
- SF less than 29%
- A decrease in the RNA LVEF by an absolute value of 10% from the previous test
- RNA LVEF less than 55%
- A decrease in the RNA LVEF with stress.

Management of anthracycline and radiation-induced cardiomyopathy should include:

- Digoxin to improve ventricular contractility
- Diuretics to decrease sodium and water retention
- Angiotensin-converting enzyme-inhibiting agents (e.g., Captopril) to decrease sodium and water retention and decrease afterload
- Enalapril to increase contractility and decrease afterload.

Prognosis of progressive late-onset heart failure is poor. Cardiac transplantation should be considered. The actuarial survival rate at 5 years after cardiac transplant is 77%.

The Children's Oncology Group provides the following guidelines for long-term follow-up of cardiac function following anthracycline and radiation exposure (available at www.Survivorshipguidelines.org). The frequency of testing is determined by the cumulative anthracycline doses and exposure to radiation (Table 32-6).

Table 32-6   Recommended Frequency of Cardiac Screening Based on Anthracycline and Heart Radiation Exposure

Recommended Frequency of Echocardiogram or MUGA Scan

Age at Treatment	Radiation with Potential Impact to the Heart	Anthracycline Dose	Recommended Frequency
	Yes	Any	Every year
1 year old	No	<200 mg/m^2	Every 2 years
		≥200 mg/m^2	Every year
	Yes	Any	Every year
1–4 years old		<100 mg/m^2	Every 5 years
	No	≥100 to <300 mg/m^2	Every 2 years
		≥300 mg/m^2	Every year
	Yes	<300 mg/m^2	Every 2 years
≥5 years old		≥300 mg/m^2	Every year
		<200 mg/m^2	Every 5 years
	No	≥200 mg/m^2 to <300 mg/m^2	Every 2 years
		≥300 mg/m^2	Every year
Any age with decrease in serial function			Every year

Adapted from: COG Survivorship Guidelines. http://www.survivorshipguidelines.org, with permission.

After completion of therapy, all patients with any amount of anthracycline exposure or thoracic radiation therapy should have an EKG and ECHO ± MUGA performed at the end of therapy. Those with normal studies at that point should be followed as per the recommended guidelines, while patients with an abnormal study either at the end of therapy or at the time of initial long-term follow-up (5 years after diagnosis) should have more frequent evaluations. It is extremely important to counsel patients to avoid tobacco smoke.

RESPIRATORY SYSTEM

Both chemotherapy and radiation therapy can cause acute and chronic lung injury. Younger children are at more risk than adolescents or adults for the development of chronic respiratory damage.

Chemotherapy

The primary offending agents include bleomycin and nitrosourea, with clinical manifestations usually occurring months after a critical cumulative dose is reached or exceeded.

Bleomycin

The critical cumulative dose of bleomycin is 400 units, after which 10% of patients experience fibrosis. This manifests as dyspnea, dry cough and rales. Radiographic findings include interstitial pneumonitis with reticular or nodular pattern and pulmonary function tests show a restrictive ventilatory defect with hypoxia, hypercapnia and chronic hyperventilation (diffusion capacity of carbon monoxide is considered to be the most sensitive test). Radiation therapy, renal insufficiency, cisplatin, cyclophosphamide, exposure to high levels of oxygen and pulmonary infections can exacerbate the effects of bleomycin.

Nitrosourea

The critical cumulative dose of nitrosourea is a dose $>1,500$ mg/m^2 in adults or >750 mg/m^2 in children. The clinical manifestations of nitrosourea toxicity are the same as bleomycin, although pulmonary fibrosis is more commonly associated with BCNU.

While bleomycin and nitrosourea are most commonly associated with long-term pulmonary toxicity, other agents such as methotrexate, 6-mercaptopurine and procarbazine have been associated with an acute hypersensitivity reaction which can result in long-term pulmonary function changes.

Cytosine arabinoside, methotrexate, ifosfamide and cyclophosphamide have been associated with noncardiogenic pulmonary edema. This complication occurs within days of the beginning of treatment and can also result in long-term pulmonary function changes.

Other host factors that can contribute to chronic pulmonary toxicity include asthma, infection, smoking and having had a history of assisted ventilation.

Radiation

Radiation therapy in children younger than 3 years of age results in increased toxicity. Radiation therapy to the lungs can cause impairment of the proliferation and maturation of alveoli, resulting in chronic respiratory insufficiency. This is thought to be consistent with a proportionate interference with the growth of both the lungs and chest wall. It also causes damage to the type II pneumocyte, which is responsible for the production of surfactant and the maintenance of patency of the alveoli. As a result of changes in surfactant production, there is a decrease in alveolar surface tension and compliance. These children exhibit decreased mean total lung volumes and DLCO (diffusion capacity of carbon monoxide) that is approximately 60% of predicted values.

The effects of direct radiation to the lungs also include damage to the endothelial cells of the capillaries resulting in alterations of perfusion and permeability of the vessel wall, likely mediated by cytokine production that stimulates septal fibroblasts increasing collagen production and pulmonary fibrosis.

The late radiation injury to the lung is characterized by the presence of progressive fibrosis of alveolar septa and obliteration of collapsed alveoli with connective tissue. In asymptomatic adolescents treated for Hodgkin lymphoma, chest radiographic findings or pulmonary function tests consistent with fibrosis have been found in over 30% of patients and these changes have been detected months to years after radiation therapy. The incidence of radiation-induced fibrosis has decreased over time due to refinements in radiation therapy.

Children who receive craniospinal radiation (either for leukemia or malignant brain tumors) also have a significant risk of developing late restrictive lung disease.

Screening and Management

It is critical to quickly identify patients developing pulmonary toxicity, obtain appropriate diagnostic tests and implement interventions to prevent the toxicity from worsening. Table 32-7 shows a scoring system for shortness of breath.

In addition to a good interval history and physical examination, important diagnostic tests for possible pulmonary toxicity include:

- Pulmonary function tests (PFT) – these should be obtained immediately following completion of therapy, at entry to long-term follow-up and as needed based on symptoms
- Chest radiograph
- CT scan of the chest based on symptoms and findings on PFTs and chest radiograph
- Quantitative ventilation–perfusion scintigraphy – blood flow to ventilated alveoli has proved most useful in quantifying radiation injury to the lungs.

Table 32-7   Shortness of Breath Scoring System

Grade	Description
0	No shortness of breath with normal activity. Shortness of breath with exertion, comparable to a well person of the same age, height and sex
1	More shortness of breath than a person of the same age while walking quietly on the level or on climbing an incline or two flights of stairs
2	More shortness of breath and unable to keep up with persons of the same age and sex while walking on the level
3	Shortness of breath while walking on the level and while performing everyday tasks at work
4	Shortness of breath while carrying out personal activities (e.g., dressing, talking, walking from one room to another)

From: McDonald S, Rubin P, Schwartz CL. Pulmonary effects of antineoplastic therapy. In: Schwartz CL, Hobbie WL, Constine LS, Ruccione KS, editors. Survivors of Childhood Cancer: Assessment and Management. St. Louis: Mosby, 1994: 177–195, with permission.

Any patient with symptoms or diagnostic testing consistent with pulmonary toxicity should be refered to a pulmonologist. In the interim, bronchodilators, expectorants, antibiotics and oxygen can be used for symptomatic relief. It is extremely important to counsel patients to avoid tobacco smoke.

CENTRAL NERVOUS SYSTEM

Understanding the risks of central nervous system (CNS) toxicity is critically important in long-term follow-up of childhood cancer survivors. Many survivors experience learning difficulties in school, which can be ameliorated with an appropriate individualized education plan (IEP) implemented by the school. A neuropsychologist familiar with the CNS effects of treatment for childhood cancer can perform a neurocognitive assessment and identify specific strengths and weaknesses in the childhood cancer survivor. These can be used to formulate interventions that can compensate for any weaknesses. Such interventions may include extra time on tests to compensate for reduced processing speed, sitting at the front of the class and having tests taken in isolation to compensate for attention deficits, or having tutoring to bolster a specific weakness. Without the knowledge of the risks of therapy, subtle learning problems may be easily overlooked, potentially leading to poor school performance, reduced self-confidence, lower education achievement, lower earning potential and a lower quality of life. Children treated before 5–6 years of age, especially those treated before 3 years of age, are at a higher risk for developing cognitive impairments than those treated after the age of 8–10 years.

Surgery

Survivors of childhood brain tumors usually have had surgery to remove the tumor. Depending on the location of the tumor and the degree of resection, almost any aspect of neurological function can be impaired. It is critically important to understand the location and extent of the surgery, as well as any perisurgical deficits that have persisted beyond therapy.

Chemotherapy

The pathogenesis of chemotherapy-induced CNS toxicity is not well understood. In addition to damage to the glial and capillary tissue, the neurotransmitter function of the brain may also be impaired. The primary offending chemotherapeutic agents leading to CNS toxicity include:

- Intrathecal chemotherapy (primarily methotrexate, cytosine arabinoside and hydrocortisone)
- High-dose IV methotrexate (crosses the blood–brain barrier)
- High-dose IV cytosine arabinoside (crosses the blood–brain barrier).

Current trends in the use of higher doses of methotrexate, as well as the increased frequency with which methotrexate is used in intrathecal therapy, have resulted in a rise of neurologic sequelae.

Radiation

Radiation to a child's brain can result in altered capillary wall permeability, resulting in alterations in cerebral blood flow, primary damage to glial cells, demyelination of glial tissue, focal white matter destruction and impaired neuronal differentiation, including dendritic formation and synaptogenesis.

Radiation to the brain over 20 Gy has been associated with pathological changes and learning deficits. Although chemotherapy or radiation therapy alone can be sufficient to induce CNS changes, the combination can have an even more profound effect. Twenty Gy cranial RT plus intrathecal methotrexate and systemic methotrexate (>40 mg/m^2 weekly) has been shown to be the most neurotoxic.

The pathologic findings as a result of radiation, intrathecal chemotherapy and systemic chemotherapy are:

*   Leukoencephalopathy
*   Mineralizing microangiopathy
*   Subacute necrotizing leukomyelopathy.

Severe leukoencephalopathy may manifest with seizures, ataxia, lethargy, slurred speech, spasticity, dysphagia, lowered IQ scores, memory impairment and confusion. Radioimaging findings include dilatation of the ventricles and subarachnoid space indicative of cerebral atrophy, white matter hypodensity by CT scan and hyperdensity on magnetic resonance imaging (MRI). With the use of current treatments, this severe form of leukoencephalopathy is seen infrequently. The subclinical form of leukoencephaly is more common, characterized by radiological abnormalities on CT scan and MRI without clinical symptoms and occurs in approximately 55–60% of patients receiving CNS prophylaxis. The more severe form is seen more often in patients treated with cranial radiation, intrathecal methotrexate and systemic methotrexate.

Mineralizing microangiopathy may manifest with seizures, electroencephalogram (EEG) abnormalities, incoordination, gait abnormalities, memory deficits, learning disabilities, decrease in IQ scores and behavioral problems. The onset of symptoms usually occurs 10 months to several years after CNS exposure and the risk of these complications increases proportionately with higher doses of IT MTX and radiation. Imaging findings include changes in the gray matter of the brain, mainly in the region of basal ganglia and less frequently in the cerebellar gray matter. Histologically, there is deposition of calcium in the small blood vessels, causing lumen occlusion by mineralized debris. There may be dystrophic calcification of the surrounding neural tissue.

Subacute necrotizing leukomyelopathy is an unusual complication occurring after the use of cranial or craniospinal irradiation combined with intrathecal methotrexate. Histologically, it shows focal myelin necrosis on the posterior and/or lateral columns of the spinal cord.

Patients with acute lymphoblastic leukemia who are younger than 5 years of age at the time of cranial irradiation are at a 1% risk to develop brain tumors within the first 10 years after therapy. Children treated under 6 years of age have an increased risk of developing high-grade gliomas and are also at risk for developing meningioma 10–20 years after cranial irradiation.

Table 32-8 lists the clinical manifestations of neurotoxicity associated with radiation therapy.

Table 32-8   Radiation-Related Neurotoxicity

Type	Timing	Etiology	Syndrome	Signs and Symptoms
Acute	During or immediately after RT	Peritumoral edema	Increased intracranial pressure	Worsening focal signs; headache, vomiting; nausea
Subacute	Weeks to 3 months	Transient Demyelination (? Oligodendroglial dysfunction)	Somnolence	Somnolence lasting days to weeks; anorexia
			Myelitis	Paresthesia down spine
			Endocrinologic	Growth failure; hypothyroidism; hypogonadism
			Neurocognitive	School difficulties; memory loss; retardation
Chronic	Months to years after RT	Vasculitis; oligodendroglial dysfunction; autoimmune etiology	Radionecrosis	Focal deficits; seizures; symptoms of increased intracranial pressure
			Large vessel occlusion	Focal deficits; seizures
			Mineralizing microangiopathy	Focal deficits; seizures; headaches
			Myelopathy	Paraparesis; quadriparesis; bowel and bladder dysfunction
			Peripheral neuropathy	Cranial nerve palsies; autonomic dysfunction weakness

From: Moore IM, Packer RJ, Karl D, Bleyer WA. Adverse effects of cancer treatment on the central nervous system. In: Schwartz CL, Hobbie WL, Constine LS, Ruccione KS, editors. Survivors of Childhood Cancer: Assessment and Management. St. Louis: Mosby, 1994: 81–95, with permission.

Screening and Management

In addition to a good interval history and physical examination, important diagnostic tests for possible CNS toxicity include:

* Neurocognitive assessment
* Neuroradiologic studies
* EEG.

Neurocognitive evaluation is essential to detect learning disability. It should be performed in all school-aged children who have received cranial irradiation, intrathecal chemotherapy, or high-dose systemic methotrexate therapy. It should be performed early in therapy to establish a baseline, at the end of therapy and at entry into long-term follow-up. If any abnormalities are noted, more frequent assessment may be needed to maintain up-to-date recommendations for the school's individual educational plan (IEP).

Neuroradiologic studies should include either CT or MRI as indicated based on physical examination, history and neurocognitive assessment. MRI is the preferred imaging modality because of its greater sensitivity in delineating white matter damage. Mild change presents as occasional punctate areas of signal abnormality, moderate as large or multiple areas of damage and severe as confluent areas of white matter damage. Patients with severe white matter damage manifest impairment in mentation, motor deficits and seizures.

EEG abnormalities are relatively nonspecific, but can be useful to identify seizure activity should that be suspected based on history or physical examination.

Management of any CNS deficits should include working directly with the school to produce a functional IEP designed to enhance the strengths and ameliorate the weaknesses identified in each child.

ENDOCRINE SYSTEM

The primary endocrine systems that can sustain injury during therapy for childhood cancer include the hypothalamic–pituitary axis, the thyroid gland, the pancreas and the gonads.

Surgery

The primary endocrine morbidities of surgery include pituitary dysfunction as a consequence of suprasellar brain tumors (i.e. germinomas and craniopharyngiomas) and gonadal dysfunction if the testes or ovaries are removed as part of therapy (i.e. for germ cell tumors).

Chemotherapy

The effects of systemic chemotherapy primarily affect the gonads. There are two aspects to gonadal function that require consideration:

- Fertility
- Hormone production.

While these can be separated in males, they cannot be easily separated in females. The primary offending agents regarding gonadal dysfunction are the alkylating agents – such as procarbazine, cyclophosphamide, ifosfamide and busulfan.

Female Gonadal Function

Female fertility and estrogen production are inexorably linked to the number of remaining oocytes. Girls are born with all of the oocytes they will have in their lifetime, approximately one million oocytes, which decline in an accelerating manner over time. When only 1,000 oocytes remain, menopause ensues. Chemotherapy exerts a downward shift in the curve, leading to acute ovarian failure if <1,000 oocytes remain after treatment or premature menopause if >1,000 oocytes remain. Without a reliable method to measure ovarian reserve, any girl treated with alkylating agents should be counseled that they may have a shortened fertility window.

Male Gonadal Function

Unlike females, fertility and hormone (testosterone) production can be separated by sensitivity to treatment. While sperm are continuously produced from spermatic stem cells, testosterone is produced in the Leydig cell. The Leydig cell tends to be quite resistant to chemotherapy, so testosterone production tends to be conserved after therapy. In order to cause permanent male infertility, chemotherapy would have to completely eliminate all spermatic stem cells, as remaining stem cells can, over time, repopulate the testes and re-establish fertility. Sperm production can be assessed through semen analysis and testosterone production can be directly measured in the serum. As Leydig cells fail, the anterior pituitary is forced to increase the production of FSH and LH to stimulate the gonads to increase testosterone production. Any male who has been found to have elevated FSH and LH should be considered to have early Leydig cell failure. Generally, doses over 10 g/m^2 of cyclophosphamide equivalents should raise concern of possible infertility and possible Leydig cell dysfunction.

Radiation

Any endocrine organ can be adversely affected by radiation. Radiation to the gonads can lead to sterility and hormone production failure as discussed above. Table 32-9 shows the effects of fractionated testicular irradiation on spermatogenesis and Leydig cell function.

Table 32-9   Effect of Fractionated Testicular Irradiation on Spermatogenesis and Leydig Cell Function

Testicular Dose in Gy (100 Rad)	Effect on Spermatogenesis	Effect on Leydig Cell Function
<0.1	No effect	No effect
0.1–0.3	Temporary oligospermia Complete recovery by 12 months	No effect
0.3–0.5	Temporary azoospermia at 4–12 months following irradiation 100% recovery by 48 months	
0.5–1.0	100% temporary azoospermia for 3–17 months from irradiation Recovery beginning at 8–26 months	Transient rise in FSH with eventual normalization
1–2	100% azoospermia from 2 months to at least 9 months Recovery beginning at 11–20 months with return of sperm counts at 30 months	Transient rise in FSH and LH No change in testosterone
2–3	100% azoospermia beginning at 1–2 months Some will suffer permanent azoospermia; others show recovery starting at 12–14 months Reduced testicular volume	Prolonged rise in FSH with some recovery Slight increase in LH No change in testosterone
3–4	100% azoospermia No recovery observed up to 40 months All have reduced testicular volume	Permanent elevation in FSH Transient rise in LH Reduced testosterone response to hCG stimulation
12	Permanent azoospermia Reduced testicular volume	Elevated FSH and LH Low testosterone Decreased or absent testosterone response to hCG stimulation Testosterone replacement may be needed to ensure pubertal changes
>24	Permanent azoospermia Reduced testicular volume	Effects more severe and profound than at 12 Gy Prepubertal testes appear moresensitive to the effects of radiation Replacement hormone treatment probably needed in all prepubertal cases

Abbreviations: FSH, follicle-stimulating hormone; LH, luteinizing hormone; HCG, human chorionic gonadotropin.
From: Leventhal BG, Halperin EC, Torano AE. The testes. In: Schwartz CL, Hobbie WL, Constine LS, Ruccione KS, editors. Survivors of Childhood Cancer: Assessment and Management. St. Louis: Mosby, 1994: 225–244, with permission.

Radiation to the neck can lead to thyroid dysfunction and radiation to the brain can damage the hypothalamic–pituitary axis.

The hypothalamic–pituitary axis plays a central role in translating neurologic and chemical signals from the brain into endocrine responses through its neuronal circuitry with the brain

Table 32-10   Anterior Pituitary Hormones, their Physiologic Functions and their Hypothalamic Regulatory Hormones

Pituitary Hormone	Functions of Pituitary Hormone	Hypothalamic Regulatory Factor
Growth hormone (GH)	Bone and soft-tissue growth through insulin-like growth hormones (ISF-1)	Growth-hormone-releasing hormones (GHRH) (+) Somatostatin (−)
Prolactin (PRL)	Induction of lactation and interruption of ovulation and menstruation during postpartum period	Dopamine (−)
Gonadotropins Luteinizing hormone (LH)	In males, LH stimulates leydig cells to produce testosterone; in females, LH is responsible for normal steroidogenesis and ovulation	Gonadotropin-releasing hormone (GnRH) (+)
Follicle-stimulating hormone (FSH)	In males, FSH stimulates spermatogenesis; in females, FSH is responsible for normal steroidogenesis and ovulation	Gonadotropin-releasing hormone (GnRH) (+)
Thyroid-stimulating hormone (TSH)	TSH regulates thyroid hormone production for thyroid gland	Thyrotropin-releasing hormone (+)
Adrenocorticotropin (ACTH)	ACTH regulates adrenal steroidogenesis	Corticotropin-releasing hormone (+)

Abbreviations: (+), stimulating effect; (−), inhibitory effect.
From: Sklar CA. Neuroendocrine complications of cancer therapy. In: Schwartz CL, Hobbie WL, Constine LS, Ruccione KS, editors. Survivors of Childhood Cancer: Assessment and Management. St. Louis: Mosby, 1994: 97–110, with permission.

involving afferent and efferent pathways. It also produces peptide hormones and biogenic amines that serve as regulators of anterior pituitary hormones. These hypothalamic factors are supplied to the anterior pituitary gland by way of the portal venous system.

Table 32-10 shows the anterior pituitary hormones, their physiologic functions and their hypothalamic regulatory hormones.

Screening and Management

Gonadal Dysfunction

In addition to a good interval history (especially menstrual and pregnancy history in females) and physical examination, important diagnostic tests for possible gonadal toxicity include:

- Serum FSH, LH and estrogen or testosterone
- Semen analysis.

For gonadal dysfunction, all patients who received alkylating agent and/or radiation to the gonad should be counseled about possible infertility and hormonal failure. Male survivors should be offered the opportunity to have a semen analysis performed and females should be counseled regarding the risk of premature menopause. All sexually mature males prior to therapy and males post-therapy who are still producing sperm but have elevated FSH and LH, should be offered sperm banking or testicular biopsy with cryopreservation. Although there is no current technique for in vitro maturation of sperm, prepubertal boys can be offered testicular biopsy with cryopreservation of testicular tissue in the hope that in the next decade the technology for in vitro maturation will be developed. For females, the only standard of care for fertility preservation remains embryo cryopreservation. This requires a partner to participate in fertilization of the oocyte. Experimental techniques are currently being clinically developed to allow for oocyte or ovarian tissue cryopreservation to allow for improved options for females receiving chemotherapy or radiation.

In addition to gonadal damage affecting fertility, the uterus and vagina can also be damaged, impacting pregnancy and delivery. Factors contributing to high-risk pregnancy and delivery after radiation to the abdomen during childhood include:

- Damage to elastic properties of the uterine musculature
- Damage to the vasculature of the uterus.

Although the rate of birth defects among childhood cancer survivors is the same as in the general population, the rate of perinatal mortality is higher than in the general population and there is a four-fold increase for low-birth-weight infants in women who have received abdominal radiation for the treatment of Wilms' tumor during their childhood. These women are also at risk for premature labor and fetal malposition.

Thyroid Dysfunction

Hypothyroidism is the most common sequela of radiotherapy to the neck. Elevated TSH levels with normal T3, T4, indicative of subclinical hypothyroidism, can be detected in up to two-thirds of patients treated with mantle field radiation greater than 2,600 cGy in Hodgkin lymphoma. However radiation dose-reduction lowered the incidence of hypothyroidism to 10–28%.

In addition to a good interval history and physical examination, important diagnostic tests for possible thyroid toxicity include:

- Annual thyroid function tests for those who received radiation to the neck, either direct or as scatter from mantle or spinal radiation
- Thyroid ultrasound to assess for possible secondary thyroid malignancies for those with a palpable nodule or abnormal thyroid function tests.

Patients with abnormal thyroid function tests should be referred to an endocrinologist for management and those with a nodule on physical examination or ultrasound should be referred to a neck surgeon for biopsy. Secondary papillary thyroid carcinoma can be treated with total thyroidectomy, radioactive iodine and TSH suppression with thyroxine.

Anterior Pituitary Dysfunction

In addition to a good interval history and physical examination, important diagnostic tests for possible anterior pituitary toxicity include:

- Close monitoring of growth velocity with the use of a growth chart
- Pubertal history and examination to ensure appropriate pubertal progression
- Serum FSH, LH and estrogen or testosterone
- First morning cortisol level if indicated by history.

Growth Hormone Deficiency

If growth failure is due to documented growth hormone (GH) deficiency, then growth hormone is appropriate treatment. Growth failure can occur due to the systemic effects of cancer or cancer treatment and not necessarily due to GH deficiency, so due diligence is required to make the diagnosis of growth hormone deficiency.

Some considerations in deciding to use growth hormone include:

- A worsened degree of scoliosis, although not an increased incidence
- It may cause benign intracranial hypertension (pseudotumor cerebri), which resolves with medical management (i.e., acetazolamide and dexamethasone) and discontinuation of GH therapy
- Children with prior neoplastic disease may have an increased risk of developing a second cancer following GH therapy, although the data are somewhat conflicting and unclear.

Luteinizing and Follicle-Stimulating Hormone Deficiency

Treatment should include:

- Estrogen and progestin therapy in females
- Androgen therapy in males.

Precocious Puberty

Treatment should include:

- Gonadotropin-releasing hormone (GnRH) analogues to suppress puberty
- GnRH plus GH for patients with coexistent GH deficiency.

Thyroid-Stimulating Hormone Deficiency

Treatment should include:

- Daily thyroxine therapy.

Adrenocorticotropin Deficiency

Treatment should include:

- Low-dose hydrocortisone therapy daily
- Stress doses of hydrocortisone during febrile illness or under anesthesia.

Hyperprolactinemia

Treatment should include:

- Bromocriptine or related dopamine agonists are used to reduce prolactin levels in young women with amenorrhea and infertility as a result of hyperprolactinemia.

Table 32-11 describes the neuroendocrine complications, their clinical manifestations, standard fractionated radiation dose limits, diagnostic studies and treatment.

Table 32-12 presents an evaluation of the hypothalamic pituitary axis using an analysis of pertinent hormonal assays.

GENITOURINARY SYSTEM

Surgery

Surgery used to treat childhood tumors affecting the kidney (e.g. Wilms' tumor and neuroblastoma) and bladder (e.g. rhabdomyosarcoma) can result in permanent damage to the genitourinary system. Other surgeries can also contribute – for example, damage to the thoracolumbar sympathetic plexus during retroperitoneal lymph node dissection can result in ejaculatory failure. Patients who have had a nephrectomy need to be counseled about the risks of trauma to their remaining single kidney (e.g. bicycle handlebar injuries or seatbelt injuries).

Chemotherapy

Ifosfamide, carboplatin and cisplatin can cause renal complications, including acute tubular dysfunction, although carboplatin is less nephrotoxic than cisplatin. The acute tubular dysfunction can be manifested by an increased excretion of potassium, phosphorus and magnesium. Ifosfamide can also cause hypophosphatemic rickets. Risk factors for ifosfamide nephrotoxicity include:

- Younger age group
- Hydronephrosis
- Prior administration of cisplatin
- Cumulative dose of ifosfamide greater than 72 g/m^2.

Table 32-11   Summary of Neuroendocrine Complications, Their Clinical Manifestations, Standard Fractionated Radiation Dose Limits to Brain Responsible for Damage, Diagnostic Studies and Treatment

Disorder	Clinical Presentation	Radiation Dose (Gy)	Diagnostic Studies	Treatment
GH deficiency	Short stature	≥18–20	Plotting growth on growth velocity chart GH stimulation test Bone age Frequent sampling for 12–24 hours GH	Recombinant GH[a]
Gonadotropin deficiency	In young child, failure to enter puberty and primary amenorrhea; in adults, infertility, sexual dysfunction and decreased libido	>30	Basal serum concentration of LH, FSH, estradiol, or testosterone GnRH stimulation test	Estrogen/ progestin (women) Depot testosterone (men)
Precocious puberty	Female 8 years of age or younger with appearance of breast and genital development and male 9 years of age or younger with testicular development Accelerated skeletal growth and premature epiphyseal fusion	>10(?)	GnRH stimulation test Estradiol or testosterone concentration Bone age Pelvic ultrasound (women) GH stimulation tests	GnRH agonists plus recombinant GH (if GH deficient)
TSH deficiency	Often subclinical	>30	Basal serum T3 uptake, T4 and TSH TRH stimulation test	L-Thyroxine
ACTH deficiency	Decreased stamina, lethargy, fasting hypoglycemia, dilutional hyponatremia	>30	Basal serum cortisol concentration Adrenal stimulation test (e.g., insulin, ACTH)	Hydrocortisone

[a]See text discussion on use of growth hormone.
Abbreviations: GH, growth hormone; TSH, thyroid-stimulating hormone; ACTH, adrenocorticotropin; LH, luteinizing hormone; FSH, follicle-stimulating hormone; GnRH, gonadotropin-releasing hormone; TRH, thyrotropin-releasing hormone.
Modified from: Sklar CA. Neuroendocrine complications of cancer therapy. In: Schwartz CL, Hobbie WL, Constine LS, Ruccione KS, editors. Survivors of Childhood Cancer, Assessment and Management. St. Louis: Mosby, 1994:97–110, with permission.

Indicators of significant Fanconi syndrome include a serum glucose less than 150 mg/dl and ratio of urine protein to urine creatinine less than 0.2, as well as serum phosphate, <3.5 mg/dl; K+, <3 mEq/l; bicarbonate, <17 mEq/l; and 1+ glycosuria.

Hemorrhagic cystitis and fibrosis of the bladder can occur often after treatment with cyclosphamide and ifosfamide, especially when the bladder is included in the radiation field.

Table 32-12   Evaluation of Hypothalamic Pituitary Axis

	Testosterone	FSH	LH	Response to GnRH	Response to hCG
Primary Leydig cell disease	nl/lo	hi	hi	—	lo
Primary disease of germinal epithelium	nl	hi	nl	—	—
Hypothalamic disease	nl/lo	nl/lo	nl/lo	nl	—
Pituitary disease	nl/lo	nl/lo	nl/lo	lo	—

Abbreviations: nl, normal; lo, low; hi, high; GnRH, gonadotropin-releasing hormone; hCG, human chorionic gonadotropin.
From: Leventhal BG, Halperin EC and Torano AE. The testes. In: Schwartz CL, Hobbie WL, Constine LS, Ruccione KS, editors. Survivors of Childhood Cancer: Assessment and Management. St. Louis: Mosby, 1994: 225–244, with permission.

Radiation

Hypoplastic kidney and renal arteriosclerosis can occur after the use of a combination of radiation therapy in a dose of 10–15 Gy with chemotherapy, or radiation therapy alone in a dose of 20–30 Gy. Nephrotic syndrome can occur after radiation therapy in a dose of 20–30 Gy of RT.

Abnormal bladder function is seen in children with pelvic rhabdomyosarcoma in whom radiation has been used. Thirty-two to 100% of these children experience dribbling and nocturnal enuresis.

Screening and Management

In addition to a good interval history (especially sexual function in boys) and physical examination, important diagnostic tests for possible genitourinary toxicity include:

- Serum BUN and creatinine annually
- Urinalysis annually
- 24-hour creatinine clearance or nuclear GFR in those with abnormal renal function.

OCULAR SYSTEM
Chemotherapy

Corticosteroid exposure may increase the risk of early cataract development, especially in conjuction with radiation to the lens.

Radiation

Several areas of the eye can be damaged by radiation. In particular:

- *Lacrimal glands:* radiation >50 Gy can result in decreased tearing or fibrosis
- *Cornea:* radiation >40 Gy can lead to ulceration, neovascularization, keratinization or edema

- *Lens:* radiation >10 Gy can cause cataracts
- *Iris:* radiation >50 can lead to neovascularization, glaucoma and atrophy
- *Retina:* radiation >50 Gy can cause infarction, exudates, hemorrhage, telangiectasia, neovascularization and macular edema
- *Optic nerve:* radiation >50 Gy can cause optic neuropathy.

Screening and Management

In addition to a good interval history and physical examination, important diagnostic tests for possible ocular toxicity include:

- An annual eye examination, including a slit lamp examination and dilation, for patients with exposure to corticosteroids or radiation to the eye.

AUDITORY SYSTEM

Chemotherapy

Cisplatin or carboplatin exposure can cause a mild to severe sensorineural hearing loss, often necessitating hearing aids. Cisplatin is significantly more ototoxic than carboplatin. As many children with cancer require frequent courses of parenteral antibiotics, it is also important to remember that aminoglycosides can contribute to hearing loss.

Radiation

Radiation exposure can cause:

- Sensorineural hearing loss can occur following radiation of more than 40 Gy
- Chronic otitis can occur following radiation of more than 40 Gy.

Screening and Management

In addition to a good interval history and physical examination, important diagnostic tests for possible audiological toxicity include:

- A hearing test or brainstem auditory evoked potential (BAER), for those too young for a behavioral hearing test, for patients with exposure to cisplatin or carboplatin or radiation to the auditory system.

GASTROINTESTINAL SYSTEM

Chemotherapy

Hepatic fibrosis/cirrhosis can occur following the use of methotrexate, actinomycin D, 6-mercaptopurine (6-MP) and 6-thioguanine (6-TG). 6-TG can also cause a mild to moderate sinusoidal obstructive syndrome of the liver.

Radiation

Radiation exposure can cause:

- Enteritis following radiation greater than 40 Gy. Actinomycin D or doxorubicin can enhance the radiation effect
- Fibrosis can occur anywhere from the esophagus to the colon following the use of radiation greater than 40 Gy
- Over 30 Gy can increase the risk of secondary colon cancer.

Screening and Management

In addition to a good interval history and physical examination, important diagnostic tests for possible gastrointestinal toxicity include:

- Liver function tests should be performed during follow-up for late effects
- Those with exposure to greater than 30 Gy of radiation should be screened for secondary colon cancer via colonoscopy starting at age 35.

It is also important to remember that administered blood transfusions prior to adequate blood donor screening for hepatitis C are at risk for chronic liver disease secondary to hepatitis C.

IMMUNOLOGIC SYSTEM

For patients who received conventional chemotherapy (i.e. not hematopoietic stem cell transplant) defects in both normal and cellular immunity generally resolve within 6 months to 1 year after cessation of chemotherapy. Although some immune deficiency may uncommonly persist for more than 2 years after completion of therapy more than half of the children have no protective antibodies to one or more previously administered vaccines or related infections. Although most of them are able to produce antibodies after reimmunization, some patients repeatedly are unable to make protective antibodies after reimmunization or despite natural disease.

In addition to the impact of chemotherapy and radiation on humoral and adaptive immunity, radiation to the mucosa (i.e. sinuses, nose, lungs) can cause defects in barrier immunity such as immotile cilia. This can lead to recurrent sino-pulmonary infections in survivors of childhood cancer.

OBESITY

Adult survivors of childhood acute lymphoblastic leukemia have an increased risk for obesity. Adult leukemia survivors are more physically inactive and have reduced exercise capacity which further increases their risk for obesity. The pathophysiology of the metabolic impacts of treatment for childhood cancer remains under investigation.

Interventions to prevent obesity and promote physical activity may decrease cardiovascular morbidity and improve the quality of life in this population. Follow-up with assessment of weight and physical activity is important in the evaluation of childhood cancer survivors.

SECOND MALIGNANT NEOPLASMS

Second malignant neoplasms are second cancers occurring in survivors of childhood cancer, caused by treatments administered for the first cancer or a heritable disposition, but otherwise unrelated to the first cancer. Second malignant neoplasms represent a significant risk to the childhood cancer survivors.

- The incidence of secondary neoplasms is 3–12% within the first 20 years from diagnosis
- Childhood cancer survivors have 10–20 times the lifetime risk of second malignant neoplasms as compared to age-matched controls
- Second malignant neoplasms are the most common cause of death in long-term survivors after recurrence of the primary cancer.

Patients with retinoblastoma, Hodgkin disease and Ewing sarcoma are at the greatest risk of developing second malignant neoplasms. Long-term survivors who have an underlying inherited susceptibility to cancer have a second malignant neoplasm rate that approaches 50%. Such inherited predispositions include:

- Retinoblastoma
- Neurofibromatosis
- Li–Fraumeni syndrome
- Familial polyposis.

Table 32-13 lists the second malignant neoplasms, their predisposing factors, signs and symptoms and recommendations for follow-up.

Table 32-13 Predisposing Factors, Signs and Symptoms and Follow-up in Second Malignant Neoplasms

Second Malignant Neoplasm	Predisposing Factor	Other	Signs and Symptoms	Recommended Follow-up
Bone or soft-tissue sarcoma	Radiotherapy (RT)	Doses >3,000 cGy; adolescents	Pain or mass in irradiated area	Radiograph, other imaging, baseline every 5 years or if symptoms arise
	Retinoblastoma	Familial and bilateral cases		
	Li-Fraumeni syndrome	Family history may be revealing; constitutional p53 mutation (?)		
	Neurofibromatosis (NF)	Diagnosis based on clinical findings; influence of RT not established		
Pineoblastoma	Retinoblastoma	Bilateral and familial cases	Change in sleep pattern	Neuroimaging baseline, then if symptoms arise
Brain tumor	RT	Younger children at greater risk	Seizures, headaches, altered mental status	
	NF			
	Nevoid basal cell carcinoma syndrome	Ionizing RT increases risk and shortens latent period		
Thyroid	RT	Younger children at greater risk	Enlarged thyroid, nodules	Ultrasound baseline, then if symptoms, ^{131}I scan
Breast	RT	Preadolescent females; interaction with family history (?)	Breast mass	Mammography at age 25, every 2 years to age 40, than every year, biopsy if mass present
Skin	RT	Ionizing RT increases risk and shortens latent period	New lesion or change in skin color or texture	Biopsy
	Nevoid basal cell carcinoma syndrome			
	Xerodema pigmentosa	Risk associated with ultraviolet light		
Leukemia	Alkylating agents	Dose response, melphalan>nitrogen mustart>cyclophosphamide; associated with chromosome 5 and 7 abnormalities	Pallor, bruising, fatigue, petechiae	Complete blood count annually
	Epipodophyllotoxins	Associated with 11q23 abnormality; schedule or dose dependent (?)		
	NF	Juvenile chronic myelogenous leukemia (JCML) most common; xanthomas; may develop monosomy 7		Bone marrow evaluation for symptoms

From: Meadows AT, Fenton JG. Follow up care of patients at risk for the development of second malignant neoplasms. In: Schwartz CL, Hobbie WL, Constine LS, Ruccione HS, editors. Survivors of Childhood Cancer: Assessment and Management. St. Louis: Mosby, 1994:319–328, with permission.

Both chemotherapy and radiation can predispose patients to second malignancies.

Chemotherapy

Most chemotherapy-induced second malignancies are leukemias, most classically acute myeloid leukemia, although acute lymphoblastic leukemia can also occur. Three classes of chemotherapeutic agents increase the risk of secondary leukemias:

*   Epipodophyllotoxins (e.g. etoposide)
*   Anthracyclines (e.g. doxorubicin)
*   Alkylating agents (e.g. cyclophosphamide).

Epipodophyllotoxins

This class of chemotherapeutics is topoisomerase II poisons. They allow the enzyme complex to cleave the double-stranded DNA, but do not allow it to re-anneal the cleaved ends. This initiates a DNA repair process called nonhomologous end joining that carries a high incidence of repair errors and therefore increases the risk of oncogenesis. As the topoisomerase II complex tends to function at the 11q23 location (the site of the MLL gene), MLL rearranged leukemias are prototypical for this agent. These leukemias are most commonly myeloid, but can also be lymphoid. The incidence is approximately 1–10% and may be dose- and schedule-dependent, although the specifics of these relationships remain under investigation.

Anthracyclines

This class of chemotherapeutics is weak topoisomerase II poisons. The consequent secondary leukemias are similar to those seen with the epipodophyllotoxins.

Alkylating Agents

Alkylating agents increase the risk of a secondary acute myeloid leukemia, usually preceded by a myelodysplastic phase. The risk has been correlated with the total dose of alkylating agents. Typical cytogenetic abnormalities include aberrations in the long arm of chromosome 5, 7, or both and the latency period is typically 3.5–5.5 years. The cumulative risk is 1% at 20 years.

The prognosis for patients with secondary acute leukemia is poor. Allogeneic hematopoietic stem cell transplantation is the treatment of choice. Radiotherapy does not contribute significantly to the development of secondary leukemias.

Radiation

Radiation therapy increases the childhood cancer survivors' risk of secondary solid tumors. The most second common tumor types include:

*   Bone sarcoma
*   Breast cancer

- Thyroid cancer
- CNS cancer (e.g. meningioma)
- Skin cancers.

The risk of developing a radiation-induced second malignancy depends on the dose, site of radiation, first cancer diagnosis, age at the time of radiation and underlying genetic risk factors. The median time to the development of a second malignant neoplasm is approximately 7 years, with a range from 1–15 years from therapy. Genetic abnormalities such as neurofibromatosis or family histories suggestive of Li–Fraumeni syndrome appear to predispose to the development of second solid neoplasms.

The risk of breast cancer following radiation increases linearly with radiation dose, until the odds ratio for developing a secondary breast cancer reaches 11-fold at 40 Gy.

Screening and Management

In addition to a good interval history and physical examination, important diagnostic tests for possible toxicity include:

- Annual complete blood count for those exposed to topoisomerase II poisons, anthracyclines or alkylating agents
- Annual dermatology checks for those exposed to radiation
- Annual thyroid function tests (with ultrasounds if abnormal) for those exposed to thyroid radiation
- Annual mammography with ultrasound or MRI for those exposed to chest/mantle radiation starting at age 25.

Patients exposed to radiation should be counseled to apply SFP 45 sunscreen liberally and to avoid tobacco smoke.

Suggested Reading

Anderson FS, Kunin-Batson AS. Neurocognitive late effects of chemotherapy in children: the past 10 years of research on brain structure and function. Pediatr Blood Cancer 2009;52(2):159–164.

Armstrong GT, Liu Q, Yasui Y, et al. Late mortality among 5-year survivors of childhood cancer: a summary from the Childhood Cancer Survivor Study. J Clin Oncol. 2009;27(14):2328–2338.

D'Angio GJ. Pediatric cancer in perspective: cure is not enough. Cancer 1975;35(3 suppl):866–870.

Diller L, Chow EJ, Gurney JG, et al. Chronic disease in the Childhood Cancer Survivor Study cohort: a review of published findings. J Clin Oncol. 2009;27(14):2339–2355.

Fish JD, Ginsberg JP. Health insurance for survivors of childhood cancer: a pre-existing problem. Pediatr Blood Cancer 2009;53(6):928–930.

Friedman DL, Meadows AT. Late effects of childhood cancer therapy. Pediatr Clin North Am. 2002; 49(5):1083–1106.

Henderson TO, Whitton J, Stovall M, et al. Secondary sarcomas in childhood cancer survivors: a report from the Childhood Cancer Survivor Study. J Natl Cancer Inst. 2007;99(4):300–308.

Johnston RJ, Wallace WH. Normal ovarian function and assessment of ovarian reserve in the survivor of childhood cancer. Pediatr Blood Cancer 2009;53(2):296–302.

Kremer LC, van Dalen EC, Offringa M, et al. Anthracycline-induced clinical heart failure in a cohort of 607 children: long-term follow-up study. J Clin Oncol. 2001;19(1):191–196.

Meadows AT. Pediatric cancer survivors: past history and future challenges. Curr Probl Cancer 2003; 27(3):112–126.

Nagarajan R, Robison LL. Pregnancy outcomes in survivors of childhood cancer. J Natl Cancer Inst Monogr. 2005;34:72–76.

Oeffinger KC, Mertens AC, Sklar CA, et al. Chronic health conditions in adult survivors of childhood cancer. N Engl J Med. 2006;355(15):1572–1582.

Park ER, Li FP, Liu Y, et al. Health insurance coverage in survivors of childhood cancer: the Childhood Cancer Survivor Study. J Clin Oncol. 2005;23(36):9187–9197.

Robison LL. Treatment-associated subsequent neoplasms among long-term survivors of childhood cancer: the experience of the Childhood Cancer Survivor Study. Pediatr Radiol. 2009;39(Suppl 1):S32–S37.

Ross L, Johansen C, Dalton SO, et al. Psychiatric hospitalizations among survivors of cancer in childhood or adolescence. N Engl J Med. 2003;349(7):650–657.

Rourke MT, Hobbie WL, Schwartz L, et al. Post-traumatic stress disorder (PTSD) in young adult survivors of childhood cancer. Pediatr Blood Cancer 2007;49(2):177–182.

Rutter MM, Rose SR. Long-term endocrine sequelae of childhood cancer. Curr Opin Pediatr. 2007; 19(4):480–487.

Shankar SM, Marina N, Hudson MM, et al. Monitoring for cardiovascular disease in survivors of childhood cancer: report from the Cardiovascular Disease Task Force of the Children's Oncology Group. Pediatrics 2008;121(2):e387–e396.

Wasilewski-Masker K, Kaste SC, Hudson MM, et al. Bone mineral density deficits in survivors of childhood cancer: long-term follow-up guidelines and review of the literature. Pediatrics 2008;121(3):e705–e713.

Weiner DJ, Maity A, Carlson CA, et al. Pulmonary function abnormalities in children treated with whole lung irradiation. Pediatr Blood Cancer 2006;46(2):222–227.

Withycombe JS, Post-White JE, Meza JL, et al. Weight patterns in children with higher risk ALL: a report from the Children's Oncology Group (COG) for CCG 1961. Pediatr Blood Cancer, 2009.

CHAPTER 33

Psychosocial Aspects of Cancer for Children and their Families

Healthcare professionals have come to recognize the psychosocial complexity of delivering care to the child and family facing cancer. The diagnosis and treatment of childhood cancer is extremely stressful and frightening for families. As with any life-threatening illness, childhood cancers interfere with the normal developmental milestones of childhood and adolescence and the ability of families to function effectively. Understanding common reactions of children and families at various stages of the illness helps the practitioner to communicate more effectively about the diagnosis, treatment and prognosis and to support the family throughout the process. While no single professional can meet a family's needs completely, collaborative and multidisciplinary healthcare team efforts can help family members enhance adaptive coping skills and mobilize their support systems to effectively engage in medical care. This chapter highlights potential challenges at each stage of the illness, reviews common psychosocial reactions and provides suggestions for building on child and family strengths, fostering adjustment and achieving the best possible quality of life for the child and family facing childhood cancer. The oncologist should provide hope and optimism unless the medical condition dictates otherwise. Since in most cases of cancer in children the prognosis is good "I expect your child will be cured" is a reasonable and comforting comment by the oncologist when medically warranted.

DIAGNOSIS

A diagnosis of childhood cancer is an acute, potentially traumatic event for a family. Preconceptions or past experiences with cancer may lead to immediate thoughts that the child will die. Even if fears about death are not at the forefront, thoughts about a long, painful and invasive treatment course may be overwhelming. The way in which the diagnosis of cancer is presented significantly influences the family's initial reactions and sets the stage for collaboration with the medical team. The meeting with the family should be held as rapidly as possible once a diagnosis is established. They should be told in advance about the purpose of this meeting so that they can involve those individuals they regard as important to the conversation. Adolescent patients may be invited and included in this meeting. If not, a separate meeting should be set up between

Manual of Pediatric Hematology and Oncology. DOI: 10.1016/B978-0-12-375154-6.00033-1

the healthcare team, parents and the patient to discuss the diagnosis and treatment in age-appropriate ways with time provided to answer the patient's questions. The oncologist must present the diagnosis of cancer in an empathic and unequivocal way, tailored to the cultural and educational characteristics of the family, patient and situation. The discussion should occur in a quiet, private area initially with the oncologist and primary team members and the multidisciplinary team approach to management explained. Information should be accurate and clearly communicated without the use of medical jargon or euphemisms. The family's reactions will determine the pace and flow of the conversation. Table 33-1 provides common reactions and suggestions for working with children at different developmental levels at the time of diagnosis and throughout cancer care. Table 33-2 provides common parental reactions and suggestions for psychosocial intervention throughout the care of the patient.

Table 33-1   Reactions to Cancer by Developmental Level and Suggestions for Psychosocial Support

Developmental Stage	Common Reactions	Suggestions for Psychosocial Support
Preschoolers	Concerns around unfamiliarity with hospital Regression (i.e. loss of developmental milestones or return to previously discarded behavior) Heightened dependence on parent (i.e. wanting to sleep with parents) Threats to body integrity (fear of mutilation) Fears of abandonment Perceive treatment as punishment or retribution for wrong doing (i.e. painful procedures) Refusal to go to bed	Address concerns about all procedures, even temperature taking, by giving a brief, honest description explaining the treatment Provide positive incentives for cooperation with procedures Use breathing exercises and developmentally appropriate distraction techniques during procedures, such as storytelling, fantasy play and puzzles. Child life specialists may be helpful in this regard The patient should be allowed to engage in medical play and to visit operating and recovery rooms if undergoing surgery to become familiar with surroundings
School age	Reaction to diagnosis may be delayed or immediate and anxiety-ridden Many children exhibit adaptive and efficient coping Younger children may display any or all reactions listed for preschoolers Psychosomatic complaints, nightmares, stoic-like acceptance and labile emotions may all be seen Child may pose difficult questions such as reasons and cause for treatment and diagnosis	School age children may like to learn about various aspects of their illness and treatment at a developmentally appropriate level They can learn how to manage stress associated with procedures by developing self-regulating skills through breathing and imagery The physician may forge a strong positive working relationship with school age children by taking time to sit and answer questions

(Continued)

Table 33-1 (Continued)

Developmental Stage	Common Reactions	Suggestions for Psychosocial Support
Adolescents	Sadness Noncompliance Feelings of being different from peers due to physical changes may lead to difficult school re-entry Lowered self-esteem and confidence levels Many exhibit adaptive and efficient coping styles Initial questions may be general (i.e. geared towards causation and prevention of disease) or more personal (i.e. individual treatment plan; disruption to social life) Experience a loss of control and independence, threatening autonomy Non-adherence may be an issue May feel self-conscious about appearance and emerging sexuality (i.e. hair loss, treatment-related delays in puberty, scars) Fear of being different from peers may interfere with school/social scene re-entry May avoid intimate relationships with peers May reconsider long-term career and family plans and expectations Concerns about fertility Feelings of isolation and vulnerability may lead to social withdrawal, hesitancy and sadness	School re-entry and other issues should be actively addressed by psychosocial staff Allow and encourage adolescents to participate in medical decisions (i.e. signing consents; when appropriate, control with scheduling, permit them to see tests, involve them in discussions of alternative treatment) Offer group support and individual counseling to all to be provided by psychosocial staff whenever possible Encourage patient to maintain active participation in daily activities Teach special skills for stress reduction such as breathing, imagery and relaxation training Family therapy may be appropriate to address issues of autonomy

Table 33-2 Parental Reactions to Cancer and Suggestions for Psychosocial Support

Development Stage	Common Reactions	Suggestions for Psychosocial Support
Diagnosis and initial treatment	Most experience shock, disbelief, fear, guilt and anxiety Often feel overwhelmed with new medical terminology and information and the need to make a rapid treatment decision May experience frequent crying, difficulty sleeping Often seek reassurance that their child "will survive"	Be factual and empathic Allow time for questions and answers Educate family on common reactions and feelings Support parents in their request for time to talk with other family members and to seek a second opinion Be sensitive to cultural beliefs and norms Provide verbal and written information

(Continued)

Table 33-2   (Continued)

Development Stage	Common Reactions	Suggestions for Psychosocial Support
		Provide guidance on ways to communicate diagnosis to the child with cancer, the sibling(s), family and friends
		Prepare parents for possible delay in emotional reaction (themselves or their child(ren))
		Help promote open communication in the family
		Encourage parents to have their child, particularly adolescents, participate in medical decisions (i.e., signing consents; when appropriate, control with scheduling, permit them to see tests)
Remission	Fear of relapse is common Financial strains may become a key issue Exhaustion as parents try to return to work and family activities while caring for the child with cancer May become more lax in following through with medical appointments and medication schedule	Monitor family communication and coping Encourage parents to maintain active participation in daily activities Teach stress reduction such as breathing, imagery and relaxation training Refer for psychosocial support services and organizations that provide financial assistance Family therapy may be appropriate to address communication, behavior or relationship issues Suggest parents monitor adherence
Recurrence	Fear, anxiety, sadness, crying, anger, difficulty sleeping An urgent need to explore all treatment options Concern how to best communicate information to the child with cancer and/or siblings	Monitor symptoms and offer counseling, particularly when symptoms interfere with ability to cope Discuss treatment options Assess family communication Encourage parents to include their child in medical decisions and when appropriate, end of life decisions Balance additional treatment with quality of life When indicated, shift focus from "cure" to symptom control or "palliative care" and/or hospice Encourage (age and cognitively appropriate) siblings to attend family conferences Offer psychosocial support to all family members
Bereavement	Most families wish to maintain a relationship with the primary oncology team	Send a personal note to the family Provide selected reading materials about bereavement to parents, siblings,

(Continued)

Table 33-2   (Continued)

Development Stage	Common Reactions	Suggestions for Psychosocial Support
	Families appreciate the opportunity to keep their child's memory alive by sharing experiences with those who knew their child during his/her illness	grandparents and extended family members, classroom teachers and others who knew the child well Be aware of "normal" grief reactions from those that are non-resolving and require counseling Offer information on how to connect with other families who have lost a child, if the family desires
Survivorship	Sense of relief along with ambivalence, fear of recurrence Anxiety about decreased contact with medical team	Normalize emotional responses to end of treatment Address realistic expectations regarding a quick "return to normal" Monitor family adjustment Assess for psychological distress (refer for counseling when indicated) Have follow up care arranged and provide education on potential late effects of cancer treatment

Despite knowing that cancer is a possibility prior to the family meeting, most family members report feeling shocked at the news. Disbelief, denial, confusion, sadness, grief, guilt, anger, helplessness, anxiety and fear are all common reactions. Parents perceive their role as providing for their children and protecting them from fear, hurt, pain and death. The diagnosis of cancer threatens that identity and sense of adequacy as a parent. Many family members report that they were unable to hear or comprehend any information from this first meeting beyond the diagnosis. Some report feeling as if they were watching the meeting happen from a distance. Severe emotional reactions are normal and should be expected. As the cancer diagnosis is accepted, anger can become a significant emotion, directed in many ways and sometimes at the treatment team. Guilt is also common and family members may pass through a period of self-blame during which they focus on transgressions they may have committed and for which they feel they are now being punished. Careful listening and supportive attention by the team is important to reassure family members that they did not cause the cancer. This includes siblings, grandparents and other relatives.

Treatment decisions need to be made quickly after the diagnosis is known. This requires the family to absorb and understand significant amounts of medical information, including weighing the risks and benefits of treatment research protocols. Families may want time to seek other medical opinions, consult with additional family members and friends and conduct their own searches for information. These efforts may be appropriate to ensure that they are doing what is best for their child and they should be encouraged to seek a second opinion from a qualified and reputable oncologist if they so desire. They should be discouraged from selectively talking to those who promote unrealistic hope or provide inaccurate information. The best way to support the family during this time is to repeat and reinforce what is known in regard to the diagnosis and treatment options, build a partnership through rapport and the encouragement of family participation in the conversation and recognize and attend to the family's level of distress while gently encouraging their competence and ability to make a good decision. It is important that multiple conversations occur with the family in which different members of the team provide the same information to ensure the family understands the information and has opportunities to both express their questions and concerns and to have them addressed.

Parents are often uncertain about how to talk with their children about the diagnosis, treatment and prognosis. Parents must talk to their children openly and honestly about the cancer and treatment that they will receive in order to increase trust within their relationship and between the child and medical team. Open communication helps their children to feel more secure and less anxious. Parents may need to be coached as to what to say and when to say it to be most effective. However, they are better informed than the healthcare team as to how their child copes with stressful information, including how much detail to provide at a given time and how soon they want to know about upcoming procedures. The healthcare team needs to support and reinforce their competence as parents.

Psychological and developmental problems in the patient or any family member add significantly to the burden of dealing with cancer and can exhaust a family's time and emotional and financial resources. Even minor symptoms can be reactive, at least in part, to the stresses and disruptions of the cancer diagnosis and treatment course. An assessment of the family's strengths and vulnerabilities, psychosocial resources and pre-existing problems can help the team anticipate the psychological adjustment of families to cancer and provide a means for quickly and efficiently providing psychosocial care based on their need. Several factors such as diagnosis of depression or anxiety, general "poor coping," or nonadherence to treatment indicate that it is appropriate to make a referral for psychological treatment. Other families seek psychological services as a "preventative" measure for their children in order to have additional support and to help guide them as parents in this arena. Potential therapeutic interventions include individual psychotherapy or cognitive-behavioral therapy (CBT) for the child, support groups and family therapy for members of the family, or a combination of approaches. Other therapeutic approaches appropriate for the early stages of treatment and beyond include art and music therapy. While not all families want or need such services, they often add a crucial dimension to comprehensive care.

TREATMENT INITIATION

Treatment initiation often provides reassurance to families because something is actively being done to help the child, but tremendous burdens are placed on the family during this time. Daily lives are disrupted due to the demands of treatment, roles and responsibilities within the family, each of which need to be renegotiated to ensure that basic needs of the family continue to be met (e.g., working to retain medical insurance and pay bills, care of healthy siblings) and highly technical medical information needs to be understood. Parents frequently report that the primary difficulty during treatment initiation was dealing with their own intense emotions. Within the first weeks following diagnosis and initiation of treatment, nearly all parents report clinically significant psychological distress in the form of anxiety, depression, or traumatic stress reactions. This distress makes it difficult for parents to establish consistently effective working relationships and open lines of communication with the medical team. It is vital to understand that parents have their child's best interests at heart but are struggling emotionally during this critical time. Furthermore, parental marital relationships are often challenged as the stress of cancer can exacerbate previous marital problems, disrupt comfortable patterns of relating, or bring to light differences in beliefs and coping styles that make working together as a parenting team difficult. Parents who have open communication with one another, are mutually supportive and trusting in their relationship can surmount these challenges.

In addition to open communication with parents, a roadmap of treatment can help the family focus on what needs to be done and provide some sense of relief, optimism and improved mood. This information needs to be presented carefully and repeatedly to the family as misunderstandings are common during this time. It is important for the treatment team to identify all involved caregivers within the family, including grandparents and to ensure that their questions are adequately answered. It is best to assess what the family understands or believes about the treatment plan and to correct erroneous perceptions. As the family comes to trust and rely on the medical team, it is also important for the physician and other members of the team to be clear about their roles and involvement in the child's care throughout the planned treatment course.

For the diagnosed child, the most significant aspects of initiating treatment are the physical effects. As treatment begins, anticipation and endurance of repeated painful procedures and side effects are the most distressing and traumatizing aspects of the cancer experience. Negotiating the impact of these physical changes on the patient's participation in normative social activities is also highly distressing. To help the child through this difficult aspect of treatment, the oncologist or member of the team should explain procedures in developmentally appropriate terms, minimize the degree to which initial procedures are painful and effectively manage chemotherapy-induced nausea and vomiting. Interventions that combine aspects of cognitive-behavioral therapy and basic behavioral strategies (e.g., preparation, desensitization, modeling, positive reinforcement) can be delivered by psychosocial specialists working as part of the multidisciplinary team. Parents can also learn to effectively use these techniques to help their children cope with procedures, boosting their self-efficacy as parents and allowing them to feel less helpless during treatment.

ILLNESS STABILIZATION

Parental distress present throughout the early stages of the illness and treatment starts to dissipate with illness stabilization. Some anxiously await the day their child is well enough to return home, especially if the initial hospitalization is long. Others find the return to home particularly stressful because they no longer have the expertise of the hospital staff monitoring and addressing the medical needs of the child. As treatment progresses and remission is established, most families develop ways of coping effectively with the cancer and its treatment, which some have described as "coming to terms" with the cancer routine. Distress decreases over time with few parents reporting clinically significant levels of anxiety and depression by one to two years post-diagnosis.

Still, there are moments of increased anxiety that surround specific stressful cancer-related events (e.g., emergency room visits, severe side effects) and symptoms of cancer-related traumatic stress (e.g., intrusive thoughts about cancer, hypervigilance and physiological arousal). This distress is normal. Parents continue to struggle to make sense of the diagnosis and experience grief and mourning related to the loss of their healthy child and prior lifestyle. Episodes of helplessness, fear and crises of confidence are common and can influence emotional reactions and behavioral patterns. Maladaptive coping may manifest as excessive concern about relapse and death, refusal to allow the child to return to everyday activities, difficulties regularly attending scheduled clinic visits, or persistent and/or escalating anxiety or depression. Interestingly, the extent of distress in response to diagnosis and treatment is not consistently related to objective measures of the severity of illness or intensity of treatment.

Regarding the child with cancer, one of the most common concerns as treatment progresses is depression. Children with cancer, particularly adolescents, may display increased sleep, loss of interest in activities and social withdrawal among other symptoms that are consistent with depressed mood. However, most research reveals that children with cancer do not evidence greater rates of clinical depression and, in fact, have frequently been noted to have better overall psychological health than the general population of children.

One of the major tasks for the child and family during the illness stabilization phase of cancer treatment is to re-establish the patterns and routines of daily life that were disrupted due to diagnosis and initiation of treatment. A new day-to-day routine needs to be established that encompasses the needs of the ill child as well as those of the well siblings and the adults in the family. Parents should be encouraged to take some time to recharge themselves and their relationships and cautioned against becoming overprotective or overindulgent with their children. Families of younger children should re-establish expectations and rules regarding bedtime, feeding, naps and playtime. Families of older children need to re-establish their child's roles and responsibilities within the family, peer relationships and school activities to the greatest extent possible. The maintenance of friendships and some level of independence from the family are important for the child and especially the adolescent, with cancer. Contact with friends can be made through short visits and phone calls, but internet-based technologies (e.g., email, internet messaging, social networking sites) are making regular contact even more convenient. In general, most children with cancer have been found to be quite resilient and to maintain good peer relationships, although they may be perceived as more socially withdrawn. Patients at highest risk for peer difficulties are those whose treatment affected the central nervous system or who have more obvious changes in physical appearance. Interventions to improve social skills can be useful for these children.

The return to school may be particularly difficult for some families. The parent may fear for the physical wellbeing of their child and the child or adolescent may feel embarrassment related to physical changes. However, school is exceptionally important to child development and for normalizing the life of the child with cancer. While still in the hospital, the child should receive school services and when home but not well enough to attend school, homebound services should be arranged. During these times, webcams to the classroom can be particularly helpful to maintain connections. Prior to returning to school, the child, classmates and school personnel need to be prepared. School reintegration programs where the family, hospital and school work together to ensure open communication and appropriate accommodations (e.g., 504 plans)[*] for the child can help the family surmount this hurdle.

Medication adherence is a vital and often unrecognized clinical issue in pediatric oncology. The clinical consequences of nonadherence can be significant, including lower survival rates and increased mortality due to relapse. Consequently, the need to anticipate, assess, monitor and treat adherence to the treatment regimen is a critical component of comprehensive clinical care.

RELAPSES AND RECURRENCES

A relapse or recurrence of a child's cancer is a crisis point for the patient and family because the threat to the child's life is renewed. In response, the family commonly

---

[*]This refers to Section 504 of the Rehabilitation Act which specifies that no one with a disability can be excluded from participating in elementary, secondary and post-secondary schooling and that modifications and accommodations have to be made so that they have the opportunity to perform at the same level as their peers.

experiences anxiety, fear, anger and sadness. For the child, the worst memories of earlier treatment may re-emerge and they must reappraise each of these in order to prepare for another round of therapy. Families gain a sense of hope when new treatment options are offered as this implies further action will be taken against the disease, but they often realize that the chances for long-term cure are slimmer after relapse. Encouraging the child and family to once again adopt a positive attitude toward treatment can be challenging for the oncology staff, especially as parents require a heightened level of emotional availability and reassurance from the team members.

Most families manage to cope adequately following relapse or recurrence. Development of a new sense of normality and stability in their lives will depend on the following:

* Treatment side effects
* The length of time the child needs to remain hospitalized or away from home and
* Concurrent stresses, such as financial pressures, career obligations and family problems.

Assessing for maladaptive coping is essential. An overly pessimistic attitude about the future may immobilize parents in their day-to-day functioning, resulting in emotional or physical withdrawal from the child, an inability to normalize the child's life, or refusal to follow through with medical care – each of which indicates the need for immediate psychosocial intervention. Similarly, an overly optimistic attitude or a desire to pursue every last medical option without concern for the child's quality of life can be problematic and should be discouraged. Individual or family sessions can help the family strengthen coping skills and provide much needed support and guidance.

Whether or not a treatment response is obtained, families often feel an urgency to search for other curative treatment options, including second opinions. Phase I trials, alternative therapies, or the choice for no further treatment interventions should be presented as options while reassuring the family that, regardless of their decision, contact with the medical team will be maintained. If medically indicated, this might be a time to shift communication from "cure" to "palliation" while carefully balancing additional treatment with quality of life for the child.

TREATMENT OUTCOMES: THE UNSUCCESSFUL COURSE

Termination of Treatment

Maintaining adequate communication with both children and their parents is of particular importance when compassionate decisions to withhold or withdraw curative and/or other life-prolonging therapies must be made. A series of meetings with the family should be held to discuss the transition in care associated with palliation. Ideally, the staff members who have been most intimately involved with the child and family during the course of the disease should be present. The child may or may not attend these meetings, depending on his or her age, developmental stage and other circumstances. If a decision is made for the child not to be present, relevant information should be communicated to the child in a developmentally appropriate manner.

Once parents understand that treatment is no longer effective, they will begin the process of accepting that their child will die while continuing to hope for cure or recovery. Parents are increasingly vulnerable during this period to the promises of nontraditional or even fraudulent healers as well as to misguided advice from Internet sites and "chat rooms." Therefore, in order to preserve a relationship that is built on trust, the physician should openly and respectfully explore all options with the family, including resuscitation status, in-hospital palliative care programs and home care with or without the support of a hospice team. Advanced discussion about resuscitation status when the physician can reassure the family that a life-threatening event is not imminent is beneficial to the child and to the parent as it is very difficult to make a nonemotionally based decision in the midst of a medical crisis. Parents usually respond best to an approach that is framed "in a worse case scenario" discussion. Once on a ventilator, the likelihood of being extubated is rare for the child with far-advanced cancer and therefore it is recommended that the phrase: "do not *attempt* resuscitation" be used (DNAR).

Palliative Care

Palliative care should ideally occur early in a child's treatment, as soon as it becomes apparent that cure will not be possible, so that the family and professional caregivers can derive maximum benefit from all that palliative care services have to offer. However, often a family's first contact with the palliative care team comes when all treatment options have been exhausted. The transition to palliative care can be much less traumatic when the family is reassured that they will not be cut off from the treatment team with whom they have developed a close relationship over the years, regardless of whether the child is at home or in the hospital. Hope should be encouraged by redirecting energies into providing as good a quality of life as possible for as long as possible, followed with as good a quality of death (absence of pain combined with the presence of loved ones) as possible. Access to care aimed at promoting optimal physical, psychological and spiritual wellbeing is of utmost importance. It is especially critical that providers are responsive and attuned to questions the child may have and to the wishes of the parents and child.

Promoting open communication between child and parents about emotional concerns, needs and desires is especially vital as the end of life nears. The child should be given honest and accurate information about end-of-life care and death, appropriate to his or her developmental age. A child's fears of separation, being alone, suffering and being forgotten are not unusual and should be openly discussed. Many children need reassurance that they will not be alone, while others wish to discuss closure activities, such as funeral arrangements, giving away of belongings and how they wish to be remembered. They need the opportunity to be involved in decision-making about how they will live the life they still have. Focusing on the quality of their child's life and death, adequate pain management, strengthening relationships with loved ones and facilitating memory building can be the greatest gift to the child and family.

As death approaches, parents often question whether they have done all they could for their child. They may describe what feels like overwhelming anxiety as they try to hold onto any semblance of control or normalcy in their relationships and day-to-day lives. Such responses are appropriate given the sequence of experiences leading to the terminal phase of illness and the physical and emotional changes that their child is undergoing. Living with a dying child is all-consuming and it has a significant impact on each family member. Fluctuations in a parents' capacity to accept the idea of their child's death are common. This enables them to emotionally be with their child in the present moment. Children, too, may talk about their funeral at one moment and soon after inquire about a future family vacation. Siblings cope similarly, gradually accepting the declining health of their brother or sister while appearing to deny the impending death. One needs to respect each family's readiness to discuss end-of-life events, as they delicately attempt to balance life issues with those related to palliative care, death and loss.

The management of the terminal phase of illness has a dramatic effect on the psychosocial recovery of the family. The medical team's participation and investment in caring for the dying child is extremely important and greatly appreciated by all families, even those who appear to demonstrate strong skills in coping and adaptation. Entering a family's world and bearing witness to a parent's, sibling's or grandparent's anger, grief and pain is emotionally demanding. When done well, being present at the time of death is also exceptionally intimate and rewarding. Opportunities for staff to reflect with other staff or professionals, especially when facing multiple losses, doubts, guilt, or emotional exhaustion, is key. The impact of burnout can be profound on the individual, the team and on the patients and families for who health care is provided.

Bereavement

The pediatric cancer experience is an intimate, long-term relationship that does not need to be severed with the child's death. Following the immediate bereavement period, families often benefit from being able to discuss their cancer journey with the primary oncology team, including a review of autopsy findings, if such was performed. This may need to occur at a location outside of the oncology ward, if parents are not ready to return. Families benefit from receiving cards and notes that remind them that their child continues to remain alive in the thoughts of those who helped care for them.

For years after the death of a child, parents try to make meaning out of the loss. Post-death adjustment is different for each family. Family members vividly recollect specific events in the course of their child's illness, some of which may be persistent and disturbing images of their child's last moments. They often have a need to review these, over and over again, until they find solace in their remembrance. Parents, siblings and others close to the child can benefit greatly from having someone available with whom to share their memories, grief and concerns. If a hospice team is not involved for bereavement counseling, a referral to a psychologist, social worker, grief counselor, or bereavement group can be helpful.

TREATMENT OUTCOMES: THE SUCCESSFUL COURSE

End of Treatment

The successful completion of therapy and discharge from treatment is a significant milestone, but one often accompanied by feelings of ambivalence within the family. The family often experiences a sense of accomplishment, relief, joy and renewed hope for the future. Yet, at the same time, they might also have underlying anxiety about the possibility of relapse or recurrence and a sense of sadness as they lose the routine of coming to the clinic and receiving support from the healthcare team. An end to the active fight against the disease can also feel unsettling. These feelings are to be expected and should be normalized for the family. Extra support and education are important for the family at this juncture. The signs and symptoms of possible relapse or recurrence should be reviewed along with remaining treatment options. The importance of monitoring for possible late effects of treatment should be explained and the family should be given a summary of the treatment received and written guidelines for self-care. Finally, the challenges that they have met and surmounted should be reviewed and their accomplishments lauded as they prepare for survivorship.

SURVIVORSHIP

With the majority of children diagnosed with cancer living 5 or more years disease-free after treatment, worries about relapse and recurrence tend to decrease as time passes from the end of treatment and concerns about possible long-term effects of treatment begin to rise. As many as two-thirds of childhood cancer survivors experience at least one late effect of treatment with one-fourth of survivors reporting severe or life-threatening late effects (see Chapter 32). Psychosocial late effects in the form of neurocognitive deficits and post-traumatic stress reactions can also emerge and warrant monitoring and intervention.

Neurocognitive deficits are one of the most widely recognized late effects of childhood cancer. While the timing of the emergence of these deficits is not well known, there is accumulating evidence that CNS-related cancers and treatments can result in deficits of memory, processing speed, visual-spatial and visual-motor skills and executive functioning.

These survivors have also been found to learn new material at slower rates than peers and to have poorer academic achievement. Children treated before the age of seven and those treated for brain tumors seem to be at the highest risk for neurocognitive late effects. To address these difficulties, the family should be encouraged to develop a collaborative working relationship with the schools and advocate for individualized educational plans to meet their child's specific needs. Specific cognitive remediation programs for childhood cancer survivors are being evaluated through research and have suggested efficacy in improving academic achievement, but these are not yet widely available. Pharmacological interventions, such as methylphenidate have been used with some success.

Most survivors of childhood cancer and their family members emerge as psychologically competent and well-functioning. Still, the experience of childhood cancer may have some long-lasting psychosocial effects. For example, the health-related quality of life of survivors, a multidimensional construct that encompasses physical functioning, psychological adjustment, social functioning and an overall sense of well-being, seems to be poorer for childhood cancer survivors compared to peers. Furthermore, subsets of survivors do experience more serious problems. There is emerging evidence that survivors of CNS-related cancers and treatments experience some difficulties in social competence and peer interactions, exhibiting greater social isolation, reporting fewer best friends and participating in fewer peer activities than other children. Also, the transition from adolescence to young adulthood is a time of psychosocial vulnerability for childhood cancer survivors. In addition to reports of increased distress during this time, milestones of adulthood (e.g., dating, employment, moving away from home, starting a family) seem to be achieved at a slower rate than for healthy peers.

A post-traumatic stress model has emerged as helpful in understanding the long-term cancer-related distress experienced by some childhood cancer survivors and their family members. Among child and adolescent survivors, rates of post-traumatic stress disorder (PTSD) are low, in the range of 5–10%. However, much higher percentages of survivors report experiencing post-traumatic stress symptoms, including intrusive thoughts about cancer, physiological arousal when reminded of cancer or their experience and a desire to avoid anything linked to their cancer experience. PTSD rates are nearly double for young adults compared to adolescent survivors of childhood cancer. Remarkably, parents and siblings seem to be more likely than survivors to report cancer-related PTSD and post-traumatic stress symptoms. Across studies, approximately 40% of mothers, 35% of fathers and one-third of siblings report moderate to severe cancer-related post-traumatic stress reactions.

Through the survivorship phase of cancer, follow-up medical appointments should routinely screen for psychosocial late effects of treatment for the survivor and the family members. Education about the cancer treatment and the possible medical late effects will help to close gaps in knowledge about the experience and potential vulnerabilities. During these conversations, it is important to address any anxiety that arises, provide anticipatory guidance about possible traumatic stress symptoms and ensure that psychological reactions do not interfere with follow-up care. When psychological symptoms or signs of poor adjustment are present, the family should be referred to an experienced clinical social worker, psychologist, or psychiatrist for consultation.

CONCLUSION

Along with utilizing the latest effective cancer treatments, comprehensive psychosocial support is an essential component of cancer care that begins with early assessment of family strengths and vulnerabilities and emphasizes the importance of the child becoming medically stable and the child and family remaining socially and emotionally intact. This chapter addresses critical elements in providing optimal care for the child with cancer, emphasizing the need for the healthcare team to join together with families to form a therapeutic alliance and promote normalcy as soon as the child is medically capable. The treatment team needs to respect the cultural differences of the family and maintain open communication with them across the trajectory of illness whether the outcome is cure or end-of-life care. No child's illness-related distress should go unrecognized and untreated and the strain on the professional staff needs to be equally recognized and addressed. In doing so, providers reduce the trauma associated with pediatric cancer and simultaneously promote hope, courage and emotional growth.

Suggested Reading

Armstrong GT, Liu Q, Yasui Y, et al. Long-term outcomes among adult survivors of childhood central nervous system malignancies in the Childhood Cancer Survivor Study. J Natl Cancer Inst. 2009;101:946–958.

Brown R, ed. Comprehensive Handbook of Childhood Cancer and Sickle Cell Disease: a Biopsychosocial Approach. New York: Oxford University Press; 2006.

Jankovic M, Spinetta J, Masera G, et al. Communicating with the dying child: An invitation to listening– a report of the SIOP working committee on psychosocial issues in pediatric oncology. Pediatric Blood Cancer. 2008;50(5):1087–1088.

Oeffinger KC, Mertens AC, Sklar CA, et al. Chronic health conditions in adult survivors of childhood cancer. N Engl J Med. 2006;355:1572–1582.

Spinetta JJ, Jankovic M, Masera G, et al. Optimal care for the child with cancer: a summary statement from the SIOP Working Committee on Psychosocial Issues in Pediatric Oncology. Pediatr Blood Cancer. 2009; 52(7):904–907.

Tonorezos ES, Oeffinger KC. Survivorship after childhood, adolescent and young adult cancer. Cancer J. 2008; 14(6):388–395.

Wiener LS, Pao M, Kazak AE, Kupst MJ, Patenaude A, eds. Quick reference for pediatric Oncology Clinicians: The Psychiatric and Psychological Dimensions of Pediatric Cancer Symptom Management. Charlottesville: IPOS Press, 2009.

Wolfe LC. End of Life Care of the Pediatric Oncology Patient (ed): Gellis Kagan's Current Pediatric Therapy 17. WB Saunders Company. 2002;1034–1037.

Zeltzer LK, Lu Q, Leisenring W, et al. Psychosocial outcomes and health-related quality of life in adult childhood cancer survivors: A report from the Childhood Cancer Survivor Study. Cancer Epidemiol Biomarkers Prev. 2008;17:435–446.

Hematological Reference Values

The normal range for most hematologic parameters in infancy and childhood is different from that in adults. Dramatic changes occur during the first few weeks of life. The recognition of variables in the pediatric age group prevents unnecessary medical and laboratory investigations.

RED CELL VALUES AND RELATED SERUM VALUES

Table A1-1   Hemoglobin Concentrations (g/dl) for Iron-Sufficient Preterm Infants[a]

| Age | Number | Birth weight (g) | |
		1,000–1,500	1,501–2,000
2 weeks	17,39	16.3 (11.7–18.4)	14.8 (11.8–19.6)
1 month	15,42	10.9 (8.7–15.2)	11.5 (8.2–15.0)
2 months	17,47	8.8 (7.1–11.5)	9.4 (8.0–11.4)
3 months	16,41	9.8 (8.9–11.2)	10.2 (9.3–11.8)
4 months	13,37	11.3 (9.1–13.1)	11.3 (9.1–13.1)
5 months	8,21	11.6 (10.2–14.3)	11.8 (10.4–13.0)
6 months	9,21	12.0 (9.4–13.8)	11.8 (10.7–12.6)

[a]These infants were admitted to the Helsinki Children's Hospital during a 15-month period. None had a complicated course during the first 2 weeks of life or had undergone an exchange transfusion. All infants were iron sufficient, as indicated by a serum ferritin greater than 10 ng/ml.
From: Lundstrom U, Siimes MA, Dallman PR. At what age does iron supplementation become necessary in low-birth weight infants. J Pediatr 1997;91:878, with permission.

Manual of Pediatric Hematology and Oncology. DOI: 10.1016/B978-0-12-375154-6.00043-4

Table A1-2   Red Cell Values on First Postnatal Day

	Gestational Age (weeks)							
	24–25 (7)[a]	26–27 (11)	28–29 (7)	30–31 (25)	32–33 (23)	34–35 (23)	36–37 (20)	Term (19)
RBC × 10⁶	4.65[b] ± 0.43	4.73 ± 0.45	4.62 ± 0.75	4.79 ± 0.74	5.0 ± 0.76	5.09 ± 0.5	5.27 ± 0.68	5.14 ± 0.7
Hb (g/dl)	19.4 ± 1.5	19.0 ± 2.5	19.3 ± 1.8	19.1 ± 2.2	18.5 ± 2.0	19.6 ± 2.1	19.2 ± 1.7	19.3 ± 2.2
Hematocrit (%)	63 ± 4	62 ± 8	60 ± 7	60 ± 8	60 ± 8	61 ± 7	64 ± 7	61 ± 7.4
MCV (fl)	135 ± 0.2	132 ± 14.4	131 ± 13.5	127 ± 12.7	123 ± 15.7	122 ± 10.0	121 ± 12.5	119 ± 9.4
Reticulocytes (%)	6.0 ± 0.5	9.6 ± 3.2	7.5 ± 2.5	5.8 ± 2.0	5.0 ± 1.9	3.9 ± 1.6	4.2 ± 1.8	3.2 ± 1.4
Weight (g)	725 ± 185	993 ± 194	1,174 ± 128	1,450 ± 232	1,816 ± 192	1,957 ± 291	2,245 ± 213	

[a]Number of infants.
[b]Mean values ± SD.
From: Zaizov R, Matoth Y. Red cell values on the first postnatal day during the last 16 weeks of gestation. Am J Hematol 1976;1:276, with permission.

Table A1-3   Mean Hematological Values in the First 2 Weeks of Life in the Term Infant

Hematological value	Cord blood	Day 1	Day 3	Day 7	Day 14
Hb (g/dl)	16.8	18.4	17.8	17.0	16.8
Hematocrit	0.53	0.58	0.55	0.54	0.52
Red cells (×10¹²/l)	5.25	5.8	5.6	5.2	5.1
MCV (fl)	107	108	99.0	98.0	96.0
MCH (pg)	34	35	33	32.5	31.5
MCHC (%)	31.7	32.5	33	33	33
Reticulocytes (%)	3–7	3–7	1–3	0–1	0–1
Nucleated RBC (mm³)	500	200	0–5	0	0
Platelets (×10⁹/l)	290	192	213	248	252

From: Oski FA, Naiman JL. Hematologic problems in the newborn. 3rd ed. Philadelphia: Saunders, 1982, with permission.

Table A1-4   Red Cell Values at Various Ages: Mean and Lower Limit of Normal (−2 SD)[a]

Age	Hemoglobin (g/dl)		Hematocrit (%)		Red Cell Count (10¹²/l)		MCV (fl)		MCH (pg)		MCHC (g/dl)		Reticulocytes	
	Mean	−2 SD	Mean	−2 SD	Mean	−2 SD	Mean	−2 SD	Mean	−2 SD	Mean	−2 SD	Mean	−2 SD
Birth (cord blood)	16.5	13.5	51	42	4.7	3.9	108	98	34	31	33	30	3.2	1.8
1–3 days (capillary)	18.5	14.5	56	45	5.3	4.0	108	95	34	31	33	29	3.0	1.5
1 week	17.5	13.5	54	42	5.1	3.9	107	88	34	28	33	28	0.5	0.1
2 weeks	16.5	12.5	51	39	4.9	3.6	105	86	34	28	33	28	0.5	0.2
1 month	14.0	10.0	43	31	4.2	3.0	104	85	34	28	33	29	0.8	0.4
2 months	11.5	9.0	35	28	3.8	2.7	96	77	30	26	33	29	1.6	0.9
3–6 months	11.5	9.5	35	29	3.8	3.1	91	74	30	25	33	30	0.7	0.4
0.5–2 years	12.0	10.5	36	33	4.5	3.7	78	70	27	23	33	30	1.0	0.2
2–6 years	12.5	11.5	37	34	4.6	3.9	81	75	27	24	34	31	1.0	0.2
6–12 years	13.5	11.5	40	35	4.6	4.0	86	77	29	25	34	31	1.0	0.2
12–18 years														
Female	14.0	12.0	41	36	4.6	4.1	90	78	30	25	34	31	1.0	0.2
Male	14.5	13.0	43	37	4.9	4.5	88	78	30	25	34	31	1.0	0.2
18–49 years														
Female	14.0	12.0	41	36	4.6	4.0	90	80	30	26	34	31	1.0	0.2
Male	15.5	13.5	47	41	5.2	4.5	90	80	30	26	34	31	1.0	0.2

[a]These data have been compiled from several sources. Emphasis is given to studies employing electronic counters and to the selection of populations that are likely to exclude individuals with iron deficiency. The mean ±2 SD can be expected to include 95% of the observations in a normal population.
From: Dallman PR. Blood and blood-forming tissue. In: Rudolph A, editor. Pediatrics. 16th ed. E. Norwalk, CT: Appleton-Cernuary-Croles, 1977, with permission.

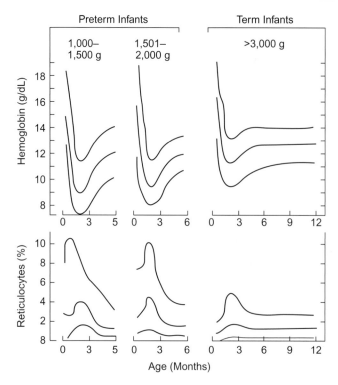

Figure A1-1 Physiologic Nadir for Term and Preterm Infants. The Mean and Range of Normal Hemoglobin and Reticulocyte Values for Term and Preterm Infants are Shown. Premature Infants Reach Nadir of Erythrocyte Production Sooner and Require Longer to Recover Than Their Term Infant Counterparts.
From: Dallman PR: Anemia of prematurity. Ann Rev Med. 1981;32:143.

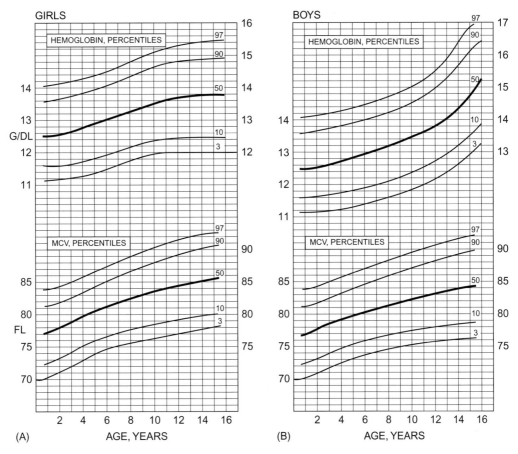

Figure A1-2 (A) Hemoglobin and MCV Percentile Curves for Girls. (B) Hemoglobin and MCV Percentile Curves for Boys.
From: Dallman PR, Siimes MA. Percentile curves for hemoglobin and red cell volume in infancy and childhood. J Pediatr. 1979;94:28, with permission.

Table A1-5   Serum Ferritin Values

Age	ng/ml
Newborn	25–200
1 month	200–600
2–5 months	50–200
6 months–15 years	7–140
Adult:	
Male	15–200
Female	12–150

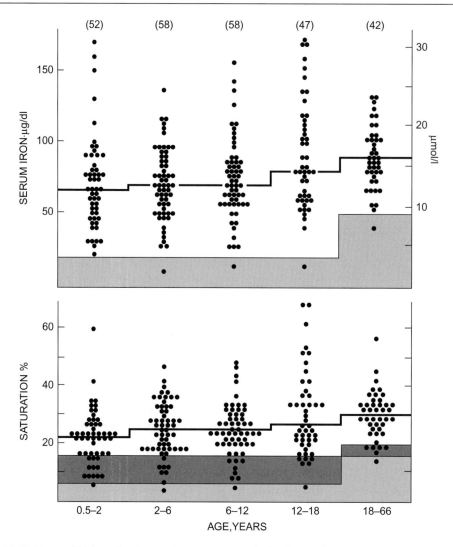

Figure A1-3 Normal Values for Serum Iron and Transferrin Saturation.
From: Koerper MA, Dallman P. Serum iron concentration and transferrin saturation in the diagnosis of iron deficiency in children: normal developmental changes. J Pediatr. 1977;91:870, with permission.

Table A1-6 Values of Serum Iron (SI), Total Iron-Binding Capacity (TIBC) and Transferrin Saturation (S%) from Infants during the First Year of Life

		Age (months)						
		0.5	1	2	4	6	9	12
SI								
Median	µmol/l	22	22	16	15	14	15	14
95% range		11–36	10–31	3–29	3–29	5–24	6–24	6–28
Median	µg/dl	120	125	87	84	77	84	78
95% range		63–201	58–172	15–159	18–164	28–135	34–135	35–155
TIBC								
Mean ± SD	µmol/l	34 ± 8	36 ± 8	44 ± 10	54 ± 7	58 ± 9	61 ± 7	64 ± 7
	µg/dl	191 ± 43	199 ± 43	246 ± 55	300 ± 39	321 ± 51	341 ± 42	358 ± 38
S%								
Median		68	63	34	27	23	25	23
95% range		30–99	35–94	21–63	7–53	10–43	10–39	10–47

Notes. These data were obtained from a group of healthy, full-term infants who were born at the Helsinki University Central Hospital. Infants received iron supplementation in formula and cereal throughout the 12-month period. Infants with hemoglobin below 10 g/dl, mean corpuscular volume of red blood cells below 71 fl, or serum ferritin below 10 ng/ml were excluded from the study. The 95% range of the transferrin saturation values indicates that the lower limit of normal is about 10% after 4 months of age.

From: Saarinen UM, Siimes MA. Serum iron and transferrin in iron deficiency. J Pediatr. 1977;91:876, with permission.

Table A1-7 Mean Serum Iron and Iron Saturation Percentage

Age (years)	Serum Iron (µg/dl)	Saturation (%)
0.5–2	68 ± 3.6 (16–120)	22 ± 1.1 (6–38)
2–6	72 ± 3.4 (20–124)	25 ± 1.2 (7–43)
6–12	73 ± 3.4 (23–123)	25 ± 1.2 (7–43)
18+	92 ± 3.8 (48–136)	30 ± 1.1 (18–46)

From: Koerper MA, Dallman PR. Serum iron concentration and transferrin saturation are lower in normal children than in adults. J Pediatr Res. 1977;11:473, with permission.

Table A1-8   Normal Serum Folic Acid Levels (ng/ml)

Folate	Age	Range	Mean ± SD
Serum folate	**Normal premature infants**		
	1–4 days	7.17–52.00	29.54 ± 0.98
	2–3 weeks	4.12–15.62	8.61 ± 0.55
	1–2 months	2.81–11.25	5.84 ± 0.35
	2–3 months	3.56–11.82	6.95 ± 0.50
	3–5 months	3.85–16.50	8.92 ± 0.86
	5–7 months	6.00–12.25	9.02 ± 0.74
	Normal children		
	1 year	3.0–35	9.3
	1–6 years	4.12–21.15	11.37 ± 0.82
	1–10 years	6.5–16.5	10.3
	Normal adults		
	20–45 years	4.50–28.00	10.29 ± 1.14
Red cell folate	Infants <1 year	74–995	277
	Children 1–11 years	96–364	215
	Adults	160–640	316
Whole blood folate	Infants <1 year	20–160	87
	Infants 1 year	31–400	86
	Infants 2–24 months	34–160	96
	Children 1–11 years	52–164	97
	Adults	50–400	195

From: Shojania A, Gross S. Folic acid deficiency and prematurity. J. Pediatr. 1964;64:323, with permission.

Table A1-9   Percentage of Hemoglobin F in the First Year of Life[a]

Age	Number Tested	Mean	2 SD	Range
1–7 days	10	74.7	5.4	61–79.6
2 weeks	13	74.9	5.7	66–88.5
1 month	11	60.2	6.3	45.7–67.3
2 months	10	45.6	10.1	29.4–60.8
3 months	10	26.6	14.5	14.8–55.9
4 months	10	17.7	6.1	9.4–28.5
5 months	10	10.4	6.7	2.3–22.4
6 months	15	6.5	3.0	2.7–13.0
8 months	11	5.1	3.6	2.3–11.9
10 months	10	2.1	0.7	1.5–3.5
12 months	10	2.6	1.5	1.3–5.0
1–14 years and adults	100	0.6	0.4	–

[a]HbF measured by alkali denaturation.
From: Schröter W, Nafz C. Diagnostic significance of hemoglobin F and A_2 levels in homo- and heterozygous beta-thalassemia during infancy. Helv Paediatr Acta. 1981;36:519.

Table A1-10 Percentage of Hemoglobin F and A₂ in Newborn and Adult

	Hemoglobin F (%)	Hemoglobin A₂(%)
Newborn	60–90	<1.0
Adult	<1.0	1.6–3.5

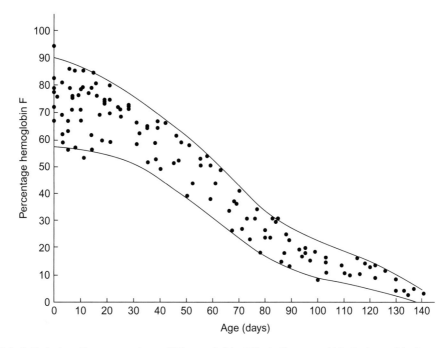

Figure A1-4 Relative Concentration of Hemoglobin F in Infants and Variation with Age.
Modified from: Garby L, Sjolin S, Vuille JC. Studies on erythro-kinetics in infancy. II The relative rate of synthesis of haemoglobin F and haemoglobin A during the first months of life. Acta Paediatr. 1962;51:245, with permission.

Table A1-11   Estimated Blood Volumes

Age	Plasma Volume (PV) (ml/kg)	Red Cell Mass (RCM) (ml/kg)	Total Blood Volume (ml/kg)	
			(From PV)	(From RCM)
Newborn	43.6	43.1	80.0	85.4
1–7 days	51–54	37.9	82.9	77.8
1–12 months	46.1	25.5	78.1	72.8
1–3 years	45.8	24.9	77.8	69.1
4–6 years	49.6	25.5	82.8	67.5
7–9 years	50.6	24.3	88.6	67.5
10–12 years	49.0	26.3	85.4	67.5
13–15 years	51.2		88.3	
16–18 years	50.1		90.2	
Adult	39–44	25–30	68–88	55–75

Modified from: Price DC, Ries C. In Handmaker H, Lowenstein JM (eds). Nuclear Medicine in Clinical Pediatrics. New York, Society of Nuclear Medicine, 1975, p. 279.

Table A1-12 Methemoglobin Levels in Normal Children[a]

	Number of Cases	Methemoglobin (g/dl)				Methemoglobin as Percentage of Total Hemoglobin				
		Number of Determinations	Mean	Range	SD	Number of Cases	Number det.	Mean	Range	SD
Prematures (birth–7 days)	29	34	0.43	(0.02–0.83)	±0.07	24	28	2.3	(0.08–4.4)	±1.26
Prematures (7–72 days)	21	29	0.31	(0.02–0.78)	±0.19	18	23	2.2	(0.02–4.7)	±1.07
Prematures (total)	50	63	0.38	(0.02–0.83)	±0.10	42	51	2.2	(0.08–4.7)	±1.10
Cook Country Hospital prematures (1–14 days)	8	8	0.52	(0.18–0.83)	±0.08	–	–	–	–	–
Newborns (1–10 days)	39	39	0.22	(0.00–0.58)	±0.17	25	30	1.5	(0.00–2.8)	±0.81
Infants (1 month–1 year)	8	8	0.14	(0.02–0.29)	±0.09	8	8	1.2	(0.17–2.4)	±0.78
Children (1–14 years)	35	35	0.11	(0.00–0.33)	±0.09	35	35	0.79	(0.00–2.4)	±0.62
Adults (14–78 years)	30	30	0.11	(0.00–0.28)	±0.09	27	27	0.82	(0.00–1.9)	±0.63

[a]The premature and full-term infants were free of known disease. None had respiratory distress or cyanosis. Analysis of milk and water ingested by these infants revealed the nitrate level less than 0.027 ppm. The premature infants routinely received vitamin C orally each day from the seventh day of life. Methemoglobin values in premature and mature infants and children. Am J Dis Child 1956;91:2, with permission.
From: Kravitz H, Elegant LD, et al. Methemoglobin values in premature and mature infants and children. Am J Dis Child 1956;91:2, with permission.

Table A1-13 Serum Erythropoietin Levels

RIA method	<5–20 mU/ml
Hemagglutination method	25–125
Bioassay method	5–18

Table A1-14   Comparison of Enzyme Activities and Glutathione Content in Newborn and Adult Red Blood Cells

Enzyme	Activity in Normal Adult RBC in IU/g Hb (Mean ± 1 SD at 37°C)	Mean Activity in Newborn RBC as Percentage of Mean (100%) Activity in Normal Adult RBC
Aldolase	3.19 ± 0.86	140
Enolase	5.39 ± 0.83	250
Glucose phosphate isomerase	60.8 ± 11.0	162
Glucose-6-phosphate dehydrogenase	8.34 ± 1.59	
WHO method	12.1 ± 2.09	174
Glutathione peroxidase	30.82 ± 4.65	56
Glyceraldehyde phosphate dehydrogenase	226 ± 41.9	170
Hexokinase	1.78 ± 0.38	239
Lactate dehydrogenase	200 ± 26.5	132
NADH-methaemoglobin reductase	19.2 ± 3.85 (at 30°C)	Increased
Phosphofructokinase	11.01 ± 2.33	97
Phosphoglycerate kinase	320 ± 36.1	165
Pyruvate kinase	15.0 ± 1.99	160
6-Phosphogluconate dehydrogenase	8.78 ± 0.78	150
Triose phosphate isomerase	211 ± 397	101
Glutathione	6,570 ± 1,040 nmol/g Hb	156

Note: The percentage activity in newborn RBC compared to mean adult (100%) values is presented with quantitative data from studies on adult RBC. Newborn data from Konrad et al., 1972; quantitative data from Beutler, 1984.
From: Hinchliffe RF, Lilleyman JS, editors. Practical paediatric haematology: a Laboratory worker's guide to blood disorders in children. New York: Wiley, 1987, with permission.

Table A1-16   Polymorphonuclear Leukocyte and Band Counts in the Newborn during the First 2 Days of Life[a]

Age (h)	Absolute Neutrophil Count (mm³)	Absolute Band Count (mm³)	B/N Ratio
0	3,500–6,000	1,300	0.14
12	8,000–15,000	1,300	0.14
24	7,000–13,000	1,300	0.14
36	5,000–9,000	700	0.11
48	3,500–5,200	700	0.11

[a]Normal values were obtained from the assessment of 3,100 separate white blood cell counts obtained from 965 infants; 513 counts were from infants considered to be completely normal at the time the count was obtained and for the preceding and subsequent 48 hours. There was no difference in the normal ranges when infants were compared by either birth weight (~2,500 g) or gestational age.
From: Manroe BL, Browne R, et al. Normal leukocyte (WBC) values in neonates. Pediatr Res. 1976;10:428, with permission.

WHITE CELL VALUES

Table A1-15   Normal Leukocyte Counts[a]

Age	Total Leukocytes		Neutrophils			Lymphocytes			Monocytes		Eosinophils	
	Mean	(Range)	Mean	(Range)	%	Mean	Range	%	Mean	%	Mean	%
Birth	18.1	(9.0–30.0)	11.0	(6.0–26.0)	61	5.5	(2.0–11.0)	31	1.1	6	0.4	2
12 hours	22.8	(13.0–38.0)	15.5	(6.0–28.0)	68	5.5	(2.0–11.0)	24	1.2	5	0.5	2
24 hours	18.9	(9.4–34.0)	11.5	(5.0–21.0)	61	5.8	(2.0–11.5)	31	1.1	6	0.5	2
1 week	12.2	(5.0–21.0)	5.5	(1.5–10.0)	45	5.0	(2.0–17.0)	41	1.1	9	0.5	4
2 weeks	11.4	(5.0–20.0)	4.5	(1.0–9.5)	40	5.5	(2.0–17.0)	48	1.0	9	0.4	3
1 month	10.8	(5.0–19.5)	3.8	(1.0–9.0)	35	6.0	(2.5–16.5)	56	0.7	7	0.3	3
6 months	11.9	(6.0–17.5)	3.8	(1.0–8.5)	32	7.3	(4.0–13.5)	61	0.6	5	0.3	3
1 year	11.4	(6.0–17.5)	3.5	(1.5–8.5)	31	7.0	(4.0–10.5)	61	0.6	5	0.3	3
2 years	10.6	(6.0–17.0)	3.5	(1.5–8.5)	33	6.3	(3.0–9.5)	59	0.5	5	0.3	3
4 years	9.1	(5.5–15.5)	3.8	(1.5–8.5)	42	4.5	(2.0–8.0)	50	0.5	5	0.3	3
6 years	8.5	(5.0–14.5)	4.3	(1.5–8.0)	51	3.5	(1.5–7.0)	42	0.4	5	0.2	3
8 years	8.3	(4.5–13.5)	4.4	(1.5–8.0)	53	3.3	(1.5–6.8)	39	0.4	4	0.2	2
10 years	8.1	(4.5–13.5)	4.4	(1.8–8.0)	54	3.1	(1.5–6.5)	38	0.4	4	0.2	2
16 years	7.8	(4.5–13.0)	4.4	(1.8–8.0)	57	2.8	(1.2–5.2)	35	0.4	5	0.2	3
21 years	7.4	(4.5–11.0)	4.4	(1.8–7.7)	59	2.5	(1.0–4.8)	34	0.3	4	0.2	3

[a]Numbers of leukocytes are in thousands per mm[3], ranges are estimates of 95% confidence limits and percentages refer to differential counts. Neutrophils include band cells at all ages and a small number of metamyelocytes and myelocytes in the first few days of life.

From: Dallman PR. Blood and blood-forming tissues. In: Rudolpg AM, editors. Pediatrics. 16th ed. E. Norwalk, CT: Appleton-Century-Crofts, 1977, with permission.

Table A1-17   Lymphocyte Subsets (25th, 50th and 75th centiles) in Children and Adults

Subset	Percentage T Cells				Percentage B Cells			Percentage Natural Killer (NK) Cells		
CD Numbers	CD3				CD19			CD16$^+$/CD56$^+$, CD3$^-$		
Reagent	Leu 4				Leu 12			Leu 11$^+$/19$^+$, Leu 4$^-$		
Percentile		P_{25}	P_{50}	P_{75}	P_{25}	P_{50}	P_{75}	P_{25}	P_{50}	P_{75}
Age group	Number									
Cord blood	24	49	55	62	14	20	23	14	19.5	30
1 day–11 months	31	55	60	67	19	25	29	11	15	19
1–6 years	54	62	64	69	21	25	28	8	11	15
7–17 years	31	64	70	74	12	16	23	8.5	11	15.5
18–70 years	300	67	72	76	11	13	16	10	14	19

Subset	Percentage T helper cells				Percentage T suppressor cells			Helper/suppressor (CD4/CD8) ratio		
CD number	CD4				CD8			CD4	CD8	
Reagent	Leu 3				Leu 2			Leu 3	Leu 2	
Percentile		P_{25}	P_{50}	P_{75}	P_{25}	P_{50}	P_{75}	P_{25}	P_{50}	P_{75}
Age group	Number									
Cord blood	24	28	35	42	26	29	33	0.80	1.15	1.75
1 day–11 months	31	35	41	48	18	23	28	1.35	1.80	2.25
1–6 years	54	32	37	40	25	32	36	1.00	1.20	1.60
7–17 years	31	33	36	40	28	31	36	1.05	1.20	1.40
18–70 years	300	38	42	46	31	35	40	1.00	1.20	1.50

Data obtained using flow cytometry. Kindly supplied by: Dr F Hulstaert, Becton Dickinson Immunocytometry Systems Medical Department Cockeysville, MD.

PLATELET VALUES

Table A1-18   Normal Platelet Counts

Subject	Platelet Count/mm³ (mean ± 1 SD)
Preterm, 27–31 weeks	275,000 ± 60,000
Preterm, 32–36 weeks	290,000 ± 70,000
Term infants	310,000 ± 68,000
Normal adult or child	300,000 ± 50,000

From: Oski FA, Naiman JL. Normal Blood values in the newborn period. In Hematologic Problems in the Newborn. Philadelphia: Saunders, 1982, with permission.

COAGULATION VALUES

Table A1-19 Reference Values for Coagulation Tests in Heathly Full-Term Infant during First 6 Months of Life

	Day 1		Day 5		Day 30		Day 90		Day 180		Adult	
	M	B	M	B	M	B	M	B	M	B	M	B
PT (s)	13.0	(10.1–15.9)[a]	12.4	(10.0–15.3)[a]	11.8	(10.0–14.3)[a]	11.9	(10.0–14.2)[a]	12.3	(10.7–13.9)[a]	12.4	(10.8–13.9)
INR	1.00	(0.53–1.62)	0.89	(0.53–1.48)	0.79	(0.53–1.26)	0.81	(0.53–1.26)	0.88	(0.61–1.17)	0.89	(0.64–1.17)
APTT (s)	42.9	(31.3–54.5)	42.6	(25.4–59.8)	40.4	(32.0–55.2)	37.1	(29.0–50.1)[a]	35.5	(28.1–42.9)[a]	33.5	(26.6–10.3)
TCT (s)	23.5	(19.0–28.3)[a]	23.1	(18.0–29.2)	24.3	(19.4–29.2)[a]	25.1	(20.5–29.7)[a]	25.5	(19.8–31.2)[a]	25.0	(19.7–30.3)
Fibrinogen (g/dl)	2.83	(1.67–3.99)[a]	3.12	(1.62–4.62)[a]	2.70	(1.62–3.78)[a]	2.43	(1.50–3.79)[a]	2.51	(1.50–3.87)[a]	2.78	(1.56–4.00)
II (units/ml)	0.48	(0.26–0.70)	0.63	(0.33–0.93)	0.68	(0.34–1.02)	0.75	(0.45–1.05)	0.88	(0.60–1.16)	1.08	(0.70–1.46)
V (units/ml)	0.72	(0.34–1.08)	0.95	(0.45–1.45)	0.98	(0.62–1.34)	0.90	(0.48–1.32)	0.91	(0.55–1.27)	1.06	(0.62–1.50)
VII (units/ ml)	0.66	(0.28–1.04)	0.89	(0.35–1.43)	0.90	(0.42–1.38)	0.91	(0.39–1.43)	0.87	(0.47–1.27)	1.05	(0.67–1.43)
VIII (units /ml)	1.00	(0.50–1.78)[a]	0.88	(0.50–1.54)[a]	0.91	(0.50–1.57)[a]	0.79	(0.50–1.25)[a]	0.73	(0.50–1.09)	0.99	(0.50–1.49)
vWF (units/ml)	1.53	(0.50–2.87)	1.40	(0.50–2.54)	1.28	(0.50–2.46)	1.18	(0.50–2.06)	1.07	(0.50–1.97)	0.92	(0.50–1.58)
IX (units/ml)	0.53	(0.15–0.91)	0.53	(0.15–0.91)	0.51	(0.21–0.81)	0.67	(0.21–1.13)	0.86	(0.36–1.36)	1.09	(0.55–1.63)
X (units/ml)	0.40	(0.12–0.68)	0.49	(0.19–0.79)	0.59	(0.31–0.87)	0.71	(0.35–1.07)	0.78	(0.38–1.18)	1.06	(0.70–1.52)
XI (units/ml)	0.38	(0.10–0.66)	0.55	(0.23–0.87)	0.53	(0.27–0.79)	0.69	(0.41–0.97)	0.86	(0.49–1.34)	0.97	(0.67–1.27)
XII (units/ml)	0.53	(0.13–0.93)	0.47	(0.11–0.83)	0.49	(0.17–0.81)	0.67	(0.25–1.09)	0.77	(0.39–1.15)	1.08	(0.52–1.64)
PK (units/ml)	0.37	(0.18–0.69)	0.48	(0.20–0.76)	0.57	(0.23–0.91)	0.73	(0.41–1.05)	0.86	(0.56–1.16)	1.12	(0.62–1.62)
HMW-K (units/ml)	0.54	(0.06–1.02)	0.74	(0.16–1.32)	0.77	(0.33–1.21)	0.82	(0.30–1.46)[a]	0.82	(0.36–1.28)[a]	0.92	(0.50–1.36)
XIII[a] (units/ml)	0.79	(0.27–1.31)	0.94	(0.44–1.44)[a]	0.93	(0.39–1.47)[a]	1.04	(0.36–1.72)[a]	1.04	(0.46–1.62)[a]	1.05	(0.55–1.55)
XIII[b] (units/ml)	0.76	(0.30–1.22)	1.06	(0.32–1.80)	1.11	(0.39–1.73)[a]	1.16	(0.48–1.84)[a]	1.10	(0.50–1.70)[a]	0.97	(0.57–1.37)

[a] Values are indistinguishable from the adult.

Abbreviations: PT, prothrombin time; INR, international normalized ratio; APTT, activated partial thromboplastin time; TCT, thrombin clotting time; VII, factor VIII procoagulant; vWF, von Willebrand's factor; PK, prekallikrein; HMW-K, high molecular weight kininogen.

All factors except fibrinogen are expressed as units per milliliter (units/ml) where pooled plasma contains 1.0 unit/ml. All values are expressed as mean (M) followed by the lower and upper boundary (B) encompassing 95% of the population. Between 40 and 77 samples were assayed for each value for the newborn. Some measurements were skewed due to a disproportionate number of high values. The lower limit, which excludes the lower 2.5% of the population, has been given.

From: Andrew M. Bailliere's Clin Hematol. 1991;4:251, with permission.

Table A1-20 Reference Values for Coagulation Tests in Heathly Premature Infants (30–36 weeks' Gestation) during First 6 Months

	Day 1 M	Day 1 B	Day 5 M	Day 5 B	Day 30 M	Day 30 B	Day 90 M	Day 90 B	Day 180 M	Day 180 B	Adult M	Adult B
PT (s)	13.0	(10.6–16.2)[b]	12.5	(10.0–15.3)[b,c]	11.8	(10.0–13.6)[b]	12.3	(10.0–14.6)[b]	12.5	(10.0–15.0)[b]	12.4	(10.8–13.9)
APTT (s)	53.6	(27.5–79.4)[c]	50.5	(26.9–74.1)[c]	44.7	(26.9–62.5)	39.5	(28.3–50.7)	37.5	(21.7–53.3)[b]	33.5	(26.6–40.3)
TCT (s)	24.8	(19.2–30.4)[b]	24.1	(18.8–29.4)[b,c,d]	24.4	(18.8–29.9)[b]	25.1	(19.4–30.8)[b]	25.2	(18.9–31.5)[b]	25.0	(19.7–30.3)
Fibrinogen (g/l)	2.43	(1.50–3.73)[b,c,d]	2.80	(1.60–4.18)[c]	2.54	(1.50–4.14)[b,c]	2.46	(1.50–3.52)[b,c]	2.28	(1.50–3.60)[d]	2.78	(1.56–4.00)
II (units/ml)	0.45	(0.20–0.77)[c]	0.57	(0.29–0.85)	0.57	(0.36–0.95)[c,d]	0.68	(0.30–1.06)	0.87	(0.51–1.23)	1.08	(0.70–1.46)
V (units/ml)	0.88	(0.41–1.44)[b,c,d]	1.00	(0.46–1.54)	1.02	(0.48–1.56)	0.99	(0.59–1.39)	1.02	(0.58–1.46)	1.06	(0.62–1.50)
VII (units/ml)	0.67	(0.21–1.13)	0.84	(0.30–1.38)	0.83	(0.21–1.45)	0.87	(0.31–1.43)	0.99	(0.47–1.51)[b]	1.05	(0.67–1.43)
VIII (units/ml)	1.11	(0.50–2.13)[b,c]	1.15	(0.53–2.05)[b,c,d]	1.11	(0.50–1.99)[b,c,d]	1.06	(0.58–1.88)[b,c,d]	0.99	(0.50–1.87)[b,c,d]	0.99	(0.50–1.49)
vWF (units/ml)	1.36	(0.78–2.10)[b]	1.33	(0.72–2.19)[c]	1.36	(0.66–2.16)[c]	1.12	(0.75–1.84)[b,c]	0.98	(0.54–1.58)[b,c]	0.92	(0.50–1.58)
IX (units/ml)	0.35	(0.19–0.65)[b,d]	0.42	(0.14–0.74)[c,d]	0.44	(0.13–0.80)[c]	0.59	(0.25–0.93)	0.81	(0.50–1.20)[c]	1.09	(0.55–1.63)
X (units/ml)	0.41	(0.11–0.71)	0.51	(0.19–0.83)	0.56	(0.20–0.92)	0.67	(0.35–0.99)	0.77	(0.35–1.19)	1.06	(0.70–1.52)
XI (units/ml)	0.30	(0.08–0.52)[c,d]	0.41	(0.13–0.69)[d]	0.43	(0.15–0.71)[d]	0.59	(0.25–0.93)[d]	0.78	(0.46–1.10)	0.97	(0.67–1.27)
XII (units/ml)	0.38	(0.10–0.66)[d]	0.39	(0.09–0.69)[d]	0.43	(0.11–0.75)	0.61	(0.15–1.07)	0.82	(0.22–1.42)	1.08	(0.52–1.64)
PK (units/ml)	0.33	(0.09–0.57)	0.45	(0.28–0.75)[c]	0.59	(0.31–0.87)	0.79	(0.37–1.21)	0.78	(0.40–1.16)	1.12	(0.62–1.62)
HMWK (units/ml)	0.49	(0.09–0.89)	0.62	(0.24–1.00)[d]	0.64	(0.16–1.12)[d]	0.78	(0.32–1.24)	0.83	(0.41–1.25)[b]	0.92	(0.50–1.36)
XIIIa (units/ml)	0.70	(0.32–1.08)	1.01	(0.57–1.45)[b]	0.99	(0.51–1.47)[b]	1.13	(0.71–1.55)[b]	1.13	(0.65–1.61)[b]	1.05	(0.55–1.55)
XIIIb (units/ml)	0.81	(0.35–1.27)	1.10	(0.68–1.58)[b]	1.07	(0.57–1.57)[b]	1.21	(0.75–1.67)	1.15	(0.67–1.63)	0.97	(0.57–1.37)
Plasminogen (CTPA, units/ml)	1.70	(1.12–2.48)[c,d]	1.91	(1.21–2.61)[d]	1.81	(1.09–2.53)	2.38	(1.58–3.18)	2.75	(1.91–3.59)[d]	3.36	(2.48–4.24)

[a] All values are give as a mean (M) followed by the lower and upper boundaries (B) encompasing 95% of the population. Between 40 and 96 samples were assayed for each value for newborns.

[b] Values indistinguishable from the adult.

[c] Measurements are skewed owing to a disproportionate number of high values. Lower limit that excludes the lower 2.5% of the population is given (B).

[d] Values differ from those of full-term infants.

From: Andrew M, Paes B, Milner R, et al. Development of the human coagulation system in the healthy premature infant. Blood 1988;72:1651, with permission.

Table A1-21   Reference Values for the Inhibitors of Coagulation in the Healthy Full-Term Infant during the First 6 Months of Life[a]

Inhibitors	Day 1 (n)	Day 5 (n)	Day 30 (n)	Day 90 (n)	Day 180 (n)	Adult (n)
AT-III (units/ml)	0.63 ± 0.12 (58)	0.67 ± 0.13 (74)	0.78 ± 0.15 (66)	0.97 ± 0.12 (60)[b]	1.04 ± 0.10 (56)[b]	1.05 ± 0.13 (28)
α_2-M (units/ml)	1.39 ± 0.22 (54)	1.48 ± 0.25 (73)	1.50 ± 0.22 (61)	1.76 ± 0.25 (55)	1.91 ± 0.21 (55)	0.86 ± 0.17 (29)
α_2-AP (units/ml)	0.85 ± 0.15 (55)	1.00 ± 0.15 (75)[b]	1.00 ± 0.12 (62)[b]	1.08 ± 0.16 (55)[b]	1.11 ± 0.14 (53)	1.02 ± 0.17 (29)
C_1E-INH (units/ml)	0.72 ± 0.18 (59)	0.90 ± 0.15 (76)[b]	0.89 ± 0.21 (63)	1.15 ± 0.22 (55)	1.41 ± 0.26 (55)[b]	1.01 ± 0.15 (29)
α_1-AT (units/ml)	0.93 ± 0.22 (57)[b]	0.89 ± 0.20 (75)[b]	0.62 ± 0.13 (61)	0.72 ± 0.15 (56)	0.77 ± 0.15 (55)	0.93 ± 0.19 (29)
HCII (units/ml)	0.43 ± 0.25 (56)	0.48 ± 0.24 (72)	0.47 ± 0.20 (58)	0.72 ± 0.37 (58)	1.20 ± 0.35 (55)	0.96 ± 0.15 (29)
Protein C (units/ml)	0.35 ± 0.09 (41)	0.42 ± 0.11 (44)	0.43 ± 0.11 (43)	0.54 ± 0.13 (44)	0.59 ± 0.11 (52)	0.96 ± 0.16 (28)
Protein S (units/ml)	0.36 ± 0.12 (40)	0.50 ± 0.14 (48)	0.63 ± 0.15 (41)	0.86 ± 0.16 (46)[b]	0.87 ± 0.16 (49)[b]	0.92 ± 0.16 (29)

[a]All values expressed in units/ml as mean ± 1 SD; *n*, numbers studied.

[b]Values indistinguishable from those of adults.

Abbreviations: AT-III, antithrombin III; α_2-M, alpha$_2$-macroglobulin; α_2-AP, alpha$_2$-antiplasmin; C_1 E-INH, C_1-esterase inhibitor; α_1-AT, alpha$_1$-antitrypsin; HCII, heparin cofactor II.

From: Andrew M, Paes B, Milner R, et al. Development of the human coagulation system in the healthy premature infant. Blood 1988;72:1651, with permission.

Table A1-22   Reference Values for Inhibitors of Coagulation Healthy Premature Infants during First 6 Months of Life[a]

	Day 1		Day 5		Day 30		Day 90		Day 180		Adult	
	M	B	M	B	M	B	M	B	M	B	M	B
AT-III (units/ml)	0.38	(0.14–0.62)[d]	0.56	(0.30–0.82)[b]	0.59	(0.37–0.81)[d]	0.83	(0.45–1.21)[d]	0.90	(0.52–1.28)[d]	1.05	(0.79–1.31)
α_2-M (units/ml)	1.10	(0.56–1.82)[c,d]	1.25	(0.71–1.77)[b]	1.38	(0.72–2.04)	1.80	(1.20–2.66)[c]	2.09	(1.10–3.21)[c]	0.86	(0.52–1.20)
α_2-AP (units/ml)	0.78	(0.40–1.16)	0.81	(0.49–1.13)[b]	0.89	(0.55–1.23)[d]	1.06	(0.64–1.48)[b]	1.15	(0.77–1.53)	1.02	(0.68–1.36)
C_1E-INH (units/ml)	0.65	(0.31–0.99)	0.83	(0.45–1.21)	0.74	(0.40–1.24)[c,d]	1.14	(0.60–1.68)	1.40	(0.96–2.04)[c]	1.01	(0.71–1.31)
α_1-AT (units/ml)	0.90	(0.36–1.44)[b]	0.94	(0.42–1.46)[d]	0.76	(0.38–1.12)[d]	0.81	(0.49–1.13)[b,d]	0.82	(0.48–1.16)[b]	0.93	(0.55–1.31)
HCII (units/ml)	0.32	(0.00–0.80)[d]	0.34	(0.00–0.69)	0.43	(0.15–0.71)	0.61	(0.20–1.11)[c]	0.89	(0.45–1.40)[b,c,d]	0.96	(0.66–1.26)
Protein C (units/ml)	0.28	(0.12–0.44)[c]	0.31	(0.11–0.51)	0.37	(0.15–0.59)[d]	0.45	(0.23–0.67)[d]	0.57	(0.31–0.83)	0.96	(0.64–1.28)
Protein S (units/ml)	0.26	(0.14–0.38)[d]	0.37	(0.13–0.61)	0.56	(0.22–0.90)	0.76	(0.40–1.12)[d]	0.82	(0.44–1.20)	0.92	(0.60–1.24)

[a]All values expressed in units/ml, were pooled plasma contains 1.0 unit/ml. All values are given a mean (M) followed by the lower and upper boundaries (B) encompassing 95% of the population. Between 40 and 75 samples were assayed for each value for the newborn.

[b]Values are indistinguishable from the adults.

[c]Measurements are skewed owing to a disproportionate number of high values. Lower limit which excludes the lower 2.5% of the population is given (B).

[d]Values differ from those of full-term infants.

From: Andrew M, Paes B, Milner R, et al. Development of the human coagulation system in the healthy premature infant. Blood 1988;72:1651, with permission.

Table A1-22a Reference Values for the Components of the Fibrinolytic System in the Healthy Full-Term Infant and the Healthy Premature infant During the First 6 Months of Life

Fibrinolytic system	Day 1		Day 5		Day 30		Day 90		Day 180		Adult	
	M	B	M	B	M	B	M	B	M	B	M	B
Healthy full-term infant												
Plasminogen (units/ml)	1.95	(1.25–2.65)	2.17	(1.41–2.93)	1.98	(1.26–2.70)	2.48	(1.74–3.22)	3.01	(2.21–3.81)	3.36	(2.48–4.24)
t-PA (ng/ml)	9.60	(5.00–18.9)	5.60	(4.00–10.0)[a]	4.10	(1.00–6.00)[a]	2.10	(1.00–5.00)[a]	2.80	(1.00–6.00)[a]	4.90	(1.40–8.40)
α_2 AP (units/ml)	0.85	(0.55–1.15)	1.00	(0.70–1.30)[a]	1.00	(0.76–1.24)[a]	1.08	(0.76–1.40)[a]	1.11	(0.83–1.39)[a]	1.02	(0.68–1.36)
PAI (units/ml)	6.40	(2.00–15.1)	2.30	(0.00–8.10)[a]	3.4	(0.00–8.80)[a]	7.20	(1.00–15.3)	8.10	(6.00–13.0)	3.6	(0.00–11.0)
Healthy premature infant												
Plasminogen (units/ml)	1.70	(1.12–2.48)[b]	1.91	(1.21–2.61)[b]	1.81	(1.09–2.53)	2.38	(1.58–3.18)	2.75	(1.91–3.59)[b]	3.36	(2.48–4.24)
t-PA (ng/ml)	8.48	(3.00–16.70)	3.97	(2.00–6.93)[a]	4.13	(2.00–7.79)[a]	3.31	(2.00–5.07)[a]	3.48	(2.00–5.85)[a]	4.96	(1.46–8.46)
α_2 AP (units/ml)	0.78	(0.40–1.16)	0.81	(0.49–1.13)[b]	0.89	(0.55–1.23)[b]	1.06	(0.64–1.48)[a]	1.15	(0.77–1.53)	1.02	(0.68–1.36)
PAI (units/ml)	5.40	(0.00–12.2)[a,b]	2.50	(0.00–7.10)[a]	4.30	(0.00–11.8)[a]	4.80	(1.00–10.2)[a,b]	4.90	(1.00–10.2)[a,b]	3.60	(0.00–11.0)

[a]Values that are indistinguishable from those of the adult.

[b]Values that are different from those of the full-term infant.

For α_2 AP, values are expressed as units per milliliter (units/ml) were pooled plasma contains 1.0 unit/ml. Plasminogen units are those recommended by the committee on Thrombolytic Agents. Values for t-PA are given as nanograms per milliliter. Values for PAI are given as units per milliliter; one unit of PAI activity is defined as the amount of PAI that inhibits one international unit of human single-chain TA. All values are given a mean (M) followed by the lower and upper boundary (B) encompassing 95% of the population.

Abbreviations: t-PA, tissue plasminogen activator; α_2 AP, α_2 antiplasmin; PAI, plasminogen activator inhibitor.

From: Andrew M, Paes B, Johnston M, et al. Development of the hemostatic system in the neonate and young infant. Am J Pediatr Hematol Oncol. 1990;12:95–104.

Table A1-23   Bleeding Time (min) in Children and Adults[a]

Subjects	Number Tested	Mean	SD	Range
Children	36	4.6	1.4	2.5–8.5
Adults	48	4.6	1.2	2.5–6.5

[a]Bleeding time performed using a template technique, with incision 6 mm long and 1 mm deep and sphygmomanometer pressure 40 mmHg. Using this technique, bleeding times up to 11.5 min have been observed in apparently healthy children tested in the author's laboratory.
From: Buchanan GR, Holtkamp CA. Prolonged bleeding time in children and young adults with hemophilia. Pediatrics 1980;66:951.

Table A1-24   Bleeding Time (min) in Newborns and Children[a]

Subjects	Number Tested	Sphygmomanometer Pressure (mmHg)	Mean	Range
Term newborn	30	30	3.4	1.9–5.8
Preterm newborn:				
<1,000 g	6	20	3.3	2.6–4.0
1,000–2,000 g	15	25	3.9	2.0–5.6
>2,000 g	5	30	3.2	2.3–5.0
Children	17	30	3.4	1.0–5.5
Adults	20	30	2.8	0.5–5.5

[a]Bleeding time performed using a template technique, with incision 5 mm long and 0.5 mm deep. All subjects had normal platelet counts.
From: Feusner JH. Normal and abnormal bleeding times in neonates and young children utilizing a fully standardized template technique. Am J Clin Pathol. 1980;74:73.

Table A1-25 Coagulation Factor Assays (Mean ± 1 SD) and Screening Tests in the Fetus and Neonate[a]

Assays of Coagulation Factors	Normal Adult Values	28–31 Weeks' Gestation	32–36 Weeks' Gestation	Term	Time at Which Values Attain Adult Norms
Fibrinogen (mg/dl)	150–400	215 ± 28 (SE) / 270 ± 85	226 ± 23 (SE) / 244 ± 55	246 ± 18 (SE) / 246 ± 55	[b]
II (%)	100	30 ± 10	35 ± 12	45 ± 15	2–12 months
V (%)	100	76 ± 7 (SE) / 90 ± 26	84 ± 9 (SE) / 72 ± 23	100 ± 5 (SE) / 98 ± 40	[b]
VII and X (%)	100	38 ± 14	40 ± 15	56 ± 16	2–12 months
VIII (%)	100	90 ± 15 (SE) / 70 ± 30	140 ± 10 (SE) / 98 ± 40	168 ± 12 (SE) / 105 ± 34	[b]
IX (%)	100	27 ± 10	NA	28 ± 8	3–9 months
XI (%)	100	5–18	NA	29–70	1–2 months
XII (%)	100	NA	30±	51 (25–70)	9–14 days
XIII					
Bioassay (%)	100	100	100	100	[b]
Quantitative (units/ml)	21 ± 5.6	5 ± 3.5	NA	11 ± 3.4	3 weeks
Prothrombin time (s)[c]	12–14	23 ±	17 (12–21)	16 (13–20)	1 week
Activated partial thromboplastin time (s)[c]	44	NA	70±	55 ± 10	2–9 months
Thrombin time (s)[c]	10	16–28	14 (11–17)	12 (10–16)	Few days

[a]Assays quoted are biologic unless otherwise specified.
[b]Adult levels attained prenatally.
[c]Values vary among laboratories, depending on reagents employed.
Abbreviations: SE, standard error; NA, not available.
From: Hathaway WE. The bleeding newborn. Semin Hematol. 1975;12:175 and from: Gross SJ, Stuart MJ. Hemostasis in the premature infant. Clin Perinatol. 1977;4:260, with permission.

Table A1-26   Vascular–platelet Interactions in the Fetus and Neonate

Vascular-platelet Ineractions	Normal Adult Values	27–31 Weeks' Gestation	32–36 Weeks' Gestation	Term	Term Infant >1–2 Months
Capillary fragility	N	Increased	N	N	N
Platelet count (10^3/mm^3)	300 ± 50	275 ± 60	290 ± 70	310 ± 68	280 ± 56
Platelet retention (%)	N	NA	NA	N or decreased	N
Platelet aggregation with ADP, epinephrine collagen	N	Abn	Abn	Abn	
Platelet aggregation with ristocetin	N	NA	NA	N or increased	N
Platelet release I (adenide nucleotides)	N	Abn	Abn	Abn	
Platelet factor 3	N	Abn	Abn	Abn	
Platelet factor 4	N	NA	NA	N	N
Bleeding time (min)	4.0 ± 1.5		4 ± 1.5	4 ± 1.5	N

Abbreviations: N, normal; Abn, abnormal; NA, not available.
From: Hathaway WE. The bleeding newborn. Semin Hematol. 1975;12:175 and from: Gross SJ, Stuart MJ. Hemostasis in the premature infant. Clin Perinatol. 1977;4:260, with permission.

BONE MARROW CELLS

Table A1-27 Bone Marrow Cell Populations of Normal Infants[a]

Cell type	Month				
	0 (n = 57)[b]	1 (n = 71)	2 (n = 48)	3 (n = 24)	4 (n = 19)
Small lymphocytes	14.42 ± 5.54	47.05 ± 9.24	42.68 ± 7.90	43.63 ± 11.83	47.06 ± 8.77
Transitional cells	1.18 ± 1.13	1.95 ± 0.94	2.38 ± 1.35	2.17 ± 1.64	1.64 ± 1.01
Proerythroblasts	0.02 ± 0.06	0.10 ± 0.14	0.13 ± 0.19	0.10 ± 0.13	0.05 ± 0.10
Basophilic erythroblasts	0.24 ± 0.25	0.34 ± 0.33	0.57 ± 0.41	0.40 ± 0.33	0.24 ± 0.24
Early erythroblasts	0.27 ± 0.26	0.44 ± 0.42	0.71 ± 0.51	0.50 ± 0.38	0.28 ± 0.30
Polychromatic erythroblasts	13.06 ± 6.78	6.90 ± 4.45	13.06 ± 3.48	10.51 ± 3.39	6.84 ± 2.58
Orthochromatic erythroblasts	0.69 ± 0.73	0.54 ± 1.88	0.66 ± 0.82	0.70 ± 0.87	0.34 ± 0.30
Extruded nuclei	0.47 ± 0.46	0.16 ± 0.17	0.26 ± 0.22	0.19 ± 0.12	0.16 ± 0.17
Late erythroblasts	14.22 ± 7.14	7.60 ± 4.84	13.99 ± 3.82	11.40 ± 3.43	7.34 ± 2.54
Early/late erythroblasts ratio[c]	1:50	1:15	1:18	1:22	1:23
Fetal erythroblasts	14.48 ± 7.24	8.04 ± 5.00	14.70 ± 3.86	11.90 ± 3.52	7.62 ± 2.56
Blood reticulocytes	4.18 ± 1.46	1.06 ± 1.13	3.39 ± 1.22	2.90 ± 0.91	1.65 ± 0.73
Neutrophils					
Promyelocytes	0.79 ± 0.91	0.76 ± 0.65	0.78 ± 0.68	0.76 ± 0.80	0.59 ± 0.51
Myelocytes	3.95 ± 2.93	2.50 ± 1.48	2.03 ± 1.14	2.24 ± 1.70	2.32 ± 1.59
Early neutrophils	4.74 ± 3.43	3.27 ± 1.94	2.81 ± 1.62	3.00 ± 2.18	2.91 ± 2.01
Metamyelocytes	19.37 ± 4.84	11.34 ± 3.59	11.27 ± 3.38	11.93 ± 13.09	6.04 ± 3.63
Bands	28.89 ± 7.56	14.10 ± 4.63	13.15 ± 4.71	14.60 ± 7.54	13.93 ± 6.13
Mature neutrophils	7.37 ± 4.64	3.64 ± 2.97	3.07 ± 2.45	3.48 ± 1.62	4.27 ± 2.69
Late neutrophils	55.63 ± 7.98	29.08 ± 6.79	27.50 ± 6.88	31.00 ± 11.17	31.30 ± 7.80
Early/late neutrophil ratio	1:12	1:9	1:9	1:9	1:11
Total neutrophils	60.37 ± 8.66	32.35 ± 7.68	30.31 ± 7.27	34.01 ± 11.95	34.21 ± 8.61
Total eosinophils	2.70 ± 1.27	2.61 ± 1.40	2.50 ± 1.22	2.54 ± 1.46	2.37 ± 4.13
Total basophils	0.12 ± 0.20	0.07 ± 0.16	0.08 ± 0.10	0.09 ± 0.09	0.11 ± 0.14
Total myeloid cells	63.19 ± 9.10	35.03 ± 8.09	32.90 ± 7.85	36.64 ± 2.26	36.69 ± 8.91
Monocytes	0.88 ± 0.85	1.01 ± 0.89	0.91 ± 0.83	0.68 ± 0.56	0.75 ± 0.75
Miscellaneous					
Megakaryocytes	0.06 ± 0.15	0.05 ± 0.09	0.10 ± 0.13	0.06 ± 0.09	0.06 ± 0.06
Plasma cells	0.00 ± 0.02	0.02 ± 0.06	0.02 ± 0.05	0.00 ± 0.02	0.01 ± 0.03
Unknown blasts	0.31 ± 0.31	0.62 ± 0.50	0.58 ± 0.50	0.63 ± 0.60	0.56 ± 0.53
Unknown cells	0.22 ± 0.34	0.21 ± 0.25	0.16 ± 0.24	0.19 ± 0.21	0.23 ± 0.25
Damaged cells	5.79 ± 2.78	5.50 ± 2.46	5.09 ± 1.78	4.75 ± 2.30	4.80 ± 2.29
Total	6.38 ± 2.84	6.39 ± 2.63	5.94 ± 1.94	5.63 ± 2.36	5.66 ± 2.30

(Continued)

Table A1-27   (Continued)

Month					
5 (*n* = 22)	6 (*n* = 22)	9 (*n* = 16)	12 (*n* = 18)	15 (*n* = 12)	18 (*n* = 19)
47.19 ± 9.93	47.55 ± 7.88	48.76 ± 8.11	47.11 ± 11.32	42.77 ± 8.94	43.55 ± 8.56
1.83 ± 0.89	2.31 ± 1.16	1.92 ± 1.39	2.32 ± 1.90	1.70 ± 0.82	1.99 ± 1.00
0.07 ± 0.10	0.09 ± 0.12	0.07 ± 0.09	0.02 ± 0.04	0.07 ± 0.12	0.08 ± 0.13
0.47 ± 0.33	0.32 ± 0.24	0.31 ± 0.24	0.30 ± 0.25	0.38 ± 0.37	0.50 ± 0.34
0.55 ± 0.36	0.41 ± 0.30	0.39 ± 0.28	0.39 ± 0.27	0.46 ± 0.36	0.59 ± 0.34
7.55 ± 2.35	7.30 ± 3.60	7.73 ± 3.39	6.83 ± 3.75	6.04 ± 1.56	6.97 ± 3.56
0.46 ± 0.51	0.38 ± 0.56	0.39 ± 0.48	0.37 ± 0.51	0.50 ± 0.65	0.44 ± 0.49
0.14 ± 0.11	0.16 ± 0.22	0.22 ± 0.25	0.23 ± 0.25	0.17 ± 0.12	0.21 ± 0.19
8.16 ± 2.58	7.85 ± 4.11	8.34 ± 3.31	7.42 ± 4.11	6.72 ± 1.80	7.62 ± 3.63
1:15	1:17	1:19	1:17	1:15	1:10
8.70 ± 2.69	8.25 ± 4.31	8.72 ± 3.34	7.81 ± 4.26	7.18 ± 1.95	8.21 ± 37.1
1.38 ± 0.65	1.74 ± 0.80	1.67 ± 0.52	1.79 ± 0.79	2.10 ± 0.91	1.84 ± 0.46
0.87 ± 0.80	0.67 ± 0.66	0.41 ± 0.34	0.69 ± 0.71	0.67 ± 0.58	0.64 ± 0.59
2.73 ± 1.82	2.22 ± 1.25	2.07 ± 1.20	2.32 ± 1.14	2.48 ± 0.94	2.49 ± 1.39
3.60 ± 2.50	2.89 ± 1.71	2.48 ± 1.46	3.02 ± 1.52	3.16 ± 1.19	3.14 ± 1.75
11.89 ± 3.24	11.02 ± 3.12	11.80 ± 3.90	11.10 ± 3.82	12.48 ± 7.45	12.42 ± 4.15
14.07 ± 5.48	14.00 ± 4.58	14.08 ± 4.53	14.02 ± 4.88	15.17 ± 4.20	14.20 ± 5.23
3.77 ± 2.44	4.85 ± 2.69	3.97 ± 2.29	5.65 ± 3.92	6.94 ± 3.88	6.31 ± 3.91
29.73 ± 7.19	29.86 ± 6.74	29.86 ± 7.36	30.77 ± 8.69	34.60 ± 7.35	32.93 ± 7.01
1:8	1:10	1:12	1:10	1:10	1:10
33.12 ± 8.34	32.75 ± 7.03	32.33 ± 7.75	33.79 ± 8.76	37.76 ± 7.32	36.06 ± 7.40
1.98 ± 0.86	2.08 ± 1.16	1.74 ± 1.08	1.92 ± 1.09	3.39 ± 1.93	2.70 ± 2.16
0.09 ± 0.13	0.10 ± 0.13	0.11 ± 0.13	0.13 ± 0.15	0.27 ± 0.37	0.10 ± 0.12
35.40 ± 8.54	34.93 ± 7.52	34.18 ± 8.13	35.83 ± 8.84	41.42 ± 7.43	38.86 ± 7.92
1.29 ± 1.06	1.21 ± 1.01	1.17 ± 0.97	1.46 ± 1.52	1.68 ± 1.09	2.12 ± 1.59
0.08 ± 0.09	0.04 ± 0.07	0.09 ± 0.12	0.05 ± 0.08	0.00 ± 0.00	0.07 ± 0.12
0.05 ± 0.11	0.03 ± 0.07	0.01 ± 0.03	0.03 ± 0.07	0.07 ± 0.12	0.06 ± 0.08
0.50 ± 0.37	0.56 ± 0.48	0.42 ± 0.50	0.37 ± 0.33	0.46 ± 0.32	0.43 ± 0.45
0.17 ± 0.22	0.10 ± 0.15	0.14 ± 0.17	0.11 ± 0.14	0.13 ± 0.18	0.20 ± 0.23
4.86 ± 1.25	5.04 ± 1.08	4.89 ± 1.60	5.34 ± 2.19	4.99 ± 1.96	5.05 ± 2.15
5.66 ± 1.41	5.78 ± 1.16	5.55 ± 1.74	5.90 ± 2.03	5.65 ± 2.02	5.81 ± 2.16

[a]Percentages of cell types (means ± Standard Derivation) in tibial bone marrow of infants from birth to 18 months of age. Data were obtained from normal American infants of black, white and Asian racial origin. The changes in the marrow during the first 18 months of postnatal life are based on differential counts of 1,000 cells classified on stained smears on each 10 serial marrow samples aspirated from the same population of infants. Criteria for including bone marrow data in this study consisted of the absence of any clinical evidence of disease, normal rate of growth and normal serum proteins and transferrin saturations.

[b]*n*, number of infants studied at each stage.

[c]Expressed in round figures for facilitating comparison. Means ± SD were calculated from values obtained in individual infants and statistical comparisons were performed.

From: Rosse C, Kraemer MJ, et al. Bone marrow cell populations of normal infants: the predominance of lymphocytes. J Lab Clin Med. 1977;89:1228, with permission.

Table A1-28 Normal Serum Vitamin E Levels (mg/dl) in Newborns[a]

Weeks	1	2	3	4	5	6	7	8	9	10
<1,500 g 28–32 weeks	0.40 [0.05][a]	0.30 [0.04]	0.25 [0.03]	0.25 [0.03]	0.25 [0.03]	0.25 [0.03]	0.25 [0.03]	0.25 [0.03]	0.35 [0.04]	0.45 [0.05]
1,500–2,000 g 32–36 weeks	0.45 [0.05]	0.40 [0.05]	0.40 [0.05]	0.45 [0.05]	0.45 [0.05]	0.45 [0.05]	0.50 [0.05]	0.50 [0.05]	0.60 [0.06]	0.70 [0.06]
2,000–2,500 g 36–40 weeks	0.50 [0.05]	0.45 [0.05]	0.50 [0.05]	0.60 [0.06]	0.70 [0.06]	0.75 [0.06]	0.75 [0.06]	0.75 [0.06]	0.75 [0.06]	0.80 [0.06]
>2,500 g term	0.55 [0.60]	0.55 [0.60]	0.55 [0.60]	0.60 [0.60]	0.75 [0.70]	0.80 [0.70]	0.85 [0.80]	0.85 [0.80]	0.85 [0.80]	0.85 [0.80]

[a]Mean ± 1 SD.
From: Klaus M, Fanaroff A. Care of the high risk neonate. 3rd ed. Philadelphia: Saunders, 1986, with permission.

Table A1-29 Levels of Urinary Homovanyllic Acid and Vanillylmandelic Acid in Different Age Groups

Age	HVA (μg/mg urinary creatinine)		VMA (μg/mg urinary creatinine)	
	Range	Mean ± SD	Range	Mean ± SD
0–3 months	11.3–35.0	22.3 ± 7.5	5.0–37.0	19.5 ± 10.3
3–12 months	8.4–44.9	27.8 ± 9.6	8.4–43.8	22.8 ± 10.6
1–2 years	12.2–31.8	19.7 ± 6.0	7.9–23.0	14.9 ± 3.9
2–5 years	3.4–32.0	15.3 ± 7.7	2.9–23.0	11.3 ± 7.6
5–10 years	6.8–23.7	12.8 ± 5.1	5.8–18.7	9.3 ± 3.7
10–15 years	3.2–13.6	8.1 ± 3.9	1.6–10.6	5.2 ± 2.5
>15 years	3.2–9.6	5.7 ± 1.9	2.8–8.3	4.5 ± 1.8

Abbreviations: HVA, homovanyllic acid; VMA, vanillylmandelic acid.
From: Tuchman M, Moriss CL, Ramnaraine ML, et al. Value of urinary homovanillic acid and vanillylmandelic acid levels in the diagnosis and management of patients with neuroblastoma: comparison with 24-hour urine collections. Pediatrics 1985;75:324, with permission.

Table A1-30   Values for Soluble Interleukin-2R (SIL-2R)

Cord Blood	267–799 units/ml
2.5–9 months	580–1712 units/ml
9.5–19.5 months	341–2337 units/ml
20–60 months	322–1207 units/ml
10+ years	80–600 units/ml
Adults	71–477 units/ml

Table A1-31   Thyroid Values

TSH, μIU/ml	0.3–5.0	Immunoradiometricassay
Total T-3, ng/dl	50–190	Radioimmunoassay
Total T-4, μg/dl	4.0–13.0	Radioimmunoassay

Biological Tumor Markers

Table A2-1

Tumor Marker	Diagnostic Significance	Normal Values
Oncofetal proteins		
α-fetoprotein in serum	Hepatoblastoma Hepatocellular carcinoma Endodermal sinus tumor Embryonal carcinoma Pancreatoblastoma	Table 27-4, average normal serum α-fetoprotein of infants at various ages
Carcinoembryonic antigen (CEA) in serum	Adenocarcinoma of colon Wilms' tumor Hepatoblastoma Hepatocellular carcinoma Germ cell tumor Pulmonary blastoma	2.5 ng/ml
Hormones		
Beta-human chorionic gonadotropin (beta-hCG) in serum	Choriocarcinoma Hepatoblastoma	5 mIU/ml
Catecholamines urinary VMA/HVA	Neuroblastoma	Appendix 1 Table A1-28
Plasma and urinary catecholamines and urinary metanephrine and VMA	Pheochromocytoma	
Testosterone serum	Leydig cell tumor (testis) Sertoli–Leydig cell (ovary)	D.O.M.
Estrogen serum	Granulosa cell tumor (ovary) Sertoli cell tumor (testis)	D.O.M.
Cortisol, aldosterone, testosterone and estrogen serum	Adrenal hyperplasia Adrenal adenoma Adrenal carcinoma Ectopic adrenal rests (liver, testis, ovary)	D.O.M.
Thyrocalcitonin serum	Medullary (C-cell) thyroid carcinoma	D.O.M.
T3, T4 serum	Hyperthyroidism Thyroid adenoma Thyroid carcinoma	Appendix 1 Table A1-31
Parathormone serum	Parathyroid hyperplasia Parathyroid adenoma	D.O.M.

(Continued)

Manual of Pediatric Hematology and Oncology. DOI: 10.1016/B978-0-12-375154-6.00044-6

Table A2-1   (Continued)

Tumor Marker	Diagnostic Significance	Normal Values
	Parathyroid carcinoma	
	Hepatoblastoma	
Vasoactive intestinal peptide (VIP)	Neuroblastoma	<75 pg/ml
	Vipoma	
Renin serum	Wilms' tumor	D.O.M.
Erythropoietin serum	Wilms' tumor	Appendix 1 Table A1-13
	Adrenal carcinoma	
	Renal carcinoma	
	Hepatoma	
Enzymes		
Neurone-specific enolase in serum	Neuroblastoma	15 ng/ml
	Primitive neuroectodermal tumors	
	Medulloblastoma	
Lactic dehydrogenase (LDH) in serum	Acute leukemias	297–537 units/l
	Non-Hodgkin's lymphoma	(depending on laboratory)
	Neuroblastoma	
	Ewing's sarcoma	
	Osteosarcoma	
	Germ cell tumor	
Specific proteins	Multiple myeloma and	See reference values
Immunoglobulins	other gammopathies	Appendix 1
Cytokines	Acute lymphoblastic leukemia	
Soluble interleukin-2 receptor (SIL-2R) in serum	Non-Hodgkin's lymphoma Hodgkin's disease	
	Hemophagocytic histiocytic syndrome	
Mucins and other glycoproteins		
CA 125	Ovarian cancer	<35 units/ml
CA 19–9	Colon cancer, pancreatic cancer	~37 units/ml (for investigation use only)
CA 15–3	Breast cancer	<31 units/ml
Other tumor markers		
5-Hydroxyindoleacetic acid in urine (5-HIAA)	Carcinoid tumor	24 hour urine: 2–8 mg/dl
Neopterins in urine	Non-Hodgkin lymphoma	7–150 µg/l
	Hemophagocytic lymphohisti-ocytosis	
Ferritin	Neuroblastoma	6 months–15 years:
	Hodgkin disease	12–113 ng/ml; 15–49
	Hepatocellular carcinoma	years: 12–156 ng/ml
	Germ cell tumor	

Abbreviations: DOM, dependent on method. The reader should consult reference laboratory or standard textbook for normal values.

Pharmacologic Properties of Commonly used Anticancer Drugs

Manual of Pediatric Hematology and Oncology. DOI: 10.1016/B978-0-12-375154-6.00045-8

Table A3-1

Drug	Synonyms	Route	Dose/m²	Schedule	Mechanism of Action	Toxicities	Antitumor Spectrum	Mechanisms of Resistance
Alkylating Agents								
Mechlorethamine	Mustargen, HN2, nitrogen mustard	i.v.	6 mg	Weekly × 2, q28 d	Alkylation; crosslinking	M, N&V, A, phlebitis, vesicant, mucositis	Hodgkin disease	↓ Transport, ↑ DNA repair, ↑ GT
Cyclophosphamide	Cytoxan, CTX	i.v.	250–1,800 mg	Daily × 1–4 d, q21–28 d	(Prodrug) alkylation; crosslinking	M, N&V, A, cystitis, water retention; cardiac (HD)	Lymphomas, leukemias, sarcomas, neuroblastoma	↑ IC catabolism, ↑ DNA repair, ↑ GT
Ifosfamide	IFOS, IFEX	p.o. i.v.	100–300 mg 1,600–2,400 mg	Daily Daily × 5, q21–28 d	(Prodrug) alkylation; crosslinking	M, N&V, A, cystitis, NT, renal; cardiac (HD)	Sarcomas, germ cell	↑ IC catabolism, ↑ DNA repair, ↑ GT
Melphalan	Alkeran, l-PAM	i.v.	10–35 mg	q21–28 d	Alkylation; crosslinking	M, N&V; mucositis and diarrhea (HD)	Rhabdomyosarcoma; sarcomas, neuroblastoma and leukemias (HD)	↓ Transport, ↑ DNA repair, ↑ GT
		p.o.	4–20 mg	Daily for 1–21 d				
		i.v.	140–220 mg	Single dose (BMT)				
Lomustine	CeeNU, CCNU	p.o.	100–150 mg	Single dose, q4–6 wk	Alkylation; crosslinking; carbamoylation	M, N&V, renal and pulmonary	Brain tumors, lymphoma, Hodgkin disease	↓ Uptake, ↑ IC catabolism, ↑ DNA repair
Carmustine	BiCNU, BCNU	i.v.	200–250 mg	Single dose, q4–6 wk	Alkylation; crosslinking; carbamoylation	M, N&V, renal and pulmonary	Brain tumors, lymphoma, Hodgkin disease	↓ Uptake, ↑ IC catabolism, ↑ DNA repair
Busulfan	Myleran	p.o.	1.8 mg	Daily	Alkylation; crosslinking	M, A, pulmonary; N&V, mucositis, NT, hepatic (HD)	CML; leukemias (BMT)	↑ DNA repair, ↑ GT
		p.o.	37.5 mg	q6h for 4 d (BMT)				
Cisplatin	Platinol, CDDP	i.v.	50–200 mg	Over 4–6 h, q21–28 d	Platination; crosslinking	M (mild), N&V, A, renal, NT, ototoxicity, HSR	Testicular and other germ cell, brain tumors, osteosarcoma, neuroblastoma	↓ Uptake, ↑ DNA repair, ↑ GT

Drug	Abbrev.	Route	Dose	Schedule	Mechanism	Toxicity	Indications	Resistance
		i.v.	20–40 mg	Daily × 5, q21–28 d				
Carboplatin	CBDCA	i.v.	400–600 mg	Single dose or daily × 2, q28 d	Platination; crosslinking	M (Plt), N&V, A, hepatic (mild), HSR	Brain tumors, germ cell neuroblastoma, sarcomas	↓ Uptake, ↑ DNA repair, ↑ GT
Oxaliplatin	Eloxatin	i.v.	85–130 mg	Single dose q21 d	Platination; crosslinking	NT	Colorectal cancer	↓ Uptake, ↑ DNA repair
		i.v.	100–175 mg	Weekly 4, q6 wk				
Dacarbazine	DTIC	i.v.	250 mg	Daily × 5, q21–28 d	(Prodrug) methylation	M (mild), N&V, flu-like syndrome, hepatic	Neuroblastoma, sarcomas, Hodgkin disease	↑ DNA repair
Temozolomide	TMZ	p.o.	200 mg	Daily × 5, q28 d	(Prodrug) methylation	M, N&V	Brain tumors	↑ DNA repair
Procarbazine	Matulan, PCZ	p.o.	100 mg	Daily for 10–14 d	(Prodrug) methylation; free radical formation	M, N&V, NT, rash, mucositis	Hodgkin disease, brain tumors	↑ DNA repair
Antimetabolites								
Methotrexate	MTX	p.o., i.m., s.c.	7.5–30.0 mg	Weekly or biweekly	Interferes with folate metabolism	M (mild), mucositis, rash, hepatic; renal, NT (HD)	Leukemia, lymphoma, osteosarcoma	↓ Transport, ↑ target enzyme, ↓ polyglutamation
		i.v.	10–33,000 mg	Bolus or CI (6–42 h)				
Mercaptopurine	Purinethol, 6-MP	p.o.	75–100 mg	Daily	(Prodrug) incorporated into DNA and RNA; blocks purine synthesis, interconversion	M, hepatic, mucositis	Leukemia (ALL, CML)	↓ Activation, ↑ IC catabolism
Thioguanine	6-TG	p.o.	75–100 mg	Daily × 5–7	(Prodrug) incorporated into DNA and RNA; blocks purine synthesis, interconversion	M, N&V, mucositis, hepatic (VOD)	Leukemia (ALL, AML)	↓ Activation, ↑ IC catabolism

(Continued)

Table A3-1 (Continued)

Drug	Synonyms	Route	Dose/m²	Schedule	Mechanism of Action	Toxicities	Antitumor Spectrum	Mechanisms of Resistance
Fludarabine phosphate	F-ara-AMP	p.o. i.v.	40–60 mg 25 mg	Daily Daily × 5	(Prodrug) incorporated into DNA; inhibits DNA polymerase, ribonucleotide reductase	M, opportunisitic infectious, neurotoxicity (high dose)	Leukemia (AML, CLL), indolent lymphomas	↓ Membrane transport, ↓ IC activation, ↑ IC catabolism
Cladribine	2-CdA	i.v.	8.9 mg	Daily × 5	(Prodrug) incorporated into DNA; inhibits DNA polymerase, ribonucleotide reductase	M, opportunisitic infectious	Leukemia (AML, CLL), indolent lymphomas	↓ Membrane transport, ↓ IC activation, ↑ IC catabolism
Cytarabine	Ara-C, cytosine arabinoside, Cytosar	i.v., s.c. i.v.	100–200 mg 3,000 mg	q12h or CI for 5–7 d q12h for 48 doses	(Prodrug) incorporated into DNA; inhibits DNA polymerase	M, N&V, mucositis, GI, flu-like syndrome; NT, ocular, skin (HD)	Leukemia, lymphoma	↓ Activation, ↓ transport, ↑ dCTP, ↑ IC catabolism
Fluorouracil	5-FU	i.v. i.v.	500 mg 800–1,200 mg	Single or daily × 5 d CI (24–120 h)	(Prodrug) inhibits thymidine synthesis; incorporated into RNA, DNA	M (bolus), mucositis, N&V, diarrhea, skin, NT, ocular, cardiac	Carcinomas, hepatic tumors	↑ IC catabolism, ↓ activation, ↑ target enzyme, altered target enzyme
Antitumor Antibiotics								
Doxorubicin	Adriamycin, ADR	i.v. i.v. i.v.	45–75 mg 20–30 mg 45–90 mg	Single, q21d Weekly CI (24–96 h)	Intercalation; DNA strand breaks (Topo II); free radical formation	M, mucositis, N&V, A, diarrhea, vesicant, cardiac (acute, chronic)	Leukemia (ALL, ANLL), lymphomas, most solid tumors	Multidrug resistance, ↓ Topo II
Daunomycin	Daunorubicin, DNR	i.v.	30–45 mg	Daily × 3 or weekly	Intercalation; DNA strand breaks	M, mucositis, N&V, diarrhea, A,	Leukemia (ALL, ANLL), lymphomas	Multidrug resistance, ↓ Topo II

		i.v.	10–15 mg	Daily or weekly × 3	(Topo II); free radical formation	vesicant, cardiac (acute, chronic)		
Idarubicin	IDA	p.o.	30–40 mg	Daily × 3	Intercalation; DNA strand breaks (Topo II); free radical formation	M, mucositis, N&V, diarrhea, A, vesicant, cardiac (acute, chronic)	Leukemia (ALL, ANLL), lymphomas	Multidrug resistance, ↓ Topo II
Mitoxantrone	Novantrone, MITO	i.v.	8–12 mg	Daily × 3–5 d	Intercalation; DNA strand breaks (Topo II)	M, mucositis, N&V, A, bluish color to urine, veins, sclerae, nails	Leukemia (ALL, ANLL), lymphomas	Multidrug resistance, ↓ Topo II
Bleomycin	Blenoxane, BLEO	i.v., i.m., s.c.	10–20 units	Weekly	Free radical–mediated DNA strand breaks	Lung, skin, fever, mucositis, alopecia, hypersensitivity, Raynaud's, N&V	Lymphoma, testicular and other germ cell	↑ IC catabolism, ↑ DNA repair
Dactinomycin	Cosmegen, ACT-D, actinomycin-D	i.v. i.v.	0.45 mg (15 µg/kg) 1.35–1.80 mg (45–60 µg/kg)	Daily × 5, q3–6 wk Single dose q3–6 wk	Intercalation; DNA strand breaks (Topo II)	M, N&V, A, mucositis, vesicant, hepatic (VOD)	Wilms', sarcomas	Multidrug resistance, ↓ Topo II
Plant Products								
Vincristine	Oncovin, VCR	i.v.	1.0–1.5 mg (maximum, 2.0 mg)	Weekly × 3–6	Mitotic inhibitor; blocks microtubule polymerization	NT, A, SIADH, hypotension, vesicant	Leukemia (ALL), lymphomas, most solid tumors	Multidrug resistance, altered tubulin subunit
Vinblastine	Velban, VLB	i.v.	3.5–6.0 mg	Weekly × 3–6	Mitotic inhibitor; blocks microtubule polymerization	M, A, mucositis, mild NT, vesicant	Histiocytosis, Hodgkin, testicular	Multidrug resistance, altered tubulin subunit
Vinorelbine	Navelbine	i.v.	30 mg	Weekly	Mitotic inhibitor; blocks microtubule polymerization	M, mild NT, A, vesicant	?	Multidrug resistance, altered tubulin subunit
Etoposide	VePesid, VP-16	i.v.	60–120 mg	Daily × 3–5, q3–6 wk	DNA strand breaks (Tolo II)	M, A, N&V, mucositis, mild NT, hypotension, HSR, secondary leukemia;	Leukemias (ALL, ANLL), lymphomas, neuroblastoma,	Multidrug resistance, ↓ or altered Topo II, ↑ DNA repair

(Continued)

Table A3-1 (Continued)

Drug	Synonyms	Route	Dose/m²	Schedule	Mechanism of Action	Toxicities	Antitumor Spectrum	Mechanisms of Resistance
		p.o.	50 mg	Daily × 21 d q4 wk		diarrhea (p.o.)	sarcomas, brain tumors	
Teniposide	VM-26	i.v.	70–180 mg	Daily × 3q 3–6 wk	DNA strand breaks (Topo II)	M, A, N&V, mucositis, mild NT, hypotension, HSR	Leukemia (ALL)	Multidrug resistance, ↓ or altered Topo II, ↑ DNA repair
Paclitaxel	Taxol	i.v.	135–250 mg	CI for 3 or 24 h, q3 wk	Mitotic inhibitor; blocks microtubule depolymerization	M, HSR, A, NT, mucositis, cardiac, EtOH poisoning	?	Multidrug resistance, altered tubulin subunits, ↑ Raf kinase
Docetaxel	Taxotere	i.v.	100–125 mg	q3 wk	Mitotic inhibitor; blocks microtubule depolymerization	M, HSR, A, NT, rash, edema, mucositis	?	Multidrug resistance, altered tubulin subunits
Topotecan	Hycamptin	i.v.	1.4–4.5 mg	Daily × 5, q3 wk	DNA strand breaks (Topo I)	M, diarrhea, mucositis, N&V, A, rash, hepatic	Neuroblastoma, rhabdomyosarcoma	↓ or altered Topo I, multidrug resistance
Irinotecan	CPT-11, Camptosar	i.v.	50 mg / 125 mg	Daily × 5, q3 wk / Weekly × 4, q6 wk	(Prodrug) DNA strand breaks (Topo I)	M, diarrhea, N&V, A, hepatic, dehydration, ileus	Rhabdomyosarcoma	↓ or altered Topo I, multidrug resistance
Miscellaneous								
Prednisone	Deltasone, PRED	p.o.	40 mg	Daily	(Prodrug) receptor-mediated lympholysis	Protean (see text)	Leukemia, lymphomas	Loss or defect in glucocorticoid receptor
Prednisolone		p.o., i.v.	40 mg	Daily	Receptor-mediated lympholysis	Protean	Leukemia, lymphomas	Loss or defect in glucocorticoid receptor
Dexamethasone	Decadron, DEX	p.o., i.v., i.m.	6 mg	Daily	Receptor-mediated lympholysis	Protean	Leukemia, lymphomas, brain tumors	Loss or defect in glucocorticoid receptor

Native asparaginase	Elspar, l-ASP	i.v., i.m.	6,000–25,000 IU	3 times per wk	Asparagine depletion; ↓ protein synthesis	Leukemia (ALL), lymphoma	HSR, coagulopathy, pancreatitis, hepatic, NT	↑ IC asparagine synthase
PEG-Asparaginase	Oncaspar, PEG-ASP	i.v., i.m.	2,500 IU	Every 1–4 wk	Asparagine depletion; ↓ protein synthesis	Leukemia (ALL), lymphoma	HSR, coagulopathy, pancreatitis, hepatic, NT	↑ IC asparagine synthase
All-*trans*-retinoic acid	ATRA, Tretinoin, Vesanoid	p.o.	45 mg	Daily for induction Daily × 7 d q28 d for maintenance	Differentiation agent	Acute promyelocytic leukemia	Retinoic acid syndrome, pseudotumor cerebri, cheilitis, conjunctivitis, dry skin, ↑ triglycerides	Mutations in *PML-RARα*
13-*cis*-retinoic acid	13cRA, Isotretinoin, Accutane	p.o.	160 mg	Daily × 14 q28 d	Differentiation agent	Minimal residual disease neuroblastoma	Cheilitis, conjunctivitis, dry mouth, xerosis, pruritis, headache, bone & joint pain, ↑ triglycerides, ↑ Ca++	
Imatinib mesylate	Gleevec, STI-571	p.o.	340 mg	Daily	Inhibitis bcr-abl tyrosine kinase	Ph+ CML	N&V, fatigue, headache, GI, hepatic, M	Mutations in bcr-abl Multidrug resistance

Abbreviations: A, alopecia; ALL, acute lymphoblastic leukemia; ANLL, acute nonlymphoblastic leukemia; BMT, bone marrow transplant; CI, continuous infusion; CML, chronic myelogenous leukemia; dCTP, deoxycytidine triphosphate; EtOH, ethanol; GI, gastrointestinal toxicity; GT, glutathione-S-transferase; HD, high dose; HSR, hypersensitivity reaction(s); IC, intracellular; M, myelosuppression; N&V, nausea and vomiting; NT, neurotoxicity; SIADH, syndrome of inappropriate antidiuretic drug hormone secretion; VOD, veno-occlusive disease; ↑, increased; ↓, decreased.

From: Pizzo, PA, Poplack DG. Principles and Practice of Pediatric Oncology 5th Ed., Philadelphia, Lippincott, Williams and Wilkins, 2006, with permission.

Index